College Algebra

Real Mathematics, Real People

SEVENTH EDITION

Ron Larson

CENGAGE
Learning®

Australia • Brazil • Japan • Korea • Mexico • Singapore • Spain • United Kingdom • United States

CENGAGE
Learning®

College Algebra: Real Mathematics, Real People, 7th Edition

College Algebra: Real Mathematics, Real People, 7th Edition
Ron Larson

For product information and technology assistance, contact us at
Cengage Learning Customer & Sales Support, 1-800-354-9706
For permission to use material from this text or product,
submit all requests online at **cengage.com/permissions**
Further permissions questions can be emailed to
permissionrequest@cengage.com

ISBN: 9781305761629

WCN: 01-100-101

Cengage Learning
20 Channel Center Street
Boston, MA 02210
USA

Cengage Learning is a leading provider of customized learning solutions with office locations around the globe, including Singapore, the United Kingdom, Australia, Mexico, Brazil, and Japan. Locate your local office at: **www.international.cengage.com/region.**

Cengage Learning products are represented in Canada by Nelson Education, Ltd.

For your lifelong learning solutions, visit **www.cengage.com/custom.**

Visit our corporate website at **www.cengage.com.**

Contents

Preface

Welcome to *College Algebra: Real Mathematics, Real People*, Seventh Edition. I am proud to present to you this new edition. As with all editions, I have been able to incorporate many useful comments from you, our user. And while much has changed in this revision, you will still find what you expect—a pedagogically sound, mathematically precise, and comprehensive textbook. In this book you will see how algebra is used by real people to solve real-life problems and make real-life decisions.

In addition to providing real and relevant mathematics, I am pleased and excited to offer you something brand new—a companion website at **LarsonPrecalculus.com.** My goal is to provide students with the tools they need to master algebra.

New To This Edition

NEW LarsonPrecalculus.com

This companion website offers multiple tools and resources to supplement your learning. Access to these features is free. View and listen to worked-out solutions of Checkpoint problems in English or Spanish, explore examples, download data sets, watch lesson videos, and much more.

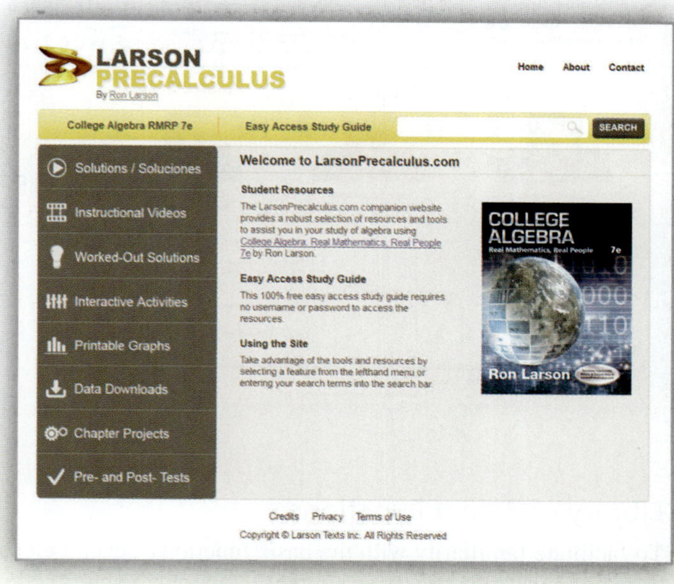

NEW Checkpoints

Accompanying every example, the Checkpoint problems encourage immediate practice and check your understanding of the concepts presented in the example. View and listen to worked-out solutions of the Checkpoint problems in English or Spanish at LarsonPrecalculus.com.

130. **HOW DO YOU SEE IT?** Decide whether the two functions shown in each graph appear to be inverse functions of each other. Explain your reasoning.

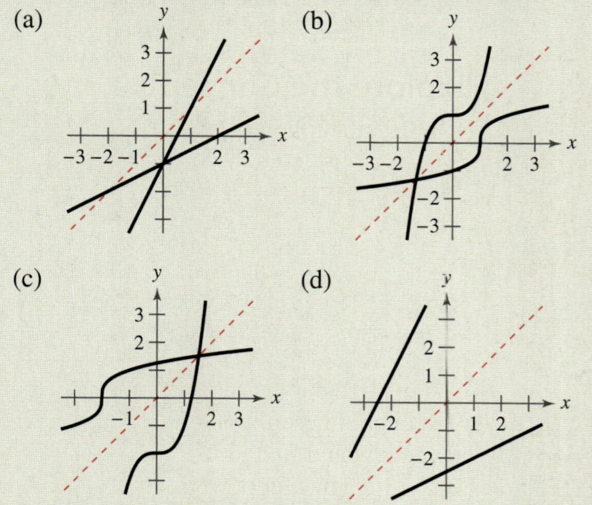

NEW How Do You See It?

The How Do You See It? feature in each section presents an exercise that you will solve by visual inspection using the concepts learned in the lesson. This exercise is excellent for classroom discussion or test preparation.

NEW Data Spreadsheets

Download these editable spreadsheets from LarsonPrecalculus.com and use the data to solve exercises.

REVISED Exercise Sets

The exercise sets have been carefully and extensively examined to ensure they are rigorous and relevant and to include all topics our users have suggested. The exercises have been **reorganized and titled** so you can better see the connections between examples and exercises. Multi-step exercises reinforce problem-solving skills and mastery of concepts by giving you the opportunity to apply the concepts in real-life situations.

REVISED Remarks

These hints and tips reinforce or expand upon concepts, help you learn how to study mathematics, address special cases, or show alternative or additional steps to a solution of an example.

Trusted Features

Calc Chat

For the past several years, an independent website—CalcChat.com—has provided free solutions to all odd-numbered problems in the text. Thousands of students have visited the site for practice and help with their homework.

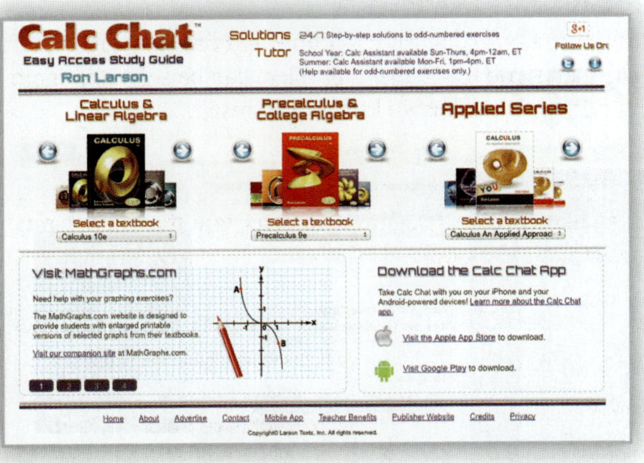

Side-By-Side Examples

Throughout the text, we present solutions to examples from multiple perspectives—algebraically, graphically, and numerically. The side-by-side format of this pedagogical feature helps you to see that a problem can be solved in more than one way and to see that different methods yield the same result. The side-by-side format also addresses different learning styles.

Why You Should Learn It Exercise

An engaging real-life application of the concepts in the section. This application exercise is typically described in the section opener as a motivator for the section.

Library of Parent Functions

To facilitate familiarity with the basic functions, several elementary and nonelementary functions have been compiled as a Library of Parent Functions. Each function is introduced at its first appearance in the text with a definition and description of basic characteristics. The Library of Parent Functions Examples are identified in the title of the example and there is a Review of Library of Parent Functions after Chapter 4. A summary of functions is presented on the inside cover of this text.

Make a Decision Exercises

The Make a Decision exercises at the end of selected sections involve in-depth applied exercises in which you will work with large, real-life data sets, often creating or analyzing models. These exercises are offered online at LarsonPrecalculus.com.

Chapter Openers

Each Chapter Opener highlights a real-life modeling problem, showing a graph of the data, a section reference, and a short description of the data.

Explore the Concept

Each Explore the Concept engages you in active discovery of mathematical concepts, strengthens critical thinking skills, and helps build intuition.

Explore the Concept

Complete the following:

$i^1 = i$	$i^7 =$
$i^2 = -1$	$i^8 =$
$i^3 = -i$	$i^9 =$
$i^4 = 1$	$i^{10} =$
$i^5 =$	$i^{11} =$
$i^6 =$	$i^{12} =$

What pattern do you see? Write a brief description of how you would find i raised to any positive integer power.

Technology Tip

Some graphing utilities will produce graphs of inequalities. For instance, you can graph $2x^2 + 5x > 12$ by setting the graphing utility to *dot* mode and entering $y = 2x^2 + 5x > 12$. Using $-10 \le x \le 10$ and $-4 \le y \le 4$, your graph should look like the graph shown below. The solution appears to be $(-\infty, -4) \cup \left(\frac{3}{2}, \infty\right)$. See Example 6 for an algebraic solution and for an alternative graphical solution.

What's Wrong?

Each What's Wrong? points out common errors made using graphing utilities.

Technology Tip

Technology Tips provide graphing calculator tips or provide alternative methods of solving a problem using a graphing utility.

Algebra of Calculus

Throughout the text, special emphasis is given to the algebraic techniques used in calculus. Algebra of Calculus examples and exercises are integrated throughout the text and are identified by the symbol $\int$.

Algebraic-Graphical-Numerical Exercises

These exercises allow you to solve a problem using multiple approaches—algebraic, graphical, and numerical. This helps you to see that a problem can be solved in more than one way and to see that different methods yield the same result.

Modeling Data Exercises

These multi-part applications that involve real-life data offer you the opportunity to generate and analyze mathematical models.

Vocabulary and Concept Check

The Vocabulary and Concept Check appears at the beginning of the exercise set for each section. Each of these checks asks fill-in-the-blank, matching, and non-computational questions designed to help you learn mathematical terminology and to test basic understanding of that section's concepts.

What you should learn

- Find the slopes of lines.
- Write linear equations given points on lines and their slopes.
- Use slope-intercept forms of linear equations to sketch lines.
- Use slope to identify parallel and perpendicular lines.

Why you should learn it

The slope of a line can be used to solve real-life problems. For instance, in Exercise 97 on page 95, you will use a linear equation to model student enrollment at Penn State University.

What you should learn/Why you should learn it

These summarize important topics in the section and why they are important in math and in life.

Chapter Summaries

The Chapter Summary includes explanations and examples of the objectives taught in the chapter.

Error Analysis Exercises

This exercise presents a sample solution that contains a common error which you are asked to identify.

ENHANCED WebAssign

Enhanced WebAssign combines exceptional algebra content with the most powerful online homework solution, WebAssign. Enhanced WebAssign engages you with immediate feedback, rich tutorial content and interactive, fully customizable eBooks (YouBook) helping you to develop a deeper conceptual understanding of the subject matter.

Instructor Resources

Complete Solutions Manual

• ISBN-13: 9781305117846

This manual contains solutions to all exercises from the text, including Chapter Review Exercises and Chapter Tests. This manual is found on the Instructors Companion Site.

Test Bank

• ISBN-13: 9781305117815

This supplement includes test forms for every chapter of the text, and is found on the instructor companion site.

Text-Specific DVDs

• ISBN-13: 9781305117761

These text-specific DVDs cover all sections of the text—providing explanations of key concepts as well as examples, exercises, and applications in a lecture-based format.

Enhanced WebAssign

Printed Access Card: 9781285858333
Instant Access Code: 9781285858319

Enhanced WebAssign combines exceptional mathematics content with the most powerful online homework solution, WebAssign. Enhanced WebAssign engages your students with immediate feedback, rich tutorial content, and an interactive, fully customizable eBook, Cengage YouBook helping students to develop a deeper conceptual understanding of the subject matter.

Instructor Companion Site

Everything you need for your course in one place! This collection of book-specific lecture and class tools is available online via www.cengage.com/login. Access and download PowerPoint presentations, images, instructor's manual, and more.

Cengage Learning Testing Powered by Cognero

• ISBN-13: 9781305258518

CLT is a flexible online system that allows you to author, edit, and manage test bank content; create multiple test versions in an instant; and deliver tests from your LMS, your classroom, or wherever you want. This is available online via www.cengage.com/login.

Student Resources

Student Solutions Manual

• ISBN-13: 9781305117754

Contains fully worked-out solutions to all of the odd-numbered exercises in the text, giving you a way to check your answers and ensure that you took the correct steps to arrive at an answer.

Enhanced WebAssign

Printed Access Card: 9781285858333
Instant Access Code: 9781285858319

Enhanced WebAssign combines exceptional mathematics content with the most powerful online homework solution, WebAssign. Enhanced WebAssign engages you with immediate feedback, rich tutorial content, and an interactive, fully customizable eBook, Cengage YouBook helping you to develop a deeper conceptual understanding of the subject matter.

CengageBrain.com

To access additional course materials, please visit www.cengagebrain.com. At the CengageBrain.com home page, search for the ISBN of your title (from the back cover of your book) using the search box at the top of the page. This will take you to the product page where these resources can be found.

Acknowledgments

I would like to thank my colleagues who have helped me develop this program. Their encouragement, criticisms, and suggestions have been invaluable to me.

Reviewers

Hugh Cornell, *University of North Florida*
Kewal Krishan, *Hudson County Community College*
Ferdinand Orock, *Hudson County Community College*
Marnie Phipps, *North Georgia College and State University*
Nancy Schendel, *Iowa Lakes Community College*
Ann Wheeler, *Texas Woman's University*

I would also like to thank the following reviewers, who have given me many useful insights to this and previous editions.

Tony Homayoon Akhlaghi, *Bellevue Community College;* Daniel D. Anderson, *University of Iowa;* Bruce Armbrust, *Lake Tahoe Community College;* Jamie Whitehead Ashby, *Texarkana College;* Teresa Barton, *Western New England College;* Kimberly Bennekin, *Georgia Perimeter College;* Charles M. Biles, *Humboldt State University;* Phyllis Barsch Bolin, *Oklahoma Christian University;* Khristo Boyadzheiv, *Ohio Northern University;* Dave Bregenzer, *Utah State University;* Anne E. Brown, *Indiana University-South Bend;* Diane Burleson, *Central Piedmont Community College;* Beth Burns, *Bowling Green State University;* Alexander Burstein, *University of Rhode Island;* Marilyn Carlson, *University of Kansas;* Victor M. Cornell, *Mesa Community College;* John Dersh, *Grand Rapids Community College;* Jennifer Dollar, *Grand Rapids Community College;* Marcia Drost, *Texas A & M University;* Cameron English, *Rio Hondo College;* Susan E. Enyart, *Otterbein College;* Patricia J. Ernst, *St. Cloud State University;* Eunice Everett, *Seminole Community College;* Kenny Fister, *Murray State University;* Susan C. Fleming, *Virginia Highlands Community College;* Jeff Frost, *Johnson County Community College;* James R. Fryxell, *College of Lake County;* Khadiga H. Gamgoum, *Northern Virginia Community College;* Nicholas E. Geller, *Collin County Community College;* Betty Givan, *Eastern Kentucky University;* Patricia K. Gramling, *Trident Technical College;* Michele Greenfield, *Middlesex County College;* Bernard Greenspan, *University of Akron;* Zenas Hartvigson, *University of Colorado at Denver;* Rodger Hergert, *Rock Valley College;* Allen Hesse, *Rochester Community College;* Rodney Holke-Farnam, *Hawkeye Community College;* Lynda Hollingsworth, *Northwest Missouri State University;* Jean M. Horn, *Northern Virginia Community College;* Spencer Hurd, *The Citadel;* Bill Huston, *Missouri Western State College;* Deborah Johnson, *Cambridge South Dorchester High School;* Francine Winston Johnson, *Howard Community College;* Luella Johnson, *State University of New York, College at Buffalo;* Susan Kellicut, *Seminole Community College;* John Kendall, *Shelby State Community College;* Donna M. Krawczyk, *University of Arizona;* Laura Lake, *Center for Advanced Technologies/ Lakewood High School;* Peter A. Lappan, *Michigan State University;* Charles G. Laws, *Cleveland State Community College;* JoAnn Lewin, *Edison Community College;* Richard J. Maher, *Loyola University;* Carl Main, *Florida College;* Marilyn McCollum, *North Carolina State University;* Judy McInerney, *Sandhills Community College;* David E. Meel, *Bowling Green University;* Beverly Michael, *University of Pittsburgh;* Wendy Morin, *Dwight D. Eisenhower High School;* Roger B. Nelsen, *Lewis and Clark College;* Stephen Nicoloff, *Paradise Valley Community College;* Jon Odell, *Richland Community College;* Paul Oswood, *Ridgewater College;* Wing M. Park, *College of Lake County;* Rupa M. Patel, *University of Portland;* Robert Pearce, *South Plains College;* David R. Peterson, *University of Central Arkansas;* Sandra Poinsett, *College of Southern Maryland;* James Pommersheim, *Reed College;* Antonio Quesada, *University of Akron;* Laura Reger, *Milwaukee Area Technical College;* Jennifer Rhinehart, *Mars Hill College;* Lila F. Roberts, *Georgia Southern University;* Keith Schwingendorf, *Purdue University North Central;* Abdallah Shuaibi, *Truman College;* George W. Shultz, *St. Petersburg Junior College;* Stephen Slack, *Kenyon College;*

Judith Smalling, *St. Petersburg Junior College;* Pamela K. M. Smith, *Fort Lewis College;* Cathryn U. Stark, *Collin County Community College;* Craig M. Steenberg, *Lewis-Clark State College;* Mary Jane Sterling, *Bradley University;* G. Bryan Stewart, *Tarrant County Junior College;* Diane Veneziale, *Burlington County College;* Mahbobeh Vezvaei, *Kent State University;* Ellen Vilas, *York Technical College;* Hayat Weiss, *Middlesex Community College;* Rich West, *Francis Marion University;* Vanessa White, *Southern University;* Howard L. Wilson, *Oregon State University;* Joel E. Wilson, *Eastern Kentucky University;* Michelle Wilson, *Franklin University;* Paul Winterbottom, *Montgomery County Community College;* Fred Worth, *Henderson State University;* Karl M. Zilm, *Lewis and Clark Community College;* Cathleen Zucco-Teveloff, *Rowan University*

I hope that you enjoy learning the mathematics presented in this text. More than that, I hope you gain a new appreciation for the relevance of mathematics to careers in science, technology, business, and medicine.

My thanks to Robert Hostetler, The Behrend College, The Pennsylvania State University, Bruce Edwards, University of Florida, and David Heyd, The Behrend College, The Pennsylvania State University, for their significant contributions to previous editions of this text.

I would also like to thank the staff of Larson Texts, Inc. who assisted in preparing the manuscript, rendering the art package, and typesetting and proofreading the pages and supplements.

On a personal level, I am grateful to my spouse, Deanna Gilbert Larson, for her love, patience, and support. Also, a special thanks goes to R. Scott O'Neil.

If you have suggestions for improving this text, please feel free to write me. Over the past two decades I have received many useful comments from both instructors and students, and I value these very much.

Ron Larson, Ph.D.
Professor of Mathematics
Penn State University
www.RonLarson.com

P Prerequisites

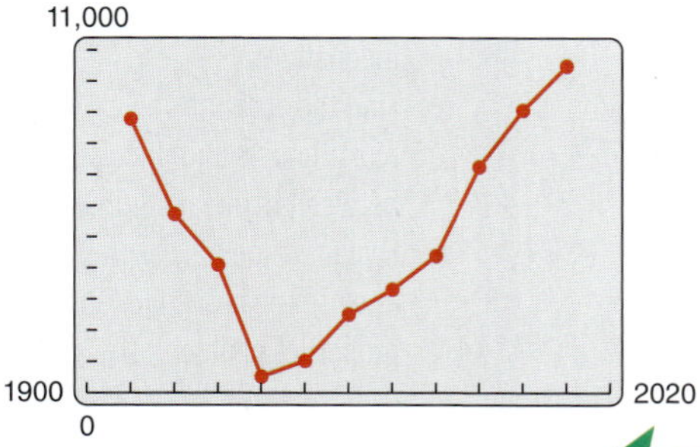

Section P.6, Example 5
U.S. Immigrants

P.1 Real Numbers

Real Numbers

Real numbers are used in everyday life to describe quantities such as age, miles per gallon, and population. Real numbers are represented by symbols such as

$$-5,\quad 9,\quad 0,\quad \tfrac{4}{3},\quad 0.666\ldots,\quad 28.21,\quad \sqrt{2},\quad \pi,\quad \text{and}\quad \sqrt[3]{-32}.$$

Here are some important **subsets** (each member of subset B is also a member of set A) of the set of real numbers.

$\{1, 2, 3, 4, \ldots\}$ Set of natural numbers

$\{0, 1, 2, 3, 4, \ldots\}$ Set of whole numbers

$\{\ldots, -3, -2, -1, 0, 1, 2, 3, \ldots\}$ Set of integers

A real number is **rational** when it can be written as the ratio p/q of two integers, where $q \neq 0$. For instance, the numbers

$$\frac{1}{3} = 0.3333\ldots = 0.\overline{3},\quad \frac{1}{8} = 0.125,\quad \text{and}\quad \frac{125}{111} = 1.126126\ldots = 1.\overline{126}$$

are rational. The decimal representation of a rational number either *repeats* $\left(\text{as in } \frac{173}{55} = 3.1\overline{45}\right)$ or *terminates* $\left(\text{as in } \frac{1}{2} = 0.5\right)$. A real number that cannot be written as the ratio of two integers is called **irrational.** Irrational numbers have infinite nonrepeating decimal representations. For instance, the numbers

$$\sqrt{2} = 1.4142135\ldots \approx 1.41$$

and

$$\pi = 3.1415926\ldots \approx 3.14$$

are irrational. (The symbol $\approx$ means "is approximately equal to.") Figure P.1 shows subsets of real numbers and their relationships to each other.

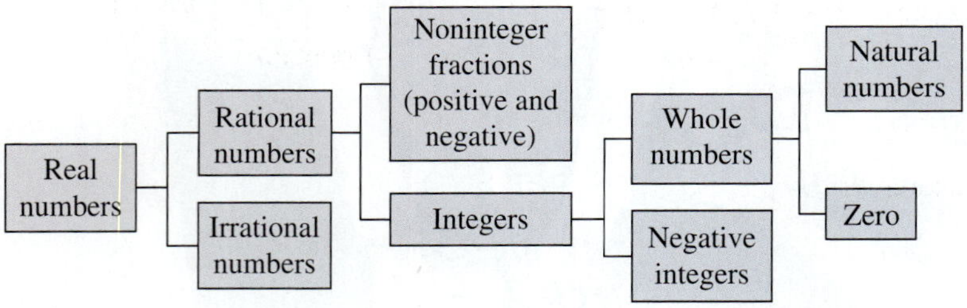

Figure P.1 *Subsets of Real Numbers*

Real numbers are represented graphically by a **real number line.** The point 0 on the real number line is the **origin.** Numbers to the right of 0 are positive and numbers to the left of 0 are negative, as shown in Figure P.2. The term **nonnegative** describes a number that is either positive or zero.

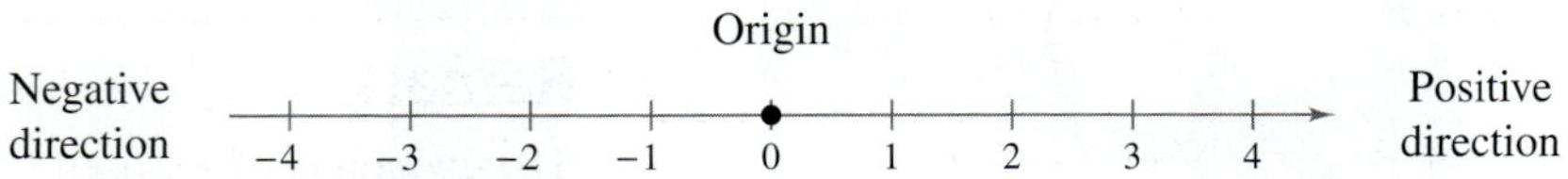

Figure P.2 *The Real Number Line*

There is a *one-to-one correspondence* between real numbers and points on the real number line. That is, every point on the real number line corresponds to exactly one real number, called its **coordinate,** and every real number corresponds to exactly one point on the real number line, as shown in Figure P.3.

What you should learn

- Represent and classify real numbers.
- Order real numbers and use inequalities.
- Find the absolute values of real numbers and the distance between two real numbers.
- Evaluate algebraic expressions and use the basic rules and properties of algebra.

Why you should learn it

Real numbers are used in every aspect of our lives, such as finding the surplus or deficit in the federal budget. See Exercises 89–94 on page 10.

Every point on the real number line corresponds to exactly one real number.

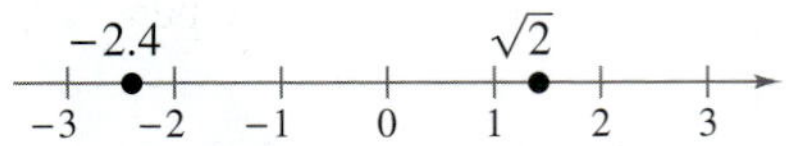

Every real number corresponds to exactly one point on the real number line.

Figure P.3 *One-to-One Correspondence*

Ordering Real Numbers

One important property of real numbers is that they are **ordered.**

> **Definition of Order on the Real Number Line**
>
> If a and b are real numbers, then a is **less than** b when $b - a$ is positive. This order is denoted by the **inequality** $a < b$. This relationship can also be described by saying that b is **greater than** a and writing $b > a$. The inequality $a \le b$ means that a is **less than or equal to** b, and the inequality $b \ge a$ means that b is **greater than or equal to** a. The symbols $<$, $>$, $\le$, and $\ge$ are **inequality symbols.**

Geometrically, this definition implies that $a < b$ if and only if a lies to the *left* of b on the real number line, as shown in Figure P.4.

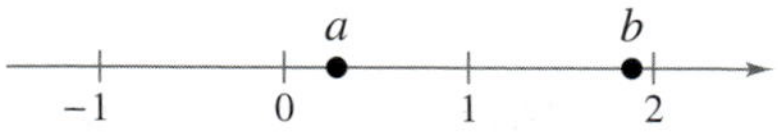

Figure P.4 *$a < b$ if and only if a lies to the left of b.*

EXAMPLE 1 Interpreting Inequalities

See LarsonPrecalculus.com for an interactive version of this type of example.

Describe the subset of real numbers that the inequality represents.

a. $x \le 2$ **b.** $x > -1$ **c.** $-2 \le x < 3$

Solution

a. The inequality $x \le 2$ denotes all real numbers less than or equal to 2, as shown in Figure P.5.

b. The inequality $x > -1$ denotes all real numbers greater than -1, as shown in Figure P.6.

c. The inequality $-2 \le x < 3$ means that $x \ge -2$ *and* $x < 3$. The "double inequality" denotes all real numbers between -2 and 3, including -2 but not including 3, as shown in Figure P.7.

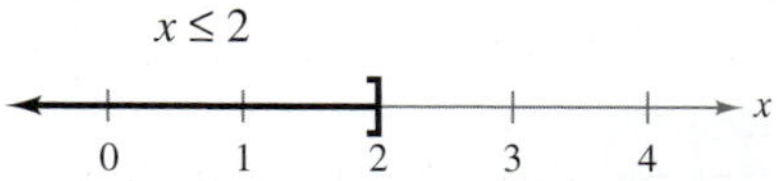

Figure P.5

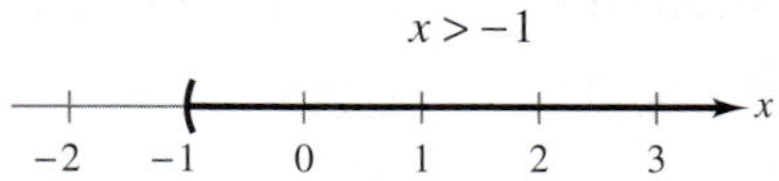

Figure P.6

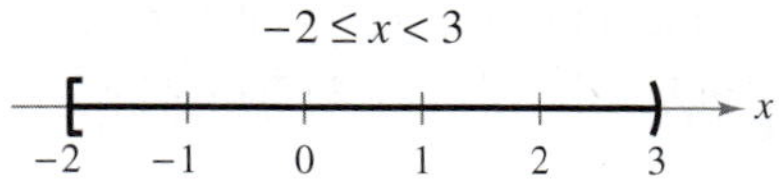

Figure P.7

✓ ***Checkpoint*** *Audio-video solution in English & Spanish at LarsonPrecalculus.com.*

Describe the subset of real numbers that the inequality represents.

a. $x > -3$ **b.** $0 < x \le 4$ ■

Inequalities can be used to describe subsets of real numbers called **intervals.** In the bounded intervals below, the real numbers a and b are the **endpoints** of each interval.

Bounded Intervals on the Real Number Line

Notation	*Interval Type*	*Inequality*	*Graph*
$[a, b]$	Closed	$a \le x \le b$	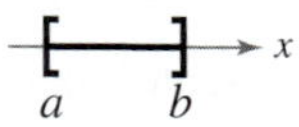
(a, b)	Open	$a < x < b$	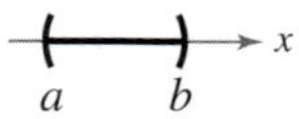
$[a, b)$		$a \le x < b$	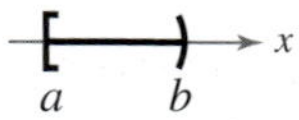
$(a, b]$		$a < x \le b$	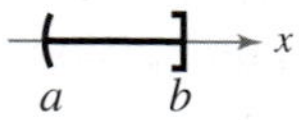

Remark

The endpoints of a closed interval are included in the interval. The endpoints of an open interval are *not* included in the interval.

The symbols ∞, **positive infinity,** and $-\infty$, **negative infinity,** do not represent real numbers. They are simply convenient symbols used to describe the unboundedness of an interval, such as $(1, \infty)$ or $(-\infty, 3]$.

Unbounded Intervals on the Real Number Line

Notation	*Interval Type*	*Inequality*	*Graph*
$[a, \infty)$		$x \geq a$	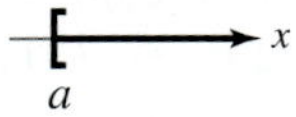
(a, ∞)	Open	$x > a$	
$(-\infty, b]$		$x \leq b$	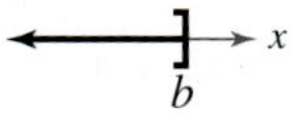
$(-\infty, b)$	Open	$x < b$	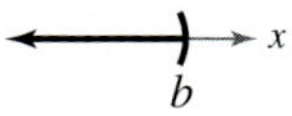
$(-\infty, \infty)$	Entire real line	$-\infty < x < \infty$	

Remark

An interval is unbounded when it continues indefinitely in one or both directions.

EXAMPLE 2 Using Inequalities to Represent Intervals

Use inequality notation to represent each of the following.

a. c is at most 2.

b. All x in the interval $(-3, 5]$

c. t is at least 4 but less than 11.

Solution

a. The statement "c is at most 2" can be represented by $c \leq 2$.

b. "All x in the interval $(-3, 5]$" can be represented by $-3 < x \leq 5$.

c. The statement "t is at least 4 but less than 11" can be represented by $4 \leq t < 11$.

✓ ***Checkpoint*** *Audio-video solution in English & Spanish at LarsonPrecalculus.com.*

Use inequality notation to represent the statement "x is greater than -2 and at most 4."

EXAMPLE 3 Interpreting Intervals

Give a verbal description of each interval.

a. $(-1, 0)$

b. $[2, \infty)$

c. $(-\infty, 0)$

Solution

a. This interval consists of all real numbers that are greater than -1 and less than 0.

b. This interval consists of all real numbers that are greater than or equal to 2.

c. This interval consists of all negative real numbers.

✓ ***Checkpoint*** *Audio-video solution in English & Spanish at LarsonPrecalculus.com.*

Give a verbal description of the interval $[-2, 5)$.

The **Law of Trichotomy** states that for any two real numbers a and b, precisely one of three relationships is possible:

$$a = b, \quad a < b, \quad \text{or} \quad a > b. \qquad \text{Law of Trichotomy}$$

Absolute Value and Distance

The **absolute value** of a real number is its *magnitude*, or the distance between the origin and the point representing the real number on the real number line.

Definition of Absolute Value

If a is a real number, then the **absolute value** of a is

$$|a| = \begin{cases} a, & a \geq 0 \\ -a, & a < 0 \end{cases}.$$

Notice from this definition that the absolute value of a real number is never negative. For instance, if $a = -5$, then $|-5| = -(-5) = 5$. The absolute value of a real number is either positive or zero. Moreover, 0 is the only real number whose absolute value is 0. So, $|0| = 0$.

Explore the Concept

Absolute value expressions can be evaluated on a graphing utility. When evaluating an expression such as $|3 - 8|$, parentheses should surround the expression, as shown below. Evaluate each expression. What can you conclude?

a. $|6|$ **b.** $|-1|$

c. $|5 - 2|$ **d.** $|2 - 5|$

EXAMPLE 4 Evaluating an Absolute Value Expression

Evaluate $\dfrac{|x|}{x}$ for (a) $x > 0$ and (b) $x < 0$.

Solution

a. If $x > 0$, then $|x| = x$ and $\dfrac{|x|}{x} = \dfrac{x}{x} = 1$.

b. If $x < 0$, then $|x| = -x$ and $\dfrac{|x|}{x} = \dfrac{-x}{x} = -1$.

✓ ***Checkpoint*** *Audio-video solution in English & Spanish at LarsonPrecalculus.com.*

Evaluate $\dfrac{|x + 3|}{x + 3}$ for (a) $x > -3$ and (b) $x < -3$. ■

Properties of Absolute Value

1. $|a| \geq 0$ **2.** $|-a| = |a|$

3. $|ab| = |a||b|$ **4.** $\left|\dfrac{a}{b}\right| = \dfrac{|a|}{|b|}, \; b \neq 0$

Absolute value can be used to define the distance between two points on the real number line. For instance, the distance between -3 and 4 is

$$|-3 - 4| = |-7| = 7$$

as shown in Figure P.8.

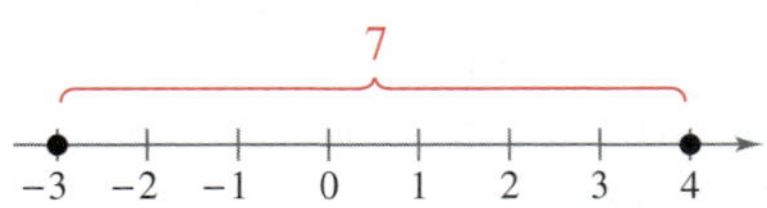

Figure P.8 *The distance between −3 and 4 is 7.*

Distance Between Two Points on the Real Number Line

Let a and b be real numbers. The **distance between a and b** is

$$d(a, b) = |b - a| = |a - b|.$$

Algebraic Expressions and the Basic Rules of Algebra

One characteristic of algebra is the use of letters to represent numbers. The letters are **variables,** and combinations of letters and numbers are **algebraic expressions.** Here are a few examples of algebraic expressions.

$$5x, \quad 2x - 3, \quad \frac{4}{x^2 + 2}, \quad 7x + y$$

Definition of an Algebraic Expression

An **algebraic expression** is a combination of letters (**variables**) and real numbers (**constants**) combined using the operations of addition, subtraction, multiplication, division, and exponentiation.

The **terms** of an algebraic expression are those parts that are separated by *addition.* For example,

$$x^2 - 5x + 8 = x^2 + (-5x) + 8$$

has three terms: x^2 and $-5x$ are the **variable terms,** and 8 is the **constant term.** The numerical factor of a term is called the **coefficient.** For instance, the coefficient of $-5x$ is -5, and the coefficient of x^2 is 1.

To **evaluate** an algebraic expression, substitute numerical values for each of the variables in the expression.

EXAMPLE 5 Evaluating Algebraic Expressions

	Expression	*Value of Variable*	*Substitute*	*Value of Expression*
a.	$-3x + 5$	$x = 3$	$-3(3) + 5$	$-9 + 5 = -4$
b.	$3x^2 + 2x - 1$	$x = -1$	$3(-1)^2 + 2(-1) - 1$	$3 - 2 - 1 = 0$
c.	$\frac{2x}{x + 1}$	$x = -3$	$\frac{2(-3)}{-3 + 1}$	$\frac{-6}{-2} = 3$

Note that you must substitute the value for *each* occurrence of the variable.

✓ Checkpoint *Audio-video solution in English & Spanish at LarsonPrecalculus.com.*

Evaluate $4x - 5$ when $x = 0$.

When an algebraic expression is evaluated, the **Substitution Principle** is used. It states, "If $a = b$, then a can be replaced by b in any expression involving a." For instance, in Example 5(a), 3 is substituted for x in the expression $-3x + 5$.

There are four arithmetic operations with real numbers: addition, multiplication, subtraction, and division, denoted by the symbols

$$+, \quad \times \text{ or } \cdot, \quad -, \quad \text{and} \quad \div \text{ or } /.$$

Of these, addition and multiplication are the two primary operations. Subtraction and division are the inverse operations of addition and multiplication, respectively.

Subtraction: Add the opposite of b.

$$a - b = a + (-b)$$

Division: Multiply by the reciprocal of b.

If $b \neq 0$, then $a/b = a\left(\frac{1}{b}\right) = \frac{a}{b}$.

In these definitions, $-b$ is the **additive inverse** (or opposite) of b, and $1/b$ is the **multiplicative inverse** (or reciprocal) of b. In the fractional form a/b, a is the **numerator** of the fraction and b is the **denominator.**

Because the properties of real numbers below are true for variables and algebraic expressions, as well as for real numbers, they are often called the **Basic Rules of Algebra.** Try to formulate a verbal description of each property. For instance, the Commutative Property of Addition states that *the order in which two real numbers are added does not affect their sum.*

Basic Rules of Algebra

Let a, b, and c be real numbers, variables, or algebraic expressions.

	Property	*Example*
Commutative Property of Addition:	$a + b = b + a$	$4x + x^2 = x^2 + 4x$
Commutative Property of Multiplication:	$ab = ba$	$(1 - x)x^2 = x^2(1 - x)$
Associative Property of Addition:	$(a + b) + c = a + (b + c)$	$(x + 5) + x^2 = x + (5 + x^2)$
Associative Property of Multiplication:	$(ab)c = a(bc)$	$(2x \cdot 3y)(8) = (2x)(3y \cdot 8)$
Distributive Properties:	$a(b + c) = ab + ac$	$3x(5 + 2x) = 3x \cdot 5 + 3x \cdot 2x$
	$(a + b)c = ac + bc$	$(y + 8)y = y \cdot y + 8 \cdot y$
Additive Identity Property:	$a + 0 = a$	$5y^2 + 0 = 5y^2$
Multiplicative Identity Property:	$a \cdot 1 = a$	$(4x^2)(1) = 4x^2$
Additive Inverse Property:	$a + (-a) = 0$	$6x^3 + (-6x^3) = 0$
Multiplicative Inverse Property:	$a \cdot \frac{1}{a} = 1, \ a \neq 0$	$(x^2 + 4)\left(\frac{1}{x^2 + 4}\right) = 1$

Because subtraction is defined as "adding the opposite," the Distributive Properties are also true for subtraction. For instance, the "subtraction form" of $a(b + c) = ab + ac$ is written as

$$a(b - c) = ab - ac.$$

Properties of Negation and Equality

Let a, b, and c be real numbers, variables, or algebraic expressions.

Property	*Example*
1. $(-1)a = -a$	$(-1)7 = -7$
2. $-(-a) = a$	$-(-6) = 6$
3. $(-a)b = -(ab) = a(-b)$	$(-5)3 = -(5 \cdot 3) = 5(-3)$
4. $(-a)(-b) = ab$	$(-2)(-x) = 2x$
5. $-(a + b) = (-a) + (-b)$	$-(x + 8) = (-x) + (-8) = -x - 8$
6. If $a = b$, then $a + c = b + c$.	$\frac{1}{2} + 3 = 0.5 + 3$
7. If $a = b$, then $ac = bc$.	$4^2(2) = 16(2)$
8. If $a \pm c = b \pm c$, then $a = b$.	$1.4 - 1 = \frac{7}{5} - 1 \implies 1.4 = \frac{7}{5}$
9. If $ac = bc$ and $c \neq 0$, then $a = b$.	$3x = 3 \cdot 4 \implies x = 4$

Remark

Be sure you see the difference between the *opposite of a number* and a *negative number.* If a is already negative, then its opposite, $-a$, is positive. For instance, if $a = -2$, then $-a = -(-2) = 2.$

Properties of Zero

Let a and b be real numbers, variables, or algebraic expressions.

1. $a + 0 = a$ and $a - 0 = a$

2. $a \cdot 0 = 0$

3. $\dfrac{0}{a} = 0, \quad a \neq 0$

4. $\dfrac{a}{0}$ is undefined.

5. Zero-Factor Property: If $ab = 0$, then $a = 0$ or $b = 0$.

The "or" in the Zero-Factor Property includes the possibility that either or both factors may be zero. This is an **inclusive or,** and it is the way the word "or" is generally used in mathematics.

Properties and Operations of Fractions

Let a, b, c, and d be real numbers, variables, or algebraic expressions such that $b \neq 0$ and $d \neq 0$.

1. Equivalent Fractions: $\dfrac{a}{b} = \dfrac{c}{d}$ if and only if $ad = bc$.

2. Rules of Signs: $-\dfrac{a}{b} = \dfrac{-a}{b} = \dfrac{a}{-b}$ and $\dfrac{-a}{-b} = \dfrac{a}{b}$

3. Generate Equivalent Fractions: $\dfrac{a}{b} = \dfrac{ac}{bc}, \quad c \neq 0$

4. Add or Subtract with Like Denominators: $\dfrac{a}{b} \pm \dfrac{c}{b} = \dfrac{a \pm c}{b}$

5. Add or Subtract with Unlike Denominators: $\dfrac{a}{b} \pm \dfrac{c}{d} = \dfrac{ad \pm bc}{bd}$

6. Multiply Fractions: $\dfrac{a}{b} \cdot \dfrac{c}{d} = \dfrac{ac}{bd}$

7. Divide Fractions: $\dfrac{a}{b} \div \dfrac{c}{d} = \dfrac{a}{b} \cdot \dfrac{d}{c} = \dfrac{ad}{bc}, \quad c \neq 0$

Remark

In Property 1, the phrase "if and only if" implies two statements. One statement is: If $a/b = c/d$, then $ad = bc$. The other statement is: If $ad = bc$, where $b \neq 0$ and $d \neq 0$, then $a/b = c/d$.

EXAMPLE 6 Properties and Operations of Fractions

a. Equivalent fractions: $\dfrac{x}{5} = \dfrac{3 \cdot x}{3 \cdot 5} = \dfrac{3x}{15}$

b. Divide fractions: $\dfrac{7}{x} \div \dfrac{3}{2} = \dfrac{7}{x} \cdot \dfrac{2}{3} = \dfrac{14}{3x}$

✓ **Checkpoint** Audio-video solution in English & Spanish at LarsonPrecalculus.com.

a. Multiply fractions: $\dfrac{3}{5} \cdot \dfrac{x}{6}$

b. Add fractions: $\dfrac{x}{10} + \dfrac{2x}{5}$

If a, b, and c are integers such that $ab = c$, then a and b are **factors** or **divisors** of c. A **prime number** is an integer that has exactly two positive factors: itself and 1. For example, 2, 3, 5, 7, and 11 are prime numbers. The numbers 4, 6, 8, 9, and 10 are **composite** because they can be written as the product of two or more prime numbers. The number 1 is neither prime nor composite. The **Fundamental Theorem of Arithmetic** states that every positive integer greater than 1 can be written as the product of prime numbers. For instance, the **prime factorization** of 24 is

$$24 = 2 \cdot 2 \cdot 2 \cdot 3.$$

P.1 Exercises

See *CalcChat.com* for tutorial help and worked-out solutions to odd-numbered exercises.
For instructions on how to use a graphing utility, see Appendix A.

Vocabulary and Concept Check

In Exercises 1–5, fill in the blank(s).

1. A real number is _______ when it can be written as the ratio $\frac{p}{q}$ of two integers, where $q \neq 0$.

2. _______ numbers have infinite nonrepeating decimal representations.

3. A _______ number is an integer with exactly two positive factors: itself and 1.

4. An algebraic expression is a combination of letters called _______ and real numbers called _______ .

5. The _______ of an algebraic expression are those parts separated by addition.

6. Is $|5 - 2| = |2 - 5|$?

In Exercises 7–12, match each property with its name.

7. Commutative Property of Addition — (a) $a \cdot 1 = a$

8. Associative Property of Multiplication — (b) $a(b + c) = ab + ac$

9. Additive Inverse Property — (c) $a + b = b + a$

10. Distributive Property — (d) $(ab)c = a(bc)$

11. Associative Property of Addition — (e) $a + (-a) = 0$

12. Multiplicative Identity Property — (f) $(a + b) + c = a + (b + c)$

Procedures and Problem Solving

Identifying Subsets of Real Numbers **In Exercises 13–18, determine which numbers are (a) natural numbers, (b) whole numbers, (c) integers, (d) rational numbers, and (e) irrational numbers.**

13. $\left\{-9, -\frac{7}{2}, 5, \frac{2}{3}, \sqrt{2}, 0, 1, -4, -1\right\}$

14. $\left\{\sqrt{5}, -7, -\frac{7}{3}, 0, 3.12, \frac{5}{4}, -2, -8, 3\right\}$

15. $\{2.01, 0.666\ldots, -13, 0.010110111\ldots, 1, -10, 20\}$

16. $\left\{2.3030030003\ldots, 0.7575, -4.63, \sqrt{10}, -2, 0.3, 8\right\}$

17. $\left\{-\pi, -\frac{1}{3}, \frac{6}{3}, \frac{1}{2}\sqrt{2}, -7.5, -2, 3, -3\right\}$

18. $\left\{25, -17, -\frac{12}{5}, \sqrt{9}, 3.12, \frac{1}{2}\pi, 6, -4, 18\right\}$

Finding the Decimal Form of a Rational Number **In Exercises 19–24, use a calculator to find the decimal form of the rational number. If it is a nonterminating decimal, write the repeating pattern.**

19. $\frac{5}{16}$ **20.** $\frac{17}{4}$

21. $\frac{41}{333}$ **22.** $\frac{3}{7}$

23. $-\frac{100}{11}$ **24.** $-\frac{218}{33}$

Writing a Decimal as a Fraction **In Exercises 25–28, use a calculator to rewrite the rational number as the ratio of two integers.**

25. 6.4 **26.** -7.5

27. -12.3 **28.** 1.87

Writing an Inequality **In Exercises 29 and 30, approximate the numbers and place the correct inequality symbol (< or >) between them.**

29.
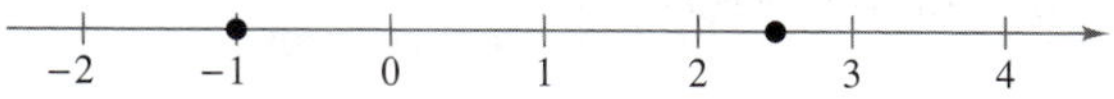

30.
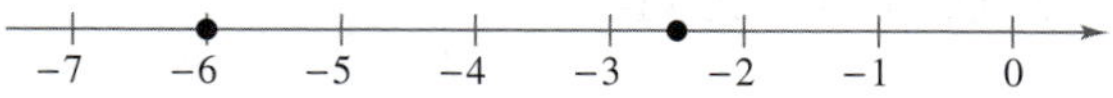

Plotting Real Numbers **In Exercises 31–36, plot the two real numbers on the real number line. Then place the correct inequality symbol (< or >) between them.**

31. $-4, 2$ **32.** $-3.5, 1$

33. $\frac{3}{2}, -\frac{7}{2}$ **34.** $-\frac{8}{7}, -\frac{3}{7}$

35. $-\frac{3}{4}, -\frac{5}{8}$ **36.** $\frac{5}{6}, \frac{2}{3}$

Interpreting Inequalities **In Exercises 37–44, (a) verbally describe the subset of real numbers represented by the inequality, (b) sketch the subset on the real number line, and (c) state whether the interval is bounded or unbounded.**

37. $x \leq 5$ **38.** $x > -3$

39. $x < 0$ **40.** $x \geq 4$

41. $-2 < x < 2$

42. $0 \leq x \leq 5$

43. $-1 \leq x < 0$

44. $-9 < x \leq -6$

Using Inequalities to Represent Intervals **In Exercises 45–52, use inequality and interval notation to represent the set.**

45. x is negative.
46. y is nonnegative.
47. z is at least 10.
48. y is no more than 25.
49. t is at least 9 and at most 24.
50. k is less than 3 but no less than -1.
51. The number of grams of radioactive material m is greater than 0 and at most 5.
52. The inflation rate r is at least 2.5% and at most 5%.

Using Interval Notation **In Exercises 53–56, use interval notation to describe the graph.**

53.
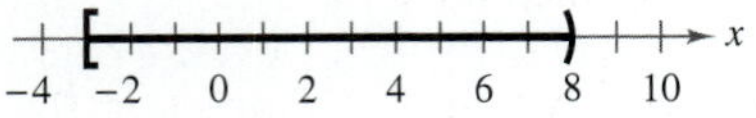

54.
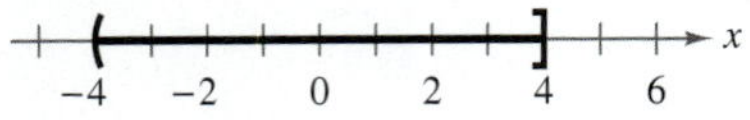

55.
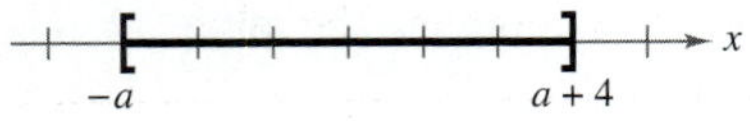

56.
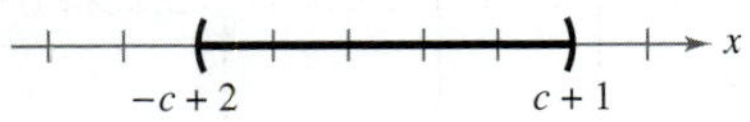

Interpreting Intervals **In Exercises 57–60, give a verbal description of the interval.**

57. $(-6, \infty)$
58. $(-\infty, 4]$
59. $(-\infty, 2]$
60. $(1, \infty)$

Evaluating the Absolute Value of a Number **In Exercises 61–66, evaluate the expression.**

61. $|-10|$
62. $|0|$
63. $-3 - |-3|$
64. $|-1| - |-2|$
65. $\dfrac{-5}{|-5|}$
66. $-3|-3|$

Evaluating an Absolute Value Expression **In Exercises 67 and 68, evaluate the expression for the given intervals of x.**

	Expression	*Intervals*	
67.	$\dfrac{\lvert x+1\rvert}{x+1}$	(a) $x > -1$	(b) $x < -1$
68.	$\dfrac{\lvert x-2\rvert}{x-2}$	(a) $x > 2$	(b) $x < 2$

Evaluating an Expression **In Exercises 69–72, evaluate the expression for the given values of x and y. Then use a graphing utility to verify your result.**

69. $|y - 4x|$ for $x = 2$ and $y = -3$
70. $|x| - 2|y|$ for $x = -2$ and $y = -1$
71. $\left|\dfrac{3x + 2y}{|x|}\right|$ for $x = 4$ and $y = 1$
72. $\left|\dfrac{3|x| - 2y}{2x + y}\right|$ for $x = -2$ and $y = -4$

Comparing Real Numbers **In Exercises 73–78, place the correct symbol (<, >, or =) between the pair of real numbers.**

73 $|-3| \;\square\; -|-3|$
74. $|-4| \;\square\; |4|$
75. $-5 \;\square\; -|5|$
76. $-|-6| \;\square\; |-6|$
77. $-|-1| \;\square\; -(-1)$
78. $-(-2) \;\square\; -2$

Finding the Distance Between Two Real Numbers **In Exercises 79–84, find the distance between a and b.**

79. $a = 126, b = 75$
80. $a = -126, b = -75$
81. $a = -\frac{5}{2}, b = \frac{9}{2}$
82. $a = \frac{1}{4}, b = \frac{11}{4}$
83. $a = \frac{16}{5}, b = \frac{112}{25}$
84. $a = \frac{7}{3}, b = -\frac{15}{8}$

Using Absolute Value Notation **In Exercises 85–88, use absolute value notation to describe the situation.**

85. The distance between x and 5 is no more than 3.
86. The distance between x and -10 is at least 6.
87. y is at least six units from 0.
88. y is at most three units from a.

Why you should learn it (p. 2) **In Exercises 89–94, use the bar graph, which shows the receipts of the federal government (in billions of dollars) for selected years from 1992 through 2012. In each exercise, you are given the expenditures of the federal government. Find the magnitude of the surplus or deficit for the year.** (Source: U.S. Office of Management and Budget)

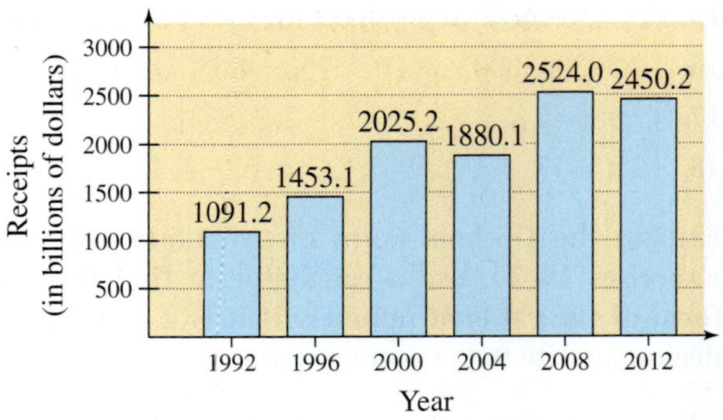

		Receipts	*Expenditures*	*|Receipts − Expenditures|*
89.	1992		\$1381.5 billion	
90.	1996		\$1560.5 billion	
91.	2000		\$1789.0 billion	
92.	2004		\$2292.8 billion	
93.	2008		\$2982.5 billion	
94.	2012		\$3537.1 billion	

Accounting **In Exercises 95–98, the accounting department of a company is checking to determine whether the actual expenses of a department differ from the budgeted expenses by more than \$500 or by more than 5%. Fill in the missing parts of the table and determine whether the actual expense passes the "budget variance test."**

	Budgeted Expense, b	*Actual Expense, a*	$\lvert a - b \rvert$	*0.05b*
95. Wages	\$112,700	\$113,356		
96. Utilities	\$9400	\$9772		
97. Taxes	\$37,600	\$37,335		
98. Insurance	\$25,800	\$25,263		

Identifying Terms and Coefficients **In Exercises 99–104, identify the terms and coefficients of the expression.**

99. $7x + 4$

100. $2x - 9$

101. $\sqrt{3}x^2 - 8x - 11$

102. $7\sqrt{5}x^2 + 3$

103. $4x^3 + \frac{x}{2} - 5$

104. $3x^4 + \frac{2x^3}{5}$

Evaluating an Algebraic Expression **In Exercises 105–108, evaluate the expression for each value of *x*. (If not possible, state the reason.)**

Expression	*Values*	
105. $2x - 5$	(a) $x = 4$	(b) $x = -\frac{1}{2}$
106. $4 - 3x$	(a) $x = 2$	(b) $x = -\frac{5}{6}$
107. $x^2 - 4$	(a) $x = 2$	(b) $x = -2$
108. $\frac{x^2}{x + 4}$	(a) $x = 1$	(b) $x = -4$

Identifying Rules of Algebra **In Exercises 109–114, identify the rule(s) of algebra illustrated by the statement.**

109. $2(x + 3) = 2x + 6$

110. $(z - 2) + 0 = z - 2$

111. $x + 9 = 9 + x$

112. $\frac{1}{h + 6}(h + 6) = 1, \quad h \neq -6$

113. $-y + (y + 10) = (-y + y) + 10 = 10$

114. $\frac{1}{7}(7 \cdot 12) = \left(\frac{1}{7} \cdot 7\right)12 = 1 \cdot 12 = 12$

Properties and Operations of Fractions **In Exercises 115–124, perform the operation(s). (Write fractional answers in simplest form.)**

115. $\frac{3}{16} + \frac{5}{16}$

116. $\frac{6}{7} - \frac{5}{7}$

117. $\frac{5}{8} - \frac{5}{12} + \frac{1}{6}$

118. $\frac{10}{11} + \frac{6}{33} - \frac{13}{66}$

119. $\frac{x}{6} + \frac{4x}{12}$

120. $\frac{2x}{5} + \frac{x}{2}$

121. $\frac{12}{x} \div \frac{1}{8}$

122. $\frac{11}{x} \div \frac{3}{4}$

123. $\left(\frac{2}{5} \div 4\right) - \left(4 \cdot \frac{3}{8}\right)$

124. $\left(\frac{3}{5} \div 3\right) - \left(6 \cdot \frac{4}{8}\right)$

125. Geography While traveling on the Pennsylvania Turnpike, you pass milepost 57 near Pittsburgh, then milepost 236 near Gettysburg. How many miles do you travel between these two mileposts?

126. Meteorology The temperature in Bismarck, North Dakota, was 60°F at noon, then 23°F at midnight. What was the change in temperature over the 12-hour period?

Conclusions

True or False? **In Exercises 127–129, determine whether the statement is true or false. Justify your answer.**

127. Every nonnegative number is positive.

128. If $a > 0$ and $b < 0$, then $ab > 0$.

129. If $a > b$, then $\frac{1}{a} > \frac{1}{b}$, where $a \neq 0$ and $b \neq 0$.

130. Think About It

(a) Use a calculator to complete the table.

n	1	0.5	0.01	0.0001	0.000001
$5/n$					

(b) Use the result from part (a) to make a conjecture about the value of $5/n$ as n approaches 0.

131. Determining the Sign of an Expression The real numbers *A*, *B*, and *C* are shown on the number line. Determine the sign of each expression.

(a) $-A$ (b) $-C$ (c) $B - A$ (d) $A - C$

132. HOW DO YOU SEE IT? Match each description with its graph. Which types of real numbers shown in Figure P.1 on page 2 may be included in a range of prices? a range of lengths? Explain.

(i)
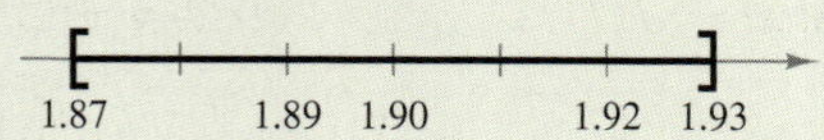

(ii)
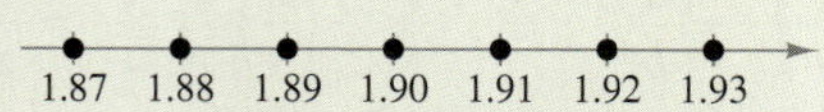

(a) The price of an item is within \$0.03 of \$1.90.

(b) The distance between the prongs of an electric plug may not differ from 1.9 centimeters by more than 0.03 centimeter.

133. Think About It Describe the real number values of u and v for which $|u + v|$ is *greater than, less than,* and *equal to* $|u| + |v|$.

134. Writing For what real numbers a is $|a| = -a$? Explain.

P.2 Exponents and Radicals

Integer Exponents

Repeated *multiplication* can be written in **exponential form.**

Repeated Multiplication	*Exponential Form*
$a \cdot a \cdot a \cdot a \cdot a$	a^5
$(-4)(-4)(-4)$	$(-4)^3$
$\frac{3}{5} \cdot \frac{3}{5} \cdot \frac{3}{5} \cdot \frac{3}{5} \cdot \frac{3}{5} \cdot \frac{3}{5}$	$\left(\frac{3}{5}\right)^6$
$(2x)(2x)(2x)(2x)$	$(2x)^4$

In general, if a is a real number, variable, or algebraic expression and n is a positive integer, then

$$a^n = \underbrace{a \cdot a \cdot a \cdots a}_{n \text{ factors}}$$

where n is the **exponent** and a is the **base.** The expression a^n is read "a to the nth **power.**" An exponent can be negative as well. Property 3 below shows how to use a negative exponent.

What you should learn

- Use properties of exponents.
- Use scientific notation to represent real numbers.
- Use properties of radicals.
- Simplify and combine radicals.
- Rationalize denominators and numerators.
- Use properties of rational exponents.

Why you should learn it

Real numbers and algebraic expressions are often written with exponents and radicals. For instance, in Exercise 139 on page 22, you will use an expression involving rational exponents to find the time required for a funnel to empty.

Properties of Exponents

Let a and b be real numbers, variables, or algebraic expressions, and let m and n be integers. (All denominators and bases are nonzero.)

Property	*Example*
1. $a^m a^n = a^{m+n}$	$3^2 \cdot 3^4 = 3^{2+4} = 3^6 = 729$
2. $\frac{a^m}{a^n} = a^{m-n}$	$\frac{x^7}{x^4} = x^{7-4} = x^3$
3. $a^{-n} = \frac{1}{a^n} = \left(\frac{1}{a}\right)^n$	$y^{-4} = \frac{1}{y^4} = \left(\frac{1}{y}\right)^4$
4. $a^0 = 1, \quad a \neq 0$	$(x^2 + 1)^0 = 1$
5. $(ab)^m = a^m b^m$	$(5x)^3 = 5^3 x^3 = 125x^3$
6. $(a^m)^n = a^{mn}$	$(y^3)^{-4} = y^{3(-4)} = y^{-12} = \frac{1}{y^{12}}$
7. $\left(\frac{a}{b}\right)^m = \frac{a^m}{b^m}$	$\left(\frac{2}{x}\right)^3 = \frac{2^3}{x^3} = \frac{8}{x^3}$
8. $\lvert a^2 \rvert = \lvert a \rvert^2 = a^2$	$\lvert(-2)^2\rvert = \lvert -2 \rvert^2 = 2^2 = 4$

It is important to recognize the difference between expressions such as $(-2)^4$ and -2^4. In $(-2)^4$, the parentheses indicate that the exponent applies to the negative sign as well as to the 2, but in $-2^4 = -(2^4)$, the exponent applies only to the 2. So,

$$(-2)^4 = 16, \quad \text{whereas} \quad -2^4 = -16.$$

It is also important to know when to use parentheses when evaluating exponential expressions using a graphing calculator. Figure P.9 shows that a graphing calculator follows the order of operations.

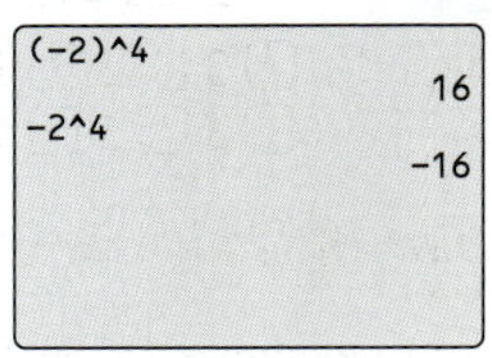

Figure P.9

The properties of exponents listed on the preceding page apply to *all* integers m and n, not just positive integers. For instance, by Property 2, you can write

$$\frac{3^4}{3^{-5}} = 3^{4-(-5)} = 3^{4+5} = 3^9.$$

EXAMPLE 1 Using Properties of Exponents

a. $(-3ab^4)(4ab^{-3}) = -12(a)(a)(b^4)(b^{-3}) = -12a^2b$

b. $(2xy^2)^3 = 2^3(x)^3(y^2)^3 = 8x^3y^6$

c. $3a(-4a^2)^0 = 3a(1) = 3a, \quad a \neq 0$

Checkpoint Audio-video solution in English & Spanish at LarsonPrecalculus.com.

Use the properties of exponents to simplify each expression.

a. $(2x^{-2}y^3)(-x^4y)$ **b.** $(4a^2b^3)^0$ **c.** $(-5z)^3(z^2)$

EXAMPLE 2 Rewriting with Positive Exponents

a. $x^{-1} = \dfrac{1}{x}$ Property 3

b. $\dfrac{1}{3x^{-2}} = \dfrac{1(x^2)}{3} = \dfrac{x^2}{3}$ The exponent -2 does not apply to 3.

c. $\dfrac{1}{(3x)^{-2}} = (3x)^2 = 9x^2$ The exponent -2 does apply to 3.

d. $\dfrac{12a^3b^{-4}}{4a^{-2}b} = \dfrac{12a^3 \cdot a^2}{4b \cdot b^4} = \dfrac{3a^5}{b^5}$ Properties 3 and 1

e. $\left(\dfrac{3x^2}{y}\right)^{-2} = \dfrac{3^{-2}(x^2)^{-2}}{y^{-2}}$ Properties 5 and 7

$= \dfrac{3^{-2}x^{-4}}{y^{-2}}$ Property 6

$= \dfrac{y^2}{3^2x^4}$ Property 3

$= \dfrac{y^2}{9x^4}$ Simplify.

Checkpoint Audio-video solution in English & Spanish at LarsonPrecalculus.com.

Rewrite each expression with positive exponents.

a. $2a^{-2}$ **b.** $\dfrac{3a^{-3}b^4}{15ab^{-1}}$ **c.** $\left(\dfrac{x}{10}\right)^{-1}$ **d.** $(-2x^2)^3(4x^3)^{-1}$

EXAMPLE 3 Calculators and Exponents

Expression	*Graphing Calculator Keystrokes*	*Display*
a. $3^{-2} + 4^{-1}$	3 [^] [(-)] 2 [+] 4 [^] [(-)] 1 [ENTER]	.3611111111
b. $\dfrac{3^5 + 1}{3^5 - 1}$	[(] 3 [^] 5 [+] 1 [)] [÷] [(] 3 [^] 5 [−] 1 [)] [ENTER]	1.008264463

Checkpoint Audio-video solution in English & Spanish at LarsonPrecalculus.com.

Use a calculator to evaluate $13^4 + 5^{-2}$.

What's Wrong?

To evaluate $(-5)^2$, you enter -5^2 on a graphing utility (see figure). You conclude that $(-5)^2 = -25$. What's wrong?

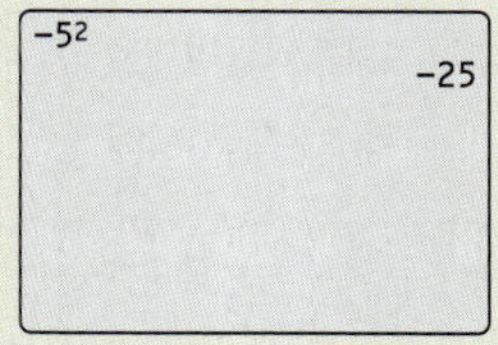

Remark

Rarely in algebra is there only one way to solve a problem. Do not be concerned when the steps you use to solve a problem are not exactly the same as the steps presented in this text. The important thing is to use steps that you understand *and*, of course, that are justified by the rules of algebra. For instance, you might prefer the following steps for Example 2(e).

$$\left(\frac{3x^2}{y}\right)^{-2} = \left(\frac{y}{3x^2}\right)^2 = \frac{y^2}{9x^4}$$

Scientific Notation

Exponents provide an efficient way of writing and computing with very large (or very small) numbers. For instance, there are about 366 billion billion gallons of water on Earth—that is, 366 followed by 18 zeros.

366,000,000,000,000,000,000

It is convenient to write such numbers in **scientific notation.** This notation has the form $\pm c \times 10^n$, where $1 \le c < 10$ and n is an integer. So, the number of gallons of water on Earth can be written in scientific notation as

$$3.66 \times 100,000,000,000,000,000,000 = 3.66 \times 10^{20}.$$

The *positive* exponent 20 indicates that the number is *large* (10 or more) and that the decimal point has been moved 20 places. A *negative* exponent indicates that the number is *small* (less than 1). For instance, the mass (in grams) of one electron is approximately

$$9.1 \times 10^{-28} = 0.00000000000000000000000000091.$$

28 decimal places

Technology Tip

The graphing calculator keystrokes given in this text may not be the same as the keystrokes for your graphing calculator. Be sure you are familiar with the use of the keys on your calculator.

EXAMPLE 4 Scientific Notation

a. $0.0000782 = 7.82 \times 10^{-5}$

b. $836,100,000 = 8.361 \times 10^8$

✓ ***Checkpoint*** *Audio-video solution in English & Spanish at LarsonPrecalculus.com.*

Write 45,850 in scientific notation.

EXAMPLE 5 Decimal Notation

a. $-9.36 \times 10^{-6} = -0.00000936$

b. $1.345 \times 10^2 = 134.5$

✓ ***Checkpoint*** *Audio-video solution in English & Spanish at LarsonPrecalculus.com.*

Write -2.718×10^{-3} in decimal notation.

EXAMPLE 6 Using Scientific Notation with a Calculator

Use a calculator to evaluate $65,000 \times 3,400,000,000$.

Solution

Because $65,000 = 6.5 \times 10^4$ and $3,400,000,000 = 3.4 \times 10^9$, you can multiply the two numbers using the following graphing calculator keystrokes.

6.5 [EE] 4 [×] 3.4 [EE] 9 [ENTER]

After entering these keystrokes, the calculator should display [2.21 E 14]. So, the product of the two numbers is

$$(6.5 \times 10^4)(3.4 \times 10^9) = 2.21 \times 10^{14} = 221,000,000,000,000.$$

✓ ***Checkpoint*** *Audio-video solution in English & Spanish at LarsonPrecalculus.com.*

Use a calculator to evaluate

$$\frac{184,000,000 - 51,000,000}{0.0048}.$$

Technology Tip

Most calculators automatically switch to scientific notation when they are showing large or small numbers that exceed the display range. Try evaluating $86,500,000 \times 6000$. A calculator that follows standard conventions should display

[5.19 11] or [5.19 E 11]

which is 5.19×10^{11}.

Radicals and Their Properties

A **square root** of a number is one of its two equal factors. For instance, 5 is a square root of 25 because 5 is one of the two equal factors of $25 = 5 \cdot 5$. In a similar way, a **cube root** of a number is one of its three equal factors, as in $125 = 5^3$.

Definition of the nth Root of a Number

Let a and b be real numbers and let $n \geq 2$ be a positive integer. If

$$a = b^n$$

then b is an ***n*th root of *a*.** If $n = 2$, then the root is a **square root.** If $n = 3$, then the root is a **cube root.**

Some numbers have more than one nth root. For example, both 5 and -5 are square roots of 25. The *principal square root* of 25, written as $\sqrt{25}$, is the positive root, 5. The **principal *n*th root** of a number is defined as follows.

Principal nth Root of a Number

Let a be a real number that has at least one nth root. The **principal *n*th root of *a*** is the nth root that has the same sign as a. It is denoted by a **radical symbol**

$$\sqrt[n]{a}. \qquad \text{Principal } n\text{th root}$$

The positive integer n is the **index** of the radical, and the number a is the **radicand.** When $n = 2$, omit the index and write $\sqrt{a}$ rather than $\sqrt[2]{a}$. (The plural of index is *indices.*)

A common misunderstanding when taking square roots of real numbers is that the square root sign implies both negative and positive roots. This is not correct. The square root sign implies only a positive root. When a negative root is needed, you must use the negative sign with the square root sign.

Incorrect: ~~$\sqrt{4} = \pm 2$~~

Correct: $\sqrt{4} = 2$

Correct: $-\sqrt{4} = -2$

EXAMPLE 7 Evaluating Expressions Involving Radicals

a. $\sqrt{36} = 6$ because $6^2 = 36$.

b. $-\sqrt{36} = -6$ because $-(\sqrt{36}) = -(\sqrt{6^2}) = -(6) = -6$.

c. $\sqrt[3]{\frac{125}{64}} = \frac{5}{4}$ because $\left(\frac{5}{4}\right)^3 = \frac{5^3}{4^3} = \frac{125}{64}$.

d. $\sqrt[5]{-32} = -2$ because $(-2)^5 = -32$.

e. $\sqrt[4]{-81}$ is not a real number because there is no real number that can be raised to the fourth power to produce -81.

✓ Checkpoint *Audio-video solution in English & Spanish at LarsonPrecalculus.com.*

Evaluate each expression (if possible).

a. $-\sqrt{144}$ **b.** $\sqrt{-144}$

c. $\sqrt{\frac{25}{64}}$ **d.** $-\sqrt[3]{\frac{8}{27}}$

Here are some generalizations about the nth roots of real numbers.

Generalizations About nth Roots of Real Numbers

Real number a	Integer n	Root(s) of a	Example
$a > 0$	$n > 0$, n is even.	$\sqrt[n]{a}, -\sqrt[n]{a}$	$\sqrt[4]{81} = 3, -\sqrt[4]{81} = -3$
$a > 0$ or $a < 0$	n is odd.	$\sqrt[n]{a}$	$\sqrt[3]{-8} = -2$
$a < 0$	n is even.	No real roots	$\sqrt{-4}$ is not a real number.
$a = 0$	n is even or odd.	$\sqrt[n]{0} = 0$	$\sqrt[5]{0} = 0$

Integers such as 1, 4, 9, 16, 25, and 36 are called **perfect squares** because they have integer square roots. Similarly, integers such as 1, 8, 27, 64, and 125 are called **perfect cubes** because they have integer cube roots.

Properties of Radicals

Let a and b be real numbers, variables, or algebraic expressions such that the indicated roots are real numbers, and let m and n be positive integers.

Property	*Example*
1. $\sqrt[n]{a^m} = \left(\sqrt[n]{a}\right)^m$	$\sqrt[3]{8^2} = \left(\sqrt[3]{8}\right)^2 = (2)^2 = 4$
2. $\sqrt[n]{a} \cdot \sqrt[n]{b} = \sqrt[n]{ab}$	$\sqrt{5} \cdot \sqrt{7} = \sqrt{5 \cdot 7} = \sqrt{35}$
3. $\dfrac{\sqrt[n]{a}}{\sqrt[n]{b}} = \sqrt[n]{\dfrac{a}{b}},\ b \neq 0$	$\dfrac{\sqrt[4]{27}}{\sqrt[4]{9}} = \sqrt[4]{\dfrac{27}{9}} = \sqrt[4]{3}$
4. $\sqrt[m]{\sqrt[n]{a}} = \sqrt[mn]{a}$	$\sqrt[3]{\sqrt{10}} = \sqrt[6]{10}$
5. $\left(\sqrt[n]{a}\right)^n = a$	$\left(\sqrt{3}\right)^2 = 3$
6. For n even, $\sqrt[n]{a^n} = \lvert a \rvert$.	$\sqrt{(-12)^2} = \lvert -12 \rvert = 12$
For n odd, $\sqrt[n]{a^n} = a$.	$\sqrt[3]{(-12)^3} = -12$

A common special case of Property 6 is

$$\sqrt{a^2} = \lvert a \rvert.$$

EXAMPLE 8 Using Properties of Radicals

Use the properties of radicals to simplify each expression.

a. $\sqrt{8} \cdot \sqrt{2}$ **b.** $\left(\sqrt[3]{5}\right)^3$ **c.** $\sqrt[3]{x^3}$ **d.** $\sqrt[6]{y^6}$

Solution

a. $\sqrt{8} \cdot \sqrt{2} = \sqrt{8 \cdot 2} = \sqrt{16} = 4$

b. $\left(\sqrt[3]{5}\right)^3 = 5$

c. $\sqrt[3]{x^3} = x$

d. $\sqrt[6]{y^6} = \lvert y \rvert$

✓ ***Checkpoint*** *Audio-video solution in English & Spanish at LarsonPrecalculus.com.*

Use the properties of radicals to simplify each expression.

a. $\dfrac{\sqrt{125}}{\sqrt{5}}$ **b.** $\sqrt[3]{125^2}$ **c.** $\sqrt[3]{x^2} \cdot \sqrt[3]{x}$ **d.** $\sqrt{\sqrt{x}}$

Technology Tip

There are three methods of evaluating radicals on most graphing calculators. For square roots, you can use the *square root key* [√]. For cube roots, you can use the *cube root key* [∛] (or menu choice). For other roots, you can use the *xth root key* [ˣ√] (or menu choice). For example, the screen below shows you how to evaluate $\sqrt{36}$, $\sqrt[3]{-8}$, and $\sqrt[5]{32}$ using one of the three methods described.

Simplifying Radicals

An expression involving radicals is in **simplest form** when the following conditions are satisfied.

1. All possible factors have been removed from the radical.
2. All fractions have radical-free denominators (accomplished by a process called *rationalizing the denominator*).
3. The index of the radical is reduced.

To simplify a radical, factor the radicand into factors whose exponents are multiples of the index. The roots of these factors are written outside the radical, and the "leftover" factors make up the new radicand.

EXAMPLE 9 Simplifying Radicals

Perfect cube — Leftover factor

a. $\sqrt[3]{24} = \sqrt[3]{8 \cdot 3} = \sqrt[3]{2^3 \cdot 3} = 2\sqrt[3]{3}$

Perfect 4th power — Leftover factor

b. $\sqrt[4]{48} = \sqrt[4]{16 \cdot 3} = \sqrt[4]{2^4 \cdot 3} = 2\sqrt[4]{3}$

c. $\sqrt{75x^3} = \sqrt{25x^2 \cdot 3x} = \sqrt{(5x)^2 \cdot 3x} = 5x\sqrt{3x}$

d. $\sqrt[3]{24a^4} = \sqrt[3]{8a^3 \cdot 3a} = \sqrt[3]{(2a)^3 \cdot 3a} = 2a\sqrt[3]{3a}$

e. $\sqrt[4]{(5x)^4} = |5x| = 5|x|$

Remark

When you simplify a radical, it is important that both expressions are defined for the same values of the variable. For instance, in Example 9(c), $\sqrt{75x^3}$ and $5x\sqrt{3x}$ are both defined only for nonnegative values of x. Similarly, in Example 9(e), $\sqrt[4]{(5x)^4}$ and $5|x|$ are both defined for all real values of x.

✓ **Checkpoint** Audio-video solution in English & Spanish at LarsonPrecalculus.com.

Simplify each radical expression.

a. $\sqrt{32}$ **b.** $\sqrt[3]{250}$ **c.** $\sqrt{24a^5}$ **d.** $\sqrt[3]{-135x^3}$

Radical expressions can be combined (added or subtracted) when they are **like radicals**—that is, when they have the same index and radicand. For instance, $\sqrt{2}$, $3\sqrt{2}$, and $\frac{1}{2}\sqrt{2}$ are like radicals, but $\sqrt{3}$ and $\sqrt{2}$ are unlike radicals. To determine whether two radicals can be combined, you should first simplify each radical.

EXAMPLE 10 Combining Radicals

a. $2\sqrt{48} - 3\sqrt{27} = 2\sqrt{16 \cdot 3} - 3\sqrt{9 \cdot 3}$ Find square factors.

$= 8\sqrt{3} - 9\sqrt{3}$ Find square roots and multiply by coefficients.

$= (8 - 9)\sqrt{3}$ Combine like radicals.

$= -\sqrt{3}$ Simplify.

b. $\sqrt[3]{16x} - \sqrt[3]{54x^4} = \sqrt[3]{8 \cdot 2x} - \sqrt[3]{27 \cdot x^3 \cdot 2x}$ Find cube factors.

$= 2\sqrt[3]{2x} - 3x\sqrt[3]{2x}$ Find cube roots.

$= (2 - 3x)\sqrt[3]{2x}$ Combine like radicals.

✓ **Checkpoint** Audio-video solution in English & Spanish at LarsonPrecalculus.com.

Simplify each radical expression.

a. $3\sqrt{8} + \sqrt{18}$

b. $\sqrt[3]{81x^5} - \sqrt[3]{24x^2}$

Rationalizing Denominators and Numerators

To rationalize a denominator or numerator of the form $a - b\sqrt{m}$ or $a + b\sqrt{m}$, multiply both numerator and denominator by a **conjugate:** $a + b\sqrt{m}$ and $a - b\sqrt{m}$ are conjugates of each other. If $a = 0$, then the rationalizing factor for $\sqrt{m}$ is itself, $\sqrt{m}$.

EXAMPLE 11 Rationalizing Denominators

a. $\dfrac{5}{2\sqrt{3}} = \dfrac{5}{2\sqrt{3}} \cdot \dfrac{\sqrt{3}}{\sqrt{3}}$ $\quad$ $\sqrt{3}$ is rationalizing factor.

$= \dfrac{5\sqrt{3}}{2(3)}$ $\quad$ Multiply.

$= \dfrac{5\sqrt{3}}{6}$ $\quad$ Simplify.

b. $\dfrac{2}{\sqrt[3]{5}} = \dfrac{2}{\sqrt[3]{5}} \cdot \dfrac{\sqrt[3]{5^2}}{\sqrt[3]{5^2}}$ $\quad$ $\sqrt[3]{5^2}$ is rationalizing factor.

$= \dfrac{2\sqrt[3]{5^2}}{\sqrt[3]{5^3}}$ $\quad$ Multiply.

$= \dfrac{2\sqrt[3]{25}}{5}$ $\quad$ Simplify.

c. $\dfrac{2}{3 + \sqrt{7}} = \dfrac{2}{3 + \sqrt{7}} \cdot \dfrac{3 - \sqrt{7}}{3 - \sqrt{7}}$ $\quad$ Multiply numerator and denominator by conjugate of denominator.

$= \dfrac{2(3 - \sqrt{7})}{(3)^2 - (\sqrt{7})^2}$ $\quad$ Find products. In denominator, $(a + b)(a - b) = a^2 - ab + ab - b^2 = a^2 - b^2$.

$= \dfrac{2(3 - \sqrt{7})}{2}$ $\quad$ Simplify denominator.

$= 3 - \sqrt{7}$ $\quad$ Divide out common factors.

Remark

Notice in Example 11(b) that the numerator and denominator are multiplied by $\sqrt[3]{5^2}$ to produce a perfect cube radicand.

✓ Checkpoint *Audio-video solution in English & Spanish at LarsonPrecalculus.com.*

Rationalize the denominator of each expression.

a. $\dfrac{5}{3\sqrt{2}}$ $\quad$ **b.** $\dfrac{1}{\sqrt[3]{25}}$ $\quad$ **c.** $\dfrac{8}{\sqrt{6} - \sqrt{2}}$

In calculus, sometimes it is necessary to rationalize the numerator of an expression.

EXAMPLE 12 Rationalizing a Numerator

$\dfrac{\sqrt{5} - \sqrt{7}}{9} = \dfrac{\sqrt{5} - \sqrt{7}}{9} \cdot \dfrac{\sqrt{5} + \sqrt{7}}{\sqrt{5} + \sqrt{7}}$ $\quad$ Multiply numerator and denominator by conjugate of numerator.

$= \dfrac{(\sqrt{5})^2 - (\sqrt{7})^2}{9(\sqrt{5} + \sqrt{7})}$ $\quad$ Find products. In numerator, $(a + b)(a - b) = a^2 - ab + ab - b^2 = a^2 - b^2$.

$= \dfrac{-2}{9(\sqrt{5} + \sqrt{7})}$ $\quad$ Simplify numerator.

Remark

Do not confuse the expression $\sqrt{5} + \sqrt{7}$ with the expression $\sqrt{5 + 7}$. In general, $\sqrt{x + y}$ does not equal $\sqrt{x} + \sqrt{y}$. Similarly, $\sqrt{x^2 + y^2}$ does not equal $x + y$.

✓ Checkpoint *Audio-video solution in English & Spanish at LarsonPrecalculus.com.*

Rationalize the numerator of $\dfrac{2 - \sqrt{2}}{3}$.

The symbol ∫ indicates an example or exercise that highlights algebraic techniques specifically used in calculus.

Rational Exponents

Definition of Rational Exponents

If a is a real number and n is a positive integer such that the principal nth root of a exists, then $a^{1/n}$ is defined as

$$a^{1/n} = \sqrt[n]{a}, \text{ where } 1/n \text{ is the } \textbf{rational exponent} \text{ of } a.$$

Moreover, if m is a positive integer that has no common factor with n, then

$$a^{m/n} = (a^{1/n})^m = \left(\sqrt[n]{a}\right)^m \quad \text{and} \quad a^{m/n} = (a^m)^{1/n} = \sqrt[n]{a^m}.$$

The numerator of a rational exponent denotes the *power* to which the base is raised, and the denominator denotes the *index* or the *root* to be taken.

Power

Index

$$b^{m/n} = \left(\sqrt[n]{b}\right)^m = \sqrt[n]{b^m}$$

When you are working with rational exponents, the properties of integer exponents still apply. For instance, $2^{1/2}2^{1/3} = 2^{(1/2)+(1/3)} = 2^{5/6}$.

Remark

Be sure you understand that the expression $b^{m/n}$ is not defined unless $\sqrt[n]{b}$ is a real number. This restriction produces some unusual results. For instance, the number $(-8)^{1/3}$ is defined because $\sqrt[3]{-8} = -2$, but the number $(-8)^{2/6}$ is undefined because $\sqrt[6]{-8}$ is not a real number.

EXAMPLE 13 Writing Exponential and Radical Forms

See LarsonPrecalculus.com for an interactive version of this type of example.

a. $\sqrt{3} = 3^{1/2}$ — Change from radical to exponential form.

b. $\sqrt{(3xy)^5} = \sqrt[2]{(3xy)^5} = (3xy)^{(5/2)}$ — Change from radical to exponential form.

c. $(x^2 + y^2)^{3/2} = \left(\sqrt{x^2 + y^2}\right)^3 = \sqrt{(x^2 + y^2)^3}$ — Change from exponential to radical form.

d. $2y^{3/4}z^{1/4} = 2(y^3z)^{1/4} = 2\sqrt[4]{y^3z}$ — Change from exponential to radical form.

e. $a^{-3/2} = \dfrac{1}{a^{3/2}} = \dfrac{1}{\sqrt{a^3}}$ — Change from exponential to radical form.

✔ ***Checkpoint*** *Audio-video solution in English & Spanish at LarsonPrecalculus.com.*

Write the expressions (a) $\sqrt[3]{27}$ and (b) $\sqrt{x^3y^5z}$ in exponential form.

Write the expressions (c) $(x^2 - 7)^{-1/2}$ and (d) $-3b^{1/3}c^{2/3}$ in radical form.

EXAMPLE 14 Simplifying with Rational Exponents

a. $(-32)^{-4/5} = \left(\sqrt[5]{-32}\right)^{-4} = (-2)^{-4} = \dfrac{1}{(-2)^4} = \dfrac{1}{16}$

b. $(-5x^{5/3})(3x^{-3/4}) = -15x^{(5/3)-(3/4)} = -15x^{11/12}, \quad x \neq 0$

c. $\sqrt[3]{\sqrt{125}} = \sqrt[6]{125} = \sqrt[6]{(5)^3} = 5^{3/6} = 5^{1/2} = \sqrt{5}$

d. $(2x - 1)^{4/3}(2x - 1)^{-1/3} = (2x - 1)^{(4/3)-(1/3)} = 2x - 1, \quad x \neq \dfrac{1}{2}$

Remark

The expression in Example 14(d) is not defined when $x = \frac{1}{2}$ because

$$\left(2 \cdot \tfrac{1}{2} - 1\right)^{-1/3} = (0)^{-1/3}$$

is not a real number.

✔ ***Checkpoint*** *Audio-video solution in English & Spanish at LarsonPrecalculus.com.*

Simplify each expression.

a. $(-125)^{-2/3}$ **b.** $(4x^2y^{3/2})(-3x^{-1/3}y^{-3/5})$

c. $\sqrt[3]{\sqrt[4]{27}}$ **d.** $(3x + 2)^{5/2}(3x + 2)^{-1/2}$

P.2 Exercises

See *CalcChat.com* for tutorial help and worked-out solutions to odd-numbered exercises. For instructions on how to use a graphing utility, see Appendix A.

Vocabulary and Concept Check

In Exercises 1–6, fill in the blank(s).

1. In the exponential form a^n, n is the _______ and a is the _______ .

2. One of the two equal factors of a number is called a _______ of the number.

3. The _______ of a number a is the nth root that has the same sign as a.

4. In the radical form $\sqrt[n]{a}$, the positive integer n is called the _______ of the radical and the number a is called the _______ .

5. Creating a radical-free denominator is known as _______ the denominator.

6. In the expression $b^{m/n}$, m denotes the _______ to which the base is raised and n denotes the _______ or root to be taken.

7. What is the conjugate of $2 + 3\sqrt{5}$?

8. When is an expression involving radicals in simplest form?

9. Is -10.678×10^3 written in scientific notation?

10. Is 64 a perfect square, a perfect cube, or both?

Procedures and Problem Solving

Evaluating an Expression **In Exercises 11–18, evaluate each expression.**

11. (a) $3 \cdot 3^3$ (b) $\dfrac{3^2}{3^4}$

12. (a) $\dfrac{5^3}{5^2}$ (b) $4^2 \cdot 4^2$

13. (a) $(4^3)^0$ (b) $\dfrac{6}{6^{-3}}$

14. (a) $24(-2)^{-5}$ (b) -7^0

15. (a) $(4 \cdot 3)^3$ (b) $(-3^2)^3$

16. (a) $(2^3 \cdot 3^2)^2$ (b) $\left(\frac{5}{8}\right)^2$

17. (a) $2^{-1} + 3^{-1}$ (b) $(3^{-1})^{-3}$

18. (a) $\dfrac{4 \cdot 3^{-2}}{2^{-2} \cdot 3^{-1}}$ (b) $3^{-1} + 2^{-2}$

Evaluating an Expression **In Exercises 19–24, evaluate the expression for the value of x.**

19. $2x^3, \quad x = -3$

20. $-3x^4, \quad x = 2$

21. $5(-x)^0, \quad x = 4$

22. $6x^0 - (6x)^0, \quad x = 7$

23. $7x^{-2}, \quad x = 2$

24. $20x^{-2} + x^{-1}, \quad x = -5$

Using Properties of Exponents **In Exercises 25–32, simplify each expression.**

25. (a) $x^3(x^2)$ (b) $x^4(3x^5)$

26. (a) $4z^4(-2z^3)$ (b) $-5x^3(4x^2)$

27. (a) $(3x)^2$ (b) $(4x^3)^0, x \neq 0$

28. (a) $6z^2(2z^5)^4$ (b) $(3x^5)^3(2x^7)^2$

29. (a) $\dfrac{7x^2}{x^3}$ (b) $\dfrac{12(x + y)^3}{9(x + y)}$

30. (a) $\dfrac{r^5}{r^9}$ (b) $\dfrac{3x^2y^4}{15(xy)^2}$

31. (a) $[(x^2y^{-2})^{-1}]^{-1}$ (b) $\left(\dfrac{a^{-2}}{b^{-2}}\right)\left(\dfrac{b}{a}\right)^3$

32. (a) $\left(\dfrac{4}{y}\right)^3\left(\dfrac{3}{y}\right)^4$ (b) $(5x^2z^6)^3(5x^2z^6)^{-3}$

Calculators and Exponents **In Exercises 33–38, use a calculator to evaluate the expression. (Round your answer to three decimal places.)**

33. $(-4)^3(5^2)$

34. $(8^{-4})(10^3)$

35. $\dfrac{3^6}{7^3}$

36. $\dfrac{4^5}{9^3}$

37. $\dfrac{4^3 - 1}{3^{-4}}$

38. $\dfrac{3^2 - 2}{4^2 - 3}$

Scientific Notation **In Exercises 39–52, write the number in scientific notation.**

39. 973.50

40. 28,022.2

41. 10,252.484

42. 525,252,118

43. -1110.25

44. $-5{,}222{,}145$

45. 0.0002485

46. 0.0000025

47. -0.0000025

48. -0.000125005

49. Land area of Earth: 57,300,000 square miles

50. Light year: 9,460,000,000,000 kilometers

51. Relative density of hydrogen: 0.0000899 gram per cubic centimeter

52. One micron (millionth of a meter): 0.000003281 foot

Decimal Notation **In Exercises 53–60, write the number in decimal notation.**

53. 1.08×10^4
54. -4.816×10^8
55. -7.65×10^{-7}
56. 5.098×10^{-10}
57. Mean SAT Mathematics score of college-bound seniors in 2013: 5.14×10^2 (Source: The College Board)
58. Core temperature of the Sun: 1.5×10^7 degrees Celsius
59. Width of a human hair: 9.0×10^{-5} meter
60. Charge of an electron: 1.6022×10^{-19} coulomb

Evaluating an Expression **In Exercises 61–66, evaluate the expression without using a calculator.**

61. $(2.0 \times 10^7)(3.0 \times 10^{-3})$
62. $(1.4 \times 10^5)(5.2 \times 10^{-2})$
63. $\dfrac{7.0 \times 10^5}{4.0 \times 10^{-3}}$
64. $\dfrac{3.0 \times 10^{-3}}{6.0 \times 10^2}$
65. $\sqrt{25 \times 10^8}$
66. $\sqrt[3]{8 \times 10^{15}}$

Using Scientific Notation with a Calculator **In Exercises 67–70, use a calculator to evaluate each expression. (Round your answer to three decimal places.)**

67. (a) $(9.3 \times 10^6)^3(6.1 \times 10^{-4})$
(b) $\dfrac{(2.414 \times 10^4)^6}{(1.68 \times 10^5)^5}$
68. (a) $750\left(1 + \dfrac{0.11}{365}\right)^{800}$
(b) $\dfrac{67{,}000{,}000 + 93{,}000{,}000}{0.0052}$
69. (a) $\sqrt{4.5 \times 10^9}$
(b) $(7.3 \times 10^4)^{1/5}$
70. (a) $(2.65 \times 10^{-4})^{1/3}$
(b) $\sqrt[4]{9.9 \times 10^6}$

Evaluating Expressions Involving Radicals **In Exercises 71–78, evaluate the expression (if possible) without using a calculator.**

71. $\sqrt{121}$
72. $\sqrt{-49}$
73. $-\sqrt[3]{-64}$
74. $\sqrt[3]{125}$
75. $\sqrt[4]{-625}$
76. $-\sqrt[7]{-128}$
77. $-\sqrt[6]{\dfrac{1}{729}}$
78. $\dfrac{\sqrt[5]{-243}}{9}$

Approximating the Value of an Expression **In Exercises 79–90, use a calculator to approximate the value of the expression. (Round your answer to three decimal places.)**

79. $\sqrt[3]{45^2}$
80. $\sqrt[5]{-27^3}$
81. $(6.1)^{-2.9}$
82. $(3.4)^{2.5}$
83. $\sqrt[4]{90} - (4.1^3)\sqrt{17}$
84. $(1.2^{-2})\sqrt{75} + 3\sqrt{8}$
85. $\dfrac{-5 + \sqrt{33}}{5}$
86. $\dfrac{\sqrt[3]{-68} + 4}{0.1}$
87. $\dfrac{3.14}{\pi} + \sqrt[3]{5}$
88. $\dfrac{\sqrt{10}}{2.5} - \pi^2$
89. $(2.8)^{-2} + 1.01 \times 10^6$
90. $2.12 \times 10^{-2} + \sqrt{15}$

Using Properties of Radicals **In Exercises 91–94, use the properties of radicals to simplify each expression.**

91. $(\sqrt[3]{20})^3$
92. $\sqrt[4]{(-3x)^4}$
93. $\sqrt{12} \cdot \sqrt{3}$
94. $\dfrac{\sqrt[3]{40x^5}}{\sqrt[3]{5x^2}}$

Simplifying a Radical Expression **In Exercises 95–102, simplify each expression.**

95. (a) $\sqrt{45}$ (b) $\sqrt[3]{\dfrac{32a^2}{b^2}}$
96. (a) $\sqrt[3]{54}$ (b) $\sqrt{32x^3y^4}$
97. (a) $\sqrt[3]{16x^5}$ (b) $\sqrt{75x^2y^{-4}}$
98. (a) $\sqrt[4]{3x^4y^2}$ (b) $\sqrt[5]{160x^8z^4}$
99. (a) $2\sqrt{50} + 12\sqrt{8}$ (b) $10\sqrt{32} - 6\sqrt{18}$
100. (a) $5\sqrt{x} - 3\sqrt{x}$ (b) $-2\sqrt{9y} + 10\sqrt{y}$
101. (a) $3\sqrt{x+1} + 10\sqrt{x+1}$
(b) $7\sqrt{80x} - 2\sqrt{125x}$
102. (a) $5\sqrt{10x^2} - \sqrt{90x^2}$ (b) $8\sqrt[3]{27x} - \frac{1}{2}\sqrt[3]{64x}$

Comparing Radical Expressions **In Exercises 103–106, complete the statement with <, =, or >.**

103. $\sqrt{\dfrac{3}{11}}$ ▭ $\dfrac{\sqrt{3}}{\sqrt{11}}$
104. $\sqrt{5} + \sqrt{3}$ ▭ $\sqrt{5+3}$
105. 5 ▭ $\sqrt{3^2 + 2^2}$
106. 5 ▭ $\sqrt{3^2 + 4^2}$

Rationalizing a Denominator **In Exercises 107–110, rationalize the denominator of the expression. Then simplify your answer.**

107. $\dfrac{1}{\sqrt{3}}$
108. $\dfrac{8}{\sqrt[3]{2}}$
109. $\dfrac{5}{\sqrt{14} - 2}$
110. $\dfrac{3}{\sqrt{5} + \sqrt{6}}$

Rationalizing a Numerator **In Exercises 111–114, rationalize the numerator of the expression. Then simplify your answer.**

111. $\dfrac{\sqrt{12}}{2}$
112. $\dfrac{\sqrt{3}}{3}$
113. $\dfrac{\sqrt{5} + \sqrt{3}}{3}$
114. $\dfrac{\sqrt{7} - 3}{4}$

The symbol $\int$ indicates an example or exercise that highlights algebraic techniques specifically used in calculus.

Reducing the Index of a Radical **In Exercises 115 and 116, reduce the index of each radical and rewrite in radical form.**

115. (a) $\sqrt[4]{3^2}$ (b) $\sqrt[6]{(x+1)^4}$

116. (a) $\sqrt[6]{x^3}$ (b) $\sqrt[4]{(3x^2)^4}$

Changing Forms **In Exercises 117–124, fill in the missing form of the expression.**

	Radical Form	*Rational Exponent Form*
117.	$\sqrt[3]{64}$	
118.		$-(144^{1/2})$
119.		$(1/32)^{1/5}$
120.	$\sqrt[3]{614.125}$	
121.		$(-243)^{1/5}$
122.	$\sqrt[3]{-216}$	
123.	$\sqrt[4]{81^3}$	
124.		$16^{5/4}$

Simplifying with Rational Exponents **In Exercises 125–132, simplify the expression.**

125. $\dfrac{(2x^2)^{3/2}}{2^{1/2}x^4}$ **126.** $\dfrac{x^{4/3}y^{2/3}}{(xy)^{1/3}}$

127. $\dfrac{x^{-3} \cdot x^{1/2}}{x^{3/2} \cdot x^{-1}}$ **128.** $\dfrac{5^{-1/2} \cdot 5x^{5/2}}{(5x)^{3/2}}$

129. $32^{-3/5}$ **130.** $\left(\frac{9}{4}\right)^{-1/2}$

131. $\left(-\frac{1}{27}\right)^{-1/3}$ **132.** $-\left(\frac{1}{125}\right)^{-4/3}$

Rewriting a Radical Expression **In Exercises 133–136, write each expression as a single radical. Then simplify your answer.**

133. $\sqrt{\sqrt{35}}$ **134.** $\sqrt{\sqrt[4]{2x}}$

135. $\sqrt{\sqrt{243(x+1)}}$ **136.** $\sqrt{\sqrt[3]{128a^7b}}$

137. Economics The table shows the 2013 estimated populations and gross domestic products (GDP) for several countries, given in scientific notation. Find the per capita GDP for each country. (Sources: U.S. Census Bureau; Central Intelligence Agency)

DATA

Country	Population	GDP (in U.S. dollars)
Brazil	2.01×10^8	2.19×10^{12}
Canada	3.46×10^7	1.83×10^{12}
Germany	8.11×10^7	3.59×10^{12}
India	1.22×10^9	1.76×10^{12}
Iran	7.99×10^7	4.12×10^{11}
Ireland	4.78×10^6	2.21×10^{11}
Mexico	1.19×10^8	1.33×10^{12}

Spreadsheet at LarsonPrecalculus.com

138. Environmental Science There were 2.51×10^8 tons of municipal waste generated in 2012. Find the number of tons for each of the categories in the graph. (Source: U.S. Environmental Protection Agency)

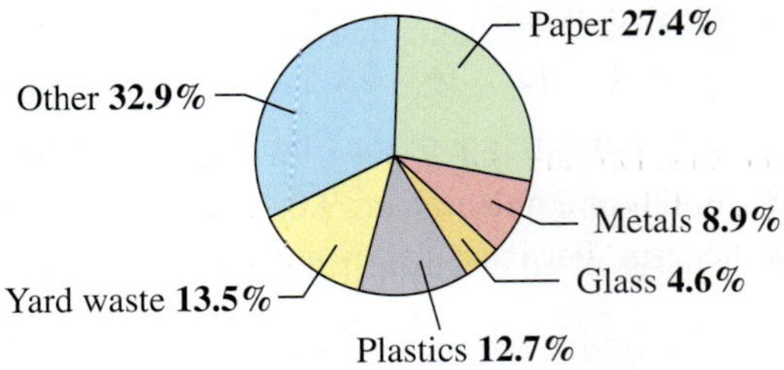

139. *Why you should learn it* *(p. 12)* A funnel is filled with hydrochloric acid to a height of h centimeters. The formula

$$t = 0.03[12^{5/2} - (12 - h)^{5/2}], \quad 0 \le h < 12$$

represents the amount of time t (in seconds) it will take for the funnel to empty. Find t for $h = 7$ centimeters.

Conclusions

True or False? **In Exercises 140–143, determine whether the statement is true or false. Justify your answer.**

140. $\dfrac{x^{k+1}}{x} = x^k$ **141.** $(a^n)^k = a^{(n^k)}$

142. $(a + b)^2 = a^2 + b^2$ **143.** $x^{-1} + y^{-1} = \dfrac{x + y}{xy}$

144. HOW DO YOU SEE IT? Package A is a cube with a volume of 500 cubic inches. Package B is a cube with a volume of 250 cubic inches. Is the length x of a side of package A greater than, less than, or equal to twice the length of a side of package B? Explain.

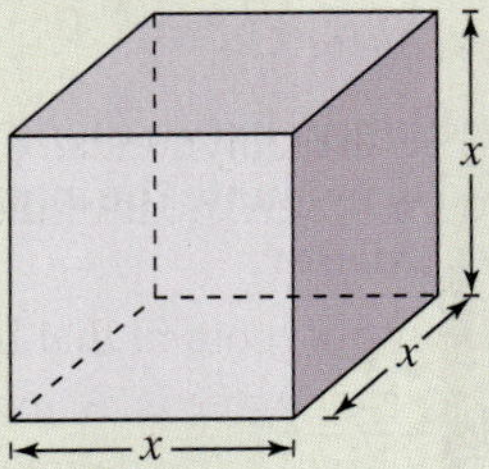

145. Think About It Verify that $a^0 = 1$, $a \neq 0$. (*Hint:* Use the property of exponents $a^m/a^n = a^{m-n}$.)

146. Exploration List all possible digits that occur in the units place of the square of a positive integer. Use that list to determine whether $\sqrt{5233}$ is an integer.

147. Think About It Square the real number $5/\sqrt{3}$ and note that the radical is eliminated from the denominator. Is this equivalent to rationalizing the denominator? Why or why not?

P.3 Polynomials and Factoring

Polynomials

The most common type of algebraic expression is the **polynomial.** Some examples are

$$2x + 5, \quad 3x^4 - 7x^2 + 2x + 4, \quad \text{and} \quad 5x^2y^2 - xy + 3.$$

The first two are *polynomials in x* and the third is a *polynomial in x and y*. The terms of a polynomial in x have the form ax^k, where a is the **coefficient** and k is the **degree** of the term. For instance, the polynomial

$$2x^3 - 5x^2 + 1 = 2x^3 + (-5)x^2 + (0)x + 1$$

has coefficients 2, -5, 0, and 1.

What you should learn

- Write polynomials in standard form.
- Add, subtract, and multiply polynomials and use special products.
- Remove common factors from polynomials.
- Factor special polynomial forms.
- Factor trinomials as the product of two binomials.
- Factor polynomials by grouping.

Why you should learn it

Polynomials can be used to model and solve real-life problems. For instance, in Exercise 152 on page 33, a polynomial is used to model the safe load on a steel beam.

Definition of a Polynomial in x

Let $a_0, a_1, a_2, \ldots, a_n$ be *real numbers* and let n be a *nonnegative integer.* A **polynomial in x** is an expression of the form

$$a_nx^n + a_{n-1}x^{n-1} + \cdots + a_1x + a_0$$

where $a_n \neq 0$. The polynomial is of **degree** n, a_n is the **leading coefficient,** and a_0 is the **constant term.**

In **standard form,** a polynomial in x is written with descending powers of x. Polynomials with one, two, and three terms are called **monomials, binomials,** and **trinomials,** respectively.

EXAMPLE 1 Writing Polynomials in Standard Form

	Polynomial	*Standard Form*	*Degree*	*Leading Coefficient*
a.	$4x^2 - 5x^7 - 2 + 3x$	$-5x^7 + 4x^2 + 3x - 2$	7	-5
b.	$4 - 9x^2$	$-9x^2 + 4$	2	-9
c.	8	$8\ (8 = 8x^0)$	0	8

✓ **Checkpoint** Audio-video solution in English & Spanish at LarsonPrecalculus.com.

Write the polynomial $6 - 7x^3 + 2x$ in standard form. Then identify the degree and leading coefficient of the polynomial.

A polynomial that has all zero coefficients is called the **zero polynomial,** denoted by 0. No degree is assigned to this particular polynomial. For polynomials in more than one variable, the degree of a *term* is the sum of the exponents of the variables in the term. The degree of the *polynomial* is the highest degree of its terms. For instance, the degree of the polynomial

$$-2x^3y^6 + 4xy - x^7y^4$$

is 11 because the sum of the exponents in the last term is the greatest. Expressions are not polynomials when a variable is underneath a radical or when a polynomial expression (with degree greater than 0) is in the denominator of a term. The following expressions are *not* polynomials.

$x^3 - \sqrt{3x} = x^3 - (3x)^{1/2}$ The exponent 1/2 is not an integer.

$x^2 + \dfrac{5}{x} = x^2 + 5x^{-1}$ The exponent -1 is not a nonnegative integer.

Operations with Polynomials and Special Products

You can add and subtract polynomials in much the same way you add and subtract real numbers. Simply add or subtract the *like terms* (terms having exactly the same variables to exactly the same powers) by adding their coefficients. For instance, $-3xy^2$ and $5xy^2$ are like terms and their sum is $-3xy^2 + 5xy^2 = (-3 + 5)xy^2 = 2xy^2$.

EXAMPLE 2 Sums and Differences of Polynomials

Perform each indicated operation.

a. $(5x^3 - 7x^2 - 3) + (x^3 + 2x^2 - x + 8)$

b. $(7x^4 - x^2 - 4x + 2) - (3x^4 - 4x^2 + 3x)$

Solution

a. $(5x^3 - 7x^2 - 3) + (x^3 + 2x^2 - x + 8)$

$= (5x^3 + x^3) + (-7x^2 + 2x^2) - x + (-3 + 8)$ Group like terms.

$= 6x^3 - 5x^2 - x + 5$ Combine like terms.

b. $(7x^4 - x^2 - 4x + 2) - (3x^4 - 4x^2 + 3x)$

$= 7x^4 - x^2 - 4x + 2 - 3x^4 + 4x^2 - 3x$ Distributive Property

$= (7x^4 - 3x^4) + (-x^2 + 4x^2) + (-4x - 3x) + 2$ Group like terms.

$= 4x^4 + 3x^2 - 7x + 2$ Combine like terms.

Remark

When a negative sign precedes an expression within parentheses, treat it like the coefficient (-1) and distribute the negative sign to each term inside the parentheses.

$-(3x^4 - 4x^2 + 3x)$

$= -3x^4 + 4x^2 - 3x$

✓ ***Checkpoint*** *Audio-video solution in English & Spanish at LarsonPrecalculus.com.*

Find the difference $(2x^3 - x + 3) - (x^2 - 2x - 3)$ and write the resulting polynomial in standard form.

To find the product of two polynomials, use the left and right Distributive Properties.

EXAMPLE 3 Multiplying Polynomials: The FOIL Method

Find the product of $3x - 2$ and $5x + 7$.

Solution

Treating $5x + 7$ as a single quantity, you can use the Distributive Property to multiply $3x - 2$ by $5x + 7$ as follows.

$$(3x - 2)(5x + 7) = 3x(5x + 7) - 2(5x + 7)$$

$$= (3x)(5x) + (3x)(7) - (2)(5x) - (2)(7)$$

$$= 15x^2 + 21x - 10x - 14$$

Product of First terms | Product of Outer terms | Product of Inner terms | Product of Last terms

$$= 15x^2 + 11x - 14$$

✓ ***Checkpoint*** *Audio-video solution in English & Spanish at LarsonPrecalculus.com.*

Use the FOIL Method to find the product of $3x - 1$ and $x - 5$.

Note that when using the **FOIL Method** (which can be used only to multiply two binomials), some of the terms in the product may be like terms that can be combined into one term.

EXAMPLE 4 The Product of Two Trinomials

Find the product of $4x^2 + x - 2$ and $-x^2 + 3x + 5$.

Solution

When multiplying two polynomials, be sure to multiply *each* term of one polynomial by *each* term of the other. A vertical format is helpful.

$4x^2 + x - 2$	Write in standard form.
$\times \quad -x^2 + 3x + 5$	Write in standard form.
$20x^2 + 5x - 10$	$5(4x^2 + x - 2)$
$12x^3 + 3x^2 - 6x$	$3x(4x^2 + x - 2)$
$-4x^4 - x^3 + 2x^2$	$-x^2(4x^2 + x - 2)$
$-4x^4 + 11x^3 + 25x^2 - x - 10$	Combine like terms.

✓ Checkpoint *Audio-video solution in English & Spanish at LarsonPrecalculus.com.*

Multiply $x^2 + 2x + 3$ by $x^2 - 2x + 3$ using a vertical arrangement.

Some binomial products have special forms that occur frequently in algebra. You do not need to memorize these formulas because you can use the Distributive Property to multiply. Becoming familiar with these formulas, however, will enable you to manipulate the algebra more quickly.

Special Products

Let u and v be real numbers, variables, or algebraic expressions.

Special Product	*Example*
Sum and Difference of Same Terms	
$(u + v)(u - v) = u^2 - v^2$	$(x + 4)(x - 4) = x^2 - 4^2 = x^2 - 16$
Square of a Binomial	
$(u + v)^2 = u^2 + 2uv + v^2$	$(x + 3)^2 = x^2 + 2(x)(3) + 3^2 = x^2 + 6x + 9$
$(u - v)^2 = u^2 - 2uv + v^2$	$(3x - 2)^2 = (3x)^2 - 2(3x)(2) + 2^2 = 9x^2 - 12x + 4$
Cube of a Binomial	
$(u + v)^3 = u^3 + 3u^2v + 3uv^2 + v^3$	$(x + 2)^3 = x^3 + 3x^2(2) + 3x(2^2) + 2^3 = x^3 + 6x^2 + 12x + 8$
$(u - v)^3 = u^3 - 3u^2v + 3uv^2 - v^3$	$(x - 1)^3 = x^3 - 3x^2(1) + 3x(1^2) - 1^3 = x^3 - 3x^2 + 3x - 1$

EXAMPLE 5 The Product of Two Trinomials

To find the product of $x + y - 2$ and $x + y + 2$, begin by grouping $x + y$ in parentheses. Then write the product of the trinomials as a special product.

Difference ↓ Sum ↓

$$\begin{aligned}(x + y - 2)(x + y + 2) &= [(x + y) - 2][(x + y) + 2] \\ &= (x + y)^2 - 2^2 \\ &= x^2 + 2xy + y^2 - 4\end{aligned}$$

✓ Checkpoint *Audio-video solution in English & Spanish at LarsonPrecalculus.com.*

Find the product of $x - 2 + 3y$ and $x - 2 - 3y$.

Factoring

The process of writing a polynomial as a product is called **factoring.** It is an important tool for solving equations and for simplifying rational expressions.

Unless noted otherwise, when you are asked to factor a polynomial, you can assume that you are looking for factors with integer coefficients. If a polynomial cannot be factored using integer coefficients, then it is **prime** or **irreducible over the integers.** For instance, the polynomial

$$x^2 - 3$$

is irreducible over the integers. Over the real numbers, this polynomial can be factored as

$$x^2 - 3 = (x + \sqrt{3})(x - \sqrt{3}).$$

A polynomial is **completely factored** when each of its factors is prime. So,

$$x^3 - x^2 + 4x - 4 = (x - 1)(x^2 + 4)$$ Completely factored

is completely factored, but

$$x^3 - x^2 - 4x + 4 = (x - 1)(x^2 - 4)$$ Not completely factored

is not completely factored. Its complete factorization is

$$x^3 - x^2 - 4x + 4 = (x - 1)(x + 2)(x - 2).$$

The simplest type of factoring involves a polynomial that can be written as the product of a monomial and another polynomial. The technique used here is the Distributive Property,

$$a(b + c) = ab + ac$$

in the *reverse* direction.

$$ab + ac = a(b + c)$$ a is a common factor.

The first step in completely factoring a polynomial is to remove (factor out) any common factors, as shown in the next example.

EXAMPLE 6 Removing Common Factors

Factor each expression.

a. $6x^3 - 4x$ **b.** $-4x^2 + 12x - 16$

c. $3x^4 + 9x^3 + 6x^2$ **d.** $(x - 2)(2x) + (x - 2)(3)$

Solution

a. $6x^3 - 4x = 2x(3x^2) - 2x(2)$ $2x$ is a common factor.

$= 2x(3x^2 - 2)$

b. $-4x^2 + 12x - 16 = -4(x^2) + (-4)(-3x) + (-4)4$ -4 is a common factor.

$= -4(x^2 - 3x + 4)$

c. $3x^4 + 9x^3 + 6x^2 = 3x^2(x^2) + 3x^2(3x) + 3x^2(2)$ $3x^2$ is a common factor.

$= 3x^2(x^2 + 3x + 2)$

$= 3x^2(x + 1)(x + 2)$

d. $(x - 2)(2x) + (x - 2)(3) = (x - 2)(2x + 3)$ $(x - 2)$ is a common factor.

✔ **Checkpoint** *Audio-video solution in English & Spanish at LarsonPrecalculus.com.*

Factor each expression.

a. $5x^3 - 15x^2$ **b.** $-3 + 6x - 12x^3$ **c.** $(x + 1)(x^2) - (x + 1)(2)$

Factoring Special Polynomial Forms

Some polynomials have special forms that arise from the special product forms on page 25. You should learn to recognize these forms so that you can factor such polynomials efficiently.

Factoring Special Polynomial Forms

Factored Form	*Example*
Difference of Two Squares	
$u^2 - v^2 = (u + v)(u - v)$	$9x^2 - 4 = (3x)^2 - 2^2 = (3x + 2)(3x - 2)$
Perfect Square Trinomial	
$u^2 + 2uv + v^2 = (u + v)^2$	$x^2 + 6x + 9 = x^2 + 2(x)(3) + 3^2 = (x + 3)^2$
$u^2 - 2uv + v^2 = (u - v)^2$	$x^2 - 6x + 9 = x^2 - 2(x)(3) + 3^2 = (x - 3)^2$
Sum or Difference of Two Cubes	
$u^3 + v^3 = (u + v)(u^2 - uv + v^2)$	$x^3 + 8 = x^3 + 2^3 = (x + 2)(x^2 - 2x + 4)$
$u^3 - v^3 = (u - v)(u^2 + uv + v^2)$	$27x^3 - 1 = (3x)^3 - 1^3 = (3x - 1)(9x^2 + 3x + 1)$

For the difference of two squares, you can think of this form as follows.

$$u^2 - v^2 = (u + v)(u - v)$$

Difference   Opposite signs

To recognize perfect square terms, look for coefficients that are squares of integers and variables raised to *even powers.*

EXAMPLE 7 Removing a Common Factor First

$$3 - 12x^2 = 3(1 - 4x^2) \qquad \text{3 is a common factor.}$$
$$= 3[1^2 - (2x)^2] \qquad \text{Write as difference of two squares.}$$
$$= 3(1 + 2x)(1 - 2x) \qquad \text{Factored form}$$

✓ ***Checkpoint*** *Audio-video solution in English & Spanish at LarsonPrecalculus.com.*

Factor $100 - 4y^2$.

Remark

In Example 7, note that the first step in factoring a polynomial is to check for a common factor. Once the common factor is removed, it is often possible to recognize patterns that were not immediately obvious.

EXAMPLE 8 Factoring the Difference of Two Squares

a. $(x + 2)^2 - y^2 = [(x + 2) + y][(x + 2) - y]$
$$= (x + 2 + y)(x + 2 - y)$$

b. $16x^4 - 81 = (4x^2)^2 - 9^2$ Write as difference of two squares.
$$= (4x^2 + 9)(4x^2 - 9)$$
$$= (4x^2 + 9)[(2x)^2 - 3^2] \qquad \text{Write as difference of two squares.}$$
$$= (4x^2 + 9)(2x + 3)(2x - 3) \qquad \text{Factored form}$$

✓ ***Checkpoint*** *Audio-video solution in English & Spanish at LarsonPrecalculus.com.*

Factor $(x - 1)^2 - 9y^4$.

A perfect square trinomial is the square of a binomial, as shown below.

$$u^2 + 2uv + v^2 = (u + v)^2 \qquad \text{or} \qquad u^2 - 2uv + v^2 = (u - v)^2$$

Like signs   Like signs

Note that the first and last terms are squares and the middle term is twice the product of u and v.

EXAMPLE 9 Factoring Perfect Square Trinomials

a. $x^2 - 10x + 25 = x^2 - 2(x)(5) + 5^2$ Rewrite in $u^2 - 2uv + v^2$ form.

$= (x - 5)^2$

b. $16x^2 + 24x + 9 = (4x)^2 + 2(4x)(3) + 3^2$ Rewrite in $u^2 + 2uv + v^2$ form.

$= (4x + 3)^2$

✓ Checkpoint *Audio-video solution in English & Spanish at LarsonPrecalculus.com.*

Factor $9x^2 - 30x + 25$.

The next two formulas show the sums and differences of cubes. Pay special attention to the signs of the terms.

Like signs   Like signs

$$u^3 + v^3 = (u + v)(u^2 - uv + v^2) \qquad u^3 - v^3 = (u - v)(u^2 + uv + v^2)$$

Unlike signs   Unlike signs

Explore the Concept

Rewrite $u^6 - v^6$ as the difference of two squares. Then find a formula for completely factoring $u^6 - v^6$. Use your formula to factor completely $x^6 - 1$ and $x^6 - 64$.

EXAMPLE 10 Factoring the Difference of Cubes

$x^3 - 27 = x^3 - 3^3$ Rewrite 27 as 3^3.

$= (x - 3)(x^2 + 3x + 9)$ Factor.

✓ Checkpoint *Audio-video solution in English & Spanish at LarsonPrecalculus.com.*

Factor $64x^3 - 1$.

EXAMPLE 11 Factoring the Sum of Cubes

a. $y^3 + 125 = y^3 + 5^3$ Rewrite 125 as 5^3.

$= (y + 5)(y^2 - 5y + 25)$ Factor.

b. $3x^3 + 192 = 3(x^3 + 64)$ 3 is a common factor.

$= 3(x^3 + 4^3)$ Rewrite 64 as 4^3.

$= 3(x + 4)(x^2 - 4x + 16)$ Factor.

✓ Checkpoint *Audio-video solution in English & Spanish at LarsonPrecalculus.com.*

Factor each expression.

a. $x^3 + 216$

b. $5y^3 + 135$

Trinomials with Binomial Factors

To factor a trinomial of the form $ax^2 + bx + c$, use the following pattern.

Factors of a

$$ax^2 + bx + c = (\square x + \square)(\square x + \square)$$

Factors of c

The goal is to find a combination of factors of a and c such that the outer and inner products add up to the middle term bx. For instance, in the trinomial $6x^2 + 17x + 5$, you can write all possible factorizations and determine which one has outer and inner products that add up to $17x$.

$$(6x + 5)(x + 1), \quad (6x + 1)(x + 5), \quad (2x + 1)(3x + 5), \quad (2x + 5)(3x + 1)$$

You can see that $(2x + 5)(3x + 1)$ is the correct factorization because the outer (O) and inner (I) products add up to $17x$.

$$(2x + 5)(3x + 1) = \overset{\text{F}}{6x^2} + \overset{\text{O}}{2x} + \overset{\text{I}}{15x} + \overset{\text{L}}{5} = 6x^2 + \overset{\text{O + I}}{17x} + 5.$$

EXAMPLE 12 Factoring a Trinomial: Leading Coefficient Is 1

Factor $x^2 - 7x + 12$.

Solution

The possible factorizations are as follows.

$$(x - 2)(x - 6), \quad (x - 1)(x - 12), \quad (x - 3)(x - 4)$$

Testing the middle term, you will find the correct factorization to be

$$x^2 - 7x + 12 = (x - 3)(x - 4). \qquad \text{O + I} = -4x + (-3x) = -7x$$

✓ Checkpoint *Audio-video solution in English & Spanish at LarsonPrecalculus.com.*

Factor $x^2 + x - 6$.

Remark

Factoring a trinomial can involve trial and error. Once you have produced the factored form, however, it is an easy matter to check your answer. For instance, you can verify the factorization in Example 12 by multiplying $(x - 3)$ by $(x - 4)$ to see that you obtain the original trinomial, $x^2 - 7x + 12$.

EXAMPLE 13 Factoring a Trinomial: Leading Coefficient Is Not 1

See LarsonPrecalculus.com for an interactive version of this type of example.

Factor $2x^2 + x - 15$.

Solution

The eight possible factorizations are as follows.

$$(2x - 1)(x + 15) \quad (2x + 1)(x - 15) \quad (2x - 5)(x + 3) \quad (2x + 5)(x - 3)$$

$$(2x - 3)(x + 5) \quad (2x + 3)(x - 5) \quad (2x - 15)(x + 1) \quad (2x + 15)(x - 1)$$

Testing the middle term, you will find the correct factorization to be

$$2x^2 + x - 15 = (2x - 5)(x + 3). \qquad \text{O + I} = 6x - 5x = x$$

✓ Checkpoint *Audio-video solution in English & Spanish at LarsonPrecalculus.com.*

Factor each trinomial.

a. $2x^2 - 5x + 3$ **b.** $12x^2 + 7x + 1$

Factoring by Grouping

Sometimes polynomials with more than three terms can be factored by a method called **factoring by grouping.**

EXAMPLE 14 Factoring by Grouping

Use factoring by grouping to factor $x^3 - 2x^2 - 3x + 6$.

Solution

$$\begin{aligned} x^3 - 2x^2 - 3x + 6 &= (x^3 - 2x^2) - (3x - 6) && \text{Group terms.} \\ &= x^2(x - 2) - 3(x - 2) && \text{Factor groups.} \\ &= (x - 2)(x^2 - 3) && (x - 2) \text{ is a common factor.} \end{aligned}$$

Remark

When grouping terms, be sure to group terms that have a common factor.

✓ **Checkpoint** Audio-video solution in English & Spanish at LarsonPrecalculus.com.

Factor $x^3 + x^2 - 5x - 5$.

Factoring a trinomial can involve quite a bit of trial and error. Some of this trial and error can be lessened by using factoring by grouping. The key to this method of factoring is knowing how to rewrite the middle term. In general, to factor a trinomial

$$ax^2 + bx + c$$

by grouping, choose factors of the product ac that add up to b and use these factors to rewrite the middle term.

EXAMPLE 15 Factoring a Trinomial by Grouping

Use factoring by grouping to factor $2x^2 + 5x - 3$.

Solution

In the trinomial $2x^2 + 5x - 3$, $a = 2$ and $c = -3$, which implies that the product ac is -6. Now, because -6 factors as $(6)(-1)$ and $6 - 1 = 5 = b$, rewrite the middle term as $5x = 6x - x$. This produces the following.

$$\begin{aligned} 2x^2 + 5x - 3 &= 2x^2 + 6x - x - 3 && \text{Rewrite middle term.} \\ &= (2x^2 + 6x) - (x + 3) && \text{Group terms.} \\ &= 2x(x + 3) - (x + 3) && \text{Factor groups.} \\ &= (x + 3)(2x - 1) && (x + 3) \text{ is a common factor.} \end{aligned}$$

So, the trinomial factors as $2x^2 + 5x - 3 = (x + 3)(2x - 1)$.

✓ **Checkpoint** Audio-video solution in English & Spanish at LarsonPrecalculus.com.

Use factoring by grouping to factor $2x^2 + 5x - 12$.

Guidelines for Factoring Polynomials

1. Factor out any common factors using the Distributive Property.
2. Factor according to one of the special polynomial forms.
3. Factor as $ax^2 + bx + c = (mx + r)(nx + s)$.
4. Factor by grouping.

P.3 Exercises

See *CalcChat.com* for tutorial help and worked-out solutions to odd-numbered exercises.
For instructions on how to use a graphing utility, see Appendix A.

Vocabulary and Concept Check

In Exercises 1–4, fill in the blank(s).

1. For the polynomial $a_nx^n + a_{n-1}x^{n-1} + \cdot \cdot \cdot + a_1x + a_0$, the degree is _______ and the leading coefficient is _______ .

2. A polynomial with one term is called a _______ .

3. The letters in "FOIL" stand for the following.
F _______ O _______ I _______ L _______

4. When a polynomial cannot be factored using integer coefficients, it is called _______ .

5. When is a polynomial completely factored?

6. List four guidelines for factoring polynomials.

Procedures and Problem Solving

Identifying Polynomials **In Exercises 7–12, match the polynomial with its description. [The polynomials are labeled (a), (b), (c), (d), (e), and (f).]**

(a) $6x$ **(b)** $1 - 4x^3$
(c) $x^3 + 2x^2 - 4x + 1$ **(d)** 7
(e) $-3x^5 + 2x^3 + x$ **(f)** $\frac{3}{4}x^4 + x^2 + 14$

7. A polynomial of degree zero

8. A trinomial of degree five

9. A binomial with leading coefficient -4

10. A monomial of positive degree

11. A trinomial with leading coefficient $\frac{3}{4}$

12. A third-degree polynomial with leading coefficient 1

Writing a Polynomial **In Exercises 13–16, write a polynomial that fits the description. (There are many correct answers.)**

13. A third-degree polynomial with leading coefficient -2

14. A fifth-degree polynomial with leading coefficient 8

15. A fourth-degree binomial with a negative leading coefficient

16. A third-degree trinomial with an even leading coefficient

Writing a Polynomial in Standard Form **In Exercises 17–22, write the polynomial in standard form. Then identify the degree and leading coefficient of the polynomial.**

17. $3x + 4x^2 + 2$ **18.** $x^2 - 4 - 3x^4$

19. $-8 + x^7$ **20.** $23 - x^3$

21. $1 - x + 6x^4 - 2x^5$ **22.** $-x^6 + 5 - 4x^5 + x^3$

Classifying an Expression **In Exercises 23–26, determine whether the expression is a polynomial. If so, write the polynomial in standard form.**

23. $y - 8y^2 + y$ **24.** $5x^4 - 2x^2 + x^{-2}$

25. $\sqrt{x^2 - x^4}$ **26.** $\dfrac{x^2 + 2x - 3}{6}$

Performing Operations with Polynomials **In Exercises 27–40, perform the operation and write the result in standard form.**

27. $(4x + 1) + (-x + 9)$ **28.** $(t^2 - 3) + (6t^2 - 4t)$

29. $(8x + 5) - (6x - 12)$ **30.** $(x^2 - 5) - (2x^2 - 3x)$

31. $(2x^3 - 9x^2 - 20) + (-2x^3 + 10x^2)$

32. $(y^3 - 6y + 3) + (5y^3 - 2y^2 + y - 10)$

33. $(15x^2 - 6) - (-8.1x^3 - 14.7x^2 - 17)$

34. $(13.6w^4 - 14w - 17.4) - (16.9w^4 - 9.2w + 13)$

35. $5z(z - 8)$ **36.** $\left(\frac{1}{6}x + 1\right)(2x^2)$

37. $\left(5 - \frac{3}{2}y\right)(-4y)$ **38.** $-7x(4 - x^3)$

39. $3x(x^2 - 2x + 1)$ **40.** $-y^2(4y^2 + 2y - 3)$

Multiplying Polynomials **In Exercises 41–66, multiply or find the special product.**

41. $(x + 3)(x + 4)$ **42.** $(x - 5)(x + 10)$

43. $(3x - 5)(2x + 1)$ **44.** $(7x - 2)(4x - 3)$

45. $(4y + 7)^2$ **46.** $(3a - 3)^2$

47. $(2 - 5x)^2$ **48.** $(5x + 8y)^2$

49. $(x - 9)(x + 9)$ **50.** $(5x + 6)(5x - 6)$

51. $(x + 2y)(x - 2y)$ **52.** $(2r^2 - 5)(2r^2 + 5)$

53. $(x + 1)^3$ **54.** $(y - 4)^3$

55. $(2x - y)^3$ **56.** $(3x + 2y)^3$

57. $\left(\frac{1}{4}x - 3\right)\left(\frac{1}{4}x + 3\right)$ **58.** $(1.5y + 0.6)(1.5y - 0.6)$

59. $\left(\frac{5}{2}x + 3\right)^2$ **60.** $(1.8y - 5)^2$

61. $(-x^2 + x - 5)(3x^2 + 4x + 1)$

62. $(x^2 + 3x + 2)(2x^2 - x + 4)$

63. $[(x + z) + 5][(x + z) - 5]$

64. $[(x - 3y) + z][(x - 3y) - z]$

65. $[(x - 3) + y]^2$ **66.** $[(x + 1) - y]^2$

Removing Common Factors In Exercises 67–74, factor out the common factor.

67. $5x - 40$
68. $4y + 20$
69. $2x^3 - 6x$
70. $3z^4 - 6z^2 + 9z$
71. $3x(x - 5) + 8(x - 5)$
72. $5(x + 1) - x(x + 1)$
73. $(5x - 4)^2 + (5x - 4)$
74. $2x(x + 3) - 3x(x + 3)^2$

Factoring the Difference of Two Squares In Exercises 75–82, factor the difference of two squares.

75. $x^2 - 36$
76. $x^2 - 81$
77. $48y^2 - 27$
78. $50 - 98z^2$
79. $4x^2 - \frac{1}{9}$
80. $\frac{25}{36}y^2 - 49$
81. $(x - 1)^2 - 4$
82. $25 - (z + 5)^2$

Factoring a Perfect Square Trinomial In Exercises 83–90, factor the perfect square trinomial.

83. $x^2 - 4x + 4$
84. $x^2 + 10x + 25$
85. $x^2 + x + \frac{1}{4}$
86. $x^2 - \frac{4}{3}x + \frac{4}{9}$
87. $4x^2 - 12x + 9$
88. $25z^2 - 10z + 1$
89. $4x^2 - \frac{4}{3}x + \frac{1}{9}$
90. $9y^2 - \frac{3}{2}y + \frac{1}{16}$

Factoring the Sum or Difference of Cubes In Exercises 91–100, factor the sum or difference of cubes.

91. $x^3 - 8$
92. $y^3 - 125$
93. $z^3 + 1$
94. $x^3 + 64$
95. $x^3 + \frac{1}{27}$
96. $w^3 - \frac{27}{216}$
97. $125v^3 - 1$
98. $343a^3 + 8$
99. $(y + 1)^3 - x^3$
100. $(2x - z)^3 + 125y^3$

Factoring a Trinomial In Exercises 101–114, factor the trinomial.

101. $x^2 + x - 2$
102. $x^2 + 6x + 8$
103. $s^2 - 5s + 6$
104. $t^2 - t - 6$
105. $20 - y - y^2$
106. $24 + 5z - z^2$
107. $3x^2 + 13x - 10$
108. $2x^2 - x - 21$
109. $5x^2 + 26x + 5$
110. $8x^2 - 45x - 18$
111. $-5u^2 - 13u + 6$
112. $-6x^2 + 23x + 4$
113. $\frac{1}{8}x^2 - \frac{1}{96}x - \frac{1}{16}$
114. $\frac{1}{81}x^2 + \frac{2}{9}x - 8$

Factoring by Grouping In Exercises 115–122, factor by grouping.

115. $x^3 - x^2 + 2x - 2$
116. $x^3 + 5x^2 - 5x - 25$
117. $x^3 - 5x^2 + x - 5$
118. $x^3 - x^2 + 3x - 3$
119. $x^2 + x - 20$
120. $b^2 - 11b + 18$
121. $6x^2 + x - 2$
122. $3x^2 + 10x + 8$

Factoring an Expression Completely In Exercises 123–148, completely factor the expression.

123. $10x^2 - 40$
124. $7z^2 - 63$
125. $y^3 - y$
126. $x^3 - 9x^2$
127. $x^2 - 2x + 1$
128. $9x^2 - 6x + 1$
129. $1 - 4x + 4x^2$
130. $16 - 6x - x^2$
131. $2x^2 + 6x - 2x^3$
132. $7y^2 + 15y - 2y^3$
133. $9x^2 + 10x + 1$
134. $13x + 6 + 5x^2$
135. $3x^3 + x^2 + 15x + 5$
136. $5 - x + 5x^2 - x^3$
137. $3u - 2u^2 + 6 - u^3$
138. $x^4 - 4x^3 + x^2 - 4x$
139. $2x^3 + x^2 - 8x - 4$
140. $3x^3 + x^2 - 27x - 9$
141. $(x^2 + 1)^2 - 4x^2$
142. $(x^2 + 8)^2 - 36x^2$
143. $3t^3 + 24$
144. $4x^3 - 32$
145. $4x(2x - 1) + 2(2x - 1)^2$
146. $5(3 - 4x)^2 - 8(3 - 4x)(5x - 1)$
147. $2(x + 1)(x - 3)^2 - 3(x + 1)^2(x - 3)$
148. $7(3x + 2)^2(1 - x)^2 + (3x + 2)(1 - x)^3$

149. MODELING DATA

After 2 years, an investment of \$1000 compounded annually at an interest rate r (in decimal form) will yield an amount of $1000(1 + r)^2$.

(a) Write this polynomial in standard form.

(b) Use a calculator to evaluate the polynomial for the values of r shown in the table.

r	1%	$1\frac{1}{2}$%	2%	$2\frac{1}{2}$%	3%
$1000(1 + r)^2$					

(c) What conclusion can you make from the table?

150. Manufacturing A shipping company is constructing an open box made by cutting squares out of the corners of a piece of cardboard that is 18 centimeters by 26 centimeters (see figure). The edge of each cut-out square is x centimeters. Find the volume of the box in terms of x. Find the volume when $x = 1$, $x = 2$, and $x = 3$.

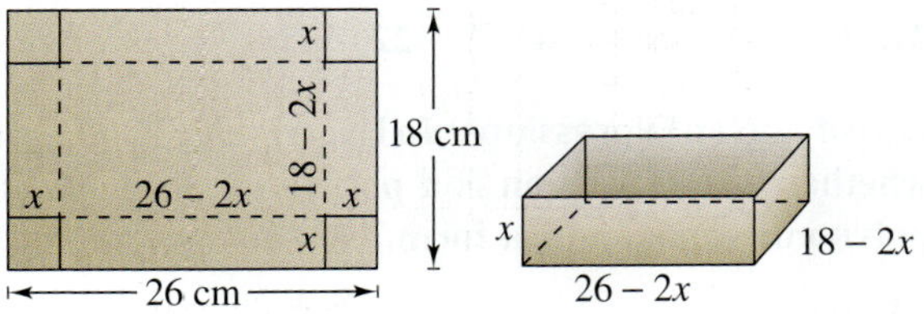

151. Public Safety The stopping distance of an automobile is the distance traveled during the driver's reaction time plus the distance traveled after the brakes are applied. In an experiment, these distances were measured (in feet) when the automobile was traveling at a speed of x miles per hour on dry, level pavement, as shown in the bar graph. The distance traveled during the reaction time R was $R = 1.1x$, and the braking distance B was $B = 0.0475x^2 - 0.001x + 0.23$.

(a) Determine the polynomial that represents the total stopping distance T.

(b) Use the result of part (a) to estimate T when $x = 30$, $x = 40$, and $x = 55$.

(c) Use the bar graph to make a statement about the total stopping distance required at increasing speeds.

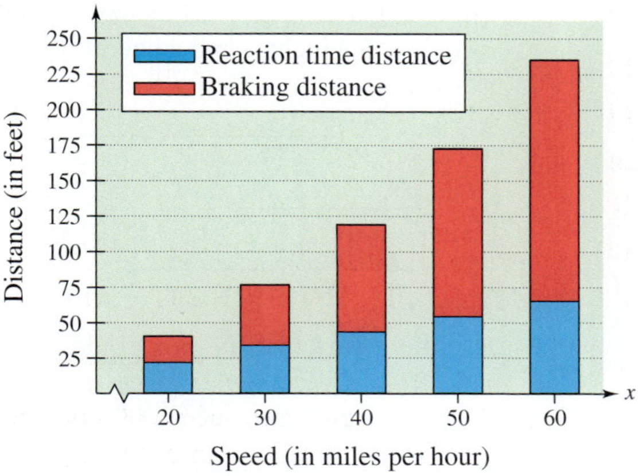

152. *Why you should learn it* *(p. 23)* A uniformly distributed load is placed on a one-inch-wide steel beam. When the span of the beam is x feet and its depth is 6 inches, the safe load S (in pounds) is approximated by

$$S_6 = (0.06x^2 - 2.42x + 38.71)^2.$$

When the depth is 8 inches, the safe load is approximated by

$$S_8 = (0.08x^2 - 3.30x + 51.93)^2.$$

(a) Use the bar graph to estimate the difference in the safe loads for these two beams when the span is 12 feet.

(b) How does the difference in safe load change as the span increases?

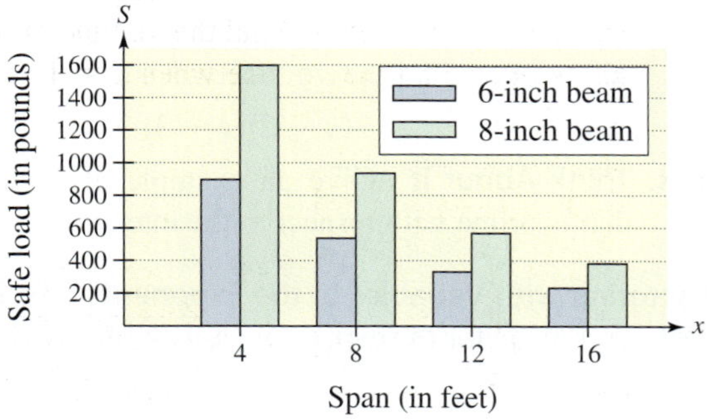

Geometric Modeling In Exercises 153 and 154, match the geometric factoring model with the correct factoring formula. For instance, the figure shown below is a factoring model for $2x^2 + 3x + 1 = (2x + 1)(x + 1)$.

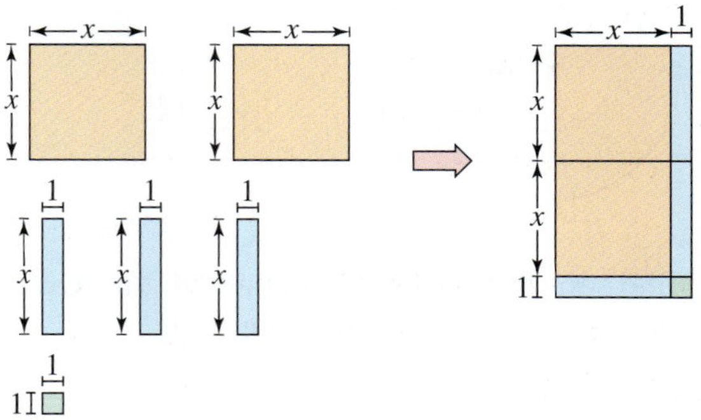

153.

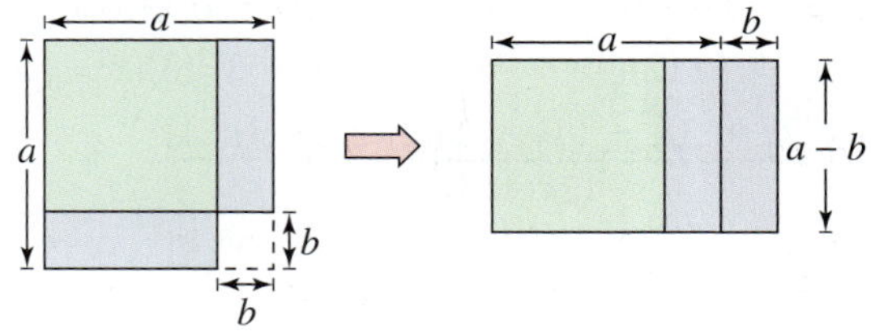

(a) $a^2 - b^2 = (a + b)(a - b)$

(b) $a^2 - 2ab + b^2 = (a - b)^2$

(c) $a^2 + 2ab + b^2 = (a + b)^2$

154.

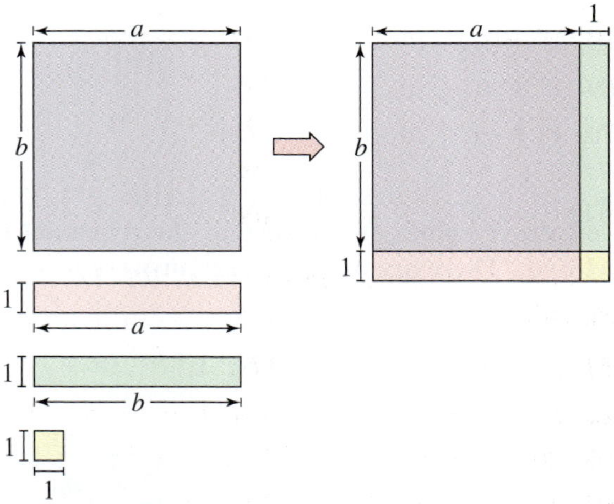

(a) $a^2 + 2a + 1 = (a + 1)^2$

(b) $ab + a + b + 1 = (a + 1)(b + 1)$

(c) $ab - a + b - 1 = (a + 1)(b - 1)$

Geometric Modeling In Exercises 155–158, draw a geometric factoring model to represent the factorization.

155. $x^2 + 3x + 2 = (x + 2)(x + 1)$

156. $x^2 + 4x + 3 = (x + 3)(x + 1)$

157. $3x^2 + 7x + 2 = (3x + 1)(x + 2)$

158. $2x^2 + 7x + 3 = (2x + 1)(x + 3)$

Geometry In Exercises 159 and 160, write an expression in factored form for the area of the shaded portion of the figure.

159.

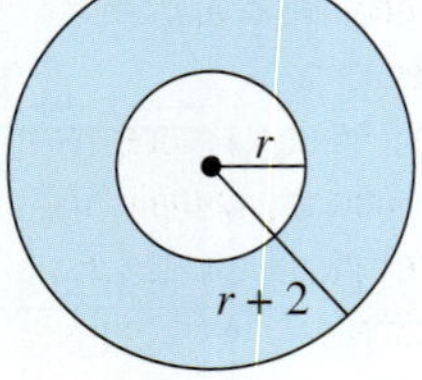

160.

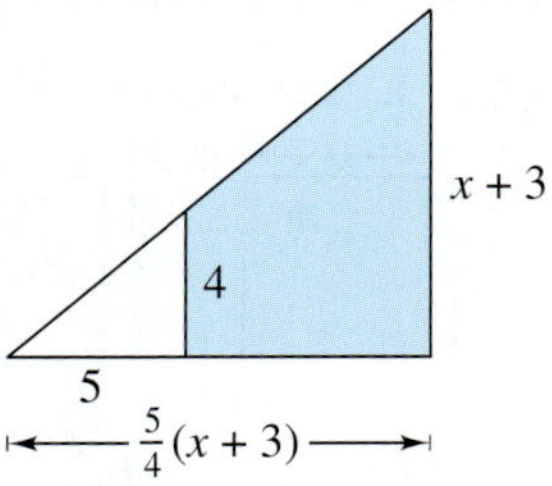

Factoring an Expression Completely In Exercises 161–166, factor the expression completely.

161. $x^4(4)(2x + 1)^3(2x) + (2x + 1)^4(4x^3)$

162. $x^3(3)(x^2 + 1)^2(2x) + (x^2 + 1)^3(3x^2)$

163. $(2x - 5)^4(3)(5x - 4)^2(5) + (5x - 4)^3(4)(2x - 5)^3(2)$

164. $(x^2 - 5)^3(2)(4x + 3)(4) + (4x + 3)^2(3)(x^2 - 5)^2(x^2)$

165. $\dfrac{4(2x + 3)^2 - (4x - 1)(2)(2x + 3)(2)}{(2x + 3)^4}$

166. $\dfrac{3(5x - 1)^3 - (3x + 1)(3)(5x - 1)^2(5)}{(5x - 1)^6}$

Classifying an Expression In Exercises 167–170, find all values of b for which the trinomial can be factored with integer coefficients.

167. $x^2 + bx - 15$

168. $x^2 + bx - 12$

169. $x^2 + bx + 50$

170. $x^2 + bx + 24$

Classifying an Expression In Exercises 171–174, find two integer values of c such that the trinomial can be factored. (There are many correct answers.)

171. $x^2 + x + c$

172. $x^2 - 9x + c$

173. $2x^2 + 5x + c$

174. $3x^2 - 10x + c$

175. Geometry The cylindrical shell shown in the figure has a volume of

$V = \pi R^2 h - \pi r^2 h.$

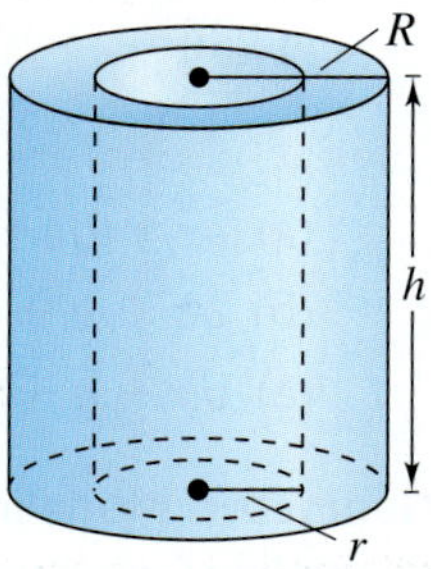

(a) Factor the expression for the volume.

(b) From the result of part (a), show that the volume is

2π(average radius)(thickness of the shell)h.

176. Chemistry The rate of change of an autocatalytic chemical reaction is $kQx - kx^2$, where Q is the amount of the original substance, x is the amount of substance formed, and k is a constant of proportionality. Factor the expression.

Conclusions

True or False? In Exercises 177–181, determine whether the statement is true or false. Justify your answer.

177. The product of two binomials is always a second-degree polynomial.

178. The product of two binomials is always a trinomial.

179. The sum of two second-degree polynomials is always a second-degree polynomial.

180. The sum of a third-degree polynomial and a fourth-degree polynomial can be a seventh-degree polynomial.

181. The expression $(3x - 6)(x + 1)$ is factored completely.

182. HOW DO YOU SEE IT? An open box has a length of $(52 - 2x)$ inches, a width of $(42 - 2x)$ inches, and a height of x inches, as shown.

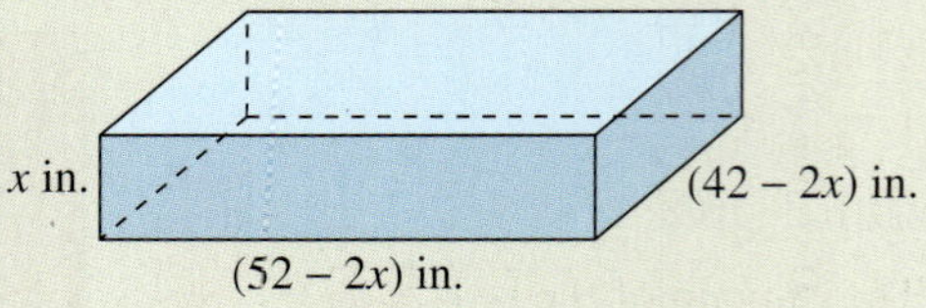

(a) Describe a way that the box could have been made from a rectangular piece of cardboard. Give the original dimensions of the cardboard.

(b) What degree is the polynomial that represents the volume of the box? Explain.

(c) Write an expression for the volume of the box in completely factored form. Use the expression to describe the possible values of x.

183. Exploration Find the degree of the product of two polynomials of degrees m and n.

184. Writing Write a paragraph explaining to a classmate why $(x + y)^2 \neq x^2 + y^2$.

185. Writing Write a paragraph explaining to a classmate a pattern that can be used to cube a binomial difference. Then use your pattern to cube the difference $(x - y)$.

186. Writing Explain what is meant when it is said that a polynomial is in factored form.

187. Error Analysis Describe the error.

$9x^2 - 9x - 54 = (3x + 6)(3x - 9)$

$= 3(x + 2)(x - 3)$

188. Think About It Give an example of a polynomial that is prime with respect to the integers.

Factoring with Variables in the Exponents In Exercises 189 and 190, factor the expression as completely as possible.

189. $x^{2n} - y^{2n}$

190. $x^{3n} + y^{3n}$

P.4 Rational Expressions

Domain of an Algebraic Expression

The set of real numbers for which an algebraic expression is defined is the **domain** of the expression. Two algebraic expressions are **equivalent** when they have the same domain and yield the same values for all numbers in their domain. For instance, the expressions $(x + 1) + (x + 2)$ and $2x + 3$ are equivalent because

$$\begin{aligned}(x + 1) + (x + 2) &= x + 1 + x + 2\\ &= x + x + 1 + 2\\ &= 2x + 3.\end{aligned}$$

What you should learn

- Find domains of algebraic expressions.
- Simplify rational expressions.
- Add, subtract, multiply, and divide rational expressions.
- Simplify complex fractions.
- Simplify expressions from calculus.

Why you should learn it

Rational expressions are useful in estimating the temperature of food as it cools. For instance, a rational expression is used in Exercise 104 on page 45 to model the temperature of food as it cools in a refrigerator set at 40°F.

EXAMPLE 1 Finding the Domain of an Algebraic Expression

Find the domain of each expression.

a. $2x^3 + 3x + 4$

b. $\sqrt{x - 2}$

c. $\dfrac{x + 2}{x - 3}$

Solution

a. The domain of the polynomial

$$2x^3 + 3x + 4$$

is the set of all real numbers. In fact, the domain of any polynomial is the set of all real numbers, unless the domain is specifically restricted.

b. The domain of the radical expression

$$\sqrt{x - 2}$$

is the set of real numbers greater than or equal to 2, because the square root of a negative number is not a real number.

c. The domain of the expression

$$\frac{x + 2}{x - 3}$$

is the set of all real numbers except $x = 3$, which would result in division by zero, which is undefined.

✓ **Checkpoint** *Audio-video solution in English & Spanish at LarsonPrecalculus.com.*

Find the domain of each expression.

a. $4x^3 + 3, \quad x \geq 0$ **b.** $\sqrt{x + 7}$ **c.** $\dfrac{1 - x}{x}$

The quotient of two algebraic expressions is a **fractional expression.** Moreover, the quotient of two *polynomials* such as

$$\frac{1}{x}, \quad \frac{2x - 1}{x + 1}, \quad \text{or} \quad \frac{x^2 - 1}{x^2 + 1}$$

is a **rational expression.**

Simplifying Rational Expressions

Recall that a fraction is in simplest form when its numerator and denominator have no factors in common aside from ± 1. To write a fraction in simplest form, divide out common factors.

$$\frac{a \cdot \cancel{c}}{b \cdot \cancel{c}} = \frac{a}{b}, \quad c \neq 0$$

The key to success in simplifying rational expressions lies in your ability to *factor* polynomials. When simplifying rational expressions, be sure to factor each polynomial completely before concluding that the numerator and denominator have no factors in common.

EXAMPLE 2 Simplifying a Rational Expression

$$\frac{x^2 + 4x - 12}{3x - 6} = \frac{(x + 6)\cancel{(x - 2)}}{3\cancel{(x - 2)}} \qquad \text{Factor completely.}$$

$$= \frac{x + 6}{3}, \quad x \neq 2 \qquad \text{Divide out common factors.}$$

Note that the original expression is undefined when $x = 2$ (because division by zero is undefined). To make the simplified expression *equivalent* to the original expression, you must restrict the domain of the simplified expression by excluding the value $x = 2$.

Remark

In Example 2, do not make the mistake of trying to simplify further by dividing out terms. Remember that to simplify fractions, divide out common *factors*, not terms.

Checkpoint *Audio-video solution in English & Spanish at LarsonPrecalculus.com.*

Write $\dfrac{4x + 12}{x^2 - 3x - 18}$ in simplest form.

It may sometimes be necessary to change the sign of a factor by factoring out (-1) to simplify a rational expression, as shown in Example 3.

EXAMPLE 3 Simplifying a Rational Expression

$$\frac{12 + x - x^2}{2x^2 - 9x + 4} = \frac{(4 - x)(3 + x)}{(2x - 1)(x - 4)} \qquad \text{Factor completely.}$$

$$= \frac{-\cancel{(x - 4)}(3 + x)}{(2x - 1)\cancel{(x - 4)}} \qquad (4 - x) = -(x - 4)$$

$$= -\frac{3 + x}{2x - 1}, \quad x \neq 4 \qquad \text{Divide out common factors.}$$

Checkpoint *Audio-video solution in English & Spanish at LarsonPrecalculus.com.*

Write $\dfrac{3x^2 - x - 2}{5 - 4x - x^2}$ in simplest form.

In this text, the domain is usually not listed with a rational expression. It is *implied* that the real numbers that make the denominator zero are excluded from the domain. Also, when performing operations with rational expressions, this text follows the convention of listing *by the simplified expression* all values of x that must be specifically excluded from the domain in order to make the domains of the simplified and original expressions agree. In Example 3, for instance, the restriction $x \neq 4$ is listed with the simplified expression to make the two domains agree. Note that the value $x = \frac{1}{2}$ is excluded from *both* domains, so it is not necessary to list this value.

Operations with Rational Expressions

To multiply or divide rational expressions, you can use the properties of fractions discussed in Section P.1. Recall that to divide fractions you invert the divisor and multiply.

EXAMPLE 4 Multiplying Rational Expressions

$$\frac{2x^2 + x - 6}{x^2 + 4x - 5} \cdot \frac{x^3 - 3x^2 + 2x}{4x^2 - 6x} = \frac{\cancel{(2x - 3)}(x + 2)}{(x + 5)\cancel{(x - 1)}} \cdot \frac{\cancel{x}(x - 2)\cancel{(x - 1)}}{2\cancel{x}\cancel{(2x - 3)}}$$

$$= \frac{(x + 2)(x - 2)}{2(x + 5)}, \quad x \neq 0, x \neq 1, x \neq \frac{3}{2}$$

Remark

In Example 4, the restrictions $x \neq 0$, $x \neq 1$, and $x \neq \frac{3}{2}$ are listed with the simplified expression in order to make the two domains agree. Note that the value $x = -5$ is excluded from both domains, so it is not necessary to list this value.

✓ Checkpoint *Audio-video solution in English & Spanish at LarsonPrecalculus.com.*

Multiply and simplify: $\frac{15x^2 + 5x}{x^3 - 3x^2 - 18x} \cdot \frac{x^2 - 2x - 15}{3x^2 - 8x - 3}$.

EXAMPLE 5 Dividing Rational Expressions

$$\frac{x^3 - 8}{x^2 - 4} \div \frac{x^2 + 2x + 4}{x^3 + 8} = \frac{x^3 - 8}{x^2 - 4} \cdot \frac{x^3 + 8}{x^2 + 2x + 4}$$ Invert and multiply.

$$= \frac{\cancel{(x - 2)}\cancel{(x^2 + 2x + 4)}}{\cancel{(x + 2)}\cancel{(x - 2)}} \cdot \frac{\cancel{(x + 2)}(x^2 - 2x + 4)}{\cancel{(x^2 + 2x + 4)}}$$

$$= x^2 - 2x + 4, \quad x \neq \pm 2$$ Divide out common factors.

✓ Checkpoint *Audio-video solution in English & Spanish at LarsonPrecalculus.com.*

Divide and simplify: $\frac{x^3 - 1}{x^2 - 1} \div \frac{x^2 + x + 1}{x^2 + 2x + 1}$.

To add or subtract rational expressions, you can use the LCD (least common denominator) method or the basic definition

$$\frac{a}{b} \pm \frac{c}{d} = \frac{ad \pm bc}{bd}, \quad b \neq 0 \text{ and } d \neq 0.$$ Basic definition

This definition provides an efficient way of adding or subtracting *two* fractions that have no common factors in their denominators.

EXAMPLE 6 Subtracting Rational Expressions

$$\frac{x}{x - 3} - \frac{2}{3x + 4} = \frac{x(3x + 4) - 2(x - 3)}{(x - 3)(3x + 4)}$$ Basic definition

$$= \frac{3x^2 + 4x - 2x + 6}{(x - 3)(3x + 4)}$$ Distributive Property

$$= \frac{3x^2 + 2x + 6}{(x - 3)(3x + 4)}$$ Combine like terms.

Remark

When subtracting rational expressions, remember to distribute the negative sign to *all* the terms in the quantity that is being subtracted.

✓ Checkpoint *Audio-video solution in English & Spanish at LarsonPrecalculus.com.*

Subtract: $\frac{x}{2x - 1} - \frac{1}{x + 2}$.

For three or more fractions, or for fractions with a repeated factor in the denominators, the LCD method works well. Recall that the least common denominator of several fractions consists of the product of all prime factors in the denominators, with each factor given the highest power of its occurrence in any denominator. Here is a numerical example.

$$\frac{1}{6}+\frac{3}{4}-\frac{2}{3}=\frac{1\cdot 2}{6\cdot 2}+\frac{3\cdot 3}{4\cdot 3}-\frac{2\cdot 4}{3\cdot 4}$$ The LCD is 12.

$$=\frac{2}{12}+\frac{9}{12}-\frac{8}{12}$$

$$=\frac{3}{12}$$

$$=\frac{1}{4}$$

Sometimes the numerator of the answer has a factor in common with the denominator. In such cases, the answer should be simplified. For instance, in the example above, $\frac{3}{12}$ was simplified to $\frac{1}{4}$.

EXAMPLE 7 Combining Rational Expressions: The LCD Method

See LarsonPrecalculus.com for an interactive version of this type of example.

Perform the operations and simplify.

$$\frac{3}{x-1}-\frac{2}{x}+\frac{x+3}{x^2-1}$$

Solution

Using the factored denominators $(x-1)$, x, and $(x+1)(x-1)$, you can see that the LCD is $x(x+1)(x-1)$.

$$\frac{3}{x-1}-\frac{2}{x}+\frac{x+3}{x^2-1}$$

$$=\frac{3}{x-1}-\frac{2}{x}+\frac{x+3}{(x+1)(x-1)}$$

$$=\frac{3(x)(x+1)}{x(x+1)(x-1)}-\frac{2(x+1)(x-1)}{x(x+1)(x-1)}+\frac{(x+3)(x)}{x(x+1)(x-1)}$$

$$=\frac{3(x)(x+1)-2(x+1)(x-1)+(x+3)(x)}{x(x+1)(x-1)}$$

$$=\frac{3x^2+3x-2x^2+2+x^2+3x}{x(x+1)(x-1)}$$ Distributive Property

$$=\frac{(3x^2-2x^2+x^2)+(3x+3x)+2}{x(x+1)(x-1)}$$ Group like terms.

$$=\frac{2x^2+6x+2}{x(x+1)(x-1)}$$ Combine like terms.

$$=\frac{2(x^2+3x+1)}{x(x+1)(x-1)}$$ Factor.

✓ Checkpoint *Audio-video solution in English & Spanish at LarsonPrecalculus.com.*

Perform the operations and simplify.

$$\frac{4}{x}-\frac{x+5}{x^2-4}+\frac{4}{x+2}$$ ■

Complex Fractions

Fractional expressions with separate fractions in the numerator, denominator, or both are called **complex fractions.** For instance,

$$\frac{\left(\frac{1}{x}\right)}{x^2 + 1}$$

and

$$\frac{\left(\frac{1}{x}\right)}{\left(\frac{1}{x^2 + 1}\right)}$$

are complex fractions. To simplify a complex fraction, combine the fractions in the numerator into a single fraction and then combine the fractions in the denominator into a single fraction. Then invert the denominator and multiply.

EXAMPLE 8 Simplifying a Complex Fraction

$$\frac{\left(\frac{2}{x} - 3\right)}{\left(1 - \frac{1}{x - 1}\right)} = \frac{\left[\frac{2 - 3(x)}{x}\right]}{\left[\frac{1(x - 1) - 1}{x - 1}\right]} \qquad \text{Combine fractions.}$$

$$= \frac{\left(\frac{2 - 3x}{x}\right)}{\left(\frac{x - 2}{x - 1}\right)} \qquad \text{Simplify.}$$

$$= \frac{2 - 3x}{x} \cdot \frac{x - 1}{x - 2} \qquad \text{Invert and multiply.}$$

$$= \frac{(2 - 3x)(x - 1)}{x(x - 2)}, \quad x \neq 1$$

✓ **Checkpoint** *Audio-video solution in English & Spanish at LarsonPrecalculus.com.*

Simplify the complex fraction $\dfrac{\left(\frac{1}{x + 2} + 1\right)}{\left(\frac{x}{3} - 1\right)}$.

In Example 8, the restriction $x \neq 1$ is listed by the final expression to make its domain agree with the domain of the original expression.

Another way to simplify a complex fraction is to multiply each term in its numerator and denominator by the LCD of all fractions in its numerator and denominator. This method is applied to the fraction in Example 8 as follows.

$$\frac{\left(\frac{2}{x} - 3\right)}{\left(1 - \frac{1}{x - 1}\right)} = \frac{\left(\frac{2}{x}\right)(x)(x - 1) - (3)(x)(x - 1)}{(1)(x)(x - 1) - \left(\frac{1}{x - 1}\right)(x)(x - 1)} \qquad \text{LCD is } x(x - 1).$$

$$= \frac{2(x - 1) - 3x(x - 1)}{x(x - 1) - x} \qquad \text{Simplify.}$$

$$= \frac{(2 - 3x)(x - 1)}{x(x - 2)}, \quad x \neq 1 \qquad \text{Factor.}$$

Simplifying Expressions from Calculus

The next four examples illustrate some methods for simplifying expressions involving negative exponents and radicals. These types of expressions occur frequently in calculus.

To simplify an expression with negative exponents, one method is to begin by factoring out the common factor with the lesser exponent. Remember that when factoring, you subtract exponents. For instance, in $3x^{-5/2} + 2x^{-3/2}$, the lesser exponent is $-\frac{5}{2}$ and the common factor is $x^{-5/2}$.

$$\begin{aligned} 3x^{-5/2} + 2x^{-3/2} &= x^{-5/2}[3(1) + 2x^{-3/2-(-5/2)}] \\ &= x^{-5/2}(3 + 2x^1) \\ &= \frac{3 + 2x}{x^{5/2}} \end{aligned}$$

EXAMPLE 9 Simplifying an Expression

Simplify $x(1 - 2x)^{-3/2} + (1 - 2x)^{-1/2}$.

Solution

Begin by factoring out the common factor with the lesser exponent.

$$\begin{aligned} x(1 - 2x)^{-3/2} + (1 - 2x)^{-1/2} &= (1 - 2x)^{-3/2}[x + (1 - 2x)^{(-1/2)-(-3/2)}] \\ &= (1 - 2x)^{-3/2}[x + (1 - 2x)^1] \\ &= \frac{1 - x}{(1 - 2x)^{3/2}} \end{aligned}$$

Checkpoint *Audio-video solution in English & Spanish at LarsonPrecalculus.com.*

Simplify $(x - 1)^{-1/3} - x(x - 1)^{-4/3}$.

A second method for simplifying an expression with negative exponents involves multiplying the numerator and denominator by another expression to eliminate the negative exponent.

EXAMPLE 10 Simplifying an Expression

Simplify $\dfrac{(4 - x^2)^{1/2} + x^2(4 - x^2)^{-1/2}}{4 - x^2}$.

Solution

$$\begin{aligned} \frac{(4 - x^2)^{1/2} + x^2(4 - x^2)^{-1/2}}{4 - x^2} &= \frac{(4 - x^2)^{1/2} + x^2(4 - x^2)^{-1/2}}{4 - x^2} \cdot \frac{(4 - x^2)^{1/2}}{(4 - x^2)^{1/2}} \\ &= \frac{(4 - x^2)^1 + x^2(4 - x^2)^0}{(4 - x^2)^{3/2}} \\ &= \frac{4 - x^2 + x^2}{(4 - x^2)^{3/2}} \\ &= \frac{4}{(4 - x^2)^{3/2}} \end{aligned}$$

Checkpoint *Audio-video solution in English & Spanish at LarsonPrecalculus.com.*

Simplify $\dfrac{x^2(x^2 - 2)^{-1/2} + (x^2 - 2)^{1/2}}{x^2 - 2}$.

EXAMPLE 11 Rewriting a Difference Quotient

The following expression from calculus is an example of a *difference quotient.*

$$\frac{\sqrt{x+h}-\sqrt{x}}{h}$$

Rewrite this expression by rationalizing its numerator.

Solution

$$\begin{aligned}\frac{\sqrt{x+h}-\sqrt{x}}{h} &= \frac{\sqrt{x+h}-\sqrt{x}}{h}\cdot\frac{\sqrt{x+h}+\sqrt{x}}{\sqrt{x+h}+\sqrt{x}}\\ &= \frac{(\sqrt{x+h})^2-(\sqrt{x})^2}{h(\sqrt{x+h}+\sqrt{x})}\\ &= \frac{x+h-x}{h(\sqrt{x+h}+\sqrt{x})}\\ &= \frac{h}{h(\sqrt{x+h}+\sqrt{x})}\\ &= \frac{1}{\sqrt{x+h}+\sqrt{x}},\quad h\neq 0\end{aligned}$$

Remark

Notice that the original expression is undefined when $h = 0$. So, you must exclude $h = 0$ from the domain of the simplified expression so that the expressions are equivalent.

✓ ***Checkpoint*** *Audio-video solution in English & Spanish at LarsonPrecalculus.com.*

Rewrite the difference quotient

$$\frac{\sqrt{9+h}-3}{h}$$

by rationalizing its numerator.

Difference quotients, like that in Example 11, occur frequently in calculus. Often, they need to be rewritten in an equivalent form that can be evaluated when $h = 0$. Note that the equivalent form is not simpler than the original form, but it has the advantage that it is defined when $h = 0$.

EXAMPLE 12 Rewriting a Difference Quotient

$$\begin{aligned}\frac{\sqrt{x-4}-\sqrt{x}}{4} &= \frac{\sqrt{x-4}-\sqrt{x}}{4}\cdot\frac{\sqrt{x-4}+\sqrt{x}}{\sqrt{x-4}+\sqrt{x}}\\ &= \frac{(\sqrt{x-4})^2-(\sqrt{x})^2}{4(\sqrt{x-4}+\sqrt{x})}\\ &= \frac{x-4-x}{4(\sqrt{x-4}+\sqrt{x})}\\ &= \frac{-4}{4(\sqrt{x-4}+\sqrt{x})}\\ &= -\frac{1}{\sqrt{x-4}+\sqrt{x}}\end{aligned}$$

✓ ***Checkpoint*** *Audio-video solution in English & Spanish at LarsonPrecalculus.com.*

Rewrite the difference quotient $\frac{\sqrt{x+2}-\sqrt{x}}{2}$ by rationalizing its numerator.

P.4 Exercises

See CalcChat.com for tutorial help and worked-out solutions to odd-numbered exercises.
For instructions on how to use a graphing utility, see Appendix A.

Vocabulary and Concept Check

In Exercises 1–4, fill in the blank.

1. The set of real numbers for which an algebraic expression is defined is the _______ of the expression.
2. The quotient of two algebraic expressions is a fractional expression, and the quotient of two polynomials is a _______ .
3. Fractional expressions with separate fractions in the numerator, denominator, or both are called _______ .
4. To simplify an expression with negative exponents, it is possible to begin by factoring out the common factor with the _______ exponent.

5. When is a rational expression in simplest form?
6. What values are excluded from the domain of a rational expression?

Procedures and Problem Solving

Finding the Domain of an Algebraic Expression **In Exercises 7–22, find the domain of the expression.**

7. $x^2 + 7x - 3$
8. $6x^2 - x - 10$
9. $5x^2 + 1,\ x > 0$
10. $9x - 4,\ x \le 0$
11. $\dfrac{x}{x + 2}$
12. $\dfrac{1 - x}{4 - x}$
13. $\dfrac{x^2 + 3x}{x^2 + 14x + 49}$
14. $\dfrac{x^2 + 2x - 8}{x^2 - 4}$
15. $\dfrac{x^2 - 2x - 3}{9x^2 - 1}$
16. $\dfrac{x^2 + 6x + 9}{x^2 - 10x + 25}$
17. $\sqrt{x + 10}$
18. $\sqrt{x - 7}$
19. $\sqrt{12 - 3x}$
20. $\sqrt{6 - 4x}$
21. $\dfrac{1}{\sqrt{x + 1}}$
22. $\dfrac{1}{\sqrt{x - 5}}$

Writing Equivalent Fractions **In Exercises 23–28, find the missing factor in the numerator such that the two fractions are equivalent.**

23. $\dfrac{5}{2x} = \dfrac{5(\quad)}{6x^2}$
24. $\dfrac{2}{3x^2} = \dfrac{2(\quad)}{3x^4}$
25. $\dfrac{3}{4} = \dfrac{3(\quad)}{4(x + 1)}$
26. $\dfrac{2}{5} = \dfrac{2(\quad)}{5(x - 3)}$
27. $\dfrac{x - 1}{4(x + 2)} = \dfrac{(x - 1)(\quad)}{4(x + 2)^2}$
28. $\dfrac{x + 3}{2(x - 1)} = \dfrac{(x + 3)(\quad)}{2(x - 1)^2}$

Simplifying a Rational Expression **In Exercises 29–46, write the rational expression in simplest form.**

29. $\dfrac{15x^2}{10x}$
30. $\dfrac{18y^2}{60y^5}$
31. $\dfrac{3xy}{x^2y + x^2}$
32. $\dfrac{2x^2y}{xy - y}$
33. $\dfrac{4y - 8y^2}{10y - 5}$
34. $\dfrac{9x^2 + 9x}{2x + 2}$
35. $\dfrac{x - 5}{10 - 2x}$
36. $\dfrac{12 - 4x}{x - 3}$
37. $\dfrac{y^2 - 16}{y + 4}$
38. $\dfrac{x^2 - 25}{5 - x}$
39. $\dfrac{x^3 + 5x^2 + 6x}{x^2 - 4}$
40. $\dfrac{x^2 + 8x - 20}{x^2 + 11x + 10}$
41. $\dfrac{y^2 - 7y + 12}{y^2 + 3y - 18}$
42. $\dfrac{-10 - x}{x^2 + 11x + 10}$
43. $\dfrac{2 - x + 2x^2 - x^3}{x - 2}$
44. $\dfrac{x^2 - 9}{x^3 + x^2 - 9x - 9}$
45. $\dfrac{z^3 - 8}{z^2 + 2z + 4}$
46. $\dfrac{y^3 - 2y^2 - 3y}{y^3 + 1}$

Comparing Rational Expressions **In Exercises 47 and 48, complete the table. What can you conclude?**

47.

x	-4	-3	-2	-1	0	1	2
$\dfrac{x^2 + 2x - 3}{x - 1}$							
$x + 3$							

48.

x	0	1	2	3	4	5	6
$\dfrac{x - 3}{x^2 - x - 6}$							
$\dfrac{1}{x + 2}$							

Multiplying or Dividing Rational Expressions In Exercises 49–56, perform the multiplication or division and simplify.

49. $\dfrac{5}{x-1} \cdot \dfrac{x-1}{25(x-2)}$

50. $\dfrac{x+13}{x^3(3-x)} \cdot \dfrac{x(x-3)}{5}$

51. $\dfrac{r}{1-r} \div \dfrac{r^2}{r^2-1}$

52. $\dfrac{4y-16}{5y+15} \div \dfrac{4-y}{2y+6}$

53. $\dfrac{t^2-t-6}{t^2+6t+9} \cdot \dfrac{t+3}{t^2-4}$

54. $\dfrac{y^3-8}{2y^3} \cdot \dfrac{4y}{y^2-5y+6}$

55. $\dfrac{3(x+y)}{4} \div \dfrac{x+y}{2}$

56. $\dfrac{2x-y}{y-1} \div \dfrac{3y-6x}{y^2-6y+5}$

Geometry In Exercises 57 and 58, find the ratio of the area of the shaded portion of the figure to the total area of the figure.

57.

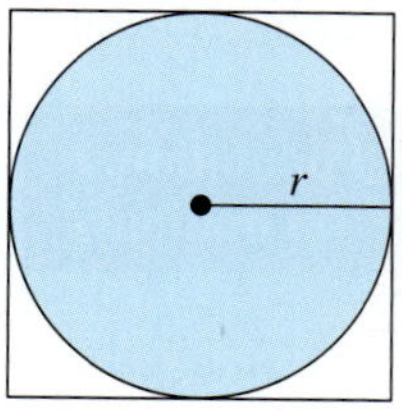

58.

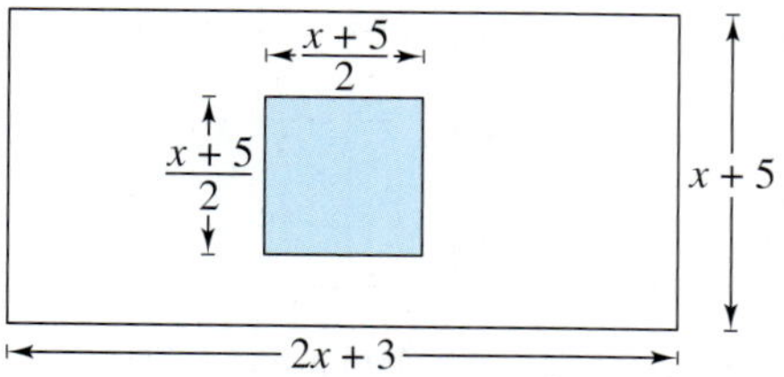

Adding or Subtracting Rational Expressions In Exercises 59–70, perform the indicated operation(s) and simplify.

59. $\dfrac{5}{x-1} + \dfrac{x}{x-1}$

60. $\dfrac{2x-1}{x+3} - \dfrac{1-x}{x+3}$

61. $\dfrac{6}{2x+1} - \dfrac{x}{x+3}$

62. $\dfrac{3}{x-1} + \dfrac{5x}{3x+4}$

63. $\dfrac{3}{x-2} + \dfrac{5}{2-x}$

64. $\dfrac{2x}{x-5} - \dfrac{5}{5-x}$

65. $\dfrac{1}{x^2-x-2} - \dfrac{x}{x^2-5x+6}$

66. $\dfrac{2}{x^2-x-2} + \dfrac{10}{x^2+2x-8}$

67. $\dfrac{2}{x+1} + \dfrac{2}{x-1} + \dfrac{1}{x^2-1}$

68. $-\dfrac{1}{x} + \dfrac{2}{x^2+1} - \dfrac{1}{x^3+x}$

69. $\dfrac{1}{x^2+x} - \dfrac{6}{x^2} + \dfrac{5}{x+1}$

70. $\dfrac{3}{x-3} - \dfrac{x}{x^2-9} - \dfrac{2}{x}$

Simplifying a Complex Fraction In Exercises 71–78, simplify the complex fraction.

71. $\dfrac{\left(\dfrac{x}{2}-1\right)}{(x-2)}$

72. $\dfrac{(x-4)}{\left(\dfrac{x}{4}-\dfrac{4}{x}\right)}$

73. $\dfrac{\left[\dfrac{x^2}{(x+1)^2}\right]}{\left[\dfrac{x}{(x+1)^3}\right]}$

74. $\dfrac{\left(\dfrac{x^2-1}{x}\right)}{\left[\dfrac{(x-1)^2}{x}\right]}$

75. $\dfrac{\left[\dfrac{1}{(x+h)^2}-\dfrac{1}{x^2}\right]}{h}$

76. $\dfrac{\left(\dfrac{x+h}{x+h+1}-\dfrac{x}{x+1}\right)}{h}$

77. $\dfrac{\left(\sqrt{x}-\dfrac{1}{2\sqrt{x}}\right)}{\sqrt{x}}$

78. $\dfrac{\left(\dfrac{t^2}{\sqrt{t^2-1}}-\sqrt{t^2-1}\right)}{t^2}$

Simplifying an Expression with Negative Exponents In Exercises 79–84, simplify the expression by removing the common factor with the lesser exponent.

79. $x^5 - 2x^{-2}$

80. $x^5 - 5x^{-3}$

81. $x^2(x^2+1)^{-5} - (x^2+1)^{-4}$

82. $2x(x-5)^{-3} - 4x^2(x-5)^{-4}$

83. $2x^2(x-1)^{1/2} - 5(x-1)^{-1/2}$

84. $4x^3(2x-1)^{3/2} - 2x(2x-1)^{-1/2}$

Simplifying an Expression with Negative Exponents In Exercises 85–90, simplify the expression.

85. $\dfrac{2x^{3/2}-x^{-1/2}}{x^2}$

86. $\dfrac{x^2(x^{-1/2})-3x^{1/2}(x^2)}{x^4}$

87. $\dfrac{-x^2(x^2+1)^{-1/2}+2x(x^2+1)^{-3/2}}{x^3}$

88. $\dfrac{x^3(4x^{-1/2})-3x^2\left(\frac{8}{3}x^{-3/2}\right)}{x^6}$

89. $\dfrac{(x^2+5)\left(\frac{1}{2}\right)(4x+3)^{-1/2}(4)-(4x+3)^{1/2}(2x)}{(x^2+5)^2}$

90. $\dfrac{(2x+1)^{1/2}(3)(x-5)^2-(x-5)^3\left(\frac{1}{2}\right)(2x+1)^{-1/2}(2)}{2x+1}$

Rewriting a Difference Quotient **In Exercises 91–96, rewrite the expression by rationalizing its numerator.**

91. $\dfrac{\sqrt{x+4}-\sqrt{x}}{4}$ **92.** $\dfrac{\sqrt{z-3}-\sqrt{z}}{3}$

93. $\dfrac{\sqrt{x+2}-\sqrt{2}}{x}$

94. $\dfrac{\sqrt{x+5}-\sqrt{5}}{x}$

95. $\dfrac{\sqrt{1-x}-1}{x}$

96. $\dfrac{\sqrt{4+x}-2}{x}$

Probability **In Exercises 97 and 98, consider an experiment in which a marble is randomly tossed into a box whose base is shown in the figure. The probability that the marble will come to rest in the shaded portion of the base is equal to the ratio of the shaded area to the total area of the figure. Find the probability.**

97.
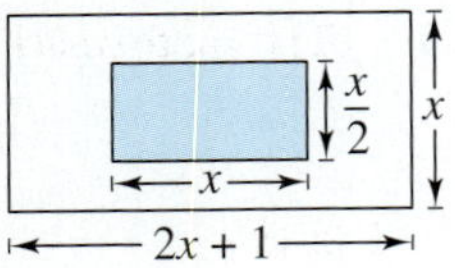

98.
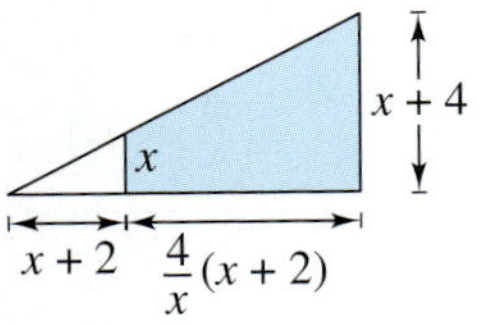

Finance **In Exercises 99 and 100, the formula that approximates the annual interest rate r of a monthly installment loan is given by**

$$r = \frac{\left[\dfrac{24(NM-P)}{N}\right]}{\left(P+\dfrac{NM}{12}\right)}$$

where N is the total number of payments, M is the monthly payment, and P is the amount financed.

99. (a) Approximate the annual interest rate for a four-year car loan of \$20,000 that has monthly payments of \$475.

(b) Simplify the expression for the annual interest rate r, and then rework part (a).

100. (a) Approximate the annual interest rate for a five-year car loan of \$28,000 that has monthly payments of \$525.

(b) Simplify the expression for the annual interest rate r, and then rework part (a).

101. Using a Rate A digital copier copies in color at a rate of 50 pages per minute.

(a) Find the time required to copy one page.

(b) Find the time required to copy x pages.

(c) Find the time required to copy 120 pages.

102. Electrical Engineering The formula for the total resistance R_T (in ohms) of a parallel circuit is given by

$$R_T = \frac{1}{\dfrac{1}{R_1}+\dfrac{1}{R_2}+\dfrac{1}{R_3}}$$

where R_1, R_2, and R_3 are the resistance values of the first, second, and third resistors, respectively.

(a) Simplify the total resistance formula.

(b) Find the total resistance in the parallel circuit when $R_1 = 6$ ohms, $R_2 = 4$ ohms, and $R_3 = 12$ ohms.

103. MODELING DATA

The table shows the number of births (in millions) and the population (in millions) of the United States from 2005 through 2012. (Sources: U.S. National Center for Health Statistics, U.S. Census Bureau)

DATA

Year	Births, B (in millions)	Population, P (in millions)
2005	4.138	295.5
2006	4.266	298.4
2007	4.316	301.2
2008	4.248	304.1
2009	4.131	306.8
2010	3.999	309.3
2011	3.954	311.6
2012	3.953	313.9

Spreadsheet at LarsonPrecalculus.com

Mathematical models for the data are

Births: $B = \dfrac{0.06815t^2 - 0.9865t + 3.948}{0.01753t^2 - 0.2530t + 1}$

and

Population: $P = 2.64t + 282.7$

where t represents the year, with $t = 5$ corresponding to 2005.

(a) Using the models, create a table to estimate the number of births and the number of people for each of the given years.

(b) Compare the estimates from part (a) with the actual data.

(c) Determine a model for the ratio of the number of births to the number of people.

(d) Use the model from part (c) to find the ratio for each of the given years. Interpret your results.

104. *Why you should learn it* (*p. 35*) When food (at room temperature) is placed in a refrigerator, the time required for the food to cool depends on the amount of food, the air circulation in the refrigerator, the original temperature of the food, and the temperature of the refrigerator. Consider the model that gives the temperature of food that is at 75°F and is placed in a 40°F refrigerator as

$$T = 10\left(\frac{4t^2 + 16t + 75}{t^2 + 4t + 10}\right)$$

where T is the temperature (in degrees Fahrenheit) and t is the time (in hours).

(a) Complete the table.

t	0	2	4	6	8	10
T						

t	12	14	16	18	20	22
T						

(b) What value of T does the mathematical model appear to be approaching?

Simplifying a Difference Quotient **In Exercises 105–108, simplify the expression.**

105. $\dfrac{(x+h)^2 - x^2}{h}$

106. $\dfrac{(x+h)^3 - x^3}{h}$

107. $\dfrac{\dfrac{1}{(x+h)^2} - \dfrac{1}{x^2}}{h}$

108. $\dfrac{\dfrac{1}{2(x+h)} - \dfrac{1}{2x}}{h}$

Simplifying an Expression **In Exercises 109 and 110, simplify the expression.**

109. $\dfrac{4}{n}\left(\dfrac{n(n+1)(2n+1)}{6}\right) + 2n\left(\dfrac{4}{n}\right)$

110. $9\left(\dfrac{3}{n}\right)\left(\dfrac{n(n+1)(2n+1)}{6}\right) - n\left(\dfrac{3}{n}\right)$

Conclusions

True or False? **In Exercises 111 and 112, determine whether the statement is true or false. Justify your answer.**

111. $\dfrac{x^{2n} - 1^{2n}}{x^n - 1^n} = x^n + 1^n$

112. $\dfrac{x^2 - 3x + 2}{x - 1} = x - 2$

113. Think About It Explain why rewriting the difference quotient

$$\frac{\sqrt{x+h} - \sqrt{x}}{h}$$

by rationalizing its numerator does not result in an expression in simplest form.

114. Think About It Is the following statement true for all nonzero real numbers a and b? Explain.

$$\frac{ax - b}{b - ax} = -1$$

Error Analysis **In Exercises 115 and 116, describe the error.**

115. $\dfrac{5x^3}{2x^3 + 4} = \dfrac{5x^3}{2x^3 + 4} = \dfrac{5}{2 + 4} = \dfrac{5}{6}$ (crossed out)

116. $\dfrac{x+4}{x+2} - \dfrac{3x-8}{x+2} = \dfrac{x + 4 - 3x - 8}{x + 2} = \dfrac{-2x - 4}{x + 2} = \dfrac{-2(x+2)}{x+2} = -2$ (crossed out)

117. Writing Write a paragraph explaining to a classmate why

$$\frac{1}{x+y} \neq \frac{1}{x} + \frac{1}{y}.$$

118. HOW DO YOU SEE IT? The mathematical model

$$P = 100\left(\frac{t^2 - t + 1}{t^2 + 1}\right), \quad t \geq 0$$

gives the percent P of the normal level of oxygen in a pond, where t is the time (in weeks) after organic waste is dumped into the pond. The bar graph shows the situation. What conclusions can you draw from the bar graph?

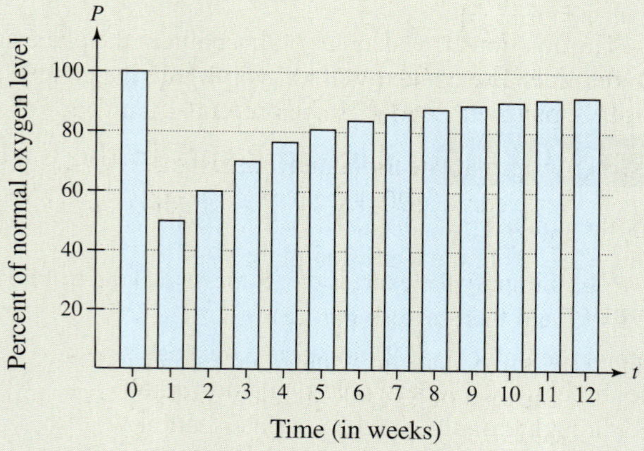

119. Exploration Simplify $2x^{-2} - x^{-4}$ by factoring out the common factor with the lesser exponent. Simplify

$$\frac{2x^{-2} - x^{-4}}{1}$$

by multiplying the numerator and denominator by an expression to eliminate the negative exponent. Discuss your results.

P.5 The Cartesian Plane

The Cartesian Plane

Just as you can represent real numbers by points on a real number line, you can represent ordered pairs of real numbers by points in a plane called the **rectangular coordinate system,** or the **Cartesian plane,** after the French mathematician René Descartes (1596–1650).

The Cartesian plane is formed by using two real number lines intersecting at right angles, as shown in Figure P.10. The horizontal real number line is usually called the ***x*-axis,** and the vertical real number line is usually called the ***y*-axis.** The point of intersection of these two axes is the **origin,** and the two axes divide the plane into four parts called **quadrants.**

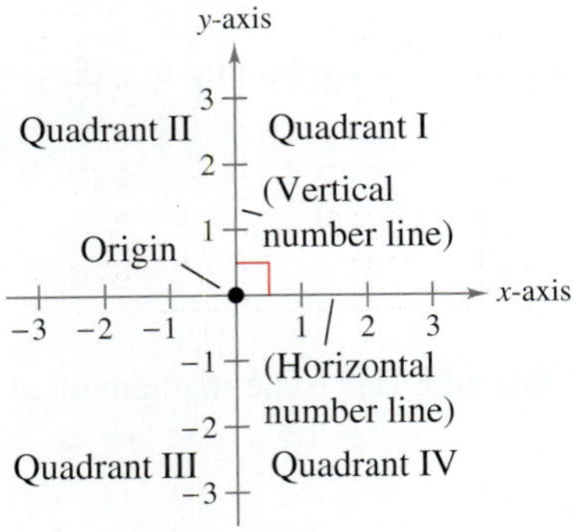

Figure P.10 ***The Cartesian Plane***

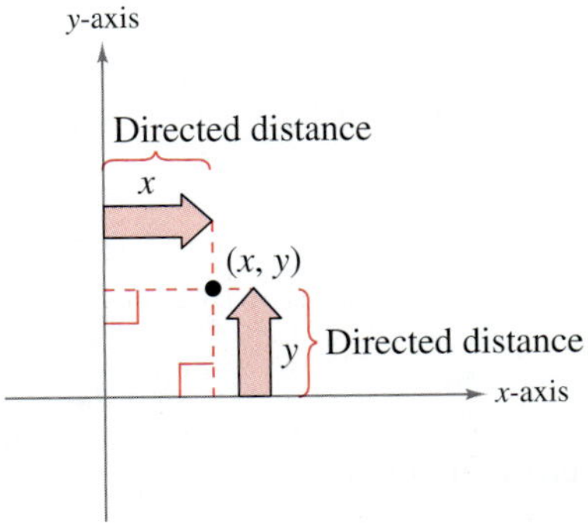

Figure P.11 ***Ordered Pair (x, y)***

Each point in the plane corresponds to an **ordered pair** (x, y) of real numbers x and y, called **coordinates** of the point. The ***x*-coordinate** represents the directed distance from the y-axis to the point, and the ***y*-coordinate** represents the directed distance from the x-axis to the point, as shown in Figure P.11.

Directed distance from y-axis → (x, y) ← Directed distance from x-axis

The notation (x, y) denotes both a point in the plane and an open interval on the real number line. The context will tell you which meaning is intended.

What you should learn

- Plot points in the Cartesian plane and sketch scatter plots.
- Use the Distance Formula to find the distance between two points.
- Use the Midpoint Formula to find the midpoint of a line segment.
- Find the equation of a circle.
- Translate points in the plane.

Why you should learn it

The Cartesian plane can be used to represent relationships between two variables. For instance, Exercise 96 on page 56 shows how to represent graphically the numbers of performers elected to the Rock and Roll Hall of Fame from 1995 through 2014.

EXAMPLE 1 Plotting Points in the Cartesian Plane

Plot the points $(-1, 2)$, $(3, 4)$, $(0, 0)$, $(3, 0)$, and $(-2, -3)$ in the Cartesian plane.

Solution

To plot the point $(-1, 2)$, imagine a vertical line through -1 on the x-axis and a horizontal line through 2 on the y-axis. The intersection of these two lines is the point $(-1, 2)$. This point is one unit to the left of the y-axis and two units up from the x-axis. The other four points can be plotted in a similar way, as shown in the figure.

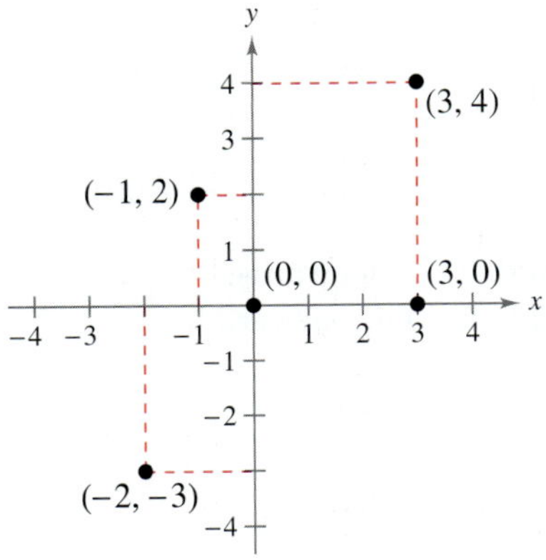

✓ Checkpoint *Audio-video solution in English & Spanish at LarsonPrecalculus.com.*

Plot the points $(-3, 2)$, $(4, -2)$, $(3, 1)$, $(0, -2)$, and $(-1, -2)$ in the Cartesian plane.

The beauty of a rectangular coordinate system is that it enables you to see relationships between two variables. It would be difficult to overestimate the importance of Descartes's introduction of coordinates in the plane. Today, his ideas are in common use in virtually every scientific and business-related field.

In the next example, data are represented graphically by points plotted in a rectangular coordinate system. This type of graph is called a **scatter plot.**

EXAMPLE 2 Sketching a Scatter Plot

The numbers of employees E (in thousands) in dentist offices in the United States from 2004 through 2013 are shown in the table, where t represents the year. Sketch a scatter plot of the data by hand. (Source: U.S. Bureau of Labor Statistics)

DATA

Year, t	Employees, E
2004	760
2005	774
2006	786
2007	808
2008	818
2009	818
2010	828
2011	844
2012	854
2013	870

Spreadsheet at LarsonPrecalculus.com

Solution

Before you sketch the scatter plot, it is helpful to represent each pair of values by an ordered pair (t, E) as follows.

(2004, 760), (2005, 774), (2006, 786), (2007, 808), (2008, 818),
(2009, 818), (2010, 828), (2011, 844), (2012, 854), (2013, 870)

To sketch a scatter plot of the data shown in the table, first draw a vertical axis to represent the number of employees (in thousands) and a horizontal axis to represent the year. Then plot the resulting points, as shown in the figure. Note that the break in the t-axis indicates that the numbers 0 through 2003 have been omitted.

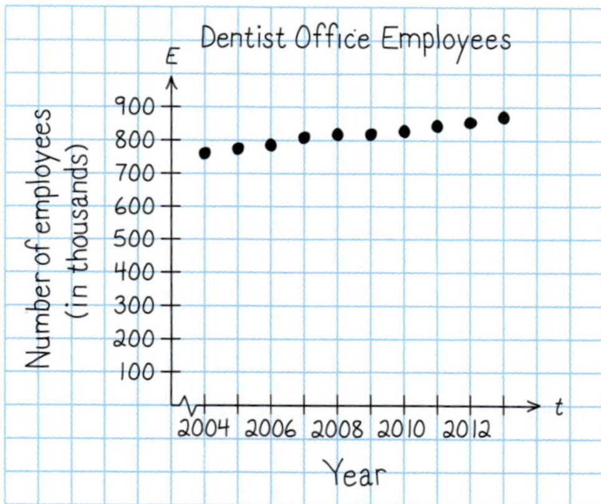

Checkpoint *Audio-video solution in English & Spanish at LarsonPrecalculus.com.*

The numbers of employees E (in thousands) in outpatient care centers in the United States from 2004 through 2013 are shown in the table, where t represents the year. Sketch a scatter plot of the data by hand. (Source: U.S. Bureau of Labor Statistics)

DATA

Year, t	Employees, E
2004	451
2005	473
2006	493
2007	512
2008	533
2009	558
2010	600
2011	621
2012	649
2013	682

Spreadsheet at LarsonPrecalculus.com

Technology Tip

You can use a graphing utility to graph the scatter plot in Example 2. For instructions on how to use the *list editor* and the *statistical plotting* feature, see Appendix A; for specific keystrokes, go to this textbook's *Companion Website.*

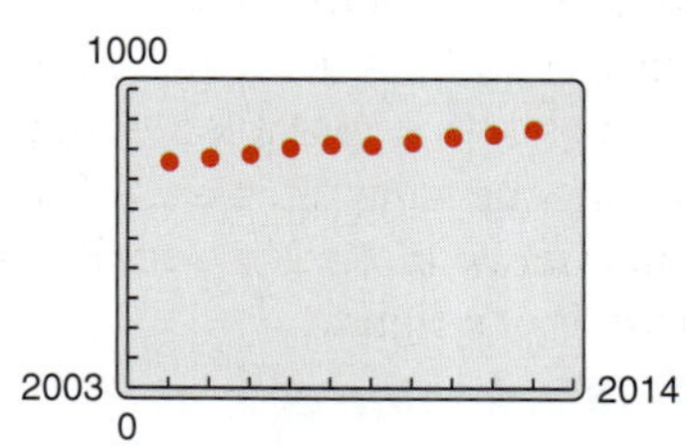

In Example 2, you could have let $t = 1$ represent the year 2004. In that case, the horizontal axis of the graph would not have been broken, and the tick marks would have been labeled 1 through 10 (instead of 2004 through 2013).

The Distance Formula

Recall from the Pythagorean Theorem that, for a right triangle with hypotenuse of length c and sides of lengths a and b, you have

$$a^2 + b^2 = c^2 \qquad \text{Pythagorean Theorem}$$

as shown in Figure P.12. (The converse is also true. That is, if $a^2 + b^2 = c^2$, then the triangle is a right triangle.)

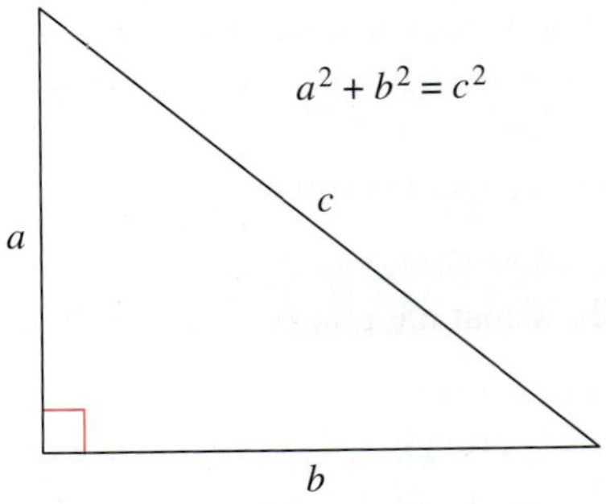

Figure P.12

Suppose you want to determine the distance d between two points (x_1, y_1) and (x_2, y_2) in the plane. Using these two points, you can form a right triangle, as shown in Figure P.13. The length of the vertical side of the triangle is $|y_2 - y_1|$ and the length of the horizontal side is $|x_2 - x_1|$. By the Pythagorean Theorem,

$$d^2 = |x_2 - x_1|^2 + |y_2 - y_1|^2$$

$$d = \sqrt{|x_2 - x_1|^2 + |y_2 - y_1|^2}$$

$$d = \sqrt{(x_2 - x_1)^2 + (y_2 - y_1)^2}.$$

This result is called the **Distance Formula.**

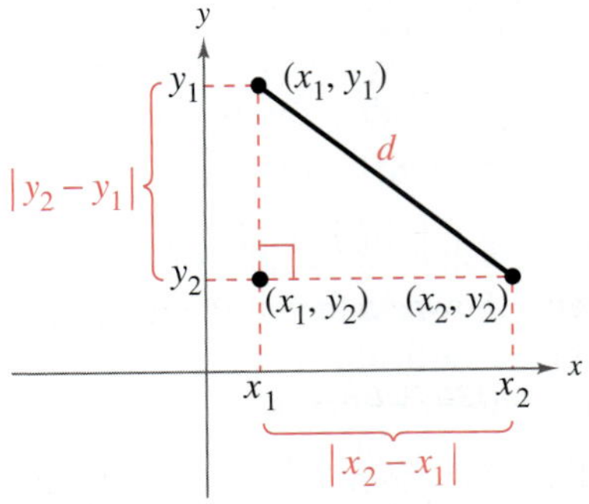

Figure P.13

The Distance Formula

The distance d between the points (x_1, y_1) and (x_2, y_2) in the plane is

$$d = \sqrt{(x_2 - x_1)^2 + (y_2 - y_1)^2}.$$

EXAMPLE 3 Finding a Distance

Find the distance between the points

$(-2, 1)$ and $(3, 4)$.

Algebraic Solution

Let $(x_1, y_1) = (-2, 1)$ and $(x_2, y_2) = (3, 4)$. Then apply the Distance Formula as follows.

$$d = \sqrt{(x_2 - x_1)^2 + (y_2 - y_1)^2} \qquad \text{Distance Formula}$$

$$= \sqrt{[3 - (-2)]^2 + (4 - 1)^2} \qquad \text{Substitute for } x_1, y_1, x_2, \text{ and } y_2.$$

$$= \sqrt{(5)^2 + (3)^2} \qquad \text{Simplify.}$$

$$= \sqrt{34} \qquad \text{Simplify.}$$

$$\approx 5.83 \qquad \text{Use a calculator.}$$

So, the distance between the points is about 5.83 units. You can use the Pythagorean Theorem to check that the distance is correct.

$$d^2 \stackrel{?}{=} 3^2 + 5^2 \qquad \text{Pythagorean Theorem}$$

$$(\sqrt{34})^2 \stackrel{?}{=} 3^2 + 5^2 \qquad \text{Substitute for } d.$$

$$34 = 34 \qquad \text{Distance checks. ✓}$$

Graphical Solution

Use centimeter graph paper to plot the points $A(-2, 1)$ and $B(3, 4)$. Carefully sketch the line segment from A to B. Then use a centimeter ruler to measure the length of the segment.

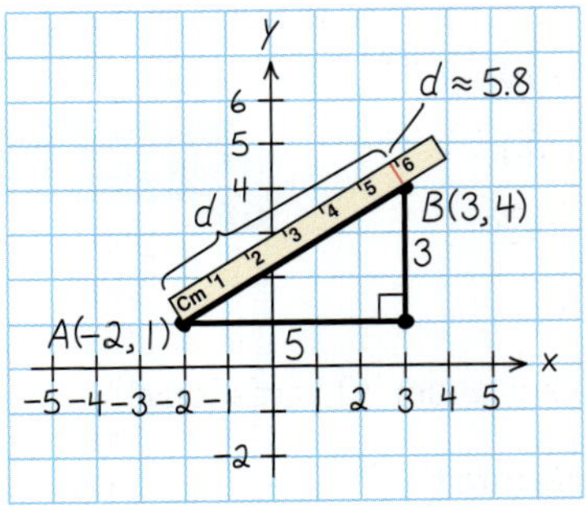

The line segment measures about 5.8 centimeters, as shown in the figure. So, the distance between the points is about 5.8 units.

✓ Checkpoint Audio-video solution in English & Spanish at LarsonPrecalculus.com.

Find the distance between the points

$(3, 1)$ and $(-3, 0)$.

When the Distance Formula is used, it does not matter which point is (x_1, y_1) and which is (x_2, y_2), because the result will be the same. For instance, in Example 3, let $(x_1, y_1) = (3, 4)$ and $(x_2, y_2) = (-2, 1)$. Then

$$d = \sqrt{(-2-3)^2 + (1-4)^2} = \sqrt{(-5)^2 + (-3)^2} = \sqrt{34} \approx 5.83.$$

EXAMPLE 4 Verifying a Right Triangle

Show that the points (2, 1), (4, 0), and (5, 7) are the vertices of a right triangle.

Solution

The three points are plotted in Figure P.14. Using the Distance Formula, you can find the lengths of the three sides as follows.

$$d_1 = \sqrt{(5-2)^2 + (7-1)^2} = \sqrt{9+36} = \sqrt{45}$$

$$d_2 = \sqrt{(4-2)^2 + (0-1)^2} = \sqrt{4+1} = \sqrt{5}$$

$$d_3 = \sqrt{(5-4)^2 + (7-0)^2} = \sqrt{1+49} = \sqrt{50}$$

Because

$$(d_1)^2 + (d_2)^2 = 45 + 5 = 50 = (d_3)^2$$

you can conclude that the triangle must be a right triangle.

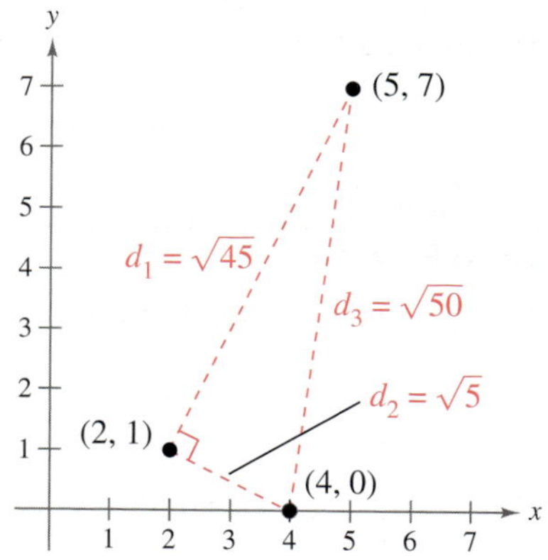

Figure P.14

Checkpoint *Audio-video solution in English & Spanish at LarsonPrecalculus.com.*

Show that the points (2, −1), (5, 5), and (6, −3) are vertices of a right triangle.

EXAMPLE 5 Finding the Length of a Pass

A football quarterback throws a pass from the 28-yard line, 40 yards from the sideline. A wide receiver catches the pass on the 5-yard line, 20 yards from the same sideline, as shown in Figure P.15. How long is the pass?

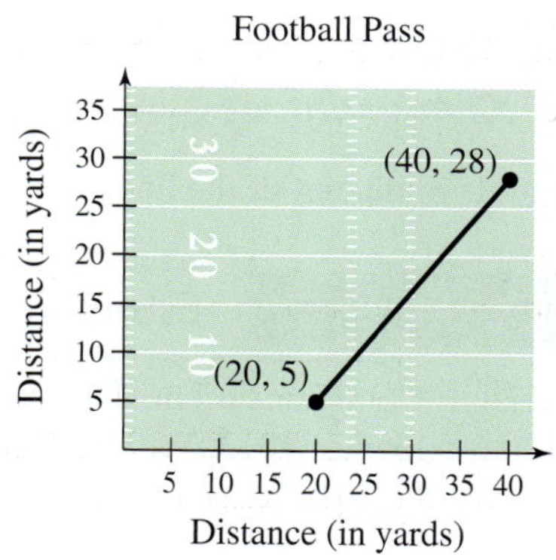

Figure P.15

Solution

You can find the length of the pass by finding the distance between the points (40, 28) and (20, 5).

$d = \sqrt{(x_2 - x_1)^2 + (y_2 - y_1)^2}$	Distance Formula
$= \sqrt{(40-20)^2 + (28-5)^2}$	Substitute for x_1, y_1, x_2, and y_2.
$= \sqrt{20^2 + 23^2}$	Simplify.
$= \sqrt{400 + 529}$	Simplify.
$= \sqrt{929}$	Simplify.
≈ 30	Use a calculator.

So, the pass is about 30 yards long.

Checkpoint *Audio-video solution in English & Spanish at LarsonPrecalculus.com.*

A football quarterback throws a pass from the 10-yard line, 10 yards from the sideline. A wide receiver catches the pass on the 32-yard line, 25 yards from the same sideline. How long is the pass?

In Example 5, the scale along the goal line does not normally appear on a football field. However, when you use coordinate geometry to solve real-life problems, you are free to place the coordinate system in any way that is convenient for the solution of the problem.

The Midpoint Formula

To find the **midpoint** of the line segment that joins two points in a coordinate plane, find the average values of the respective coordinates of the two endpoints using the **Midpoint Formula.**

> **The Midpoint Formula** (See the proof on page 72.)
>
> The midpoint of the line segment joining the points (x_1, y_1) and (x_2, y_2) is given by the Midpoint Formula
>
> $$\text{Midpoint} = \left(\frac{x_1 + x_2}{2}, \frac{y_1 + y_2}{2}\right).$$

EXAMPLE 6 Finding a Line Segment's Midpoint

Find the midpoint of the line segment joining the points $(-5, -3)$ and $(9, 3)$.

Solution

Let $(x_1, y_1) = (-5, -3)$ and $(x_2, y_2) = (9, 3)$.

$$\text{Midpoint} = \left(\frac{x_1 + x_2}{2}, \frac{y_1 + y_2}{2}\right) \qquad \text{Midpoint Formula}$$

$$= \left(\frac{-5 + 9}{2}, \frac{-3 + 3}{2}\right) \qquad \text{Substitute for } x_1, y_1, x_2, \text{ and } y_2.$$

$$= (2, 0) \qquad \text{Simplify.}$$

The midpoint of the line segment is $(2, 0)$, as shown in Figure P.16.

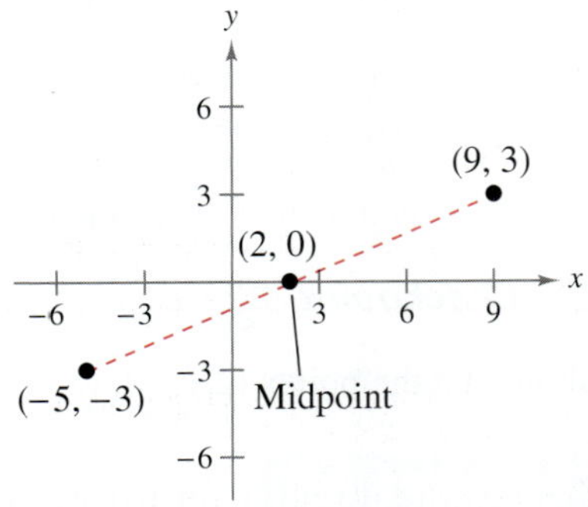

Figure P.16

Checkpoint Audio-video solution in English & Spanish at LarsonPrecalculus.com.

Find the midpoint of the line segment joining the points $(-2, 8)$ and $(4, -10)$.

EXAMPLE 7 Estimating Annual Revenues

Verizon Communications had annual revenues of \$110.9 billion in 2011 and \$120.6 billion in 2013. Without knowing any additional information, what would you estimate the 2012 revenue to have been? (Source: Verizon Communications)

Solution

One solution to the problem is to assume that revenue followed a linear pattern. With this assumption, you can estimate the 2012 revenue by finding the midpoint of the line segment connecting the points (2011, 110.9) and (2013, 120.6).

$$\text{Midpoint} = \left(\frac{2011 + 2013}{2}, \frac{110.9 + 120.6}{2}\right)$$

$$= (2012, 115.75)$$

So, you would estimate the 2012 revenue to have been about \$115.75 billion, as shown in Figure P.17. (The actual 2012 revenue was \$115.8 billion.)

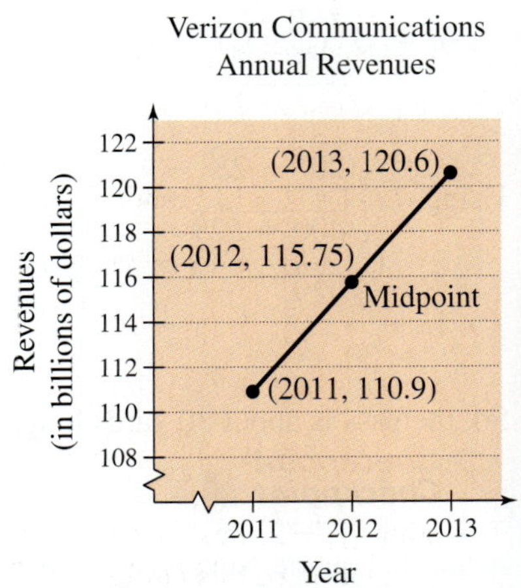

Figure P.17

Checkpoint Audio-video solution in English & Spanish at LarsonPrecalculus.com.

Google, Inc. had annual revenues of approximately \$37.9 billion in 2011 and \$59.8 billion in 2013. Without knowing any additional information, what would you estimate the 2012 revenue to have been? (Source: Google, Inc.)

The Equation of a Circle

The Distance Formula provides a convenient way to define circles. A circle of radius r with center at the point (h, k) is shown in Figure P.18. The point (x, y) is on this circle if and only if its distance from the center (h, k) is r. This means that a **circle** in the plane consists of all points (x, y) that are a given positive distance r from a fixed point (h, k). Using the Distance Formula, you can express this relationship by saying that the point (x, y) lies on the circle if and only if

$$\sqrt{(x-h)^2+(y-k)^2}=r.$$

By squaring each side of this equation, you obtain the **standard form of the equation of a circle.**

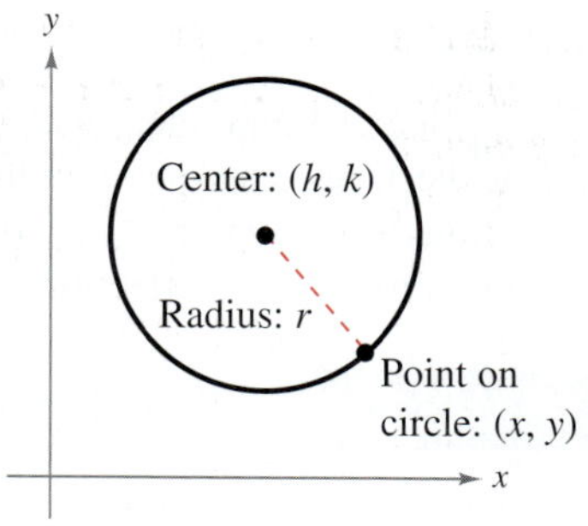

Figure P. 18

Standard Form of the Equation of a Circle

The **standard form of the equation of a circle** is

$$(x-h)^2+(y-k)^2=r^2. \quad \text{Circle with center at } (h, k)$$

The point (h, k) is the **center** of the circle, and the positive number r is the **radius** of the circle. The standard form of the equation of a circle whose center is the origin, $(h, k) = (0, 0)$, is

$$x^2+y^2=r^2. \quad \text{Circle with center at origin}$$

$(1, -3)$

EXAMPLE 8 Writing an Equation of a Circle

The point $(3, 4)$ lies on a circle whose center is at $(-1, 2)$, as shown in Figure P.19. Write the standard form of the equation of this circle.

Solution

The radius r of the circle is the distance between $(-1, 2)$, and $(3, 4)$.

$$\begin{aligned} r &= \sqrt{(x-h)^2+(y-k)^2} && \text{Distance Formula} \\ &= \sqrt{[3-(-1)]^2+(4-2)^2} && \text{Substitute for } x, y, h, \text{ and } k. \\ &= \sqrt{4^2+2^2} && \text{Simplify.} \\ &= \sqrt{16+4} && \text{Simplify.} \\ &= \sqrt{20} && \text{Radius} \end{aligned}$$

Using $(h, k) = (-1, 2)$ and $r = \sqrt{20}$, the equation of the circle is

$$\begin{aligned} (x-h)^2+(y-k)^2 &= r^2 && \text{Equation of circle} \\ [x-(-1)]^2+(y-2)^2 &= \left(\sqrt{20}\right)^2 && \text{Substitute for } h, k, \text{ and } r. \\ (x+1)^2+(y-2)^2 &= 20. && \text{Standard form} \end{aligned}$$

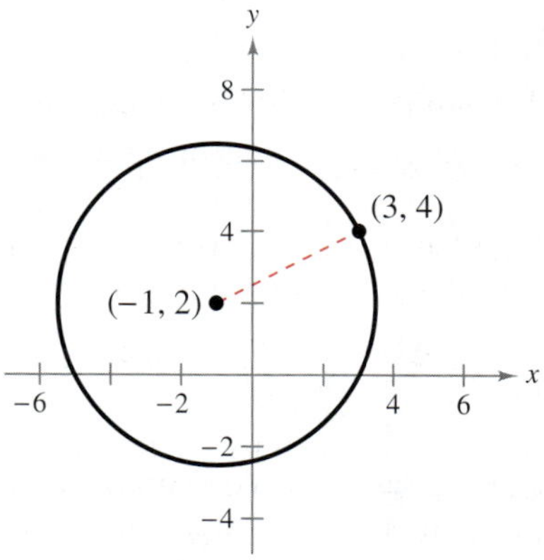

Figure P.19

✔ **Checkpoint** *Audio-video solution in English & Spanish at LarsonPrecalculus.com.*

The point $(1, -2)$ lies on a circle whose center is at $(-3, -5)$. Write the standard form of the equation of this circle.

Be careful when you are finding h and k from the standard form of the equation of a circle. For instance, to find h and k from the equation of the circle in Example 8, rewrite the quantities $(x + 1)^2$ and $(y - 2)^2$ using subtraction.

$$(x+1)^2=[x-(-1)]^2 \quad \text{and} \quad (y-2)^2=[y-(2)]^2$$

So, $h = -1$ and $k = 2$.

Application

Much of computer graphics, including the computer-generated goldfish tessellation shown at the right, consists of transformations of points in a coordinate plane. One type of transformation, a translation, is illustrated in Example 9. Other types of transformations include reflections, rotations, and stretches.

EXAMPLE 9 Translating Points in the Plane

See LarsonPrecalculus.com for an interactive version of this type of example.

The triangle in Figure P.20 has vertices at the points $(-1, 2)$, $(1, -4)$, and $(2, 3)$. Shift the triangle three units to the right and two units upward and find the vertices of the shifted triangle, as shown in Figure P.21.

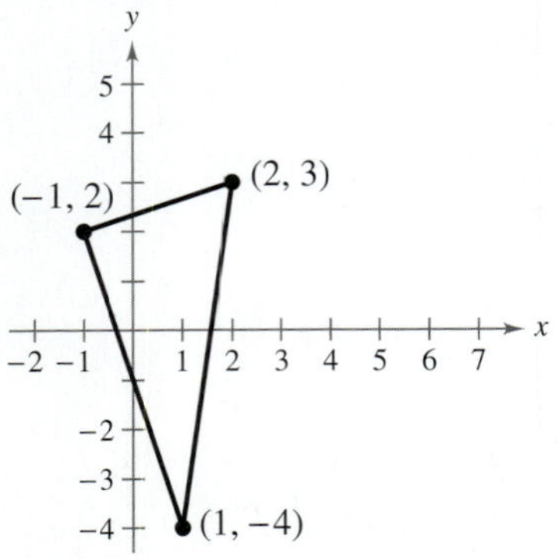

Figure P.20

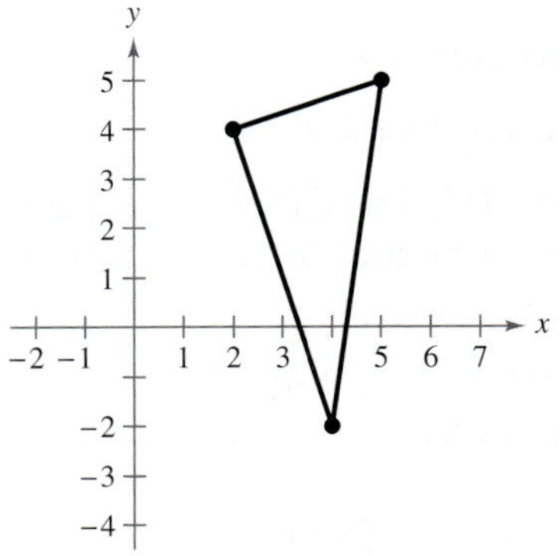

Figure P.21

Solution

To shift the vertices three units to the right, add 3 to each of the x-coordinates. To shift the vertices two units upward, add 2 to each of the y-coordinates.

Original Point	*Translated Point*
$(-1, 2)$	$(-1 + 3, 2 + 2) = (2, 4)$
$(1, -4)$	$(1 + 3, -4 + 2) = (4, -2)$
$(2, 3)$	$(2 + 3, 3 + 2) = (5, 5)$

Plotting the translated points and sketching the line segments between them produces the shifted triangle shown in Figure P.21.

Checkpoint Audio-video solution in English & Spanish at LarsonPrecalculus.com.

Find the vertices of the parallelogram shown in Figure P.22 after translating it two units to the left and four units down.

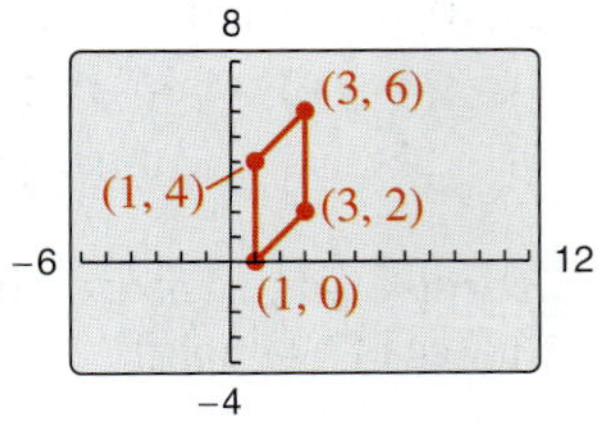

Figure P.22

Example 9 shows how to translate points in a coordinate plane. The following transformed points are related to the original points as follows.

Original Point	*Transformed Point*	
(x, y)	$(-x, y)$	$(-x, y)$ is a reflection of the original point in the y-axis.
(x, y)	$(x, -y)$	$(x, -y)$ is a reflection of the original point in the x-axis.
(x, y)	$(-x, -y)$	$(-x, -y)$ is a reflection of the original point through the origin.

The figures provided with Example 9 were not really essential to the solution. Nevertheless, it is strongly recommended that you develop the habit of including sketches with your solutions because they serve as useful problem-solving tools.

P.5 Exercises

See CalcChat.com for tutorial help and worked-out solutions to odd-numbered exercises.
For instructions on how to use a graphing utility, see Appendix A.

Vocabulary and Concept Check

In Exercises 1–4, fill in the blank(s).

1. An ordered pair of real numbers can be represented in a plane called the rectangular coordinate system or the _______ plane.

2. The _______ is a result derived from the Pythagorean Theorem.

3. Finding the average values of the respective coordinates of the two endpoints of a line segment in a coordinate plane is also known as using the _______ .

4. The standard form of the equation of a circle is _______ , where the point (h, k) is the _______ of the circle and the positive number r is the _______ of the circle.

In Exercises 5–10, match each term with its definition.

5. x-axis (a) point of intersection of vertical axis and horizontal axis

6. y-axis (b) directed distance from the x-axis

7. origin (c) horizontal real number line

8. quadrants (d) four regions of the coordinate plane

9. x-coordinate (e) directed distance from the y-axis

10. y-coordinate (f) vertical real number line

Procedures and Problem Solving

Approximating Coordinates of Points **In Exercises 11 and 12, approximate the coordinates of the points.**

11.

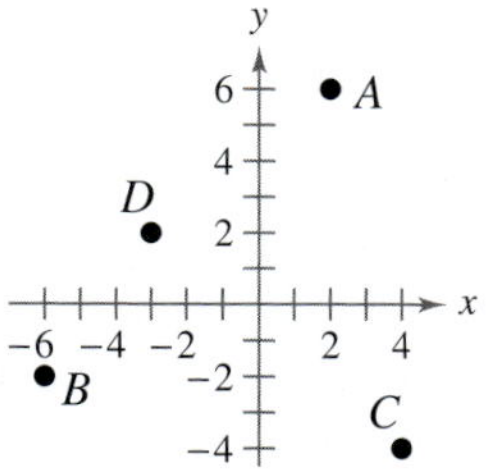

12.

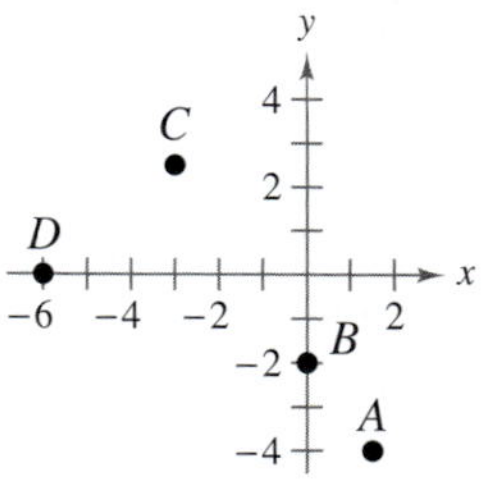

Plotting Points in the Cartesian Plane **In Exercises 13–16, plot the points in the Cartesian plane.**

13. $(-4, 2), (-3, -6), (0, 5), (1, -4)$

14. $(4, -2), (0, 0), (-4, 0), (-5, -5)$

15. $(3, 8), (0.5, -1), (5, -6), (-2, -2.5)$

16. $\left(1, -\frac{1}{2}\right), \left(-\frac{5}{2}, 2\right), (3, -3), \left(\frac{3}{2}, 1\right)$

Finding Coordinates of Points **In Exercises 17–20, find the coordinates of the point.**

17. The point is located five units to the left of the y-axis and four units above the x-axis.

18. The point is located three units below the x-axis and two units to the right of the y-axis.

19. The point is on the y-axis and six units below the x-axis.

20. The point is on the x-axis and 11 units to the left of the y-axis.

Determining Quadrants **In Exercises 21–30, determine the quadrant(s) in which (x, y) is located so that the condition(s) is (are) satisfied.**

21. $x > 0$ and $y < 0$

22. $x < 0$ and $y < 0$

23. $x = -4$ and $y > 0$

24. $x > 2$ and $y = 3$

25. $y < -5$

26. $x > 4$

27. $x < 0$ and $-y > 0$

28. $-x > 0$ and $y < 0$

29. $xy > 0$

30. $xy < 0$

31. **Accounting** The table shows the sales y (in millions of dollars) for Apple for the years 2002 through 2013. Sketch a scatter plot of the data. (Source: Apple Inc.)

DATA

Year	Sales, y (in millions of dollars)
2002	5,742
2003	6,207
2004	8,279
2005	13,931
2006	19,315
2007	24,006
2008	32,479
2009	36,537
2010	65,255
2011	108,249
2012	156,508
2013	170,910

Spreadsheet at LarsonPrecalculus.com

32. Meteorology The table shows the lowest temperature on record y (in degrees Fahrenheit) in Flagstaff, Arizona, for each month x, where $x = 1$ represents January. Sketch a scatter plot of the data. (Source: National Oceanic and Atmospheric Administration)

DATA

Month, x	Temperature, y
1	-22
2	-23
3	-16
4	-2
5	14
6	22
7	32
8	24
9	23
10	-2
11	-13
12	-23

Spreadsheet at LarsonPrecalculus.com

Finding a Distance **In Exercises 33–42, find the distance between the points algebraically and confirm graphically by using centimeter graph paper and a centimeter ruler.**

33. $(6, -3), (6, 5)$

34. $(-11, 4), (-1, 4)$

35. $(-2, 6), (3, -6)$

36. $(8, 5), (0, 20)$

37. $(2, 6), (-5, 5)$

38. $(-3, -7), (1, -15)$

39. $\left(\frac{1}{2}, \frac{4}{3}\right), (2, -1)$

40. $\left(-\frac{2}{3}, 3\right), \left(-1, \frac{5}{4}\right)$

41. $(-4.2, 3.1), (-12.5, 4.8)$

42. $(9.5, -2.6), (-3.9, 8.2)$

Verifying a Right Triangle **In Exercises 43–46, (a) find the length of each side of the right triangle and (b) show that these lengths satisfy the Pythagorean Theorem.**

43.

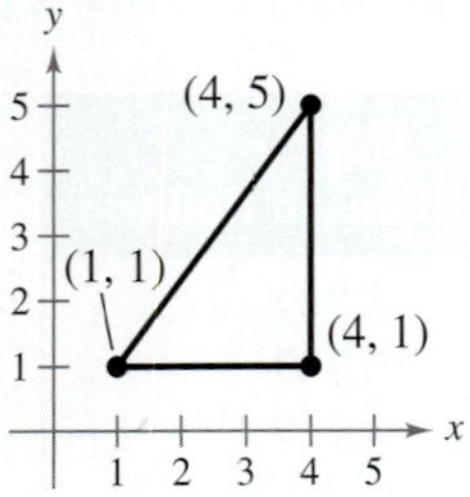

44.

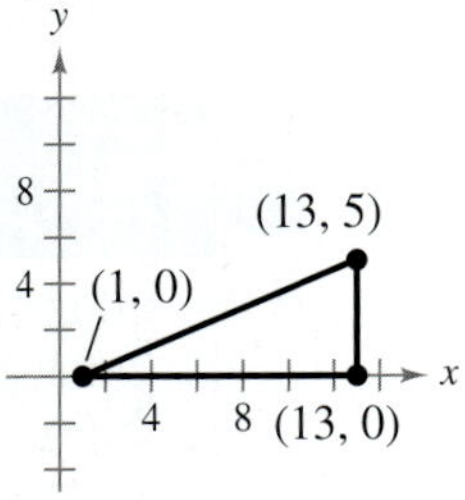

45.

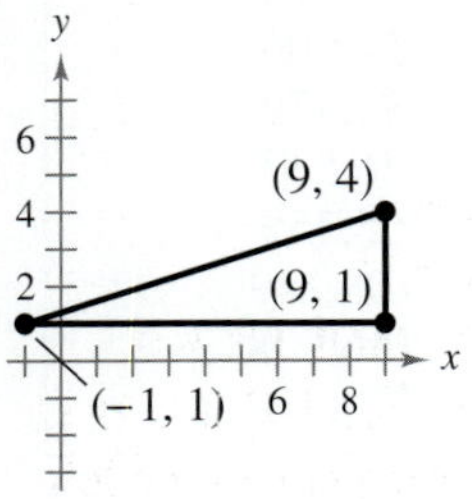

46.

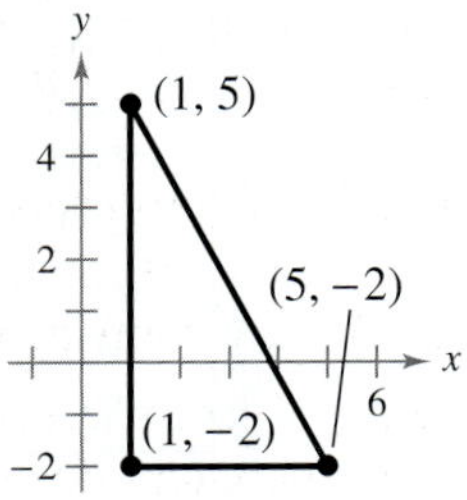

Verifying a Polygon **In Exercises 47–54, show that the points form the vertices of the polygon.**

47. Right triangle: $(4, 0), (2, 1), (-1, -5)$

48. Right triangle: $(-1, 3), (3, 5), (5, 1)$

49. Isosceles triangle: $(1, -3), (3, 2), (-2, 4)$

50. Isosceles triangle: $(2, 3), (4, 9), (-2, 7)$

51. Parallelogram: $(2, 5), (0, 9), (-2, 0), (0, -4)$

52. Parallelogram: $(0, 1), (3, 7), (4, 4), (1, -2)$

53. Rectangle: $(-5, 6), (0, 8), (-3, 1), (2, 3)$ (*Hint:* Show that the diagonals are of equal length.)

54. Rectangle: $(2, 4), (3, 1), (1, 2), (4, 3)$ (*Hint:* Show that the diagonals are of equal length.)

55. Physical Education In a football game, a quarterback throws a pass from the 15-yard line, 10 yards from the sideline, as shown in the figure. The pass is caught on the 40-yard line, 45 yards from the same sideline. How long is the pass?

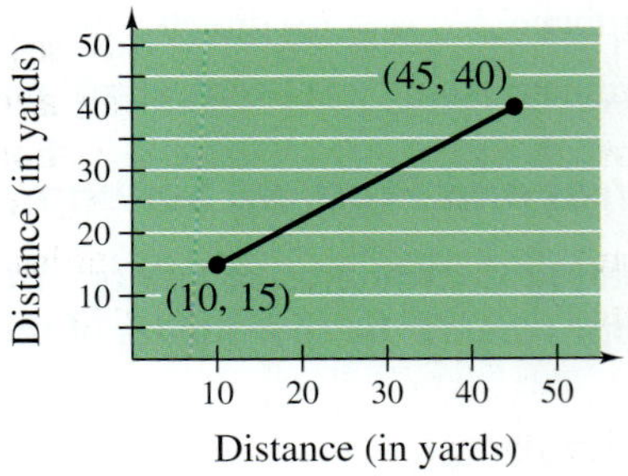

56. Physical Education A major league baseball diamond is a square with 90-foot sides. In the figure, home plate is at the origin and the first base line lies on the positive x-axis. The right fielder fields the ball at the point (300, 25). How far does the right fielder have to throw the ball to get a runner out at home plate? How far does the right fielder have to throw the ball to get a runner out at third base? (Round your answers to one decimal place.)

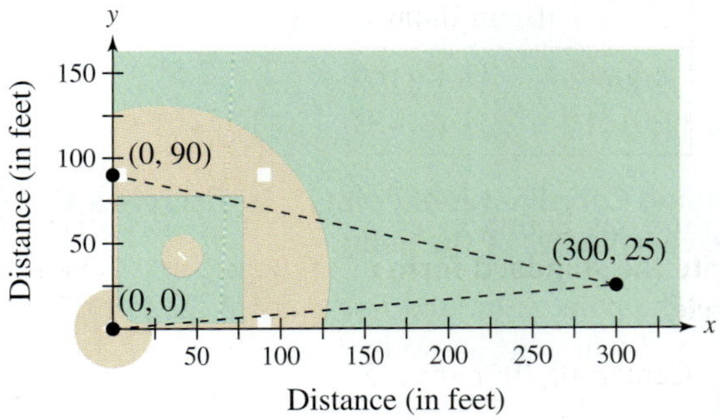

57. Aviation A jet plane flies from Naples, Italy, in a straight line to Rome, Italy, which is 120 kilometers north and 150 kilometers west of Naples. How far does the plane fly?

58. **Geography** Oklahoma City, Oklahoma, is about 514 miles east of Albuquerque, New Mexico, and about 360 miles north of Austin, Texas. Find the approximate distance between Albuquerque and Austin.

Finding a Line Segment's Midpoint **In Exercises 59–68, (a) plot the points and (b) find the midpoint of the line segment joining the points.**

59. $(0, 0), (8, 6)$

60. $(1, 12), (9, 0)$

61. $(-1, 2), (5, 4)$

62. $(2, 10), (10, 2)$

63. $(-4, 10), (4, -5)$

64. $(-7, -4), (2, 8)$

65. $\left(\frac{1}{2}, 1\right), \left(-\frac{5}{2}, \frac{4}{3}\right)$

66. $\left(-\frac{1}{3}, -\frac{1}{3}\right), \left(-\frac{1}{6}, -\frac{1}{2}\right)$

67. $(6.2, 5.4), (-3.7, 1.8)$

68. $(-16.8, 12.3), (5.6, 4.9)$

Estimating Annual Revenues **In Exercises 69 and 70, use the Midpoint Formula to estimate the annual revenues (in millions of dollars) for Texas Roadhouse and Papa John's in 2011. The revenues for the two companies in 2009 and 2013 are shown in the tables. Assume that the revenues followed a linear pattern.** (Sources: Texas Roadhouse, Inc.; Papa John's International)

69. Texas Roadhouse

Year	Annual revenue (in millions of dollars)
2009	942
2013	1423

70. Papa John's Intl.

Year	Annual revenue (in millions of dollars)
2009	1106
2013	1439

Writing an Equation of a Circle **In Exercises 71–84, write the standard form of the equation of the specified circle.**

71. Center: $(0, 0)$; radius: 5

72. Center: $(0, 0)$; radius: 6

73. Center: $(2, -1)$; radius: 4

74. Center: $(-5, 3)$; radius: 2

75. Center: $(-1, 2)$; solution point: $(0, 0)$

76. Center: $(3, -2)$; solution point: $(-1, 1)$

77. Endpoints of a diameter: $(-4, -1), (4, 1)$

78. Endpoints of a diameter: $(0, 0), (6, 8)$

79. Center: $(-2, 1)$; tangent to the x-axis

80. Center: $(3, -2)$; tangent to the y-axis

81. The circle inscribed in the square with vertices $(7, -2)$, $(-1, -2)$, $(-1, -10)$, and $(7, -10)$

82. The circle inscribed in the square with vertices $(-12, 10)$, $(8, 10)$, $(8, -10)$, and $(-12, -10)$

83.

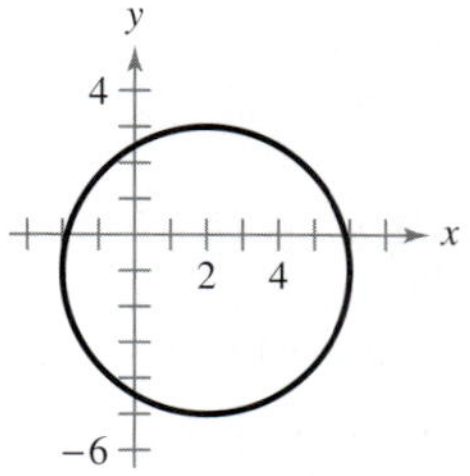

84.

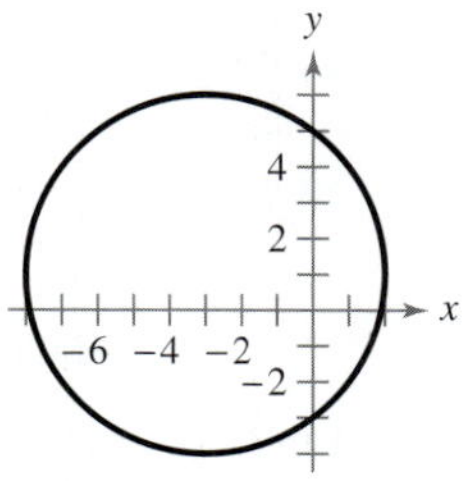

Sketching a Circle **In Exercises 85–90, find the center and radius, and sketch the circle.**

85. $x^2 + y^2 = 25$

86. $x^2 + y^2 = 64$

87. $(x - 6)^2 + y^2 = 9$

88. $(x + 1)^2 + (y - 3)^2 = 4$

89. $\left(x - \frac{1}{3}\right)^2 + \left(y + \frac{2}{3}\right)^2 = \frac{16}{9}$

90. $(x + 4.5)^2 + (y - 0.5)^2 = 2.25$

Translating Points in the Plane **In Exercises 91–94, the polygon is shifted to a new position in the plane. Find the coordinates of the vertices of the polygon in the new position.**

91.

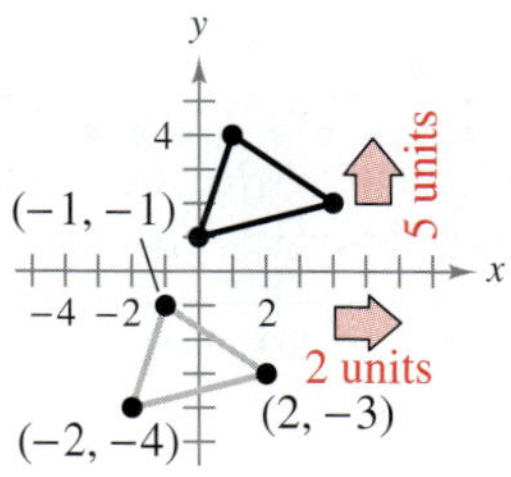

92.

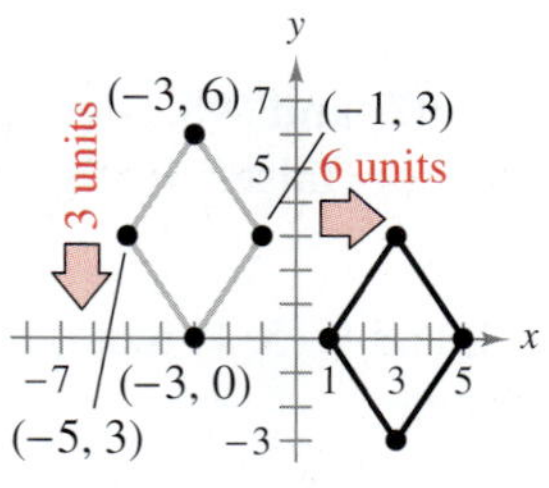

93. Original coordinates of vertices:
$(0, 2), (-3, 5), (-5, 2), (-2, -1)$
Shift: three units upward, one unit to the left

94. Original coordinates of vertices: $(1, -1), (3, 2), (1, -2)$
Shift: two units downward, three units to the left

95. Education The scatter plot shows the mathematics entrance test scores x and the final examination scores y in an algebra course for a sample of 10 students.

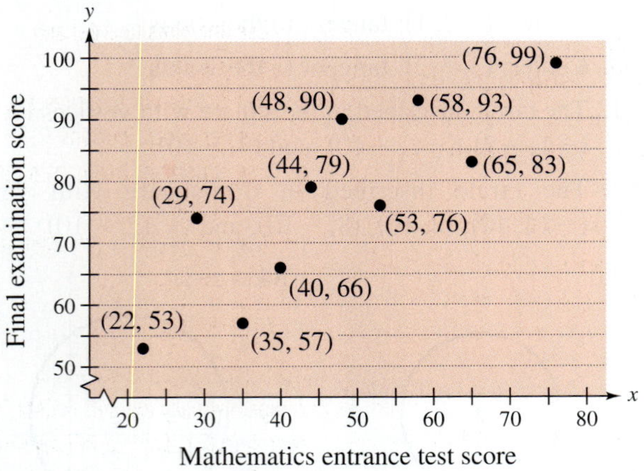

(a) Find the entrance exam score of any student with a final exam score in the 80s.

(b) Does a higher entrance exam score necessarily imply a higher final exam score? Explain.

96. ***Why you should learn it*** *(p. 46)* The graph shows the numbers of performers who were elected to the Rock and Roll Hall of Fame from 1995 through 2014. Describe any trends or irregularities in the data. From these trends, predict the number of performers elected in 2015. (Source: rockhall.com)

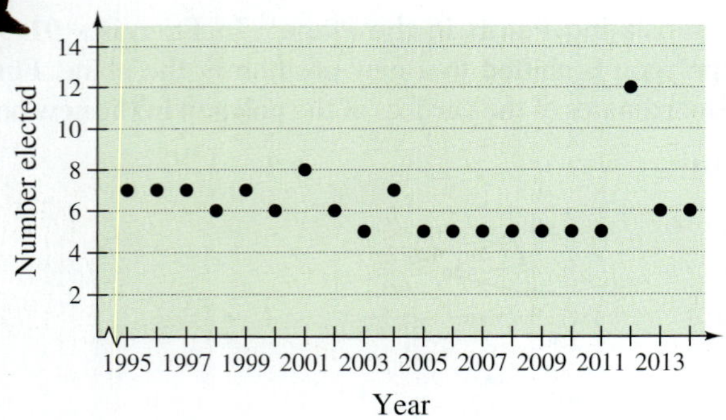

Conclusions

True or False? In Exercises 97 and 98, determine whether the statement is true or false. Justify your answer.

97. The points $(-8, 4)$, $(2, 11)$, and $(-5, 1)$ represent the vertices of an isosceles triangle.

98. If four points represent the vertices of a polygon, and the four sides are equal, then the polygon must be a square.

99. Think About It What is the y-coordinate of any point on the x-axis? What is the x-coordinate of any point on the y-axis?

100. Exploration A line segment has (x_1, y_1) as one endpoint and (x_m, y_m) as its midpoint. Find the other endpoint (x_2, y_2) of the line segment in terms of x_1, y_1, x_m, and y_m. Use the result to find the coordinates of the endpoint of a line segment when the coordinates of the other endpoint and midpoint are, respectively,

(a) $(1, -2), (4, -1)$.

(b) $(-5, 11), (2, 4)$.

101. Exploration Use the Midpoint Formula three times to find the three points that divide the line segment joining (x_1, y_1) and (x_2, y_2) into four parts. Use the result to find the points that divide the line segment joining the given points into four equal parts.

(a) $(1, -2), (4, -1)$ (b) $(-2, -3), (0, 0)$

102. HOW DO YOU SEE IT? Use the plot of the point (x_0, y_0) in the figure. Match the transformation of the point with the correct plot. Explain your reasoning. [The plots are labeled (i), (ii), (iii), and (iv).]

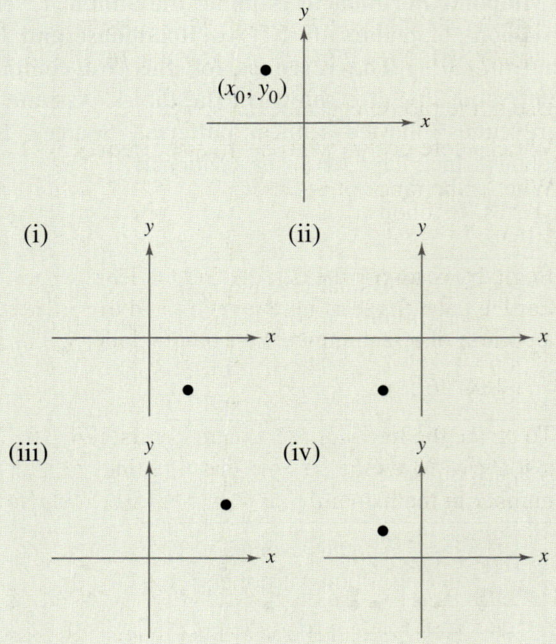

(a) $(x_0, -y_0)$ (b) $(-2x_0, y_0)$

(c) $\left(x_0, \frac{1}{2}y_0\right)$ (d) $(-x_0, -y_0)$

103. Proof Prove that the diagonals of the parallelogram in the figure intersect at their midpoints.

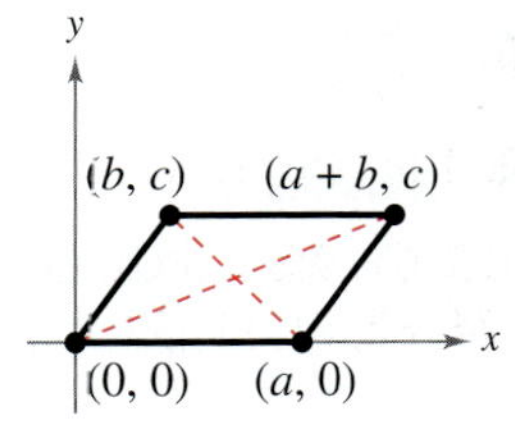

P.6 Representing Data Graphically

Line Plots

Statistics is the branch of mathematics that studies techniques for collecting, organizing, and interpreting data. In this section, you will study several ways to organize data. The first is a **line plot,** which uses a portion of a real number line to order numbers. Line plots are especially useful for ordering small sets of numbers (about 50 or less) by hand.

Many statistical measures can be obtained from a line plot. Two such measures are the *frequency* and *range* of the data. The **frequency** measures the number of times a value occurs in a data set. The **range** is the difference between the greatest and least data values. For example, consider the data values

20, 21, 21, 25, 32.

The frequency of 21 in the data set is 2 because 21 occurs twice. The range is 12 because the difference between the greatest and least data values is $32 - 20 = 12$.

What you should learn

- Use line plots to order and analyze data.
- Use histograms to represent frequency distributions.
- Use bar graphs to represent and analyze data.
- Use line graphs to represent and analyze data.

Why you should learn it

Double bar graphs allow you to compare visually two sets of data over time. For example, in Exercises 15 and 16 on page 63, you are asked to estimate the difference in tuition between public and private institutions of higher education.

EXAMPLE 1 Constructing a Line Plot

The scores from an economics class of 30 students are listed. The scores are for a 100-point exam. *(Spreadsheet at LarsonPrecalculus.com)*

DATA 93, 70, 76, 67, 86, 93, 82, 78, 83, 86, 64, 78, 76, 66, 83, 83, 96, 74, 69, 76, 64, 74, 79, 76, 88, 76, 81, 82, 74, 70

a. Use a line plot to organize the scores.

b. Which score occurs with the greatest frequency?

c. What is the range of the scores?

Solution

a. Begin by scanning the data to find the least and greatest numbers. For the data, the least number is 64 and the greatest is 96. Next, draw a portion of a real number line that includes the interval

$[64, 96]$.

To create the line plot, start with the first number, 93, and record a • above 93 on the number line. Continue recording a • for each number in the list until you obtain the line plot shown in the figure.

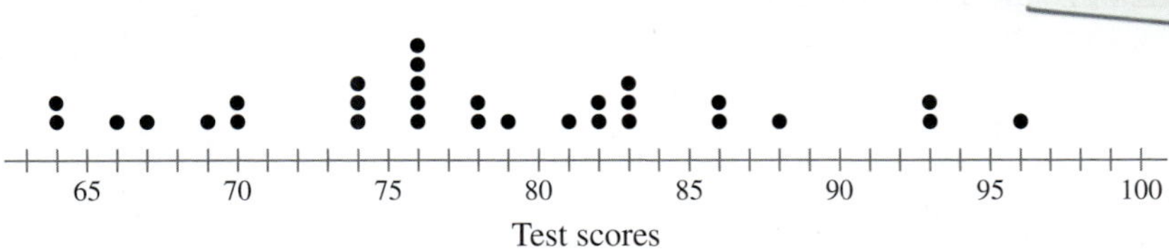

b. From the line plot, you can see that 76 occurs with the greatest frequency.

c. Because the range is the difference between the greatest and least data values, the range of scores is

$96 - 64 = 32.$

✓ **Checkpoint** *Audio-video solution in English & Spanish at LarsonPrecalculus.com.*

Rework Example 1 for the test scores listed. *(Spreadsheet at LarsonPrecalculus.com)*

DATA 68, 73, 67, 95, 71, 82, 85, 74, 82, 61, 87, 92, 78, 74, 64, 71, 74, 82, 71, 83, 92, 82, 78, 72, 82, 64, 85, 67, 71, 62

Histograms and Frequency Distributions

When you want to organize large sets of data, it is useful to group the data into intervals and plot the frequency of the data in each interval. A **frequency distribution** can be used to construct a **histogram.** A histogram uses a portion of a real number line as its horizontal axis. The bars of a histogram are not separated by spaces.

EXAMPLE 2 Constructing a Histogram

The table at the right shows the percent of the resident population of each state and the District of Columbia who were at least 65 years old in 2012. Construct a frequency distribution and a histogram for the data. (Source: U.S. Census Bureau)

DATA

AK	8.5	MT	15.8
AL	14.5	NC	13.8
AR	15.0	ND	14.4
AZ	14.8	NE	13.8
CA	12.1	NH	14.7
CO	11.8	NJ	14.1
CT	14.8	NM	14.1
DC	11.4	NV	13.0
DE	15.4	NY	14.1
FL	18.2	OH	14.8
GA	11.5	OK	14.1
HI	15.1	OR	14.9
IA	15.3	PA	16.0
ID	13.2	RI	15.1
IL	13.2	SC	14.7
IN	13.6	SD	14.5
KS	13.7	TN	14.3
KY	14.0	TX	10.9
LA	13.0	UT	9.5
MA	14.5	VA	13.0
MD	13.0	VT	15.7
ME	17.0	WA	13.2
MI	14.6	WI	14.4
MN	13.6	WV	16.8
MO	14.7	WY	13.0
MS	13.5		

Spreadsheet at LarsonPrecalculus.com

Solution

To begin constructing a frequency distribution, you must first decide on the number of intervals. There are several ways to group the data. However, because the least number is 8.5 and the greatest is 18.2, it seems that six intervals would be appropriate. The first would be the interval

$[8, 10)$

the second would be

$[10, 12)$

and so on. By tallying the data into the six intervals, you obtain the frequency distribution shown below.

Interval	*Tally*
$[8, 10)$	||
$[10, 12)$	||||
$[12, 14)$	||||/ ||||/ ||||/
$[14, 16)$	||||/ ||||/ ||||/ ||||/ ||||/ |
$[16, 18)$	|||
$[18, 20)$	|

You can construct the histogram by drawing a vertical axis to represent the number of states and a horizontal axis to represent the percent of the population 65 and older. Then, for each interval, draw a vertical bar whose height is the total tally, as shown in Figure P.23. Note that the break in the horizontal axis indicates that the numbers 0 through 7 have been omitted.

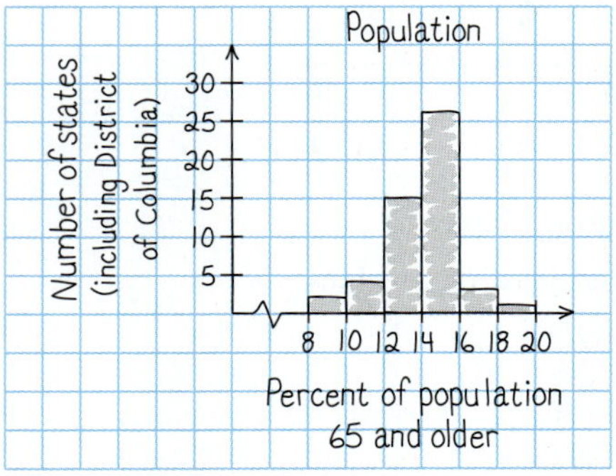

Figure P.23

✓ Checkpoint Audio-video solution in English & Spanish at LarsonPrecalculus.com.

Construct a frequency distribution and a histogram for the data in Example 1.

Bar Graphs

A **bar graph** is similar to a histogram, except that the bars can be either horizontal or vertical, and the labels of the bars are not necessarily numbers. For instance, the labels of the bars can be the months of a year, names of cities, or types of vehicles. Another difference between a bar graph and a histogram is that the bars in a bar graph are usually separated by spaces.

EXAMPLE 3 Constructing a Bar Graph

The table shows the numbers of Wal-Mart stores from 2008 through 2013. Construct a bar graph for the data. What can you conclude? (Source: Wal-Mart Stores, Inc.)

DATA

Year	Number of stores
2008	7,720
2009	8,416
2010	8,970
2011	10,130
2012	10,773
2013	10,942

Spreadsheet at LarsonPrecalculus.com

Solution

To create a bar graph, begin by drawing a vertical axis to represent the number of stores and a horizontal axis to represent the year. The bar graph is shown below.

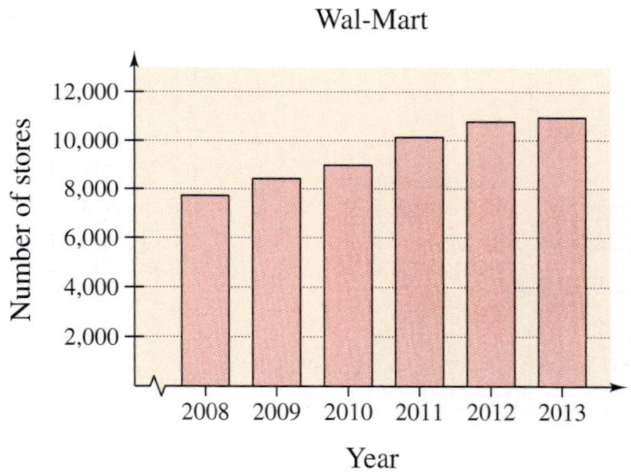

From the graph, you can see that the number of stores has steadily increased from 2008 to 2013.

Checkpoint Audio-video solution in English & Spanish at LarsonPrecalculus.com.

The table shows the numbers of Dollar General stores from 2008 through 2013. Construct a bar graph for the data. What can you conclude? (Source: Dollar General Corporation)

DATA

Year	Number of stores
2008	8,362
2009	8,828
2010	9,372
2011	9,937
2012	10,506
2013	11,132

Spreadsheet at LarsonPrecalculus.com

EXAMPLE 4 Constructing a Double Bar Graph

The table shows the percents of associate degrees awarded to males and females for selected fields of study in the United States in 2012. Construct a double bar graph for the data. (Source: U.S. National Center for Education Statistics)

DATA

Field of Study	% Female	% Male
Agriculture and Natural Resources	34.8	65.2
Biological and Biomedical Sciences	66.6	33.4
Accounting	75.1	24.9
Business/Commerce	58.6	41.4
Communication/Journalism	53.5	46.5
Education	87.2	12.8
Engineering	13.3	86.7
Family and Consumer Sciences	96.1	3.9
Emergency Medical Technician	29.4	70.6
Legal Professions	86.2	13.8
Mathematics and Statistics	31.3	68.7
Physical Sciences	43.3	56.7
Precision Production	6.1	93.9
Psychology	76.8	23.2
Social Sciences	65.1	34.9

Spreadsheet at LarsonPrecalculus.com

Solution

For the data, a horizontal bar graph seems to be appropriate. This makes it easier to label and read the bars, as shown below.

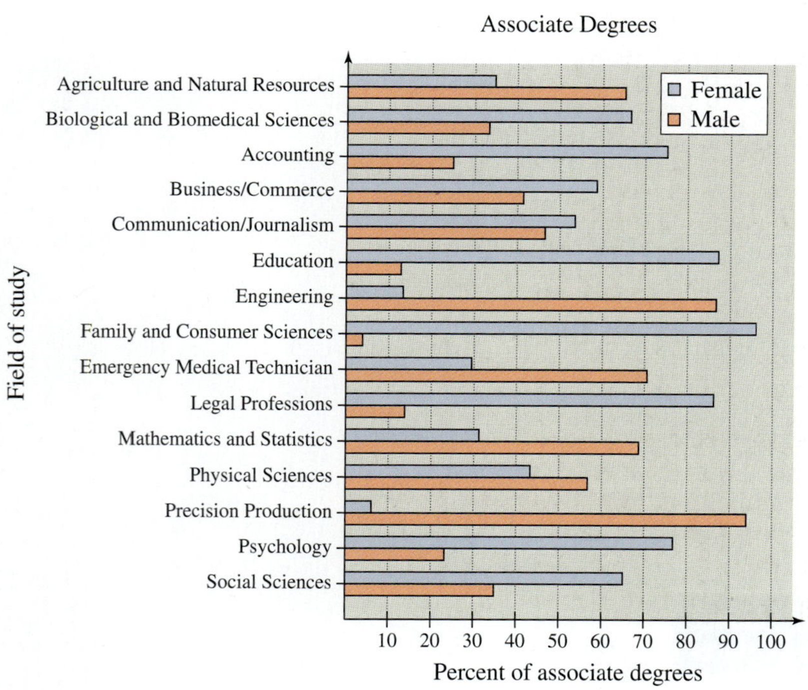

✓ **Checkpoint** *Audio-video solution in English & Spanish at LarsonPrecalculus.com.*

Construct a double bar graph for the numbers of Wal-Mart and Dollar General stores in Example 3 and the Checkpoint after Example 3, respectively.

Line Graphs

A **line graph** is similar to a standard coordinate graph. Line graphs are usually used to show trends over periods of time.

EXAMPLE 5 Constructing a Line Graph

See LarsonPrecalculus.com for an interactive version of this type of example.

The table at the right shows the number of immigrants (in thousands) legally entering the United States for each decade from 1901 through 2010. Construct a line graph for the data. What can you conclude? (Source: U.S. Department of Homeland Security)

Decade	Number (in thousands)
1901–1910	8,795
1911–1920	5,736
1921–1930	4,107
1931–1940	528
1941–1950	1,035
1951–1960	2,515
1961–1970	3,322
1971–1980	4,399
1981–1990	7,256
1991–2000	9,081
2001–2010	10,501

Solution

Begin by drawing a vertical axis to represent the number of immigrants in thousands. Then label the horizontal axis with decades and plot the points shown in the table. Finally, connect the points with line segments, as shown in Figure P.24. From the line graph, you can see that the number of immigrants hit a low point during the depression of the 1930s. Since then, the number has steadily increased.

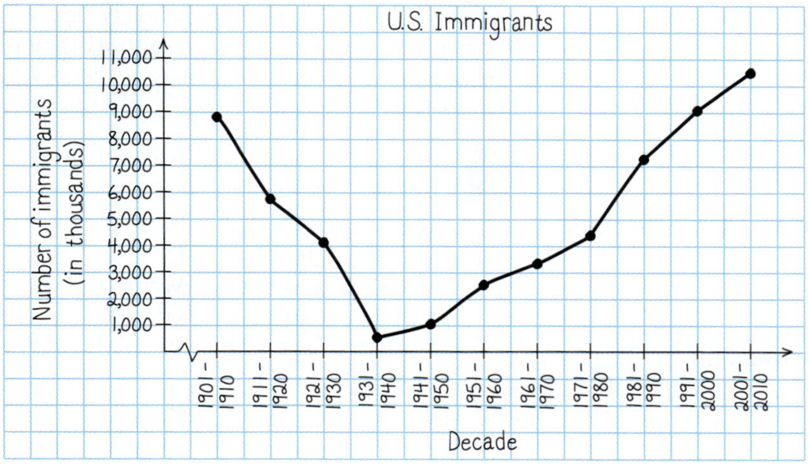

Figure P.24

You can use a graphing utility to check your sketch, as shown in Figure P.25.

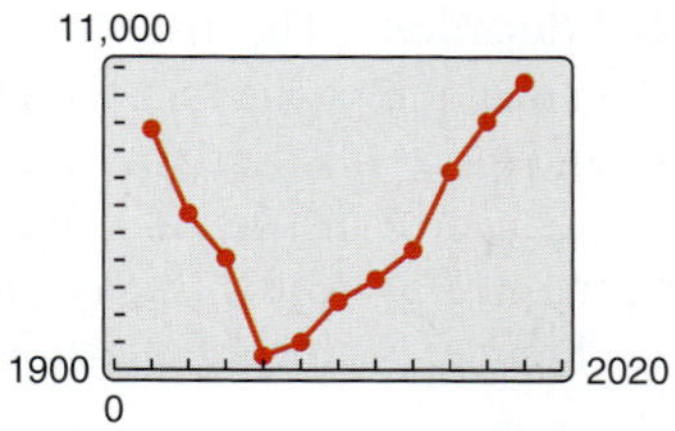

Figure P.25

✓ ***Checkpoint*** *Audio-video solution in English & Spanish at LarsonPrecalculus.com.*

Construct a line graph for the data in Example 3.

Technology Tip

You can use the *statistical plotting* feature of a graphing utility to create different types of graphs, such as line graphs. For instructions on how to use the *statistical plotting* feature, see Appendix A; for specific keystrokes, go to this textbook's *Companion Website*.

P.6 Exercises

See *CalcChat.com* for tutorial help and worked-out solutions to odd-numbered exercises.
For instructions on how to use a graphing utility, see Appendix A.

Vocabulary and Concept Check

In Exercises 1 and 2, fill in the blank.

1. _______ are useful for ordering small sets of numbers by hand.
2. _______ show trends over periods of time.

In Exercises 3–6, match the data display with its name.

3. Line plot
4. Bar graph
5. Histogram
6. Line graph

(a) 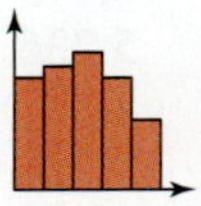

(b) 

(c) 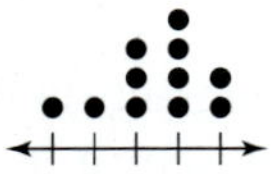

(d) 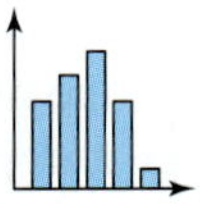

Procedures and Problem Solving

7. **Economics** The line plot shows a sample of average prices of unleaded regular gasoline in 25 different cities.

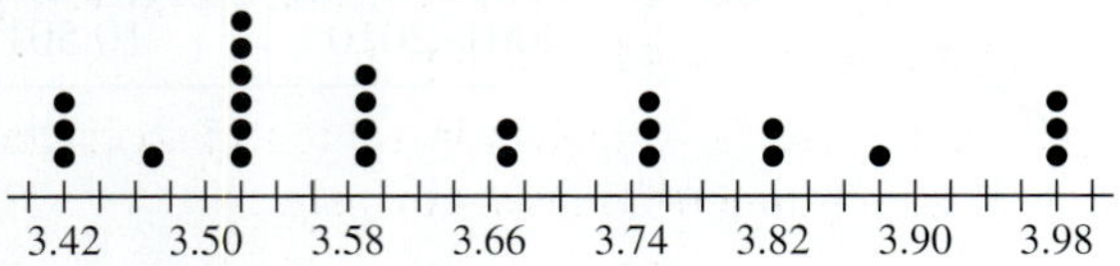

(a) What price occurred with the greatest frequency?
(b) What is the range of prices?

8. **Waste Management** The line plot shows the weights (to the nearest hundred pounds) of municipal waste hauled by a garbage truck in 30 trips to a landfill.

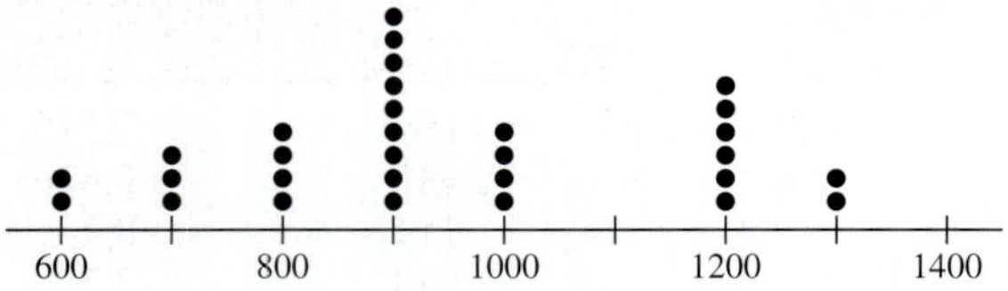

(a) What weight occurred with the greatest frequency?
(b) What is the range of weights?

Education In Exercises 9 and 10, use the following scores from an algebra class of 30 students. The scores are for one 25-point quiz and one 100-point exam.

Quiz **20, 15, 14, 20, 16, 19, 10, 21, 24, 15, 15, 14, 15, 21, 19, 15, 20, 18, 18, 22, 18, 16, 18, 19, 21, 19, 16, 20, 14, 12**

Exam **77, 100, 77, 70, 83, 89, 87, 85, 81, 84, 81, 78, 89, 78, 88, 85, 90, 92, 75, 81, 85, 100, 98, 81, 78, 75, 85, 89, 82, 75**

9. Construct a line plot for the quiz. Which score(s) occurred with the greatest frequency?

10. Construct a line plot for the exam. Which score(s) occurred with the greatest frequency?

11. **Demographics** The list shows the percents of individuals living below the poverty level in the 50 states in 2012. Use a frequency distribution and a histogram to organize the data. *(Spreadsheet at LarsonPrecalculus.com)* (Source: U.S. Census Bureau)

DATA

AK 10.8	AL 15.8	AR 19.4	AZ 18.1
CA 16.4	CO 12.5	CT 10.2	DE 13.6
FL 15.1	GA 18.3	HI 13.0	IA 10.4
ID 15.1	IL 13.4	IN 15.4	KS 14.2
KY 16.9	LA 21.1	MA 10.9	MD 9.6
ME 13.1	MI 14.3	MN 10.0	MO 15.3
MS 19.7	MT 15.0	NC 16.3	ND 10.7
NE 11.2	NH 7.9	NJ 10.4	NM 21.3
NV 15.6	NY 16.6	OH 15.2	OK 16.0
OR 13.9	PA 13.2	RI 13.5	SC 17.8
SD 13.7	TN 17.4	TX 17.2	UT 11.0
VA 11.0	VT 11.4	WA 12.1	WI 12.2
WV 17.1	WY 10.2		

12. **Education** The list shows the numbers of students enrolled in public schools (in thousands) in the 50 states and the District of Columbia in 2012. Use a frequency distribution and a histogram to organize the data. *(Spreadsheet at LarsonPrecalculus.com)* (Source: National Education Association)

DATA

AK 128	AL 736	AR 472	AZ 1070
CA 6185	CO 869	CT 550	DC 76
DE 131	FL 2681	GA 1703	HI 177
IA 501	ID 294	IL 2084	IN 1031
KS 486	KY 659	LA 708	MA 954
MD 860	ME 185	MI 1544	MN 843
MO 907	MS 493	MT 143	NC 1488
ND 99	NE 303	NH 189	NJ 1358
NM 334	NV 478	NY 2590	OH 1868
OK 673	OR 564	PA 1739	RI 134
SC 683	SD 127	TN 980	TX 5059
UT 601	VA 1266	VT 82	WA 1053
WI 872	WV 282	WY 91	

13. Meterology The table shows the normal monthly precipitation (in inches) in Seattle, Washington. Construct a bar graph for the data. Write a brief statement regarding the precipitation in Seattle, Washington, throughout the year. (Source: National Oceanic and Atmospheric Administration)

Month	Precipitation
January	5.47
February	3.52
March	3.85
April	2.79
May	2.01
June	1.57
July	0.85
August	0.94
September	1.30
October	3.38
November	5.98
December	6.06

Spreadsheet at LarsonPrecalculus.com

14. Business The table shows the revenues (in billions of dollars) for Costco Wholesale stores in the years from 1997 through 2013. Construct a bar graph for the data. Write a brief statement regarding the revenue of Costco Wholesale stores over time. (Source: Costco Wholesale Corporation)

Year	Revenue (in billions of dollars)
1997	21.874
1998	24.270
1999	27.456
2000	32.164
2001	34.797
2002	38.762
2003	42.546
2004	48.107
2005	52.935
2006	60.151
2007	64.909
2008	72.483
2009	71.449
2010	77.946
2011	88.915
2012	99.137
2013	105.156

Spreadsheet at LarsonPrecalculus.com

Why you should learn it *(p. 57)* **In Exercises 15 and 16, the double bar graph shows the mean tuitions (in dollars) charged by public and private institutions of higher education in the United States from 2008 through 2013.** (Source: U.S. National Center for Education Statistics)

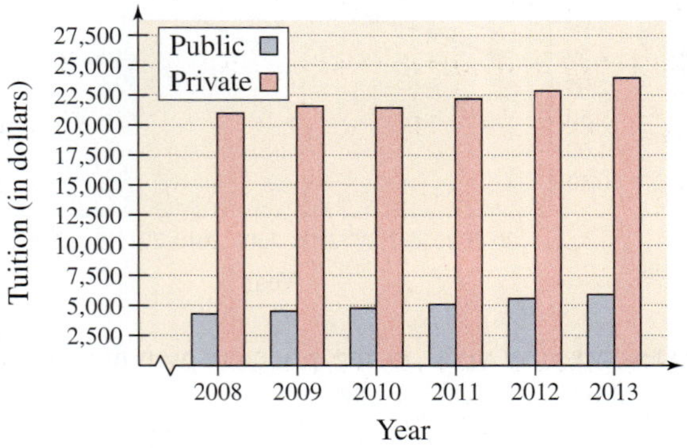

15. Approximate the difference in tuition charges for public and private schools for each year.

16. Approximate the increase in tuition charges for each type of institution from year to year.

17. Education The table shows the total college enrollments (in thousands) for women and men in the United States from 2006 through 2012. Construct a double bar graph for the data. (Source: U.S. National Center for Education Statistics)

Year	Women (in thousands)	Men (in thousands)
2006	10,184	7,575
2007	10,432	7,816
2008	10,914	8,189
2009	11,658	8,770
2010	11,971	9,045
2011	11,968	9,026
2012	11,724	8,919

Spreadsheet at LarsonPrecalculus.com

18. Demographics The table shows the populations (in millions) of the five most populous cities in the United States in 1990 and 2012. Construct a double bar graph for the data. (Source: U.S. Census Bureau)

City	1990 population (in millions)	2012 population (in millions)
New York, NY	7.32	8.34
Los Angeles, CA	3.49	3.86
Chicago, IL	2.78	2.71
Houston, TX	1.63	2.16
Philadelphia, PA	1.59	1.55

Spreadsheet at LarsonPrecalculus.com

Economics In Exercises 19–22, use the line graph, which shows the average prices of a gallon of premium unleaded gasoline from 2005 through 2013. (Source: U.S. Energy Information Administration)

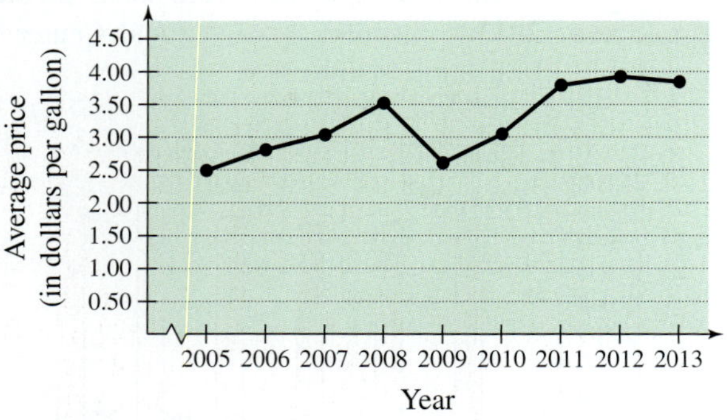

19. Describe the trend in the price of premium unleaded gasoline from 2005 to 2008.

20. Describe the trend in the price of premium unleaded gasoline from 2008 to 2009.

21. Approximate the percent decrease in the price per gallon of premium unleaded gasoline from 2008 to 2010.

22. Approximate the percent increase in the price per gallon of premium unleaded gasoline from 2009 to 2013.

Agriculture In Exercises 23–26, use the line graph, which shows the average retail price (in dollars) of one dozen Grade A large eggs in the United States for each month in 2013. (Source: U.S. Bureau of Labor Statistics)

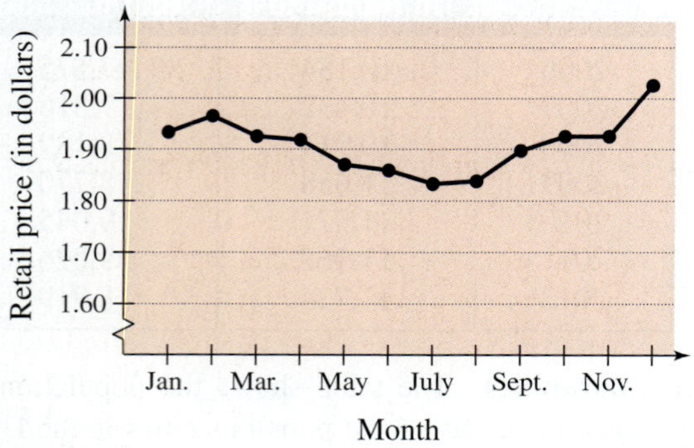

23. What is the highest price of one dozen Grade A large eggs shown in the graph? When did this price occur?

24. What was the difference between the highest price and the lowest price of one dozen Grade A large eggs in 2013?

25. Determine when the price of eggs showed the greatest rate of increase from one month to the next.

26. Describe any trends shown by the line graph. Then predict the average price of eggs in February of 2014. Are you confident that your prediction is within $0.10 of the actual average retail price? Explain your reasoning.

27. Human Resources The table shows the percents of wives in two-income families who earned more than their husbands in the United States from 2000 through 2012. Construct a line graph for the data. Write a brief statement describing what the graph reveals. (Source: U.S. Bureau of Labor Statistics)

DATA

Year	Percent
2000	23.3
2001	24.1
2002	25.0
2003	25.2
2004	25.3
2005	25.5
2006	25.7
2007	25.9
2008	26.6
2009	28.9
2010	29.2
2011	28.1
2012	29.0

Spreadsheet at LarsonPrecalculus.com

28. Economics The table shows the trade deficits (the differences between imports and exports) of the United States from 2004 through 2013. Construct a line graph for the data. Write a brief statement describing what the graph reveals. (Source: U.S. Census Bureau)

DATA

Year	Trade deficit (in billions of dollars)
2004	655
2005	772
2006	828
2007	809
2008	816
2009	504
2010	635
2011	725
2012	731
2013	689

Spreadsheet at LarsonPrecalculus.com

29. Marketing The table shows the costs (in millions of dollars) of a 30-second television advertisement during the Super Bowl from 1999 through 2014. Use a graphing utility to construct a line graph for the data. (Source: superbowl-ads.com)

DATA

Year	Cost (in millions of dollars)
1999	1.60
2000	2.10
2001	2.05
2002	1.90
2003	2.10
2004	2.25
2005	2.40
2006	2.50
2007	2.60
2008	2.70
2009	3.00
2010	2.65
2011	3.00
2012	3.50
2013	3.80
2014	4.00

Spreadsheet at LarsonPrecalculus.com

30. Networking The list shows the percent of households in each of the 50 states and the District of Columbia with Internet access in 2012. Use a graphing utility to organize the data in the graph of your choice. Explain your choice of graph. *(Spreadsheet at LarsonPrecalculus.com)* (Source: U.S. Census Bureau)

DATA

AK 83.1	AL 72.0	AR 74.8
AZ 73.7	CA 81.3	CO 85.7
CT 84.0	DC 77.5	DE 78.2
FL 81.5	GA 79.2	HI 85.0
IA 80.2	ID 86.8	IL 82.7
IN 77.7	KS 80.7	KY 74.4
LA 67.7	MA 86.0	MD 83.6
ME 84.2	MI 80.8	MN 87.4
MO 75.4	MS 64.8	MT 78.3
NC 77.7	ND 84.0	NE 76.1
NH 87.7	NJ 83.7	NM 73.9
NV 82.8	NY 81.0	OH 75.6
OK 74.8	OR 87.9	PA 80.8
RI 82.2	SC 73.7	SD 78.3
TN 74.3	TX 72.6	UT 85.2
VA 77.2	VT 85.7	WA 85.1
WI 80.7	WV 70.6	WY 82.9

31. Athletics The table shows the numbers of participants (in thousands) in high school athletic programs in the United States from 2002 through 2013. Organize the data in an appropriate display. Explain your choice of graph. (Source: National Federation of State High School Associations)

DATA

Year	Female athletes (in thousands)	Male athletes (in thousands)
2002	2807	3961
2003	2856	3989
2004	2865	4038
2005	2908	4110
2006	2953	4207
2007	3022	4321
2008	3057	4372
2009	3114	4423
2010	3173	4456
2011	3174	4494
2012	3208	4485
2013	3223	4491

Spreadsheet at LarsonPrecalculus.com

Conclusions

32. Writing Describe the differences between a bar graph and a histogram.

33. Writing Describe the differences between a line plot and a line graph.

34. HOW DO YOU SEE IT? The graphs shown below represent the same data points.

(a) Which of the two graphs is misleading and why?

(b) Discuss other ways in which graphs can be misleading.

(c) Why would it be beneficial for someone to use a misleading graph?

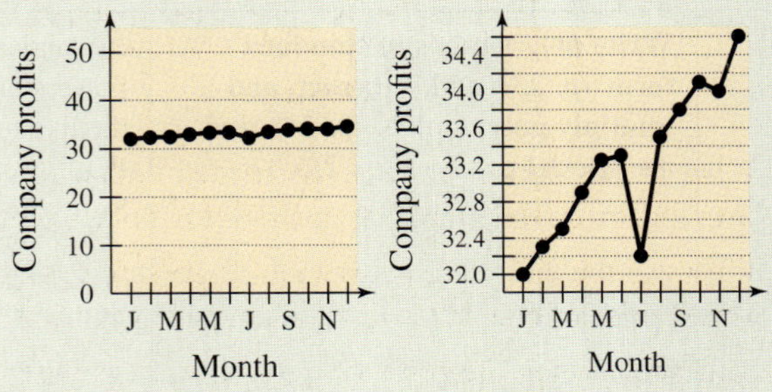

35. Think About It How can you decide which type of graph to use when you are organizing data?

P Chapter Summary

	What did you learn?	Explanation and Examples	Review Exercises
P.1	**Represent and classify real numbers** ***(p. 2).***	**Real numbers:** set of all rational and irrational numbers **Rational numbers:** real numbers that can be written as the ratio of two integers **Irrational numbers:** real numbers that cannot be written as the ratio of two integers	1, 2
	Order real numbers and use inequalities ***(p. 3).***	$a < b$: a is less than b. $a > b$: a is greater than b. $a \leq b$: a is less than or equal to b. $a \geq b$: a is greater than or equal to b.	3–8
	Find the absolute values of real numbers and the distance between two real numbers ***(p. 5).***	**Absolute value of a:** $\lvert a\rvert = \begin{cases} a, & a \geq 0 \\ -a, & a < 0 \end{cases}$ **Distance between a and b:** $d(a, b) = \lvert b - a\rvert = \lvert a - b\rvert$	9–12
	Evaluate algebraic expressions ***(p. 6)*** **and use the basic rules and properties of algebra** ***(p. 7).***	To evaluate an algebraic expression, substitute numerical values for each of the variables in the expression.	13–22
P.2	**Use properties of exponents** ***(p. 12).***	**1.** $a^m a^n = a^{m+n}$ **2.** $a^m/a^n = a^{m-n}$ **3.** $a^{-n} = (1/a)^n$ **4.** $a^0 = 1, a \neq 0$ **5.** $(ab)^m = a^m b^m$ **6.** $(a^m)^n = a^{mn}$ **7.** $(a/b)^m = a^m/b^m$ **8.** $\lvert a^2\rvert = a^2$	23–26
	Use scientific notation to represent real numbers ***(p. 14).***	A number written in scientific notation has the form $\pm c \times 10^n$, where $1 \leq c < 10$ and n is an integer.	27–34
	Use properties of radicals ***(p. 15)*** **to simplify and combine radicals** ***(p. 17).***	**1.** $\sqrt[n]{a^m} = (\sqrt[n]{a})^m$ **2.** $\sqrt[n]{a} \cdot \sqrt[n]{b} = \sqrt[n]{ab}$ **3.** $\sqrt[n]{a}/\sqrt[n]{b} = \sqrt[n]{a/b},\ b \neq 0$ **4.** $\sqrt[m]{\sqrt[n]{a}} = \sqrt[mn]{a}$ **5.** $(\sqrt[n]{a})^n = a$ **6.** **n even:** $\sqrt[n]{a^n} = \lvert a\rvert$, **$n$ odd:** $\sqrt[n]{a^n} = a$	35–50
	Rationalize denominators and numerators ***(p. 18).***	To rationalize a denominator or numerator of the form $a - b\sqrt{m}$ or $a + b\sqrt{m}$, multiply both numerator and denominator by a conjugate.	51–54
	Use properties of rational exponents ***(p. 19).***	If a is a real number and n is a positive integer such that the principal nth root of a exists, then $a^{1/n}$ is defined as $a^{1/n} = \sqrt[n]{a}$, where $1/n$ is the rational exponent of a.	55–58
P.3	**Write polynomials in standard form** ***(p. 23)*****, add, subtract, and multiply polynomials** ***(p. 24)*****, and use special products** ***(p. 25).***	In standard form, a polynomial in x is written with descending powers of x. To add and subtract polynomials, add or subtract the like terms. To find the product of two binomials, use the FOIL Method. **Sum and Difference of Same Terms:** $(u + v)(u - v) = u^2 - v^2$ **Square of a Binomial:** $(u + v)^2 = u^2 + 2uv + v^2$ $(u - v)^2 = u^2 - 2uv + v^2$ **Cube of a Binomial:** $(u + v)^3 = u^3 + 3u^2v + 3uv^2 + v^3$ $(u - v)^3 = u^3 - 3u^2v + 3uv^2 - v^3$	59–76

	What did you learn?	Explanation and Examples	Review Exercises
P.3	**Remove common factors from polynomials** *(p. 26).*	The process of writing a polynomial as a product is called factoring. Removing (factoring out) any common factors is the first step in completely factoring a polynomial.	77–82
	Factor special polynomial forms *(p. 27).*	**Difference of Two Squares:** $u^2 - v^2 = (u + v)(u - v)$ **Perfect Square Trinomial:** $u^2 + 2uv + v^2 = (u + v)^2$ $u^2 - 2uv + v^2 = (u - v)^2$ **Sum or Difference of Two Cubes:** $u^3 + v^3 = (u + v)(u^2 - uv + v^2)$ $u^3 - v^3 = (u - v)(u^2 + uv + v^2)$	83–88
	Factor trinomials as the product of two binomials *(p. 29).*	$ax^2 + bx + c = (\square x + \square)(\square x + \square)$ Factors of a — Factors of c	89–92
	Factor polynomials by grouping *(p. 30).*	Polynomials with more than three terms can sometimes be factored by a method called factoring by grouping. (See Example 14.)	93–96
P.4	**Find domains of algebraic expressions** *(p. 35).*	The set of real numbers for which an algebraic expression is defined is the domain of the expression.	97–100
	Simplify rational expressions *(p. 36).*	When simplifying rational expressions, be sure to factor each polynomial completely before concluding that the numerator and denominator have no factors in common.	101–104
	Add, subtract, multiply, and divide rational expressions *(p. 37).*	To add or subtract, use the LCD method or the basic definition $\frac{a}{b} \pm \frac{c}{d} = \frac{ad \pm bc}{bd}$, $b \neq 0$, $d \neq 0$. To multiply or divide, use properties of fractions.	105–112
	Simplify complex fractions *(p. 39)*, **and simplify expressions from calculus** *(p. 40).*	To simplify a complex fraction, combine the fractions in the numerator into a single fraction and then combine the fractions in the denominator into a single fraction. Then invert the denominator and multiply.	113–116
P.5	**Plot points in the Cartesian plane and sketch scatter plots** *(p. 46).*	For an ordered pair (x, y), the x-coordinate is the directed distance from the y-axis to the point, and the y-coordinate is the directed distance from the x-axis to the point.	117–122
	Use the Distance Formula *(p. 48)* **and the Midpoint Formula** *(p. 50).*	**Distance Formula:** $d = \sqrt{(x_2 - x_1)^2 + (y_2 - y_1)^2}$ **Midpoint Formula:** Midpoint $= \left(\frac{x_1 + x_2}{2}, \frac{y_1 + y_2}{2}\right)$	123–126
	Find the equation of a circle *(p. 51)*, **and translate points in the plane** *(p. 52).*	The standard form of the equation of a circle is $(x - h)^2 + (y - k)^2 = r^2$.	127–130
P.6	**Use line plots to order and analyze data** *(p. 57)*, **histograms to represent frequency distributions** *(p. 58)*, **and bar graphs** *(p. 59)* **and line graphs** *(p. 61)* **to represent and analyze data.**	A line plot uses a portion of a real number line to order numbers. A frequency distribution and a histogram can be used to organize and display a large set of data. A bar graph is similar to a histogram, except that the bars can be either horizontal or vertical, the labels are not necessarily numbers, and the bars are usually separated by spaces. A line graph is usually used to show trends over a period of time.	131–134

P Review Exercises

See CalcChat.com for tutorial help and worked-out solutions to odd-numbered exercises. For instructions on how to use a graphing utility, see Appendix A.

P.1

Identifying Subsets of Real Numbers In Exercises 1 and 2, determine which numbers are (a) natural numbers, (b) whole numbers, (c) integers, (d) rational numbers, and (e) irrational numbers.

1. $\{11, -14, -\frac{8}{9}, \frac{5}{2}, \sqrt{6}, 0.4\}$
2. $\{\sqrt{15}, -22, -\frac{10}{3}, 0, 5.2, \frac{3}{7}\}$

Writing an Inequality In Exercises 3 and 4, use a calculator to find the decimal form of each rational number. If it is a nonterminating decimal, write the repeating pattern. Then plot the numbers on the real number line and place the correct inequality symbol (< or >) between them.

3. (a) $\frac{5}{6}$ (b) $\frac{7}{8}$
4. (a) $\frac{1}{3}$ (b) $\frac{9}{25}$

Interpreting Inequalities In Exercises 5–8, (a) verbally describe the subset of real numbers represented by the inequality, (b) sketch the subset on the real number line, and (c) state whether the interval is bounded or unbounded.

5. $x \geq -6$
6. $x < 1$
7. $-4 \leq x \leq 0$
8. $7 \leq x < 10$

Finding the Distance Between Two Real Numbers In Exercises 9 and 10, find the distance between a and b.

9. $a = -74, \quad b = 48$
10. $a = -123, \quad b = -9$

Using Absolute Value Notation In Exercises 11 and 12, use absolute value notation to describe the situation.

11. The distance between x and 7 is at least 6.
12. The distance between y and -30 is less than 5.

Evaluating an Algebraic Expression In Exercises 13–16, evaluate the expression for each value of x. (If not possible, state the reason.)

	Expression	*Values*	
13.	$9x - 2$	(a) $x = -1$	(b) $x = 3$
14.	$x^2 - 11x + 24$	(a) $x = -2$	(b) $x = 2$
15.	$\dfrac{-2x + 3}{x}$	(a) $x = 0$	(b) $x = 6$
16.	$\dfrac{4x}{x - 1}$	(a) $x = -1$	(b) $x = 1$

Identifying Rules of Algebra In Exercises 17–22, identify the rule of algebra illustrated by the statement.

17. $2x + (3x - 10) = (2x + 3x) - 10$
18. $4(t + 2) = 4 \cdot t + 4 \cdot 2$
19. $0 + (a - 5) = a - 5$
20. $(t^2 + 1) + 3 = 3 + (t^2 + 1)$
21. $\dfrac{2}{y + 4} \cdot \dfrac{y + 4}{2} = 1, \ y \neq -4$
22. $1 \cdot (3x + 4) = 3x + 4$

P.2

Using Properties of Exponents In Exercises 23–26, simplify each expression.

23. (a) $(-2z)^3$ (b) $(a^2b^4)(3ab^{-2})$
24. (a) $\dfrac{(4y^2)^3}{y^2}$ (b) $\dfrac{40(b - 3)^3}{75(b - 3)^5}$
25. (a) $\dfrac{36u^0v^{-3}}{12u^{-2}v}$ (b) $\dfrac{3^{-4}m^{-1}n^{-3}}{9^{-2}mn^{-3}}$
26. (a) $(a^4b^{-3}c^0)^{-1}a^2$ (b) $\left(\dfrac{y^{-2}}{x}\right)^{-1}\left(\dfrac{x^2}{y^{-2}}\right)$

Scientific Notation In Exercises 27–30, write the number in scientific notation.

27. 2,585,000,000
28. $-3{,}250{,}000$
29. -0.000000125
30. 0.00000008064

Decimal Notation In Exercises 31–34, write the number in decimal notation.

31. 1.28×10^5
32. -4.002×10^2
33. 1.80×10^{-5}
34. -4.02×10^{-2}

Using Properties of Radicals In Exercises 35–38, use the properties of radicals to simplify the expression.

35. $(\sqrt[4]{78})^4$
36. $\sqrt{(5x)^2}$
37. $\sqrt[5]{8} \cdot \sqrt[5]{4}$
38. $\sqrt[3]{\sqrt{xy}}$

Simplifying a Radical Expression In Exercises 39–50, simplify the expression.

39. $\sqrt{25a^3}$
40. $\sqrt[5]{64x^6}$
41. $\sqrt{\frac{81}{144}}$
42. $\sqrt[3]{\frac{125}{216}}$
43. $\sqrt{\dfrac{75x^2}{y^4}}$
44. $\sqrt[3]{\dfrac{2x^3}{27}}$
45. $\sqrt{48} - \sqrt{27}$
46. $3\sqrt{32} + 4\sqrt{98}$
47. $8\sqrt{3x} - 5\sqrt{3x}$
48. $-11\sqrt{36y} - 6\sqrt{y}$
49. $\sqrt{8x^3} + \sqrt{2x}$
50. $3\sqrt{14x^2} - \sqrt{56x^2}$

Rationalizing a Denominator In Exercises 51 and 52, rationalize the denominator of the expression. Then simplify your answer.

51. $\dfrac{1}{3 - \sqrt{5}}$
52. $\dfrac{1}{\sqrt{x} - 1}$

Rationalizing a Numerator In Exercises 53 and 54, rationalize the numerator of the expression. Then simplify your answer.

53. $\frac{\sqrt{20}}{4}$

54. $\frac{\sqrt{2} - \sqrt{11}}{3}$

Simplifying with Rational Exponents In Exercises 55–58, simplify the expression.

55. $64^{5/2}$

56. $64^{-2/3}$

57. $(-3x^{-1/6})(-2x^{1/2})$

58. $(x - 1)^{1/3}(x - 1)^{-1/4}$

P.3

Writing a Polynomial in Standard Form In Exercises 59 and 60, write the polynomial in standard form. Then identify the degree and leading coefficient of the polynomial.

59. $15x^2 - 2x^5 + 3x^3 + 5 - x^4$

60. $-4x^4 + x^2 - 10 - x + x^3$

Performing Operations with Polynomials In Exercises 61–66, perform the operation and write the result in standard form.

61. $(3x^2 + 2x) - (1 - 5x)$

62. $(8y^2 + 2y) + (3y - 8)$

63. $(2x^3 - 5x^2 + 10x - 7) + (4x^2 - 7x - 2)$

64. $(6x^4 - 4x^3 - x + 3 - 20x^2) - (16 + 9x^4 - 11x^2)$

65. $-2a(a^2 + a - 3)$

66. $(y^2 - 4y)(y^3)$

Multiplying Polynomials In Exercises 67–74, multiply or find the special product.

67. $(x + 4)(x + 9)$

68. $(z + 1)(5z - 6)$

69. $(x + 8)(x - 8)$

70. $(7x - 4)^2$

71. $(x - 4)^3$

72. $(2x + 1)^3$

73. $[(m - 7) + n][(m - 7) - n]$

74. $[(x - y) - 4][(x - y) + 4]$

75. **Geometry** Use the area model to write two different expressions for the area. Then equate the two expressions and name the algebraic property that is illustrated.

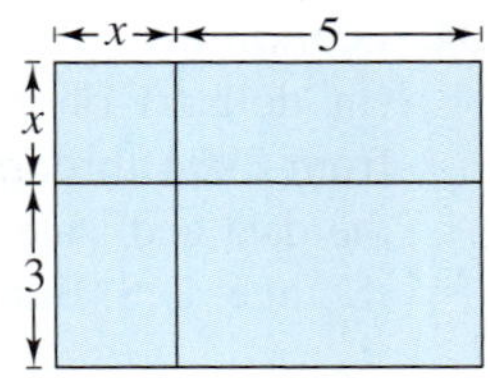

76. **Compound Interest** After 2 years, an investment of \$2500 compounded annually at an interest rate r will yield an amount of $2500(1 + r)^2$. Write this polynomial in standard form.

Removing Common Factors In Exercises 77–82, factor out the common factor.

77. $7x + 35$

78. $4b - 12$

79. $2x^3 + 18x^2 - 4x$

80. $-6x^4 - 3x^3 + 12x$

81. $x(x - 3) + 4(x - 3)$

82. $8(2 - y) - (2 - y)^2$

Factoring Polynomials In Exercises 83–92, factor the expression.

83. $x^2 - 169$

84. $9x^2 - \frac{1}{25}$

85. $x^2 + 6x + 9$

86. $4x^2 - 4x + 1$

87. $x^3 + 216$

88. $64x^3 - 27$

89. $x^2 - 6x - 27$

90. $x^2 - 9x + 14$

91. $2x^2 + 21x + 10$

92. $3x^2 + 14x + 8$

Factoring by Grouping In Exercises 93–96, factor by grouping.

93. $x^3 - 4x^2 - 3x + 12$

94. $x^3 - 6x^2 - x + 6$

95. $2x^2 - x - 15$

96. $6x^2 + x - 12$

P.4

Finding the Domain of an Algebraic Expression In Exercises 97–100, find the domain of the expression.

97. $-5x^2 - x - 1$

98. $7x^4 + 8, \quad x < 0$

99. $\frac{4}{2x - 3}$

100. $\sqrt{x + 12}$

Simplifying a Rational Expression In Exercises 101–104, write the rational expression in simplest form.

101. $\frac{4x^2}{4x^3 + 28x}$

102. $\frac{6xy}{xy + 2x}$

103. $\frac{x^2 - x - 30}{x^2 - 25}$

104. $\frac{x^2 - 9x + 18}{8x - 48}$

Performing Operations with Rational Expressions In Exercises 105–112, perform the operations and simplify your answer.

105. $\frac{x^2 - 4}{x^4 - 2x^2 - 8} \cdot \frac{x^2 + 2}{x^2}$

106. $\frac{2x - 1}{x + 1} \cdot \frac{x^2 - 1}{2x^2 - 7x + 3}$

107. $\frac{x^2(5x - 6)}{2x + 3} \div \frac{5x}{2x + 3}$

108. $\frac{4x - 6}{(x - 1)^2} \div \frac{2x^2 - 3x}{x^2 + 2x - 3}$

109. $x - 1 + \frac{1}{x + 2} + \frac{1}{x - 1}$

110. $2x + \frac{3}{2(x - 4)} - \frac{1}{2(x + 2)}$

111. $\frac{1}{x} - \frac{x - 1}{x^2 + 1}$

112. $\frac{1}{x - 1} + \frac{1 - x}{x^2 + x + 1}$

Simplifying a Complex Fraction In Exercises 113 and 114, simplify the complex fraction.

113. $\frac{\left(\frac{1}{x} - \frac{1}{y}\right)}{(x^2 - y^2)}$

114. $\frac{\left(\frac{1}{2x - 3} - \frac{1}{2x + 3}\right)}{\left(\frac{1}{2x} - \frac{1}{2x + 3}\right)}$

Simplifying an Expression with Negative Exponents In Exercises 115 and 116, simplify the expression.

115. $x^3(2x^2 + 1)^{-4} + x(2x^2 + 1)^{-3}$

116. $\dfrac{x^2(5x^{-3/2}) - 2x^3(x^{-1/2})}{x^3}$

P.5

Plotting Points in the Cartesian Plane In Exercises 117–120, plot the point in the Cartesian plane and determine the quadrant in which it is located.

117. $(8, -3)$
118. $(-4, -9)$
119. $\left(-\frac{5}{2}, 10\right)$
120. $(6.5, 0.5)$

Accounting In Exercises 121 and 122, use the table, which shows the revenues (in millions of U.S. dollars) of the Fédération Internationale de Football Association (FIFA) for the years 2008 through 2013 (Source: FIFA)

Year	Revenue (in millions of U.S. dollars)
2008	957
2009	1059
2010	1291
2011	1070
2012	1166
2013	1386

Spreadsheet at LarsonPrecalculus.com

121. Sketch a scatter plot of the data.

122. Write a brief statement regarding FIFA's revenues over time.

Using the Distance and Midpoint Formulas In Exercises 123–126, (a) plot the points, (b) find the distance between the points, and (c) find the midpoint of the line segment joining the points.

123. $(-3, 8), (1, 5)$
124. $(-12, 5), (4, -7)$
125. $(5.6, 0), (0, 4.2)$
126. $(3.8, 2.6), (-1.2, -9.4)$

Writing an Equation of a Circle In Exercises 127 and 128, write the standard form of the equation of the specified circle.

127. Center: $(3, -1)$; solution point: $(-5, 1)$

128. Endpoints of a diameter: $(-4, 6), (10, -2)$

Translating Points in the Plane In Exercises 129 and 130, the polygon is shifted to a new position in the plane. Find the new coordinates of the vertices.

129. Original coordinates of vertices:
$(4, 8), (6, 8), (4, 3), (6, 3)$
Shift: three units downward, two units to the left

130. Original coordinates of vertices:
$(0, 1), (3, 3), (0, 5), (-3, 3)$
Shift: five units upward, four units to the right

P.6

131. Merchandising Use a line plot to organize the following sample of prices (in dollars) of running shoes. Which price occurred with the greatest frequency? *(Spreadsheet at LarsonPrecalculus.com)*

DATA 100, 65, 67, 88, 69, 60, 100, 100, 88, 79, 99, 75, 65, 89, 68, 74, 100, 66, 81, 95, 75, 69, 85, 91, 71

132. Athletics The list shows the average numbers of points per game in the series for the 27 players who played in the 2014 NBA finals. Use a frequency distribution and a histogram to organize the data. *(Spreadsheet at LarsonPrecalculus.com)* (Source: ESPN)

DATA 8.6, 1.0, 9.0, 14.0, 28.2, 1.3, 2.6, 2.8, 0.0, 3.2, 9.8, 0.0, 15.2, 4.4, 15.4, 0.7, 17.8, 2.0, 1.3, 18.0, 4.6, 10.2, 6.2, 1.7, 14.4, 9.2, 6.2

133. Meteorology The normal daily maximum and minimum temperatures (in °F) for several months for the city of Nashville, Tennessee, are shown in the table. Construct a double bar graph for the data. (Source: National Oceanic and Atmospheric Administration)

Month	Maximum	Minimum
January	46.9	28.4
February	51.8	31.6
March	61.0	39.0
April	70.5	47.5
May	78.2	56.8
June	86.0	65.4

Spreadsheet at LarsonPrecalculus.com

134. Economics The table shows the average retail prices (in dollars) of a pound of white all-purpose flour from 2004 through 2013. Construct a line graph for the data and state what information the graph reveals. (Source: U.S. Bureau of Labor Statistics)

Year	Retail price (in dollars)
2004	0.30
2005	0.32
2006	0.33
2007	0.36
2008	0.51
2009	0.50
2010	0.48
2011	0.52
2012	0.52
2013	0.52

Spreadsheet at LarsonPrecalculus.com

P Chapter Test

See CalcChat.com for tutorial help and worked-out solutions to odd-numbered exercises.
For instructions on how to use a graphing utility, see Appendix A.

Take this test as you would take a test in class. After you are finished, check your work against the answers given in the back of the book.

1. Use $<$ or $>$ to show the relationship between $-\frac{10}{3}$ and $-|-4|$.
2. Find the distance between the real numbers -16 and 38.
3. Identify the rule of algebra illustrated by $5 \cdot (1 - x) \cdot 2 = 5 \cdot 2 \cdot (1 - x)$
4. Evaluate each expression without using a calculator.

   (a) $\left(\frac{3^2}{2}\right)^{-3}$ (b) $\sqrt{5} \cdot \sqrt{125}$ (c) $\frac{5.4 \times 10^8}{3 \times 10^3}$ (d) $(3 \times 10^4)^3$

In Exercises 5 and 6, simplify each expression.

5. (a) $3z^2(2z^3)^2$ (b) $(u - 2)^{-4}(u - 2)^{-3}$ (c) $\left(\frac{x^{-2}y^2}{3}\right)^{-1}$
6. (a) $9z\sqrt{8z} - 3\sqrt{2z^3}$ (b) $-5\sqrt{16y} + 10\sqrt{y}$ (c) $\sqrt[3]{\frac{16}{v^5}}$
7. Write the polynomial $3 - 2x^5 + 3x^3 - x^4$ in standard form. Identify the degree and leading coefficient.

In Exercises 8–11, perform the operations and simplify.

8. $(x^2 + 3) - [3x + (8 - x^2)]$
9. $(2x - 5)(4x^2 + 6)$
10. $\frac{8x}{x - 3} + \frac{24}{3 - x}$
11. $\frac{\left(\frac{2}{x} - \frac{2}{x + 1}\right)}{\left(\frac{4}{x^2 - 1}\right)}$

In Exercises 12–14, find the special product.

12. $(x + \sqrt{5})(x - \sqrt{5})$
13. $(x - 2)^3$
14. $[(x + y) - z][(x + y) + z]$

In Exercises 15–17, factor the expression completely.

15. $2x^4 - 3x^3 - 2x^2$
16. $x^3 + 2x^2 - 4x - 8$
17. $8x^3 - 64$
18. Find the domain of each expression: (a) $\frac{x + 3}{x^2 - 16}$ and (b) $\sqrt{7 - x}$.
19. Rationalize each denominator: (a) $\frac{16}{\sqrt[3]{16}}$, (b) $\frac{6}{1 - \sqrt{3}}$, and (c) $\frac{1}{\sqrt{x + 2} - \sqrt{2}}$.
20. Write an expression for the area of the shaded region in the figure at the right and simplify the result.

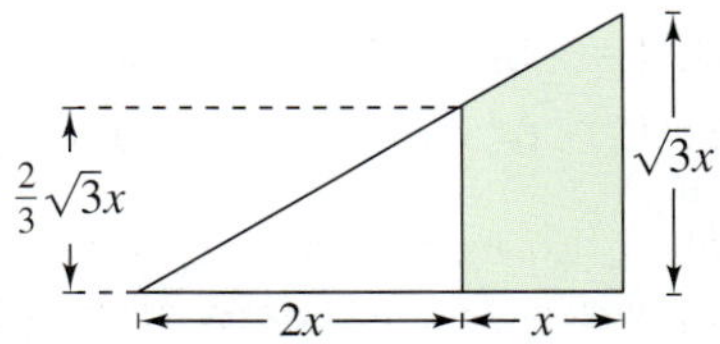

Figure for 20

21. Plot the points $(-2, 5)$ and $(6, 0)$. Find the coordinates of the midpoint of the line segment joining the points and the distance between the points.
22. The endpoints of the diameter of a circle are $(-3, 4)$ and $(1, -8)$. (a) Find the coordinates of the endpoints when the circle is shifted 5 units to the left. (b) Write the standard form of the equation of the circle following the translation in part (a).
23. The numbers (in millions) of votes cast for the Democratic candidates for president in 1988, 1992, 1996, 2000, 2004, 2008, and 2012 were 41.8, 44.9, 47.4, 51.0, 59.0, 69.5, and 65.9, respectively. Construct a bar graph for the data. (Source: U.S. Federal Election Commission)

Proofs in Mathematics

What does the word *proof* mean to you? In mathematics, the word *proof* means a valid argument. When you are proving a statement or theorem, you must use facts, definitions, and accepted properties in a logical order. You can also use previously proved theorems in your proof. For instance, the Distance Formula is used in the proof of the Midpoint Formula below. There are several different proof methods, which you will see in later chapters.

The Midpoint Formula (p. 50)

The midpoint of the line segment joining the points (x_1, y_1) and (x_2, y_2) is given by the Midpoint Formula

$$\text{Midpoint} = \left(\frac{x_1 + x_2}{2}, \frac{y_1 + y_2}{2}\right).$$

The Cartesian Plane

The Cartesian plane was named after the French mathematician René Descartes (1596–1650). While Descartes was lying in bed, he noticed a fly buzzing around on the square ceiling tiles. He discovered that the position of the fly could be described by which ceiling tile the fly landed on. This led to the development of the Cartesian plane. Descartes felt that a coordinate plane could be used to facilitate description of the positions of objects.

Proof

Using the figure, you must show that $d_1 = d_2$ and $d_1 + d_2 = d_3$.

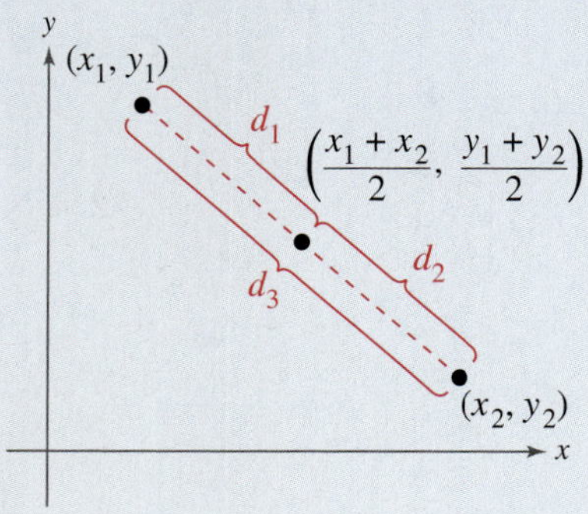

By the Distance Formula, you obtain

$$d_1 = \sqrt{\left(\frac{x_1 + x_2}{2} - x_1\right)^2 + \left(\frac{y_1 + y_2}{2} - y_1\right)^2}$$

$$= \frac{1}{2}\sqrt{(x_2 - x_1)^2 + (y_2 - y_1)^2}$$

for d_1,

$$d_2 = \sqrt{\left(x_2 - \frac{x_1 + x_2}{2}\right)^2 + \left(y_2 - \frac{y_1 + y_2}{2}\right)^2}$$

$$= \frac{1}{2}\sqrt{(x_2 - x_1)^2 + (y_2 - y_1)^2}$$

for d_2, and

$$d_3 = \sqrt{(x_2 - x_1)^2 + (y_2 - y_1)^2}$$

for d_3. So, it follows that $d_1 = d_2$ and $d_1 + d_2 = d_3$. ■

1 Functions and Their Graphs

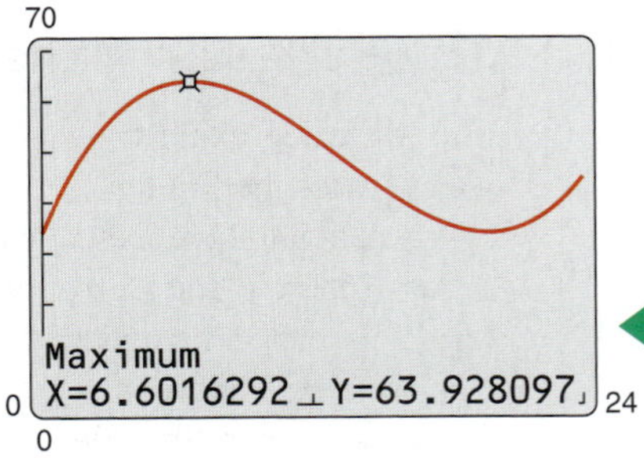

Section 1.4, Example 7
Maximum Temperature

Introduction to Library of Parent Functions

In Chapter 1, you will be introduced to the concept of a *function*. As you proceed through the text, you will see that functions play a primary role in modeling real-life situations.

There are three basic types of functions that have proven to be the most important in modeling real-life situations. These functions are algebraic functions, exponential and logarithmic functions, and trigonometric and inverse trigonometric functions. These three types of functions are referred to as the *elementary functions*, though they are often placed in the two categories of *algebraic functions* and *transcendental functions*. Each time a new type of function is studied in detail in this text, it will be highlighted in a box similar to those shown below. The graphs of these functions are shown on the inside covers of this text.

Algebraic Functions

These functions are formed by applying algebraic operations to the linear function $f(x) = x$.

Name	*Function*	*Location*
Linear	$f(x) = x$	Section 1.2
Quadratic	$f(x) = x^2$	Section 3.1
Cubic	$f(x) = x^3$	Section 3.2
Rational	$f(x) = \frac{1}{x}$	Section 3.6
Square root	$f(x) = \sqrt{x}$	Section 1.3

Transcendental Functions

These functions cannot be formed from the linear function by using algebraic operations.

Name	*Function*	*Location*
Exponential	$f(x) = a^x, a > 0, a \neq 1$	Section 4.1
Logarithmic	$f(x) = \log_a x, x > 0, a > 0, a \neq 1$	Section 4.2
Trigonometric	$f(x) = \sin x$	
(Not covered in	$f(x) = \cos x$	
this text)	$f(x) = \tan x$	
	$f(x) = \csc x$	
	$f(x) = \sec x$	
	$f(x) = \cot x$	
Inverse trigonometric	$f(x) = \arcsin x$	
(Not covered in	$f(x) = \arccos x$	
this text)	$f(x) = \arctan x$	

Nonelementary Functions

Some useful nonelementary functions include the following.

Name	*Function*	*Location*
Absolute value	$f(x) = \|x\|$	Section 1.3
Greatest integer	$f(x) = [\![x]\!]$	Section 1.4

1.1 Graphs of Equations

The Graph of an Equation

News magazines often show graphs comparing the rate of inflation, the federal deficit, or the unemployment rate to the time of year. Businesses use graphs to report monthly sales statistics. Such graphs provide geometric pictures of the way one quantity changes with respect to another. Frequently, the relationship between two quantities is expressed as an **equation.** This section introduces the basic procedure for determining the geometric picture associated with an equation.

For an equation in the variables x and y, a point (a, b) is a **solution point** when substitution of a for x and b for y satisfies the equation. Most equations have *infinitely many* solution points. For example, the equation

$$3x + y = 5$$

has solution points

$$(0, 5), \quad (1, 2), \quad (2, -1), \quad (3, -4)$$

and so on. The set of all solution points of an equation is the **graph of the equation.**

What you should learn

- Sketch graphs of equations by point plotting.
- Graph equations using a graphing utility.
- Use graphs of equations to solve real-life problems.

Why you should learn it

The graph of an equation can help you see relationships between real-life quantities. For example, in Exercise 64 on page 83, a graph can be used to analyze the life expectancies of children born in the United States.

EXAMPLE 1 Determining Solution Points

Determine whether (a) $(2, 13)$ and (b) $(-1, -3)$ lie on the graph of

$$y = 10x - 7.$$

Solution

a. $y = 10x - 7$ — Write original equation.

$13 \stackrel{?}{=} 10(2) - 7$ — Substitute 2 for x and 13 for y.

$13 = 13$ — $(2, 13)$ is a solution. ✓

The point $(2, 13)$ *does* lie on the graph of $y = 10x - 7$ because it is a solution point of the equation.

b. $y = 10x - 7$ — Write original equation.

$-3 \stackrel{?}{=} 10(-1) - 7$ — Substitute -1 for x and -3 for y.

$-3 \neq -17$ — $(-1, -3)$ is not a solution.

The point $(-1, -3)$ *does not* lie on the graph of $y = 10x - 7$ because it is not a solution point of the equation.

✓ ***Checkpoint*** *Audio-video solution in English & Spanish at LarsonPrecalculus.com.*

Determine whether (a) $(3, -5)$ and (b) $(-2, 26)$ lie on the graph of $y = 14 - 6x$. ■

Remark

When evaluating an expression or an equation, remember to follow the Basic Rules of Algebra. To review the Basic Rules of Algebra, see Section P.1.

The basic technique used for sketching the graph of an equation is the point-plotting method.

Sketching the Graph of an Equation by Point Plotting

1. If possible, rewrite the equation so that one of the variables is isolated on one side of the equation.
2. Make a table of values showing several solution points.
3. Plot these points in a rectangular coordinate system.
4. Connect the points with a smooth curve or line.

EXAMPLE 2 Sketching a Graph by Point Plotting

Use point plotting and graph paper to sketch the graph of $3x + y = 6$.

Solution

In this case, you can isolate the variable y.

$y = 6 - 3x$ Solve equation for y.

Using negative and positive values of x, and $x = 0$, you can obtain the following table of values (solution points).

x	-1	0	1	2	3
$y = 6 - 3x$	9	6	3	0	-3
Solution point	$(-1, 9)$	$(0, 6)$	$(1, 3)$	$(2, 0)$	$(3, -3)$

Next, plot the solution points and connect them, as shown in Figure 1.1. It appears that the points lie on a line. You will study lines extensively in Section 1.2.

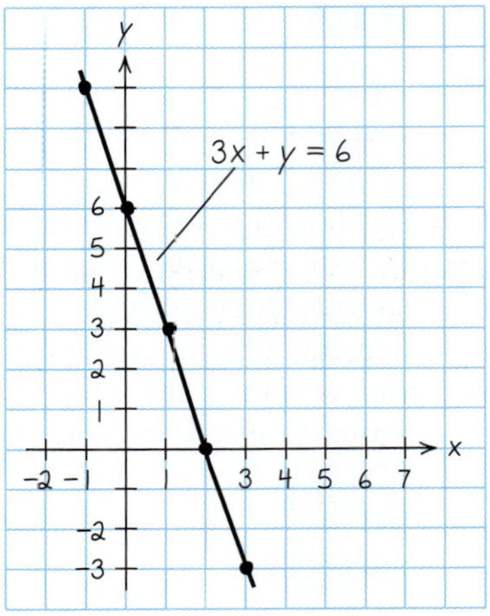

Figure 1.1

Checkpoint Audio-video solution in English & Spanish at LarsonPrecalculus.com.

Sketch the graph of $y - 2x = 1$.

The points at which a graph touches or crosses an axis are called the **intercepts** of the graph. For instance, in Example 2 the point $(0, 6)$ is the y-intercept of the graph because the graph crosses the y-axis at that point. The point $(2, 0)$ is the x-intercept of the graph because the graph crosses the x-axis at that point.

EXAMPLE 3 Sketching a Graph by Point Plotting

See LarsonPrecalculus.com for an interactive version of this type of example.

Use point plotting and graph paper to sketch the graph of

$y = x^2 - 2.$

Solution

Because the equation is already solved for y, make a table of values by choosing several convenient values of x and calculating the corresponding values of y.

x	-2	-1	0	1	2	3
$y = x^2 - 2$	2	-1	-2	-1	2	7
Solution point	$(-2, 2)$	$(-1, -1)$	$(0, -2)$	$(1, -1)$	$(2, 2)$	$(3, 7)$

Next, plot the solution points, as shown in Figure 1.2(a). Finally, connect the points with a smooth curve, as shown in Figure 1.2(b). This graph is called a *parabola*. You will study parabolas in Section 3.1.

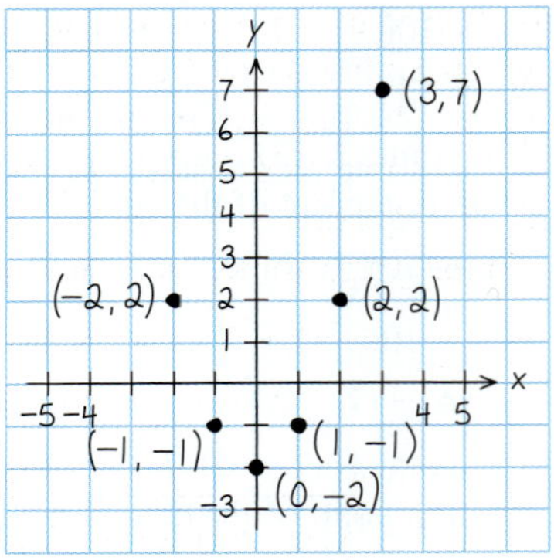

(a)

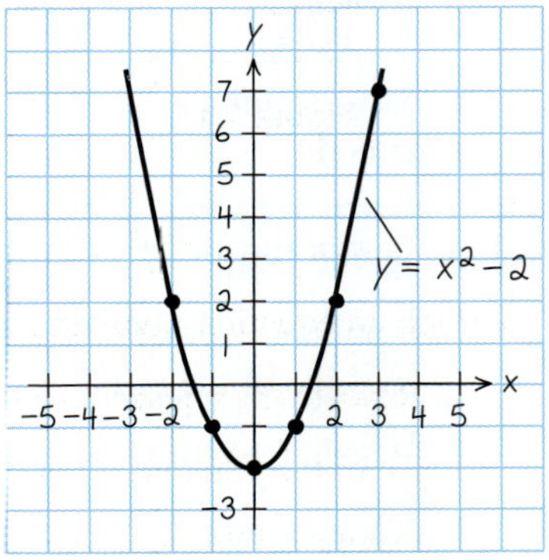

(b)

Figure 1.2

Checkpoint Audio-video solution in English & Spanish at LarsonPrecalculus.com.

Sketch the graph of $y = 1 - x^2$.

In this text, you will study two basic ways to create graphs: *by hand* and *using a graphing utility*. For instance, the graphs in Figures 1.1 and 1.2 were sketched by hand, and the graph in Figure 1.5 (on the next page) was created using a graphing utility.

Using a Graphing Utility

One of the disadvantages of the point-plotting method is that to get a good idea about the shape of a graph, you need to plot *many* points. With only a few points, you could misrepresent the graph of an equation. For instance, consider the equation

$$y = \frac{1}{30}x(x^4 - 10x^2 + 39).$$

When you plot the points $(-3, -3)$, $(-1, -1)$, $(0, 0)$, $(1, 1)$, and $(3, 3)$, as shown in Figure 1.3(a), you might think that the graph of the equation is a line. This is not correct. By plotting several more points and connecting the points with a smooth curve, you can see that the actual graph is not a line, as shown in Figure 1.3(b).

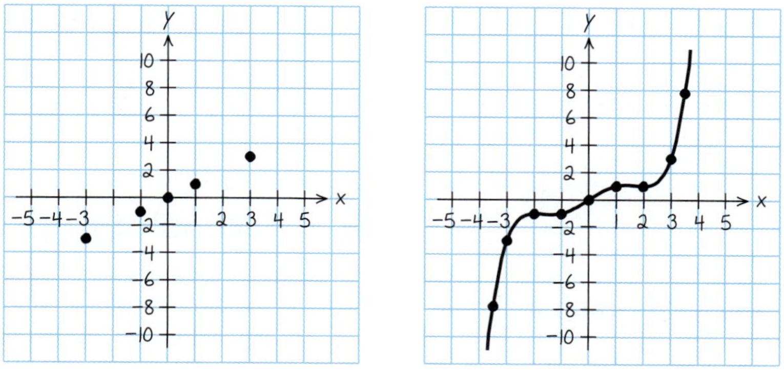

(a) (b)

Figure 1.3

From this, you can see that the point-plotting method leaves you with a dilemma. This method can be very inaccurate when only a few points are plotted, and it is very time-consuming to plot a dozen (or more) points. Technology can help solve this dilemma. Plotting several (even several hundred) points in a rectangular coordinate system is something that a computer or calculator can do easily. For instance, you can enter the equation $y = \frac{1}{30}x(x^4 - 10x^2 + 39)$ in a graphing utility (see Figure 1.4) to obtain the graph shown in Figure 1.5.

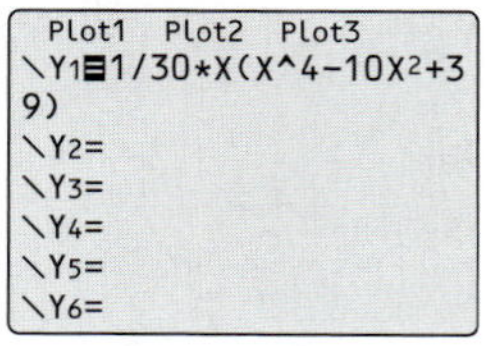

Figure 1.4

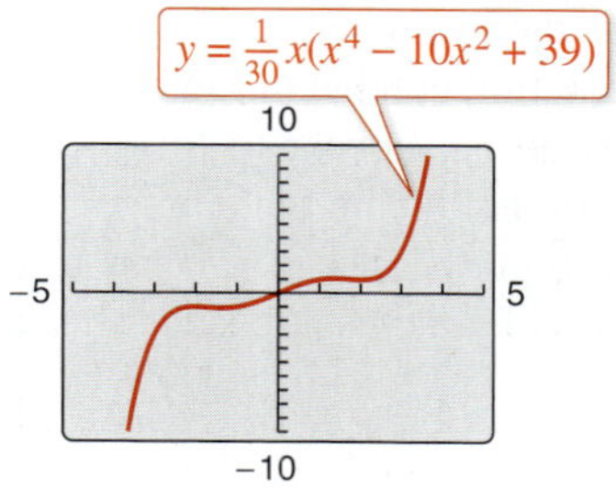

Figure 1.5

Using a Graphing Utility to Graph an Equation

To graph an equation involving x and y on a graphing utility, do the following.

1. Rewrite the equation so that y is isolated on the left side.
2. Enter the equation in the graphing utility.
3. Determine a *viewing window* that shows all important features of the graph.
4. Graph the equation.

Technology Tip

Many graphing utilities are capable of creating a table of values such as the following, which shows some points of the graph in Figure 1.3(b). For instructions on how to use the *table* feature, see Appendix A; for specific keystrokes, go to this textbook's *Companion Website*.

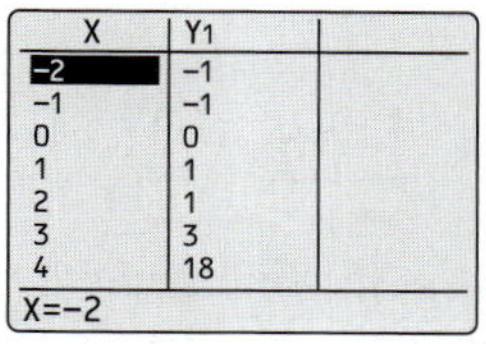

X	Y1
−2	−1
−1	−1
0	0
1	1
2	1
3	3
4	18

X=−2

Technology Tip

By choosing different viewing windows for a graph, it is possible to obtain very different impressions of the graph's shape. For instance, Figure 1.6 shows a different viewing window for the graph of the equation in Figure 1.5. Note how Figure 1.6 does not show *all* of the important features of the graph as does Figure 1.5. For instructions on how to set up a viewing window, see Appendix A; for specific keystrokes, go to this textbook's *Companion Website*.

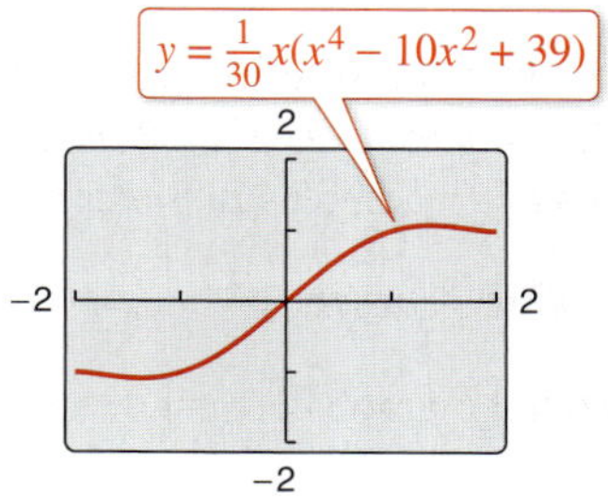

Figure 1.6

EXAMPLE 4 Using a Graphing Utility to Graph an Equation

To graph

$$y = -0.5x^3 + 2x$$

enter the equation in a graphing utility. Then use a standard viewing window (see Figure 1.7) to obtain the graph shown in Figure 1.8.

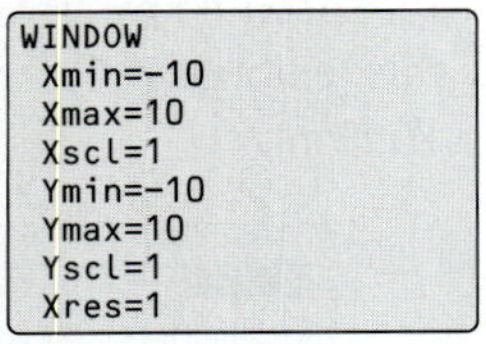

Figure 1.7

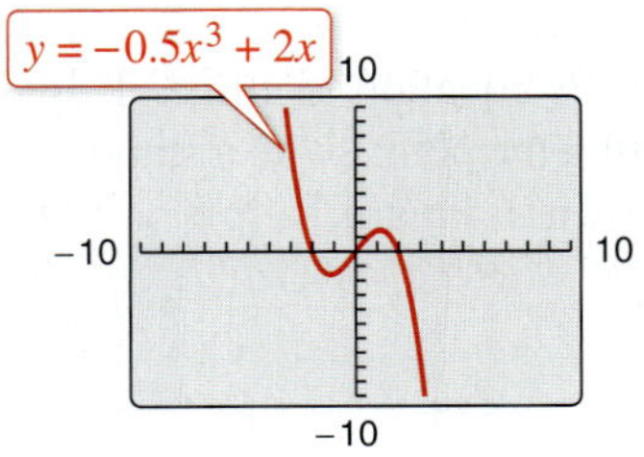

Figure 1.8

✓ ***Checkpoint*** *Audio-video solution in English & Spanish at LarsonPrecalculus.com.*

Use a graphing utility to graph $y = (x - 1)^2 - 3$.

Technology Tip

The standard viewing window on many graphing utilities does not give a true geometric perspective because the screen is rectangular, which distorts the image. That is, perpendicular lines will not appear to be perpendicular, and circles will not appear to be circular. To overcome this, you can use a *square setting*, as demonstrated in Example 5.

EXAMPLE 5 Using a Graphing Utility to Graph a Circle

Use a graphing utility to graph $x^2 + y^2 = 9$.

Solution

The graph of $x^2 + y^2 = 9$ is a circle whose center is the origin and whose radius is 3. To graph the equation, begin by solving the equation for y.

$x^2 + y^2 = 9$ — Write original equation.

$y^2 = 9 - x^2$ — Subtract x^2 from each side.

$y = \pm\sqrt{9 - x^2}$ — Take the square root of each side.

Remember that when you take the square root of a variable expression, you must account for both the positive and negative solutions. The graph of $y = \sqrt{9 - x^2}$ is the upper semicircle. The graph of $y = -\sqrt{9 - x^2}$ is the lower semicircle. Enter *both* equations in your graphing utility and generate the resulting graphs. In Figure 1.9, note that for a standard viewing window, the two graphs do not appear to form a circle. You can overcome this problem by using a *square setting*, in which the horizontal and vertical tick marks have equal spacing, as shown in Figure 1.10. On many graphing utilities, a square setting can be obtained when the ratio of y to x is 2 to 3. For instance, in Figure 1.10, the ratio of y to x is

$$\frac{Y_{max} - Y_{min}}{X_{max} - X_{min}} = \frac{4 - (-4)}{6 - (-6)} = \frac{8}{12} = \frac{2}{3}.$$

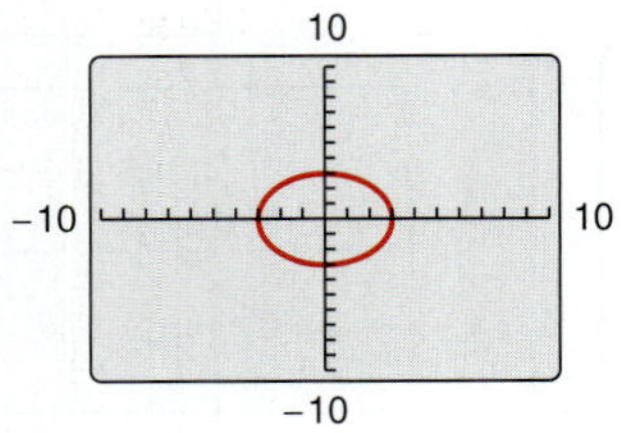

Figure 1.9

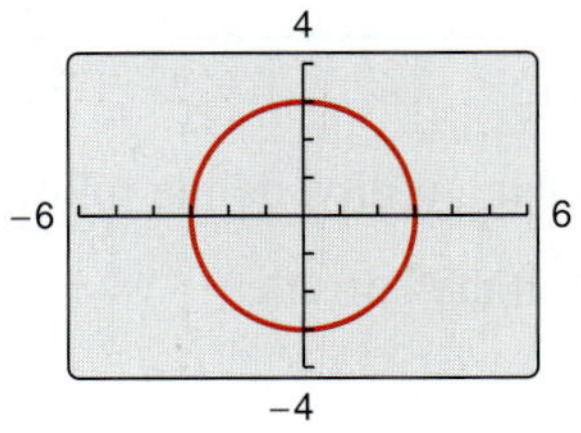

Figure 1.10

✓ ***Checkpoint*** *Audio-video solution in English & Spanish at LarsonPrecalculus.com.*

Use a graphing utility to graph $x^2 + y^2 = 4$.

Applications

Throughout this course, you will learn that there are many ways to approach a problem. Three common approaches are illustrated in Examples 6 and 7.

An Algebraic Approach: Use the rules of algebra.

A Graphical Approach: Draw and use a graph.

A Numerical Approach: Construct and use a table.

You should develop the habit of using at least two approaches to solve every problem in order to build your intuition and to check that your answer is reasonable.

The following two applications show how to develop mathematical models to represent real-world situations. You will see that both a graphing utility and algebra can be used to understand and solve the problems posed.

EXAMPLE 6 Running a Marathon

A runner runs at a constant rate of 4.8 miles per hour. The verbal model and algebraic equation relating distance run and elapsed time are as follows.

Verbal Model: Distance = Rate · Time

Equation: $d = 4.8t$

a. Determine how far the runner can run in 3.1 hours.

b. Determine how long it will take the runner to run a 26.2-mile marathon.

Algebraic Solution

a. To begin, find how far the runner can run in 3.1 hours by substituting 3.1 for t in the equation.

$d = 4.8t$ — Write original equation.

$= 4.8(3.1)$ — Substitute 3.1 for t.

≈ 14.9 — Use a calculator.

So, the runner can run about 14.9 miles in 3.1 hours. Use estimation to check your answer. Because 4.8 is about 5 and 3.1 is about 3, the distance is about $5(3) = 15$. So, 14.9 is reasonable.

b. You can find how long it will take to run a 26.2-mile marathon as follows. (For help with solving linear equations, see Appendix D at this textbook's *Companion Website*.)

$d = 4.8t$ — Write original equation.

$26.2 = 4.8t$ — Substitute 26.2 for d.

$\frac{26.2}{4.8} = t$ — Divide each side by 4.9.

$5.5 \approx t$ — Use a calculator.

So, it will take the runner about 5.5 hours to run 26.2 miles.

Graphical Solution

a. Use a graphing utility to graph $d = 4.8t$. (Represent d by y and t by x.) Choose a viewing window that shows the graph at $x = 3.1$.

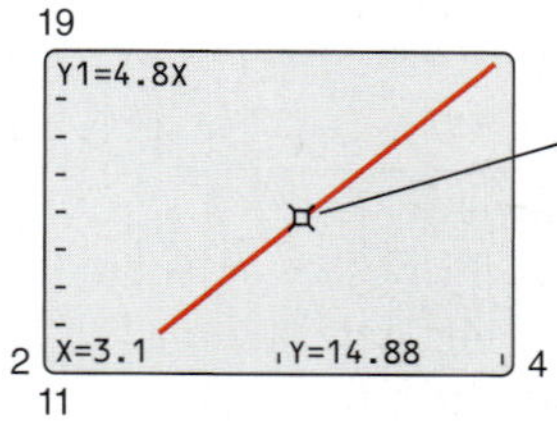

Use the *value* feature to estimate that when $x = 3.1$, the distance is $y \approx 14.9$ miles.

b. Adjust the viewing window so that it shows the graph at $y = 26.2$.

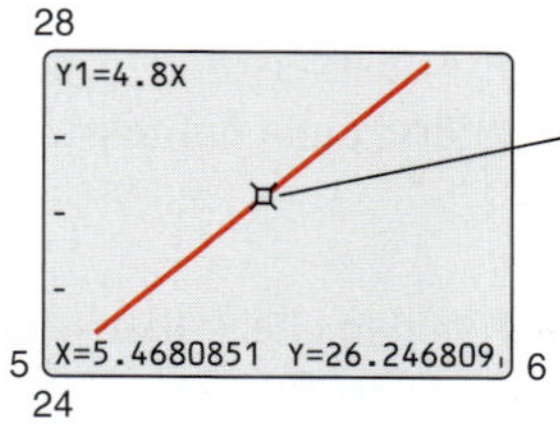

Use the *zoom* and *trace* features to estimate that when $y \approx 26.2$, the time is $x \approx 5.5$ hours.

✓ **Checkpoint** Audio-video solution in English & Spanish at LarsonPrecalculus.com.

Rework Example 6 when the runner runs at a constant rate of 5.2 miles per hour. (The equation relating distance run and elapsed time is $d = 5.2t$.)

EXAMPLE 7 Monthly Wage

You receive a monthly salary of \$2000 plus a commission of 10% of sales. The verbal model and algebraic equation relating wages, salary, and commission are as follows.

Verbal Model: Wages = Salary + Commission on sales

Equation: $y = 2000 + 0.1x$

a. Sales are \$1480 in August. What are your wages for that month?

b. You receive \$2225 for September. What are your sales for that month?

Numerical Solution

a. Enter $y = 2000 + 0.1x$ in a graphing utility. Then use the *table* feature of the graphing utility to create a table. Start the table at $x = 1400$ with a table step of 10.

X	Y1
1420	2142
1430	2143
1440	2144
1450	2145
1460	2146
1470	2147
1480	2148

Y1=2148

When $x = 1480$, the wages are $y = \$2148$.

b. Start the table at $x = 2000$ with a table step of 100.

X	Y1
2000	2200
2100	2210
2200	2220
2300	2230
2400	2240
2500	2250
2600	2260

X=2000

From the table, you can see that wages of \$2225 result from sales between \$2200 and \$2300.

You can improve the estimate by starting the table at $x = 2200$ with a table step of 10.

X	Y1
2200	2220
2210	2221
2220	2222
2230	2223
2240	2224
2250	2225
2260	2226

X=2250

From the table, you can see that wages of \$2225 result from sales of \$2250.

Graphical Solution

a. Use a graphing utility to graph $y = 2000 + 0.1x$. Choose a viewing window that shows the graph at $x = 1480$.

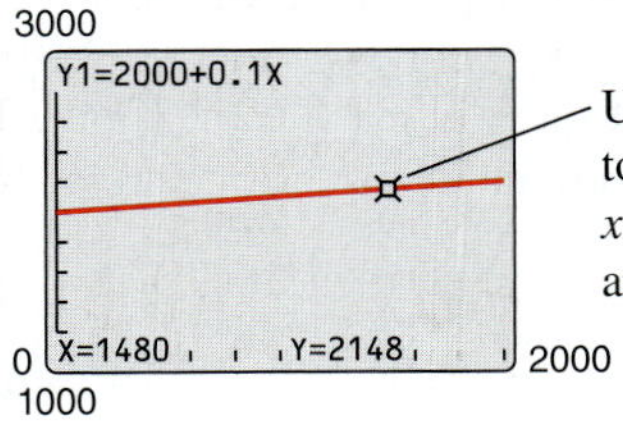

Use the *value* feature to estimate that when $x = 1480$, the wages are $y = \$2148$.

b. Use the graphing utility to find the value along the x-axis (sales) that corresponds to a y-value of 2225 (wages). Adjust the viewing window so that it shows the graph at $y = 2225$.

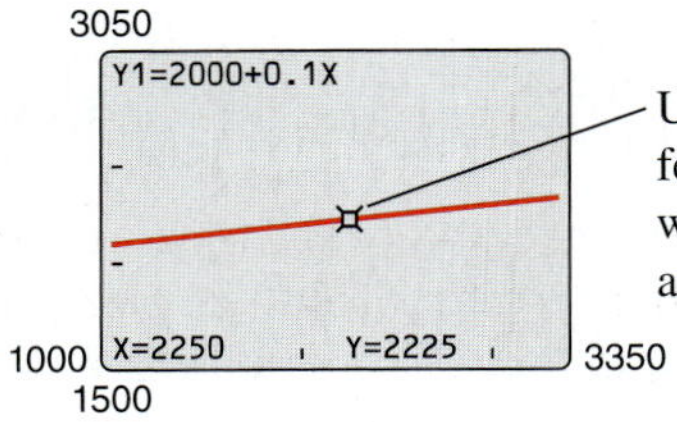

Use the *zoom* and *trace* features to estimate that when $y = 2225$, the sales are $x = \$2250$.

Checkpoint Audio-video solution in English & Spanish at LarsonPrecalculus.com.

Rework Example 7 when the monthly salary is \$2075 plus a commission of 8% of sales. (The equation relating wages, salary, and commission is $y = 2075 + 0.08x$.)

Remember to use a different approach to check that your answer is reasonable. For instance, to check the numerical solution to Example 7, use a graphical approach as shown above or use an algebraic approach as follows.

a. Substitute 1480 for x in the original equation and solve for y.

$$y = 2000 + 0.1(1480) = \$2148$$

b. Substitute 2225 for y in the original equation and solve for x.

$$2225 = 2000 + 0.1x \quad \Rightarrow \quad x = \$2250$$

1.1 Exercises

See *CalcChat.com* for tutorial help and worked-out solutions to odd-numbered exercises. For instructions on how to use a graphing utility, see Appendix A.

Vocabulary and Concept Check

In Exercises 1 and 2, fill in the blank.

1. For an equation in x and y, if substitution of a for x and b for y satisfies the equation, then the point (a, b) is a _______ .
2. The set of all solution points of an equation is the _______ of the equation.
3. Name three common approaches you can use to solve problems mathematically.
4. List the steps for sketching the graph of an equation by point plotting.

Procedures and Problem Solving

Determining Solution Points **In Exercises 5–12, determine whether each point lies on the graph of the equation.**

Equation	*Points*	
5. $y = \sqrt{x + 4}$	(a) $(0, 2)$	(b) $(12, 4)$
6. $y = \sqrt{5 - x}$	(a) $(1, 2)$	(b) $(5, 0)$
7. $y = 4 - \lvert x - 2\rvert$	(a) $(1, 5)$	(b) $(1.2, 3.2)$
8. $y = \lvert x - 1\rvert + 2$	(a) $(2, 1)$	(b) $(3.2, 4.2)$
9. $2x - y - 3 = 0$	(a) $(1, 2)$	(b) $(1, -1)$
10. $x^2 + y^2 = 20$	(a) $(3, -2)$	(b) $(-4, 2)$
11. $y = x^2 - 3x + 2$	(a) $\left(\frac{5}{2}, \frac{3}{4}\right)$	(b) $(-2, 8)$
12. $y = \frac{1}{3}x^3 - 2x^2$	(a) $\left(2, -\frac{16}{3}\right)$	(b) $(-3, 9)$

Sketching a Graph by Point Plotting **In Exercises 13–16, complete the table. Use the resulting solution points to sketch the graph of the equation. Use a graphing utility to verify the graph.**

13. $3x - 2y = 2$

x	-2	0	$\frac{2}{3}$	1	2
y					
Solution point					

14. $-4x + 2y = 10$

x	-3	$-\frac{5}{2}$	0	1	2
y					
Solution point					

15. $2x + y = x^2$

x	-1	0	1	2	3
y					
Solution point					

16. $6x - 2y = -2x^2$

x	-4	-3	-2	0	1
y					
Solution point					

Matching an Equation with Its Graph **In Exercises 17–22, match the equation with its graph. [The graphs are labeled (a), (b), (c), (d), (e), and (f).]**

(a)

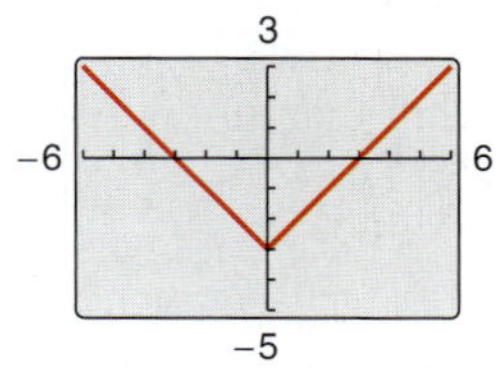

(b)

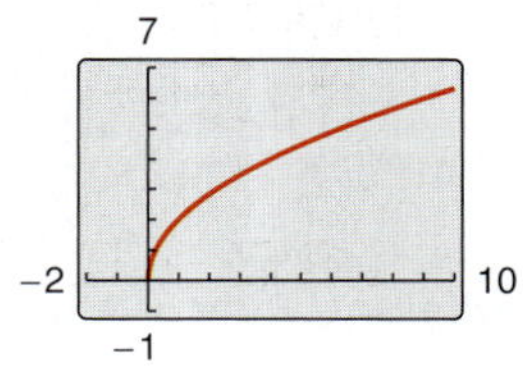

(c)

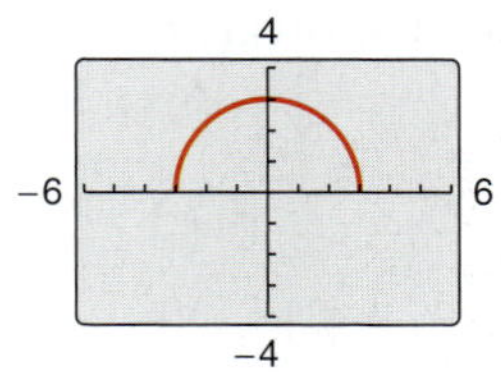

(d)

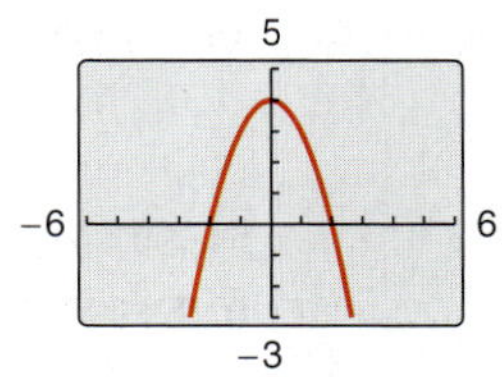

(e)

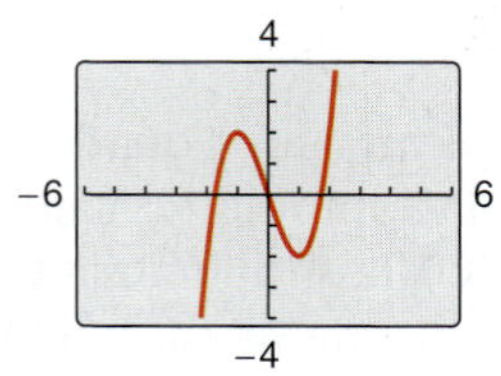

(f)

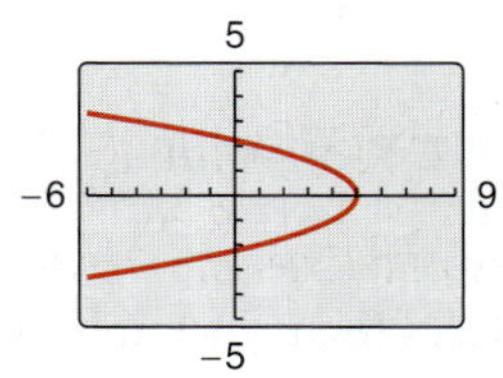

17. $y = 2\sqrt{x}$
18. $y = 4 - x^2$
19. $y = \sqrt{9 - x^2}$
20. $y = \lvert x\rvert - 3$
21. $y = x^3 - 3x$
22. $x = 5 - y^2$

Sketching the Graph of an Equation **In Exercises 23–30, sketch the graph of the equation.**

23. $y = 2 - x^2$
24. $y = x^3 - 3$
25. $y = \sqrt{x - 3}$
26. $y = \sqrt{1 - x}$
27. $y = \lvert x - 2\rvert$
28. $y = 4 - \lvert x\rvert$
29. $x = y^2 - 1$
30. $x = y^2 + 4$

Using a Graphing Utility to Graph an Equation In Exercises 31–44, use a graphing utility to graph the equation. Use a standard viewing window. Approximate any x- or y-intercepts of the graph.

31. $y = 5 - \frac{3}{2}x$

32. $y = \frac{2}{3}x - 1$

33. $y = |x + 2| - 3$

34. $y = -|x - 3| + 1$

35. $y = \dfrac{2x}{x - 1}$

36. $y = \dfrac{10}{x^2 + 2}$

37. $y = x\sqrt{x + 3}$

38. $y = (6 - x)\sqrt{x}$

39. $y = \sqrt[3]{x - 8}$

40. $y = \sqrt[3]{x + 1}$

41. $x^2 - y = 4x - 3$

42. $2y - x^2 + 8 = 2x$

43. $y - 4x = x^2(x - 4)$

44. $x^3 + y = 1$

Describing the Viewing Window of a Graphing Utility In Exercises 45 and 46, describe the viewing window of the graph shown.

45. $y = -10x + 50$

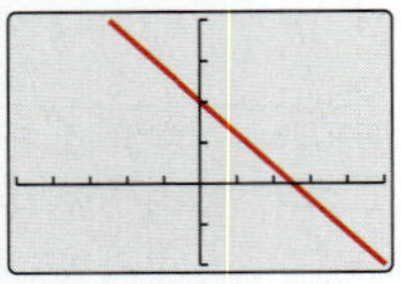

46. $y = \sqrt{x + 2} - 1$

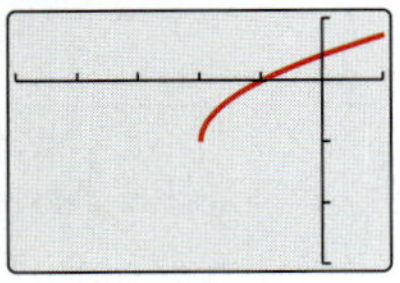

Verifying a Rule of Algebra In Exercises 47–50, explain how to use a graphing utility to verify that $y_1 = y_2$. Identify the rule of algebra that is illustrated.

47. $y_1 = \frac{1}{4}(x^2 - 8)$
$y_2 = \frac{1}{4}x^2 - 2$

48. $y_1 = \frac{1}{2}x + (x + 1)$
$y_2 = \frac{3}{2}x + 1$

49. $y_1 = \dfrac{1}{5}[10(x^2 - 1)]$
$y_2 = 2(x^2 - 1)$

50. $y_1 = (x^2 + 3) \cdot \dfrac{1}{x^2 + 3}$
$y_2 = 1$

Using a Graphing Utility to Graph an Equation In Exercises 51–54, use a graphing utility to graph the equation. Use the *trace* feature of the graphing utility to approximate the unknown coordinate of each solution point accurate to two decimal places. (*Hint:* You may need to use the *zoom* feature of the graphing utility to obtain the required accuracy.)

51. $y = \sqrt{5 - x}$
(a) $(3, y)$
(b) $(x, 3)$

52. $y = x^2(x - 3)$
(a) $(-1, y)$
(b) $(x, 6)$

53. $y = x^5 - 5x$
(a) $(-0.5, y)$
(b) $(x, -2)$

54. $y = |x^2 - 6x + 5|$
(a) $(2, y)$
(b) $(x, 1.5)$

Using a Graphing Utility to Graph a Circle In Exercises 55–58, solve for y and use a graphing utility to graph each of the resulting equations in the same viewing window. (Adjust the viewing window so that the circle appears circular.)

55. $x^2 + y^2 = 16$

56. $x^2 + y^2 = 36$

57. $(x - 1)^2 + (y - 2)^2 = 49$

58. $(x - 3)^2 + (y - 1)^2 = 25$

Determining Solution Points In Exercises 59 and 60, determine which point lies on the graph of the circle. (There may be more than one correct answer.)

59. $(x - 1)^2 + (y - 2)^2 = 25$
(a) $(1, 3)$ (b) $(-2, 6)$
(c) $(5, -1)$ (d) $(0, 2 + 2\sqrt{6})$

60. $(x + 2)^2 + (y - 3)^2 = 25$
(a) $(-2, 3)$ (b) $(0, 0)$
(c) $(1, -1)$ (d) $(-1, 3 - 2\sqrt{6})$

61. MODELING DATA

A hospital purchases a new magnetic resonance imaging (MRI) machine for \$500,000. The depreciated value (reduced value) y after t years is $y = 500{,}000 - 47{,}000t$, for $0 \le t \le 9$.

(a) Use the constraints of the model and a graphing utility to graph the equation using an appropriate viewing window.

(b) Use the *value* feature of the graphing utility to determine the value of y when $t = 5.8$. Verify your answer algebraically.

(c) Use the *zoom* and *trace* features of the graphing utility to determine the value of t when $y = 156{,}900$. Verify your answer algebraically.

62. MODELING DATA

You buy a personal watercraft for \$8250. The depreciated value y after t years is $y = 8250 - 689t$, for $0 \le t \le 10$.

(a) Use the constraints of the model and a graphing utility to graph the equation using an appropriate viewing window.

(b) Use the *zoom* and *trace* features of the graphing utility to determine the value of t when $y = 5545.25$. Verify your answer algebraically.

(c) Use the *value* feature of the graphing utility to determine the value of y when $t = 5.5$. Verify your answer algebraically.

63. MODELING DATA

The table shows the numbers of new single-family houses y (in thousands) in the western United States from 2006 through 2013. (Source: U.S. Census Bureau)

Year	New houses, y (in thousands)
2006	415
2007	294
2008	190
2009	118
2010	103
2011	91
2012	101
2013	129

Spreadsheet at LarsonPrecalculus.com

A model that represents the data is given by

$$y = 13.42t^2 - 294.1t + 1692, \quad 6 \leq t \leq 13$$

where t represents the year, with $t = 6$ corresponding to 2006.

(a) Use the model and the *table* feature of a graphing utility to find the numbers of new houses from 2006 through 2013. How well does the model fit the data? Explain.

(b) Use the graphing utility to graph the data from the table and the model in the same viewing window. How well does the model fit the data? Explain.

(c) Use the model to estimate the numbers of new houses in 2015 and 2017. Do the values seem reasonable? Explain.

(d) Use the *zoom* and *trace* features of the graphing utility to determine during which year(s) the number of new houses was approximately 100,000.

64. ***Why you should learn it*** *(p. 75)* The table shows the life expectancy y of a child (at birth) in the United States for each of the selected years from 1940 through 2010. (Source: U.S. National Center for Health Statistics)

Year	Life expectancy, y
1940	62.9
1950	68.2
1960	69.7
1970	70.8
1980	73.7
1990	75.4
2000	76.8
2010	78.7

Spreadsheet at LarsonPrecalculus.com

A model that represents the data is given by

$$y = \frac{63.6 + 0.97t}{1 + 0.01t}, \quad 0 \leq t \leq 70$$

where t is the time in years, with $t = 0$ corresponding to 1940.

(a) Use a graphing utility to graph the data from the table above and the model in the same viewing window. How well does the model fit the data? Explain.

(b) Find the y-intercept of the graph of the model. What does it represent in the context of the problem?

(c) Use the *zoom* and *trace* features of the graphing utility to determine the year when the life expectancy was 70.1. Verify your answer algebraically.

(d) Determine the life expectancy in 1978 both graphically and algebraically.

Conclusions

True or False? **In Exercises 65 and 66, determine whether the statement is true or false. Justify your answer.**

65. A parabola can have only one x-intercept.

66. The graph of a linear equation can have either no x-intercepts or only one x-intercept.

67. Writing Your employer offers you a choice of wage scales: a monthly salary of \$3000 plus commission of 7% of sales or a salary of \$3400 plus a 5% commission. Write a short paragraph discussing how you would choose your option. At what sales level would the options yield the same salary?

68. **HOW DO YOU SEE IT?** Use the graph of $y = (x - 1)^2 - 4$ to answer each question.

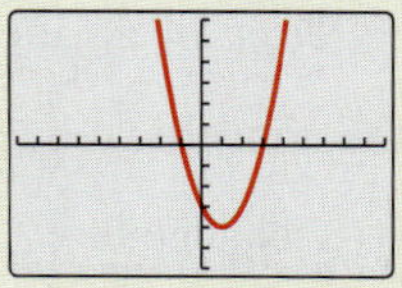

(a) Describe the viewing window of the graph.

(b) Approximate any x- or y-intercepts of the graph.

(c) Determine whether $(-4, 1)$ and $(2, -3)$ are solution points of the equation. Explain.

Cumulative Mixed Review

Performing Operations with Polynomials **In Exercises 69 and 70, perform the operation and write the result in standard form.**

69. $(9x - 4) + (2x^2 - x + 15)$

70. $(3x^2 - 5)(-x^2 + 1)$

1.2 Lines in the Plane

The Slope of a Line

In this section, you will study lines and their equations. The **slope** of a nonvertical line represents the number of units the line rises or falls vertically for each unit of horizontal change from left to right. For instance, consider the two points

(x_1, y_1) and (x_2, y_2)

on the line shown in Figure 1.11.

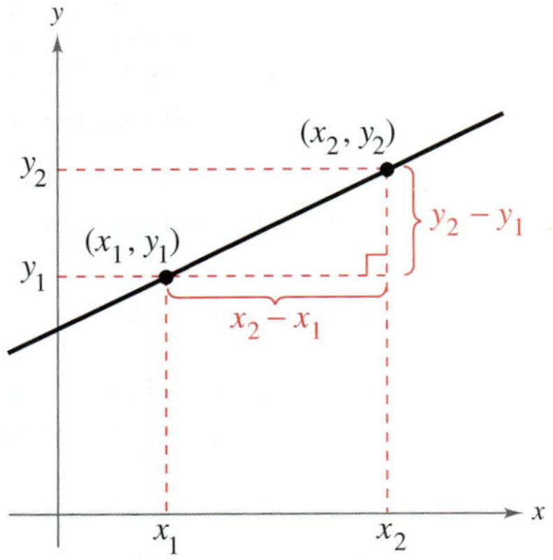

Figure 1.11

As you move from left to right along this line, a change of $(y_2 - y_1)$ units in the vertical direction corresponds to a change of $(x_2 - x_1)$ units in the horizontal direction. That is,

$y_2 - y_1 =$ the change in y

and

$x_2 - x_1 =$ the change in x.

The slope of the line is given by the ratio of these two changes.

What you should learn

- Find the slopes of lines.
- Write linear equations given points on lines and their slopes.
- Use slope-intercept forms of linear equations to sketch lines.
- Use slope to identify parallel and perpendicular lines.

Why you should learn it

The slope of a line can be used to solve real-life problems. For instance, in Exercise 97 on page 95, you will use a linear equation to model student enrollment at Penn State University.

Definition of the Slope of a Line

The **slope** m of the nonvertical line through (x_1, y_1) and (x_2, y_2) is

$$m = \frac{y_2 - y_1}{x_2 - x_1} = \frac{\text{change in } y}{\text{change in } x}$$

where $x_1 \neq x_2$.

When this formula for slope is used, the *order of subtraction* is important. Given two points on a line, you are free to label either one of them as (x_1, y_1) and the other as (x_2, y_2). Once you have done this, however, you must form the numerator and denominator using the same order of subtraction.

$$m = \frac{y_2 - y_1}{x_2 - x_1} \qquad m = \frac{y_1 - y_2}{x_1 - x_2} \qquad m = \frac{y_2 - y_1}{x_1 - x_2}$$

Correct Correct Incorrect

Throughout this text, the term *line* always means a *straight* line.

EXAMPLE 1 Finding the Slope of a Line

Find the slope of the line passing through each pair of points.

a. $(-2, 0)$ and $(3, 1)$ **b.** $(-1, 2)$ and $(2, 2)$ **c.** $(0, 4)$ and $(1, -1)$

Solution

a. $m = \dfrac{y_2 - y_1}{x_2 - x_1} = \dfrac{1 - 0}{3 - (-2)} = \dfrac{1}{3 + 2} = \dfrac{1}{5}$

Difference in y-values (numerator); Difference in x-values (denominator)

b. $m = \dfrac{2 - 2}{2 - (-1)} = \dfrac{0}{3} = 0$

c. $m = \dfrac{-1 - 4}{1 - 0} = \dfrac{-5}{1} = -5$

The graphs of the three lines are shown in Figure 1.12. Note that the *square setting* gives the correct "steepness" of the lines.

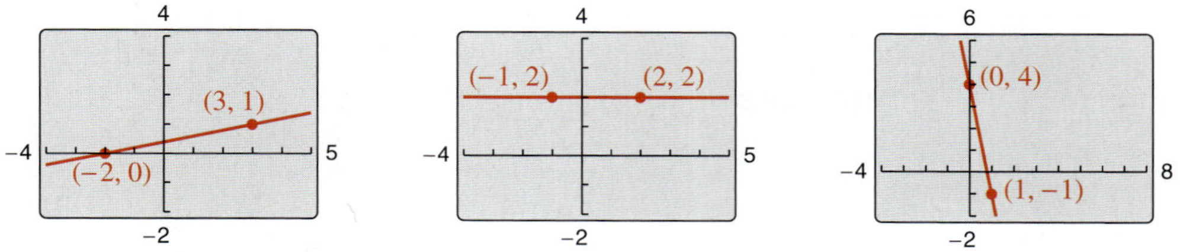

(a) (b) (c)

Figure 1.12

✓ Checkpoint *Audio-video solution in English & Spanish at LarsonPrecalculus.com.*

Find the slope of the line passing through each pair of points.

a. $(-5, -6)$ and $(2, 8)$ **b.** $(4, 2)$ and $(2, 5)$ **c.** $(0, -1)$ and $(3, -1)$

The definition of slope does not apply to vertical lines. For instance, consider the points $(3, 4)$ and $(3, 1)$ on the vertical line shown in Figure 1.13. Applying the formula for slope, you obtain

$$m = \frac{4 - 1}{3 - 3} = \frac{3}{0}. \quad \text{Undefined}$$

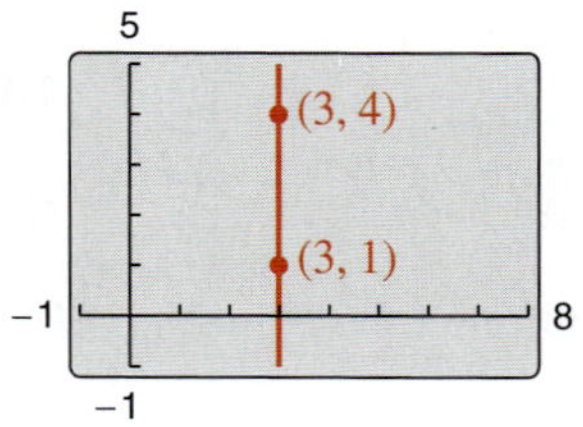

Figure 1.13

Because division by zero is undefined, the slope of a vertical line is undefined.

From the lines shown in Figures 1.12 and 1.13, you can make the following generalizations about the slope of a line.

The Slope of a Line

1. A line with positive slope $(m > 0)$ *rises* from left to right.
2. A line with negative slope $(m < 0)$ *falls* from left to right.
3. A line with zero slope $(m = 0)$ is *horizontal.*
4. A line with undefined slope is *vertical.*

Explore the Concept

Use a graphing utility to compare the slopes of the lines $y = 0.5x$, $y = x$, $y = 2x$, and $y = 4x$. What do you observe about these lines? Compare the slopes of the lines $y = -0.5x$, $y = -x$, $y = -2x$, and $y = -4x$. What do you observe about these lines? (*Hint:* Use a *square setting* to obtain a true geometric perspective.)

The Point-Slope Form of the Equation of a Line

When you know the slope of a line *and* you also know the coordinates of one point on the line, you can find an equation of the line. For instance, in Figure 1.14, let (x_1, y_1) be a point on the line whose slope is m. When (x, y) is any *other* point on the line, it follows that

$$\frac{y - y_1}{x - x_1} = m.$$

This equation in the variables x and y can be rewritten in the **point-slope form** of the equation of a line.

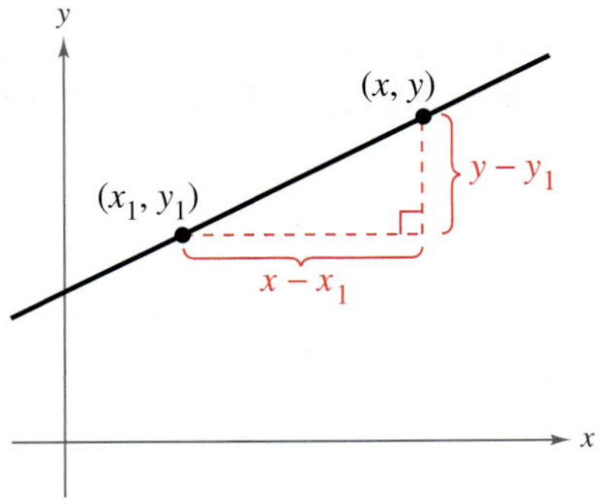

Figure 1.14

> **Point-Slope Form of the Equation of a Line**
>
> The **point-slope form** of the equation of the line that passes through the point (x_1, y_1) and has a slope of m is
>
> $$y - y_1 = m(x - x_1).$$

EXAMPLE 2 The Point-Slope Form of the Equation of a Line

Find an equation of the line that passes through the point

$(1, -2)$

and has a slope of 3.

Solution

$y - y_1 = m(x - x_1)$	Point-slope form
$y - (-2) = 3(x - 1)$	Substitute for y_1, m, and x_1.
$y + 2 = 3x - 3$	Simplify.
$y = 3x - 5$	Solve for y.

The line is shown in Figure 1.15.

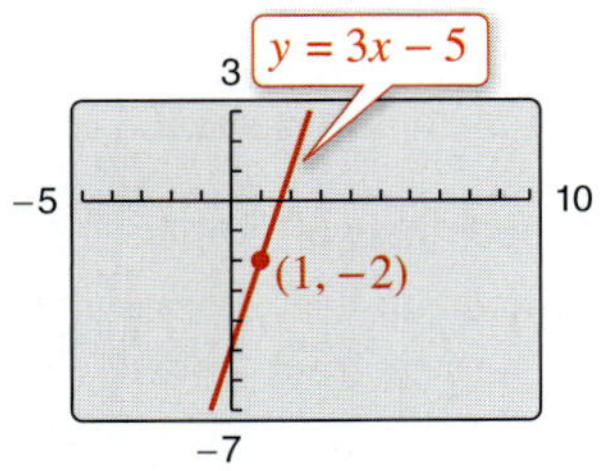

Figure 1.15

✓ **Checkpoint** *Audio-video solution in English & Spanish at LarsonPrecalculus.com.*

Find an equation of the line that passes through the point $(3, -7)$ and has a slope of 2.

The point-slope form can be used to find an equation of a nonvertical line passing through two points

(x_1, y_1) and (x_2, y_2).

First, find the slope of the line.

$$m = \frac{y_2 - y_1}{x_2 - x_1}, \quad x_1 \neq x_2$$

Then use the point-slope form to obtain the equation

$$y - y_1 = \frac{y_2 - y_1}{x_2 - x_1}(x - x_1).$$

This is sometimes called the **two-point form** of the equation of a line.

> **Remark**
>
> When you find an equation of the line that passes through two given points, you need to substitute the coordinates of only one of the points into the point-slope form. It does not matter which point you choose because both points will yield the same result.

EXAMPLE 3 A Linear Model for Profits Prediction

In 2011, Tyson Foods had sales of \$32.266 billion, and in 2012, sales were \$33.278 billion. Write a linear equation giving the sales y in terms of the year x. Then use the equation to predict the sales for 2013. (Source: Tyson Foods, Inc.)

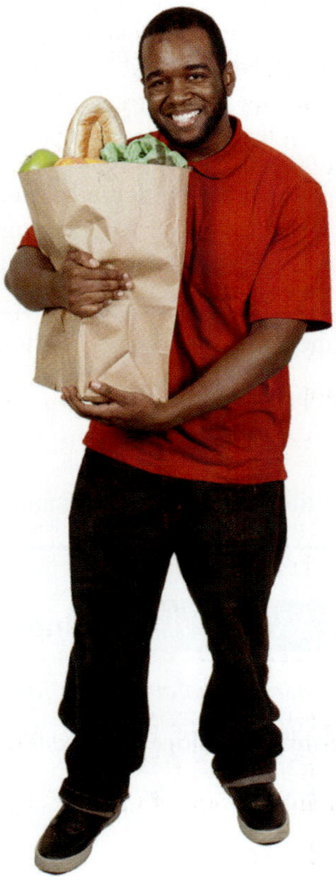

Solution

Let $x = 0$ represent 2000. In Figure 1.16, let $(11, 32.266)$ and $(12, 33.278)$ be two points on the line representing the sales. The slope of this line is

$$m = \frac{33.278 - 32.266}{12 - 11} = 1.012.$$

Next, use the point-slope form to find the equation of the line.

$$y - 32.266 = 1.012(x - 11)$$

$$y = 1.012x + 21.134$$

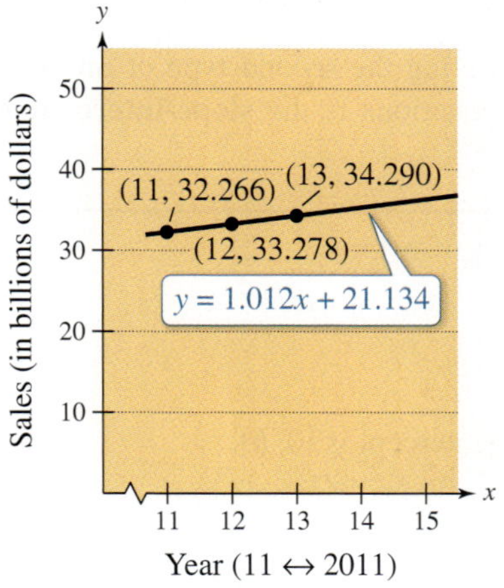

Figure 1.16

Now, using this equation, you can predict the 2013 sales ($x = 13$) to be

$$y = 1.012(13) + 21.134 = 13.156 + 21.134 = \$34.290 \text{ billion.}$$

(In this case, the prediction is quite good—the actual sales in 2013 were \$34.374 billion.)

✓ **Checkpoint** *Audio-video solution in English & Spanish at LarsonPrecalculus.com.*

In 2012, Apple had sales of \$156.508 billion, and in 2013, sales were \$170.910 billion. Write a linear equation giving the sales y in terms of the year x. Then use the equation to predict the sales for 2014. (Source: Apple, Inc.)

Library of Parent Functions: Linear Function

In the next section, you will be introduced to the precise meaning of the term *function*. The simplest type of function is the *parent linear function*

$$f(x) = x.$$

As its name implies, the graph of the parent linear function is a line. The basic characteristics of the parent linear function are summarized below and on the inside cover of this text. (Note that some of the terms below will be defined later in the text.)

Graph of $f(x) = x$
Domain: $(-\infty, \infty)$
Range: $(-\infty, \infty)$
Intercept: $(0, 0)$
Increasing

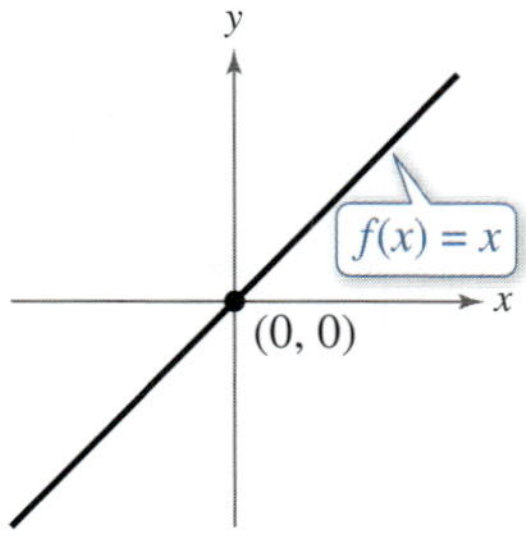

The function $f(x) = x$ is also referred to as the *identity function*. Later in this text, you will learn that the graph of the linear function $f(x) = mx + b$ is a line with slope m and y-intercept $(0, b)$. When $m = 0$, $f(x) = b$ is called a *constant function* and its graph is a horizontal line.

Sketching Graphs of Lines

Many problems in coordinate geometry can be classified in two categories.

1. Given a graph (or parts of it), find its equation.
2. Given an equation, sketch its graph.

For lines, the first problem can be solved by using the point-slope form. This formula, however, is not particularly useful for solving the second type of problem. The form that is better suited to graphing linear equations is the **slope-intercept form** of the equation of a line, $y = mx + b$.

Slope-Intercept Form of the Equation of a Line

The graph of the equation

$$y = mx + b$$

is a line whose slope is m and whose y-intercept is $(0, b)$.

EXAMPLE 4 Using the Slope-Intercept Form

See LarsonPrecalculus.com for an interactive version of this type of example.

Determine the slope and y-intercept of each linear equation. Then describe its graph.

a. $x + y = 2$ **b.** $y = 2$

Algebraic Solution

a. Begin by writing the equation in slope-intercept form.

$x + y = 2$ Write original equation.

$y = 2 - x$ Subtract x from each side.

$y = -x + 2$ Write in slope-intercept form.

From the slope-intercept form of the equation, the slope is -1 and the y-intercept is

$(0, 2)$.

Because the slope is negative, you know that the graph of the equation is a line that falls one unit for every unit it moves to the right.

b. By writing the equation $y = 2$ in slope-intercept form

$y = (0)x + 2$

you can see that the slope is 0 and the y-intercept is

$(0, 2)$.

A zero slope implies that the line is horizontal.

Graphical Solution

a.

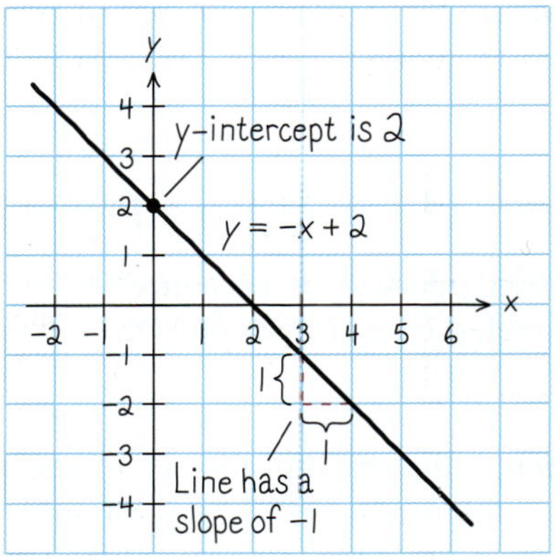

b.

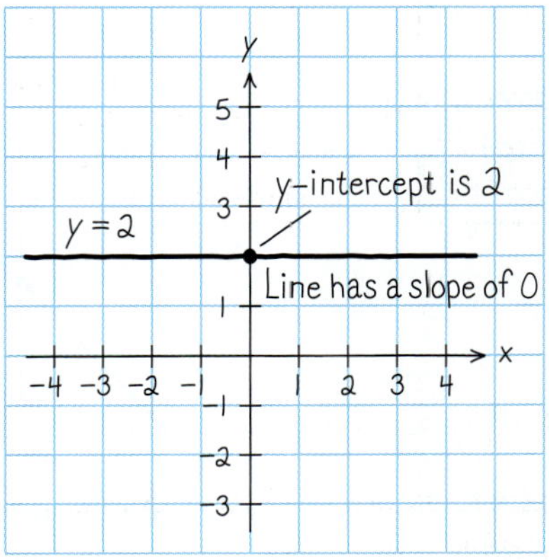

✔ **Checkpoint** *Audio-video solution in English & Spanish at LarsonPrecalculus.com.*

Determine the slope and y-intercept of $x - 2y = 4$. Then describe its graph.

From the slope-intercept form of the equation of a line, you can see that a horizontal line ($m = 0$) has an equation of the form $y = b$. This is consistent with the fact that each point on a horizontal line through $(0, b)$ has a y-coordinate of b. Similarly, each point on a vertical line through $(a, 0)$ has an x-coordinate of a. So, a vertical line has an equation of the form $x = a$. This equation cannot be written in slope-intercept form because the slope of a vertical line is undefined. However, every line has an equation that can be written in the **general form**

$$Ax + By + C = 0 \qquad \text{General form of the equation of a line}$$

where A and B are not *both* zero.

Summary of Equations of Lines

1. General form: $Ax + By + C = 0$
2. Vertical line: $x = a$
3. Horizontal line: $y = b$
4. Slope-intercept form: $y = mx + b$
5. Point-slope form: $y - y_1 = m(x - x_1)$

EXAMPLE 5 Different Viewing Windows

When a graphing utility is used to graph a line, it is important to realize that the line may not visually appear to have the slope indicated by its equation. This occurs because of the viewing window used for the graph. For instance, Figure 1.17 shows graphs of $y = 2x + 1$ produced on a graphing utility using three different viewing windows. Notice that the slopes in Figures 1.17(a) and (b) do not visually appear to be equal to 2. When you use a *square setting*, as in Figure 1.17(c), the slope visually appears to be 2.

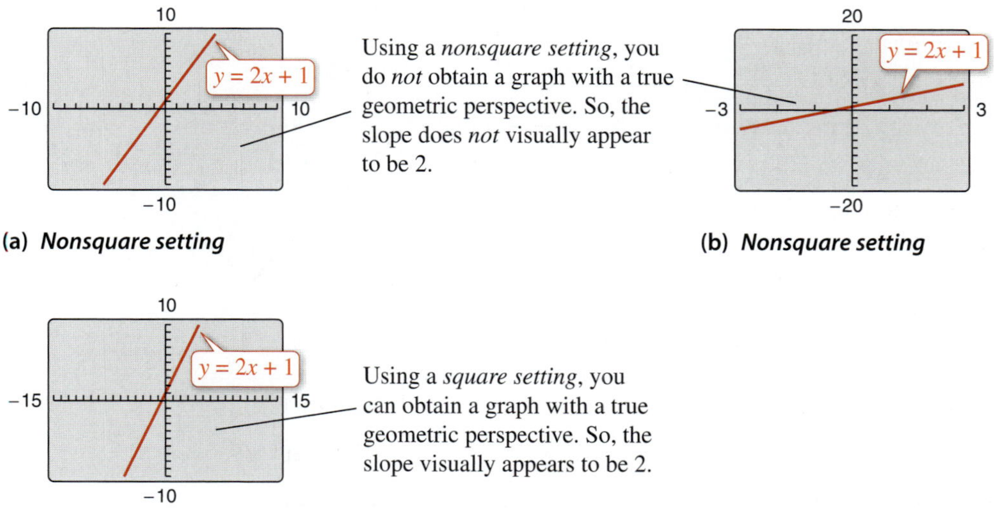

(a) *Nonsquare setting*

(b) *Nonsquare setting*

(c) *Square setting*

Figure 1.17

✓ *Checkpoint* *Audio-video solution in English & Spanish at LarsonPrecalculus.com.*

Use a graphing utility to graph $y = 0.5x - 3$ using each viewing window. Describe the difference in the graphs.

a. Xmin = -5, Xmax = 10, Xscl = 1, Ymin = -1, Ymax = 10, Yscl = 1

b. Xmin = -2, Xmax = 10, Xscl = 1, Ymin = -4, Ymax = 1, Yscl = 1

c. Xmin = -5, Xmax = 10, Xscl = 1, Ymin = -7, Ymax = 3, Yscl = 1

Parallel and Perpendicular Lines

The slope of a line is a convenient tool for determining whether two lines are parallel or perpendicular.

> **Parallel Lines**
>
> Two distinct nonvertical lines are **parallel** if and only if their slopes are equal. That is, $m_1 = m_2$.

Explore the Concept

Graph the lines $y_1 = \frac{1}{2}x + 1$ and $y_2 = -2x + 1$ in the same viewing window. What do you observe?

Graph the lines $y_1 = 2x + 1$, $y_2 = 2x$, and $y_3 = 2x - 1$ in the same viewing window. What do you observe?

EXAMPLE 6 Equations of Parallel Lines

Find the slope-intercept form of the equation of the line that passes through the point $(2, -1)$ and is parallel to the line $2x - 3y = 5$.

Solution

Begin by writing the equation of the line in slope-intercept form.

$$2x - 3y = 5 \quad \text{Write original equation.}$$

$$-2x + 3y = -5 \quad \text{Multiply by } -1.$$

$$3y = 2x - 5 \quad \text{Add } 2x \text{ to each side.}$$

$$y = \frac{2}{3}x - \frac{5}{3} \quad \text{Write in slope-intercept form.}$$

Therefore, the given line has a slope of

$m = \frac{2}{3}$.

Any line parallel to the given line must also have a slope of $\frac{2}{3}$. So, the line through $(2, -1)$ has the following equation.

$$y - y_1 = m(x - x_1) \quad \text{Point-slope form}$$

$$y - (-1) = \frac{2}{3}(x - 2) \quad \text{Substitute for } y_1, m, \text{ and } x_1.$$

$$y + 1 = \frac{2}{3}x - \frac{4}{3} \quad \text{Simplify.}$$

$$y = \frac{2}{3}x - \frac{7}{3} \quad \text{Write in slope-intercept form.}$$

Notice the similarity between the slope-intercept form of the original equation and the slope-intercept form of the parallel equation. The graphs of both equations are shown in Figure 1.18.

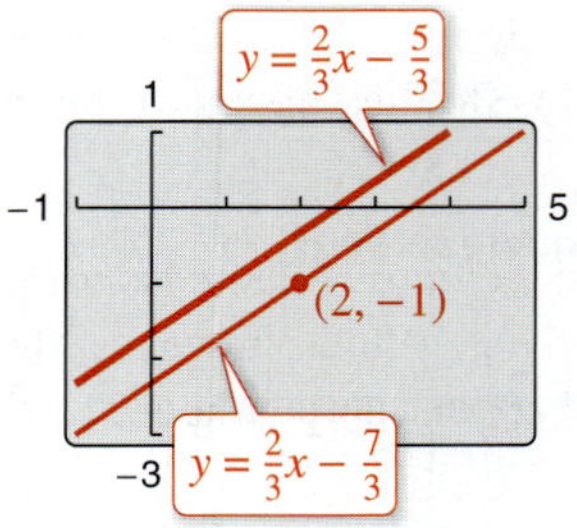

Figure 1.18

✓ **Checkpoint** *Audio-video solution in English & Spanish at LarsonPrecalculus.com.*

Find the slope-intercept form of the equation of the line that passes through the point $(-4, 1)$ and is parallel to the line $5x - 3y = 8$.

> **Perpendicular Lines**
>
> Two nonvertical lines are **perpendicular** if and only if their slopes are negative reciprocals of each other. That is,
>
> $$m_1 = -\frac{1}{m_2}.$$

EXAMPLE 7 Equations of Perpendicular Lines

Find the slope-intercept form of the equation of the line that passes through the point $(2, -1)$ and is perpendicular to the line $2x - 3y = 5$.

Solution

From Example 6, you know that the equation can be written in the slope-intercept form $y = \frac{2}{3}x - \frac{5}{3}$. You can see that the line has a slope of $\frac{2}{3}$. So, any line perpendicular to this line must have a slope of $-\frac{3}{2}$ (because $-\frac{3}{2}$ is the negative reciprocal of $\frac{2}{3}$). So, the line through the point $(2, -1)$ has the following equation.

$$y - (-1) = -\tfrac{3}{2}(x - 2) \quad \text{Write in point-slope form.}$$

$$y + 1 = -\tfrac{3}{2}x + 3 \quad \text{Simplify.}$$

$$y = -\tfrac{3}{2}x + 2 \quad \text{Write in slope-intercept form.}$$

The graphs of both equations are shown in the figure.

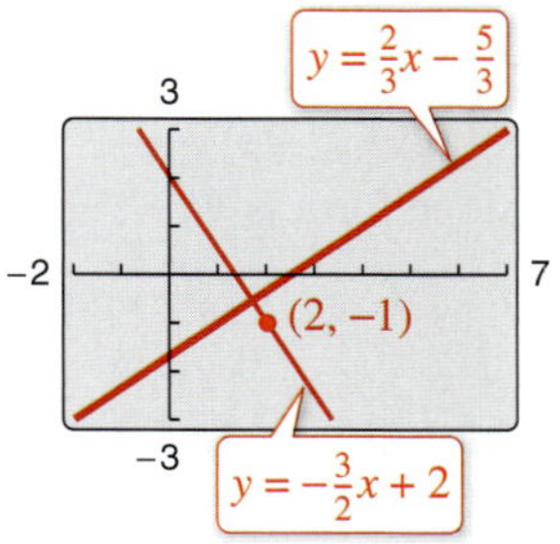

✓ Checkpoint *Audio-video solution in English & Spanish at LarsonPrecalculus.com.*

Find the slope-intercept form of the equation of the line that passes through the point $(-4, 1)$ and is perpendicular to the line $5x - 3y = 8$.

EXAMPLE 8 Graphs of Perpendicular Lines

Use a graphing utility to graph the lines $y = x + 1$ and $y = -x + 3$ in the same viewing window. The lines are perpendicular (they have slopes of $m_1 = 1$ and $m_2 = -1$). Do they appear to be perpendicular on the display?

Solution

When the viewing window is nonsquare, as in Figure 1.19, the two lines will not appear perpendicular. When, however, the viewing window is square, as in Figure 1.20, the lines will appear perpendicular.

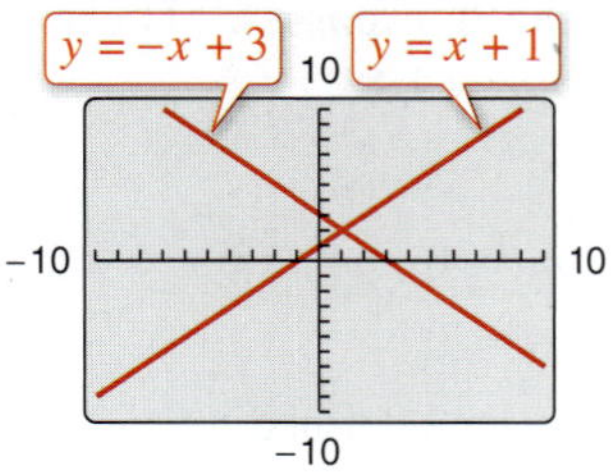

Figure 1.19

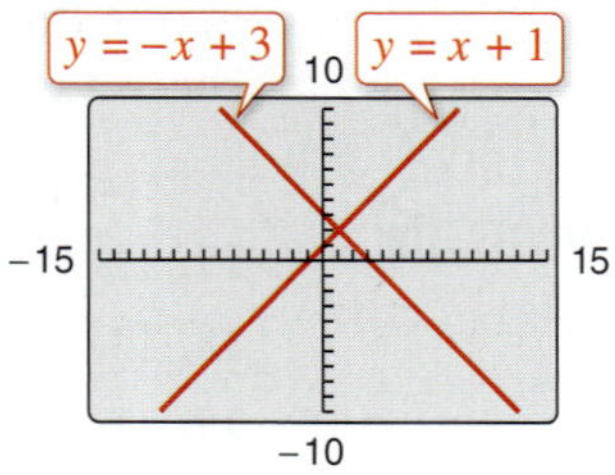

Figure 1.20

✓ Checkpoint *Audio-video solution in English & Spanish at LarsonPrecalculus.com.*

Identify any relationships that exist among the lines $y = 2x$, $y = -2x$, and $y = \frac{1}{2}x$. Then use a graphing utility to graph the three equations in the same viewing window. Adjust the viewing window so that each slope appears visually correct. Use the slopes of the lines to verify your results.

What's Wrong?

You use a graphing utility to graph $y_1 = 1.5x$ and $y_2 = -1.5x + 5$, as shown in the figure. You use the graph to conclude that the lines are perpendicular. What's wrong?

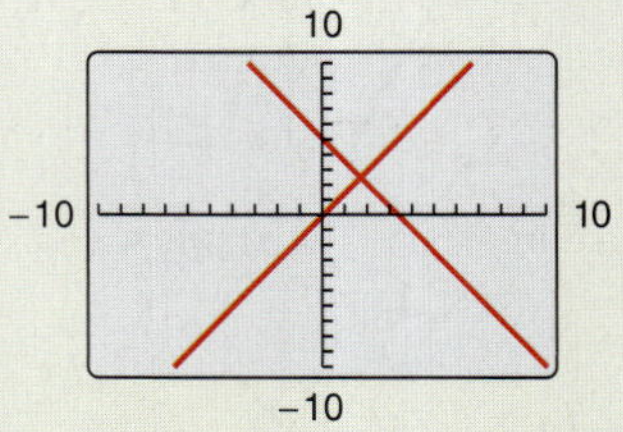

1.2 Exercises

See *CalcChat.com* for tutorial help and worked-out solutions to odd-numbered exercises.
For instructions on how to use a graphing utility, see Appendix A.

Vocabulary and Concept Check

1. Match each equation with its form.

(a) $Ax + By + C = 0$ (i) vertical line
(b) $x = a$ (ii) slope-intercept form
(c) $y = b$ (iii) general form
(d) $y = mx + b$ (iv) point-slope form
(e) $y - y_1 = m(x - x_1)$ (v) horizontal line

In Exercises 2 and 3, fill in the blank.

2. For a line, the ratio of the change in y to the change in x is called the _______ of the line.

3. Two lines are _______ if and only if their slopes are equal.

4. What is the relationship between two lines whose slopes are -3 and $\frac{1}{3}$?

5. What is the slope of a line that is perpendicular to the line represented by $x = 3$?

6. Give the coordinates of a point on the line whose equation in point-slope form is $y - (-1) = \frac{1}{4}(x - 8)$.

Procedures and Problem Solving

Using Slope **In Exercises 7 and 8, identify the line that has the indicated slope.**

7. (a) $m = \frac{2}{3}$ (b) m is undefined. (c) $m = -2$

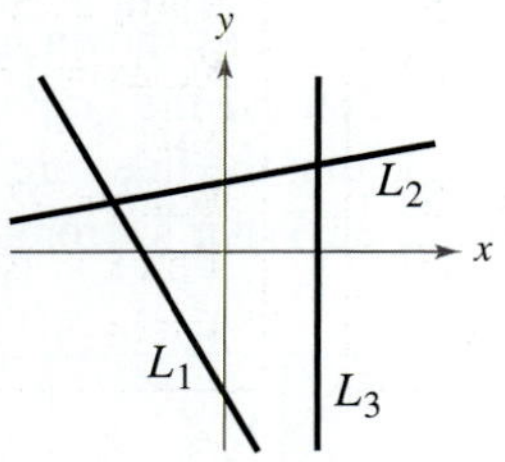

8. (a) $m = 0$ (b) $m = -\frac{3}{4}$ (c) $m = 1$

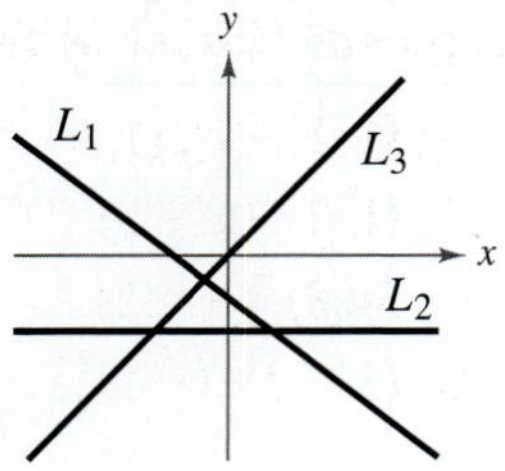

Estimating Slope **In Exercises 9 and 10, estimate the slope of the line.**

9.

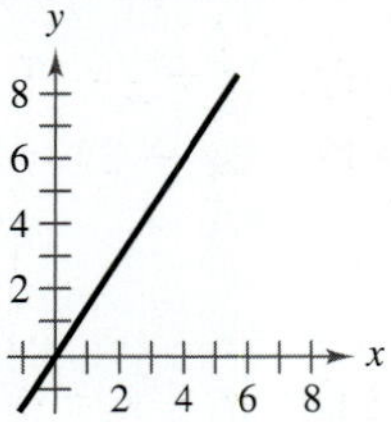

10.

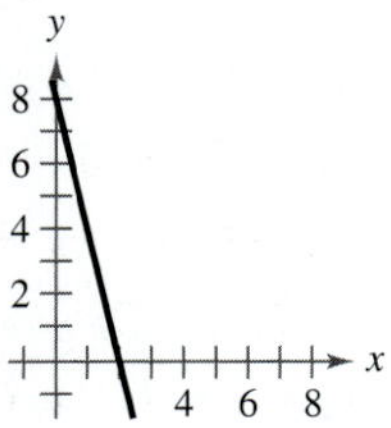

Sketching Lines **In Exercises 11 and 12, sketch the lines through the point with the indicated slopes on the same set of coordinate axes.**

	Point	*Slopes*			
11.	(2, 3)	(a) 0	(b) 1	(c) 2	(d) -3
12.	$(-4, 1)$	(a) 4	(b) -2	(c) $\frac{1}{2}$	(d) Undefined

Finding the Slope of a Line **In Exercises 13–16, find the slope of the line passing through the pair of points. Then use a graphing utility to plot the points and use the *draw* feature to graph the line segment connecting the two points. (Use a *square setting*.)**

13. $(0, -10), (-4, 0)$
14. $(2, 4), (4, -4)$
15. $(-6, -1), (-6, 4)$
16. $(4, 9), (6, 12)$

Using Slope **In Exercises 17–24, use the point on the line and the slope of the line to find three additional points through which the line passes. (There are many correct answers.)**

	Point	*Slope*
17.	(2, 1)	$m = 0$
18.	$(3, -2)$	$m = 0$
19.	(1, 5)	m is undefined.
20.	$(-4, 1)$	m is undefined.
21.	$(0, -9)$	$m = -2$
22.	$(-5, 4)$	$m = 4$
23.	$(7, -2)$	$m = \frac{1}{2}$
24.	$(-1, -6)$	$m = -\frac{1}{3}$

The Point-Slope Form of the Equation of a Line **In Exercises 25–32, find an equation of the line that passes through the given point and has the indicated slope. Sketch the line by hand. Use a graphing utility to verify your sketch, if possible.**

25. $(0, -2), \quad m = 3$

26. $(-3, 6), \quad m = -3$

27. $(2, -3), \quad m = -\frac{1}{2}$

28. $(-2, -5), \quad m = \frac{3}{4}$

29. $(6, -1), \quad m$ is undefined.

30. $(-10, 4), \quad m$ is undefined.

31. $\left(-\frac{1}{2}, \frac{3}{2}\right), \quad m = 0$

32. $(2.3, -8.5), \quad m = 0$

33. Finance The median player salary for the New York Yankees was \$1.5 million in 2007 and \$1.7 million in 2013. Write a linear equation giving the median salary y in terms of the year x. Then use the equation to predict the median salary in 2020.

34. Finance The median player salary for the Dallas Cowboys was \$348,000 in 2004 and \$555,000 in 2013. Write a linear equation giving the median salary y in terms of the year x. Then use the equation to predict the median salary in 2019.

Using the Slope-Intercept Form **In Exercises 35–42, determine the slope and y-intercept (if possible) of the linear equation. Then describe its graph.**

35. $2x - 3y = 9$

36. $3x + 4y = 1$

37. $2x - 5y + 10 = 0$

38. $4x - 3y - 9 = 0$

39. $x = -6$

40. $y = 12$

41. $3y + 2 = 0$

42. $2x - 5 = 0$

Using the Slope-Intercept Form **In Exercises 43–48, (a) find the slope and y-intercept (if possible) of the equation of the line algebraically, and (b) sketch the line by hand. Use a graphing utility to verify your answers to parts (a) and (b).**

43. $5x - y + 3 = 0$

44. $2x + 3y - 9 = 0$

45. $5x - 2 = 0$

46. $3x + 7 = 0$

47. $3y + 5 = 0$

48. $-11 - 4y = 0$

Finding the Slope-Intercept Form **In Exercises 49 and 50, find the slope-intercept form of the equation of the line shown.**

49.

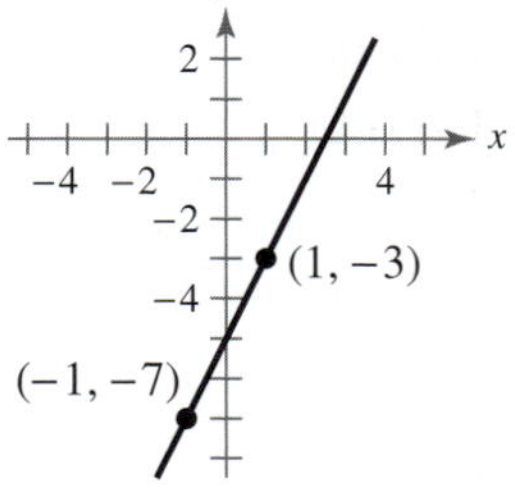

50.

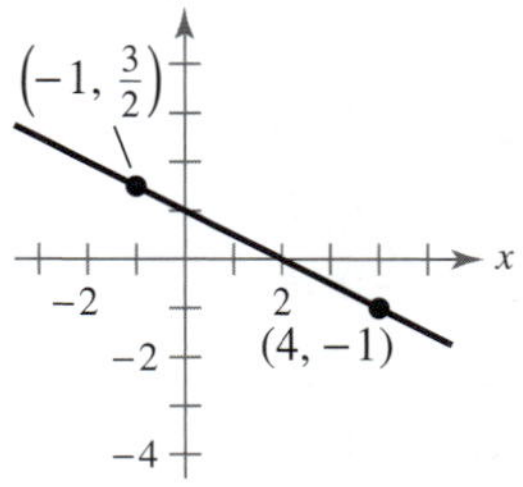

Finding the Slope-Intercept Form **In Exercises 51–60, write an equation of the line that passes through the points. Use the slope-intercept form (if possible). If not possible, explain why and use the general form. Use a graphing utility to graph the line (if possible).**

51. $(5, -1), (-5, 5)$

52. $(4, 3), (-4, -4)$

53. $(-8, 1), (-8, 7)$

54. $(-1, 6), (5, 6)$

55. $\left(2, \frac{1}{2}\right), \left(\frac{1}{2}, \frac{5}{4}\right)$

56. $(1, 1), \left(6, -\frac{2}{3}\right)$

57. $\left(-\frac{1}{10}, -\frac{3}{5}\right), \left(\frac{9}{10}, -\frac{9}{5}\right)$

58. $\left(\frac{3}{4}, \frac{3}{2}\right), \left(-\frac{4}{3}, \frac{7}{4}\right)$

59. $(1, 0.6), (-2, -0.6)$

60. $(-8, 0.6), (2, -2.4)$

Different Viewing Windows **In Exercises 61 and 62, use a graphing utility to graph the equation using each viewing window. Describe the differences in the graphs.**

61. $y = 0.25x - 2$

Xmin = -1	Xmin = -5	Xmin = -5
Xmax = 9	Xmax = 10	Xmax = 10
Xscl = 1	Xscl = 1	Xscl = 1
Ymin = -5	Ymin = -3	Ymin = -5
Ymax = 4	Ymax = 4	Ymax = 5
Yscl = 1	Yscl = 1	Yscl = 1

62. $y = -8x + 5$

Xmin = -5	Xmin = -5	Xmin = -5
Xmax = 5	Xmax = 10	Xmax = 13
Xscl = 1	Xscl = 1	Xscl = 1
Ymin = -10	Ymin = -80	Ymin = -2
Ymax = 10	Ymax = 80	Ymax = 10
Yscl = 1	Yscl = 20	Yscl = 1

Parallel and Perpendicular Lines **In Exercises 63–66, determine whether the lines L_1 and L_2 passing through the pairs of points are parallel, perpendicular, or neither.**

63. L_1: $(0, -1), (5, 9)$
L_2: $(0, 3), (4, 1)$

64. L_1: $(-2, -1), (1, 5)$
L_2: $(1, 3), (5, -5)$

65. L_1: $(3, 6), (-6, 0)$
L_2: $(0, -1), \left(5, \frac{7}{3}\right)$

66. L_1: $(4, 8), (-4, 2)$
L_2: $(3, -5), \left(-1, \frac{1}{3}\right)$

Equations of Parallel and Perpendicular Lines **In Exercises 67–76, write the slope-intercept forms of the equations of the lines through the given point (a) parallel to the given line and (b) perpendicular to the given line.**

67. $(2, 1), \quad 4x - 2y = 3$

68. $(-3, 2), \quad x + y = 7$

69. $\left(-\frac{2}{3}, \frac{7}{8}\right), \quad 3x + 4y = 7$

70. $\left(\frac{2}{5}, -1\right), \quad 3x - 2y = 6$

71. $(-3.9, -1.4), \quad 6x + 5y = 9$

72. $(-1.2, 2.4), \quad 5x + 4y = 1$

73. $(3, -2), \quad x - 4 = 0$

74. $(3, -1), \quad y - 2 = 0$

75. $(-5, 1), \quad y + 2 = 0$

76. $(-2, 4), \quad x + 5 = 0$

Equations of Parallel Lines In Exercises 77 and 78, the lines are parallel. Find the slope-intercept form of the equation of line y_2.

77.

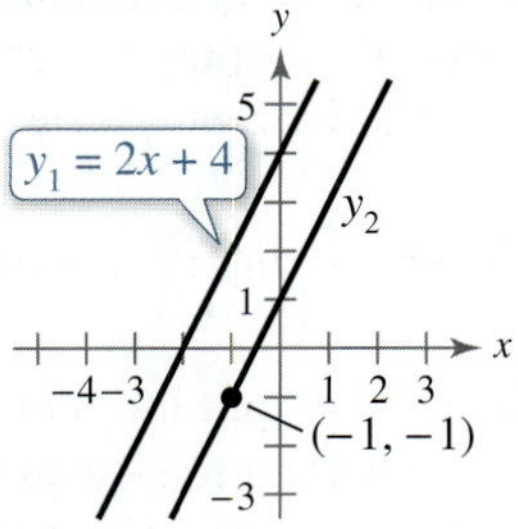

78.

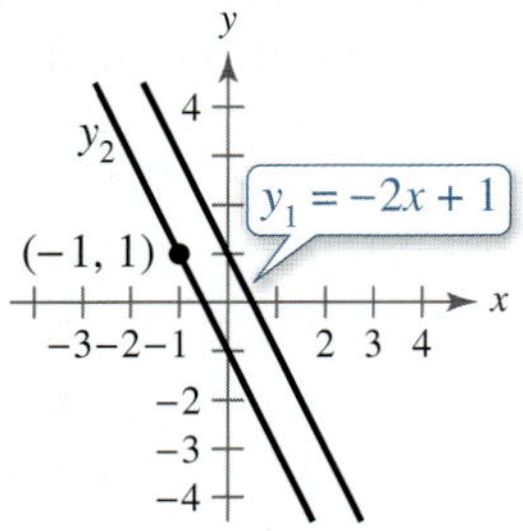

Equations of Perpendicular Lines In Exercises 79 and 80, the lines are perpendicular. Find the slope-intercept form of the equation of line y_2.

79.

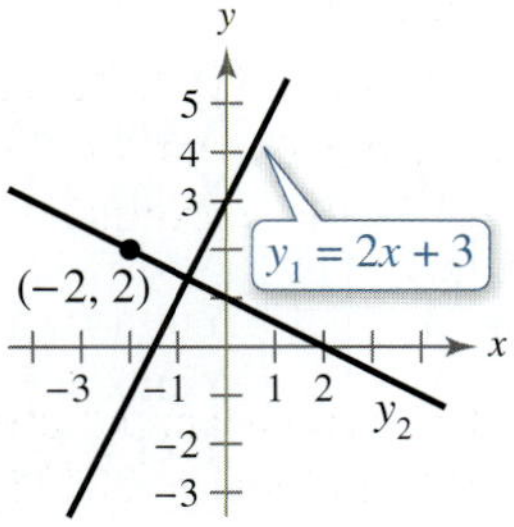

80.

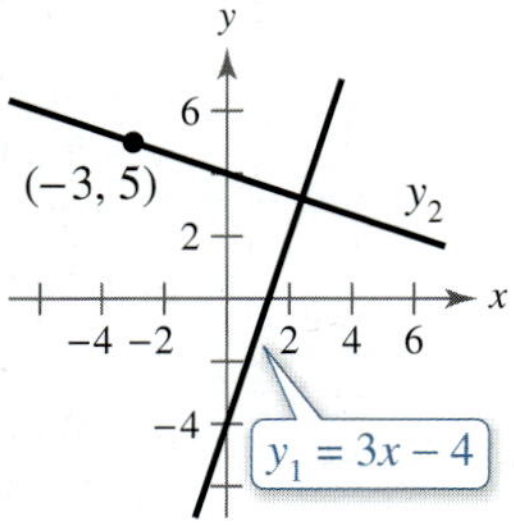

Graphs of Parallel and Perpendicular Lines In Exercises 81–84, identify any relationships that exist among the lines, and then use a graphing utility to graph the three equations in the same viewing window. Adjust the viewing window so that each slope appears visually correct. Use the slopes of the lines to verify your results.

81. (a) $y = 4x$ (b) $y = -4x$ (c) $y = \frac{1}{4}x$

82. (a) $y = \frac{2}{3}x$ (b) $y = -\frac{3}{2}x$ (c) $y = \frac{2}{3}x + 2$

83. (a) $y = -\frac{1}{2}x$ (b) $y = -\frac{1}{2}x + 3$ (c) $y = 2x - 4$

84. (a) $y = x - 8$ (b) $y = x + 1$ (c) $y = -x + 3$

85. Architectural Design The rise-to-run ratio of the roof of a house determines the steepness of the roof. The rise-to-run ratio of the roof in the figure is 3 to 4. Determine the maximum height in the attic of the house if the house is 32 feet wide.

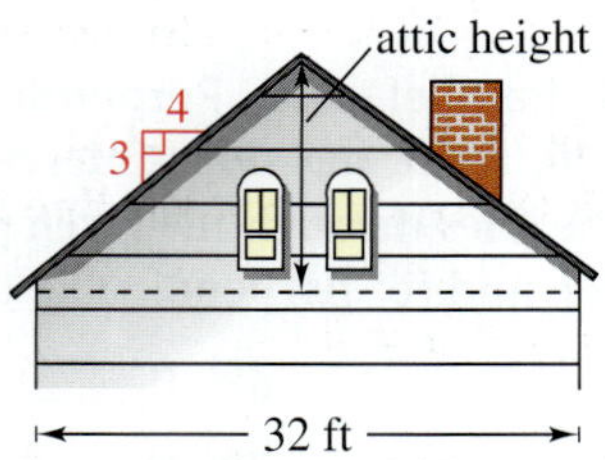

86. Highway Engineering When driving down a mountain road, you notice warning signs indicating that it is a "12% grade." This means that the slope of the road is $-\frac{12}{100}$. Approximate the amount of horizontal change in your position if you note from elevation markers that you have descended 2000 feet vertically.

87. MODELING DATA

The graph shows the sales y (in billions of dollars) of the Coca-Cola Company each year x from 2005 through 2012, where $x = 5$ represents 2005. (Source: Coca-Cola Company)

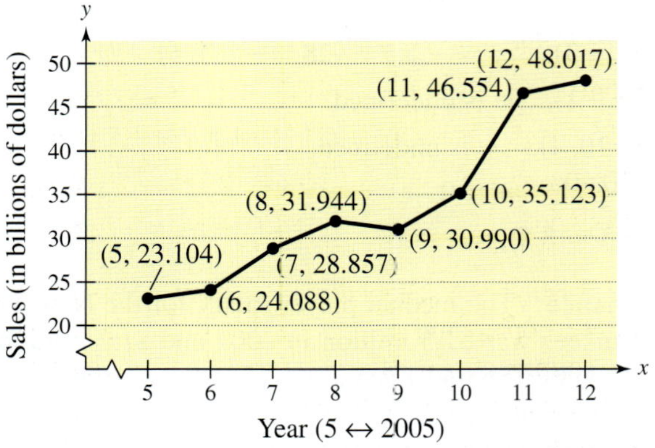

(a) Use the slopes to determine the years in which the sales showed the greatest increase and greatest decrease.

(b) Find the equation of the line between the years 2005 and 2012.

(c) Interpret the meaning of the slope of the line from part (b) in the context of the problem.

(d) Use the equation from part (b) to estimate the sales of the Coca-Cola Company in 2017. Do you think this is an accurate estimate? Explain.

88. MODELING DATA

The table shows the profits y (in millions of dollars) for Buffalo Wild Wings for each year x from 2007 through 2013, where $x = 7$ represents 2007. (Source: Buffalo Wild Wings, Inc.)

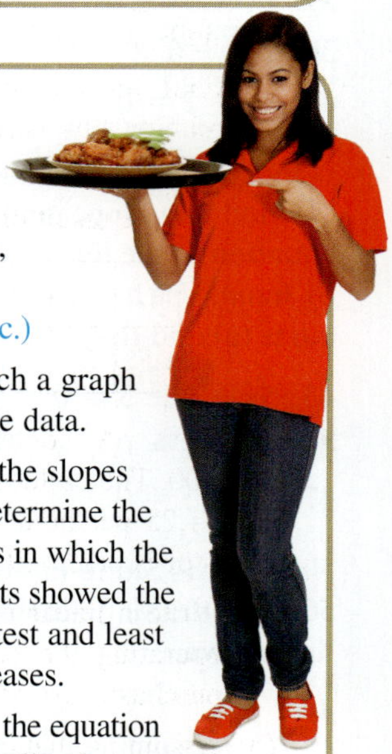

DATA

Year, x	Profits, y
7	19.7
8	24.4
9	30.7
10	38.4
11	50.4
12	57.3
13	71.6

Spreadsheet at LarsonPrecalculus.com

(a) Sketch a graph of the data.

(b) Use the slopes to determine the years in which the profits showed the greatest and least increases.

(c) Find the equation of the line between the years 2007 and 2013.

(d) Interpret the meaning of the slope of the line from part (c) in the context of the problem.

(e) Use the equation from part (c) to estimate the profit for Buffalo Wild Wings in 2017. Do you think this is an accurate estimate? Explain.

Using a Rate of Change to Write an Equation **In Exercises 89–92, you are given the dollar value of a product in 2015 and the rate at which the value of the product is expected to change during the next 5 years. Write a linear equation that gives the dollar value V of the product in terms of the year t. (Let $t = 15$ represent 2015.)**

	2015 Value	*Rate*
89.	\$2540	\$125 increase per year
90.	\$156	\$5.50 increase per year
91.	\$20,400	\$2000 decrease per year
92.	\$245,000	\$5600 decrease per year

93. Accounting A school district purchases a high-volume printer, copier, and scanner for \$25,000. After 10 years, the equipment will have to be replaced. Its value at that time is expected to be \$2000.

(a) Write a linear equation giving the value V of the equipment for each year t during its 10 years of use.

(b) Use a graphing utility to graph the linear equation representing the depreciation of the equipment, and use the *value* or *trace* feature to complete the table. Verify your answers algebraically by using the equation you found in part (a).

t	0	1	2	3	4	5	6	7	8	9	10
V											

94. Meterology Recall that water freezes at 0°C (32°F) and boils at 100°C (212°F).

(a) Find an equation of the line that shows the relationship between the temperature in degrees Celsius C and degrees Fahrenheit F.

(b) Use the result of part (a) to complete the table.

C		−10°	10°			177°
F	0°			68°	90°	

95. Business A contractor purchases a bulldozer for \$36,500. The bulldozer requires an average expenditure of \$11.25 per hour for fuel and maintenance, and the operator is paid \$19.50 per hour.

(a) Write a linear equation giving the total cost C of operating the bulldozer for t hours. (Include the purchase cost of the bulldozer.)

(b) Assuming that customers are charged \$80 per hour of bulldozer use, write an equation for the revenue R derived from t hours of use.

(c) Use the profit formula $(P = R - C)$ to write an equation for the profit gained from t hours of use.

(d) Use the result of part (c) to find the break-even point (the number of hours the bulldozer must be used to gain a profit of 0 dollars).

96. Real Estate A real estate office handles an apartment complex with 50 units. When the rent per unit is \$580 per month, all 50 units are occupied. However, when the rent is \$625 per month, the average number of occupied units drops to 47. Assume that the relationship between the monthly rent p and the demand x is linear.

(a) Write an equation of the line giving the demand x in terms of the rent p.

(b) Use a graphing utility to graph the demand equation and use the *trace* feature to estimate the number of units occupied when the rent is \$655. Verify your answer algebraically.

(c) Use the demand equation to predict the number of units occupied when the rent is lowered to \$595. Verify your answer graphically.

97. ***Why you should learn it*** *(p. 84)* In 1994, Penn State University had an enrollment of 73,500 students. By 2013, the enrollment had increased to 98,097. (Source: Penn State Fact Book)

(a) What was the average annual change in enrollment from 1994 to 2013?

(b) Use the average annual change in enrollment to estimate the enrollments in 1996, 2006, and 2011.

(c) Write an equation of a line that represents the given data. What is its slope? Interpret the slope in the context of the problem.

98. Writing Using the results of Exercise 97, write a short paragraph discussing the concepts of *slope* and *average rate of change*.

Conclusions

True or False? **In Exercises 99 and 100, determine whether the statement is true or false. Justify your answer.**

99. The line through $(-8, 2)$ and $(-1, 4)$ and the line through $(0, -4)$ and $(-7, 7)$ are parallel.

100. If the points $(10, -3)$ and $(2, -9)$ lie on the same line, then the point $\left(-12, -\frac{37}{2}\right)$ also lies on that line.

Exploration **In Exercises 101–104, use a graphing utility to graph the equation of the line in the form**

$$\frac{x}{a} + \frac{y}{b} = 1, \quad a \neq 0, b \neq 0.$$

Use the graphs to make a conjecture about what a and b represent. Verify your conjecture.

101. $\dfrac{x}{7} + \dfrac{y}{-3} = 1$

102. $\dfrac{x}{-6} + \dfrac{y}{2} = 1$

103. $\dfrac{x}{4} + \dfrac{y}{-\frac{2}{3}} = 1$

104. $\dfrac{x}{\frac{1}{2}} + \dfrac{y}{5} = 1$

Using Intercepts **In Exercises 105–108, use the results of Exercises 101–104 to write an equation of the line that passes through the points.**

105. x-intercept: $(2, 0)$
y-intercept: $(0, 9)$

106. x-intercept: $(-5, 0)$
y-intercept: $(0, -4)$

107. x-intercept: $\left(-\frac{1}{6}, 0\right)$
y-intercept: $\left(0, -\frac{2}{3}\right)$

108. x-intercept: $\left(\frac{3}{4}, 0\right)$
y-intercept: $\left(0, \frac{4}{3}\right)$

Think About It **In Exercises 109 and 110, determine which equation(s) may be represented by the graphs shown. (There may be more than one correct answer.)**

109.

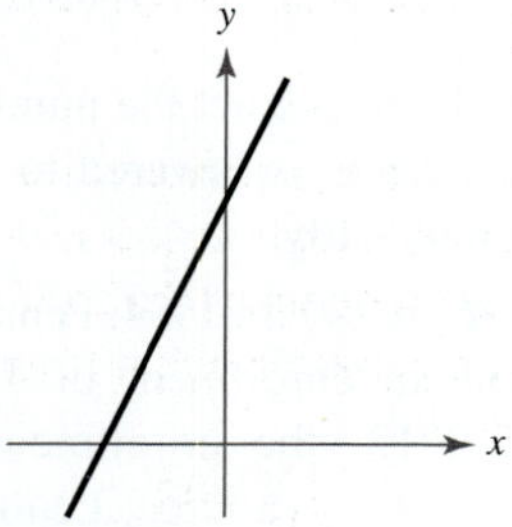

(a) $2x - y = -10$
(b) $2x + y = 10$
(c) $x - 2y = 10$
(d) $x + 2y = 10$

110.

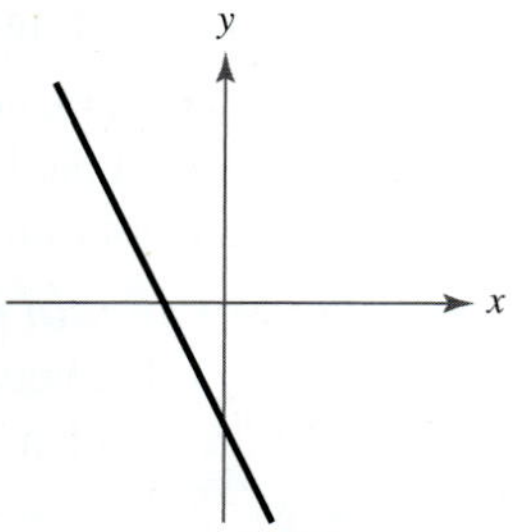

(a) $2x + y = 5$
(b) $2x + y = -5$
(c) $x - 2y = 5$
(d) $x - 2y = -5$

Think About It **In Exercises 111 and 112, determine which pair of equations may be represented by the graphs shown.**

111.

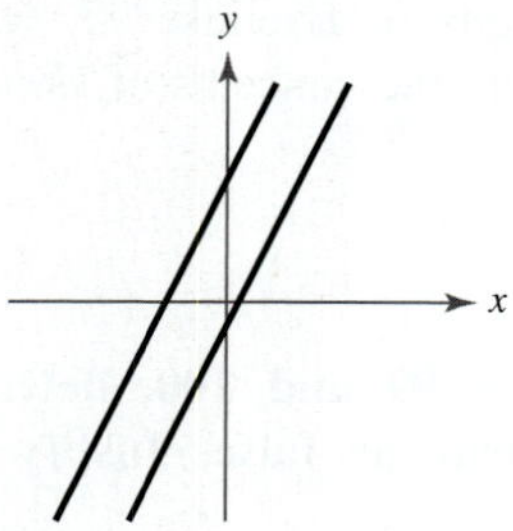

(a) $2x - y = 5$
$2x - y = 1$
(b) $2x + y = -5$
$2x + y = 1$
(c) $2x - y = -5$
$2x - y = 1$
(d) $x - 2y = -5$
$x - 2y = -1$

112.

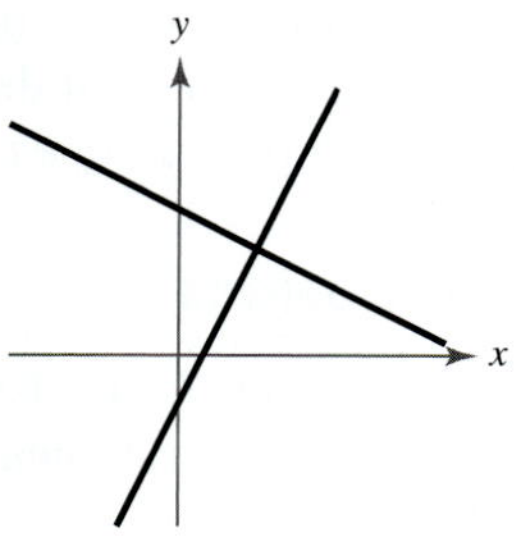

(a) $2x - y = 2$
$x + 2y = 12$
(b) $x - y = 1$
$x + y = 6$
(c) $2x + y = 2$
$x - 2y = 12$
(d) $x - 2y = 2$
$x + 2y = 12$

113. **Think About It** Does every line have both an x-intercept and a y-intercept? Explain.

114. **Think About It** Can every line be written in slope-intercept form? Explain.

115. **Think About It** Does every line have an infinite number of lines that are parallel to it? Explain.

116. **HOW DO YOU SEE IT?** Match the description with its graph. Determine the slope and y-intercept of each graph and interpret their meaning in the context of the problem. [The graphs are labeled (i), (ii), (iii), and (iv).]

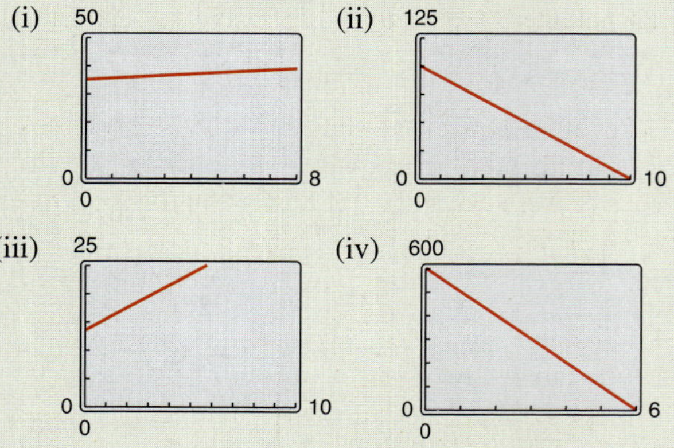

(a) You are paying \$10 per week to repay a \$100 loan.
(b) An employee is paid \$13.50 per hour plus \$2 for each unit produced per hour.
(c) A sales representative receives \$35 per day for food plus \$0.50 for each mile traveled.
(d) A tablet computer that was purchased for \$600 depreciates \$100 per year.

Cumulative Mixed Review

Identifying Polynomials **In Exercises 117–122, determine whether the expression is a polynomial. If it is, write the polynomial in standard form.**

117. $x + 20$

118. $3x - 10x^2 + 1$

119. $4x^2 + x^{-1} - 3$

120. $2x^2 - 2x^4 - x^3 + \sqrt{2}$

121. $\dfrac{x^2 + 3x + 4}{x^2 - 9}$

122. $\sqrt{x^2 + 7x + 6}$

Factoring Trinomials **In Exercises 123–126, factor the trinomial.**

123. $x^2 - 6x - 27$

124. $x^2 + 11x + 28$

125. $2x^2 + 11x - 40$

126. $3x^2 - 16x + 5$

127. ***Make a Decision*** To work an extended application analyzing the numbers of bachelor's degrees earned by women in the United States from 2001 through 2012, visit this textbook's website at *LarsonPrecalulus.com.* (Data Source: National Center for Education Statistics)

The *Make a Decision* exercise indicates a multipart exercise using large data sets. Visit this textbook's website at *LarsonPrecalculus.com.*

1.3 Functions

Introduction to Functions

Many everyday phenomena involve two quantities that are related to each other by some rule of correspondence. The mathematical term for such a rule of correspondence is a **relation.** Here are two examples.

1. The simple interest I earned on an investment of \$1000 for 1 year is related to the annual interest rate r by the formula $I = 1000r$.
2. The area A of a circle is related to its radius r by the formula $A = \pi r^2$.

Not all relations have simple mathematical formulas. For instance, you can match NFL starting quarterbacks with touchdown passes and time of day with temperature. In each of these cases, there is some relation that matches each item from one set with exactly one item from a different set. Such a relation is called a **function.**

What you should learn

- Determine whether a relation between two variables represents a function.
- Use function notation and evaluate functions.
- Find the domains of functions.
- Use functions to model and solve real-life problems.
- Evaluate difference quotients.

Why you should learn it

Many natural phenomena can be modeled by functions, such as the force of water against the face of a dam, explored in Exercise 80 on page 108.

Definition of a Function

A **function** f from a set A to a set B is a relation that assigns to each element x in the set A exactly one element y in the set B. The set A is the **domain** (or set of inputs) of the function f, and the set B contains the **range** (or set of outputs).

To help understand this definition, look at the function that relates the time of day to the temperature in Figure 1.21.

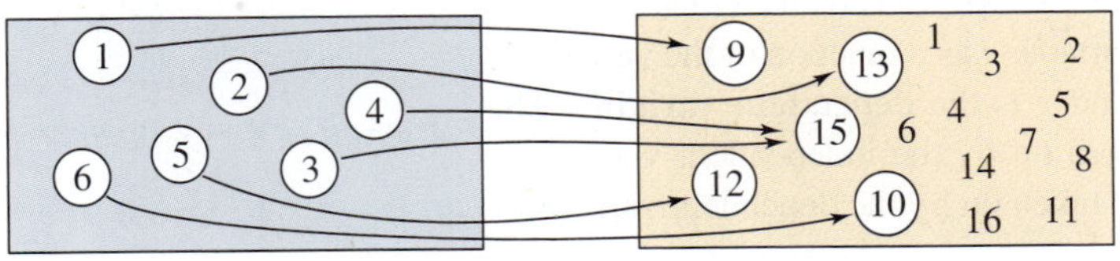

Figure 1.21

This function can be represented by the ordered pairs

$$\{(1, 9^\circ), (2, 13^\circ), (3, 15^\circ), (4, 15^\circ), (5, 12^\circ), (6, 10^\circ)\}.$$

In each ordered pair, the first coordinate (x-value) is the **input** and the second coordinate (y-value) is the **output.**

Characteristics of a Function from Set A to Set B

1. Each element of A must be matched with an element of B.
2. Some elements of B may not be matched with any element of A.
3. Two or more elements of A may be matched with the same element of B.
4. An element of A (the domain) cannot be matched with two different elements of B.

To determine whether or not a relation is a function, you must decide whether each input value is matched with exactly one output value. When any input value is matched with two or more output values, the relation is not a function.

EXAMPLE 1 Testing for Functions

Determine whether each relation represents y as a function of x.

a.

Input, x	2	2	3	4	5
Output, y	11	10	8	5	1

b.

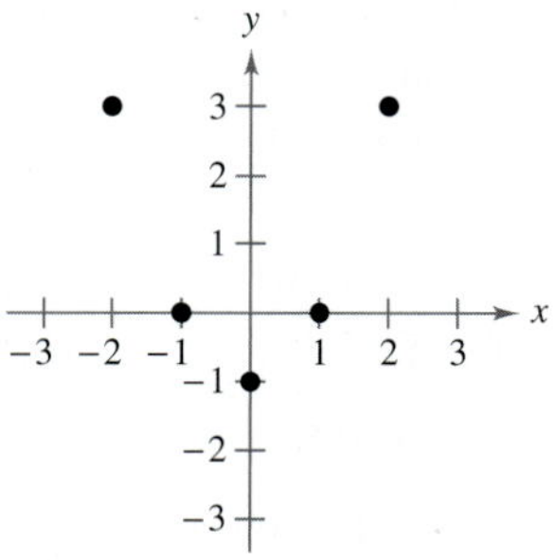

Solution

a. This table *does not* describe y as a function of x. The input value 2 is matched with two different y-values.

b. The graph *does* describe y as a function of x. Each input value is matched with exactly one output value.

✓ Checkpoint *Audio-video solution in English & Spanish at LarsonPrecalculus.com.*

Determine whether the relation represents y as a function of x.

Input, x	0	1	2	3	4
Output, y	−4	−2	0	2	4

In algebra, it is common to represent functions by equations or formulas involving two variables. For instance, $y = x^2$ represents the variable y as a function of the variable x. In this equation, x is the **independent variable** and y is the **dependent variable.** The domain of the function is the set of all values taken on by the independent variable x, and the range of the function is the set of all values taken on by the dependent variable y.

EXAMPLE 2 Testing for Functions Represented Algebraically

See LarsonPrecalculus.com for an interactive version of this type of example.

Determine whether each equation represents y as a function of x.

a. $x^2 + y = 1$ **b.** $-x + y^2 = 1$

Solution

To determine whether y is a function of x, try to solve for y in terms of x.

a. $x^2 + y = 1$ Write original equation.

$y = 1 - x^2$ Solve for y.

Each value of x corresponds to exactly one value of y. So, y is a function of x.

b. $-x + y^2 = 1$ Write original equation.

$y^2 = 1 + x$ Add x to each side.

$y = \pm\sqrt{1 + x}$ Solve for y.

The $\pm$ indicates that for a given value of x, there correspond two values of y. For instance, when $x = 3$, $y = 2$ or $y = -2$. So, y is not a function of x.

✓ Checkpoint *Audio-video solution in English & Spanish at LarsonPrecalculus.com.*

Determine whether each equation represents y as a function of x.

a. $x^2 + y^2 = 8$ **b.** $y - 4x^2 = 36$

Explore the Concept

Use a graphing utility to graph $x^2 + y = 1$. Then use the graph to write a convincing argument that each x-value corresponds to at most one y-value.

Use a graphing utility to graph $-x + y^2 = 1$. (*Hint:* You will need to use two equations.) Does the graph represent y as a function of x? Explain.

Function Notation

When an equation is used to represent a function, it is convenient to name the function so that it can be referenced easily. For example, you know that the equation $y = 1 - x^2$ describes y as a function of x. Suppose you give this function the name "f." Then you can use the following **function notation.**

Input	*Output*	*Equation*
x	$f(x)$	$f(x) = 1 - x^2$

The symbol $f(x)$ is read as the *value of f at x* or simply *f of x*. The symbol $f(x)$ corresponds to the y-value for a given x. So, you can write $y = f(x)$. Keep in mind that f is the *name* of the function, whereas $f(x)$ is the *output value* of the function at the *input value* x. In function notation, the *input* is the independent variable and the *output* is the dependent variable. For instance, the function $f(x) = 3 - 2x$ has *function values* denoted by $f(-1)$, $f(0)$, and so on. To find these values, substitute the specified input values into the given equation.

For $x = -1$, $\quad f(-1) = 3 - 2(-1) = 3 + 2 = 5.$

For $x = 0$, $\quad f(0) = 3 - 2(0) = 3 - 0 = 3.$

Although f is often used as a convenient function name and x is often used as the independent variable, you can use other letters. For instance,

$$f(x) = x^2 - 4x + 7, \quad f(t) = t^2 - 4t + 7, \quad \text{and} \quad g(s) = s^2 - 4s + 7$$

all define the same function. In fact, the role of the independent variable is that of a "placeholder." Consequently, the function could be written as

$$f(\quad) = (\quad)^2 - 4(\quad) + 7.$$

EXAMPLE 3 Evaluating a Function

Let $g(x) = -x^2 + 4x + 1$. Find each value of the function.

a. $g(2)$ **b.** $g(t)$ **c.** $g(x + 2)$

Solution

a. Replacing x with 2 in $g(x) = -x^2 + 4x + 1$ yields the following.

$$g(2) = -(2)^2 + 4(2) + 1 = -4 + 8 + 1 = 5$$

b. Replacing x with t yields the following.

$$g(t) = -(t)^2 + 4(t) + 1 = -t^2 + 4t + 1$$

c. Replacing x with $x + 2$ yields the following.

$g(x + 2) = -(x + 2)^2 + 4(x + 2) + 1$ Substitute $x + 2$ for x.

$= -(x^2 + 4x + 4) + 4x + 8 + 1$ Multiply.

$= -x^2 - 4x - 4 + 4x + 8 + 1$ Distributive Property

$= -x^2 + 5$ Simplify.

✓ *Checkpoint* *Audio-video solution in English & Spanish at LarsonPrecalculus.com.*

Let $f(x) = 10 - 3x^2$. Find each function value.

a. $f(2)$ **b.** $f(-4)$ **c.** $f(x - 1)$

In part (c) of Example 3, note that $g(x + 2)$ is not equal to $g(x) + g(2)$. In general, $g(u + v) \neq g(u) + g(v)$.

Library of Parent Functions: Absolute Value Function

The *parent absolute value function* given by

$$f(x) = |x|$$

can be written as a piecewise-defined function. The basic characteristics of the parent absolute value function are summarized below and on the inside cover of this text.

Graph of $f(x) = |x| = \begin{cases} x, & x \geq 0 \\ -x, & x < 0 \end{cases}$

Domain: $(-\infty, \infty)$
Range: $[0, \infty)$
Intercept: $(0, 0)$
Decreasing on $(-\infty, 0)$
Increasing on $(0, \infty)$

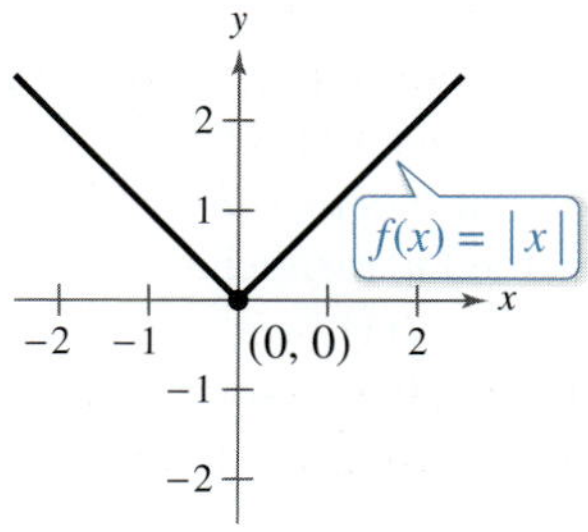

A function defined by two or more equations over a specified domain is called a **piecewise-defined function.**

EXAMPLE 4 A Piecewise-Defined Function

Evaluate the function when $x = -1, 0$, and 1.

$$f(x) = \begin{cases} x^2 + 1, & x < 0 \\ x - 1, & x \geq 0 \end{cases}$$

Solution

Because $x = -1$ is less than 0, use $f(x) = x^2 + 1$ to obtain

$f(-1) = (-1)^2 + 1$ Substitute -1 for x.

$= 2.$ Simplify.

For $x = 0$, use $f(x) = x - 1$ to obtain

$f(0) = 0 - 1$ Substitute 0 for x.

$= -1.$ Simplify.

For $x = 1$, use $f(x) = x - 1$ to obtain

$f(1) = 1 - 1$ Substitute 1 for x.

$= 0.$ Simplify.

The graph of f is shown in the figure.

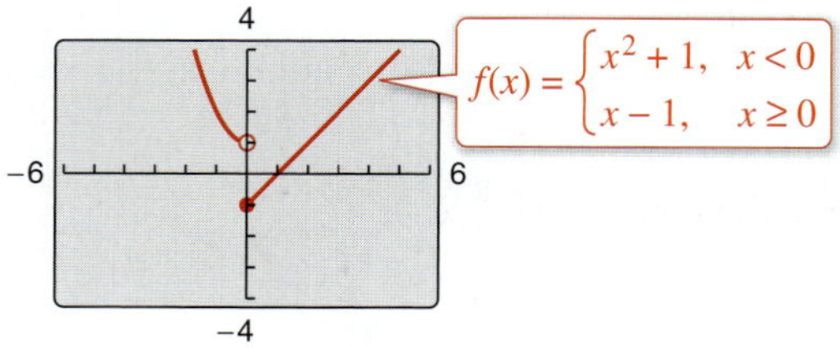

Checkpoint *Audio-video solution in English & Spanish at LarsonPrecalculus.com.*

Evaluate the function given in Example 4 when $x = -2, 2$, and 3.

Technology Tip

Most graphing utilities can graph piecewise-defined functions. For instructions on how to enter a piecewise-defined function into your graphing utility, consult your user's manual. You may find it helpful to set your graphing utility to *dot mode* before graphing such functions.

The Domain of a Function

The domain of a function can be described explicitly or it can be *implied* by the expression used to define the function. The **implied domain** is the set of all real numbers for which the expression is defined. For instance, the function

$$f(x) = \frac{1}{x^2 - 4}$$ Domain excludes x-values that result in division by zero.

has an implied domain that consists of all real numbers x other than $x = \pm 2$. These two values are excluded from the domain because division by zero is undefined. Another common type of implied domain is that used to avoid even roots of negative numbers. For example, the function

$$(x) = \sqrt{x}$$ Domain excludes x-values that result in even roots of negative numbers.

is defined only for $x \geq 0$. So, its implied domain is the interval $[0, \infty)$. In general, the domain of a function *excludes* values that would cause division by zero *or* result in the even root of a negative number.

Explore the Concept

Use a graphing utility to graph $y = \sqrt{4 - x^2}$. What is the domain of this function? Then graph $y = \sqrt{x^2 - 4}$. What is the domain of this function? Do the domains of these two functions overlap? If so, for what values?

Library of Parent Functions: Square Root Function

Radical functions arise from the use of rational exponents. The most common radical function is the *parent square root* function given by $f(x) = \sqrt{x}$. The basic characteristics of the parent square root function are summarized below and on the inside cover of this text.

Graph of $f(x) = \sqrt{x}$

Domain: $[0, \infty)$
Range: $[0, \infty)$
Intercept: $(0, 0)$
Increasing on $(0, \infty)$

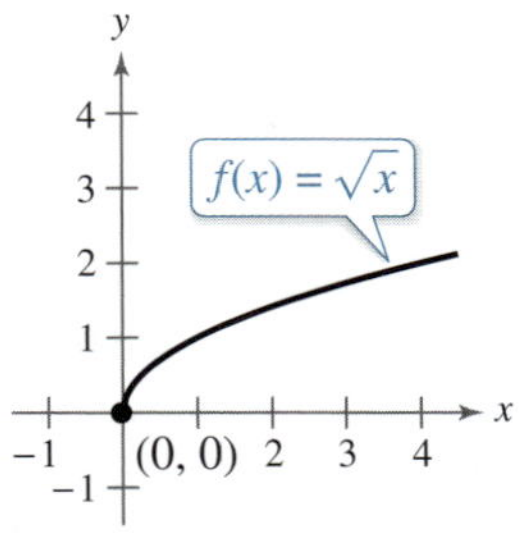

Remark

Because the square root function is not defined for $x < 0$, you must be careful when analyzing the domains of complicated functions involving the square root symbol.

EXAMPLE 5 Finding the Domain of a Function

Find the domain of each function.

a. f: $\{(-3, 0), (-1, 4), (0, 2), (2, 2), (4, -1)\}$

b. $g(x) = -3x^2 + 4x + 5$

c. $h(x) = \dfrac{1}{x + 5}$

Solution

a. The domain of f consists of all first coordinates in the set of ordered pairs.

$$\text{Domain} = \{-3, -1, 0, 2, 4\}$$

b. The domain of g is the set of all *real* numbers.

c. Excluding x-values that yield zero in the denominator, the domain of h is the set of all real numbers x except $x = -5$.

✓ ***Checkpoint*** *Audio-video solution in English & Spanish at LarsonPrecalculus.com.*

Find the domain of each function.

a. f: $\{(-2, 2), (-1, 1), (0, 3), (1, 1), (2, 2)\}$ **b.** $g(x) = \dfrac{1}{3 - x}$

EXAMPLE 6 Finding the Domain of a Function

Find the domain of each function.

a. Volume of a sphere: $V = \frac{4}{3}\pi r^3$ **b.** $k(x) = \sqrt{4 - 3x}$

Solution

a. Because this function represents the volume of a sphere, the values of the radius r must be positive (see Figure 1.22). So, the domain is the set of all real numbers r such that $r > 0$.

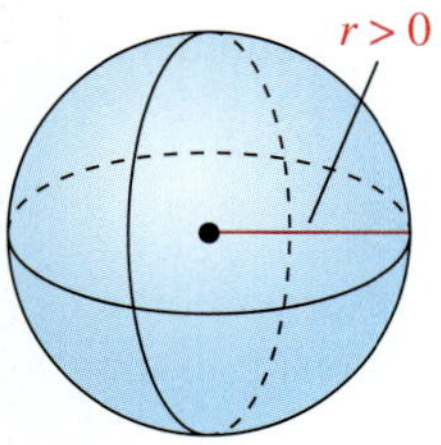

Figure 1.22

b. This function is defined only for x-values for which $4 - 3x \geq 0$. By solving this inequality, you will find that the domain of k is all real numbers that are less than or equal to $\frac{4}{3}$.

✓ ***Checkpoint*** *Audio-video solution in English & Spanish at LarsonPrecalculus.com.*

Find the domain of each function.

a. Circumference of a circle: $C = 2\pi r$ **b.** $h(x) = \sqrt{x - 16}$

In Example 6(a), note that the *domain of a function may be implied by the physical context.* For instance, from the equation $V = \frac{4}{3}\pi r^3$, you would have no reason to restrict r to positive values, but the physical context implies that a sphere cannot have a negative or zero radius.

For some functions, it may be easier to find the domain and range of the function by examining its graph.

EXAMPLE 7 Finding the Domain and Range of a Function

Use a graphing utility to find the domain and range of the function $f(x) = \sqrt{9 - x^2}$.

Solution

Graph the function as $y = \sqrt{9 - x^2}$, as shown in Figure 1.23. Using the *trace* feature of a graphing utility, you can determine that the x-values extend from -3 to 3 and the y-values extend from 0 to 3. So, the domain of the function f is all real numbers such that $-3 \leq x \leq 3$, and the range of f is all real numbers such that $0 \leq y \leq 3$.

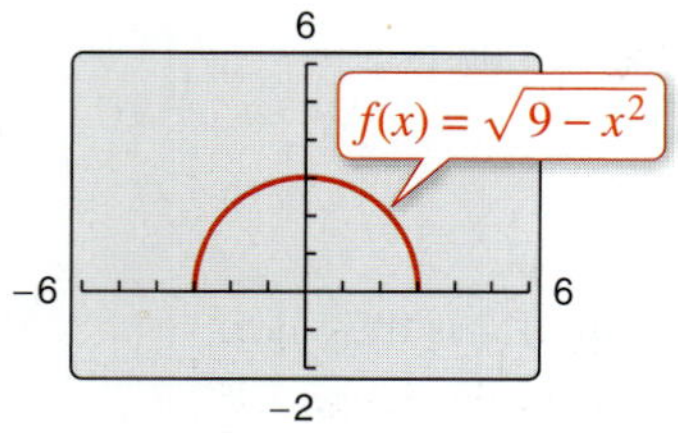

Figure 1.23

Remark

Be sure you see that the *range* of a function is not the same as the use of *range* relating to the viewing window of a graphing utility.

✓ ***Checkpoint*** *Audio-video solution in English & Spanish at LarsonPrecalculus.com.*

Use a graphing utility to find the domain and range of the function $f(x) = \sqrt{4 - x^2}$.

Applications

EXAMPLE 8 Interior Design Services Employees

The number N (in thousands) of employees in the interior design services industry in the United States decreased in a linear pattern from 2007 through 2010 (see Figure 1.24). In 2011, the number rose and increased through 2012 in a different linear pattern. These two patterns can be approximated by the function

$$N(t) = \begin{cases} -4.64t + 76.2, & 7 \le t \le 10 \\ 0.90t + 20.0, & 11 \le t \le 12 \end{cases}$$

where t represents the year, with $t = 7$ corresponding to 2007. Use this function to approximate the number of employees for the years 2007 and 2011. (Source: U.S. Bureau of Labor Statistics)

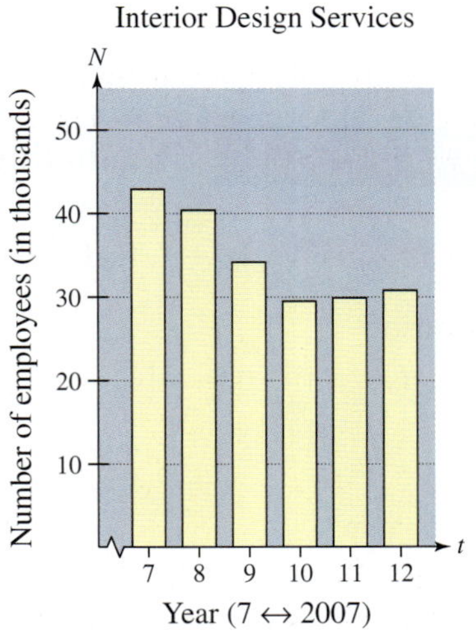

Figure 1.24

Solution

For 2007, $t = 7$, so use $N(t) = -4.64t + 76.2$.

$$N(7) = -4.64(7) + 76.2 = -32.48 + 76.2 = 43.72 \text{ thousand employees}$$

For 2011, $t = 11$, so use $N(t) = 0.90t + 20.0$.

$$N(11) = 0.90(11) + 20.0 = 9.9 + 20.0 = 29.9 \text{ thousand employees}$$

Checkpoint *Audio-video solution in English & Spanish at LarsonPrecalculus.com.*

Use the function in Example 8 to approximate the number of employees for the years 2010 and 2012.

EXAMPLE 9 The Path of a Baseball

A baseball is hit at a point 3 feet above the ground at a velocity of 100 feet per second and an angle of 45°. The path of the baseball is given by $f(x) = -0.0032x^2 + x + 3$, where $f(x)$ is the height of the baseball (in feet) and x is the horizontal distance from home plate (in feet). Will the baseball clear a 10-foot fence located 300 feet from home plate?

Algebraic Solution

The height of the baseball is a function of the horizontal distance from home plate. When $x = 300$, you can find the height of the baseball as follows.

$$f(x) = -0.0032x^2 + x + 3 \quad \text{Write original function.}$$

$$f(300) = -0.0032(300)^2 + 300 + 3 \quad \text{Substitute 300 for } x.$$

$$= 15 \quad \text{Simplify.}$$

When $x = 300$, the height of the baseball is 15 feet. So, the baseball will clear a 10-foot fence.

Graphical Solution

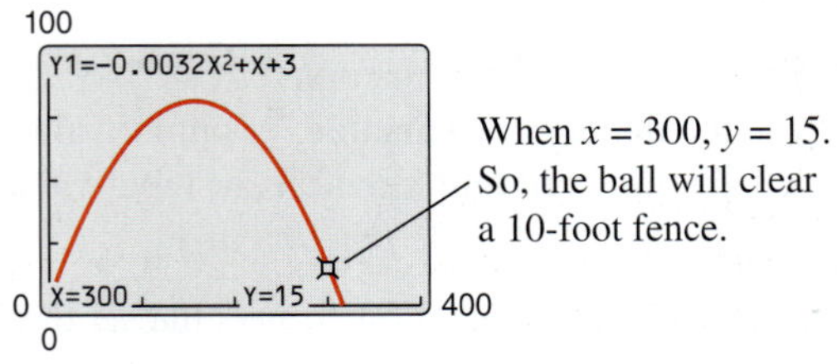

Checkpoint *Audio-video solution in English & Spanish at LarsonPrecalculus.com.*

A second baseman throws a baseball toward the first baseman 60 feet away. The path of the baseball is given by $f(x) = -0.004x^2 + 0.3x + 6$, where $f(x)$ is the height of the baseball (in feet) and x is the horizontal distance from the second baseman (in feet). The first baseman can reach 8 feet high. Can the first baseman catch the baseball without jumping?

Difference Quotients

One of the basic definitions in calculus employs the ratio

$$\frac{f(x+h)-f(x)}{h}, \quad h \neq 0.$$

This ratio is called a **difference quotient,** as illustrated in Example 10.

EXAMPLE 10 Evaluating a Difference Quotient

For $f(x) = x^2 - 4x + 7$, find $\dfrac{f(x+h)-f(x)}{h}$.

Solution

$$\frac{f(x+h)-f(x)}{h} = \frac{[(x+h)^2 - 4(x+h) + 7] - (x^2 - 4x + 7)}{h}$$

$$= \frac{x^2 + 2xh + h^2 - 4x - 4h + 7 - x^2 + 4x - 7}{h}$$

$$= \frac{2xh + h^2 - 4h}{h}$$

$$= \frac{h(2x + h - 4)}{h}$$

$$= 2x + h - 4, \quad h \neq 0$$

Remark

Notice in Example 10 that h cannot be zero in the original expression. Therefore, you must restrict the domain of the simplified expression by listing $h \neq 0$ so that the simplified expression is equivalent to the original expression.

✓ ***Checkpoint*** *Audio-video solution in English & Spanish at LarsonPrecalculus.com.*

For $f(x) = x^2 + 2x - 3$, evaluate the difference quotient in Example 10.

Summary of Function Terminology

Function: A **function** is a relationship between two variables such that to each value of the independent variable there corresponds exactly one value of the dependent variable.

Function Notation: $y = f(x)$

- f is the *name* of the function.
- y is the **dependent variable,** or output value.
- x is the **independent variable,** or input value.
- $f(x)$ is the *value of the function at x.*

Domain: The **domain** of a function is the set of all values (inputs) of the independent variable for which the function is defined. If x is in the domain of f, then f is said to be *defined* at x. If x is not in the domain of f, then f is said to be *undefined* at x.

Range: The **range** of a function is the set of all values (outputs) assumed by the dependent variable (that is, the set of all function values).

Implied Domain: If f is defined by an algebraic expression and the domain is not specified, then the **implied domain** consists of all real numbers for which the expression is defined.

The symbol $\int$ indicates an example or exercise that highlights algebraic techniques specifically used in calculus.

1.3 Exercises

See *CalcChat.com* for tutorial help and worked-out solutions to odd-numbered exercises.
For instructions on how to use a graphing utility, see Appendix A.

Vocabulary and Concept Check

In Exercises 1 and 2, fill in the blanks.

1. A relation that assigns to each element x from a set of inputs, or _______ , exactly one element y in a set of outputs, or _______ , is called a _______ .

2. For an equation that represents y as a function of x, the _______ variable is the set of all x in the domain, and the _______ variable is the set of all y in the range.

3. Can the ordered pairs (3, 0) and (3, 5) represent a function?

4. To find $g(x + 1)$, what do you substitute for x in the function $g(x) = 3x - 2$?

5. Does the domain of the function $f(x) = \sqrt{1 + x}$ include $x = -2$?

6. Is the domain of a piecewise-defined function *implied* or *explicitly described*?

Procedures and Problem Solving

Testing for Functions **In Exercises 7–10, does the relation describe a function? Explain your reasoning.**

7. *Domain* *Range*

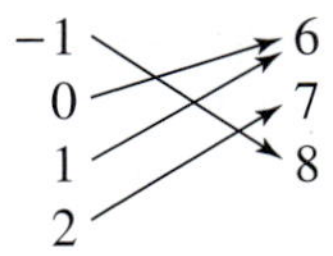

8. *Domain* *Range*

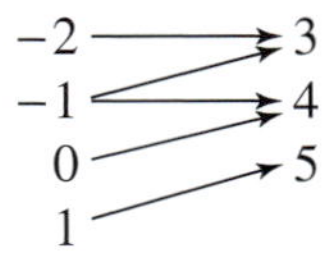

9. *Domain* *Range*

10. *Domain* (State) *Range* (Electoral votes 2010–2012)

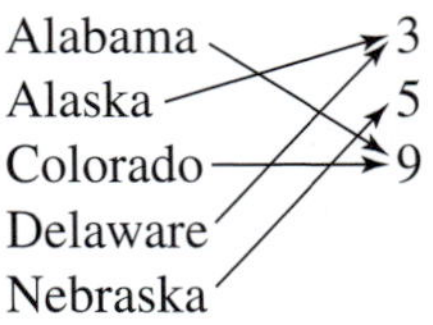

Testing for Functions **In Exercises 11–14, determine whether the relation represents y as a function of x. Explain your reasoning.**

11.

Input, x	−3	−1	0	1	3
Output, y	−9	−1	0	1	−9

12.

Input, x	0	1	2	1	0
Output, y	−4	−2	0	2	4

13.

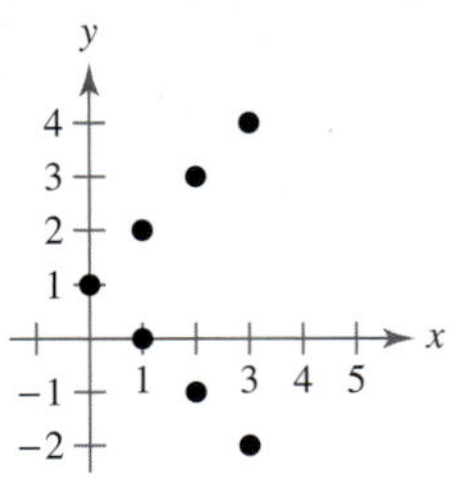

14.

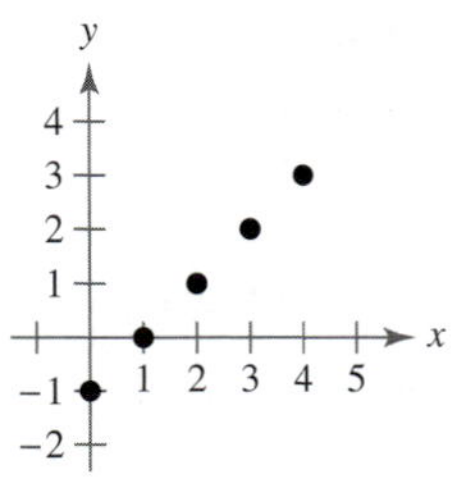

Testing for Functions **In Exercises 15 and 16, which sets of ordered pairs represent functions from A to B? Explain.**

15. $A = \{0, 1, 2, 3\}$ and $B = \{-2, -1, 0, 1, 2\}$

(a) $\{(0, 1), (1, -2), (2, 0), (3, 2)\}$

(b) $\{(0, -1), (2, 2), (1, -2), (3, 0), (1, 1)\}$

(c) $\{(1, 0), (-2, 3), (-1, 3), (0, 0)\}$

16. $A = \{a, b, c\}$ and $B = \{0, 1, 2, 3\}$

(a) $\{(a, 1), (c, 2), (c, 3), (b, 3)\}$

(b) $\{(a, 1), (b, 2), (c, 3)\}$

(c) $\{(1, a), (0, a), (2, c), (3, b)\}$

Pharmacology **In Exercises 17 and 18, use the graph, which shows the average prices of name brand and generic drug prescriptions in the United States.** (Source: National Association of Chain Drug Stores)

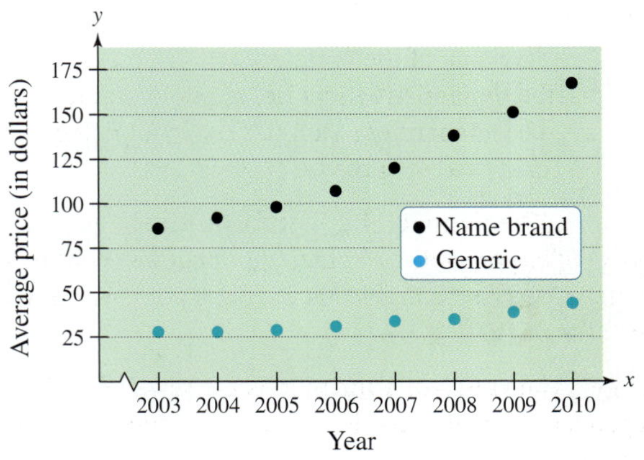

17. Is the average price of a name brand prescription a function of the year? Is the average price of a generic prescription a function of the year? Explain.

18. Let $b(t)$ and $g(t)$ represent the average prices of name brand and generic prescriptions, respectively, in year t. Find $b(2009)$ and $g(2006)$.

Testing for Functions Represented Algebraically **In Exercises 19–30, determine whether the equation represents *y* as a function of *x*.**

19. $x^2 + y^2 = 4$
20. $x = y^2 + 1$
21. $y = \sqrt{x^2 - 1}$
22. $y = \sqrt{x + 5}$
23. $2x + 3y = 4$
24. $x = -y + 5$
25. $y^2 = x^2 - 1$
26. $x + y^2 = 3$
27. $y = |4 - x|$
28. $|y| = 3 - 2x$
29. $x = -7$
30. $y = 8$

Evaluating a Function **In Exercises 31–46, evaluate the function at each specified value of the independent variable and simplify.**

31. $f(t) = 3t + 1$
(a) $f(2)$ (b) $f(-4)$ (c) $f(t + 2)$

32. $g(y) = 7 - 3y$
(a) $g(0)$ (b) $g\left(\frac{7}{3}\right)$ (c) $g(s + 5)$

33. $h(t) = t^2 - 2t$
(a) $h(2)$ (b) $h(1.5)$ (c) $h(x - 4)$

34. $V(r) = \frac{4}{3}\pi r^3$
(a) $V(3)$ (b) $V\left(\frac{3}{2}\right)$ (c) $V(2r)$

35. $f(y) = 3 - \sqrt{y}$
(a) $f(4)$ (b) $f(0.25)$ (c) $f(4x^2)$

36. $f(x) = \sqrt{x + 8} + 2$
(a) $f(-4)$ (b) $f(8)$ (c) $f(x - 8)$

37. $q(x) = \dfrac{1}{x^2 - 9}$
(a) $q(-3)$ (b) $q(2)$ (c) $q(y + 3)$

38. $q(t) = \dfrac{2t^2 + 3}{t^2}$
(a) $q(2)$ (b) $q(0)$ (c) $q(-x)$

39. $f(x) = \dfrac{|x|}{x}$
(a) $f(9)$ (b) $f(-9)$ (c) $f(t)$

40. $f(x) = |x| + 4$
(a) $f(5)$ (b) $f(-5)$ (c) $f(t)$

41. $f(x) = \begin{cases} 2x + 1, & x < 0 \\ 2x + 2, & x \geq 0 \end{cases}$
(a) $f(-1)$ (b) $f(0)$ (c) $f(2)$

42. $f(x) = \begin{cases} 2x + 5, & x \leq 0 \\ 2 - x, & x > 0 \end{cases}$
(a) $f(-2)$ (b) $f(0)$ (c) $f(1)$

43. $f(x) = \begin{cases} x^2 + 2, & x \leq 1 \\ 2x^2 + 2, & x > 1 \end{cases}$
(a) $f(-2)$ (b) $f(1)$ (c) $f(2)$

44. $f(x) = \begin{cases} x^2 - 4, & x \leq 0 \\ 1 - 2x^2, & x > 0 \end{cases}$
(a) $f(-2)$ (b) $f(0)$ (c) $f(1)$

45. $f(x) = \begin{cases} x + 2, & x < 0 \\ 4, & 0 \leq x < 2 \\ x^2 + 1, & x \geq 2 \end{cases}$
(a) $f(-2)$ (b) $f(0)$ (c) $f(2)$

46. $f(x) = \begin{cases} 5 - 2x, & x < 0 \\ 5, & 0 \leq x < 1 \\ 4x + 1, & x \geq 1 \end{cases}$
(a) $f(-4)$ (b) $f(0)$ (c) $f(1)$

Evaluating a Function **In Exercises 47–50, assume that the domain of *f* is the set $A = \{-2, -1, 0, 1, 2\}$. Determine the set of ordered pairs representing the function *f*.**

47. $f(x) = (x - 1)^2$
48. $f(x) = x^2 - 3$
49. $f(x) = |x| + 2$
50. $f(x) = |x + 1|$

Evaluating a Function **In Exercises 51 and 52, complete the table.**

51. $h(t) = \frac{1}{2}|t + 3|$

t	-5	-4	-3	-2	-1
$h(t)$					

52. $f(s) = \dfrac{|s - 2|}{s - 2}$

s	0	1	$\frac{3}{2}$	$\frac{5}{2}$	4
$f(s)$					

Finding the Inputs That Have Outputs of Zero **In Exercises 53–56, find all values of *x* such that $f(x) = 0$.**

53. $f(x) = 15 - 3x$
54. $f(x) = 5x + 1$
55. $f(x) = \dfrac{9x - 4}{5}$
56. $f(x) = \dfrac{2x - 3}{7}$

Finding the Domain of a Function **In Exercises 57–66, find the domain of the function.**

57. $f(x) = 5x^2 + 2x - 1$
58. $g(x) = 1 - 2x^2$
59. $h(t) = \dfrac{4}{t}$
60. $s(y) = \dfrac{3y}{y + 5}$
61. $f(x) = \sqrt[3]{x - 4}$
62. $f(x) = \sqrt[4]{x^2 + 3x}$
63. $g(x) = \dfrac{1}{x} - \dfrac{3}{x + 2}$
64. $h(x) = \dfrac{10}{x^2 - 2x}$
65. $g(y) = \dfrac{y + 2}{\sqrt{y - 10}}$
66. $f(x) = \dfrac{\sqrt{x + 6}}{6 + x}$

Finding the Domain and Range of a Function **In Exercises 67–70, use a graphing utility to graph the function. Find the domain and range of the function.**

67. $f(x) = \sqrt{16 - x^2}$

68. $f(x) = \sqrt{x^2 + 1}$

69. $g(x) = |2x + 3|$

70. $g(x) = |3x - 5|$

71. Geometry Write the area A of a circle as a function of its circumference C.

72. Geometry Write the area A of an equilateral triangle as a function of the length s of its sides.

73. Exploration An open box of maximum volume is to be made from a square piece of material, 24 centimeters on a side, by cutting equal squares from the corners and turning up the sides. (See figure.)

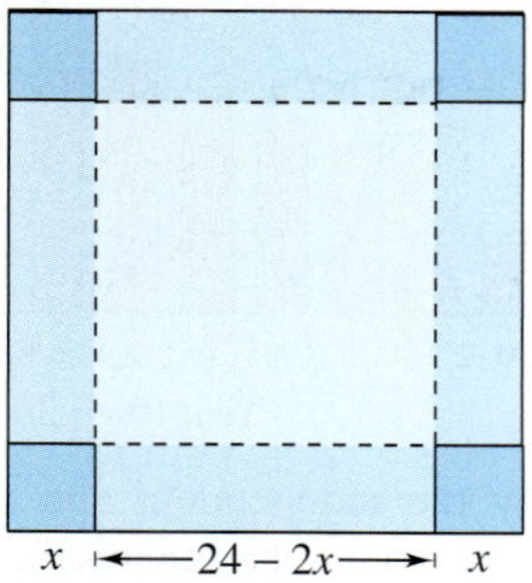

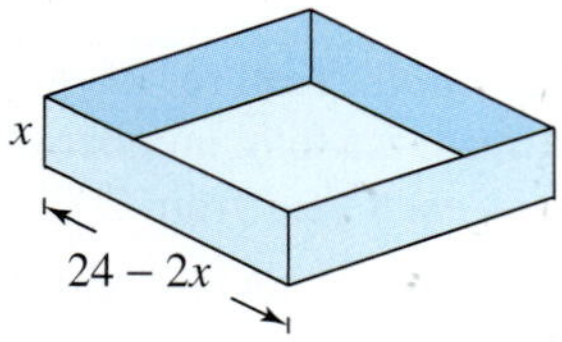

(a) The table shows the volume V (in cubic centimeters) of the box for various heights x (in centimeters). Use the table to estimate the maximum volume.

Height, x	Volume, V
1	484
2	800
3	972
4	1024
5	980
6	864

(b) Plot the points (x, V) from the table in part (a). Does the relation defined by the ordered pairs represent V as a function of x?

(c) If V is a function of x, write the function and determine its domain.

(d) Use a graphing utility to plot the points from the table in part (a) with the function from part (c). How closely does the function represent the data? Explain.

74. Geometry A right triangle is formed in the first quadrant by the x- and y-axes and a line through the point (2, 1), as shown in the figure. Write the area A of the triangle as a function of x and determine the domain of the function.

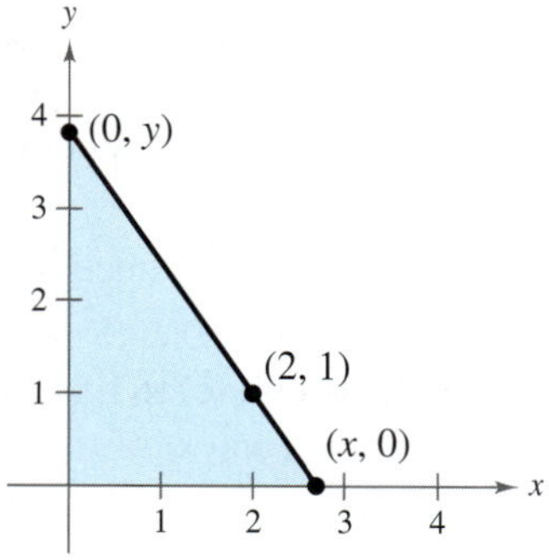

75. Geometry A rectangle is bounded by the x-axis and the semicircle $y = \sqrt{36 - x^2}$, as shown in the figure. Write the area A of the rectangle as a function of x and determine the domain of the function.

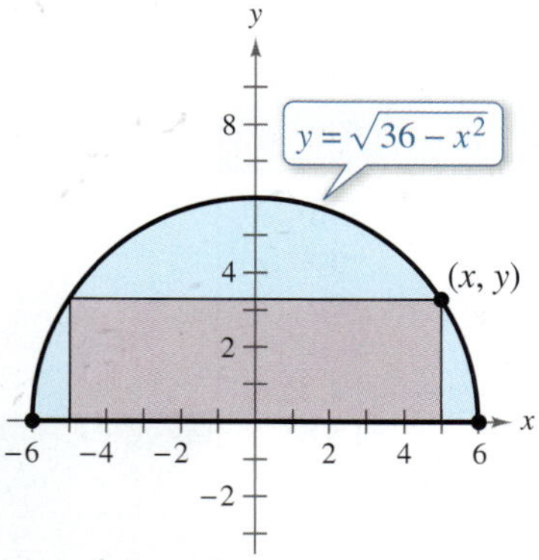

76. Geometry A rectangular package to be sent by the U.S. Postal Service can have a maximum combined length and girth (perimeter of a cross section) of 108 inches. (See figure.)

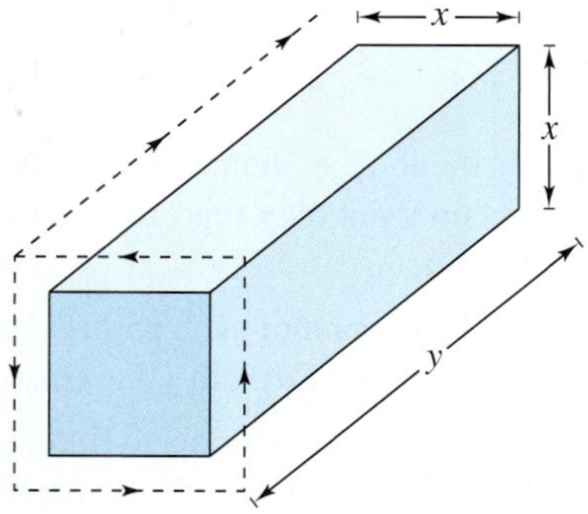

(a) Write the volume V of the package as a function of x. What is the domain of the function?

(b) Use a graphing utility to graph the function. Be sure to use an appropriate viewing window.

(c) What dimensions will maximize the volume of the package? Explain.

77. Business A company produces a product for which the variable cost is \$68.75 per unit and the fixed costs are \$248,000. The product sells for \$99.99. Let x be the number of units produced and sold.

(a) The total cost for a business is the sum of the variable cost and the fixed costs. Write the total cost C as a function of the number of units produced.

(b) Write the revenue R as a function of the number of units sold.

(c) Write the profit P as a function of the number of units sold. (*Note*: $P = R - C$.)

(d) Use the model in part (c) to find $P(20{,}000)$. Interpret your result in the context of the situation.

(e) Use the model in part (c) to find $P(0)$. Interpret your result in the context of the situation.

78. MODELING DATA

The table shows the revenue y (in thousands of dollars) of a landscaping business for each month of 2015, with $x = 1$ representing January.

Month, x	Revenue, y
1	5.2
2	5.6
3	6.6
4	8.3
5	11.5
6	15.8
7	12.8
8	10.1
9	8.6
10	6.9
11	4.5
12	2.7

The mathematical model below represents the data.

$$f(x) = \begin{cases} -1.97x + 26.3 \\ 0.505x^2 - 1.47x + 6.3 \end{cases}$$

(a) Identify the independent and dependent variables and explain what they represent in the context of the problem.

(b) What is the domain of each part of the piecewise-defined function? Explain your reasoning.

(c) Use the mathematical model to find $f(5)$. Interpret your result in the context of the problem.

(d) Use the mathematical model to find $f(11)$. Interpret your result in the context of the problem.

(e) How do the values obtained from the models in parts (c) and (d) compare with the actual data values?

79. Civil Engineering The total numbers n (in millions) of miles for all public roadways in the United States from 2000 through 2011 can be approximated by the model

$$n(t) = \begin{cases} 0.0050t^2 + 0.005t + 3.95, & 0 \le t \le 2 \\ 0.013t + 3.95, & 2 < t \le 11 \end{cases}$$

where t represents the year, with $t = 0$ corresponding to 2000. The actual numbers are shown in the bar graph. (Source: U.S. Federal Highway Administration)

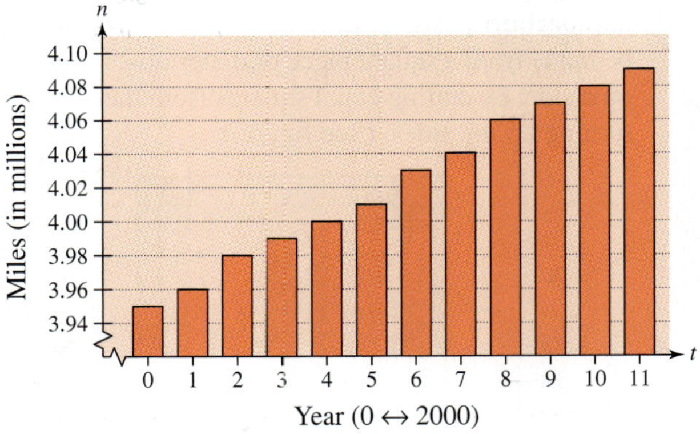

(a) Identify the independent and dependent variables and explain what they represent in the context of the problem.

(b) Use the *table* feature of a graphing utility to approximate the total number of miles for all public roadways each year from 2000 through 2011.

(c) Compare the values in part (b) with the actual values shown in the bar graph. How well does the model fit the data?

(d) Do you think the piecewise-defined function could be used to predict the total number of miles for all public roadways for years outside the domain? Explain your reasoning.

80. ***Why you should learn it*** *(p. 97)* The force F (in tons) of water against the face of a dam is estimated by the function

$$F(y) = 149.76\sqrt{10}y^{5/2}$$

where y is the depth of the water (in feet).

(a) Complete the table. What can you conclude?

y	5	10	20	30	40
$F(y)$					

(b) Use a graphing utility to graph the function. Describe your viewing window.

(c) Use the table to approximate the depth at which the force against the dam is 1,000,000 tons. Verify your answer graphically. How could you find a better estimate?

81. **Projectile Motion** The height y (in feet) of a baseball thrown by a child is

$$y = -0.1x^2 + 3x + 6$$

where x is the horizontal distance (in feet) from where the ball was thrown. Will the ball fly over the glove of another child 30 feet away trying to catch the ball? Explain. (Assume that the child who is trying to catch the ball holds a baseball glove at a height of 5 feet.)

82. **Business** The graph shows the sales (in millions of dollars) of Green Mountain Coffee Roasters from 2005 through 2013. Let $f(x)$ represent the sales in year x. (Source: Green Mountain Coffee Roasters, Inc.)

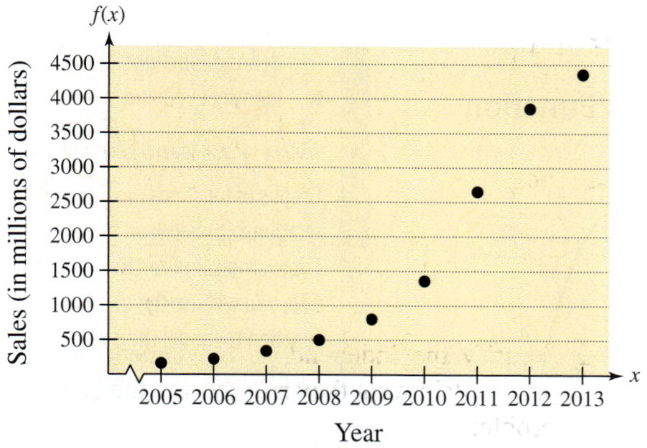

(a) Find $\dfrac{f(2013) - f(2005)}{2013 - 2005}$ and interpret the result in the context of the problem.

(b) An approximate model for the function is

$$S(t) = 90.442t^2 - 1075.25t + 3332.5, \quad 5 \le t \le 13$$

where S is the sales (in millions of dollars) and $t = 5$ represents 2005. Complete the table and compare the results with the data in the graph.

t	5	6	7	8	9	10	11	12	13
$S(t)$									

Evaluating a Difference Quotient **In Exercises 83–86, find the difference quotient and simplify your answer.**

83. $f(x) = 2x, \quad \dfrac{f(x + c) - f(x)}{c}, \quad c \neq 0$

84. $g(x) = 3x - 1, \quad \dfrac{g(x + h) - g(x)}{h}, \quad h \neq 0$

85. $f(x) = x^2 - x + 1, \quad \dfrac{f(2 + h) - f(2)}{h}, \quad h \neq 0$

86. $f(x) = x^3 + x, \quad \dfrac{f(x + h) - f(x)}{h}, \quad h \neq 0$

Conclusions

True or False? **In Exercises 87 and 88, determine whether the statement is true or false. Justify your answer.**

87. The domain of the function $f(x) = x^4 - 1$ is $(-\infty, \infty)$, and the range of f is $(0, \infty)$.

88. The set of ordered pairs $\{(-8, -2), (-6, 0), (-4, 0), (-2, 2), (0, 4), (2, -2)\}$ represents a function.

Think About It **In Exercises 89 and 90, write a square root function for the graph shown. Then identify the domain and range of the function.**

89.

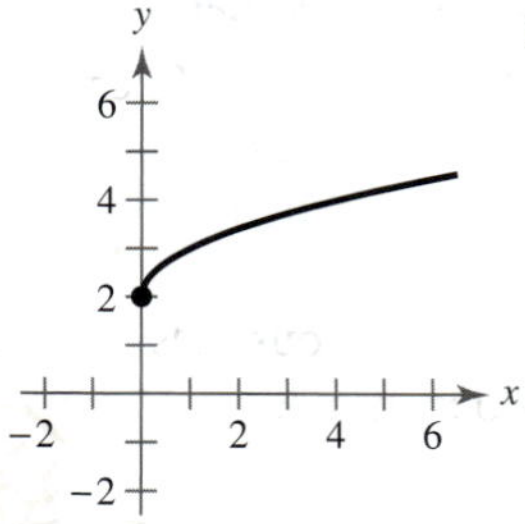

90.

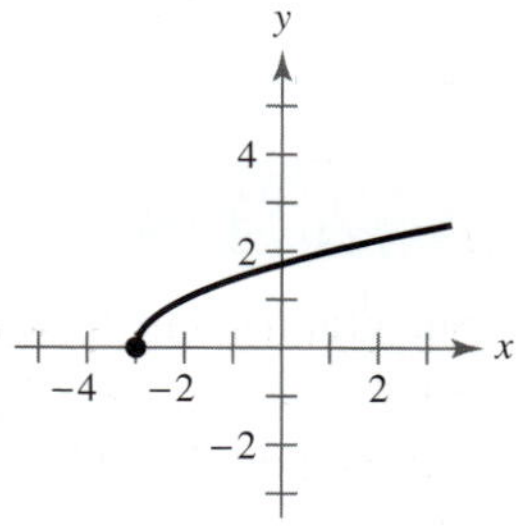

91. **Think About It** Given $f(x) = x^2$, is f the independent variable? Why or why not?

92. **HOW DO YOU SEE IT?** The graph represents the height h of a projectile after t seconds.

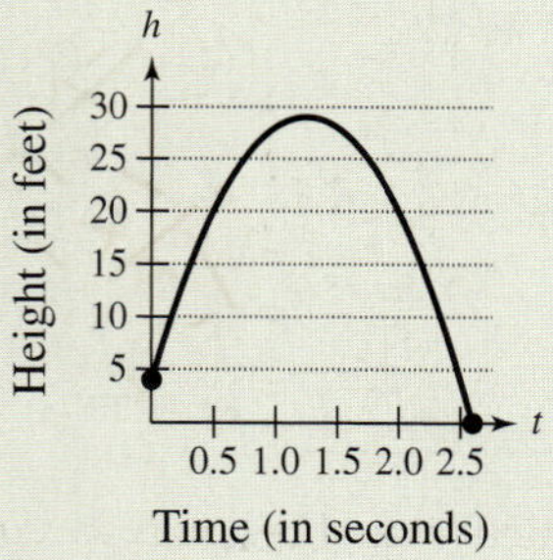

(a) Explain why h is a function of t.

(b) Approximate the height of the projectile after 0.5 second and after 1.25 seconds.

(c) Approximate the domain of h.

(d) Is t a function of h? Explain.

Cumulative Mixed Review

Operations with Rational Expressions **In Exercises 93–96, perform the operation and simplify.**

93. $12 - \dfrac{4}{x + 2}$

94. $\dfrac{3}{x^2 + x - 20} + \dfrac{2x}{x^2 + 4x - 5}$

95. $\dfrac{x^5}{2x^3 + 4x^2} \cdot \dfrac{4x + 8}{3x}$

96. $\dfrac{x + 7}{2(x - 9)} \div \dfrac{x - 7}{2(x - 9)}$

The symbol $\int$ indicates an example or exercise that highlights algebraic techniques specifically used in calculus.

1.4 Graphs of Functions

The Graph of a Function

In Section 1.3, some functions were represented graphically by points on a graph in a coordinate plane in which the input values are represented by the horizontal axis and the output values are represented by the vertical axis. The **graph of a function** f is the collection of ordered pairs $(x, f(x))$ such that x is in the domain of f. As you study this section, remember the geometric interpretations of x and $f(x)$.

$x =$ the directed distance from the y-axis

$f(x) =$ the directed distance from the x-axis

Example 1 shows how to use the graph of a function to find the domain and range of the function.

What you should learn

- Find the domains and ranges of functions and use the Vertical Line Test for functions.
- Determine intervals on which functions are increasing, decreasing, or constant.
- Determine relative maximum and relative minimum values of functions.
- Identify and graph step functions and other piecewise-defined functions.
- Identify even and odd functions.

Why you should learn it

Graphs of functions provide visual relationships between two variables. For example, in Exercise 92 on page 120, you will use the graph of a step function to model the cost of sending a package.

EXAMPLE 1 Finding the Domain and Range of a Function

Use the graph of the function f to find (a) the domain of f, (b) the function values $f(-1)$ and $f(2)$, and (c) the range of f.

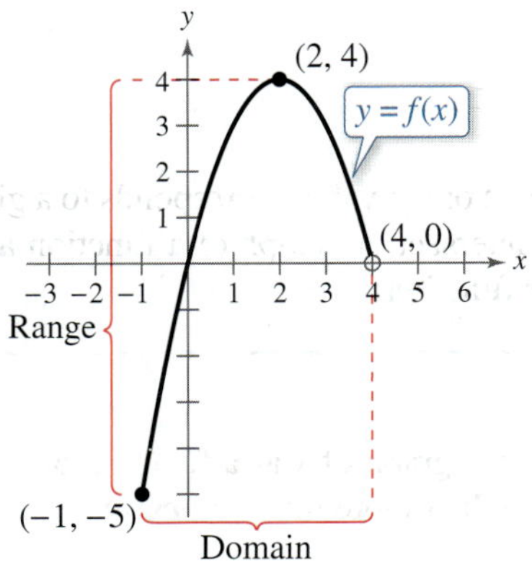

Solution

a. The closed dot at $(-1, -5)$ indicates that $x = -1$ is in the domain of f, whereas the open dot at $(4, 0)$ indicates that $x = 4$ is not in the domain. So, the domain of f is all x in the interval $[-1, 4)$.

b. Because $(-1, -5)$ is a point on the graph of f, it follows that

$$f(-1) = -5.$$

Similarly, because $(2, 4)$ is a point on the graph of f, it follows that

$$f(2) = 4.$$

c. Because the graph does not extend below $f(-1) = -5$ or above $f(2) = 4$, the range of f is the interval $[-5, 4]$.

✓ Checkpoint Audio-video solution in English & Spanish at LarsonPrecalculus.com.

Use the graph of the function f to find (a) the domain of f, (b) the function values $f(0)$ and $f(3)$, and (c) the range of f.

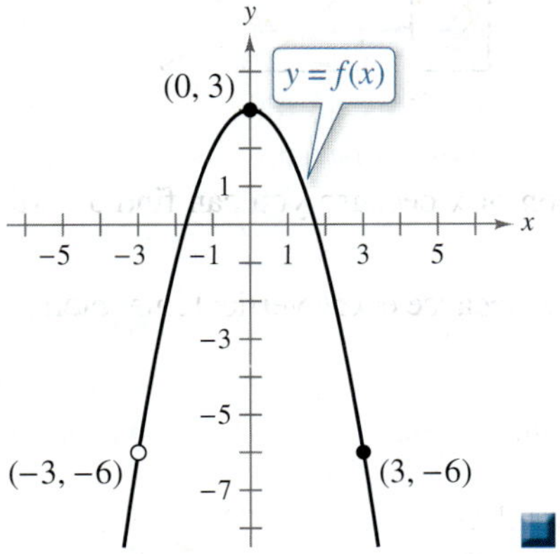

The use of dots (open or closed) at the extreme left and right points of a graph indicates that the graph does not extend beyond these points. When no such dots are shown, assume that the graph extends beyond these points.

EXAMPLE 2 Finding the Domain and Range of a Function

Find the domain and range of $f(x) = \sqrt{x - 4}$.

Algebraic Solution

Because the expression under a radical cannot be negative, the domain of $f(x) = \sqrt{x - 4}$ is the set of all real numbers such that $x - 4 \geq 0$. Solve this linear inequality for x as follows. (For help with solving linear inequalities, see Appendix D at this textbook's *Companion Website.*)

$x - 4 \geq 0$ Write original inequality.

$x \geq 4$ Add 4 to each side.

So, the domain is the set of all real numbers greater than or equal to 4. Because the value of a radical expression is never negative, the range of $f(x) = \sqrt{x - 4}$ is the set of all nonnegative real numbers.

Graphical Solution

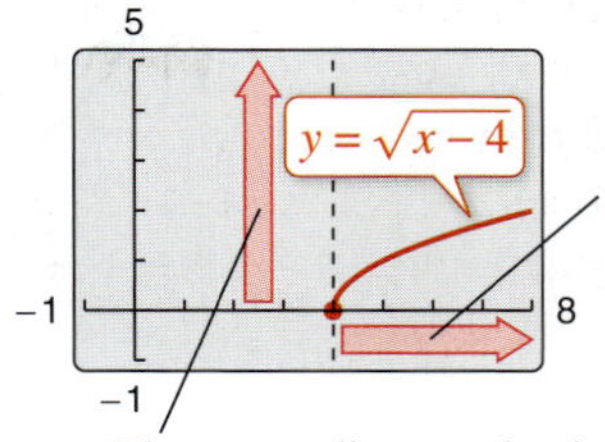

The x-coordinates of points on the graph extend from 4 to the right. So, the domain is the set of all real numbers greater than or equal to 4.

The y-coordinates of points on the graph extend from 0 upwards. So, the range is the set of all nonnegative real numbers.

✓ Checkpoint *Audio-video solution in English & Spanish at LarsonPrecalculus.com.*

Find the domain and range of $f(x) = \sqrt{x - 1}$.

By the definition of a function, at most one y-value corresponds to a given x-value. It follows, then, that a vertical line can intersect the graph of a function at most once. This leads to the **Vertical Line Test** for functions.

> **Vertical Line Test for Functions**
>
> A set of points in a coordinate plane is the graph of y as a function of x if and only if no vertical line intersects the graph at more than one point.

EXAMPLE 3 Vertical Line Test for Functions

Use the Vertical Line Test to decide whether each graph represents y as a function of x.

a.

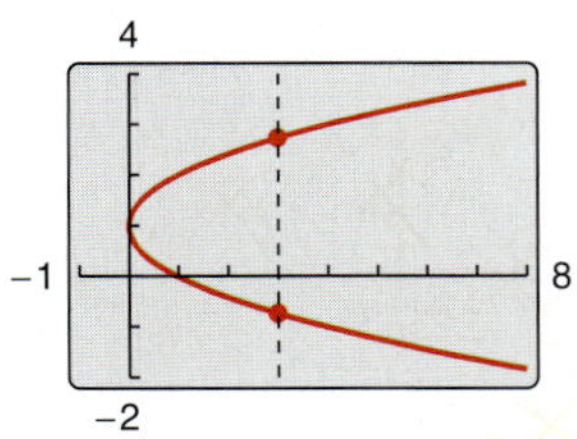

b.

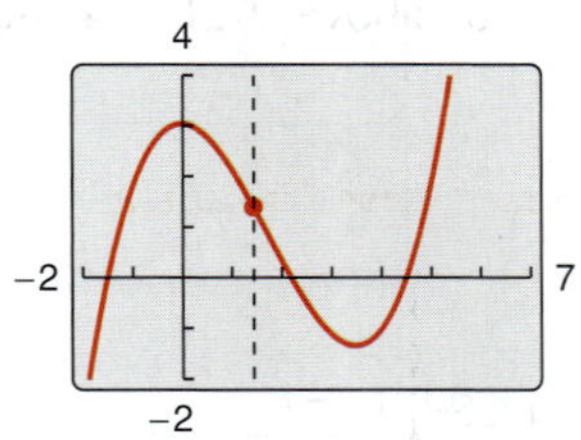

Solution

a. This is *not* a graph of y as a function of x because you can find a vertical line that intersects the graph twice.

b. This *is* a graph of y as a function of x because every vertical line intersects the graph at most once.

✓ Checkpoint *Audio-video solution in English & Spanish at LarsonPrecalculus.com.*

Use the Vertical Line Test to decide whether the graph at the right represents y as a function of x.

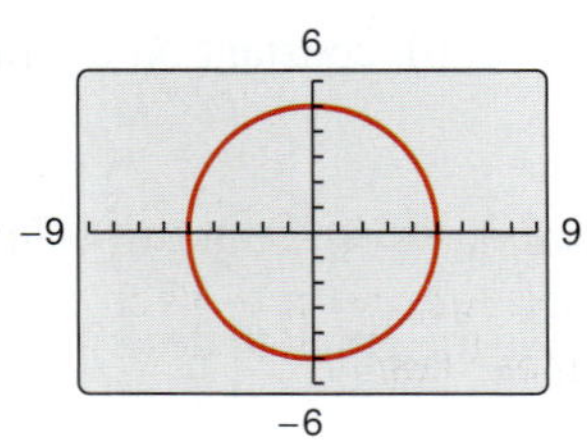

Technology Tip

Most graphing utilities are designed to graph functions of x more easily than other types of equations. For instance, the graph shown in Example 3(a) represents the equation $x - (y - 1)^2 = 0$. To use a graphing utility to duplicate this graph, you must first solve the equation for y to obtain $y = 1 \pm \sqrt{x}$, and then graph the two equations $y_1 = 1 + \sqrt{x}$ and $y_2 = 1 - \sqrt{x}$ in the same viewing window.

Increasing and Decreasing Functions

The more you know about the graph of a function, the more you know about the function itself. Consider the graph shown in Figure 1.25. Moving from *left to right,* this graph falls from $x = -2$ to $x = 0$, is constant from $x = 0$ to $x = 2$, and rises from $x = 2$ to $x = 4$.

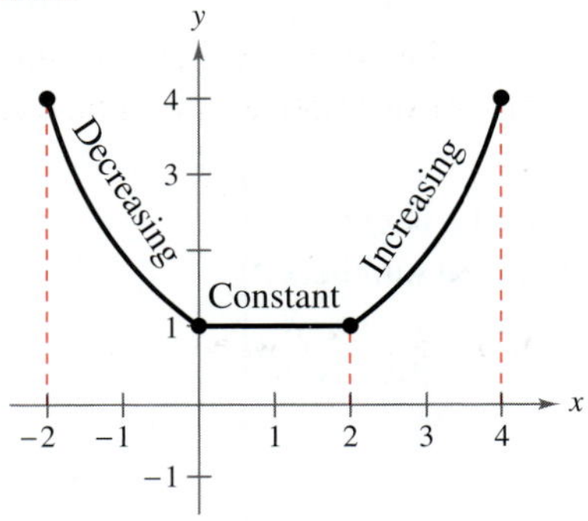

Figure 1.25

> **Increasing, Decreasing, and Constant Functions**
>
> A function f is **increasing** on an interval when, for any x_1 and x_2 in the interval,
>
> $x_1 < x_2$ implies $f(x_1) < f(x_2)$.
>
> A function f is **decreasing** on an interval when, for any x_1 and x_2 in the interval,
>
> $x_1 < x_2$ implies $f(x_1) > f(x_2)$.
>
> A function f is **constant** on an interval when, for any x_1 and x_2 in the interval,
>
> $f(x_1) = f(x_2)$.

EXAMPLE 4 Increasing and Decreasing Functions

In Figure 1.26, determine the open intervals on which each function is increasing, decreasing, or constant.

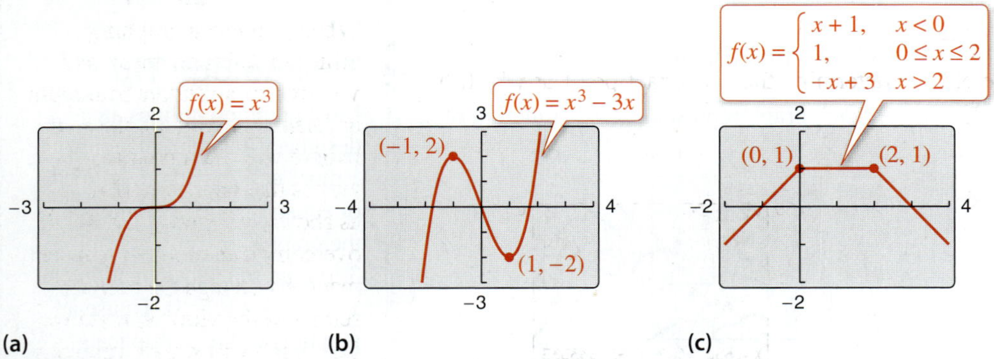

Figure 1.26

Solution

a. Although it might appear that there is an interval in which this function is constant, you can see that if $x_1 < x_2$, then $(x_1)^3 < (x_2)^3$, which implies that $f(x_1) < f(x_2)$. So, the function is increasing over the entire real line.

b. This function is increasing on the interval $(-\infty, -1)$, decreasing on the interval $(-1, 1)$, and increasing on the interval $(1, \infty)$.

c. This function is increasing on the interval $(-\infty, 0)$, constant on the interval $(0, 2)$, and decreasing on the interval $(2, \infty)$.

✔ ***Checkpoint*** *Audio-video solution in English & Spanish at LarsonPrecalculus.com.*

Graph the function $f(x) = x^3 + 3x^2 - 1$. Then use the graph to describe the increasing and decreasing behavior of the function.

Relative Minimum and Maximum Values

The points at which a function changes its increasing, decreasing, or constant behavior are helpful in determining the relative maximum or relative minimum values of the function.

Definition of Relative Minimum and Relative Maximum

A function value $f(a)$ is called a **relative minimum** of f when there exists an interval (x_1, x_2) that contains a such that

$$x_1 < x < x_2 \quad \text{implies} \quad f(a) \le f(x).$$

A function value $f(a)$ is called a **relative maximum** of f when there exists an interval (x_1, x_2) that contains a such that

$$x_1 < x < x_2 \quad \text{implies} \quad f(a) \ge f(x).$$

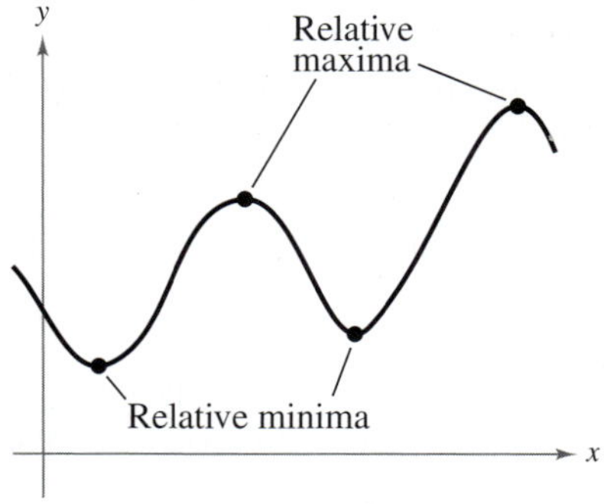

Figure 1.27

Figure 1.27 shows several different examples of relative minima and relative maxima. In Section 3.1, you will study a technique for finding the *exact points* at which a second-degree polynomial function has a relative minimum or relative maximum. For the time being, however, you can use a graphing utility to find reasonable approximations of these points.

EXAMPLE 5 Approximating a Relative Minimum

Use a graphing utility to approximate the relative minimum of the function given by $f(x) = 3x^2 - 4x - 2$.

Solution

The graph of f is shown in Figure 1.28. By using the *zoom* and *trace* features of a graphing utility, you can estimate that the function has a relative minimum at the point

$(0.67, -3.33)$. See Figure 1.29.

Later, in Section 3.1, you will be able to determine that the exact point at which the relative minimum occurs is $\left(\frac{2}{3}, -\frac{10}{3}\right)$.

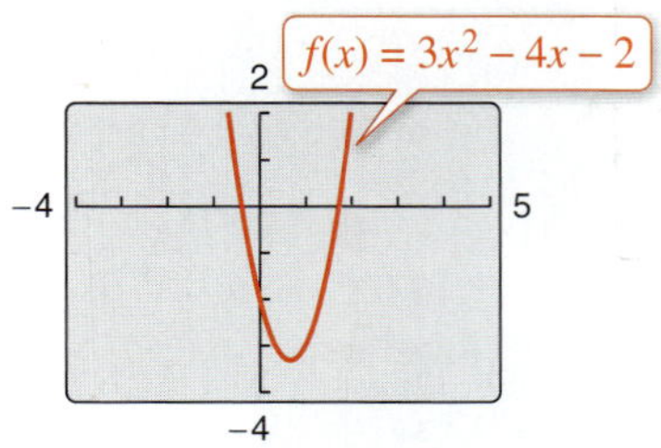

Figure 1.28

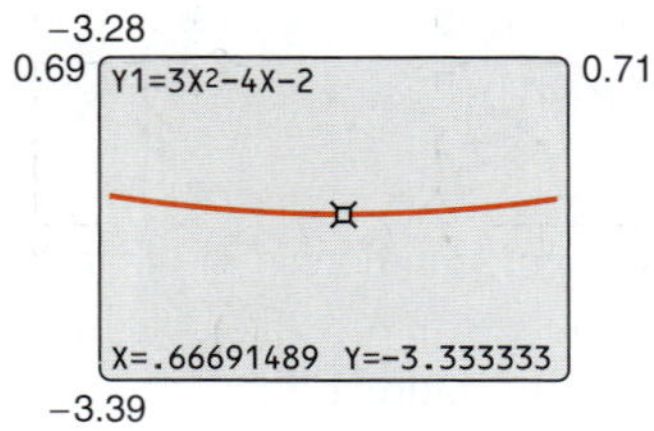

Figure 1.29

Technology Tip

When you use a graphing utility to estimate the x- and y-values of a relative minimum or relative maximum, the *zoom* feature will often produce graphs that are nearly flat, as shown in Figure 1.29. To overcome this problem, you can manually change the vertical setting of the viewing window. The graph will stretch vertically when the values of Ymin and Ymax are closer together.

✓ ***Checkpoint*** *Audio-video solution in English & Spanish at LarsonPrecalculus.com.*

Use a graphing utility to approximate the relative maximum of the function

$f(x) = -4x^2 - 7x + 3.$

You can also use the *table* feature of a graphing utility to numerically approximate the relative minimum of the function in Example 5. Using a table that begins at 0.6 and increments the value of x by 0.01, you can approximate that the minimum of f occurs at the point $(0.67, -3.33)$. Some graphing utilities have built-in programs that will find minimum or maximum values, as shown in Example 6.

EXAMPLE 6 Approximating Relative Minima and Maxima

Use a graphing utility to approximate the relative minimum and relative maximum of the function $f(x) = -x^3 + x$.

Solution

By using the *minimum* and *maximum* features of the graphing utility, you can estimate that the function has a relative minimum at the point

$(-0.58, -0.38)$ See Figure 1.30.

and a relative maximum at the point

$(0.58, 0.38)$. See Figure 1.31.

If you take a course in calculus, you will learn a technique for finding the exact points at which this function has a relative minimum and a relative maximum.

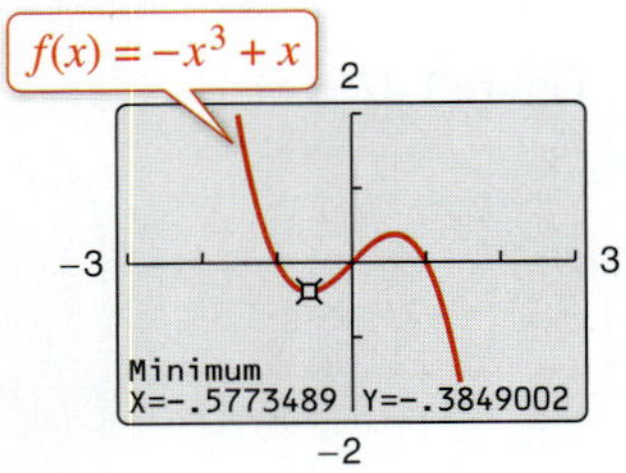

Figure 1.30

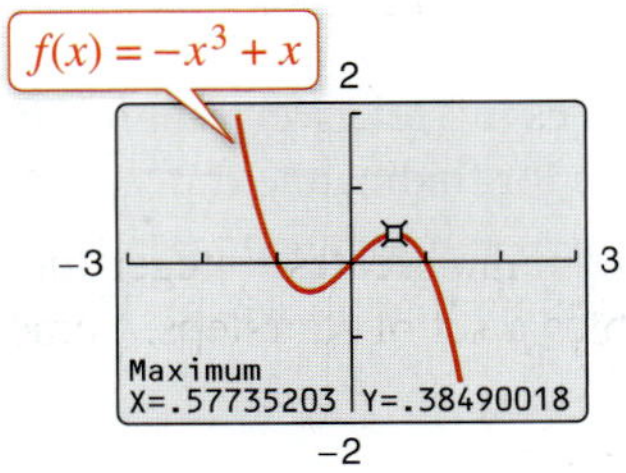

Figure 1.31

Checkpoint *Audio-video solution in English & Spanish at LarsonPrecalculus.com.*

Use a graphing utility to approximate the relative minimum and relative maximum of the function $f(x) = 2x^3 + 3x^2 - 12x$.

EXAMPLE 7 Temperature

During a 24-hour period, the temperature y (in degrees Fahrenheit) of a certain city can be approximated by the model

$$y = 0.026x^3 - 1.03x^2 + 10.2x + 34, \quad 0 \le x \le 24$$

where x represents the time of day, with $x = 0$ corresponding to 6 A.M. Approximate the maximum temperature during this 24-hour period.

Solution

Using the *maximum* feature of a graphing utility, you can determine that the maximum temperature during the 24-hour period was approximately 64°F. This temperature occurred at about 12:36 P.M. ($x \approx 6.6$), as shown in Figure 1.32.

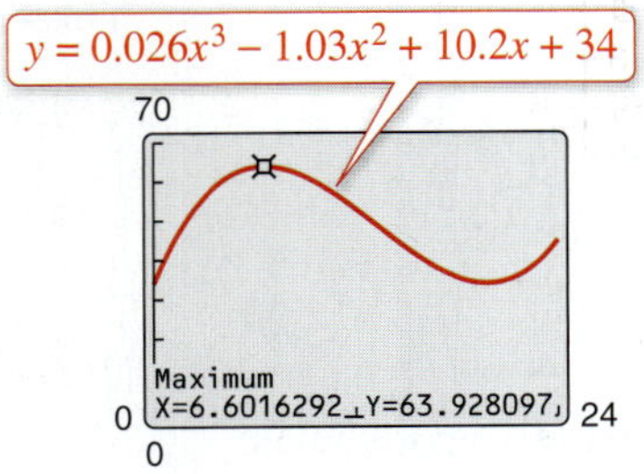

Figure 1.32

Checkpoint *Audio-video solution in English & Spanish at LarsonPrecalculus.com.*

In Example 7, approximate the minimum temperature during the 24-hour period.

Step Functions and Piecewise-Defined Functions

Library of Parent Functions: Greatest Integer Function

The *greatest integer function*, denoted by $[\![x]\!]$ and defined as the greatest integer less than or equal to x, has an infinite number of breaks or steps—one at each integer value in its domain. The basic characteristics of the greatest integer function are summarized below.

Graph of $f(x) = [\![x]\!]$

Domain: $(-\infty, \infty)$

Range: the set of integers

x-intercepts: in the interval $[0, 1)$

y-intercept: $(0, 0)$

Constant between each pair of consecutive integers

Jumps vertically one unit at each integer value

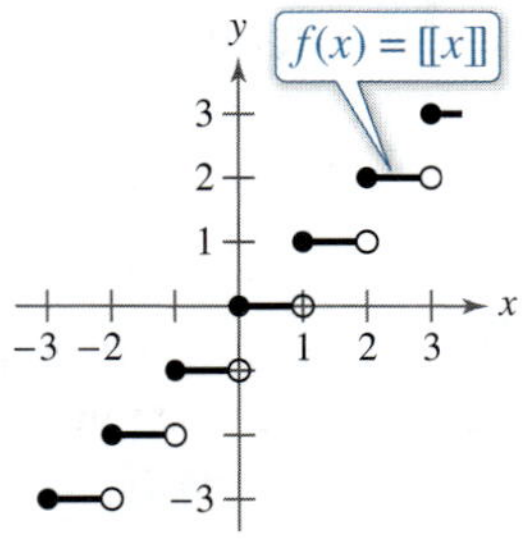

Technology Tip

Most graphing utilities display graphs in *connected mode*, which works well for graphs that do not have breaks. For graphs that do have breaks, such as the greatest integer function, it is better to use *dot mode*. Graph the greatest integer function [often called Int (x)] in *connected* and *dot modes*, and compare the two results.

Because of the vertical jumps described above, the greatest integer function is an example of a **step function** whose graph resembles a set of stairsteps. Some values of the greatest integer function are as follows.

$$[\![-1]\!] = (\text{greatest integer} \leq -1) = -1$$

$$[\![-\tfrac{1}{2}]\!] = \left(\text{greatest integer} \leq -\tfrac{1}{2}\right) = -1$$

$$[\![\tfrac{1}{10}]\!] = \left(\text{greatest integer} \leq \tfrac{1}{10}\right) = 0$$

$$[\![1.5]\!] = (\text{greatest integer} \leq 1.5) = 1$$

In Section 1.3, you learned that a piecewise-defined function is a function that is defined by two or more equations over a specified domain. To sketch the graph of a piecewise-defined function, you need to sketch the graph of each equation on the appropriate portion of the domain.

EXAMPLE 8 Sketching a Piecewise-Defined Function

Sketch the graph of

$$f(x) = \begin{cases} 2x + 3, & x \leq 1 \\ -x + 4, & x > 1 \end{cases}$$

by hand.

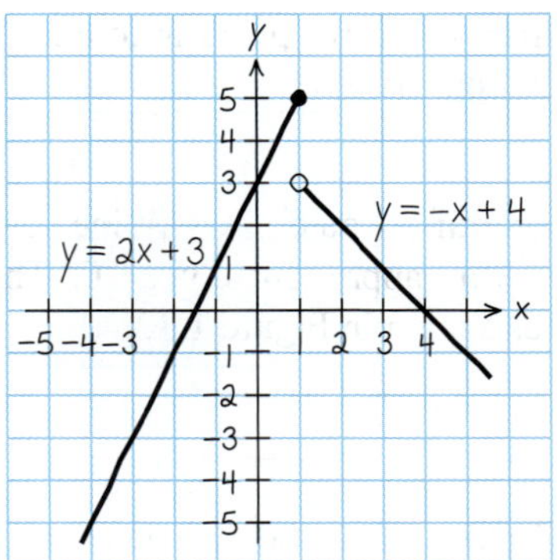

Solution

This piecewise-defined function is composed of two linear functions. At and to the left of $x = 1$, the graph is the line $y = 2x + 3$. To the right of $x = 1$, the graph is the line $y = -x + 4$, as shown in the figure. Notice that the point $(1, 5)$ is a solid dot and the point $(1, 3)$ is an open dot. This is because $f(1) = 5$.

Checkpoint *Audio-video solution in English & Spanish at LarsonPrecalculus.com.*

Sketch the graph of $f(x) = \begin{cases} -\frac{1}{2}x - 6, & x \leq -4 \\ x + 5, & x > -4 \end{cases}$ by hand.

What's Wrong?

You use a graphing utility to graph

$$f(x) = \begin{cases} x^2 + 1, & x \leq 0 \\ 4 - x, & x > 0 \end{cases}$$

by letting $y_1 = x^2 + 1$ and $y_2 = 4 - x$, as shown in the figure. You conclude that this is the graph of f. What's wrong?

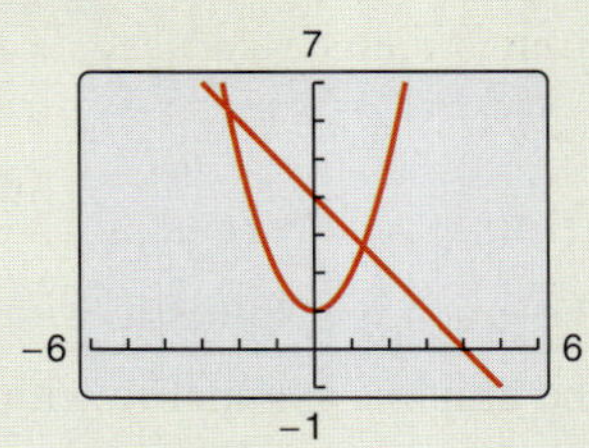

Even and Odd Functions

A graph has *symmetry with respect to the y-axis* if whenever (x, y) is on the graph, then so is the point $(-x, y)$. A graph has *symmetry with respect to the origin* if whenever (x, y) is on the graph, then so is the point $(-x, -y)$. A graph has *symmetry with respect to the x-axis* if whenever (x, y) is on the graph, then so is the point $(x, -y)$. A function whose graph is symmetric with respect to the y-axis is an **even function.** A function whose graph is symmetric with respect to the origin is an **odd function.** A graph that is symmetric with respect to the x-axis is not the graph of a function (except for the graph of $y = 0$). These three types of symmetry are illustrated in Figure 1.33.

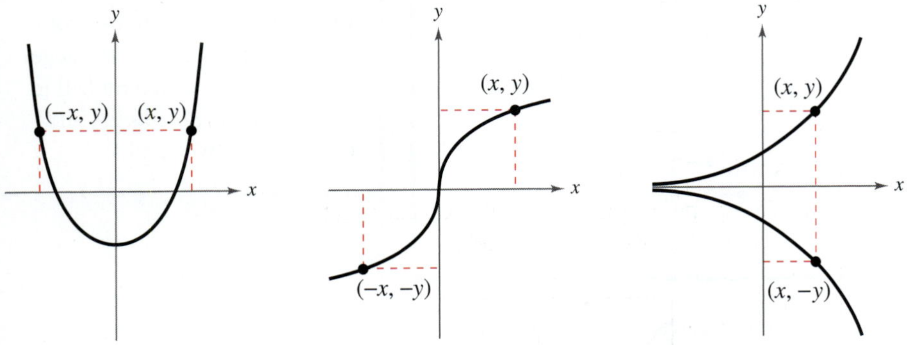

Symmetric to y-axis *Even function*

Symmetric to origin *Odd function*

Symmetric to x-axis *Not a function*

Figure 1.33

> ### Explore the Concept
>
> Graph each function with a graphing utility. Determine whether the function is *even*, *odd*, or *neither*.
>
> $f(x) = x^2 - x^4$
>
> $g(x) = 2x^3 + 1$
>
> $h(x) = x^5 - 2x^3 + x$
>
> $j(x) = 2 - x^6 - x^8$
>
> $k(x) = x^5 - 2x^4 + x - 2$
>
> $p(x) = x^9 + 3x^5 - x^3 + x$
>
> What do you notice about the equations of functions that are (a) odd and (b) even? Describe a way to identify a function as (c) odd, (d) even, or (e) neither odd nor even by inspecting the equation.

EXAMPLE 9 Even and Odd Functions

See LarsonCollege Algebra.com for an interactive version of this type of example.

For each graph, determine whether the function is even, odd, or neither.

a.

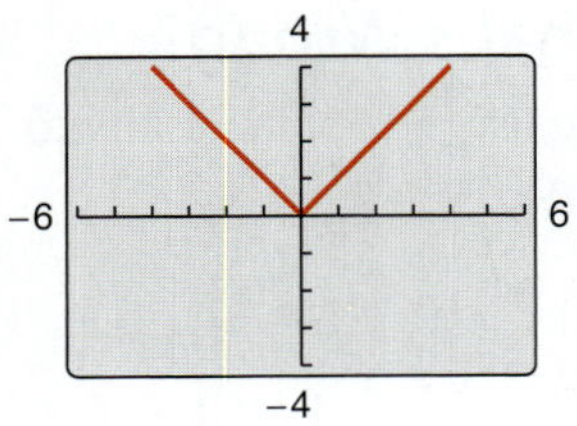

b.

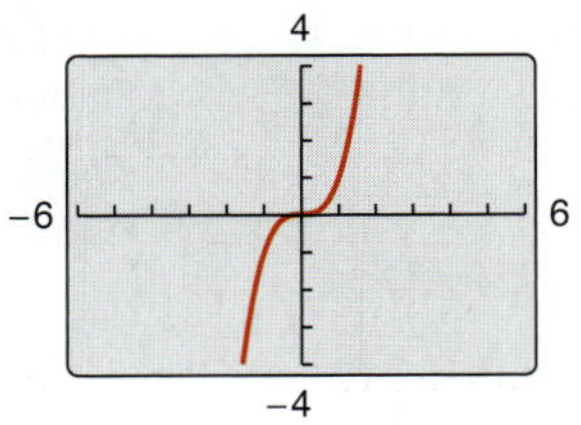

c.

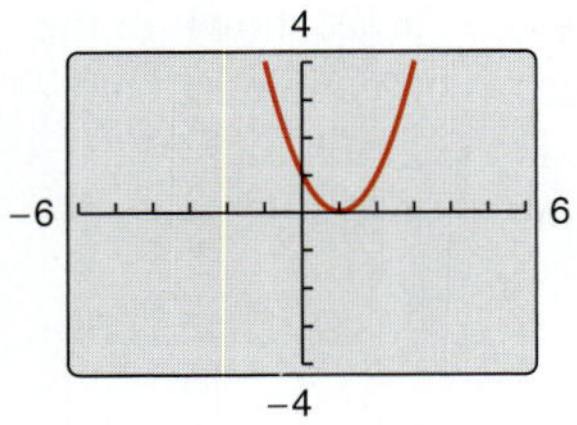

d.

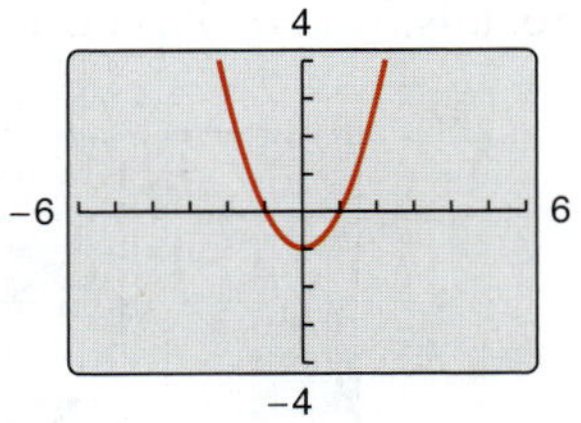

Solution

a. The graph is symmetric with respect to the y-axis. So, the function is even.

b. The graph is symmetric with respect to the origin. So, the function is odd.

c. The graph is neither symmetric with respect to the origin nor with respect to the y-axis. So, the function is neither even nor odd.

d. The graph is symmetric with respect to the y-axis. So, the function is even.

✓ Checkpoint *Audio-video solution in English & Spanish at LarsonPrecalculus.com.*

Use a graphing utility to graph $f(x) = x^2 - 4$ and determine whether it is even, odd, or neither.

Test for Even and Odd Functions

A function f is **even** when, for each x in the domain of f, $f(-x) = f(x)$.
A function f is **odd** when, for each x in the domain of f, $f(-x) = -f(x)$.

EXAMPLE 10 Even and Odd Functions

Determine whether each function is even, odd, or neither.

a. $g(x) = x^3 - x$

b. $h(x) = x^2 + 1$

c. $f(x) = x^3 - 1$

Algebraic Solution

a. This function is odd because

$$\begin{aligned} g(-x) &= (-x)^3 - (-x) \\ &= -x^3 + x \\ &= -(x^3 - x) \\ &= -g(x). \end{aligned}$$

b. This function is even because

$$\begin{aligned} h(-x) &= (-x)^2 + 1 \\ &= x^2 + 1 \\ &= h(x). \end{aligned}$$

c. Substituting $-x$ for x produces

$$\begin{aligned} f(-x) &= (-x)^3 - 1 \\ &= -x^3 - 1. \end{aligned}$$

Because

$$f(x) = x^3 - 1$$

and

$$-f(x) = -x^3 + 1$$

you can conclude that

$$f(-x) \neq f(x)$$

and

$$f(-x) \neq -f(x).$$

So, the function is neither even nor odd.

Graphical Solution

a. The graph is symmetric with respect to the origin. So, this function is odd.

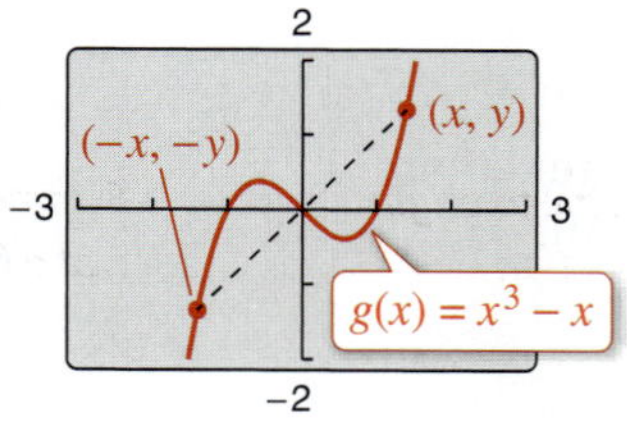

b. The graph is symmetric with respect to the y-axis. So, this function is even.

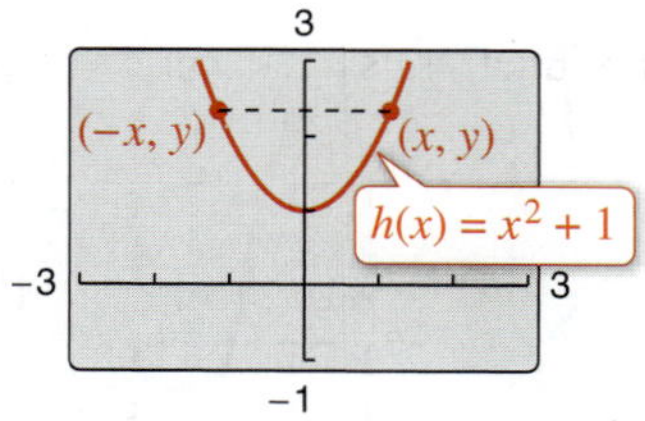

c. The graph is neither symmetric with respect to the origin nor with respect to the y-axis. So, this function is neither even nor odd.

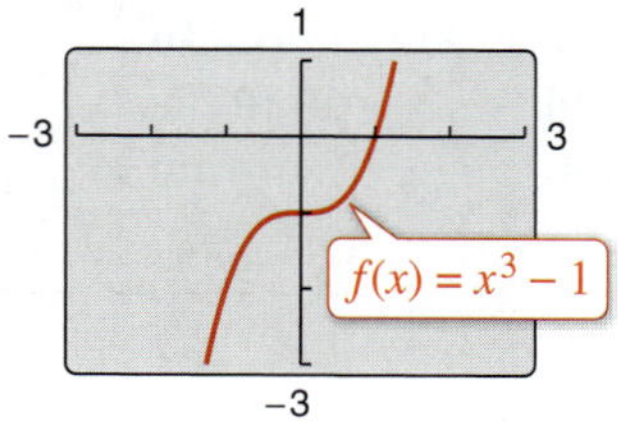

✓ Checkpoint Audio-video solution in English & Spanish at LarsonPrecalculus.com.

Determine whether the function is even, odd, or neither. Then describe the symmetry.

a. $f(x) = 5 - 3x$ **b.** $g(x) = x^4 - x^2 - 1$ **c.** $h(x) = 2x^3 + 3x$

To help visualize symmetry with respect to the origin, place a pin at the origin of a graph and rotate the graph 180°. If the result after rotation coincides with the original graph, then the graph is symmetric with respect to the origin.

1.4 Exercises

See CalcChat.com for tutorial help and worked-out solutions to odd-numbered exercises.
For instructions on how to use a graphing utility, see Appendix A.

Vocabulary and Concept Check

In Exercises 1 and 2, fill in the blank.

1. A function f is _______ on an interval when, for any x_1 and x_2 in the interval, $x_1 < x_2$ implies $f(x_1) > f(x_2)$.
2. A function f is _______ when, for each x in the domain of f, $f(-x) = f(x)$.

3. The graph of a function f is the segment from $(1, 2)$ to $(4, 5)$, including the endpoints. What is the domain of f?
4. A vertical line intersects a graph twice. Does the graph represent a function?
5. Let f be a function such that $f(2) \geq f(x)$ for all values of x in the interval $(0, 3)$. Does $f(2)$ represent a relative minimum or a relative maximum?
6. Given $f(x) = [\![x]\!]$, in what interval does $f(x) = 5$?

Procedures and Problem Solving

Finding the Domain and Range of a Function **In Exercises 7–10, use the graph of the function to find the domain and range of f. Then find $f(0)$.**

7.

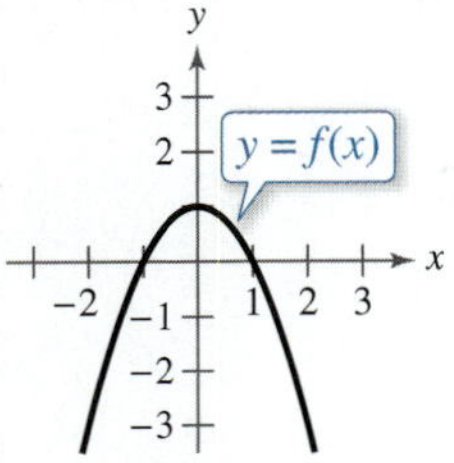

8.

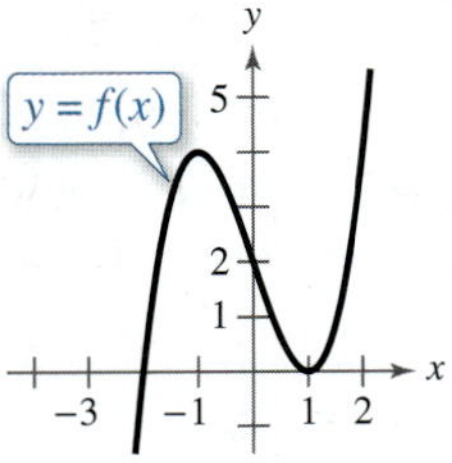

9.

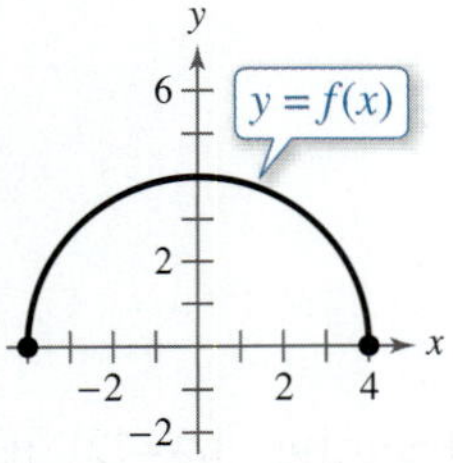

10.

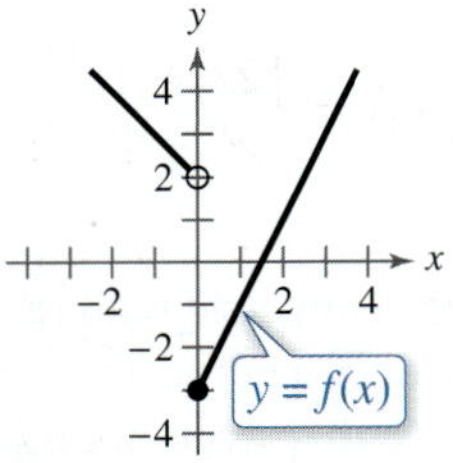

Finding the Domain and Range of a Function **In Exercises 11–16, use a graphing utility to graph the function and estimate its domain and range. Then find the domain and range algebraically.**

11. $f(x) = -2x^2 + 3$
12. $f(x) = x^2 - 1$
13. $f(x) = \sqrt{x + 2}$
14. $h(t) = \sqrt{4 - t^2}$
15. $f(x) = |x + 3|$
16. $f(x) = -\frac{1}{4}|x - 5|$

Analyzing a Graph **In Exercises 17 and 18, use the graph of the function to answer the questions.**

(a) Determine the domain of the function.

(b) Determine the range of the function.

(c) Find the value(s) of x for which $f(x) = 0$.

(d) What are the values of x from part (c) referred to graphically?

(e) Find $f(0)$, if possible.

(f) What is the value from part (e) referred to graphically?

(g) What is the value of f at $x = 1$? What are the coordinates of the point?

(h) What is the value of f at $x = -1$? What are the coordinates of the point?

(i) The coordinates of the point on the graph of f at which $x = -3$ can be labeled $(-3, f(-3))$, or $(-3, \quad)$.

17.

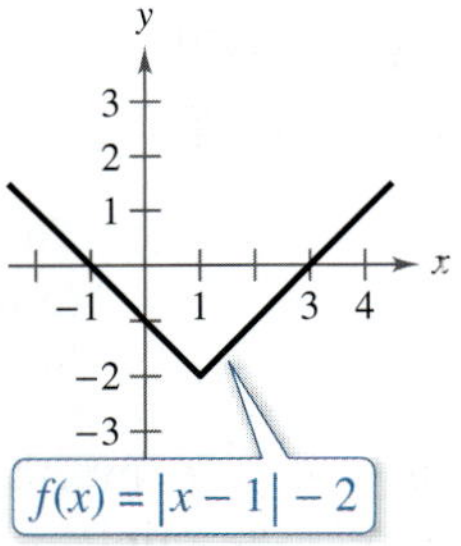

18.

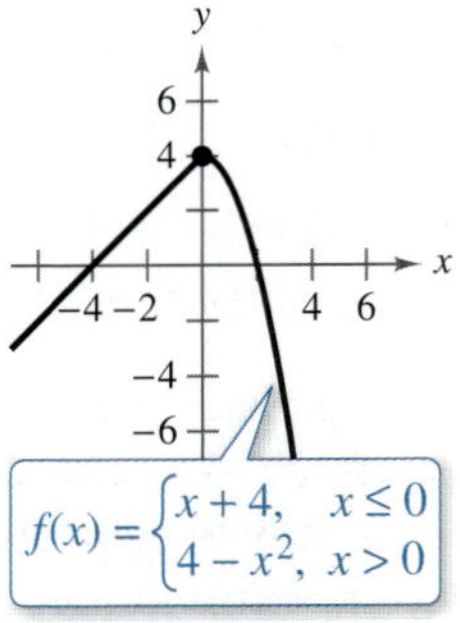

Vertical Line Test for Functions **In Exercises 19–22, use the Vertical Line Test to determine whether y is a function of x. Describe how you can use a graphing utility to produce the given graph.**

19. $y = \frac{1}{2}x^2$

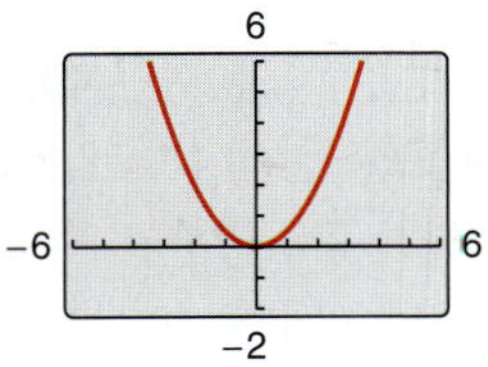

20. $x - y^2 = 1$

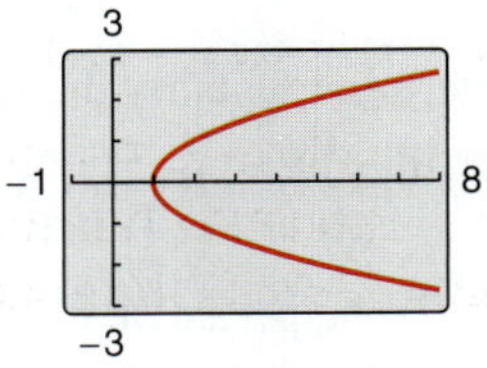

21. $0.25x^2 + y^2 = 1$

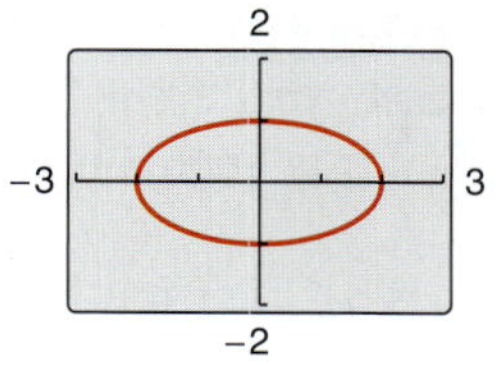

22. $x^2 = 2xy - 1$

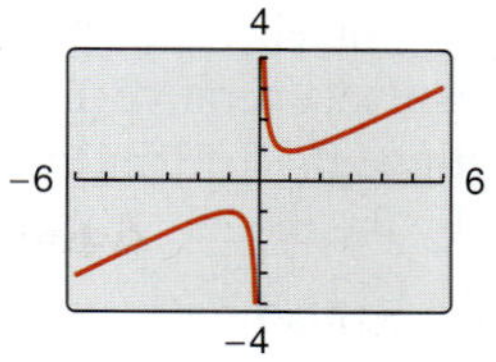

Increasing and Decreasing Functions **In Exercises 23–26, determine the open intervals on which the function is increasing, decreasing, or constant.**

23. $f(x) = \frac{3}{2}x$

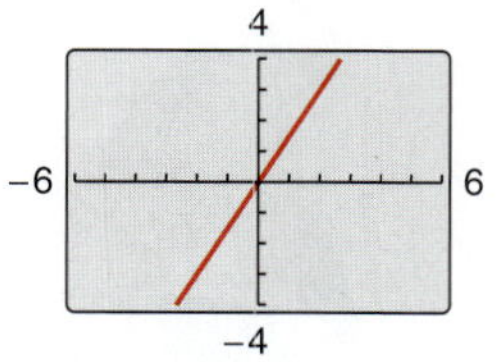

24. $f(x) = x^2 - 4x$

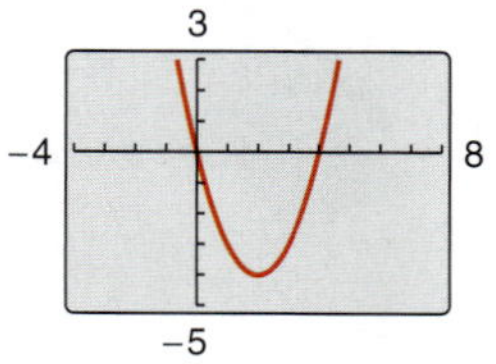

25. $f(x) = x^3 - 3x^2 + 2$

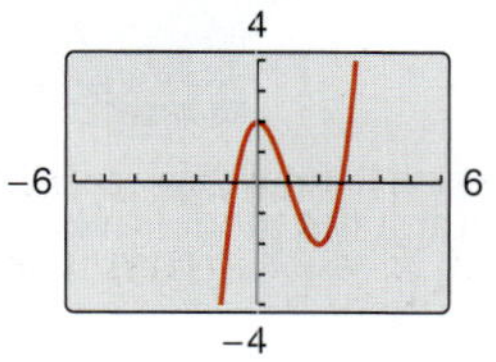

26. $f(x) = \sqrt{x^2 - 1}$

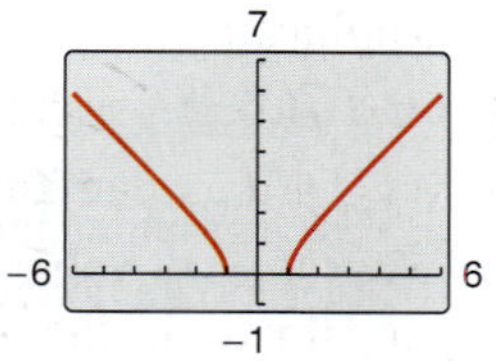

Increasing and Decreasing Functions **In Exercises 27–34, (a) use a graphing utility to graph the function and (b) determine the open intervals on which the function is increasing, decreasing, or constant.**

27. $f(x) = 3$ **28.** $f(x) = x$

29. $f(x) = x^{2/3}$ **30.** $f(x) = -x^{3/4}$

31. $f(x) = x\sqrt{x + 3}$ **32.** $f(x) = x\sqrt{3 - x}$

33. $f(x) = |x + 1| + |x - 1|$

34. $f(x) = -|x + 4| - |x + 1|$

Approximating Relative Minima and Maxima **In Exercises 35–46, use a graphing utility to graph the function and to approximate any relative minimum or relative maximum values of the function.**

35. $f(x) = x^2 - 6x$ **36.** $f(x) = 3x^2 - 2x - 5$

37. $y = -2x^3 - x^2 + 14x$ **38.** $y = x^3 - 6x^2 + 15$

39. $h(x) = (x - 1)\sqrt{x}$ **40.** $g(x) = x\sqrt{4 - x}$

41. $f(x) = x^2 - 4x - 5$ **42.** $f(x) = 3x^2 - 12x$

43. $f(x) = x^3 - 3x$ **44.** $f(x) = -x^3 + 3x^2$

45. $f(x) = 3x^2 - 6x + 1$ **46.** $f(x) = 8x - 4x^2$

Library of Parent Functions **In Exercises 47–52, sketch the graph of the function by hand. Then use a graphing utility to verify the graph.**

47. $f(x) = [\![x]\!] + 2$ **48.** $f(x) = [\![x]\!] - 3$

49. $f(x) = [\![x - 1]\!] - 2$ **50.** $f(x) = [\![x + 2]\!] + 1$

51. $f(x) = 2[\![x]\!]$ **52.** $f(x) = [\![4x]\!]$

Describing a Step Function **In Exercises 53 and 54, use a graphing utility to graph the function. State the domain and range of the function. Describe the pattern of the graph.**

53. $s(x) = 2\left(\frac{1}{4}x - [\![\frac{1}{4}x]\!]\right)$

54. $g(x) = 2\left(\frac{1}{4}x - [\![\frac{1}{4}x]\!]\right)^2$

Sketching a Piecewise-Defined Function **In Exercises 55–62, sketch the graph of the piecewise-defined function by hand.**

55. $f(x) = \begin{cases} 2x + 3, & x < 0 \\ 3 - x, & x \ge 0 \end{cases}$

56. $f(x) = \begin{cases} x + 6, & x \le -4 \\ 3x - 4, & x > -4 \end{cases}$

57. $f(x) = \begin{cases} \sqrt{4 + x}, & x < 0 \\ \sqrt{4 - x}, & x \ge 0 \end{cases}$

58. $f(x) = \begin{cases} 1 - (x - 1)^2, & x \le 2 \\ \sqrt{x - 2}, & x > 2 \end{cases}$

59. $f(x) = \begin{cases} x + 3, & x \le 0 \\ 3, & 0 < x \le 2 \\ 2x - 1, & x > 2 \end{cases}$

60. $g(x) = \begin{cases} x + 5, & x \le -3 \\ 5, & -3 < x < 1 \\ 5x - 4, & x \ge 1 \end{cases}$

61. $f(x) = \begin{cases} 2x + 1, & x \le -1 \\ x^2 - 2, & x > -1 \end{cases}$

62. $h(x) = \begin{cases} 3 + x, & x < 0 \\ x^2 + 1, & x \ge 0 \end{cases}$

Even and Odd Functions **In Exercises 63–72, use a graphing utility to graph the function and determine whether it is even, odd, or neither.**

63. $f(x) = 5$ **64.** $f(x) = -9$

65. $f(x) = 3x - 2$ **66.** $f(x) = 4 - 5x$

67. $h(x) = x^2 + 6$ **68.** $f(x) = -x^2 - 8$

69. $f(x) = \sqrt{1 - x}$ **70.** $g(t) = \sqrt[3]{t - 1}$

71. $f(x) = |x + 2|$ **72.** $f(x) = -|x - 5|$

Think About It **In Exercises 73–78, find the coordinates of a second point on the graph of a function f if the given point is on the graph and the function is (a) even and (b) odd.**

73. $\left(\frac{3}{2}, 4\right)$ **74.** $\left(-\frac{5}{3}, -7\right)$

75. $(-2, -9)$ **76.** $(5, -1)$

77. $(x, -y)$ **78.** $(2a, 2c)$

Algebraic-Graphical-Numerical **In Exercises 79–86, determine whether the function is even, odd, or neither (a) algebraically, (b) graphically by using a graphing utility to graph the function, and (c) numerically by using the *table* feature of the graphing utility to compare $f(x)$ and $f(-x)$ for several values of x.**

79. $f(t) = t^2 + 2t - 3$

80. $f(x) = x^6 - 2x^2 + 3$

81. $g(x) = x^3 - 5x$

82. $h(x) = x^5 - 4x^3$

83. $f(x) = x\sqrt{1 - x^2}$

84. $f(x) = x\sqrt{x + 5}$

85. $g(s) = 4s^{2/3}$

86. $f(s) = 4s^{3/5}$

Finding the Intervals Where a Function is Positive **In Exercises 87–90, graph the function and determine the interval(s) (if any) on the real axis for which $f(x) \ge 0$. Use a graphing utility to verify your results.**

87. $f(x) = 4 - x$

88. $f(x) = 4x + 8$

89. $f(x) = x^2 - 9$

90. $f(x) = x^2 - 4x$

91. Business The cost of parking in a metered lot is \$1.00 for the first hour and \$0.50 for each additional hour or portion of an hour.

(a) A customer needs a model for the cost C of parking in the metered lot for t hours. Which of the following is the appropriate model?

$C_1(t) = 1 + 0.50[\![t - 1]\!]$

$C_2(t) = 1 - 0.50[\![-(t - 1)]\!]$

(b) Use a graphing utility to graph the appropriate model. Estimate the cost of parking in the metered lot for 7 hours and 10 minutes.

92. ***Why you should learn it*** *(p. 110)* The cost of sending an overnight package from New York to Atlanta is \$23.20 for a package weighing up to but not including 1 pound and \$2.00 for each additional pound or portion of a pound. Use the greatest integer function to create a model for the cost C of overnight delivery of a package weighing x pounds, where $x > 0$. Sketch the graph of the function.

Using the Graph of a Function **In Exercises 93 and 94, write the height h of the rectangle as a function of x.**

93.

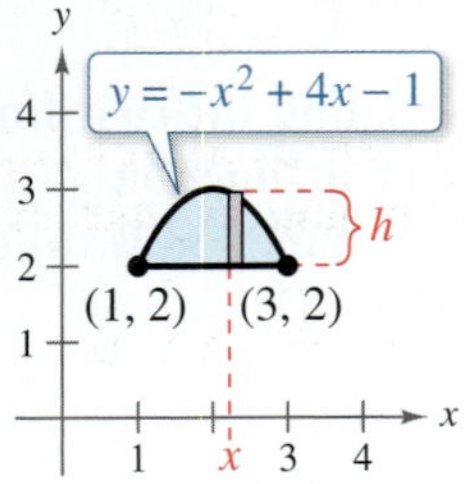

94.

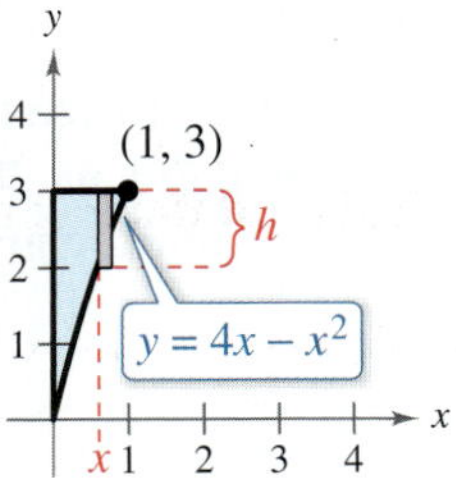

95. MODELING DATA

The number N (in thousands) of existing condominiums and cooperative homes sold each year from 2010 through 2013 in the United States is approximated by the model

$N = -24.83t^3 + 906t^2 - 10{,}928.2t + 44{,}114,$

$10 \le t \le 13$

where t represents the year, with $t = 10$ corresponding to 2010. (Source: National Association of Realtors)

(a) Use a graphing utility to graph the model over the appropriate domain.

(b) Use the graph from part (a) to determine during which years the number of cooperative homes and condos was increasing. During which years was the number decreasing?

(c) Approximate the minimum number of cooperative homes and condos sold from 2010 through 2013.

96. Mechanical Engineering The intake pipe of a 100-gallon tank has a flow rate of 10 gallons per minute, and two drain pipes have a flow rate of 5 gallons per minute each. The graph shows the volume V of fluid in the tank as a function of time t. Determine in which pipes the fluid is flowing in specific subintervals of the one-hour interval of time shown on the graph. (There are many correct answers.)

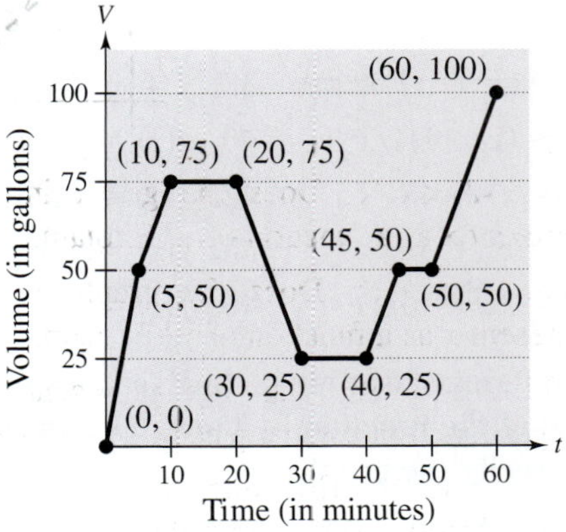

Conclusions

True or False? **In Exercises 97 and 98, determine whether the statement is true or false. Justify your answer.**

97. A function with a square root cannot have a domain that is the set of all real numbers.

98. It is possible for an odd function to have the interval $[0, \infty)$ as its domain.

Think About It **In Exercises 99–104, match the graph of the function with the description that best fits the situation.**

(a) The air temperature at a beach on a sunny day

(b) The height of a football kicked in a field goal attempt

(c) The number of children in a family over time

(d) The population of California as a function of time

(e) The depth of the tide at a beach over a 24-hour period

(f) The number of cupcakes on a tray at a party

99.

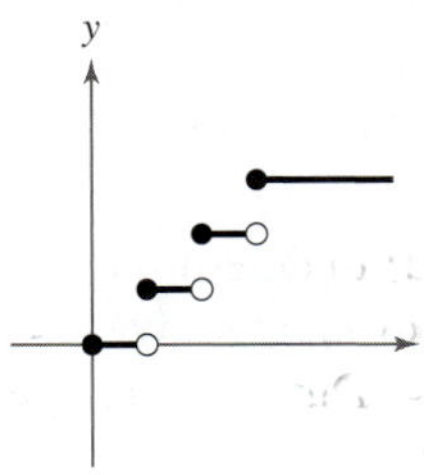

100.

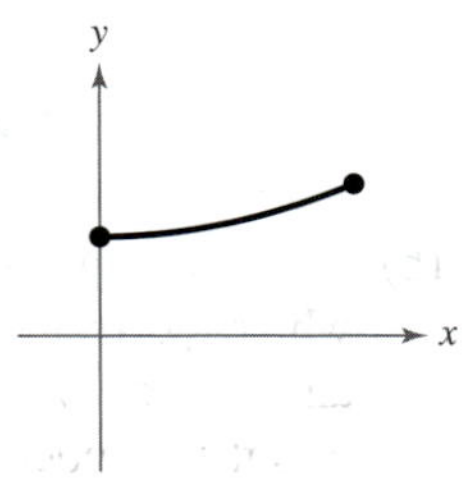

101.

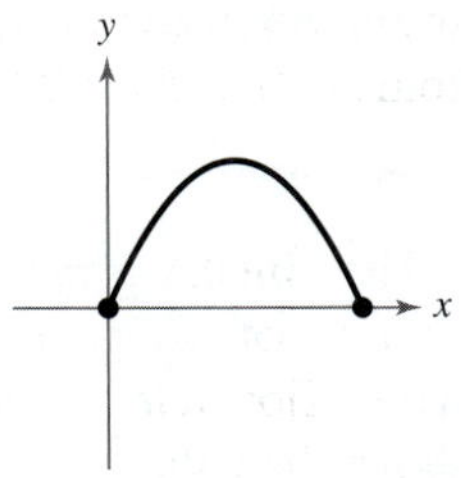

102.

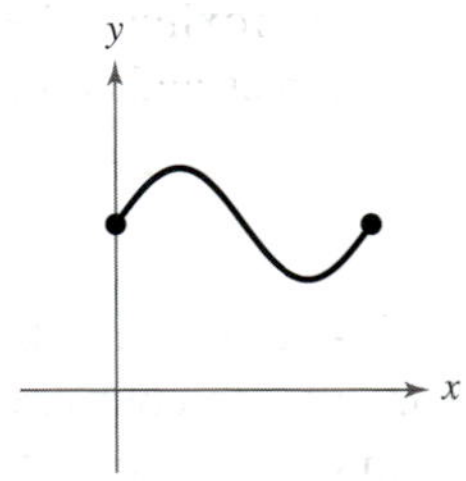

103.

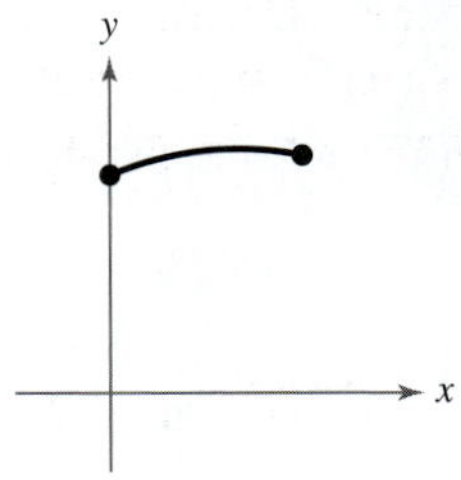

104.

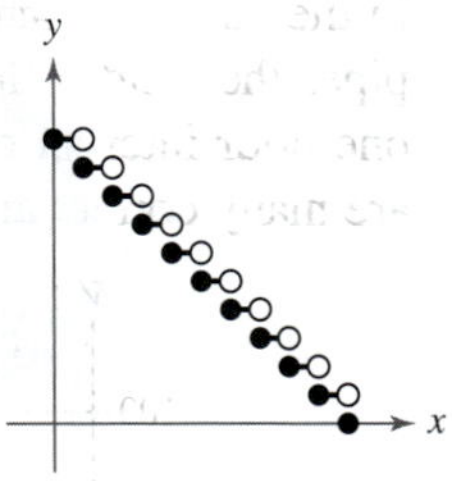

105. **Think About It** Does the graph in Exercise 19 represent x as a function of y? Explain.

106. **Think About It** Does the graph in Exercise 21 represent x as a function of y? Explain.

107. **Think About It** Can you represent the greatest integer function using a piecewise-defined function?

108. **Think About It** How does the graph of the greatest integer function differ from the graph of a line with a slope of zero?

109. **Think About It** Let f be an even function. Determine whether g is even, odd, or neither. Explain.

(a) $g(x) = -f(x)$ (b) $g(x) = f(-x)$

(c) $g(x) = f(x) - 2$ (d) $g(x) = -f(x + 3)$

110. **HOW DO YOU SEE IT?** Half of the graph of an odd function is shown.

(a) Sketch a complete graph of the function.

(b) Find the domain and range of the function.

(c) Identify the open intervals on which the function is increasing, decreasing, or constant.

(d) Find any relative minimum and relative maximum values of the function.

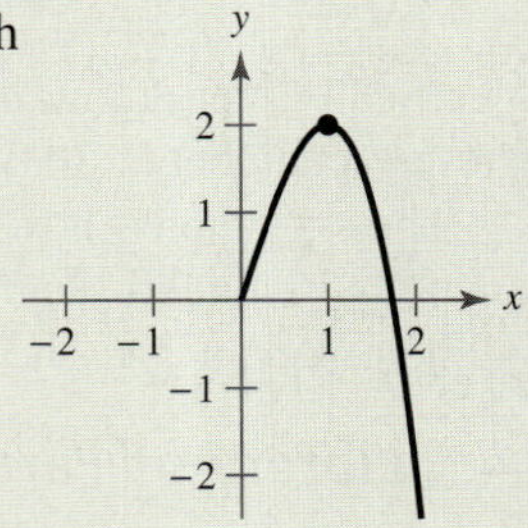

111. **Proof** Prove that a function of the following form is odd.

$$y = a_{2n+1}x^{2n+1} + a_{2n-1}x^{2n-1} + \cdots + a_3x^3 + a_1x$$

112. **Proof** Prove that a function of the following form is even.

$$y = a_{2n}x^{2n} + a_{2n-2}x^{2n-2} + \cdots + a_2x^2 + a_0$$

Cumulative Mixed Review

Identifying Terms and Coefficients **In Exercises 113–116, identify the terms. Then identify the coefficients of the variable terms of the expression.**

113. $-2x^2 + 11x + 3$ **114.** $10 + 3x$

115. $\frac{x}{3} - 5x^2 + x^3$ **116.** $7x^4 + \sqrt{2}x^2 - x$

Evaluating a Function **In Exercises 117 and 118, evaluate the function at each specified value of the independent variable and simplify.**

117. $f(x) = -x^2 - x + 3$

(a) $f(4)$ (b) $f(-5)$ (c) $f(x - 2)$

118. $f(x) = x\sqrt{x - 3}$

(a) $f(3)$ (b) $f(12)$ (c) $f(6)$

Evaluating a Difference Quotient **In Exercises 119 and 120, find the difference quotient and simplify your answer.**

119. $f(x) = x^2 - 2x + 9, \frac{f(3 + h) - f(3)}{h}, h \neq 0$

120. $f(x) = 5 + 6x - x^2, \frac{f(6 + h) - f(6)}{h}, h \neq 0$

1.5 Shifting, Reflecting, and Stretching Graphs

Summary of Graphs of Parent Functions

One of the goals of this text is to enable you to build your intuition for the basic shapes of the graphs of different types of functions. For instance, from your study of lines in Section 1.2, you can determine the basic shape of the graph of the parent linear function

$$f(x) = x.$$

Specifically, you know that the graph of this function is a line whose slope is 1 and whose y-intercept is $(0, 0)$.

The six graphs shown in Figure 1.34 represent the most commonly used types of functions in algebra. Familiarity with the basic characteristics of these parent graphs will help you analyze the shapes of more complicated graphs.

Library of Parent Functions: Commonly Used Functions

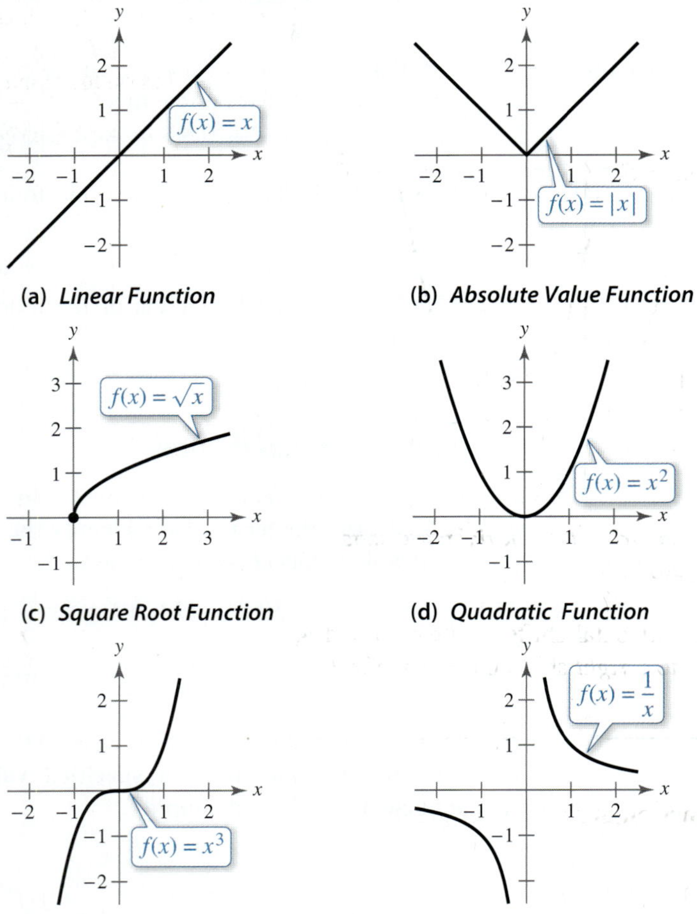

(a) *Linear Function* (b) *Absolute Value Function* (c) *Square Root Function* (d) *Quadratic Function* (e) *Cubic Function* (f) *Rational Function*

Figure 1.34

Throughout this section, you will discover how many complicated graphs are derived by shifting, stretching, shrinking, or reflecting the parent graphs shown above. Shifts, stretches, shrinks, and reflections are called *transformations*. Many graphs of functions can be created from combinations of these transformations.

What you should learn

- Recognize graphs of parent functions.
- Use vertical and horizontal shifts to sketch graphs of functions.
- Use reflections to sketch graphs of functions.
- Use nonrigid transformations to sketch graphs of functions.

Why you should learn it

Recognizing the graphs of parent functions and knowing how to shift, reflect, and stretch graphs of functions can help you sketch or describe the graphs of a wide variety of simple functions. For example, in Exercise 66 on page 130, you are asked to describe a transformation that produces the graph of a model for the sales of the WD-40 Company.

Vertical and Horizontal Shifts

Many functions have graphs that are simple transformations of the graphs of parent functions summarized in Figure 1.24. For example, you can obtain the graph of

$$h(x) = x^2 + 2$$

by shifting the graph of $f(x) = x^2$ two units *upward*, as shown in Figure 1.35. In function notation, h and f are related as follows.

$$h(x) = x^2 + 2$$
$$= f(x) + 2 \qquad \text{Upward shift of two units}$$

Similarly, you can obtain the graph of

$$g(x) = (x - 2)^2$$

by shifting the graph of $f(x) = x^2$ two units to the *right*, as shown in Figure 1.36. In this case, the functions g and f have the following relationship.

$$g(x) = (x - 2)^2$$
$$= f(x - 2) \qquad \text{Right shift of two units}$$

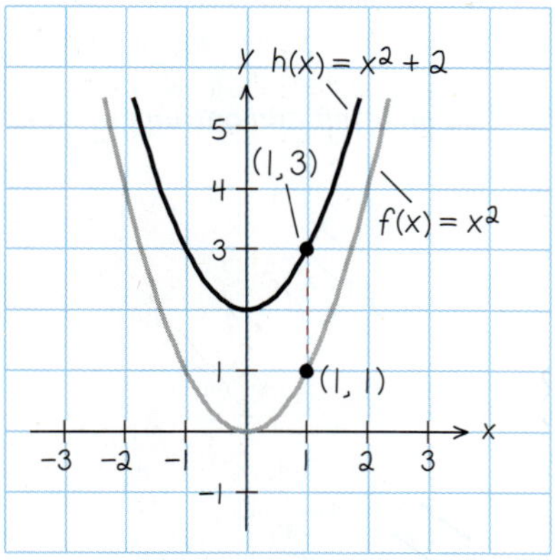

Vertical shift upward: two units
Figure 1.35

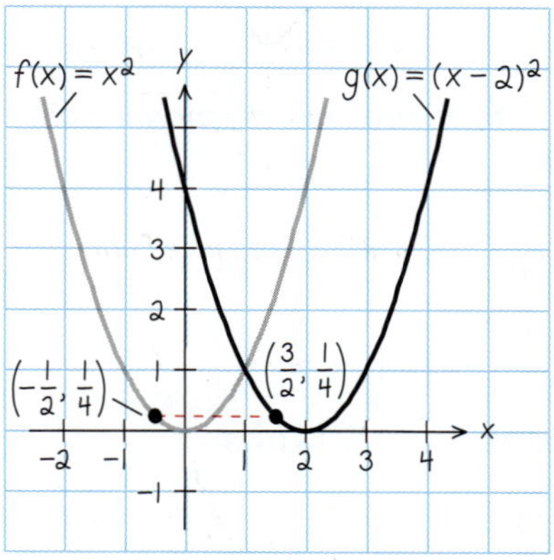

Horizontal shift to the right: two units
Figure 1.36

Explore the Concept

Use a graphing utility to display (in the same viewing window) the graphs of $y = x^2 + c$, where $c = -2, 0, 2,$ and 4. Use the results to describe the effect that c has on the graph.

Use a graphing utility to display (in the same viewing window) the graphs of $y = (x + c)^2$, where $c = -2, 0, 2,$ and 4. Use the results to describe the effect that c has on the graph.

The following list summarizes vertical and horizontal shifts. In items 3 and 4, be sure you see that $h(x) = f(x - c)$ corresponds to a *right* shift and $h(x) = f(x + c)$ corresponds to a *left* shift for $c > 0$.

Vertical and Horizontal Shifts

Let c be a positive real number. **Vertical and horizontal shifts** in the graph of $y = f(x)$ are represented as follows.

1. Vertical shift c units *upward*: $h(x) = f(x) + c$

2. Vertical shift c units *downward*: $h(x) = f(x) - c$

3. Horizontal shift c units to the *right*: $h(x) = f(x - c)$

4. Horizontal shift c units to the *left*: $h(x) = f(x + c)$

Some graphs can be obtained from combinations of vertical and horizontal shifts, as shown in Example 1(c) on the next page. Vertical and horizontal shifts generate a *family of functions*, each with a graph that has the same shape but at a different location in the plane.

EXAMPLE 1 Shifts in the Graph of a Function

a. To use the graph of $f(x) = x^3$ to sketch the graph of $g(x) = x^3 - 1$, shift the graph of f one unit downward.

b. To use the graph of $f(x) = x^3$ to sketch the graph of $h(x) = (x - 1)^3$, shift the graph of f one unit to the right.

c. To use the graph of $f(x) = x^3$ to sketch the graph of $k(x) = (x + 2)^3 + 1$, shift the graph of f two units to the left and then one unit upward.

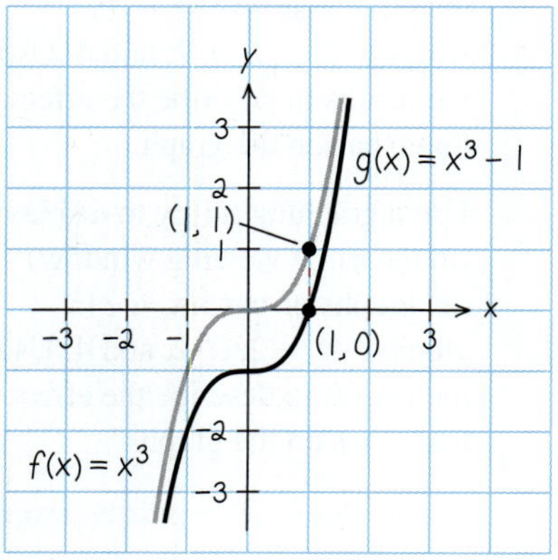

(a) ***Vertical shift: one unit downward***

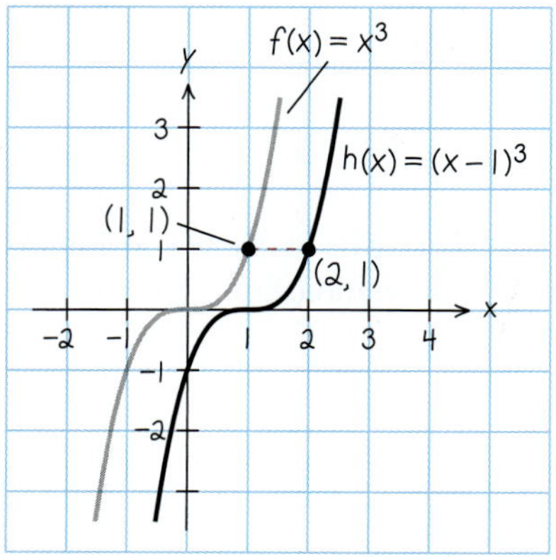

(b) ***Horizontal shift: one unit right***

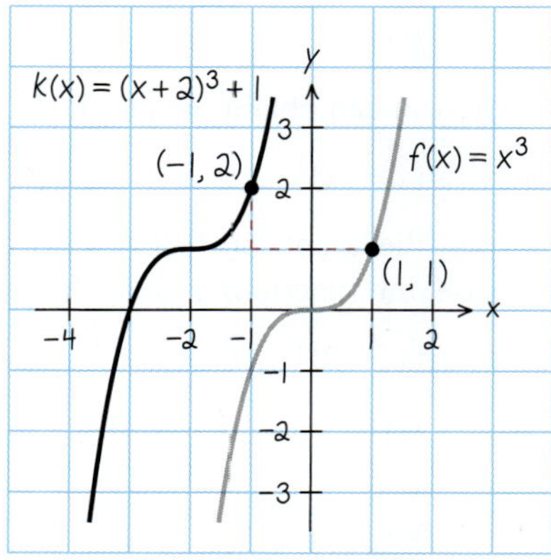

(c) ***Two units left and one unit upward***

✔ ***Checkpoint*** *Audio-video solution in English & Spanish at LarsonPrecalculus.com.*

Use the graph of $f(x) = x^3$ to sketch the graph of each function.

a. $h(x) = x^3 + 5$ **b.** $g(x) = (x - 3)^3 + 2$

EXAMPLE 2 Writing Equations from Graphs

Each graph is a transformation of the graph of $f(x) = x^2$. Write an equation for each function.

a.

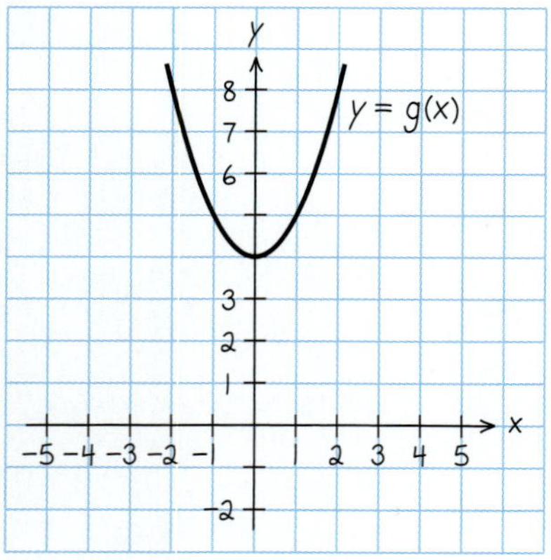

b.

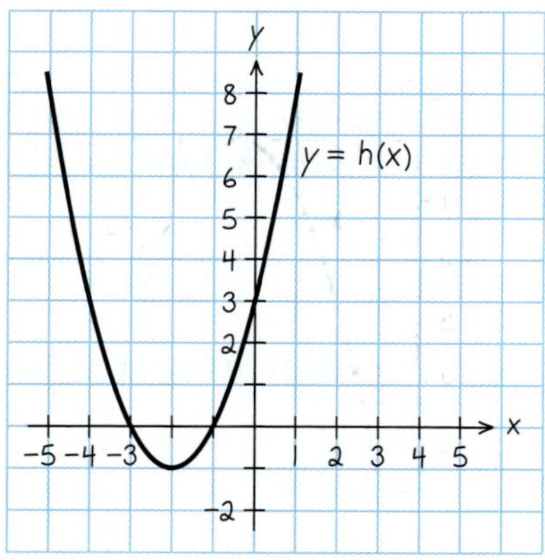

Solution

a. The graph of g is a vertical shift of four units upward of the graph of $f(x) = x^2$. So, the equation for g is $g(x) = x^2 + 4$.

b. The graph of h is a horizontal shift of two units to the left, and a vertical shift of one unit downward, of the graph of $f(x) = x^2$. So, the equation for h is $h(x) = (x + 2)^2 - 1$.

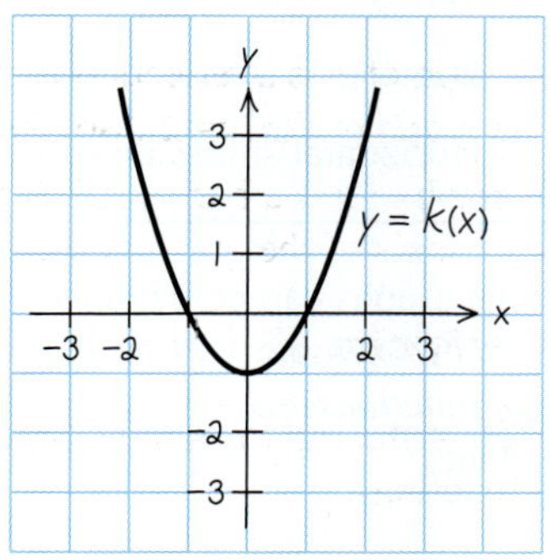

✔ ***Checkpoint*** *Audio-video solution in English & Spanish at LarsonPrecalculus.com.*

The graph at the right is a transformation of the graph of $f(x) = x^2$. Write an equation for the function.

Reflecting Graphs

Another common type of transformation is called a **reflection.** For instance, when you consider the x-axis to be a mirror, the graph of $h(x) = -x^2$ is the mirror image (or reflection) of the graph of $f(x) = x^2$, as shown in Figure 1.37.

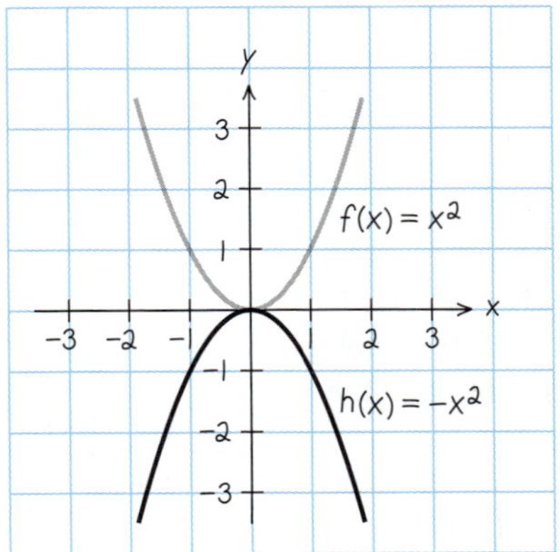

Figure 1.37

Explore the Concept

Compare the graph of each function with the graph of $f(x) = x^2$ by using a graphing utility to graph the function and f in the same viewing window. Describe the transformation.

a. $g(x) = -x^2$

b. $h(x) = (-x)^2$

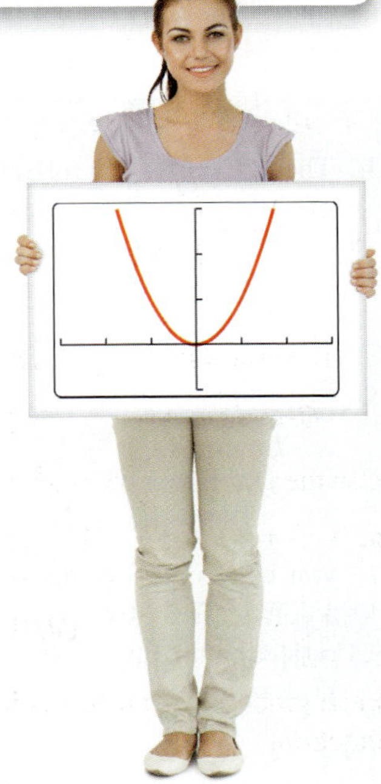

Reflections in the Coordinate Axes

Reflections in the coordinate axes of the graph of $y = f(x)$ are represented as follows.

1. Reflection in the x-axis: $h(x) = -f(x)$
2. Reflection in the y-axis: $h(x) = f(-x)$

EXAMPLE 3 Writing Equations from Graphs

Each graph is a transformation of the graph of $f(x) = x^2$. Write an equation for each function.

a.

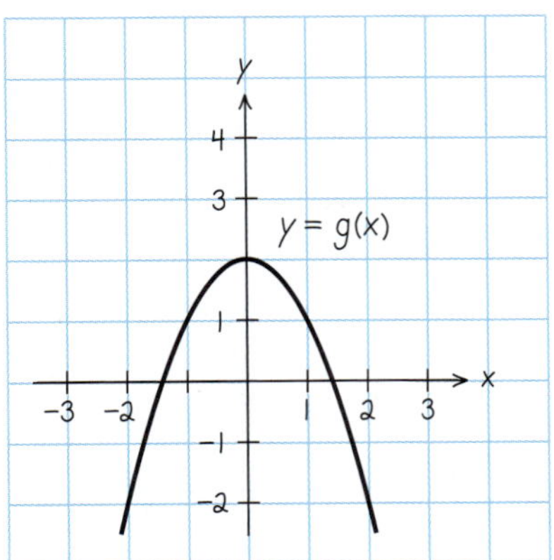

b.

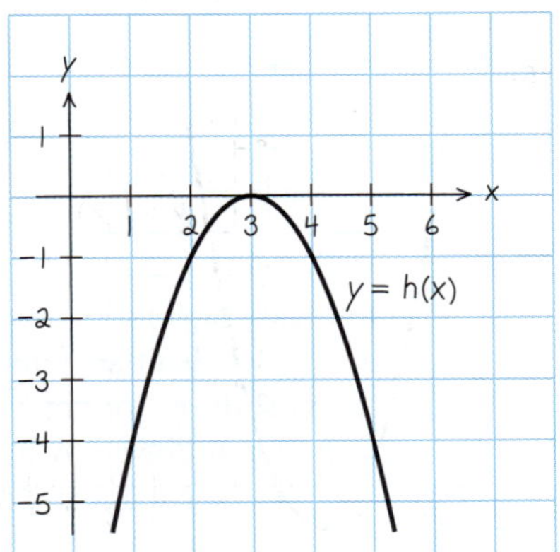

Solution

a. The graph of g is a reflection in the x-axis *followed by* an upward shift of two units of the graph of $f(x) = x^2$. So, the equation for g is $g(x) = -x^2 + 2$.

b. The graph of h is a horizontal shift of three units to the right *followed by* a reflection in the x-axis of the graph of $f(x) = x^2$. So, the equation for h is $h(x) = -(x - 3)^2$.

Checkpoint Audio-video solution in English & Spanish at LarsonPrecalculus.com.

The graph at the right is a transformation of the graph of $f(x) = x^2$. Write an equation for the function.

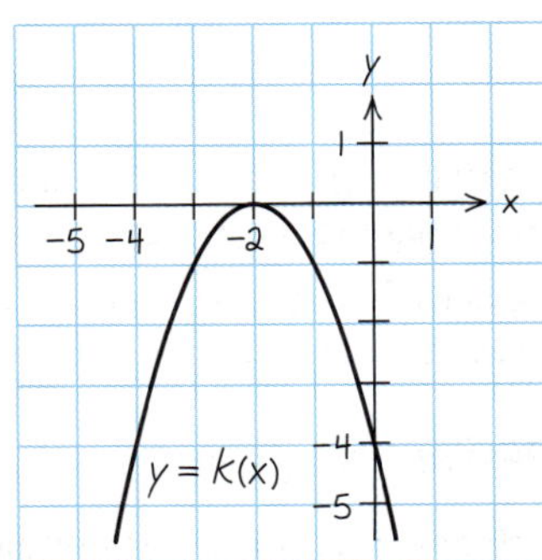

EXAMPLE 4 Reflections and Shifts

Compare the graph of each function with the graph of

$f(x) = \sqrt{x}.$

a. $g(x) = -\sqrt{x}$

b. $h(x) = \sqrt{-x}$

c. $k(x) = -\sqrt{x+2}$

Algebraic Solution

a. Relative to the graph of $f(x) = \sqrt{x}$, the graph of g is a reflection in the x-axis because

$$g(x) = -\sqrt{x}$$
$$= -f(x).$$

b. The graph of h is a reflection of the graph of $f(x) = \sqrt{x}$ in the y-axis because

$$h(x) = \sqrt{-x}$$
$$= f(-x).$$

c. From the equation

$$k(x) = -\sqrt{x+2}$$
$$= -f(x+2)$$

you can conclude that the graph of k is a left shift of two units, followed by a reflection in the x-axis, of the graph of $f(x) = \sqrt{x}$.

Graphical Solution

a. From the graph in Figure 1.38, you can see that the graph of g is a reflection of the graph of f in the x-axis. Note that the domain of g is $x \geq 0$.

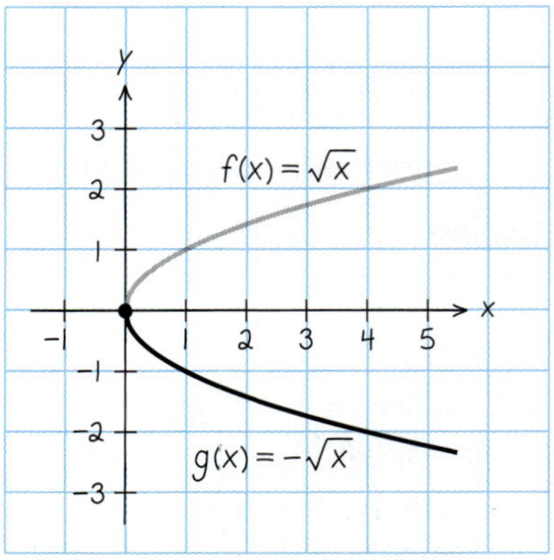

Figure 1.38

b. From the graph in Figure 1.39, you can see that the graph of h is a reflection of the graph of f in the y-axis. Note that the domain of h is $x \leq 0$.

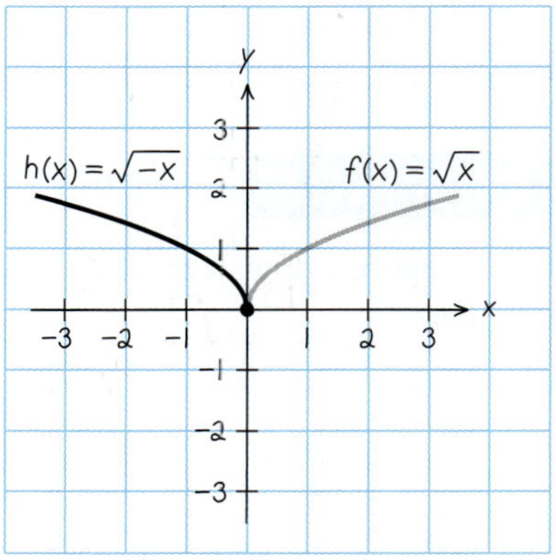

Figure 1.39

c. From the graph in Figure 1.40, you can see that the graph of k is a left shift of two units of the graph of f, followed by a reflection in the x-axis. Note that the domain of k is $x \geq -2$.

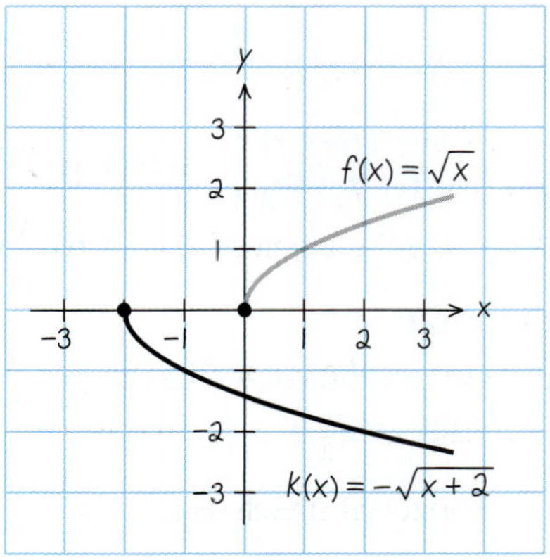

Figure 1.40

Checkpoint *Audio-video solution in English & Spanish at LarsonPrecalculus.com.*

Compare the graph of $g(x) = -|x|$ with the graph of $f(x) = |x|$.

Nonrigid Transformations

Horizontal shifts, vertical shifts, and reflections are called **rigid transformations** because the basic shape of the graph is unchanged. These transformations change only the *position* of the graph in the coordinate plane. **Nonrigid transformations** are those that cause a *distortion*—a change in the shape of the original graph. For instance, a nonrigid transformation of the graph of $y = f(x)$ is represented by $g(x) = cf(x)$, where the transformation is a **vertical stretch** when $c > 1$ and a **vertical shrink** when $0 < c < 1$. Another nonrigid transformation of the graph of $y = f(x)$ is represented by $h(x) = f(cx)$, where the transformation is a **horizontal shrink** when $c > 1$ and a **horizontal stretch** when $0 < c < 1$.

EXAMPLE 5 Nonrigid Transformations

See LarsonPrecalculus.com for an interactive version of this type of example.

Compare the graphs of (a) $h(x) = 3|x|$ and (b) $g(x) = \frac{1}{3}|x|$ with the graph of $f(x) = |x|$.

Solution

a. Relative to the graph of $f(x) = |x|$, the graph of $h(x) = 3|x| = 3f(x)$ is a vertical stretch (each y-value is multiplied by 3) of the graph of f. (See Figure 1.41.)

b. Similarly, the graph of $g(x) = \frac{1}{3}|x| = \frac{1}{3}f(x)$ is a vertical shrink $\left(\text{each } y\text{-value is multiplied by } \frac{1}{3}\right)$ of the graph of f. (See Figure 1.42.)

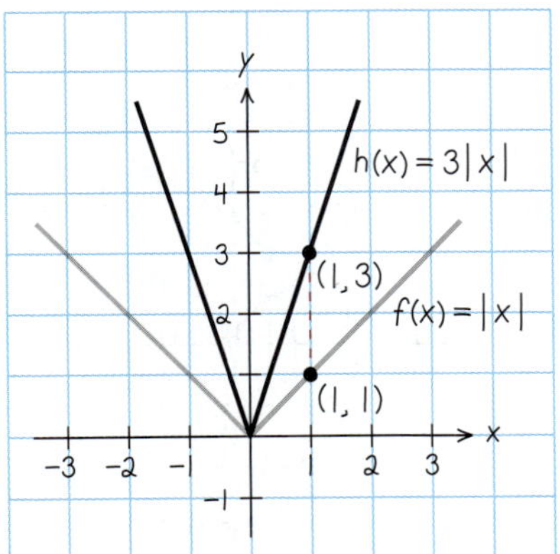

Figure 1.41

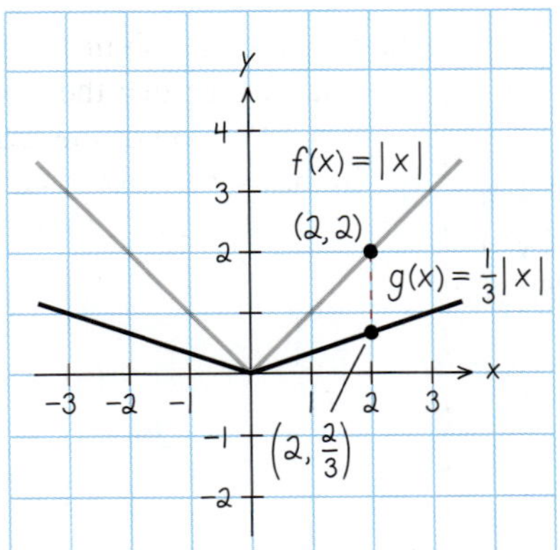

Figure 1.42

✔ **Checkpoint** *Audio-video solution in English & Spanish at LarsonPrecalculus.com.*

Compare the graphs of (a) $g(x) = 4x^2$ and (b) $h(x) = \frac{1}{4}x^2$ with the graph of $f(x) = x^2$.

EXAMPLE 6 Nonrigid Transformations

Compare the graph of $h(x) = f\left(\frac{1}{2}x\right)$ with the graph of $f(x) = 2 - x^3$.

Solution

Relative to the graph of $f(x) = 2 - x^3$, the graph of

$$h(x) = f\left(\tfrac{1}{2}x\right) = 2 - \left(\tfrac{1}{2}x\right)^3 = 2 - \tfrac{1}{8}x^3$$

is a horizontal stretch (each x-value is multiplied by 2) of the graph of f. (See Figure 1.43.)

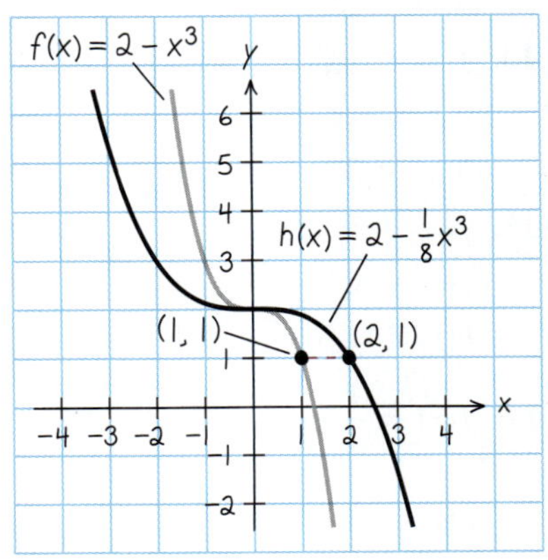

Figure 1.43

✔ **Checkpoint** *Audio-video solution in English & Spanish at LarsonPrecalculus.com.*

Compare the graphs of (a) $g(x) = f(2x)$ and (b) $h(x) = f\left(\frac{1}{2}x\right)$ with the graph of $f(x) = x^2 + 3$.

1.5 Exercises

See *CalcChat.com* for tutorial help and worked-out solutions to odd-numbered exercises. For instructions on how to use a graphing utility, see Appendix A.

Vocabulary and Concept Check

1. Name three types of rigid transformations.

2. Match the rigid transformation of $y = f(x)$ with the correct representation, where $c > 0$.

(a) $h(x) = f(x) + c$ (i) horizontal shift c units to the left

(b) $h(x) = f(x) - c$ (ii) vertical shift c units upward

(c) $h(x) = f(x - c)$ (iii) horizontal shift c units to the right

(d) $h(x) = f(x + c)$ (iv) vertical shift c units downward

In Exercises 3 and 4, fill in the blanks.

3. A reflection in the x-axis of $y = f(x)$ is represented by $h(x) =$ _______ , while a reflection in the y-axis of $y = f(x)$ is represented by $h(x) =$ _______ .

4. A nonrigid transformation of $y = f(x)$ represented by $cf(x)$ is a vertical stretch when _______ and a vertical shrink when _______ .

Procedures and Problem Solving

Sketching Transformations **In Exercises 5–18, sketch the graphs of the three functions by hand on the same rectangular coordinate system. Verify your results with a graphing utility.**

5. $f(x) = x$
$g(x) = x - 4$
$h(x) = 3x$

6. $f(x) = \frac{1}{2}x$
$g(x) = \frac{1}{2}x + 2$
$h(x) = 4(x - 2)$

7. $f(x) = x^2$
$g(x) = x^2 + 2$
$h(x) = (x - 2)^2$

8. $f(x) = x^2$
$g(x) = 3x^2$
$h(x) = (x + 2)^2 + 1$

9. $f(x) = -x^2$
$g(x) = -x^2 + 1$
$h(x) = -(x + 3)^2$

10. $f(x) = (x - 2)^2$
$g(x) = (x + 2)^2 + 2$
$h(x) = -(x - 2)^2 - 1$

11. $f(x) = x^2$
$g(x) = \frac{1}{2}x^2$
$h(x) = (2x)^2$

12. $f(x) = x^2$
$g(x) = \frac{1}{4}x^2 + 2$
$h(x) = -\frac{1}{4}x^2$

13. $f(x) = |x|$
$g(x) = |x| - 1$
$h(x) = 3|x - 3|$

14. $f(x) = |x|$
$g(x) = |2x|$
$h(x) = -2|x + 2| - 1$

15. $f(x) = -\sqrt{x}$
$g(x) = \sqrt{x + 1}$
$h(x) = \sqrt{x - 2} + 1$

16. $f(x) = \sqrt{x}$
$g(x) = \frac{1}{2}\sqrt{x}$
$h(x) = -\sqrt{x - 4}$

17. $f(x) = \dfrac{1}{x}$
$g(x) = \dfrac{1}{x} - 2$
$h(x) = \dfrac{1}{x - 1} + 2$

18. $f(x) = \dfrac{1}{x}$
$g(x) = \dfrac{1}{x} - 4$
$h(x) = \dfrac{1}{x + 3} - 1$

Sketching Transformations **In Exercises 19 and 20, use the graph of f to sketch each graph. To print an enlarged copy of the graph, go to *MathGraphs.com*.**

19. (a) $y = f(x) + 2$
(b) $y = -f(x)$
(c) $y = f(x - 2)$
(d) $y = f(x + 3)$
(e) $y = 2f(x)$
(f) $y = f(-x)$
(g) $y = f\left(\frac{1}{2}x\right)$

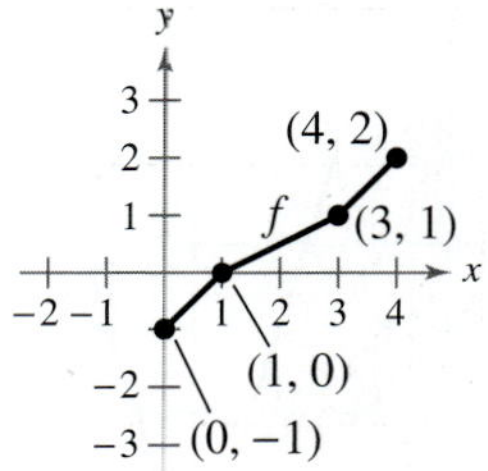

20. (a) $y = f(x) - 1$
(b) $y = f(x + 2)$
(c) $y = f(x - 1)$
(d) $y = -f(x - 2)$
(e) $y = f(-x)$
(f) $y = \frac{1}{2}f(x)$
(g) $y = f(2x)$

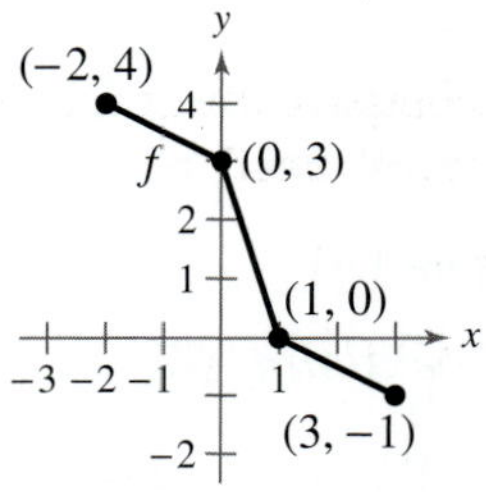

Error Analysis **In Exercises 21 and 22, describe the error in graphing the function.**

21. $f(x) = (x + 1)^2$

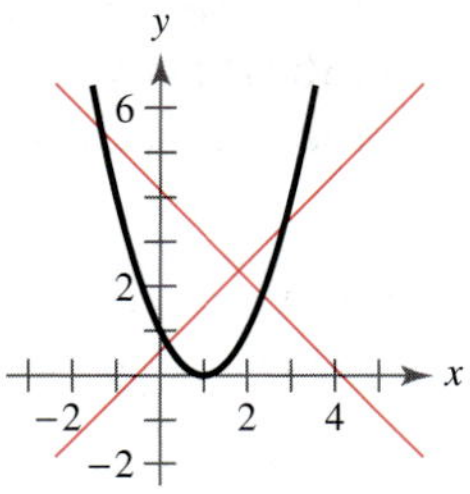

22. $f(x) = (x - 1)^2$

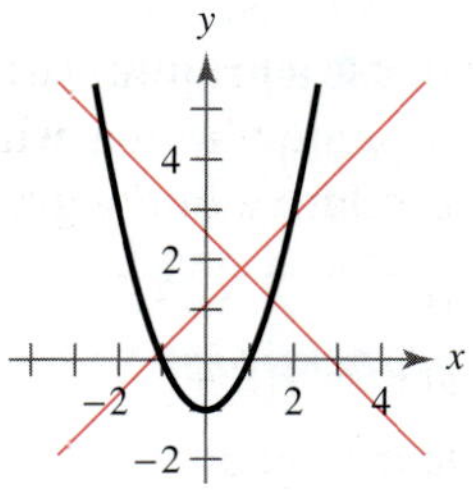

Library of Parent Functions **In Exercises 23–28, compare the graph of the function with the graph of its parent function.**

23. $y = \sqrt{x} + 2$ **24.** $y = \dfrac{1}{x} - 5$

25. $y = (x - 4)^3$ **26.** $y = |x + 5|$

27. $y = x^2 - 2$ **28.** $y = \sqrt{x - 2}$

Library of Parent Functions **In Exercises 29–34, identify the parent function and describe the transformation shown in the graph. Write an equation for the graphed function.**

29.

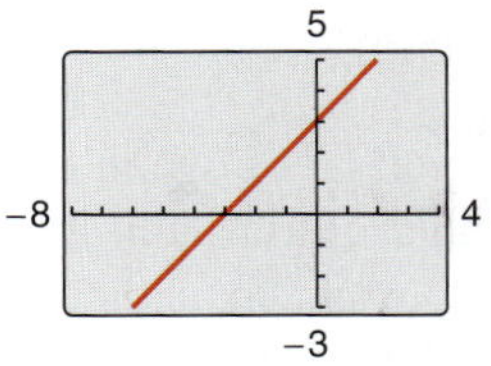

30.

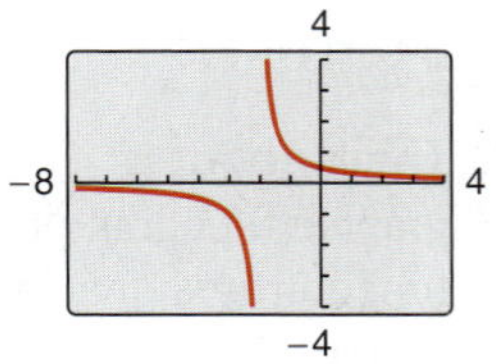

31.

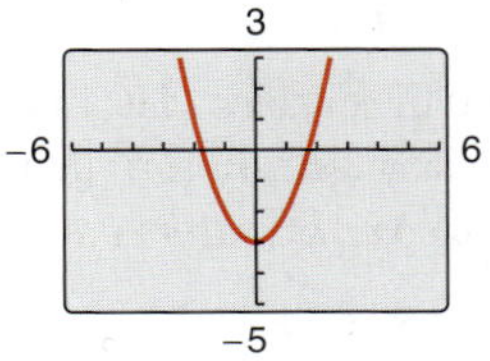

32.

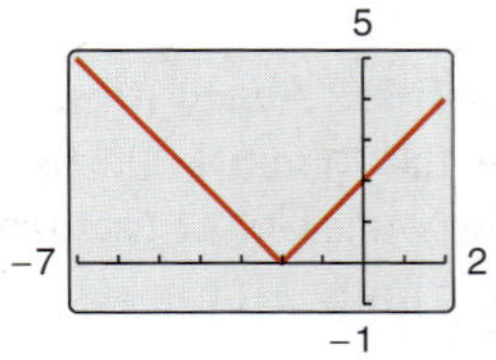

33.

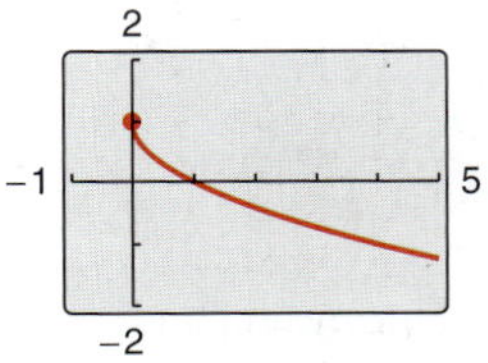

34.

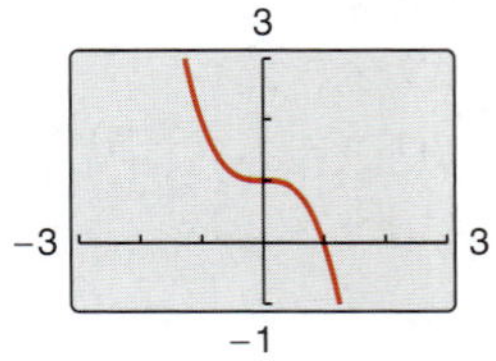

Rigid and Nonrigid Transformations **In Exercises 35–46, compare the graph of the function with the graph of its parent function.**

35. $y = -x$ **36.** $y = |-x|$

37. $y = (-x)^2$ **38.** $y = -x^3$

39. $y = \dfrac{1}{-x}$ **40.** $y = -\dfrac{1}{x}$

41. $h(x) = 4|x|$ **42.** $p(x) = \frac{1}{2}x^2$

43. $g(x) = \frac{1}{4}x^3$ **44.** $y = 2\sqrt{x}$

45. $f(x) = \sqrt{4x}$ **46.** $y = \left|\frac{1}{2}x\right|$

Rigid and Nonrigid Transformations **In Exercises 47–50, use a graphing utility to graph the three functions in the same viewing window. Describe the graphs of *g* and *h* relative to the graph of *f*.**

47. $f(x) = x^3 - 3x^2$
$g(x) = f(x + 2)$
$h(x) = \frac{1}{2}f(x)$

48. $f(x) = x^3 - 3x^2 + 2$
$g(x) = f(x - 1)$
$h(x) = f(3x)$

49. $f(x) = x^3 - 3x^2$
$g(x) = -\frac{1}{3}f(x)$
$h(x) = f(-x)$

50. $f(x) = x^3 - 3x^2 + 2$
$g(x) = -f(x)$
$h(x) = f(2x)$

Describing Transformations **In Exercises 51–64, *g* is related to one of the six parent functions on page 122. (a) Identify the parent function *f*. (b) Describe the sequence of transformations from *f* to *g*. (c) Sketch the graph of *g* by hand. (d) Use function notation to write *g* in terms of the parent function *f*.**

51. $g(x) = 2 - (x + 5)^2$ **52.** $g(x) = (x - 10)^2 + 5$

53. $g(x) = 3 + 2(x - 4)^2$ **54.** $g(x) = -\frac{1}{4}(x + 2)^2 - 2$

55. $g(x) = \frac{1}{3}(x - 2)^3$ **56.** $g(x) = \frac{1}{2}(x + 1)^3$

57. $g(x) = (x - 1)^3 + 2$

58. $g(x) = -(x + 3)^3 - 10$

59. $g(x) = \dfrac{1}{x + 8} - 9$ **60.** $g(x) = \dfrac{1}{x - 7} + 4$

61. $g(x) = -2|x - 1| - 4$ **62.** $g(x) = \frac{1}{2}|x - 2| - 3$

63. $g(x) = -\frac{1}{2}\sqrt{x + 3} - 1$ **64.** $g(x) = -3\sqrt{x + 1} - 6$

65. MODELING DATA

The numbers N (in millions) of households in the United States from 2000 through 2013 are given by the ordered pairs of the form $(t, N(t))$, where $t = 0$ represents 2000. A model for the data is

$$N(t) = -0.03(t - 26.17)^2 + 126.5.$$

(Source: U.S. Census Bureau)

(0, 104.7)
(1, 108.2)
(2, 109.3)
(3, 111.3)
(4, 112.0)
(5, 113.3)
(6, 114.4)
(7, 116.0)
(8, 116.8)
(9, 117.2)
(10, 117.5)
(11, 119.9)
(12, 121.1)
(13, 122.5)

(a) Describe the transformation of the parent function $f(t) = t^2$.

(b) Use a graphing utility to graph the model and the data in the same viewing window.

(c) Rewrite the function so that $t = 0$ represents 2009. Explain how you got your answer.

66. ***Why you should learn it*** *(p. 122)* The depreciation D (in millions of dollars) of the WD-40 Company assets from 2009 through 2013 can be approximated by the function

$$D(t) = 1.9\sqrt{t + 3.7}$$

where $t = 0$ represents 2009. (Source: WD-40 Company)

(a) Describe the transformation of the parent function $f(t) = \sqrt{t}$.

(b) Use a graphing utility to graph the model over the interval $0 \le t \le 4$.

(c) According to the model, in what year will the depreciation of WD-40 assets be approximately 6 million dollars?

(d) Rewrite the function so that $t = 0$ represents 2011. Explain how you got your answer.

Conclusions

True or False? **In Exercises 67 and 68, determine whether the statement is true or false. Justify your answer.**

67. The graph of $y = f(-x)$ is a reflection of the graph of $y = f(x)$ in the x-axis.

68. The graphs of $f(x) = |x| + 6$ and $f(x) = |-x| + 6$ are identical.

Exploration **In Exercises 69–72, use the fact that the graph of $y = f(x)$ has x-intercepts at $x = 2$ and $x = -3$ to find the x-intercepts of the given graph. If not possible, state the reason.**

69. $y = f(-x)$

70. $y = 2f(x)$

71. $y = f(x) + 2$

72. $y = f(x - 3)$

Library of Parent Functions **In Exercises 73–76, determine which equation(s) may be represented by the graph shown. (There may be more than one correct answer.)**

73.

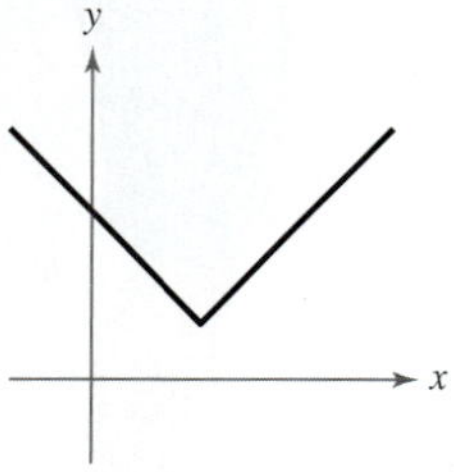

(a) $f(x) = |x + 2| + 1$
(b) $f(x) = |x - 1| + 2$
(c) $f(x) = |x - 2| + 1$
(d) $f(x) = 2 + |x - 2|$
(e) $f(x) = |(x - 2) + 1|$
(f) $f(x) = 1 - |x - 2|$

74.

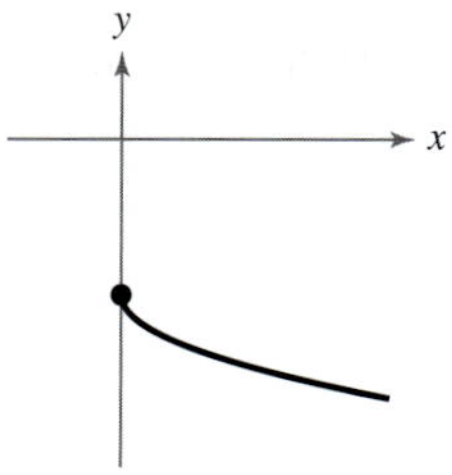

(a) $f(x) = -\sqrt{x} - 4$
(b) $f(x) = -4 - \sqrt{x}$
(c) $f(x) = -4 - \sqrt{-x}$
(d) $f(x) = \sqrt{-x} - 4$
(e) $f(x) = \sqrt{-x} + 4$
(f) $f(x) = \sqrt{x} - 4$

75.

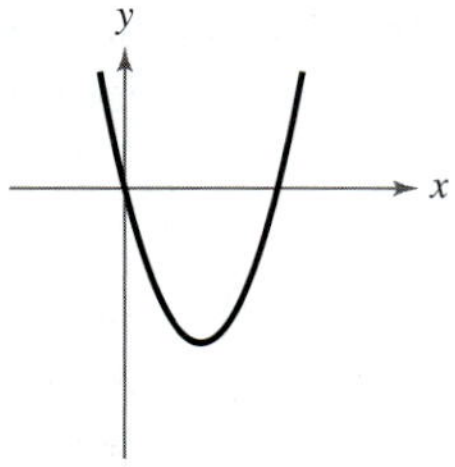

(a) $f(x) = (x - 2)^2 - 2$
(b) $f(x) = (x + 4)^2 - 4$
(c) $f(x) = (x - 2)^2 - 4$
(d) $f(x) = (x + 2)^2 - 4$
(e) $f(x) = 4 - (x - 2)^2$
(f) $f(x) = 4 - (x + 2)^2$

76.

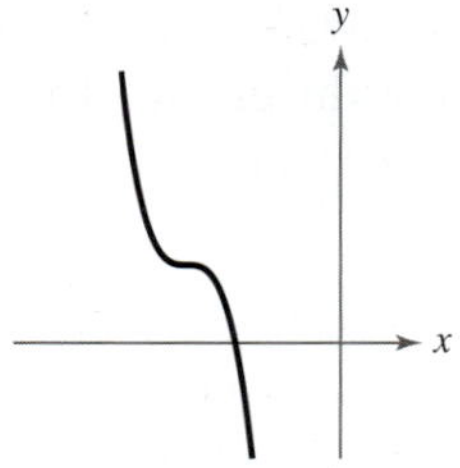

(a) $f(x) = -(x - 4)^3 + 2$
(b) $f(x) = -(x + 4)^3 + 2$
(c) $f(x) = -(x - 2)^3 + 4$
(d) $f(x) = (-x - 4)^3 + 2$
(e) $f(x) = (x + 4)^3 + 2$
(f) $f(x) = (-x + 4)^3 + 2$

77. **Think About It** You can use either of two methods to graph a function: plotting points, or translating a parent function as shown in this section. Which method do you prefer to use for each function? Explain.

(a) $f(x) = 3x^2 - 4x + 1$ (b) $f(x) = 2(x - 1)^2 - 6$

78. **Think About It** The graph of $y = f(x)$ passes through the points $(0, 1)$, $(1, 2)$, and $(2, 3)$. Find the corresponding points on the graph of $y = f(x + 2) - 1$.

79. **Think About It** Compare the graph of $g(x) = ax^2$ with the graph of $f(x) = x^2$ when (a) $0 < a < 1$ and (b) $a > 1$.

80. **HOW DO YOU SEE IT?** Use the graph of $y = f(x)$ to find the intervals on which each of the graphs in (a)–(c) is increasing and decreasing. If not possible, then state the reason.

(a) $y = -f(x)$
(b) $y = f(x) - 3$
(c) $y = f(x - 1)$

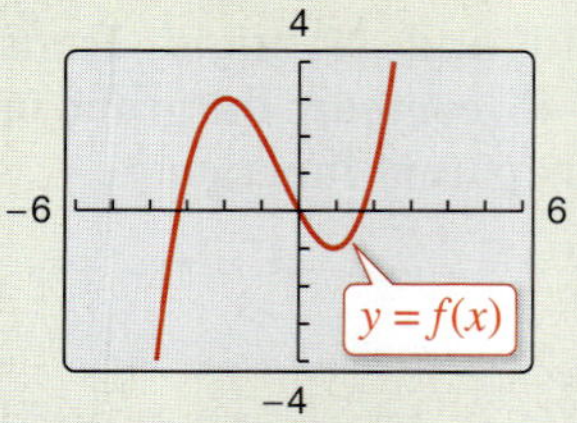

Cumulative Mixed Review

Parallel and Perpendicular Lines **In Exercises 81 and 82, determine whether the lines L_1 and L_2 passing through the pairs of points are parallel, perpendicular, or neither.**

81. L_1: $(-2, -2), (2, 10)$
L_2: $(-1, 3), (3, 9)$

82. L_1: $(-1, -7), (4, 3)$
L_2: $(1, 5), (-2, -7)$

Finding the Domain of a Function **In Exercises 83–86, find the domain of the function.**

83. $f(x) = \dfrac{4}{9 - x}$

84. $f(x) = \dfrac{\sqrt{x - 5}}{x - 7}$

85. $f(x) = \sqrt{100 - x^2}$

86. $f(x) = \sqrt[3]{16 - x^2}$

1.6 Combinations of Functions

Arithmetic Combinations of Functions

Just as two real numbers can be combined by the operations of addition, subtraction, multiplication, and division to form other real numbers, two *functions* can be combined to create new functions. When $f(x) = 2x - 3$ and $g(x) = x^2 - 1$, you can form the sum, difference, product, and quotient of f and g as follows.

$$f(x) + g(x) = (2x - 3) + (x^2 - 1) = x^2 + 2x - 4 \quad \text{Sum}$$

$$f(x) - g(x) = (2x - 3) - (x^2 - 1) = -x^2 + 2x - 2 \quad \text{Difference}$$

$$f(x) \cdot g(x) = (2x - 3)(x^2 - 1) = 2x^3 - 3x^2 - 2x + 3 \quad \text{Product}$$

$$\frac{f(x)}{g(x)} = \frac{2x - 3}{x^2 - 1}, \quad x \neq \pm 1 \quad \text{Quotient}$$

The domain of an **arithmetic combination** of functions f and g consists of all real numbers that are common to the domains of f and g. In the case of the quotient

$$\frac{f(x)}{g(x)}$$

there is the further restriction that $g(x) \neq 0$.

What you should learn

- Add, subtract, multiply, and divide functions.
- Find compositions of one function with another function.
- Use combinations of functions to model and solve real-life problems.

Why you should learn it

You can model some situations by combining functions. For instance, in Exercise 79 on page 138, you will model the stopping distance of a car by combining the driver's reaction time with the car's braking distance.

Sum, Difference, Product, and Quotient of Functions

Let f and g be two functions with overlapping domains. Then, for all x common to both domains, the sum, difference, product, and quotient of f and g are defined as follows.

1. **Sum:** $(f + g)(x) = f(x) + g(x)$
2. **Difference:** $(f - g)(x) = f(x) - g(x)$
3. **Product:** $(fg)(x) = f(x) \cdot g(x)$
4. **Quotient:** $\left(\frac{f}{g}\right)(x) = \frac{f(x)}{g(x)}, \quad g(x) \neq 0$

EXAMPLE 1 Finding the Sum of Two Functions

Given $f(x) = 2x + 1$ and $g(x) = x^2 + 2x - 1$, find $(f + g)(x)$. Then evaluate the sum when $x = 2$.

Solution

$$\begin{aligned}(f + g)(x) &= f(x) + g(x) \\ &= (2x + 1) + (x^2 + 2x - 1) \\ &= x^2 + 4x\end{aligned}$$

When $x = 2$, the value of this sum is $(f + g)(2) = 2^2 + 4(2) = 12$.

Checkpoint *Audio-video solution in English & Spanish at LarsonPrecalculus.com.*

Given $f(x) = x^2$ and $g(x) = 1 - x$, find $(f + g)(x)$. Then evaluate the sum when $x = 2$.

EXAMPLE 2 Finding the Difference of Two Functions

Given $f(x) = 2x + 1$ and $g(x) = x^2 + 2x - 1$, find $(f - g)(x)$. Then evaluate the difference when $x = 2$.

Algebraic Solution

The difference of the functions f and g is

$$(f - g)(x) = f(x) - g(x)$$
$$= (2x + 1) - (x^2 + 2x - 1)$$
$$= -x^2 + 2.$$

When $x = 2$, the value of this difference is

$$(f - g)(2) = -(2)^2 + 2$$
$$= -2.$$

Graphical Solution

Enter $y_1 = f(x)$, $y_2 = g(x)$, and $y_3 = f(x) - g(x)$. Then graph the difference of the two functions, y_3.

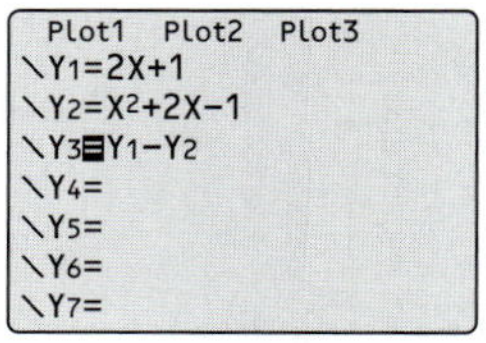

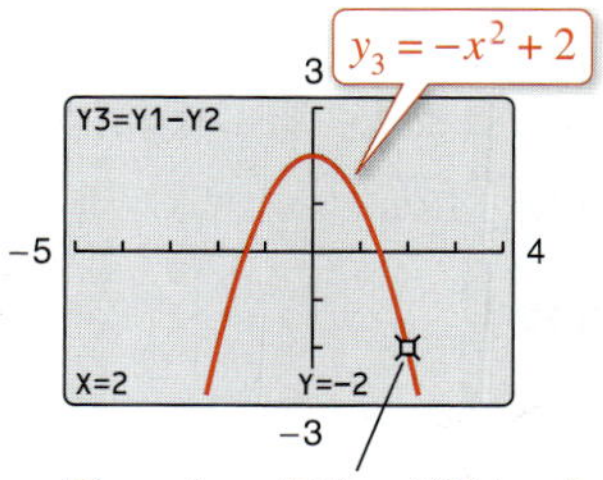

The value of $(f - g)(2)$ is -2.

✓ Checkpoint Audio-video solution in English & Spanish at LarsonPrecalculus.com.

Given $f(x) = x^2$ and $g(x) = 1 - x$, find $(f - g)(x)$. Then evaluate the difference when $x = 3$.

EXAMPLE 3 Finding the Product of Two Functions

Given $f(x) = x^2$ and $g(x) = x - 3$, find $(fg)(x)$. Then evaluate the product when $x = 4$.

Solution

$$(fg)(x) = f(x)g(x) = (x^2)(x - 3) = x^3 - 3x^2$$

When $x = 4$, the value of this product is $(fg)(4) = 4^3 - 3(4)^2 = 16$.

✓ Checkpoint Audio-video solution in English & Spanish at LarsonPrecalculus.com.

Given $f(x) = x^2$ and $g(x) = 1 - x$, find $(fg)(x)$. Then evaluate the product when $x = 3$.

In Examples 1–3, both f and g have domains that consist of all real numbers. So, the domains of $(f + g)$ and $(f - g)$ are also the set of all real numbers. Remember to consider any restrictions on the domains of f or g when forming the sum, difference, product, or quotient of f and g. For instance, the domain of $f(x) = 1/x$ is all $x \neq 0$, and the domain of $g(x) = \sqrt{x}$ is $[0, \infty)$. This implies that the domain of $(f + g)$ is $(0, \infty)$.

EXAMPLE 4 Finding the Quotient of Two Functions

Given $f(x) = \sqrt{x}$ and $g(x) = \sqrt{4 - x^2}$, find $(f/g)(x)$. Then find the domain of f/g.

Solution

$$\left(\frac{f}{g}\right)(x) = \frac{f(x)}{g(x)} = \frac{\sqrt{x}}{\sqrt{4 - x^2}}$$

The domain of f is $[0, \infty)$ and the domain of g is $[-2, 2]$. The intersection of these domains is $[0, 2]$. So, the domain of f/g is $[0, 2)$.

✓ Checkpoint Audio-video solution in English & Spanish at LarsonPrecalculus.com.

Find $(f/g)(x)$ and $(g/f)(x)$ for the functions $f(x) = \sqrt{x - 3}$ and $g(x) = \sqrt{16 - x^2}$. Then find the domains of f/g and g/f.

Compositions of Functions

Another way of combining two functions is to form the **composition** of one with the other. For instance, when $f(x) = x^2$ and $g(x) = x + 1$, the composition of f with g is

$$\begin{aligned} f(g(x)) &= f(x + 1) \\ &= (x + 1)^2. \end{aligned}$$

This composition is denoted as $f \circ g$ and is read as "f composed with g."

Definition of Composition of Two Functions

The **composition** of the function f with the function g is

$$(f \circ g)(x) = f(g(x)).$$

The domain of $f \circ g$ is the set of all x in the domain of g such that $g(x)$ is in the domain of f. (See Figure 1.44.)

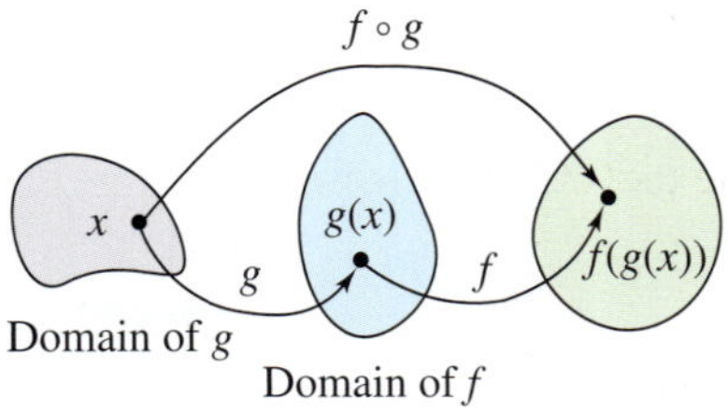

Figure 1.44

EXAMPLE 5 Forming the Composition of f with g

Find $(f \circ g)(x)$ for $f(x) = \sqrt{x}, x \geq 0$, and $g(x) = x - 1, x \geq 1$. If possible, find $(f \circ g)(2)$ and $(f \circ g)(0)$.

Solution

The composition of f with g is

$$\begin{aligned} (f \circ g)(x) &= f(g(x)) && \text{Definition of } f \circ g \\ &= f(x - 1) && \text{Definition of } g(x) \\ &= \sqrt{x - 1},\ x \geq 1. && \text{Definition of } f(x) \end{aligned}$$

The domain of $(f \circ g)$ is $[1, \infty)$ (see Figure 1.45). So,

$$(f \circ g)(2) = \sqrt{2 - 1} = 1$$

is defined, but $(f \circ g)(0)$ is not defined because 0 is not in the domain of $f \circ g$.

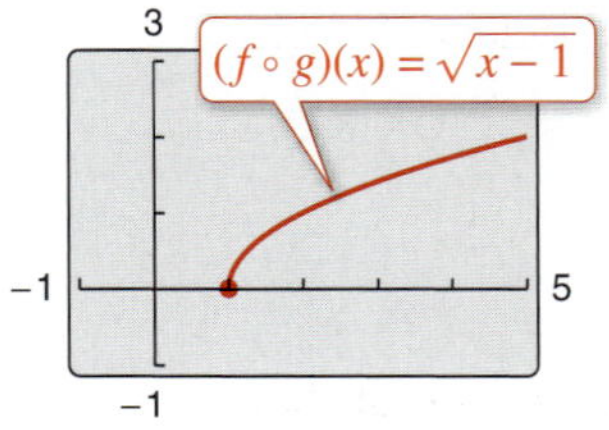

Figure 1.45

Checkpoint Audio-video solution in English & Spanish at LarsonPrecalculus.com.

Find $(f \circ g)(x)$ for $f(x) = x^2$ and $g(x) = x - 1$. If possible, find $(f \circ g)(0)$.

Explore the Concept

Let $f(x) = x + 2$ and $g(x) = 4 - x^2$. Are the compositions $f \circ g$ and $g \circ f$ equal? You can use your graphing utility to answer this question by entering and graphing the following functions.

$$y_1 = (4 - x^2) + 2$$

$$y_2 = 4 - (x + 2)^2$$

What do you observe? Which function represents $f \circ g$ and which represents $g \circ f$?

The composition of f with g is generally not the same as the composition of g with f. This is illustrated in Example 6.

EXAMPLE 6 Compositions of Functions

See LarsonPrecalculus.com for an interactive version of this type of example.

Given $f(x) = x + 2$ and $g(x) = 4 - x^2$, evaluate (a) $(f \circ g)(x)$ and (b) $(g \circ f)(x)$ when $x = 0$ and 1.

Algebraic Solution

a. $(f \circ g)(x) = f(g(x))$ Definition of $f \circ g$

$= f(4 - x^2)$ Definition of $g(x)$

$= (4 - x^2) + 2$ Definition of $f(x)$

$= -x^2 + 6$

$(f \circ g)(0) = -0^2 + 6 = 6$

$(f \circ g)(1) = -1^2 + 6 = 5$

b. $(g \circ f)(x) = g(f(x))$ Definition of $g \circ f$

$= g(x + 2)$ Definition of $f(x)$

$= 4 - (x + 2)^2$ Definition of $g(x)$

$= 4 - (x^2 + 4x + 4)$

$= -x^2 - 4x$

$(g \circ f)(0) = -0^2 - 4(0) = 0$

$(g \circ f)(1) = -1^2 - 4(1) = -5$

Note that $(f \circ g) \neq g \circ f$.

Numerical Solution

a. and b. Enter $y_1 = f(x)$, $y_2 = g(x)$, $y_3(f \circ g)(x)$, and $y_4 = (g \circ f)(x)$. Then use the *table* feature to find the desired function values.

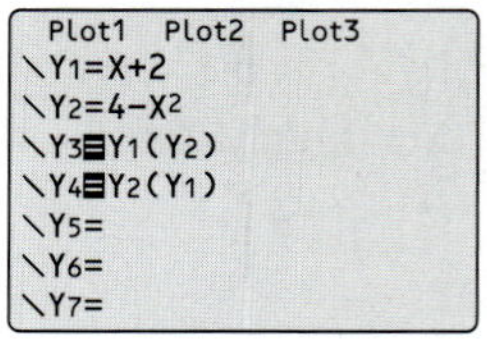

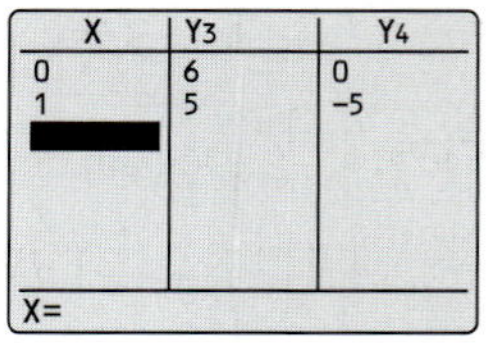

From the table, you can see that $f \circ g \neq g \circ f$.

✓ Checkpoint *Audio-video solution in English & Spanish at LarsonPrecalculus.com.*

Given $f(x) = 2x + 5$ and $g(x) = 4x^2 + 1$, find the following.

a. $(f \circ g)(x)$ **b.** $(g \circ f)(x)$ **c.** $(f \circ g)\left(-\frac{1}{2}\right)$

EXAMPLE 7 Finding the Domain of a Composite Function

Find the domain of $f \circ g$ for the functions $f(x) = x^2 - 9$ and $g(x) = \sqrt{9 - x^2}$.

Algebraic Solution

The composition of the functions is as follows.

$$(f \circ g)(x) = f(g(x))$$
$$= f\left(\sqrt{9 - x^2}\right)$$
$$= \left(\sqrt{9 - x^2}\right)^2 - 9$$
$$= 9 - x^2 - 9$$
$$= -x^2$$

From this, it might appear that the domain of the composition is the set of all real numbers. This, however, is not true. Because the domain of f is the set of all real numbers and the domain of g is $[-3, 3]$, the domain of $f \circ g$ is $[-3, 3]$.

Graphical Solution

The x-coordinates of points on the graph extend from -3 to 3. So, the domain of $f \circ g$ is $[-3, 3]$.

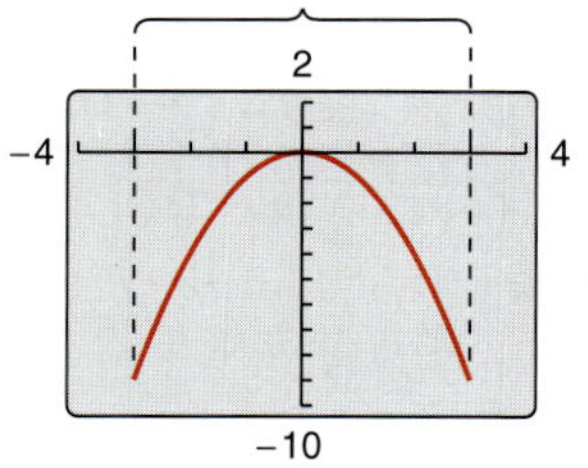

✓ Checkpoint *Audio-video solution in English & Spanish at LarsonPrecalculus.com.*

Find the domain of $(f \circ g)(x)$ for the functions $f(x) = \sqrt{x}$ and $g(x) = x^2 + 4$.

EXAMPLE 8 A Case in Which $f \circ g = g \circ f$

Given $f(x) = 2x + 3$ and $g(x) = \frac{1}{2}(x - 3)$, find (a) $(f \circ g)(x)$ and (b) $(g \circ f)(x)$.

Solution

a. $(f \circ g)(x) = f(g(x))$

$$= f\left(\frac{1}{2}(x - 3)\right)$$
$$= 2\left[\frac{1}{2}(x - 3)\right] + 3$$
$$= x - 3 + 3$$
$$= x$$

b. $(g \circ f)(x) = g(f(x))$

$$= g(2x + 3)$$
$$= \frac{1}{2}[(2x + 3) - 3]$$
$$= \frac{1}{2}(2x)$$
$$= x$$

Checkpoint Audio-video solution in English & Spanish at LarsonPrecalculus.com.

Given $f(x) = x^{1/3}$ and $g(x) = x^6$, find (a) $(f \circ g)(x)$ and (b) $(g \circ f)(x)$.

In Examples 5–8, you formed the composition of two given functions. In calculus, it is also important to be able to identify two functions that make up a given composite function. Basically, to "decompose" a composite function, look for an "inner" and an "outer" function.

EXAMPLE 9 Identifying a Composite Function

Write the function

$$h(x) = \frac{1}{(x - 2)^2}$$

as a composition of two functions.

Solution

One way to write h as a composition of two functions is to take the inner function to be $g(x) = x - 2$ and the outer function to be

$$f(x) = \frac{1}{x^2} = x^{-2}.$$

Then you can write

$$h(x) = \frac{1}{(x - 2)^2} = (x - 2)^{-2} = f(x - 2) = f(g(x)).$$

Checkpoint Audio-video solution in English & Spanish at LarsonPrecalculus.com.

Write the function

$$h(x) = \frac{\sqrt[3]{8 - x}}{5}$$

as a composition of two functions.

Remark

In Example 8, note that the two composite functions $f \circ g$ and $g \circ f$ are equal, and both represent the identity function. That is, $(f \circ g)(x) = x$ and $(g \circ f)(x) = x$. You will study this special case in the next section.

Explore the Concept

Write each function as a composition of two functions.

a. $h(x) = |x^3 - 2|$

b. $r(x) = |x^3| - 2$

What do you notice about the inner and outer functions?

Application

EXAMPLE 10 Bacteria Count

The number N of bacteria in a refrigerated petri dish is given by

$$N(T) = 20T^2 - 80T + 500, \quad 2 \le T \le 14$$

where T is the temperature of the petri dish (in degrees Celsius). When the petri dish is removed from refrigeration, the temperature of the petri dish is given by

$$T(t) = 4t + 2, \quad 0 \le t \le 3$$

where t is the time (in hours).

a. Find the composition $N(T(t))$ and interpret its meaning in context.

b. Find the number of bacteria in the petri dish when $t = 2$ hours.

c. Find the time when the bacteria count reaches 2000.

Microbiologist

Solution

a.
$$\begin{aligned} N(T(t)) &= 20(4t + 2)^2 - 80(4t + 2) + 500 \\ &= 20(16t^2 + 16t + 4) - 320t - 160 + 500 \\ &= 320t^2 + 320t + 80 - 320t - 160 + 500 \\ &= 320t^2 + 420 \end{aligned}$$

The composite function $N(T(t))$ represents the number of bacteria as a function of the amount of time the petri dish has been out of refrigeration.

b. When $t = 2$, the number of bacteria is

$$N = 320(2)^2 + 420 = 1280 + 420 = 1700.$$

c. The bacteria count will reach $N = 2000$ when $320t^2 + 420 = 2000$. You can solve this equation for t algebraically as follows.

$$\begin{aligned} 320t^2 + 420 &= 2000 \\ 320t^2 &= 1580 \\ t^2 &= \frac{79}{16} \\ t &= \frac{\sqrt{79}}{4} \quad \Rightarrow \quad t \approx 2.22 \text{ hours} \end{aligned}$$

So, the count will reach 2000 when $t \approx 2.22$ hours. Note that the negative value is rejected because it is not in the domain of the composite function. To confirm your solution, graph the equation $N = 320t^2 + 420$, as shown in Figure 1.46. Then use the *zoom* and *trace* features to approximate $N = 2000$ when $t \approx 2.22$, as shown in Figure 1.47.

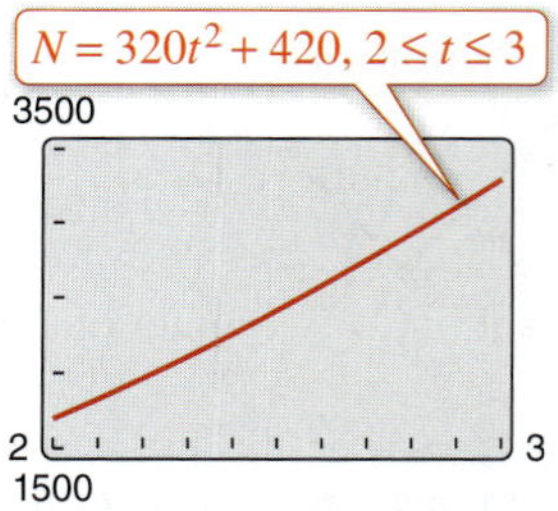

Figure 1.46

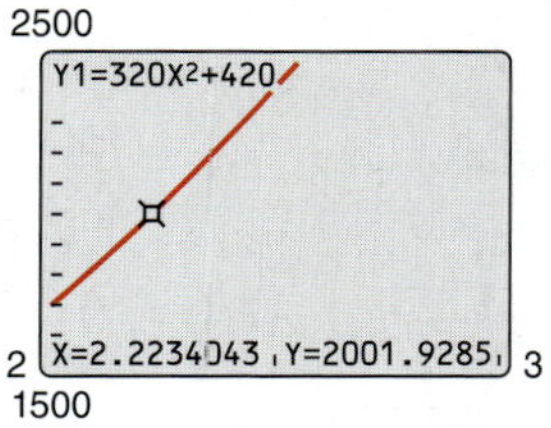

Figure 1.47

✓ Checkpoint *Audio-video solution in English & Spanish at LarsonPrecalculus.com.*

The number N of bacteria in a refrigerated food is given by

$$N(T) = 8T^2 - 14T + 200, \quad 2 \le T \le 12$$

where T is the temperature of the food in degrees Celsius. When the food is removed from refrigeration, the temperature of the food is given by

$$T(t) = 2t + 2, \quad 0 \le t \le 5$$

where t is the time in hours. Find (a) $(N \circ T)(t)$ and (b) the time when the bacteria count reaches 1000.

1.6 Exercises

See *CalcChat.com* for tutorial help and worked-out solutions to odd-numbered exercises.
For instructions on how to use a graphing utility, see Appendix A.

Vocabulary and Concept Check

In Exercises 1–4, fill in the blank(s).

1. Two functions f and g can be combined by the arithmetic operations of _______ , _______ , _______ , and _______ to create new functions.
2. The _______ of the function f with the function g is $(f \circ g)(x) = f(g(x))$.
3. The domain of $f \circ g$ is the set of all x in the domain of g such that _______ is in the domain of f.
4. To decompose a composite function, look for an _______ and an _______ function.
5. Given $f(x) = x^2 + 1$ and $(fg)(x) = 2x(x^2 + 1)$, what is $g(x)$?
6. Given $(f \circ g)(x) = f(x^2 + 1)$, what is $g(x)$?

Procedures and Problem Solving

Graphing the Sum of Two Functions **In Exercises 7–10, use the graphs of f and g to graph $h(x) = (f + g)(x)$. To print an enlarged copy of the graph, go to *MathGraphs.com*.**

7.

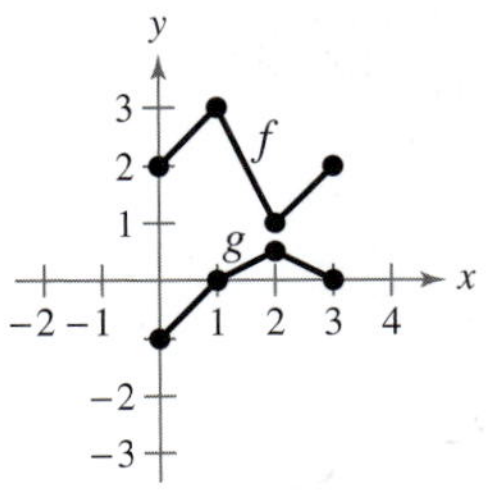

8.

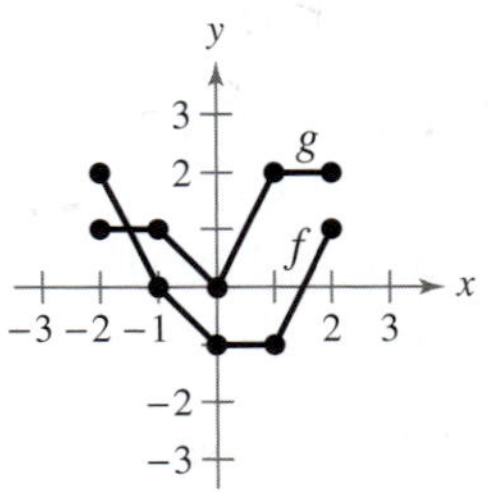

9.

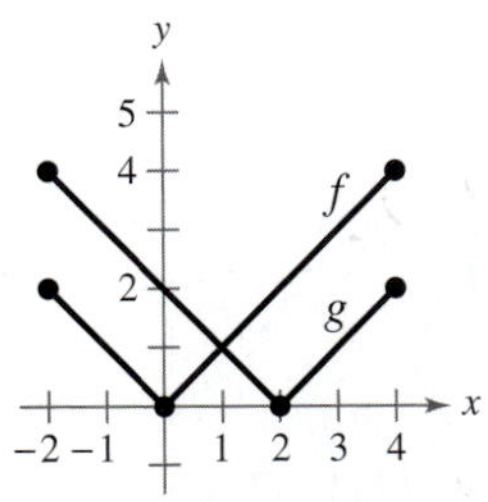

10.

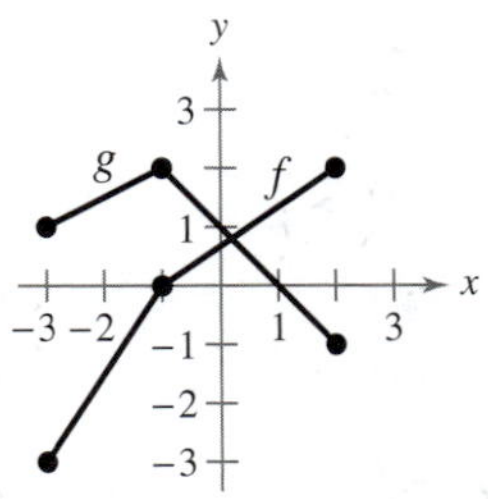

Finding Arithmetic Combinations of Functions **In Exercises 11–18, find (a) $(f + g)(x)$, (b) $(f - g)(x)$, (c) $(fg)(x)$, and (d) $(f/g)(x)$. What is the domain of f/g?**

11. $f(x) = x + 3, \quad g(x) = x - 3$
12. $f(x) = 2x - 5, \quad g(x) = 1 - x$
13. $f(x) = 3x^2, \quad g(x) = 6 - 5x$
14. $f(x) = 2x + 5, \quad g(x) = x^2 - 9$
15. $f(x) = x^2 + 5, \quad g(x) = \sqrt{1 - x}$
16. $f(x) = \sqrt{x^2 - 4}, \quad g(x) = \dfrac{x^2}{x^2 + 1}$
17. $f(x) = \dfrac{1}{x}, \quad g(x) = \dfrac{1}{x^2}$
18. $f(x) = \dfrac{x}{x + 1}, \quad g(x) = \dfrac{1}{x^3}$

Evaluating an Arithmetic Combination of Functions **In Exercises 19–32, evaluate the indicated function for $f(x) = x^2 - 1$ and $g(x) = x - 2$ algebraically. If possible, use a graphing utility to verify your answer.**

19. $(f + g)(3)$
20. $(f - g)(-2)$
21. $(f - g)(0)$
22. $(f + g)(1)$
23. $(fg)(-6)$
24. $(fg)(4)$
25. $(f/g)(-5)$
26. $(f/g)(0)$
27. $(f - g)(t + 1)$
28. $(f + g)(t - 3)$
29. $(fg)(-5t)$
30. $(fg)(3t^2)$
31. $(f/g)(t - 4)$
32. $(f/g)(t + 2)$

Graphing an Arithmetic Combination of Functions **In Exercises 33–36, use a graphing utility to graph the functions f, g, and h in the same viewing window.**

33. $f(x) = \frac{1}{2}x, \quad g(x) = x - 1, \quad h(x) = f(x) + g(x)$
34. $f(x) = \frac{1}{3}x, \quad g(x) = -x + 4, \quad h(x) = f(x) - g(x)$
35. $f(x) = x^2, \quad g(x) = -2x + 5, \quad h(x) = f(x) \cdot g(x)$
36. $f(x) = 4 - x^2, \quad g(x) = x, \quad h(x) = f(x)/g(x)$

Graphing a Sum of Functions **In Exercises 37–40, use a graphing utility to graph f, g, and $f + g$ in the same viewing window. Which function contributes most to the magnitude of the sum when $0 \le x \le 2$? Which function contributes most to the magnitude of the sum when $x > 6$?**

37. $f(x) = 3x, \quad g(x) = -\dfrac{x^3}{10}$
38. $f(x) = \dfrac{x}{2}, \quad g(x) = \sqrt{x}$
39. $f(x) = 3x + 2, \quad g(x) = -\sqrt{x + 5}$
40. $f(x) = x^2 - \frac{1}{2}, \quad g(x) = -3x^2 - 1$

Compositions of Functions **In Exercises 41–44, find (a) $f \circ g$, (b) $g \circ f$, and, if possible, (c) $(f \circ g)(0)$.**

41. $f(x) = 2x^2, \quad g(x) = x + 4$

42. $f(x) = \sqrt[3]{x - 1}, \quad g(x) = x^3 + 1$

43. $f(x) = 3x + 5, \quad g(x) = 5 - x$

44. $f(x) = x^3, \quad g(x) = \dfrac{1}{x}$

Finding the Domain of a Composite Function **In Exercises 45–54, determine the domains of (a) f, (b) g, and (c) $f \circ g$. Use a graphing utility to verify your results.**

45. $f(x) = \sqrt{x - 7}, \quad g(x) = 4x^2$

46. $f(x) = \sqrt{x + 3}, \quad g(x) = \dfrac{x}{2}$

47. $f(x) = x^2 + 1, \quad g(x) = \sqrt{x}$

48. $f(x) = x^{1/4}, \quad g(x) = x^4$

49. $f(x) = \dfrac{1}{x}, \quad g(x) = \dfrac{1}{x + 3}$

50. $f(x) = \dfrac{1}{x}, \quad g(x) = \dfrac{1}{2x}$

51. $f(x) = |x - 4|, \quad g(x) = 3 - x$

52. $f(x) = \dfrac{2}{|x|}, \quad g(x) = x - 5$

53. $f(x) = x + 2, \quad g(x) = \dfrac{1}{x^2 - 4}$

54. $f(x) = \dfrac{3}{x^2 - 1}, \quad g(x) = x + 1$

Determining Whether $f \circ g = g \circ f$ **In Exercises 55–60, (a) find $f \circ g$, $g \circ f$, and the domain of $f \circ g$. (b) Use a graphing utility to graph $f \circ g$ and $g \circ f$. Determine whether $f \circ g = g \circ f$.**

55. $f(x) = \sqrt{x + 4}, \quad g(x) = x^2$

56. $f(x) = \sqrt[3]{x + 1}, \quad g(x) = x^3 - 1$

57. $f(x) = \frac{1}{3}x - 3, \quad g(x) = 3x + 9$

58. $f(x) = \sqrt{x}, \quad g(x) = \sqrt{x}$

59. $f(x) = x^{2/3}, \quad g(x) = x^6$

60. $f(x) = |x|, \quad g(x) = -x^2 + 1$

Determining Whether $f \circ g = g \circ f$ **In Exercises 61–66, (a) find $(f \circ g)(x)$ and $(g \circ f)(x)$, (b) determine algebraically whether $(f \circ g)(x) = (g \circ f)(x)$, and (c) use a graphing utility to complete a table of values for the two compositions to confirm your answer to part (b).**

61. $f(x) = 5x + 4, \quad g(x) = \frac{1}{5}(x - 4)$

62. $f(x) = \frac{1}{4}(x - 1), \quad g(x) = 4x + 1$

63. $f(x) = \sqrt{x + 6}, \quad g(x) = x^2 - 5$

64. $f(x) = x^3 - 4, \quad g(x) = \sqrt[3]{x + 10}$

65. $f(x) = |x|, \quad g(x) = 2x^3$

66. $f(x) = \dfrac{6}{3x - 5}, \quad g(x) = -x$

Evaluating Combinations of Functions **In Exercises 67–70, use the graphs of f and g to evaluate the functions.**

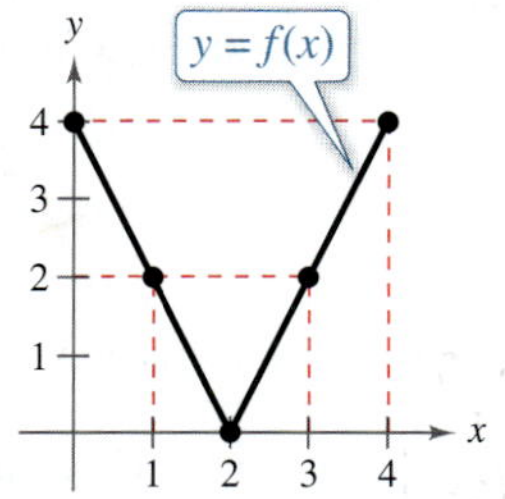

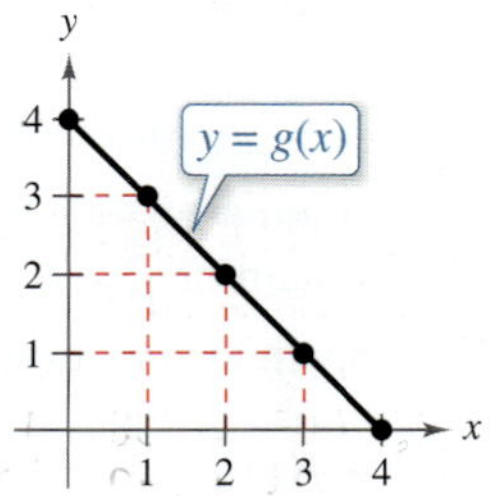

67. (a) $(f + g)(3)$ (b) $(f/g)(2)$

68. (a) $(f - g)(1)$ (b) $(fg)(4)$

69. (a) $(f \circ g)(3)$ (b) $(g \circ f)(2)$

70. (a) $(f \circ g)(1)$ (b) $(g \circ f)(3)$

Identifying a Composite Function **In Exercises 71–78, find two functions f and g such that $(f \circ g)(x) = h(x)$. (There are many correct answers.)**

71. $h(x) = (2x + 1)^2$

72. $h(x) = (1 - x)^3$

73. $h(x) = \sqrt[3]{x^2 - 4}$

74. $h(x) = \sqrt{9 - x}$

75. $h(x) = \dfrac{1}{x + 2}$

76. $h(x) = \dfrac{4}{(5x + 2)^2}$

77. $h(x) = (x + 4)^2 + 2(x + 4)$

78. $h(x) = (x + 3)^{3/2} + 4(x + 3)^{1/2}$

79. ***Why you should learn it*** *(p. 131)* The research and development department of an automobile manufacturer has determined that when required to stop quickly to avoid an accident, the distance (in feet) a car travels during the driver's reaction time is given by

$R(x) = \frac{3}{4}x$

where x is the speed of the car in miles per hour. The distance (in feet) traveled while the driver is braking is given by

$B(x) = \frac{1}{15}x^2.$

(a) Find the function that represents the total stopping distance T.

(b) Use a graphing utility to graph the functions R, B, and T in the same viewing window for $0 \le x \le 60$.

(c) Which function contributes most to the magnitude of the sum at higher speeds? Explain.

80. MODELING DATA

The table shows the total amounts (in billions of dollars) of health consumption expenditures in the United States (including Puerto Rico) for the years 2002 through 2012. The variables y_1, y_2, and y_3 represent out-of-pocket payments, insurance premiums, and other types of payments, respectively. (Source: U.S. Centers for Medicare and Medicaid Services)

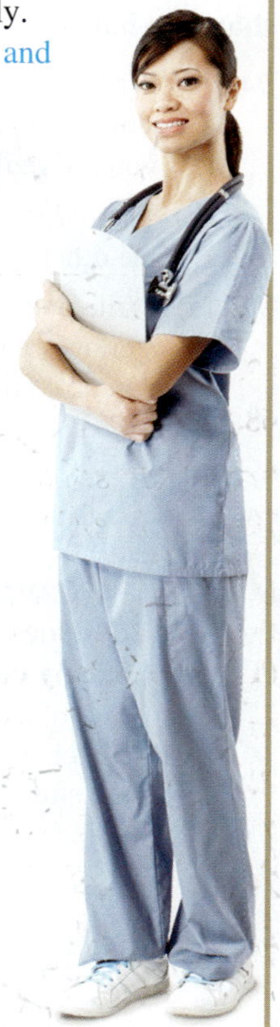

DATA

Year	y_1	y_2	y_3
2002	222	1122	140
2003	238	1223	153
2004	252	1322	159
2005	267	1417	168
2006	277	1521	177
2007	294	1612	187
2008	301	1703	182
2009	301	1799	185
2010	306	1874	195
2011	316	1943	202
2012	328	2014	216

Spreadsheet at LarsonPrecalculus.com

The data are approximated by the following models, where t represents the year, with $t = 2$ corresponding to 2002.

$$y_1 = -0.62t^2 + 18.7t + 188$$

$$y_2 = -1.86t^2 + 116.4t + 890$$

$$y_3 = -0.12t^2 + 8.3t + 128$$

(a) Use the models and the *table* feature of a graphing utility to create a table showing the values of y_1, y_2, and y_3 for each year from 2002 through 2012. Compare these models with the original data. Are the models a good fit? Explain.

(b) Use the graphing utility to graph y_1, y_2, y_3, and $y_T = y_1 + y_2 + y_3$ in the same viewing window. What does the function y_T represent?

81. Geometry A square concrete foundation was prepared as a base for a large cylindrical gasoline tank (see figure).

(a) Write the radius r of the tank as a function of the length x of the sides of the square.

(b) Write the area A of the circular base of the tank as a function of the radius r.

(c) Find and interpret $(A \circ r)(x)$.

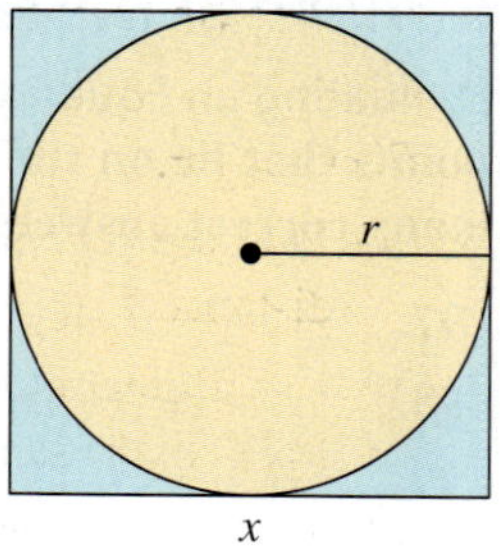

82. Geometry A pebble is dropped into a calm pond, causing ripples in the form of concentric circles. The radius (in feet) of the outermost ripple is given by $r(t) = 0.6t$, where t is the time (in seconds) after the pebble strikes the water. The area of the circle is given by $A(r) = \pi r^2$. Find and interpret $(A \circ r)(t)$.

83. Business A company owns two retail stores. The annual sales (in thousands of dollars) of the stores each year from 2009 through 2015 can be approximated by the models

$$S_1 = 973 + 1.3t^2 \quad \text{and} \quad S_2 = 349 + 72.4t$$

where t is the year, with $t = 9$ corresponding to 2009.

(a) Write a function T that represents the total annual sales of the two stores.

(b) Use a graphing utility to graph S_1, S_2, and T in the same viewing window.

84. Business The annual cost C (in thousands of dollars) and revenue R (in thousands of dollars) for a company each year from 2009 through 2015 can be approximated by the models

$$C = 254 - 9t + 1.1t^2 \quad \text{and} \quad R = 341 + 3.2t$$

where t is the year, with $t = 9$ corresponding to 2009.

(a) Write a function P that represents the annual profits of the company.

(b) Use a graphing utility to graph C, R, and P in the same viewing window.

85. Biology The number of bacteria in a refrigerated food product is given by

$$N(T) = 10T^2 - 20T + 600, \quad 1 \le T \le 20$$

where T is the temperature of the food in degrees Celsius. When the food is removed from the refrigerator, the temperature of the food is given by

$$T(t) = 2t + 1$$

where t is the time in hours.

(a) Find the composite function $N(T(t))$ or $(N \circ T)(t)$ and interpret its meaning in the context of the situation.

(b) Find $(N \circ T)(12)$ and interpret its meaning.

(c) Find the time when the bacteria count reaches 1200.

86. Environmental Science The spread of a contaminant is increasing in a circular pattern on the surface of a lake. The radius of the contaminant can be modeled by $r(t) = 5.25\sqrt{t}$, where r is the radius in meters and t is time in hours since contamination.

(a) Find a function that gives the area A of the circular leak in terms of the time t since the spread began.

(b) Find the size of the contaminated area after 36 hours.

(c) Find when the size of the contaminated area is 6250 square meters.

87. **Air Traffic Control** An air traffic controller spots two planes flying at the same altitude. Their flight paths form a right angle at point P. One plane is 150 miles from point P and is moving at 450 miles per hour. The other plane is 200 miles from point P and is moving at 450 miles per hour. Write the distance s between the planes as a function of time t.

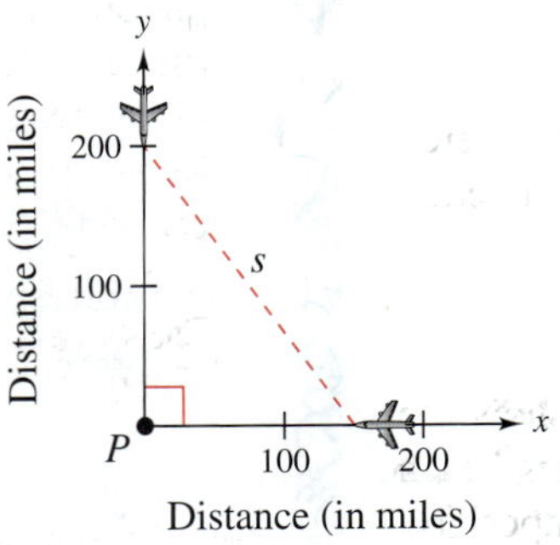

88. **Marketing** The suggested retail price of a new car is p dollars. The dealership advertised a factory rebate of \$2000 and a 9% discount.

(a) Write a function R in terms of p giving the cost of the car after receiving the rebate from the factory.

(b) Write a function S in terms of p giving the cost of the car after receiving the dealership discount.

(c) Form the composite functions $(R \circ S)(p)$ and $(S \circ R)(p)$ and interpret each.

(d) Find $(R \circ S)(24{,}795)$ and $(S \circ R)(24{,}795)$. Which yields the lower cost for the car? Explain.

Conclusions

True or False? **In Exercises 89 and 90, determine whether the statement is true or false. Justify your answer.**

89. A function that represents the graph of $f(x) = x^2$ shifted three units to the right is $f(g(x))$, where $g(x) = x + 3$.

90. Given two functions f and g, you can calculate $(f \circ g)(x)$ if and only if the range of g is a subset of the domain of f.

91. **Exploration** The function in Example 9 can be decomposed in other ways. For which of the following pairs of functions is $h(x) = \dfrac{1}{(x-2)^2}$ equal to $f(g(x))$?

(a) $g(x) = \dfrac{1}{x-2}$ and $f(x) = x^2$

(b) $g(x) = x^2$ and $f(x) = \dfrac{1}{x-2}$

(c) $g(x) = (x-2)^2$ and $f(x) = \dfrac{1}{x}$

92. **Proof** Prove that the product of two odd functions is an even function, and that the product of two even functions is an even function.

93. **Proof** Use examples to hypothesize whether the product of an odd function and an even function is even or odd. Then prove your hypothesis.

Exploration **In Exercises 94 and 95, three siblings are of three different ages. The oldest is twice the age of the middle sibling, and the middle sibling is six years older than one-half the age of the youngest.**

94. (a) Write a composite function that gives the oldest sibling's age in terms of the youngest. Explain how you arrived at your answer.

(b) The oldest sibling is 16 years old. Find the ages of the other two siblings.

95. (a) Write a composite function that gives the youngest sibling's age in terms of the oldest. Explain how you arrived at your answer.

(b) The youngest sibling is two years old. Find the ages of the other two siblings.

96. **HOW DO YOU SEE IT?** The graphs labeled L_1, L_2, L_3, and L_4 represent four different pricing discounts, where p is the original price (in dollars) and S is the sale price (in dollars). Match each function with its graph. Describe the situations in parts (c) and (d).

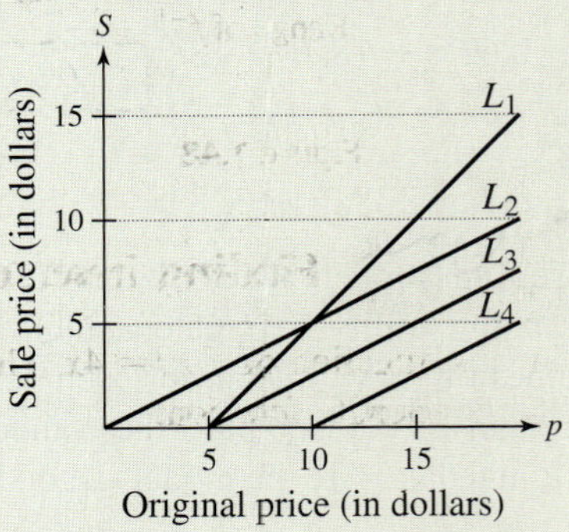

(a) $f(p)$: A 50% discount is applied.

(b) $g(p)$: A \$5 discount is applied.

(c) $(g \circ f)(p)$

(d) $(f \circ g)(p)$

Cumulative Mixed Review

Evaluating an Equation **In Exercises 97–100, find three points that lie on the graph of the equation. (There are many correct answers.)**

97. $y = -x^2 + x - 5$

98. $y = \frac{1}{5}x^3 - 4x^2 + 1$

99. $x^2 + y^2 = 49$

100. $y = \dfrac{x}{x^2 - 5}$

1.7 Inverse Functions

Inverse Functions

Recall from Section 1.3 that a function can be represented by a set of ordered pairs. For instance, the function $f(x) = x + 4$ from the set $A = \{1, 2, 3, 4\}$ to the set $B = \{5, 6, 7, 8\}$ can be written as follows.

$$f(x) = x + 4: \{(1, 5), (2, 6), (3, 7), (4, 8)\}$$

In this case, by interchanging the first and second coordinates of each of these ordered pairs, you can form the **inverse function** of f, which is denoted by f^{-1}. It is a function from the set B to the set A, and can be written as follows.

$$f^{-1}(x) = x - 4: \{(5, 1), (6, 2), (7, 3), (8, 4)\}$$

Note that the domain of f is equal to the range of f^{-1}, and vice versa, as shown in Figure 1.48. Also note that the functions f and f^{-1} have the effect of "undoing" each other. In other words, when you form the composition of f with f^{-1} or the composition of f^{-1} with f, you obtain the identity function.

$$f(f^{-1}(x)) = f(x - 4) = (x - 4) + 4 = x$$

$$f^{-1}(f(x)) = f^{-1}(x + 4) = (x + 4) - 4 = x$$

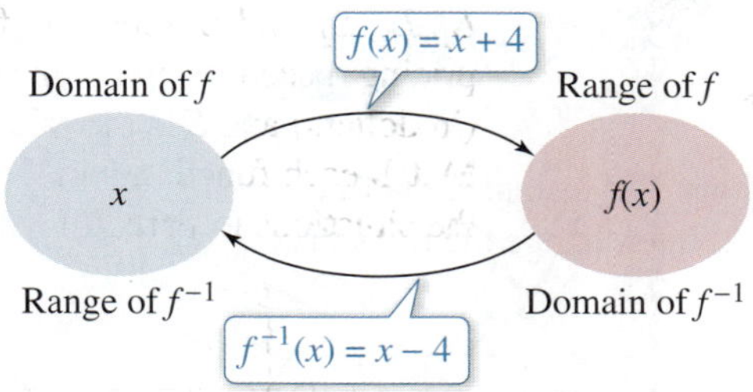

Figure 1.48

What you should learn

- Find inverse functions informally and verify that two functions are inverse functions of each other.
- Use graphs of functions to decide whether functions have inverse functions.
- Determine whether functions are one-to-one.
- Find inverse functions algebraically.

Why you should learn it

Inverse functions can be helpful in further exploring how two variables relate to each other. For example, in Exercise 115 on page 150, you will use inverse functions to find the European shoe sizes from the corresponding U.S. shoe sizes.

EXAMPLE 1 Finding Inverse Functions Informally

Find the inverse function of $f(x) = 4x$. Then verify that both $f(f^{-1}(x))$ and $f^{-1}(f(x))$ are equal to the identity function.

Solution

The function f *multiplies* each input by 4. To "undo" this function, you need to *divide* each input by 4. So, the inverse function of $f(x) = 4x$ is given by

$$f^{-1}(x) = \frac{x}{4}.$$

You can verify that both $f(f^{-1}(x))$ and $f^{-1}(f(x))$ are equal to the identity function as follows.

$$f(f^{-1}(x)) = f\left(\frac{x}{4}\right) = 4\left(\frac{x}{4}\right) = x$$

$$f^{-1}(f(x)) = f^{-1}(4x) = \frac{4x}{4} = x$$

✓ Checkpoint *Audio-video solution in English & Spanish at LarsonPrecalculus.com.*

Find the inverse function of $f(x) = \frac{1}{5}x$. Then verify that both $f(f^{-1}(x))$ and $f^{-1}(f(x))$ are equal to the identity function.

Do not be confused by the use of the exponent -1 to denote the inverse function f^{-1}. In this text, whenever f^{-1} is written, it always refers to the inverse function of the function f and not to the reciprocal of $f(x)$, which is given by

$$\frac{1}{f(x)}.$$

EXAMPLE 2 Finding Inverse Functions Informally

Find the inverse function of $f(x) = x - 6$. Then verify that both $f(f^{-1}(x))$ and $f^{-1}(f(x))$ are equal to the identity function.

Solution

The function f *subtracts* 6 from each input. To "undo" this function, you need to *add* 6 to each input. So, the inverse function of $f(x) = x - 6$ is given by $f^{-1}(x) = x + 6$. You can verify that both $f(f^{-1}(x))$ and $f^{-1}(f(x))$ are equal to the identity function as follows.

$$f(f^{-1}(x)) = f(x + 6) = (x + 6) - 6 = x$$

$$f^{-1}(f(x)) = f^{-1}(x - 6) = (x - 6) + 6 = x$$

Checkpoint *Audio-video solution in English & Spanish at LarsonPrecalculus.com.*

Find the inverse function of $f(x) = x + 7$. Then verify that both $f(f^{-1}(x))$ and $f^{-1}(f(x))$ are equal to the identity function.

A table of values can help you understand inverse functions. For instance, the first table below shows several values of the function in Example 2. Interchange the rows of this table to obtain values of the inverse function.

x	-2	-1	0	1	2
$f(x)$	-8	-7	-6	-5	-4

x	-8	-7	-6	-5	-4
$f^{-1}(x)$	-2	-1	0	1	2

In the table at the left, each output is 6 less than the input, and in the table at the right, each output is 6 more than the input.

The formal definition of an inverse function is as follows.

Definition of Inverse Function

Let f and g be two functions such that

$f(g(x)) = x$ for every x in the domain of g

and

$g(f(x)) = x$ for every x in the domain of f.

Under these conditions, the function g is the **inverse function** of the function f. The function g is denoted by f^{-1} (read "f-inverse"). So,

$$f(f^{-1}(x)) = x \quad \text{and} \quad f^{-1}(f(x)) = x.$$

The domain of f must be equal to the range of f^{-1}, and the range of f must be equal to the domain of f^{-1}.

If the function g is the inverse function of the function f, then it must also be true that the function f is the inverse function of the function g. For this reason, you can say that the functions f and g are *inverse functions of each other*.

EXAMPLE 3 Verifying Inverse Functions Algebraically

Show that the functions are inverse functions of each other.

$$f(x) = 2x^3 - 1 \quad \text{and} \quad g(x) = \sqrt[3]{\frac{x+1}{2}}$$

Solution

$$\begin{aligned} f(g(x)) &= f\left(\sqrt[3]{\frac{x+1}{2}}\right) \\ &= 2\left(\sqrt[3]{\frac{x+1}{2}}\right)^3 - 1 \\ &= 2\left(\frac{x+1}{2}\right) - 1 \\ &= x + 1 - 1 \\ &= x \end{aligned}$$

$$\begin{aligned} g(f(x)) &= g(2x^3 - 1) \\ &= \sqrt[3]{\frac{(2x^3 - 1) + 1}{2}} \\ &= \sqrt[3]{\frac{2x^3}{2}} \\ &= \sqrt[3]{x^3} \\ &= x \end{aligned}$$

Technology Tip

Most graphing utilities can graph $y = x^{1/3}$ in two ways:

$y_1 = x \wedge (1/3)$ or

$y_1 = \sqrt[3]{x}$.

On some graphing utilities, you may not be able to obtain the complete graph of $y = x^{2/3}$ by entering $y_1 = x \wedge (2/3)$. If not, you should use

$y_1 = (x \wedge (1/3))^2$ or

$y_1 = \sqrt[3]{x^2}$.

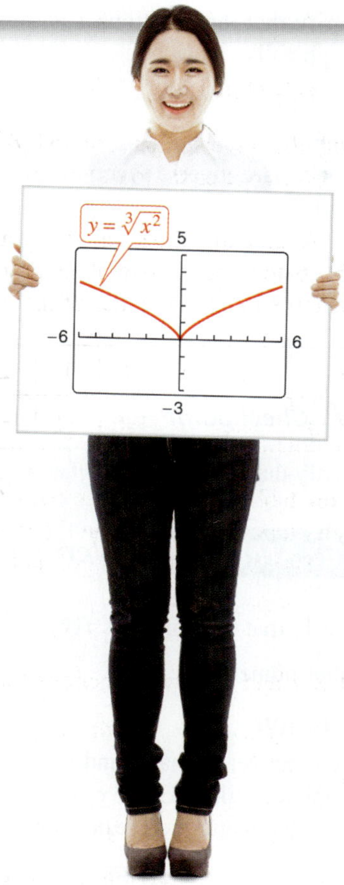

✓ Checkpoint *Audio-video solution in English & Spanish at LarsonPrecalculus.com.*

Show that $f(x) = x^5$ and $g(x) = \sqrt[5]{x}$ are inverse functions of each other.

EXAMPLE 4 Verifying Inverse Functions Algebraically

Which of the functions is the inverse function of $f(x) = \dfrac{5}{x-2}$?

$$g(x) = \frac{x-2}{5} \quad \text{or} \quad h(x) = \frac{5}{x} + 2$$

Solution

By forming the composition of f with g, you have

$$f(g(x)) = f\left(\frac{x-2}{5}\right) = \frac{5}{\left(\frac{x-2}{5}\right) - 2} = \frac{25}{x - 12} \neq x.$$

Because this composition is not equal to the identity function x, it follows that g *is not* the inverse function of f. By forming the composition of f with h, you have

$$f(h(x)) = f\left(\frac{5}{x} + 2\right) = \frac{5}{\left(\frac{5}{x} + 2\right) - 2} = \frac{5}{\left(\frac{5}{x}\right)} = x.$$

So, it appears that h is the inverse function of f. You can confirm this by showing that the composition of h with f is also equal to the identity function.

$$h(f(x)) = h\left(\frac{5}{x-2}\right) = \frac{5}{\left(\frac{5}{x-2}\right)} + 2 = x - 2 + 2 = x$$

✓ Checkpoint *Audio-video solution in English & Spanish at LarsonPrecalculus.com.*

Which of the functions is the inverse function of $f(x) = \dfrac{x-4}{7}$?

$$g(x) = 7x + 4 \qquad h(x) = \frac{7}{x-4}$$

The Graph of an Inverse Function

The graphs of a function f and its inverse function f^{-1} are related to each other in the following way. If the point

(a, b)

lies on the graph of f, then the point

(b, a)

must lie on the graph of f^{-1}, and vice versa. This means that the graph of f^{-1} is a reflection of the graph of f in the line $y = x$, as shown in Figure 1.49.

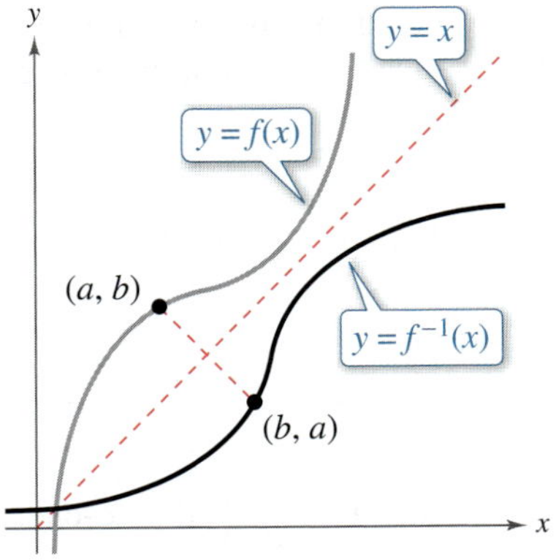

Figure 1.49

Technology Tip

Many graphing utilities have a built-in feature for drawing an inverse function. For instructions on how to use the *draw inverse* feature, see Appendix A; for specific keystrokes, go to this textbook's *Companion Website.*

EXAMPLE 5 Verifying Inverse Functions Graphically

Verify that the functions f and g from Example 3 are inverse functions of each other graphically.

Solution

From the figure, it appears that f and g are inverse functions of each other.

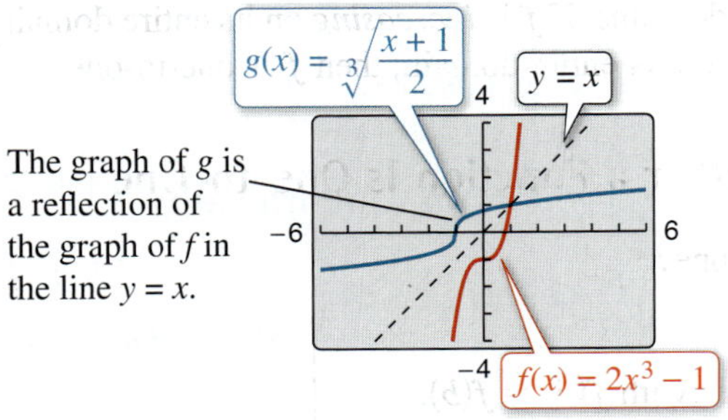

✓ Checkpoint Audio-video solution in English & Spanish at LarsonPrecalculus.com.

Verify that $f(x) = x^5$ and $g(x) = \sqrt[5]{x}$ are inverse functions of each other graphically.

EXAMPLE 6 Verifying Inverse Functions Numerically

Verify that the functions $f(x) = \dfrac{x - 5}{2}$ and $g(x) = 2x + 5$ are inverse functions of each other numerically.

Solution

You can verify that f and g are inverse functions of each other *numerically* by using a graphing utility. Enter $y_1 = f(x)$, $y_2 = g(x)$, $y_3 = f(g(x))$, and $y_4 = g(f(x))$. Then use the *table* feature to create a table.

```
Plot1  Plot2  Plot3
\Y1=(X-5)/2
\Y2=2X+5
\Y3=Y1(Y2)
\Y4=Y2(Y1)
\Y5=
\Y6=
\Y7=
```

X	Y3	Y4
-2	-2	-2
-1	-1	-1
0	0	0
1	1	1
2	2	2
3	3	3
4	4	4

X=-2

Note that the entries for x, y_3, and y_4 are the same. So, it appears that $f(g(x)) = x$ and $g(f(x)) = x$, and that f and g are inverse functions of each other.

✓ Checkpoint Audio-video solution in English & Spanish at LarsonPrecalculus.com.

Verify that $f(x) = x^5$ and $g(x) = \sqrt[5]{x}$ are inverse functions of each other numerically. ■

The Existence of an Inverse Function

To have an inverse function, a function must be **one-to-one,** which means that no two elements in the domain of f correspond to the same element in the range of f.

Definition of a One-to-One Function

A function f is **one-to-one** when, for a and b in its domain, $f(a) = f(b)$ implies that $a = b$.

Existence of an Inverse Function

A function f has an inverse function f^{-1} if and only if f is one-to-one.

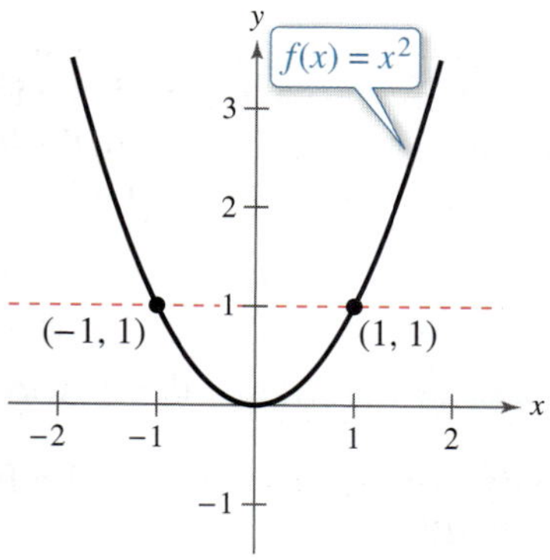

$f(x) = x^2$ is not one-to-one.
Figure 1.50

From its graph, it is easy to tell whether a function of x is one-to-one. Simply check to see that every horizontal line intersects the graph of the function at most once. This is called the **Horizontal Line Test.** For instance, Figure 1.50 shows the graph of $f(x) = x^2$. On the graph, you can find a horizontal line that intersects the graph twice. So, f is not one-to-one and does not have an inverse function.

Two special types of functions that pass the Horizontal Line Test are those that are increasing or decreasing on their entire domains. If f is *increasing* on its entire domain, then f is one-to-one. If f is *decreasing* on its entire domain, then f is one-to-one.

EXAMPLE 7 Testing Whether a Function Is One-to-One

Is the function $f(x) = \sqrt{x} + 1$ one-to-one?

Algebraic Solution

Let a and b be nonnegative real numbers with $f(a) = f(b)$.

$$\sqrt{a} + 1 = \sqrt{b} + 1 \quad \text{Set } f(a) = f(b).$$
$$\sqrt{a} = \sqrt{b}$$
$$a = b$$

So, $f(a) = f(b)$ implies that $a = b$. You can conclude that f *is* one-to-one and *does* have an inverse function.

Graphical Solution

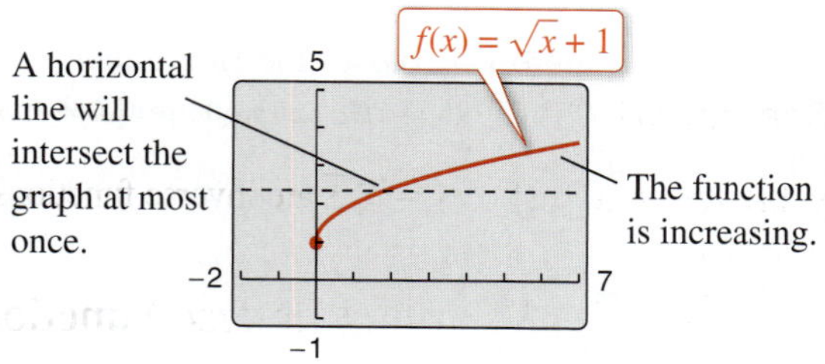

From the figure, you can conclude that f *is* one-to-one and *does* have an inverse function.

✓ **Checkpoint** *Audio-video solution in English & Spanish at LarsonPrecalculus.com.*

Is the function $f(x) = \sqrt[3]{x}$ one-to-one?

EXAMPLE 8 Testing Whether a Function Is One-to-One

See LarsonPrecalculus.com for an interactive version of this type of example.

To determine whether $f(x) = x^2 - x$ is one-to-one, note that $f(-1) = (-1)^2 - (-1) = 2$ and $f(2) = 2^2 - 2 = 2$. Because you have two inputs matched with the same output, f *is not* one-to-one and *does not* have an inverse function. You can confirm this graphically by noticing that the horizontal line $y = 2$ intersects the graph of f twice, as shown in Figure 1.51.

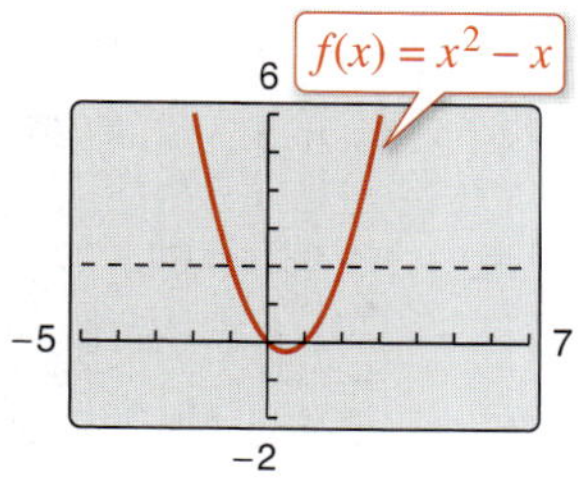

Figure 1.51

✓ **Checkpoint** *Audio-video solution in English & Spanish at LarsonPrecalculus.com.*

Is the function $f(x) = |x|$ one-to-one?

Finding Inverse Functions Algebraically

For relatively simple functions (such as the ones in Examples 1 and 2), you can find inverse functions by inspection. For more complicated functions, however, it is best to use the following guidelines.

Finding an Inverse Function

1. Use the Horizontal Line Test to decide whether f has an inverse function.
2. In the equation for $f(x)$, replace $f(x)$ with y.
3. Interchange the roles of x and y, and solve for y.
4. Replace y with $f^{-1}(x)$ in the new equation.
5. Verify that f and f^{-1} are inverse functions of each other by showing that the domain of f is equal to the range of f^{-1}, the range of f is equal to the domain of f^{-1}, and $f(f^{-1}(x)) = x$ and $f^{-1}(f(x)) = x$.

The key step in these guidelines is Step 3—interchanging the roles of x and y. This step corresponds to the fact that inverse functions have ordered pairs with the coordinates reversed.

What's Wrong?

You use a graphing utility to graph $y_1 = x^2$ and then use the *draw inverse* feature to conclude that $f(x) = x^2$ has an inverse function (see figure). What's wrong?

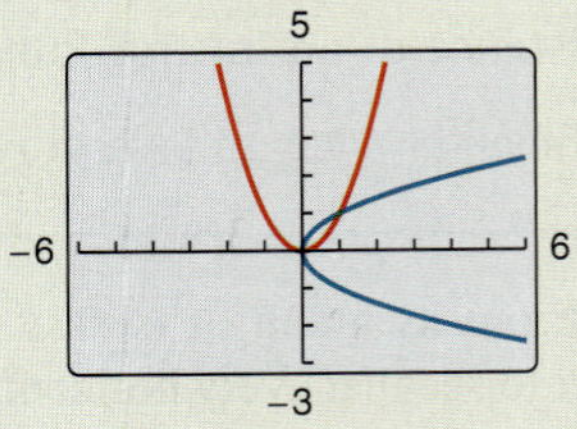

EXAMPLE 9 Finding an Inverse Function Algebraically

Find the inverse function of

$$f(x) = \frac{5 - 3x}{2}.$$

Solution

The graph of f in Figure 1.52 passes the Horizontal Line Test. So, you know that f is one-to-one and has an inverse function.

$$f(x) = \frac{5 - 3x}{2}$$ Write original function.

$$y = \frac{5 - 3x}{2}$$ Replace $f(x)$ with y.

$$x = \frac{5 - 3y}{2}$$ Interchange x and y.

$$2x = 5 - 3y$$ Multiply each side by 2.

$$3y = 5 - 2x$$ Isolate the y-term.

$$y = \frac{5 - 2x}{3}$$ Solve for y.

$$f^{-1}(x) = \frac{5 - 2x}{3}$$ Replace y with $f^{-1}(x)$.

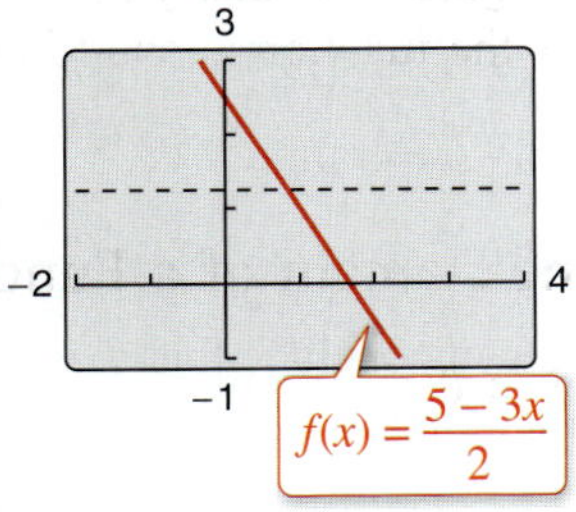

Figure 1.52

The domains and ranges of f and f^{-1} consist of all real numbers. Verify that $f(f^{-1}(x)) = x$ and $f^{-1}(f(x)) = x$.

✓ Checkpoint *Audio-video solution in English & Spanish at LarsonPrecalculus.com.*

Find the inverse function of $f(x) = 2x - 3$.

EXAMPLE 10 Finding an Inverse Function Algebraically

Find the inverse function of $f(x) = \sqrt{2x - 3}$.

Solution

The graph of f in Figure 1.53 passes the Horizontal Line Test. So, you know that f is one-to-one and has an inverse function.

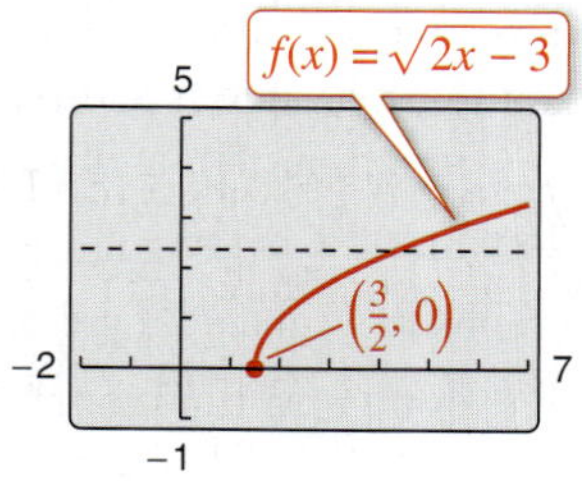

Figure 1.53

$f(x) = \sqrt{2x - 3}$ Write original function.

$y = \sqrt{2x - 3}$ Replace $f(x)$ with y.

$x = \sqrt{2y - 3}$ Interchange x and y.

$x^2 = 2y - 3$ Square each side.

$2y = x^2 + 3$ Isolate y.

$y = \dfrac{x^2 + 3}{2}$ Solve for y.

$f^{-1}(x) = \dfrac{x^2 + 3}{2}, \; x \geq 0$ Replace y with $f^{-1}(x)$.

The range of f is the interval $[0, \infty)$, which implies that the domain of f^{-1} is the interval $[0, \infty)$. Moreover, the domain of f is the interval $\left[\frac{3}{2}, \infty\right)$, which implies that the range of f^{-1} is the interval $\left[\frac{3}{2}, \infty\right)$. Verify that $f(f^{-1}(x)) = x$ and $f^{-1}(f(x)) = x$.

✓ **Checkpoint** *Audio-video solution in English & Spanish at LarsonPrecalculus.com.*

Find the inverse function of $f(x) = \sqrt[3]{x + 10}$.

A function f with an implied domain of all real numbers may not pass the Horizontal Line Test. In this case, the domain of f may be restricted so that f does have an inverse function. Recall from Figure 1.50 that $f(x) = x^2$ is not one-to-one. By restricting the domain of f to $x \geq 0$, the function is one-to-one and does have an inverse function (see Figure 1.54).

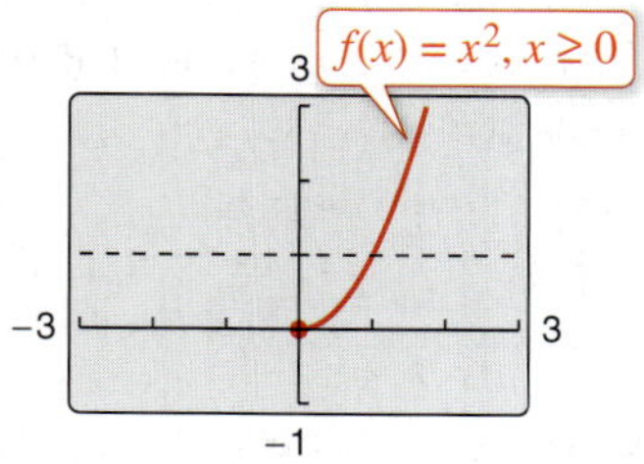

Figure 1.54

EXAMPLE 11 Restricting the Domain of a Function

Find the inverse function of $f(x) = x^2, x \geq 0$.

Solution

The graph of f in Figure 1.54 passes the Horizontal Line Test. So, you know that f is one-to-one and has an inverse function.

$f(x) = x^2, \; x \geq 0$ Write original function with restricted domain.

$y = x^2$ Replace $f(x)$ with y.

$x = y^2, \; y \geq 0$ Interchange x and y.

$y = \sqrt{x}$ Solve for y. Extract positive square root $(y \geq 0)$.

$f^{-1}(x) = \sqrt{x}$ Replace y with $f^{-1}(x)$.

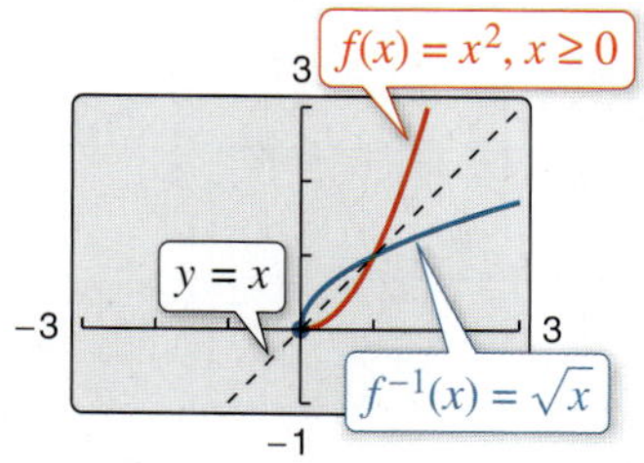

Figure 1.55

The graph of f^{-1} in Figure 1.55 is the reflection of the graph of f in the line $y = x$. The domains and ranges of f and f^{-1} consist of all nonnegative real numbers. Verify that $f(f^{-1}(x)) = x$ and $f^{-1}(f(x)) = x$.

✓ **Checkpoint** *Audio-video solution in English & Spanish at LarsonPrecalculus.com.*

Find the inverse function of $f(x) = (x - 1)^2, x \geq 1$.

1.7 Exercises

See *CalcChat.com* for tutorial help and worked-out solutions to odd-numbered exercises.
For instructions on how to use a graphing utility, see Appendix A.

Vocabulary and Concept Check

In Exercises 1–4, fill in the blank(s).

1. If f and g are functions such that $f(g(x)) = x$ and $g(f(x)) = x$, then the function g is the _______ function of f, and is denoted by _______ .

2. The domain of f is the _______ of f^{-1}, and the _______ of f^{-1} is the range of f.

3. The graphs of f and f^{-1} are reflections of each other in the line _______ .

4. To have an inverse function, a function f must be _______ ; that is, $f(a) = f(b)$ implies $a = b$.

5. How many times can a horizontal line intersect the graph of a function that is one-to-one?

6. Can (1, 4) and (2, 4) be two ordered pairs of a one-to-one function?

Procedures and Problem Solving

Finding Inverse Functions Informally **In Exercises 7–14, find the inverse function of f informally. Verify that $f(f^{-1}(x)) = x$ and $f^{-1}(f(x)) = x$.**

7. $f(x) = 6x$
8. $f(x) = \frac{1}{3}x$
9. $f(x) = x + 11$
10. $f(x) = x + 3$
11. $f(x) = (x - 1)/2$
12. $f(x) = 4(x - 1)$
13. $f(x) = \sqrt[3]{x}$
14. $f(x) = x^7$

Identifying Graphs of Inverse Functions **In Exercises 15–18, match the graph of the function with the graph of its inverse function. [The graphs of the inverse functions are labeled (a), (b), (c), and (d).]**

(a)
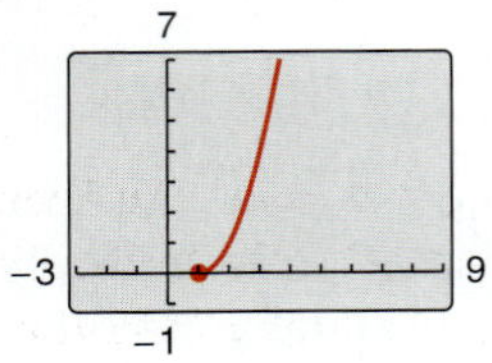

(b)
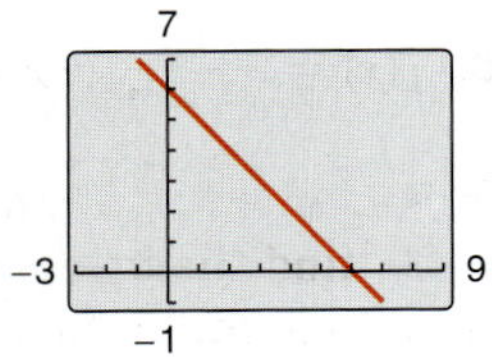

(c)
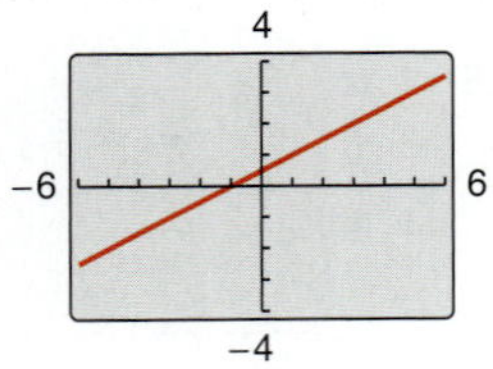

(d)
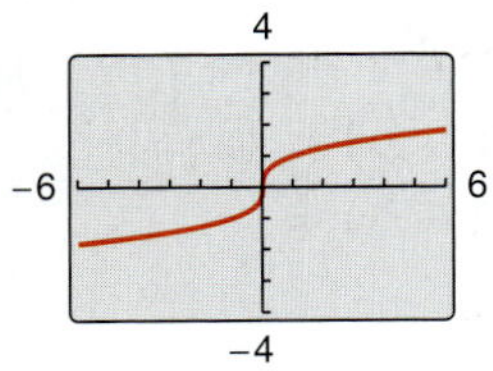

15.
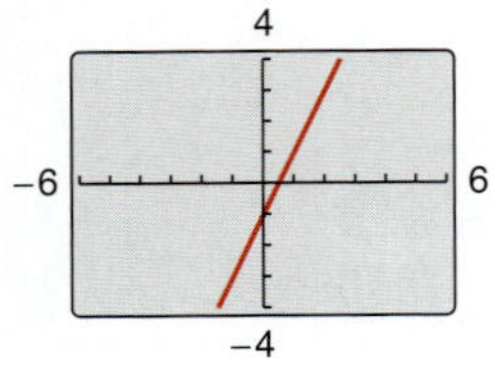

16.
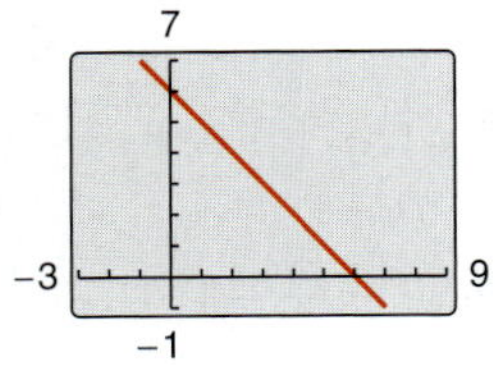

17.
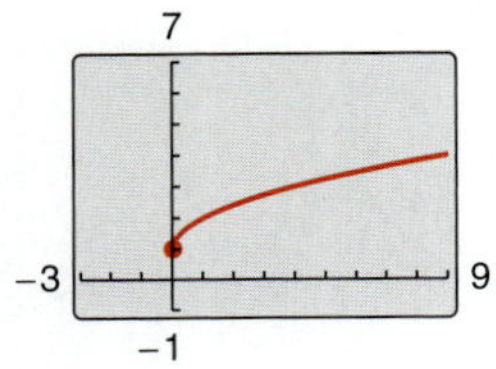

18.
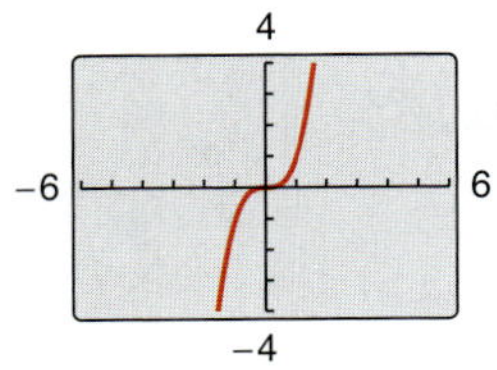

Verifying Inverse Functions Algebraically **In Exercises 19–24, show that f and g are inverse functions algebraically. Use a graphing utility to graph f and g in the same viewing window. Describe the relationship between the graphs.**

19. $f(x) = x^3, \quad g(x) = \sqrt[3]{x}$
20. $f(x) = \dfrac{1}{x}, \quad g(x) = \dfrac{1}{x}$

21. $f(x) = \sqrt{x - 4}; \quad g(x) = x^2 + 4, \quad x \geq 0$

22. $f(x) = 9 - x^2, x \geq 0; \quad g(x) = \sqrt{9 - x}$

23. $f(x) = 1 - x^3, \quad g(x) = \sqrt[3]{1 - x}$

24. $f(x) = \dfrac{1}{1 + x}, x \geq 0; \quad g(x) = \dfrac{1 - x}{x}, 0 < x \leq 1$

Algebraic-Graphical-Numerical **In Exercises 25–34, show that f and g are inverse functions (a) algebraically, (b) graphically, and (c) numerically.**

25. $f(x) = -\dfrac{7}{2}x - 3, \quad g(x) = -\dfrac{2x + 6}{7}$

26. $f(x) = \dfrac{x - 9}{4}, \quad g(x) = 4x + 9$

27. $f(x) = x^3 + 5, \quad g(x) = \sqrt[3]{x - 5}$

28. $f(x) = \dfrac{x^3}{2}, \quad g(x) = \sqrt[3]{2x}$

29. $f(x) = -\sqrt{x - 8}; \quad g(x) = 8 + x^2, \quad x \leq 0$

30. $f(x) = \sqrt[4]{3x - 10}; \quad g(x) = \dfrac{x^4 + 10}{3}, \quad x \geq 0$

31. $f(x) = 2x, \quad g(x) = \dfrac{x}{2}$

32. $f(x) = -3x + 5, \quad g(x) = -\dfrac{x - 5}{3}$

33. $f(x) = \dfrac{x - 1}{x + 5}, \quad g(x) = -\dfrac{5x + 1}{x - 1}$

34. $f(x) = \dfrac{x + 3}{x - 2}, \quad g(x) = \dfrac{2x + 3}{x - 1}$

Identifying Whether Functions Have Inverses In Exercises 35–38, does the function have an inverse? Explain.

35. *Domain* → *Range*

1 can → \$1
6 cans → \$5
12 cans → \$9
24 cans → \$16

36. *Domain* → *Range*

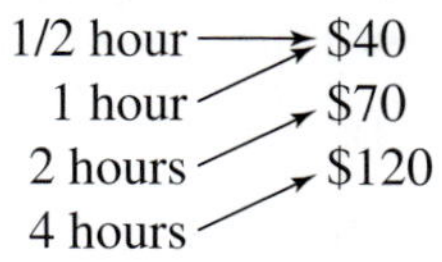

37. $\{(-3, 6), (-1, 5), (0, 6)\}$

38. $\{(2, 4), (3, 7), (7, 2)\}$

Recognizing One-to-One Functions In Exercises 39–44, determine whether the graph is that of a function. If so, determine whether the function is one-to-one.

39.

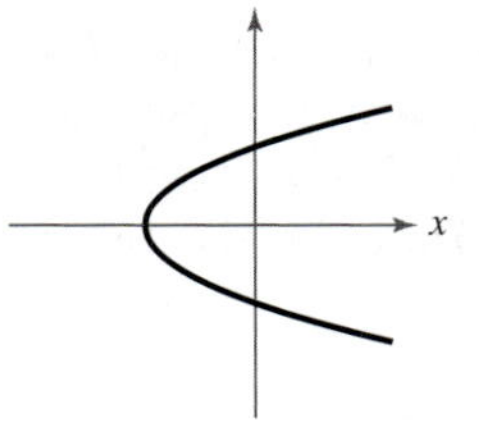

40.

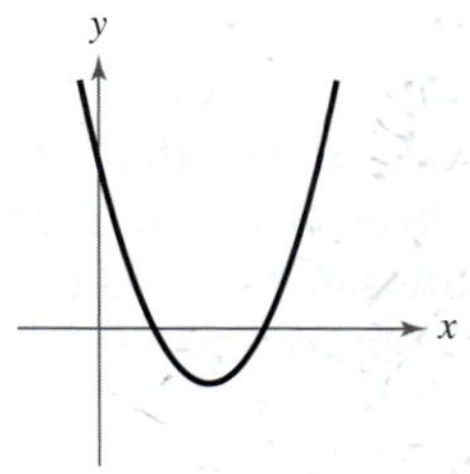

41.

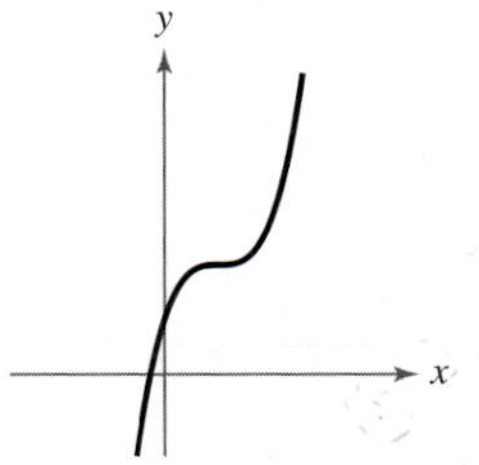

42.

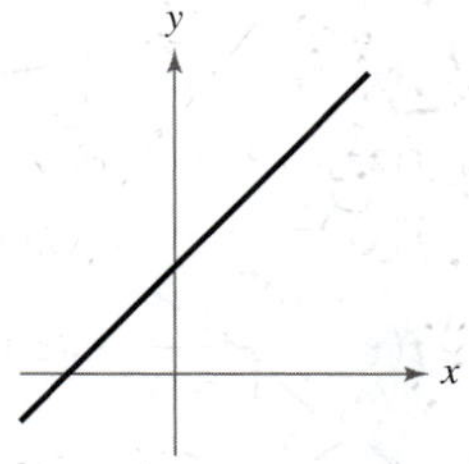

43.

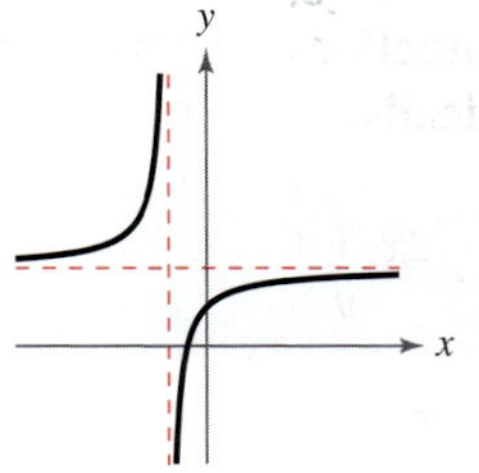

44.

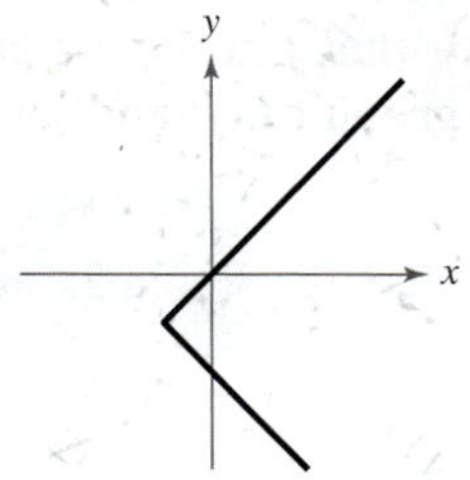

Using the Horizontal Line Test In Exercises 45–56, use a graphing utility to graph the function and use the Horizontal Line Test to determine whether the function is one-to-one and so has an inverse function.

45. $f(x) = 3 - \frac{1}{2}x$

46. $f(x) = 2x^{1/3} + 5$

47. $h(x) = \dfrac{x^2}{x^2 + 1}$

48. $g(x) = \dfrac{4 - x}{6x^2}$

49. $h(x) = \sqrt{16 - x^2}$

50. $f(x) = -2x\sqrt{16 - x^2}$

51. $f(x) = 10$

52. $f(x) = -0.65$

53. $g(x) = (x + 5)^3$

54. $f(x) = x^5 - 7$

55. $h(x) = |x| - |x - 4|$

56. $f(x) = -\dfrac{|x^2 - 9|}{|x^2 + 7|}$

Analyzing a Piecewise-Defined Function In Exercises 57 and 58, sketch the graph of the piecewise-defined function by hand and use the graph to determine whether an inverse function exists.

57. $f(x) = \begin{cases} x^2, & 0 \le x \le 1 \\ x, & x > 1 \end{cases}$

58. $f(x) = \begin{cases} (x - 2)^3, & x < 3 \\ (x - 4)^2, & x \ge 3 \end{cases}$

Testing Whether a Function Is One-to-One In Exercises 59–70, determine algebraically whether the function is one-to-one. Verify your answer graphically. If the function is one-to-one, find its inverse.

59. $f(x) = x^4$

60. $g(x) = x^2 - x^4$

61. $f(x) = \dfrac{3x + 4}{5}$

62. $f(x) = 3x + 5$

63. $f(x) = \dfrac{1}{x^2}$

64. $h(x) = \dfrac{4}{x^2}$

65. $f(x) = (x + 3)^2, \quad x \ge -3$

66. $q(x) = (x - 5)^2, \quad x \le 5$

67. $f(x) = \sqrt{2x + 3}$

68. $f(x) = \sqrt{x - 2}$

69. $f(x) = |x - 2|, \quad x \le 2$

70. $f(x) = \dfrac{x^2}{x^2 + 1}, \quad x \ge 0$

Finding an Inverse Function Algebraically In Exercises 71–80, find the inverse function of f algebraically. Use a graphing utility to graph both f and f^{-1} in the same viewing window. Describe the relationship between the graphs.

71. $f(x) = 4x - 9$

72. $f(x) = 3x$

73. $f(x) = x^5$

74. $f(x) = x^3 + 1$

75. $f(x) = x^4, \quad x \le 0$

76. $f(x) = x^2, \quad x \ge 0$

77. $f(x) = \sqrt{4 - x^2}, \quad 0 \le x \le 2$

78. $f(x) = \sqrt{16 - x^2}, \quad -4 \le x \le 0$

79. $f(x) = \dfrac{4}{x^3}$

80. $f(x) = \dfrac{6}{\sqrt{x}}$

Think About It **In Exercises 81–90, restrict the domain of the function f so that the function is one-to-one and has an inverse function. Then find the inverse function f^{-1}. State the domains and ranges of f and f^{-1}. Explain your results. (There are many correct answers.)**

81. $f(x) = (x - 2)^2$

82. $f(x) = x^4 + 1$

83. $f(x) = |x + 2|$

84. $f(x) = |x - 2|$

85. $f(x) = (x + 3)^2$

86. $f(x) = (x - 4)^2$

87. $f(x) = -2x^2 - 5$

88. $f(x) = \frac{1}{2}x^2 + 1$

89. $f(x) = |x - 4| + 1$

90. $f(x) = -|x - 1| - 2$

Using the Properties of Inverse Functions **In Exercises 91 and 92, use the graph of the function f to complete the table and sketch the graph of f^{-1}.**

91.

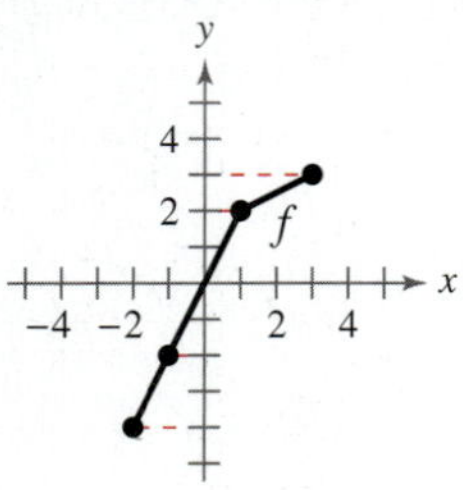

x	$f^{-1}(x)$
-4	
-2	
2	
3	

92.

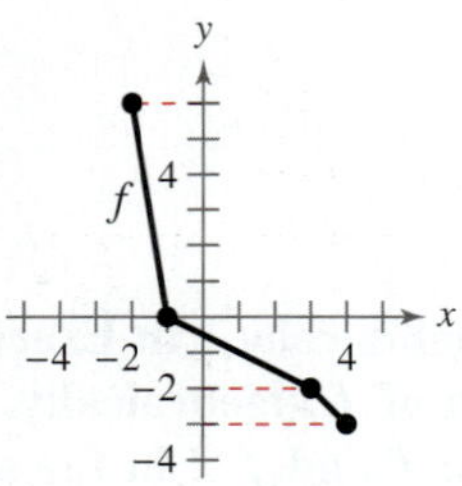

x	$f^{-1}(x)$
-3	
-2	
0	
6	

Using Graphs to Evaluate a Function **In Exercises 93–100, use the graphs of $y = f(x)$ and $y = g(x)$ to evaluate the function.**

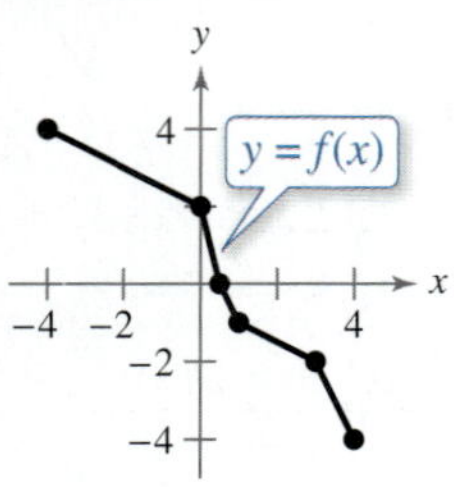

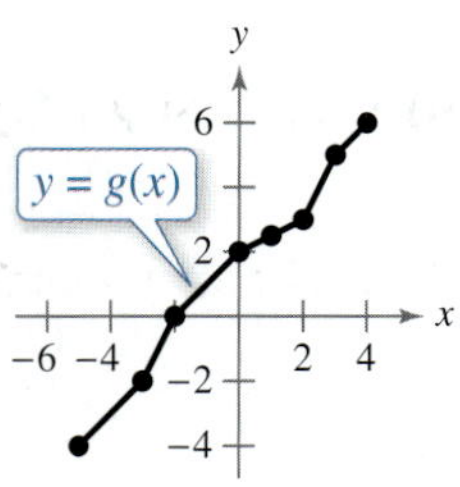

93. $f^{-1}(-4)$

94. $g^{-1}(0)$

95. $(f \circ g)(2)$

96. $g(f(-4))$

97. $f^{-1}(g(0))$

98. $(g^{-1} \circ f)(3)$

99. $(g \circ f^{-1})(2)$

100. $(f^{-1} \circ g^{-1})(6)$

Using the *Draw Inverse* Feature **In Exercises 101–104, (a) use a graphing utility to graph the function f, (b) use the *draw inverse* feature of the graphing utility to draw the inverse relation of the function, and (c) determine whether the inverse relation is an inverse function. Explain your reasoning.**

101. $f(x) = x^3 + x + 1$

102. $f(x) = x\sqrt{4 - x^2}$

103. $f(x) = \dfrac{3x^2}{x^2 + 1}$

104. $f(x) = \dfrac{4x}{\sqrt{x^2 + 15}}$

Evaluating a Composition of Functions **In Exercises 105–110, use the functions $f(x) = \frac{1}{8}x - 3$ and $g(x) = x^3$ to find the indicated value or function.**

105. $(f^{-1} \circ g^{-1})(1)$

106. $(g^{-1} \circ f^{-1})(-3)$

107. $(f^{-1} \circ f^{-1})(-6)$

108. $(g^{-1} \circ g^{-1})(4)$

109. $f^{-1} \circ g^{-1}$

110. $g^{-1} \circ f^{-1}$

Finding a Composition of Functions **In Exercises 111–114, use the functions $f(x) = x + 4$ and $g(x) = 2x - 5$ to find the specified function.**

111. $g^{-1} \circ f^{-1}$

112. $f^{-1} \circ g^{-1}$

113. $(f \circ g)^{-1}$

114. $(g \circ f)^{-1}$

115. ***Why you should learn it*** *(p. 141)* The table shows men's shoe sizes in the United States and the corresponding European shoe sizes. Let $y = f(x)$ represent the function that gives the men's European shoe size in terms of x, the men's U.S. size.

Men's U.S. shoe size	Men's European shoe size
8	41
9	42
10	43
11	44
12	45
13	46

(a) Is f one-to-one? Explain.

(b) Find $f(11)$.

(c) Find $f^{-1}(43)$, if possible.

(d) Find $f(f^{-1}(41))$.

(e) Find $f^{-1}(f(12))$.

116. Fashion Design Let $y = g(x)$ represent the function that gives the women's European shoe size in terms of x, the women's U.S. size. A women's U.S. size 6 shoe corresponds to a European size 37. Find $g^{-1}(g(6))$.

117. Military Science You can encode and decode messages using functions and their inverses. To code a message, first translate the letters to numbers using 1 for "A," 2 for "B," and so on. Use 0 for a space. So, "A ball" becomes

1 0 2 1 12 12.

Then, use a one-to-one function to convert to coded numbers. Using $f(x) = 2x - 1$, "A ball" becomes

1 −1 3 1 23 23.

(a) Encode "Call me later" using the function $f(x) = 5x + 4$.

(b) Find the inverse function of $f(x) = 5x + 4$ and use it to decode 119 44 9 104 4 104 49 69 29.

118. Production Management Your wage is $12.00 per hour plus $0.55 for each unit produced per hour. So, your hourly wage y in terms of the number of units produced x is $y = 12 + 0.55x$.

(a) Find the inverse function. What does each variable in the inverse function represent?

(b) Use a graphing utility to graph the function and its inverse function.

(c) Use the *trace* feature of the graphing utility to find the hourly wage when 9 units are produced per hour.

(d) Use the *trace* feature of the graphing utility to find the number of units produced per hour when your hourly wage is $21.35.

Conclusions

True or False? **In Exercises 119 and 120, determine whether the statement is true or false. Justify your answer.**

119. If f is an even function, then f^{-1} exists.

120. If the inverse function of f exists, and the graph of f has a y-intercept, then the y-intercept of f is an x-intercept of f^{-1}.

Think About It **In Exercises 121–124, determine whether the situation could be represented by a one-to-one function. If so, write a statement that describes the inverse function.**

121. The number of miles n a marathon runner has completed in terms of the time t in hours

122. The population p of a town in terms of the year t from 2005 through 2015 given that the population was greatest in 2012

123. The depth of the tide d at a beach in terms of the time t over a 24-hour period

124. The height h in inches of a human child from age 2 to age 14 in terms of his or her age n in years

125. Writing Describe the relationship between the graph of a function f and the graph of its inverse function f^{-1}.

126. Think About It The domain of a one-to-one function f is $[0, 9]$ and the range is $[-3, 3]$. Find the domain and range of f^{-1}.

127. Think About It The function $f(x) = \frac{9}{5}x + 32$ can be used to convert a temperature of x degrees Celsius to its corresponding temperature in degrees Fahrenheit.

(a) Using the expression for f, make a conceptual argument to show that f has an inverse function.

(b) What does $f^{-1}(50)$ represent?

128. Think About It A function f is increasing over its entire domain. Does f have an inverse function? Explain.

129. Think About It Describe a type of function that is *not* one-to-one on any interval of its domain.

130. HOW DO YOU SEE IT? Decide whether the two functions shown in each graph appear to be inverse functions of each other. Explain your reasoning.

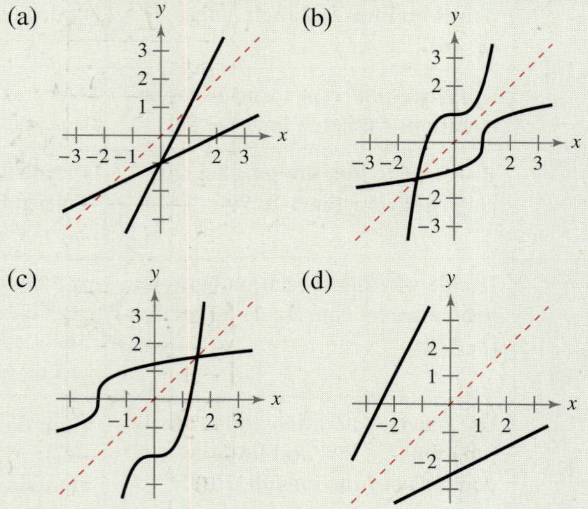

131. Proof Prove that if f and g are one-to-one functions, then $(f \circ g)^{-1}(x) = (g^{-1} \circ f^{-1})(x)$.

132. Proof Prove that if f is a one-to-one odd function, then f^{-1} is an odd function.

Cumulative Mixed Review

Simplifying a Rational Expression **In Exercises 133–136, write the rational expression in simplest form.**

133. $\dfrac{27x^3}{3x^2}$

134. $\dfrac{5x^2y^2 + 25x^2y}{xy + 5x}$

135. $\dfrac{x^2 - 36}{6 - x}$

136. $\dfrac{x^2 + 3x - 40}{x^2 - 3x - 10}$

Testing for Functions **In Exercises 137–140, determine whether the equation represents y as a function of x.**

137. $x = 5$

138. $y - 7 = -3$

139. $x^2 + y = 5$

140. $x - y^2 = 0$

1 Chapter Summary

What did you learn?		Explanation and Examples	Review Exercises
1.1	**Sketch graphs of equations by point plotting *(p. 75).***	**1.** If possible, rewrite the equation so that one of the variables is isolated on one side of the equation. **2.** Make a table of values showing several solution points. **3.** Plot these points in a rectangular coordinate system. **4.** Connect the points with a smooth curve or line.	1, 2
	Graph equations using a graphing utility *(p. 77)*, and use graphs of equations to solve real-life problems *(p. 79).*	**1.** Rewrite the equation so that y is isolated on the left side. **2.** Enter the equation in the graphing utility. **3.** Determine a *viewing window* that shows all important features of the graph. **4.** Graph the equation.	3–8
1.2	**Find the slopes of lines *(p. 84).***	**Slope m of nonvertical line:** $m = \dfrac{y_2 - y_1}{x_2 - x_1},\ x_1 \neq x_2$	9–16
	Write linear equations given points on lines and their slopes *(p. 86).*	The point-slope form of the equation of the line that passes through the point (x_1, y_1) and has a slope of m is $y - y_1 = m(x - x_1)$.	17–22
	Use slope-intercept forms of linear equations to sketch lines *(p. 88).*	The graph of the equation $y = mx + b$ is a line whose slope is m and whose y-intercept is $(0, b)$.	23–30
	Use slope to identify parallel and perpendicular lines *(p. 90).*	**Parallel lines:** Slopes are equal. **Perpendicular lines:** Slopes are negative reciprocals of each other.	31, 32
1.3	**Decide whether a relation between two variables represents a function *(p. 97).***	A function f from a set A to a set B is a relation that assigns to each element x in the set A exactly one element y in the set B. The set A is the domain (or set of inputs) of the function f, and the set B contains the range (or set of outputs).	33–40
	Use function notation and evaluate functions *(p. 99)*, and find the domains of functions *(p. 101).*	**Equation:** $f(x) = 5 - x^2$ **$f(2)$:** $f(2) = 5 - 2^2 = 1$ **Domain of $f(x) = 5 - x^2$:** All real numbers	41–46
	Use functions to model and solve real-life problems *(p. 103).*	A function can be used to model the number of interior design services employees in the United States. (See Example 8.)	47, 48
	Evaluate difference quotients *(p. 104).*	**Difference quotient:** $\dfrac{f(x + h) - f(x)}{h},\ h \neq 0$	49, 50
1.4	**Find the domains and ranges of functions *(p. 110).***	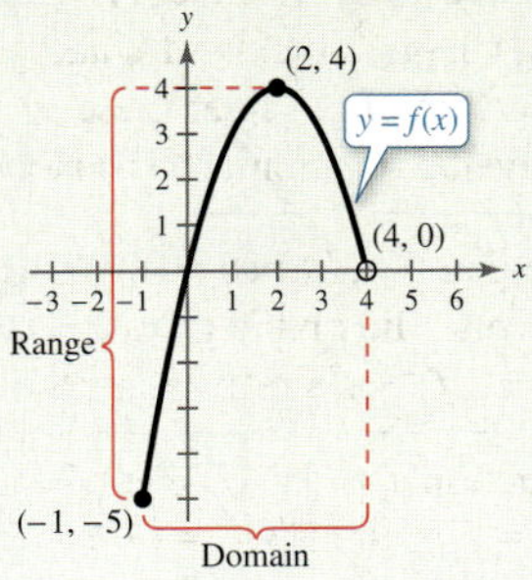	51–58
	Use the Vertical Line Test for functions *(p. 111).*	A set of points in a coordinate plane is the graph of y as a function of x if and only if no *vertical* line intersects the graph at more than one point.	59, 60

	What did you learn?	Explanation and Examples	Review Exercises
1.4	**Determine intervals on which functions are increasing, decreasing, or constant *(p. 112).***	A function f is increasing on an interval when, for any x_1 and x_2 in the interval, $x_1 < x_2$ implies $f(x_1) < f(x_2)$. A function f is decreasing on an interval when, for any x_1 and x_2 in the interval, $x_1 < x_2$ implies $f(x_1) > f(x_2)$. A function f is constant on an interval when, for any x_1 and x_2 in the interval, $f(x_1) = f(x_2)$.	61–64
	Determine relative maximum and relative minimum values of functions *(p. 113).*	A function value $f(a)$ is called a relative minimum of f when there exists an interval (x_1, x_2) that contains a such that $x_1 < x < x_2$ implies $f(a) \le f(x)$. A function value $f(a)$ is called a relative maximum of f when there exists an interval (x_1, x_2) that contains a such that $x_1 < x < x_2$ implies $f(a) \ge f(x)$.	65, 66
	Identify and graph step functions and other piecewise-defined functions *(p. 115)*, and identify even and odd functions *(p. 116).*	**Greatest integer:** $f(x) = [\![x]\!]$ **Even:** For each x in the domain of f, $f(-x) = f(x)$. **Odd:** For each x in the domain of f, $f(-x) = -f(x)$.	67–78
1.5	**Recognize graphs of parent functions *(p. 122).***	**Linear:** $f(x) = x$; **Quadratic:** $f(x) = x^2$; **Cubic:** $f(x) = x^3$; **Absolute value:** $f(x) = \|x\|$; **Square root:** $f(x) = \sqrt{x}$; **Rational:** $f(x) = 1/x$ (See Figure 1.36, page 122.)	79–82
	Use vertical and horizontal shifts *(p. 123)*, reflections *(p. 125)*, and nonrigid transformations *(p. 127)* to graph functions.	**Vertical shifts:** $h(x) = f(x) + c$ or $h(x) = f(x) - c$ **Horizontal shifts:** $h(x) = f(x - c)$ or $h(x) = f(x + c)$ **Reflection in the x-axis:** $h(x) = -f(x)$ **Reflection in the y-axis:** $h(x) = f(-x)$ **Nonrigid transformations:** $h(x) = cf(x)$ or $h(x) = f(cx)$	83–94
1.6	**Add, subtract, multiply, and divide functions *(p. 131)*, find the compositions of functions *(p. 133)*, and write a function as a composition of two functions *(p. 135).***	$(f + g)(x) = f(x) + g(x)$ $\quad (f - g)(x) = f(x) - g(x)$ $(fg)(x) = f(x) \cdot g(x)$ $\quad (f/g)(x) = f(x)/g(x),\ g(x) \neq 0$ **Composition of functions:** $(f \circ g)(x) = f(g(x))$	95–108
	Use combinations of functions to model and solve real-life problems *(p. 136).*	A composite function can be used to represent the number of bacteria in a petri dish as a function of the amount of time the petri dish has been out of refrigeration. (See Example 10.)	109, 110
1.7	**Find inverse functions informally and verify that two functions are inverse functions of each other *(p. 141).***	Let f and g be two functions such that $f(g(x)) = x$ for every x in the domain of g and $g(f(x)) = x$ for every x in the domain of f. Under these conditions, the function g is the inverse function of the function f.	111–114
	Use graphs of functions to decide whether functions have inverse functions *(p. 144).*	If the point (a, b) lies on the graph of f, then the point (b, a) must lie on the graph of f^{-1}, and vice versa. In short, f^{-1} is a reflection of f in the line $y = x$.	115, 116
	Determine whether functions are one-to-one *(p. 145).*	A function f is one-to-one when, for a and b in its domain, $f(a) = f(b)$ implies $a = b$.	117–120
	Find inverse functions algebraically *(p. 146).*	To find inverse functions, replace $f(x)$ by y, interchange the roles of x and y, and solve for y. Replace y by $f^{-1}(x)$.	121–128

1 Review Exercises

See CalcChat.com for tutorial help and worked-out solutions to odd-numbered exercises. For instructions on how to use a graphing utility, see Appendix A.

1.1

Sketching a Graph by Point Plotting **In Exercises 1 and 2, complete the table. Use the resulting solution points to sketch the graph of the equation. Use a graphing utility to verify the graph.**

1. $y = -\frac{1}{2}x + 2$

x	-2	0	2	3	4
y					
Solution point					

2. $y = x^2 - 6x$

x	-1	0	3	4	6
y					
Solution point					

Using a Graphing Utility to Graph an Equation **In Exercises 3–6, use a graphing utility to graph the equation. Approximate any x- or y-intercepts.**

3. $y = \frac{1}{4}x^4 - 9x^2$

4. $y = \frac{1}{2}x^3 - 8x$

5. $y = x\sqrt{9 - x^2}$

6. $y = x^2\sqrt{x + 4}$

7. Economics You purchase a compact car for \$17,500. The depreciated value y after t years is

$y = 17{,}500 - 1400t, \quad 0 \le t \le 6.$

(a) Use the constraints of the model to determine an appropriate viewing window.

(b) Use a graphing utility to graph the equation.

(c) Use the *zoom* and *trace* features of the graphing utility to determine the value of t when $y = \$11{,}900$.

8. Business The revenues R (in millions of dollars) for Under Armour from 2004 through 2013 can be approximated by the model

$R = 23.540t^2 - 177.56t + 595.2, \quad 4 \le t \le 13$

where t represents the year, with $t = 4$ corresponding to 2004. (Source: Under Armour, Inc.)

(a) Use a graphing utility to graph the model.

(b) Use the *table* feature of the graphing utility to approximate the revenues for Under Armour from 2004 through 2013.

(c) Use the *zoom* and *trace* features of the graphing utility to predict the first year that the revenues for Under Armour exceed \$5 billion. Confirm your result algebraically.

1.2

Finding the Slope of a Line **In Exercises 9–16, plot the two points and find the slope of the line passing through the points.**

9. $(-3, 2), (8, 2)$

10. $(3, -1), (-3, -1)$

11. $(-5, -1), (-5, 9)$

12. $(8, -1), (8, 2)$

13. $\left(\frac{3}{2}, 1\right), \left(5, \frac{5}{2}\right)$

14. $\left(-\frac{3}{4}, \frac{5}{6}\right), \left(\frac{1}{2}, -\frac{5}{2}\right)$

15. $(-4.5, 6), (2.1, 3)$

16. $(-2.7, -6.3), (0, 1.8)$

The Point-Slope Form of the Equation of a Line **In Exercises 17–22, (a) use the point on the line and the slope of the line to find an equation of the line, and (b) find three additional points through which the line passes. (There are many correct answers.)**

17. $(2, -1), m = \frac{1}{4}$

18. $(-3, 5), m = -\frac{3}{2}$

19. $(0, -5), m = -\frac{3}{2}$

20. $(0, 1), m = \frac{4}{5}$

21. $(-2, 6), m = 0$

22. $(-8, 8), m = 0$

Finding the Slope-Intercept Form **In Exercises 23–30, write an equation of the line that passes through the points. Use the slope-intercept form, if possible. If not possible, explain why. Use a graphing utility to graph the line (if possible).**

23. $(2, -1), (4, -1)$

24. $(0, 0), (0, 10)$

25. $\left(\frac{5}{6}, -1\right), \left(\frac{5}{6}, 3\right)$

26. $\left(7, \frac{4}{3}\right), \left(9, \frac{4}{3}\right)$

27. $(-1, 0), (6, 2)$

28. $(1, 6), (4, 2)$

29. $(3, -1), (-3, 2)$

30. $\left(-\frac{5}{2}, 1\right), \left(-4, \frac{2}{9}\right)$

Equations of Parallel and Perpendicular Lines **In Exercises 31 and 32, write the slope-intercept forms of the equations of the lines through the given point (a) parallel to the given line and (b) perpendicular to the given line. Verify your result with a graphing utility (use a *square setting*).**

	Point	*Line*
31.	$(3, -2)$	$5x - 4y = 8$
32.	$(-8, 3)$	$2x + 3y = 5$

1.3

Testing for Functions **In Exercises 33 and 34, which set of ordered pairs represents a function from A to B? Explain.**

33. $A = \{10, 20, 30, 40\}$ and $B = \{0, 2, 4, 6\}$

(a) $\{(20, 4), (40, 0), (20, 6), (30, 2)\}$

(b) $\{(10, 4), (20, 4), (30, 4), (40, 4)\}$

34. $A = \{u, v, w\}$ and $B = \{-2, -1, 0, 1, 2\}$

(a) $\{(u, -2), (v, 2), (w, 1)\}$

(b) $\{(w, -2), (v, 0), (w, 2)\}$

Testing for Functions Represented Algebraically **In Exercises 35–40, determine whether the equation represents y as a function of x.**

35. $16x^2 - y^2 = 0$
36. $x^3 + y^2 = 64$
37. $2x - y - 3 = 0$
38. $2x + y = 10$
39. $y = \sqrt{1 - x}$
40. $y = \sqrt{x^2 + 4}$

Evaluating a Function **In Exercises 41 and 42, evaluate the function at each specified value of the independent variable, and simplify.**

41. $f(x) = x^2 + 1$
    (a) $f(1)$ (b) $f(-3)$
    (c) $f(b^3)$ (d) $f(x - 1)$
42. $g(x) = \sqrt{x^2 + 1}$
    (a) $g(-1)$ (b) $g(3)$
    (c) $g(3x)$ (d) $g(x + 2)$

Finding the Domain of a Function **In Exercises 43–46, find the domain of the function.**

43. $f(x) = \dfrac{x - 1}{x + 2}$
44. $f(x) = \dfrac{x^2}{x^2 + 1}$
45. $f(x) = \sqrt{25 - x^2}$
46. $f(x) = \sqrt{x^2 - 16}$

47. **Industrial Engineering** A hand tool manufacturer produces a product for which the variable cost is \$5.25 per unit and the fixed costs are \$17,500. The company sells the product for \$8.43 and can sell all that it produces.
    (a) Write the total cost C as a function of x, the number of units produced.
    (b) Write the profit P as a function of x.

48. **Education** The numbers n (in millions) of students enrolled in public schools in the United States from 2005 through 2012 can be approximated by

$$n(t) = \begin{cases} -0.35t^3 + 6.95t^2 - 44.8t + 158, & 5 \le t \le 8 \\ -0.35t^2 + 7.7t + 26, & 8 < t \le 12 \end{cases}$$

where t is the year, with $t = 5$ corresponding to 2005. (Source: U.S. Census Bureau)
    (a) Use the *table* feature of a graphing utility to approximate the enrollments from 2005 through 2012.
    (b) Use the graphing utility to graph the model and estimate the enrollments for the years 2013 through 2017. Do the values seem reasonable? Explain.

Evaluating a Difference Quotient **In Exercises 49 and 50, find the difference quotient $\dfrac{f(x+h) - f(x)}{h}$ for the given function and simplify your answer.**

49. $f(x) = 2x^2 + 3x - 1$
50. $f(x) = x^2 - 3x + 5$

1.4

Finding the Domain and Range of a Function **In Exercises 51–58, use a graphing utility to graph the function and estimate its domain and range. Then find the domain and range algebraically.**

51. $f(x) = 3 - 2x^2$
52. $f(x) = 2x^2 + 5$
53. $f(x) = \sqrt{x + 3} + 4$
54. $f(x) = 2 - \sqrt{x - 5}$
55. $h(x) = \sqrt{36 - x^2}$
56. $f(x) = \sqrt{x^2 - 9}$
57. $f(x) = |x - 5| + 2$
58. $f(x) = |x + 1| - 3$

Vertical Line Test for Functions **In Exercises 59 and 60, use the Vertical Line Test to determine whether y is a function of x. Describe how to enter the equation into a graphing utility to produce the given graph.**

59. $y - 4x = x^2$
60. $3x + y^2 - 2 = 0$

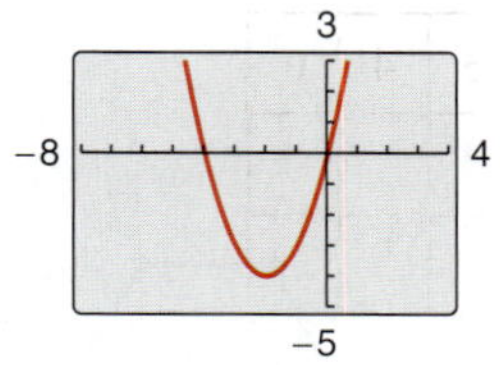

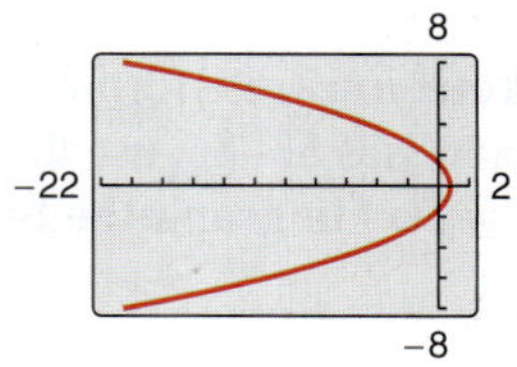

Increasing and Decreasing Functions **In Exercises 61–64, (a) use a graphing utility to graph the function and (b) determine the open intervals on which the function is increasing, decreasing, or constant.**

61. $f(x) = x^3 - 3x$
62. $f(x) = \sqrt{x^2 - 9}$
63. $f(x) = x\sqrt{x - 6}$
64. $f(x) = \dfrac{|x + 8|}{2}$

Approximating Relative Minima and Maxima **In Exercises 65 and 66, use a graphing utility to approximate (to two decimal places) any relative minimum or relative maximum values of the function.**

65. $f(x) = (x^2 - 4)^2$
66. $f(x) = x^3 - 4x^2 - 1$

Sketching Graphs **In Exercises 67–70, sketch the graph of the function by hand.**

67. $f(x) = \begin{cases} 3x + 5, & x < 0 \\ x - 4, & x \ge 0 \end{cases}$
68. $f(x) = \begin{cases} \frac{1}{2}x + 3, & x < 0 \\ 4 - x^2, & x \ge 0 \end{cases}$
69. $f(x) = [\![x]\!] - 3$
70. $f(x) = [\![x + 2]\!]$

Even and Odd Functions **In Exercises 71–78, determine algebraically whether the function is even, odd, or neither. Verify your answer using a graphing utility.**

71. $f(x) = x^2 + 6$
72. $f(x) = x^2 - x - 1$
73. $f(x) = (x^2 - 8)^2$
74. $f(x) = 2x^3 - x$
75. $f(x) = 3x^{5/2}$
76. $f(x) = 3x^{2/5}$
77. $f(x) = 2x\sqrt{4 - x^2}$
78. $f(x) = x\sqrt{x^2 - 1}$

1.5

Library of Parent Functions **In Exercises 79–82, identify the parent function and describe the transformation shown in the graph. Write an equation for the graphed function.**

79.

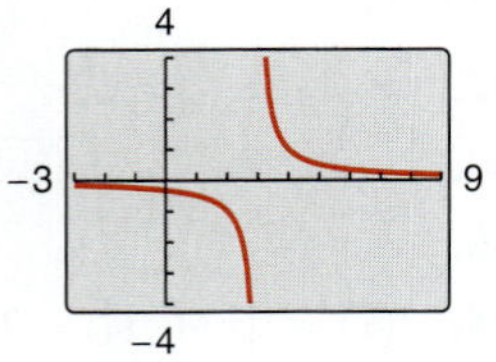

80.

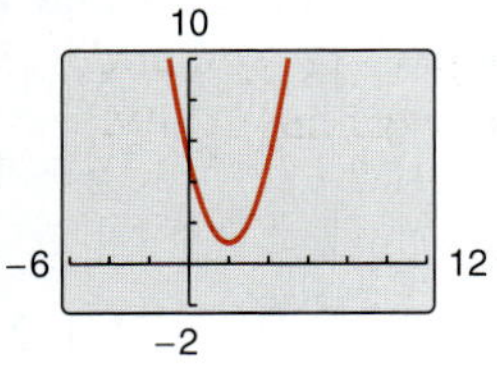

81.

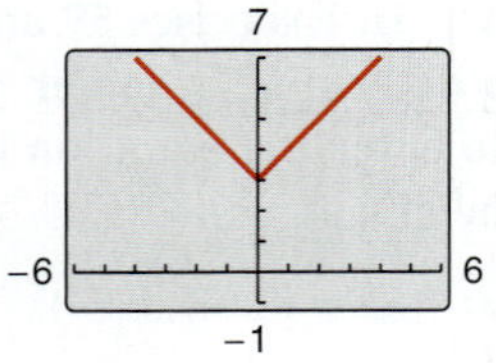

82.

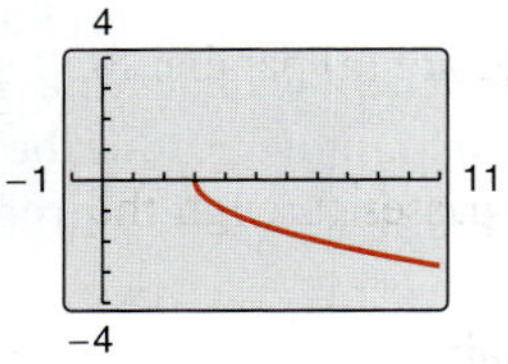

Sketching Transformations **In Exercises 83–86, use the graph of $y = f(x)$ to graph the function.**

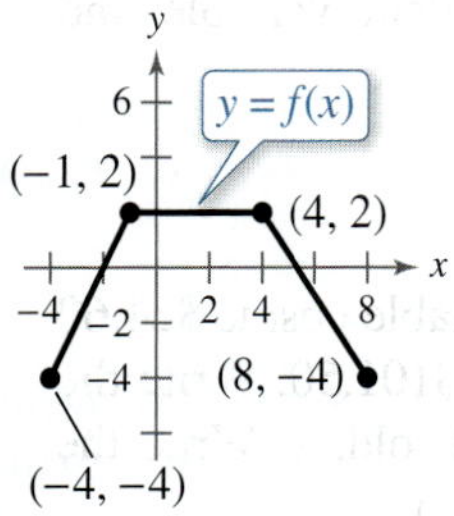

83. $y = f(-x)$

84. $y = -f(x)$

85. $y = f(x) + 2$

86. $y = f(x - 1)$

Describing Transformations **In Exercises 87–94, h is related to one of the six parent functions on page 122. (a) Identify the parent function f. (b) Describe the sequence of transformations from f to h. (c) Sketch the graph of h by hand. (d) Use function notation to write h in terms of the parent function f.**

87. $h(x) = \dfrac{1}{x} - 6$

88. $h(x) = -\dfrac{1}{x} + 3$

89. $h(x) = (x - 2)^3 + 5$

90. $h(x) = (-x)^2 - 8$

91. $h(x) = -\sqrt{x} - 6$

92. $h(x) = \sqrt{x - 1} + 4$

93. $h(x) = |3x| + 9$

94. $h(x) = |2x + 8| - 1$

1.6

Evaluating a Combination of Functions **In Exercises 95–102, let $f(x) = 3 - 2x$, $g(x) = \sqrt{x}$, and $h(x) = 3x^2 + 2$, and find the indicated values.**

95. $(f - g)(4)$

96. $(f + h)(5)$

97. $(f + g)(25)$

98. $(g - h)(1)$

99. $(fh)(1)$

100. $(g/h)(1)$

101. $(h \circ g)(5)$

102. $(g \circ f)(-3)$

Identifying a Composite Function **In Exercises 103–108, find two functions f and g such that $(f \circ g)(x) = h(x)$. (There are many correct answers.)**

103. $h(x) = (x + 3)^2$

104. $h(x) = (1 - 2x)^3$

105. $h(x) = \sqrt{4x + 2}$

106. $h(x) = \sqrt[3]{(x + 2)^2}$

107. $h(x) = \dfrac{4}{x + 2}$

108. $h(x) = \dfrac{6}{(3x + 1)^3}$

Education **In Exercises 109 and 110, the numbers (in thousands) of students taking the SAT (y_1) and the ACT (y_2) for the years 2008 through 2013 can be modeled by**

$$y_1 = -5.824t^3 + 182.05t^2 - 1832.8t + 7515 \text{ and}$$

$$y_2 = 3.398t^3 - 103.63t^2 + 1106.5t - 2543$$

where t represents the year, with $t = 8$ corresponding to 2008. (Source: College Entrance Examination Board and ACT, Inc.)

109. Use a graphing utility to graph y_1, y_2, and $y_1 + y_2$ in the same viewing window.

110. Use the model $y_1 + y_2$ to estimate the total number of students taking the SAT and ACT in 2017.

1.7

Finding Inverse Functions Informally **In Exercises 111–114, find the inverse function of f informally. Verify that $f(f^{-1}(x)) = x$ and $f^{-1}(f(x)) = x$.**

111. $f(x) = 6x$

112. $f(x) = x + 5$

113. $f(x) = \dfrac{1}{2}x + 3$

114. $f(x) = \dfrac{x - 4}{5}$

Algebraic-Graphical-Numerical **In Exercises 115 and 116, show that f and g are inverse functions (a) algebraically, (b) graphically, and (c) numerically.**

115. $f(x) = 3 - 4x$, $g(x) = \dfrac{3 - x}{4}$

116. $f(x) = \sqrt{x + 1}$; $g(x) = x^2 - 1$, $x \geq 0$

Using the Horizontal Line Test **In Exercises 117–120, use a graphing utility to graph the function and use the Horizontal Line Test to determine whether the function is one-to-one and an inverse function exists.**

117. $f(x) = \frac{1}{2}x - 3$

118. $f(x) = (x - 1)^2$

119. $h(t) = (t + 5)^{2/3}$

120. $g(x) = \dfrac{5}{x - 4}$

Finding an Inverse Function Algebraically **In Exercises 121–128, find the inverse function of f algebraically.**

121. $f(x) = \dfrac{1}{2}x - 5$

122. $f(x) = \dfrac{7x + 3}{8}$

123. $f(x) = -5x^3 - 3$

124. $f(x) = 5x^3 + 2$

125. $f(x) = \sqrt{x + 10}$

126. $f(x) = 4\sqrt{6 - x}$

127. $f(x) = \frac{1}{4}x^2 + 1, \quad x \geq 0$

128. $f(x) = 5 - \frac{1}{9}x^2, \quad x \leq 0$

1 Chapter Test

See *CalcChat.com* for tutorial help and worked-out solutions to odd-numbered exercises.
For instructions on how to use a graphing utility, see Appendix A.

Take this test as you would take a test in class. After you are finished, check your work against the answers given in the back of the book.

In Exercises 1–6, use the point-plotting method to graph the equation by hand and identify any x- and y-intercepts. Verify your results using a graphing utility.

1. $y = 2|x| - 1$ **2.** $y = -\frac{2}{3}x + 3$

3. $y = 2x^2 - 4x$ **4.** $y = x^3 - x$

5. $y = -x^2 + 9$ **6.** $y = \sqrt{x - 2}$

7. Find equations of the lines that pass through the point $(0, 4)$ and are (a) parallel to and (b) perpendicular to the line $5x + 2y = 3$.

8. Find the slope-intercept form of the equation of the line that passes through the points $(2, -1)$ and $(-3, 4)$.

9. Does the graph at the right represent y as a function of x? Explain.

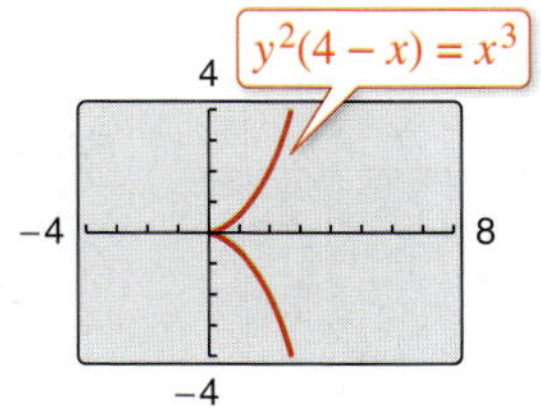

Figure for 9

10. Evaluate $f(x) = |x + 2| - 15$ at each value of the independent variable and simplify.

(a) $f(-8)$ (b) $f(14)$ (c) $f(t - 6)$

11. Find the domain of $f(x) = 10 - \sqrt{3 - x}$.

12. An electronics company produces a car stereo for which the variable cost is \$24.60 per unit and the fixed costs are \$25,000. The product sells for \$101.50. Write the total cost C as a function of the number of units produced and sold, x. Write the profit P as a function of the number of units produced and sold, x.

In Exercises 13 and 14, determine algebraically whether the function is even, odd, or neither.

13. $f(x) = 2x^3 - 3x$ **14.** $f(x) = 3x^4 + 5x^2$

In Exercises 15 and 16, determine the open intervals on which the function is increasing, decreasing, or constant.

15. $h(x) = \frac{1}{4}x^4 - 2x^2$ **16.** $g(t) = |t + 2| - |t - 2|$

In Exercises 17 and 18, use a graphing utility to graph the functions and to approximate (to two decimal places) any relative minimum or relative maximum values of the function.

17. $f(x) = -x^3 - 5x^2 + 12$ **18.** $f(x) = x^5 - 3x^3 + 4$

In Exercises 19–21, (a) identify the parent function f, (b) describe the sequence of transformations from f to g, and (c) sketch the graph of g.

19. $g(x) = -2(x - 5)^3 + 3$ **20.** $g(x) = \sqrt{-7x - 14}$ **21.** $g(x) = 4|-x| - 7$

22. Use the functions $f(x) = x^2$ and $g(x) = \sqrt{2 - x}$ to find the specified function and its domain.

(a) $(f - g)(x)$ (b) $\left(\frac{f}{g}\right)(x)$ (c) $(f \circ g)(x)$ (d) $(g \circ f)(x)$

In Exercises 23–25, determine whether the function has an inverse function, and if so, find the inverse function.

23. $f(x) = x^3 + 8$ **24.** $f(x) = x^2 + 6$ **25.** $f(x) = \dfrac{3x\sqrt{x}}{8}$

Proofs in Mathematics

Conditional Statements

Many theorems are written in the **if-then form** "if p, then q," which is denoted by

$p \rightarrow q$ Conditional statement

where p is the **hypothesis** and q is the **conclusion.** Here are some other ways to express the conditional statement $p \rightarrow q$.

p implies q. p, only if q. p is sufficient for q.

Conditional statements can be either true or false. The conditional statement $p \rightarrow q$ is false only when p is true and q is false. To show that a conditional statement is true, you must prove that the conclusion follows for all cases that fulfill the hypothesis. To show that a conditional statement is false, you need to describe only a single **counterexample** that shows that the statement is not always true.

For instance, $x = -4$ is a counterexample that shows that the following statement is false.

If $x^2 = 16$, then $x = 4$.

The hypothesis "$x^2 = 16$" is true because $(-4)^2 = 16$. However, the conclusion "$x = 4$" is false. This implies that the given conditional statement is false.

For the conditional statement $p \rightarrow q$, there are three important associated conditional statements.

1. The **converse** of $p \rightarrow q$: $q \rightarrow p$
2. The **inverse** of $p \rightarrow q$: $\sim p \rightarrow \sim q$
3. The **contrapositive** of $p \rightarrow q$: $\sim q \rightarrow \sim p$

The symbol $\sim$ means the **negation** of a statement. For instance, the negation of "The engine is running" is "The engine is not running."

EXAMPLE 1 Writing the Converse, Inverse, and Contrapositive

Write the converse, inverse, and contrapositive of the conditional statement "If I get a B on my test, then I will pass the course."

Solution

a. *Converse:* If I pass the course, then I got a B on my test.

b. *Inverse:* If I do not get a B on my test, then I will not pass the course.

c. *Contrapositive:* If I do not pass the course, then I did not get a B on my test. ■

In the example above, notice that neither the converse nor the inverse is logically equivalent to the original conditional statement. On the other hand, the contrapositive is logically equivalent to the original conditional statement.

2 Solving Equations and Inequalities

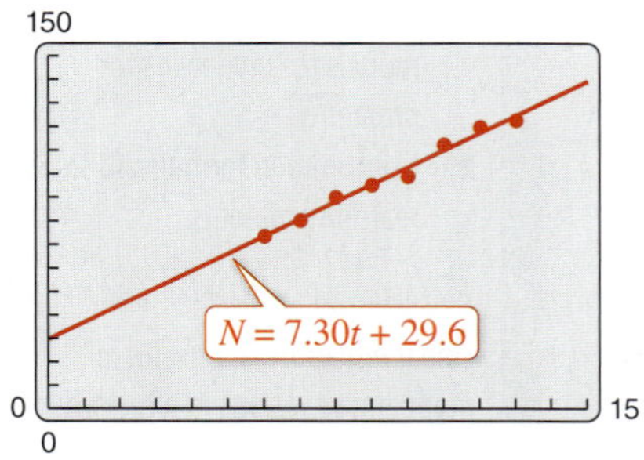

Section 2.7, Example 4
e-Filed Tax Returns

2.1 Linear Equations and Problem Solving

Equations and Solutions of Equations

An **equation** in x is a statement that two algebraic expressions are equal. For example,

$$3x - 5 = 7, \quad x^2 - x - 6 = 0, \quad \text{and} \quad \sqrt{2x} = 4$$

are equations. To **solve** an equation in x means to find all values of x for which the equation is true. Such values are **solutions.** For instance, $x = 4$ is a solution of the equation $3x - 5 = 7$, because $3(4) - 5 = 7$ is a true statement.

The solutions of an equation depend on the kinds of numbers being considered. For instance, in the set of rational numbers, $x^2 = 10$ has no solution because there is no rational number whose square is 10. In the set of real numbers, however, the equation has two solutions: $x = \sqrt{10}$ and $x = -\sqrt{10}$.

An equation can be classified as an *identity*, a *conditional*, or a *contradiction*, as shown in the following table.

Equation	*Definition*	*Example*
Identity	An equation that is true for *every* real number in the domain of the variable	$\frac{x}{3x^2} = \frac{1}{3x}$ is a true statement for any nonzero real value of x.
Conditional	An equation that is true for *some* (but not all) of the real numbers in the domain of the variable	$x^2 - 9 = 0$ is a true statement for $x = 3$ and $x = -3$, but not for any other real values.
Contradiction	An equation that is *false* for every real number in the domain of the variable	$2x + 1 = 2x - 3$ is a false statement for any real value of x.

A **linear equation in one variable x** is an equation that can be written in the standard form $ax + b = 0$, where a and b are real numbers, with $a \neq 0$. (For a review of solving one- and two-step linear equations, see Appendix D at this textbook's *Companion Website*.)

To solve an equation involving fractional expressions, find the least common denominator (LCD) of all terms in the equation and multiply every term by this LCD. This procedure clears the equation of fractions.

EXAMPLE 1 Solving an Equation Involving Fractions

$$\frac{x}{3} + \frac{3x}{4} = 2 \qquad \text{Original equation}$$

$$(12)\frac{x}{3} + (12)\frac{3x}{4} = (12)2 \qquad \text{Multiply each term by the LCD of 12.}$$

$$4x + 9x = 24 \qquad \text{Divide out and multiply.}$$

$$13x = 24 \qquad \text{Combine like terms.}$$

$$x = \frac{24}{13} \qquad \text{Divide each side by 13.}$$

✓ Checkpoint *Audio-video solution in English & Spanish at LarsonPrecalculus.com.*

Solve $\frac{4x}{9} - \frac{1}{3} = x + \frac{5}{3}$.

What you should learn

- Solve equations involving fractional expressions.
- Write and use mathematical models to solve real-life problems.
- Use common formulas to solve real-life problems.

Why you should learn it

Linear equations are useful for modeling situations in which you need to find missing information. For instance, Exercise 60 on page 168 shows how to use a linear equation to determine the height of a flagpole by measuring its shadow.

After solving an equation, you should check each solution in the original equation. For instance, you can check the solution to Example 1 as follows.

$$\frac{x}{3} + \frac{3x}{4} = 2 \qquad \text{Write original equation.}$$

$$\frac{\frac{24}{13}}{3} + \frac{3\left(\frac{24}{13}\right)}{4} \stackrel{?}{=} 2 \qquad \text{Substitute } \tfrac{24}{13} \text{ for } x.$$

$$\frac{8}{13} + \frac{18}{13} \stackrel{?}{=} 2 \qquad \text{Simplify.}$$

$$2 = 2 \qquad \text{Solution checks. ✓}$$

When multiplying or dividing an equation by a *variable* expression, it is possible to introduce an **extraneous solution**—one that does not satisfy the original equation. The next example demonstrates the importance of checking your solution when you have multiplied or divided by a variable expression.

EXAMPLE 2 An Equation with an Extraneous Solution

See LarsonPrecalculus.com for an interactive version of this type of example.

Solve $\dfrac{1}{x-2} = \dfrac{3}{x+2} - \dfrac{6x}{x^2-4}$.

Solution

The LCD is $x^2 - 4 = (x+2)(x-2)$. Multiply each term by the LCD.

$$\frac{1}{x-2}(x+2)(x-2) = \frac{3}{x+2}(x+2)(x-2) - \frac{6x}{x^2-4}(x+2)(x-2)$$

$$x + 2 = 3(x-2) - 6x, \quad x \neq \pm 2$$

$$x + 2 = 3x - 6 - 6x$$

$$4x = -8$$

$$x = -2 \qquad \text{Extraneous solution}$$

In the original equation, $x = -2$ yields a denominator of zero. So, $x = -2$ is an extraneous solution, and the original equation has *no solution*. To check this result graphically, graph

$$y_1 = \frac{1}{x-2} \quad \text{and} \quad y_2 = \frac{3}{x+2} - \frac{6x}{x^2-4}$$

in the same viewing window, as shown in the figure.

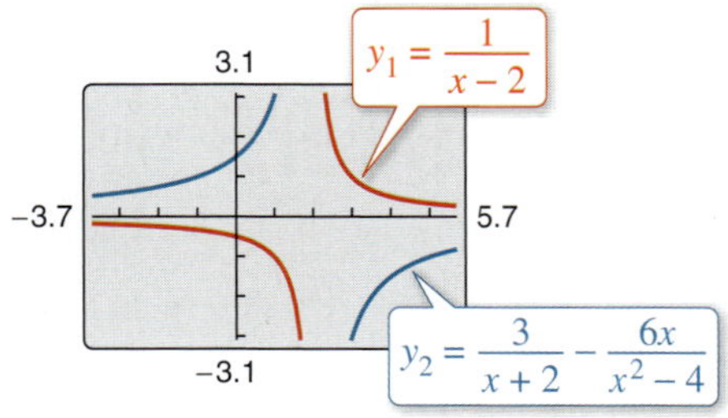

The graphs of the equations do not appear to intersect. This means that there is no x-value for which the left side of the equation y_1 is equal to the right side of the equation y_2. So, the equation appears to have *no solution*.

✓ Checkpoint *Audio-video solution in English & Spanish at LarsonPrecalculus.com.*

Solve $\dfrac{3x}{x-4} = 5 + \dfrac{12}{x-4}$.

What's Wrong?

To approximate the solution of $0.9x + 1 = 0.91x$, you use a graphing utility to graph $y_1 = 0.9x + 1$ and $y_2 = 0.91x$, as shown in the figure. You use the graph to conclude that the equation has no solution because the lines do not intersect. What's wrong?

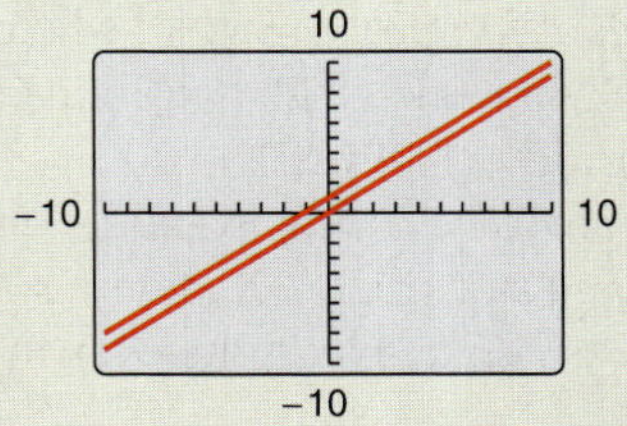

Using Mathematical Models to Solve Problems

One of the primary goals of this text is to learn how algebra can be used to solve problems that occur in real-life situations. This procedure, introduced in Chapter 1, is called **mathematical modeling.**

A good approach to mathematical modeling is to use two stages. Begin by using the verbal description of the problem to form a *verbal model*. Then, after assigning labels to the quantities in the verbal model, form a *mathematical model* or an *algebraic equation*.

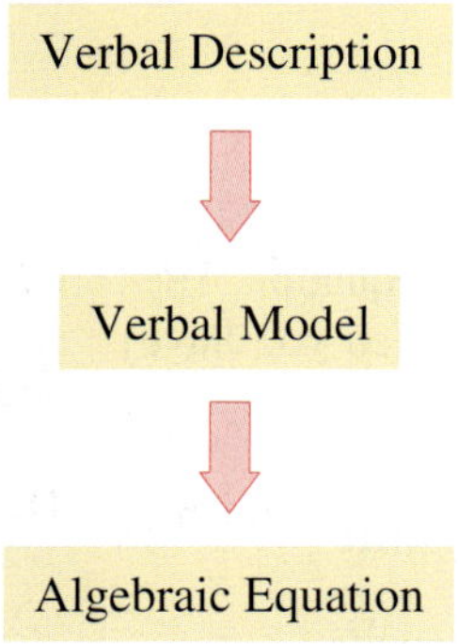

Remark

Students sometimes say that although a solution looks easy when it is worked out in class, they don't see where to begin when solving a problem alone. Keep in mind that no one—not even great mathematicians—can expect to look at every mathematical problem and know immediately where to begin. Many problems involve some trial and error before a solution is found. To make algebra work for you, put in a lot of time, expect to try solution methods that end up not working, and learn from both your successes and your failures.

EXAMPLE 3 Finding the Dimensions of a Room

A rectangular kitchen is twice as long as it is wide, and its perimeter is 84 feet. Find the dimensions of the kitchen.

Solution

For this problem, it helps to draw a diagram, as shown in Figure 2.1.

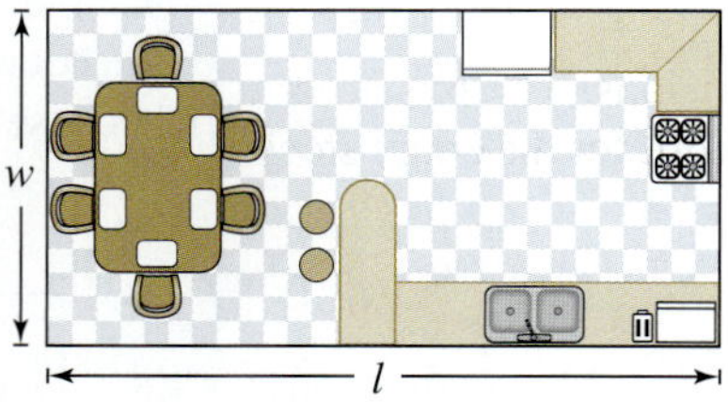

Figure 2.1

Verbal Model: $2 \cdot$ Length $+ 2 \cdot$ Width $=$ Perimeter

Labels: Perimeter $= 84$ (feet)

Width $= w$ (feet)

Length $= l = 2w$ (feet)

Equation: $2(2w) + 2w = 84$ Write equation.

$6w = 84$ Combine like terms.

$w = 14$ Divide each side by 6.

Because the length is twice the width, you have

$l = 2w = 2(14) = 28.$

So, the dimensions of the room are 14 feet by 28 feet.

Checkpoint *Audio-video solution in English & Spanish at LarsonPrecalculus.com.*

A rectangular family room is 3 times as long as it is wide, and its perimeter is 112 feet. Find the dimensions of the family room.

EXAMPLE 4 Height of a Building

You measure the shadow cast by the Aon Center in Chicago, Illinois, and find it to be 142 feet long, as shown in Figure 2.2. Then you measure the shadow cast by a 48-inch post and find it to be 6 inches long. Determine the building's height.

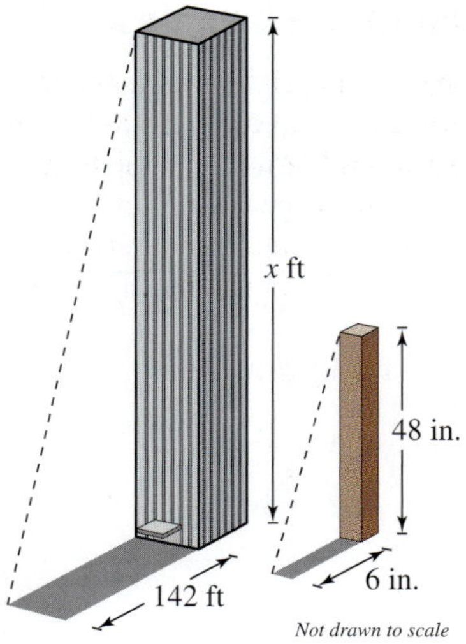

Figure 2.2

Solution

To solve this problem, use the result from geometry that states that the ratios of corresponding sides of similar triangles are equal.

Verbal Model: $\dfrac{\text{Height of building}}{\text{Length of building's shadow}} = \dfrac{\text{Height of post}}{\text{Length of post's shadow}}$

Labels: Height of building $= x$ (feet)

Length of building's shadow $= 142$ (feet)

Height of post $= 48$ (inches)

Length of post's shadow $= 6$ (inches)

Equation: $\dfrac{x}{142} = \dfrac{48}{6}$ ⇨ $x = 1136$

So, the Aon Center is 1136 feet high.

✓ Checkpoint *Audio-video solution in English & Spanish at LarsonPrecalculus.com.*

The shadow cast by a building is 55 feet long. The shadow cast by a 4-foot post is 1.8 feet long. Determine the building's height.

EXAMPLE 5 An Inventory Problem

A store has \$30,000 of inventory in 32-inch and 46-inch televisions. The profit on a 32-inch set is 22% and the profit on a 46-inch set is 40%. The profit for the entire stock is 35%. How much was invested in each type of television?

Store Manager

Solution

Verbal Model: Profit from 32-inch sets + Profit from 46-inch sets = Total profit

Labels: Inventory of 32-inch sets $= x$ (dollars)

Inventory of 46-inch sets $= 30{,}000 - x$ (dollars)

Profit from 32-inch sets $= 0.22x$ (dollars)

Profit from 46-inch sets $= 0.40(30{,}000 - x)$ (dollars)

Total profit $= 0.35(30{,}000) = 10{,}500$ (dollars)

Equation:
$$0.22x + 0.40(30{,}000 - x) = 10{,}500$$
$$0.22x + 12{,}000 - 0.40x = 10{,}500$$
$$-0.18x = -1500$$
$$x \approx 8333.33$$

So, \$8333.33 was invested in 32-inch sets and $30{,}000 - x$, or \$21,666.67, was invested in 46-inch sets.

Remark

Notice in the solution to Example 5 that each percent is expressed as a decimal. For instance, 22% is written as 0.22.

✓ Checkpoint *Audio-video solution in English & Spanish at LarsonPrecalculus.com.*

Rework Example 5 when the store has \$50,000 worth of inventory.

Common Formulas

Many common types of geometric, scientific, and investment problems use ready-made equations called **formulas.** Knowing these formulas will help you translate and solve a wide variety of real-life applications.

Common Formulas for Area A, Perimeter P, Circumference C, and Volume V

Square

$A = s^2$

$P = 4s$

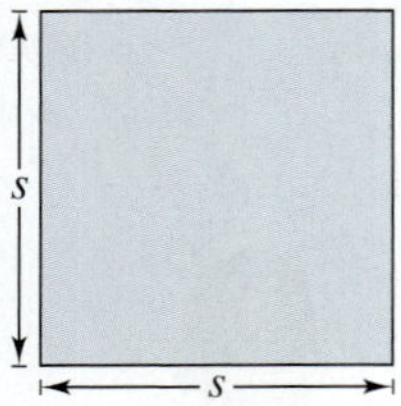

Rectangle

$A = lw$

$P = 2l + 2w$

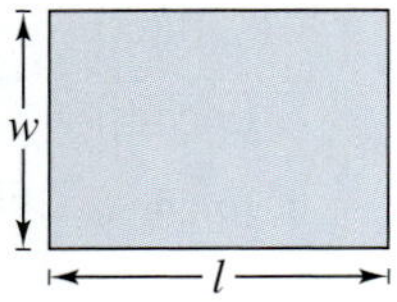

Circle

$A = \pi r^2$

$C = 2\pi r$

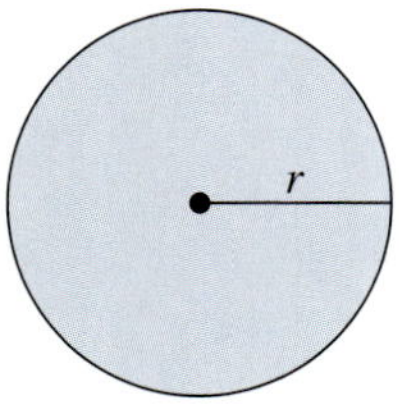

Triangle

$A = \frac{1}{2}bh$

$P = a + b + c$

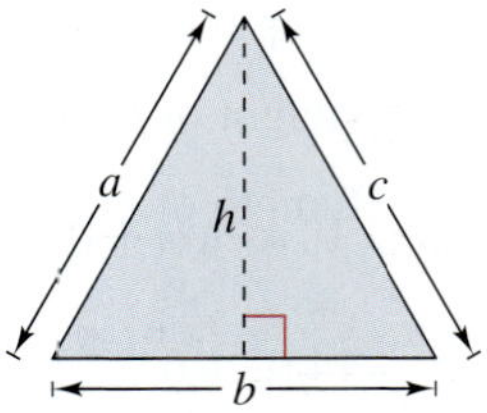

Cube

$V = s^3$

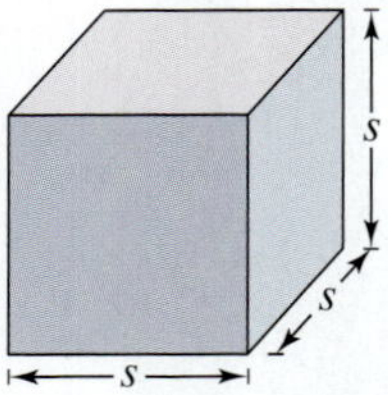

Rectangular Solid

$V = lwh$

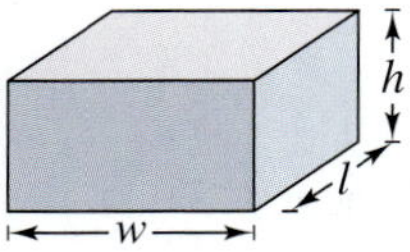

Right Circular Cylinder

$V = \pi r^2 h$

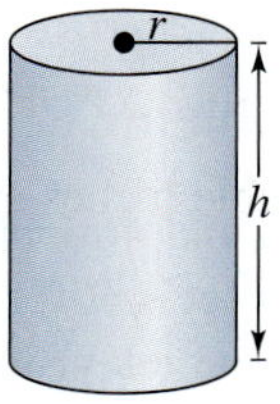

Sphere

$V = \frac{4}{3}\pi r^3$

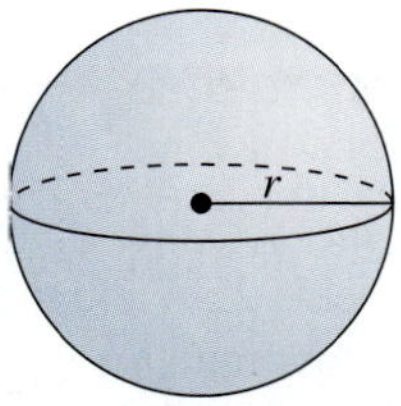

Miscellaneous Common Formulas

Temperature: $F = \frac{9}{5}C + 32$

F = degrees Fahrenheit
C = degrees Celsius

Simple Interest: $I = Prt$

I = interest
P = principal (original deposit)
r = annual interest rate
t = time in years

Compound Interest: $A = P\left(1 + \frac{r}{n}\right)^{nt}$

A = balance
P = principal (original deposit)
r = annual interest rate
n = compoundings (number of times interest is calculated) per year
t = time in years

Distance: $d = rt$

d = distance traveled
r = rate
t = time

When working with applied problems, you may find it helpful to rewrite a common formula. For instance, the formula for the perimeter of a rectangle, $P = 2l + 2w$, can be solved for w as $w = \frac{1}{2}(P - 2l)$.

EXAMPLE 6 Using a Formula

A cylindrical can has a volume of 600 cubic centimeters and a radius of 4 centimeters, as shown in Figure 2.3. Find the height of the can.

Figure 2.3

Solution

The formula for the volume of a cylinder is

$$V = \pi r^2 h. \qquad \text{Volume of a cylinder}$$

To find the height of the can, solve for h.

$$h = \frac{V}{\pi r^2}$$

Then, using $V = 600$ and $r = 4$, find the height.

$$h = \frac{600}{\pi(4)^2} = \frac{600}{16\pi} \approx 11.94$$

The height of the can is about 11.94 centimeters. You can use unit analysis to check that your answer is reasonable.

$$\frac{600 \text{ cm}^3}{16\pi \text{ cm}^2} \approx 11.94 \text{ cm}$$

✓ **Checkpoint** *Audio-video solution in English & Spanish at LarsonPrecalculus.com.*

A cylindrical container has a volume of 84 cubic inches and a radius of 3 inches. Find the height of the container.

EXAMPLE 7 Using a Formula

The average daily temperature in a city is 64.4°F. What is the city's average daily temperature in degrees Celsius?

Solution

First solve for C in the formula for temperature. Then use $F = 64.4$ to find the temperature in degrees Celsius.

$$F = \frac{9}{5}C + 32 \qquad \text{Formula for temperature}$$

$$F - 32 = \frac{9}{5}C \qquad \text{Subtract 32 from each side.}$$

$$\frac{5}{9}(F - 32) = C \qquad \text{Multiply each side by } \tfrac{5}{9}.$$

$$\frac{5}{9}(64.4 - 32) = C \qquad \text{Substitute 64.4 for } F.$$

$$18 = C \qquad \text{Simplify.}$$

The average daily temperature in the city is 18°C.

✓ **Checkpoint** *Audio-video solution in English & Spanish at LarsonPrecalculus.com.*

You jog at a steady speed of 6 miles per hour. How long will it take you to jog 5 miles?

Remark

Once you have rewritten the formula for temperature, you can easily find other Celsius values. Simply substitute other Fahrenheit values and evaluate.

2.1 Exercises

See *CalcChat.com* for tutorial help and worked-out solutions to odd-numbered exercises.
For instructions on how to use a graphing utility, see Appendix A.

Vocabulary and Concept Check

In Exercises 1–4, fill in the blank.

1. A(n) _______ is a statement that two algebraic expressions are equal.

2. A linear equation in one variable is an equation that can be written in the standard form _______ .

3. When solving an equation, it is possible to introduce a(n) _______ solution, which is a value that does not satisfy the original equation.

4. Many real-life problems can be solved using ready-made equations called _______ .

5. Is the equation $x + 1 = 3$ an identity, a conditional equation, or a contradiction?

6. How can you clear the equation $\frac{x}{2} + 1 = \frac{1}{4}$ of fractions?

Procedures and Problem Solving

Checking Solutions of an Equation **In Exercises 7–10, determine whether each value of x is a solution of the equation.**

Equation	*Values*	
7. $\frac{5}{2x} - \frac{4}{x} = 3$	(a) $x = -\frac{1}{2}$	(b) $x = 4$
	(c) $x = 0$	(d) $x = \frac{1}{4}$
8. $\frac{x}{2} + \frac{6x}{7} = \frac{19}{14}$	(a) $x = -2$	(b) $x = 1$
	(c) $x = \frac{1}{2}$	(d) $x = 7$
9. $\frac{\sqrt{x+4}}{6} + 3 = 4$	(a) $x = -3$	(b) $x = 0$
	(c) $x = 21$	(d) $x = 32$
10. $\frac{\sqrt[3]{x-8}}{3} = -1$	(a) $x = 4$	(b) $x = 0$
	(c) $x = -19$	(d) $x = 16$

Classifying Equations **In Exercises 11–18, determine whether the equation is an identity, a conditional equation, or a contradiction.**

11. $2(x - 1) = 2x - 2$

12. $-5(x - 1) = -5(x + 1)$

13. $(x + 3)(x - 5) = x^2 - 2(x + 7)$

14. $x^2 - 8x + 5 = (x - 4)^2 - 11$

15. $(x + 6)^2 = (x + 8)(x + 2)$

16. $(x + 1)(x - 5) = (x + 3)(x - 1)$

17. $3 + \frac{1}{x + 1} = \frac{4x}{x + 1}$

18. $\frac{5}{x} + \frac{3}{x} = 24$

Solving an Equation Involving Fractions **In Exercises 19–22, solve the equation using two methods. Then explain which method you prefer.**

19. $\frac{3x}{8} - \frac{4x}{3} = 4$

20. $\frac{3z}{8} - \frac{z}{10} = 6$

21. $\frac{2x}{5} + 5x = \frac{4}{3}$

22. $\frac{4y}{3} - 2y = \frac{16}{5}$

Solving Equations **In Exercises 23–42, solve the equation (if possible).**

23. $3x - 5 = 2x + 7$

24. $5x + 3 = 6 - 2x$

25. $3(y - 5) = 3 + 5y$

26. $4(z - 3) + 3z = 1 + 8z$

27. $\frac{x}{5} - \frac{x}{2} = 3$

28. $\frac{3x}{4} + \frac{x}{2} = -5$

29. $\frac{5x - 4}{5x + 4} = \frac{2}{3}$

30. $\frac{10x + 3}{5x + 6} = \frac{1}{2}$

31. $\frac{2}{5}(z - 4) + \frac{3z}{10} = 4z$

32. $\frac{3x}{2} + \frac{1}{4}(x - 2) = 10$

33. $\frac{17 + y}{y} + \frac{32 + y}{y} = 100$

34. $\frac{x - 11}{x} = \frac{x - 9}{x} + 2$

35. $\frac{1}{x - 3} + \frac{1}{x + 3} = \frac{10}{x^2 - 9}$

36. $\frac{1}{x - 2} + \frac{3}{x + 3} = \frac{4}{x^2 + x - 6}$

37. $\frac{1}{x} + \frac{2}{x - 5} = 0$

38. $3 = 2 + \frac{2}{z + 2}$

39. $\frac{2}{(x - 4)(x - 2)} = \frac{1}{x - 4} + \frac{2}{x - 2}$

40. $\frac{2}{x(x - 2)} + \frac{5}{x} = \frac{1}{x - 2}$

41. $\dfrac{3}{x^2 - 3x} + \dfrac{4}{x} = \dfrac{1}{x - 3}$ **42.** $\dfrac{6}{x} - \dfrac{2}{x + 3} = \dfrac{3(x + 5)}{x(x + 3)}$

Solving for a Variable **In Exercises 43–48, solve for the indicated variable.**

43. ***Area of a Triangle***

Solve for h: $A = \frac{1}{2}bh$

44. ***Area of a Trapezoid***

Solve for b: $A = \frac{1}{2}(a + b)h$

45. ***Investment at Compound Interest***

Solve for P: $A = P\left(1 + \dfrac{r}{n}\right)^{nt}$

46. ***Investment at Simple Interest***

Solve for r: $A = P + Prt$

47. ***Volume of a Right Circular Cylinder***

Solve for h: $V = \pi r^2 h$

48. ***Volume of a Right Circular Cone***

Solve for h: $V = \frac{1}{3}\pi r^2 h$

MODELING DATA

In Exercises 49 and 50, use the following information. The relationship between the length of an adult's femur (thigh bone) and the height of the adult can be approximated by the linear equations

$\mathbf{y = 0.386x - 19.20}$ **Female**

$\mathbf{y = 0.442x - 29.37}$ **Male**

where y is the length of the femur in centimeters and x is the height of the adult in centimeters. (See figure.)

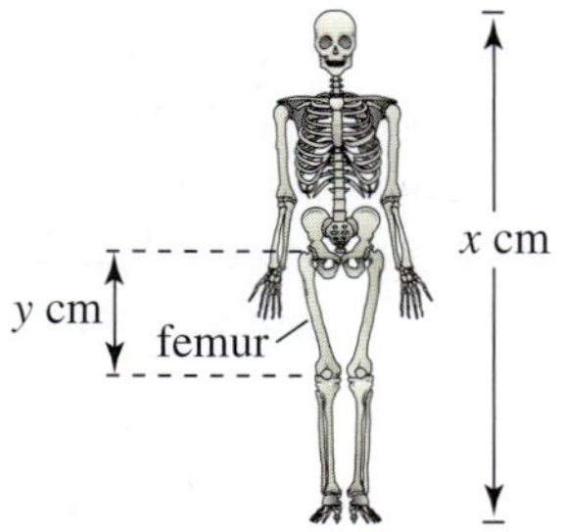

49. An anthropologist discovers a femur belonging to an adult human female. The bone is 43 centimeters long. Estimate the height of the female.

50. From the foot bones of an adult human male, an anthropologist estimates that the person's height was 175 centimeters. A few feet away from the site where the foot bones were discovered, the anthropologist discovers a male adult femur that is 48 centimeters long. Is it likely that both the foot bones and the thigh bone came from the same person?

51. **Interior Design** A room is 1.5 times as long as it is wide, and its perimeter is 25 meters.

(a) Draw a diagram that gives a visual representation of the problem. Identify the length as l and the width as w.

(b) Write l in terms of w and write an equation for the perimeter in terms of w.

(c) Find the dimensions of the room.

52. **Woodworking** A picture frame has a total perimeter of 3 meters. The height of the frame is $\frac{2}{3}$ times its width.

(a) Draw a diagram that gives a visual representation of the problem. Identify the width as w and the height as h.

(b) Write h in terms of w and write an equation for the perimeter in terms of w.

(c) Find the dimensions of the picture frame.

53. **Education** To get an A in a course, you must have an average of at least 90 on four tests of 100 points each. The scores on your first three tests were 93, 91, and 84.

(a) Write a verbal model for the test average for the course.

(b) What is the least you can score on the fourth test to get an A in the course?

54. **Business** A store generates Monday through Thursday sales of \$150, \$125, \$75, and \$180. What sales on Friday would give a weekday average of \$150?

55. **Travel Time** A salesperson is driving from the office to a client, a distance of about 250 kilometers. After 30 minutes, the salesperson passes a town that is 50 kilometers from the office. Assuming the salesperson continues at the same constant speed, how long will it take to drive from the office to the client?

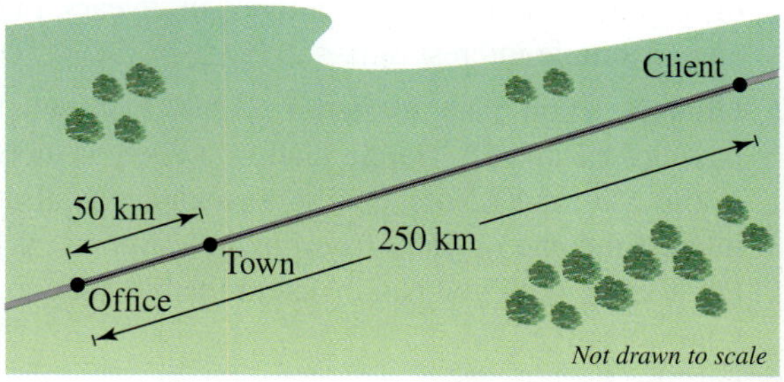

56. **Travel Time** On the first part of a 336-mile trip, a salesperson averaged 58 miles per hour. The salesperson averaged only 52 miles per hour on the last part of the trip because of an increased volume of traffic. The total time of the trip was 6 hours. Find the amount of time at each of the two speeds.

57. **Average Speed** A truck driver traveled at an average speed of 55 miles per hour on a 200-mile trip to pick up a load of freight. On the return trip (with the truck fully loaded), the average speed was 40 miles per hour. Find the average speed for the round trip.

58. Geography You are driving on a Canadian freeway to a town that is 300 kilometers from your home. After 30 minutes, you pass a freeway exit that you know is 50 kilometers from your home. Assuming that you continue at the same constant speed, how long will it take for the entire trip?

59. Dendrology To determine the height of a pine tree, you measure the shadow cast by the tree and find it to be 20 feet long. Then you measure the shadow cast by a 36-inch-tall oak sapling and find it to be 24 inches long (see figure). Estimate the height of the pine tree.

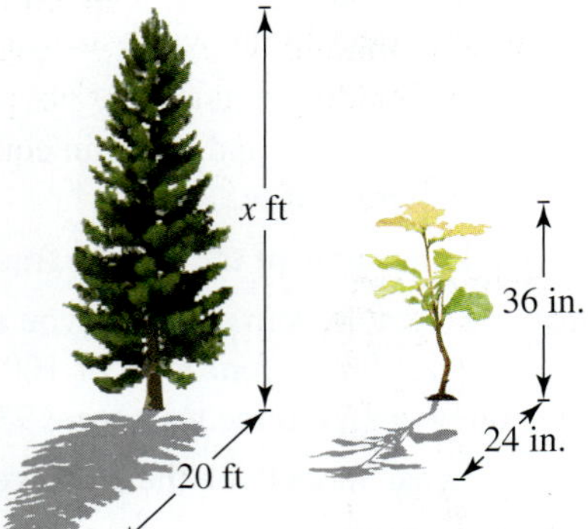

Not drawn to scale

60. *Why you should learn it* *(p. 160)* A person who is 6 feet tall walks away from a flagpole toward the tip of the shadow of the flagpole. When the person is 30 feet from the flagpole, the tips of the person's shadow and the shadow cast by the flagpole coincide at a point 5 feet in front of the person.

(a) Draw a diagram that illustrates the problem. Let h represent the height of the flagpole.

(b) Find the height of the flagpole.

61. Finance A certificate of deposit with an initial deposit of \$8000 accumulates \$200 interest in 2 years. Find the annual simple interest rate.

62. Finance You plan to invest \$12,000 in two funds paying $4\frac{1}{2}\%$ and 5% simple interest. (There is more risk in the 5% fund.) Your goal is to obtain a total annual interest income of \$560 from the investments. What is the least amount you can invest in the 5% fund to meet your objective?

63. Merchandising A grocer mixes peanuts that cost \$2.50 per pound and walnuts that cost \$8.00 per pound to make 100 pounds of a mixture that costs \$5.25 per pound. How much of each kind of nut is put into the mixture?

64. Forestry A forester mixes gasoline and oil to make 2 gallons of mixture for a two-cycle chainsaw engine. This mixture is 32 parts gasoline and 1 part oil. How much gasoline must be added to bring the mixture to 40 parts gasoline and 1 part oil?

65. Retail Management A store has \$40,000 of inventory in notebook computers and tablet computers. The profit on a notebook computer is 20% and the profit on a tablet computer is 25%. The profit for the entire stock is 24%. How much is invested in notebook computers and how much in tablet computers?

66. Retail Management A store has \$4500 of inventory in 8×10 picture frames and 5×7 picture frames. The profit on an 8×10 frame is 25% and the profit on a 5×7 frame is 22%. The profit on the entire stock is 24%. How much is invested in the 8×10 picture frames and how much in the 5×7 picture frames?

67. Sailing A triangular sail has an area of 182.25 square feet. The sail has a base of 13.5 feet. Find the height of the sail.

68. Geometry The figure shows three squares. The perimeter of square I is 20 inches and the perimeter of square II is 32 inches. Find the area of square III.

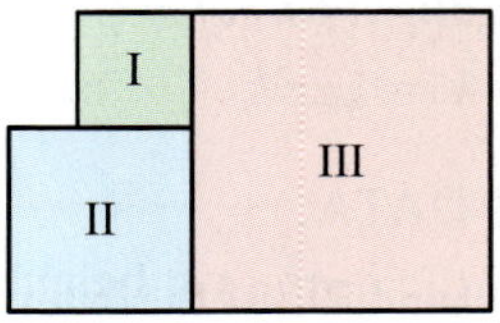

69. Geometry The volume of a rectangular package is 2304 cubic inches. The length of the package is 3 times its width, and the height is $1\frac{1}{2}$ times its width.

(a) Draw a diagram that illustrates the problem. Label the height, width, and length accordingly.

(b) Find the dimensions of the package. Use a graphing utility to verify your result.

70. Geometry The volume of a globe is about 6255 cubic centimeters. Use a graphing utility to find the radius of the globe. Round your result to two decimal places.

71. Meteorology The line graph shows the temperatures (in degrees Fahrenheit) on a summer day in Cleveland, Ohio, from 10:00 A.M. to 6:00 P.M. Create a new line graph showing the temperatures throughout the day in degrees Celsius.

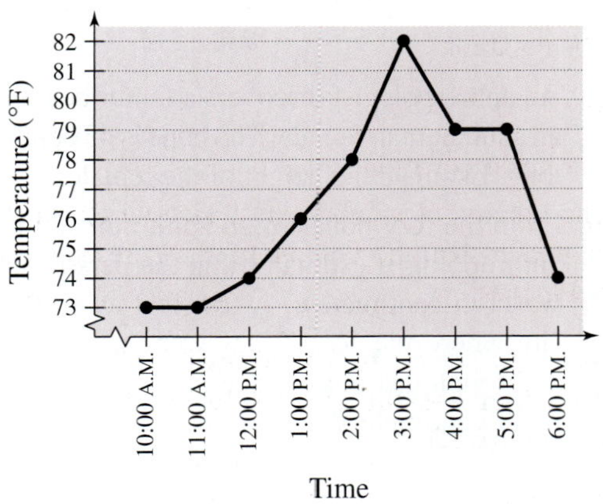

72. **Meteorology** The average July 2013 temperature in the contiguous United States was 74.3°F. What was the average temperature in degrees Celsius? (Source: NOAA)

73. **Aviation** An executive flew in the corporate jet to a meeting in a city 1500 kilometers away. After traveling the same amount of time on the return flight, the pilot mentioned that they still had 300 kilometers to go. The air speed of the plane was 600 kilometers per hour. How fast was the wind blowing? (Assume that the wind direction was parallel to the flight path and constant all day.)

74. **Parks and Recreation** A gondola tower in an amusement park casts a shadow that is 80 feet long, while a sign that is 4 feet tall casts a shadow that is $3\frac{1}{2}$ feet long. Draw a diagram for the situation. Then find the height of the tower.

Physics **In Exercises 75 and 76, you have a uniform beam of length L with a fulcrum x feet from one end. Objects with weights W_1 and W_2 are placed at opposite ends of the beam (see figure). The beam will balance when**

$$W_1x = W_2(L - x).$$

Find x such that the beam will balance.

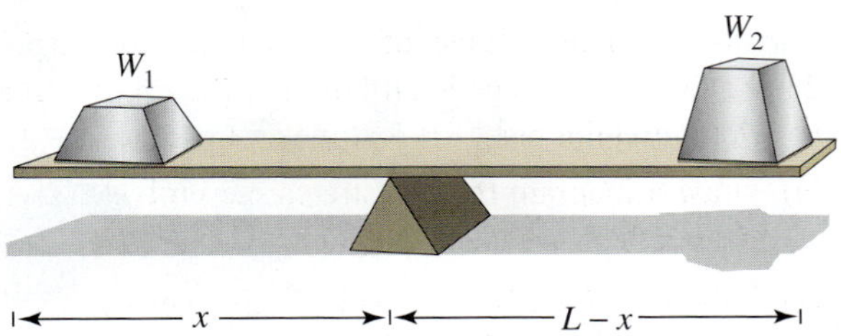

75. Two children weighing 50 pounds (W_1) and 75 pounds (W_2) are going to play on a seesaw that is 10 feet long.

76. A person weighing 200 pounds (W_1) is attempting to move a 550-pound rock (W_2) with a bar that is 5 feet long.

Conclusions

True or False? **In Exercises 77 and 78, determine whether the statement is true or false. Justify your answer.**

77. The equation

$$x(3 - x) = 10$$

is a linear equation.

78. The volume of a cube with a side length of 9.5 inches is greater than the volume of a sphere with a radius of 5.9 inches.

Writing Equations **In Exercises 79–82, write a linear equation that has the given solution. (There are many correct answers.)**

79. $x = -3$ **80.** $x = 0$

81. $x = \frac{1}{4}$ **82.** $x = -2.5$

83. **Error Analysis** Describe the error in solving the equation.

$$\frac{1}{x+1} + \frac{1}{x-1} = \frac{2}{(x+1)(x-1)}$$

$$(x - 1) + (x + 1) = 2$$

$$2x = 2$$

$$x = 1$$

The solution is $x = 1$.

84. **HOW DO YOU SEE IT?** To determine the height of a building, you measure the shadows cast by the building and a four-foot post. (See figure.)

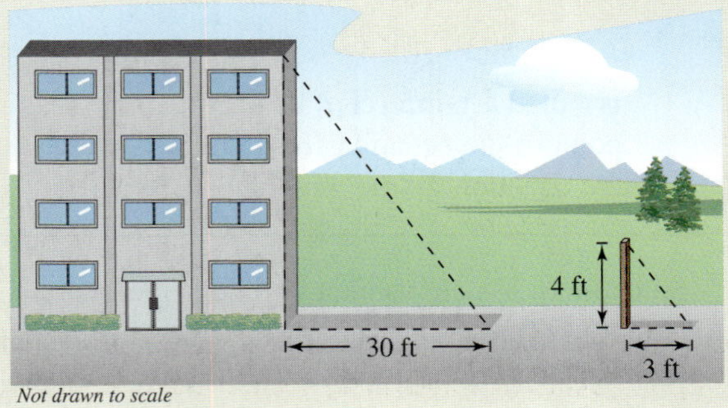

(a) Write a verbal model for the situation.

(b) Translate your verbal model into an algebraic equation.

85. **Writing** Consider the equation

$$\frac{6}{(x-3)(x-1)} = \frac{3}{x-3} + \frac{4}{x-1}.$$

Without performing any calculations, explain how to clear this equation of fractions. Is it possible that this process will introduce an extraneous solution? If so, describe two ways to determine whether a solution is extraneous.

86. **Think About It** Find c such that $x = 2$ is a solution of the linear equation $5x + 2c = 12 + 4x - 2c$.

Cumulative Mixed Review

Sketching a Graph **In Exercises 87–92, sketch the graph of the equation by hand. Verify using a graphing utility.**

87. $y = 5 - 4x$ **88.** $y = \dfrac{3x - 5}{2} + 2$

89. $y = (x - 3)^2 - 7$ **90.** $y = 4 - \frac{1}{3}x^2$

91. $y = -\frac{1}{2}|x + 4| - 1$ **92.** $y = |x - 2| + 10$

2.2 Solving Equations Graphically

Intercepts, Zeros, and Solutions

In Section 1.1, you learned that the intercepts of a graph are the points at which the graph intersects the x- or y-axis.

Definitions of Intercepts

1. The point

$(a, 0)$

is called an ***x*-intercept** of the graph of an equation when it is a solution point of the equation. To find the x-intercept(s), set y equal to 0 and solve the equation for x.

2. The point

$(0, b)$

is called a ***y*-intercept** of the graph of an equation when it is a solution point of the equation. To find the y-intercept(s), set x equal to 0 and solve the equation for y.

What you should learn

- Find x- and y-intercepts of graphs of equations.
- Find solutions of equations graphically.
- Find the points of intersection of two graphs.

Why you should learn it

Because some real-life problems involve equations that are difficult to solve algebraically, it is helpful to use a graphing utility to approximate the solutions of such equations. For instance, Exercise 85 on page 178 shows how to find the intersection point of two equations to determine the year during which the population of one state exceeded the population of another state.

Sometimes it is convenient to denote the x-intercept as simply the x-coordinate of the point $(a, 0)$ rather than the point itself. Likewise, you can denote the y-intercept as the y-coordinate of the point $(0, b)$. Unless it is necessary to make a distinction, the term *intercept* will refer to either the point or the coordinate.

It is possible for a graph to have no intercepts, one intercept, or several intercepts. For instance, consider the four graphs shown in Figure 2.4.

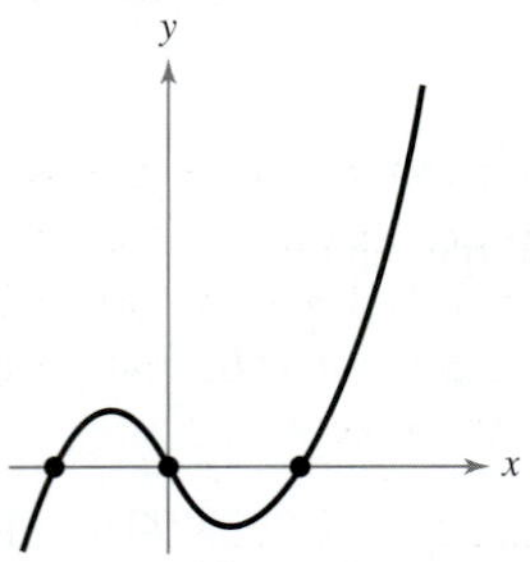

Three x-Intercepts
One y-Intercept

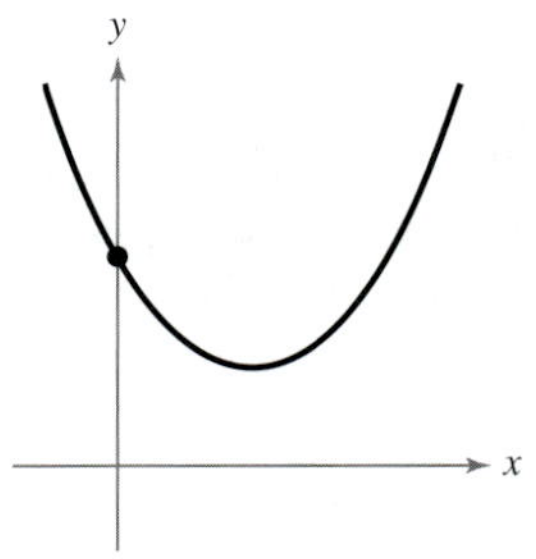

No x-Intercepts
One y-Intercept

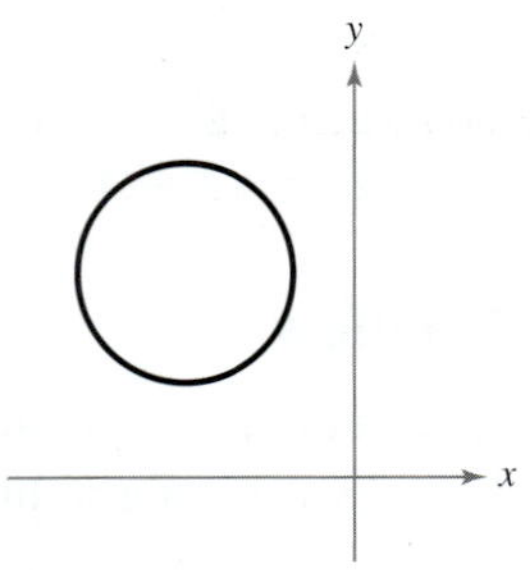

No Intercepts

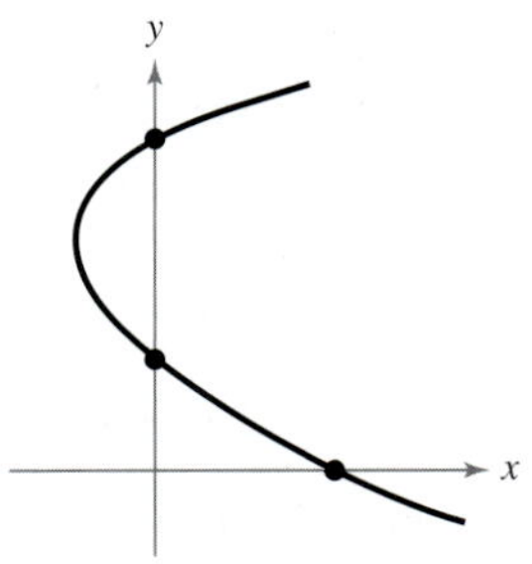

One x-Intercept
Two y-Intercepts

Figure 2.4

As you study this section, you will see the connection among x-intercepts, zeros of functions, and solutions of equations.

EXAMPLE 1 Finding x- and y-Intercepts

Find the x- and y-intercepts of the graph of $2x + 3y = 5$.

Solution

To find the x-intercept, let $y = 0$ and solve for x. This produces

$$2x = 5 \quad \Rightarrow \quad x = \frac{5}{2}$$

which implies that the graph has one x-intercept at $\left(\frac{5}{2}, 0\right)$. To find the y-intercept, let $x = 0$ and solve for y. This produces

$$3y = 5 \quad \Rightarrow \quad y = \frac{5}{3}$$

which implies that the graph has one y-intercept at $\left(0, \frac{5}{3}\right)$. See Figure 2.5.

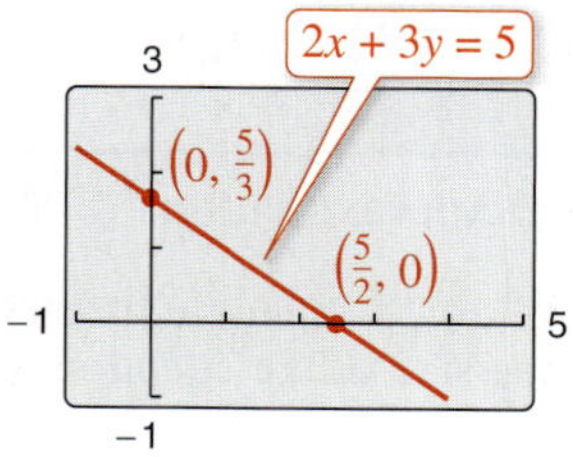

Figure 2.5

✓ ***Checkpoint*** *Audio-video solution in English & Spanish at LarsonPrecalculus.com.*

Find the x- and y-intercepts of the graph of $y = 2x - 3$.

A **zero** of a function $y = f(x)$ is a number a such that $f(a) = 0$. So, to find the zeros of a function, you must solve the equation $f(x) = 0$.

The concepts of x-intercepts, zeros of functions, and solutions of equations are closely related. In fact, the following statements are equivalent when a is a real number.

1. The point $(a, 0)$ is an *x-intercept* of the graph of $y = f(x)$.
2. The number a is a *zero* of the function f.
3. The number a is a *solution* of the equation $f(x) = 0$.

EXAMPLE 2 Verifying Zeros of Functions

Verify that the real numbers -2 and 3 are zeros of $f(x) = x^2 - x - 6$.

Algebraic Solution

To verify that -2 is a zero of f, check that $f(-2) = 0$.

$f(x) = x^2 - x - 6$ — Write original function.

$f(-2) = (-2)^2 - (-2) - 6$ — Substitute -2 for x.

$= 4 + 2 - 6$ — Simplify.

$= 0$ — -2 is a zero. ✓

To verify that 3 is a zero of f, check that $f(3) = 0$.

$f(x) = x^2 - x - 6$ — Write original function.

$f(3) = (3)^2 - 3 - 6$ — Substitute 3 for x.

$= 9 - 3 - 6$ — Simplify.

$= 0$ — 3 is a zero. ✓

Graphical Solution

Use a graphing utility to graph $y = x^2 - x - 6$. From the figure shown, it appears that the graph of f has x-intercepts when $x = -2$ and $x = 3$. So, -2 and 3 are zeros of f. Use the *zero* or *root* feature of the graphing utility to verify these zeros.

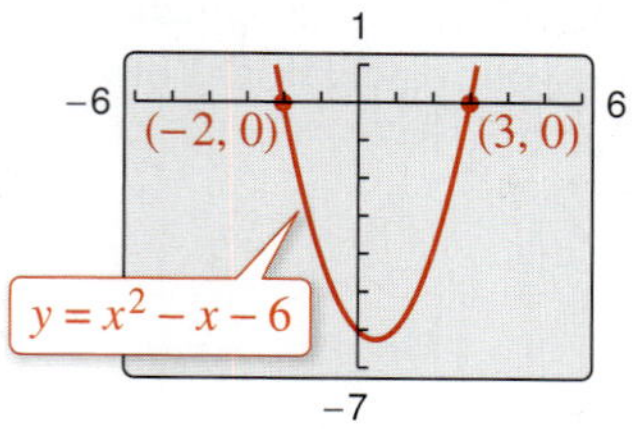

✓ ***Checkpoint*** *Audio-video solution in English & Spanish at LarsonPrecalculus.com.*

Verify both algebraically and graphically that $x = 4$ is a zero of $f(x) = 24 - 6x$.

The close connection among x-intercepts, zeros, and solutions is crucial to your study of algebra. You can take advantage of this connection in two ways. Use your algebraic "equation-solving skills" to find the x-intercepts of a graph and your "graphing skills" to approximate the solutions of an equation.

Finding Solutions Graphically

Polynomial equations of degree 1 or 2 can be solved in relatively straightforward ways. Solving polynomial equations of higher degrees can, however, be quite difficult, especially when you rely only on algebraic techniques. For such equations, a graphing utility can be very helpful.

Graphical Approximations of Solutions of an Equation

1. Write the equation in *general form*, $f(x) = 0$, with the nonzero terms on one side of the equation and zero on the other side.
2. Use a graphing utility to graph the function $y = f(x)$. Be sure the viewing window shows all the relevant features of the graph.
3. Use the *zero* or *root* feature or the *zoom* and *trace* features of the graphing utility to approximate the x-intercepts of the graph of f.

In Chapter 3 you will learn techniques for determining the number of solutions of a polynomial equation. For now, you should know that a polynomial equation of degree n cannot have more than n different solutions.

EXAMPLE 3 Finding Solutions Graphically

Use a graphing utility to approximate any solutions of

$$2x^3 - 3x + 2 = 0.$$

Solution

Graph the function $y = 2x^3 - 3x + 2$. You can see from the graph shown at the right that there is one x-intercept. It lies between -2 and -1 and is approximately -1.5. By using the *zero* or *root* feature of a graphing utility, you can improve the approximation. Choose a left bound of $x = -2$ (see Figure 2.6) and a right bound of $x = -1$ (see Figure 2.7). To two-decimal-place accuracy, the solution is $x \approx -1.48$, as shown in Figure 2.8. Check this approximation on your calculator. You will find that the value of y is

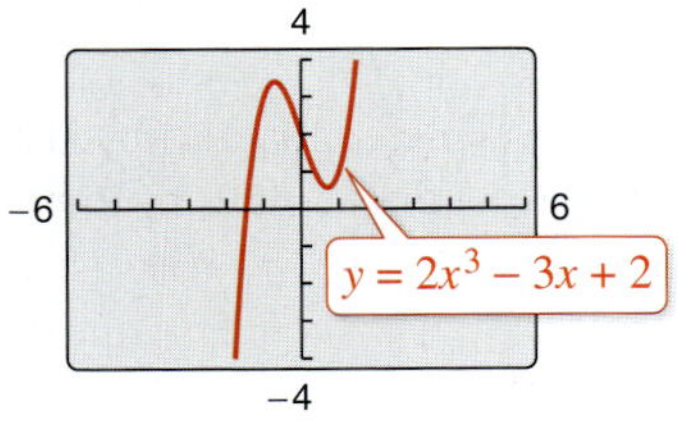

$$y = 2(-1.48)^3 - 3(-1.48) + 2$$
$$\approx -0.04.$$

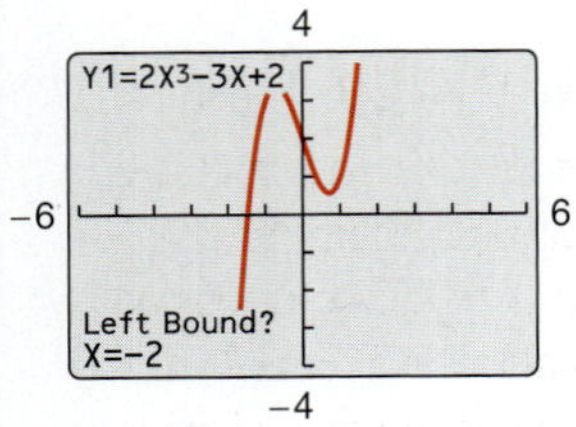

Figure 2.6

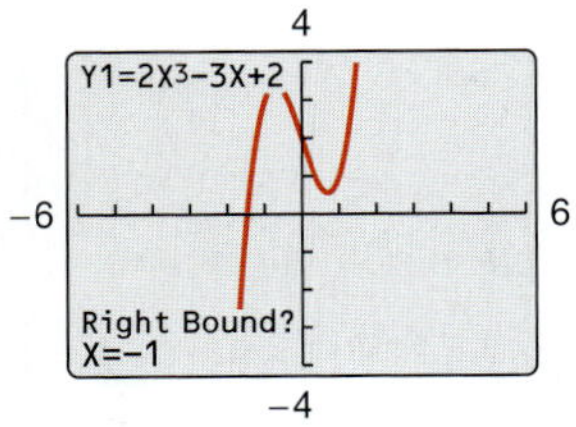

Figure 2.7

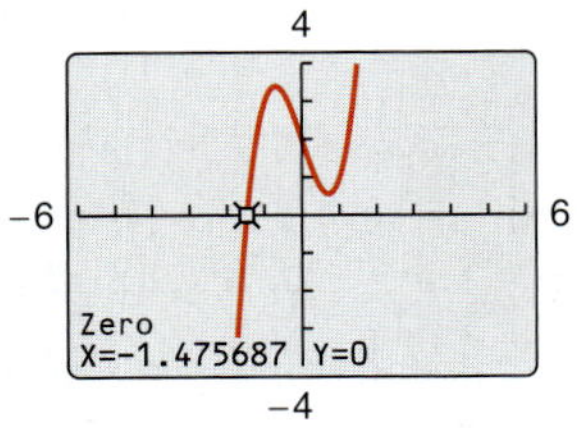

Figure 2.8

Checkpoint *Audio-video solution in English & Spanish at LarsonPrecalculus.com.*

Use a graphing utility to approximate any solutions of $-x^3 + x + 3 = 0$.

You can also use a graphing utility's *zoom* and *trace* features to approximate the solution of an equation. Here are some suggestions for using the *zoom-in* feature of a graphing utility.

1. With each successive zoom-in, adjust the x-scale (if necessary) so that the resulting viewing window shows at least one scale mark on each side of the solution.
2. The accuracy of the approximation will always be such that the error is less than the distance between two scale marks. For instance, to approximate the zero to the nearest hundredth, set the x-scale to 0.01. To approximate the zero to the nearest thousandth, set the x-scale to 0.001.
3. The graphing utility's *trace* feature can sometimes be used to add one more decimal place of accuracy without changing the viewing window.

Unless stated otherwise, all real solutions in this text will be approximated with an error of *at most* 0.01.

EXAMPLE 4 Approximating Solutions Graphically

See LarsonPrecalculus.com for an interactive version of this type of example.

Use a graphing utility to approximate any solutions of

$$x^2 + 3 = 5x.$$

Solution

In general form, this equation is

$$x^2 - 5x + 3 = 0. \qquad \text{Equation in general form}$$

So, you can begin by graphing

$$y = x^2 - 5x + 3 \qquad \text{Function to be graphed}$$

as shown in Figure 2.9. This graph has two x-intercepts, and by using the *zoom* and *trace* features you can approximate the corresponding solutions to be

$$x \approx 0.70 \text{ and } x \approx 4.30$$

as shown in Figures 2.10 and 2.11, respectively.

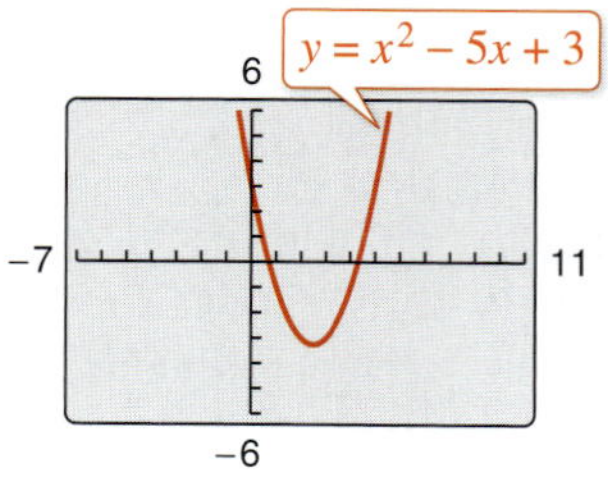

Figure 2.9

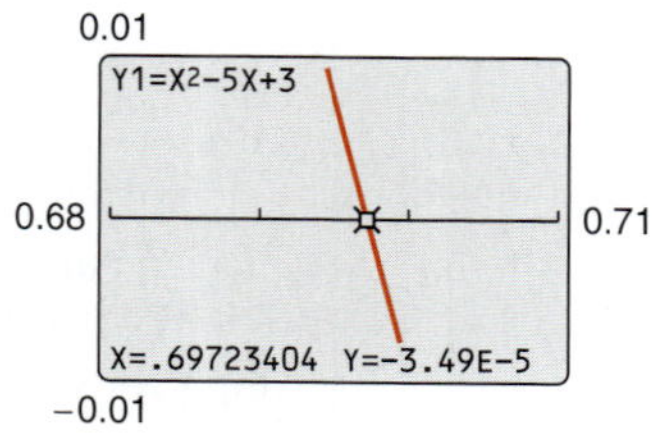

Figure 2.10

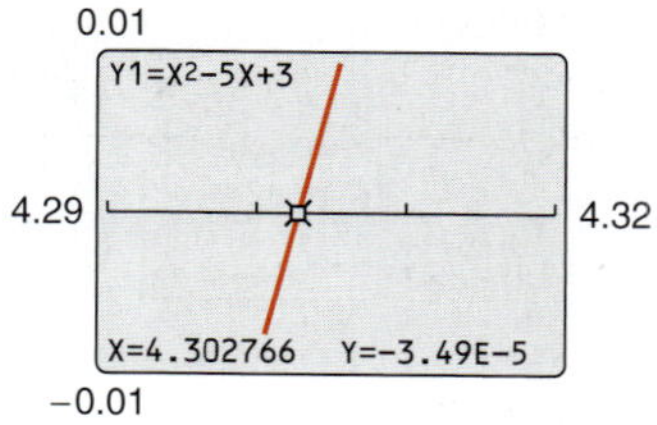

Figure 2.11

Technology Tip

Use the *zero* or *root* feature of a graphing utility to approximate the solutions of the equation in Example 4 to see that it yields a similar result.

Checkpoint *Audio-video solution in English & Spanish at LarsonPrecalculus.com.*

Use a graphing utility to approximate any solutions of $3x^4 = 3x^3 + 3$.

Points of Intersection of Two Graphs

An ordered pair that is a solution of two different equations is called a **point of intersection** of the graphs of the two equations. For instance, in Figure 2.12 you can see that the graphs of the following equations have two points of intersection.

$y = x + 2$ Equation 1

$y = x^2 - 2x - 2$ Equation 2

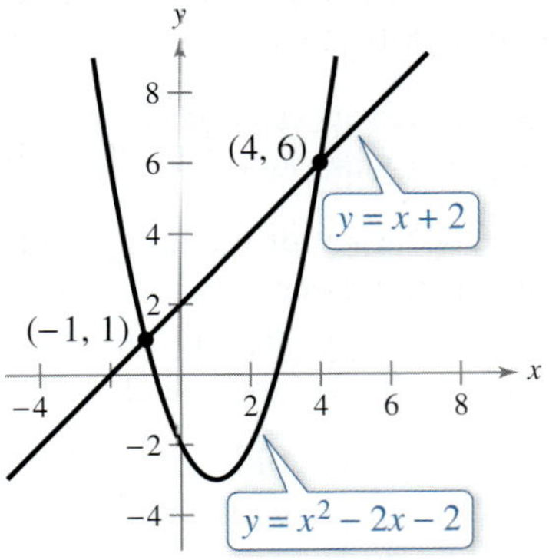

Figure 2.12

The point $(-1, 1)$ is a solution of both equations, and the point $(4, 6)$ is a solution of both equations. To check this algebraically, substitute

$x = -1$ and $x = 4$

into each equation.

Check that $(-1, 1)$ *is a solution.*

Equation 1: $y = -1 + 2 = 1$ Solution checks. ✓

Equation 2: $y = (-1)^2 - 2(-1) - 2 = 1$ Solution checks. ✓

Check that $(4, 6)$ *is a solution.*

Equation 1: $y = 4 + 2 = 6$ Solution checks. ✓

Equation 2: $y = (4)^2 - 2(4) - 2 = 6$ Solution checks. ✓

To find the points of intersection of the graphs of two equations, solve each equation for y (or x) and set the two results equal to each other. The resulting equation will be an equation in one variable, which can be solved using standard procedures, as shown in Example 5.

EXAMPLE 5 Finding Points of Intersection

Find any points of intersection of the graphs of

$2x - 3y = -2$ and $4x - y = 6$.

Algebraic Solution

To begin, solve each equation for y to obtain

$$y = \frac{2}{3}x + \frac{2}{3} \quad \text{and} \quad y = 4x - 6.$$

Next, set the two expressions for y equal to each other and solve the resulting equation for x, as follows.

$\frac{2}{3}x + \frac{2}{3} = 4x - 6$ Equate expressions for y.

$2x + 2 = 12x - 18$ Multiply each side by 3.

$-10x = -20$ Subtract $12x$ and 2 from each side.

$x = 2$ Divide each side by -10.

When $x = 2$, the y-value of each of the original equations is 2. So, the point of intersection is $(2, 2)$.

Graphical Solution

To begin, solve each equation for y to obtain

$$y_1 = \frac{2}{3}x + \frac{2}{3} \quad \text{and} \quad y_2 = 4x - 6.$$

Then use a graphing utility to graph both equations in the same viewing window, as shown in the figure. Use the *intersect* feature to approximate the point of intersection.

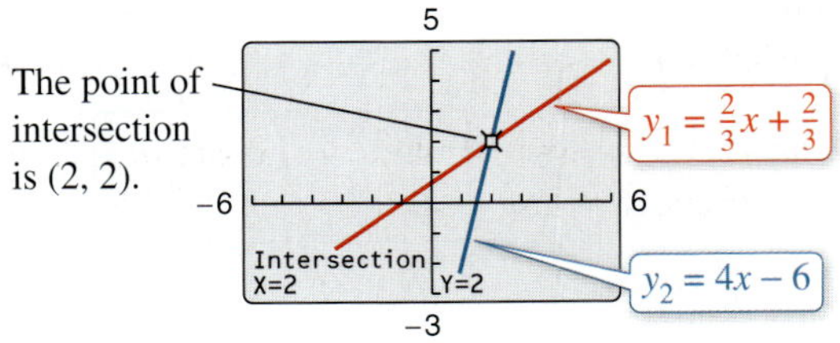

✓ **Checkpoint** *Audio-video solution in English & Spanish at LarsonPrecalculus.com.*

Find any points of intersection of the graphs of

$x + 4y = 2$ and $x - y = 0$.

Another way to approximate the points of intersection of two graphs is to graph both equations with a graphing utility and use the *zoom* and *trace* features to find the point or points at which the two graphs intersect.

EXAMPLE 6 Approximating Points of Intersection Graphically

Use a graphing utility to approximate any points of intersection of the graphs of the following equations.

$y = x^2 - 3x - 4$ Equation 1 (quadratic function)

$y = x^3 + 3x^2 - 2x - 1$ Equation 2 (cubic function)

Solution

Begin by using the graphing utility to graph both functions, as shown in Figure 2.13. From this display, you can see that the two graphs have only one point of intersection. Then, using the *zoom* and *trace* features, approximate the point of intersection to be

$(-2.17, 7.25)$ Point of intersection

as shown in Figure 2.14.

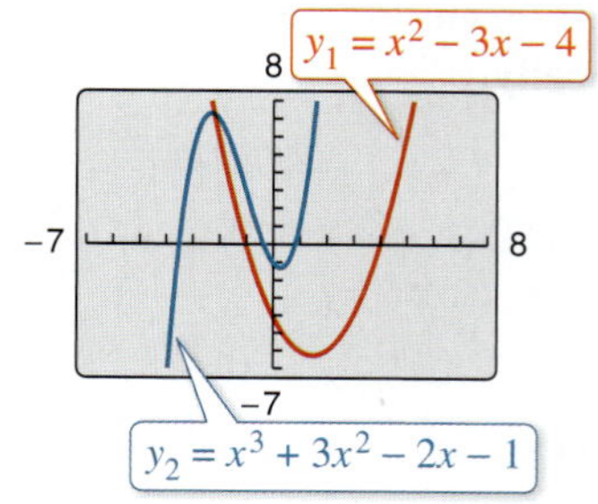

Figure 2.13

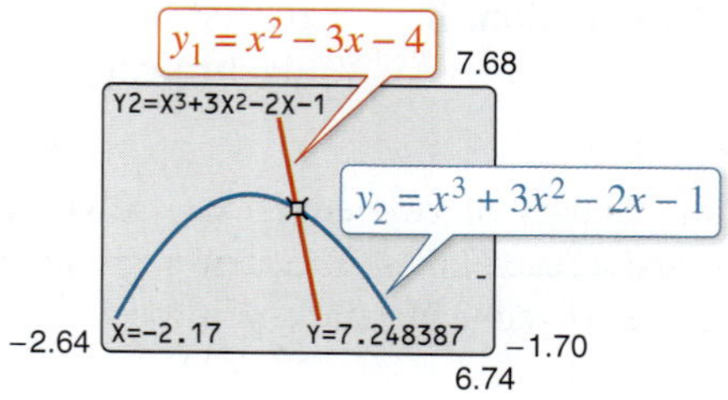

Figure 2.14

To test the reasonableness of this approximation, you can evaluate both functions at $x = -2.17$.

Quadratic Function

$y = (-2.17)^2 - 3(-2.17) - 4 \approx 7.22$

Cubic Function

$y = (-2.17)^3 + 3(-2.17)^2 - 2(-2.17) - 1 \approx 7.25$

Because both functions yield approximately the same y-value, you can conclude that the approximate coordinates of the point of intersection are

$x \approx -2.17$ and $y \approx 7.25$.

✓ Checkpoint *Audio-video solution in English & Spanish at LarsonPrecalculus.com.*

Use a graphing utility to approximate any points of intersection of the graphs of the equations $y = 3x^2$ and $y = 2x^4 - 4x^2$.

Technology Tip

Use the *intersect* feature of a graphing utility to approximate the point of intersection in Example 6 to see that it yields a similar result.

Technology Tip

The table shows some points on the graphs of the equations in Example 5. You can find any points of intersection of the graphs by finding all values of x for which y_1 and y_2 are equal.

2.2 Exercises

See *CalcChat.com* for tutorial help and worked-out solutions to odd-numbered exercises.
For instructions on how to use a graphing utility, see Appendix A.

Vocabulary and Concept Check

In Exercises 1 and 2, fill in the blank(s).

1. The points $(a, 0)$ and $(0, b)$ are called the _______ and _______, respectively, of the graph of an equation.
2. A _______ of a function is a number a such that $f(a) = 0$.

In Exercises 3–6, use the figure to answer the questions.

3. What are the x-intercepts of the graph of $y = f(x)$?
4. What is the y-intercept of the graph of $y = g(x)$?
5. What are the zero(s) of the function f?
6. What are the solutions of the equation $f(x) = g(x)$?

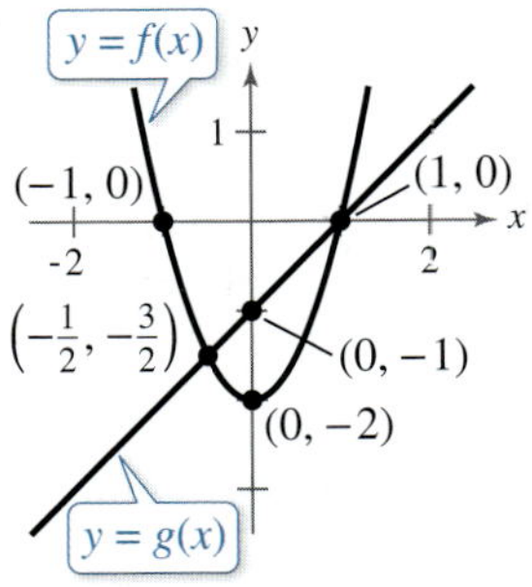

Figure for Exercises 3–6

Procedures and Problem Solving

Finding x- and y-Intercepts **In Exercises 7–16, find the x- and y-intercepts of the graph of the equation, if possible.**

7. $y = x - 5$

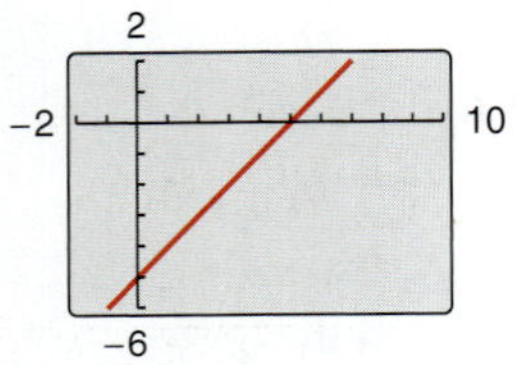

8. $y = -\frac{3}{4}x - 3$

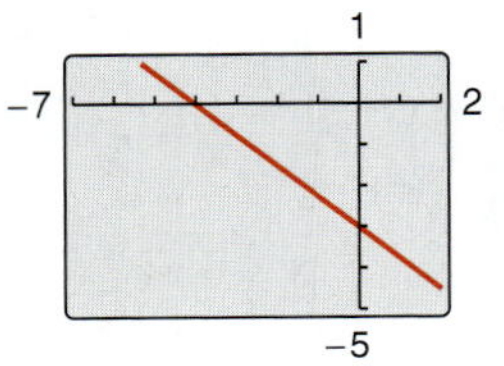

9. $y = x^2 + 2x + 2$

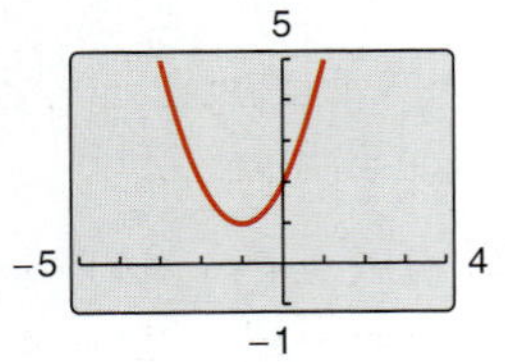

10. $y = 4 - x^2$

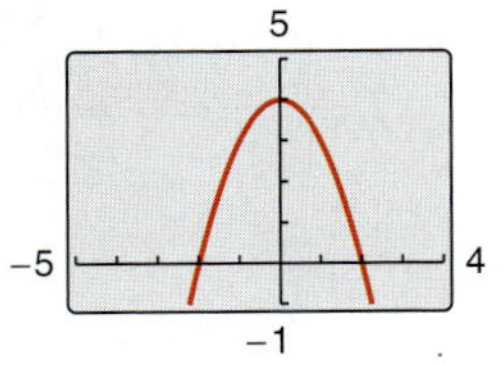

11. $y = x\sqrt{x + 2}$

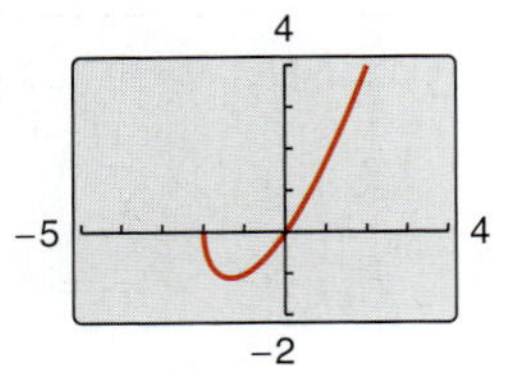

12. $y = -\frac{1}{2}x\sqrt{x + 3} + 1$

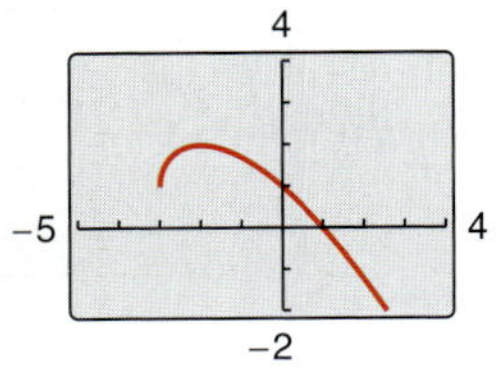

13. $y = \dfrac{4x - 8}{x}$
14. $y = \dfrac{3x - 1}{4x}$
15. $xy - 2y - x + 1 = 0$
16. $xy - x + 4y = 3$

Approximating x- and y-Intercepts **In Exercises 17–20, use a graphing utility to graph the equation and approximate any x- and y-intercepts. Verify your results algebraically.**

17. $y = 3(x - 2) - 5$
18. $y = 4(x + 3) - 2$
19. $y = 20 - (3x - 10)$
20. $y = 10 + 2(x - 2)$

Verifying Zeros of Functions **In Exercises 21–26, the zero(s) of the function are given. Verify the zero(s) both algebraically and graphically.**

	Function	*Zero(s)*
21.	$f(x) = 4(3 - x)$	$x = 3$
22.	$f(x) = 3(x - 5) + 9$	$x = 2$
23.	$f(x) = x^3 - 6x^2 + 5x$	$x = 0, 5, 1$
24.	$f(x) = x^3 - 9x^2 + 18x$	$x = 0, 3, 6$
25.	$f(x) = \dfrac{x + 1}{2} - \dfrac{x - 2}{7} + 1$	$x = -5$
26.	$f(x) = x - 3 - \dfrac{10}{x}$	$x = -2, 5$

Finding Solutions of an Equation Algebraically **In Exercises 27–40, solve the equation algebraically. Then write the equation in the form $f(x) = 0$ and use a graphing utility to verify the algebraic solution.**

27. $2.7x - 0.4x = 1.2$
28. $3.6x - 8.2 = 0.5x$
29. $12(x + 2) = 15(x - 4) - 3$
30. $1200 = 300 + 2(x - 500)$
31. $\dfrac{3x}{2} + \dfrac{1}{4}(x + 2) = 10$
32. $\dfrac{2x}{3} + \dfrac{1}{2}(x - 5) = 6$
33. $0.60x + 0.40(100 - x) = 1.2$
34. $0.75x + 0.2(80 - x) = 3.9$
35. $\dfrac{x - 3}{3} = \dfrac{3x - 5}{2}$
36. $\dfrac{x - 3}{25} = \dfrac{x - 5}{12}$
37. $\dfrac{x - 5}{4} + \dfrac{x}{2} = 10$
38. $\dfrac{x - 5}{10} - \dfrac{x - 3}{5} = 1$
39. $(x + 2)^2 = x^2 - 6x + 1$
40. $(x + 1)^2 + 2(x - 2) = (x + 1)(x - 2)$

Finding Solutions of an Equation Graphically In Exercises 41–60, use a graphing utility to approximate any solutions of the equation. [Remember to write the equation in the form $f(x) = 0$.]

41. $x^3 + x + 4 = 0$

42. $2x^3 + x + 4 = 0$

43. $\frac{1}{4}(x^2 - 10x + 17) = 0$

44. $-\frac{1}{2}(x^2 - 6x + 6) = 0$

45. $2x^3 - x^2 - 18x + 9 = 0$

46. $4x^3 + 12x^2 - 26x - 24 = 0$

47. $x^5 = 3x^3 - 3$

48. $x^5 = 3 + 2x^3$

49. $\dfrac{2}{x+2} = 3$

50. $\dfrac{1}{x-3} = -2$

51. $\dfrac{5}{x} = 1 + \dfrac{3}{x+2}$

52. $\dfrac{3}{x} + 1 = \dfrac{3}{x-1}$

53. $-|x + 1| = -3$

54. $-|x - 2| = -6$

55. $|3x - 2| - 1 = 4$

56. $|4x + 1| + 2 = 8$

57. $\sqrt{x-2} = 3$

58. $\sqrt{x-4} = 8$

59. $2 - \sqrt{x+5} = 1$

60. $8 - \sqrt{x+9} = 6$

61. Exploration

(a) Use a graphing utility to complete the table. Determine the interval in which the solution to the equation $3.2x - 5.8 = 0$ is located. Explain your reasoning.

x	-1	0	1	2	3	4
$3.2x - 5.8$						

(b) Use the graphing utility to complete the table. Determine the interval in which the solution to the equation $3.2x - 5.8 = 0$ is located. Explain how this process can be used to approximate the solution to any desired degree of accuracy. Then use the graphing utility to verify graphically the solution to $3.2x - 5.8 = 0$.

x	1.5	1.6	1.7	1.8	1.9	2
$3.2x - 5.8$						

62. Exploration Use the procedure in Exercise 61 to approximate the solution of the equation

$0.3(x - 1.8) - 1 = 0$

accurate to two decimal places.

Finding Points of Intersection Algebraically In Exercises 63–70, determine any point(s) of intersection algebraically. Then verify your result numerically by creating a table of values for each function.

63. $y = 6 - x$
$y = 3x - 2$

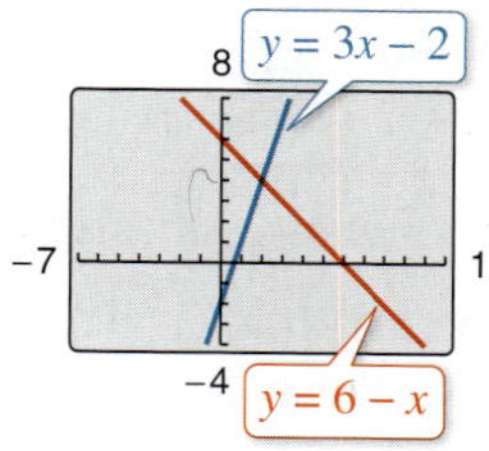

64. $y = 2x - 3$
$y = 9 - x$

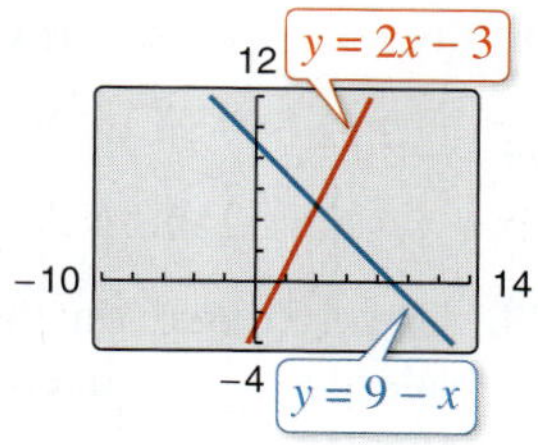

65. $2x + y = 6$
$-x + y = 0$

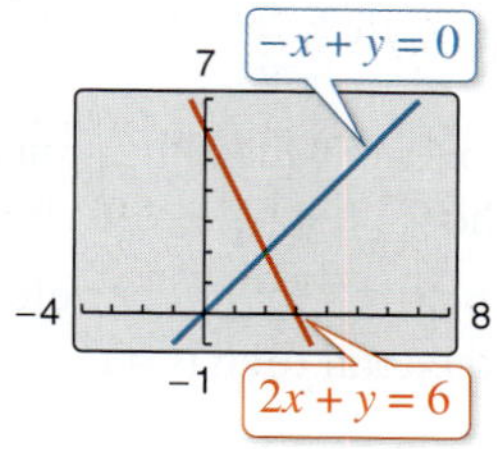

66. $x - y = -4$
$x + 2y = 5$

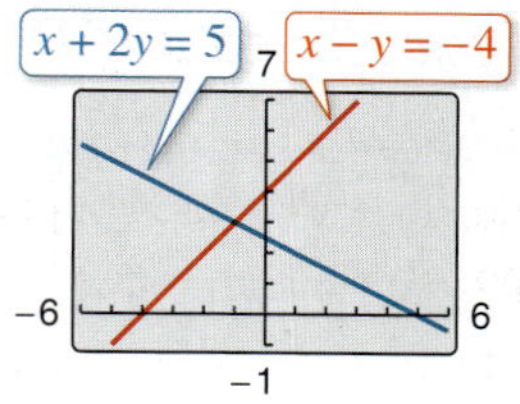

67. $x - y = 10$
$x + 2y = 4$

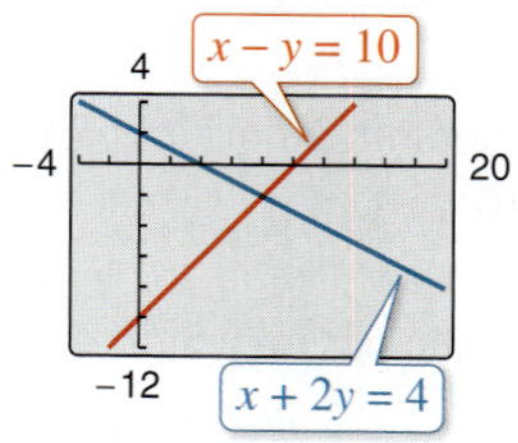

68. $4x - y = 4$
$x - 4y = 1$

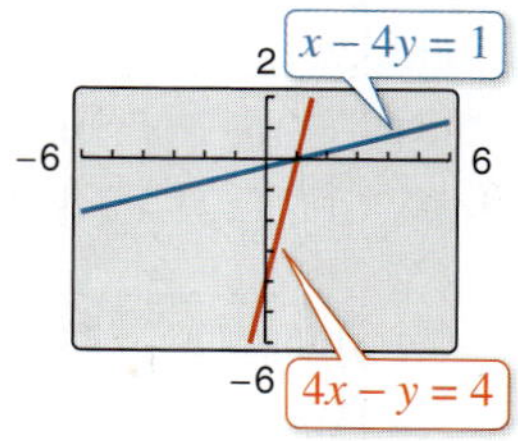

69. $y = x^2 - x + 1$
$y = x^2 + 2x + 4$

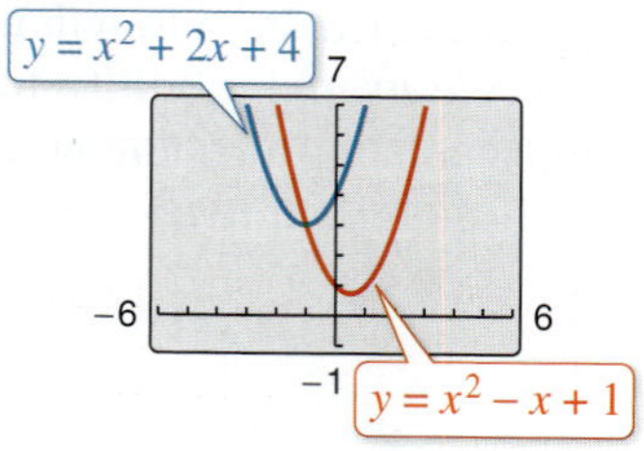

70. $y = -x^2 + 3x + 1$
$y = -x^2 - 2x - 4$

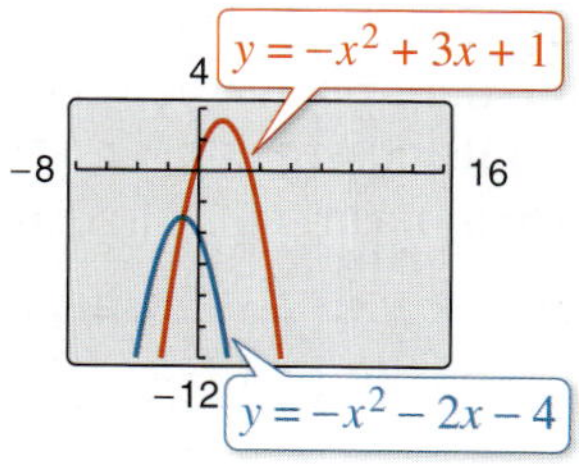

Approximating Points of Intersection Graphically In Exercises 71–76, use a graphing utility to approximate any points of intersection of the graphs of the equations. Check your results algebraically.

71. $y = 9 - 2x$
$y = x - 3$

72. $x - 3y = -3$
$5x - 2y = 11$

73. $y = x$
$y = 2x - x^2$

74. $y = 4 - x^2$
$y = 2x - 1$

75. $x^3 - y = 3$
$2x + y = 5$

76. $y = 2x^2$
$y = x^4 - 2x^2$

Comparing Methods **In Exercises 77 and 78, evaluate the expression in two ways. (a) Calculate entirely on your calculator by storing intermediate results and then rounding the final answer to two decimal places. (b) Round both the numerator and denominator to two decimal places before dividing, and then round the final answer to two decimal places. Does the method in part (b) decrease the accuracy? Explain.**

77. $\dfrac{1 + 0.73205}{1 - 0.73205}$ **78.** $\dfrac{1 + 0.86603}{1 - 0.86603}$

79. Travel Time On the first part of a 280-mile trip, a salesperson averaged 63 miles per hour. The salesperson averaged only 54 miles per hour on the last part of the trip because of an increased volume of traffic.

(a) Write the total time t for the trip as a function of the distance x traveled at an average speed of 63 miles per hour.

(b) Use a graphing utility to graph the time function. What is the domain of the function?

(c) Approximate the number of miles traveled at 63 miles per hour when the total time is 4 hours and 45 minutes.

80. Chemistry A 55-gallon barrel contains a mixture with a concentration of 33% sodium chloride. You remove x gallons of this mixture and replace it with 100% sodium chloride.

(a) Write the amount A of sodium chloride in the final mixture as a function of x.

(b) Use a graphing utility to graph the concentration function. What is the domain of the function?

(c) Approximate (accurate to one decimal place) the value of x when the final mixture is 60% sodium chloride.

Geometry **In Exercises 81 and 82, (a) write a function for the area of the region, (b) use a graphing utility to graph the function, and (c) approximate the value of x when the area of the region is 180 square units.**

81.

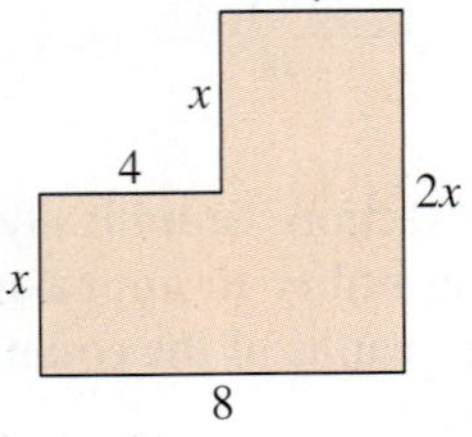

82.

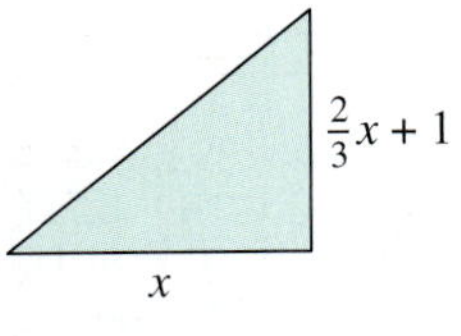

83. Accounting The following information describes a possible negative income tax for a family consisting of two adults and two children. The plan would guarantee the poor a minimum income while encouraging a family to increase its private income ($0 \le x \le 20{,}000$). (A *subsidy* is a grant of money.)

Family's earned income: $I = x$

Subsidy: $S = 10{,}000 - \frac{1}{2}x$

Total income: $T = I + S$

(a) Write the total income T in terms of x.

(b) Use a graphing utility to find the earned income x when the subsidy is \$6600. Verify your answer algebraically.

(c) Use the graphing utility to find the earned income x when the total income is \$13,800. Verify your answer algebraically.

(d) Find the subsidy S graphically when the total income is \$12,500.

84. Payroll The median weekly earnings y (in dollars) of full-time workers in the United States from 2000 through 2012 can be modeled by $y = 16.9t + 574$, $0 \le t \le 12$, where t is the year, with $t = 0$ corresponding to 2000. (Source: U.S. Bureau of Labor Statistics)

(a) Find algebraically and interpret the y-intercept of the model.

(b) What is the slope of the model and what does it tell you about the median weekly earnings of full-time workers in the United States?

(c) Do you think the model can be used to predict the median weekly earnings of full-time workers in the United States for years beyond 2012? If so, for what time period? Explain.

(d) Explain, both algebraically and graphically, how you could find when the median weekly earnings of full-time workers reaches \$800.

85. ***Why you should learn it*** *(p. 170)* The populations (in thousands) of Maryland M and Wisconsin W from 2001 through 2013 can be modeled by

$M = 43.4t + 5355$, $1 \le t \le 13$

$W = 28.4t + 5398$, $1 \le t \le 13$

where t represents the year, with $t = 1$ corresponding to 2001. (Source: U.S. Census Bureau)

(a) Use a graphing utility to graph each model in the same viewing window over the appropriate domain. Approximate the point of intersection. Round your result to one decimal place. Explain the meaning of the coordinates of the point.

(b) Find the point of intersection algebraically. Round your result to one decimal place. What does the point of intersection represent?

(c) Explain the meaning of the slopes of both models and what they tell you about the population growth rates.

(d) Use the models to estimate the population of each state in 2016. Do the values seem reasonable? Explain.

86. Design and Construction Consider the swimming pool in the figure. (When finding its volume, use the fact that the volume is the area of the region on the vertical sidewall times the width of the pool.)

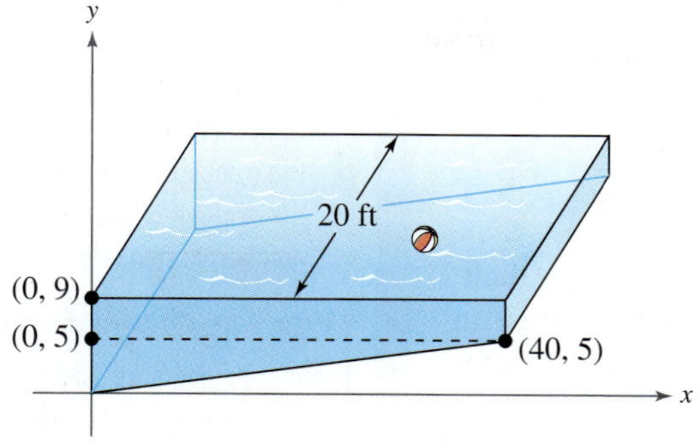

(a) Find the volume of the pool.

(b) How many gallons of water are in the pool? (There are about 7.48 gallons of water in 1 cubic foot.)

(c) Find an equation of the line representing the base of the pool.

(d) The depth of the water at the deep end of the pool is d feet. Show that the volume of water is

$$V(d) = \begin{cases} 80d^2, & 0 \le d \le 5 \\ 800d - 2000, & 5 < d \le 9 \end{cases}.$$

(e) Graph the volume function.

(f) Use a graphing utility to complete the table.

d	3	5	7	9
V				

(g) Approximate the depth of the water at the deep end when the volume is 4800 cubic feet.

Conclusions

True or False? **In Exercises 87 and 88, determine whether the statement is true or false. Justify your answer.**

87. To find the y-intercept of a graph, let $x = 0$ and solve the equation for y.

88. Every linear equation has at least one y-intercept or x-intercept.

89. Writing You graphically approximate the solution of the equation

$$\frac{x}{x-1} - \frac{99}{100} = 0$$

to be $x = -99.1$. Substituting this value for x produces

$$\frac{-99.1}{-99.1-1} - \frac{99}{100} = 0.00000999 = 9.99 \times 10^{-6}.$$

Is -99.1 a good approximation of the solution? Write a short paragraph explaining why or why not.

90. HOW DO YOU SEE IT? Use the table to solve each linear equation where

$$y_1 = f(x) \quad \text{and} \quad y_2 = g(x).$$

X	Y1	Y2
-2	-15	0
-1	-12	2
0	-9	4
1	-6	6
2	-3	8
3	0	10
4	3	12
X=-2		

(a) $f(x) = 0$

(b) $g(x) = 0$

(c) $g(x) = -f(x)$

(d) $f(x) = -6g(x)$

91. Algebraic-Graphical-Numerical For each of the following, find the answer algebraically, numerically, and graphically.

(a) Find the x- and y-intercepts of the graph of

$y = 2x + 2.$

(b) Verify that the real numbers -1 and 1 are zero(s) of the function

$f(x) = x^2 - 1.$

(c) Find the points of intersection of the graphs of

$y = 2x + 2$

and

$y = x^2 - 1.$

Cumulative Mixed Review

Rationalizing a Denominator **In Exercises 92–95, rationalize the denominator.**

92. $\dfrac{12}{5\sqrt{3}}$

93. $\dfrac{10}{\sqrt{14} - 2}$

94. $\dfrac{3}{8 + \sqrt{11}}$

95. $\dfrac{14}{3\sqrt{10} - 1}$

Multiplying Polynomials **In Exercises 96–99, find the product.**

96. $(x - 6)(3x - 5)$

97. $(3x + 13)(4x - 7)$

98. $(2x - 9)(2x + 9)$

99. $(4x + 1)^2$

2.3 Complex Numbers

What you should learn

- Use the imaginary unit i to write complex numbers.
- Add, subtract, and multiply complex numbers.
- Use complex conjugates to write the quotient of two complex numbers in standard form.

Why you should learn it

Complex numbers are used to model numerous aspects of the natural world, such as the impedance of an electrical circuit, as shown in Exercise 74 on page 185.

The Imaginary Unit i

Some quadratic equations have no real solutions. For instance, the quadratic equation $x^2 + 1 = 0$ has no real solution because there is no real number x that can be squared to produce -1. To overcome this deficiency, mathematicians created an expanded system of numbers using the **imaginary unit i,** defined as

$$i = \sqrt{-1} \quad \text{Imaginary unit}$$

where $i^2 = -1$. By adding real numbers to real multiples of this imaginary unit, you obtain the set of **complex numbers.** Each complex number can be written in the **standard form** $a + bi$. For instance, the standard form of the complex number $\sqrt{-9} - 5$ is $-5 + 3i$ because

$$\begin{aligned}\sqrt{-9} - 5 &= \sqrt{3^2(-1)} - 5\\ &= 3\sqrt{-1} - 5\\ &= 3i - 5\\ &= -5 + 3i.\end{aligned}$$

In the standard form $a + bi$, the real number a is called the **real part** of the **complex number** $a + bi$, and the number bi (where b is a real number) is called the **imaginary part** of the complex number.

Definition of a Complex Number

If a and b are real numbers, then the number $a + bi$ is a **complex number,** and it is said to be written in **standard form.** If $b = 0$, then the number $a + bi = a$ is a real number. If $b \neq 0$, then the number $a + bi$ is called an **imaginary number.** A number of the form bi, where $b \neq 0$, is called a **pure imaginary number.**

The set of real numbers is a subset of the set of complex numbers, as shown in Figure 2.15. This is true because every real number a can be written as a complex number using $b = 0$. That is, for every real number a, you can write $a = a + 0i$.

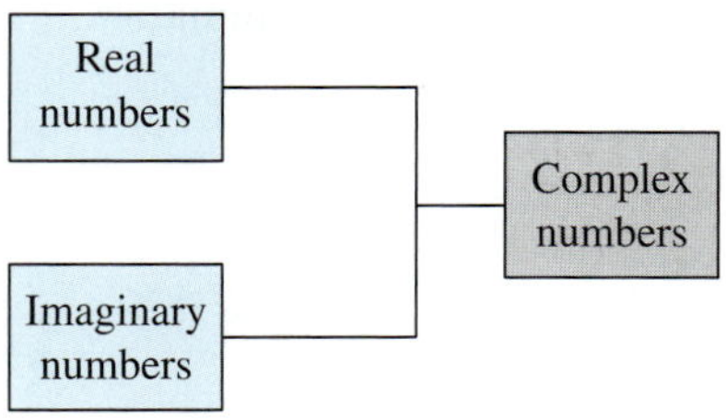

Figure 2.15

Equality of Complex Numbers

Two complex numbers $a + bi$ and $c + di$, written in standard form, are equal to each other

$$a + bi = c + di \quad \text{Equality of two complex numbers}$$

if and only if $a = c$ and $b = d$.

Operations with Complex Numbers

To add (or subtract) two complex numbers, you add (or subtract) the real and imaginary parts of the numbers separately.

Addition and Subtraction of Complex Numbers

If $a + bi$ and $c + di$ are two complex numbers written in standard form, then their sum and difference are defined as follows.

Sum: $(a + bi) + (c + di) = (a + c) + (b + d)i$

Difference: $(a + bi) - (c + di) = (a - c) + (b - d)i$

The **additive identity** in the complex number system is zero (the same as in the real number system). Furthermore, the **additive inverse** of the complex number $a + bi$ is

$-(a + bi) = -a - bi.$ Additive inverse

So, you have $(a + bi) + (-a - bi) = 0 + 0i = 0$.

EXAMPLE 1 Adding and Subtracting Complex Numbers

a. $(3 - i) + (2 + 3i) = 3 - i + 2 + 3i$ Remove parentheses.

$= 3 + 2 - i + 3i$ Group like terms.

$= (3 + 2) + (-1 + 3)i$

$= 5 + 2i$ Write in standard form.

b. $\sqrt{-4} + \left(-4 - \sqrt{-4}\right) = 2i + (-4 - 2i)$ Write in *i*-form.

$= 2i - 4 - 2i$ Remove parentheses.

$= -4 + 2i - 2i$ Group like terms.

$= -4$ Write in standard form.

c. $3 - (-2 + 3i) + (-5 + i) = 3 + 2 - 3i - 5 + i$

$= 3 + 2 - 5 - 3i + i$

$= 0 - 2i$

$= -2i$

d. $(3 + 2i) + (4 - i) - (7 + i) = 3 + 2i + 4 - i - 7 - i$

$= 3 + 4 - 7 + 2i - i - i$

$= 0 + 0i$

$= 0$

✓ Checkpoint Audio-video solution in English & Spanish at LarsonPrecalculus.com.

Perform each operation and write the result in standard form.

a. $(7 + 3i) + (5 - 4i)$

b. $\left(-1 + \sqrt{-8}\right) + \left(8 - \sqrt{-50}\right)$

c. $2i + (-3 - 4i) - (-3 - 3i)$

d. $(5 - 3i) + (3 + 5i) - (8 + 2i)$

In Examples 1(b) and 1(d), note that the sum of complex numbers can be a real number.

Many of the properties of real numbers are valid for complex numbers as well. Here are some examples.

Associative Properties of Addition and Multiplication

Commutative Properties of Addition and Multiplication

Distributive Property of Multiplication over Addition

Notice how these properties are used when two complex numbers are multiplied.

$(a + bi)(c + di) = a(c + di) + bi(c + di)$	Distributive Property
$= ac + (ad)i + (bc)i + (bd)i^2$	Distributive Property
$= ac + (ad)i + (bc)i + (bd)(-1)$	$i^2 = -1$
$= ac - bd + (ad)i + (bc)i$	Commutative Property
$= (ac - bd) + (ad + bc)i$	Distributive Property

The procedure above is similar to multiplying two polynomials and combining like terms, as in the FOIL Method discussed in Section P.3.

Explore the Concept

Complete the following:

$i^1 = i$	$i^7 =$
$i^2 = -1$	$i^8 =$
$i^3 = -i$	$i^9 =$
$i^4 = 1$	$i^{10} =$
$i^5 =$	$i^{11} =$
$i^6 =$	$i^{12} =$

What pattern do you see? Write a brief description of how you would find i raised to any positive integer power.

EXAMPLE 2 Multiplying Complex Numbers

See LarsonPrecalculus.com for an interactive version of this type of example.

a. $\sqrt{-4} \cdot \sqrt{-16} = (2i)(4i)$ — Write each factor in i-form.

$= 8i^2$ — Multiply.

$= 8(-1)$ — $i^2 = -1$

$= -8$ — Simplify.

b. $(2 - i)(4 + 3i) = 8 + 6i - 4i - 3i^2$ — FOIL Method

$= 8 + 6i - 4i - 3(-1)$ — $i^2 = -1$

$= 8 + 3 + 6i - 4i$ — Group like terms.

$= 11 + 2i$ — Write in standard form.

c. $(3 + 2i)(3 - 2i) = 9 - 6i + 6i - 4i^2$ — FOIL Method

$= 9 - 4(-1)$ — $i^2 = -1$

$= 9 + 4$ — Simplify.

$= 13$ — Write in standard form.

d. $4i(-1 + 5i) = 4i(-1) + 4i(5i)$ — Distributive Property

$= -4i + 20i^2$ — Simplify.

$= -4i + 20(-1)$ — $i^2 = -1$

$= -20 - 4i$ — Write in standard form.

e. $(3 + 2i)^2 = 9 + 12i + 4i^2$ — $(a + b)^2 = a^2 + 2ab + b^2$

$= 9 + 12i + 4(-1)$ — $i^2 = -1$

$= 9 - 4 + 12i$ — Group like terms.

$= 5 + 12i$ — Write in standard form.

Remark

Before you perform operations with complex numbers, be sure to rewrite the terms or factors in i-form first and then proceed with the operations, as shown in Example 2(a).

✓ Checkpoint *Audio-video solution in English & Spanish at LarsonPrecalculus.com.*

Perform each operation and write the result in standard form.

a. $\sqrt{-9} \cdot \sqrt{-25}$

b. $(2 - 4i)(3 + 3i)$

c. $(4 + 2i)^2$

Complex Conjugates

Notice in Example 2(c) that the product of two complex numbers can be a real number. This occurs with pairs of complex numbers of the forms $a + bi$ and $a - bi$, called **complex conjugates.**

$$\begin{aligned}(a + bi)(a - bi) &= a^2 - abi + abi - b^2i^2 \\ &= a^2 - b^2(-1) \\ &= a^2 + b^2\end{aligned}$$

EXAMPLE 3 Multiplying Conjugates

Multiply each complex number by its complex conjugate.

a. $1 + i$ **b.** $4 - 3i$

Solution

a. The complex conjugate of $1 + i$ is $1 - i$.

$$(1 + i)(1 - i) = 1^2 - i^2 = 1 - (-1) = 2$$

b. The complex conjugate of $4 - 3i$ is $4 + 3i$.

$$(4 - 3i)(4 + 3i) = 4^2 - (3i)^2 = 16 - 9i^2 = 16 - 9(-1) = 25$$

✓ ***Checkpoint*** *Audio-video solution in English & Spanish at LarsonPrecalculus.com.*

Multiply each complex number by its complex conjugate.

a. $3 + 6i$ **b.** $2 - 5i$

To write the quotient of $a + bi$ and $c + di$ in standard form, where c and d are not both zero, multiply the numerator and denominator by the complex conjugate of the *denominator* to obtain

$$\frac{a + bi}{c + di} = \frac{a + bi}{c + di}\left(\frac{c - di}{c - di}\right)$$ Multiply numerator and denominator by complex conjugate of denominator.

$$= \frac{ac + bd}{c^2 + d^2} + \left(\frac{bc - ad}{c^2 + d^2}\right)i.$$ Standard form

Technology Tip

Some graphing utilities can perform operations with complex numbers. For instance, on some graphing utilities, to divide $2 + 3i$ by $4 - 2i$, use the following keystrokes.

(2 + 3 i) ÷
(4 − 2 i) ENTER

EXAMPLE 4 Quotient of Complex Numbers in Standard Form

$$\frac{2 + 3i}{4 - 2i} = \frac{2 + 3i}{4 - 2i}\left(\frac{4 + 2i}{4 + 2i}\right)$$ Multiply numerator and denominator by complex conjugate of denominator.

$$= \frac{8 + 4i + 12i + 6i^2}{16 - 4i^2}$$ Expand.

$$= \frac{8 - 6 + 16i}{16 + 4}$$ $i^2 = -1$

$$= \frac{2 + 16i}{20}$$ Simplify.

$$= \frac{1}{10} + \frac{4}{5}i$$ Write in standard form.

✓ ***Checkpoint*** *Audio-video solution in English & Spanish at LarsonPrecalculus.com.*

Write $\dfrac{2 + i}{2 - i}$ in standard form.

2.3 Exercises

See *CalcChat.com* for tutorial help and worked-out solutions to odd-numbered exercises.
For instructions on how to use a graphing utility, see Appendix A.

Vocabulary and Concept Check

1. Match the type of complex number with its definition.

(a) real number (i) $a + bi, a = 0, b \neq 0$

(b) imaginary number (ii) $a + bi, b = 0$

(c) pure imaginary number (iii) $a + bi, a \neq 0, b \neq 0$

In Exercises 2 and 3, fill in the blanks.

2. The imaginary unit i is defined as $i =$ _______ , where $i^2 =$ _______ .

3. The set of real multiples of the imaginary unit i combined with the set of real numbers is called the set of _______ numbers, which are written in the standard form _______ .

4. What method for multiplying two polynomials can you use when multiplying two complex numbers?

5. What is the additive inverse of the complex number $2 - 4i$?

6. What is the complex conjugate of the complex number $2 - 4i$?

Procedures and Problem Solving

Equality of Complex Numbers **In Exercises 7–10, find real numbers a and b such that the equation is true.**

7. $a + bi = -9 + 4i$ **8.** $a + bi = 12 + 5i$

9. $(a - 1) + (b + 3)i = 5 + 8i$

10. $(a + 6) + 2bi = 6 - 5i$

Writing a Complex Number in Standard Form **In Exercises 11–20, write the complex number in standard form.**

11. $4 + \sqrt{-9}$ **12.** $7 - \sqrt{-25}$

13. 12 **14.** -3

15. $-8i - i^2$ **16.** $2i^2 - 6i$

17. $(\sqrt{-16})^2 + 5$ **18.** $-i - (\sqrt{-23})^2$

19. $\sqrt{-0.09}$ **20.** $\sqrt{-0.0004}$

Adding and Subtracting Complex Numbers **In Exercises 21–30, perform the addition or subtraction and write the result in standard form.**

21. $(4 + i) - (7 - 2i)$ **22.** $(11 - 2i) - (-3 + 6i)$

23. $13i - (14 - 7i)$

24. $22 + (-5 + 8i) - 9i$

25. $\left(\frac{3}{2} + \frac{5}{2}i\right) + \left(\frac{5}{3} + \frac{11}{3}i\right)$

26. $\left(\frac{3}{4} + \frac{7}{5}i\right) - \left(\frac{5}{6} - \frac{1}{6}i\right)$

27. $(1.6 + 3.2i) + (-5.8 + 4.3i)$

28. $-(-3.7 - 12.8i) - (6.1 - 16.3i)$

29. $(5 + \sqrt{-27}) - (-12 + \sqrt{-48})$

30. $(7 + \sqrt{-18}) + (3 + \sqrt{-32})$

Multiplying Complex Numbers **In Exercises 31–46, perform the operation and write the result in standard form.**

31. $\sqrt{-6} \cdot \sqrt{-2}$ **32.** $\sqrt{-5} \cdot \sqrt{-10}$

33. $(\sqrt{-10})^2$ **34.** $(\sqrt{-75})^2$

35. $4(3 + 5i)$ **36.** $-6(5 - 3i)$

37. $(1 + i)(3 - 2i)$ **38.** $(6 - 2i)(2 - 3i)$

39. $4i(8 + 5i)$ **40.** $-3i(6 - i)$

41. $(\sqrt{14} + \sqrt{10}i)(\sqrt{14} - \sqrt{10}i)$

42. $(\sqrt{3} + \sqrt{15}i)(\sqrt{3} - \sqrt{15}i)$

43. $(6 + 7i)^2$ **44.** $(5 - 4i)^2$

45. $(4 + 5i)^2 - (4 - 5i)^2$ **46.** $(1 - 2i)^2 - (1 + 2i)^2$

Multiplying Conjugates **In Exercises 47–54, write the complex conjugate of the complex number. Then multiply the number by its complex conjugate.**

47. $6 - 2i$ **48.** $3 + 5i$

49. $-1 + \sqrt{7}i$ **50.** $-4 - \sqrt{3}i$

51. $\sqrt{-29}$ **52.** $\sqrt{-10}$

53. $9 - \sqrt{6}i$ **54.** $-8 + \sqrt{15}i$

Writing a Quotient of Complex Numbers in Standard Form **In Exercises 55–62, write the quotient in standard form.**

55. $\dfrac{6}{i}$ **56.** $-\dfrac{5}{2i}$

57. $\dfrac{2}{4 - 5i}$ **58.** $\dfrac{3}{1 - i}$

59. $\dfrac{3 - i}{3 + i}$ **60.** $\dfrac{8 - 7i}{1 - 2i}$

61. $\dfrac{i}{(4 - 5i)^2}$ **62.** $\dfrac{5i}{(2 + 3i)^2}$

Adding or Subtracting Quotients of Complex Numbers **In Exercises 63–66, perform the operation and write the result in standard form.**

63. $\dfrac{2}{1 + i} - \dfrac{3}{1 - i}$ **64.** $\dfrac{2i}{2 + i} + \dfrac{5}{2 - i}$

65. $\dfrac{i}{3 - 2i} + \dfrac{2i}{3 + 8i}$ **66.** $\dfrac{1 + i}{i} - \dfrac{3}{4 - i}$

Expressions Involving Powers of *i* **In Exercises 67–72, simplify the complex number and write it in standard form.**

67. $-6i^3 + i^2$ **68.** $4i^2 - 2i^3$

69. $(\sqrt{-75})^3$ **70.** $(\sqrt{-2})^6$

71. $\dfrac{1}{i^3}$ **72.** $\dfrac{1}{(2i)^3}$

73. **Powers of *i*** Use the results of the Explore the Concept feature on page 182 to find each power of i.

(a) i^{20} (b) i^{45} (c) i^{67} (d) i^{114}

74. ***Why you should learn it*** *(p. 180)* The opposition to current in an electrical circuit is called its impedance. The impedance z in a parallel circuit with two pathways satisfies the equation

$$\frac{1}{z} = \frac{1}{z_1} + \frac{1}{z_2}$$

where z_1 is the impedance (in ohms) of pathway 1, and z_2 is the impedance (in ohms) of pathway 2. Use the table to determine the impedance of each parallel circuit. (*Hint:* You can find the impedance of each pathway in a parallel circuit by adding the impedances of all components in the pathway.)

	Resistor	Inductor	Capacitor
Symbol	$a\,\Omega$	$b\,\Omega$	$c\,\Omega$
Impedance	a	bi	$-ci$

(a)

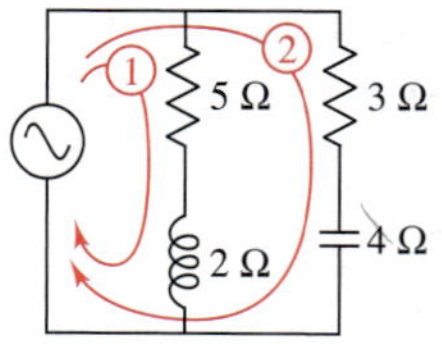

(b)

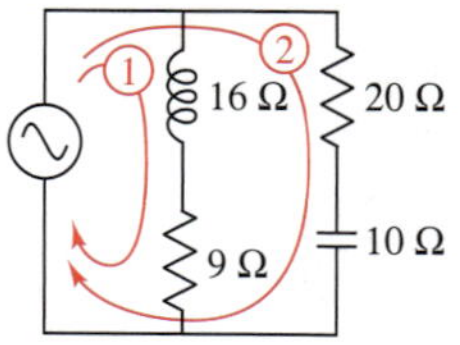

Conclusions

True or False? **In Exercises 75–79, determine whether the statement is true or false. Justify your answer.**

75. No complex number is equal to its complex conjugate.

76. $i^{44} + i^{150} - i^{74} - i^{109} + i^{61} = -1$

77. The sum of two imaginary numbers is always an imaginary number.

78. The conjugate of the product of two complex numbers is equal to the product of the conjugates of the two complex numbers.

79. The conjugate of the sum of two complex numbers is equal to the sum of the conjugates of the two complex numbers.

80. **HOW DO YOU SEE IT?** The coordinate system shown below is called the *complex plane*. In the complex plane, the point that corresponds to the complex number $a + bi$ is (a, b). Match each complex number with its corresponding point.

(i) 2 (ii) $2i$ (iii) $-2 + i$ (iv) $1 - 2i$

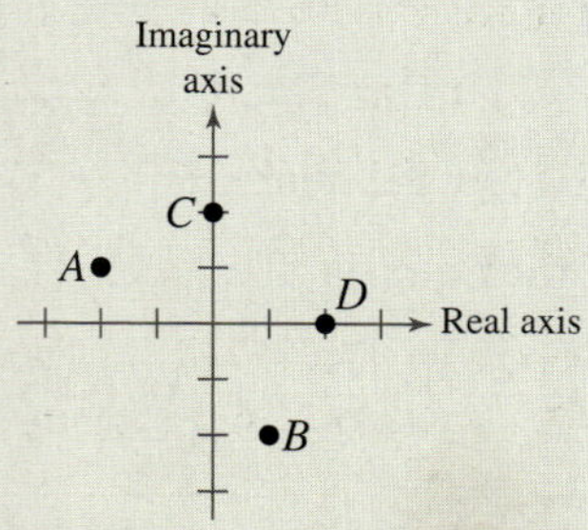

81. **Error Analysis** Describe the error.

~~$\sqrt{-6}\sqrt{-6} = \sqrt{(-6)(-6)} = \sqrt{36} = 6$~~

82. **Exploration** Consider the binomials $x + 5$ and $2x - 1$ and the complex numbers $1 + 5i$ and $2 - i$.

(a) Find the sum of the binomials and the sum of the complex numbers. Describe the similarities and differences in your results.

(b) Find the product of the binomials and the product of the complex numbers. Describe the similarities and differences in your results.

(c) Explain why the products in part (b) are not related in the same way as the sums in part (a).

Cumulative Mixed Review

Multiplying Polynomials **In Exercises 83–86, perform the operation and write the result in standard form.**

83. $(4x - 5)(4x + 5)$ **84.** $(x + 2)^3$

85. $\left(3x - \frac{1}{2}\right)(x + 4)$ **86.** $(2x - 5)^2$

2.4 Solving Quadratic Equations Algebraically

What you should learn

- Solve quadratic equations by factoring.
- Solve quadratic equations by extracting square roots.
- Solve quadratic equations by completing the square.
- Use the Quadratic Formula to solve quadratic equations.
- Use quadratic equations to model and solve real-life problems.

Why you should learn it

Quadratic equations can help you model and solve real-life problems. For instance, in Exercise 100 on page 198, you will use a quadratic equation to model the total public debt in the United States.

Quadratic Equations

A **quadratic equation in** x is an equation that can be written in the general form

$$ax^2 + bx + c = 0$$

where a, b, and c are real numbers with $a \neq 0$. A quadratic equation in x is also known as a **second-degree polynomial equation in** x. You should be familiar with the following four methods for solving quadratic equations.

Solving a Quadratic Equation

Factoring: If $ab = 0$, then $a = 0$ or $b = 0$. Zero-Factor Property

Example:

$$x^2 - x - 6 = 0$$
$$(x - 3)(x + 2) = 0$$
$$x - 3 = 0 \Rightarrow x = 3$$
$$x + 2 = 0 \Rightarrow x = -2$$

Extracting Square Roots: If $u^2 = c$, then $u = \pm\sqrt{c}$.

Example: $(x + 3)^2 = 16$

$$x + 3 = \pm 4$$
$$x = -3 \pm 4$$
$$x = 1 \quad \text{or} \quad x = -7$$

Completing the Square: If $x^2 + bx = c$, then

$$x^2 + bx + \left(\frac{b}{2}\right)^2 = c + \left(\frac{b}{2}\right)^2$$
$$\left(x + \frac{b}{2}\right)^2 = c + \frac{b^2}{4}.$$

Example:

$$x^2 + 6x = 5$$
$$x^2 + 6x + 3^2 = 5 + 3^2$$
$$(x + 3)^2 = 14$$
$$x + 3 = \pm\sqrt{14}$$
$$x = -3 \pm \sqrt{14}$$

Quadratic Formula: If $ax^2 + bx + c = 0$, then $x = \dfrac{-b \pm \sqrt{b^2 - 4ac}}{2a}$.

Example: $2x^2 + 3x - 1 = 0$

$$x = \frac{-3 \pm \sqrt{3^2 - 4(2)(-1)}}{2(2)}$$
$$x = \frac{-3 \pm \sqrt{17}}{4}$$

Note that you can solve every quadratic equation by completing the square or by using the Quadratic Formula.

Technology Tip

Try programming the Quadratic Formula into a computer or graphing calculator. Programs for several graphing calculator models can be found at this textbook's *Companion Website*.

EXAMPLE 1 Solving Quadratic Equations by Factoring

Solve each quadratic equation by factoring.

a. $6x^2 = 3x$

b. $9x^2 - 6x = -1$

Solution

a.

$6x^2 = 3x$	Write original equation.
$6x^2 - 3x = 0$	Write in general form.
$3x(2x - 1) = 0$	Factor.
$3x = 0 \implies x = 0$	Set 1st factor equal to 0.
$2x - 1 = 0 \implies x = \frac{1}{2}$	Set 2nd factor equal to 0.

b.

$9x^2 - 6x = -1$	Write original equation.
$9x^2 - 6x + 1 = 0$	Write in general form.
$(3x - 1)^2 = 0$	Factor.
$3x - 1 = 0 \implies x = \frac{1}{3}$	Set repeated factor equal to 0.

Throughout the text, when solving equations, be sure to check your solutions either *algebraically* by substituting in the original equation or *graphically*.

Check

a.

$6x^2 = 3x$	Write original equation.
$6(0)^2 \stackrel{?}{=} 3(0)$	Substitute 0 for x.
$0 = 0$	Solution checks. ✓
$6(\frac{1}{2})^2 \stackrel{?}{=} 3(\frac{1}{2})$	Substitute $\frac{1}{2}$ for x.
$\frac{6}{4} = \frac{3}{2}$	Solution checks. ✓

b.

$9x^2 - 6x = -1$	Write original equation.
$9(\frac{1}{3})^2 - 6(\frac{1}{3}) \stackrel{?}{=} -1$	Substitute $\frac{1}{3}$ for x.
$1 - 2 \stackrel{?}{=} -1$	Simplify.
$-1 = -1$	Solution checks. ✓

Similarly, you can check your solutions graphically using a graphing utility, as shown in Figure 2.16.

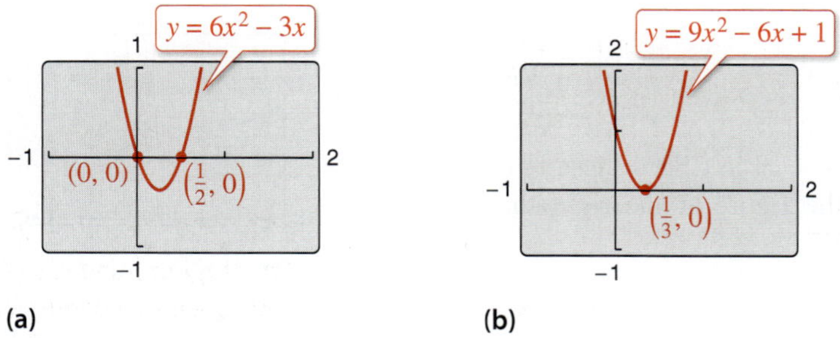

Figure 2.16

✓ Checkpoint *Audio-video solution in English & Spanish at LarsonPrecalculus.com.*

Solve $2x^2 - 3x + 1 = 6$ by factoring.

Extracting Square Roots

EXAMPLE 2 Extracting Square Roots

Solve each quadratic equation by extracting square roots.

a. $4x^2 = 12$ **b.** $(x - 3)^2 = 7$ **c.** $(2x - 1)^2 = -9$

Solution

a. $4x^2 = 12$ Write original equation.

$x^2 = 3$ Divide each side by 4.

$x = \pm\sqrt{3}$ Extract square roots.

This equation has two solutions: $x = \pm\sqrt{3} \approx \pm 1.73$. You can check your solutions graphically, as shown in the figure.

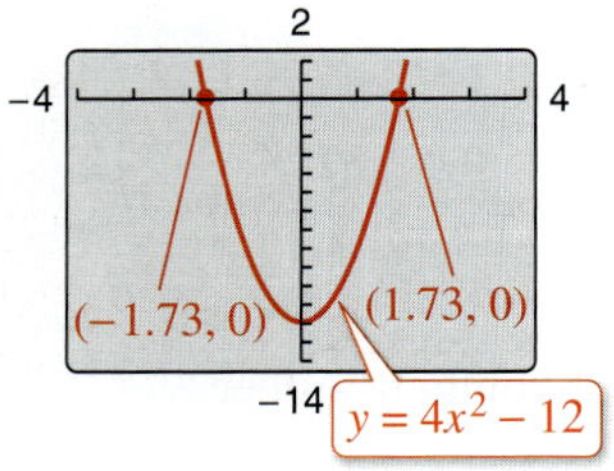

Remark

When you take the square root of a variable expression, you must account for both positive and negative solutions.

b. $(x - 3)^2 = 7$ Write original equation.

$x - 3 = \pm\sqrt{7}$ Extract square roots.

$x = 3 \pm \sqrt{7}$ Add 3 to each side.

This equation has two solutions: $x = 3 + \sqrt{7} \approx 5.65$ and $x = 3 - \sqrt{7} \approx 0.35$. You can check your solutions graphically, as shown in the figure.

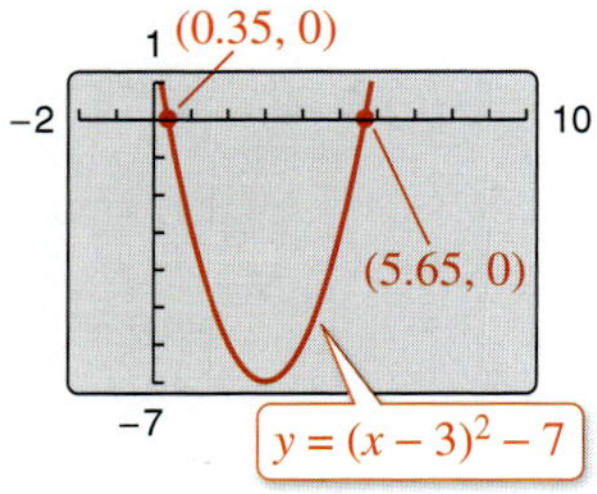

c. $(2x - 1)^2 = -9$ Write original equation.

$2x - 1 = \pm\sqrt{-9}$ Extract square roots.

$2x - 1 = \pm 3i$ Write in *i*-form.

$x = \frac{1}{2} \pm \frac{3}{2}i$ Solve for *x*.

This equation has two imaginary solutions: $x = \frac{1}{2} \pm \frac{3}{2}i$. Check these in the original equation. Notice from the graph shown in the figure at the right that you can conclude that the equation has no real solution.

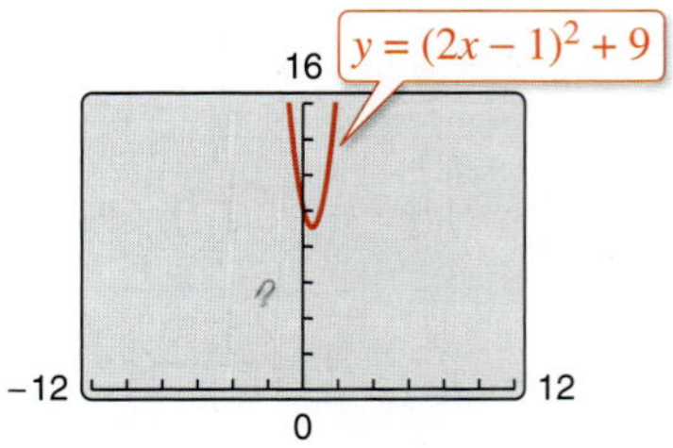

Checkpoint Audio-video solution in English & Spanish at LarsonPrecalculus.com.

Solve each equation by extracting square roots.

a. $3x^2 = 36$

b. $(x - 1)^2 = 10$

c. $(3x - 2)^2 = -6$

Completing the Square

Completing the square is best suited for quadratic equations in general form $ax^2 + bx + c = 0$ with $a = 1$ and b an even number (see page 186). When the leading coefficient of the quadratic is not 1, divide each side of the equation by this coefficient before completing the square (see Examples 4 and 5).

EXAMPLE 3 Completing the Square: Leading Coefficient Is 1

Solve $x^2 + 2x - 6 = 0$ by completing the square.

Solution

$x^2 + 2x - 6 = 0$ Write original equation.

$x^2 + 2x = 6$ Add 6 to each side.

$x^2 + 2x + 1^2 = 6 + 1^2$ Add 1^2 to each side.

(Half of 2)^2

$(x + 1)^2 = 7$ Simplify.

$x + 1 = \pm\sqrt{7}$ Extract square roots.

$x = -1 \pm \sqrt{7}$ Subtract 1 from each side.

Using a calculator, the two solutions are $x \approx 1.65$ and $x \approx -3.65$, which agree with the graphical solutions shown in Figure 2.17.

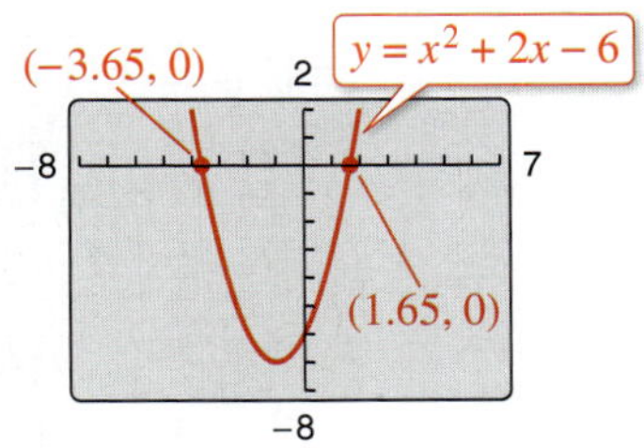

Figure 2.17

✓ Checkpoint Audio-video solution in English & Spanish at LarsonPrecalculus.com.

Solve $x^2 - 4x - 1 = 0$ by completing the square.

EXAMPLE 4 Completing the Square: Leading Coefficient Is Not 1

$2x^2 + 8x + 3 = 0$ Original equation

$2x^2 + 8x = -3$ Subtract 3 from each side.

$x^2 + 4x = -\frac{3}{2}$ Divide each side by 2.

$x^2 + 4x + 2^2 = -\frac{3}{2} + 2^2$ Add 2^2 to each side.

(Half of 4)^2

$(x + 2)^2 = \frac{5}{2}$ Simplify.

$x + 2 = \pm\sqrt{\frac{5}{2}}$ Extract square roots.

$x + 2 = \pm\frac{\sqrt{10}}{2}$ Rationalize denominator.

$x = -2 \pm \frac{\sqrt{10}}{2}$ Subtract 2 from each side.

Using a calculator, the two solutions are $x \approx -0.42$ and $x \approx -3.58$, which agree with the graphical solutions shown in Figure 2.18.

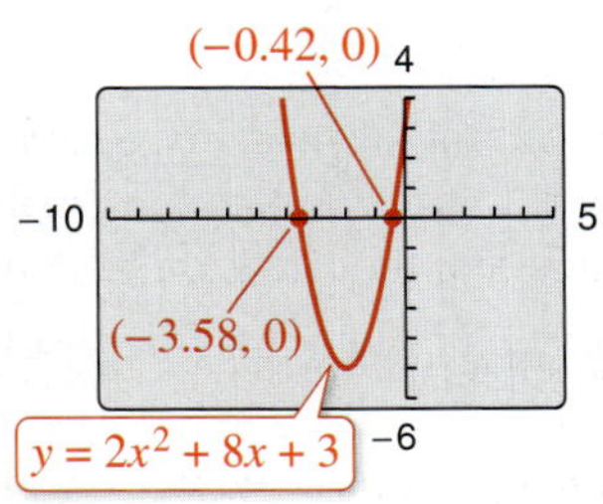

Figure 2.18

✓ Checkpoint Audio-video solution in English & Spanish at LarsonPrecalculus.com.

Solve $2x^2 - 4x + 1 = 0$ by completing the square.

EXAMPLE 5 Completing the Square: Leading Coefficient Is Not 1

Solve $3x^2 - 4x - 5 = 0$ by completing the square.

Solution

$$3x^2 - 4x - 5 = 0 \qquad \text{Write original equation.}$$

$$3x^2 - 4x = 5 \qquad \text{Add 5 to each side.}$$

$$x^2 - \frac{4}{3}x = \frac{5}{3} \qquad \text{Divide each side by 3.}$$

$$x^2 - \frac{4}{3}x + \left(-\frac{2}{3}\right)^2 = \frac{5}{3} + \left(-\frac{2}{3}\right)^2 \qquad \text{Add } \left(-\tfrac{2}{3}\right)^2 \text{ to each side.}$$

$$\left(\text{Half of } -\frac{4}{3}\right)^2$$

$$\left(x - \frac{2}{3}\right)^2 = \frac{19}{9} \qquad \text{Simplify.}$$

$$x - \frac{2}{3} = \pm\frac{\sqrt{19}}{3} \qquad \text{Extract square roots.}$$

$$x = \frac{2}{3} \pm \frac{\sqrt{19}}{3} \qquad \text{Add } \tfrac{2}{3} \text{ to each side.}$$

Using a calculator, the two solutions are

$x \approx 2.12$ and $x \approx -0.79$

which agree with the graphical solutions shown in Figure 2.19.

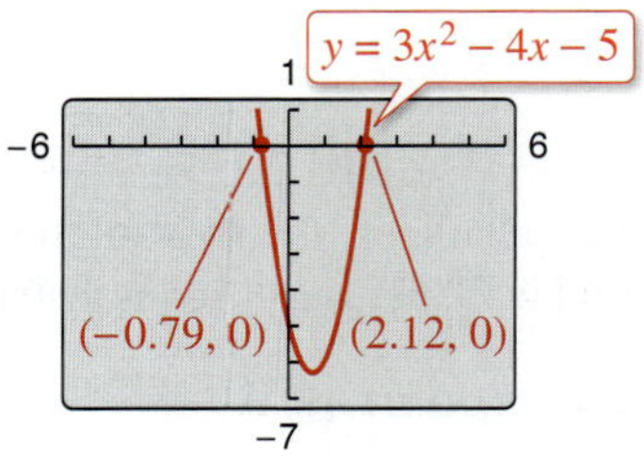

Figure 2.19

Checkpoint *Audio-video solution in English & Spanish at LarsonPrecalculus.com.*

Solve $3x^2 - 10x - 2 = 0$ by completing the square.

EXAMPLE 6 Completing the Square: No Real Solution

Solve $x^2 - 4x + 8 = 0$ by completing the square.

Solution

$$x^2 - 4x + 8 = 0 \qquad \text{Write original equation.}$$

$$x^2 - 4x = -8 \qquad \text{Subtract 8 from each side.}$$

$$x^2 - 4x + (-2)^2 = -8 + (-2)^2 \qquad \text{Add } (-2)^2 \text{ to each side.}$$

$$(\text{Half of } -4)^2$$

$$(x - 2)^2 = -4 \qquad \text{Simplify.}$$

$$x - 2 = \pm 2i \qquad \text{Extract square roots.}$$

$$x = 2 \pm 2i \qquad \text{Add 2 to each side.}$$

This equation has two imaginary solutions:

$2 + 2i$ and $2 - 2i$.

Check these in the original equation.

Checkpoint *Audio-video solution in English & Spanish at LarsonPrecalculus.com.*

Solve $x^2 + 2x + 17 = 0$ by completing the square.

The Quadratic Formula

The Quadratic Formula is derived by completing the square for a general quadratic equation.

$$ax^2 + bx + c = 0 \qquad \text{Quadratic equation in general form, } a \neq 0$$

$$ax^2 + bx = -c \qquad \text{Subtract } c \text{ from each side.}$$

$$x^2 + \frac{b}{a}x = -\frac{c}{a} \qquad \text{Divide each side by } a.$$

$$x^2 + \frac{b}{a}x + \left(\frac{b}{2a}\right)^2 = -\frac{c}{a} + \left(\frac{b}{2a}\right)^2 \qquad \text{Complete the square.}$$

$\left(\text{Half of } \frac{b}{a}\right)^2$

$$\left(x + \frac{b}{2a}\right)^2 = \frac{b^2 - 4ac}{4a^2} \qquad \text{Simplify.}$$

$$x + \frac{b}{2a} = \pm\sqrt{\frac{b^2 - 4ac}{4a^2}} \qquad \text{Extract square roots.}$$

$$x = -\frac{b}{2a} \pm \frac{\sqrt{b^2 - 4ac}}{2|a|} \qquad \text{Solutions}$$

Because $\pm 2|a|$ represents the same numbers as $\pm 2a$, the formula simplifies to

$$x = \frac{-b \pm \sqrt{b^2 - 4ac}}{2a}.$$

Explore the Concept

Use a graphing utility to graph the three quadratic equations

$$y_1 = x^2 - 2x$$

$$y_2 = x^2 - 2x + 1$$

$$y_3 = x^2 - 2x + 2$$

in the same viewing window. Compute the *discriminant* $b^2 - 4ac$ for each equation and discuss the relationship between the discriminant and the number of zeros of the quadratic function.

EXAMPLE 7 Quadratic Formula: Two Distinct Solutions

See LarsonPrecalculus.com for an interactive version of this type of example.

Use the Quadratic Formula to solve

$x^2 + 3x = 9.$

Algebraic Solution

$$x^2 + 3x = 9 \qquad \text{Write original equation.}$$

$$x^2 + 3x - 9 = 0 \qquad \text{Write in general form.}$$

$$x = \frac{-b \pm \sqrt{b^2 - 4ac}}{2a} \qquad \text{Quadratic Formula}$$

$$x = \frac{-3 \pm \sqrt{3^2 - 4(1)(-9)}}{2(1)} \qquad \text{Substitute 3 for } b, \text{ 1 for } a, \text{ and } -9 \text{ for } c.$$

$$x = \frac{-3 \pm \sqrt{45}}{2} \qquad \text{Simplify.}$$

$$x = \frac{-3 \pm 3\sqrt{5}}{2} \qquad \text{Simplify radical.}$$

$$x \approx 1.85 \text{ or } -4.85 \qquad \text{Solutions}$$

The equation has two solutions: $x \approx 1.85$ and $x \approx -4.85$. Check these solutions in the original equation.

Graphical Solution

Use a graphing utility to graph $y_1 = x^2 + 3x$ and $y_2 = 9$ in the same viewing window. Use the *intersect* feature of the graphing utility to determine that the graphs intersect when $x \approx 1.85$ and $x \approx -4.85$, as shown in the figure. The x-coordinates of the points of intersection are the solutions of the equation $x^2 + 3x = 9$.

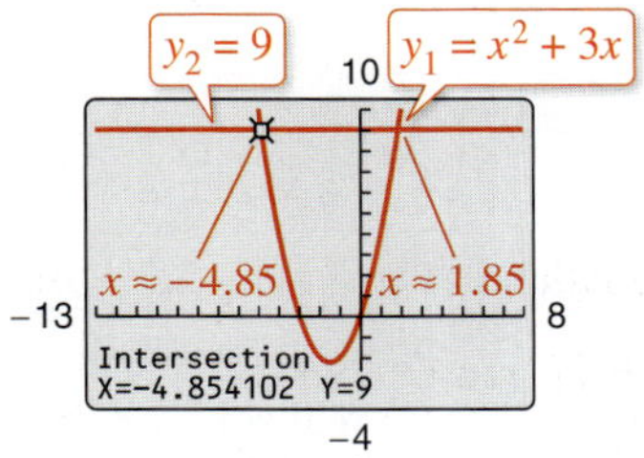

✓ Checkpoint *Audio-video solution in English & Spanish at LarsonPrecalculus.com.*

Use the Quadratic Formula to solve $3x^2 + 2x - 10 = 0.$

EXAMPLE 8 Quadratic Formula: One Repeated Solution

Use the Quadratic Formula to solve

$8x^2 - 24x + 18 = 0.$

Algebraic Solution

This equation has a common factor of 2. You can simplify the equation by dividing each side of the equation by 2.

$8x^2 - 24x + 18 = 0$ — Write original equation.

$4x^2 - 12x + 9 = 0$ — Divide each side by 2.

$$x = \frac{-b \pm \sqrt{b^2 - 4ac}}{2a}$$ Quadratic Formula

$$x = \frac{-(-12) \pm \sqrt{(-12)^2 - 4(4)(9)}}{2(4)}$$

$$x = \frac{12 \pm \sqrt{0}}{8}$$ Simplify.

$$x = \frac{3}{2}$$ Repeated solution

The equation has only one solution: $x = \frac{3}{2}$. Check this solution in the original equation.

Graphical Solution

Use a graphing utility to graph

$y = 8x^2 - 24x + 18.$

Use the *zero* feature of the graphing utility to approximate any values of x for which the function is equal to zero. From the figure, it appears that the function is equal to zero when $x = 1.5 = \frac{3}{2}$. This is the only solution of the equation $8x^2 - 24x + 18 = 0$.

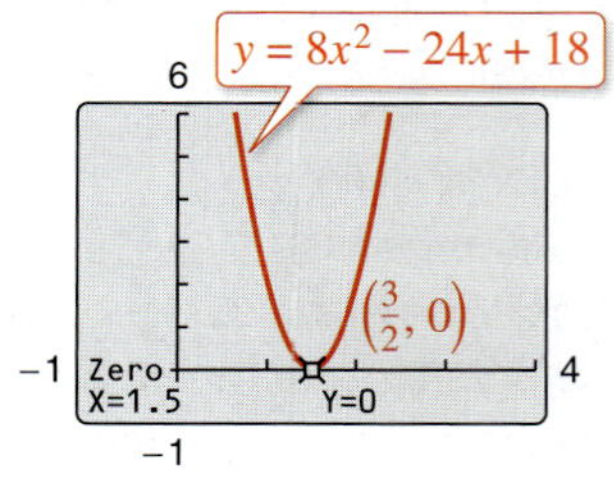

Checkpoint Audio-video solution in English & Spanish at LarsonPrecalculus.com.

Use the Quadratic Formula to solve $3x^2 - 6x + 3 = 0$.

EXAMPLE 9 Quadratic Formula: No Real Solution

Use the Quadratic Formula to solve

$3x^2 - 2x + 5 = 0.$

Algebraic Solution

This equation is in general form, so use the Quadratic Formula with $a = 3$, $b = -2$, and $c = 5$.

$$x = \frac{-b \pm \sqrt{b^2 - 4ac}}{2a}$$ Quadratic Formula

$$x = \frac{-(-2) \pm \sqrt{(-2)^2 - 4(3)(5)}}{2(3)}$$ Substitute -2 for b, 3 for a, and 5 for c.

$$x = \frac{2 \pm \sqrt{-56}}{6}$$ Simplify.

$$x = \frac{2 \pm 2\sqrt{14}i}{6}$$ Simplify radical.

$$x = \frac{1}{3} \pm \frac{\sqrt{14}}{3}i$$ Solutions

The equation has no real solution, but it has two imaginary solutions:

$x = \frac{1}{3}(1 + \sqrt{14}i)$ and $x = \frac{1}{3}(1 - \sqrt{14}i)$.

Graphical Solution

Use a graphing utility to graph

$y = 3x^2 - 2x + 5.$

Note in the figure that the graph of the function appears to have no x-intercept. From this, you can conclude that the equation $3x^2 - 2x + 5 = 0$ has no real solution. You can solve the equation algebraically to find the imaginary solutions.

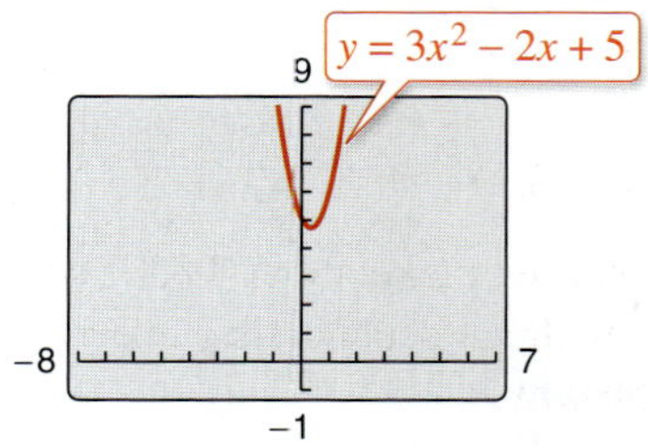

Checkpoint Audio-video solution in English & Spanish at LarsonPrecalculus.com.

Use the Quadratic Formula to solve $x^2 - 2x + 2 = 0$.

Applications

A common application of quadratic equations involves an object that is falling (or projected into the air). The general equation that gives the height of such an object is called a **position equation,** and on Earth's surface it has the form

$$s = -16t^2 + v_0t + s_0.$$

In this equation, s represents the height of the object (in feet), v_0 represents the initial velocity of the object (in feet per second), s_0 represents the initial height of the object (in feet), and t represents the time (in seconds). Note that this position equation ignores air resistance, so it should only be used to model falling objects that have little air resistance and that fall short distances.

Remark

In the position equation $s = -16t^2 + v_0t + s_0$, the initial velocity v_0 is positive when the object is rising and negative when the object is falling.

EXAMPLE 10 Falling Time

A construction worker accidentally drops a wrench from a height of 235 feet (see Figure 2.20) and yells, "Look out below!" Could a person at ground level hear this warning in time to get out of the way? (*Note:* The speed of sound is about 1100 feet per second.)

Solution

Because sound travels at about 1100 feet per second, and the distance to the ground is 235 feet, it follows that a person at ground level hears the warning within 1 second of the time the wrench is dropped. To set up a mathematical model for the height of the wrench, use the position equation

$$s = -16t^2 + v_0t + s_0.$$

Because the object is dropped rather than thrown, the initial velocity is $v_0 = 0$ feet per second. So, with an initial height of $s_0 = 235$ feet, you have the model

$$s = -16t^2 + (0)t + 235 = -16t^2 + 235.$$

After falling for 1 second, the height of the wrench is

$$-16(1)^2 + 235 = 219 \text{ feet.}$$

After falling for 2 seconds, the height of the wrench is

$$-16(2)^2 + 235 = 171 \text{ feet.}$$

To find the number of seconds it takes the wrench to hit the ground, let the height s be zero and solve the equation for t.

$s = -16t^2 + 235$ — Write position equation.

$0 = -16t^2 + 235$ — Substitute 0 for s.

$16t^2 = 235$ — Add $16t^2$ to each side.

$t^2 = \dfrac{235}{16}$ — Divide each side by 16.

$t = \dfrac{\sqrt{235}}{4}$ — Extract positive square root.

$t \approx 3.83$ — Use a calculator.

The wrench will take about 3.83 seconds to hit the ground. So, a person who hears the warning within 1 second after the wrench is dropped will have almost 3 seconds to get out of the way.

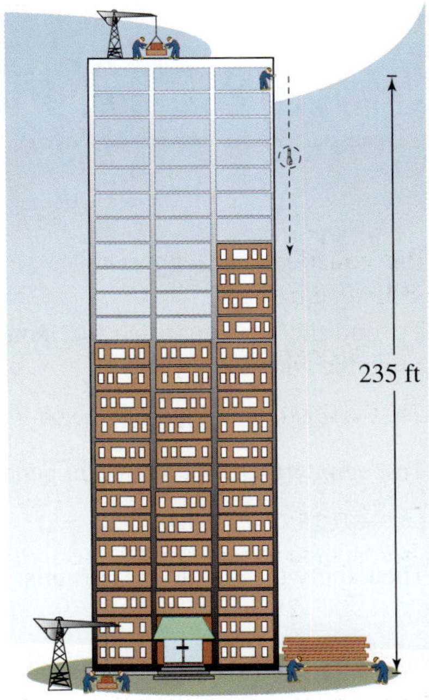

Figure 2.20

Checkpoint *Audio-video solution in English & Spanish at LarsonPrecalculus.com.*

You drop a rock from a height of 196 feet. How long does it take the rock to hit the ground?

EXAMPLE 11 Quadratic Modeling: Internet Users

From 1995 through 2014, the estimated numbers I (in millions) of U.S. adults who use the Internet can be modeled by

$$I = -0.25t^2 + 16.9t - 50, \quad 5 \le t \le 24$$

where t represents the year, with $t = 5$ corresponding to 1995 (see figure). According to the model, in which year did the number of U.S. adults who use the Internet reach 94 million? (Sources: Pew Research Center, U.S. Census Bureau)

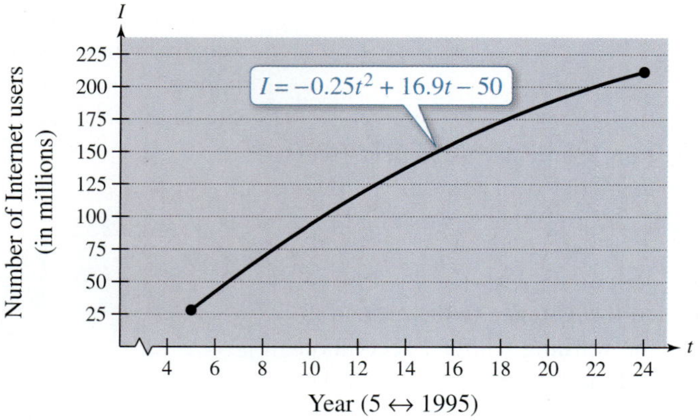

Solution

To find the year in which the number of Internet users reached 94 million, you can solve the equation

$$-0.25t^2 + 16.9t - 50 = 94.$$

To begin, write the equation in general form.

$$-0.25t^2 + 16.9t - 144 = 0$$

Then apply the Quadratic Formula.

$$t = \frac{-b \pm \sqrt{b^2 - 4ac}}{2a}$$

$$t = \frac{-16.9 \pm \sqrt{16.9^2 - 4(-0.25)(-144)}}{2(-0.25)}$$

$$t = \frac{-16.9 \pm \sqrt{141.61}}{-0.5} \quad \Rightarrow \quad t = 10 \text{ or } 57.6$$

The domain of the model is $5 \le t \le 24$, so choose $t = 10$. Because $t = 5$ corresponds to 1995, it follows that $t = 10$ must correspond to 2000. So, the number of U.S. adults who use the Internet reached 94 million during 2000.

✓ Checkpoint *Audio-video solution in English & Spanish at LarsonPrecalculus.com.*

According to the model in Example 11, in which year did the number of U.S. adults who use the Internet reach 188 million?

Technology Tip

You can solve Example 11 with your graphing utility by graphing $y_1 = -0.25t^2 + 16.9t - 50$ and $y_2 = 94$ in the same viewing window and finding their point of intersection. You should obtain $t = 10$.

Another type of application that often involves a quadratic equation is one dealing with the hypotenuse of a right triangle. These types of applications often use the Pythagorean Theorem, which states that $a^2 + b^2 = c^2$, where a and b are the legs of a right triangle and c is the hypotenuse, as indicated in Figure 2.21.

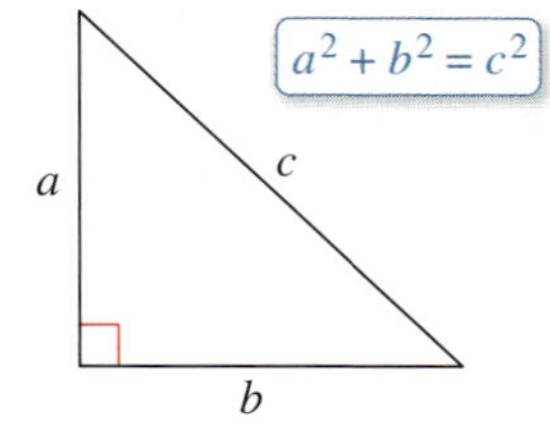

Figure 2.21

EXAMPLE 12 An Application Involving the Pythagorean Theorem

An L-shaped sidewalk from the athletic center to the library on a college campus is shown in the figure. The sidewalk was constructed so that the length of one sidewalk forming the L is twice as long as the other. The length of the diagonal sidewalk that cuts across the grounds between the two buildings is 102 feet. How many feet does a person save by walking on the diagonal sidewalk?

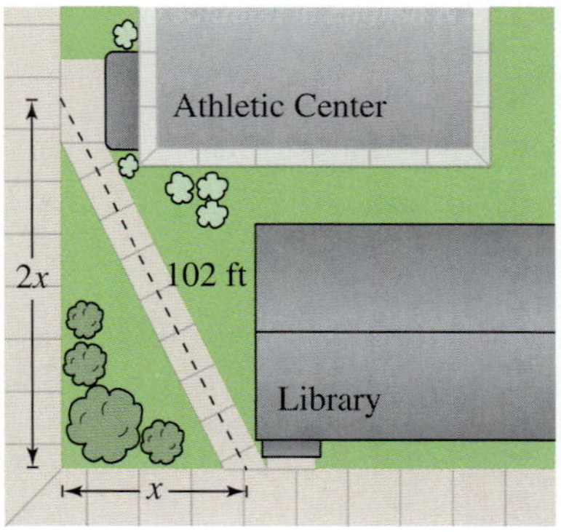

Solution

Using the Pythagorean Theorem, you have

$$a^2 + b^2 = c^2 \qquad \text{Pythagorean Theorem}$$

$$x^2 + (2x)^2 = 102^2 \qquad \text{Substitute for } a, b, \text{ and } c.$$

$$5x^2 = 10{,}404 \qquad \text{Simplify.}$$

$$x^2 = 2080.8 \qquad \text{Divide each side by 5.}$$

$$x = \sqrt{2080.8}. \qquad \text{Extract positive square root.}$$

The total distance covered by walking on the L-shaped sidewalk is

$$x + 2x = 3x = 3\sqrt{2080.8} \approx 136.85 \text{ feet.}$$

Walking on the diagonal sidewalk saves a person about

$$136.85 - 102 = 34.85 \text{ feet.}$$

✓ **Checkpoint** *Audio-video solution in English & Spanish at LarsonPrecalculus.com.*

In Example 12, how many feet does a person save by walking on the diagonal sidewalk when the length of one sidewalk forming the L is three times as long as the other?

2.4 Exercises

See *CalcChat.com* for tutorial help and worked-out solutions to odd-numbered exercises. For instructions on how to use a graphing utility, see Appendix A.

Vocabulary and Concept Check

In Exercises 1 and 2, fill in the blank.

1. An equation of the form $ax^2 + bx + c = 0$, where a, b, and c are real numbers and $a \neq 0$, is a _______ , or a second-degree polynomial equation in x.
2. The part of the Quadratic Formula $b^2 - 4ac$, known as the _______ , determines the type of solutions of a quadratic equation.

3. List four methods that can be used to solve a quadratic equation.
4. What does the equation $s = -16t^2 + v_0t + s_0$ represent? What do v_0 and s_0 represent?

Procedures and Problem Solving

Writing a Quadratic Equation in General Form **In Exercises 5–8, write the quadratic equation in general form. Do not solve the equation.**

5. $2x^2 = 3 - 5x$
6. $x^2 = 25x + 26$
7. $\frac{1}{5}(3x^2 - 10) = 12x$
8. $x(x + 2) = 3x^2 + 1$

Solving a Quadratic Equation by Factoring **In Exercises 9–20, solve the quadratic equation by factoring. Check your solutions in the original equation.**

9. $15x^2 + 5x = 0$
10. $9x^2 - 21x = 0$
11. $x^2 - 10x + 21 = 0$
12. $x^2 - 10x + 9 = 0$
13. $x^2 - 8x + 16 = 0$
14. $4x^2 + 12x + 9 = 0$
15. $3x^2 = 8 - 2x$
16. $2x^2 = 19x + 33$
17. $-x^2 - 11x = 28$
18. $-x^2 - 11x = 30$
19. $(x + a)^2 - b^2 = 0$
20. $x^2 + 2ax + a^2 = 0$

Extracting Square Roots **In Exercises 21–32, solve the equation by extracting square roots. List both the exact solutions and the decimal solutions rounded to the nearest hundredth.**

21. $x^2 = 49$
22. $x^2 = 144$
23. $3x^2 = 81$
24. $9x^2 = 36$
25. $x^2 + 12 = 112$
26. $x^2 - 3 = 78$
27. $(x - 12)^2 = 16$
28. $(x - 5)^2 = 25$
29. $(3x - 1)^2 + 6 = 0$
30. $(2x + 3)^2 + 25 = 0$
31. $(x - 7)^2 = (x + 3)^2$
32. $(x + 5)^2 = (x + 4)^2$

Completing the Square **In Exercises 33–46, solve the quadratic equation by completing the square. Verify your answer graphically.**

33. $x^2 + 4x - 32 = 0$
34. $x^2 - 2x - 3 = 0$
35. $x^2 - 6x + 2 = 0$
36. $x^2 + 8x + 14 = 0$
37. $x^2 - 4x + 13 = 0$
38. $x^2 - 6x + 34 = 0$
39. $x^2 + 8x + 32 = 0$
40. $x^2 + 18x + 117 = 0$
41. $-6 + 2x - x^2 = 0$
42. $-x^2 + 6x - 16 = 0$
43. $9x^2 - 18x + 3 = 0$
44. $4x^2 - 16x - 5 = 0$
45. $2x^2 + 5x - 8 = 0$
46. $9x^2 - 12x - 14 = 0$

Graphing a Quadratic Equation **In Exercises 47–50, (a) use a graphing utility to graph the equation, (b) use the graph to approximate any x-intercepts of the graph, and (c) verify your results algebraically.**

47. $y = (x + 3)^2 - 4$
48. $y = 1 - (x - 2)^2$
49. $y = -4x^2 + 4x + 3$
50. $y = x^2 + 3x - 4$

Using a Graphing Utility to Find Solutions **In Exercises 51–56, use a graphing utility to determine the number of real solutions of the quadratic equation.**

51. $x^2 - 4x + 4 = 0$
52. $2x^2 - x - 1 = 0$
53. $\frac{4}{7}x^2 - 8x + 28 = 0$
54. $\frac{1}{3}x^2 - 5x + 25 = 0$
55. $-0.2x^2 + 1.2x - 8 = 0$
56. $9 + 2.4x - 8.3x^2 = 0$

Using the Quadratic Formula **In Exercises 57–68, use the Quadratic Formula to solve the equation. Use a graphing utility to verify your solutions graphically.**

57. $x^2 - 9x + 19 = 0$
58. $x^2 - 10x + 22 = 0$
59. $x^2 + 3x = -8$
60. $x^2 + 16 = -5x$
61. $4x = 8 - x^2$
62. $8x = 4 - x^2$
63. $20x^2 - 20x + 5 = 0$
64. $9x^2 - 18x + 9 = 0$
65. $16x^2 + 24x + 9 = 0$
66. $9x^2 + 30x + 25 = 0$
67. $4x^2 + 16x + 17 = 0$
68. $9x^2 - 6x + 37 = 0$

Solving a Quadratic Equation **In Exercises 69–76, solve the equation using any convenient method.**

69. $x^2 - 3x - 4 = 0$
70. $11x^2 + 33x = 0$
71. $(x + 3)^2 = 81$
72. $(x - 1)^2 = -1$
73. $x^2 - 2x = -\frac{13}{4}$
74. $x^2 + 4x = -\frac{19}{4}$
75. $5x^2 = 3x + 1$
76. $4x^2 = 7x + 3$

Finding *x*-Intercepts Algebraically **In Exercises 77–80, find algebraically the *x*-intercept(s), if any, of the graph of the equation.**

77.
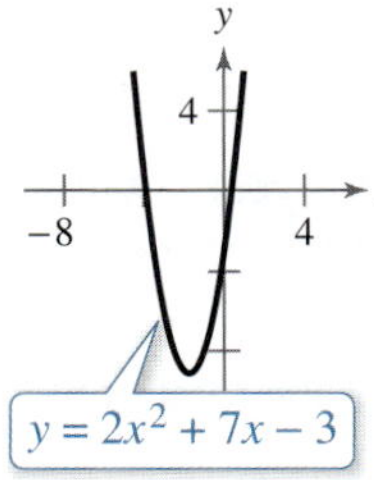

78.
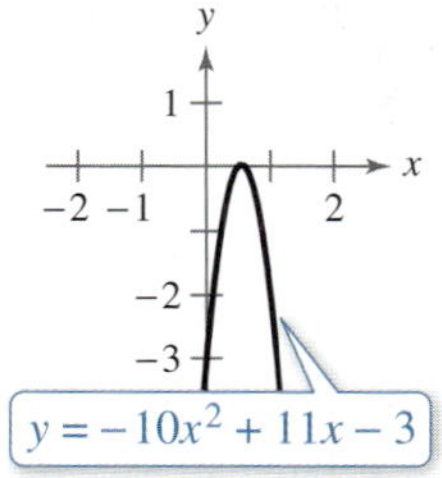

79.
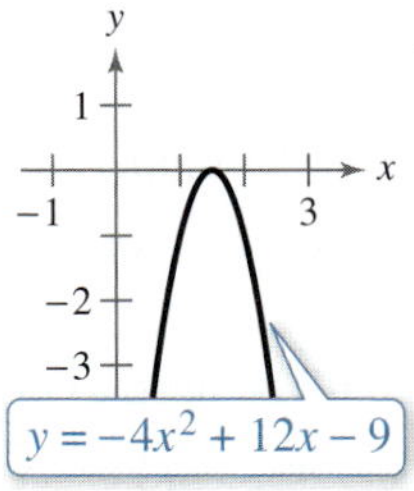

80.
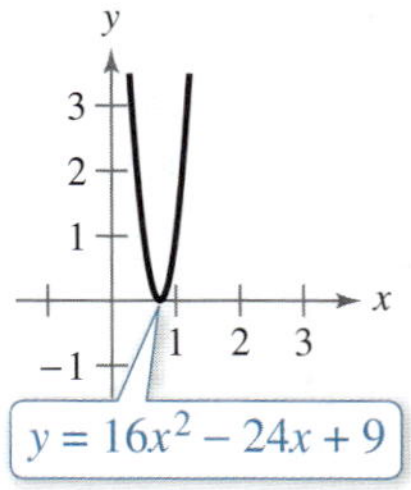

Writing Quadratic Equations **In Exercises 81–90, find two quadratic equations having the given solutions. (There are many correct answers.)**

81. $-6, 5$
82. $-2, 1$
83. $-\frac{7}{3}, \frac{6}{7}$
84. $-\frac{2}{3}, \frac{4}{3}$
85. $5\sqrt{3}, -5\sqrt{3}$
86. $2\sqrt{5}, -2\sqrt{5}$
87. $1 + 2\sqrt{3}, 1 - 2\sqrt{3}$
88. $2 + 3\sqrt{5}, 2 - 3\sqrt{5}$
89. $2 + i, 2 - i$
90. $3 + 4i, 3 - 4i$

91. Architecture The floor of a one-story building is 14 feet longer than it is wide. The building has 1632 square feet of floor space.

(a) Draw a diagram that gives a visual representation of the floor space. Represent the width as w and show the length in terms of w.

(b) Write a quadratic equation for the area of the floor in terms of w.

(c) Find the length and width of the building floor.

92. Design and Construction An above-ground swimming pool with a square base is to be constructed such that the surface area of the pool is 561 square feet. The height of the pool is to be 4 feet (see figure). What should the dimensions of the base be? (*Hint:* The surface area is $S = x^2 + 4xh$.)

93. MODELING DATA

A gardener has 100 meters of fencing to enclose two adjacent rectangular gardens, as shown in the figure.

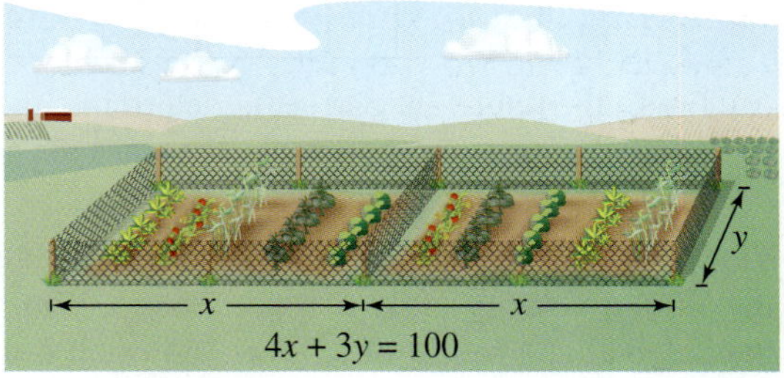

(a) Write the area A of the enclosed region as a function of x. Determine the domain of the function.

(b) Use a graphing utility to generate additional rows of the table. Use the table to estimate the dimensions that will produce a maximum area.

x	y	Area
2	$\frac{92}{3}$	$\frac{368}{3} \approx 123$
4	28	224

(c) Use the graphing utility to graph the area function, and use the graph to estimate the dimensions that will produce a maximum area.

(d) Use the graph to approximate the dimensions such that the enclosed area is 350 square meters.

(e) Find the required dimensions of part (d) algebraically.

94. **Geometry** An open box is to be made from a square piece of material by cutting four-centimeter squares from each corner and turning up the sides (see figure). The volume of the finished box is to be 576 cubic centimeters. Find the size of the original piece of material.

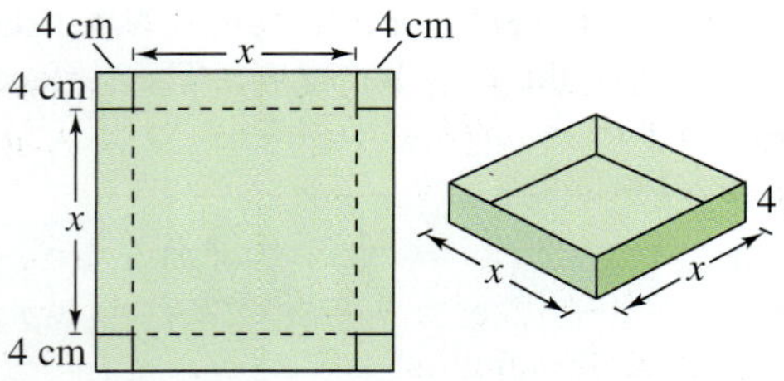

Using the Position Equation **In Exercises 95–98, use the position equation given on page 193 as the model for the problem.**

95. **Falling Object** At 2080 feet tall, the Tokyo Skytree in Tokyo, Japan, is the world's tallest tower. An object is dropped from the top of the tower.

(a) Find the position equation $s = -16t^2 + v_0t + s_0$.

(b) Complete the table.

t	0	2	4	6	8	10	12
s							

(c) From the table in part (b), determine the time interval during which the object reaches the ground. Find the time algebraically.

96. **Vertical Motion** You throw a baseball straight up into the air at a velocity of 45 feet per second. You release the baseball at a height of 5.5 feet and catch it when it falls back to a height of 6 feet.

(a) Use the position equation to write a mathematical model for the height of the baseball.

(b) Find the height of the baseball after 0.5 second.

(c) How many seconds is the baseball in the air? (Use a graphing utility to verify your answer.)

97. **Projectile Motion** A cargo plane flying at 8000 feet over level terrain drops a 500-pound supply package.

(a) How long will it take the package to strike the ground?

(b) The plane is flying at 600 miles per hour. How far will the package travel horizontally during its descent?

98. **Wildlife Management** A game commission plane, flying at 1100 feet, drops oral rabies vaccination pellets into a game preserve.

(a) Use the position equation to write a mathematical model for the height of the vaccination pellets.

(b) The plane is flying at 95 miles per hour. How far will the pellets travel horizontally during their descent?

99. **Education** The average salaries S (in thousands of dollars) of secondary classroom teachers in the United States from 2005 through 2013 can be approximated by the model

$$S = -0.143t^2 + 3.73t + 32.5, \quad 5 \le t \le 13$$

where t represents the year, with $t = 5$ corresponding to 2005. (Source: National Education Association)

(a) Determine algebraically when the average salary of a secondary classroom teacher was $50,000.

(b) Verify your answer to part (a) by creating a table of values for the model.

(c) Use a graphing utility to graph the model.

(d) Use the model to determine when the average salary reached $55,500.

(e) Do you believe the model could be used to predict the average salaries for years beyond 2013? Explain your reasoning.

100. ***Why you should learn it*** *(p. 186)* The total public debt D (in trillions of dollars) in the United States from 2005 through 2014 can be approximated by the model

$$D = 0.051t^2 + 0.20t + 5.0, \quad 5 \le t \le 14$$

where t represents the year, with $t = 5$ corresponding to 2005. (Source: U.S. Department of Treasury)

(a) Determine algebraically when the total public debt reached $10 trillion.

(b) Verify your answer to part (a) by creating a table of values for the model.

(c) Use a graphing utility to graph the model.

(d) Use the model to predict when the total public debt will reach $20 trillion.

(e) Do you believe the model could be used to predict the total public debt for years beyond 2014? Explain your reasoning.

101. **Biology** The metabolic rate of an ectothermic organism increases with increasing temperature within a certain range. Experimental data for oxygen consumption C (in microliters per gram per hour) of a beetle at certain temperatures yielded the model

$$C = 0.45x^2 - 1.73x + 52.65, \quad 10 \le x \le 25$$

where x is the air temperature (in degrees Celsius).

(a) Use a graphing utility to graph the consumption model over the specified domain.

(b) Use the graph to approximate the air temperature resulting in oxygen consumption of 150 microliters per gram per hour.

(c) When the temperature is increased from 10°C to 20°C, the oxygen consumption will be increased by approximately what factor?

102. Mechanical Engineering The fuel efficiency F (in miles per gallon) of a car is approximated by

$$F = -0.0191s^2 + 1.639s + 2.20, \quad 5 \le s \le 65$$

where s is the average speed of the car (in miles per hour).

(a) Use a graphing utility to graph the function over the specified domain.

(b) Use the graph to determine the greatest fuel efficiency of the car. How fast should the car travel?

(c) When the average speed of the car is increased from 20 miles per hour to 30 miles per hour, the fuel efficiency will be increased by approximately what factor?

103. Boating A winch is used to tow a boat to a dock. The rope is attached to the boat at a point 15 feet below the level of the winch. (See figure.)

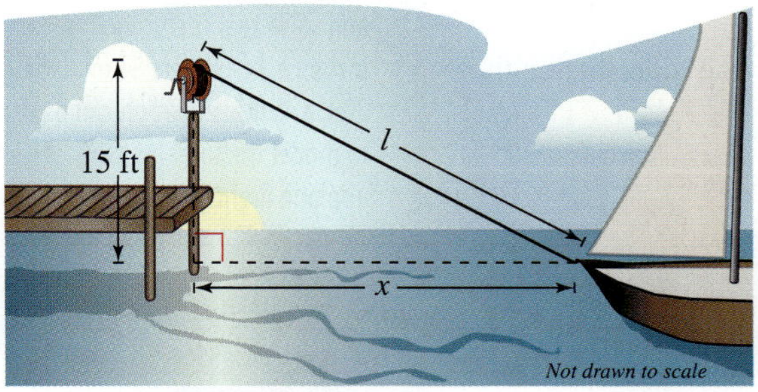

(a) Use the Pythagorean Theorem to write an equation giving the relationship between l and x.

(b) Find the distance from the boat to the dock when there is 75 feet of rope out.

104. Aviation Two planes leave simultaneously from Chicago's O'Hare Airport, one flying due north and the other due east (see figure). The northbound plane is flying 50 miles per hour faster than the eastbound plane. After 3 hours, the planes are 2440 miles apart. Find the speed of each plane.

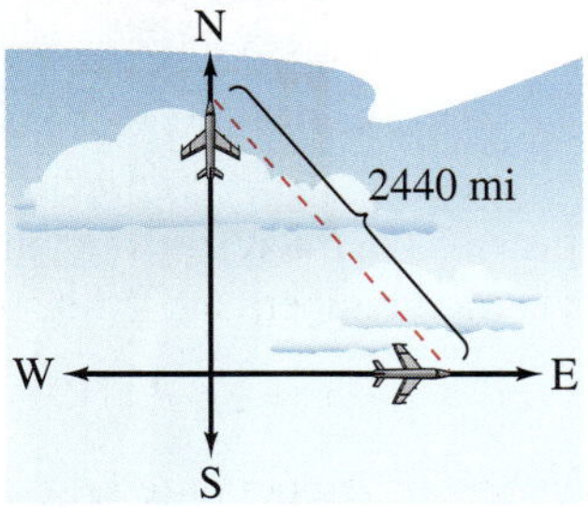

Conclusions

True or False? **In Exercises 105–107, determine whether the statement is true or false. Justify your answer.**

105. The quadratic equation $-3x^2 - x = 10$ has two real solutions.

106. If $(2x - 3)(x + 5) = 8$, then $2x - 3 = 8$ or $x + 5 = 8$.

107. A quadratic equation with real coefficients can have one real solution and one imaginary solution.

108. Exploration Given that a and b are nonzero real numbers, determine the solutions of the equations.

(a) $ax^2 + bx = 0$ (b) $ax^2 - ax = 0$

109. Proof Given that the solutions of a quadratic equation are $x = \left(-b \pm \sqrt{b^2 - 4ac}\right)/(2a)$, show that the sum of the solutions is $S = -b/a$.

110. Proof Given that the solutions of a quadratic equation are $x = \left(-b \pm \sqrt{b^2 - 4ac}\right)/(2a)$, show that the product of the solutions is $P = c/a$.

111. Think About It Match the equation with a method you would use to solve it. Explain your reasoning. (Use each method once and do not solve the equations.)

(a) $3x^2 + 5x - 11 = 0$ (i) Factoring

(b) $x^2 + 10x = 3$ (ii) Extracting square roots

(c) $x^2 - 16x + 64 = 0$ (iii) Completing the square

(d) $x^2 - 15 = 0$ (iv) Quadratic Formula

112. HOW DO YOU SEE IT? Match the quadratic equation with its graph. Explain how you found your result. [The graphs are labeled (i) and (ii).]

(i)

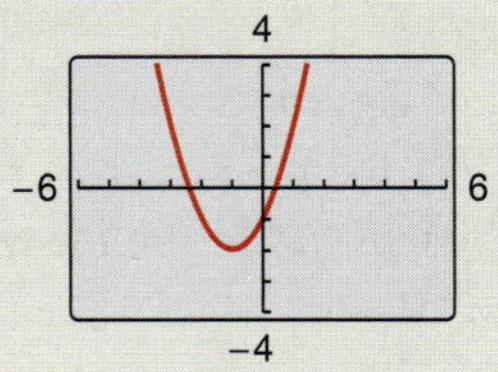

(ii)

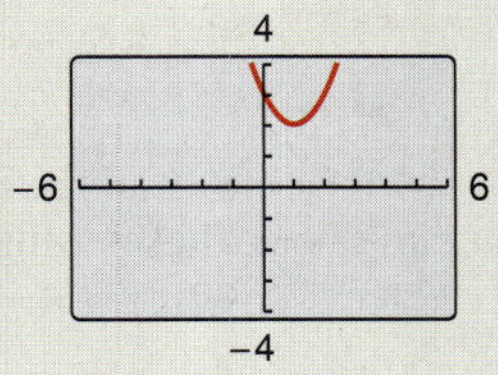

(a) $0 = (x - 1)^2 + 2$ (b) $0 = (x + 1)^2 - 2$

Cumulative Mixed Review

Factoring an Expression **In Exercises 113–116, completely factor the expression over the real numbers.**

113. $x^5 - 27x^2$

114. $x^3 - 5x^2 - 14x$

115. $x^3 + 5x^2 - 2x - 10$

116. $5(x + 5)x^{1/3} + 4x^{4/3}$

---

117. ***Make a Decision*** To work an extended application analyzing the population of the United States, visit this textbook's website at *LarsonPrecalculus.com.* (Data Source: U.S. Census Bureau)

2.5 Solving Other Types of Equations Algebraically

What you should learn

- Solve polynomial equations of degree three or greater.
- Solve equations involving radicals.
- Solve equations involving fractions or absolute values.
- Use polynomial equations and equations involving radicals to model and solve real-life problems.

Why you should learn it

Polynomial equations, radical equations, and absolute value equations can be used to model and solve real-life problems. For instance, in Exercise 84 on page 208, a radical equation can be used to model the total monthly cost of airplane flights between Chicago and Denver.

Polynomial Equations

In this section, the techniques for solving equations are extended to nonlinear and nonquadratic equations. At this point in the text, you have four basic algebraic methods for solving nonlinear equations—*factoring, extracting square roots, completing the square,* and the *Quadratic Formula.* The main goal of this section is to learn to *rewrite* nonlinear equations in a form to which you can apply one of these methods.

Example 1 shows how to use factoring to solve a **polynomial equation,** which is an equation that can be written in the general form

$$a_nx^n + a_{n-1}x^{n-1} + \cdot \cdot \cdot + a_2x^2 + a_1x + a_0 = 0.$$

EXAMPLE 1 Solving a Polynomial Equation by Factoring

Solve $3x^4 = 48x^2$.

Solution

First write the polynomial equation in general form with zero on one side, factor the other side, and then set each factor equal to zero and solve.

$3x^4 = 48x^2$	Write original equation.
$3x^4 - 48x^2 = 0$	Write in general form.
$3x^2(x^2 - 16) = 0$	Factor out common factor.
$3x^2(x + 4)(x - 4) = 0$	Factor completely.
$3x^2 = 0 \Rightarrow x = 0$	Set 1st factor equal to 0.
$x + 4 = 0 \Rightarrow x = -4$	Set 2nd factor equal to 0.
$x - 4 = 0 \Rightarrow x = 4$	Set 3rd factor equal to 0.

You can check these solutions by substituting in the original equation, as follows.

Check

$3(0)^4 \stackrel{?}{=} 48(0)^2$	Substitute 0 for x.
$0 = 0$	0 checks. ✓
$3(-4)^4 \stackrel{?}{=} 48(-4)^2$	Substitute -4 for x.
$768 = 768$	-4 checks. ✓
$3(4)^4 \stackrel{?}{=} 48(4)^2$	Substitute 4 for x.
$768 = 768$	4 checks. ✓

So, you can conclude that the solutions are $x = 0$, $x = -4$, and $x = 4$.

✓ **Checkpoint** *Audio-video solution in English & Spanish at LarsonPrecalculus.com.*

Solve $9x^4 - 12x^2 = 0$.

A common mistake that is made in solving an equation such as that in Example 1 is to divide each side of the equation by the variable factor x^2. This loses the solution $x = 0$. When solving an equation, always write the equation in general form, then factor the equation and set each factor equal to zero. Do not divide each side of an equation by a variable factor in an attempt to simplify the equation.

EXAMPLE 2 Solving a Polynomial Equation by Factoring

Solve $2x^3 - 6x^2 + 6x - 18 = 0$.

Solution

This equation has a common factor of 2. You can simplify the equation by first dividing each side of the equation by 2.

$2x^3 - 6x^2 + 6x - 18 = 0$		Write original equation.
$x^3 - 3x^2 + 3x - 9 = 0$		Divide each side by 2.
$x^2(x - 3) + 3(x - 3) = 0$		Factor by grouping.
$(x - 3)(x^2 + 3) = 0$		Distributive Property
$x - 3 = 0$	$\Rightarrow x = 3$	Set 1st factor equal to 0.
$x^2 + 3 = 0$	$\Rightarrow x = \pm\sqrt{3}i$	Set 2nd factor equal to 0.

The solutions are $x = 3$, $x = \sqrt{3}i$, and $x = -\sqrt{3}i$. Check these solutions in the original equation. Figure 2.22 verifies the real solution $x = 3$ graphically.

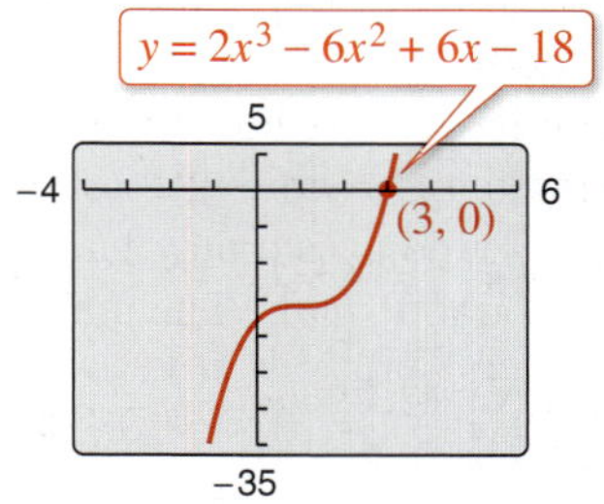

Figure 2.22

✓ Checkpoint *Audio-video solution in English & Spanish at LarsonPrecalculus.com.*

Solve each equation.

a. $x^3 - 5x^2 - 2x + 10 = 0$ **b.** $6x^3 - 27x^2 - 54x = 0$

Occasionally, mathematical models involve equations that are of **quadratic type.** In general, an equation is of quadratic type when it can be written in the form $au^2 + bu + c = 0$, where $a \neq 0$ and u is an algebraic expression.

EXAMPLE 3 Solving an Equation of Quadratic Type

Solve $x^4 - 3x^2 + 2 = 0$.

Solution

This equation is of quadratic type with $u = x^2$.

$x^4 - 3x^2 + 2 = 0$		Write original equation.
$(x^2)^2 - 3(x^2) + 2 = 0$		Write in quadratic form.
$u^2 - 3u + 2 = 0$		$u = x^2$
$(u - 1)(u - 2) = 0$		Factor.
$u - 1 = 0$	$\Rightarrow u = 1$	Set 1st factor equal to 0.
$u - 2 = 0$	$\Rightarrow u = 2$	Set 2nd factor equal to 0.

Next, replace u with x^2 and solve for x.

$u = 1 \Rightarrow x^2 = 1 \Rightarrow x = \pm 1$

$u = 2 \Rightarrow x^2 = 2 \Rightarrow x = \pm\sqrt{2}$

The solutions are $x = -1$, $x = 1$, $x = \sqrt{2}$, and $x = -\sqrt{2}$. Check these solutions in the original equation. Figure 2.23 verifies the solutions graphically.

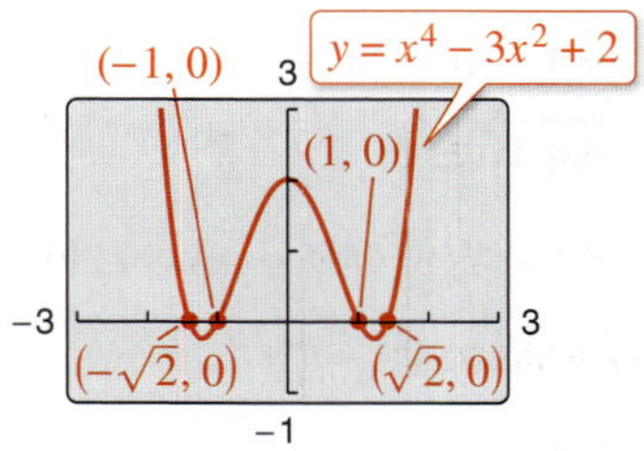

Figure 2.23

✓ Checkpoint *Audio-video solution in English & Spanish at LarsonPrecalculus.com.*

Solve each equation.

a. $x^4 - 7x^2 + 12 = 0$ **b.** $9x^4 - 37x^2 + 4 = 0$

Equations Involving Radicals

An equation involving a radical expression can often be cleared of radicals by raising each side of the equation to an appropriate power. When using this procedure, remember to check for extraneous solutions.

EXAMPLE 4 Solving Equations Involving Radicals

a. $\sqrt{2x+7} - x = 2$ — Original equation

$\sqrt{2x+7} = x + 2$ — Isolate radical.

$2x + 7 = x^2 + 4x + 4$ — Square each side.

$x^2 + 2x - 3 = 0$ — Write in general form.

$(x+3)(x-1) = 0$ — Factor.

$x + 3 = 0 \Rightarrow x = -3$ — Set 1st factor equal to 0.

$x - 1 = 0 \Rightarrow x = 1$ — Set 2nd factor equal to 0.

By checking these values, you can determine that the only solution is $x = 1$. Figure 2.24 verifies the solution graphically.

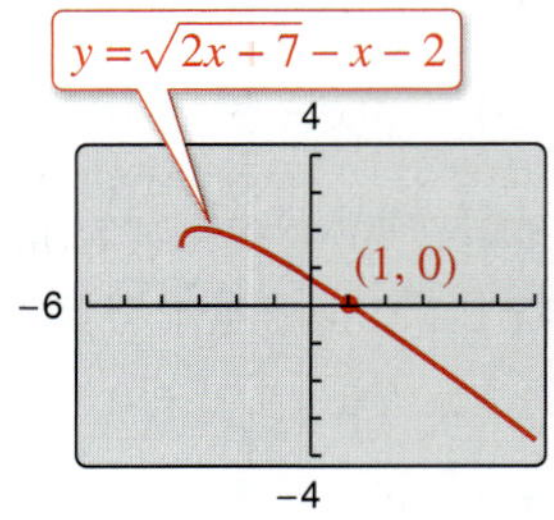

Figure 2.24

b. $\sqrt{2x+6} - \sqrt{x+4} = 1$ — Original equation

$\sqrt{2x+6} = 1 + \sqrt{x+4}$ — Isolate $\sqrt{2x+6}$.

$2x + 6 = 1 + 2\sqrt{x+4} + (x+4)$ — Square each side.

$x + 1 = 2\sqrt{x+4}$ — Isolate $2\sqrt{x+4}$.

$x^2 + 2x + 1 = 4(x+4)$ — Square each side.

$x^2 - 2x - 15 = 0$ — Write in general form.

$(x-5)(x+3) = 0$ — Factor.

$x - 5 = 0 \Rightarrow x = 5$ — Set 1st factor equal to 0.

$x + 3 = 0 \Rightarrow x = -3$ — Set 2nd factor equal to 0.

By checking these values, you can determine that the only solution is $x = 5$. Figure 2.25 verifies the solution graphically.

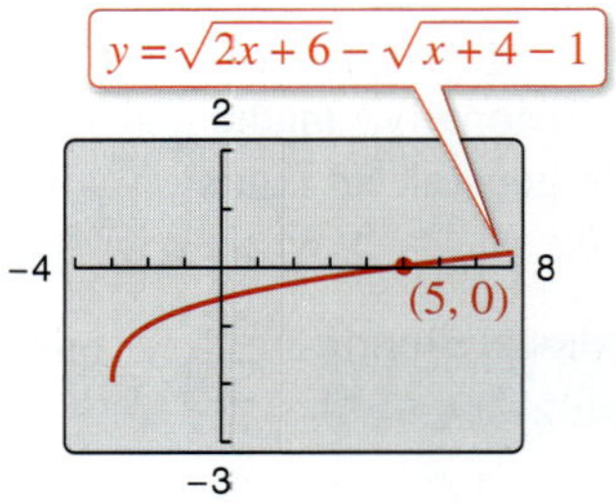

Figure 2.25

✓ **Checkpoint** *Audio-video solution in English & Spanish at LarsonPrecalculus.com.*

Solve $-\sqrt{40 - 9x} + 2 = x$.

EXAMPLE 5 Solving an Equation Involving a Rational Exponent

Solve $(x+1)^{2/3} = 4$.

Solution

$(x+1)^{2/3} = 4$ — Write original equation.

$\sqrt[3]{(x+1)^2} = 4$ — Rewrite in radical form.

$(x+1)^2 = 64$ — Cube each side.

$x + 1 = \pm 8$ — Extract square roots.

$x = 7, x = -9$ — Subtract 1 from each side.

The solutions are $x = 7$ and $x = -9$. Check these in the original equation. Figure 2.26 verifies the solutions graphically.

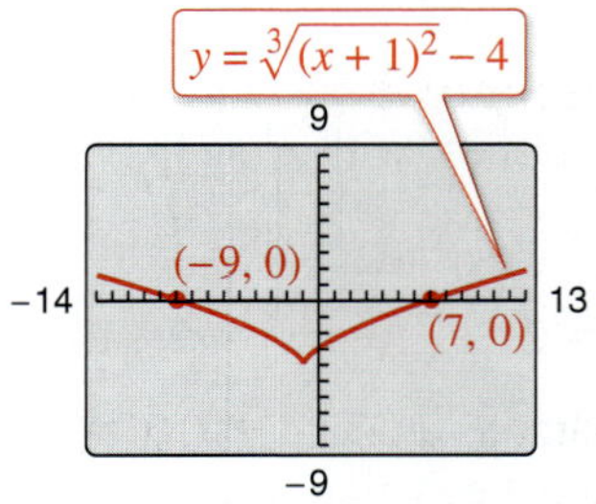

Figure 2.26

✓ **Checkpoint** *Audio-video solution in English & Spanish at LarsonPrecalculus.com.*

Solve $(x-5)^{2/3} = 16$.

Equations Involving Fractions or Absolute Values

As demonstrated in Section 2.1, you can solve an equation involving fractions algebraically by multiplying each side of the equation by the least common denominator of all terms in the equation to clear the equation of fractions.

EXAMPLE 6 Solving an Equation Involving Fractions

Solve $\frac{2}{x} = \frac{3}{x-2} - 1$.

Solution

For this equation, the least common denominator of the three terms is $x(x-2)$, so you can begin by multiplying each term of the equation by this expression.

$$\frac{2}{x} = \frac{3}{x-2} - 1 \quad \text{Write original equation.}$$

$$x(x-2)\frac{2}{x} = x(x-2)\frac{3}{x-2} - x(x-2)(1) \quad \text{Multiply each term by the LCD.}$$

$$2(x-2) = 3x - x(x-2), \quad x \neq 0, 2 \quad \text{Simplify.}$$

$$2x - 4 = -x^2 + 5x \quad \text{Simplify.}$$

$$x^2 - 3x - 4 = 0 \quad \text{Write in general form.}$$

$$(x-4)(x+1) = 0 \quad \text{Factor.}$$

$$x - 4 = 0 \implies x = 4 \quad \text{Set 1st factor equal to 0.}$$

$$x + 1 = 0 \implies x = -1 \quad \text{Set 2nd factor equal to 0.}$$

Check these solutions in the original equation as follows.

Check $x = 4$

$$\frac{2}{x} = \frac{3}{x-2} - 1$$

$$\frac{2}{4} \stackrel{?}{=} \frac{3}{4-2} - 1$$

$$\frac{1}{2} \stackrel{?}{=} \frac{3}{2} - 1$$

$$\frac{1}{2} = \frac{1}{2} \checkmark$$

Check $x = -1$

$$\frac{2}{x} = \frac{3}{x-2} - 1$$

$$\frac{2}{-1} \stackrel{?}{=} \frac{3}{-1-2} - 1$$

$$-2 \stackrel{?}{=} -1 - 1$$

$$-2 = -2 \checkmark$$

So, the solutions are $x = 4$ and $x = -1$.

✓ Checkpoint *Audio-video solution in English & Spanish at LarsonPrecalculus.com.*

Solve $\frac{4}{x} + \frac{2}{x+3} = -3$.

Explore the Concept

Using *dot* mode, graph the equations

$$y_1 = \frac{2}{x} \quad \text{and} \quad y_2 = \frac{3}{x-2} - 1$$

in the same viewing window. How many times do the graphs of the equations intersect? What does this tell you about the solution to Example 6?

EXAMPLE 7 Solving an Equation Involving an Absolute Value

See LarsonPrecalculus.com for an interactive version of this type of example.

Solve $|x^2 - 3x| = -4x + 6$.

Solution

To solve an equation involving an absolute value algebraically, you must consider the fact that the expression inside the absolute value symbols can be positive or negative. This results in *two* separate equations, each of which must be solved.

First Equation

$x^2 - 3x = -4x + 6$ Use positive expression.

$x^2 + x - 6 = 0$ Write in general form.

$(x + 3)(x - 2) = 0$ Factor.

$x + 3 = 0 \Rightarrow x = -3$ Set 1st factor equal to 0.

$x - 2 = 0 \Rightarrow x = 2$ Set 2nd factor equal to 0.

Second Equation

$-(x^2 - 3x) = -4x + 6$ Use negative expression.

$x^2 - 7x + 6 = 0$ Write in general form.

$(x - 1)(x - 6) = 0$ Factor.

$x - 1 = 0 \Rightarrow x = 1$ Set 1st factor equal to 0.

$x - 6 = 0 \Rightarrow x = 6$ Set 2nd factor equal to 0.

Check

$|(-3)^2 - 3(-3)| \stackrel{?}{=} -4(-3) + 6$ Substitute -3 for x.

$18 = 18$ -3 checks. ✓

$|2^2 - 3(2)| \stackrel{?}{=} -4(2) + 6$ Substitute 2 for x.

$2 \neq -2$ 2 does not check.

$|1^2 - 3(1)| \stackrel{?}{=} -4(1) + 6$ Substitute 1 for x.

$2 = 2$ 1 checks. ✓

$|6^2 - 3(6)| \stackrel{?}{=} -4(6) + 6$ Substitute 6 for x.

$18 \neq -18$ 6 does not check.

The solutions are $x = -3$ and $x = 1$. Figure 2.27 shows the solutions graphically.

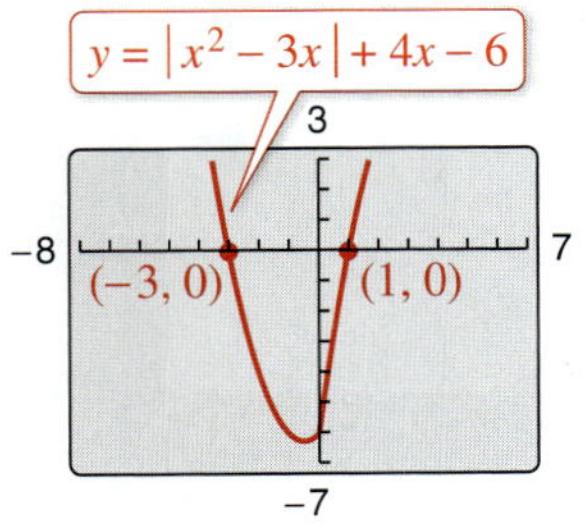

Figure 2.27

Explore the Concept

In Figure 2.27, the graph of

$$y = |x^2 - 3x| + 4x - 6$$

appears to be a straight line to the right of the y-axis. Is it? Explain your reasoning.

✓ ***Checkpoint*** *Audio-video solution in English & Spanish at LarsonPrecalculus.com.*

Solve $|x^2 + 4x| = 5x + 12$.

Applications

It would be impossible to categorize the many different types of applications that involve nonlinear and nonquadratic models. However, from the few examples and exercises that are given, you will gain some appreciation for the variety of applications that can occur.

EXAMPLE 8 Reduced Rates

The athletic booster club of a university chartered a bus for an away football game at a cost of \$960. To lower the bus fare per person, the club invited nonmembers to go along. After five nonmembers joined the trip, the fare per person decreased by \$9.60. How many club members are going on the trip?

Solution

Verbal Model: Cost per person · Number of people = Cost of trip

Labels:
Cost of trip $= 960$ (dollars)
Number of athletic booster club members $= x$ (people)
Number of people $= x + 5$ (people)
Original cost per member $= \dfrac{960}{x}$ (dollars per person)
Cost per person $= \dfrac{960}{x} - 9.60$ (dollars per person)

Equation:

$$\left(\frac{960}{x} - 9.60\right)(x + 5) = 960$$

$$\left(\frac{960 - 9.6x}{x}\right)(x + 5) = 960 \qquad \text{Write } \left(\frac{960}{x} - 9.60\right) \text{ as a fraction.}$$

$$(960 - 9.6x)(x + 5) = 960x, \quad x \neq 0 \qquad \text{Multiply each side by } x.$$

$$960x + 4800 - 9.6x^2 - 48x = 960x \qquad \text{Multiply.}$$

$$-9.6x^2 - 48x + 4800 = 0 \qquad \text{Subtract } 960x \text{ from each side.}$$

$$x^2 + 5x - 500 = 0 \qquad \text{Divide each side by } -9.6.$$

$$(x + 25)(x - 20) = 0 \qquad \text{Factor.}$$

$$x + 25 = 0 \implies x = -25$$

$$x - 20 = 0 \implies x = 20$$

Choosing the positive value of x, you can conclude that 20 club members are going on the trip. Check this in the original statement of the problem, as follows.

$$\left(\frac{960}{20} - 9.60\right)(20 + 5) \stackrel{?}{=} 960 \qquad \text{Substitute 20 for } x.$$

$$(48 - 9.60)25 \stackrel{?}{=} 960 \qquad \text{Simplify.}$$

$$960 = 960 \qquad \text{20 checks ✓}$$

✓ Checkpoint *Audio-video solution in English & Spanish at LarsonPrecalculus.com.*

A high school charters a bus for \$560 to take a group of students to an observatory. When 8 more students join the trip, the cost per student decreases by \$3.50. How many students were in the original group?

Interest in a savings account is calculated by one of three basic methods: simple interest, interest compounded n times per year, and interest compounded continuously. The next example uses the formula for interest that is compounded n times per year.

$$A = P\left(1 + \frac{r}{n}\right)^{nt}$$

In this formula, A is the balance in the account, P is the principal (or original deposit), r is the annual interest rate (in decimal form), n is the number of compoundings per year, and t is the time in years.

Remark

In Chapter 4, you will study a derivation of this formula for interest compounded continuously.

EXAMPLE 9 Compound Interest

When you were born, your grandparents deposited \$5000 in a long-term investment in which the interest was compounded quarterly. On your 25th birthday, the value of the investment is \$25,062.59. What is the annual interest rate for this investment?

Solution

Formula: $A = P\left(1 + \dfrac{r}{n}\right)^{nt}$

Labels:	Balance $= A = 25{,}062.59$	(dollars)
	Principal $= P = 5000$	(dollars)
	Time $= t = 25$	(years)
	Compoundings per year $= n = 4$	(compoundings per year)
	Annual interest rate $= r$	(percent in decimal form)

Equation: $25{,}062.59 = 5000\left(1 + \dfrac{r}{4}\right)^{4(25)}$

$\dfrac{25{,}062.59}{5000} = \left(1 + \dfrac{r}{4}\right)^{100}$ Divide each side by 5000.

$5.0125 \approx \left(1 + \dfrac{r}{4}\right)^{100}$ Use a calculator.

$(5.0125)^{1/100} \approx 1 + \dfrac{r}{4}$ Raise each side to reciprocal power.

$1.01625 \approx 1 + \dfrac{r}{4}$ Use a calculator.

$0.01625 \approx \dfrac{r}{4}$ Subtract 1 from each side.

$0.065 \approx r$ Multiply each side by 4.

The annual interest rate is about 0.065, or 6.5%. Check this in the original statement of the problem. Note that you can use a graphing utility to check your answer, as shown below.

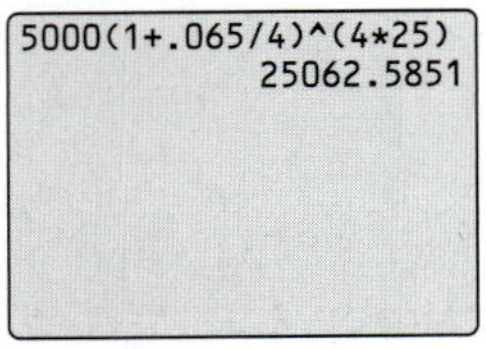

Checkpoint *Audio-video solution in English & Spanish at LarsonPrecalculus.com.*

A deposit of \$2500 in a mutual fund reaches a balance of \$3544.06 after 5 years. What annual interest rate on a certificate of deposit compounded monthly would yield an equivalent return?

2.5 Exercises

See CalcChat.com for tutorial help and worked-out solutions to odd-numbered exercises.
For instructions on how to use a graphing utility, see Appendix A.

Vocabulary and Concept Check

In Exercises 1 and 2, fill in the blank.

1. The general form of a _______ equation is $a_nx^n + a_{n-1}x^{n-1} + \cdots + a_2x^2 + a_1x + a_0 = 0$.
2. To clear the equation $\frac{4}{x} + 5 = \frac{6}{x-3}$ of fractions, multiply each side of the equation by the least common denominator _______ .
3. Describe the step needed to remove the radical from the equation $\sqrt{x+2} = x$.
4. Is the equation $x^4 - 2x + 4 = 0$ of quadratic type?

Procedures and Problem Solving

Solving a Polynomial Equation by Factoring **In Exercises 5–12, find all solutions of the equation algebraically. Use a graphing utility to verify the solutions graphically.**

5. $4x^4 - 16x^2 = 0$
6. $8x^4 - 18x^2 = 0$
7. $7x^3 + 63x = 0$
8. $x^3 + 512 = 0$
9. $5x^3 + 30x^2 + 45x = 0$
10. $9x^4 - 24x^3 + 16x^2 = 0$
11. $x^3 + 5 = 5x^2 + x$
12. $x^4 - 2x^3 = 16 + 8x - 4x^3$

Solving an Equation of Quadratic Type **In Exercises 13–16, find all solutions of the equation algebraically. Check your solutions.**

13. $x^4 - 4x^2 + 3 = 0$
14. $x^4 - 5x^2 - 36 = 0$
15. $36t^4 + 29t^2 - 7 = 0$
16. $4x^4 - 65x^2 + 16 = 0$

Solving an Equation Involving Radicals **In Exercises 17–36, find all solutions of the equation algebraically. Check your solutions.**

17. $3\sqrt{x} - 10 = 0$
18. $3\sqrt{x} - 6 = 0$
19. $\sqrt{x-10} - 4 = 0$
20. $\sqrt{2x+5} + 3 = 0$
21. $\sqrt[3]{6x} + 9 = 0$
22. $2\sqrt[3]{x} - 10 = 0$
23. $\sqrt[3]{2x+1} + 8 = 0$
24. $\sqrt[3]{4x-3} + 2 = 0$
25. $\sqrt{5x-26} + 4 = x$
26. $x - \sqrt{8x-31} = 5$
27. $\sqrt{x+1} - 3x = 1$
28. $\sqrt{x+5} - 2x = 3$
29. $\sqrt{x+1} = \sqrt{3x+1}$
30. $\sqrt{x+5} = \sqrt{2x-5}$
31. $2x + 9\sqrt{x} - 5 = 0$
32. $6x - 7\sqrt{x} - 3 = 0$
33. $\sqrt{x} - \sqrt{x-5} = 1$
34. $\sqrt{x} + \sqrt{x-20} = 10$
35. $3\sqrt{x-5} + \sqrt{x-1} = 0$
36. $4\sqrt{x-3} - \sqrt{6x-17} = 3$

Solving an Equation Involving Rational Exponents **In Exercises 37–46, find all solutions of the equation algebraically. Check your solutions.**

37. $3x^{1/3} + 2x^{2/3} = 5$
38. $9t^{2/3} + 24t^{1/3} = -16$
39. $(x+6)^{3/2} = 1$
40. $(x-1)^{3/2} = 8$
41. $(x-9)^{2/3} = 25$
42. $(x-7)^{2/3} = 9$
43. $(x^2 - 5x - 2)^{1/3} = -2$
44. $(x^2 - x - 22)^{4/3} = 16$
45. $3x(x-1)^{1/2} + 2(x-1)^{3/2} = 0$
46. $4x^2(x-1)^{1/3} + 6x(x-1)^{4/3} = 0$

Solving an Equation Involving Fractions **In Exercises 47–58, find all solutions of the equation. Check your solutions.**

47. $x = \frac{3}{x} + \frac{1}{2}$
48. $\frac{4}{x} - \frac{5}{3} = \frac{x}{6}$
49. $\frac{20-x}{x} = x$
50. $4x + 1 = \frac{3}{x}$
51. $\frac{1}{x} - \frac{1}{x+1} = 3$
52. $\frac{4}{x+1} - \frac{3}{x+2} = 1$
53. $\frac{1}{t^2} + \frac{8}{t} + 15 = 0$
54. $6 - \frac{1}{x} - \frac{1}{x^2} = 0$
55. $\frac{x-2}{x} - \frac{1}{x+2} = 0$
56. $\frac{x}{x^2-4} + \frac{1}{x+2} = 3$
57. $6\left(\frac{s}{s+1}\right)^2 + 5\left(\frac{s}{s+1}\right) - 6 = 0$
58. $8\left(\frac{t}{t-1}\right)^2 - 2\left(\frac{t}{t-1}\right) - 3 = 0$

Solving an Equation Involving an Absolute Value **In Exercises 59–64, find all solutions of the equation algebraically. Check your solutions.**

59. $|2x - 5| = 11$
60. $|3x + 2| = 7$
61. $|x| = x^2 + x - 24$
62. $|x^2 - 6x| = 3x + 18$
63. $|x + 1| = x^2 - 5$
64. $|x - 15| = x^2 - 15x$

Graphical Analysis **In Exercises 65–76, (a) use a graphing utility to graph the equation, (b) use the graph to approximate any x-intercepts of the graph, (c) set $y = 0$ and solve the resulting equation, and (d) compare the result of part (c) with the x-intercepts of the graph.**

65. $y = x^3 - 2x^2 - 3x$

66. $y = 2x^4 - 15x^3 + 18x^2$

67. $y = x^4 - 10x^2 + 9$

68. $y = x^4 - 29x^2 + 100$

69. $y = \sqrt{11x - 30} - x$

70. $y = 2x - \sqrt{15 - 4x}$

71. $y = 3x - 3\sqrt{x} - 4$

72. $y = \sqrt{7x + 36} - \sqrt{5x + 16} - 2$

73. $y = \dfrac{1}{x} - \dfrac{4}{x - 1} - 1$

74. $y = x - 5 + \dfrac{7}{x + 3}$

75. $y = x + \dfrac{9}{x + 1} - 5$

76. $y = 2x + \dfrac{8}{x - 5} - 2$

77. $y = |x + 1| - 2$

78. $y = |x - 2| - 3$

79. Reduced Rates A college charters a bus for \$1700 to take a group of students to see a Broadway production. When 6 more students join the trip, the cost per student decreases by \$7.50. How many students were in the original group?

80. Reduced Rates Three students are planning to rent an apartment for a year and share equally in the cost. By adding a fourth person, each person saves \$75 a month. How much is the monthly rent?

81. Finance A deposit of \$7500 in a mutual fund reaches a balance of \$11,752.45 after 10 years. What annual interest rate on a certificate of deposit compounded monthly would yield an equivalent return?

82. Finance Twenty years ago, your parents deposited \$3000 in a long-term investment in which interest was compounded biannually. Today, the value of the investment is \$8055.19. What is the annual interest rate for this investment?

83. Physics The temperature T (in degrees Fahrenheit) of saturated steam increases as pressure increases. This relationship is approximated by the model

$$T = 75.82 - 2.11x + 43.51\sqrt{x}, \quad 5 \le x \le 40$$

where x is the absolute pressure (in pounds per square inch). Approximate the pressure for saturated steam at a temperature of 212°F.

84. ***Why you should learn it*** *(p. 200)* An airline offers daily flights between Chicago and Denver. The total monthly cost C (in millions of dollars) of these flights is modeled by

$$C = \sqrt{0.2x + 1}$$

where x is the number of passengers (in thousands). The total cost of the flights for June is 2.5 million dollars. How many passengers flew in June?

85. Electrical Energy A power station is on one side of a river that is $\frac{3}{4}$ mile wide, and a factory is 8 miles downstream on the other side of the river, as shown in the figure. It costs \$24 per foot to run power lines over land and \$30 per foot to run them under water.

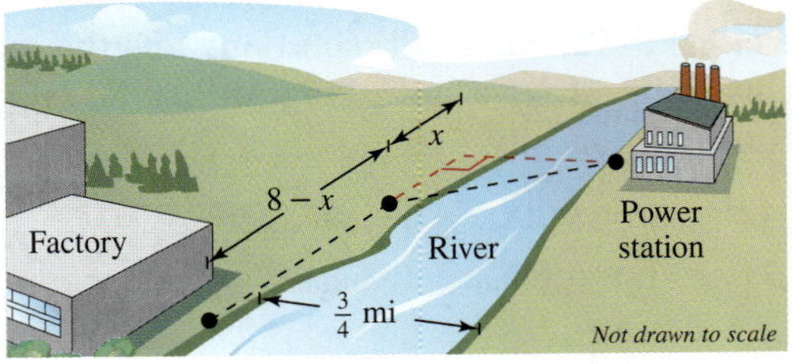

(a) Write the total cost C of running power lines in terms of x.

(b) Find the total cost when $x = 3$.

(c) Find the length x when $C = \$1,098,662.40$.

(d) Use a graphing utility to graph the equation from part (a).

(e) Use your graph from part (d) to find the value of x that minimizes the cost.

86. Meteorology A meteorologist is positioned 100 feet from the point at which a weather balloon is launched. When the balloon is at height h, the distance d (in feet) between the meteorologist and the balloon is given by

$$d = \sqrt{100^2 + h^2}.$$

(a) Use a graphing utility to graph the equation. Use the *trace* feature to approximate the value of h when $d = 200$.

(b) Complete the table. Use the table to approximate the value of h when $d = 200$.

h	160	165	170	175	180	185
d						

(c) Find h algebraically when $d = 200$.

(d) Compare the results of each method. In each case, what information did you gain that wasn't revealed by another solution method?

87. **Algebraic-Graphical-Numerical** The numbers of crimes (in millions) committed in the United States from 2008 through 2012 can be approximated by the model

$$C = \sqrt{1.49145t^2 - 35.034t + 309.6},\quad 8 \le t \le 12$$

where t is the year, with $t = 8$ corresponding to 2008. (Source: Federal Bureau of Investigation)

(a) Use the *table* feature of a graphing utility to estimate the number of crimes committed in the U.S. each year from 2008 through 2012.

(b) According to the table, when was the first year that the number of crimes committed fell below 11 million?

(c) Find the answer to part (b) algebraically.

(d) Use the graphing utility to graph the model and find the answer to part (b).

88. **MODELING DATA**

The table shows the average salaries S (in thousands of dollars) for baseball players in Major League Baseball from 2001 through 2012. (Source: CBS Broadcasting, Inc.)

Year	Salary, S (in thousands of dollars)
2001	2264.4
2002	2383.2
2003	2555.5
2004	2486.6
2005	2632.7
2006	2866.5
2007	2944.6
2008	3154.8
2009	3240.2
2010	3297.8
2011	3305.4
2012	3440.0

Spreadsheet at LarsonPrecalculus.com

(a) Use a graphing utility to create a scatter plot of the data, where t represents the year, with $t = 1$ corresponding to 2001.

(b) A model that approximates the data is

$$S = 110.386t - \frac{16.4988}{t} + 2167.7,\quad 1 \le t \le 12$$

Graph the model and the data in the same viewing window. Is the model a good fit for the data?

(c) Use the model to predict the first year in which the average MLB player salary exceeds $4 million.

Conclusions

True or False? **In Exercises 89 and 90, determine whether the statement is true or false. Justify your answer.**

89. An equation can never have more than one extraneous solution.

90. When solving an absolute value equation, you always have to check more than one solution.

91. **Think About It** Find x such that the distance between each set of points is 13.

(a) $(1, 2), (x, -10)$ (b) $(-8, 0), (x, 5)$

92. **Think About It** Find a and b in the equation $x + \sqrt{x - a} = b$ when the solution is $x = 20$. (There are many correct answers.)

93. **Error Analysis** Describe the error.

$$x^3 - 25x = 0$$
$$x^3 = 25x$$
$$x^2 = 25$$
$$x = \pm 5$$

94. **HOW DO YOU SEE IT?** The figure shows a glass cube partially filled with water.

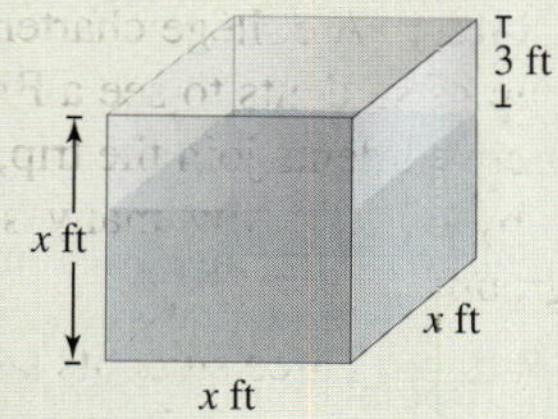

(a) What does the expression $x^2(x - 3)$ represent?

(b) Given $x^2(x - 3) = 320$, explain how you can find the capacity of the cube.

Cumulative Mixed Review

Operations with Rational Expressions **In Exercises 95–98, simplify the expression.**

95. $\dfrac{8}{3x} + \dfrac{3}{2x}$

96. $\dfrac{2}{x^2 - 4} - \dfrac{1}{x^2 - 3x + 2}$

97. $\dfrac{2}{z + 2} - \left(3 - \dfrac{2}{z}\right)$

98. $25y^2 \div \dfrac{xy}{5}$

Solving a Quadratic Equation **In Exercises 99 and 100, find all real solutions of the equation.**

99. $x^2 - 22x + 121 = 0$

100. $x(x - 20) + 3(x - 20) = 0$

2.6 Solving Inequalities Algebraically and Graphically

Properties of Inequalities

Simple inequalities were reviewed in Section P.1. There, the inequality symbols

$<, \quad \leq, \quad >, \quad \text{and} \quad \geq$ Inequality symbols

were used to compare two numbers and to denote subsets of real numbers. For instance, the simple inequality $x \geq 3$ denotes all real numbers x that are greater than or equal to 3.

In this section, you will study inequalities that contain more involved statements such as

$$5x - 7 > 3x + 9$$

and

$$-3 \leq 6x - 1 < 3.$$

As with an equation, you **solve an inequality** in the variable x by finding all values of x for which the inequality is true. These values are **solutions** of the inequality and are said to **satisfy** the inequality. For instance, the number 9 is a solution of the first inequality listed above because

$$5(9) - 7 > 3(9) + 9$$
$$45 - 7 > 27 + 9$$
$$38 > 36.$$

On the other hand, the number 7 is not a solution because

$$5(7) - 7 \ngtr 3(7) + 9$$
$$35 - 7 \ngtr 21 + 9$$
$$28 \ngtr 30.$$

The set of all real numbers that are solutions of an inequality is the **solution set** of the inequality.

The set of all points on the real number line that represent the solution set is the **graph of the inequality.** Graphs of many types of inequalities consist of intervals on the real number line.

The procedures for solving linear inequalities in one variable are much like those for solving linear equations. To isolate the variable, you can make use of the **properties of inequalities.** These properties are similar to the properties of equality, but there are two important exceptions. When each side of an inequality is multiplied or divided by a negative number, *the direction of the inequality symbol must be reversed* in order to maintain a true statement. Here is an example.

$-2 < 5$ Original inequality

$(-3)(-2) > (-3)(5)$ Multiply each side by -3 and reverse the inequality symbol.

$6 > -15$ Simplify.

Two inequalities that have the same solution set are **equivalent inequalities.** instance, the inequalities

$$x + 2 < 5 \quad \text{and} \quad x < 3$$

are equivalent. To obtain the second inequality from the first, you can subtract 2 from each side of the inequality. The properties listed at the top of the next page describe operations that can be used to create equivalent inequalities.

What you should learn

- Use properties of inequalities to solve linear inequalities.
- Solve inequalities involving absolute values.
- Solve polynomial inequalities.
- Solve rational inequalities.
- Use inequalities to model and solve real-life problems.

Why you should learn it

An inequality can be used to determine when a real-life quantity exceeds a given level. For instance, Exercises 99–102 on page 222 show how to use linear inequalities to determine when the number of Bed Bath & Beyond stores exceeds the number of Williams-Sonoma stores.

Properties of Inequalities

Let a, b, c, and d be real numbers.

1. *Transitive Property*

   $a < b$ and $b < c \Rightarrow a < c$

2. *Addition of Inequalities*

   $a < b$ and $c < d \Rightarrow a + c < b + d$

3. *Addition of a Constant*

   $a < b \Rightarrow a + c < b + c$

4. *Multiplying by a Constant*

   For $c > 0$, $a < b \Rightarrow ac < bc$

   For $c < 0$, $a < b \Rightarrow ac > bc$ Reverse the inequality symbol.

Each of the properties above is true when the symbol $<$ is replaced by $\leq$ and $>$ is replaced by $\geq$. For instance, another form of Property 3 is as follows.

$$a \leq b \Rightarrow a + c \leq b + c$$

The simplest type of inequality to solve is a **linear inequality** in one variable, such as $x + 3 > 4$. (For help with solving one-step linear inequalities, see Appendix D at this textbook's *Companion Website.*)

EXAMPLE 1 Solving a Linear Inequality

Solve $5x - 7 > 3x + 9$.

Solution

$5x - 7 > 3x + 9$	Write original inequality.
$2x - 7 > 9$	Subtract $3x$ from each side.
$2x > 16$	Add 7 to each side.
$x > 8$	Divide each side by 2.

The solution set is all real numbers that are greater than 8, which is denoted by $(8, \infty)$. The graph of this solution set is shown below. Note that a parenthesis at 8 on the real number line indicates that 8 is *not* part of the solution set.

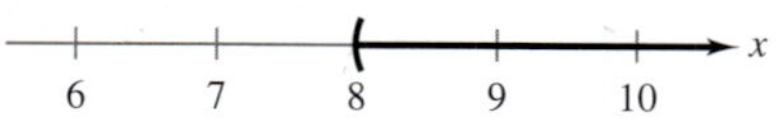

Solution Interval: **$(8, \infty)$**

Checkpoint *Audio-video solution in English & Spanish at LarsonPrecalculus.com.*

Solve $7x - 3 \leq 2x + 7$. Then graph the solution set.

Note that the four inequalities forming the solution steps of Example 1 are all *equivalent* in the sense that each has the same solution set.

Checking the solution set of an inequality is not as simple as checking the solution(s) of an equation because there are simply too many x-values to substitute into the original inequality. However, you can get an indication of the validity of the solution set by substituting a few convenient values of x. For instance, in Example 1, try substituting $x = 6$ and $x = 10$ into the original inequality.

Explore the Concept

Use a graphing utility to graph $f(x) = 5x - 7$ and $g(x) = 3x + 9$ in the same viewing window. (Use $-1 \leq x \leq 15$ and $-5 \leq y \leq 50$.) For which values of x does the graph of f lie above the graph of g? Explain how the answer to this question can be used to solve the inequality in Example 1.

EXAMPLE 2 Solving an Inequality

Solve $1 - \frac{3}{2}x \geq x - 4$.

Algebraic Solution

$1 - \frac{3}{2}x \geq x - 4$ Write original inequality.

$2 - 3x \geq 2x - 8$ Multiply each side by the LCD.

$2 - 5x \geq -8$ Subtract $2x$ from each side.

$-5x \geq -10$ Subtract 2 from each side.

$x \leq 2$ Divide each side by -5 and reverse the inequality symbol.

The solution set is all real numbers that are less than or equal to 2, which is denoted by $(-\infty, 2]$. The graph of this solution set is shown below. Note that a bracket at 2 on the number line indicates that 2 *is* part of the solution set.

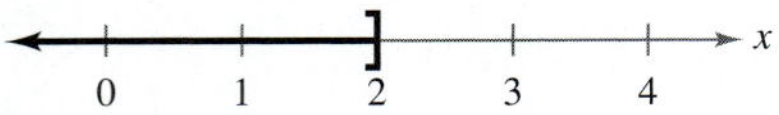

Solution Interval: $(-\infty, 2]$

Graphical Solution

Use a graphing utility to graph $y_1 = 1 - \frac{3}{2}x$ and $y_2 = x - 4$ in the same viewing window, as shown below.

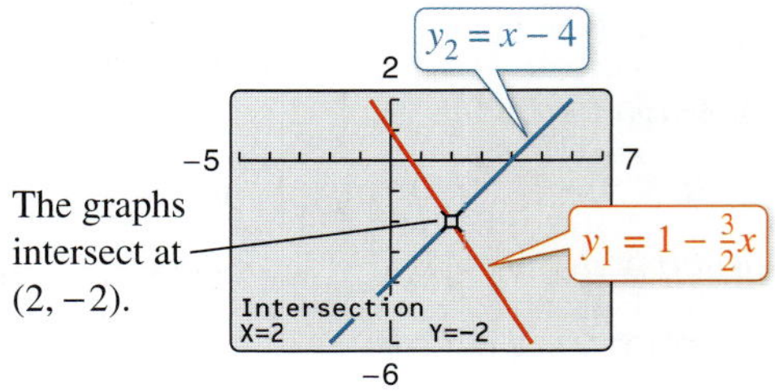

The graph of y_1 lies above the graph of y_2 to the left of their point of intersection, $(2, -2)$, which implies that $y_1 \geq y_2$ for all $x \leq 2$.

✓ Checkpoint *Audio-video solution in English & Spanish at LarsonPrecalculus.com.*

Solve $2 - \frac{5}{3}x > x - 6$ (a) algebraically and (b) graphically.

Sometimes it is possible to write two inequalities as a **double inequality,** as demonstrated in Example 3.

EXAMPLE 3 Solving a Double Inequality

Solve $-3 \leq 6x - 1$ and $6x - 1 < 3$.

Algebraic Solution

$-3 \leq 6x - 1 < 3$ Write as a double inequality.

$-3 + 1 \leq 6x - 1 + 1 < 3 + 1$ Add 1 to each part.

$-2 \leq 6x < 4$ Simplify.

$\frac{-2}{6} \leq \frac{6x}{6} < \frac{4}{6}$ Divide each part by 6.

$-\frac{1}{3} \leq x < \frac{2}{3}$ Simplify.

The solution set is all real numbers that are greater than or equal to $-\frac{1}{3}$ *and* less than $\frac{2}{3}$. The interval notation for this solution set is $\left[-\frac{1}{3}, \frac{2}{3}\right)$. The graph of this solution set is shown below.

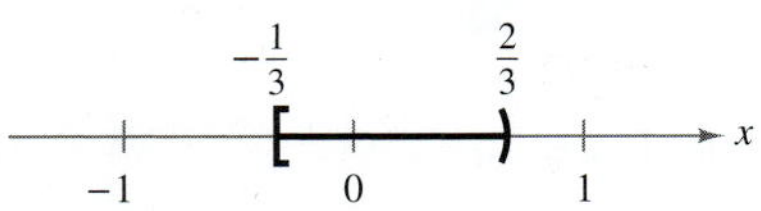

Solution Interval: $\left[-\frac{1}{3}, \frac{2}{3}\right)$

Graphical Solution

Use a graphing utility to graph $y_1 = 6x - 1$, $y_2 = -3$, and $y_3 = 3$ in the same viewing window, as shown below.

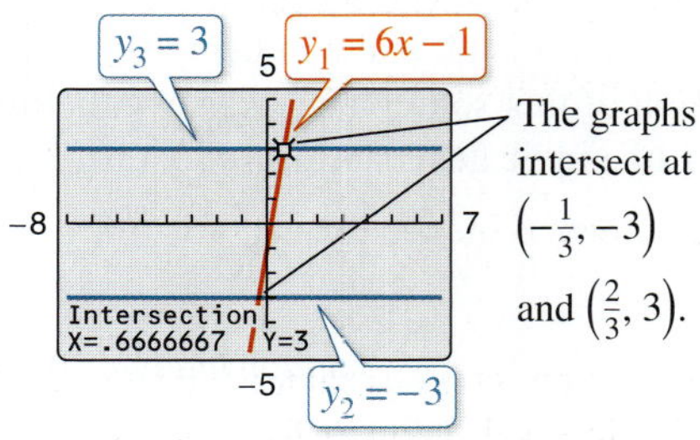

The graph of y_1 lies above the graph of y_2 to the right of $\left(-\frac{1}{3}, -3\right)$ *and* the graph of y_1 lies below the graph of y_3 to the left of $\left(\frac{2}{3}, 3\right)$. This implies that $y_2 \leq y_1 < y_3$ when $-\frac{1}{3} \leq x < \frac{2}{3}$.

✓ Checkpoint *Audio-video solution in English & Spanish at LarsonPrecalculus.com.*

Solve $1 < 2x + 7 < 11$. Then graph the solution set.

Inequalities Involving Absolute Values

Solving an Absolute Value Inequality

Let x be a variable or an algebraic expression and let a be a positive real number.

1. The solutions of $|x| < a$ are all values of x that lie between $-a$ and a.

   $|x| < a$ if and only if $-a < x < a$. Double inequality

2. The solutions of $|x| > a$ are all values of x that are less than $-a$ or greater than a.

   $|x| > a$ if and only if $x < -a$ or $x > a$. Compound inequality

These rules are also valid when $<$ is replaced by $\leq$ and $>$ is replaced by $\geq$.

EXAMPLE 4 Solving Absolute Value Inequalities

Solve each inequality.

a. $|x - 5| < 2$

b. $|x - 5| > 2$

Algebraic Solution

a. $|x - 5| < 2$ Write original inequality.

$-2 < x - 5 < 2$ Write double inequality.

$3 < x < 7$ Add 5 to each part.

The solution set is all real numbers that are greater than 3 *and* less than 7. The interval notation for this solution set is $(3, 7)$. The graph of this solution set is shown below.

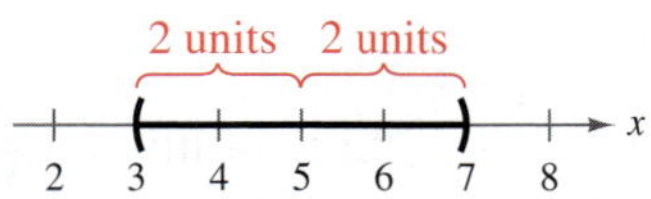

b. The absolute value inequality $|x - 5| > 2$ is equivalent to the following compound inequality: $x - 5 < -2$ *or* $x - 5 > 2$.

Solve first inequality: $x - 5 < -2$ Write first inequality.

$x < 3$ Add 5 to each side.

Solve second inequality: $x - 5 > 2$ Write second inequality.

$x > 7$ Add 5 to each side.

The solution set is all real numbers that are less than 3 *or* greater than 7. The interval notation for this solution set is $(-\infty, 3) \cup (7, \infty)$. The symbol $\cup$ is called a *union* symbol and is used to denote the combining of two sets. The graph of this solution set is shown below.

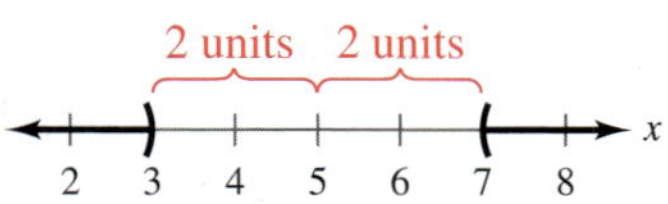

Graphical Solution

Use a graphing utility to graph

$$y_1 = |x - 5| \quad \text{and} \quad y_2 = 2$$

in the same viewing window, as shown below.

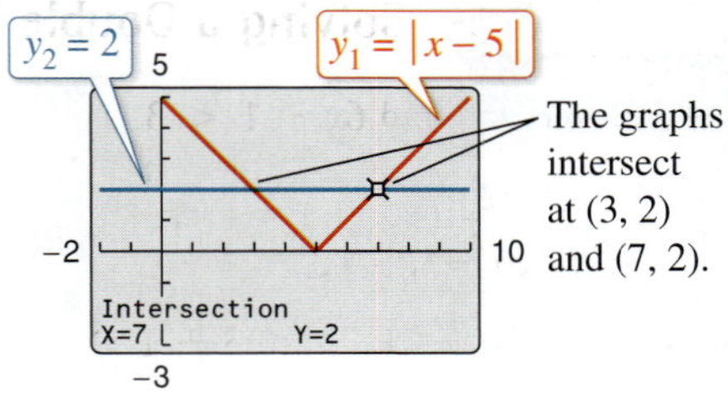

a. You can see that the graph of y_1 lies below the graph of y_2 when

$$3 < x < 7.$$

This implies that the solution set is all real numbers greater than 3 *and* less than 7.

b. You can see that the graph of y_1 lies above the graph of y_2 when

$$x < 3$$

or when

$$x > 7.$$

This implies that the solution set is all real numbers that are less than 3 *or* greater than 7.

✔ **Checkpoint** Audio-video solution in English & Spanish at LarsonPrecalculus.com.

Solve $|x - 20| \leq 4$. Then graph the solution set.

Polynomial Inequalities

To solve a polynomial inequality such as $x^2 - 2x - 3 < 0$, use the fact that a polynomial can change signs only at its zeros (the x-values that make the polynomial equal to zero). Between two consecutive zeros, a polynomial must be entirely positive or entirely negative. This means that when the real zeros of a polynomial are put in order, they divide the real number line into intervals in which the polynomial has no sign changes. These zeros are the **key numbers** of the inequality, and the resulting open intervals are the **test intervals** for the inequality. For instance, the polynomial above factors as $x^2 - 2x - 3 = (x + 1)(x - 3)$ and has two zeros, $x = -1$ and $x = 3$, which divide the real number line into three test intervals: $(-\infty, -1)$, $(-1, 3)$, and $(3, \infty)$. To solve the inequality $x^2 - 2x - 3 < 0$, you need to test only one value in each test interval.

Finding Test Intervals for a Polynomial

To determine the intervals on which the values of a polynomial are entirely negative or entirely positive, use the following steps.

1. Find all real zeros of the polynomial, and arrange the zeros in increasing order. These zeros are the key numbers of the polynomial.
2. Use the key numbers to determine the test intervals.
3. Choose one representative x-value in each test interval and evaluate the polynomial at that value. If the value of the polynomial is negative, then the polynomial will have negative values for *every* x-value in the interval. If the value of the polynomial is positive, then the polynomial will have positive values for *every* x-value in the interval.

Technology Tip

Some graphing utilities will produce graphs of inequalities. For instance, you can graph $2x^2 + 5x > 12$ by setting the graphing utility to *dot* mode and entering $y = 2x^2 + 5x > 12$. Using $-10 \le x \le 10$ and $-4 \le y \le 4$, your graph should look like the graph shown below. The solution appears to be $(-\infty, -4) \cup \left(\frac{3}{2}, \infty\right)$. See Example 6 for an algebraic solution and for an alternative graphical solution.

EXAMPLE 5 Investigating Polynomial Behavior

To determine the intervals on which $x^2 - 3$ is entirely negative and those on which it is entirely positive, factor the quadratic as $x^2 - 3 = (x + \sqrt{3})(x - \sqrt{3})$. The key numbers occur at $x = -\sqrt{3}$ and $x = \sqrt{3}$. So, the test intervals for the quadratic are

$$(-\infty, -\sqrt{3}),\quad (-\sqrt{3}, \sqrt{3}),\quad \text{and} \quad (\sqrt{3}, \infty).$$

In each test interval, choose a representative x-value and evaluate the polynomial, shown in the table.

Interval	x-Value	Value of Polynomial	Sign of Polynomial
$(-\infty, -\sqrt{3})$	$x = -3$	$(-3)^2 - 3 = 6$	Positive
$(-\sqrt{3}, \sqrt{3})$	$x = 0$	$(0)^2 - 3 = -3$	Negative
$(\sqrt{3}, \infty)$	$x = 5$	$(5)^2 - 3 = 22$	Positive

The polynomial has negative values for every x in the interval $(-\sqrt{3}, \sqrt{3})$ and positive values for every x in the intervals $(-\infty, -\sqrt{3})$ and $(\sqrt{3}, \infty)$. In Figure 2.27, the graph of $y = x^2 - 3$ confirms this result.

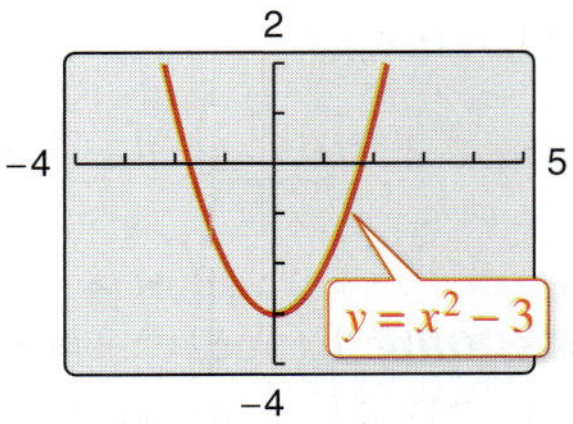

Figure 2.27

Checkpoint Audio-video solution in English & Spanish at LarsonPrecalculus.com.

Determine the intervals on which $x^2 - 2x - 3$ is entirely negative and those on which it is entirely positive.

To determine the test intervals for a polynomial inequality, the inequality must first be written in general form with the polynomial on one side and zero on the other.

EXAMPLE 6 Solving a Polynomial Inequality

See LarsonPrecalculus.com for an interactive version of this type of example.

Solve $2x^2 + 5x > 12$.

Algebraic Solution

$2x^2 + 5x - 12 > 0$ Write inequality in general form.

$(x + 4)(2x - 3) > 0$ Factor.

Key Numbers: $x = -4, x = \frac{3}{2}$

Test Intervals: $(-\infty, -4), \left(-4, \frac{3}{2}\right), \left(\frac{3}{2}, \infty\right)$

Test: Is $(x + 4)(2x - 3) > 0$?

After testing these intervals, you can see that the polynomial $2x^2 + 5x - 12$ is positive on the open intervals $(-\infty, -4)$ and $\left(\frac{3}{2}, \infty\right)$. So, the solution set of the inequality is

$$(-\infty, -4) \cup \left(\tfrac{3}{2}, \infty\right).$$

Graphical Solution

First write the polynomial inequality $2x^2 + 5x > 12$ as $2x^2 + 5x - 12 > 0$. Then use a graphing utility to graph $y = 2x^2 + 5x - 12$. In the figure, you can see that the graph is *above* the x-axis when x is less than -4 *or* when x is greater than $\frac{3}{2}$. So, you can graphically approximate the solution set to be $(-\infty, -4) \cup \left(\frac{3}{2}, \infty\right)$.

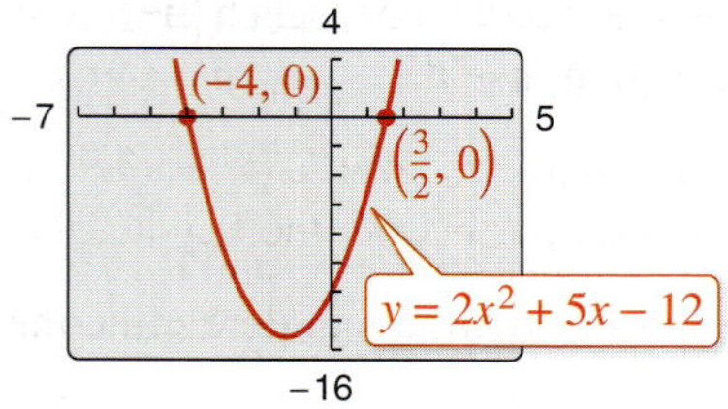

✔ **Checkpoint** *Audio-video solution in English & Spanish at LarsonPrecalculus.com.*

Solve $2x^2 + 3x < 5$ (a) algebraically and (b) graphically.

EXAMPLE 7 Solving a Polynomial Inequality

$2x^3 - 3x^2 - 32x + 48 > 0$ Original inequality

$x^2(2x - 3) - 16(2x - 3) > 0$ Factor by grouping.

$(x^2 - 16)(2x - 3) > 0$ Distributive Property

$(x - 4)(x + 4)(2x - 3) > 0$ Factor difference of two squares.

The key numbers are $x = -4$, $x = \frac{3}{2}$, and $x = 4$; and the test intervals are $(-\infty, -4)$, $\left(-4, \frac{3}{2}\right)$, $\left(\frac{3}{2}, 4\right)$, and $(4, \infty)$.

Interval	*x-Value*	*Polynomial Value*	*Conclusion*
$(-\infty, -4)$	$x = -5$	$2(-5)^3 - 3(-5)^2 - 32(-5) + 48 = -117$	Negative
$\left(-4, \frac{3}{2}\right)$	$x = 0$	$2(0)^3 - 3(0)^2 - 32(0) + 48 = 48$	Positive
$\left(\frac{3}{2}, 4\right)$	$x = 2$	$2(2)^3 - 3(2)^2 - 32(2) + 48 = -12$	Negative
$(4, \infty)$	$x = 5$	$2(5)^3 - 3(5)^2 - 32(5) + 48 = 63$	Positive

From this you can conclude that the polynomial is positive on the open intervals $\left(-4, \frac{3}{2}\right)$ and $(4, \infty)$. So, the solution set is $\left(-4, \frac{3}{2}\right) \cup (4, \infty)$.

✔ **Checkpoint** *Audio-video solution in English & Spanish at LarsonPrecalculus.com.*

Solve $3x^3 - x^2 - 12x > -4$. Then graph the solution set.

When solving a polynomial inequality, be sure to account for the particular type of inequality symbol given in the inequality. For instance, in Example 7, the original inequality contained a "greater than" symbol and the solution consisted of two open intervals. If the original inequality had been $2x^3 - 3x^2 - 32x + 48 \geq 0$, the solution would have consisted of the closed interval $\left[-4, \frac{3}{2}\right]$ and the interval $[4, \infty)$.

EXAMPLE 8 Unusual Solution Sets

a. The solution set of

$$x^2 + 2x + 4 > 0$$

consists of the entire set of real numbers, $(-\infty, \infty)$. In other words, the value of the quadratic $x^2 + 2x + 4$ is positive for every real value of x, as shown in the figure at the right. (Note that this quadratic inequality has *no* key numbers. In such a case, there is only one test interval—the entire real number line.)

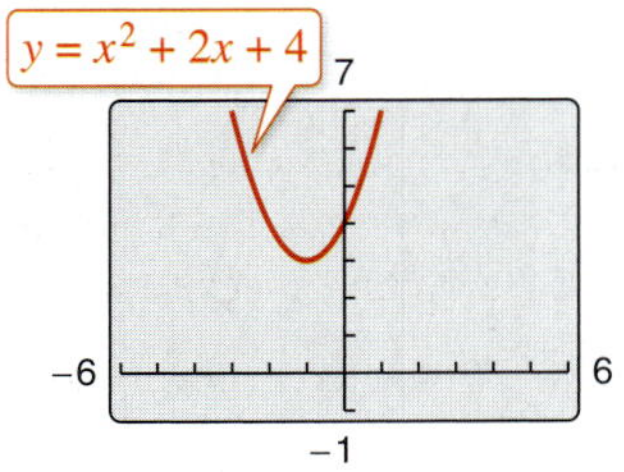

b. The solution set of

$$x^2 + 2x + 1 \le 0$$

consists of the single real number $\{-1\}$, because the quadratic

$$x^2 + 2x + 1$$

has one key number, $x = -1$, and it is the only value that satisfies the inequality, as shown in the figure at the right.

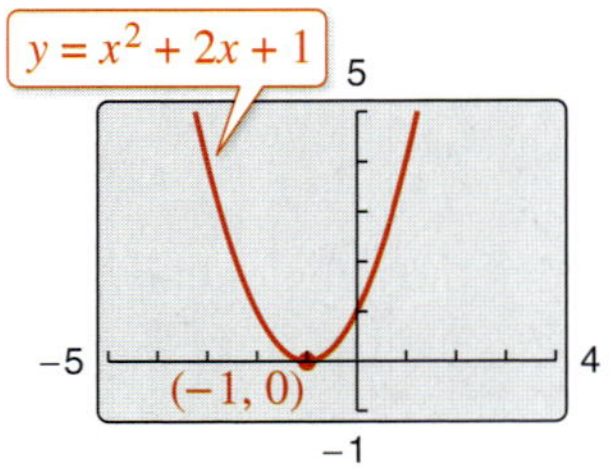

c. The solution set of

$$x^2 + 3x + 5 < 0$$

is empty. In other words, the quadratic

$$x^2 + 3x + 5$$

is not less than zero for any value of x, as shown in the figure at the right.

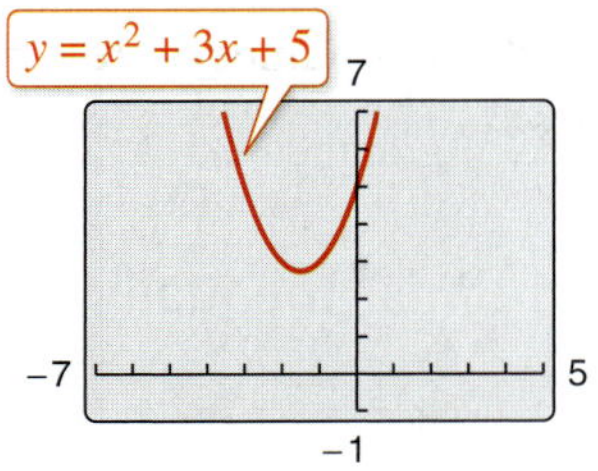

d. The solution set of

$$x^2 - 4x + 4 > 0$$

consists of all real numbers *except* the number 2. In interval notation, this solution set can be written as $(-\infty, 2) \cup (2, \infty)$. The graph of $y = x^2 - 4x + 4$ lies above the x-axis except at $x = 2$, where it touches the x-axis, as shown in the figure at the right.

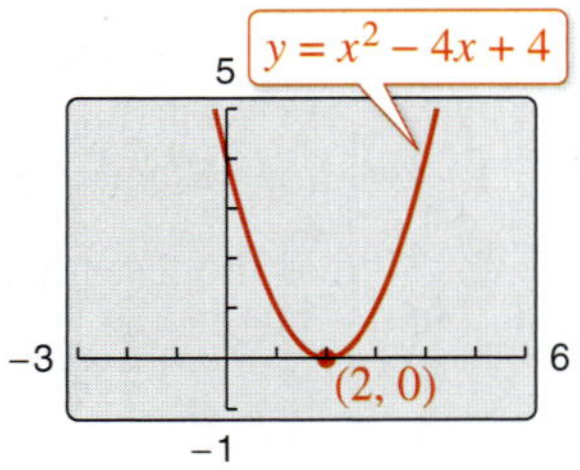

✓ **Checkpoint** Audio-video solution in English & Spanish at LarsonPrecalculus.com.

What is unusual about the solution set of each inequality?

a. $x^2 + 6x + 9 < 0$ **b.** $x^2 + 4x + 4 \le 0$

c. $x^2 - 6x + 9 > 0$ **d.** $x^2 - 2x + 1 \ge 0$

Technology Tip

One of the advantages of technology is that you can solve complicated polynomial inequalities that might be difficult, or even impossible, to factor. For instance, you could use a graphing utility to approximate the solution of the inequality

$$x^3 - 0.2x^2 - 3.16x + 1.4 < 0.$$

Rational Inequalities

The concepts of key numbers and test intervals can be extended to inequalities involving rational expressions. To do this, use the fact that the value of a rational expression can change sign only at its *zeros* (the x-values for which its numerator is zero) and its *undefined values* (the x-values for which its denominator is zero). These two types of numbers make up the *key numbers* of a rational inequality. When solving a rational inequality, begin by writing the inequality in general form with the rational expression on one side and zero on the other.

EXAMPLE 9 Solving a Rational Inequality

Solve $\frac{2x-7}{x-5} \le 3$.

Algebraic Solution

$$\frac{2x-7}{x-5} - 3 \le 0 \qquad \text{Write in general form.}$$

$$\frac{2x-7-3x+15}{x-5} \le 0 \qquad \text{Write as single fraction.}$$

$$\frac{-x+8}{x-5} \le 0 \qquad \text{Simplify.}$$

Now, in standard form you can see that the key numbers are $x = 5$ and $x = 8$, and you can proceed as follows.

Key Numbers: $x = 5, x = 8$

Test Intervals: $(-\infty, 5), (5, 8), (8, \infty)$

Test: Is $\frac{-x+8}{x-5} \le 0$?

Interval	*x-Value*	*Polynomial Value*	*Conclusion*
$(-\infty, 5)$	$x = 0$	$\frac{-0+8}{0-5} = -\frac{8}{5}$	Negative
$(5, 8)$	$x = 6$	$\frac{-6+8}{6-5} = 2$	Positive
$(8, \infty)$	$x = 9$	$\frac{-9+8}{9-5} = -\frac{1}{4}$	Negative

By testing these intervals, you can determine that the rational expression $(-x+8)/(x-5)$ is negative in the open intervals $(-\infty, 5)$ and $(8, \infty)$. Moreover, because

$$\frac{-x+8}{x-5} = 0$$

when $x = 8$, you can conclude that the solution set of the inequality is $(-\infty, 5) \cup [8, \infty)$.

Graphical Solution

Use a graphing utility to graph

$$y_1 = \frac{2x-7}{x-5} \quad \text{and} \quad y_2 = 3$$

in the same viewing window, as shown below.

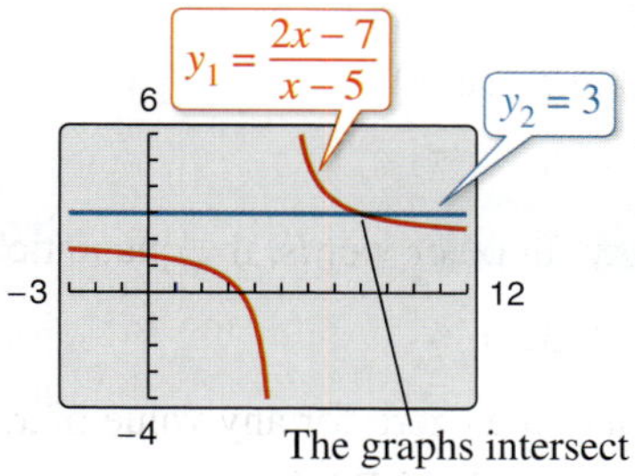

The graph of y_1 lies on or below the graph of y_2 on the intervals $(-\infty, 5)$ and $[8, \infty)$. So, you can graphically estimate the solution set to be all real numbers less than 5 *or* greater than or equal to 8.

Checkpoint Audio-video solution in English & Spanish at LarsonPrecalculus.com.

Solve $\frac{4x-1}{x-6} > 3$.

Note in Example 9 that $x = 5$ is not included in the solution set because the inequality is undefined when $x = 5$.

Applications

In Section 1.3, you studied the *implied domain* of a function, the set of all x-values for which the function is defined. A common type of implied domain is used to avoid even roots of negative numbers, as shown in Example 10.

EXAMPLE 10 Finding the Domain of an Expression

Find the domain of $\sqrt{64 - 4x^2}$.

Solution

Because $\sqrt{64 - 4x^2}$ is defined only when $64 - 4x^2$ is nonnegative, the domain is given by $64 - 4x^2 \geq 0$.

$64 - 4x^2 \geq 0$ Write in general form.

$16 - x^2 \geq 0$ Divide each side by 4.

$(4 - x)(4 + x) \geq 0$ Factor.

The inequality has two key numbers: $x = -4$ and $x = 4$. A test shows that $64 - 4x^2 \geq 0$ in the *closed interval* $[-4, 4]$. The graph of $y = \sqrt{64 - 4x^2}$, shown in Figure 2.29, confirms that the domain is $[-4, 4]$.

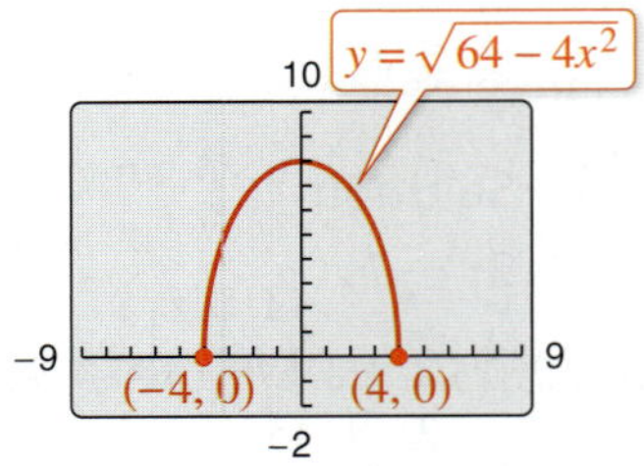

Figure 2.29

Checkpoint Audio-video solution in English & Spanish at LarsonPrecalculus.com.

Find the domain of $\sqrt{x^2 - 7x + 10}$.

EXAMPLE 11 Height of a Projectile

A projectile is fired straight upward from ground level with an initial velocity of 384 feet per second. During what time period will the height of the projectile exceed 2000 feet?

Solution

In Section 2.4, you saw that the position of an object moving vertically can be modeled by the *position equation*

$$s = -16t^2 + v_0 t + s_0$$

where s is the height in feet and t is the time in seconds. In this case, $s_0 = 0$ and $v_0 = 384$. So, you need to solve the inequality $-16t^2 + 384t > 2000$. Using a graphing utility, graph $y_1 = -16t^2 + 384t$ and $y_2 = 2000$, as shown in Figure 2.30. From the graph, you can determine that $-16t^2 + 384t > 2000$ for t between approximately 7.6 and 16.4. You can verify this result algebraically.

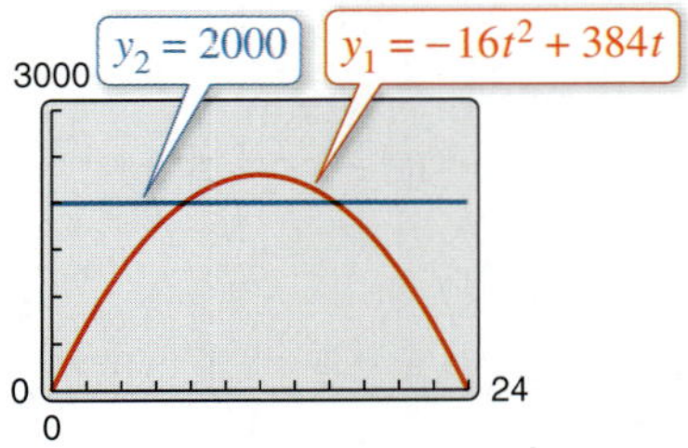

Figure 2.30

$-16t^2 + 384t > 2000$ Write original inequality.

$t^2 - 24t < -125$ Divide by -16 and reverse inequality symbol.

$t^2 - 24t + 125 < 0$ Write in general form.

By the Quadratic Formula, the key numbers are $t = 12 - \sqrt{19}$ and $t = 12 + \sqrt{19}$, or approximately 7.64 and 16.36. A test will verify that the height of the projectile will exceed 2000 feet when $7.64 < t < 16.36$—that is, during the time interval (7.64, 16.36) seconds.

Checkpoint Audio-video solution in English & Spanish at LarsonPrecalculus.com.

A projectile is fired straight upward from ground level with an initial velocity of 208 feet per second. During what time period will its height exceed 640 feet? ■

2.6 Exercises

See *CalcChat.com* for tutorial help and worked-out solutions to odd-numbered exercises. For instructions on how to use a graphing utility, see Appendix A.

Vocabulary and Concept Check

In Exercises 1–4, fill in the blank(s).

1. It is sometimes possible to write two inequalities as one inequality, called a _______ inequality.
2. The solutions of $|x| \le a$ are those values of x such that _______ .
3. The solutions of $|x| \ge a$ are those values of x such that _______ or _______ .
4. The key numbers of a rational inequality are its _______ and its _______ .

5. Are the inequalities $x - 4 < 5$ and $x > 9$ equivalent?
6. Which property of inequalities is shown below?

   $a < b$ and $b < c$ $\Rightarrow$ $a < c$

Procedures and Problem Solving

Matching an Inequality with Its Graph **In Exercises 7–12, match the inequality with its graph. [The graphs are labeled (a), (b), (c), (d), (e), and (f).]**

(a) 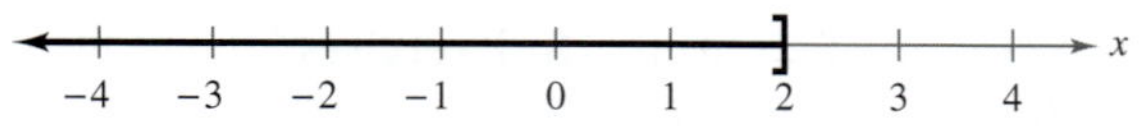

(b) 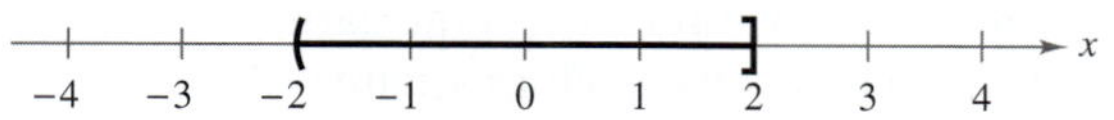

(c) 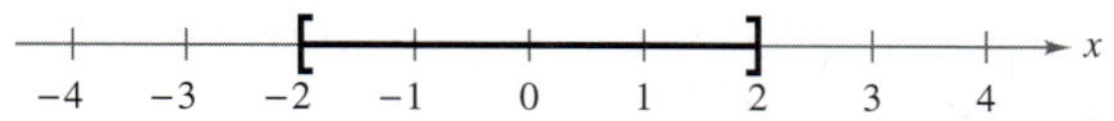

(d) 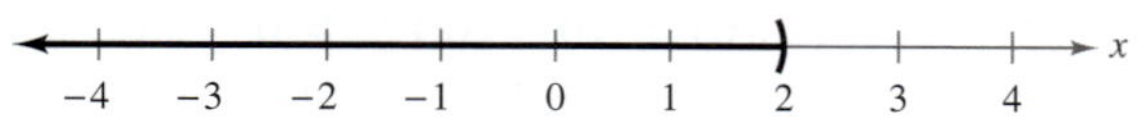

(e) 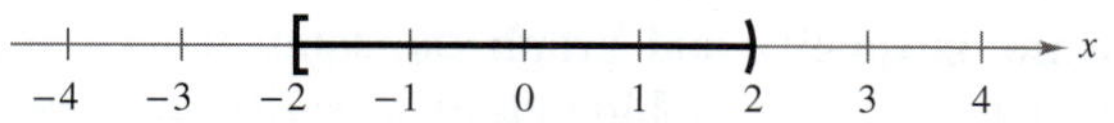

(f) 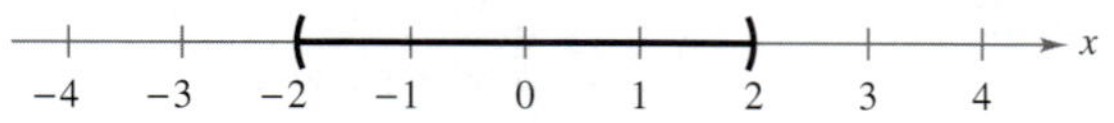

7. $x < 2$
8. $x \le 2$
9. $-2 < x < 2$
10. $-2 < x \le 2$
11. $-2 \le x < 2$
12. $-2 \le x \le 2$

Determining Solutions of an Inequality **In Exercises 13–16, determine whether each value of x is a solution of the inequality.**

Inequality	*Values*	
13. $5x - 12 > 0$	(a) $x = 3$	(b) $x = -3$
	(c) $x = \frac{5}{2}$	(d) $x = \frac{3}{2}$
14. $-5 < 2x - 1 \le 1$	(a) 2	(b) -2
	(c) 0	(d) $-\frac{1}{2}$
15. $-1 < \dfrac{3 - x}{2} \le 1$	(a) $x = -1$	(b) $x = \sqrt{5}$
	(c) $x = 1$	(d) $x = 5$
16. $\lvert x - 10 \rvert \ge 3$	(a) $x = 13$	(b) $x = -1$
	(c) $x = 14$	(d) $x = 8$

Solving an Inequality **In Exercises 17–30, solve the inequality and sketch the solution on the real number line. Use a graphing utility to verify your solution graphically.**

17. $6x > 42$
18. $-10x \le 40$
19. $4x + 7 < 3 + 2x$
20. $3x + 1 \ge 2 + x$
21. $2(1 - x) < 3x + 7$
22. $2x + 7 < 3(x - 4)$
23. $\frac{3}{4}x - 6 \le x - 7$
24. $3 + \frac{2}{7}x > x - 2$
25. $1 \le 2x + 3 \le 9$
26. $-8 \le -3x + 5 < 13$
27. $-8 \le 1 - 3(x - 2) < 13$
28. $0 \le 2 - 3(x + 1) < 20$
29. $-4 < \dfrac{2x - 3}{3} < 4$
30. $0 \le \dfrac{x + 3}{2} < 5$

Approximating a Solution **In Exercises 31–34, use a graphing utility to approximate the solution.**

31. $5 - 2x \ge 1$
32. $20 < 6x - 1$
33. $3(x + 1) < x + 7$
34. $4(x - 3) > 8 - x$

Approximating Solutions **In Exercises 35–38, use a graphing utility to graph the equation and graphically approximate the values of x that satisfy the specified inequalities. Then solve each inequality algebraically.**

Equation	*Inequalities*	
35. $y = 2x - 3$	(a) $y \ge 1$	(b) $y \le 0$
36. $y = \frac{2}{3}x + 1$	(a) $y \le 5$	(b) $y \ge 0$
37. $y = -3x + 8$	(a) $-1 \le y \le 3$	(b) $y \le 0$
38. $y = -\frac{1}{2}x + 2$	(a) $0 \le y \le 3$	(b) $y \ge 0$

Solving an Absolute Value Inequality In Exercises 39–46, solve the inequality and sketch the solution on the real number line. Use a graphing utility to verify your solutions graphically.

39. $|5x| > 10$

40. $\left|\frac{x}{2}\right| \leq 1$

41. $|x - 7| \leq 6$

42. $|x - 20| > 4$

43. $\left|\frac{x-3}{2}\right| \geq 5$

44. $|x + 14| + 3 \geq 17$

45. $10|1 - x| < 5$

46. $3|4 - 5x| < 9$

Approximating Solutions In Exercises 47 and 48, use a graphing utility to graph the equation and graphically approximate the values of x that satisfy the specified inequalities. Then solve each inequality algebraically.

	Equation	*Inequalities*	
47.	$y = \|x - 3\|$	(a) $y \leq 2$	(b) $y \geq 4$
48.	$y = \left\|\frac{1}{2}x + 1\right\|$	(a) $y \leq 4$	(b) $y \geq 1$

Using Absolute Value Notation In Exercises 49–56, use absolute value notation to define the interval (or pair of intervals) on the real number line.

49.
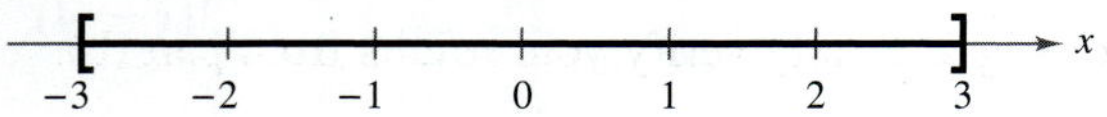

50.
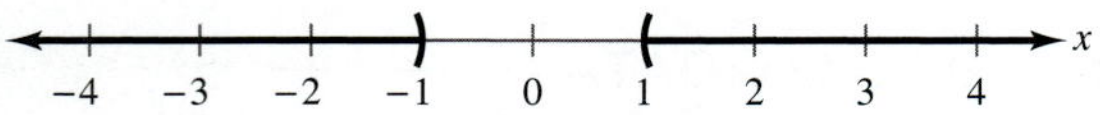

51.
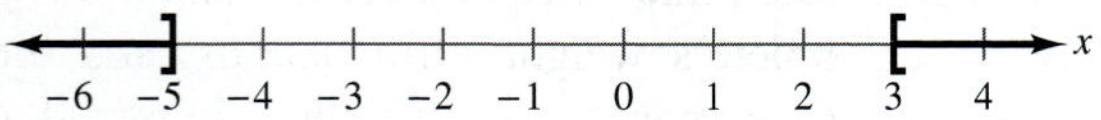

52.
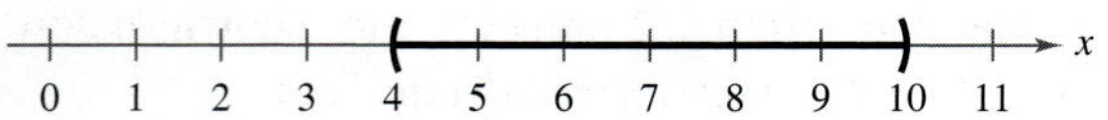

53. All real numbers less than 10 units from 6

54. All real numbers no more than 8 units from -5

55. All real numbers more than 3 units from -1

56. All real numbers at least 5 units from 3

Investigating Polynomial Behavior In Exercises 57–62, determine the intervals on which the polynomial is entirely negative and those on which it is entirely positive.

57. $x^2 - 4x - 5$

58. $x^2 - 3x - 4$

59. $2x^2 - 4x - 3$

60. $-2x^2 + x + 5$

61. $-x^2 + 6x - 10$

62. $3x^2 + 8x + 6$

Solving a Polynomial Inequality In Exercises 63–76, solve the inequality and graph the solution on the real number line. Use a graphing utility to verify your solution graphically.

63. $x^2 + 4x + 4 \geq 9$

64. $x^2 - 6x + 9 < 16$

65. $(x + 2)^2 < 25$

66. $(x - 3)^2 \geq 1$

67. $x^3 - 4x^2 \geq 0$

68. $x^5 - 3x^4 \leq 0$

69. $2x^3 + 5x^2 > 6x + 9$

70. $2x^3 + 3x^2 < 11x + 6$

71. $x^3 - 3x^2 - x > -3$

72. $2x^3 + 13x^2 - 8x - 46 \geq 6$

73. $3x^2 - 11x + 16 \leq 0$

74. $4x^2 + 12x + 9 \leq 0$

75. $4x^2 - 4x + 1 > 0$

76. $x^2 + 3x + 8 > 0$

Using Graphs to Find Solutions In Exercises 77 and 78, use the graph of the function to solve the equation or inequality.

(a) $f(x) = g(x)$ (b) $f(x) \geq g(x)$ (c) $f(x) > g(x)$

77.
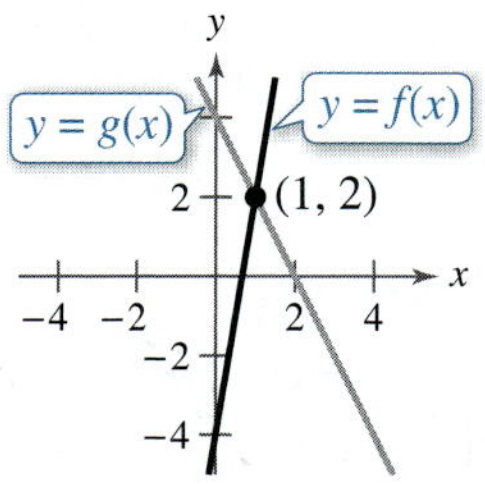

78.
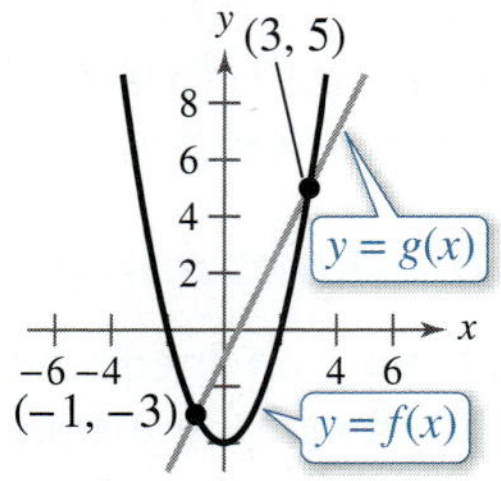

Approximating Solutions In Exercises 79 and 80, use a graphing utility to graph the equation and graphically approximate the values of x that satisfy the specified inequalities. Then solve each inequality algebraically.

	Equation	*Inequalities*	
79.	$y = -x^2 + 2x + 3$	(a) $y \leq 0$	(b) $y \geq 3$
80.	$y = x^3 - x^2 - 16x + 16$	(a) $y \leq 0$	(b) $y \geq 36$

Solving a Rational Inequality In Exercises 81–84, solve the inequality and graph the solution on the real number line. Use a graphing utility to verify your solution graphically.

81. $\frac{1}{x} - x > 0$

82. $\frac{1}{x} - 4 < 0$

83. $\frac{x+6}{x+1} - 2 \leq 0$

84. $\frac{x+12}{x+2} - 3 \geq 0$

Approximating Solutions In Exercises 85 and 86, use a graphing utility to graph the equation and graphically approximate the values of x that satisfy the specified inequalities. Then solve each inequality algebraically.

	Equation	*Inequalities*	
85.	$y = \frac{3x}{x-2}$	(a) $y \leq 0$	(b) $y \geq 6$
86.	$y = \frac{5x}{x^2+4}$	(a) $y \geq 1$	(b) $y \leq 0$

Finding the Domain of an Expression **In Exercises 87–92, find the domain of x in the expression.**

87. $\sqrt{x-5}$

88. $\sqrt{6x+15}$

89. $\sqrt{-x^2+x+12}$

90. $\sqrt{2x^2-8}$

91. $\sqrt[4]{3x^2-20x-7}$

92. $\sqrt[4]{2x^2+4x+3}$

93. MODELING DATA

The graph models the population P (in thousands) of Sacramento, California, from 2003 through 2012, where t is the year, with $t = 3$ corresponding to 2003. Also shown is the line $y = 450$. Use the graphs of the model and the horizontal line to write an equation or an inequality that could be solved to answer the question. Then answer the question. (Source: U.S. Census Bureau)

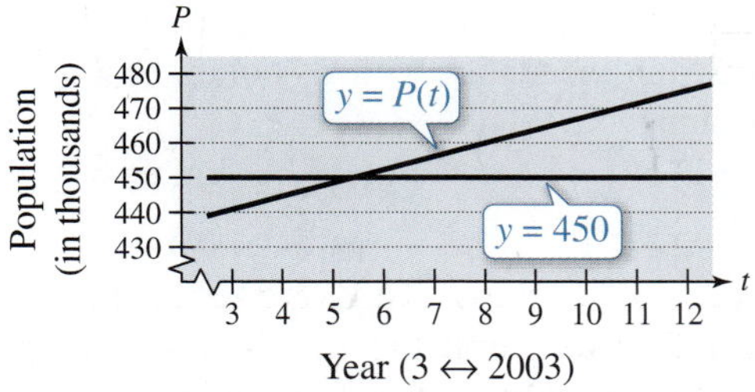

(a) In what year did the population of Sacramento reach 450,000?

(b) During what time period was the population of Sacramento less than 450,000? greater than 450,000?

94. MODELING DATA

The graph models the population P (in thousands) of Cleveland, Ohio, from 2003 through 2012, where t is the year, with $t = 3$ corresponding to 2003. Also shown is the line $y = 400$. Use the graphs of the model and the horizontal line to write an equation or an inequality that could be solved to answer the question. Then answer the question. (Source: U.S. Census Bureau)

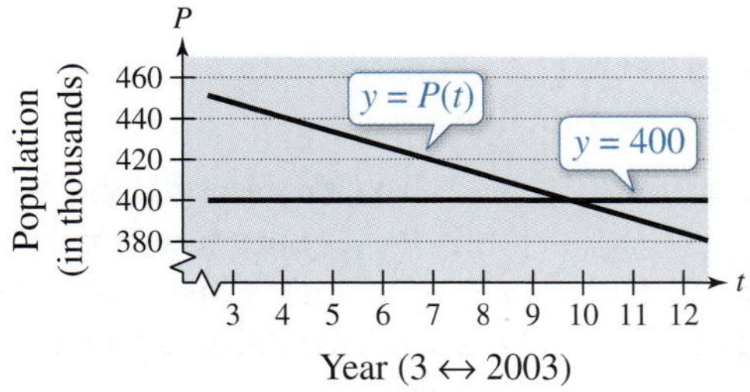

(a) In what year did the population of Cleveland reach 400,000?

(b) During what time period was the population of Cleveland less than 400,000? greater than 400,000?

95. Height of a Projectile A projectile is fired straight upward from ground level with an initial velocity of 160 feet per second.

(a) At what instant will it be back at ground level?

(b) When will the height exceed 384 feet?

96. Height of a Projectile A projectile is fired straight upward from ground level with an initial velocity of 128 feet per second.

(a) At what instant will it be back at ground level?

(b) When will the height be less than 128 feet?

97. MODELING DATA

The numbers D of doctorate degrees (in thousands) awarded to female students from 1991 through 2012 in the United States can be approximated by the model

$$D = 0.0743t^2 + 0.628t + 42.61, \quad 0 \le t \le 22$$

where t is the year, with $t = 1$ corresponding to 1991. (Source: U.S. National Center for Education Statistics)

(a) Use a graphing utility to graph the model.

(b) Use the *zoom* and *trace* features to find when the number of degrees was between 50 and 60 thousand.

(c) Algebraically verify your results from part (b).

98. MODELING DATA

You want to determine whether there is a relationship between an athlete's weight x (in pounds) and the athlete's maximum bench-press weight y (in pounds). Sample data from 12 athletes are shown below. *(Spreadsheet at LarsonPrecalculus.com)*

DATA (165, 170), (184, 185), (150, 200), (210, 255), (196, 205), (240, 295), (202, 190), (170, 175), (185, 195), (190, 185), (230, 250), (160, 150)

(a) Use a graphing utility to plot the data.

(b) A model for the data is

$y = 1.3x - 36.$

Use the graphing utility to graph the equation in the same viewing window used in part (a).

(c) Use the graph to estimate the values of x that predict a maximum bench-press weight of at least 200 pounds.

(d) Use the graph to write a statement about the accuracy of the model. If you think the graph indicates that an athlete's weight is not a good indicator of the athlete's maximum bench-press weight, list other factors that might influence an individual's maximum bench-press weight.

Why you should learn it *(p. 210)* **In Exercises 99–102, use the models below, which approximate the numbers of Bed Bath & Beyond stores B and Williams-Sonoma stores W for the years 2000 through 2013, where t is the year, with $t = 0$ corresponding to 2000.** (Sources: Bed Bath & Beyond, Inc.; Williams-Sonoma, Inc.)

Bed Bath & Beyond:

$$B = 86.5t + 342, \quad 0 \le t \le 13$$

Williams-Sonoma:

$$W = -2.92t^2 + 52.0t + 381, \quad 0 \le t \le 13$$

99. Solve the inequality $B(t) \ge 900$. Explain what the solution of the inequality represents.

100. Solve the inequality $W(t) \le 600$. Explain what the solution of the inequality represents.

101. Solve the equation $B(t) = W(t)$. Explain what the solution of the equation represents.

102. Solve the inequality $B(t) \ge W(t)$. Explain what the solution of the inequality represents.

Music **In Exercises 103–106, use the following information. Michael Kasha, of Florida State University, used physics and mathematics to design a classical guitar. He used the model for the frequency of the vibrations on a circular plate**

$$v = \frac{2.6t}{d^2}\sqrt{\frac{E}{\rho}}$$

where v is the frequency (in vibrations per second), t is the plate thickness (in millimeters), d is the diameter of the plate, E is the elasticity of the plate material, and ρ is the density of the plate material. For fixed values of d, E, and ρ, the graph of the equation is a line, as shown in the figure.

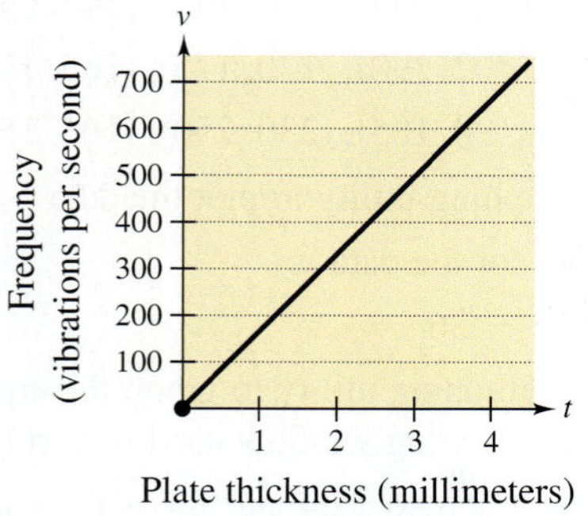

103. Estimate the frequency when the plate thickness is 2 millimeters.

104. Estimate the plate thickness when the frequency is 600 vibrations per second.

105. Approximate the interval for the plate thickness when the frequency is between 200 and 400 vibrations per second.

106. Approximate the interval for the frequency when the plate thickness is less than 3 millimeters.

Conclusions

True or False? **In Exercises 107–109, determine whether the statement is true or false. Justify your answer.**

107. If $-10 \le x \le 8$, then $-10 \ge -x$ and $-x \ge -8$.

108. The solution set of the inequality $\frac{3}{2}x^2 + 3x + 6 \ge 0$ is the entire set of real numbers.

109. The domain of $\sqrt[3]{6 - x}$ is $(-\infty, 6]$.

110. **Proof** The arithmetic mean of a and b is given by $(a + b)/2$. Order the statements of the proof to show that if $a < b$, then $a < (a - b)/2 < b$.

(i) $a < \dfrac{a + b}{2} < b$ (ii) $2a < 2b$

(iii) $2a < a + b < 2b$ (iv) $a < b$

111. **Think About It** Without performing any calculations, match the inequality with its solution. Explain your reasoning.

(a) $2x \le -6$ (i) $x \le -8$ or $x \ge 4$

(b) $-2x \le 6$ (ii) $x \ge -3$

(c) $|x + 2| \le 6$ (iii) $-8 \le x \le 4$

(d) $|x + 2| \ge 6$ (iv) $x \le -3$

112. **HOW DO YOU SEE IT?** Consider the polynomial $(x - a)(x - b)$ and the real number line shown below.

(a) Identify the points on the line at which the polynomial is zero.

(b) Determine the intervals on which the polynomial is entirely negative and those on which it is entirely positive. Explain your reasoning.

(c) At what x-values does the polynomial change signs?

Cumulative Mixed Review

Finding the Inverse of a Function **In Exercises 113–116, find the inverse function.**

113. $y = 12x$ **114.** $y = 5x + 8$

115. $y = x^3 + 7$ **116.** $y = \sqrt[3]{x - 7}$

117. ***Make a Decision*** To work an extended application analyzing the number of lung transplants in the United States, visit this textbook's website at *LarsonPrecalculus.com.* (Data Source: U.S. Department of Health and Human Services)

2.7 Linear Models and Scatter Plots

Scatter Plots and Correlation

Many real-life situations involve finding relationships between two variables, such as the year and the population of the United States. In a typical situation, data are collected and written as a set of ordered pairs. The graph of such a set, called a *scatter plot*, was discussed briefly in Section P.5.

What you should learn

- Construct scatter plots and interpret correlation.
- Use scatter plots and a graphing utility to find linear models for data.

Why you should learn it

Real-life data often follow a linear pattern. For instance, in Exercise 27 on page 231, you will find a linear model for the winning times in the women's 400-meter freestyle Olympic swimming event.

EXAMPLE 1 Constructing a Scatter Plot

The populations P (in millions) of the United States from 2008 through 2013 are shown in the table. Construct a scatter plot of the data. (Source: U.S. Census Bureau)

DATA

Year	Population, P (in millions)
2008	304.1
2009	306.8
2010	309.3
2011	311.6
2012	313.9
2013	316.1

Spreadsheet at LarsonPrecalculus.com

Solution

Begin by representing the data with a set of ordered pairs. Let t represent the year, with $t = 8$ corresponding to 2008.

(8, 304.1), (9, 306.8), (10, 309.3),
(11, 311.6), (12, 313.9), (13, 316.1)

Then plot each point in a coordinate plane, as shown in the figure.

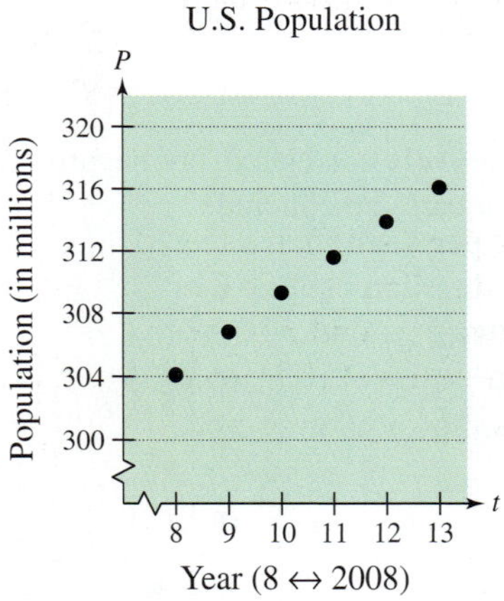

Checkpoint Audio-video solution in English & Spanish at LarsonPrecalculus.com.

The median sales prices (in thousands of dollars) of new homes sold in a neighborhood from 2006 through 2013 are given by the following ordered pairs. *(Spreadsheet at LarsonPrecalculus.com)* Construct a scatter plot of the data.

DATA (2006, 179.4), (2007, 185.4), (2008, 191.0), (2009, 196.7),
(2010, 202.6), (2011, 208.7), (2012, 214.9), (2013, 221.4)

From the scatter plot in Example 1, it appears that the points describe a relationship that is nearly linear. The relationship is not *exactly* linear because the population did not increase by precisely the same amount each year.

A mathematical equation that approximates the relationship between t and P is a *mathematical model*. When developing a mathematical model to describe a set of data, you strive for two (often conflicting) goals—accuracy and simplicity. For the data above, a linear model of the form $P = at + b$ (where a and b are constants) appears to be best. It is simple and relatively accurate.

Consider a collection of ordered pairs of the form (x, y). If y tends to increase as x increases, then the collection is said to have a **positive correlation.** If y tends to decrease as x increases, then the collection is said to have a **negative correlation.** Figure 2.31 shows three examples: one with a positive correlation, one with a negative correlation, and one with no (discernible) correlation.

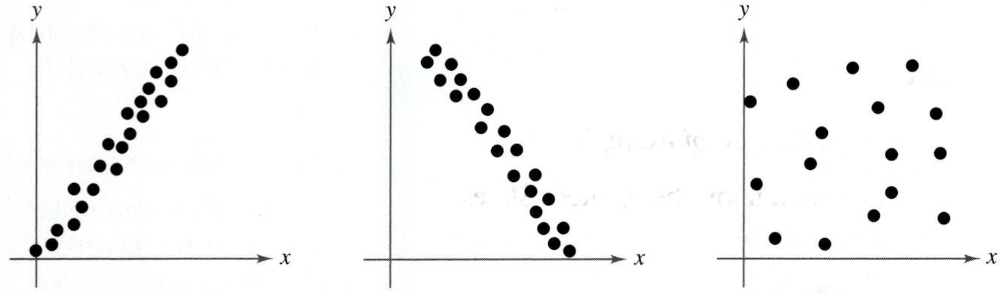

Positive Correlation ***Negative Correlation*** ***No Correlation***

Figure 2.31

EXAMPLE 2 Interpreting Correlation

On a Friday, 22 students in a class were asked to record the numbers of hours they spent studying for a test on Monday. The results are shown below. The first coordinate is the number of hours and the second coordinate is the score obtained on the test. *(Spreadsheet at LarsonPrecalculus.com)*

DATA *Study Hours:* (0, 40), (1, 41), (2, 51), (3, 58), (3, 49), (4, 48), (4, 64), (5, 55), (5, 69), (5, 58), (5, 75), (6, 68), (6, 63), (6, 93), (7, 84), (7, 67), (8, 90), (8, 76), (9, 95), (9, 72), (9, 85), (10, 98)

a. Construct a scatter plot of the data.

b. Determine whether the points are positively correlated, are negatively correlated, or have no discernible correlation. What can you conclude?

Solution

a. The scatter plot is shown in the figure.

b. The scatter plot relating study hours and test scores has a positive correlation. This means that the more a student studied, the higher his or her score tended to be.

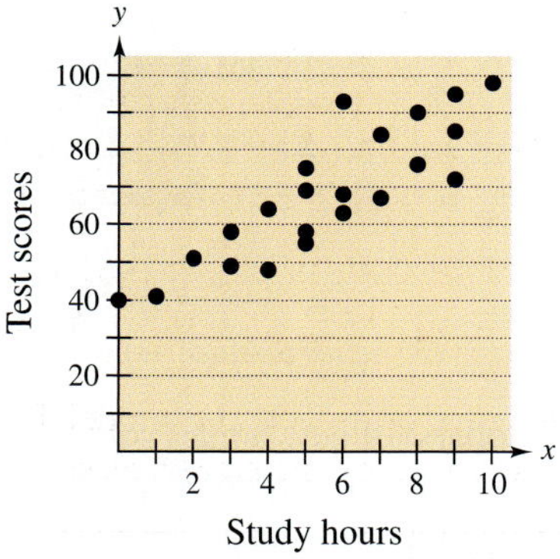

Checkpoint *Audio-video solution in English & Spanish at LarsonPrecalculus.com.*

The students in Example 2 also recorded the numbers of hours they spent watching television. The results are shown below. The first coordinate is the number of hours and the second coordinate is the score obtained on the test. *(Spreadsheet at LarsonPrecalculus.com)*

DATA *TV Hours:* (0, 98), (1, 85), (2, 72), (2, 90), (3, 67), (3, 93), (3, 95), (4, 68), (4, 84), (5, 76), (7, 75), (7, 58), (9, 63), (9, 69), (11, 55), (12, 58), (14, 64), (16, 48), (17, 51), (18, 41), (19, 49), (20, 40)

a. Construct a scatter plot of the data.

b. Determine whether the points are positively correlated, are negatively correlated, or have no discernible correlation. What can you conclude?

Fitting a Line to Data

Finding a linear model to represent the relationship described by a scatter plot is called **fitting a line to data.** You can do this graphically by simply sketching the line that appears to fit the points, finding two points on the line, and then finding the equation of the line that passes through the two points.

EXAMPLE 3 Fitting a Line to Data

See LarsonPrecalculus.com for an interactive version of this type of example.

Find a linear model that relates the year to the population of the United States. (See Example 1.)

DATA

Year	U.S. Population, P (in millions)
2008	304.1
2009	306.8
2010	309.3
2011	311.6
2012	313.9
2013	316.1

Spreadsheet at LarsonPrecalculus.com

U.S. Population

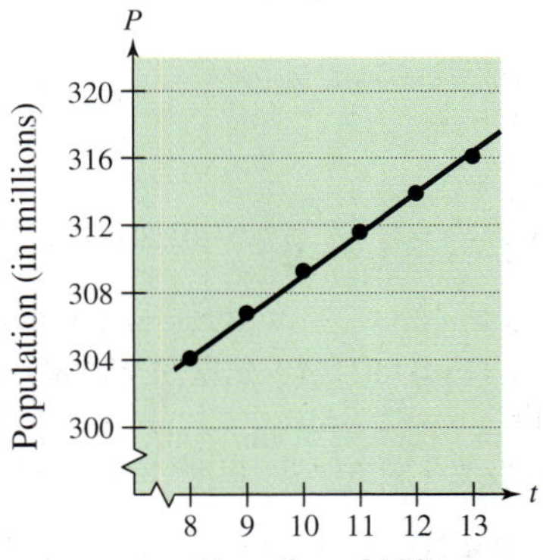

Solution

Let t represent the year, with $t = 8$ corresponding to 2008. After plotting the data in the table, draw the line that you think best represents the data, as shown in the figure. Two points that lie on this line are (8, 304.1) and (12, 313.9). Using the point-slope form, you can find the equation of the line to be

$$P = 2.45(t - 8) + 304.1$$

$$= 2.45t + 284.5. \quad \text{Linear model}$$

Checkpoint *Audio-video solution in English & Spanish at LarsonPrecalculus.com.*

Find a linear model that relates the year to the median sales prices (in thousands of dollars) of new homes sold in a neighborhood from 2006 through 2013. (See Checkpoint for Example 1.) *(Spreadsheet at LarsonPrecalculus.com)*

DATA (2006, 179.4), (2007, 185.4), (2008, 191.0), (2009, 196.7), (2010, 202.6), (2011, 208.7), (2012, 214.9), (2013, 221.4)

After finding a model, you can measure how well the model fits the data by comparing the actual values with the values given by the model, as shown in the table.

	t	8	9	10	11	12	13
Actual	P	304.1	306.8	309.3	311.6	313.9	316.1
Model	P	304.1	306.6	309.0	311.5	313.9	316.4

The sum of the squares of the differences between the actual values and the model values is called the **sum of the squared differences.** The model that has the least sum is called the **least squares regression line** for the data. For the model in Example 3, the sum of the squared differences is 0.23. The least squares regression line for the data is

$$P = 2.39t + 285.2. \quad \text{Best-fitting linear model}$$

Its sum of squared differences is 0.14. For more on the least squares regression line, see Appendix B.2 at this textbook's *Companion Website.*

Remark

The model in Example 3 is based on the two data points chosen. When different points are chosen, the model may change somewhat. For instance, when you choose (9, 306.8) and (12, 313.9), the new model is

$$P = 2.37(t - 9) + 306.8$$

$$= 2.37t + 285.5.$$

Another way to find a linear model to represent the relationship described by a scatter plot is to enter the data points into a graphing utility and use the *linear regression* feature. This method is demonstrated in Example 4.

> **Technology Tip**
>
> For instructions on how to use the *linear regression* feature, see Appendix A; for specific keystrokes, go to this textbook's *Companion Website*.

EXAMPLE 4 A Mathematical Model

The table shows the numbers N (in millions) of tax returns made through e-file from 2006 through 2013. (Source: Internal Revenue Service)

DATA

Year	Number of tax returns made through e-file, N (in millions)
2006	73.3
2007	80.0
2008	89.9
2009	95.0
2010	98.7
2011	112.2
2012	119.6
2013	122.5

Spreadsheet at LarsonPrecalculus.com

a. Use the *regression* feature of a graphing utility to find a linear model for the data. Let t represent the year, with $t = 6$ corresponding to 2006.

b. How closely does the model represent the data?

Graphical Solution

a. Use the *linear regression* feature of a graphing utility to obtain the model shown in the figure. You can approximate the model to be $N = 7.30t + 29.6$.

b. Graph the actual data and the model. From the figure, it appears that the model is a good fit for the actual data.

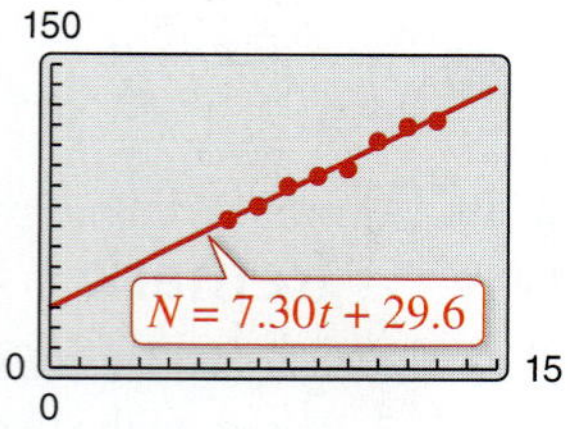

Numerical Solution

a. Using the *linear regression* feature of a graphing utility, you can find that a linear model for the data is $N = 7.30t + 29.6$.

b. You can see how well the model fits the data by comparing the actual values of N with the values of N given by the model, which are labeled N^* in the table below. From the table, you can see that the model appears to be a good fit for the actual data.

Year	N	N^*
2006	73.3	73.4
2007	80.0	80.7
2008	89.9	88.0
2009	95.0	95.3
2010	98.7	102.6
2011	112.2	109.9
2012	119.6	117.2
2013	122.5	124.5

✓ **Checkpoint** Audio-video solution in English & Spanish at LarsonPrecalculus.com.

The numbers N (in millions) of people unemployed in the U.S. from 2007 through 2013 are given by the following ordered pairs. *(Spreadsheet at LarsonPrecalculus.com)* (Source: U.S. Bureau of Labor Statistics)

DATA (2007, 7.078), (2008, 8.924), (2009, 14.265), (2010, 14.825), (2011, 13.747), (2012, 12.506), (2013, 11.460)

a. Use the *regression* feature of a graphing utility to find a linear model for the data. Let t represent the year, with $t = 7$ corresponding to 2007.

b. How closely does the model represent the data?

When you use the *regression* feature of a graphing calculator or computer program to find a linear model for data, you will notice that the program may also output an "r-value." For instance, the r-value from Example 4 was $r \approx 0.992$. This r-value is the **correlation coefficient** of the data and gives a measure of how well the model fits the data. The correlation coefficient r varies between -1 and 1. Basically, the closer $|r|$ is to 1, the better the points can be described by a line. Three examples are shown in Figure 2.32.

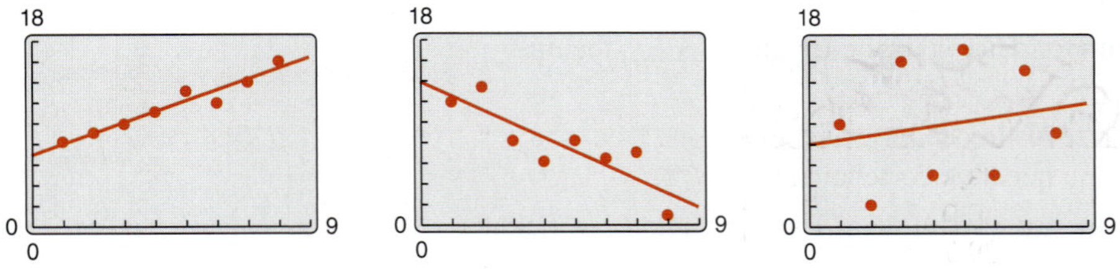

$r = 0.972$ $\qquad r = -0.856$ $\qquad r = 0.190$

Figure 2.32

Technology Tip

For some calculators, the *diagnostics on* feature must be selected before the *regression* feature is used in order to see the value of the correlation coefficient r. To learn how to use this feature, consult your user's manual.

EXAMPLE 5 Finding a Least Squares Regression Line

The ordered pairs (w, h) represent the shoe sizes w and the heights h (in inches) of 25 men. Use the *regression* feature of a graphing utility to find the least squares regression line for the data.

(10.0, 70.5)	(10.5, 71.0)	(9.5, 69.0)	(11.0, 72.0)	(12.0, 74.0)
(8.5, 67.0)	(9.0, 68.5)	(13.0, 76.0)	(10.5, 71.5)	(10.5, 70.5)
(10.0, 71.0)	(9.5, 70.0)	(10.0, 71.0)	(10.5, 71.0)	(11.0, 71.5)
(12.0, 73.5)	(12.5, 75.0)	(11.0, 72.0)	(9.0, 68.0)	(10.0, 70.0)
(13.0, 75.5)	(10.5, 72.0)	(10.5, 71.0)	(11.0, 73.0)	(8.5, 67.5)

Solution

After entering the data into a graphing utility (see Figure 2.33), you obtain the model shown in Figure 2.34. So, the least squares regression line for the data is

$$h = 1.84w + 51.9.$$

In Figure 2.35, this line is plotted with the data. Note that the plot does not have 25 points because some of the ordered pairs graph as the same point. The correlation coefficient for this model is $r \approx 0.981$, which implies that the model is a good fit for the data.

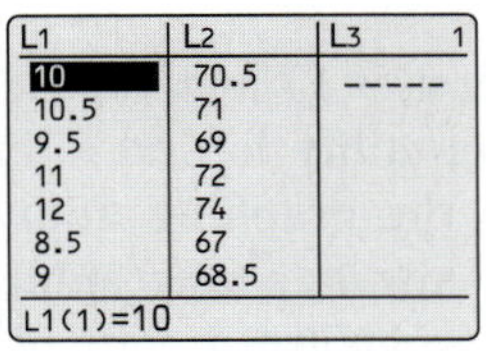

Figure 2.33

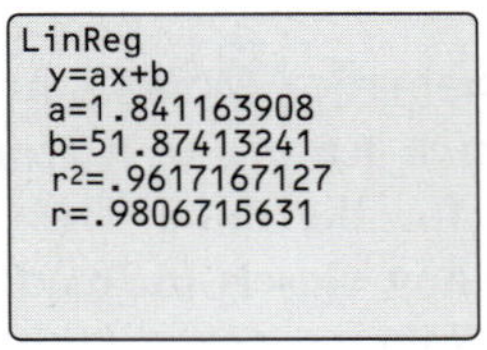

Figure 2.34

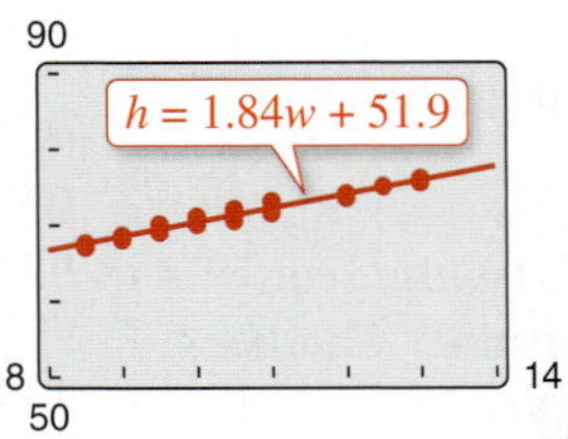

Figure 2.35

✓ **Checkpoint** Audio-video solution in English & Spanish at LarsonPrecalculus.com.

The ordered pairs (g, e) represent the gross domestic products g (in millions of U.S. dollars) and the carbon dioxide emissions e (in millions of metric tons) for seven countries. *(Spreadsheet at LarsonPrecalculus.com)* Use the *regression* feature of a graphing utility to find the least squares regression line for the data. (Sources: World Bank and U.S. Energy Information Administration)

DATA (14.99, 5491), (5.87, 1181), (3.6, 748), (2.77, 374), (2.48, 475), (2.45, 497), (2.19, 401)

2.7 Exercises

See CalcChat.com for tutorial help and worked-out solutions to odd-numbered exercises.
For instructions on how to use a graphing utility, see Appendix A.

Vocabulary and Concept Check

In Exercises 1 and 2, fill in the blank.

1. Consider a collection of ordered pairs of the form (x, y). If y tends to increase as x increases, then the collection is said to have a _______ correlation.

2. To find the least squares regression line for data, you can use the _______ feature of a graphing utility.

3. In a collection of ordered pairs (x, y), y tends to decrease as x increases. Does the collection have a positive correlation or a negative correlation?

4. You find the least squares regression line for a set of data. The correlation coefficient is 0.114. Is the model a good fit?

Procedures and Problem Solving

5. Constructing a Scatter Plot The following ordered pairs give the years of experience x for 15 sales representatives and the monthly sales y (in thousands of dollars).

(1.5, 41.7), (1.0, 32.4), (0.3, 19.2), (3.0, 48.4), (4.0, 51.2), (0.5, 28.5), (2.5, 50.4), (1.8, 35.5), (2.0, 36.0), (1.5, 40.0), (3.5, 50.3), (4.0, 55.2), (0.5, 29.1), (2.2, 43.2), (2.0, 41.6)

(a) Create a scatter plot of the data.

(b) Does the relationship between x and y appear to be approximately linear? Explain.

6. Constructing a Scatter Plot The following ordered pairs give the scores on two consecutive 15-point quizzes for a class of 18 students.

(7, 13), (9, 7), (14, 14), (15, 15), (10, 15), (9, 7), (14, 11), (14, 15), (8, 10), (9, 10), (15, 9), (10, 11), (11, 14), (7, 14), (11, 10), (14, 11), (10, 15), (9, 6)

(a) Create a scatter plot of the data.

(b) Does the relationship between consecutive quiz scores appear to be approximately linear? If not, give some possible explanations.

Interpreting Correlation **In Exercises 7–10, the scatter plot of a set of data is shown. Determine whether the points are positively correlated, are negatively correlated, or have no discernible correlation.**

7.

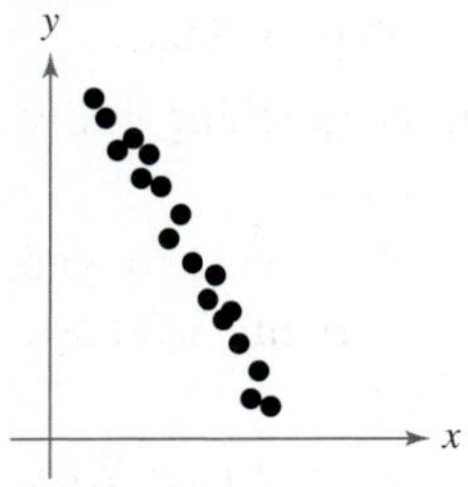

8.

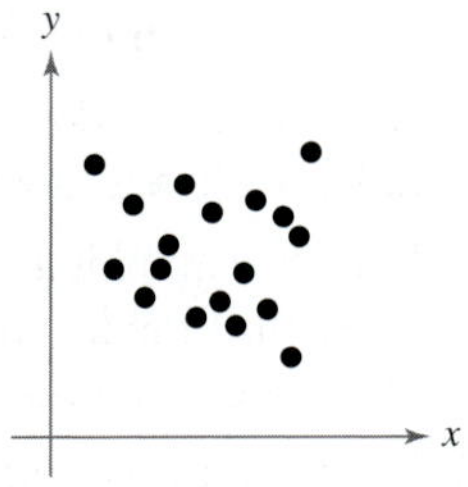

9.

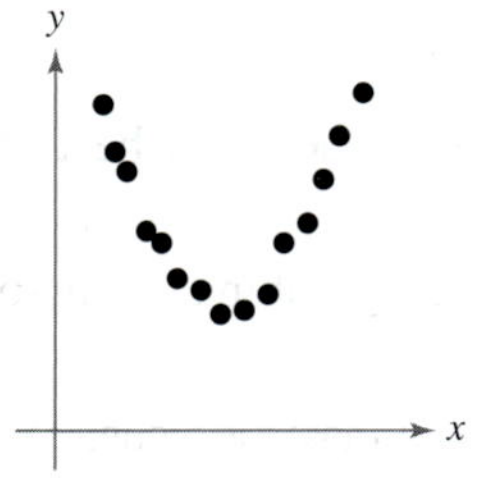

10.

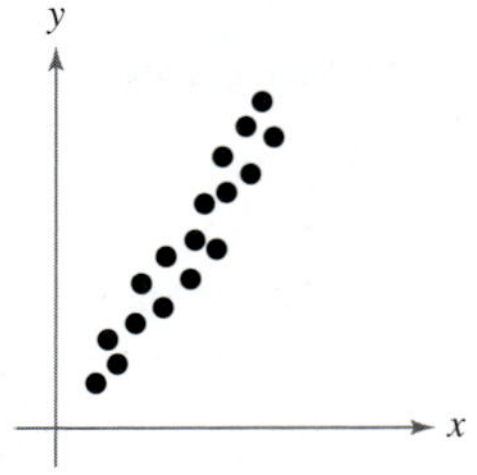

Fitting a Line to Data **In Exercises 11–16, (a) create a scatter plot of the data, (b) draw a line of fit that passes through two of the points, and (c) use the two points to find an equation of the line.**

11. $(-3, -3), (3, 4), (1, 1), (3, 2), (4, 4), (-1, -1)$

12. $(-2, 3), (-2, 4), (-1, 2), (1, -2), (0, 0), (0, 1)$

13. $(0, 2), (-2, 1), (3, 3), (1, 3), (4, 4)$

14. $(3, 2), (2, 3), (1, 5), (4, 0), (5, 0)$

15. $(0, 7), (3, 2), (6, 0), (4, 3), (2, 5)$

16. $(3, 4), (2, 2), (5, 6), (1, 1), (0, 2)$

A Mathematical Model **In Exercises 17 and 18, use the *regression* feature of a graphing utility to find a linear model for the data. Then use the graphing utility to decide how closely the model fits the data (a) graphically and (b) numerically. To print an enlarged copy of the graph, go to *MathGraphs.com*.**

17.

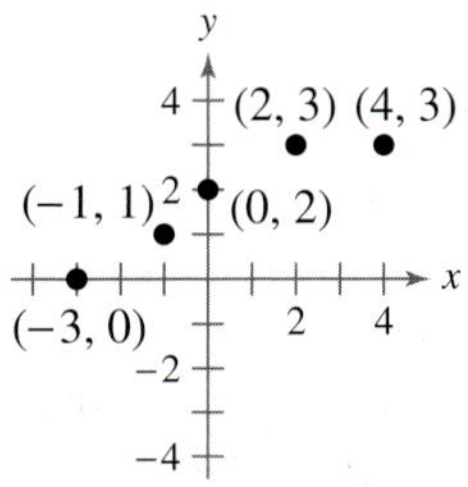

18.

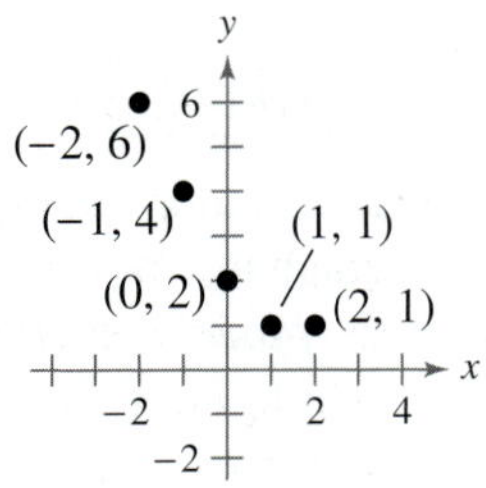

19. MODELING DATA

Hooke's Law states that the force F required to compress or stretch a spring (within its elastic limits) is proportional to the distance d that the spring is compressed or stretched from its original length. That is, $F = kd$, where k is the measure of the stiffness of the spring and is called the *spring constant*. The table shows the elongation d in centimeters of a spring when a force of F kilograms is applied.

Force, F	Elongation, d
20	1.4
40	2.5
60	4.0
80	5.3
100	6.6

(a) Sketch a scatter plot of the data.

(b) Find the equation of the line that seems to best fit the data.

(c) Use the *regression* feature of a graphing utility to find a linear model for the data.

(d) Use the model from part (c) to estimate the elongation of the spring when a force of 55 kilograms is applied.

20. MODELING DATA

The numbers of subscribers S (in millions) to wireless networks from 2006 through 2012 are shown in the table. (Source: CTIA–The Wireless Association)

DATA

Year	Subscribers, S (in millions)
2006	233.0
2007	255.4
2008	270.3
2009	285.6
2010	296.3
2011	316.0
2012	326.5

Spreadsheet at LarsonPrecalculus.com

(a) Use a graphing utility to create a scatter plot of the data, with $t = 6$ corresponding to 2006.

(b) Use the *regression* feature of the graphing utility to find a linear model for the data.

(c) Use the graphing utility to plot the data and graph the model in the same viewing window. Is the model a good fit? Explain.

(d) Use the model to predict the number of subscribers in 2020. Is your answer reasonable? Explain.

21. MODELING DATA

The total enterprise values V (in millions of dollars) for the Pittsburgh Penguins from 2008 through 2012 are shown in the table. (Source: Forbes)

Year	Total enterprise value, V (in millions of dollars)
2008	195
2009	222
2010	235
2011	264
2012	288

(a) Use a graphing utility to create a scatter plot of the data, with $t = 8$ corresponding to 2008.

(b) Use the *regression* feature of the graphing utility to find a linear model for the data.

(c) Use the graphing utility to plot the data and graph the model in the same viewing window. Is the model a good fit? Explain.

(d) Use the model to predict the values in 2015 and 2020. Do the results seem reasonable? Explain.

(e) What is the slope of your model? What does it tell you about the enterprise value?

22. MODELING DATA

The mean salaries S (in thousands of dollars) of public school teachers in the United States from 2005 through 2011 are shown in the table. (Source: National Education Association)

DATA

Year	Mean salary, S (in thousands of dollars)
2005	47.5
2006	49.1
2007	51.1
2008	52.8
2009	54.3
2010	55.5
2011	57.2

Spreadsheet at LarsonPrecalculus.com

(a) Use a graphing utility to create a scatter plot of the data, with $t = 5$ corresponding to 2005.

(b) Use the *regression* feature of the graphing utility to find a linear model for the data.

(c) Use the graphing utility to plot the data and graph the model in the same viewing window. Is the model a good fit? Explain.

(d) Use the model to predict the mean salaries in 2020 and 2022. Do the results seem reasonable? Explain.

23. MODELING DATA

The populations P (in thousands) of New Jersey from 2008 through 2013 are shown in the table. (Source: U.S. Census Bureau)

DATA

Year	Population, P (in thousands)
2008	8711
2009	8756
2010	8792
2011	8821
2012	8868
2013	8899

Spreadsheet at LarsonPrecalculus.com

(a) Use a graphing utility to create a scatter plot of the data, with $t = 8$ corresponding to 2008.

(b) Use the *regression* feature of the graphing utility to find a linear model for the data.

(c) Use the graphing utility to plot the data and graph the model in the same viewing window.

(d) Create a table showing the actual values of P and the values of P given by the model. How closely does the model fit the data?

(e) Use the model to predict the population of New Jersey in 2050. Does the result seem reasonable? Explain.

24. MODELING DATA

The populations P (in thousands) of Wyoming from 2009 through 2013 are shown in the table. (Source: U.S. Census Bureau)

Year	Population, P (in thousands)
2009	560
2010	564
2011	568
2012	577
2013	583

(a) Use a graphing utility to create a scatter plot of the data, with $t = 9$ corresponding to 2009.

(b) Use the *regression* feature of the graphing utility to find a linear model for the data.

(c) Use the graphing utility to plot the data and graph the model in the same viewing window.

(d) Create a table showing the actual values of P and the values of P given by the model. How closely does the model fit the data?

(e) Use the model to predict the population of Wyoming in 2050. Does the result seem reasonable? Explain.

25. MODELING DATA

The table shows the advertising expenditures x and sales volumes y for a company for nine randomly selected months. Both are measured in thousands of dollars.

DATA

Month	Advertising expenditures, x	Sales volume, y
1	2.4	215
2	1.6	182
3	1.4	172
4	2.3	214
5	2.0	201
6	2.6	230
7	1.4	176
8	1.6	177
9	2.0	196

Spreadsheet at LarsonPrecalculus.com

(a) Use the *regression* feature of a graphing utility to find a linear model for the data.

(b) Use the graphing utility to plot the data and graph the model in the same viewing window.

(c) Interpret the slope of the model in the context of the problem.

(d) Use the model to estimate sales for advertising expenditures of \$1500.

26. MODELING DATA

The table shows the numbers T of stores owned by the Target Corporation from 2006 through 2012. (Source: Target Corp.)

DATA

Year	Number of stores, T
2006	1488
2007	1591
2008	1682
2009	1740
2010	1750
2011	1763
2012	1778

Spreadsheet at LarsonPrecalculus.com

(a) Use a graphing utility to make a scatter plot of the data, with $t = 6$ corresponding to 2006. Identify two sets of points in the scatter plot that are approximately linear.

(b) Use the *regression* feature of the graphing utility to find a linear model for each set of points.

(c) Write a piecewise-defined model for the data. Use the graphing utility to graph the piecewise-defined model.

(d) Describe a scenario that could be the cause of the break in the data.

27. ***Why you should learn it*** *(p. 223)* The following ordered pairs (t, T) represent the Olympic year t and the winning time T (in minutes) in the women's 400-meter freestyle swimming event. *(Spreadsheet at LarsonPrecalculus.com)* (Source: International Olympic Committee)

DATA

(1956, 4.91)	(1988, 4.06)
(1960, 4.84)	(1992, 4.12)
(1964, 4.72)	(1996, 4.12)
(1968, 4.53)	(2000, 4.10)
(1972, 4.32)	(2004, 4.09)
(1976, 4.16)	(2008, 4.05)
(1980, 4.15)	(2012, 4.02)
(1984, 4.12)	

(a) Use the *regression* feature of a graphing utility to find a linear model for the data and to identify the correlation coefficient. Let t represent the year, with $t = 0$ corresponding to 1950.

(b) What information is given by the sign of the slope of the model?

(c) Use the graphing utility to plot the data and graph the model in the same viewing window.

(d) Create a table showing the actual values of y and the values of y given by the model. How closely does the model fit the data?

(e) How can you use the value of the correlation coefficient to help answer the question in part (d)?

(f) Would you use the model to predict the winning times in the future? Explain.

28. MODELING DATA

In a study, 60 colts were measured every 14 days from birth. The ordered pairs (d, l) represent the average length l (in centimeters) of the 60 colts d days after birth. *(Spreadsheet at LarsonPrecalculus.com)* (Source: American Society of Animal Science)

DATA

(14, 81.2)	(28, 87.1)	(42, 93.7)
(56, 98.3)	(70, 102.4)	(84, 106.2)
(98, 110.0)		

(a) Use the *regression* feature of a graphing utility to find a linear model for the data and to identify the correlation coefficient.

(b) According to the correlation coefficient, does the model represent the data well? Explain.

(c) Use the graphing utility to plot the data and graph the model in the same viewing window. How closely does the model fit the data?

(d) Use the model to predict the average length of a colt 112 days after birth.

Conclusions

True or False? **In Exercises 29 and 30, determine whether the statement is true or false. Justify your answer.**

29. A linear regression model with a positive correlation will have a slope that is greater than 0.

30. When the correlation coefficient for a linear regression model is close to -1, the regression line is a poor fit for the data.

31. Writing Use your school's library, the Internet, or some other reference source to locate data that you think describe a linear relationship. Create a scatter plot of the data and find the least squares regression line that represents the points. Interpret the slope and y-intercept in the context of the data. Write a summary of your findings.

32. **HOW DO YOU SEE IT?** Each graphing utility screen below shows the least squares regression line for a set of data. The equations and r-values for the models are given.

$y = 0.68x + 2.7$
$y = 0.41x + 2.7$
$y = -0.62x + 10.0$
$r = 0.973$
$r = -0.986$
$r = 0.624$

(i)

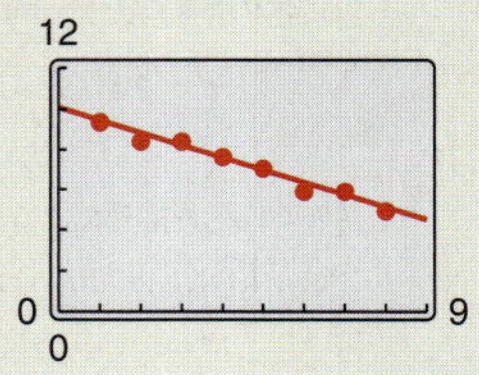

(ii)

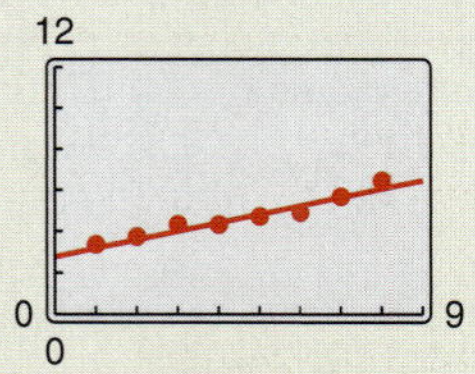

(iii)

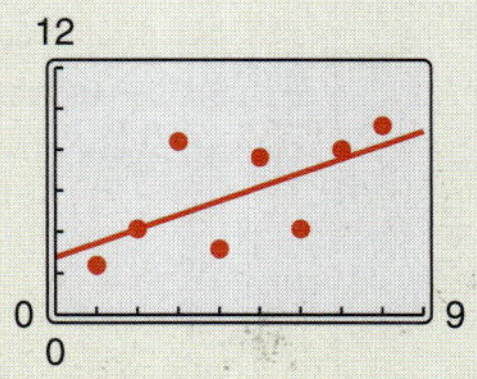

(a) Determine the equation and correlation coefficient (r-value) that represent each graph. Explain how you found your answers.

(b) According to the correlation coefficients, which model is the best fit for its data? Explain.

Cumulative Mixed Review

Evaluating a Function **In Exercises 33 and 34, evaluate the function at each value of the independent variable and simplify.**

33. $f(x) = 2x^2 - 3x + 5$
(a) $f(-1)$
(b) $f(w + 2)$

34. $g(x) = 5x^2 - 6x + 1$
(a) $g(-2)$
(b) $g(z - 2)$

Solving Equations **In Exercises 35–38, solve the equation algebraically. Check your solution graphically.**

35. $6x + 1 = -9x - 8$

36. $3(x - 3) = 7x + 2$

37. $8x^2 - 10x - 3 = 0$

38. $10x^2 - 23x - 5 = 0$

2 Chapter Summary

	What did you learn?	Explanation and Examples	Review Exercises
2.1	**Solve equations involving fractional expressions *(p. 160)*.**	To solve an equation involving fractional expressions, find the least common denominator (LCD) of all terms in the equation and multiply every term by this LCD.	1–12
	Write and use mathematical models to solve real-life problems *(p. 162)*.	To form a mathematical model, begin by using a verbal description of the problem to form a verbal model. Then, after assigning labels to the quantities in the verbal model, write the algebraic equation.	13–16
	Use common formulas to solve real-life problems *(p. 164)*.	Many common types of geometric, scientific, and investment problems use ready-made equations called formulas. Knowing these formulas will help you translate and solve a wide variety of real-life applications. (See page 164.)	17, 18
2.2	**Find x- and y-intercepts of graphs of equations *(p. 170)*.**	The point $(a, 0)$ is an x-intercept and the point $(0, b)$ is a y-intercept of the graph of $y = f(x)$.	19–22
	Find solutions of equations graphically *(p. 172)*.	**1.** Write the equation in general form, $f(x) = 0$, with the nonzero terms on one side of the equation and zero on the other side. **2.** Use a graphing utility to graph the function $y = f(x)$. Be sure the viewing window shows all the relevant features of the graph. **3.** Use the *zero* or *root* feature or the *zoom* and *trace* features of the graphing utility to approximate the x-intercepts of the graph of f.	23–28
	Find the points of intersection of two graphs *(p. 174)*.	To find the points of intersection of the graphs of two equations, solve each equation for y (or x) and set the two results equal to each other. The resulting equation will be an equation in one variable, which can be solved using standard procedures.	29–32
2.3	**Use the imaginary unit i to write complex numbers *(p. 180)*.**	The imaginary unit i is defined as $i = \sqrt{-1}$. If a and b are real numbers, the number $a + bi$ is a complex number, and it is written in standard form.	33–36
	Add, subtract, and multiply complex numbers *(p. 181)*.	**Sum:** $(a + bi) + (c + di) = (a + c) + (b + d)i$ **Difference:** $(a + bi) - (c + di) = (a - c) + (b - d)i$	37–48
	Use complex conjugates to write the quotient of two complex numbers in standard form *(p. 183)*.	Complex numbers of the forms $a + bi$ and $a - bi$ are complex conjugates. To write $(a + bi)/(c + di)$ in standard form, multiply by $(c - di)/(c - di)$.	49–52
2.4	**Solve quadratic equations by factoring *(p. 186)*, extracting square roots *(p. 188)*, completing the square *(p. 189)*, and using the Quadratic Formula *(p. 191)*.**	**Factoring:** If $ab = 0$, then $a = 0$ or $b = 0$. **Extracting Square Roots:** If $u^2 = c$, then $u = \pm\sqrt{c}$. **Completing the Square:** If $x^2 + bx = c$, then $x^2 + bx + \left(\frac{b}{2}\right)^2 = c + \left(\frac{b}{2}\right)^2$ $\left(x + \frac{b}{2}\right)^2 = c + \frac{b^2}{4}$. **Quadratic Formula:** If $ax^2 + bx + c = 0$, then $x = \frac{-b \pm \sqrt{b^2 - 4ac}}{2a}$.	53–78

	What did you learn?	Explanation and Examples	Review Exercises
2.4	**Use quadratic equations to model and solve real-life problems *(p. 193).***	A quadratic equation can be used to model the numbers of U.S. adults who use the Internet from 1995 to 2014. (See Example 11.)	79, 80
2.5	**Solve polynomial equations of degree three or greater *(p. 200).***	To solve a polynomial equation, factor if possible. Then use the methods used in solving linear and quadratic equations.	81–88
	Solve equations involving radicals *(p. 202).*	To solve an equation involving a radical, isolate the radical on one side of the equation and raise each side to an appropriate power.	89–98
	Solve equations involving fractions or absolute values *(p. 203)*, and use polynomial equations and equations involving radicals to model and solve real-life problems *(p. 205).*	• To solve an equation with a fraction, multiply each term by the LCD, then solve the resulting equation. • To solve an equation involving an absolute value, isolate the absolute value term on one side of the equation. Then set up two equations, one where the absolute value term is positive and one where the absolute value term is negative. Solve both equations.	99–116
2.6	**Use properties of inequalities to solve linear inequalities *(p. 210).***	**1. Transitive Property** $a < b$ and $b < c$ $\Rightarrow$ $a < c$ **2. Addition of Inequalities** $a < b$ and $c < d$ $\Rightarrow$ $a + c < b + d$ **3. Addition of a Constant** $a < b$ $\Rightarrow$ $a + c < b + c$ **4. Multiplying by a Constant** For $c > 0$, $a < b$ $\Rightarrow$ $ac < bc$ For $c < 0$, $a < b$ $\Rightarrow$ $ac > bc$ Each of these properties is true when the symbol $<$ is replaced by $\leq$ and $>$ is replaced by $\geq$.	117–122
	Solve inequalities involving absolute values *(p. 213).*	To solve an inequality involving an absolute value, rewrite the inequality as a double inequality or as a compound inequality.	123–128
	Solve polynomial inequalities *(p. 214).*	To solve a polynomial inequality, write the polynomial in general form, find all the real zeros (key numbers) of the polynomial, and test the intervals bounded by the key numbers to determine the intervals that are solutions of the polynomial inequality.	129–134
	Solve rational inequalities *(p. 217).*	To solve a rational inequality, find the x-values for which the rational expression is 0 or undefined (key numbers) and test the intervals bounded by the key numbers to determine the intervals that are solutions of the rational inequality.	135–138
	Use inequalities to model and solve real-life problems *(p. 218).*	A polynomial inequality can be used to model the position of a projectile to estimate the time period when the height of the projectile will exceed 2000 feet. (See Example 11.)	139–142
2.7	**Construct scatter plots *(p. 223)* and interpret correlation *(p. 224).***	A scatter plot is a graphical representation of data written as a set of ordered pairs.	143, 144
	Use scatter plots *(p. 225)* and a graphing utility *(p. 226)* to find linear models for data.	The best-fitting linear model can be found using the *linear regression* feature of a graphing utility or a computer program.	145, 146

2 Review Exercises

See *CalcChat.com* for tutorial help and worked-out solutions to odd-numbered exercises. For instructions on how to use a graphing utility, see Appendix A.

2.1

Checking Solutions of an Equation **In Exercises 1 and 2, determine whether each value of x is a solution of the equation.**

Equation	*Values*	
1. $6 + \dfrac{3}{x-4} = 5$	(a) $x = \frac{11}{2}$	(b) $x = 0$
	(c) $x = -2$	(d) $x = 1$
2. $6 + \dfrac{2}{x+3} = \dfrac{6x+1}{3}$	(a) $x = -3$	(b) $x = 3$
	(c) $x = 0$	(d) $x = -\frac{2}{3}$

Solving Equations **In Exercises 3–12, solve the equation (if possible). Then use a graphing utility to verify your solution.**

3. $\dfrac{x}{18} + \dfrac{x}{10} = 7$

4. $\dfrac{x}{2} + \dfrac{x}{7} = 9$

5. $\dfrac{5}{x-2} = \dfrac{13}{2x-3}$

6. $\dfrac{10}{x+1} = \dfrac{12}{3x-2}$

7. $14 + \dfrac{2}{x-1} = 10$

8. $10 + \dfrac{2}{x-1} = 4$

9. $6 - \dfrac{4}{x} = 6 + \dfrac{5}{x}$

10. $2 - \dfrac{1}{x^2} = 2 + \dfrac{4}{x^2}$

11. $\dfrac{9x}{3x-1} - \dfrac{4}{3x+1} = 3$

12. $\dfrac{5}{x-5} + \dfrac{1}{x+5} = \dfrac{2}{x^2-25}$

13. Retail Management In October, a greeting card company's total profit was 12% more than it was in September. The total profit for the two months was \$689,000. Find the profit for each month.

14. Chemistry A car radiator contains 10 liters of a 30% antifreeze solution. How many liters will have to be replaced with pure antifreeze so that the resulting solution is 50% antifreeze?

15. Dendrology To obtain the height of a tree, you measure the tree's shadow and find that it is 8 meters long. You also measure the shadow of a two-meter lamppost and find that it is 75 centimeters long.

(a) Draw a diagram that illustrates the problem. Let h represent the height of the tree.

(b) Find the height of the tree in meters.

16. Finance You invest \$12,000 in a fund paying $2\frac{1}{2}\%$ simple interest and \$10,000 in a fund with a variable interest rate. At the end of the year, you are notified that the total interest for both funds is \$870. Find the equivalent simple interest rate on the variable-rate fund.

17. Meteorology The average daily temperature for the month of January in Juneau, Alaska, is 28.3°F. What is Juneau's average daily temperature for the month of January in degrees Celsius? (Source: National Oceanic and Atmospheric Administration)

18. Physical Education A basketball and a baseball have circumferences of 30 inches and $9\frac{1}{4}$ inches, respectively. Find the volume of each.

2.2

Finding *x*- and *y*-Intercepts **In Exercises 19–22, find the x- and y-intercepts of the graph of the equation.**

19. $-x + y = 3$

20. $x - 5y = 20$

21. $y = x^2 - 9x + 8$

22. $y = 25 - x^2$

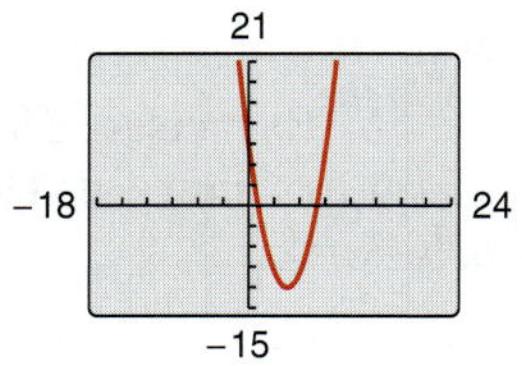

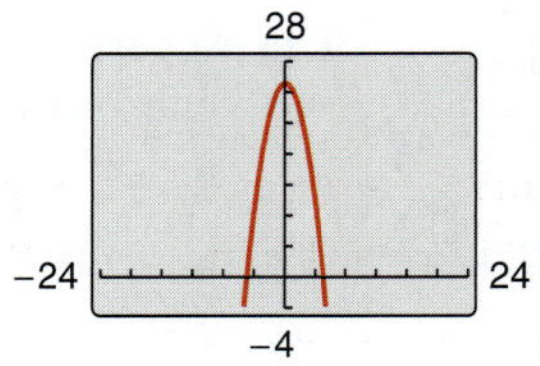

Finding Real Solutions of an Equation Graphically **In Exercises 23–28, use a graphing utility to approximate the real solutions of the equation. [Remember to write the equation in the form $f(x) = 0$.]**

23. $5(x-2) - 1 = 0$

24. $12 - 5(x-7) = 0$

25. $3x^2 + 3 = 5x$

26. $\frac{1}{3}x^3 + 4 = 2x$

27. $0.3x^4 = 5x + 2$

28. $0.8x^4 = 0.5x - 5$

Approximating Points of Intersection Graphically **In Exercises 29–32, use a graphing utility to approximate any points of intersection of the graphs of the equations. Check your results algebraically.**

29. $3x + 5y = -7$
$-x - 2y = 3$

30. $x - y = 3$
$2x + y = 12$

31. $4x^2 + 2y = 14$
$2x + y = 3$

32. $y = -x + 7$
$y = 2x^3 - x + 9$

2.3

Writing a Complex Number in Standard Form **In Exercises 33–36, write the complex number in standard form.**

33. $3 - \sqrt{-16}$

34. $\sqrt{-50} + 8$

35. $-i + 4i^2$

36. $7i - 9i^2$

Operations with Complex Numbers **In Exercises 37–48, perform the operation(s) and write the result in standard form.**

37. $(2 + 13i) + (6 - 5i)$

38. $\left(\frac{1}{2}+\frac{\sqrt{3}}{4}i\right)-\left(\frac{1}{2}-\frac{\sqrt{3}}{4}i\right)$

39. $5i(13-8i)$

40. $(1+6i)(5-2i)$

41. $(\sqrt{-16}+3)(\sqrt{-25}-2)$

42. $(5-\sqrt{-4})(5+\sqrt{-4})$

43. $\sqrt{-9}+3+\sqrt{-36}$

44. $7-\sqrt{-81}+\sqrt{-49}$

45. $(10-8i)(2-3i)$

46. $i(6+i)(3-2i)$

47. $(3+7i)^2+(3-7i)^2$

48. $(4-i)^2-(4+i)^2$

Writing a Quotient of Complex Numbers in Standard Form **In Exercises 49–52, write the quotient in standard form.**

49. $\frac{6+i}{i}$

50. $\frac{4}{-3i}$

51. $\frac{3+2i}{5+i}$

52. $\frac{1-7i}{2+3i}$

2.4

Solving a Quadratic Equation **In Exercises 53–78, solve the equation using any convenient method. Use a graphing utility to verify your solution(s).**

53. $9x^2 = 49$

54. $8x = 2x^2$

55. $6x = 3x^2$

56. $16x^2 = 25$

57. $x^2 - 7x - 8 = 0$

58. $x^2 + 3x - 18 = 0$

59. $(x+4)^2 = 18$

60. $(x+1)^2 = 24$

61. $(3x-1)^2 + 4 = 0$

62. $(5x-3)^2 + 16 = 0$

63. $x^2 + 4x - 9 = 0$

64. $x^2 - 6x - 5 = 0$

65. $x^2 - 12x + 30 = 0$

66. $x^2 + 6x - 3 = 0$

67. $x^2 - 3x = 28$

68. $x^2 + 3x = 40$

69. $x^2 - 10x = 9$

70. $x^2 + 8x = 7$

71. $2x^2 + 9x - 5 = 0$

72. $4x^2 + x - 5 = 0$

73. $-x^2 - x + 15 = 0$

74. $-x^2 - 3x + 2 = 0$

75. $x^2 + 4x + 10 = 0$

76. $x^2 + 6x - 1 = 0$

77. $2x^2 - 6x + 21 = 0$

78. $2x^2 - 8x + 11 = 0$

79. **Medicine** The average costs per day C (in dollars) for hospital care from 2000 through 2011 in the U.S. can be approximated by the model

$$C = -0.54t^2 + 82.6t + 1136, \quad 0 \le t \le 11$$

where t is the year, with $t = 0$ corresponding to 2000. (Source: Health Forum)

(a) Use a graphing utility to graph the model in an appropriate viewing window.

(b) Use the *zoom* and *trace* features of the graphing utility to estimate when the cost per day reached \$1800.

(c) Algebraically find when the cost per day reached \$1800.

(d) Do you believe the model could be used to predict the cost per day for years beyond 2011? Explain.

80. **Business** The revenues R (in millions of dollars) for Priceline Group from 2004 through 2013 can be approximated by the model

$$R = 86.727t^2 - 839.83t + 2967.9, \quad 4 \le t \le 13$$

where t is the year, with $t = 4$ corresponding to 2004. (Source: Priceline Group)

(a) Use a graphing utility to graph the model in an appropriate viewing window.

(b) Use the *zoom* and *trace* features of the graphing utility to estimate when the revenue reached 4 billion dollars.

(c) Algebraically find when the revenue reached 4 billion dollars.

(d) Use the model to predict when the revenue will reach 10 billion dollars.

2.5

Solving Equations **In Exercises 81–108, find all solutions of the equation algebraically. Use a graphing utility to verify the solutions graphically.**

81. $3x^3 - 26x^2 + 16x = 0$

82. $36x^3 - x = 0$

83. $5x^4 - 12x^3 = 0$

84. $4x^3 - 6x^2 = 0$

85. $x^4 - x^2 - 12 = 0$

86. $x^4 - 4x^2 - 5 = 0$

87. $2x^4 - 22x^2 = -56$

88. $3x^4 + 18x^2 = -24$

89. $\sqrt{x+4} = 3$

90. $\sqrt{x-2} - 8 = 0$

91. $2x - 5\sqrt{x} + 3 = 0$

92. $\sqrt{3x-2} = 4 - x$

93. $\sqrt{2x+3} + \sqrt{x-2} = 2$

94. $5\sqrt{x} - \sqrt{x-1} = 6$

95. $(x-1)^{2/3} - 25 = 0$

96. $(x+2)^{3/4} = 27$

97. $(x+4)^{1/2} + 5x(x+4)^{3/2} = 0$

98. $8x^2(x^2-4)^{1/3} + (x^2-4)^{4/3} = 0$

99. $\frac{x}{8} + \frac{3}{8} = \frac{1}{2x}$

100. $\frac{3x}{2} = \frac{1}{x} - \frac{5}{2}$

101. $\frac{5}{x} = 1 + \frac{3}{x+2}$

102. $\frac{6}{x} + \frac{8}{x+5} = 3$

103. $3 + \frac{2}{x} = \frac{16}{x^2}$

104. $\frac{2x}{x^2-1} - \frac{3}{x+1} = 1$

105. $|x-5| = 10$

106. $|2x+3| = 7$

107. $|x^2-3| = 2x$

108. $|x^2-6| = x$

109. Reduced Rates A group of investors agree to share equally in the cost of a \$240,000 apartment complex. If two more investors join the group, then each person's share of the cost will decrease by \$20,000. How many investors are presently in the group?

110. Reduced Rates A college charters a bus for \$1700 to take a group to a museum. When six more students join the trip, the cost per student drops by \$7.50. How many students were in the original group?

111. Aviation An airplane makes a commuter flight of 145 miles from Portland, Oregon, to Seattle, Washington. The return trip to Portland is 12 minutes shorter because the average speed is 40 miles per hour faster. What is the average speed of the airplane on the return trip?

112. Business You drove 56 miles one way on a service call. On the return trip, your average speed was 8 miles per hour greater and the trip took 10 fewer minutes. What was your average speed on the return trip?

113. Finance A deposit of \$1000 in a mutual fund reaches a balance of \$1196.95 after 6 years. What annual interest rate on a certificate of deposit compounded monthly would yield an equivalent return?

114. Finance A deposit of \$1500 in a mutual fund reaches a balance of \$2465.43 after 10 years. What annual interest rate on a certificate of deposit compounded quarterly would yield an equivalent return?

115. MODELING DATA

The numbers of students S (in millions) enrolled in private, nonprofit colleges in the United States from 2000 through 2012 can be modeled by the equation

$$S = \sqrt{0.51049t + 9.5287}, \quad 0 \le t \le 12$$

where t is the year, with $t = 0$ corresponding to 2000. (Source: U.S. National Center for Educational Statistics)

(a) Use the *table* feature of a graphing utility to find the number of students enrolled for each year from 2000 through 2012.

(b) Use the graphing utility to graph the model in an appropriate viewing window.

(c) Use the *zoom* and *trace* features of the graphing utility to find when enrollment reached 3.5 million.

(d) Algebraically confirm your approximation in part (c).

(e) Use the model to predict when the enrollment will reach 6 million. Does this answer seem reasonable?

(f) Do you believe the enrollment population will ever reach 7 million? Explain your reasoning.

116. MODELING DATA

The populations P (in millions) of New York City from 2003 through 2012 can be modeled by the equation

$$P = \sqrt{0.13296t^2 - 1.4650t + 68.243}, \quad 3 \le t \le 12$$

where t is the year, with $t = 3$ corresponding to 2003. (Source: U.S. Census Bureau)

(a) Use the *table* feature of a graphing utility to find the population of New York City for each year from 2003 through 2012.

(b) Use the graphing utility to graph the model in an appropriate viewing window.

(c) Use the *zoom* and *trace* features of the graphing utility to find when the population reached 8.2 million.

(d) Algebraically confirm your approximation in part (c).

(e) Use the model to predict when the population will reach 8.5 million. Does this answer seem reasonable?

(f) Do you believe the population will ever reach 15 million? Explain your reasoning.

2.6

Solving an Inequality **In Exercises 117–138, solve the inequality and sketch the solution on the real number line. Use a graphing utility to verify your solution graphically.**

117. $8x - 3 < 6x + 15$

118. $9x - 8 \le 7x + 16$

119. $\frac{1}{2}(3 - x) > \frac{1}{3}(2 - 3x)$

120. $4(5 - 2x) \ge \frac{1}{2}(8 - x)$

121. $-2 < -x + 7 \le 10$

122. $-6 \le 3 - 2(x - 5) < 14$

123. $|x| \le 4$

124. $|x - 2| < 1$

125. $\left|x - \frac{3}{2}\right| > \frac{3}{2}$

126. $|x - 3| \ge 4$

127. $4|3 - 2x| \le 16$

128. $|x + 9| + 7 > 19$

129. $x^2 - 2x > 3$

130. $x^2 - 6x - 27 < 0$

131. $4x^2 - 23x \le 6$

132. $6x^2 + 5x \ge 4$

133. $x^3 - 16x \ge 0$

134. $12x^3 - 20x^2 < 0$

135. $\frac{1}{x} + 3 > 0$

136. $\frac{9}{x} \ge x$

137. $\frac{3x + 8}{x - 3} \le 4$

138. $\frac{x + 8}{x + 5} - 2 < 0$

Finding the Domain of an Expression **In Exercises 139 and 140, find the domain of x in the expression.**

139. $\sqrt{16 - x^2}$

140. $\sqrt[4]{x^2 - 5x - 14}$

141. Sales You stop at a gas station to buy 15 gallons of 87-octane gasoline at $3.69 a gallon. The gas pump is accurate to within $\frac{1}{10}$ of a gallon. How much might you be overcharged or undercharged?

142. Meteorology An electronic device is to be operated in an environment with relative humidity h in the interval defined by $|h - 50| \le 30$. What are the minimum and maximum relative humidities for the operation of this device?

2.7

143. Education The following ordered pairs give the entrance exam scores x and the grade-point averages y after 1 year of college for 10 students. *(Spreadsheet at LarsonPrecalculus.com)*

DATA (75, 2.3), (82, 3.0), (90, 3.6), (65, 2.0), (70, 2.1), (88, 3.5), (93, 3.9), (69, 2.0), (80, 2.8), (85, 3.3)

(a) Create a scatter plot of the data.

(b) Does the relationship between x and y appear to be approximately linear? Explain.

144. Industrial Engineering A machine part was tested by bending it x centimeters 10 times per minute until it failed (y equals the time to failure in hours). The results are given as the following ordered pairs. *(Spreadsheet at LarsonPrecalculus.com)*

DATA (3, 61), (6, 56), (9, 53), (12, 55), (15, 48), (18, 35), (21, 36), (24, 33), (27, 44), (30, 23)

(a) Create a scatter plot of the data.

(b) Does the relationship between x and y appear to be approximately linear? If not, give some possible explanations.

145. MODELING DATA

In an experiment, students measured the speed s (in meters per second) of a ball t seconds after it was released. The results are shown in the table.

Time, t	Speed, s
0	0
1	11.0
2	19.4
3	29.2
4	39.4

(a) Sketch a scatter plot of the data.

(b) Find the equation of the line that seems to best fit the data.

(c) Use the *regression* feature of a graphing utility to find a linear model for the data and to identify the correlation coefficient.

(d) Use the model from part (c) to estimate the speed of the ball after 2.5 seconds.

146. MODELING DATA

The following ordered pairs (x, y) represent the Olympic year x and the winning time y (in minutes) in the men's 400-meter freestyle swimming event. *(Spreadsheet at LarsonPrecalculus.com)* (Source: International Olympic Committee)

DATA
(1968, 4.150) (1984, 3.854) (2000, 3.677)
(1972, 4.005) (1988, 3.783) (2004, 3.718)
(1976, 3.866) (1992, 3.750) (2008, 3.698)
(1980, 3.855) (1996, 3.800) (2012, 3.669)

(a) Use the *regression* feature of a graphing utility to find a linear model for the data. Let x represent the year, with $x = 8$ corresponding to 1968.

(b) Use the graphing utility to create a scatter plot of the data. Graph the model in the same viewing window.

(c) Is the model a good fit for the data? Explain.

(d) Is the model appropriate for predicting the winning times in future Olympics? Explain.

Conclusions

True or False? **In Exercises 147–149, determine whether the statement is true or false. Justify your answer.**

147. The graph of a function may have two distinct y-intercepts.

148. The sum of two complex numbers cannot be a real number.

149. The sign of the slope of a regression line is always positive.

150. Writing In your own words, explain the difference between an identity and a conditional equation.

151. Writing Describe the relationship among the x-intercepts of a graph, the zeros of a function, and the solutions of an equation.

152. Think About It Consider the linear equation $ax + b = 0$.

(a) What is the sign of the solution when $ab > 0$?

(b) What is the sign of the solution when $ab < 0$?

153. Error Analysis Describe the error.

$$\sqrt{-8}\sqrt{-8} = \sqrt{(-8)(-8)} = \sqrt{64} = 8$$

154. Error Analysis Describe the error.

$$-i\left(\sqrt{-4} - 1\right) = -i(4i - 1)$$
$$= -4i^2 + i$$
$$= 4 + i$$

155. Finding Powers of i Write each of the powers of i as i, $-i$, 1, or -1.

(a) i^{40} (b) i^{25} (c) i^{50} (d) i^{67}

2 Chapter Test

See *CalcChat.com* for tutorial help and worked-out solutions to odd-numbered exercises.
For instructions on how to use a graphing utility, see Appendix A.

Take this test as you would take a test in class. After you are finished, check your work against the answers given in the back of the book.

In Exercises 1 and 2, solve the equation (if possible). Then use a graphing utility to verify your solution.

1. $\frac{12}{x} - 7 = -\frac{27}{x} + 6$

2. $\frac{4}{3x - 2} - \frac{9x}{3x + 2} = -3$

In Exercises 3–6, perform the operation(s) and write the result in standard form.

3. $(-8 - 3i) + (-1 - 15i)$

4. $(10 + \sqrt{-20}) - (4 - \sqrt{-14})$

5. $(2 + i)(6 - i)$

6. $(4 + 3i)^2 - (5 + i)^2$

In Exercises 7 and 8, write the quotient in standard form.

7. $\frac{8 + 5i}{i}$

8. $(2i - 1) \div (3i + 2)$

In Exercises 9–12, use a graphing utility to approximate any solutions of the equation. [Remember to write the equation in the form $f(x) = 0$.]

9. $3x^2 - 6 = 0$

10. $8x^2 - 2 = 0$

11. $x^3 + 5x = 4x^2$

12. $x = x^3$

In Exercises 13–16, solve the equation using any convenient method. Use a graphing utility to verify the solutions graphically.

13. $x^2 - 15x + 56 = 0$

14. $x^2 + 12x - 2 = 0$

15. $4x^2 - 81 = 0$

16. $5x^2 + 7x + 6 = 0$

In Exercises 17–20, find all solutions of the equation algebraically. Use a graphing utility to verify the solutions graphically.

17. $3x^3 - 4x^2 - 12x + 16 = 0$

18. $x + \sqrt{22 - 3x} = 6$

19. $(x^2 + 6)^{2/3} = 16$

20. $|8x - 1| = 21$

In Exercises 21–24, solve the inequality and sketch the solution on the real number line. Use a graphing utility to verify your solution graphically.

21. $6x - 1 > 3x - 10$

22. $2|x - 8| < 10$

23. $6x^2 + 5x + 1 \geq 0$

24. $\frac{8 - 5x}{2 + 3x} \leq -2$

25. A projectile is fired straight upward from ground level with an initial velocity of 224 feet per second. During what time period will the height of the projectile exceed 350 feet?

26. The table shows the average monthly costs C of expanded basic cable television from 2004 through 2012, where t represents the year, with $t = 4$ corresponding to 2004. Use the *regression* feature of a graphing utility to find a linear model for the data. Use the model to estimate the year in which the average monthly cost will reach \$85. (Source: Federal Communications Commission)

DATA

Year, t	Average monthly cost, C (in dollars)
4	41.04
5	43.04
6	45.26
7	47.27
8	49.65
9	52.37
10	54.44
11	57.46
12	61.63

Spreadsheet at LarsonPrecalculus.com

Table for 26

P–2 Cumulative Test

See *CalcChat.com* for tutorial help and worked-out solutions to odd-numbered exercises. For instructions on how to use a graphing utility, see Appendix A.

Take this test to review the material in Chapters P–2. After you are finished, check your work against the answers given in the back of the book.

In Exercises 1–3, simplify the expression.

1. $\dfrac{14x^2y^{-3}}{32x^{-1}y^2}$ **2.** $8\sqrt{60} - 2\sqrt{135} - \sqrt{15}$ **3.** $\sqrt{28x^4y^3}$

In Exercises 4–6, perform the operation and simplify the result.

4. $4x - [2x + 5(2 - x)]$ **5.** $(x - 2)(x^2 + x - 3)$ **6.** $\dfrac{2}{x + 3} - \dfrac{1}{x + 1}$

In Exercises 7–9, factor the expression completely.

7. $36 - (x - 4)^2$ **8.** $x - 5x^2 - 6x^3$ **9.** $54 - 16x^3$

10. Find the midpoint of the line segment connecting the points $\left(-\frac{7}{2}, 4\right)$ and $\left(\frac{5}{2}, -8\right)$. Then find the distance between the points.

11. Write the standard form of the equation of a circle with center $\left(-\frac{1}{2}, -8\right)$ and a radius of 4.

In Exercises 12–14, use point plotting to sketch the graph of the equation.

12. $x - 3y + 12 = 0$ **13.** $y = x^2 - 9$ **14.** $y = \sqrt{4 - x}$

In Exercises 15–17, (a) write the general form of the equation of the line that satisfies the given conditions and (b) find three additional points through which the line passes.

15. The line contains the points $(-5, 8)$ and $(-1, 4)$.

16. The line contains the point $\left(-\frac{1}{2}, 1\right)$ and has a slope of -2.

17. The line has an undefined slope and contains the point $\left(-\frac{3}{7}, \frac{1}{8}\right)$.

18. Find the equation of the line that passes through the point $(2, 3)$ and is (a) parallel to and (b) perpendicular to the line $6x - y = 4$.

In Exercises 19 and 20, evaluate the function at each specified value of the independent variable and simplify.

19. $f(x) = \dfrac{x}{x - 2}$

(a) $f(5)$ (b) $f(2)$ (c) $f(5 + 4s)$

20. $f(x) = \begin{cases} 3x - 8, & x < 0 \\ x^2 + 4, & x \ge 0 \end{cases}$

(a) $f(-8)$ (b) $f(0)$ (c) $f(4)$

In Exercises 21–24, find the domain of the function.

21. $f(x) = (x + 2)(3x - 4)$

22. $f(t) = \sqrt{5 + 7t}$

23. $g(s) = \sqrt{9 - s^2}$

24. $h(x) = \dfrac{4}{5x + 2}$

25. Determine whether the function given by $g(x) = 3x - x^3$ is even, odd, or neither.

26. Does the graph at the right represent y as a function of x? Explain.

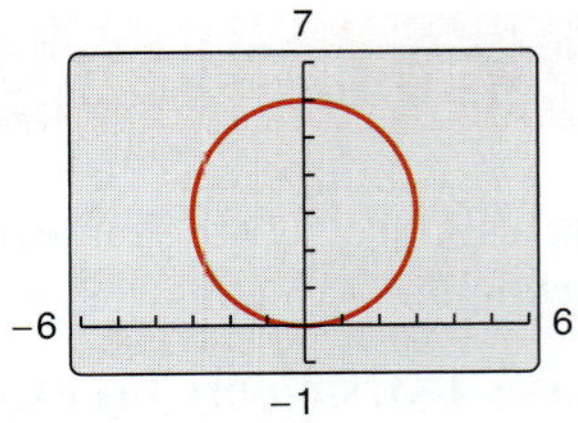

Figure for 26

27. Use a graphing utility to graph the function $f(x) = 2|x - 5| - |x + 5|$. Then determine the open intervals over which the function is increasing, decreasing, or constant.

28. Compare the graph of each function with the graph of $f(x) = \sqrt{x}$.

(a) $r(x) = \frac{1}{2}\sqrt{x}$

(b) $h(x) = \sqrt{x} + 2$

(c) $g(x) = -\sqrt{x + 2}$

In Exercises 29–32, evaluate the indicated function for $f(x) = x^2 + x$ and $g(x) = 3x - 2$.

29. $(f + g)(x)$ **30.** $(g - f)(x)$

31. $(g \circ f)(x)$ **32.** $(fg)(x)$

In Exercises 33–35, determine whether the function has an inverse function, and if so, find the inverse function.

33. $f(x) = -5x + 4$ **34.** $f(x) = (x - 1)^2$ **35.** $f(x) = \sqrt[3]{x} - 2$

In Exercises 36–39, solve the equation algebraically. Then write the equation in the form $f(x) = 0$ and use a graphing utility to verify the algebraic solution.

36. $4x^3 - 12x^2 + 8x = 0$ **37.** $\dfrac{5}{x} = \dfrac{10}{x - 3}$

38. $|3x + 4| - 2 = 0$ **39.** $\sqrt{x^2 + 1} + x - 9 = 0$

In Exercises 40–43, solve the inequality and sketch the solution on the real number line. Use a graphing utility to verify your solution graphically.

40. $\dfrac{x}{5} - 6 \le -\dfrac{x}{2} + 6$ **41.** $2x^2 + x \ge 15$

42. $|7 + 8x| > 5$ **43.** $\dfrac{2(x - 2)}{x + 1} \le 0$

44. A soccer ball has a volume of about 370.7 cubic inches. Find the radius of the soccer ball (accurate to three decimal places).

45. A rectangular plot of land with a perimeter of 546 feet has a width of x.

(a) Write the area A of the plot as a function of x.

(b) Use a graphing utility to graph the area function. What is the domain of the function?

(c) Approximate the dimensions of the plot when the area is 15,000 square feet.

46. The table shows the net profits P (in millions of dollars) for McDonald's from 2004 through 2013. (Source: McDonald's Corp.)

(a) Construct a bar graph for the data. Write a brief statement regarding the profit of McDonald's over time.

(b) Use the *regression* feature of a graphing utility to find a linear model for the data and to identify the correlation coefficient. Let t represent the year, with $t = 4$ corresponding to 2004.

(c) Use the graphing utility to plot the data and graph the model in the same viewing window.

(d) Use the model to predict the profits for McDonald's in 2015 and 2018.

(e) In your opinion, is the model appropriate for predicting future profits? Explain.

DATA

Year, t	Profit, P
2004	2458.6
2005	2509.8
2006	2873.0
2007	3522.6
2008	4201.1
2009	4405.5
2010	4961.9
2011	5503.1
2012	5464.8
2013	5585.9

Spreadsheet at LarsonPrecalculus.com

Table for 46

Proofs in Mathematics

Biconditional Statements

Recall from the Proofs in Mathematics in Chapter 1 that a conditional statement is a statement of the form "if p, then q." A statement of the form "p if and only if q" is called a **biconditional statement.** A biconditional statement, denoted by

$p \leftrightarrow q$ Biconditional statement

is the conjunction of the conditional statement $p \rightarrow q$ and its converse $q \rightarrow p$.

A biconditional statement can be either true or false. To be true, *both* the conditional statement and its converse must be true.

EXAMPLE 1 Analyzing a Biconditional Statement

Consider the statement

$x = 3$ if and only if $x^2 = 9$.

a. Is the statement a biconditional statement?

b. Is the statement true?

Solution

a. The statement is a biconditional statement because it is of the form "p if and only if q."

b. The statement can be rewritten as the following conditional statement and its converse.

Conditional statement: If $x = 3$, then $x^2 = 9$.

Converse: If $x^2 = 9$, then $x = 3$.

The first of these statements is true, but the second is false because x could also equal -3. So, the biconditional statement is false. ■

Knowing how to use biconditional statements is an important tool for reasoning in mathematics.

EXAMPLE 2 Analyzing a Biconditional Statement

Determine whether the biconditional statement is true or false. If it is false, provide a counterexample.

A number is divisible by 5 if and only if it ends in 0.

Solution

The biconditional statement can be rewritten as the following conditional statement and its converse.

Conditional statement: If a number is divisible by 5, then it ends in 0.

Converse: If a number ends in 0, then it is divisible by 5.

The conditional statement is false. A counterexample is the number 15, which is divisible by 5 but does not end in 0. So, the biconditional statement is false. ■

Progressive Summary (Chapters P–2)

This chart outlines the topics that have been covered so far in this text. Progressive Summary charts appear after Chapters 2, 4, and 7. In each Progressive Summary, new topics encountered for the first time appear in blue.

Algebraic Functions

Polynomial, Rational, Radical

Rewriting

Polynomial form ↔ Factored form
Operations with polynomials
Rationalize denominators
Simplify rational expressions
Exponent form ↔ Radical form
Operations with complex numbers

Solving

Equation	*Strategy*
Linear	Isolate variable
Quadratic	Factor, set to zero Extract square roots Complete the square Quadratic Formula
Polynomial	Factor, set to zero
Rational	Multiply by LCD
Radical	Isolate, raise to power
Absolute value	Isolate, form two equations

Analyzing

Graphically	*Algebraically*
Intercepts	Domain, Range
Symmetry	Transformations
Slope	Composition

Numerically
Table of values

Transcendental Functions

Rewriting

Solving

Analyzing

Other Topics

Rewriting

Solving

Analyzing

3 Polynomial and Rational Functions

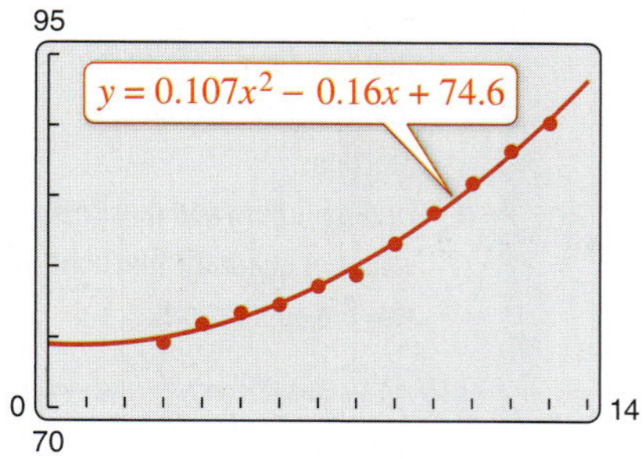

Section 3.7, Example 4
People Not in the
U.S. Labor Force

3.1 Quadratic Functions

The Graph of a Quadratic Function

In this and the next section, you will study the graphs of polynomial functions.

Definition of Polynomial Function

Let n be a nonnegative integer and let $a_n, a_{n-1}, \ldots a_2, a_1, a_0$ be real numbers with $a_n \neq 0$. The function given by

$$f(x) = a_n x^n + a_{n-1}x^{n-1} + \cdots + a_2x^2 + a_1x + a_0$$

is called a **polynomial function of x with degree n.**

Polynomial functions are classified by degree. For instance, the polynomial function

$$f(x) = a, \quad a \neq 0 \qquad \text{Constant function}$$

has degree 0 and is called a **constant function.** In Chapter 1, you learned that the graph of this type of function is a horizontal line. The polynomial function

$$f(x) = mx + b, \quad m \neq 0 \qquad \text{Linear function}$$

has degree 1 and is called a **linear function.** You learned in Chapter 1 that the graph of $f(x) = mx + b$ is a line whose slope is m and whose y-intercept is $(0, b)$. In this section, you will study second-degree polynomial functions, which are called **quadratic functions.**

Definition of Quadratic Function

Let a, b, and c be real numbers with $a \neq 0$. The function given by

$$f(x) = ax^2 + bx + c \qquad \text{Quadratic function}$$

is called a **quadratic function.**

Often real-life data can be modeled by quadratic functions. For instance, the table at the right shows the height h (in feet) of a projectile fired from a height of 6 feet with an initial velocity of 256 feet per second at any time t (in seconds). A quadratic model for the data in the table is

$$h(t) = -16t^2 + 256t + 6 \quad \text{for} \quad 0 \leq t \leq 16.$$

Time, (in seconds) t	Height, (in feet) h
0	6
2	454
4	774
6	966
8	1030
10	966
12	774
14	454
16	6

The graph of a quadratic function is a special type of U-shaped curve called a **parabola.** Parabolas occur in many real-life applications, especially those involving reflective properties, such as satellite dishes or flashlight reflectors. You will study these properties in a later chapter.

All parabolas are symmetric with respect to a line called the **axis of symmetry,** or simply the **axis** of the parabola. The point where the axis intersects the parabola is called the **vertex** of the parabola. These and other basic characteristics of quadratic functions are summarized on the next page.

What you should learn

- Analyze graphs of quadratic functions.
- Write quadratic functions in standard form and use the results to sketch graphs of functions.
- Find minimum and maximum values of quadratic functions in real-life applications.

Why you should learn it

Quadratic functions can be used to model the design of a room. For instance, Exercise 63 on page 251 shows how the size of an indoor fitness room with a running track can be modeled.

Basic Characteristics of Quadratic Functions

Graph of $f(x) = ax^2, a > 0$
Domain: $(-\infty, \infty)$
Range: $[0, \infty)$
Intercept: $(0, 0)$
Decreasing on $(-\infty, 0)$
Increasing on $(0, \infty)$
Even function
Axis of symmetry: $x = 0$
Relative minimum or vertex: $(0, 0)$

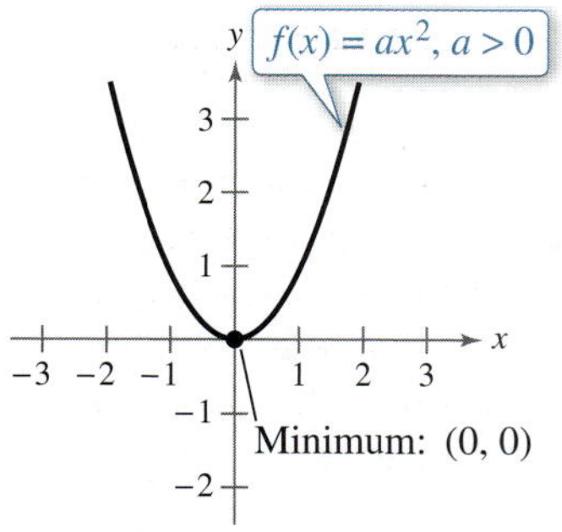

Graph of $f(x) = ax^2, a < 0$
Domain: $(-\infty, \infty)$
Range: $(-\infty, 0]$
Intercept: $(0, 0)$
Increasing on $(-\infty, 0)$
Decreasing on $(0, \infty)$
Even function
Axis of symmetry: $x = 0$
Relative maximum or vertex: $(0, 0)$

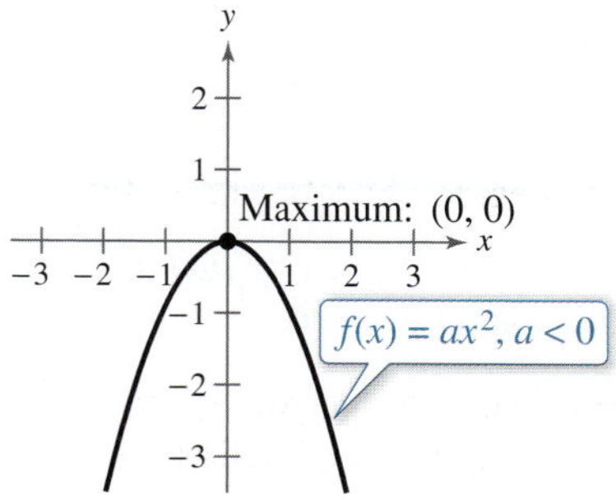

For the general quadratic form $f(x) = ax^2 + bx + c$, when the leading coefficient a is positive, the parabola opens upward; and when the leading coefficient a is negative, the parabola opens downward. Later in this section you will learn ways to find the coordinates of the vertex of a parabola.

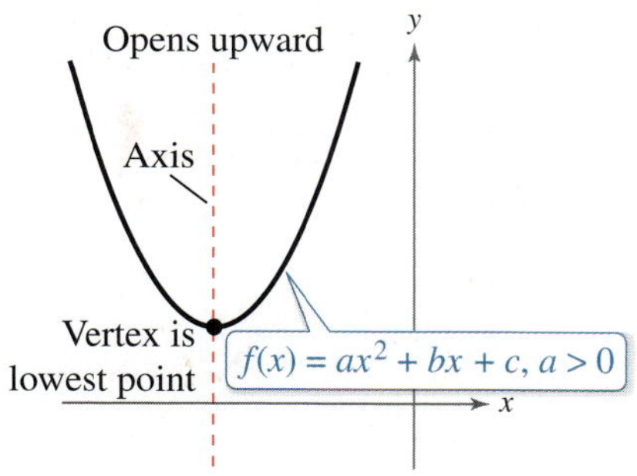

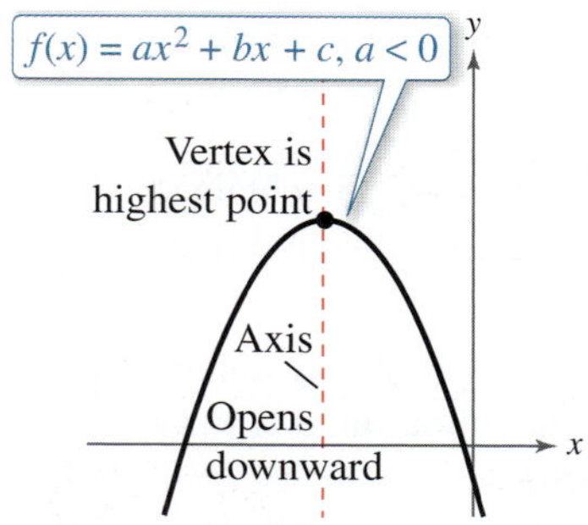

When sketching the graph of $f(x) = ax^2$, it is helpful to use the graph of $y = x^2$ as a reference, as discussed in Section 1.5. There you saw that when $a > 1$, the graph of $y = af(x)$ is a vertical stretch of the graph of $y = f(x)$. When $0 < a < 1$, the graph of $y = af(x)$ is a vertical shrink of the graph of $y = f(x)$. Notice in Figure 3.1 that the coefficient a determines how widely the parabola given by $f(x) = ax^2$ opens. When $|a|$ is small, the parabola opens more widely than when $|a|$ is large.

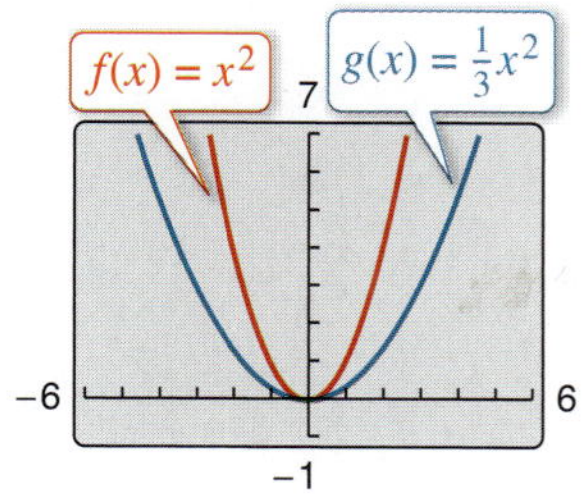

Vertical shrink

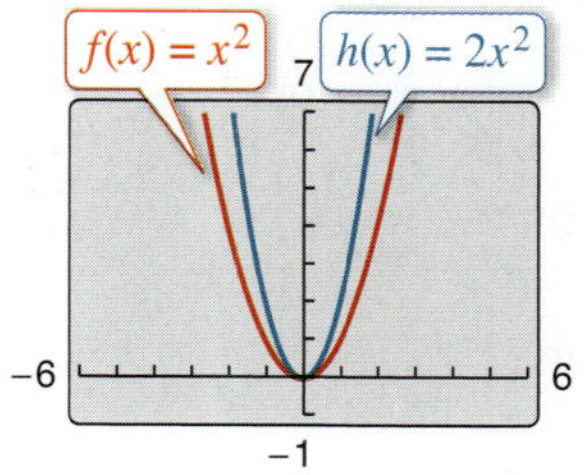

Vertical stretch

Figure 3.1

Library of Parent Functions: Quadratic Functions

The *parent quadratic function* is $f(x) = x^2$, also known as the *squaring function*. The basic characteristics of the parent quadratic function are summarized below and on the inside cover of this text.

Graph of $f(x) = x^2$
Domain: $(-\infty, \infty)$
Range: $[0, \infty)$
Intercept: $(0, 0)$
Decreasing on $(-\infty, 0)$
Increasing on $(0, \infty)$
Even function
Axis of symmetry: $x = 0$
Relative minimum or vertex: $(0, 0)$

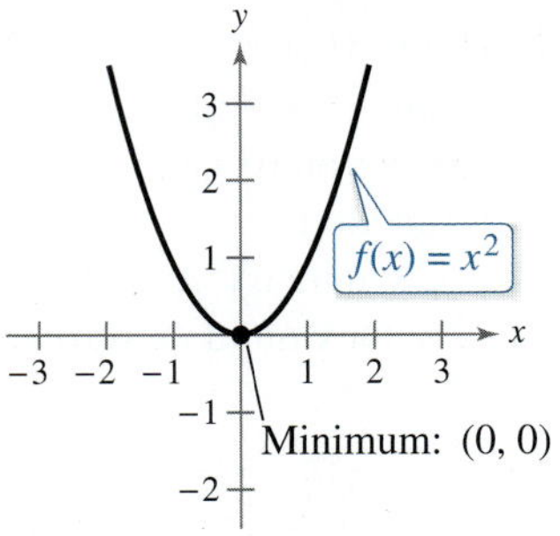

Recall from Section 1.5 that the graphs of $y = f(x \pm c)$, $y = f(x) \pm c$, $y = -f(x)$, and $y = f(-x)$ are rigid transformations of the graph of $y = f(x)$.

$y = f(x \pm c)$	Horizontal shift	$y = -f(x)$	Reflection in x-axis
$y = f(x) \pm c$	Vertical shift	$y = f(-x)$	Reflection in y-axis

EXAMPLE 1 Library of Parent Functions: $f(x) = x^2$

See LarsonPrecalculus.com for an interactive version of this type of example.

Sketch the graph of each function by hand and compare it with the graph of $f(x) = x^2$.

a. $g(x) = -x^2 + 1$ **b.** $h(x) = (x + 2)^2 - 3$

Solution

a. With respect to the graph of $f(x) = x^2$, the graph of g is obtained by a *reflection* in the x-axis and a vertical shift one unit *upward*, as shown in Figure 3.2. Confirm this with a graphing utility.

b. With respect to the graph of $f(x) = x^2$, the graph of h is obtained by a horizontal shift two units *to the left* and a vertical shift three units *downward*, as shown in Figure 3.3. Confirm this with a graphing utility.

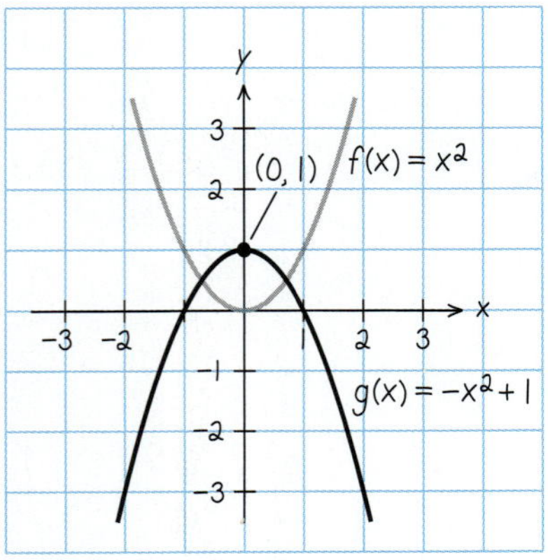

Figure 3.2

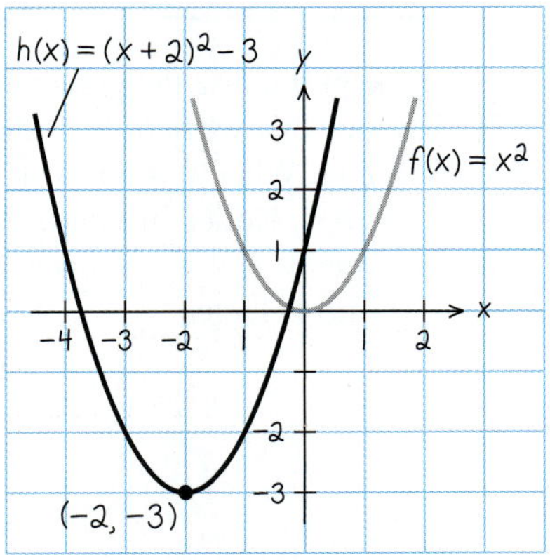

Figure 3.3

✓ **Checkpoint** Audio-video solution in English & Spanish at LarsonPrecalculus.com.

Sketch the graph of each function by hand and compare it with the graph of $f(x) = x^2$.

a. $g(x) = x^2 - 4$ **b.** $h(x) = (x - 3)^2 + 1$

The Standard Form of a Quadratic Function

The equation in Example 1(b) is written in the **standard form**

$$f(x) = a(x - h)^2 + k.$$

This form is especially convenient for sketching a parabola because it identifies the vertex of the parabola as (h, k).

> **Standard Form of a Quadratic Function**
>
> The quadratic function given by
>
> $$f(x) = a(x - h)^2 + k, \quad a \neq 0$$
>
> is in **standard form.** The graph of f is a parabola whose axis is the vertical line $x = h$ and whose vertex is the point (h, k). When $a > 0$, the parabola opens upward, and when $a < 0$, the parabola opens downward.

EXAMPLE 2 Identifying the Vertex of a Quadratic Function

Describe the graph of

$$f(x) = 2x^2 + 8x + 7$$

and identify the vertex.

Solution

Write the quadratic function in standard form by completing the square. Recall that the first step is to factor out any coefficient of x^2 that is not 1.

$f(x) = 2x^2 + 8x + 7$	Write original function.
$= (2x^2 + 8x) + 7$	Group x-terms.
$= 2(x^2 + 4x) + 7$	Factor 2 out of x-terms.
$= 2(x^2 + 4x + 4 - 4) + 7$ (with $4 = (4/2)^2$)	Add and subtract $(4/2)^2 = 4$ within parentheses to complete the square.
$= 2(x^2 + 4x + 4) - 2(4) + 7$	Regroup terms.
$= 2(x + 2)^2 - 1$	Write in standard form.

From the standard form, you can see that the graph of f is a parabola that opens upward with vertex

$$(-2, -1)$$

as shown in the figure. This corresponds to a left shift of two units and a downward shift of one unit relative to the graph of

$$y = 2x^2.$$

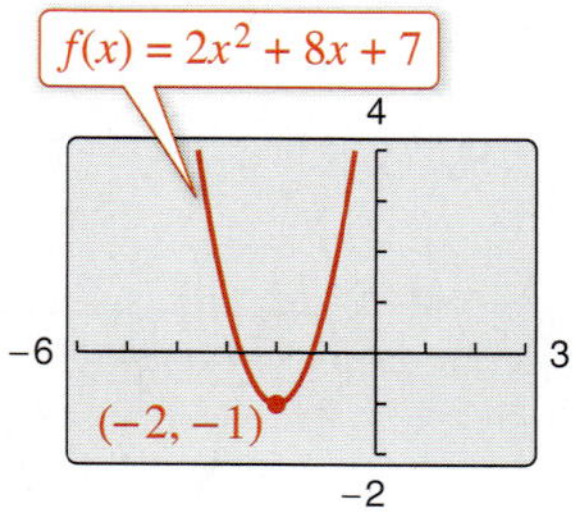

Checkpoint Audio-video solution in English & Spanish at LarsonPrecalculus.com.

Describe the graph of $f(x) = 3x^2 - 6x + 4$ and identify the vertex.

To find the x-intercepts of the graph of $f(x) = ax^2 + bx + c$, solve the equation $ax^2 + bx + c = 0$. When $ax^2 + bx + c$ does not factor, you can use the Quadratic Formula to find the x-intercepts, or a graphing utility to approximate the x-intercepts. Remember, however, that a parabola may not have x-intercepts.

Explore the Concept

Use a graphing utility to graph $y = ax^2$ with $a = -2, -1, -0.5, 0.5, 1,$ and 2. How does changing the value of a affect the graph?

Use a graphing utility to graph $y = (x - h)^2$ with $h = -4, -2, 2,$ and 4. How does changing the value of h affect the graph?

Use a graphing utility to graph $y = x^2 + k$ with $k = -4, -2, 2,$ and 4. How does changing the value of k affect the graph?

EXAMPLE 3 Identifying x-Intercepts of a Quadratic Function

Describe the graph of $f(x) = -x^2 + 6x - 8$ and identify any x-intercepts.

Solution

$f(x) = -x^2 + 6x - 8$ — Write original function.

$= -(x^2 - 6x) - 8$ — Factor -1 out of x-terms.

$= -(x^2 - 6x + 9 - 9) - 8$ — Add and subtract $(-6/2)^2 = 9$ within parentheses.

$(-6/2)^2$

$= -(x^2 - 6x + 9) - (-9) - 8$ — Regroup terms.

$= -(x - 3)^2 + 1$ — Write in standard form.

The graph of f is a parabola that opens downward with vertex $(3, 1)$, as shown in Figure 3.4. The x-intercepts are determined as follows.

$-(x^2 - 6x + 8) = 0$ — Factor out -1.

$-(x - 2)(x - 4) = 0$ — Factor.

$x - 2 = 0 \Rightarrow x = 2$ — Set 1st factor equal to 0.

$x - 4 = 0 \Rightarrow x = 4$ — Set 2nd factor equal to 0.

So, the x-intercepts are $(2, 0)$ and $(4, 0)$, as shown in Figure 3.4.

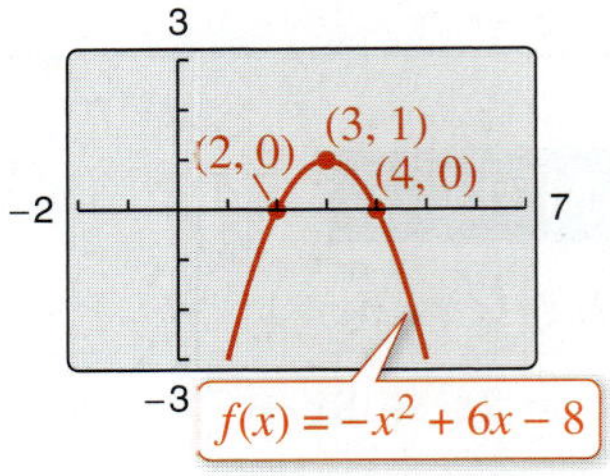

Figure 3.4

✓ **Checkpoint** *Audio-video solution in English & Spanish at LarsonPrecalculus.com.*

Describe the graph of $f(x) = x^2 - 4x + 3$ and identify any x-intercepts.

EXAMPLE 4 Writing the Equation of a Parabola

Write the standard form of the equation of the parabola whose vertex is $(1, 2)$ and that passes through the point $(3, -6)$.

Solution

Because the vertex of the parabola is $(h, k) = (1, 2)$, the equation has the form

$f(x) = a(x - 1)^2 + 2.$ — Substitute for h and k in standard form.

Because the parabola passes through the point $(3, -6)$, it follows that $f(3) = -6$. So, you obtain

$f(x) = a(x - 1)^2 + 2$ — Write in standard form.

$-6 = a(3 - 1)^2 + 2$ — Substitute -6 for $f(x)$ and 3 for x.

$-6 = 4a + 2$ — Simplify.

$-8 = 4a$ — Subtract 2 from each side.

$-2 = a.$ — Divide each side by 4.

The equation in standard form is $f(x) = -2(x - 1)^2 + 2$. You can confirm this answer by graphing f with a graphing utility, as shown in Figure 3.5. Use the *zoom* and *trace* features or the *maximum* and *value* features to confirm that its vertex is $(1, 2)$ and that it passes through the point $(3, -6)$.

Remark

In Example 4, there are infinitely many different parabolas that have a vertex at $(1, 2)$. Of these, however, the only one that passes through the point $(3, -6)$ is the one given by

$$f(x) = -2(x - 1)^2 + 2.$$

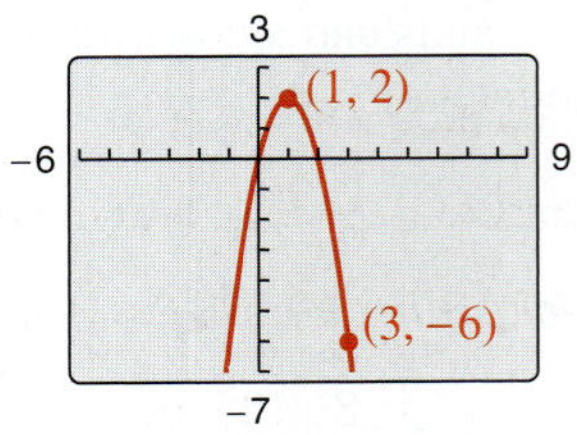

Figure 3.5

✓ **Checkpoint** *Audio-video solution in English & Spanish at LarsonPrecalculus.com.*

Write the standard form of the equation of the parabola whose vertex is $(-4, 11)$ and that passes through the point $(-6, 15)$.

Finding Minimum and Maximum Values

Many applications involve finding the maximum or minimum value of a quadratic function. By completing the square of the quadratic function $f(x) = ax^2 + bx + c$, you can rewrite the function in standard form.

$$f(x) = a\left(x + \frac{b}{2a}\right)^2 + \left(c - \frac{b^2}{4a}\right) \qquad \text{Standard form}$$

So, the vertex of the graph of f is $\left(-\frac{b}{2a}, f\left(-\frac{b}{2a}\right)\right)$, which implies the following.

Minimum and Maximum Values of Quadratic Functions

Consider the function $f(x) = ax^2 + bx + c$ with vertex $\left(-\frac{b}{2a}, f\left(-\frac{b}{2a}\right)\right)$.

1. If $a > 0$, then f has a *minimum* at $x = -\frac{b}{2a}$.

   The minimum value is $f\left(-\frac{b}{2a}\right)$.

2. If $a < 0$, then f has a *maximum* at $x = -\frac{b}{2a}$.

   The maximum value is $f\left(-\frac{b}{2a}\right)$.

EXAMPLE 5 The Maximum Height of a Projectile

The path of a baseball is given by the function $f(x) = -0.0032x^2 + x + 3$, where $f(x)$ is the height of the baseball (in feet) and x is the horizontal distance from home plate (in feet). What is the maximum height reached by the baseball?

Algebraic Solution

For this quadratic function, you have

$$f(x) = ax^2 + bx + c = -0.0032x^2 + x + 3$$

which implies that $a = -0.0032$ and $b = 1$. Because the function has a maximum when $x = -b/(2a)$, you can conclude that the baseball reaches its maximum height when it is x feet from home plate, where x is

$$x = -\frac{b}{2a}$$

$$= -\frac{1}{2(-0.0032)}$$

$$= 156.25 \text{ feet.}$$

At this distance, the maximum height is

$$f(156.25) = -0.0032(156.25)^2 + 156.25 + 3$$

$$= 81.125 \text{ feet.}$$

Graphical Solution

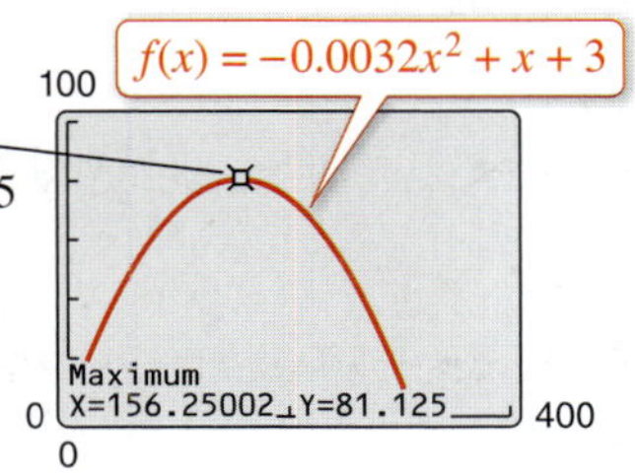

The maximum height is $y \approx 81.125$ feet at $x \approx 156.25$ feet.

✓ Checkpoint Audio-video solution in English & Spanish at LarsonPrecalculus.com.

Rework Example 5 when the path of the baseball is given by the function

$$f(x) = -0.007x^2 + x + 4.$$

3.1 Exercises

See *CalcChat.com* for tutorial help and worked-out solutions to odd-numbered exercises.
For instructions on how to use a graphing utility, see Appendix A.

Vocabulary and Concept Check

In Exercises 1 and 2, fill in the blanks.

1. A polynomial function with degree n and leading coefficient a_n is a function of the form $f(x) = a_nx^n + a_{n-1}x^{n-1} + \cdot \cdot \cdot + a_2x^2 + a_1x + a_0$, $a_n \neq 0$, where n is a ________ and $a_n, a_{n-1}, \ldots, a_2, a_1, a_0$ are ________ numbers.

2. A ________ function is a second-degree polynomial function, and its graph is called a ________ .

3. Is the quadratic function $f(x) = (x - 2)^2 + 3$ written in standard form? Identify the vertex of the graph of f.

4. Does the graph of the quadratic function $f(x) = -3x^2 + 5x + 2$ have a relative minimum value at its vertex?

Procedures and Problem Solving

Graphs of Quadratic Functions **In Exercises 5–8, match the quadratic function with its graph. [The graphs are labeled (a), (b), (c), and (d).]**

(a)

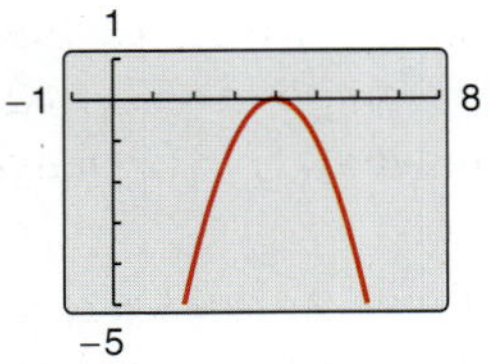

(b)

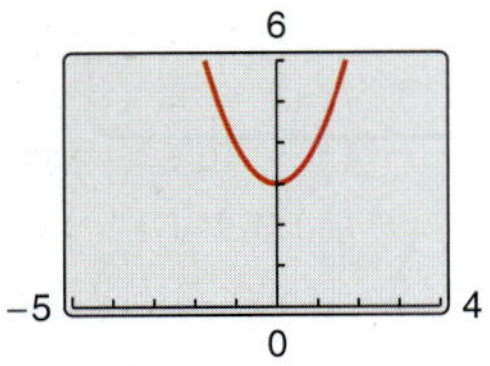

(c)

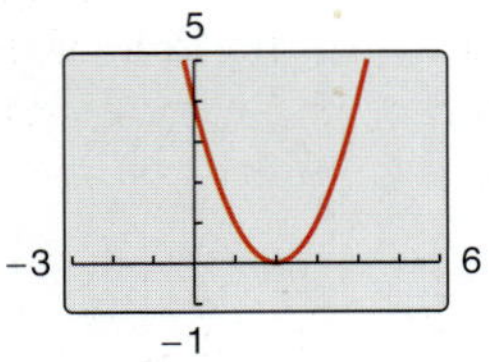

(d)

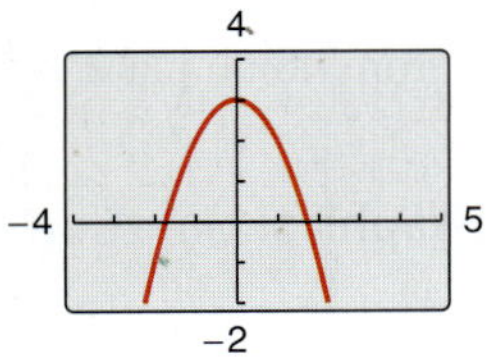

5. $f(x) = (x - 2)^2$ **6.** $f(x) = 3 - x^2$

7. $f(x) = x^2 + 3$ **8.** $f(x) = -(x - 4)^2$

Library of Parent Functions **In Exercises 9–16, sketch the graph of the function and compare it with the graph of $y = x^2$.**

9. $y = -x^2$ **10.** $y = x^2 - 1$

11. $y = (x + 3)^2$ **12.** $y = -(x + 3)^2 - 1$

13. $y = (x + 1)^2$ **14.** $y = -x^2 + 2$

15. $y = (x - 3)^2$ **16.** $y = -(x - 3)^2 + 1$

Identifying the Vertex of a Quadratic Function **In Exercises 17–30, describe the graph of the function and identify the vertex. Use a graphing utility to verify your results.**

17. $f(x) = 20 - x^2$ **18.** $f(x) = x^2 + 8$

19. $f(x) = \frac{1}{2}x^2 - 5$ **20.** $f(x) = -6 - \frac{1}{4}x^2$

21. $f(x) = (x + 3)^2 - 4$

22. $f(x) = (x - 7)^2 + 2$

23. $h(x) = x^2 - 2x + 1$

24. $g(x) = x^2 + 16x + 64$

25. $f(x) = x^2 - x + \frac{5}{4}$

26. $f(x) = x^2 + 3x + \frac{1}{4}$

27. $f(x) = -x^2 + 2x + 5$

28. $f(x) = -x^2 - 4x + 1$

29. $h(x) = 4x^2 - 4x + 21$

30. $f(x) = 2x^2 - x + 1$

Identifying *x*-Intercepts of a Quadratic Function **In Exercises 31–36, describe the graph of the quadratic function. Identify the vertex and *x*-intercept(s). Use a graphing utility to verify your results.**

31. $g(x) = x^2 + 8x + 11$

32. $f(x) = x^2 + 10x + 14$

33. $f(x) = -(x^2 - 2x - 15)$

34. $f(x) = -(x^2 + 3x - 4)$

35. $f(x) = -2x^2 + 16x - 31$

36. $f(x) = -4x^2 + 24x - 41$

Writing the Equation of a Parabola in Standard Form **In Exercises 37 and 38, write an equation of the parabola in standard form. Use a graphing utility to graph the equation and verify your result.**

37.

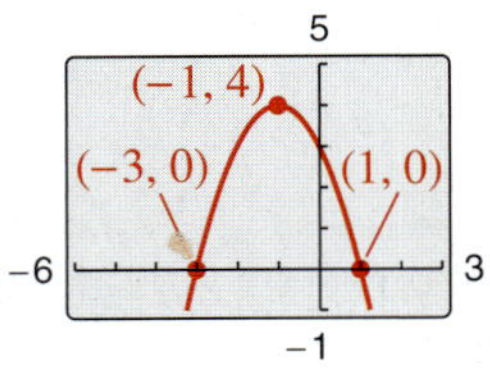

38.

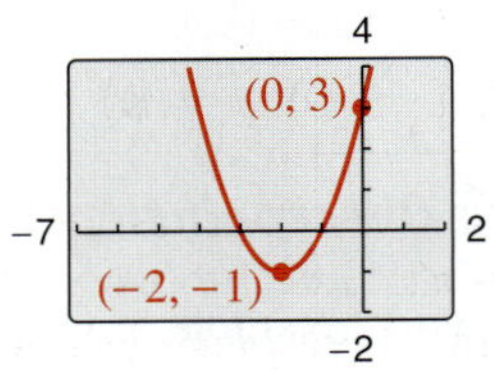

Writing the Equation of a Parabola in Standard Form **In Exercises 39–44, write the standard form of the quadratic function that has the indicated vertex and whose graph passes through the given point. Use a graphing utility to verify your result.**

39. Vertex: $(-2, 5)$;	Point: $(0, 9)$
40. Vertex: $(4, 1)$;	Point: $(6, -7)$
41. Vertex: $(1, -2)$;	Point: $(-1, 14)$
42. Vertex: $(-4, -1)$;	Point: $(-2, 4)$
43. Vertex: $\left(\frac{1}{2}, 1\right)$;	Point: $\left(-2, -\frac{21}{5}\right)$
44. Vertex: $\left(-\frac{1}{4}, -1\right)$;	Point: $\left(0, -\frac{17}{16}\right)$

Using a Graph to Identify *x*-Intercepts **In Exercises 45–48, determine the *x*-intercept(s) of the graph visually. Then find the *x*-intercept(s) algebraically to verify your answer.**

45.

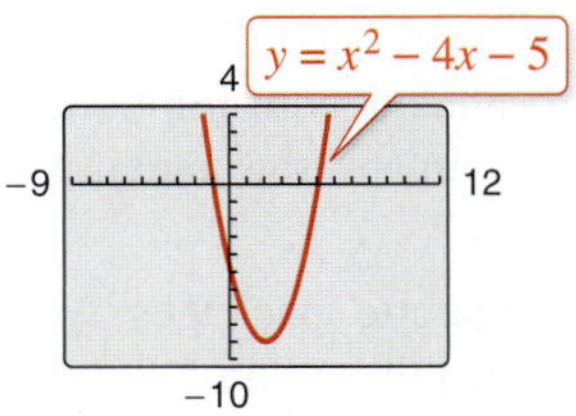

46.

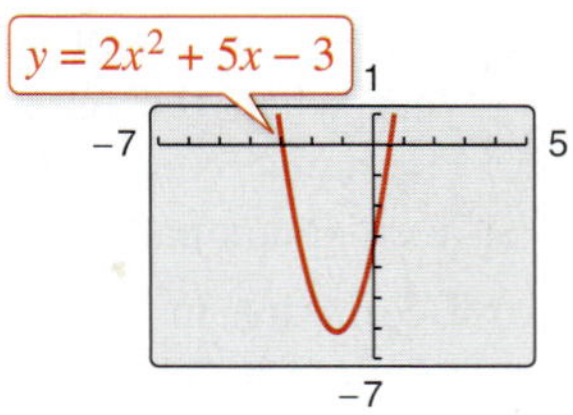

47.

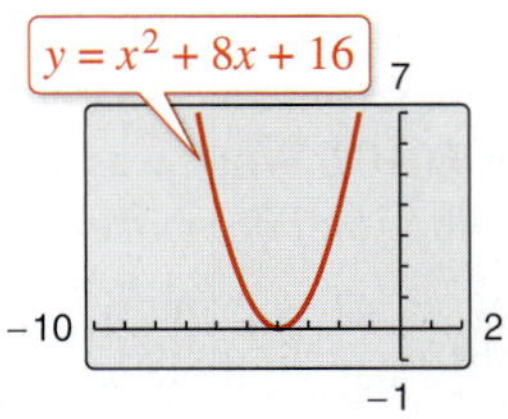

48.

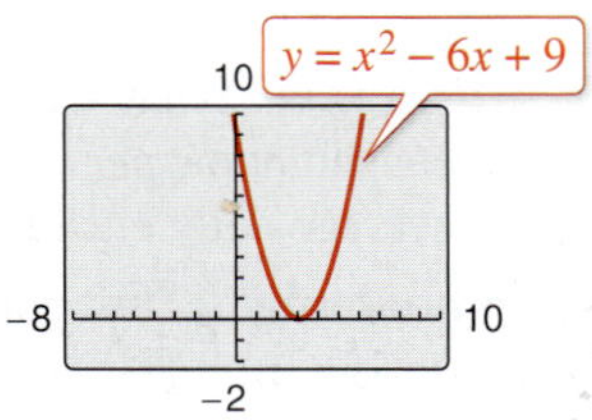

Graphing to Identify *x*-Intercepts **In Exercises 49–54, use a graphing utility to graph the quadratic function and find the *x*-intercepts of the graph. Then find the *x*-intercepts algebraically to verify your answer.**

49. $y = x^2 - 4x$	**50.** $y = -2x^2 + 10x$
51. $y = 2x^2 - 7x - 30$	**52.** $y = 4x^2 + 25x - 21$
53. $y = -\frac{1}{2}(x^2 - 6x - 7)$	**54.** $y = \frac{7}{10}(x^2 + 12x - 45)$

Using the *x*-Intercepts to Write Equations **In Exercises 55–58, find two quadratic functions, one that opens upward and one that opens downward, whose graphs have the given *x*-intercepts. (There are many correct answers.)**

55. $(-1, 0), (3, 0)$	**56.** $(0, 0), (10, 0)$
57. $(-3, 0), \left(-\frac{1}{2}, 0\right)$	**58.** $\left(-\frac{5}{2}, 0\right), (2, 0)$

Maximizing a Product of Two Numbers **In Exercises 59–62, find the two positive real numbers with the given sum whose product is a maximum.**

59. The sum is 110. **60.** The sum is 66.

61. The sum of the first and twice the second is 24.

62. The sum of the first and three times the second is 42.

63. ***Why you should learn it*** *(p. 244)* An indoor physical fitness room consists of a rectangular region with a semicircle on each end. The perimeter of the room is to be a 200-meter single-lane running track.

(a) Draw a diagram that illustrates the problem. Let x and y represent the length and width of the rectangular region, respectively.

(b) Determine the radius of the semicircular ends of the track. Determine the distance, in terms of y, around the inside edge of the two semicircular parts of the track.

(c) Use the result of part (b) to write an equation, in terms of x and y, for the distance traveled in one lap around the track. Solve for y.

(d) Use the result of part (c) to write the area A of the rectangular region as a function of x.

(e) Use a graphing utility to graph the area function from part (d). Use the graph to approximate the dimensions that will produce a rectangle of maximum area.

64. **Algebraic-Graphical-Numerical** A child-care center has 200 feet of fencing to enclose two adjacent rectangular safe play areas (see figure). Use the following methods to determine the dimensions that will produce a maximum enclosed area.

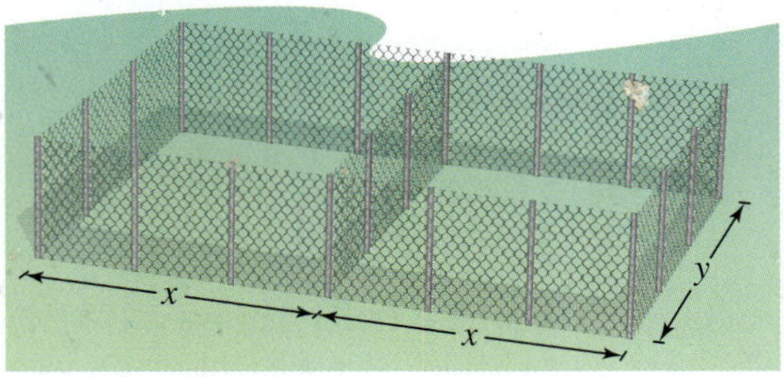

(a) Write the total area A of the play areas as a function of x.

(b) Use the *table* feature of a graphing utility to create a table showing possible values of x and the corresponding total area A of the play areas. Use the table to estimate the dimensions that will produce the maximum enclosed area.

(c) Use the graphing utility to graph the area function. Use the graph to approximate the dimensions that will produce the maximum enclosed area.

(d) Write the area function in standard form to find algebraically the dimensions that will produce the maximum enclosed area.

(e) Compare your results from parts (b), (c), and (d).

65. Height of a Projectile The height y (in feet) of a punted football is approximated by

$$y = -\frac{16}{2025}x^2 + \frac{9}{5}x + \frac{3}{2}$$

where x is the horizontal distance (in feet) from where the football is punted. (See figure.)

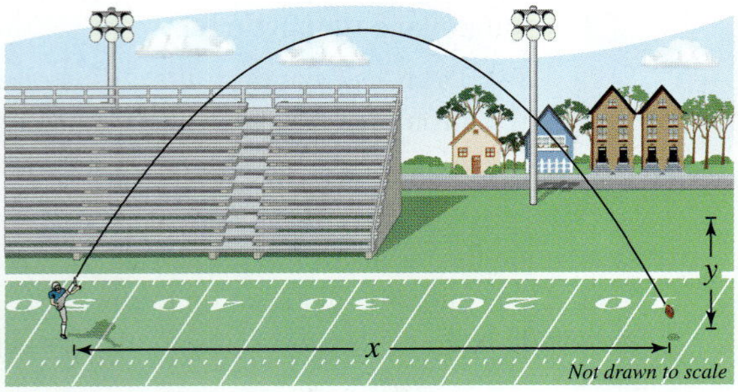

(a) Use a graphing utility to graph the path of the football.

(b) How high is the football when it is punted? (*Hint:* Find y when $x = 0$.)

(c) What is the maximum height of the football?

(d) How far from the punter does the football strike the ground?

66. Physics The path of a diver is approximated by

$$y = -\frac{4}{9}x^2 + \frac{24}{9}x + 12$$

where y is the height (in feet) and x is the horizontal distance (in feet) from the end of the diving board (see figure). What is the maximum height of the diver?

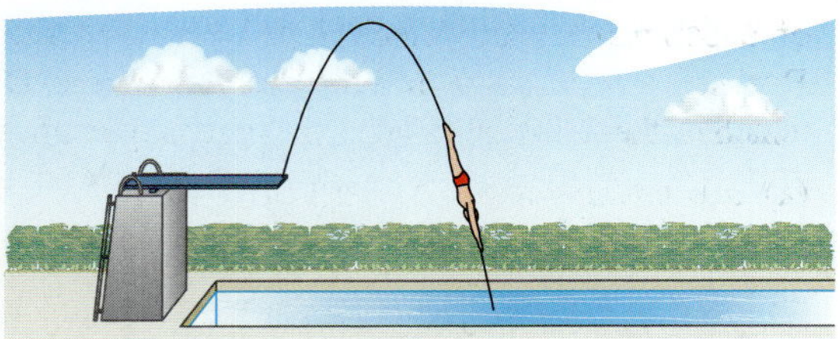

67. Geometry You have a steel wire that is 100 inches long. To make a sign holder, you bend the wire x inches from each end to form two right angles. To use the sign holder, you insert each end 6 inches into the ground. (See figure.)

(a) Write a function for the rectangular area A enclosed by the sign holder in terms of x.

(b) Use the *table* feature of a graphing utility to determine the value of x that maximizes the rectangular area enclosed by the sign holder.

68. Economics The monthly revenue R (in thousands of dollars) from the sales of a digital picture frame is approximated by $R(p) = -10p^2 + 1580p$, where p is the price per unit (in dollars).

(a) Find the monthly revenues for unit prices of \$50, \$70, and \$90.

(b) Find the unit price that will yield a maximum monthly revenue.

(c) What is the maximum monthly revenue?

69. Public Health For selected years from 1955 through 2010, the annual per capita consumption C of cigarettes by Americans (ages 18 and older) can be modeled by

$$C(t) = -1.39t^2 + 36.5t + 3871, \quad 5 \le t \le 60$$

where t is the year, with $t = 5$ corresponding to 1955. (Sources: Centers for Disease Control and Prevention and U.S. Census Bureau)

(a) Use a graphing utility to graph the model.

(b) Use the graph of the model to approximate the year when the maximum annual consumption of cigarettes occurred. Approximate the maximum average annual consumption.

(c) Beginning in 1966, all cigarette packages were required by law to carry a health warning. Do you think the warning had any effect? Explain.

(d) In 2010, the U.S. population (ages 18 and older) was 234,564,000. Of those, about 45,271,000 were smokers. What was the average annual cigarette consumption *per smoker* in 2010? What was the average daily cigarette consumption *per smoker*?

70. Demography The population P of Germany (in thousands) from 2000 through 2013 can be modeled by

$$P(t) = -14.82t^2 + 95.9t + 82{,}276, \quad 0 \le t \le 13$$

where t is the year, with $t = 0$ corresponding to 2000. (Source: U.S. Census Bureau)

(a) According to the model, in what year did Germany have its greatest population? What was the population?

(b) According to the model, what will Germany's population be in the year 2075? Is this result reasonable? Explain.

Conclusions

True or False? **In Exercises 71–74, determine whether the statement is true or false. Justify your answer.**

71. The function $f(x) = -12x^2 - 1$ has no x-intercepts.

72. The function $f(x) = a(x - 5)^2$ has exactly one x-intercept for any nonzero value of a.

73. The functions $f(x) = 3x^2 + 6x + 7$ and $g(x) = 3x^2 + 6x - 1$ have the same vertex.

74. The graphs of $f(x) = -4x^2 - 10x + 7$ and $g(x) = 12x^2 + 30x + 1$ have the same axis of symmetry.

Library of Parent Functions **In Exercises 75 and 76, determine which equation(s) may be represented by the graph shown. (There may be more than one correct answer.)**

75. (a) $f(x) = -(x - 4)^2 + 2$
(b) $f(x) = -(x + 2)^2 + 4$
(c) $f(x) = -(x + 2)^2 - 4$
(d) $f(x) = -x^2 - 4x - 8$
(e) $f(x) = -(x - 2)^2 - 4$
(f) $f(x) = -x^2 + 4x - 8$

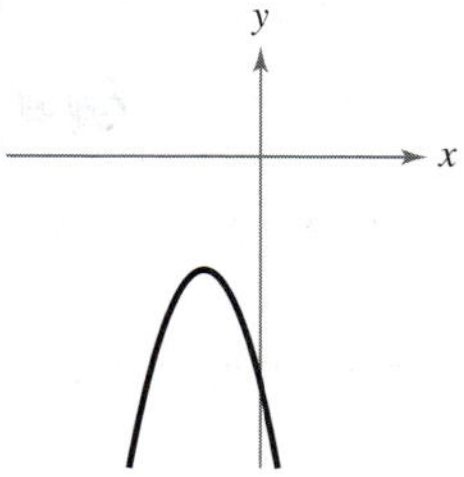

76. (a) $f(x) = (x - 1)^2 + 3$
(b) $f(x) = (x + 1)^2 + 3$
(c) $f(x) = (x - 3)^2 + 1$
(d) $f(x) = x^2 + 2x + 4$
(e) $f(x) = (x + 3)^2 + 1$
(f) $f(x) = x^2 + 6x + 10$

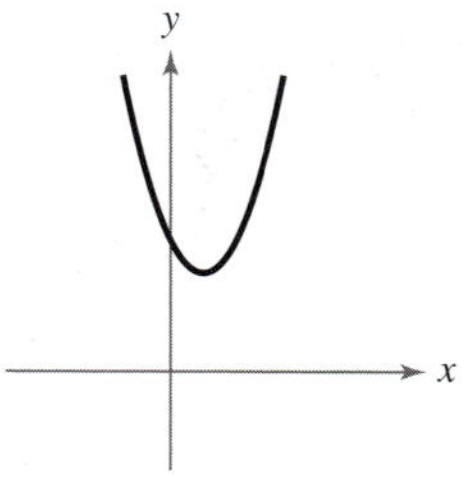

Describing Parabolas **In Exercises 77–80, let z represent a positive real number. Describe how the family of parabolas represented by the given function compares with the graph of $g(x) = x^2$.**

77. $f(x) = (x - z)^2$
78. $f(x) = x^2 - z$
79. $f(x) = z(x - 3)^2$
80. $f(x) = zx^2 + 4$

Think About It **In Exercises 81–84, find the value of b such that the function has the given maximum or minimum value.**

81. $f(x) = -x^2 + bx - 75$; Maximum value: 25
82. $f(x) = -x^2 + bx - 16$; Maximum value: 48
83. $f(x) = x^2 + bx + 26$; Minimum value: 10
84. $f(x) = x^2 + bx - 25$; Minimum value: -50

85. **Proof** Let x and y be two positive real numbers whose sum is S. Show that the maximum product of x and y occurs when x and y are both equal to $S/2$.

86. **Proof** Assume that the function $f(x) = ax^2 + bx + c$, $a \neq 0$, has two real zeros. Show that the x-coordinate of the vertex of the graph is the average of the zeros of f. (*Hint:* Use the Quadratic Formula.)

87. **Writing** The parabola in the figure has an equation of the form $y = ax^2 + bx - 4$. Find the equation of this parabola two different ways, by hand and with technology. Write a paragraph describing the methods you used and comparing the results.

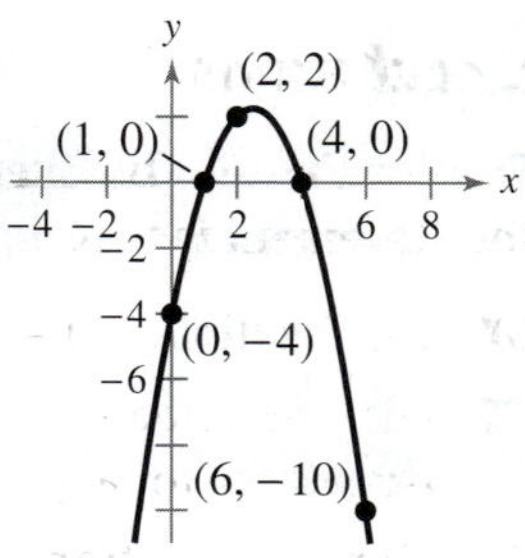

88. **HOW DO YOU SEE IT?** The graph shows a quadratic function of the form $R(t) = at^2 + bt + c$, which represents the yearly revenues for a company, where $R(t)$ is the revenue in year t.

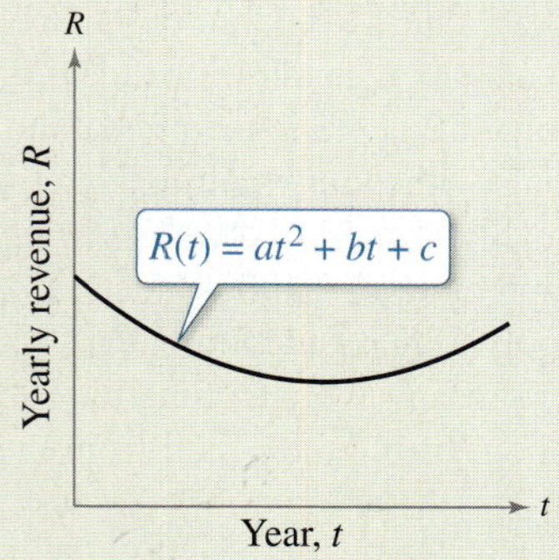

(a) Is the value of a positive, negative, or zero?
(b) Write an expression in terms of a and b that represents the year t when the company had the least revenue.
(c) The company made the same yearly revenues in 2004 and 2014. Estimate the year in which the company had the least revenue.
(d) Assume that the model is still valid today. Are the yearly revenues currently increasing, decreasing, or constant? Explain.

89. **Think About It** The annual profit P (in dollars) of a company is modeled by a function of the form $P = at^2 + bt + c$, where t represents the year. Discuss which of the following models the company might prefer.
(a) a is positive and $t \geq -b/(2a)$.
(b) a is positive and $t \leq -b/(2a)$.
(c) a is negative and $t \geq -b/(2a)$.
(d) a is negative and $t \leq -b/(2a)$.

Cumulative Mixed Review

Finding Points of Intersection **In Exercises 90–93, determine algebraically any point(s) of intersection of the graphs of the equations. Verify your results using the *intersect* feature of a graphing utility.**

90. $x + y = 8$
$-\frac{2}{3}x + y = 6$

91. $y = 3x - 10$
$y = \frac{1}{4}x + 1$

92. $y = 9 - x^2$
$y = x + 3$

93. $y = x^3 + 2x - 1$
$y = -2x + 15$

94. ***Make a Decision*** To work an extended application analyzing the heights of a softball after it has been dropped, visit this textbook's website at *LarsonPrecalculus.com.*

3.2 Polynomial Functions of Higher Degree

Graphs of Polynomial Functions

At this point, you should be able to sketch accurate graphs of polynomial functions of degrees 0, 1, and 2. The graphs of polynomial functions of degree greater than 2 are more difficult to sketch by hand. In this section, however, you will learn how to recognize some of the basic features of the graphs of polynomial functions. Using these features along with point plotting, intercepts, and symmetry, you should be able to make reasonably accurate sketches *by hand*.

The graph of a polynomial function is **continuous.** Essentially, this means that the graph of a polynomial function has no breaks, holes, or gaps, as shown in Figure 3.6. Informally, you can say that a function is continuous when its graph can be drawn with a pencil without lifting the pencil from the paper.

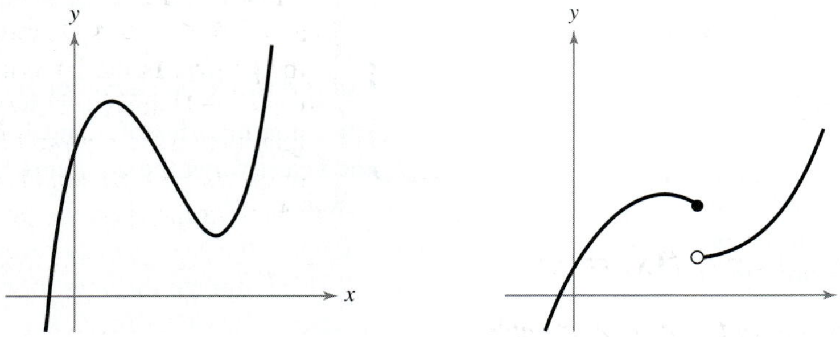

(a) Polynomial functions have continuous graphs.

(b) Functions with graphs that are not continuous are not polynomial functions.

Figure 3.6

Another feature of the graph of a polynomial function is that it has only smooth, rounded turns, as shown in Figure 3.7(a). It cannot have a sharp turn, such as the one shown in Figure 3.7(b).

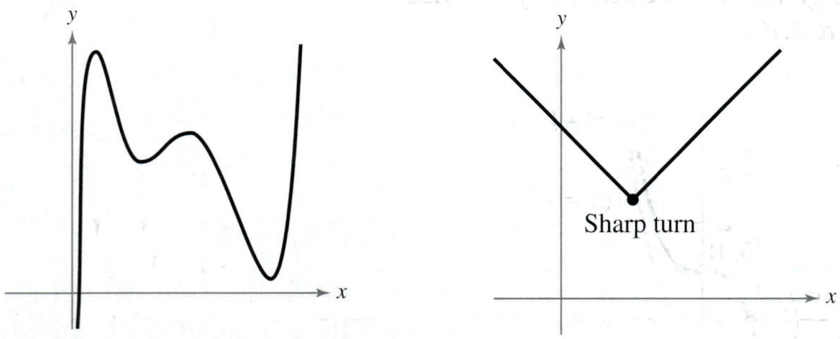

(a) Polynomial functions have graphs with smooth, rounded turns.

(b) Functions with graphs that have sharp turns are not polynomial functions.

Figure 3.7

The graphs of polynomial functions of degree 1 are lines, and those of functions of degree 2 are parabolas. The graphs of all polynomial functions are smooth and continuous. A polynomial function of degree n has the form

$$f(x) = a_nx^n + a_{n-1}x^{n-1} + \cdot \cdot \cdot + a_2x^2 + a_1x + a_0$$

where n is a positive integer and $a_n \neq 0$.

What you should learn

- Use transformations to sketch graphs of polynomial functions.
- Use the Leading Coefficient Test to determine the end behavior of graphs of polynomial functions.
- Find and use zeros of polynomial functions as sketching aids.
- Use the Intermediate Value Theorem to help locate zeros of polynomial functions.

Why you should learn it

You can use polynomial functions to model various aspects of nature, such as the growth of a red oak tree, as shown in Exercise 112 on page 264.

The polynomial functions that have the simplest graphs are monomials of the form $f(x) = x^n$, where n is an integer greater than zero. The greater the value of n, the flatter the graph near the origin. When n is even, the graph is similar to the graph of $f(x) = x^2$ and touches the x-axis at the x-intercept. When n is odd, the graph is similar to the graph of $f(x) = x^3$ and crosses the x-axis at the x-intercept. Polynomial functions of the form $f(x) = x^n$ are often referred to as **power functions.**

Library of Parent Functions: Cubic Function

The basic characteristics of the *parent cubic function* $f(x) = x^3$ are summarized below and on the inside cover of this text.

Graph of $f(x) = x^3$
Domain: $(-\infty, \infty)$
Range: $(-\infty, \infty)$
Intercept: $(0, 0)$
Increasing on $(-\infty, \infty)$
Odd function
Origin symmetry

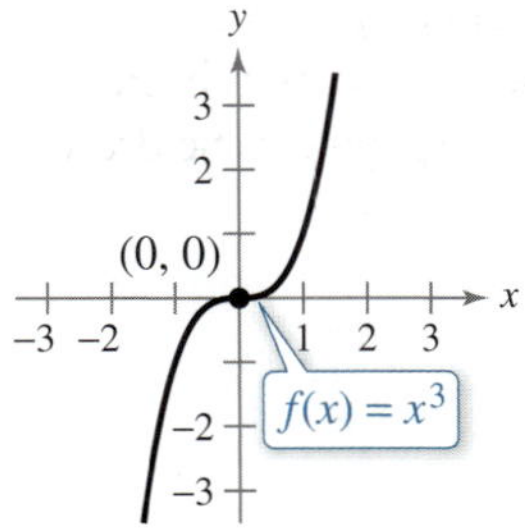

Explore the Concept

Use a graphing utility to graph $y = x^n$ for $n = 2$, 4, and 8. (Use the viewing window $-1.5 \le x \le 1.5$ and $-1 \le y \le 6$.) Compare the graphs. In the interval $(-1, 1)$, which graph is on the bottom? Outside the interval $(-1, 1)$, which graph is on the bottom?

Use a graphing utility to graph $y = x^n$ for $n = 3$, 5, and 7. (Use the viewing window $-1.5 \le x \le 1.5$ and $-4 \le y \le 4$.) Compare the graphs. In the intervals $(-\infty, -1)$ and $(0, 1)$, which graph is on the bottom? In the intervals $(-1, 0)$ and $(1, \infty)$, which graph is on the bottom?

EXAMPLE 1 Library of Parent Functions: $f(x) = x^3$

See LarsonPrecalculus.com for an interactive version of this type of example.

Sketch the graphs of (a) $g(x) = -x^3$, (b) $h(x) = x^3 + 1$, and (c) $k(x) = (x - 1)^3$ by hand.

Solution

a. With respect to the graph of $f(x) = x^3$, the graph of g is obtained by a *reflection* in the x-axis, as shown in Figure 3.8.

b. With respect to the graph of $f(x) = x^3$, the graph of h is obtained by a vertical shift one unit *upward*, as shown in Figure 3.9.

c. With respect to the graph of $f(x) = x^3$, the graph of k is obtained by a horizontal shift one unit *to the right*, as shown in Figure 3.10.

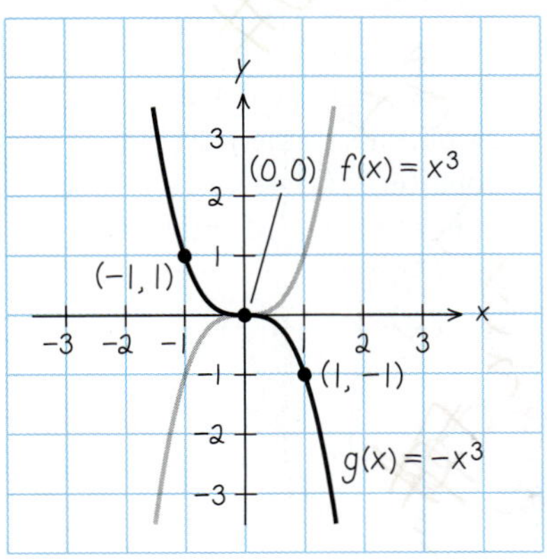

Figure 3.8

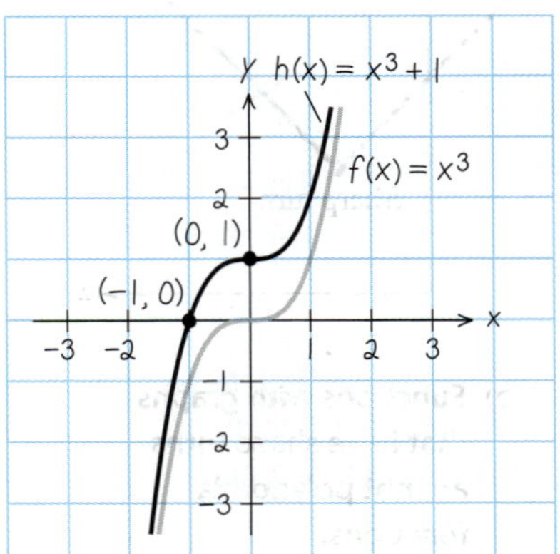

Figure 3.9

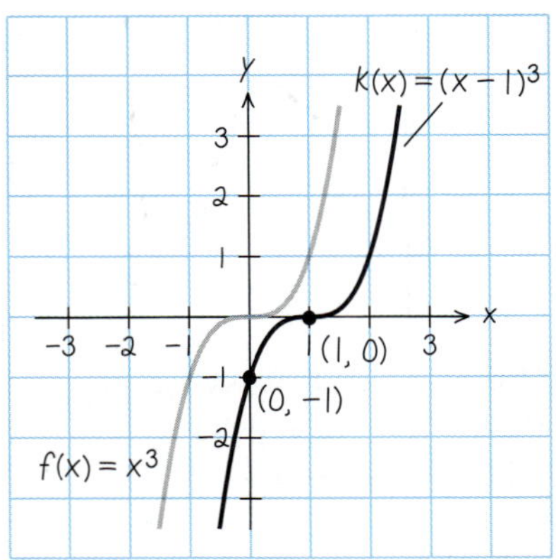

Figure 3.10

✓ **Checkpoint** Audio-video solution in English & Spanish at LarsonPrecalculus.com.

Sketch the graphs of (a) $g(x) = -x^3 + 2$, (b) $h(x) = x^3 - 5$, and (c) $k(x) = (x + 2)^3$ by hand. ■

The Leading Coefficient Test

In Example 1, note that all three graphs eventually rise or fall without bound as x moves to the right. Whether the graph of a polynomial eventually rises or falls can be determined by the polynomial function's degree (even or odd) and by its leading coefficient, as indicated in the **Leading Coefficient Test.**

Leading Coefficient Test

As x moves without bound to the left or to the right, the graph of the polynomial function

$$f(x) = a_n x^n + \cdot \cdot \cdot + a_1 x + a_0, \quad a_n \neq 0$$

eventually rises or falls in the following manner.

1. When n is odd:

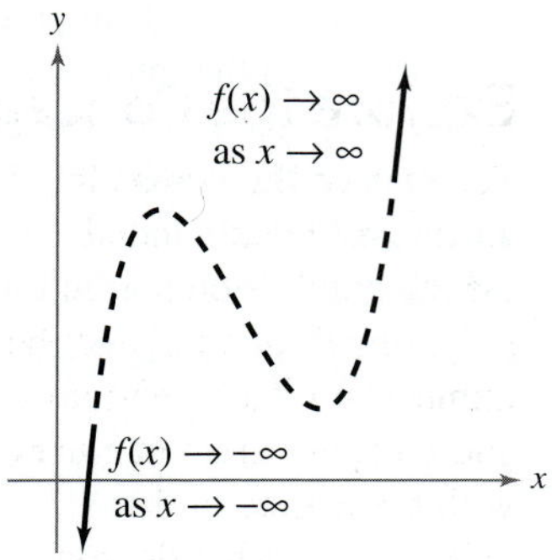

If the leading coefficient is positive ($a_n > 0$), then the graph falls to the left and rises to the right.

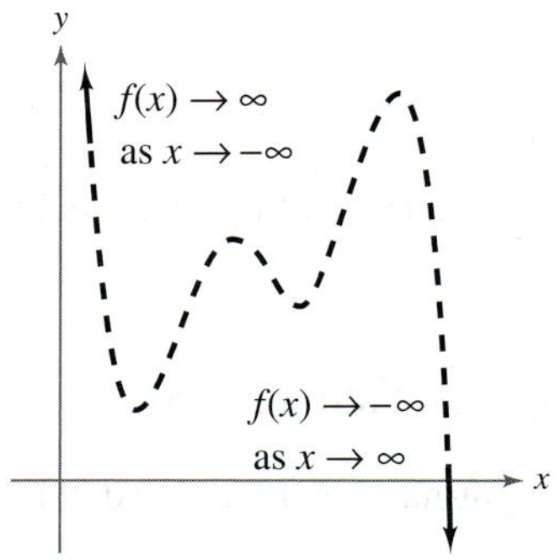

If the leading coefficient is negative ($a_n < 0$), then the graph rises to the left and falls to the right.

2. When n is even:

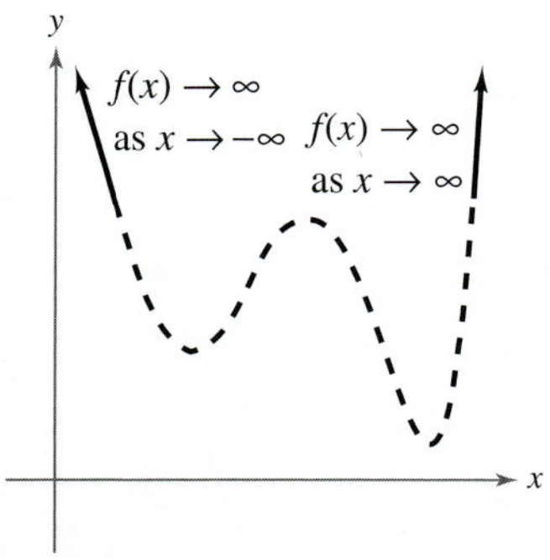

If the leading coefficient is positive ($a_n > 0$), then the graph rises to the left and right.

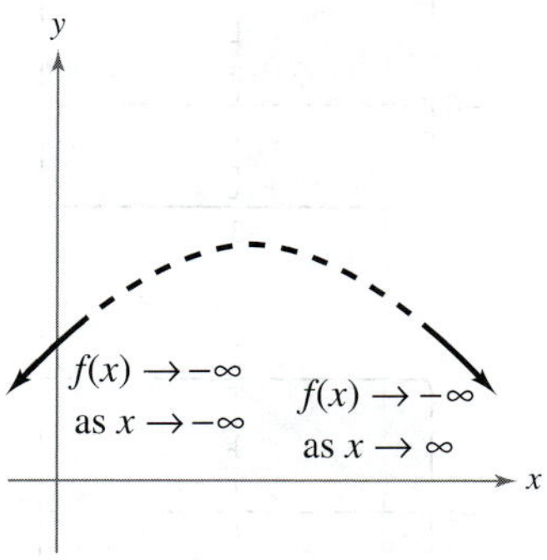

If the leading coefficient is negative ($a_n < 0$), then the graph falls to the left and right.

Note that the dashed portions of the graphs indicate that the test determines only the right-hand and left-hand behavior of the graph.

As you continue to study polynomial functions and their graphs, you will notice that the degree of a polynomial plays an important role in determining other characteristics of the polynomial function and its graph.

Explore the Concept

For each function, identify the degree of the function and whether the degree of the function is even or odd. Identify the leading coefficient and whether the leading coefficient is positive or negative. Use a graphing utility to graph each function. Describe the relationship between the degree and sign of the leading coefficient of the function, and the right- and left-hand behavior of the graph of the function.

a. $y = x^3 - 2x^2 - x + 1$

b. $y = 2x^5 + 2x^2 - 5x + 1$

c. $y = -2x^5 - x^2 + 5x + 3$

d. $y = -x^3 + 5x - 2$

e. $y = 2x^2 + 3x - 4$

f. $y = x^4 - 3x^2 + 2x - 1$

g. $y = -x^2 + 3x + 2$

h. $y = -x^6 - x^2 - 5x + 4$

Remark

The notation "$f(x) \to -\infty$ as $x \to -\infty$" indicates that the graph falls to the left. The notation "$f(x) \to \infty$ as $x \to \infty$" indicates that the graph rises to the right.

EXAMPLE 2 Applying the Leading Coefficient Test

Use the Leading Coefficient Test to describe the right-hand and left-hand behavior of the graph of $f(x) = -x^3 + 4x$.

Solution

Because the degree is odd and the leading coefficient is negative, the graph rises to the left and falls to the right, as shown in the figure.

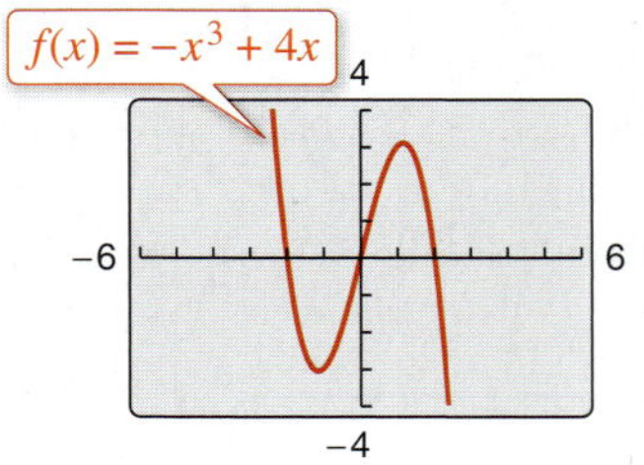

✓ Checkpoint *Audio-video solution in English & Spanish at LarsonPrecalculus.com.*

Use the Leading Coefficient Test to describe the right-hand and left-hand behavior of the graph of $f(x) = 2x^3 - 3x^2 + 5$.

EXAMPLE 3 Applying the Leading Coefficient Test

Use the Leading Coefficient Test to describe the right-hand and left-hand behavior of the graph of each polynomial function.

a. $f(x) = x^4 - 5x^2 + 4$ **b.** $f(x) = x^5 - x$

Solution

a. Because the degree is even and the leading coefficient is positive, the graph rises to the left and right, as shown in the figure.

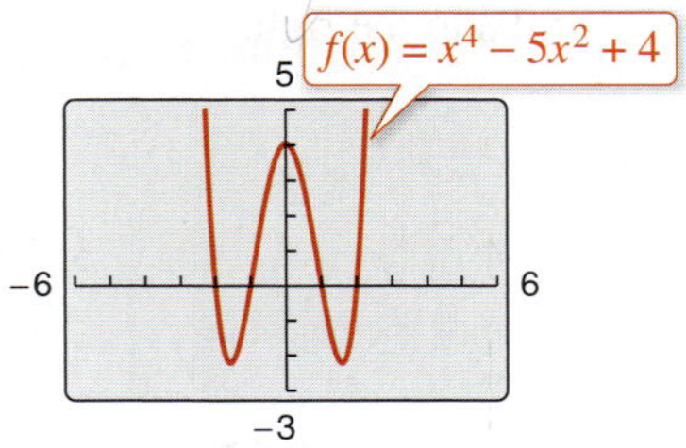

b. Because the degree is odd and the leading coefficient is positive, the graph falls to the left and rises to the right, as shown in the figure.

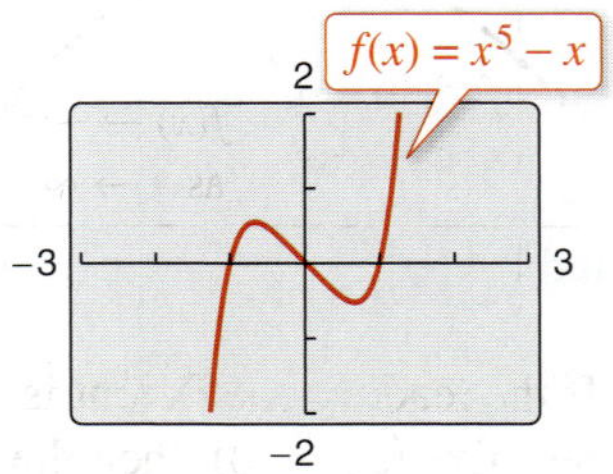

✓ Checkpoint *Audio-video solution in English & Spanish at LarsonPrecalculus.com.*

Use the Leading Coefficient Test to describe the right-hand and left-hand behavior of the graph of each polynomial function.

a. $f(x) = -x^4 + 2x^2 + 6$ **b.** $f(x) = -x^5 + 3x^4 - x$

In Examples 2 and 3, note that the Leading Coefficient Test tells you only whether the graph *eventually* rises or falls to the right or left. Other characteristics of the graph, such as intercepts and minimum and maximum points, must be determined by other tests.

Explore the Concept

For each of the graphs in Examples 2 and 3, count the number of zeros of the polynomial function and the number of relative extrema, and compare these numbers with the degree of the polynomial. What do you observe?

Zeros of Polynomial Functions

It can be shown that for a polynomial function f of degree n, the following statements are true.

1. The function f has at most n real zeros. (You will study this result in detail in Section 3.4 on the Fundamental Theorem of Algebra.)
2. The graph of f has at most $n - 1$ relative **extrema** (relative minima or maxima).

Recall that a zero of a function f is a number x for which $f(x) = 0$. Finding the zeros of polynomial functions is one of the most important problems in algebra. You have already seen that there is a strong interplay between graphical and algebraic approaches to this problem. Sometimes you can use information about the graph of a function to help find its zeros. In other cases, you can use information about the zeros of a function to find a good viewing window.

Real Zeros of Polynomial Functions

When f is a polynomial function and a is a real number, the following statements are equivalent.

1. $x = a$ is a *zero* of the function f.
2. $x = a$ is a *solution* of the polynomial equation $f(x) = 0$.
3. $(x - a)$ is a *factor* of the polynomial $f(x)$.
4. $(a, 0)$ is an *x-intercept* of the graph of f.

Finding zeros of polynomial functions is closely related to factoring and finding x-intercepts, as demonstrated in Examples 4, 5, and 6.

EXAMPLE 4 Finding Zeros of a Polynomial Function

Find all real zeros of $f(x) = x^3 - x^2 - 2x$.

Algebraic Solution

$f(x) = x^3 - x^2 - 2x$ — Write original function.

$0 = x^3 - x^2 - 2x$ — Substitute 0 for $f(x)$.

$0 = x(x^2 - x - 2)$ — Remove common monomial factor.

$0 = x(x - 2)(x + 1)$ — Factor completely.

So, the real zeros are $x = 0$, $x = 2$, and $x = -1$, and the corresponding x-intercepts are $(0, 0)$, $(2, 0)$, and $(-1, 0)$.

Check

$(0)^3 - (0)^2 - 2(0) = 0$ — $x = 0$ is a zero. ✓

$(2)^3 - (2)^2 - 2(2) = 0$ — $x = 2$ is a zero. ✓

$(-1)^3 - (-1)^2 - 2(-1) = 0$ — $x = -1$ is a zero. ✓

Graphical Solution

The graph of f has the x-intercepts

$(0, 0)$, $(2, 0)$, and $(-1, 0)$

as shown in the figure. So, the real zeros of f are

$x = 0$, $x = 2$, and $x = -1$.

Use the *zero* or *root* feature of a graphing utility to verify these zeros.

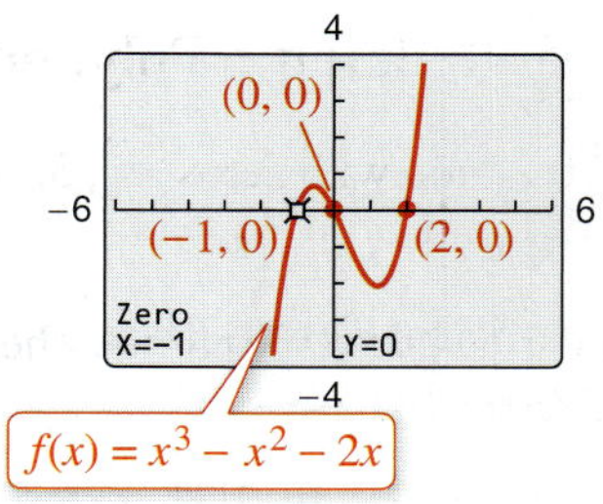

✓ Checkpoint Audio-video solution in English & Spanish at LarsonPrecalculus.com.

Find all real zeros of $f(x) = x^3 + x^2 - 6x$.

EXAMPLE 5 Analyzing a Polynomial Function

Find all real zeros and relative extrema of $f(x) = -2x^4 + 2x^2$.

Solution

$0 = -2x^4 + 2x^2$ — Substitute 0 for $f(x)$.

$0 = -2x^2(x^2 - 1)$ — Remove common monomial factor.

$0 = -2x^2(x - 1)(x + 1)$ — Factor completely.

So, the real zeros are $x = 0$, $x = 1$, and $x = -1$, and the corresponding x-intercepts are $(0, 0)$, $(1, 0)$, and $(-1, 0)$, as shown in Figure 3.11. Using the *minimum* and *maximum* features of a graphing utility, you can approximate the three relative extrema to be $(-0.71, 0.5)$, $(0, 0)$, and $(0.71, 0.5)$.

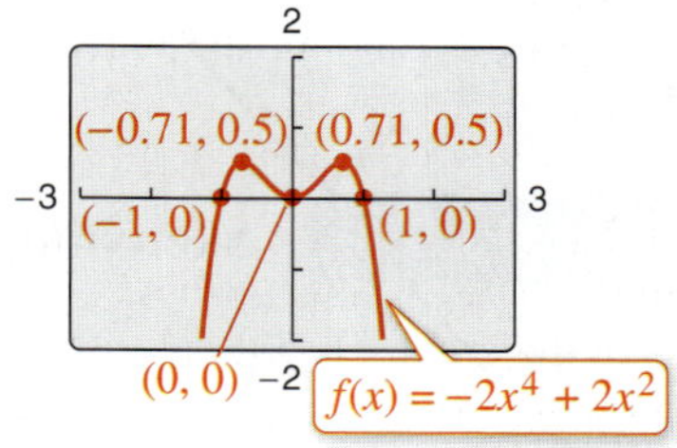

Figure 3.11

✓ Checkpoint *Audio-video solution in English & Spanish at LarsonPrecalculus.com.*

Find all real zeros and relative extrema of $f(x) = x^3 + 2x^2 - 3x$.

Repeated Zeros

For a polynomial function, a factor of $(x - a)^k$, $k > 1$, yields a **repeated zero** $x = a$ of **multiplicity** k.

1. If k is odd, then the graph *crosses* the x-axis at $x = a$.
2. If k is even, then the graph *touches* the x-axis (but does not cross the x-axis) at $x = a$.

Remark

In Example 5, note that because k is even, the factor $-2x^2$ yields the repeated zero $x = 0$. The graph touches (but does not cross) the x-axis at $x = 0$, as shown in Figure 3.11.

EXAMPLE 6 Analyzing a Polynomial Function

To find all real zeros of $f(x) = x^5 - 3x^3 - x^2 - 4x - 1$, use the *zero* feature of a graphing utility. From Figure 3.12, the zeros are $x \approx -1.86$, $x \approx -0.25$, and $x \approx 2.11$. Note that this fifth-degree polynomial factors as

$$f(x) = x^5 - 3x^3 - x^2 - 4x - 1 = (x^2 + 1)(x^3 - 4x - 1).$$

The three zeros obtained above are the zeros of the cubic factor $x^3 - 4x - 1$. The quadratic factor $x^2 + 1$ has no real zeros but does have two *imaginary* zeros. You will learn more about imaginary zeros in Section 3.4.

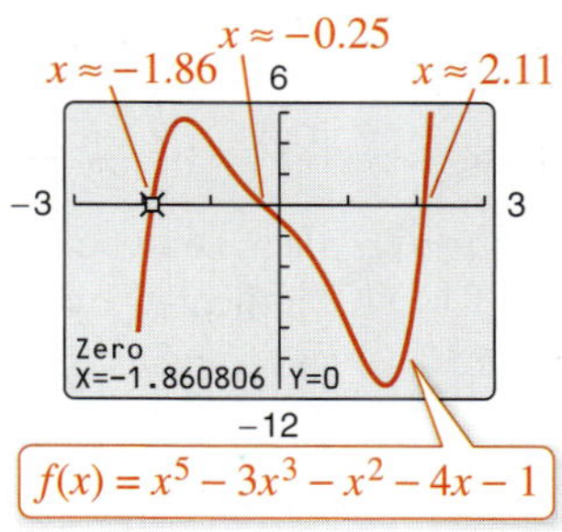

Figure 3.12

✓ Checkpoint *Audio-video solution in English & Spanish at LarsonPrecalculus.com.*

Find all real zeros of $f(x) = x^5 - 2x^4 + x^3 - 2x^2$.

EXAMPLE 7 Finding a Polynomial Function with Given Zeros

Find a polynomial function with zeros $-\frac{1}{2}$, 3, and 3. (There are many correct solutions.)

Solution

Note that the zero $x = -\frac{1}{2}$ corresponds to either $\left(x + \frac{1}{2}\right)$ or $(2x + 1)$. To avoid fractions, choose the second factor and write

$$f(x) = (2x + 1)(x - 3)^2 = (2x + 1)(x^2 - 6x + 9) = 2x^3 - 11x^2 + 12x + 9.$$

✓ Checkpoint *Audio-video solution in English & Spanish at LarsonPrecalculus.com.*

Find a polynomial function with zeros -2, -1, 1, and 2. (There are many correct solutions.)

Note in Example 7 that there are many polynomial functions with the indicated zeros. In fact, multiplying the function by any real number does not change the zeros of the function. For instance, multiply the function from Example 7 by $\frac{1}{2}$ to obtain

$$f(x) = x^3 - \tfrac{11}{2}x^2 + 6x + \tfrac{9}{2}.$$

Then find the zeros of the function. You will obtain the zeros $-\frac{1}{2}$, 3, and 3, as given in Example 7.

EXAMPLE 8 Sketching the Graph of a Polynomial Function

Sketch the graph of $f(x) = -2x^3 + 6x^2 - \frac{9}{2}x$.

Solution

1. *Apply the Leading Coefficient Test.* Because the leading coefficient is negative and the degree is odd, you know that the graph eventually rises to the left and falls to the right (see Figure 3.13).

2. *Find the Real Zeros of the Polynomial.* By factoring

$$\begin{aligned} f(x) &= -2x^3 + 6x^2 - \tfrac{9}{2}x \\ &= -\tfrac{1}{2}x(4x^2 - 12x + 9) \\ &= -\tfrac{1}{2}x(2x - 3)^2 \end{aligned}$$

you can see that the real zeros of f are $x = 0$ (of odd multiplicity 1) and $x = \frac{3}{2}$ (of even multiplicity 2). So, the x-intercepts occur at $(0, 0)$ and $\left(\frac{3}{2}, 0\right)$. Add these points to your graph, as shown in Figure 3.13.

3. *Plot a Few Additional Points.* To sketch the graph, find a few additional points, as shown in the table. Then plot the points (see Figure 3.14).

x	-0.5	0.5	1	2
$f(x)$	4	-1	-0.5	-1

4. *Draw the Graph.* Draw a continuous curve through the points, as shown in Figure 3.14. As indicated by the multiplicities of the zeros, the graph crosses the x-axis at $(0, 0)$ and touches (but does not cross) the x-axis at $\left(\frac{3}{2}, 0\right)$.

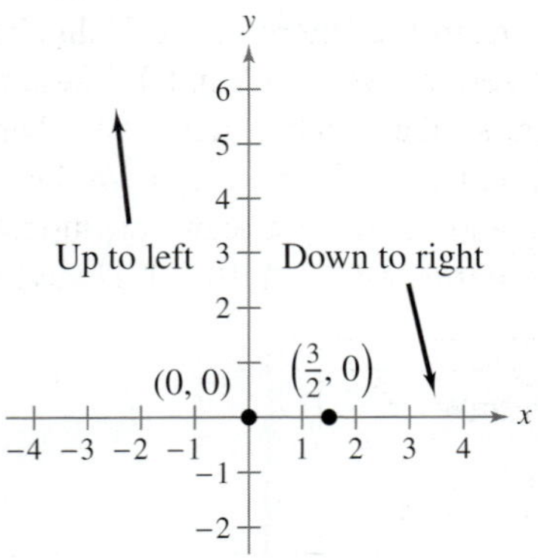

Figure 3.13

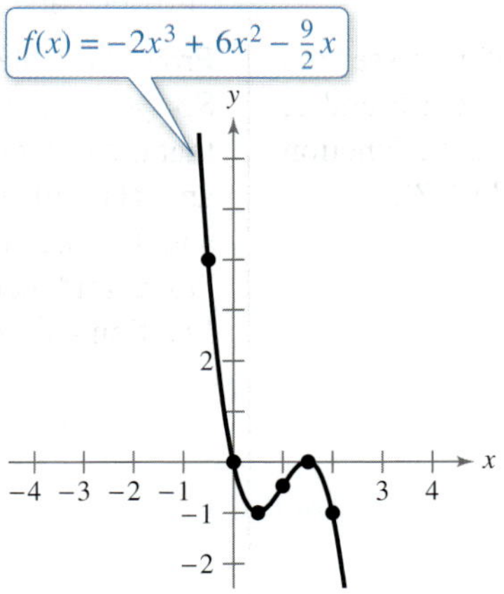

Figure 3.14

Checkpoint *Audio-video solution in English & Spanish at LarsonPrecalculus.com.*

Sketch the graph of $f(x) = -\frac{1}{4}x^4 + \frac{3}{2}x^3 - \frac{9}{4}x^2$.

Remark

Observe in Example 8 that the sign of $f(x)$ is positive to the left of and negative to the right of the zero $x = 0$. Similarly, the sign of $f(x)$ is negative to the left and to the right of the zero $x = \frac{3}{2}$. This suggests that if a zero of a polynomial function is of *odd* multiplicity, then the sign of $f(x)$ changes from one side of the zero to the other side. If a zero is of *even* multiplicity, then the sign of $f(x)$ does not change from one side of the zero to the other side. The following table helps to illustrate this result.

x	-0.5	0	0.5
$f(x)$	4	0	-1
Sign	$+$		$-$

x	1	$\frac{3}{2}$	2
$f(x)$	-0.5	0	-1
Sign	$-$		$-$

This sign analysis may be helpful in graphing polynomial functions.

The Intermediate Value Theorem

The **Intermediate Value Theorem** implies that if $(a, f(a))$ and $(b, f(b))$ are two points on the graph of a polynomial function such that $f(a) \neq f(b)$, then for any number d between $f(a)$ and $f(b)$, there must be a number c between a and b such that $f(c) = d$. (See figure shown at the right.)

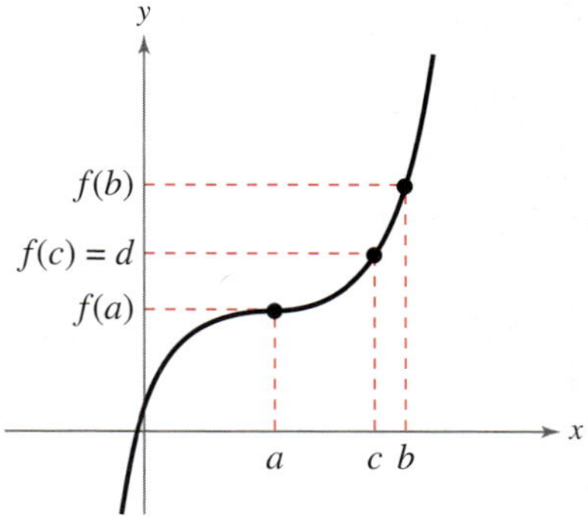

> **Intermediate Value Theorem**
>
> Let a and b be real numbers such that $a < b$. If f is a polynomial function such that $f(a) \neq f(b)$, then in the interval $[a, b]$, f takes on every value between $f(a)$ and $f(b)$.

This theorem helps you locate the real zeros of a polynomial function in the following way. If you can find a value $x = a$ at which a polynomial function is positive, and another value $x = b$ at which it is negative, then you can conclude that the function has at least one real zero between these two values. For example, the function $f(x) = x^3 + x^2 + 1$ is negative when $x = -2$ and positive when $x = -1$. So, it follows from the Intermediate Value Theorem that f must have a real zero somewhere between -2 and -1, as shown in the figure. By continuing this line of reasoning, you can approximate any real zeros of a polynomial function to any desired accuracy.

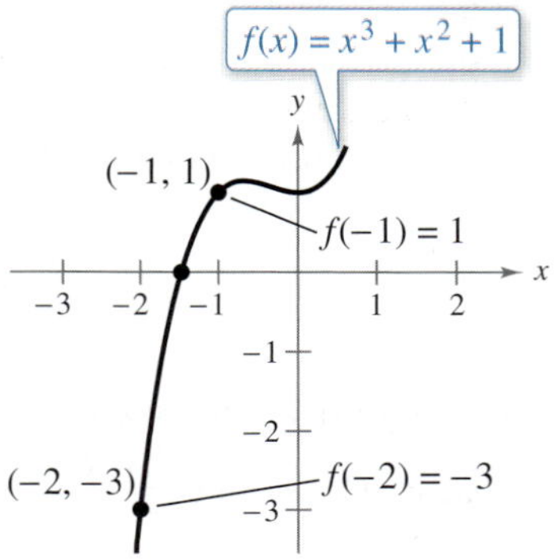

The function f must have a real zero somewhere between -2 and -1.

EXAMPLE 9 Approximating the Zeros of a Function

Find three intervals of length 1 in which the polynomial

$$f(x) = 12x^3 - 32x^2 + 3x + 5$$

is guaranteed to have a zero.

Graphical Solution

From the figure, you can see that the graph of f crosses the x-axis three times—between -1 and 0, between 0 and 1, and between 2 and 3. So, you can conclude that the function has zeros in the intervals $(-1, 0)$, $(0, 1)$, and $(2, 3)$.

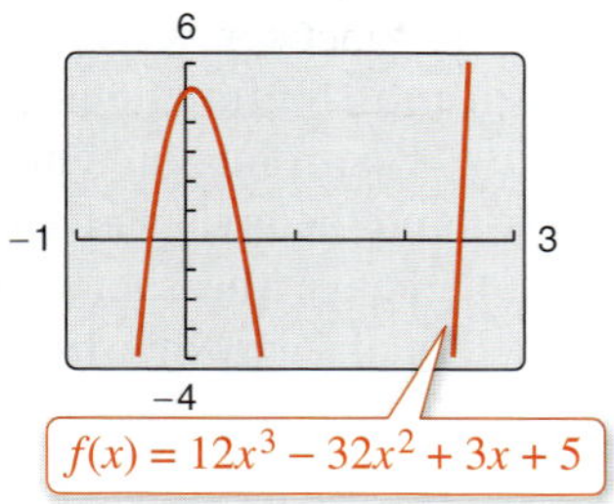

Numerical Solution

From the table, you can see that $f(-1)$ and $f(0)$ differ in sign. So, you can conclude from the Intermediate Value Theorem that the function has a zero between -1 and 0. Similarly, $f(0)$ and $f(1)$ differ in sign, so the function has a zero between 0 and 1. Likewise, $f(2)$ and $f(3)$ differ in sign, so the function has a zero between 2 and 3. So, you can conclude that the function has zeros in the intervals $(-1, 0)$, $(0, 1)$, and $(2, 3)$.

X	Y1
−2	−225
−1	−42
0	5
1	−12
2	−21
3	50
4	273

X=−1

Checkpoint Audio-video solution in English & Spanish at LarsonPrecalculus.com.

Find three intervals of length 1 in which the polynomial $f(x) = x^3 - 4x^2 + 1$ is guaranteed to have a zero.

3.2 Exercises

See *CalcChat.com* for tutorial help and worked-out solutions to odd-numbered exercises.
For instructions on how to use a graphing utility, see Appendix A.

Vocabulary and Concept Check

In Exercises 1–4, fill in the blank(s).

1. The graph of a polynomial function is _______ , so it has no breaks, holes, or gaps.

2. A polynomial function of degree n has at most _______ real zeros and at most _______ relative extrema.

3. When $x = a$ is a zero of a polynomial function f, the following statements are true.

(a) $x = a$ is a _______ of the polynomial equation $f(x) = 0$.

(b) _______ is a factor of the polynomial $f(x)$.

(c) The point _______ is an x-intercept of the graph of f.

4. If a zero of a polynomial function f is of even multiplicity, then the graph of f _______ the x-axis, and if the zero is of odd multiplicity, then the graph of f _______ the x-axis.

For Exercises 5–8, the graph shows the right-hand and left-hand behavior of a polynomial function f.

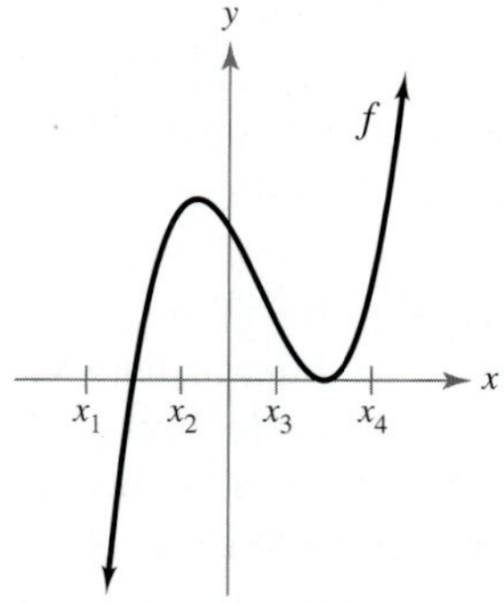

5. Can f be a fourth-degree polynomial function?

6. Can the leading coefficient of f be negative?

7. The graph shows that $f(x_1) < 0$. What other information shown in the graph allows you to apply the Intermediate Value Theorem to guarantee that f has a zero in the interval $[x_1, x_2]$?

8. Is the repeated zero of f in the interval $[x_3, x_4]$ of even or odd multiplicity?

Procedures and Problem Solving

Identifying Graphs of Polynomial Functions **In Exercises 9–16, match the polynomial function with its graph. [The graphs are labeled (a) through (h).]**

(a)
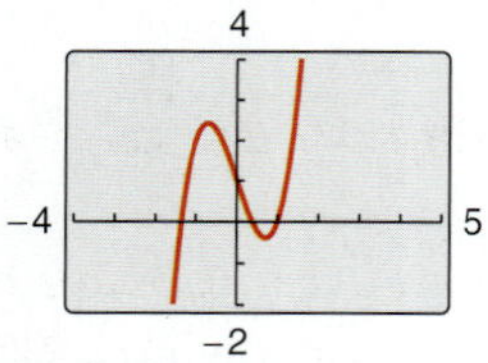

(b)
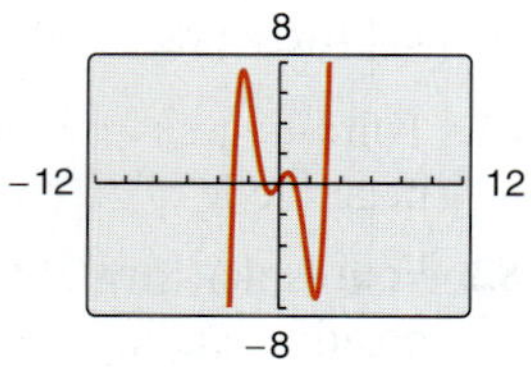

(c)
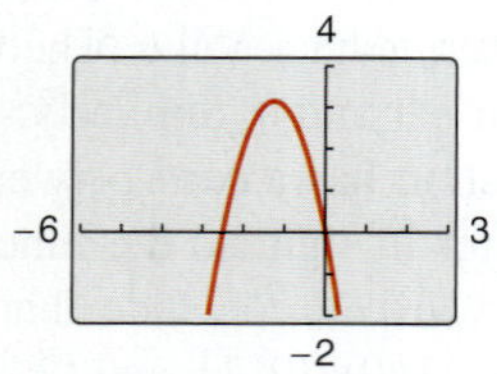

(d)
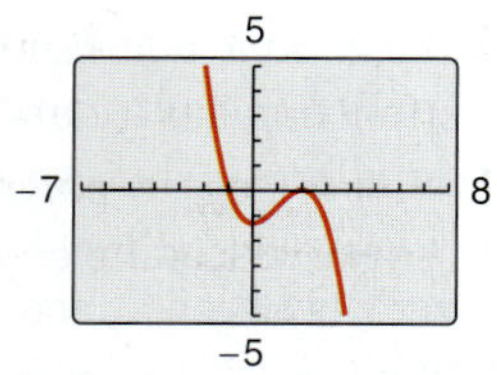

(e)
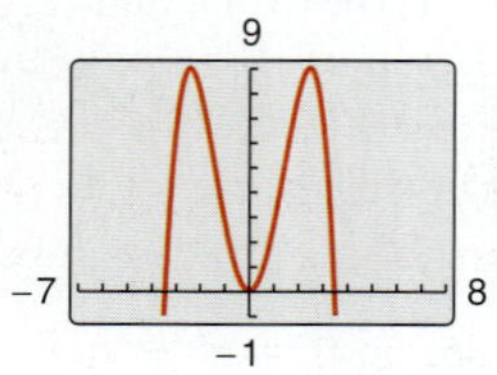

(f)
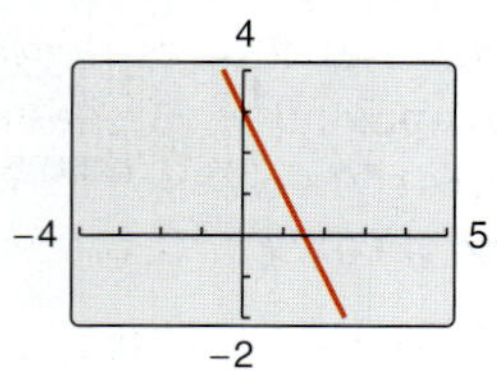

(g)
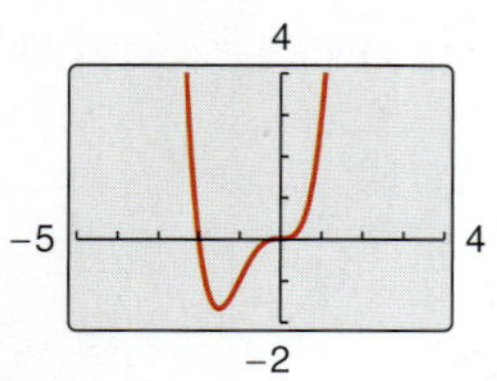

(h)
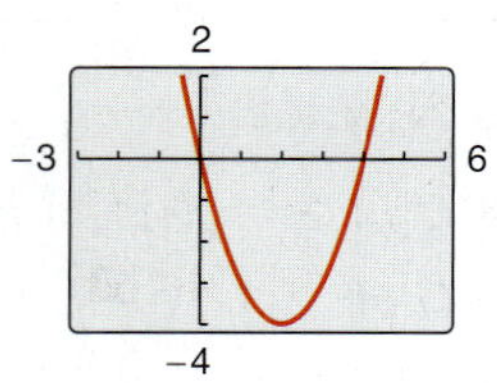

9. $f(x) = -2x + 3$

10. $f(x) = x^2 - 4x$

11. $f(x) = -2x^2 - 5x$

12. $f(x) = 2x^3 - 3x + 1$

13. $f(x) = -\frac{1}{4}x^4 + 3x^2$

14. $f(x) = -\frac{1}{3}x^3 + x^2 - \frac{4}{3}$

15. $f(x) = x^4 + 2x^3$

16. $f(x) = \frac{1}{5}x^5 - 2x^3 + \frac{9}{5}x$

Library of Parent Functions **In Exercises 17–22, sketch the graph of $f(x) = x^3$ and the graph of the function g. Describe the transformation from f to g.**

17. $g(x) = (x - 3)^3$

18. $g(x) = x^3 - 3$

19. $g(x) = -x^3 + 4$

20. $g(x) = (x - 2)^3 - 3$

21. $g(x) = -(x - 3)^3$

22. $g(x) = (x + 4)^3 + 1$

Comparing End Behavior **In Exercises 23–28, use a graphing utility to graph the functions f and g in the same viewing window. Zoom out far enough to see the right-hand and left-hand behavior of each graph. Do the graphs of f and g have the same right-hand and left-hand behavior? Explain why or why not.**

23. $f(x) = 3x^3 - 9x + 1, \quad g(x) = 3x^3$

24. $f(x) = -\frac{1}{3}(x^3 - 3x + 2), \quad g(x) = -\frac{1}{3}x^3$

25. $f(x) = -(x^4 - 4x^3 + 16x), \quad g(x) = -x^4$

26. $f(x) = 3x^4 - 6x^2, \quad g(x) = 3x^4$

27. $f(x) = -2x^3 + 4x^2 - 1, \quad g(x) = 2x^3$

28. $f(x) = -(x^4 - 6x^2 - x + 10), \quad g(x) = x^4$

Applying the Leading Coefficient Test In Exercises 29–36, use the Leading Coefficient Test to describe the right-hand and left-hand behavior of the graph of the polynomial function. Use a graphing utility to verify your results.

29. $f(x) = 2x^4 - 3x + 1$

30. $h(x) = 1 - x^6$

31. $g(x) = 5 - \frac{7}{2}x - 3x^2$

32. $f(x) = \frac{1}{3}x^3 + 5x$

33. $f(x) = \dfrac{6x^5 - 2x^4 + 4x^2 - 5x}{3}$

34. $f(x) = \dfrac{3x^7 - 2x^5 + 5x^3 + 6x^2}{4}$

35. $h(t) = -\frac{2}{3}(t^2 - 5t + 3)$

36. $f(s) = -\frac{7}{8}(s^3 + 5s^2 - 7s + 1)$

Finding Zeros of a Polynomial Function In Exercises 37–48, (a) find the zeros algebraically, (b) use a graphing utility to graph the function, and (c) use the graph to approximate any zeros and compare them with those from part (a).

37. $f(x) = 3x^2 - 12x + 3$

38. $g(x) = 5x^2 - 10x - 5$

39. $g(t) = \frac{1}{2}t^4 - \frac{1}{2}$

40. $y = \frac{1}{4}x^3(x^2 - 9)$

41. $f(x) = x^5 + x^3 - 6x$

42. $g(t) = t^5 - 6t^3 + 9t$

43. $f(x) = 2x^4 - 2x^2 - 40$

44. $f(x) = 5x^4 + 15x^2 + 10$

45. $f(x) = x^3 - 4x^2 - 25x + 100$

46. $y = 4x^3 + 4x^2 - 7x + 2$

47. $y = 4x^3 - 20x^2 + 25x$

48. $y = x^5 - 5x^3 + 4x$

Finding Zeros and Their Multiplicities In Exercises 49–58, find all the real zeros of the polynomial function. Determine the multiplicity of each zero. Use a graphing utility to verify your results.

49. $f(x) = x^2 - 25$

50. $f(x) = 49 - x^2$

51. $h(t) = t^2 - 6t + 9$

52. $f(x) = x^2 + 10x + 25$

53. $f(x) = x^2 + x - 2$

54. $f(x) = 2x^2 - 14x + 24$

55. $f(t) = t^3 - 4t^2 + 4t$

56. $f(x) = x^4 - x^3 - 20x^2$

57. $f(x) = \frac{1}{2}x^2 + \frac{5}{2}x - \frac{3}{2}$

58. $f(x) = \frac{5}{3}x^2 + \frac{8}{3}x - \frac{4}{3}$

Analyzing a Polynomial Function In Exercises 59–64, use a graphing utility to graph the function and approximate (accurate to three decimal places) any real zeros and relative extrema.

59. $f(x) = 2x^4 - 6x^2 + 1$

60. $f(x) = -\frac{3}{8}x^4 - x^3 + 2x^2 + 5$

61. $f(x) = x^5 + 3x^3 - x + 6$

62. $f(x) = -3x^3 - 4x^2 + x - 3$

63. $f(x) = -2x^4 + 5x^2 - x - 1$

64. $f(x) = 3x^5 - 2x^2 - x + 1$

Finding a Polynomial Function with Given Zeros In Exercises 65–74, find a polynomial function that has the given zeros. (There are many correct answers.)

65. $0, 7$

66. $-2, 5$

67. $0, -2, -4$

68. $0, 1, 6$

69. $4, -3, 3, 0$

70. $-2, -1, 0, 1, 2$

71. $1 + \sqrt{2}, 1 - \sqrt{2}$

72. $4 + \sqrt{3}, 4 - \sqrt{3}$

73. $2, 2 + \sqrt{5}, 2 - \sqrt{5}$

74. $3, 2 + \sqrt{7}, 2 - \sqrt{7}$

Finding a Polynomial Function with Given Zeros In Exercises 75–80, find a polynomial function with the given zeros, multiplicities, and degree. (There are many correct answers.)

75. Zero: -2, multiplicity: 2
Zero: -1, multiplicity: 1
Degree: 3

76. Zero: 3, multiplicity: 1
Zero: 2, multiplicity: 3
Degree: 4

77. Zero: -4, multiplicity: 2
Zero: 3, multiplicity: 2
Degree: 4

78. Zero: 5, multiplicity: 3
Zero: 0, multiplicity: 2
Degree: 5

79. Zero: -1, multiplicity: 2
Zero: -2, multiplicity: 1
Degree: 3
Rises to the left,
Falls to the right

80. Zero: 1, multiplicity: 2
Zero: 4, multiplicity: 2
Degree: 4
Falls to the left,
Falls to the right

Sketching a Polynomial with Given Conditions In Exercises 81–84, sketch the graph of a polynomial function that satisfies the given conditions. If not possible, explain your reasoning. (There are many correct answers.)

81. Third-degree polynomial with two real zeros and a negative leading coefficient

82. Fourth-degree polynomial with three real zeros and a positive leading coefficient

83. Fifth-degree polynomial with three real zeros and a positive leading coefficient

84. Fourth-degree polynomial with two real zeros and a negative leading coefficient

Sketching the Graph of a Polynomial Function In Exercises 85–94, sketch the graph of the function by (a) applying the Leading Coefficient Test, (b) finding the zeros of the polynomial, (c) plotting sufficient solution points, and (d) drawing a continuous curve through the points.

85. $f(x) = x^3 - 9x$

86. $g(x) = x^4 - 4x^2$

87. $f(x) = x^3 - 3x^2$

88. $f(x) = 3x^3 - 24x^2$

89. $f(x) = -x^4 + 9x^2 - 20$

90. $f(x) = -x^6 + 7x^3 + 8$

91. $f(x) = x^3 + 3x^2 - 9x - 27$

92. $h(x) = x^5 - 4x^3 + 8x^2 - 32$

93. $g(t) = -\frac{1}{4}t^4 + 2t^2 - 4$

94. $g(x) = \frac{1}{10}(x^4 - 4x^3 - 2x^2 + 12x + 9)$

Approximating the Zeros of a Function **In Exercises 95–100, (a) use the Intermediate Value Theorem and a graphing utility to find graphically any intervals of length 1 in which the polynomial function is guaranteed to have a zero, and (b) use the *zero* or *root* feature of the graphing utility to approximate the real zeros of the function. Verify your answers in part (a) by using the *table* feature of the graphing utility.**

95. $f(x) = x^3 - 3x^2 + 3$ **96.** $f(x) = -2x^3 - 6x^2 + 3$

97. $g(x) = 3x^4 + 4x^3 - 3$ **98.** $h(x) = x^4 - 10x^2 + 2$

99. $f(x) = x^4 - 3x^3 - 4x - 3$

100. $f(x) = x^3 - 4x^2 - 2x + 10$

Identifying Symmetry and x-Intercepts **In Exercises 101–108, use a graphing utility to graph the function. Identify any symmetry with respect to the *x*-axis, *y*-axis, or origin. Determine the number of *x*-intercepts of the graph.**

101. $f(x) = x^2(x + 6)$ **102.** $h(x) = x^3(x - 3)^2$

103. $g(t) = -\frac{1}{2}(t - 4)^2(t + 4)^2$

104. $g(x) = \frac{1}{8}(x + 1)^2(x - 3)^3$

105. $f(x) = x^3 - 4x$ **106.** $f(x) = x^4 - 2x^2$

107. $g(x) = \frac{1}{5}(x + 1)^2(x - 3)(2x - 9)$

108. $h(x) = \frac{1}{5}(x + 2)^2(3x - 5)^2$

109. Geometry An open box is to be made from a square piece of material 36 centimeters on a side by cutting equal squares with sides of length x from the corners and turning up the sides (see figure).

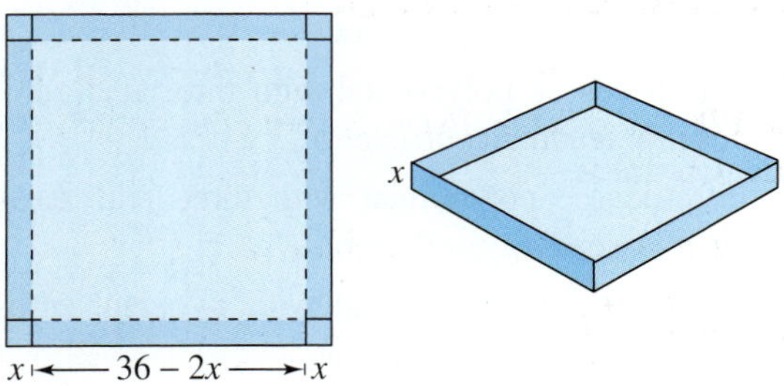

(a) Verify that the volume of the box is given by the function $V(x) = x(36 - 2x)^2$.

(b) Determine the domain of the function V.

(c) Use the *table* feature of a graphing utility to create a table that shows various box heights x and the corresponding volumes V. Use the table to estimate a range of dimensions within which the maximum volume is produced.

(d) Use the graphing utility to graph V and use the range of dimensions from part (c) to find the x-value for which $V(x)$ is maximum.

110. Geometry An open box with locking tabs is to be made from a square piece of material 24 inches on a side. This is done by cutting equal squares from the corners and folding along the dashed lines, as shown in the figure.

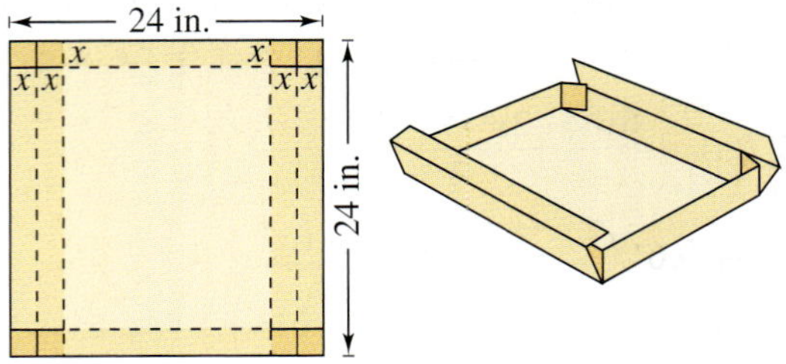

(a) Verify that the volume of the box is given by the function $V(x) = 8x(6 - x)(12 - x)$.

(b) Determine the domain of the function V.

(c) Sketch the graph of the function and estimate the value of x for which $V(x)$ is maximum.

111. Marketing The total revenue R (in millions of dollars) for a company is related to its advertising expense by the function

$$R = 0.00001(-x^3 + 600x^2), \quad 0 \le x \le 400$$

where x is the amount spent on advertising (in tens of thousands of dollars). Use the graph of the function shown in the figure to estimate the point on the graph at which the function is increasing most rapidly. This point is called the **point of diminishing returns** because any expense above this amount will yield less return per dollar invested in advertising.

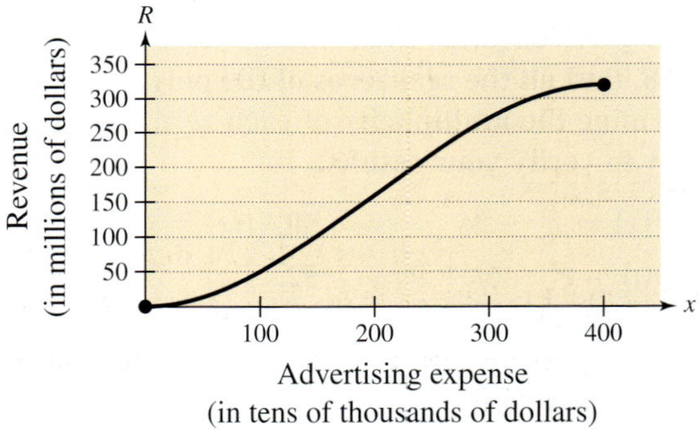

112. ***Why you should learn it*** *(p. 254)* The growth of a red oak tree is approximated by the function

$$G = -0.003t^3 + 0.137t^2 + 0.458t - 0.839$$

where G is the height of the tree (in feet) and $t (2 \le t \le 32)$ is its age (in years). Use a graphing utility to graph the function and estimate the age of the tree when it is growing most rapidly. This point is called the **point of diminishing returns** because the increase in growth will be less with each additional year. (*Hint:* Use a viewing window in which $0 \le x \le 35$ and $0 \le y \le 60$.)

113. MODELING DATA

The U.S. production of crude oil y_1 (in quadrillions of British thermal units) and of solar and photovoltaic energy y_2 (in trillions of British thermal units) are shown in the table for the years 2004 through 2013, where t represents the year, with $t = 4$ corresponding to 2004. These data can be approximated by the models

$$y_1 = 0.00281t^4 - 0.0850t^3 + 1.027t^2 - 5.71t + 22.7$$

and

$$y_2 = 0.618t^3 - 10.80t^2 + 66.2t - 71.$$

(Source: Energy Information Administration)

DATA

Year, t	y_1	y_2
4	11.6	63
5	11.0	63
6	10.8	68
7	10.7	76
8	10.6	89
9	11.3	98
10	11.6	126
11	12.0	171
12	13.8	234
13	15.7	321

Spreadsheet at LarsonPrecalculus.com

(a) Use a graphing utility to plot the data and graph the model for y_1 in the same viewing window. How closely does the model represent the data?

(b) Extend the viewing window of the graphing utility to show the right-hand behavior of the model y_1. Would you use the model to estimate the production of crude oil in 2015? in 2020? Explain.

(c) Repeat parts (a) and (b) for y_2.

Conclusions

True or False? **In Exercises 114–120, determine whether the statement is true or false. Justify your answer.**

114. It is possible for a sixth-degree polynomial to have only one zero.

115. It is possible for a fifth-degree polynomial to have no real zeros.

116. It is possible for a polynomial with an even degree to have a range of $(-\infty, \infty)$.

117. The graph of the function $f(x) = x^6 - x^7$ rises to the left and falls to the right.

118. The graph of the function $f(x) = 2x(x - 1)^2(x + 3)^3$ crosses the x-axis at $x = 1$.

119. The graph of the function $f(x) = 2x(x - 1)^2(x + 3)^3$ touches, but does not cross, the x-axis.

120. The graph of the function $f(x) = 2x(x - 1)^2(x + 3)^3$ rises to the left and falls to the right.

121. Exploration Use a graphing utility to graph

$$y_1 = x + 2 \quad \text{and} \quad y_2 = (x + 2)(x - 1).$$

Predict the shape of the graph of

$$y_3 = (x + 2)(x - 1)(x - 3).$$

Use the graphing utility to verify your answer.

122. HOW DO YOU SEE IT? For each graph, describe a polynomial function that could represent the graph. (Indicate the degree of the function and the sign of its leading coefficient.)

(a)

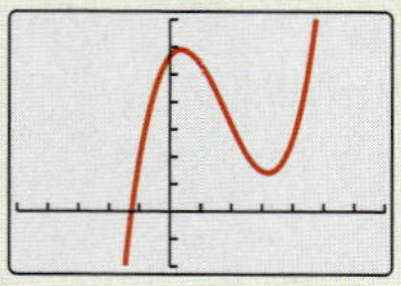

(b)

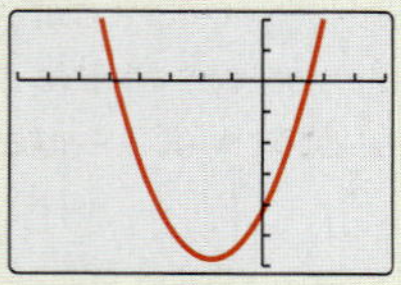

(c)

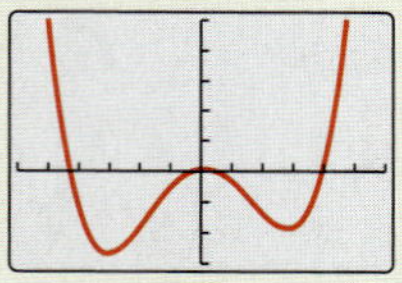

(d)

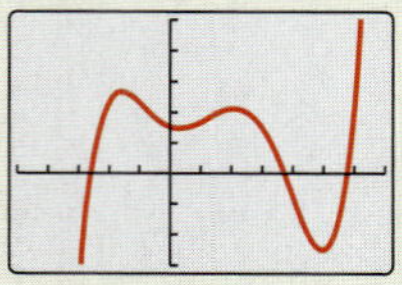

Cumulative Mixed Review

Evaluating Combinations of Functions **In Exercises 123–128, let $f(x) = 14x - 3$ and $g(x) = 8x^2$. Find the indicated value.**

123. $(f + g)(-4)$

124. $(g - f)(3)$

125. $(fg)\left(-\frac{4}{7}\right)$

126. $\left(\frac{f}{g}\right)(-1.5)$

127. $(f \circ g)(-1)$

128. $(g \circ f)(0)$

Solving Inequalities **In Exercises 129–132, solve the inequality and sketch the solution on the real number line. Use a graphing utility to verify your solution graphically.**

129. $3(x - 5) < 4x - 7$

130. $2x^2 - x \geq 1$

131. $\dfrac{5x - 2}{x - 7} \leq 4$

132. $|x + 8| - 1 \geq 15$

3.3 Real Zeros of Polynomial Functions

Long Division of Polynomials

In the graph of $f(x) = 6x^3 - 19x^2 + 16x - 4$ shown in Figure 3.15, it appears that $x = 2$ is a zero of f. Because $f(2) = 0$, you know that $x = 2$ is a zero of the polynomial function f, and that $(x - 2)$ is a factor of $f(x)$. This means that there exists a second-degree polynomial $q(x)$ such that $f(x) = (x - 2) \cdot q(x)$. To find $q(x)$, you can use **long division of polynomials.**

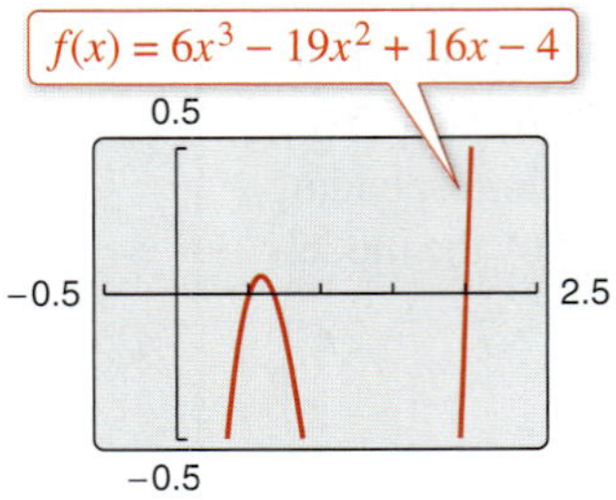

Figure 3.15

What you should learn

- Use long division to divide polynomials by other polynomials.
- Use synthetic division to divide polynomials by binomials of the form $(x - k)$.
- Use the Remainder and Factor Theorems.
- Use the Rational Zero Test to determine possible rational zeros of polynomial functions.
- Use Descartes's Rule of Signs and the Upper and Lower Bound Rules to find zeros of polynomials.

Why you should learn it

The Remainder Theorem can be used to determine the number of employees in education and health services in the United States in a given year based on a polynomial model, as shown in Exercise 106 on page 280.

EXAMPLE 1 Long Division of Polynomials

Divide the polynomial $6x^3 - 19x^2 + 16x - 4$ by $x - 2$, and use the result to factor the polynomial completely.

Solution

Think $\frac{6x^3}{x} = 6x^2$.

Think $\frac{-7x^2}{x} = -7x$.

Think $\frac{2x}{x} = 2$.

$$
\begin{array}{r|l}
 & 6x^2 - 7x + 2 \\
x - 2 & 6x^3 - 19x^2 + 16x - 4 \\
 & \underline{6x^3 - 12x^2} \quad \text{Multiply: } 6x^2(x - 2). \\
 & -7x^2 + 16x \quad \text{Subtract.} \\
 & \underline{-7x^2 + 14x} \quad \text{Multiply: } -7x(x - 2). \\
 & 2x - 4 \quad \text{Subtract.} \\
 & \underline{2x - 4} \quad \text{Multiply: } 2(x - 2). \\
 & 0 \quad \text{Subtract.}
\end{array}
$$

You can see that

$$6x^3 - 19x^2 + 16x - 4 = (x - 2)(6x^2 - 7x + 2)$$
$$= (x - 2)(2x - 1)(3x - 2).$$

Note that this factorization agrees with the graph of f (see Figure 3.15) in that the three x-intercepts occur at $x = 2$, $x = \frac{1}{2}$, and $x = \frac{2}{3}$.

Checkpoint Audio-video solution in English & Spanish at LarsonPrecalculus.com.

Divide the polynomial $3x^2 + 19x + 28$ by $x + 4$, and use the result to factor the polynomial completely.

Note that in Example 1, the division process requires $-7x^2 + 14x$ to be subtracted from $-7x^2 + 16x$. So, it is implied that

$$\begin{array}{r} -7x^2 + 16x \\ \underline{-(-7x^2 + 14x)} \end{array} = \begin{array}{r} -7x^2 + 16x \\ \underline{7x^2 - 14x} \end{array}$$

and instead is written simply as

$$\begin{array}{r} -7x^2 + 16x \\ \underline{-7x^2 + 14x} \\ 2x. \end{array}$$

In Example 1, $x - 2$ is a factor of the polynomial

$$6x^3 - 19x^2 + 16x - 4$$

and the long division process produces a remainder of zero. Often, long division will produce a nonzero remainder. For instance, when you divide $x^2 + 3x + 5$ by $x + 1$, you obtain the following.

$$\begin{array}{rl} & \;x + 2 \quad \leftarrow \text{Quotient} \\ \text{Divisor} \rightarrow \quad x + 1 & \overline{)\, x^2 + 3x + 5} \quad \leftarrow \text{Dividend} \\ & \underline{\;x^2 + x\phantom{{}+5}} \\ & \phantom{x^2 +{}} 2x + 5 \\ & \phantom{x^2 +{}} \underline{2x + 2} \\ & \phantom{x^2 + 2x + {}} 3 \quad \leftarrow \text{Remainder} \end{array}$$

In fractional form, you can write this result as follows.

$$\underbrace{\overbrace{x^2 + 3x + 5}^{\text{Dividend}}}_{} \over \underbrace{x + 1}_{\text{Divisor}}$$

$$\frac{\overbrace{x^2 + 3x + 5}^{\text{Dividend}}}{\underbrace{x + 1}_{\text{Divisor}}} = \overbrace{x + 2}^{\text{Quotient}} + \frac{\overset{\text{Remainder}}{3}}{\underbrace{x + 1}_{\text{Divisor}}}$$

This implies that

$$x^2 + 3x + 5 = (x + 1)(x + 2) + 3 \qquad \text{Multiply each side by } (x + 1).$$

which illustrates the following theorem, called the **Division Algorithm.**

The Division Algorithm

If $f(x)$ and $d(x)$ are polynomials such that $d(x) \neq 0$, and the degree of $d(x)$ is less than or equal to the degree of $f(x)$, then there exist unique polynomials $q(x)$ and $r(x)$ such that

$$\underset{\text{Dividend}}{f(x)} = \underset{\text{Divisor}}{d(x)}\,\underset{\text{Quotient}}{q(x)} + \underset{\text{Remainder}}{r(x)}$$

where $r(x) = 0$ *or* the degree of $r(x)$ is less than the degree of $d(x)$. If the remainder $r(x)$ is zero, then $d(x)$ *divides evenly* into $f(x)$.

The Division Algorithm can also be written as

$$\frac{f(x)}{d(x)} = q(x) + \frac{r(x)}{d(x)}.$$

In the Division Algorithm, the rational expression $f(x)/d(x)$ is **improper** because the degree of $f(x)$ is greater than or equal to the degree of $d(x)$. On the other hand, the rational expression $r(x)/d(x)$ is **proper** because the degree of $r(x)$ is less than the degree of $d(x)$.

Before you apply the Division Algorithm, follow these steps.

1. Write the dividend and divisor in descending powers of the variable.

2. Insert placeholders with zero coefficients for missing powers of the variable.

Note how these steps are applied in the next two examples.

EXAMPLE 2 Long Division of Polynomials

Divide $8x^3 - 1$ by $2x - 1$.

Solution

Because there is no x^2-term or x-term in the dividend, you need to line up the subtraction by using zero coefficients (or leaving spaces) for the missing terms.

$$
\begin{array}{r l}
 & \quad 4x^2 + 2x + 1 \\
2x - 1 \,\big) & 8x^3 + 0x^2 + 0x - 1 \\
 & \underline{8x^3 - 4x^2} \\
 & \qquad 4x^2 + 0x \\
 & \qquad \underline{4x^2 - 2x} \\
 & \qquad\qquad 2x - 1 \\
 & \qquad\qquad \underline{2x - 1} \\
 & \qquad\qquad\qquad 0
\end{array}
$$

So, $2x - 1$ divides evenly into $8x^3 - 1$, and you can write

$$\frac{8x^3 - 1}{2x - 1} = 4x^2 + 2x + 1, \quad x \neq \frac{1}{2}.$$

You can check this result by multiplying.

$$
\begin{aligned}
(2x - 1)(4x^2 + 2x + 1) &= 8x^3 + 4x^2 + 2x - 4x^2 - 2x - 1 \\
&= 8x^3 - 1
\end{aligned}
$$

✔ **Checkpoint** *Audio-video solution in English & Spanish at LarsonPrecalculus.com.*

Divide $x^3 - 2x^2 - 9$ by $x - 3$.

In each of the long division examples presented so far, the divisor has been a first-degree polynomial. The long division algorithm works just as well with polynomial divisors of degree two or more, as shown in Example 3.

EXAMPLE 3 Long Division of Polynomials

Divide $-2 + 3x - 5x^2 + 4x^3 + 2x^4$ by $x^2 + 2x - 3$.

Solution

Begin by writing the dividend in descending powers of x.

$$
\begin{array}{r l}
 & \qquad\qquad 2x^2 \qquad\quad + 1 \\
x^2 + 2x - 3 \,\big) & 2x^4 + 4x^3 - 5x^2 + 3x - 2 \\
 & \underline{2x^4 + 4x^3 - 6x^2} \\
 & \qquad\qquad\quad x^2 + 3x - 2 \\
 & \qquad\qquad\quad \underline{x^2 + 2x - 3} \\
 & \qquad\qquad\qquad\quad x + 1
\end{array}
$$

Note that the first subtraction eliminated two terms from the dividend. When this happens, the quotient skips a term. You can write the result as

$$\frac{2x^4 + 4x^3 - 5x^2 + 3x - 2}{x^2 + 2x - 3} = 2x^2 + 1 + \frac{x + 1}{x^2 + 2x - 3}.$$

✔ **Checkpoint** *Audio-video solution in English & Spanish at LarsonPrecalculus.com.*

Divide $-x^3 + 9x + 6x^4 - x^2 - 3$ by $1 + 3x$.

Synthetic Division

There is a nice shortcut for long division of polynomials when dividing by divisors of the form

$$x - k.$$

The shortcut is called **synthetic division.** The pattern for synthetic division of a cubic polynomial is summarized as follows. (The pattern for higher-degree polynomials is similar.)

Synthetic Division (of a Cubic Polynomial)

To divide $ax^3 + bx^2 + cx + d$ by $x - k$, use the following pattern.

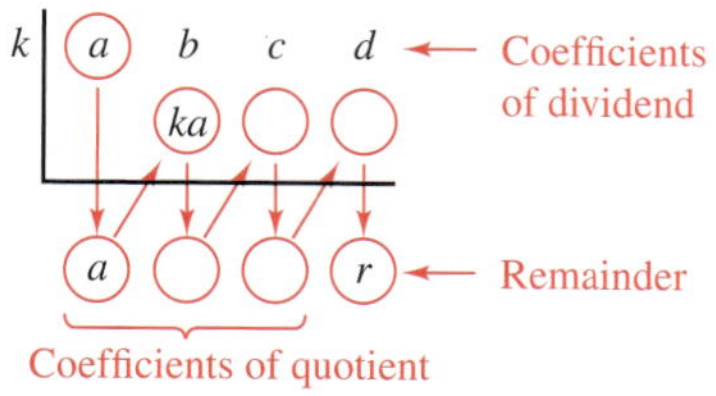

Vertical pattern: Add terms.
Diagonal pattern: Multiply by k.

This algorithm for synthetic division works *only* for divisors of the form $x - k$. Remember that

$$x + k = x - (-k).$$

EXAMPLE 4 Using Synthetic Division

Use synthetic division to divide

$$x^4 - 10x^2 - 2x + 4 \quad \text{by} \quad x + 3.$$

Solution

You should set up the array as follows. Note that a zero is included for the missing term in the dividend.

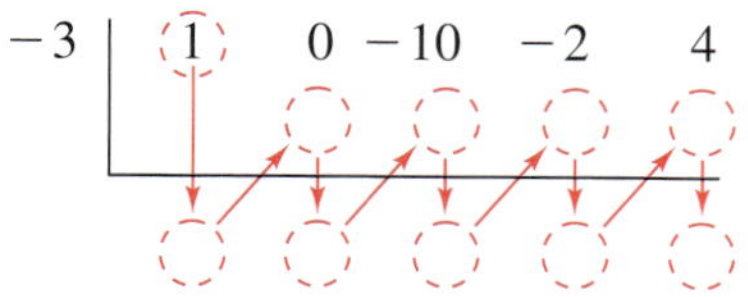

Then, use the synthetic division pattern by adding terms in columns and multiplying the results by -3.

Divisor: $x + 3$ Dividend: $x^4 - 10x^2 - 2x + 4$

-3	1	0	-10	-2	4
		-3	9	3	-3
	1	-3	-1	1	1

Remainder: 1

Quotient: $x^3 - 3x^2 - x + 1$

So, you have

$$\frac{x^4 - 10x^2 - 2x + 4}{x + 3} = x^3 - 3x^2 - x + 1 + \frac{1}{x + 3}.$$

Checkpoint *Audio-video solution in English & Spanish at LarsonPrecalculus.com.*

Use synthetic division to divide $5x^3 + 8x^2 - x + 6$ by $x + 2$.

Explore the Concept

Evaluate the polynomial $x^4 - 10x^2 - 2x + 4$ when $x = -3$. What do you observe?

The Remainder and Factor Theorems

The remainder obtained in the synthetic division process has an important interpretation, as described in the **Remainder Theorem.**

The Remainder Theorem (See the proof on page 321.)

If a polynomial $f(x)$ is divided by $x - k$, then the remainder is

$$r = f(k).$$

The Remainder Theorem tells you that synthetic division can be used to evaluate a polynomial function. That is, to evaluate a polynomial $f(x)$ when $x = k$, divide $f(x)$ by $x - k$. The remainder will be $f(k)$.

EXAMPLE 5 Using the Remainder Theorem

Use the Remainder Theorem to evaluate

$$f(x) = 3x^3 + 8x^2 + 5x - 7$$

when $x = -2$.

Solution

Using synthetic division, you obtain the following.

-2	3	8	5	-7
		-6	-4	-2
	3	2	1	-9

Because the remainder is $r = -9$, you can conclude that

$$f(-2) = -9. \qquad r = f(k)$$

This means that $(-2, -9)$ is a point on the graph of f. You can check this by substituting $x = -2$ in the original function.

Check

$$f(-2) = 3(-2)^3 + 8(-2)^2 + 5(-2) - 7$$
$$= 3(-8) + 8(4) - 10 - 7$$
$$= -24 + 32 - 10 - 7$$
$$= -9$$

✓ ***Checkpoint*** *Audio-video solution in English & Spanish at LarsonPrecalculus.com.*

Use the Remainder Theorem to find each function value given

$$f(x) = 4x^3 + 10x^2 - 3x - 8.$$

a. $f(-1)$ **b.** $f(4)$

c. $f(\frac{1}{2})$ **d.** $f(-3)$

Another important theorem is the **Factor Theorem.** This theorem states that you can test whether a polynomial has $(x - k)$ as a factor by evaluating the polynomial at $x = k$. If the result is 0, then $(x - k)$ is a factor.

The Factor Theorem (See the proof on page 321.)

A polynomial $f(x)$ has a factor $(x - k)$ if and only if $f(k) = 0$.

EXAMPLE 6 Factoring a Polynomial: Repeated Division

Show that $(x - 2)$ and $(x + 3)$ are factors of

$$f(x) = 2x^4 + 7x^3 - 4x^2 - 27x - 18.$$

Then find the remaining factors of $f(x)$.

Algebraic Solution

Using synthetic division with the factor $(x - 2)$, you obtain the following.

2	2	7	−4	−27	−18
		4	22	36	18
	2	11	18	9	0

→ 0 remainder; $(x - 2)$ is a factor.

Take the result of this division and perform synthetic division again using the factor $(x + 3)$.

−3	2	11	18	9
		−6	−15	−9
	2	5	3	0

$2x^2 + 5x + 3$ → 0 remainder; $(x + 3)$ is a factor.

Because the resulting quadratic factors as

$$2x^2 + 5x + 3 = (2x + 3)(x + 1)$$

the complete factorization of $f(x)$ is

$$f(x) = (x - 2)(x + 3)(2x + 3)(x + 1).$$

Graphical Solution

From the graph of

$$f(x) = 2x^4 + 7x^3 - 4x^2 - 27x - 18$$

you can see that there are four x-intercepts (see Figure 3.16). These occur at $x = -3$, $x = -\frac{3}{2}$, $x = -1$, and $x = 2$. (Check this algebraically.) This implies that $(x + 3)$, $\left(x + \frac{3}{2}\right)$, $(x + 1)$, and $(x - 2)$ are factors of $f(x)$. [Note that $\left(x + \frac{3}{2}\right)$ and $(2x + 3)$ are equivalent factors because they both yield the same zero, $x = -\frac{3}{2}$.]

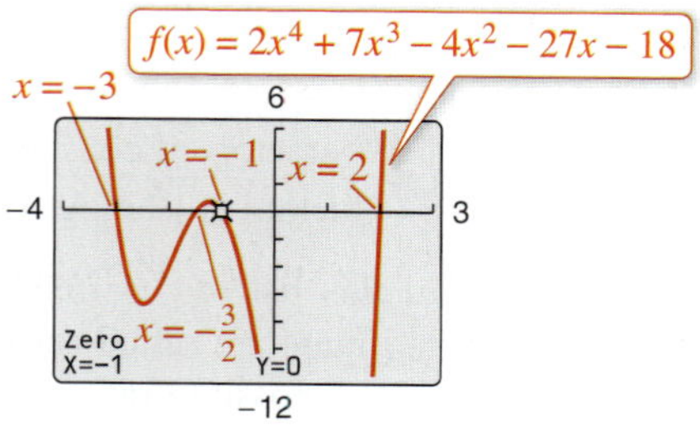

Figure 3.16

✓ **Checkpoint** Audio-video solution in English & Spanish at LarsonPrecalculus.com.

Show that $(x + 3)$ is a factor of $f(x) = x^3 - 19x - 30$. Then find the remaining factors of $f(x)$.

Note in Example 6 that the complete factorization of $f(x)$ implies that f has four real zeros:

$$x = 2, \quad x = -3, \quad x = -\frac{3}{2}, \quad \text{and} \quad x = -1.$$

This is confirmed by the graph of f, which is shown in Figure 3.16.

Using the Remainder in Synthetic Division

In summary, the remainder r, obtained in the synthetic division of a polynomial $f(x)$ by $x - k$, provides the following information.

1. The remainder r gives the *value* of f at $x = k$. That is, $r = f(k)$.
2. If $r = 0$, then $(x - k)$ is a *factor* of $f(x)$.
3. If $r = 0$, then $(k, 0)$ is an *x-intercept* of the graph of f.

Throughout this text, the importance of developing several problem-solving strategies is emphasized. In the exercises for this section, try using more than one strategy to solve several of the exercises. For instance, when you find that $x - k$ divides evenly into $f(x)$, try sketching the graph of f. You should find that $(k, 0)$ is an x-intercept of the graph.

The Rational Zero Test

The **Rational Zero Test** relates the possible rational zeros of a polynomial (having integer coefficients) to the leading coefficient and to the constant term of the polynomial.

The Rational Zero Test

If the polynomial

$$f(x) = a_n x^n + a_{n-1} x^{n-1} + \cdot \cdot \cdot + a_2 x^2 + a_1 x + a_0$$

has integer coefficients, then every rational zero of f has the form

$$\text{Rational zero} = \frac{p}{q}$$

where p and q have no common factors other than 1, p is a factor of the constant term a_0, and q is a factor of the leading coefficient a_n.

Remark

Use a graphing utility to graph the polynomial

$$y = x^3 - 53x^2 + 103x - 51$$

in a standard viewing window. From the graph alone, it appears that there is only one zero. From the Leading Coefficient Test, you know that because the degree of the polynomial is odd and the leading coefficient is positive, the graph falls to the left and rises to the right. So, the function must have another zero. From the Rational Zero Test, you know that ± 51 might be zeros of the function. When you zoom out several times, you will see a more complete picture of the graph. Your graph should confirm that $x = 51$ is a zero of f.

To use the Rational Zero Test, first list all rational numbers whose numerators are factors of the constant term and whose denominators are factors of the leading coefficient.

$$\text{Possible rational zeros} = \frac{\text{factors of constant term}}{\text{factors of leading coefficient}}$$

Now that you have formed this list of *possible rational zeros,* use a trial-and-error method to determine which, if any, are actual zeros of the polynomial. Note that when the leading coefficient is 1, the possible rational zeros are simply the factors of the constant term. This case is illustrated in Example 7.

EXAMPLE 7 Rational Zero Test with Leading Coefficient of 1

Find the rational zeros of $f(x) = x^3 + x + 1$.

Solution

Because the leading coefficient is 1, the possible rational zeros are simply the factors of the constant term.

Possible rational zeros: ± 1

By testing these possible zeros, you can see that neither works.

$$f(1) = (1)^3 + 1 + 1 = 3$$

$$f(-1) = (-1)^3 + (-1) + 1 = -1$$

So, you can conclude that the polynomial has *no* rational zeros. Note from the graph of f shown below that f does have one real zero between -1 and 0. However, by the Rational Zero Test, you know that this real zero is *not* a rational number.

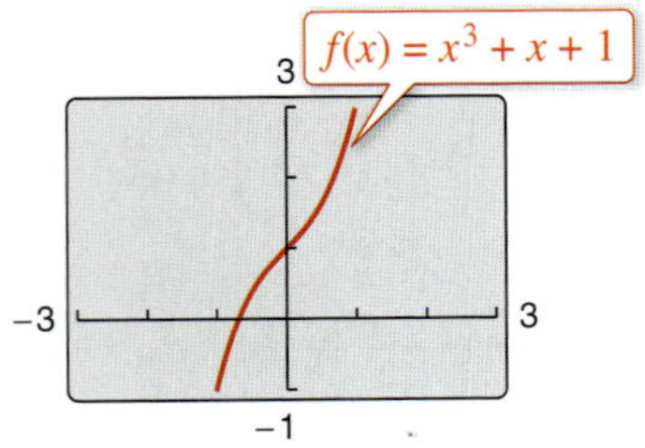

Checkpoint *Audio-video solution in English & Spanish at LarsonPrecalculus.com.*

Find the rational zeros of $f(x) = x^3 - 5x^2 + 2x + 8$.

When the leading coefficient of a polynomial is not 1, the list of possible rational zeros can increase dramatically. In such cases, the search can be shortened in several ways.

1. A graphing utility can be used to speed up the calculations.
2. A graph, drawn either by hand or with a graphing utility, can give good estimates of the locations of the zeros.
3. The Intermediate Value Theorem, along with a table generated by a graphing utility, can give approximations of zeros.
4. The Factor Theorem and synthetic division can be used to test the possible rational zeros.

Finding the first zero is often the most difficult part. After that, the search is simplified by working with the lower-degree polynomial obtained in synthetic division, as shown in Example 8.

EXAMPLE 8 Using the Rational Zero Test

Find the rational zeros of

$$f(x) = 2x^3 + 3x^2 - 8x + 3.$$

Solution

The leading coefficient is 2, and the constant term is 3.

Possible rational zeros: $\dfrac{\text{Factors of 3}}{\text{Factors of 2}} = \dfrac{\pm1, \pm3}{\pm1, \pm2} = \pm1, \pm3, \pm\dfrac{1}{2}, \pm\dfrac{3}{2}$

By synthetic division, you can determine that $x = 1$ is a rational zero.

$$\begin{array}{r|rrrr} 1 & 2 & 3 & -8 & 3 \\ & & 2 & 5 & -3 \\ \hline & 2 & 5 & -3 & 0 \end{array}$$

So, $f(x)$ factors as

$$\begin{aligned} f(x) &= (x - 1)(2x^2 + 5x - 3) \\ &= (x - 1)(2x - 1)(x + 3) \end{aligned}$$

and you can conclude that the rational zeros of f are $x = 1$, $x = \frac{1}{2}$, and $x = -3$, as shown in the figure.

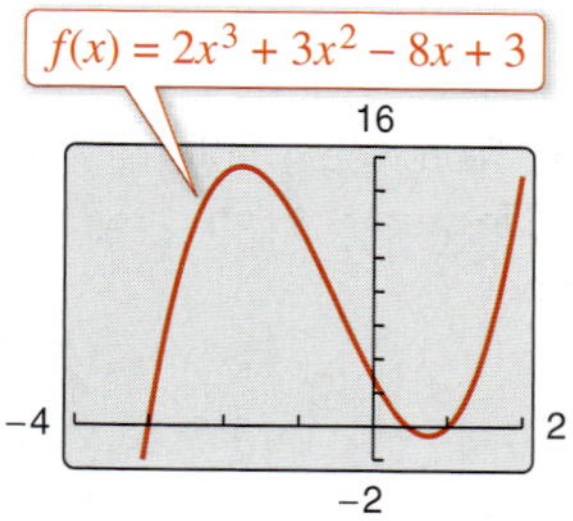

✓ **Checkpoint** *Audio-video solution in English & Spanish at LarsonPrecalculus.com.*

Find the rational zeros of $f(x) = 2x^4 - 9x^3 - 18x^2 + 71x - 30$.

Remember that when you try to find the rational zeros of a polynomial function with many possible rational zeros, as in Example 8, you must use trial and error. There is no quick algebraic method to determine which of the possibilities is an actual zero; however, sketching a graph may be helpful.

Other Tests for Zeros of Polynomials

You know that an nth-degree polynomial function can have *at most n* real zeros. Of course, many nth-degree polynomials do not have that many real zeros. For instance, $f(x) = x^2 + 1$ has no real zeros, and $f(x) = x^3 + 1$ has only one real zero. The following theorem, called **Descartes's Rule of Signs,** sheds more light on the number of real zeros of a polynomial.

> **Decartes's Rule of Signs**
>
> Let $f(x) = a_nx^n + a_{n-1}x^{n-1} + \cdot \cdot \cdot + a_2x^2 + a_1x + a_0$ be a polynomial with real coefficients and $a_0 \neq 0$.
>
> 1. The number of *positive real zeros* of f is either equal to the number of variations in sign of $f(x)$ or less than that number by an even integer.
> 2. The number of *negative real zeros* of f is either equal to the number of variations in sign of $f(-x)$ or less than that number by an even integer.

A **variation in sign** means that two consecutive coefficients have opposite signs. Missing terms (those with zero coefficients) can be ignored.

When using Descartes's Rule of Signs, a zero of multiplicity k should be counted as k zeros. For instance, the polynomial $x^3 - 3x + 2$ has two variations in sign, and so has either two positive or no positive real zeros. Because

$$x^3 - 3x + 2 = (x - 1)(x - 1)(x + 2)$$

you can see that the two positive real zeros are $x = 1$ of multiplicity 2.

EXAMPLE 9 Using Descartes's Rule of Signs

Determine the possible numbers of positive and negative real zeros of

$$f(x) = 3x^3 - 5x^2 + 6x - 4.$$

Solution

The original polynomial has *three* variations in sign.

+ to −   + to −

$$f(x) = 3x^3 - 5x^2 + 6x - 4$$

− to +

The polynomial

$$f(-x) = 3(-x)^3 - 5(-x)^2 + 6(-x) - 4$$
$$= -3x^3 - 5x^2 - 6x - 4$$

has no variations in sign. So, from Descartes's Rule of Signs, the polynomial $f(x) = 3x^3 - 5x^2 + 6x - 4$ has either three positive real zeros or one positive real zero, and has no negative real zeros. By using the *trace* feature of a graphing utility, you can see that the function has only one real zero (it is a positive number near $x = 1$), as shown in the figure.

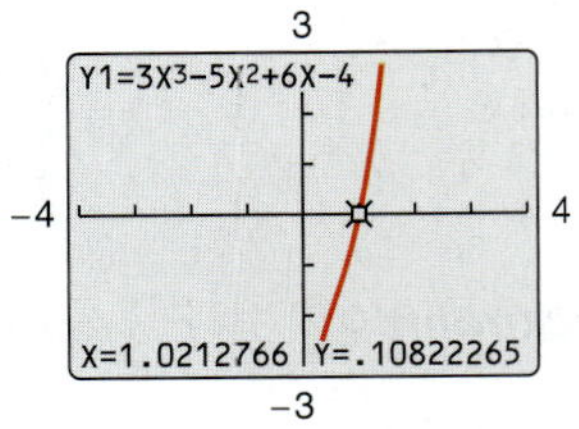

Checkpoint *Audio-video solution in English & Spanish at LarsonPrecalculus.com.*

Determine the possible numbers of positive and negative real zeros of

$$f(x) = -2x^3 + 5x^2 - x + 8.$$

Another test for zeros of a polynomial function is related to the sign pattern in the last row of the synthetic division array. This test can give you an upper or lower bound for the real zeros of f, which can help you eliminate possible real zeros. A real number c is an **upper bound** for the real zeros of f when no zeros are greater than c. Similarly, c is a **lower bound** when no real zeros of f are less than c.

Upper and Lower Bound Rules

Let $f(x)$ be a polynomial with real coefficients and a positive leading coefficient. Suppose $f(x)$ is divided by $x - c$, using synthetic division.

1. If $c > 0$ and each number in the last row is either positive or zero, then c is an **upper bound** for the real zeros of f.
2. If $c < 0$ and the numbers in the last row are alternately positive and negative (zero entries count as positive or negative), then c is a **lower bound** for the real zeros of f.

EXAMPLE 10 Finding the Zeros of a Polynomial Function

Find all real zeros of $f(x) = 6x^3 - 4x^2 + 3x - 2$.

Solution

The possible real zeros are as follows.

$$\frac{\text{Factors of } -2}{\text{Factors of } 6} = \frac{\pm 1, \pm 2}{\pm 1, \pm 2, \pm 3, \pm 6} = \pm 1, \pm \frac{1}{2}, \pm \frac{1}{3}, \pm \frac{1}{6}, \pm \frac{2}{3}, \pm 2$$

The original polynomial $f(x)$ has three variations in sign. The polynomial

$$\begin{aligned} f(-x) &= 6(-x)^3 - 4(-x)^2 + 3(-x) - 2 \\ &= -6x^3 - 4x^2 - 3x - 2 \end{aligned}$$

has no variations in sign. So, you can apply Descartes's Rule of Signs to conclude that there are either three positive real zeros or one positive real zero, and no negative real zeros. Trying $x = 1$ produces the following.

$$\begin{array}{r|rrrr} 1 & 6 & -4 & 3 & -2 \\ & & 6 & 2 & 5 \\ \hline & 6 & 2 & 5 & 3 \end{array}$$

So, $x = 1$ is not a zero, but because the last row has all positive entries, you know that $x = 1$ is an upper bound for the real zeros. Therefore, you can restrict the search to zeros between 0 and 1. By trial and error, you can determine that $x = \frac{2}{3}$ is a zero. So,

$$f(x) = \left(x - \frac{2}{3}\right)(6x^2 + 3).$$

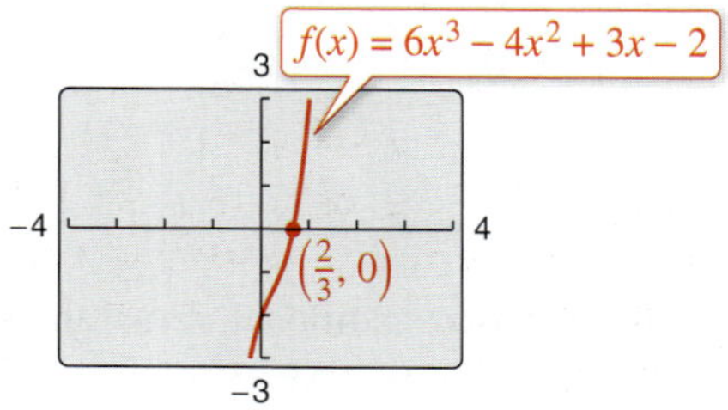

Because $6x^2 + 3$ has no real zeros, it follows that $x = \frac{2}{3}$ is the only real zero, as shown in the figure.

Checkpoint Audio-video solution in English & Spanish at LarsonPrecalculus.com.

Find all real zeros of

$$f(x) = 8x^3 - 4x^2 + 6x - 3.$$

Explore the Concept

Use a graphing utility to graph the polynomial

$$y_1 = 6x^3 - 4x^2 + 3x - 2.$$

Notice that the graph intersects the x-axis at the point $\left(\frac{2}{3}, 0\right)$. How does this information relate to the real zero found in Example 10? Use the graphing utility to graph

$$y_2 = x^4 - 5x^3 + 3x^2 + x.$$

How many times does the graph intersect the x-axis? How many real zeros does y_2 have?

Here are two additional hints that can help you find the real zeros of a polynomial.

1. When the terms of $f(x)$ have a common monomial factor, it should be factored out before applying the tests in this section. For instance, by writing

$$f(x) = x^4 - 5x^3 + 3x^2 + x = x(x^3 - 5x^2 + 3x + 1)$$

you can see that $x = 0$ is a zero of f and that the remaining zeros can be obtained by analyzing the cubic factor.

2. When you are able to find all but two zeros of f, you can always use the Quadratic Formula on the remaining quadratic factor. For instance, after writing

$$f(x) = x^4 - 5x^3 + 3x^2 + x = x(x - 1)(x^2 - 4x - 1)$$

you can apply the Quadratic Formula to $x^2 - 4x - 1$ to conclude that the two remaining zeros are $x = 2 + \sqrt{5}$ and $x = 2 - \sqrt{5}$.

Note how these hints are applied in the next example.

Explore the Concept

Use a graphing utility to graph the polynomial

$$y = x^3 + 4.8x^2 - 127x + 309$$

in a standard viewing window. From the graph, what do the real zeros appear to be? Discuss how the mathematical tools of this section might help you realize that the graph does not show all the important features of the polynomial function. Now use the *zoom* feature to find all the zeros of this function.

EXAMPLE 11 Finding the Zeros of a Polynomial Function

See LarsonPrecalculus.com for an interactive version of this type of example.

Find all real zeros of $f(x) = 10x^4 - 15x^3 - 16x^2 + 12x$.

Solution

Remove the common monomial factor x to write

$$f(x) = 10x^4 - 15x^3 - 16x^2 + 12x = x(10x^3 - 15x^2 - 16x + 12).$$

So, $x = 0$ is a zero of f. You can find the remaining zeros of f by analyzing the cubic factor. Because the leading coefficient is 10 and the constant term is 12, there is a long list of possible rational zeros.

Possible rational zeros:

$$\frac{\text{Factors of }12}{\text{Factors of }10} = \frac{\pm1, \pm2, \pm3, \pm4, \pm6, \pm12}{\pm1, \pm2, \pm5, \pm10}$$

With so many possibilities (32, in fact), it is worth your time to use a graphing utility to focus on just a few. By using the *trace* feature of a graphing utility, it looks like three reasonable choices are $x = -\frac{6}{5}$, $x = \frac{1}{2}$, and $x = 2$ (see figure). Synthetic division shows that only $x = 2$ works. (You could also use the Factor Theorem to test these choices.)

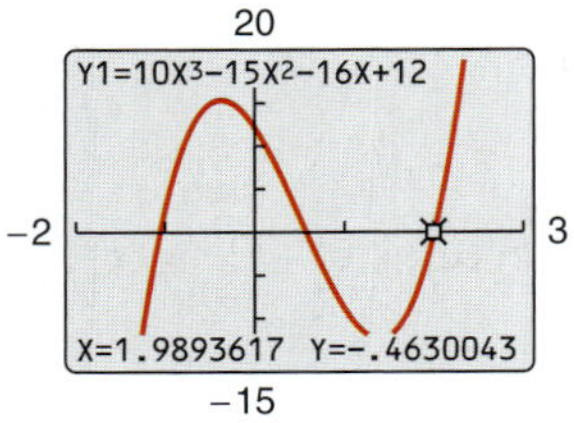

$$\begin{array}{r|rrrr} 2 & 10 & -15 & -16 & 12 \\ & & 20 & 10 & -12 \\ \hline & 10 & 5 & -6 & 0 \end{array}$$

So, $x = 2$ is one zero and you have

$$f(x) = x(x - 2)(10x^2 + 5x - 6).$$

Using the Quadratic Formula, you find that the two additional zeros are irrational numbers.

$$x = \frac{-5 + \sqrt{265}}{20} \approx 0.56 \quad \text{and} \quad x = \frac{-5 - \sqrt{265}}{20} \approx -1.06$$

Checkpoint *Audio-video solution in English & Spanish at LarsonPrecalculus.com.*

Find all real zeros of $f(x) = 3x^4 - 14x^2 - 4x$.

3.3 Exercises

See *CalcChat.com* for tutorial help and worked-out solutions to odd-numbered exercises.
For instructions on how to use a graphing utility, see Appendix A.

Vocabulary and Concept Check

1. Two forms of the Division Algorithm are shown below. Identify and label each part.

$$f(x) = d(x)q(x) + r(x) \qquad \frac{f(x)}{d(x)} = q(x) + \frac{r(x)}{d(x)}$$

In Exercises 2–5, fill in the blank(s).

2. The rational expression $p(x)/q(x)$ is called _______ when the degree of the numerator is greater than or equal to that of the denominator.

3. Every rational zero of a polynomial function with integer coefficients has the form p/q, where p is a factor of the _______ and q is a factor of the _______ .

4. The theorem that can be used to determine the possible numbers of positive real zeros and negative real zeros of a function is called _______ of _______ .

5. A real number c is a(n) _______ bound for the real zeros of f when no zeros are greater than c, and is a(n) _______ bound when no real zeros of f are less than c.

6. How many negative real zeros are possible for a polynomial function f, given that $f(-x)$ has 5 variations in sign?

7. You divide the polynomial $f(x)$ by $(x - 4)$ and obtain a remainder of 7. What is $f(4)$?

8. What value should you write in the circle to check whether $(x - 2)$ is a factor of $f(x) = x^3 + 6x^2 - 5x + 3$?

○ | 1 6 −5 3

Procedures and Problem Solving

Long Division of Polynomials **In Exercises 9–12, use long division to divide and use the result to factor the dividend completely.**

9. $(x^2 + 5x + 6) \div (x + 3)$

10. $(5x^2 - 17x - 12) \div (x - 4)$

11. $(x^3 + 5x^2 - 12x - 36) \div (x + 2)$

12. $(2x^3 - 3x^2 - 50x + 75) \div (2x - 3)$

Long Division of Polynomials **In Exercises 13–22, use long division to divide.**

13. $(x^3 - 4x^2 - 17x + 6) \div (x - 3)$

14. $(4x^3 - 7x^2 - 11x + 5) \div (4x + 5)$

15. $(7x^3 + 3) \div (x + 2)$

16. $(8x^4 - 5) \div (2x + 1)$

17. $(5x - 1 + 10x^3 - 2x^2) \div (2x^2 + 1)$

18. $(1 + 3x^2 + x^4) \div (3 - 2x + x^2)$

19. $(x^3 - 9) \div (x^2 + 1)$

20. $(x^5 + 7) \div (x^3 - 1)$

21. $\dfrac{2x^3 - 4x^2 - 15x + 5}{(x - 1)^2}$

22. $\dfrac{x^4}{(x - 1)^3}$

Using Synthetic Division **In Exercises 23–32, use synthetic division to divide.**

23. $(3x^3 - 17x^2 + 15x - 25) \div (x - 5)$

24. $(5x^3 + 18x^2 + 7x - 6) \div (x + 3)$

25. $(6x^3 + 7x^2 - x + 26) \div (x - 3)$

26. $(2x^3 + 14x^2 - 20x + 7) \div (x + 6)$

27. $(9x^3 - 18x^2 - 16x + 32) \div (x - 2)$

28. $(5x^3 + 6x + 8) \div (x + 2)$

29. $(x^3 + 512) \div (x + 8)$

30. $(x^3 - 729) \div (x - 9)$

31. $\dfrac{4x^3 + 16x^2 - 23x - 15}{x + \frac{1}{2}}$

32. $\dfrac{3x^3 - 4x^2 + 5}{x - \frac{3}{2}}$

Verifying Quotients **In Exercises 33–36, use a graphing utility to graph the two equations in the same viewing window. Use the graphs to verify that the expressions are equivalent. Verify the results algebraically.**

33. $y_1 = \dfrac{x^2}{x + 2}, \quad y_2 = x - 2 + \dfrac{4}{x + 2}$

34. $y_1 = \dfrac{x^2 + 2x - 1}{x + 3}, \quad y_2 = x - 1 + \dfrac{2}{x + 3}$

35. $y_1 = \dfrac{x^4 - 3x^2 - 1}{x^2 + 5}, \quad y_2 = x^2 - 8 + \dfrac{39}{x^2 + 5}$

36. $y_1 = \dfrac{x^4 + x^2 - 1}{x^2 + 1}, \quad y_2 = x^2 - \dfrac{1}{x^2 + 1}$

Verifying the Remainder Theorem **In Exercises 37–42, write the function in the form $f(x) = (x - k)q(x) + r(x)$ for the given value of k. Use a graphing utility to demonstrate that $f(k) = r$.**

	Function	*Value of k*
37.	$f(x) = x^3 - x^2 - 14x + 11$	$k = 4$
38.	$f(x) = 15x^4 + 10x^3 - 6x^2 + 14$	$k = -\frac{2}{3}$

Function	*Value of k*
39. $f(x) = x^3 + 3x^2 - 2x - 14$	$k = \sqrt{2}$
40. $f(x) = x^3 + 2x^2 - 5x - 4$	$k = -\sqrt{5}$
41. $f(x) = 4x^3 - 6x^2 - 12x - 4$	$k = 1 - \sqrt{3}$
42. $f(x) = -3x^3 + 8x^2 + 10x - 8$	$k = 2 + \sqrt{2}$

Using the Remainder Theorem **In Exercises 43–46, use the Remainder Theorem and synthetic division to evaluate the function at each given value. Use a graphing utility to verify your results.**

43. $f(x) = 2x^3 - 7x + 3$
(a) $f(1)$ (b) $f(-2)$ (c) $f(\frac{1}{2})$ (d) $f(2)$

44. $g(x) = 2x^6 + 3x^4 - x^2 + 3$
(a) $g(2)$ (b) $g(1)$ (c) $g(3)$ (d) $g(-1)$

45. $h(x) = x^3 - 5x^2 - 7x + 4$
(a) $h(3)$ (b) $h(2)$ (c) $h(-2)$ (d) $h(-5)$

46. $f(x) = 4x^4 - 16x^3 + 7x^2 + 20$
(a) $f(1)$ (b) $f(-2)$ (c) $f(5)$ (d) $f(-10)$

Using the Factor Theorem **In Exercises 47–52, use synthetic division to show that x is a solution of the third-degree polynomial equation, and use the result to factor the polynomial completely. List all the real solutions of the equation.**

Polynomial Equation	*Value of x*
47. $x^3 - 13x - 12 = 0$	$x = 4$
48. $x^3 - 31x + 30 = 0$	$x = -6$
49. $2x^3 - 17x^2 + 12x + 63 = 0$	$x = -\frac{3}{2}$
50. $60x^3 - 89x^2 + 41x - 6 = 0$	$x = \frac{1}{3}$
51. $x^3 + 2x^2 - 3x - 6 = 0$	$x = \sqrt{3}$
52. $x^3 - x^2 - 13x - 3 = 0$	$x = 2 - \sqrt{5}$

Factoring a Polynomial **In Exercises 53–58, (a) verify the given factor(s) of the function f, (b) find the remaining factors of f, (c) use your results to write the complete factorization of f, and (d) list all real zeros of f. Confirm your results by using a graphing utility to graph the function.**

Function	*Factor(s)*
53. $f(x) = 2x^3 + x^2 - 5x + 2$	$(x + 2)$
54. $f(x) = 3x^3 + 2x^2 - 19x + 6$	$(x + 3)$
55. $f(x) = x^4 - 4x^3 - 15x^2 + 58x - 40$	$(x - 5), (x + 4)$
56. $f(x) = 8x^4 - 14x^3 - 71x^2 - 10x + 24$	$(x + 2), (x - 4)$
57. $f(x) = 6x^3 + 41x^2 - 9x - 14$	$(2x + 1)$
58. $f(x) = 2x^3 - x^2 - 10x + 5$	$(2x - 1)$

Using the Rational Zero Test **In Exercises 59–62, use the Rational Zero Test to list all possible rational zeros of f. Then find the rational zeros.**

59. $f(x) = x^3 + 3x^2 - x - 3$

60. $f(x) = x^3 - 4x^2 - 4x + 16$

61. $f(x) = 2x^4 - 17x^3 + 35x^2 + 9x - 45$

62. $f(x) = 4x^5 - 8x^4 - 5x^3 + 10x^2 + x - 2$

Using Descartes's Rule of Signs **In Exercises 63–66, use Descartes's Rule of Signs to determine the possible numbers of positive and negative real zeros of the function.**

63. $f(x) = 2x^4 - x^3 + 6x^2 - x + 5$

64. $f(x) = 3x^4 + 5x^3 - 6x^2 + 8x - 3$

65. $g(x) = 4x^3 - 5x + 8$

66. $g(x) = 2x^3 - 4x^2 - 5$

Finding the Zeros of a Polynomial Function **In Exercises 67–72, (a) use Descartes's Rule of Signs to determine the possible numbers of positive and negative real zeros of f, (b) list the possible rational zeros of f, (c) use a graphing utility to graph f so that some of the possible zeros in parts (a) and (b) can be disregarded, and (d) determine all the real zeros of f.**

67. $f(x) = x^3 + x^2 - 4x - 4$

68. $f(x) = -3x^3 + 20x^2 - 36x + 16$

69. $f(x) = -2x^4 + 13x^3 - 21x^2 + 2x + 8$

70. $f(x) = 4x^4 - 17x^2 + 4$

71. $f(x) = 32x^3 - 52x^2 + 17x + 3$

72. $f(x) = x^4 - x^3 - 29x^2 - x - 30$

Finding the Zeros of a Polynomial Function **In Exercises 73–76, use synthetic division to verify the upper and lower bounds of the real zeros of f. Then find all real zeros of the function.**

73. $f(x) = x^4 - 4x^3 + 15$
Upper bound: $x = 4$
Lower bound: $x = -1$

74. $f(x) = 2x^3 - 3x^2 - 12x + 8$
Upper bound: $x = 4$
Lower bound: $x = -3$

75. $f(x) = x^4 - 4x^3 + 16x - 16$
Upper bound: $x = 5$
Lower bound: $x = -3$

76. $f(x) = 2x^4 - 8x + 3$
Upper bound: $x = 3$
Lower bound: $x = -4$

Occasionally, throughout this text, you will be asked to round to a place value rather than to a number of decimal places.

Rewriting to Use the Rational Zero Test **In Exercises 77–80, find the rational zeros of the polynomial function.**

77. $P(x) = x^4 - \frac{25}{4}x^2 + 9 = \frac{1}{4}(4x^4 - 25x^2 + 36)$

78. $f(x) = x^3 - \frac{3}{2}x^2 - \frac{23}{2}x + 6 = \frac{1}{2}(2x^3 - 3x^2 - 23x + 12)$

79. $f(x) = x^3 - \frac{1}{4}x^2 - x + \frac{1}{4} = \frac{1}{4}(4x^3 - x^2 - 4x + 1)$

80. $f(z) = z^3 + \frac{11}{6}z^2 - \frac{1}{2}z - \frac{1}{3} = \frac{1}{6}(6z^3 + 11z^2 - 3z - 2)$

A Cubic Polynomial with Two Terms **In Exercises 81–84, match the cubic function with the correct number of rational and irrational zeros.**

(a) Rational zeros: 0; Irrational zeros: 1

(b) Rational zeros: 3; Irrational zeros: 0

(c) Rational zeros: 1; Irrational zeros: 2

(d) Rational zeros: 1; Irrational zeros: 0

81. $f(x) = x^3 - 1$ **82.** $f(x) = x^3 - 2$

83. $f(x) = x^3 - x$ **84.** $f(x) = x^3 - 2x$

Using a Graph to Help Find Zeros **In Exercises 85–88, the graph of $y = f(x)$ is shown. Use the graph as an aid to find all real zeros of the function.**

85. $y = 2x^4 - 9x^3 + 5x^2 + 3x - 1$

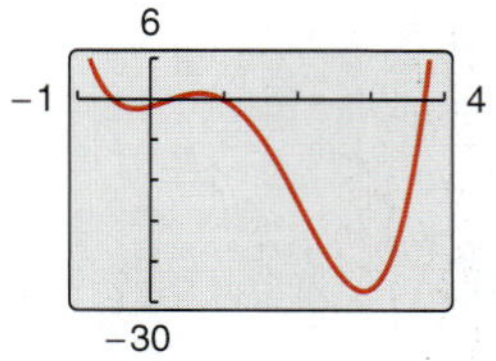

86. $y = x^4 - 5x^3 - 7x^2 + 13x - 2$

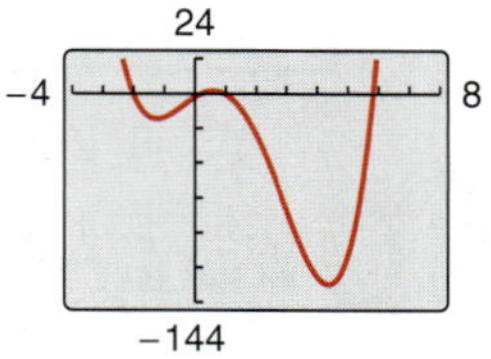

87. $y = -2x^4 + 17x^3 - 3x^2 - 25x - 3$

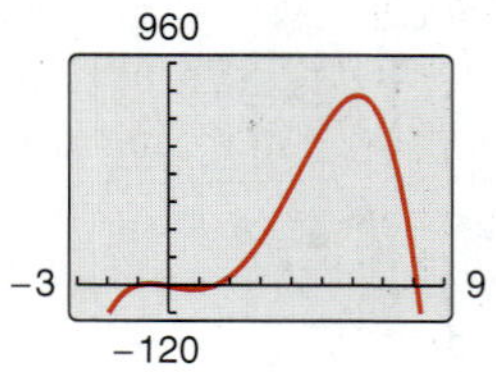

88. $y = -x^4 + 5x^3 - 10x - 4$

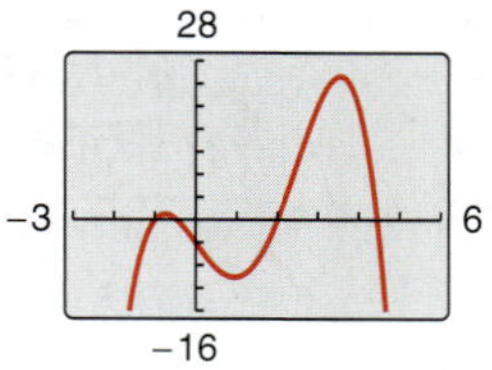

Finding the Zeros of a Polynomial Function **In Exercises 89–100, find all real zeros of the polynomial function.**

89. $f(x) = 5x^4 + 9x^3 - 19x^2 - 3x$

90. $g(x) = 4x^4 - 11x^3 - 22x^2 + 8x$

91. $f(z) = z^4 - z^3 - 2z - 4$

92. $f(x) = 4x^3 + 7x^2 - 11x - 18$

93. $g(y) = 2y^4 + 7y^3 - 26y^2 + 23y - 6$

94. $h(x) = x^5 - x^4 - 3x^3 + 5x^2 - 2x$

95. $f(x) = 4x^4 - 55x^2 - 45x + 36$

96. $z(x) = 6x^4 + 33x^3 - 69x + 30$

97. $g(x) = 8x^4 + 28x^3 + 9x^2 - 9x$

98. $h(x) = x^5 + 5x^4 - 5x^3 - 15x^2 - 6x$

99. $f(x) = 8x^5 + 6x^4 - 37x^3 - 36x^2 + 29x + 30$

100. $g(x) = 4x^5 + 8x^4 - 15x^3 - 23x^2 + 11x + 15$

Using a Rational Zero **In Exercises 101–104, (a) use the *zero* or *root* feature of a graphing utility to approximate (accurate to the nearest thousandth) the zeros of the function, (b) determine one of the exact zeros and use synthetic division to verify your result, and (c) factor the polynomial completely.**

101. $h(t) = t^3 - 2t^2 - 7t + 2$

102. $f(s) = s^3 - 12s^2 + 40s - 24$

103. $h(x) = x^5 - 7x^4 + 10x^3 + 14x^2 - 24x$

104. $g(x) = 6x^4 - 11x^3 - 51x^2 + 99x - 27$

105. MODELING DATA

The table shows the numbers S of cellular phone subscriptions per 100 people in the United States from 1995 through 2012. (Source: International Telecommunications Union)

Year	Subscriptions per 100 people, S
1995	12.6
1996	16.2
1997	20.1
1998	24.9
1999	30.6
2000	38.5
2001	44.7
2002	48.9
2003	54.9
2004	62.6
2005	68.3
2006	76.3
2007	82.1
2008	85.2
2009	88.6
2010	91.3
2011	94.7
2012	95.5

Spreadsheet at LarsonPrecalculus.com

The data can be approximated by the model

$S = -0.0223t^3 + 0.825t^2 - 3.58t + 12.6,$
$5 \le t \le 22$

where t represents the year, with $t = 5$ corresponding to 1995.

(a) Use a graphing utility to plot the data and graph the model in the same viewing window.

(b) How well does the model fit the data?

(c) Use the Remainder Theorem to evaluate the model for the year 2020. Is the value reasonable? Explain.

106. ***Why you should learn it*** *(p. 266)* The numbers of employees E (in thousands) in education and health services in the United States from 1960 through 2013 are approximated by $E = -0.088t^3 + 10.77t^2 + 14.6t + 3197$, $0 \leq t \leq 53$, where t is the year, with $t = 0$ corresponding to 1960. (Source: U.S. Bureau of Labor Statistics)

(a) Use a graphing utility to graph the model over the domain.

(b) Estimate the number of employees in education and health services in 1960. Use the Remainder Theorem to estimate the number in 2010.

(c) Is this a good model for making predictions in future years? Explain.

107. Geometry A rectangular package sent by a delivery service can have a maximum combined length and girth (perimeter of a cross section) of 120 inches (see figure).

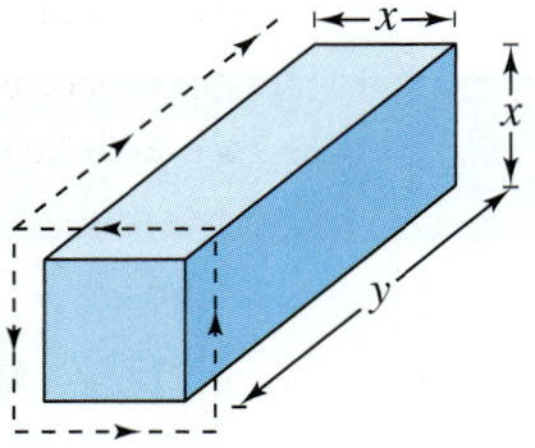

(a) Show that the volume of the package is given by the function $V(x) = 4x^2(30 - x)$.

(b) Use a graphing utility to graph the function and approximate the dimensions of the package that yield a maximum volume.

(c) Find values of x such that $V = 13{,}500$. Which of these values is a physical impossibility in the construction of the package? Explain.

108. Environmental Science The number of parts per million of nitric oxide emissions y from a car engine is approximated by $y = -5.05x^3 + 3857x - 38{,}411.25$, $13 \leq x \leq 18$, where x is the air-fuel ratio.

(a) Use a graphing utility to graph the model.

(b) There are two air-fuel ratios that produce 2400 parts per million of nitric oxide. One is $x = 15$. Use the graph to approximate the other.

(c) Find the second air-fuel ratio from part (b) algebraically. (*Hint:* Use the known value of $x = 15$ and synthetic division.)

Conclusions

True or False? **In Exercises 109 and 110, determine whether the statement is true or false. Justify your answer.**

109. If $(7x + 4)$ is a factor of some polynomial function f, then $\frac{4}{7}$ is a zero of f.

110. The value $x = \frac{1}{7}$ is a zero of the polynomial function $f(x) = 3x^5 - 2x^4 + x^3 - 16x^2 + 3x - 8$.

Think About It **In Exercises 111 and 112, the graph of a cubic polynomial function $y = f(x)$ with integer zeros is shown. Find the factored form of f.**

111.

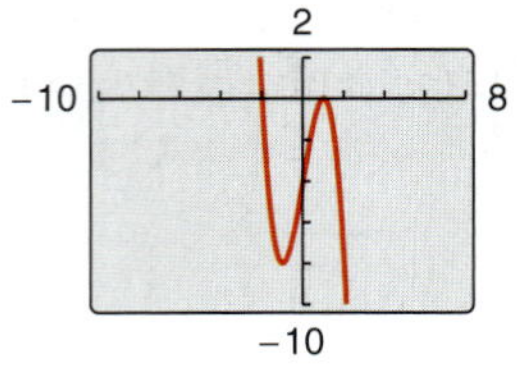

112.

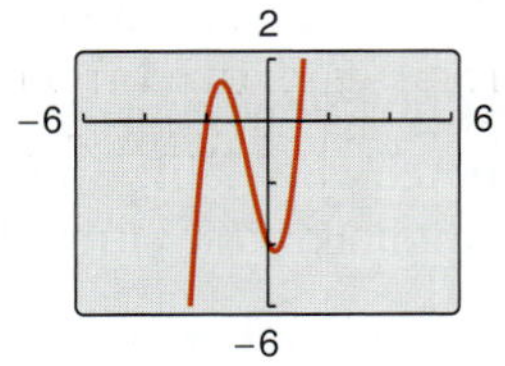

113. Think About It Let $y = f(x)$ be a fourth-degree polynomial with leading coefficient $a = -1$ and $f(\pm 1) = f(\pm 2) = 0$. Find the factored form of f.

114. Think About It Find the value of k such that $x - 3$ is a factor of $x^3 - kx^2 + 2kx - 12$.

115. Writing Complete each polynomial division. Write a brief description of the pattern that you obtain, and use your result to find a formula for the polynomial division $(x^n - 1)/(x - 1)$. Create a numerical example to test your formula.

(a) $\dfrac{x^2 - 1}{x - 1} = \square$ (b) $\dfrac{x^3 - 1}{x - 1} = \square$

(c) $\dfrac{x^4 - 1}{x - 1} = \square$

116. HOW DO YOU SEE IT? A graph of $y = f(x)$ is shown, where

$f(x) = 2x^5 - 3x^4 + x^3 - 8x^2 + 5x + 3$ and

$f(-x) = -2x^5 - 3x^4 - x^3 - 8x^2 - 5x + 3$.

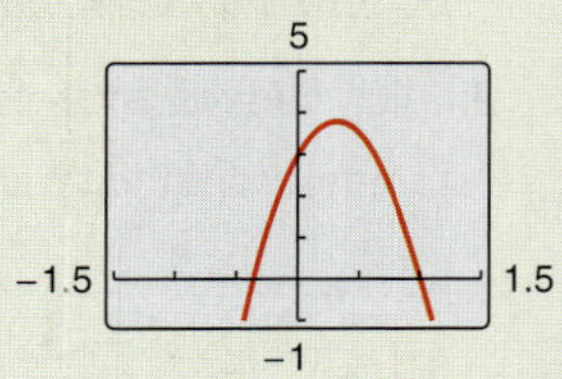

(a) How many negative real zeros does f have? Explain.

(b) How many positive real zeros are *possible* for f? Explain. What does this tell you about the eventual right-hand behavior of the graph?

(c) Is $x = -\frac{1}{3}$ a possible rational zero of f? Explain.

(d) Explain how to check whether $\left(x - \frac{3}{2}\right)$ is a factor of f and whether $x = \frac{3}{2}$ is an upper bound for the real zeros of f.

Cumulative Mixed Review

Solving a Quadratic Equation **In Exercises 117–120, use any convenient method to solve the quadratic equation.**

117. $4x^2 - 17 = 0$

118. $25x^2 - 1 = 0$

119. $3x^2 - 11x - 20 = 0$

120. $6x^2 + 4x - 3 = 0$

3.4 The Fundamental Theorem of Algebra

The Fundamental Theorem of Algebra

You know that an nth-degree polynomial can have at most n real zeros. In the complex number system, this statement can be improved. That is, in the complex number system, every nth-degree polynomial function has *precisely n zeros*. This important result is derived from the **Fundamental Theorem of Algebra,** first proved by the German mathematician Carl Friedrich Gauss (1777–1855).

The Fundamental Theorem of Algebra

If $f(x)$ is a polynomial of degree n, where $n > 0$, then f has at least one zero in the complex number system.

Using the Fundamental Theorem of Algebra and the equivalence of zeros and factors, you obtain the **Linear Factorization Theorem.**

Linear Factorization Theorem (See the proof on page 322.)

If $f(x)$ is a polynomial of degree n, where $n > 0$, then f has precisely n linear factors

$$f(x) = a_n(x - c_1)(x - c_2) \cdot \cdot \cdot (x - c_n)$$

where $c_1, c_2, \ldots, c_n$ are complex numbers.

Note that the Fundamental Theorem of Algebra and the Linear Factorization Theorem tell you only that the zeros or factors of a polynomial exist, not how to find them. Such theorems are called *existence theorems.* To find the zeros of a polynomial function, you still must rely on other techniques.

What you should learn

- Use the Fundamental Theorem of Algebra to determine the number of zeros of a polynomial function.
- Find all zeros of polynomial functions.
- Find conjugate pairs of complex zeros.
- Find zeros of polynomials by factoring.

Why you should learn it

Being able to find zeros of polynomial functions is an important part of modeling real-life problems. For instance, Exercise 69 on page 287 shows how to determine whether a football kicked with a given velocity can reach a certain height.

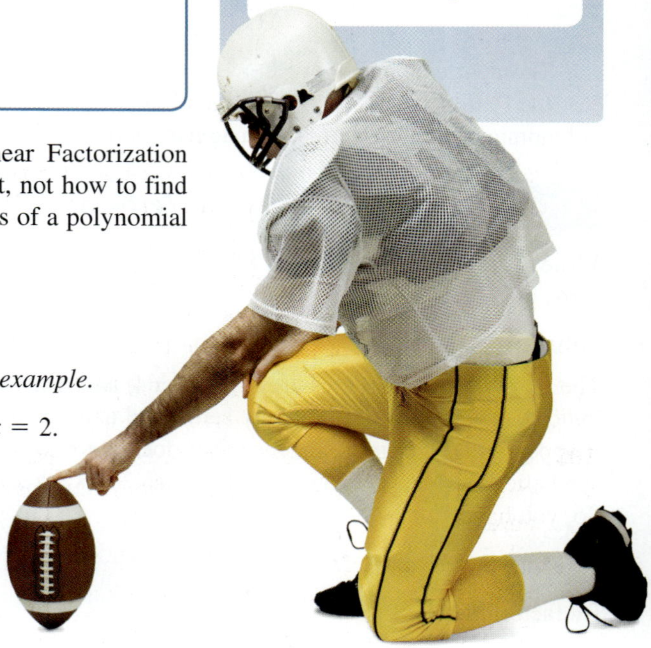

EXAMPLE 1 Zeros of Polynomial Functions

See LarsonPrecalculus.com for an interactive version of this type of example.

a. The first-degree polynomial $f(x) = x - 2$ has exactly *one* zero: $x = 2$.

b. Counting multiplicity, the second-degree polynomial function

$$f(x) = x^2 - 6x + 9$$
$$= (x - 3)(x - 3)$$

has exactly *two* zeros: $x = 3$ and $x = 3$. (This is called a *repeated zero.*)

c. The third-degree polynomial function

$$f(x) = x^3 + 4x = x(x^2 + 4) = x(x - 2i)(x + 2i)$$

has exactly *three* zeros: $x = 0$, $x = 2i$, and $x = -2i$.

d. The fifth-degree polynomial function

$$f(x) = x^5 + 9x^3 = x^3(x^2 + 9) = x^3(x - 3i)(x + 3i)$$

has exactly *five* zeros: $x = 0$, $x = 0$, $x = 0$, $x = 3i$, and $x = -3i$.

Checkpoint *Audio-video solution in English & Spanish at LarsonPrecalculus.com.*

Determine the number of zeros of the polynomial function $f(x) = x^4 - 1$.

Finding Zeros of a Polynomial Function

Remember that the n zeros of a polynomial function can be real or imaginary, and they may be repeated. Examples 2 and 3 illustrate several cases.

EXAMPLE 2 Complex Zeros of a Polynomial Function

Confirm that the third-degree polynomial function $f(x) = x^3 + 4x$ has exactly three zeros: $x = 0$, $x = 2i$, and $x = -2i$.

Solution

Factor the polynomial completely as $x(x - 2i)(x + 2i)$. So, the zeros are

$$x(x - 2i)(x + 2i) = 0$$

$x = 0$ — Set 1st factor equal to 0.

$x - 2i = 0 \Rightarrow x = 2i$ — Set 2nd factor equal to 0.

$x + 2i = 0 \Rightarrow x = -2i.$ — Set 3rd factor equal to 0.

In Figure 3.17, only the real zero $x = 0$ appears as an x-intercept.

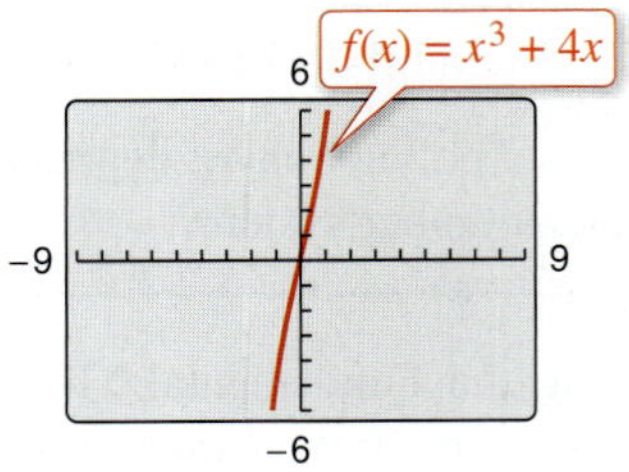

Figure 3.17

✓ Checkpoint Audio-video solution in English & Spanish at LarsonPrecalculus.com.

Confirm that the fourth-degree polynomial function $f(x) = x^4 - 1$ has exactly four zeros: $x = 1$, $x = -1$, $x = i$, and $x = -i$.

The next example shows how to use the methods described in Sections 3.2 and 3.3 (the Rational Zero Test, synthetic division, and factoring) to find all the zeros of a polynomial function, including imaginary zeros.

EXAMPLE 3 Finding the Zeros of a Polynomial Function

Write $f(x) = x^5 + x^3 + 2x^2 - 12x + 8$ as the product of linear factors, and list all the zeros of f.

Solution

The possible rational zeros are ± 1, ± 2, ± 4, and ± 8. The graph shown in Figure 3.18 indicates that 1 and -2 are likely zeros, and that 1 is possibly a repeated zero because it appears that the graph touches (but does not cross) the x-axis at this point. Using synthetic division, you can determine that -2 is a zero and 1 is a repeated zero of f. So, you have

$$f(x) = x^5 + x^3 + 2x^2 - 12x + 8 = (x - 1)(x - 1)(x + 2)(x^2 + 4).$$

By factoring $x^2 + 4$ as

$$x^2 - (-4) = \left(x - \sqrt{-4}\right)\left(x + \sqrt{-4}\right) = (x - 2i)(x + 2i)$$

you obtain

$$f(x) = (x - 1)(x - 1)(x + 2)(x - 2i)(x + 2i)$$

which gives the following five zeros of f.

$$x = 1, x = 1, x = -2, x = 2i, \quad \text{and} \quad x = -2i$$

Note from the graph of f shown in Figure 3.18 that the *real* zeros are the only ones that appear as x-intercepts.

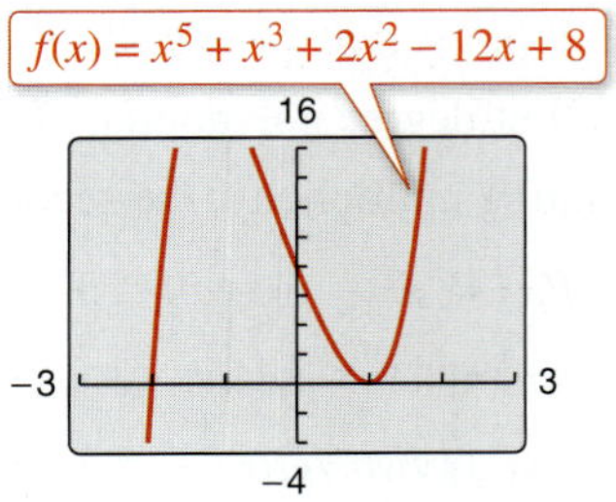

Figure 3.18

✓ Checkpoint Audio-video solution in English & Spanish at LarsonPrecalculus.com.

Write $f(x) = x^4 - 256$ as the product of linear factors, and list all the zeros of f.

Conjugate Pairs

In Example 3, note that the two complex zeros $2i$ and $-2i$ are **conjugates.** That is, they are of the forms $a + bi$ and $a - bi$.

Complex Zeros Occur in Conjugate Pairs

Let f be a polynomial function that has *real coefficients.* If $a + bi$, where $b \neq 0$, is a zero of the function, then the conjugate $a - bi$ is also a zero of the function.

Be sure you see that this result is true only when the polynomial function has *real coefficients.* For instance, the result applies to the function $f(x) = x^2 + 1$, but not to the function $g(x) = x - i$.

EXAMPLE 4 Finding a Polynomial with Given Zeros

Find a *fourth-degree* polynomial function with real coefficients that has -1, -1, and $3i$ as zeros.

Solution

Because $3i$ is a zero *and* the polynomial is stated to have real coefficients, you know that the conjugate $-3i$ must also be a zero. So, from the Linear Factorization Theorem, $f(x)$ can be written as

$$f(x) = a(x + 1)(x + 1)(x - 3i)(x + 3i).$$

For simplicity, let $a = 1$ to obtain

$$f(x) = (x^2 + 2x + 1)(x^2 + 9) = x^4 + 2x^3 + 10x^2 + 18x + 9.$$

✓ Checkpoint *Audio-video solution in English & Spanish at LarsonPrecalculus.com.*

Find a *fourth-degree* polynomial function with real coefficients that has 2, -2, and $7i$ as zeros.

EXAMPLE 5 Finding a Polynomial with Given Zeros

Find a *cubic* polynomial function f with real coefficients that has 2 and $1-i$ as zeros, and $f(1) = 3$.

Solution

Because $1 - i$ is a zero of f, the conjugate $1 + i$ must also be a zero.

$$\begin{aligned} f(x) &= a(x - 2)[x - (1 - i)][x - (1 + i)] \\ &= a(x - 2)[x^2 - x(1 + i) - x(1 - i) + 1 - i^2] \\ &= a(x - 2)(x^2 - 2x + 2) \\ &= a(x^3 - 4x^2 + 6x - 4) \end{aligned}$$

To find the value of a, use the fact that $f(1) = 3$ to obtain

$$a[(1)^3 - 4(1)^2 + 6(1) - 4] = 3.$$

So, $a = -3$ and you can conclude that

$$f(x) = -3(x^3 - 4x^2 + 6x - 4) = -3x^3 + 12x^2 - 18x + 12.$$

✓ Checkpoint *Audio-video solution in English & Spanish at LarsonPrecalculus.com.*

Find a *quartic* polynomial function f with real coefficients that has 1, -2, and $2i$ as zeros, and $f(-1) = 10$.

Factoring a Polynomial

The Linear Factorization Theorem states that you can write any nth-degree polynomial as the product of n linear factors.

$$f(x) = a_n(x - c_1)(x - c_2)(x - c_3) \cdot \cdot \cdot (x - c_n)$$

This result, however, includes the possibility that some of the values of c_i are imaginary. The next theorem states that even when you do not want to get involved with "imaginary factors," you can still write $f(x)$ as the product of linear and quadratic factors.

Factors of a Polynomial (See the proof on page 322.)

Every polynomial of degree $n > 0$ with real coefficients can be written as the product of linear and quadratic factors with real coefficients, where the quadratic factors have no real zeros.

Remark

Recall that irrational and rational numbers are subsets of the set of real numbers, and the real numbers are a subset of the set of complex numbers.

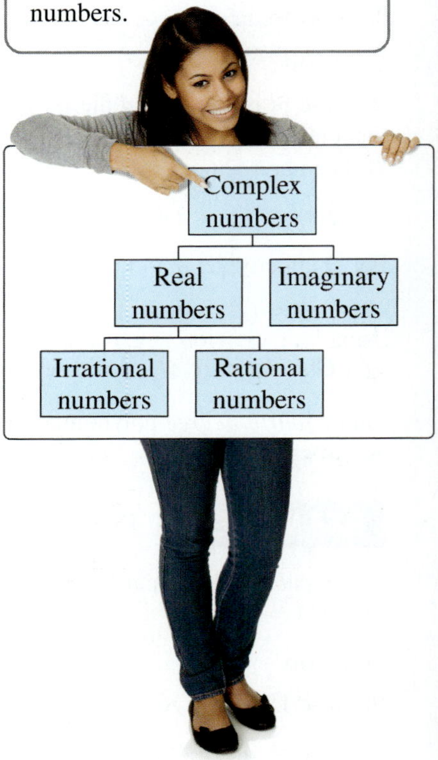

A quadratic factor with no real zeros is said to be **prime** or **irreducible over the reals.** Be sure you see that this is not the same as being *irreducible over the rationals.* For example, the quadratic

$$x^2 + 1 = (x - i)(x + i)$$

is irreducible over the reals (and therefore over the rationals). On the other hand, the quadratic

$$x^2 - 2 = \left(x - \sqrt{2}\right)\left(x + \sqrt{2}\right)$$

is irreducible over the rationals, but *reducible* over the reals.

EXAMPLE 6 Factoring a Polynomial

Write the polynomial $f(x) = x^4 - x^2 - 20$

a. as the product of factors that are irreducible over the *rationals*,

b. as the product of linear factors and quadratic factors that are irreducible over the *reals*, and

c. in completely factored form.

Solution

a. Begin by factoring the polynomial as the product of two quadratic polynomials.

$$x^4 - x^2 - 20 = (x^2 - 5)(x^2 + 4)$$

Both of these factors are irreducible over the rationals.

b. By factoring over the reals, you have

$$x^4 - x^2 - 20 = \left(x + \sqrt{5}\right)\left(x - \sqrt{5}\right)(x^2 + 4)$$

where the quadratic factor is irreducible over the reals.

c. In completely factored form, you have

$$x^4 - x^2 - 20 = \left(x + \sqrt{5}\right)\left(x - \sqrt{5}\right)(x - 2i)(x + 2i).$$

✓ ***Checkpoint*** *Audio-video solution in English & Spanish at LarsonPrecalculus.com.*

Write the polynomial $f(x) = x^4 - 2x^2 - 3$ (a) as the product of factors that are irreducible over the *rationals*, (b) as the product of linear factors and quadratic factors that are irreducible over the *reals*, and (c) in completely factored form.

In Example 6, notice from the completely factored form that the *fourth*-degree polynomial has *four* zeros.

Throughout this chapter, the results and theorems have been stated in terms of zeros of polynomial functions. Be sure you see that the same results could have been stated in terms of solutions of polynomial equations. This is true because the zeros of the polynomial function

$$f(x) = a_n x^n + a_{n-1}x^{n-1} + \cdot \cdot \cdot + a_2 x^2 + a_1 x + a_0$$

are precisely the solutions of the polynomial equation

$$a_n x^n + a_{n-1}x^{n-1} + \cdot \cdot \cdot + a_2 x^2 + a_1 x + a_0 = 0.$$

EXAMPLE 7 Finding the Zeros of a Polynomial Function

Find all the zeros of

$$f(x) = x^4 - 3x^3 + 6x^2 + 2x - 60$$

given that $1 + 3i$ is a zero of f.

Algebraic Solution

Because complex zeros occur in conjugate pairs, you know that $1 - 3i$ is also a zero of f. This means that both

$$x - (1 + 3i) \quad \text{and} \quad x - (1 - 3i)$$

are factors of f. Multiplying these two factors produces

$$\begin{aligned}[x - (1 + 3i)][x - (1 - 3i)] &= [(x - 1) - 3i][(x - 1) + 3i] \\ &= (x - 1)^2 - 9i^2 \\ &= x^2 - 2x + 10.\end{aligned}$$

Using long division, you can divide $x^2 - 2x + 10$ into f to obtain the following.

$$\begin{array}{r} x^2 - x - 6 \\ x^2 - 2x + 10 \overline{\big) x^4 - 3x^3 + 6x^2 + 2x - 60} \\ \underline{x^4 - 2x^3 + 10x^2} \\ -x^3 - 4x^2 + 2x \\ \underline{-x^3 + 2x^2 - 10x} \\ -6x^2 + 12x - 60 \\ \underline{-6x^2 + 12x - 60} \\ 0 \end{array}$$

So, you have

$$\begin{aligned} f(x) &= (x^2 - 2x + 10)(x^2 - x - 6) \\ &= (x^2 - 2x + 10)(x - 3)(x + 2) \end{aligned}$$

and you can conclude that the zeros of f are

$$x = 1 + 3i, x = 1 - 3i, x = 3, \text{ and } x = -2.$$

Graphical Solution

Using the *zero* feature of a graphing utility, you can conclude that $x = -2$ and $x = 3$ are zeros of f. (See figure.)

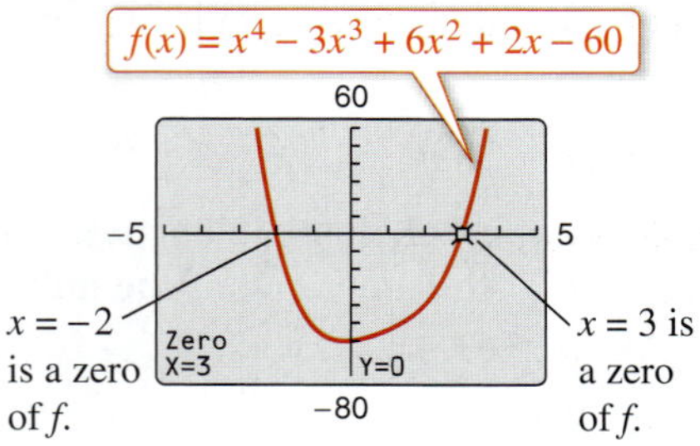

Because $1 + 3i$ is a zero of f, you know that the conjugate $1 - 3i$ must also be a zero. So, you can conclude that the zeros of f are

$$x = 1 + 3i, x = 1 - 3i, x = 3, \text{ and } x = -2.$$

Checkpoint *Audio-video solution in English & Spanish at LarsonPrecalculus.com.*

Find all the zeros of $f(x) = 3x^3 - 2x^2 + 48x - 32$ given that $4i$ is a zero of f.

In Example 7, if you were not told that $1 + 3i$ is a zero of f, you could still find all the zeros of the function by using synthetic division to find the real zeros -2 and 3. Then, you could factor the polynomial as $(x + 2)(x - 3)(x^2 - 2x + 10)$. Finally, by using the Quadratic Formula, you could determine that the zeros are $x = 1 + 3i$, $x = 1 - 3i$, $x = 3$, and $x = -2$.

3.4 Exercises

See *CalcChat.com* for tutorial help and worked-out solutions to odd-numbered exercises. For instructions on how to use a graphing utility, see Appendix A.

Vocabulary and Concept Check

In Exercises 1 and 2, fill in the blanks.

1. The _______ of _______ states that if $f(x)$ is a polynomial of degree n $(n > 0)$, then f has at least one zero in the complex number system.

2. A quadratic factor that cannot be factored as a product of linear factors containing real numbers is said to be _______ over the _______ .

3. How many linear factors does a polynomial function f of degree n have, where $n > 0$?

4. Three of the zeros of a fourth-degree polynomial function f are -1, 3, and $2i$. What is the other zero of f?

Procedures and Problem Solving

Zeros of a Polynomial Function **In Exercises 5–8, match the function with its exact number of zeros.**

5. $f(x) = -2x^4 + 32$ (a) 1 zero

6. $f(x) = x^5 - x^3$ (b) 3 zeros

7. $f(x) = x^3 + 3x^2 + 2x$ (c) 4 zeros

8. $f(x) = x - 14$ (d) 5 zeros

Complex Zeros of a Polynomial Function **In Exercises 9–12, confirm that the function has the indicated zeros.**

9. $f(x) = x^2 + 5;\ -\sqrt{5}i,\ \sqrt{5}i$

10. $f(x) = x^3 + 9x;\ 0,\ -3i,\ 3i$

11. $f(x) = 3x^4 - 48;\ -2,\ 2,\ -2i,\ 2i$

12. $f(x) = 2x^5 - 2x;\ 0,\ 1,\ -1\ -i,\ i$

Comparing the Zeros and the x-Intercepts **In Exercises 13–16, find all the zeros of the function. Is there a relationship between the number of real zeros and the number of x-intercepts of the graph? Explain.**

13. $f(x) = x^3 - 4x^2 + x - 4$

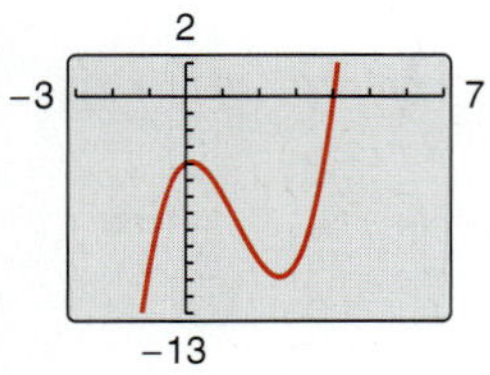

14. $f(x) = x^3 - 4x^2 - 4x + 16$

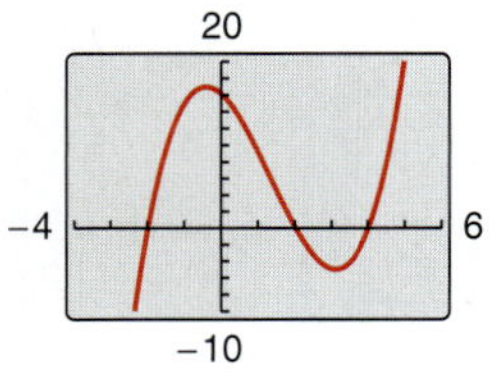

15. $f(x) = x^4 + 4x^2 + 4$

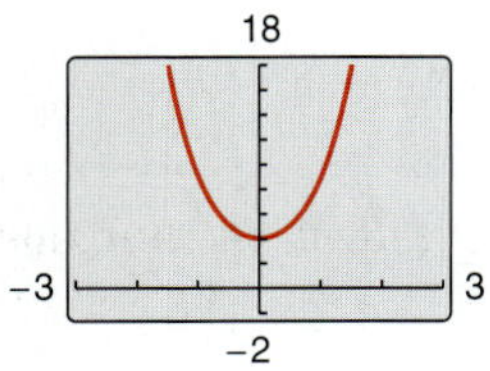

16. $f(x) = x^4 - 3x^2 - 4$

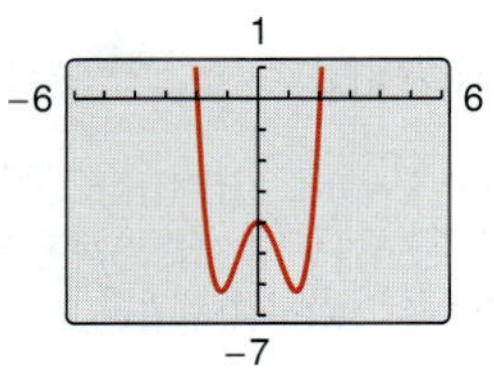

Finding the Zeros of a Polynomial Function **In Exercises 17–36, find all the zeros of the function and write the polynomial as a product of linear factors. Use a graphing utility to verify your results graphically. (If possible, use the graphing utility to verify the imaginary zeros.)**

17. $h(x) = x^2 - 4x + 1$

18. $g(x) = x^2 + 10x + 23$

19. $f(x) = x^2 - 12x + 26$

20. $f(x) = x^2 + 6x - 2$

21. $f(x) = x^2 + 25$

22. $f(x) = x^2 + 36$

23. $f(x) = 16x^4 - 81$

24. $f(y) = 81y^4 - 625$

25. $f(z) = z^2 - z + 56$

26. $h(x) = x^2 - 4x - 3$

27. $f(x) = x^4 + 10x^2 + 9$

28. $f(x) = x^4 + 29x^2 + 100$

29. $f(x) = 3x^3 - 5x^2 + 48x - 80$

30. $f(x) = 3x^3 - 2x^2 + 75x - 50$

31. $f(t) = t^3 - 3t^2 - 15t + 125$

32. $f(x) = x^3 + 11x^2 + 39x + 29$

33. $f(x) = 5x^3 - 9x^2 + 28x + 6$

34. $f(s) = 3s^3 - 4s^2 + 8s + 8$

35. $g(x) = x^4 - 4x^3 + 8x^2 - 16x + 16$

36. $h(x) = x^4 + 6x^3 + 10x^2 + 6x + 9$

Using the Zeros to Find x-Intercepts **In Exercises 37–44, (a) find all zeros of the function, (b) write the polynomial as a product of linear factors, and (c) use your factorization to determine the x-intercepts of the graph of the function. Use a graphing utility to verify that the real zeros are the only x-intercepts.**

37. $f(x) = x^2 - 14x + 46$

38. $f(x) = x^2 - 12x + 34$

39. $f(x) = 2x^3 - 3x^2 + 8x - 12$

40. $f(x) = 2x^3 - 5x^2 + 18x - 45$

41. $f(x) = x^3 - 11x + 150$

42. $f(x) = x^3 + 10x^2 + 33x + 34$

43. $f(x) = x^4 + 25x^2 + 144$

44. $f(x) = x^4 - 8x^3 + 17x^2 - 8x + 16$

Finding a Polynomial with Given Zeros **In Exercises 45–50, find a polynomial function with real coefficients that has the given zeros. (There are many correct answers.)**

45. $5, i, -i$

46. $3, 4i, -4i$

47. $1, 1, 2i$

48. $-3, -3, i$

49. $0, -5, 1 + \sqrt{2}i$

50. $0, 4, 1 + \sqrt{2}i$

Finding a Polynomial with Given Zeros **In Exercises 51–54, a polynomial function f with real coefficients has the given degree, zeros, and solution point. Write the function (a) in completely factored form and (b) in polynomial form.**

	Degree	*Zeros*	*Solution Point*
51.	4	$-1, 2, i$	$f(1) = 8$
52.	4	$1, -4, \sqrt{3}i$	$f(0) = -6$
53.	3	$-1, 2 + \sqrt{5}i$	$f(2) = 45$
54.	3	$-2, 2 + 2\sqrt{2}i$	$f(-1) = -34$

Factoring a Polynomial **In Exercises 55–58, write the polynomial (a) as the product of factors that are irreducible over the *rationals,* (b) as the product of linear and quadratic factors that are irreducible over the *reals,* and (c) in completely factored form.**

55. $f(x) = x^4 - 6x^2 - 7$

56. $f(x) = x^4 + 6x^2 - 27$

57. $f(x) = x^4 - 2x^3 - 3x^2 + 12x - 18$

(*Hint:* One factor is $x^2 - 6$.)

58. $f(x) = x^4 - 3x^3 - x^2 - 12x - 20$

(*Hint:* One factor is $x^2 + 4$.)

Finding the Zeros of a Polynomial Function **In Exercises 59–64, use the given zero to find all the zeros of the function.**

	Function	*Zero*
59.	$f(x) = 2x^3 + 3x^2 + 50x + 75$	$5i$
60.	$f(x) = x^3 + x^2 + 9x + 9$	$3i$
61.	$g(x) = x^3 - 7x^2 - x + 87$	$5 + 2i$
62.	$g(x) = 4x^3 + 23x^2 + 34x - 10$	$-3 + i$
63.	$h(x) = 3x^3 - 4x^2 + 8x + 8$	$1 - \sqrt{3}i$
64.	$f(x) = 25x^3 - 55x^2 - 54x - 18$	$\frac{1}{5}(-2 + \sqrt{2}i)$

Using a Graph to Locate the Real Zeros **In Exercises 65–68, (a) use a graphing utility to find the real zeros of the function, and then (b) use the real zeros to find the exact values of the imaginary zeros.**

65. $f(x) = x^4 + 3x^3 - 5x^2 - 21x + 22$

66. $f(x) = x^3 + 4x^2 + 14x + 20$

67. $h(x) = 8x^3 - 14x^2 + 18x - 9$

68. $f(x) = 25x^3 - 55x^2 - 54x - 18$

69. ***Why you should learn it*** *(p. 281)* A football is kicked off the ground with an initial upward velocity of 48 feet per second. The football's height h (in feet) is given by

$$h(t) = -16t^2 + 48t, \quad 0 \le t \le 3$$

where t is the time (in seconds). Does the football reach a height of 50 feet? Explain.

70. Marketing The demand equation for a microwave is $p = 140 - 0.001x$, where p is the unit price (in dollars) of the microwave and x is the number of units produced and sold. The cost equation for the microwave is $C = 40x + 150{,}000$, where C is the total cost (in dollars) and x is the number of units produced. The total profit P obtained by producing and selling x units is given by $P = R - C = xp - C$. Is there a price p that yields a profit of \$3 million? Explain.

Conclusions

True or False? **In Exercises 71 and 72, decide whether the statement is true or false. Justify your answer.**

71. It is possible for a third-degree polynomial function with integer coefficients to have no real zeros.

72. If $[x + (4 + 3i)]$ is a factor of a polynomial function f with real coefficients, then $[x - (4 + 3i)]$ is also a factor of f.

73. Writing Compile a list of all the various techniques for factoring a polynomial that have been covered so far in the text. Give an example illustrating each technique and write a paragraph discussing when the use of each technique is appropriate.

74. HOW DO YOU SEE IT? Describe a translation of the graph that will result in a function with (a) four distinct real zeros, (b) two real zeros, each of multiplicity 2, (c) two real zeros and two imaginary zeros, and (d) four imaginary zeros.

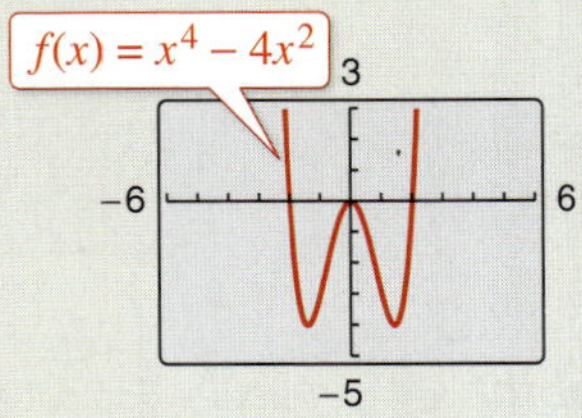

Cumulative Mixed Review

Identifying the Vertex of a Quadratic Function **In Exercises 75–78, describe the graph of the function and identify the vertex.**

75. $f(x) = x^2 - 7x - 8$

76. $f(x) = -x^2 + x + 6$

77. $f(x) = 6x^2 + 5x - 6$

78. $f(x) = 4x^2 + 2x - 12$

3.5 Rational Functions and Asymptotes

Introduction to Rational Functions

A **rational function** can be written in the form

$$f(x) = \frac{N(x)}{D(x)}$$

where $N(x)$ and $D(x)$ are polynomials and $D(x)$ is not the zero polynomial.

In general, the *domain* of a rational function of x includes all real numbers except x-values that make the denominator zero. Much of the discussion of rational functions will focus on their graphical behavior near these x-values.

EXAMPLE 1 Finding the Domain of a Rational Function

See LarsonPrecalculus.com for an interactive version of this type of example.

Find the domain of $f(x) = 1/x$ and discuss the behavior of f near any excluded x-values.

Solution

Because the denominator is zero when $x = 0$, the domain of f is all real numbers except $x = 0$. To determine the behavior of f near this excluded value, evaluate $f(x)$ to the left and right of $x = 0$, as indicated in the following tables.

x	-1	-0.5	-0.1	-0.01	-0.001	$\rightarrow 0$
$f(x)$	-1	-2	-10	-100	-1000	$\rightarrow -\infty$

x	$0 \leftarrow$	0.001	0.01	0.1	0.5	1
$f(x)$	$\infty \leftarrow$	1000	100	10	2	1

From the table, note that as x approaches 0 *from the left*, $f(x)$ decreases without bound. In contrast, as x approaches 0 *from the right*, $f(x)$ increases without bound. Because $f(x)$ decreases without bound from the left and increases without bound from the right, you can conclude that f is not continuous. The graph of f is shown in the figure.

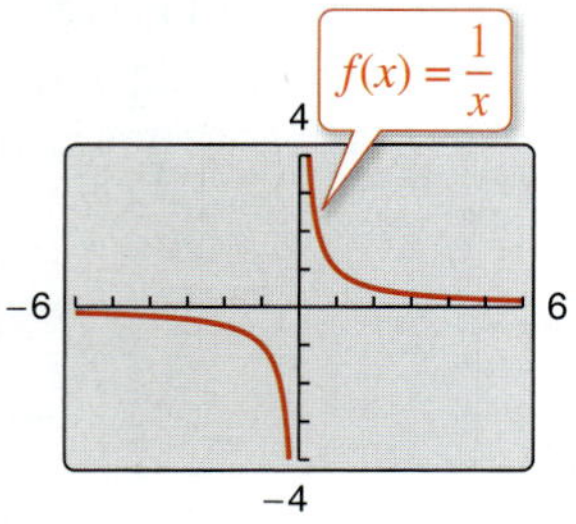

Checkpoint *Audio-video solution in English & Spanish at LarsonPrecalculus.com.*

Find the domain of $f(x) = \dfrac{3x}{x - 1}$ and discuss the behavior of f near any excluded x-values.

Explore the Concept

Use the *table* and *trace* features of a graphing utility to verify that the function $f(x) = 1/x$ in Example 1 is not continuous.

What you should learn

- Find the domains of rational functions.
- Find vertical and horizontal asymptotes of graphs of rational functions.
- Use rational functions to model and solve real-life problems.

Why you should learn it

Rational functions are convenient in modeling a wide variety of real-life problems, such as environmental scenarios. For instance, Exercise 49 on page 296 shows how to determine the cost of supplying recycling bins to the population of a rural township.

Vertical and Horizontal Asymptotes

In Example 1, the behavior of f near $x = 0$ is denoted as follows.

$$\underbrace{f(x) \to -\infty \text{ as } x \to 0^-}_{f(x) \text{ decreases without bound as } x \text{ approaches 0 from the left.}} \qquad \underbrace{f(x) \to \infty \text{ as } x \to 0^+}_{f(x) \text{ increases without bound as } x \text{ approaches 0 from the right.}}$$

The line $x = 0$ is a **vertical asymptote** of the graph of f, as shown in Figure 3.19. From this figure, you can see that the graph of f also has a **horizontal asymptote**—the line $y = 0$. This means the values of $f(x) = 1/x$ approach zero as x increases or decreases without bound.

$$\underbrace{f(x) \to 0 \text{ as } x \to -\infty}_{f(x) \text{ approaches 0 as } x \text{ decreases without bound.}} \qquad \underbrace{f(x) \to 0 \text{ as } x \to \infty}_{f(x) \text{ approaches 0 as } x \text{ increases without bound.}}$$

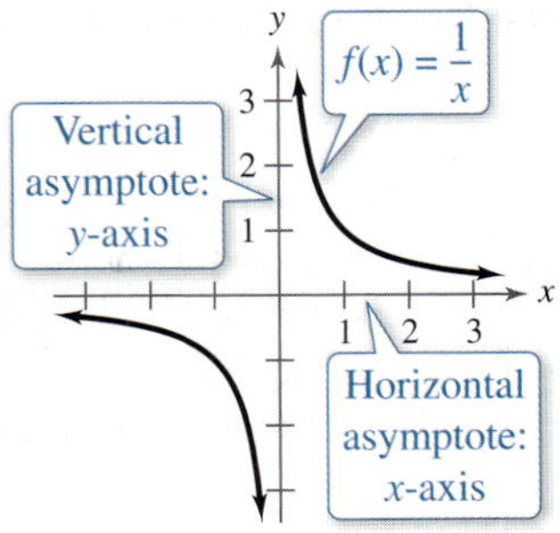

Figure 3.19

Definition of Vertical and Horizontal Asymptotes

1. The line $x = a$ is a **vertical asymptote** of the graph of f when

$$f(x) \to \infty \quad \text{or} \quad f(x) \to -\infty$$

as $x \to a$, either from the right or from the left.

2. The line $y = b$ is a **horizontal asymptote** of the graph of f when

$$f(x) \to b$$

as $x \to \infty$ or $x \to -\infty$.

Figure 3.20 shows the vertical and horizontal asymptotes of the graphs of three rational functions. Note in Figure 3.20 that eventually (as $x \to \infty$ or $x \to -\infty$) the distance between the horizontal asymptote and the points on the graph must approach zero. [The graphs shown in Figures 3.19 and 3.20(a) are **hyperbolas.** You will study hyperbolas in Section 10.3.]

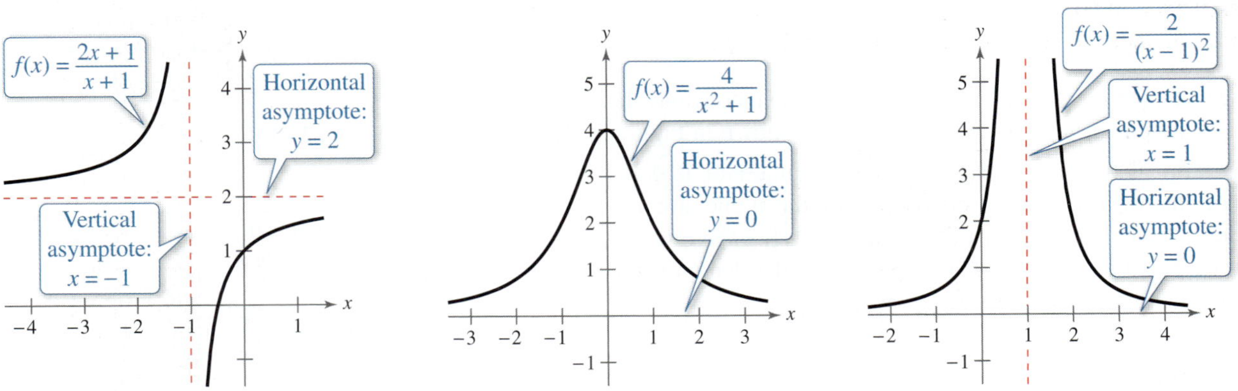

(a) (b) (c)

Figure 3.20

Explore the Concept

Use a table of values to determine whether the functions in Figure 3.20 are continuous. When the graph of a function has an asymptote, can you conclude that the function is not continuous? Explain.

Vertical and Horizontal Asymptotes of a Rational Function

Let f be the rational function

$$f(x) = \frac{N(x)}{D(x)} = \frac{a_n x^n + a_{n-1}x^{n-1} + \cdots + a_1 x + a_0}{b_m x^m + b_{m-1}x^{m-1} + \cdots + b_1 x + b_0}$$

where $N(x)$ and $D(x)$ have no common factors.

1. The graph of f has *vertical* asymptotes at the zeros of the denominator, $D(x)$.

2. The graph of f has at most one *horizontal* asymptote determined by comparing the degrees of $N(x)$ and $D(x)$.

a. If $n < m$, then the graph of f has the line $y = 0$ (the x-axis) as a horizontal asymptote.

b. If $n = m$, then the graph of f has the line

$$y = \frac{a_n}{b_m}$$

as a horizontal asymptote, where a_n is the leading coefficient of the numerator and b_m is the leading coefficient of the denominator.

c. If $n > m$, then the graph of f has no horizontal asymptote.

What's Wrong?

You use a graphing utility to graph

$$y_1 = \frac{2x^3 + 1000x^2 + x}{x^3 + 1000x^2 + x + 1000}$$

as shown in the figure. You use the graph to conclude that the graph of y_1 has the line $y = 1$ as a horizontal asymptote. What's wrong?

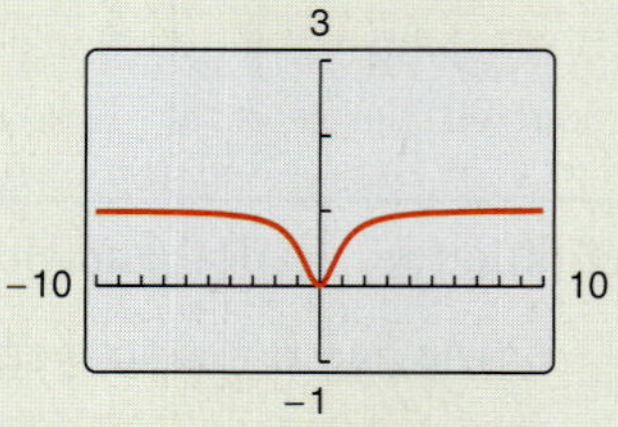

EXAMPLE 2 Finding Vertical and Horizontal Asymptotes

Find all asymptotes of the graph of each rational function.

a. $f(x) = \dfrac{2x}{3x^2 + 1}$ **b.** $f(x) = \dfrac{2x^2}{x^2 - 1}$

Solution

a. For this rational function, the degree of the numerator is *less than* the degree of the denominator, so the graph has the line $y = 0$ as a horizontal asymptote. To find any vertical asymptotes, set the denominator equal to zero and solve the resulting equation for x.

$$3x^2 + 1 = 0 \qquad \text{Set denominator equal to zero.}$$

Because this equation has no real solutions, you can conclude that the graph has no vertical asymptote. The graph of the function is shown in Figure 3.21.

b. For this rational function, the degree of the numerator is *equal* to the degree of the denominator. The leading coefficient of the numerator is 2 and the leading coefficient of the denominator is 1, so the graph has the line $y = 2$ as a horizontal asymptote. To find any vertical asymptotes, set the denominator equal to zero and solve the resulting equation for x.

$$x^2 - 1 = 0 \qquad \text{Set denominator equal to zero.}$$

$$(x + 1)(x - 1) = 0 \qquad \text{Factor.}$$

$$x + 1 = 0 \Rightarrow x = -1 \qquad \text{Set 1st factor equal to 0.}$$

$$x - 1 = 0 \Rightarrow x = 1 \qquad \text{Set 2nd factor equal to 0.}$$

This equation has two real solutions, $x = -1$ and $x = 1$, so the graph has the lines $x = -1$ and $x = 1$ as vertical asymptotes, as shown in Figure 3.22.

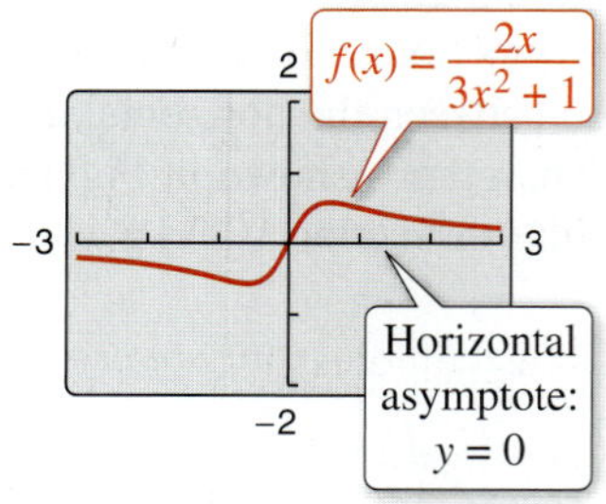

Figure 3.21

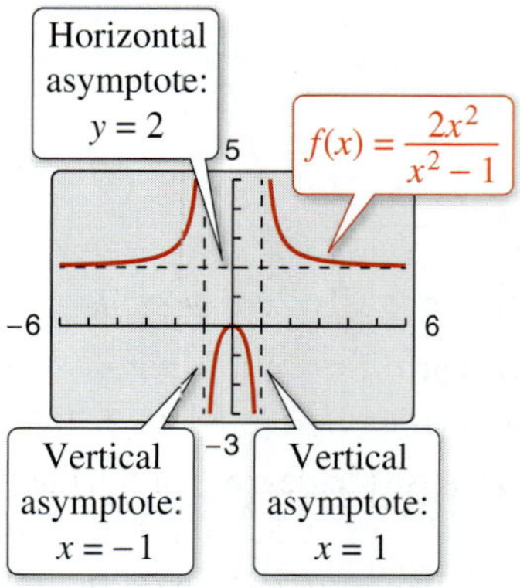

Figure 3.22

✔ ***Checkpoint*** *Audio-video solution in English & Spanish at LarsonPrecalculus.com.*

Find all vertical and horizontal asymptotes of the graph of $f(x) = \dfrac{5x^2}{x^2 - 1}$.

Values for which a rational function is undefined (the denominator is zero) result in a vertical asymptote or a hole in the graph, as shown in Example 3.

EXAMPLE 3 Finding Asymptotes and Holes

Find all asymptotes and holes in the graph of

$$f(x) = \frac{x^2 + x - 2}{x^2 - x - 6}.$$

Solution

For this rational function, the degree of the numerator is *equal* to the degree of the denominator. The leading coefficients of the numerator and denominator are both 1, so the graph has the line $y = 1$ as a horizontal asymptote. To find any vertical asymptotes, first factor the numerator and denominator as follows.

$$f(x) = \frac{x^2 + x - 2}{x^2 - x - 6} = \frac{(x - 1)\cancel{(x + 2)}}{\cancel{(x + 2)}(x - 3)} = \frac{x - 1}{x - 3}, \quad x \neq -2$$

By setting the denominator $x - 3$ (of the simplified function) equal to zero, you can determine that the graph has the line $x = 3$ as a vertical asymptote, as shown in the figure. To find any holes in the graph, note that the function is undefined at $x = -2$ and $x = 3$. Because $x = -2$ is not a vertical asymptote of the function, there is a hole in the graph at $x = -2$. To find the y-coordinate of the hole, substitute $x = -2$ into the simplified form of the function.

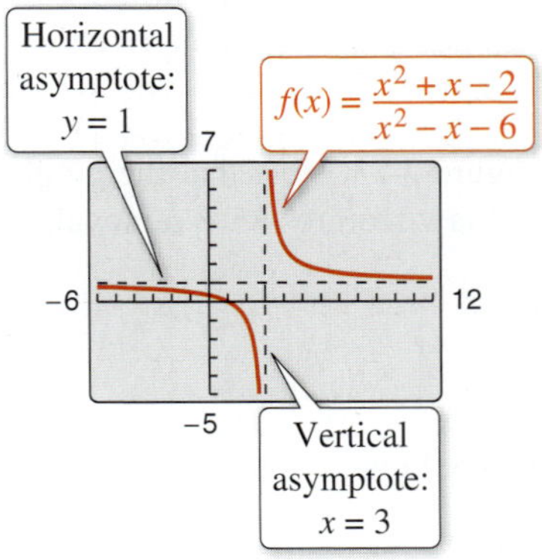

$$y = \frac{x - 1}{x - 3} = \frac{-2 - 1}{-2 - 3} = \frac{3}{5}$$

So, the graph of the rational function has a hole at $\left(-2, \frac{3}{5}\right)$.

Checkpoint Audio-video solution in English & Spanish at LarsonPrecalculus.com.

Find all asymptotes and holes in the graph of $f(x) = \dfrac{x^2 - 25}{x^2 + 5x}$.

EXAMPLE 4 Finding a Function's Domain and Asymptotes

For the function f, find (a) the domain of f, (b) the vertical asymptote of f, and (c) the horizontal asymptote of f.

$$f(x) = \frac{3x^3 + 7x^2 + 2}{-2x^3 + 16}$$

Solution

a. Because the denominator is zero when $-2x^3 + 16 = 0$, solve this equation to determine that the domain of f is all real numbers except $x = 2$.

b. Because the denominator of f has a zero at $x = 2$, and 2 is not a zero of the numerator, the graph of f has the vertical asymptote $x = 2$.

c. Because the degrees of the numerator and denominator are the same, and the leading coefficient of the numerator is 3 and the leading coefficient of the denominator is -2, the horizontal asymptote of f is $y = -\frac{3}{2}$.

Checkpoint Audio-video solution in English & Spanish at LarsonPrecalculus.com.

Repeat Example 4 using the function $f(x) = \dfrac{3x^2 + x - 5}{x^2 + 1}$.

Technology Tip

Graphing utilities are limited in their resolution and therefore may not show a break or hole in the graph. You can use the *table* feature of a graphing utility to verify the values of x at which a function is not defined. Try doing this for the function in Example 3.

Application

There are many examples of asymptotic behavior in real life. For instance, Example 5 shows how a vertical asymptote can be used to analyze the cost of removing pollutants from smokestack emissions.

EXAMPLE 5 Cost-Benefit Model

A utility company burns coal to generate electricity. The cost C (in dollars) of removing $p\%$ of the smokestack pollutants is given by

$$C = \frac{80{,}000p}{100 - p}$$

for $0 \le p < 100$. Use a graphing utility to graph this function. You are a member of a state legislature that is considering a law that would require utility companies to remove 90% of the pollutants from their smokestack emissions. The current law requires 85% removal. How much additional cost would the utility company incur as a result of the new law?

Solution

The graph of this function is shown in Figure 3.23. Note that the graph has a vertical asymptote at $p = 100$. Because the current law requires 85% removal, the current cost to the utility company is

$$C = \frac{80{,}000(85)}{100 - 85} \approx \$453{,}333.$$ Evaluate C when $p = 85$.

The cost to remove 90% of the pollutants would be

$$C = \frac{80{,}000(90)}{100 - 90} \approx \$720{,}000.$$ Evaluate C when $p = 90$.

So, the new law would require the utility company to spend an additional

$$720{,}000 - 453{,}333 = \$266{,}667.$$ Subtract 85% removal cost from 90% removal cost.

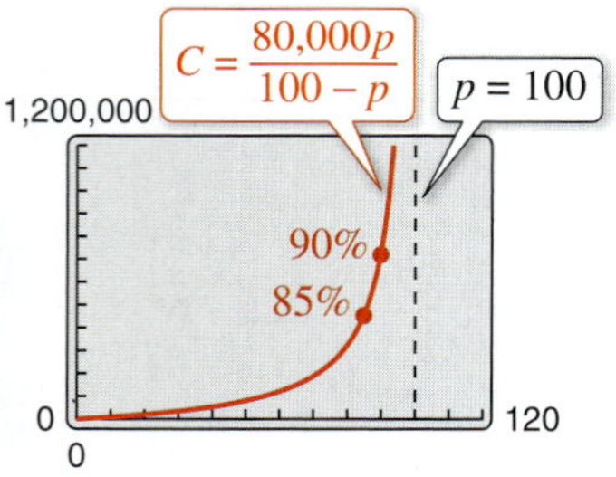

Figure 3.23

Explore the Concept

The *table* feature of a graphing utility can be used to estimate vertical and horizontal asymptotes of rational functions. Use the *table* feature to find any vertical or horizontal asymptotes of

$$f(x) = \frac{2x}{x + 1}.$$

Write a statement explaining how you found the asymptote(s) using the table.

✓ Checkpoint *Audio-video solution in English & Spanish at LarsonPrecalculus.com.*

The cost C (in millions of dollars) of removing $p\%$ of the industrial and municipal pollutants discharged into a river is given by

$$C = \frac{255p}{100 - p}, \quad 0 \le p < 100.$$

a. Find the costs of removing 20%, 45%, and 80% of the pollutants.

b. According to the model, would it be possible to remove 100% of the pollutants? Explain.

3.5 Exercises

See *CalcChat.com* for tutorial help and worked-out solutions to odd-numbered exercises. For instructions on how to use a graphing utility, see Appendix A.

Vocabulary and Concept Check

In Exercises 1 and 2, fill in the blank.

1. Functions of the form $f(x) = N(x)/D(x)$, where $N(x)$ and $D(x)$ are polynomials and $D(x)$ is not the zero polynomial, are called _______ .

2. If $f(x) \to \pm\infty$ as $x \to a$ from the left (or right), then $x = a$ is a _______ of the graph of f.

3. What feature of the graph of $y = \dfrac{9}{x-3}$ can you find by solving $x - 3 = 0$?

4. Is $y = \dfrac{2}{3}$ a horizontal asymptote of the function $f(x) = \dfrac{2x}{3x^2 - 5}$?

Procedures and Problem Solving

Finding the Domain of a Rational Function **In Exercises 5–10, (a) find the domain of the function, (b) complete each table, and (c) discuss the behavior of f near any excluded x-values.**

x	$f(x)$
0.5	
0.9	
0.99	
0.999	

x	$f(x)$
1.5	
1.1	
1.01	
1.001	

x	$f(x)$
5	
10	
100	
1000	

x	$f(x)$
−5	
−10	
−100	
−1000	

5. $f(x) = \dfrac{1}{x-1}$

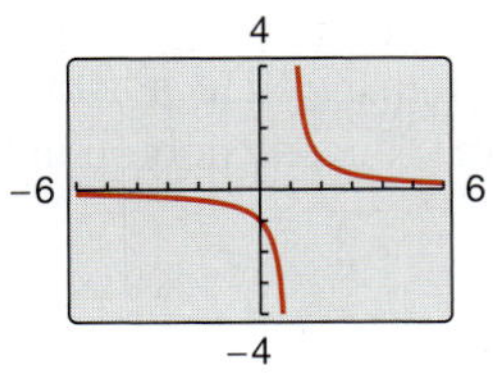

6. $f(x) = \dfrac{5x}{x-1}$

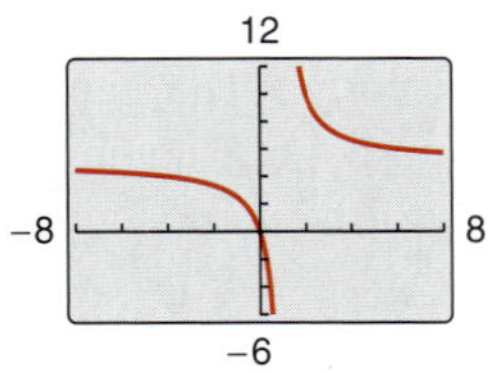

7. $f(x) = \dfrac{3x}{|x-1|}$

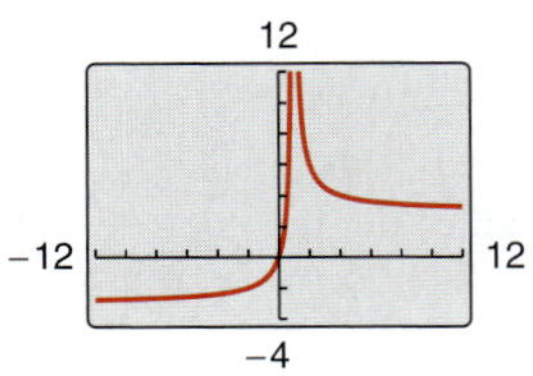

8. $f(x) = \dfrac{3}{|x-1|}$

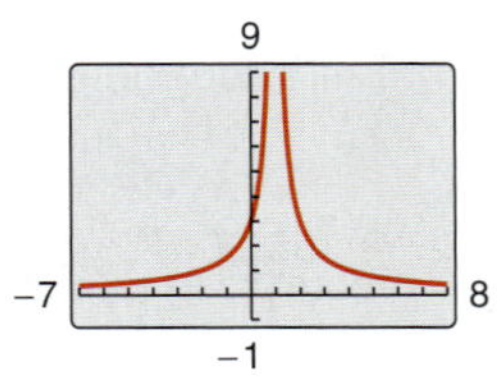

9. $f(x) = \dfrac{3x^2}{x^2 - 1}$

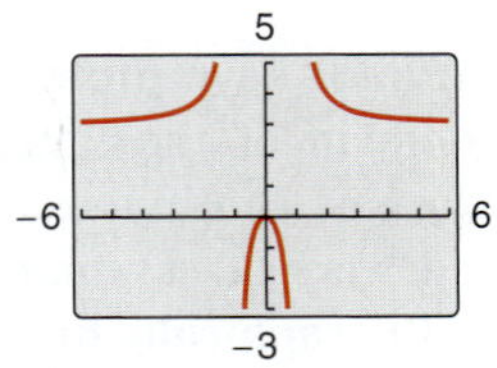

10. $f(x) = \dfrac{4x}{x^2 - 1}$

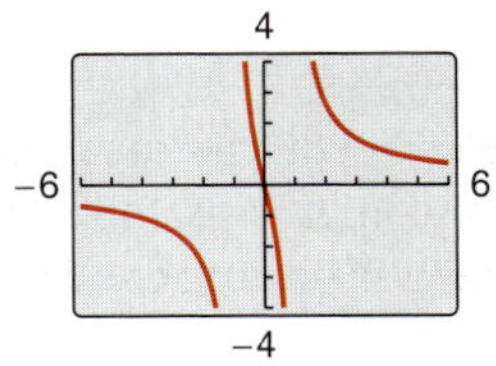

Identifying Graphs of Rational Functions **In Exercises 11–16, match the function with its graph. [The graphs are labeled (a), (b), (c), (d), (e), and (f).]**

(a)

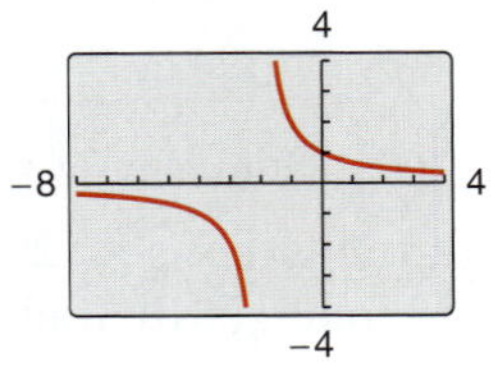

(b)

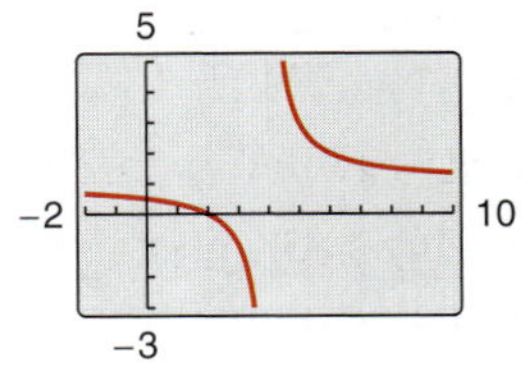

(c)

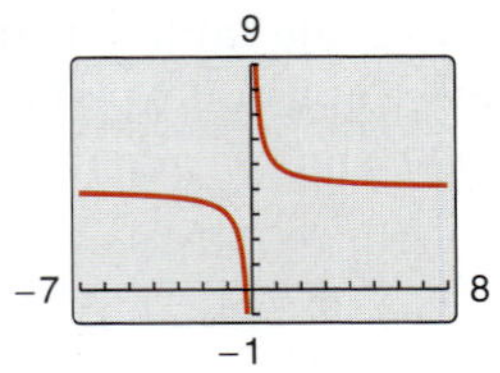

(d)

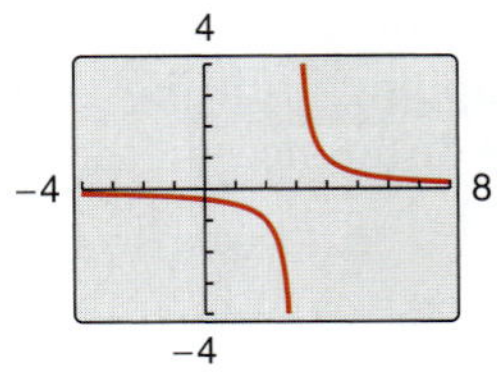

(e)

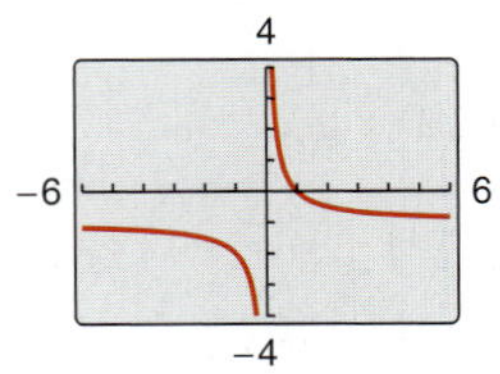

(f)

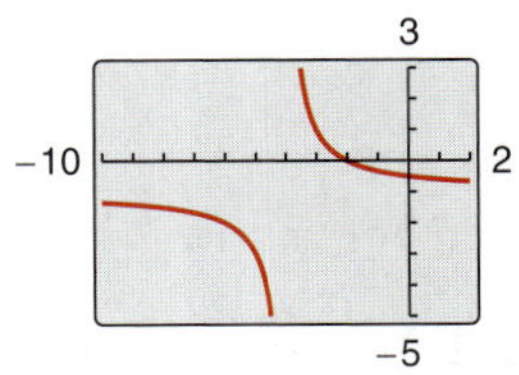

11. $f(x) = \dfrac{2}{x+2}$

12. $f(x) = \dfrac{1}{x-3}$

13. $f(x) = \dfrac{4x+1}{x}$

14. $f(x) = \dfrac{1-x}{x}$

15. $f(x) = \dfrac{x-2}{x-4}$

16. $f(x) = -\dfrac{x+2}{x+4}$

Finding Vertical and Horizontal Asymptotes In Exercises 17–20, find any asymptotes of the graph of the rational function. Verify your answers by using a graphing utility to graph the function.

17. $f(x) = \dfrac{1}{x^2}$

18. $f(x) = \dfrac{3}{(x-2)^3}$

19. $f(x) = \dfrac{2x^2}{x^2 + x - 6}$

20. $f(x) = \dfrac{x^2 - 4x}{x^2 - 4}$

Finding Asymptotes and Holes In Exercises 21–24, find any asymptotes and holes in the graph of the rational function. Verify your answers by using a graphing utility.

21. $f(x) = \dfrac{x(2 + x)}{2x - x^2}$

22. $f(x) = \dfrac{x^2 + 2x + 1}{2x^2 - x - 3}$

23. $f(x) = \dfrac{x^2 - 16}{x^2 + 8x}$

24. $f(x) = \dfrac{3 - 14x - 5x^2}{3 + 7x + 2x^2}$

Finding a Function's Domain and Asymptotes In Exercises 25–28, (a) find the domain of the function, (b) decide whether the function is continuous, and (c) identify any horizontal and vertical asymptotes. Verify your answer to part (a) both graphically by using a graphing utility and numerically by creating a table of values.

25. $f(x) = \dfrac{5x^2 - 2x - 6}{x^2 + 4}$

26. $f(x) = \dfrac{3x^2 + 1}{x^2 + x + 9}$

27. $f(x) = \dfrac{x^2 + 3x - 4}{-x^3 + 27}$

28. $f(x) = \dfrac{4x^3 - x^2 + 3}{3x^3 + 24}$

Algebraic-Graphical-Numerical In Exercises 29–32, (a) determine the domains of f and g, (b) find any vertical asymptotes and holes in the graphs of f and g, (c) compare f and g by completing the table, (d) use a graphing utility to graph f and g, and (e) explain why the differences in the domains of f and g are not shown in their graphs.

29. $f(x) = \dfrac{x^2 - 16}{x - 4}$, $g(x) = x + 4$

x	1	2	3	4	5	6	7
$f(x)$							
$g(x)$							

30. $f(x) = \dfrac{x^2 - 9}{x - 3}$, $g(x) = x + 3$

x	0	1	2	3	4	5	6
$f(x)$							
$g(x)$							

31. $f(x) = \dfrac{x^2 - 1}{x^2 - 2x - 3}$, $g(x) = \dfrac{x - 1}{x - 3}$

x	-2	-1	0	1	2	3	4
$f(x)$							
$g(x)$							

32. $f(x) = \dfrac{x^2 - 4}{x^2 - 3x + 2}$, $g(x) = \dfrac{x + 2}{x - 1}$

x	-3	-2	-1	0	1	2	3
$f(x)$							
$g(x)$							

Exploration In Exercises 33–36, determine the value that the function f approaches as the magnitude of x increases. Is $f(x)$ greater than or less than this value when x is positive and large in magnitude? What about when x is negative and large in magnitude?

33. $f(x) = 4 - \dfrac{1}{x}$

34. $f(x) = 2 + \dfrac{1}{x - 3}$

35. $f(x) = \dfrac{2x - 1}{x - 3}$

36. $f(x) = \dfrac{2x - 1}{x^2 + 1}$

Finding the Zeros of a Rational Function In Exercises 37–44, find the zeros (if any) of the rational function. Use a graphing utility to verify your answer.

37. $f(x) = \dfrac{x^2 - 9}{x^2 + 5}$

38. $h(x) = \dfrac{x^3 + 8}{x^2 - 11}$

39. $g(x) = 1 + \dfrac{6}{x - 3}$

40. $f(x) = 3 - \dfrac{-12}{x^2 + 2}$

41. $h(x) = \dfrac{x^2 - x - 20}{x^2 + 7}$

42. $g(x) = \dfrac{x^2 - 8x + 12}{x^2 + 4}$

43. $f(x) = \dfrac{x^2 + 4x - 21}{x^2 - 4x + 3}$

44. $h(x) = \dfrac{2x^2 + 11x + 5}{3x^2 + 13x - 10}$

45. **Biology** The game commission introduces 100 deer into newly acquired state game lands. The population N of the herd is given by

$$N = \frac{20(5 + 3t)}{1 + 0.04t}, \quad t \geq 0$$

where t is the time in years.

(a) Use a graphing utility to graph the model.

(b) Find the populations when $t = 5$, $t = 10$, and $t = 25$.

(c) What is the limiting size of the herd as time increases? Explain.

46. MODELING DATA

The endpoints of the interval over which distinct vision is possible are called the *near point* and *far point* of the eye (see figure). With increasing age, these points normally change. The table shows the approximate near points y (in inches) for various ages x (in years).

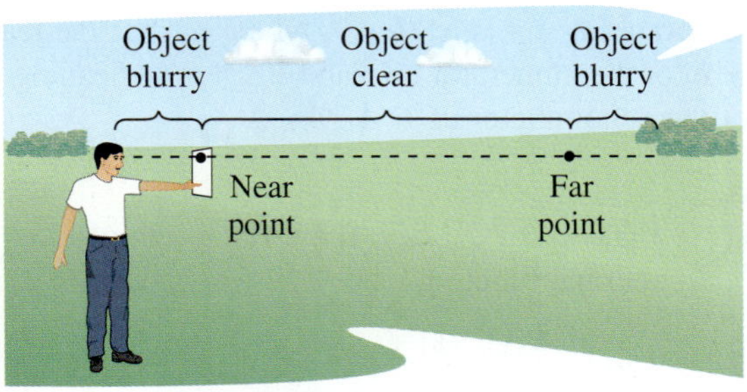

Age, x	Near point, y
16	3.0
32	4.7
44	9.8
50	19.7
60	39.7

(a) Find a rational model for the data. Take the reciprocals of the near points to generate the points

$$\left(x, \frac{1}{y}\right).$$

Use the *regression* feature of a graphing utility to find a linear model for the data. The resulting line has the form

$$\frac{1}{y} = ax + b.$$

Solve for y.

(b) Use the *table* feature of the graphing utility to create a table showing the predicted near point based on the model for each of the ages in the original table.

(c) Do you think the model can be used to predict the near point for a person who is 70 years old? Explain.

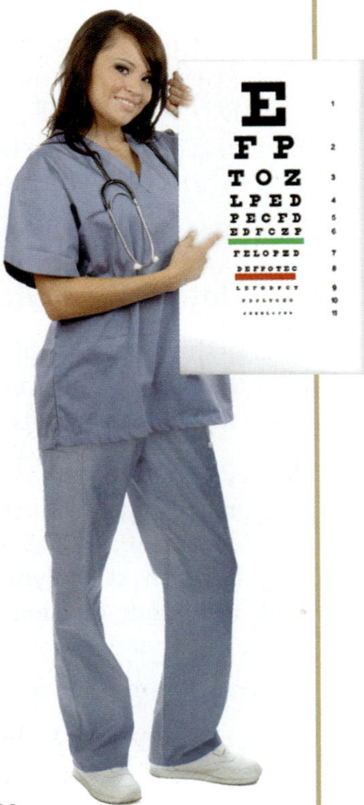

47. Physics Consider a physics laboratory experiment designed to determine an unknown mass. A flexible metal meter stick is clamped to a table with 50 centimeters overhanging the edge (see figure). Known masses M ranging from 200 grams to 2000 grams are attached to the end of the meter stick. For each mass, the meter stick is displaced vertically and then allowed to oscillate. The average time t (in seconds) of one oscillation for each mass is recorded in the table.

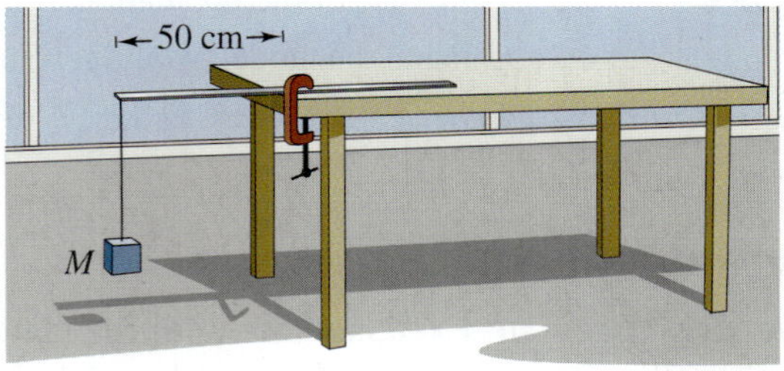

DATA

Mass, M	Time, t
200	0.450
400	0.597
600	0.712
800	0.831
1000	0.906
1200	1.003
1400	1.088
1600	1.126
1800	1.218
2000	1.338

Spreadsheet at LarsonPrecalculus.com

A model for the data is given by

$$t = \frac{38M + 16{,}965}{10(M + 5000)}.$$

(a) Use the *table* feature of a graphing utility to create a table showing the estimated time based on the model for each of the masses shown in the table. What can you conclude?

(b) Use the model to approximate the mass of an object when the average time for one oscillation is 1.056 seconds.

48. Business The sales S (in thousands of units) of a tablet computer during the nth week after the tablet is released are given by

$$S = \frac{150n}{n^2 + 100}, \quad n \geq 0.$$

(a) Use a graphing utility to graph the sales function.

(b) Find the sales in week 5, week 10, and week 20.

(c) According to this model, will sales ever drop to zero units? Explain.

49. ***Why you should learn it*** *(p. 288)* The cost C (in dollars) of supplying recycling bins to $p\%$ of the population of a rural township is given by

$$C = \frac{25{,}000p}{100 - p}, \quad 0 \le p < 100.$$

(a) Use a graphing utility to graph the cost function.

(b) Find the costs of supplying bins to 15%, 50%, and 90% of the population.

(c) According to the model, would it be possible to supply bins to 100% of the population? Explain.

Conclusions

True or False? **In Exercises 50 and 51, determine whether the statement is true or false. Justify your answer.**

50. A rational function can have infinitely many vertical asymptotes.

51. A rational function must have at least one vertical asymptote.

52. **Writing a Rational Function** Write a rational function f that has the specified characteristics. (There are many correct answers.)

(a) Vertical asymptote: $x = 2$
Horizontal asymptote: $y = 0$
Zero: $x = 1$

(b) Vertical asymptote: $x = -1$
Horizontal asymptote: $y = 0$
Zero: $x = 2$

(c) Vertical asymptotes: $x = -2, x = 1$
Horizontal asymptote: $y = 2$
Zeros: $x = 3, x = -3$

(d) Vertical asymptotes: $x = -1, x = 2$
Horizontal asymptote: $y = -2$
Zeros: $x = -2, x = 3$

(e) Vertical asymptotes: $x = 0, x = \pm 3$
Horizontal asymptote: $y = 3$
Zeros: $x = -1, x = 1, x = 2$

53. **Think About It** A real zero of the numerator of a rational function f is $x = c$. Must $x = c$ also be a zero of f? Explain.

54. **Think About It** When the graph of a rational function f has a vertical asymptote at $x = 4$, can f have a common factor of $(x - 4)$ in the numerator and denominator? Explain.

55. **Exploration** Use a graphing utility to compare the graphs of y_1 and y_2.

$$y_1 = \frac{3x^3 - 5x^2 + 4x - 5}{2x^2 - 6x + 7}, \quad y_2 = \frac{3x^3}{2x^2}$$

Start with a viewing window of $-5 \le x \le 5$ and $-10 \le y \le 10$, and then zoom out. Make a conjecture about how the graph of a rational function f is related to the graph of $y = a_nx^n/b_mx^m$, where a_nx^n is the leading term of the numerator of f and b_mx^m is the leading term of the denominator of f.

56. **HOW DO YOU SEE IT?** The graph of a rational function

$$f(x) = \frac{N(x)}{D(x)}$$

is shown below. Determine which of the statements about the function is false. Justify your answer.

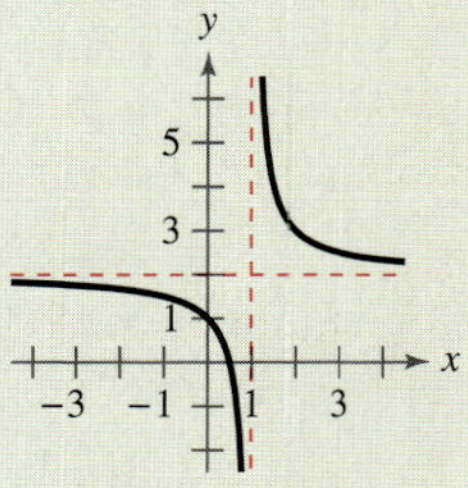

(a) $D(1) = 0$.

(b) The degrees of $N(x)$ and $D(x)$ are equal.

(c) The ratio of the leading coefficients of $N(x)$ and $D(x)$ is 1.

Cumulative Mixed Review

Finding the Equation of a Line **In Exercises 57–60, write the general form of the equation of the line that passes through the points.**

57. $(3, 2), (0, -1)$

58. $(-6, 1), (4, -5)$

59. $(2, 7), (3, 10)$

60. $(0, 0), (-9, 4)$

Long Division of Polynomials **In Exercises 61–64, divide using long division.**

61. $(x^2 + 5x + 6) \div (x - 4)$

62. $(x^2 - 10x + 15) \div (x - 3)$

63. $(2x^4 + x^2 - 11) \div (x^2 + 5)$

64. $(4x^5 + 3x^3 - 10) \div (2x + 3)$

3.6 Graphs of Rational Functions

The Graph of a Rational Function

To sketch the graph of a rational function, use the following guidelines.

Guidelines for Graphing Rational Functions

Let $f(x) = N(x)/D(x)$, where $N(x)$ and $D(x)$ are polynomials and $D(x)$ is not the zero polynomial.

1. Simplify f, if possible. Any restrictions on the domain of f not in the simplified function should be listed.
2. Find and plot the y-intercept (if any) by evaluating $f(0)$.
3. Find the zeros of the numerator (if any) by setting the numerator equal to zero. Then plot the corresponding x-intercepts.
4. Find the zeros of the denominator (if any) by setting the denominator equal to zero. Then sketch the corresponding vertical asymptotes using dashed vertical lines and plot the corresponding holes using open circles.
5. Find and sketch any other asymptotes of the graph using dashed lines.
6. Plot at least one point *between* and one point *beyond* each x-intercept and vertical asymptote.
7. Use smooth curves to complete the graph between and beyond the vertical asymptotes, excluding any points where f is not defined.

When graphing simple rational functions, testing for symmetry can be useful. For instance, the graph of $f(x) = 1/x$ is symmetrical with respect to the origin, and the graph of $g(x) = 1/x^2$ is symmetrical with respect to the y-axis.

What you should learn

- Analyze and sketch graphs of rational functions.
- Sketch graphs of rational functions that have slant asymptotes.
- Use graphs of rational functions to model and solve real-life problems.

Why you should learn it

The graph of a rational function provides a good indication of the behavior of a mathematical model. Exercise 89 on page 305 models the concentration of a chemical in the bloodstream after injection.

Technology Tip

Some graphing utilities have difficulty graphing rational functions that have vertical asymptotes. Often, the utility will connect parts of the graph that are not supposed to be connected. Notice that the graph in Figure 3.24(a) should consist of two *unconnected* portions—one to the left of $x = 2$ and the other to the right of $x = 2$. To eliminate this problem, you can try changing the *mode* of the graphing utility to *dot mode*. The problem with this mode is that the graph is then represented as a collection of dots rather than as a smooth curve, as shown in Figure 3.24(b).

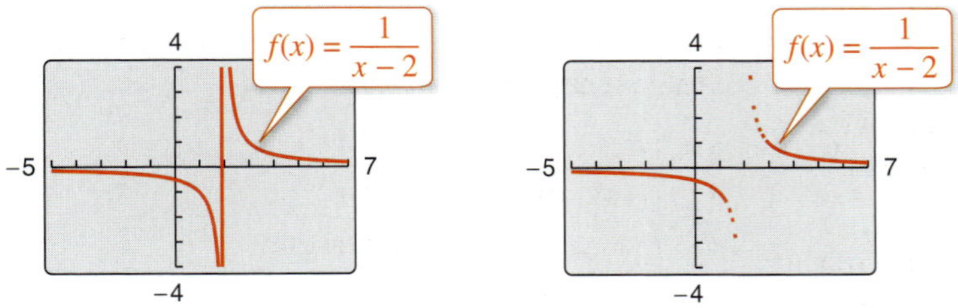

(a) Connected mode (b) Dot mode

Figure 3.24

Library of Parent Functions: Rational Function

The simplest type of rational function is the *parent rational function* $f(x) = 1/x$, also known as the *reciprocal function.* The basic characteristics of the parent rational function are summarized below and on the inside cover of this text.

Graph of $f(x) = \dfrac{1}{x}$

Domain: $(-\infty, 0) \cup (0, \infty)$
Range: $(-\infty, 0) \cup (0, \infty)$
No intercepts
Decreasing on $(-\infty, 0)$ and $(0, \infty)$
Odd function
Origin symmetry
Vertical asymptote: y-axis
Horizontal asymptote: x-axis

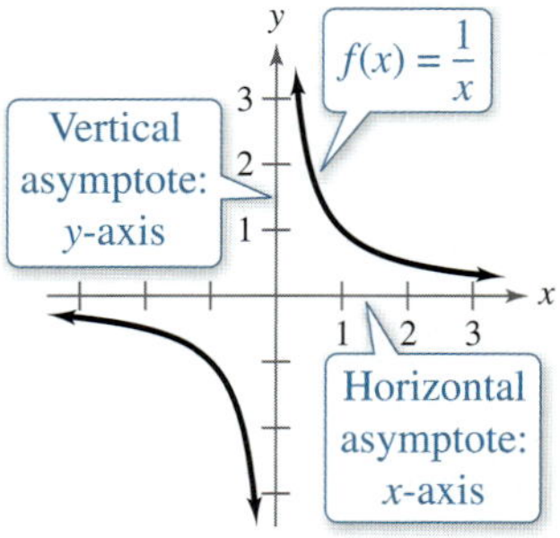

Explore the Concept

Use a graphing utility to graph

$$f(x) = 1 + \frac{1}{x - \frac{1}{x}}.$$

Set the graphing utility to *dot* mode and use a decimal viewing window. Use the *trace* feature to find three "holes" or "breaks" in the graph. Do all three holes or breaks represent zeros of the denominator

$$x - \frac{1}{x}?$$

Explain.

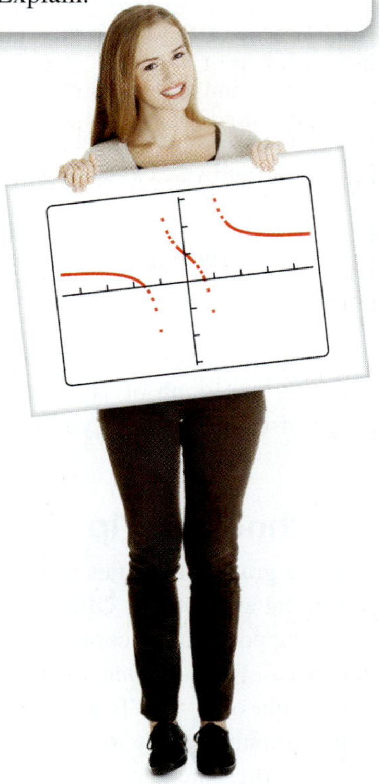

EXAMPLE 1 Library of Parent Functions: $f(x) = 1/x$

See LarsonPrecalculus.com for an interactive version of this type of example.

Sketch the graph of each function by hand and compare it with the graph of $f(x) = 1/x$.

a. $g(x) = \dfrac{-1}{x + 2}$

b. $h(x) = \dfrac{1}{x - 1} + 3$

Solution

a. With respect to the graph of $f(x) = 1/x$, the graph of g is obtained by a *reflection* in the y-axis and a horizontal shift two units *to the left*, as shown in Figure 3.25. Confirm this with a graphing utility.

b. With respect to the graph of $f(x) = 1/x$, the graph of h is obtained by a horizontal shift one unit *to the right* and a vertical shift three units *upward*, as shown in Figure 3.26. Confirm this with a graphing utility.

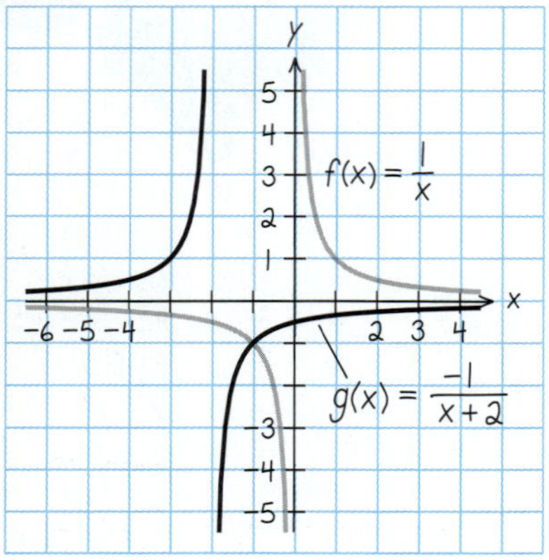

Figure 3.25

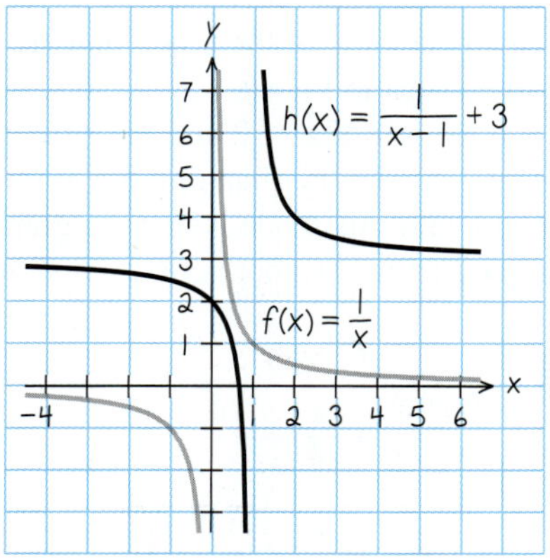

Figure 3.26

Checkpoint *Audio-video solution in English & Spanish at LarsonPrecalculus.com.*

Sketch the graph of each function by hand and compare it with the graph of $f(x) = 1/x$.

a. $g(x) = \dfrac{1}{x - 4}$ **b.** $h(x) = \dfrac{1}{x + 3} - 1$

In Examples 2–6, note that the vertical asymptotes are included in the tables of additional points. This is done to emphasize numerically the behavior of the graph of the function.

EXAMPLE 2 Sketching the Graph of a Rational Function

Sketch the graph of $g(x) = \dfrac{3}{x-2}$ by hand.

Solution

y-intercept: $\left(0, -\frac{3}{2}\right)$, because $g(0) = -\frac{3}{2}$

x-intercept: None, because $3 \neq 0$

Vertical asymptote: $x = 2$, zero of denominator

Horizontal asymptote: $y = 0$, because degree of $N(x) <$ degree of $D(x)$

Additional points:

x	-4	1	2	3	5
$g(x)$	-0.5	-3	Undefined	3	1

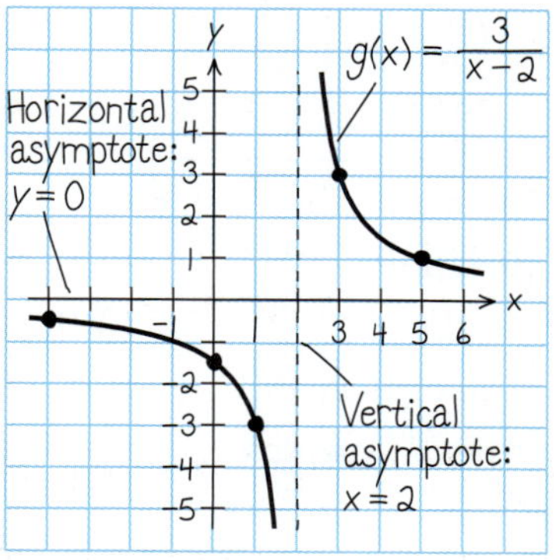

Figure 3.27

By plotting the intercept, asymptotes, and a few additional points, you can obtain the graph shown in Figure 3.27. Confirm this with a graphing utility.

✓ Checkpoint *Audio-video solution in English & Spanish at LarsonPrecalculus.com.*

Sketch the graph of $f(x) = \dfrac{1}{x+3}$ by hand.

Note that the graph of g in Example 2 is a vertical stretch and a right shift of the graph of

$$f(x) = \frac{1}{x}$$

because

$$g(x) = \frac{3}{x-2} = 3\left(\frac{1}{x-2}\right) = 3f(x-2).$$

EXAMPLE 3 Sketching the Graph of a Rational Function

Sketch the graph of $f(x) = \dfrac{2x-1}{x}$ by hand.

Solution

y-intercept: None, because $x = 0$ is not in the domain

x-intercept: $\left(\frac{1}{2}, 0\right)$, because $2x - 1 = 0$ when $x = \frac{1}{2}$

Vertical asymptote: $x = 0$, zero of denominator

Horizontal asymptote: $y = 2$, because degree of $N(x) =$ degree of $D(x)$

Additional points:

x	-4	-1	0	$\frac{1}{4}$	4
$f(x)$	2.25	3	Undefined	-2	1.75

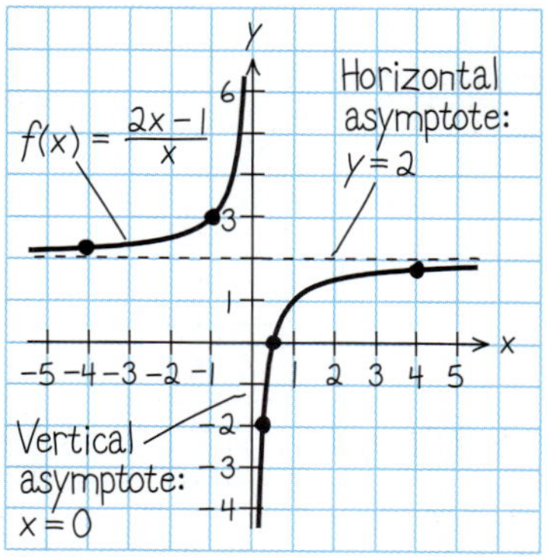

Figure 3.28

By plotting the intercept, asymptotes, and a few additional points, you can obtain the graph shown in Figure 3.28. Confirm this with a graphing utility.

✓ Checkpoint *Audio-video solution in English & Spanish at LarsonPrecalculus.com.*

Sketch the graph of $f(x) = \dfrac{2x+3}{x+1}$ by hand.

EXAMPLE 4 Sketching the Graph of a Rational Function

Sketch the graph of $f(x) = \dfrac{x}{x^2 - x - 2}$.

Solution

Factor the denominator to determine the zeros of the denominator.

$$f(x) = \frac{x}{x^2 - x - 2}$$

$$= \frac{x}{(x + 1)(x - 2)}$$

y-intercept: (0, 0), because $f(0) = 0$

x-intercept: (0, 0)

Vertical asymptotes: $x = -1, x = 2$, zeros of denominator

Horizontal asymptote: $y = 0$, because degree of $N(x) <$ degree of $D(x)$

Additional points:

x	-3	-1	-0.5	1	2	3
$f(x)$	-0.3	Undefined	0.4	-0.5	Undefined	0.75

The graph is shown in Figure 3.29.

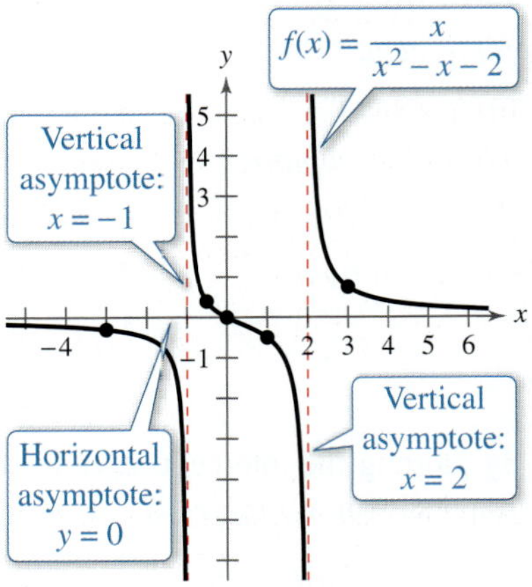

Figure 3.29

✓ ***Checkpoint*** *Audio-video solution in English & Spanish at LarsonPrecalculus.com.*

Sketch the graph of $f(x) = \dfrac{3x}{x^2 + x - 2}$.

EXAMPLE 5 Sketching the Graph of a Rational Function

Sketch the graph of $f(x) = \dfrac{x^2 - 9}{x^2 - 2x - 3}$.

Solution

By factoring the numerator and denominator, you have

$$f(x) = \frac{x^2 - 9}{x^2 - 2x - 3} = \frac{\cancel{(x - 3)}(x + 3)}{\cancel{(x - 3)}(x + 1)} = \frac{x + 3}{x + 1}, \quad x \neq 3.$$

y-intercept: (0, 3), because $f(0) = 3$

x-intercept: $(-3, 0)$, because $x + 3 = 0$ when $x = -3$

Vertical asymptote: $x = -1$, zero of (simplified) denominator

Hole: $\left(3, \frac{3}{2}\right)$, f is not defined at $x = 3$

Horizontal asymptote: $y = 1$, because degree of $N(x) =$ degree of $D(x)$

Additional points:

x	-5	-2	-1	-0.5	1	3	4
$f(x)$	0.5	-1	Undefined	5	2	Undefined	1.4

The graph is shown in Figure 3.30.

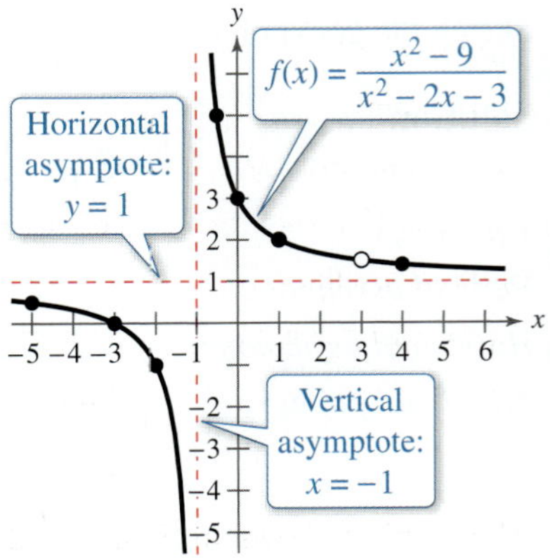

Figure 3.30 *Hole at x = 3*

✓ ***Checkpoint*** *Audio-video solution in English & Spanish at LarsonPrecalculus.com.*

Sketch the graph of $f(x) = \dfrac{x^2 - 4}{x^2 - x - 6}$.

Slant Asymptotes

Consider a rational function whose denominator is of degree 1 or greater. If the degree of the numerator is exactly *one more* than the degree of the denominator, then the graph of the function has a **slant** (or **oblique**) **asymptote.** For example, the graph of

$$f(x) = \frac{x^2 - x}{x + 1}$$

has a slant asymptote, as shown in Figure 3.31. To find the equation of a slant asymptote, use long division. For instance, by dividing $x + 1$ into $x^2 - x$, you have

$$f(x) = \frac{x^2 - x}{x + 1} = \underbrace{x - 2}_{\text{Slant asymptote } (y = x - 2)} + \frac{2}{x + 1}.$$

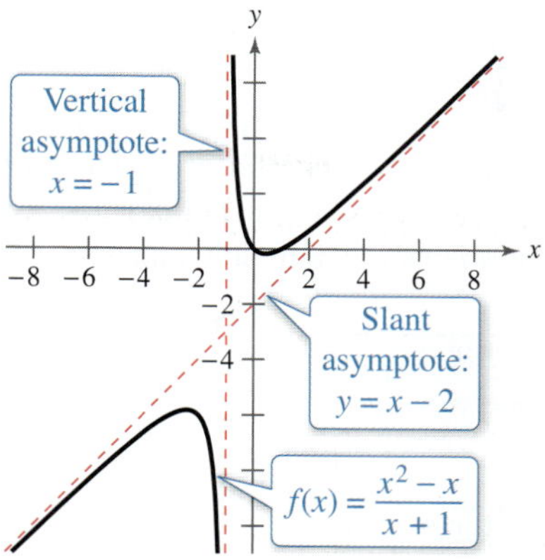

Figure 3.31

As x increases or decreases without bound, the remainder term $2/(x + 1)$ approaches 0, so the graph of f approaches the line $y = x - 2$, as shown in Figure 3.31.

Explore the Concept

Do you think it is possible for the graph of a rational function to cross its horizontal asymptote or its slant asymptote? Use the graphs of the following functions to investigate this question. Write a summary of your conclusion. Explain your reasoning.

$$f(x) = \frac{x}{x^2 + 1}$$

$$g(x) = \frac{2x}{3x^2 - 2x + 1}$$

$$h(x) = \frac{x^3}{x^2 + 1}$$

EXAMPLE 6 A Rational Function with a Slant Asymptote

Sketch the graph of $f(x) = \dfrac{x^2 - x - 2}{x - 1}$.

Solution

First, write $f(x)$ in two different ways. Factoring the numerator

$$f(x) = \frac{x^2 - x - 2}{x - 1} = \frac{(x - 2)(x + 1)}{x - 1}$$

enables you to recognize the x-intercepts. Long division

$$f(x) = \frac{x^2 - x - 2}{x - 1} = x - \frac{2}{x - 1}$$

enables you to recognize that the line $y = x$ is a slant asymptote of the graph.

y-intercept: (0, 2), because $f(0) = 2$

x-intercepts: $(-1, 0)$ and $(2, 0)$

Vertical asymptote: $x = 1$, zero of denominator

Horizontal asymptote: None, because degree of $N(x)$ > degree of $D(x)$

Slant asymptote: $y = x$

Additional points:

x	-2	0.5	1	1.5	3
$f(x)$	-1.33	4.5	Undefined	-2.5	2

The graph is shown in Figure 3.32.

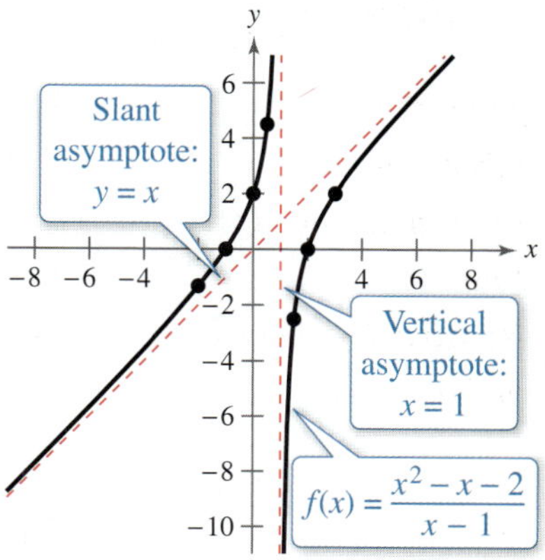

Figure 3.32

✔ **Checkpoint** Audio-video solution in English & Spanish at LarsonPrecalculus.com.

Sketch the graph of $f(x) = \dfrac{3x^2 + 1}{x}$.

Application

EXAMPLE 7 Publishing

A rectangular page is designed to contain 48 square inches of print. The margins on each side of the page are $1\frac{1}{2}$ inches wide. The margins at the top and bottom are each 1 inch deep. What should the dimensions of the page be so that the minimum amount of paper is used?

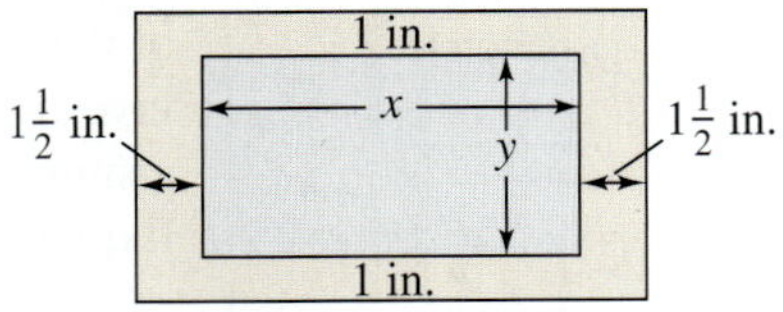

Figure 3.33

Graphical Solution

Let A be the area to be minimized. From Figure 3.33, you can write

$$A = (x + 3)(y + 2).$$

The printed area inside the margins is modeled by $48 = xy$ or $y = 48/x$. To find the minimum area, rewrite the equation for A in terms of just one variable by substituting $48/x$ for y.

$$A = (x + 3)\left(\frac{48}{x} + 2\right) = \frac{(x + 3)(48 + 2x)}{x}, \quad x > 0$$

The graph of this rational function is shown in Figure 3.34. Because x represents the width of the printed area, you need consider only the portion of the graph for which x is positive. Using the *minimum* feature of a graphing utility, you can approximate the minimum value of A to occur when $x \approx 8.5$ inches. The corresponding value of y is $48/8.5 \approx 5.6$ inches. So, the dimensions should be

$$x + 3 \approx 11.5 \text{ inches by } y + 2 \approx 7.6 \text{ inches.}$$

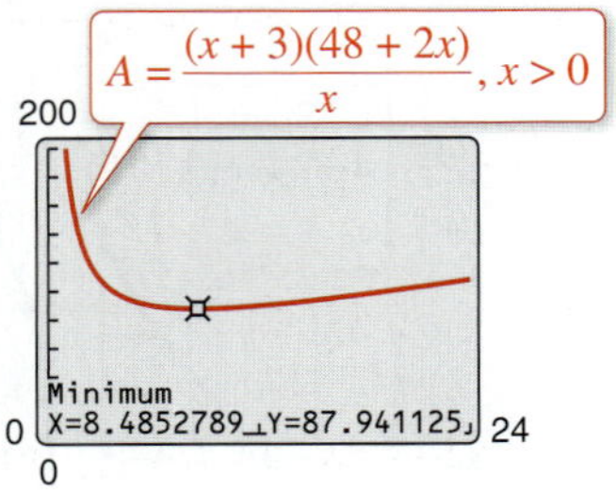

Figure 3.34

Numerical Solution

Let A be the area to be minimized. From Figure 3.33, you can write

$$A = (x + 3)(y + 2).$$

The printed area inside the margins is modeled by $48 = xy$ or $y = 48/x$. To find the minimum area, rewrite the equation for A in terms of just one variable by substituting $48/x$ for y.

$$A = (x + 3)\left(\frac{48}{x} + 2\right) = \frac{(x + 3)(48 + 2x)}{x}, \quad x > 0$$

Use the *table* feature of a graphing utility to create a table of values for the function

$$y_1 = \frac{(x + 3)(48 + 2x)}{x}$$

beginning at $x = 1$. From the table, you can see that the minimum value of y_1 occurs when x is somewhere between 8 and 9, as shown in Figure 3.35. To approximate the minimum value of y_1 to one decimal place, change the table to begin at $x = 8$ and set the table step to 0.1. The minimum value of y_1 occurs when $x \approx 8.5$, as shown in Figure 3.36. The corresponding value of y is $48/8.5 \approx 5.6$ inches. So, the dimensions should be

$$x + 3 \approx 11.5 \text{ inches by } y + 2 \approx 7.6 \text{ inches.}$$

X	Y1
6	90
7	88.571
8	88
9	88
10	88.4
11	89.091
12	90

X=8

Figure 3.35

X	Y1
8.2	87.961
8.3	87.949
8.4	87.943
8.5	87.941
8.6	87.944
8.7	87.952
8.8	87.964

X=8.5

Figure 3.36

✓ ***Checkpoint*** *Audio-video solution in English & Spanish at LarsonPrecalculus.com.*

Rework Example 7 when the margins on each side are 2 inches wide and the page contains 40 square inches of print.

If you go on to take a course in calculus, you will learn an analytic technique for finding the exact value of x that produces a minimum area in Example 7. In this case, that value is $x = 6\sqrt{2} \approx 8.485$.

3.6 Exercises

See *CalcChat.com* for tutorial help and worked-out solutions to odd-numbered exercises. For instructions on how to use a graphing utility, see Appendix A.

Vocabulary and Concept Check

In Exercises 1 and 2, fill in the blank(s).

1. For the rational function $f(x) = N(x)/D(x)$, if the degree of $N(x)$ is exactly one more than the degree of $D(x)$, then the graph of f has a _______ (or oblique) _______ .

2. The graph of $f(x) = 1/x$ has a _______ asymptote at $x = 0$.

3. Does the graph of $f(x) = \dfrac{x^3 - 1}{x^2 + 2}$ have a slant asymptote?

4. Using long division, you find that $f(x) = \dfrac{x^2 + 1}{x + 1} = x - 1 + \dfrac{2}{x + 1}$. What is the slant asymptote of the graph of f?

Procedures and Problem Solving

Library of Parent Functions: $f(x) = 1/x$ **In Exercises 5–8, sketch the graph of the function g and describe how the graph is related to the graph of $f(x) = 1/x$.**

5. $g(x) = \dfrac{-1}{x} + 2$ **6.** $g(x) = \dfrac{1}{x - 6}$

7. $g(x) = \dfrac{1}{x - 3} - 1$ **8.** $g(x) = \dfrac{-1}{x + 2} - 4$

Describing a Transformation of $f(x) = 2/x$ **In Exercises 9–12, use a graphing utility to graph $f(x) = 2/x$ and the function g in the same viewing window. Describe the relationship between the two graphs.**

9. $g(x) = f(x) + 1$ **10.** $g(x) = f(x - 1)$

11. $g(x) = -f(x)$ **12.** $g(x) = \frac{1}{2}f(x + 2)$

Describing a Transformation of $f(x) = 3/x^2$ **In Exercises 13–16, use a graphing utility to graph $f(x) = 3/x^2$ and the function g in the same viewing window. Describe the relationship between the two graphs.**

13. $g(x) = f(x) - 2$ **14.** $g(x) = -2f(x)$

15. $g(x) = f(x - 2)$ **16.** $g(x) = \frac{1}{4}f(x)$

Sketching the Graph of a Rational Function **In Exercises 17–32, sketch the graph of the rational function by hand. As sketching aids, check for intercepts, vertical asymptotes, horizontal asymptotes, and holes. Use a graphing utility to verify your graph.**

17. $f(x) = \dfrac{1}{x + 2}$ **18.** $f(x) = \dfrac{1}{x - 6}$

19. $C(x) = \dfrac{5 + 2x}{1 + x}$ **20.** $P(x) = \dfrac{1 - 3x}{1 - x}$

21. $f(t) = \dfrac{1 - 2t}{t}$ **22.** $g(x) = \dfrac{1}{x + 2} + 2$

23. $f(x) = \dfrac{x^2}{x^2 - 4}$ **24.** $g(x) = \dfrac{x}{x^2 - 9}$

25. $g(x) = \dfrac{4(x + 1)}{x(x - 4)}$ **26.** $h(x) = \dfrac{2}{x^2(x - 3)}$

27. $f(x) = \dfrac{3x}{x^2 - x - 2}$ **28.** $f(x) = \dfrac{2x}{x^2 + x - 2}$

29. $f(x) = \dfrac{x^2 + 3x}{x^2 + x - 6}$ **30.** $g(x) = \dfrac{5(x + 4)}{x^2 + x - 12}$

31. $f(x) = \dfrac{x^2 - 1}{x + 1}$ **32.** $f(x) = \dfrac{x^2 - 16}{x - 4}$

Finding the Domain and Asymptotes **In Exercises 33–42, use a graphing utility to graph the function. Determine its domain and identify any vertical or horizontal asymptotes.**

33. $f(x) = \dfrac{2 + x}{1 - x}$ **34.** $f(x) = \dfrac{3 - x}{2 - x}$

35. $g(x) = \dfrac{3x - 4}{-x}$ **36.** $h(x) = \dfrac{2x - 1}{x + 5}$

37. $g(x) = \dfrac{5}{x^2 + 1}$ **38.** $g(x) = -\dfrac{x}{(x - 2)^2}$

39. $f(x) = \dfrac{x + 1}{x^2 - x - 6}$ **40.** $f(x) = \dfrac{x + 4}{x^2 + x - 6}$

41. $f(x) = \dfrac{20x}{x^2 + 1} - \dfrac{1}{x}$ **42.** $f(x) = 5\left(\dfrac{1}{x - 4} - \dfrac{1}{x + 2}\right)$

Exploration **In Exercises 43–48, use a graphing utility to graph the function. What do you observe about its asymptotes?**

43. $h(x) = \dfrac{6x}{\sqrt{x^2 + 1}}$ **44.** $f(x) = -\dfrac{x}{\sqrt{9 + x^2}}$

45. $g(x) = \dfrac{4|x - 2|}{x + 1}$ **46.** $f(x) = -\dfrac{8|3 + x|}{x - 2}$

47. $f(x) = \dfrac{4(x - 1)^2}{x^2 - 4x + 5}$ **48.** $g(x) = \dfrac{3x^4 - 5x + 3}{x^4 + 1}$

A Rational Function with a Slant Asymptote **In Exercises 49–56, sketch the graph of the rational function by hand. As sketching aids, check for intercepts, vertical asymptotes, and slant asymptotes.**

49. $f(x) = \dfrac{2x^2 + 1}{x}$

50. $g(x) = \dfrac{1 - x^2}{x}$

51. $h(x) = \dfrac{x^2}{x - 1}$

52. $f(x) = \dfrac{x^3}{x^2 - 1}$

53. $g(x) = \dfrac{x^3}{2x^2 - 8}$

54. $f(x) = \dfrac{x^3}{x^2 + 4}$

55. $f(x) = \dfrac{x^3 + 2x^2 + 4}{2x^2 + 1}$

56. $f(x) = \dfrac{2x^2 - 5x + 5}{x - 2}$

Finding the *x*-Intercepts **In Exercises 57–60, use the graph to estimate any *x*-intercepts of the rational function. Set $y = 0$ and solve the resulting equation to confirm your result.**

57. $y = \dfrac{x + 1}{x - 3}$

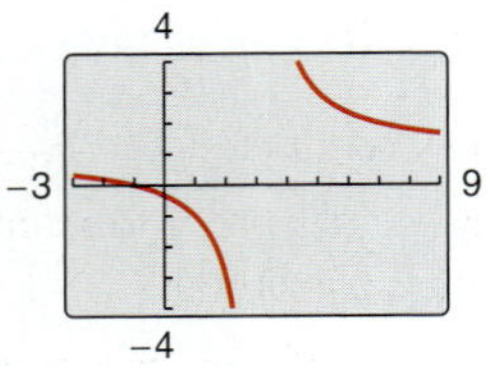

58. $y = \dfrac{2x}{x - 3}$

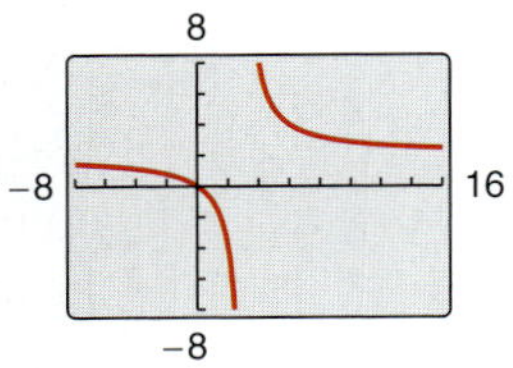

59. $y = \dfrac{1}{x} - x$

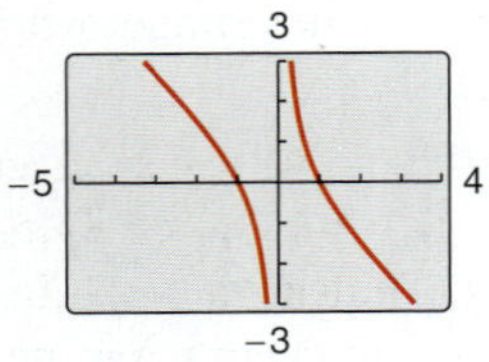

60. $y = x - 3 + \dfrac{2}{x}$

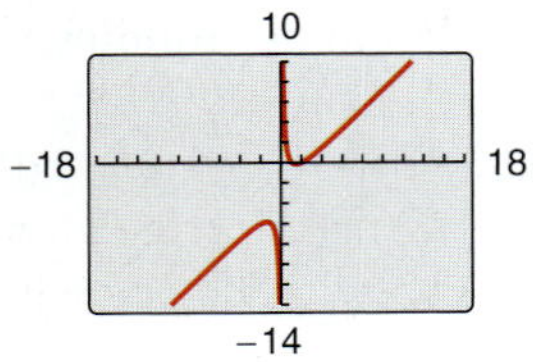

Finding the Domain and Asymptotes **In Exercises 61–64, use a graphing utility to graph the rational function. Determine the domain of the function and identify any asymptotes.**

61. $y = \dfrac{2x^2 + x}{x + 1}$

62. $y = \dfrac{x^2 + 5x + 8}{x + 3}$

63. $y = \dfrac{1 + 3x^2 - x^3}{x^2}$

64. $y = \dfrac{12 - 2x - x^2}{2(4 + x)}$

Finding Asymptotes and Holes **In Exercises 65–70, find all vertical asymptotes, horizontal asymptotes, slant asymptotes, and holes in the graph of the function. Then use a graphing utility to verify your results.**

65. $f(x) = \dfrac{x^2 - 5x + 4}{x^2 - 4}$

66. $f(x) = \dfrac{x^2 - 2x - 8}{x^2 - 9}$

67. $f(x) = \dfrac{2x^2 - 5x + 2}{2x^2 - x - 6}$

68. $f(x) = \dfrac{3x^2 - 8x + 4}{2x^2 - 3x - 2}$

69. $f(x) = \dfrac{2x^3 - x^2 - 2x + 1}{x^2 + 3x + 2}$

70. $f(x) = \dfrac{2x^3 + x^2 - 8x - 4}{x^2 - 3x + 2}$

Finding *x*-Intercepts Graphically **In Exercises 71–82, use a graphing utility to graph the function and determine any *x*-intercepts. Set $y = 0$ and solve the resulting equation to confirm your result.**

71. $y = \dfrac{1}{x + 5} + \dfrac{4}{x}$

72. $y = \dfrac{1}{x - 2} - \dfrac{5}{x}$

73. $y = \dfrac{2}{x + 2} - \dfrac{3}{x - 1}$

74. $y = \dfrac{6}{x + 3} - \dfrac{1}{x + 4}$

75. $y = x - \dfrac{2}{x + 1}$

76. $y = 2x - \dfrac{8}{x}$

77. $y = x + 2 - \dfrac{1}{x + 1}$

78. $y = 2x - 1 + \dfrac{1}{x - 2}$

79. $y = x + 1 + \dfrac{2}{x - 1}$

80. $y = x + 2 + \dfrac{2}{x + 2}$

81. $y = x + 3 - \dfrac{2}{2x - 1}$

82. $y = x - 1 - \dfrac{2}{2x - 3}$

83. Chemistry A 1000-liter tank contains 50 liters of a 25% brine solution. You add x liters of a 75% brine solution to the tank.

(a) Show that the concentration C (the ratio of brine to the total solution) of the final mixture is given by

$$C = \frac{3x + 50}{4(x + 50)}.$$

(b) Determine the domain of the function based on the physical constraints of the problem.

(c) Use a graphing utility to graph the function. As the tank is filled, what happens to the rate at which the concentration of brine increases? What percent does the concentration of brine appear to approach?

84. Geometry A rectangular region of length x and width y has an area of 500 square meters.

(a) Write the width y as a function of x.

(b) Determine the domain of the function based on the physical constraints of the problem.

(c) Sketch a graph of the function and determine the width of the rectangle when $x = 30$ meters.

85. Publishing A page that is x inches wide and y inches high contains 30 square inches of print (see figure). The margins at the top and bottom are 2 inches deep and the margins on each side are 1 inch wide.

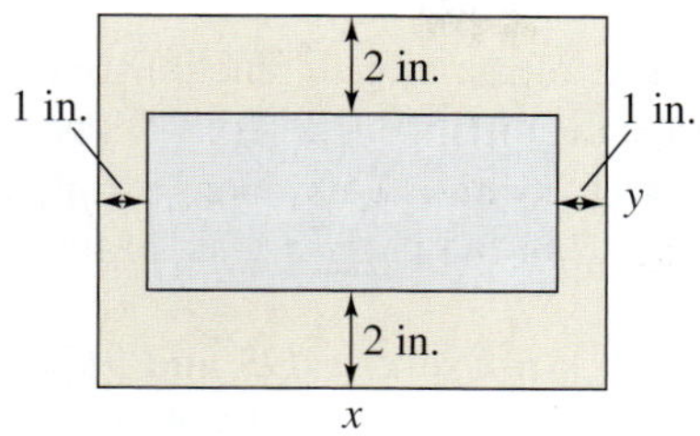

(a) Show that the total area A of the page is given by

$$A = \frac{2x(2x + 11)}{x - 2}.$$

(b) Determine the domain of the function based on the physical constraints of the problem.

(c) Use a graphing utility to graph the area function and approximate the page size such that the minimum amount of paper will be used. Verify your answer numerically using the *table* feature of the graphing utility.

86. Geometry A right triangle is formed in the first quadrant by the x-axis, the y-axis, and a line segment through the point (3, 2). (See figure.)

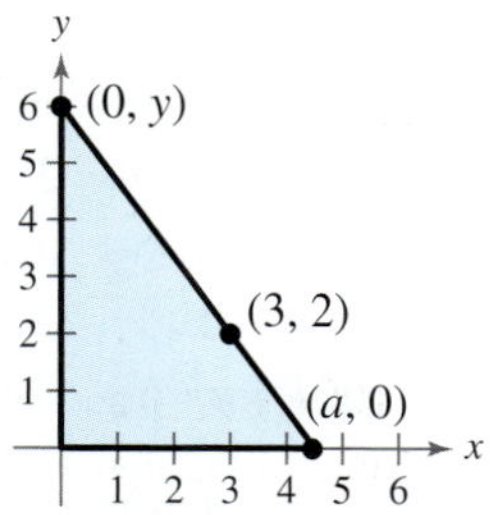

(a) Show that an equation of the line segment is given by

$$y = \frac{2(a - x)}{a - 3}, \quad 0 \le x \le a.$$

(b) Show that the area of the triangle is given by

$$A = \frac{a^2}{a - 3}.$$

(c) Use a graphing utility to graph the area function and estimate the value of a that yields a minimum area. Estimate the minimum area. Verify your answer numerically using the *table* feature of the graphing utility.

87. Cost Management The ordering and transportation cost C (in thousands of dollars) for the components used in manufacturing a product is given by

$$C = 100\left(\frac{200}{x^2} + \frac{x}{x + 30}\right), \quad x \ge 1$$

where x is the order size (in hundreds). Use a graphing utility to graph the cost function. From the graph, estimate the order size that minimizes cost.

88. Cost Management The cost C of producing x units of a product is given by $C = 0.2x^2 + 10x + 5$, and the average cost per unit is given by

$$\overline{C} = \frac{C}{x} = \frac{0.2x^2 + 10x + 5}{x}, \quad x > 0.$$

Sketch the graph of the average cost function, and estimate the number of units that should be produced to minimize the average cost per unit.

89. *Why you should learn it* *(p. 297)* The concentration C of a chemical in the bloodstream t hours after injection into muscle tissue is given by

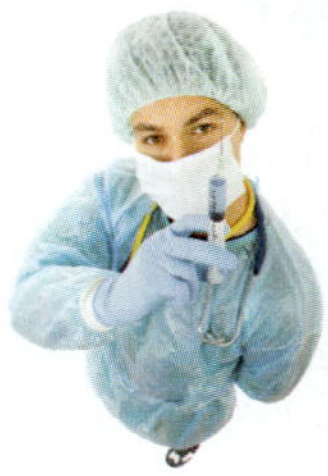

$$C = \frac{3t^2 + t}{t^3 + 50}, \quad t \ge 0.$$

(a) Determine the horizontal asymptote of the function and interpret its meaning in the context of the problem.

(b) Use a graphing utility to graph the function and approximate the time when the bloodstream concentration is greatest.

(c) Use the graphing utility to determine when the concentration is less than 0.345.

90. Algebraic-Graphical-Numerical A driver averaged 50 miles per hour on the round trip between Baltimore, Maryland, and Philadelphia, Pennsylvania, 100 miles away. The average speeds for going and returning were x and y miles per hour, respectively.

(a) Show that $y = \dfrac{25x}{x - 25}$.

(b) Determine the vertical and horizontal asymptotes of the function.

(c) Use a graphing utility to complete the table. What do you observe?

x	30	35	40	45	50	55	60
y							

(d) Use the graphing utility to graph the function.

(e) Is it possible to average 20 miles per hour in one direction and still average 50 miles per hour on the round trip? Explain.

91. MODELING DATA

Data are recorded at 225 monitoring sites throughout the United States to study national trends in air quality. The table shows the mean amount A of carbon monoxide (in parts per million) recorded at these sites in each year from 2003 through 2012. (Source: EPA)

DATA

Year	Amount, A (in parts per million)
2003	2.8
2004	2.6
2005	2.3
2006	2.2
2007	2.0
2008	1.8
2009	1.8
2010	1.6
2011	1.6
2012	1.5

Spreadsheet at LarsonPrecalculus.com

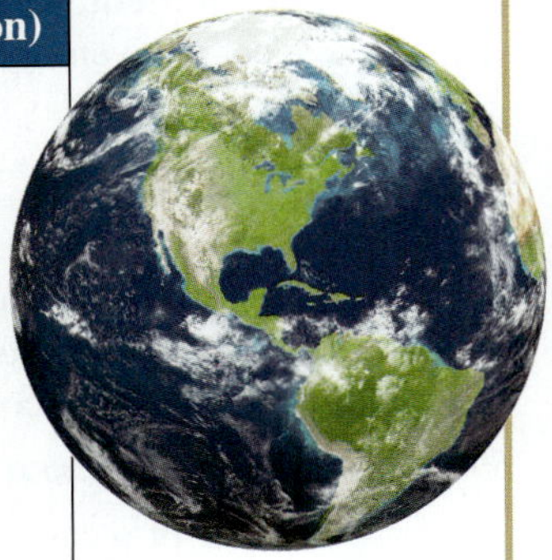

(a) Use the *regression* feature of a graphing utility to find a linear model for the data. Let $t = 3$ represent 2003. Use the graphing utility to plot the data and graph the model in the same viewing window.

(b) Find a rational model for the data. Take the reciprocal of A to generate the points $(t, 1/A)$. Use the *regression* feature of the graphing utility to find a linear model for these data. The resulting line has the form $1/A = at + b$. Solve for A. Use the graphing utility to plot the data and graph the rational model in the same viewing window.

(c) Which model do you prefer? Why?

92. HOW DO YOU SEE IT? A herd of elk is released onto state game lands. The graph shows the expected population P of the herd, where t is the time (in years) since the initial number of elk were released.

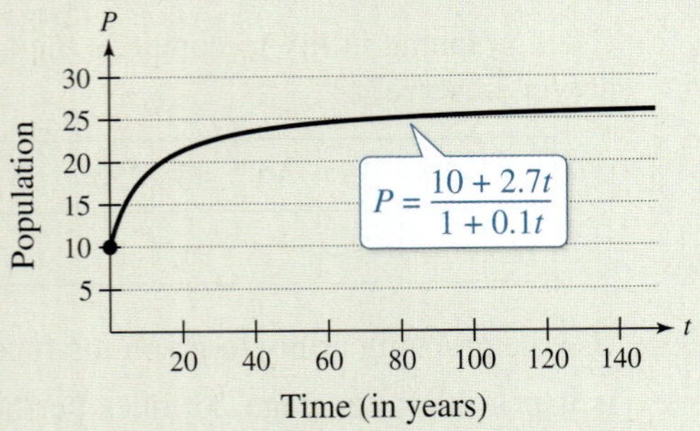

(a) Determine the domain of the function. Explain.

(b) Find the initial number of elk in the herd.

(c) Is there a limit to the size of the herd? If so, what is the expected population?

Conclusions

True or False? **In Exercises 93 and 94, determine whether the statement is true or false. Justify your answer.**

93. The graph of a rational function is continuous only when the denominator is a constant polynomial.

94. The graph of a rational function can never cross one of its asymptotes.

Think About It **In Exercises 95 and 96, use a graphing utility to graph the function. Explain why there is no vertical asymptote when a superficial examination of the function might indicate that there should be one.**

95. $h(x) = \dfrac{6 - 2x}{3 - x}$ **96.** $g(x) = \dfrac{x^2 + x - 2}{x - 1}$

97. Writing Write a set of guidelines for finding all the asymptotes of a rational function given that the degree of the numerator is not more than 1 greater than the degree of the denominator.

98. Writing a Rational Function Write a rational function that has the specified characteristics. (There are many correct answers.)

(a) Vertical asymptote: $x = -2$
Slant asymptote: $y = x + 1$
Zero of the function: $x = 2$

(b) Vertical asymptote: $x = -4$
Slant asymptote: $y = x - 2$
Zero of the function: $x = 3$

Cumulative Mixed Review

Simplifying Exponential Expressions **In Exercises 99–102, simplify the expression.**

99. $\left(\dfrac{x}{8}\right)^{-3}$ **100.** $(4x^2)^{-2}$

101. $\dfrac{3^{7/6}}{3^{1/6}}$ **102.** $\dfrac{(x^{-2})(x^{1/2})}{(x^{-1})(x^{5/2})}$

Finding the Domain and Range of a Function **In Exercises 103–106, use a graphing utility to graph the function and find its domain and range.**

103. $f(x) = \sqrt{6 + x^2}$ **104.** $f(x) = \sqrt{121 - x^2}$

105. $f(x) = -|x + 9|$ **106.** $f(x) = -x^2 + 9$

107. *Make a Decision* To work an extended application analyzing the median sales prices of existing one-family homes, visit this textbook's website at *LarsonPrecalculus.com.* (Data Source: National Association of Realtors)

3.7 Quadratic Models

Classifying Scatter Plots

In real life, many relationships between two variables are parabolic, as in Section 3.1, Example 5. A scatter plot can be used to give you an idea of which type of model will best fit a set of data.

What you should learn

- Classify scatter plots.
- Use scatter plots and a graphing utility to find quadratic models for data.
- Choose a model that best fits a set of data.

Why you should learn it

Many real-life situations can be modeled by quadratic equations. For instance, in Exercise 17 on page 311, a quadratic equation is used to model the monthly precipitation for San Francisco, California.

EXAMPLE 1 Classifying Scatter Plots

See LarsonPrecalculus.com for an interactive version of this type of example.

Decide whether each set of data could be better modeled by a linear model,

$$y = ax + b$$

a quadratic model,

$$y = ax^2 + bc + c$$

or neither.

a. (0.9, 1.7), (1.2, 2.0), (1.3, 1.9), (1.4, 2.1), (1.6, 2.5), (1.8, 2.8), (2.1, 3.0), (2.5, 3.4), (2.9, 3.7), (3.2, 3.9), (3.3, 4.1), (3.6, 4.4), (4.0, 4.7), (4.2, 4.8), (4.3, 5.0)

b. (0.9, 3.2), (1.2, 4.0), (1.3, 4.1), (1.4, 4.4), (1.6, 5.1), (1.8, 6.0), (2.1, 7.6), (2.5, 9.8), (2.9, 12.4), (3.2, 14.3), (3.3, 15.2), (3.6, 18.1), (4.0, 22.7), (4.2, 24.9), (4.3, 27.2)

c. (0.9, 1.2), (1.2, 6.5), (1.3, 9.3), (1.4, 11.6), (1.6, 15.2), (1.8, 16.9), (2.1, 14.7), (2.5, 8.1), (2.9, 3.7), (3.2, 5.8), (3.3, 7.1), (3.6, 11.5), (4.0, 20.2), (4.2, 23.7), (4.3, 26.9)

Solution

a. Begin by entering the data into a graphing utility. Then display the scatter plot, as shown in Figure 3.37. From the scatter plot, it appears the data follow a linear pattern. So, the data can be better modeled by a linear function.

b. Enter the data into a graphing utility and then display the scatter plot (see Figure 3.38). From the scatter plot, it appears the data follow a parabolic pattern. So, the data can be better modeled by a quadratic function.

c. Enter the data into a graphing utility and then display the scatter plot (see Figure 3.39). From the scatter plot, it appears the data do not follow either a linear or a parabolic pattern. So, the data cannot be modeled by either a linear function or a quadratic function.

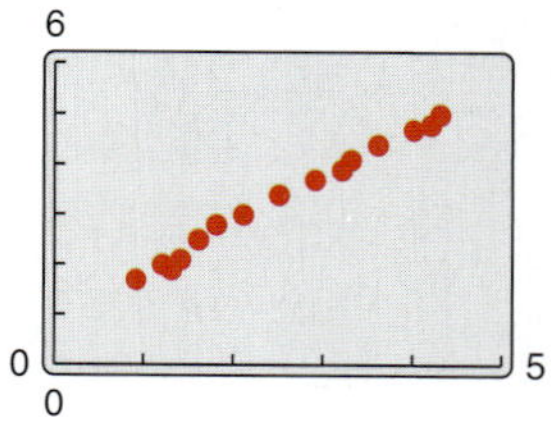

Figure 3.37

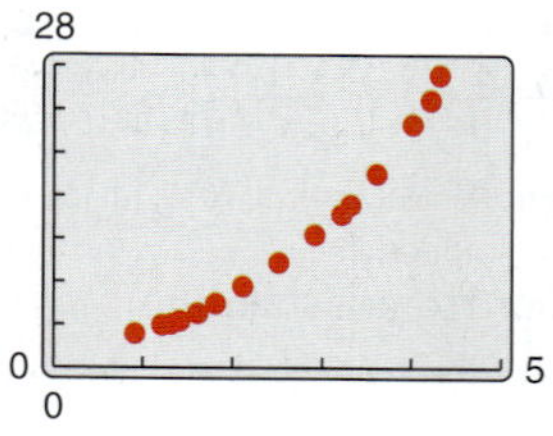

Figure 3.38

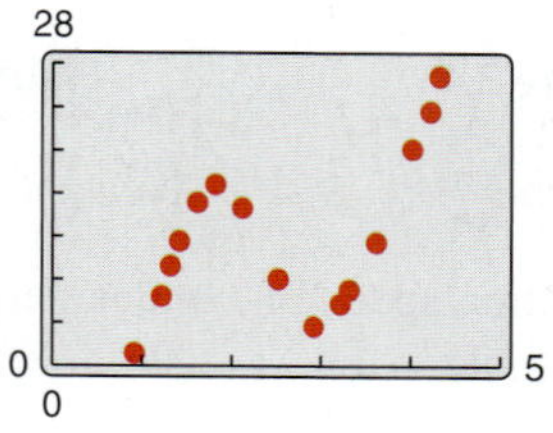

Figure 3.39

✓ ***Checkpoint*** *Audio-video solution in English & Spanish at LarsonPrecalculus.com.*

Decide whether the data could be better modeled by a linear model, $y = ax + b$, a quadratic model, $y = ax^2 + bx + c$, or neither.

(0, 3480), (5, 2235), (10, 1250), (15, 565), (20, 150), (25, 12), (30, 145), (35, 575), (40, 1275), (45, 2225), (50, 3500), (55, 5010)

Fitting a Quadratic Model to Data

In Section 2.7, you created scatter plots of data and used a graphing utility to find the least squares regression lines for the data. You can use a similar procedure to find a model for nonlinear data. Once you have used a scatter plot to determine the type of model that would best fit a set of data, there are several ways that you can actually find the model. Each method is best used with a computer or calculator, rather than with hand calculations.

EXAMPLE 2 Fitting a Quadratic Model to Data

A study was done to compare the speed x (in miles per hour) with the mileage y (in miles per gallon) of an automobile. The results are shown in the table.

DATA

Speed, x	Mileage, y
15	22.3
20	25.5
25	27.5
30	29.0
35	28.7
40	29.9
45	30.4
50	30.2
55	30.0
60	28.8
65	27.4
70	25.3
75	23.3

Spreadsheet at LarsonPrecalculus.com

a. Use a graphing utility to create a scatter plot of the data.

b. Use the *regression* feature of the graphing utility to find a model that best fits the data.

c. Approximate the speed at which the mileage is the greatest.

Solution

a. Begin by entering the data into a graphing utility and displaying the scatter plot, as shown in Figure 3.40. From the scatter plot, you can see that the data appear to follow a parabolic pattern.

b. Using the *regression* feature of the graphing utility, you can find the quadratic model, as shown in Figure 3.41. So, the quadratic equation that best fits the data is given by

$$y = -0.0082x^2 + 0.75x + 13.5. \quad \text{Quadratic model}$$

c. Graph the data and the model in the same viewing window, as shown in Figure 3.42. Use the *maximum* feature or the *zoom* and *trace* features of the graphing utility to approximate the speed at which the mileage is greatest. You should obtain a maximum of approximately (46, 31), as shown in Figure 3.42. So, the speed at which the mileage is greatest is about 46 miles per hour.

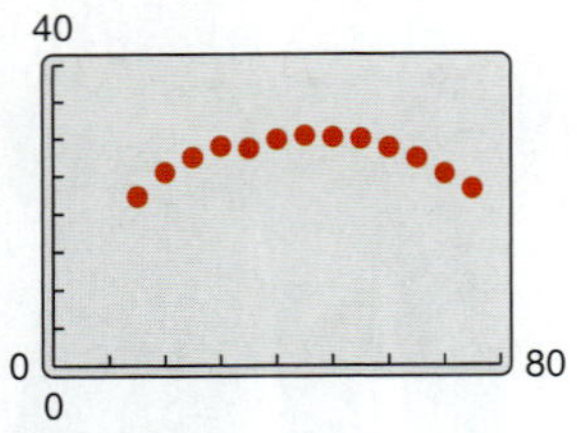

Figure 3.40

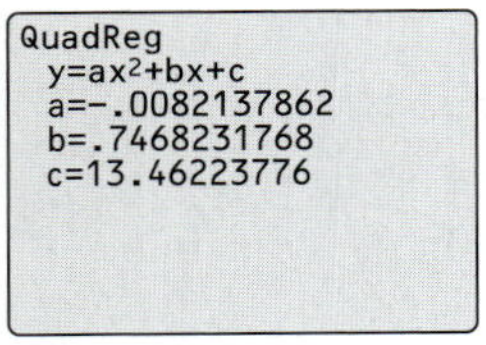

Figure 3.41

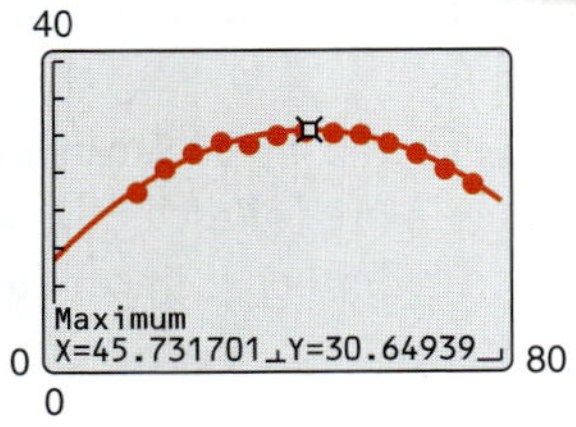

Figure 3.42

✓ ***Checkpoint*** *Audio-video solution in English & Spanish at LarsonPrecalculus.com.*

The time y (in seconds) required to attain a speed of x miles per hour from a standing start for an automobile is shown in the table.

DATA

Speed, x	0	20	30	40	50	60	70	80
Time, y	0	1.4	2.6	3.8	4.9	6.3	8.0	9.9

Spreadsheet at LarsonPrecalculus.com

a. Use a graphing utility to create a scatter plot of the data.

b. Use the *regression* feature of the graphing utility to find a model that best fits the data.

c. Use the model to estimate how long it takes the automobile, from a standing start, to reach a speed of 55 miles per hour.

Supertrooper | Dreamstime

EXAMPLE 3 Fitting a Quadratic Model to Data

A basketball is dropped from a height of about 5.25 feet. The height of the basketball is recorded 23 times at intervals of about 0.02 second. The results are shown in the table. Use a graphing utility to find a model that best fits the data. Then use the model to predict the time when the basketball will hit the ground.

Time, x	Height, y
0.0	5.23594
0.02	5.20353
0.04	5.16031
0.06	5.09910
0.08	5.02707
0.099996	4.95146
0.119996	4.85062
0.139992	4.74979
0.159988	4.63096
0.179988	4.50132
0.199984	4.35728
0.219984	4.19523
0.23998	4.02958
0.25993	3.84593
0.27998	3.65507
0.299976	3.44981
0.319972	3.23375
0.339961	3.01048
0.359961	2.76921
0.379951	2.52074
0.399941	2.25786
0.419941	1.98058
0.439941	1.63488

Spreadsheet at LarsonPrecalculus.com

Solution

Begin by entering the data into a graphing utility and displaying the scatter plot, as shown in Figure 3.43. From the scatter plot, you can see that the data show a parabolic trend. So, using the *regression* feature of the graphing utility, you can find the quadratic model, as shown in Figure 3.44. The quadratic model that best fits the data is given by $y = -15.449x^2 - 1.30x + 5.2$.

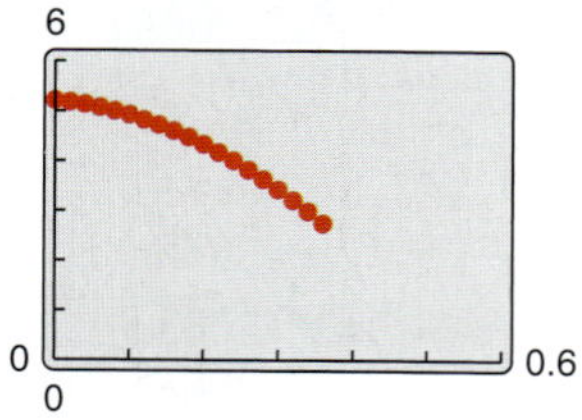

Figure 3.43

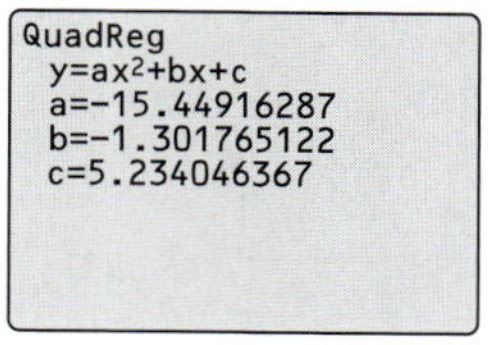

Figure 3.44

You can graph the data and the model in the same viewing window to see that the model fits the data well, as shown in the next figure.

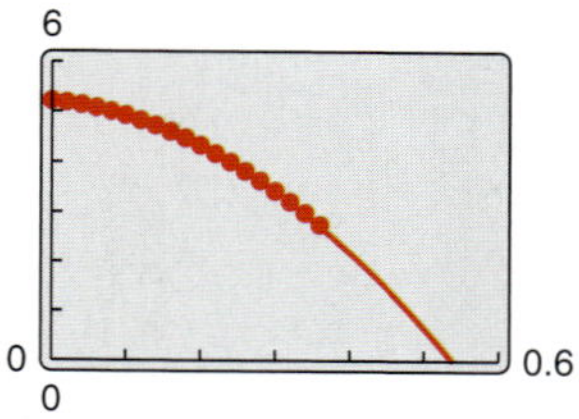

Using this model, you can predict the time when the basketball will hit the ground by substituting 0 for y and solving the resulting equation for x.

$$y = -15.449x^2 - 1.30x + 5.2 \quad \text{Write original model.}$$

$$0 = -15.449x^2 - 1.30x + 5.2 \quad \text{Substitute 0 for } y.$$

$$x = \frac{-b \pm \sqrt{b^2 - 4ac}}{2a} \quad \text{Quadratic Formula}$$

$$= \frac{-(-1.30) \pm \sqrt{(-1.30)^2 - 4(-15.449)(5.2)}}{2(-15.449)} \quad \text{Substitute for } a, b, \text{ and } c.$$

$$\approx 0.54 \quad \text{Choose positive solution.}$$

So, the solution is about 0.54 second. In other words, the basketball will continue to fall for about $0.54 - 0.44 = 0.1$ second more before hitting the ground.

✓ **Checkpoint** *Audio-video solution in English & Spanish at LarsonPrecalculus.com.*

The table shows the annual sales y (in millions of dollars) of a department store chain. Use a graphing utility to find a model that best fits the data. Then use the model to estimate the first year when the annual sales will be less than $190 million.

Year, x	1	2	3	4	5	6
Sales, y	221	222	220	219	216	211

Spreadsheet at LarsonPrecalculus.com

Choosing a Model

Sometimes it is not easy to distinguish from a scatter plot which type of model will best fit the data. You should first find several models for the data, using the *Library of Parent Functions*, and then choose the model that best fits the data by comparing the y-values of each model with the actual y-values.

EXAMPLE 4 Choosing a Model

The table shows the numbers y (in millions) of people ages 16 and older that were not in the U.S. labor force from 2003 through 2013. Use the *regression* feature of a graphing utility to find a linear model and a quadratic model for the data. Determine which model better fits the data. (Source: U.S. Department of Labor)

DATA

Year	Number of people unemployed (in millions), y
2003	74.7
2004	76.0
2005	76.8
2006	77.4
2007	78.7
2008	79.5
2009	81.7
2010	83.9
2011	86.0
2012	88.3
2013	90.3

Spreadsheet at LarsonPrecalculus.com

Solution

Let x represent the year, with $x = 3$ corresponding to 2003. Begin by entering the data into a graphing utility. Using the *regression* feature, a linear model for the data is

$$y = 1.55x + 68.8$$

and a quadratic model for the data is

$$y = 0.107x^2 - 0.16x + 74.6.$$

Plot the data and the linear model in the same viewing window, as shown in Figure 3.45. Then plot the data and the quadratic model in the same viewing window, as shown in Figure 3.46. To determine which model fits the data better, compare the y-values given by each model with the actual y-values. The model whose y-values are closest to the actual values is the better fit. In this case, the better-fitting model is the quadratic model.

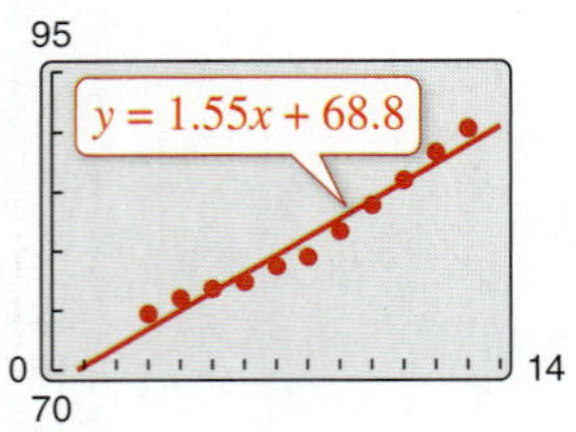

Figure 3.45

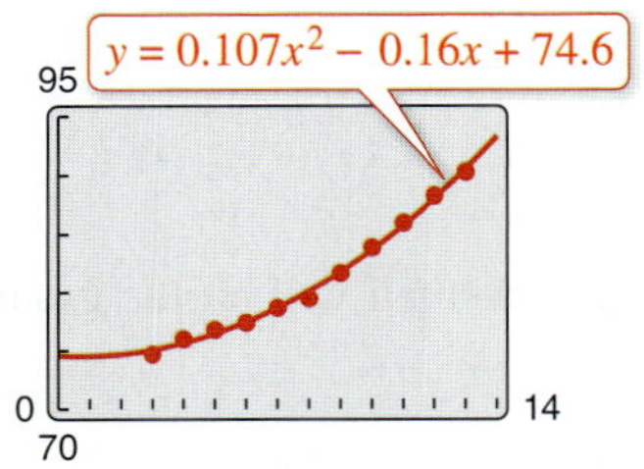

Figure 3.46

Checkpoint Audio-video solution in English & Spanish at LarsonPrecalculus.com.

The table shows the numbers y (in thousands) of full-size, alternative fueled vehicles in use from 2005 through 2011. Use the *regression* feature of a graphing utility to find a linear model and a quadratic model for the data. Determine which model better fits the data. (Source: U.S. Energy Information Administration)

DATA

Year	Number of alternative fueled vehicles (in thousands), y
2005	19.2
2006	31.3
2007	44.9
2008	59.8
2009	64.2
2010	72.1
2011	81.3

Spreadsheet at LarsonPrecalculus.com

Technology Tip

When you use the *regression* feature of a graphing utility, the program may output an "r^2-value." This r^2-value is the **coefficient of determination** of the data and gives a measure of how well the model fits the data. The coefficient of determination for the linear model in Example 4 is $r^2 \approx 0.9608$, and the coefficient of determination for the quadratic model is $r^2 \approx 0.9962$. Because the coefficient of determination for the quadratic model is closer to 1, the quadratic model better fits the data.

3.7 Exercises

See *CalcChat.com* for tutorial help and worked-out solutions to odd-numbered exercises.
For instructions on how to use a graphing utility, see Appendix A.

Vocabulary and Concept Check

1. What type of model best represents data that follow a parabolic pattern?
2. Which coefficient of determination indicates a better model for a set of data, $r^2 = 0.0365$ or $r^2 = 0.9688$?

Procedures and Problem Solving

Classifying Scatter Plots **In Exercises 3–8, determine whether the scatter plot could best be modeled by a linear model, a quadratic model, or neither.**

3.
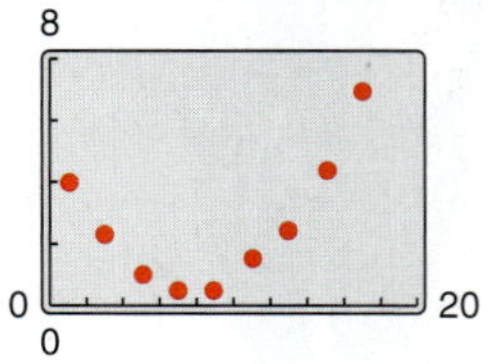

4.
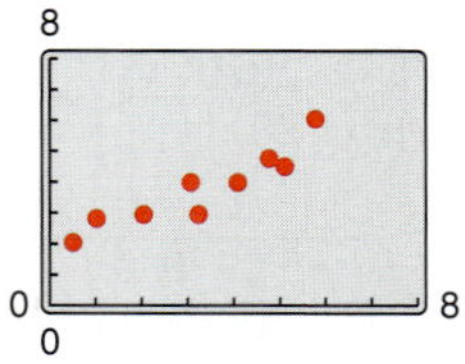

5.
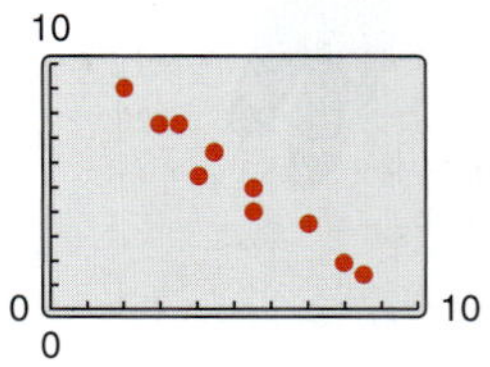

6.
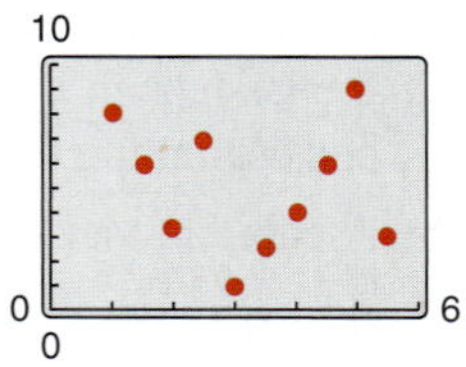

7.
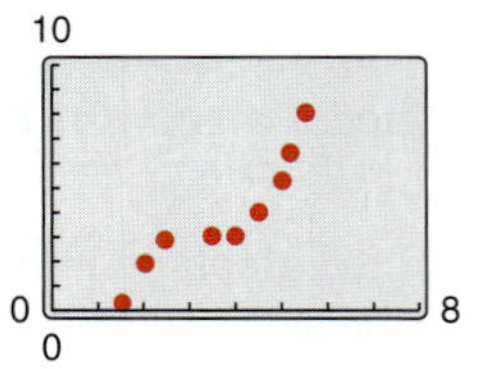

8.
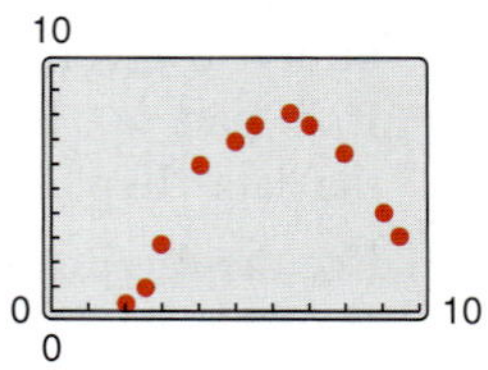

Choosing a Model **In Exercises 9–16, (a) use a graphing utility to create a scatter plot of the data, (b) determine whether the data could be better modeled by a linear model or a quadratic model, (c) use the *regression* feature of the graphing utility to find a model for the data, (d) use the graphing utility to graph the model with the scatter plot from part (a), and (e) create a table comparing the original data with the data given by the model.**

9. (0, 2.1), (1, 2.4), (2, 2.5), (3, 2.8), (4, 2.9), (5, 3.0), (6, 3.0), (7, 3.2), (8, 3.4), (9, 3.5), (10, 3.6)
10. (−2, 11.0), (−1, 10.7), (0, 10.4), (1, 10.3), (2, 10.1), (3, 9.9), (4, 9.6), (5, 9.4), (6, 9.4), (7, 9.2), (8, 9.0)
11. (0, 2795), (5, 1590), (10, 650), (15, −30), (20, −450), (25, −615), (30, −520), (35, −55), (40, 625), (45, 1630), (50, 2845), (55, 4350)
12. (0, 6140), (2, 6815), (4, 7335), (6, 7710), (8, 7915), (10, 7590), (12, 7975), (14, 7700), (16, 7325), (18, 6820), (20, 6125), (22, 5325)
13. (1, 4.0), (2, 6.5), (3, 8.8), (4, 10.6), (5, 13.9), (6, 15.0), (7, 17.5), (8, 20.1), (9, 24.0), (10, 27.1)
14. (−6, 10.7), (−4, 9.0), (−2, 7.0), (0, 5.4), (2, 3.5), (4, 1.7), (6, −0.1), (8, −1.8), (10, −3.6), (12, −5.3)
15. (0, 587), (5, 551), (10, 512), (15, 478), (20, 436), (25, 430), (30, 424), (35, 420), (40, 423), (45, 429), (50, 444)
16. (2, 34.3), (3, 33.8), (4, 32.6), (5, 30.1), (6, 27.8), (7, 22.5), (8, 19.1), (9, 14.8), (10, 9.4), (11, 3.7), (12, −1.6)

17. ***Why you should learn it*** *(p. 307)* The table shows the monthly normal precipitation P (in inches) for San Francisco, California. (Source: The Weather Channel)

DATA

Month	Precipitation, P
January	4.50
February	4.61
March	3.26
April	1.46
May	0.70
June	0.16
July	0.00
August	0.06
September	0.21
October	1.12
November	3.16
December	4.56

Spreadsheet at LarsonPrecalculus.com

(a) Use a graphing utility to create a scatter plot of the data. Let t represent the month, with $t = 1$ corresponding to January.

(b) Use the *regression* feature of the graphing utility to find a quadratic model for the data and identify the coefficient of determination.

(c) Use the graphing utility to graph the model with the scatter plot from part (a).

(d) Use the graph from part (c) to determine in which month the normal precipitation in San Francisco is the least.

(e) Use the table to determine the month in which the normal precipitation in San Francisco is the least. Compare your answer with that of part (d).

18. MODELING DATA

The table shows the annual sales S (in billions of dollars) of pharmacies and drug stores in the United States from 2007 through 2012. (Source: U.S. Census Bureau)

Spreadsheet at LarsonPrecalculus.com

Year	Sales, S (in billions of dollars)
2007	202.3
2008	211.0
2009	217.6
2010	222.8
2011	231.5
2012	233.4

(a) Use a graphing utility to create a scatter plot of the data. Let t represent the year, with $t = 7$ corresponding to 2007.

(b) Use the *regression* feature of the graphing utility to find a quadratic model for the data.

(c) Use the graphing utility to graph the model with the scatter plot from part (a).

(d) Use the model to estimate the first year when the annual sales of pharmacies and drug stores will be less than \$200 billion. Is this a good model for predicting future sales? Explain.

19. MODELING DATA

The table shows the percents P of U.S. households with Internet access from 2007 through 2012. (Source: U.S. Census Bureau)

Spreadsheet at LarsonPrecalculus.com

Year	Percent, P
2007	54.7
2008	61.7
2009	68.7
2010	71.1
2011	71.7
2012	74.8

(a) Use a graphing utility to create a scatter plot of the data. Let t represent the year, with $t = 7$ corresponding to 2007.

(b) Use the *regression* feature of the graphing utility to find a quadratic model for the data.

(c) Use the graphing utility to graph the model with the scatter plot from part (a).

(d) According to the model, in what year will the percent of U.S. households with Internet access be less than 60%? Is this a good model for making future predictions? Explain.

20. MODELING DATA

The table shows the estimated average numbers of hours H that adults in the United States spent reading newspapers each year from 2003 through 2012. (Source: Statista)

Spreadsheet at LarsonPrecalculus.com

Year	Hours, H
2003	198
2004	195
2005	191
2006	182
2007	176
2008	169
2009	158
2010	155
2011	152
2012	150

(a) Use a graphing utility to create a scatter plot of the data. Let t represent the year, with $t = 3$ corresponding to 2003.

(b) A cubic model for the data is

$$H = 0.131t^3 - 2.81t^2 + 12.1t + 183$$

which has an r^2-value of 0.9962. Use the graphing utility to graph the model with the scatter plot from part (a). Is the cubic model a good fit for the data? Explain.

(c) Use the *regression* feature of the graphing utility to find a quadratic model for the data and identify the coefficient of determination.

(d) Use the graphing utility to graph the quadratic model with the scatter plot from part (a). Is the quadratic model a good fit for the data? Explain.

(e) Which model is a better fit for the data? Explain.

(f) A consumer research company makes projections about the average numbers of hours H^* that adults spent reading newspapers each year from 2013 through 2015. The company's projections are shown in the table. Use the models from parts (b) and (c) to *predict* the average numbers of hours for 2013 through 2015. Explain why your values may differ from those in the table.

Year	2013	2014	2015
H^*	145	143	139

21. MODELING DATA

The table shows the numbers of U.S. households with televisions (in millions) from 2000 through 2012. (Source: The Nielsen Company)

DATA

Year	Televisions, T (in millions)
2000	100.8
2001	102.2
2002	105.5
2003	106.7
2004	108.4
2005	109.6
2006	110.2
2007	111.4
2008	112.8
2009	114.5
2010	114.9
2011	115.9
2012	114.7

Spreadsheet at LarsonPrecalculus.com

(a) Use a graphing utility to create a scatter plot of the data. Let t represent the year, with $t = 0$ corresponding to 2000.

(b) Use the *regression* feature of the graphing utility to find a linear model for the data and identify the coefficient of determination.

(c) Use the graphing utility to graph the linear model with the scatter plot from part (a).

(d) Use the *regression* feature of the graphing utility to find a quadratic model for the data and identify the coefficient of determination.

(e) Use the graphing utility to graph the quadratic model with the scatter plot from part (a).

(f) Which model is a better fit for the data? Explain.

(g) Use each model to approximate the year when the number of households with televisions will reach 120 million, if possible.

Conclusions

True or False? **In Exercises 22–24, determine whether the statement is true or false. Justify your answer.**

22. The graph of a quadratic model with a negative leading coefficient will have a maximum value at its vertex.

23. The graph of a quadratic model with a positive leading coefficient will have a minimum value at its vertex.

24. Data that are positively correlated are always better modeled by a linear equation than by a quadratic equation.

25. Writing Explain why the parabola shown in the figure is not a good fit for the data.

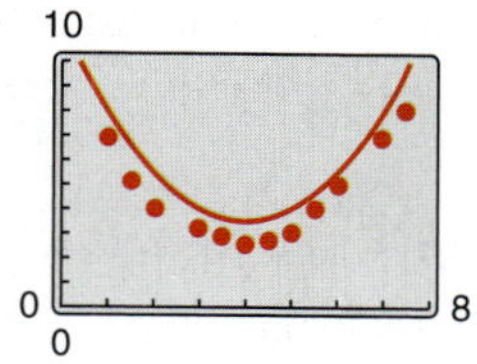

26. HOW DO YOU SEE IT? The r^2-values representing the coefficients of determination for the least squares linear model and the least squares quadratic model for the data shown are given below. Which is which? Explain your reasoning.

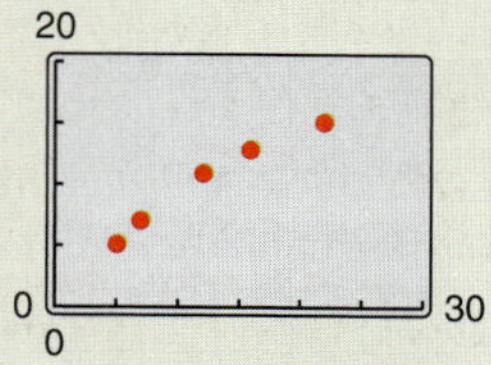

$r^2 \approx 0.9995$

$r^2 \approx 0.9782$

Cumulative Mixed Review

Compositions of Functions **In Exercises 27–30, find (a) $f \circ g$ and (b) $g \circ f$.**

27. $f(x) = 2x - 1,\ g(x) = x^2 + 3$

28. $f(x) = 5x + 8,\ g(x) = 2x^2 - 1$

29. $f(x) = x^3 - 1,\ g(x) = \sqrt[3]{x + 1}$

30. $f(x) = \sqrt[3]{x + 5},\ g(x) = x^3 - 5$

Testing Whether a Function is One-to-One **In Exercises 31–34, determine algebraically whether the function is one-to-one. If it is, find its inverse function. Verify your answer graphically.**

31. $f(x) = 2x + 5$

32. $f(x) = \dfrac{x - 4}{5}$

33. $f(x) = x^2 + 5,\ x \ge 0$

34. $f(x) = 2x^2 - 3,\ x \ge 0$

Multiplying Complex Conjugates **In Exercises 35–38, write the complex conjugate of the complex number. Then multiply the number by its complex conjugate.**

35. $1 - 3i$

36. $-2 + 4i$

37. $-5i$

38. $8i$

3 Chapter Summary

	What did you learn?	Explanation and Examples	Review Exercises
3.1	**Analyze graphs of quadratic functions *(p. 244).***	Let a, b, and c be real numbers with $a \neq 0$. The function $f(x) = ax^2 + bx + c$ is called a quadratic function. Its graph is a "U-shaped" curve called a parabola.	1–6
	Write quadratic functions in standard form and use the results to sketch graphs of functions *(p. 247).*	The quadratic function $f(x) = a(x - h)^2 + k$, $a \neq 0$, is in standard form. The graph of f is a parabola whose axis is the vertical line $x = h$ and whose vertex is the point (h, k). The parabola opens upward when $a > 0$ and opens downward when $a < 0$.	7–12
	Find minimum and maximum values of quadratic functions in real-life applications *(p. 249).*	Consider $f(x) = ax^2 + bx + c$ with vertex $\left(-\frac{b}{2a}, f\left(-\frac{b}{2a}\right)\right)$. If $a > 0$, then f has a *minimum* at $x = -b/(2a)$. If $a < 0$, then f has a *maximum* at $x = -b/(2a)$.	13, 14
3.2	**Use transformations to sketch graphs of polynomial functions *(p. 254).***	The graph of a polynomial function is continuous (no breaks, holes, or gaps) and has only smooth, rounded turns.	15–20
	Use the Leading Coefficient Test to determine the end behavior of graphs of polynomial functions *(p. 256).*	Consider $f(x) = a_nx^n + \cdot \cdot \cdot + a_1x + a_0$, $a_n \neq 0$. ***n* is odd:** If $a_n > 0$, then the graph falls to the left and rises to the right. If $a_n < 0$, then the graph rises to the left and falls to the right. ***n* is even:** If $a_n > 0$, then the graph rises to the left and right. If $a_n < 0$, then the graph falls to the left and right.	21, 22
	Find and use zeros of polynomial functions as sketching aids *(p. 258).*	If f is a polynomial function and a is a real number, the following are equivalent: (1) $x = a$ is a *zero* of the function f, (2) $x = a$ is a *solution* of the polynomial equation $f(x) = 0$, (3) $(x - a)$ is a *factor* of the polynomial $f(x)$, and (4) $(a, 0)$ is an *x-intercept* of the graph of f.	23–34
	Use the Intermediate Value Theorem to help locate zeros of polynomial functions *(p. 261).*	Let a and b be real numbers such that $a < b$. If f is a polynomial function such that $f(a) \neq f(b)$, then, in $[a, b]$, f takes on every value between $f(a)$ and $f(b)$.	35–38
3.3	**Use long division to divide polynomials by other polynomials *(p. 266).***	Dividend → $\frac{x^2 + 3x + 5}{x + 1} = x + 2 + \frac{3}{x + 1}$ ← Divisor Divisor → (denominator $x + 1$); Quotient: $x + 2$; Remainder: 3	39–44
	Use synthetic division to divide polynomials by binomials of the form $x - k$ *(p. 269).*	Divisor: $x + 3$ Dividend: $x^4 - 10x^2 - 2x + 4$ -3 | $1 \quad 0 \quad -10 \quad -2 \quad 4$ | $\quad\; -3 \quad 9 \quad 3 \quad -3$ $1 \quad -3 \quad -1 \quad 1 \quad (1)$ ← Remainder: 1 Quotient: $x^3 - 3x^2 - x + 1$	45–50
	Use the Remainder Theorem and the Factor Theorem *(p. 270).*	**The Remainder Theorem:** If a polynomial $f(x)$ is divided by $x - k$, then the remainder is $r = f(k)$. **The Factor Theorem:** A polynomial $f(x)$ has a factor $(x - k)$ if and only if $f(k) = 0$.	51–54
	Use the Rational Zero Test to determine possible rational zeros of polynomial functions *(p. 272).*	The Rational Zero Test relates the possible rational zeros of a polynomial to the leading coefficient and to the constant term of the polynomial.	55, 56

	What did you learn?	Explanation and Examples	Review Exercises
3.3	**Use Descartes's Rule of Signs *(p. 274)* and the Upper and Lower Bound Rules *(p. 275)* to find zeros of polynomials.**	**Descartes's Rule of Signs** Let $f(x) = a_n x^n + a_{n-1}x^{n-1} + \cdots + a_2x^2 + a_1x + a_0$ be a polynomial with real coefficients and $a_0 \neq 0$. **1.** The number of *positive real zeros* of f is either equal to the number of variations in sign of $f(x)$ or less than that number by an even integer. **2.** The number of *negative real zeros* of f is either equal to the number of variations in sign of $f(-x)$ or less than that number by an even integer.	57–64
3.4	**Use the Fundamental Theorem of Algebra to determine the number of zeros of a polynomial function *(p. 281)*.**	**The Fundamental Theorem of Algebra** If $f(x)$ is a polynomial of degree n, where $n > 0$, then f has at least one zero in the complex number system. **Linear Factorization Theorem** If $f(x)$ is a polynomial of degree n, where $n > 0$, then f has precisely n linear factors $f(x) = a_n(x - c_1)(x - c_2) \cdots (x - c_n)$ where $c_1, c_2, \ldots, c_n$ are complex numbers.	65–68
	Find all zeros of polynomial functions *(p. 282)* and find conjugate pairs of complex zeros *(p. 283)*.	**Complex Zeros Occur in Conjugate Pairs** Let f be a polynomial function that has real coefficients. If $a + bi$ $(b \neq 0)$ is a zero of the function, the conjugate $a - bi$ is also a zero of the function.	69–86
	Find zeros of polynomials by factoring *(p. 284)*.	Every polynomial of degree $n > 0$ with real coefficients can be written as the product of linear and quadratic factors with real coefficients, where the quadratic factors have no real zeros.	87–90
3.5	**Find the domains *(p. 288)* and vertical and horizontal asymptotes *(p. 289)* of rational functions.**	The domain of a rational function of x includes all real numbers except x-values that make the denominator zero.	91–102
	Use rational functions to model and solve real-life problems *(p. 292)*.	A rational function can be used to model the cost of removing a given percent of smokestack pollutants at a utility company that burns coal. (See Example 5.)	103, 104
3.6	**Analyze and sketch graphs of rational functions *(p. 297)*, including functions with slant asymptotes *(p. 301)*.**	Consider a rational function whose denominator is of degree 1 or greater. If the degree of the numerator is exactly *one more* than the degree of the denominator, then the graph of the function has a slant asymptote.	105–118
	Use rational functions to model and solve real-life problems *(p. 302)*.	A rational function can be used to model the area of a page. The model can be used to determine the dimensions of the page that use the minimum amount of paper. (See Example 7.)	119, 120
3.7	**Classify scatter plots *(p. 307)*, find quadratic models for data *(p. 308)*, and choose a model that best fits a set of data *(p. 310)*.**	Sometimes it is not easy to distinguish from a scatter plot which type of model will best fit the data. You should first find several models for the data and then choose the model that best fits the data by comparing the y-values of each model with the actual y-values.	121–123

3 Review Exercises

See CalcChat.com for tutorial help and worked-out solutions to odd-numbered exercises. For instructions on how to use a graphing utility, see Appendix A.

3.1

Library of Parent Functions **In Exercises 1–6, sketch the graph of each function and compare it with the graph of $y = x^2$.**

1. $y = x^2 + 1$
2. $y = -x^2 + 5$
3. $y = (x - 6)^2$
4. $y = -(x + 1)^2$
5. $y = -(x + 5)^2 - 1$
6. $y = -(x - 2)^2 + 2$

Identifying the Vertex of a Quadratic Function **In Exercises 7–10, describe the graph of the function and identify the vertex. Then, sketch the graph of the function. Identify any x-intercepts.**

7. $f(x) = \left(x + \frac{3}{2}\right)^2 + 1$
8. $f(x) = (x - 4)^2 - 4$
9. $f(x) = \frac{1}{3}(x^2 + 5x - 4)$
10. $f(x) = 3x^2 - 12x + 11$

Writing the Equation of a Parabola in Standard Form **In Exercises 11 and 12, write the standard form of the quadratic function that has the indicated vertex and whose graph passes through the given point. Use a graphing utility to verify your result.**

11. Vertex: $(-3, 4)$ Point: $(0, -5)$
12. Vertex: $(2, -1)$ Point: $(5, 2)$

13. **Geometry** A rectangle is inscribed in the region bounded by the x-axis, the y-axis, and the graph of $x + 2y - 8 = 0$, as shown in the figure.

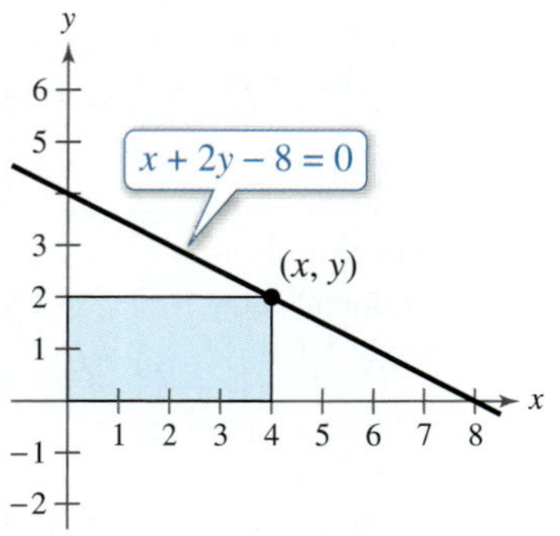

(a) Write the area A of the rectangle as a function of x. Determine the domain of the function in the context of the problem.

(b) Use a graphing utility to graph the area function. Use the graph to approximate the dimensions that will produce a maximum area.

(c) Write the area function in standard form to find algebraically the dimensions that will produce a maximum area. Compare your results with your answer from part (b).

14. **Physical Education** A college has 1500 feet of portable rink boards to form three adjacent outdoor ice rinks, as shown in the figure. Determine the dimensions that will produce the maximum total area of ice surface.

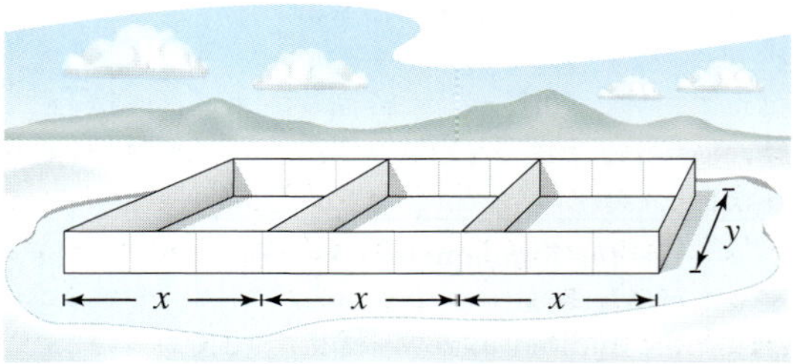

3.2

Library of Parent Functions **In Exercises 15–18, sketch the graph of $f(x) = x^3$ and the graph of the function g. Describe the transformation from f to g.**

15. $g(x) = x^3 - 2$
16. $g(x) = (x - 5)^3$
17. $g(x) = -(x + 4)^3$
18. $g(x) = -(x - 2)^3 + 6$

Comparing End Behavior **In Exercises 19 and 20, use a graphing utility to graph the functions f and g in the same viewing window. Zoom out far enough to see the right-hand and left-hand behavior of each graph. Do the graphs of f and g have the same right-hand and left-hand behavior? Explain why or why not.**

19. $f(x) = \frac{1}{2}x^3 - 2x + 1,\quad g(x) = \frac{1}{2}x^3$
20. $f(x) = -x^4 + 2x^3,\quad g(x) = -x^4$

Applying the Leading Coefficient Test **In Exercises 21 and 22, use the Leading Coefficient Test to describe the right-hand and left-hand behavior of the graph of the polynomial function.**

21. $f(x) = x^2 - x - 6$
22. $g(x) = -3x^3 + 4x^2 - 1$

Finding Zeros of a Polynomial Function **In Exercises 23–28, (a) find the zeros algebraically, (b) use a graphing utility to graph the function, and (c) use the graph to approximate any zeros and compare them with those in part (a).**

23. $g(x) = x^4 - x^3 - 2x^2$
24. $h(x) = -2x^3 - x^2 + x$
25. $f(t) = t^3 - 3t$
26. $f(x) = -(x + 6)^3 - 8$
27. $f(x) = x(x + 3)^2$
28. $f(t) = t^4 - 4t^2$

Finding a Polynomial Function with Given Zeros **In Exercises 29–32, find a polynomial function that has the given zeros. (There are many correct answers.)**

29. $4, -2, -2$
30. $-1, 0, 3, 5$
31. $3, 2 - \sqrt{3}, 2 + \sqrt{3}$
32. $-7, 4 - \sqrt{6}, 4 + \sqrt{6}$

Sketching the Graph of a Polynomial Function In Exercises 33 and 34, sketch the graph of the function by (a) applying the Leading Coefficient Test, (b) finding the zeros of the polynomial, (c) plotting sufficient solution points, and (d) drawing a continuous curve through the points.

33. $f(x) = x^4 - 2x^3 - 12x^2 + 18x + 27$
34. $f(x) = 18 + 27x - 2x^2 - 3x^3$

Approximating the Zeros of a Function In Exercises 35–38, (a) use the Intermediate Value Theorem and a graphing utility to find graphically any intervals of length 1 in which the polynomial function is guaranteed to have a zero, and (b) use the *zero* or *root* feature of the graphing utility to approximate the real zeros of the function. Verify your results in part (a) by using the *table* feature of the graphing utility.

35. $f(x) = x^3 + 2x^2 - x - 1$
36. $f(x) = x^4 - 6x^2 - 4$
37. $f(x) = 0.24x^3 - 2.6x - 1.4$
38. $f(x) = 2x^4 + \frac{7}{2}x^3 - 2$

3.3

Long Division of Polynomials In Exercises 39–44, use long division to divide.

39. $\dfrac{24x^2 - x - 8}{3x - 2}$
40. $\dfrac{4x^2 + 7}{3x - 2}$
41. $\dfrac{x^4 - 3x^2 + 2}{x^2 - 1}$
42. $\dfrac{3x^4 + x^2 - 1}{x^2 - 1}$
43. $(5x^3 - 13x^2 - x + 2) \div (x^2 - 3x + 1)$
44. $\dfrac{6x^4 + 10x^3 + 13x^2 - 5x + 2}{2x^2 - 1}$

Using Synthetic Division In Exercises 45–50, use synthetic division to divide.

45. $(3x^3 - 10x^2 + 12x - 22) \div (x - 4)$
46. $(2x^3 + 6x^2 - 14x + 9) \div (x - 1)$
47. $(0.25x^4 - 4x^3) \div (x + 2)$
48. $(0.1x^3 + 0.3x^2 - 0.5) \div (x - 5)$
49. $(6x^4 - 4x^3 - 27x^2 + 18x) \div \left(x - \frac{2}{3}\right)$
50. $(2x^3 + 2x^2 - x + 2) \div \left(x - \frac{1}{2}\right)$

Using the Remainder Theorem In Exercises 51 and 52, use the Remainder Theorem and synthetic division to evaluate the function at each given value. Use a graphing utility to verify your results.

51. $f(x) = x^4 + 10x^3 - 24x^2 + 20x + 44$
    (a) $f(-3)$ (b) $f(-2)$
52. $g(t) = 2t^5 - 5t^4 - 8t + 20$
    (a) $g(-4)$ (b) $g(\sqrt{2})$

Factoring a Polynomial In Exercises 53 and 54, (a) verify the given factor(s) of the function f, (b) find the remaining factors of f, (c) use your results to write the complete factorization of f, and (d) list all real zeros of f. Confirm your results by using a graphing utility to graph the function.

	Function	*Factor(s)*
53.	$f(x) = x^3 + 4x^2 - 25x - 28$	$(x - 4)$
54.	$f(x) = x^4 - 4x^3 - 7x^2 + 22x + 24$	$(x + 2), (x - 3)$

Using the Rational Zero Test In Exercises 55 and 56, use the Rational Zero Test to list all possible rational zeros of f. Use a graphing utility to verify that all the zeros of f are contained in the list.

55. $f(x) = 4x^3 - 11x^2 + 10x - 3$
56. $f(x) = 10x^3 + 21x^2 - x - 6$

Using Descartes's Rule of Signs In Exercises 57 and 58, use Descartes's Rule of Signs to determine the possible numbers of positive and negative real zeros of the function.

57. $g(x) = 5x^3 - 6x + 9$
58. $f(x) = 2x^5 - 3x^2 + 2x - 1$

Finding the Zeros of a Polynomial Function In Exercises 59 and 60, use synthetic division to verify the upper and lower bounds of the real zeros of f. Then find all real zeros of the function.

59. $f(x) = 4x^3 - 3x^2 + 4x - 3$
    Upper bound: $x = 1$; Lower bound: $x = -\frac{1}{4}$
60. $f(x) = 2x^3 - 5x^2 - 14x + 8$
    Upper bound: $x = 8$; Lower bound: $x = -4$

Finding the Zeros of a Polynomial Function In Exercises 61–64, find all real zeros of the polynomial function.

61. $f(x) = x^3 - 4x^2 + x + 6$
62. $f(x) = x^3 + x^2 - 28x - 10$
63. $f(x) = 6x^4 - 25x^3 + 14x^2 + 27x - 18$
64. $f(x) = 5x^4 + 126x^2 + 25$

3.4

Zeros of a Polynomial Function In Exercises 65–68, confirm that the function has the indicated zero(s).

65. $f(x) = x^2 + 6x + 9$; Repeated zero: -3
66. $f(x) = x^2 - 10x + 25$; Repeated zero: 5
67. $f(x) = x^3 + 16x$; $0, -4i, 4i$
68. $f(x) = x^3 + 144x$; $0, -12i, 12i$

Using the Factored Form of a Function In Exercises 69 and 70, find all the zeros of the function.

69. $f(x) = -4x(x + 3)$
70. $f(x) = (x - 8)^3(x + 2)$

Finding the Zeros of a Polynomial Function **In Exercises 71–76, find all the zeros of the function and write the polynomial as a product of linear factors. Verify your results by using a graphing utility to graph the function.**

71. $h(x) = x^3 - 7x^2 + 18x - 24$

72. $f(x) = 2x^3 - 5x^2 - 9x + 40$

73. $f(x) = 2x^4 - 5x^3 + 10x - 12$

74. $g(x) = 3x^4 - 4x^3 + 7x^2 + 10x - 4$

75. $f(x) = x^5 + x^4 + 5x^3 + 5x^2$

76. $f(x) = x^5 - 5x^3 + 4x$

Using the Zeros to Find the x-Intercepts **In Exercises 77–82, (a) find all the zeros of the function, (b) write the polynomial as a product of linear factors, and (c) use your factorization to determine the x-intercepts of the graph of the function. Use a graphing utility to verify that the real zeros are the only x-intercepts.**

77. $f(x) = x^3 - 4x^2 + 6x - 4$

78. $f(x) = x^3 - 5x^2 - 7x + 51$

79. $f(x) = -3x^3 - 19x^2 - 4x + 12$

80. $f(x) = 2x^3 - 9x^2 + 22x - 30$

81. $f(x) = x^4 + 34x^2 + 225$

82. $f(x) = x^4 + 10x^3 + 26x^2 + 10x + 25$

Finding a Polynomial with Given Zeros **In Exercises 83–86, find a polynomial function with real coefficients that has the given zeros. (There are many correct answers.)**

83. $5, 3i$

84. $-6, -i$

85. $-2, -2 - 4i$

86. $1, -5 + \sqrt{2}i$

Factoring a Polynomial **In Exercises 87 and 88, write the polynomial (a) as the product of factors that are irreducible over the *rationals,* (b) as the product of linear and quadratic factors that are irreducible over the *reals,* and (c) in completely factored form.**

87. $f(x) = x^4 - 2x^3 + 8x^2 - 18x - 9$

(*Hint:* One factor is $x^2 + 9$.)

88. $f(x) = x^4 - 4x^3 + 3x^2 + 8x - 16$

(*Hint:* One factor is $x^2 - x - 4$.)

Finding the Zeros of a Polynomial Function **In Exercises 89 and 90, use the given zero to find all the zeros of the function.**

	Function	*Zero*
89.	$f(x) = x^3 + 3x^2 + 4x + 12$	$-2i$
90.	$f(x) = 2x^3 - 7x^2 + 14x + 9$	$2 + \sqrt{5}i$

3.5

Finding a Function's Domain and Asymptotes **In Exercises 91–102, (a) find the domain of the function, (b) decide whether the function is continuous, and (c) identify any horizontal and vertical asymptotes.**

91. $f(x) = \dfrac{2 - x}{x + 3}$

92. $f(x) = \dfrac{4x}{x - 8}$

93. $f(x) = \dfrac{2}{x^2 - 3x - 18}$

94. $f(x) = \dfrac{2x^2 + 3}{x^2 + x + 3}$

95. $f(x) = \dfrac{7 + x}{7 - x}$

96. $f(x) = \dfrac{6x}{x^2 - 1}$

97. $f(x) = \dfrac{4x^2}{2x^2 - 3}$

98. $f(x) = \dfrac{3x^2 - 11x - 4}{x^2 + 2}$

99. $f(x) = \dfrac{2x - 10}{x^2 - 2x - 15}$

100. $f(x) = \dfrac{4 - x}{x^3 + 6x^2}$

101. $f(x) = \dfrac{3x^2 - 15}{x^3 - 5x^2 - 24x}$

102. $f(x) = \dfrac{x^2 + 3x + 2}{x^3 - 4x^2}$

103. Criminology The cost C (in millions of dollars) for the U.S. government to seize $p\%$ of an illegal drug as it enters the country is given by

$$C = \frac{528p}{100 - p}, \quad 0 \le p < 100.$$

(a) Find the costs of seizing 25%, 50%, and 75% of the illegal drug.

(b) Use a graphing utility to graph the function. Be sure to choose an appropriate viewing window. Explain why you chose the values you used in your viewing window.

(c) According to this model, would it be possible to seize 100% of the drug? Explain.

104. Biology A biology class performs an experiment comparing the quantity of food consumed by a certain kind of moth with the quantity supplied. The model for the experimental data is given by

$$y = \frac{1.568x - 0.001}{6.360x + 1}, \quad x > 0$$

where x is the quantity (in milligrams) of food supplied and y is the quantity (in milligrams) eaten (see figure). At what level of consumption will the moth become satiated?

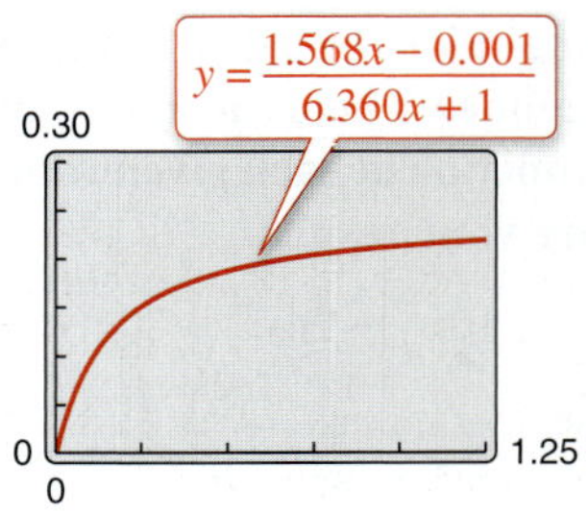

3.6

Finding Asymptotes and Holes **In Exercises 105–108, find all of the vertical, horizontal, and slant asymptotes, and any holes in the graph of the function. Then use a graphing utility to verify your results.**

105. $f(x) = \dfrac{x^2 - 5x + 4}{x^2 - 1}$ **106.** $f(x) = \dfrac{2x^2 - 7x + 3}{2x^2 - 3x - 9}$

107. $f(x) = \dfrac{3x^2 + 5x - 2}{x + 1}$ **108.** $f(x) = \dfrac{2x^2 + 5x + 3}{x - 2}$

Sketching the Graph of a Rational Function **In Exercises 109–118, sketch the graph of the rational function by hand. As sketching aids, check for intercepts, vertical asymptotes, horizontal asymptotes, slant asymptotes, and holes.**

109. $f(x) = \dfrac{1}{x} + 3$ **110.** $f(x) = \dfrac{-1}{x + 2}$

111. $f(x) = \dfrac{2x - 1}{x - 5}$ **112.** $f(x) = \dfrac{x - 3}{x - 2}$

113. $f(x) = \dfrac{2}{(x + 1)^2}$ **114.** $f(x) = \dfrac{4}{(x - 1)^2}$

115. $f(x) = \dfrac{2x^2}{x^2 - 4}$ **116.** $f(x) = \dfrac{5x}{x^2 + 1}$

117. $f(x) = \dfrac{x^2 - x + 1}{x - 3}$

118. $f(x) = \dfrac{2x^2 + 7x + 3}{x + 1}$

119. Biology A parks and wildlife commission releases 80,000 fish into a lake. After t years, the population N of the fish (in thousands) is given by

$$N = \frac{20(4 + 3t)}{1 + 0.05t}, \quad t \geq 0.$$

(a) Use a graphing utility to graph the function and find the populations when $t = 5$, $t = 10$, and $t = 25$.

(b) What is the maximum number of fish in the lake as time passes? Explain your reasoning.

120. Publishing A page that is x inches wide and y inches high contains 30 square inches of print. The top and bottom margins are 2 inches deep and the margins on each side are 2 inches wide.

(a) Draw a diagram that illustrates the problem.

(b) Show that the total area A of the page is given by

$$A = \frac{2x(2x + 7)}{x - 4}.$$

(c) Determine the domain of the function based on the physical constraints of the problem.

(d) Use a graphing utility to graph the area function and approximate the page size such that the minimum amount of paper will be used.

3.7

Classifying Scatter Plots **In Exercises 121 and 122, determine whether the scatter plot could best be modeled by a linear model, a quadratic model, or neither.**

121.

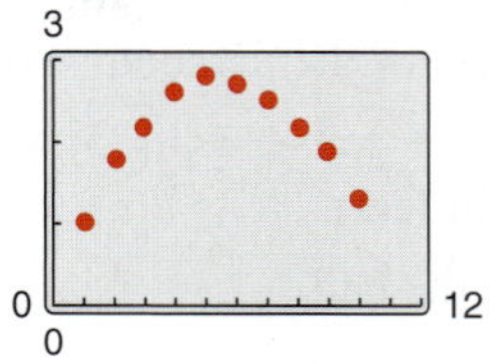

122.

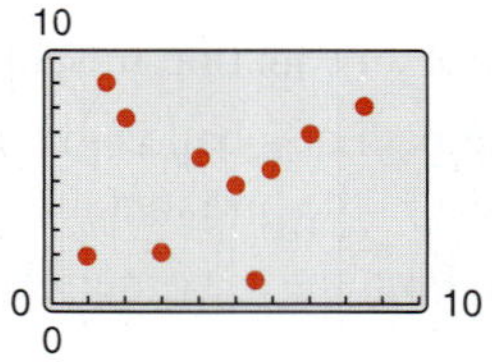

123. MODELING DATA

The table shows the numbers of commercial FM radio stations S in the United States from 2004 through 2013. (Source: Federal Communications Commission)

DATA

Year	FM Stations, S
2004	6218
2005	6231
2006	6266
2007	6309
2008	6427
2009	6479
2010	6526
2011	6542
2012	6598
2013	6612

Spreadsheet at LarsonPrecalculus.com

(a) Use a graphing utility to create a scatter plot of the data. Let t represent the year, with $t = 4$ corresponding to 2004.

(b) A cubic model for the data is

$$S = -1.088t^3 + 26.61t^2 - 150.9t + 6459.$$

Use the graphing utility to graph this model with the scatter plot from part (a).

(c) Use the *regression* feature of the graphing utility to find a quadratic model for the data. Then graph the model with the scatter plot from part (a).

(d) Which model is a better fit for the data? Explain.

(e) Use the model you chose in part (d) to predict the number of commercial FM radio stations in 2015.

Conclusions

True or False? **In Exercises 124 and 125, determine whether the statement is true or false. Justify your answer.**

124. A fourth-degree polynomial with real coefficients can have -5, $-8i$, $4i$, and 5 as its zeros.

125. The sum of two complex numbers cannot be a real number.

3 Chapter Test

See CalcChat.com for tutorial help and worked-out solutions to odd-numbered exercises. For instructions on how to use a graphing utility, see Appendix A.

Take this test as you would take a test in class. After you are finished, check your work against the answers given in the back of the book.

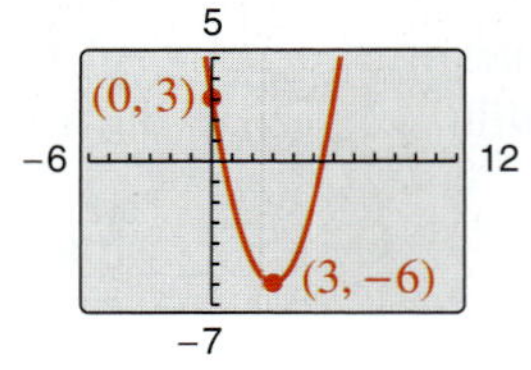

Figure for 2

1. Identify the vertex and intercepts of the graph of $y = x^2 + 5x + 6$.

2. Write an equation in standard form of the parabola shown at the right.

3. The path of a ball is given by $y = -\frac{1}{20}x^2 + 3x + 5$, where y is the height (in feet) and x is the horizontal distance (in feet).

(a) Find the maximum height of the ball.

(b) Which term represents the height at which the ball was thrown? Does changing this term change the maximum height of the ball? Explain.

4. Find all the real zeros of $f(x) = 4x^3 - 12x^2 + 9x$. Determine the multiplicity of each zero.

5. Sketch the graph of the function $f(x) = -x^3 + 7x + 6$.

6. Divide using long division: $(2x^3 + 5x - 3) \div (x^2 + 2)$.

7. Divide using synthetic division: $(2x^4 - 5x^2 - 3) \div (x - 2)$.

8. Use synthetic division to evaluate $f(-2)$ for $f(x) = 3x^4 - 6x^2 + 5x - 1$.

In Exercises 9 and 10, list all the possible rational zeros of the function. Use a graphing utility to graph the function and find all the rational zeros.

9. $g(t) = 2t^4 - 3t^3 + 16t - 24$

10. $h(x) = 3x^5 + 2x^4 - 3x - 2$

11. Find all the zeros of the function $f(x) = x^3 - 7x^2 + 11x + 19$ and write the polynomial as a product of linear factors.

In Exercises 12–14, find a polynomial with real coefficients that has the given zeros. (There are many correct answers.)

12. $0, 2, 2 + i$ **13.** $1 - \sqrt{3}i, 2, 2$ **14.** $0, 1 + i$

In Exercises 15–17, sketch the graph of the rational function. As sketching aids, check for intercepts, vertical asymptotes, horizontal asymptotes, and slant asymptotes.

15. $h(x) = \dfrac{4}{x^2} - 1$ **16.** $g(x) = \dfrac{x^2 + 2}{x - 1}$ **17.** $f(x) = \dfrac{2x^2 + 9}{5x^2 + 2}$

18. The table shows the amounts A (in billions of dollars) spent on national defense by the United States for the years 2007 through 2013. (Source: U.S. Office of Management and Budget)

(a) Use a graphing utility to create a scatter plot of the data. Let t represent the year, with $t = 7$ corresponding to 2007.

(b) Use the *regression* feature of the graphing utility to find a quadratic model for the data.

(c) Use the graphing utility to graph the quadratic model with the scatter plot from part (a). Is the quadratic model a good fit for the data?

(d) Use the model to estimate the amounts spent on national defense in 2015 and 2020.

(e) Do you believe the model is useful for predicting the amounts spent on national defense for years beyond 2013? Explain.

DATA

Year	National defense, A (in billions of dollars)
2007	551.3
2008	616.1
2009	661.0
2010	693.5
2011	705.6
2012	677.9
2013	633.4

Spreadsheet at LarsonPrecalculus.com

Table for 18

Proofs in Mathematics

These two pages contain proofs of four important theorems about polynomial functions. The first two theorems are from Section 3.3, and the second two theorems are from Section 3.4.

The Remainder Theorem (p. 270)

If a polynomial $f(x)$ is divided by $x - k$, then the remainder is

$$r = f(k).$$

Proof

From the Division Algorithm, you have

$$f(x) = (x - k)q(x) + r(x)$$

and because either $r(x) = 0$ or the degree of $r(x)$ is less than the degree of $x - k$, you know that $r(x)$ must be a constant. That is, $r(x) = r$. Now, by evaluating $f(x)$ at $x = k$, you have

$$\begin{aligned} f(k) &= (k - k)q(k) + r \\ &= (0)q(k) + r \\ &= r. \end{aligned}$$

To be successful in algebra, it is important that you understand the connection among the *factors* of a polynomial, the *zeros* of a polynomial function, and the *solutions* or *roots* of a polynomial equation. The Factor Theorem is the basis for this connection.

The Factor Theorem (p. 270)

A polynomial $f(x)$ has a factor $(x - k)$ if and only if $f(k) = 0$.

Proof

Using the Division Algorithm with the factor $(x - k)$, you have

$$f(x) = (x - k)q(x) + r(x).$$

By the Remainder Theorem, $r(x) = r = f(k)$, and you have

$$f(x) = (x - k)q(x) + f(k)$$

where $q(x)$ is a polynomial of lesser degree than $f(x)$. If $f(k) = 0$, then

$$f(x) = (x - k)q(x)$$

and you see that $(x - k)$ is a factor of $f(x)$. Conversely, if $(x - k)$ is a factor of $f(x)$, then division of $f(x)$ by $(x - k)$ yields a remainder of 0. So, by the Remainder Theorem, you have $f(k) = 0$.

Linear Factorization Theorem (p. 281)

If $f(x)$ is a polynomial of degree n, where $n > 0$, then f has precisely n linear factors

$$f(x) = a_n(x - c_1)(x - c_2) \cdot \cdot \cdot (x - c_n)$$

where $c_1, c_2, \ldots, c_n$ are complex numbers.

Proof

Using the Fundamental Theorem of Algebra, you know that f must have at least one zero, c_1. Consequently, $(x - c_1)$ is a factor of $f(x)$, and you have

$$f(x) = (x - c_1)f_1(x).$$

If the degree of $f_1(x)$ is greater than zero, then apply the Fundamental Theorem again to conclude that f_1 must have a zero c_2, which implies that

$$f(x) = (x - c_1)(x - c_2)f_2(x).$$

It is clear that the degree of $f_1(x)$ is $n - 1$, that the degree of $f_2(x)$ is $n - 2$, and that you can repeatedly apply the Fundamental Theorem n times until you obtain

$$f(x) = a_n(x - c_1)(x - c_2) \cdot \cdot \cdot (x - c_n)$$

where a_n is the leading coefficient of the polynomial $f(x)$.

Factors of a Polynomial (p. 284)

Every polynomial of degree $n > 0$ with real coefficients can be written as the product of linear and quadratic factors with real coefficients, where the quadratic factors have no real zeros.

Proof

To begin, you use the Linear Factorization Theorem to conclude that $f(x)$ can be *completely* factored in the form

$$f(x) = d(x - c_1)(x - c_2)(x - c_3) \ldots (x - c_n).$$

If each c_i is real, then there is nothing more to prove. If any c_i is complex ($c_i = a + bi$, $b \neq 0$), then, because the coefficients of $f(x)$ are real, you know that the conjugate $c_j = a - bi$ is also a zero. By multiplying the corresponding factors, you obtain

$$\begin{aligned}(x - c_i)(x - c_j) &= [x - (a + bi)][x - (a - bi)] \\ &= x^2 - 2ax + (a^2 + b^2)\end{aligned}$$

where each coefficient is real.

The Fundamental Theorem of Algebra

The Linear Factorization Theorem is closely related to the Fundamental Theorem of Algebra. The Fundamental Theorem of Algebra has a long and interesting history. In the early work with polynomial equations, the Fundamental Theorem of Algebra was thought to have been not true, because imaginary solutions were not considered. In fact, in the very early work by mathematicians such as Abu al-Khwarizmi (c. 800 A.D.), negative solutions were also not considered.

Once imaginary numbers were accepted, several mathematicians attempted to give a general proof of the Fundamental Theorem of Algebra. These included Gottfried von Leibniz (1702), Jean d'Alembert (1746), Leonhard Euler (1749), Joseph-Louis Lagrange (1772), and Pierre Simon Laplace (1795). The mathematician usually credited with the first correct proof of the Fundamental Theorem of Algebra is Carl Friedrich Gauss, who published the proof in his doctoral thesis in 1799.

4 Exponential and Logarithmic Functions

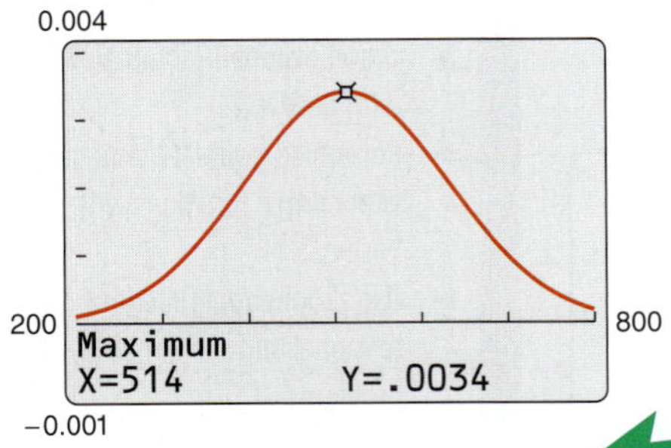

Section 4.5, Example 4
SAT Scores

4.1 Exponential Functions and Their Graphs

Exponential Functions

So far, this text has dealt mainly with **algebraic functions,** which include polynomial functions and rational functions. In this chapter, you will study two types of nonalgebraic functions—*exponential functions* and *logarithmic functions*. These functions are examples of **transcendental functions.**

> **Definition of Exponential Function**
>
> The **exponential function f with base a** is denoted by
>
> $f(x) = a^x$
>
> where $a > 0$, $a \neq 1$, and x is any real number.

Note that in the definition of an exponential function, the base $a = 1$ is excluded because it yields

$f(x) = 1^x = 1.$ Constant function

This is a constant function, not an exponential function.

You have already evaluated a^x for integer and rational values of x. For example, you know that

$4^3 = 64$ and $4^{1/2} = 2.$

However, to evaluate 4^x for any real number x, you need to interpret forms with *irrational* exponents. For the purposes of this text, it is sufficient to think of $a^{\sqrt{2}}$ (where $\sqrt{2} \approx 1.41421356$) as the number that has the successively closer approximations

$a^{1.4}, a^{1.41}, a^{1.414}, a^{1.4142}, a^{1.41421}, \ldots.$

Example 1 shows how to use a calculator to evaluate exponential functions.

What you should learn

- Recognize and evaluate exponential functions with base a.
- Graph exponential functions with base a.
- Recognize, evaluate, and graph exponential functions with base e.
- Use exponential functions to model and solve real-life problems.

Why you should learn it

Exponential functions are useful in modeling data that represent quantities that increase or decrease quickly. For instance, Exercise 82 on page 335 shows how an exponential function is used to model the depreciation of a new vehicle.

EXAMPLE 1 Evaluating Exponential Functions

Use a calculator to evaluate each function at the indicated value of x.

	Function	*Value*
a.	$f(x) = 2^x$	$x = -3.1$
b.	$f(x) = 2^{-x}$	$x = \pi$
c.	$f(x) = 0.6^x$	$x = \frac{3}{2}$
d.	$f(x) = 1.05^{2x}$	$x = 12$

Solution

	Function Value	*Graphing Calculator Keystrokes*	*Display*
a.	$f(-3.1) = 2^{-3.1}$	2 [^] [(-)] 3.1 [ENTER]	0.1166291
b.	$f(\pi) = 2^{-\pi}$	2 [^] [(-)] π [ENTER]	0.1133147
c.	$f(\frac{3}{2}) = (0.6)^{3/2}$	.6 [^] [(] 3 [÷] 2 [)] [ENTER]	0.4647580
d.	$f(12) = (1.05)^{2(12)}$	1.05 [^] [(] 2 [×] 12 [)] [ENTER]	3.2250999

✔ **Checkpoint** Audio-video solution in English & Spanish at LarsonPrecalculus.com.

Use a calculator to evaluate $f(x) = 8^{-x}$ at $x = \sqrt{2}$.

> **Technology Tip**
>
> When evaluating exponential functions with a calculator, remember to enclose fractional exponents in parentheses. Because the calculator follows the order of operations, parentheses are crucial in order to obtain the correct result.

Graphs of Exponential Functions

The graphs of all exponential functions have similar characteristics, as shown in Examples 2, 3, and 4.

EXAMPLE 2 Graphs of $y = a^x$

In the same coordinate plane, sketch the graphs of $f(x) = 2^x$ and $g(x) = 4^x$.

Solution

The table below lists some values for each function. By plotting these points and connecting them with smooth curves, you obtain the graphs shown in Figure 4.1. Note that both graphs are increasing. Moreover, the graph of $g(x) = 4^x$ is increasing more rapidly than the graph of $f(x) = 2^x$.

x	-3	-2	-1	0	1	2
2^x	$\frac{1}{8}$	$\frac{1}{4}$	$\frac{1}{2}$	1	2	4
4^x	$\frac{1}{64}$	$\frac{1}{16}$	$\frac{1}{4}$	1	4	16

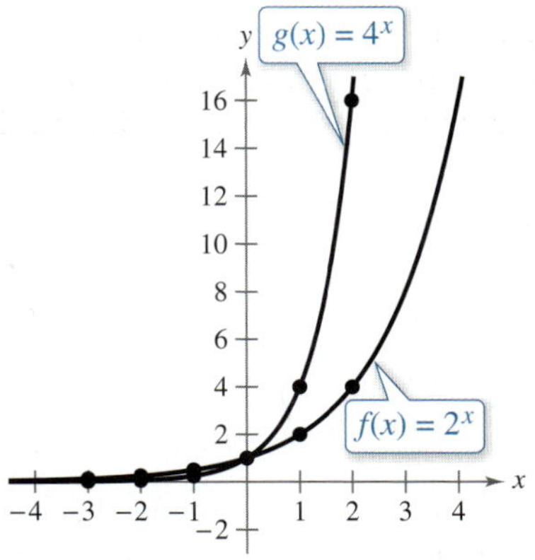

Figure 4.1

Checkpoint Audio-video solution in English & Spanish at LarsonPrecalculus.com.

In the same coordinate plane, sketch the graphs of $f(x) = 3^x$ and $g(x) = 9^x$.

EXAMPLE 3 Graphs of $y = a^{-x}$

In the same coordinate plane, sketch the graphs of $F(x) = 2^{-x}$ and $G(x) = 4^{-x}$.

Solution

The table below lists some values for each function. By plotting these points and connecting them with smooth curves, you obtain the graphs shown in Figure 4.2. Note that both graphs are decreasing. Moreover, the graph of $G(x) = 4^{-x}$ is decreasing more rapidly than the graph of $F(x) = 2^{-x}$.

x	-2	-1	0	1	2	3
2^{-x}	4	2	1	$\frac{1}{2}$	$\frac{1}{4}$	$\frac{1}{8}$
4^{-x}	16	4	1	$\frac{1}{4}$	$\frac{1}{16}$	$\frac{1}{64}$

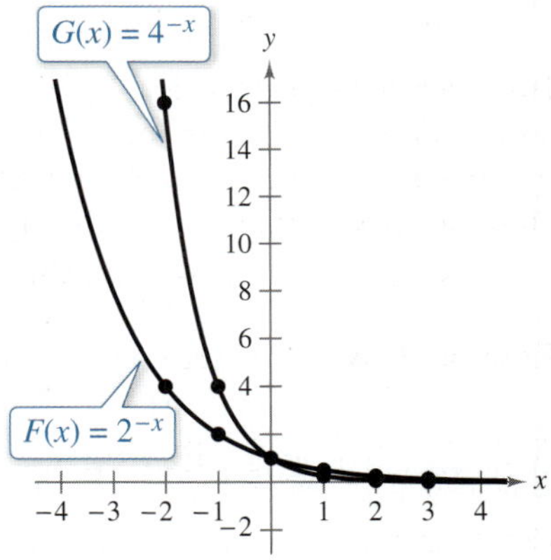

Figure 4.2

Checkpoint Audio-video solution in English & Spanish at LarsonPrecalculus.com.

In the same coordinate plane, sketch the graphs of $f(x) = 3^{-x}$ and $g(x) = 9^{-x}$.

The properties of exponents can also be applied to real-number exponents. For review, these properties are listed below.

1. $a^x a^y = a^{x+y}$ **2.** $\dfrac{a^x}{a^y} = a^{x-y}$ **3.** $a^{-x} = \dfrac{1}{a^x} = \left(\dfrac{1}{a}\right)^x$ **4.** $a^0 = 1$

5. $(ab)^x = a^x b^x$ **6.** $(a^x)^y = a^{xy}$ **7.** $\left(\dfrac{a}{b}\right)^x = \dfrac{a^x}{b^x}$ **8.** $|a^2| = |a|^2 = a^2$

In Example 3, note that the functions $F(x) = 2^{-x}$ and $G(x) = 4^{-x}$ can be rewritten with positive exponents as

$$F(x) = 2^{-x} = \left(\frac{1}{2}\right)^x \quad \text{and} \quad G(x) = 4^{-x} = \left(\frac{1}{4}\right)^x.$$

Comparing the functions in Examples 2 and 3, observe that

$F(x) = 2^{-x} = f(-x)$ and $G(x) = 4^{-x} = g(-x)$.

Consequently, the graph of F is a reflection (in the y-axis) of the graph of f, as shown in Figure 4.3. The graphs of G and g have the same relationship, as shown in Figure 4.4.

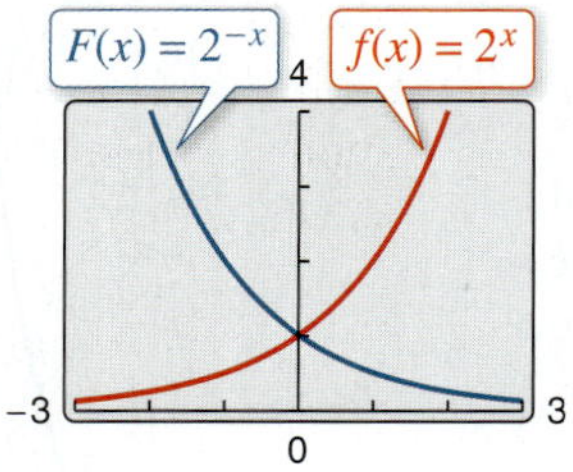

Figure 4.3

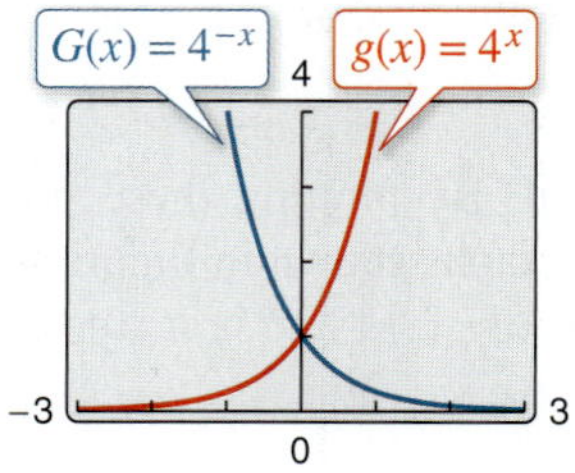

Figure 4.4

The graphs in Figures 4.3 and 4.4 are typical of the graphs of the exponential functions

$f(x) = a^x$ and $f(x) = a^{-x}$.

They have one y-intercept and one horizontal asymptote (the x-axis), and they are continuous. The basic characteristics of these exponential functions are summarized below.

Library of Parent Functions: Exponential Function

The *parent exponential function*

$f(x) = a^x, a > 0, a \neq 1$

is different from all the functions you have studied so far because the variable x is an *exponent*. A distinguishing characteristic of an exponential function is its rapid increase as x increases (for $a > 1$). Many real-life phenomena with patterns of rapid growth (or decline) can be modeled by exponential functions. The basic characteristics of the exponential function are summarized below and on the inside cover of this text.

Graph of $f(x) = a^x, a > 1$
Domain: $(-\infty, \infty)$
Range: $(0, \infty)$
Intercept: $(0, 1)$
Increasing on $(-\infty, \infty)$
x-axis is a horizontal asymptote
$(a^x \to 0$ as $x \to -\infty)$
Continuous

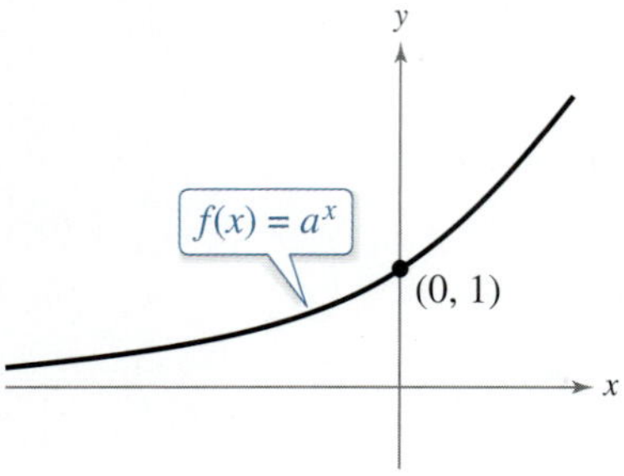

Graph of $f(x) = a^{-x}, a > 1$
Domain: $(-\infty, \infty)$
Range: $(0, \infty)$
Intercept: $(0, 1)$
Decreasing on $(-\infty, \infty)$
x-axis is a horizontal asymptote
$(a^{-x} \to 0$ as $x \to \infty)$
Continuous

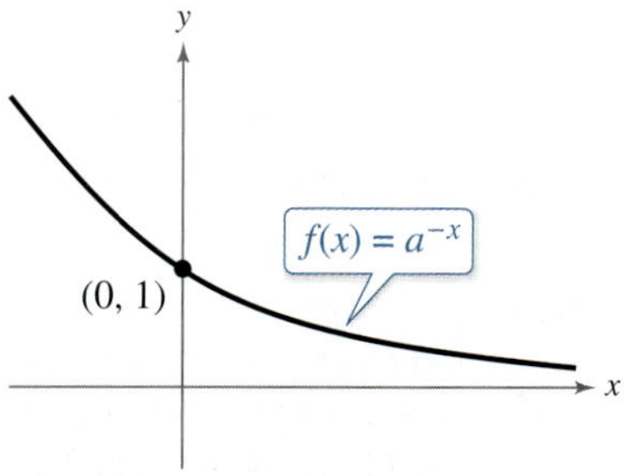

Explore the Concept

Use a graphing utility to graph $y = a^x$ for $a = 3, 5,$ and 7 in the same viewing window. (Use a viewing window in which $-2 \leq x \leq 1$ and $0 \leq y \leq 2$.) How do the graphs compare with each other? Which graph is on the top in the interval $(-\infty, 0)$? Which is on the bottom? Which graph is on the top in the interval $(0, \infty)$? Which is on the bottom? Repeat this experiment with the graphs of $y = b^x$ for $b = \frac{1}{3}, \frac{1}{5},$ and $\frac{1}{7}$. (Use a viewing window in which $-1 \leq x \leq 2$ and $0 \leq y \leq 2$.) What can you conclude about the shape of the graph of $y = b^x$ and the value of b?

In the next example, notice how the graph of $y = a^x$ can be used to sketch the graphs of functions of the form $f(x) = b \pm a^{x+c}$.

EXAMPLE 4 Library of Parent Functions: $f(x) = a^x$

See LarsonPrecalculus.com for an interactive version of this type of example.

Each of the following graphs is a transformation of the graph of $f(x) = 3^x$.

a. Because $g(x) = 3^{x+1} = f(x + 1)$, the graph of g can be obtained by shifting the graph of f one unit to the *left*, as shown in Figure 4.5.

b. Because $h(x) = 3^x - 2 = f(x) - 2$, the graph of h can be obtained by shifting the graph of f *downward* two units, as shown in Figure 4.6.

c. Because $k(x) = -3^x = -f(x)$, the graph of k can be obtained by *reflecting* the graph of f in the x-axis, as shown in Figure 4.7.

d. Because $j(x) = 3^{-x} = f(-x)$, the graph of j can be obtained by *reflecting* the graph of f in the y-axis, as shown in Figure 4.8.

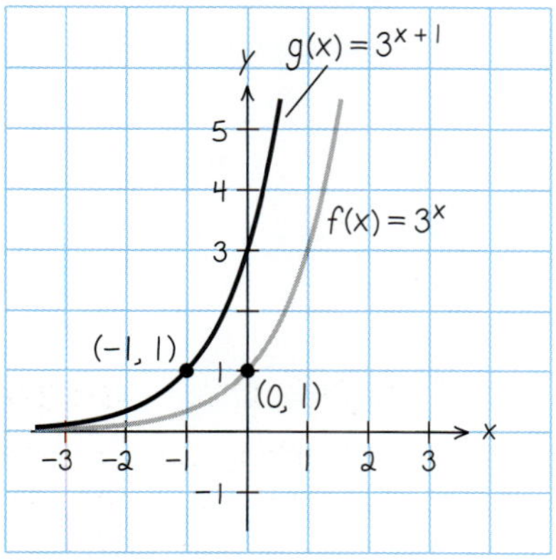

Figure 4.5

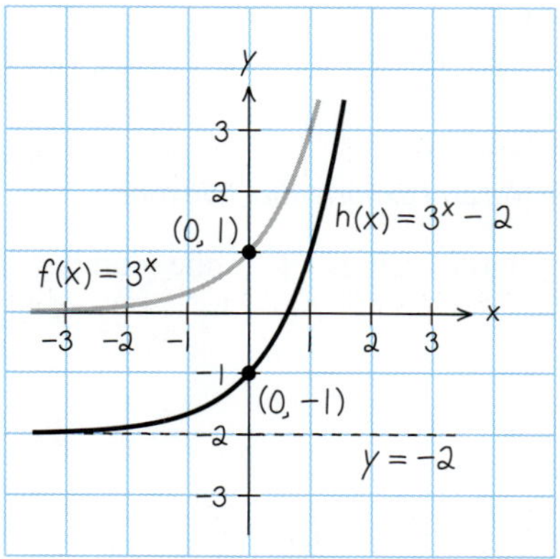

Figure 4.6

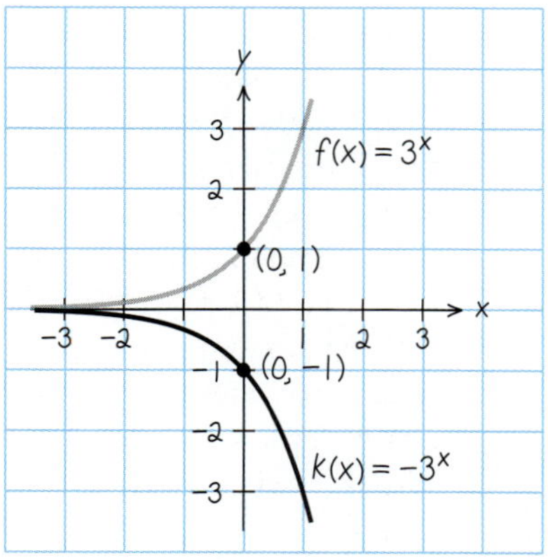

Figure 4.7

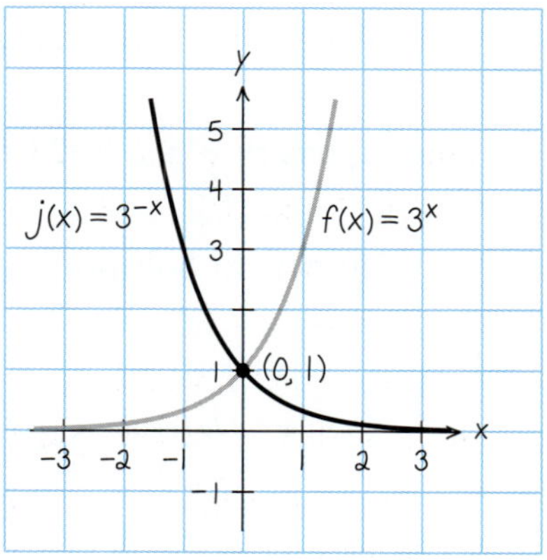

Figure 4.8

✓ Checkpoint Audio-video solution in English & Spanish at LarsonPrecalculus.com.

Use the graph of $f(x) = 4^x$ to describe the transformation that yields the graph of each function.

a. $g(x) = 4^{x-2}$ **b.** $h(x) = 4^x + 3$ **c.** $k(x) = 4^{-x} - 3$

Notice that the transformations in Figures 4.5, 4.7, and 4.8 keep the x-axis ($y = 0$) as a horizontal asymptote, but the transformation in Figure 4.6 yields a new horizontal asymptote of $y = -2$. Also, be sure to note how the y-intercept is affected by each transformation.

What's Wrong?

You use a graphing utility to graph $f(x) = 3^x$ and $g(x) = 3^{x+2}$, as shown in the figure. You use the graph to conclude that the graph of g can be obtained by shifting the graph of f upward two units. What's wrong?

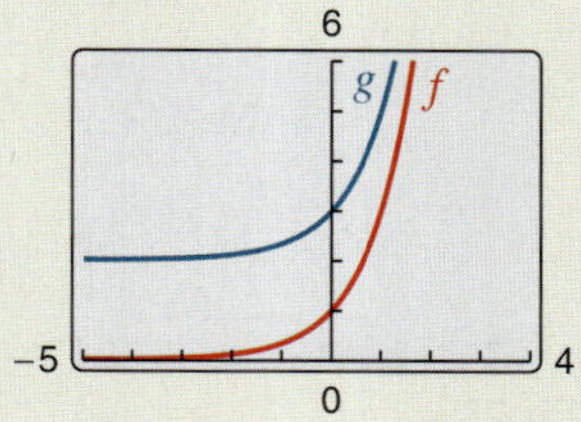

Explore the Concept

The following table shows some points on the graphs in Figure 4.5, where $Y_1 = f(x)$ and $Y_2 = g(x)$. Explain how you can use the table to describe the transformation.

X	Y1	Y2
−3	.03704	.11111
−2	.11111	.33333
−1	.33333	1
0	1	3
1	3	9
2	9	27
3	27	81

X=−3

The Natural Base e

For many applications, the convenient choice for a base is the irrational number

$$e = 2.71828128 \ldots.$$

This number is called the **natural base.** The function $f(x) = e^x$ is called the **natural exponential function,** and its graph is shown in Figure 4.9. The graph of the natural exponential function has the same basic characteristics as the graph of the function $f(x) = a^x$ (see page 326). Be sure you see that for the natural exponential function $f(x) = e^x$, e is the constant 2.718281828 . . . , whereas x is the variable.

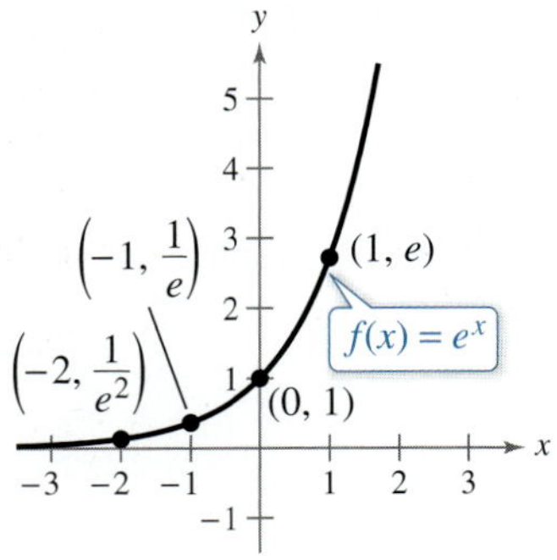

Figure 4.9 ***The Natural Exponential Function***

In Example 5, you will see that, for large values of x, the number e can be approximated by the expression

$$\left(1 + \frac{1}{x}\right)^x.$$

EXAMPLE 5 Approximation of the Number e

Evaluate the expression

$$\left(1 + \frac{1}{x}\right)^x$$

for several large values of x to see that the values approach $e \approx 2.718281828$ as x increases without bound.

Graphical Solution

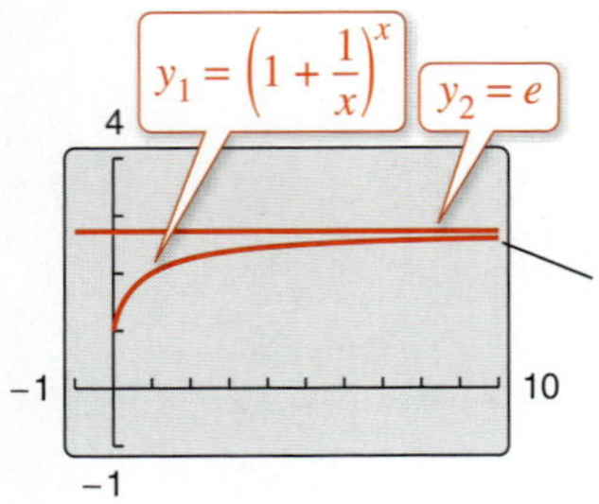

As x increases, the graph of y_1 gets closer and closer to the graph of the line $y_2 = e$.

Numerical Solution

Enter $y_1 = [1 + (1/x)]^x$.

Use the *table* feature (in *ask* mode) to evaluate y_1 for increasing values of x.

X	Y1
10	2.5937
100	2.7048
1000	2.7169
10000	2.7181
100000	2.7183
1E6	2.7183

X=2.71828046932

From the table, it seems reasonable to conclude that

$$\left(1 + \frac{1}{x}\right)^x \to e \text{ as } x \to \infty.$$

Checkpoint *Audio-video solution in English & Spanish at LarsonPrecalculus.com.*

Evaluate the expression $\left(1 + \frac{2}{x}\right)^x$ for several large values of x to see that the values approach $e^2 \approx 7.389056099$.

Explore the Concept

Use your graphing utility to graph the functions

$$y_1 = 2^x$$
$$y_2 = e^x$$
$$y_3 = 3^x$$

in the same viewing window. From the relative positions of these graphs, make a guess as to the value of the real number e. Then try to find a number a such that the graphs of $y_2 = e^x$ and $y_4 = a^x$ are as close to each other as possible.

EXAMPLE 6 Evaluating the Natural Exponential Function

Use a calculator to evaluate the function $f(x) = e^x$ at each indicated value of x.

a. $x = -2$

b. $x = 0.25$

c. $x = -0.4$

d. $x = \frac{2}{3}$

Solution

	Function Value	*Graphing Calculator Keystrokes*	*Display*
a.	$f(-2) = e^{-2}$	[e^x] [(–)] 2 [ENTER]	0.1353353
b.	$f(0.25) = e^{0.25}$	[e^x] .25 [ENTER]	1.2840254
c.	$f(-0.4) = e^{-0.4}$	[e^x] [(–)] .4 [ENTER]	0.6703200
d.	$f(\frac{2}{3}) = e^{2/3}$	[e^x] [(] 2 [÷] 3 [)] [ENTER]	1.9477340

✓ ***Checkpoint*** *Audio-video solution in English & Spanish at LarsonPrecalculus.com.*

Use a calculator to evaluate the function $f(x) = e^x$ at each value of x.

a. $x = 0.3$ **b.** $x = -1.2$ **c.** $x = 6.2$

EXAMPLE 7 Graphing Natural Exponential Functions

Sketch the graphs of $f(x) = 2e^{0.24x}$ and $g(x) = \frac{1}{2}e^{-0.58x}$.

Solution

To sketch these two graphs, you can use a calculator to construct a table of values, as shown below.

x	-3	-2	-1	0	1	2	3
$f(x)$	0.974	1.238	1.573	2.000	2.542	3.232	4.109
$g(x)$	2.849	1.595	0.893	0.500	0.280	0.157	0.088

After constructing the table, plot the points and connect them with smooth curves. Note that the graph in Figure 4.10 is increasing, whereas the graph in Figure 4.11 is decreasing. Use a graphing calculator to verify these graphs.

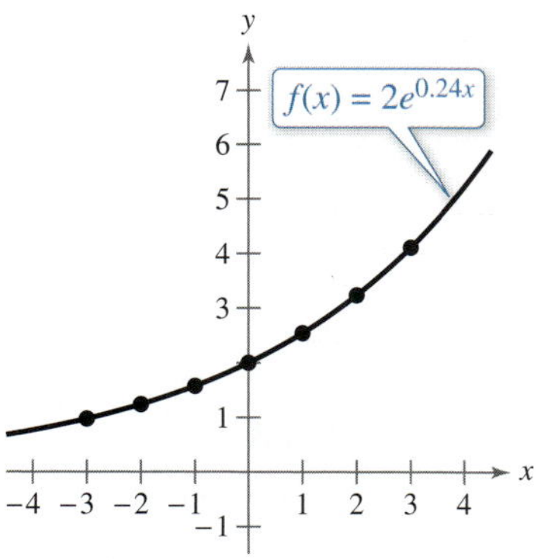

Figure 4.10

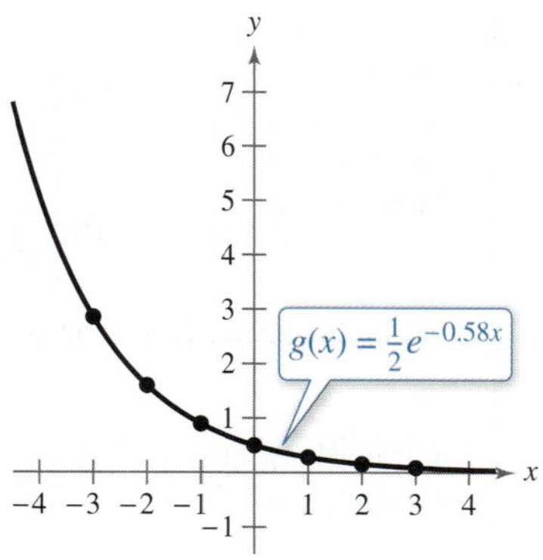

Figure 4.11

✓ ***Checkpoint*** *Audio-video solution in English & Spanish at LarsonPrecalculus.com.*

Sketch the graph of $f(x) = 5e^{0.17x}$.

Explore the Concept

Use a graphing utility to graph $y = (1 + x)^{1/x}$. Describe the behavior of the graph near $x = 0$. Is there a y-intercept? How does the behavior of the graph near $x = 0$ relate to the result of Example 5? Use the *table* feature of the graphing utility to create a table that shows values of y for values of x near $x = 0$ to help you describe the behavior of the graph near this point.

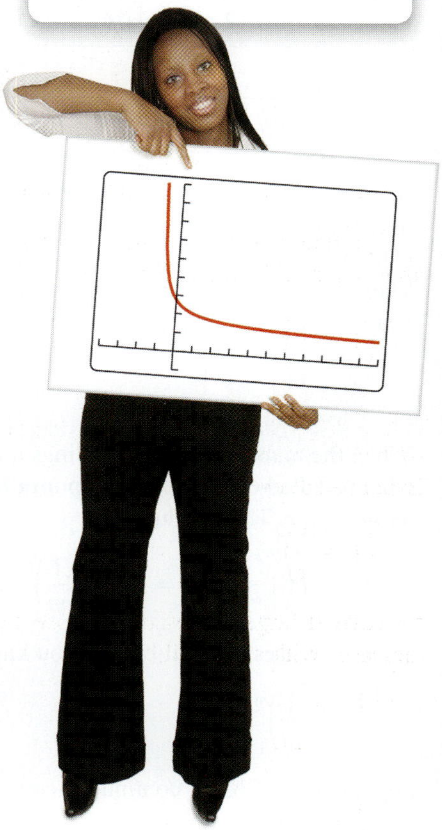

Applications

One of the most familiar examples of exponential growth is an investment earning *continuously compounded interest*. Suppose a principal P is invested at an annual interest rate r, compounded once a year. If the interest is added to the principal at the end of the year, then the new balance P_1 is

$$P_1 = P + Pr = P(1 + r).$$

This pattern of multiplying the previous principal by $1 + r$ is then repeated each successive year, as shown in the table.

Time in years	Balance after each compounding
0	$P = P$
1	$P_1 = P(1 + r)$
2	$P_2 = P_1(1 + r) = P(1 + r)(1 + r) = P(1 + r)^2$
⋮	⋮
t	$P_t = P(1 + r)^t$

To accommodate more frequent (quarterly, monthly, or daily) compounding of interest, let n be the number of compoundings per year and let t be the number of years. (The product nt represents the total number of times the interest will be compounded.) Then the interest rate per compounding period is r/n, and the account balance after t years is

$$A = P\left(1 + \frac{r}{n}\right)^{nt}.$$ Amount (balance) with n compoundings per year

When the number of compoundings n increases without bound, the process approaches what is called **continuous compounding.** In the formula for n compoundings per year, let $m = n/r$. This produces

$$A = P\left(1 + \frac{r}{n}\right)^{nt} = P\left(1 + \frac{1}{m}\right)^{mrt} = P\left[\left(1 + \frac{1}{m}\right)^{m}\right]^{rt}.$$

As m increases without bound, you know from Example 5 that

$$\left(1 + \frac{1}{m}\right)^{m}$$

approaches e. So, for continuous compounding, it follows that

$$P\left[\left(1 + \frac{1}{m}\right)^{m}\right]^{rt} \Rightarrow P[e]^{rt}$$

and you can write $A = Pe^{rt}$. This result is part of the reason that e is the "natural" choice for a base of an exponential function.

Formulas for Compound Interest

After t years, the balance A in an account with principal P and annual interest rate r (in decimal form) is given by the following formulas.

1. For n compoundings per year: $A = P\left(1 + \dfrac{r}{n}\right)^{nt}$

2. For continuous compounding: $A = Pe^{rt}$

Explore the Concept

Use the formula

$$A = P\left(1 + \frac{r}{n}\right)^{nt}$$

to calculate the amount in an account when $P = \$3000$, $r = 6\%$, $t = 10$ years, and the interest is compounded (a) by the day, (b) by the hour, (c) by the minute, and (d) by the second. Does increasing the number of compoundings per year result in unlimited growth of the amount in the account? Explain.

Remark

The interest rate r in the formulas for compound interest should be written as a decimal. For example, an interest rate of 2.5% would be written as $r = 0.025$.

EXAMPLE 8 Finding the Balance for Compound Interest

A total of \$9000 is invested at an annual interest rate of 2.5%, compounded annually. Find the balance in the account after 5 years.

Algebraic Solution

In this case, $P = 9000$, $r = 2.5\% = 0.025$, $n = 1$, and $t = 5$. Using the formula for compound interest with n compoundings per year, you have

$$A = P\left(1 + \frac{r}{n}\right)^{nt} \quad \text{Formula for compound interest}$$

$$= 9000\left(1 + \frac{0.025}{1}\right)^{1(5)} \quad \text{Substitute for } P, r, n, \text{ and } t.$$

$$= 9000(1.025)^5 \quad \text{Simplify.}$$

$$\approx \$10{,}182.67. \quad \text{Use a calculator.}$$

So, the balance in the account after 5 years is \$10,182.67.

Graphical Solution

Substitute the values for P, r, and n into the formula for compound interest with n compoundings per year and simplify to obtain $A = 9000(1.025)^t$. Use a graphing utility to graph $A = 9000(1.025)^t$. Then use the *value* feature to approximate the value of A when $t = 5$, as shown in the figure.

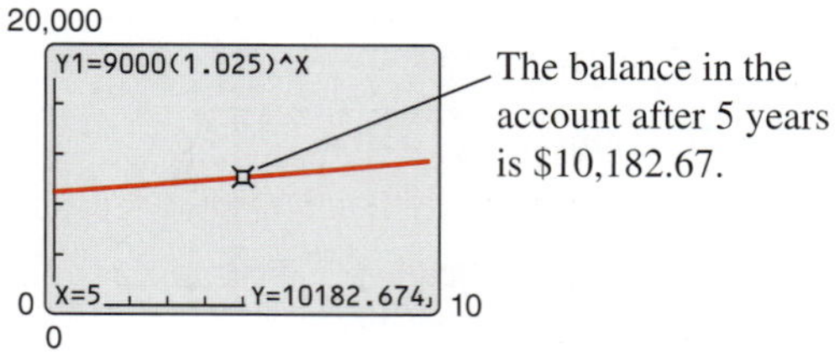

✓ Checkpoint Audio-video solution in English & Spanish at LarsonPrecalculus.com.

For the account in Example 8, find the balance after 10 years.

EXAMPLE 9 Finding Compound Interest

A total of \$12,000 is invested at an annual interest rate of 3%. Find the balance after 4 years when the interest is compounded (a) quarterly and (b) continuously.

Solution

a. For quarterly compoundings, $n = 4$. So, after 4 years at 3%, the balance is

$$A = P\left(1 + \frac{r}{n}\right)^{nt} \quad \text{Formula for compound interest}$$

$$= 12{,}000\left(1 + \frac{0.03}{4}\right)^{4(4)} \quad \text{Substitute for } P, r, n, \text{ and } t.$$

$$\approx \$13{,}523.91. \quad \text{Use a calculator.}$$

b. For continuous compounding, the balance is

$$A = Pe^{rt} \quad \text{Formula for continuous compounding}$$

$$= 12{,}000e^{0.03(4)} \quad \text{Substitute for } P, r, \text{ and } t.$$

$$\approx \$13{,}529.96. \quad \text{Use a calculator.}$$

Note that a continuous-compounding account yields more than a quarterly-compounding account.

Financial Analyst

✓ Checkpoint Audio-video solution in English & Spanish at LarsonPrecalculus.com.

You invest \$6000 at an annual rate of 4%. Find the balance after 7 years when the interest is compounded (a) quarterly, (b) monthly, and (c) continuously.

Example 9 illustrates the following general rule. For a given principal, interest rate, and time, the more often the interest is compounded per year, the greater the balance will be. Moreover, the balance obtained by continuous compounding is greater than the balance obtained by compounding n times per year.

EXAMPLE 10 Radioactive Decay

Let y represent a mass, in grams, of radioactive strontium (^{90}Sr), whose half-life is about 29 years. The quantity of strontium present after t years is $y = 10(\frac{1}{2})^{t/29}$.

a. What is the initial mass (when $t = 0$)?

b. How much of the initial mass is present after 80 years?

Algebraic Solution

a. $y = 10\left(\frac{1}{2}\right)^{t/29}$ Write original equation.

$= 10\left(\frac{1}{2}\right)^{0/29}$ Substitute 0 for t.

$= 10$ Simplify.

So, the initial mass is 10 grams.

b. $y = 10\left(\frac{1}{2}\right)^{t/29}$ Write original equation.

$= 10\left(\frac{1}{2}\right)^{80/29}$ Substitute 80 for t.

$\approx 10\left(\frac{1}{2}\right)^{2.759}$ Simplify.

≈ 1.48 Use a calculator.

So, about 1.48 grams are present after 80 years.

Graphical Solution

a.

When $t = 0$, $y = 10$. So, the initial mass is 10 grams.

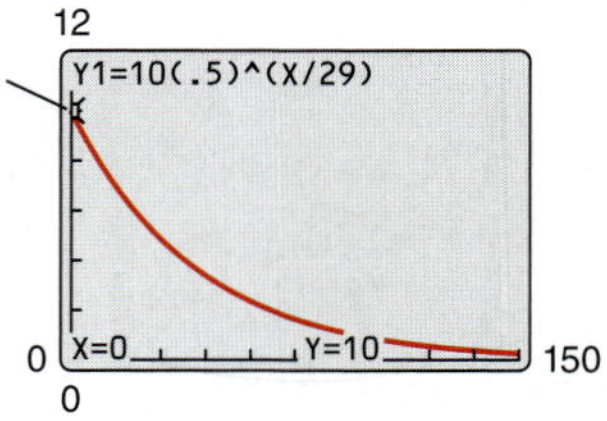

b.

When $t = 80$, $y \approx 1.48$. So, about 1.48 grams are present after 80 years.

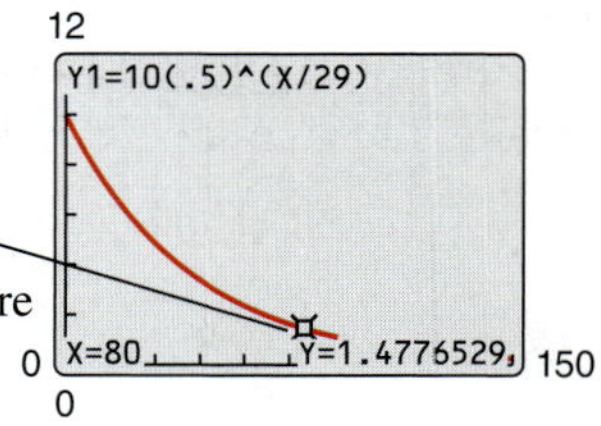

Checkpoint Audio-video solution in English & Spanish at LarsonPrecalculus.com.

In Example 10, how much of the initial mass is present after 160 years?

EXAMPLE 11 Population Growth

The approximate number of fruit flies in an experimental population after t hours is given by $Q(t) = 20e^{0.03t}$, where $t \geq 0$.

a. Find the initial number of fruit flies in the population.

b. How large is the population of fruit flies after 72 hours?

c. Graph Q.

Solution

a. To find the initial population, evaluate $Q(t)$ when $t = 0$.

$$Q(0) = 20e^{0.03(0)} = 20e^{0} = 20(1) = 20 \text{ flies}$$

b. After 72 hours, the population size is

$$Q(72) = 20e^{0.03(72)} = 20e^{2.16} \approx 173 \text{ flies.}$$

c. The graph of Q is shown in the figure.

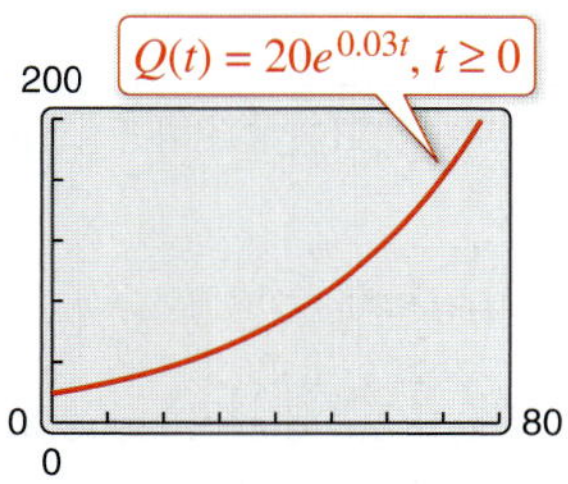

Checkpoint Audio-video solution in English & Spanish at LarsonPrecalculus.com.

Rework Example 11 when the approximate number of fruit flies in the experimental population after t hours is given by $Q(t) = 10e^{0.02t}$.

4.1 Exercises

See *CalcChat.com* for tutorial help and worked-out solutions to odd-numbered exercises. For instructions on how to use a graphing utility, see Appendix A.

Vocabulary and Concept Check

In Exercises 1 and 2, fill in the blank(s).

1. Exponential and logarithmic functions are examples of nonalgebraic functions, also called _______ functions.

2. The exponential function $f(x) = e^x$ is called the _______ function, and the base e is called the _______ base.

3. What type of transformation of the graph of $f(x) = 5^x$ is the graph of $f(x + 1)$?

4. The formula $A = Pe^{rt}$ gives the balance A of an account earning what type of interest?

Procedures and Problem Solving

Evaluating Exponential Functions **In Exercises 5–8, use a calculator to evaluate the function at the indicated value of x. Round your result to three decimal places.**

Function	*Value*
5. $f(x) = 3.4^x$	$x = 6.8$
6. $f(x) = 1.2^x$	$x = \frac{1}{3}$
7. $g(x) = 5^x$	$x = -\pi$
8. $h(x) = 8.6^{-3x}$	$x = -\sqrt{2}$

Graphs of $y = a^x$ and $y = a^{-x}$ **In Exercises 9–16, graph the exponential function by hand. Identify any asymptotes and intercepts and determine whether the graph of the function is increasing or decreasing.**

9. $g(x) = 5^x$ **10.** $f(x) = 10^x$

11. $f(x) = 5^{-x}$ **12.** $h(x) = 10^{-x}$

13. $h(x) = \left(\frac{5}{4}\right)^x$ **14.** $g(x) = \left(\frac{3}{2}\right)^x$

15. $g(x) = \left(\frac{5}{4}\right)^{-x}$ **16.** $f(x) = \left(\frac{3}{2}\right)^{-x}$

Library of Parent Functions **In Exercises 17–20, use the graph of $y = 2^x$ to match the function with its graph. [The graphs are labeled (a), (b), (c), and (d).]**

(a)

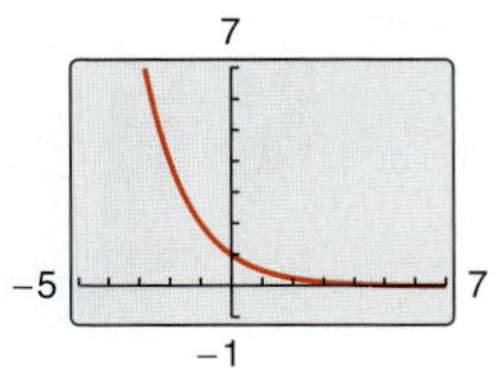

(b)

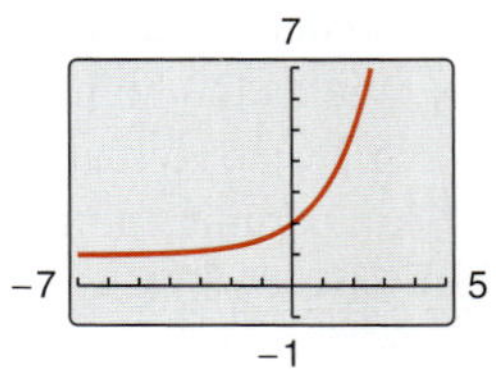

(c)

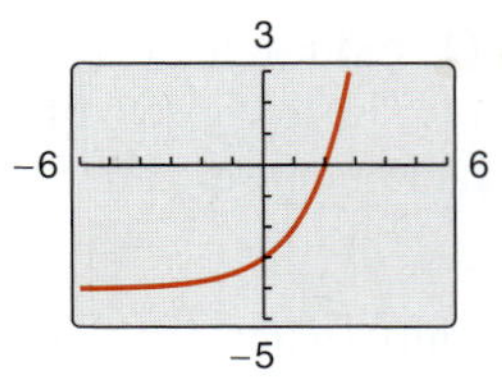

(d)

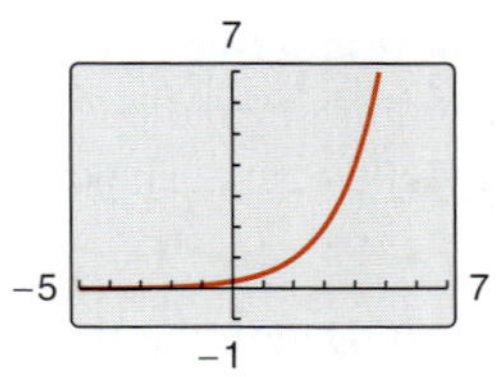

17. $f(x) = 2^{x-2}$ **18.** $f(x) = 2^{-x}$

19. $f(x) = 2^x - 4$ **20.** $f(x) = 2^x + 1$

Library of Parent Functions **In Exercises 21–30, use the graph of f to describe the transformation that yields the graph of g. Then sketch the graphs of f and g by hand.**

21. $f(x) = 3^x, \quad g(x) = 3^{x-5}$

22. $f(x) = 10^x, \quad g(x) = 10^{x+4}$

23. $f(x) = 6^x, \quad g(x) = 6^x + 5$

24. $f(x) = -2^x, \quad g(x) = 5 - 2^x$

25. $f(x) = \left(\frac{3}{5}\right)^x, \quad g(x) = -\left(\frac{3}{5}\right)^{x+4}$

26. $f(x) = 0.3^x, \quad g(x) = -0.3^x + 5$

27. $f(x) = 0.4^x, \quad g(x) = 0.4^{x-2} - 3$

28. $f(x) = \left(\frac{2}{3}\right)^x, \quad g(x) = \left(\frac{2}{3}\right)^{x+2} - 3$

29. $f(x) = \left(\frac{1}{4}\right)^x, \quad g(x) = \left(\frac{1}{4}\right)^{-x} + 2$

30. $f(x) = \left(\frac{1}{2}\right)^x, \quad g(x) = \left(\frac{1}{2}\right)^{-(x+4)}$

Approximation of a Power with Base e **In Exercises 31 and 32, show that the value of $f(x)$ approaches the value of $g(x)$ as x increases without bound (a) graphically and (b) numerically.**

31. $f(x) = 1 + \left(\frac{0.5}{x}\right)^x, \quad g(x) = e^{0.5}$

32. $f(x) = 1 + \left(\frac{3}{x}\right)^x, \quad g(x) = e^3$

Evaluating the Natural Exponential Function **In Exercises 33–36, use a calculator to evaluate the function at the indicated value of x. Round your result to the nearest thousandth.**

Function	*Value*
33. $f(x) = e^x$	$x = 9.2$
34. $f(x) = e^{-x}$	$x = -\frac{3}{4}$
35. $g(x) = 50e^{4x}$	$x = 0.02$
36. $h(x) = -5.5e^{-x}$	$x = 200$

Graphing an Exponential Function **In Exercises 37–52, use a graphing utility to construct a table of values for the function. Then sketch the graph of the function. Identify any asymptotes of the graph.**

37. $f(x) = \left(\frac{5}{2}\right)^x$

38. $f(x) = \left(\frac{5}{2}\right)^{-x}$

39. $f(x) = 6^x$

40. $f(x) = 2^{x-1}$

41. $f(x) = 3^{x+2}$

42. $y = 2^{-x^2}$

43. $y = 3^{x-2} + 1$

44. $y = 4^{x+1} - 2$

45. $f(x) = e^{-x}$

46. $s(t) = 3e^{-0.2t}$

47. $f(x) = 3e^{x+4}$

48. $f(x) = 2e^{x-3}$

49. $f(x) = 2 + e^{x-5}$

50. $g(x) = e^{x+1} + 2$

51. $s(t) = 2e^{-0.12t}$

52. $g(x) = e^{0.5x} - 1$

Finding Asymptotes **In Exercises 53–56, use a graphing utility to (a) graph the function and (b) find any asymptotes numerically by creating a table of values for the function.**

53. $f(x) = \dfrac{8}{1 + e^{-0.5x}}$

54. $g(x) = \dfrac{8}{1 + e^{-0.5/x}}$

55. $f(x) = -\dfrac{6}{2 - e^{0.2x}}$

56. $f(x) = \dfrac{6}{2 - e^{0.2/x}}$

Finding Points of Intersection **In Exercises 57–60, use a graphing utility to find the point(s) of intersection, if any, of the graphs of the functions. Round your result to three decimal places.**

57. $y = 20e^{0.05x}$
$y = 1500$

58. $y = 100e^{0.01x}$
$y = 12{,}500$

59. $y = 2.5e^{0.3x}$
$y = 0.2$

60. $y = 3.2e^{0.5x}$
$y = 0.9$

Approximating Relative Extrema **In Exercises 61–64, (a) use a graphing utility to graph the function, (b) use the graph to find the open intervals on which the function is increasing and decreasing, and (c) approximate any relative maximum or minimum values.**

61. $f(x) = x^2e^{-x}$

62. $f(x) = 2x^2e^{x+1}$

63. $f(x) = x^3e^x$

64. $f(x) = x^3e^{-x+2}$

Finding the Balance for Compound Interest **In Exercises 65–68, complete the table to determine the balance A for \$2500 invested at rate r for t years and compounded n times per year.**

n	1	2	4	12	365	Continuous
A						

65. $r = 2\%, t = 10$ years

66. $r = 6\%, t = 10$ years

67. $r = 4\%, t = 20$ years

68. $r = 3\%, t = 40$ years

Finding the Balance for Compound Interest **In Exercises 69–72, complete the table to determine the balance A for \$12,000 invested at rate r for t years, compounded continuously.**

t	1	10	20	30	40	50
A						

69. $r = 4\%$

70. $r = 6\%$

71. $r = 3.5\%$

72. $r = 2.5\%$

Finding the Amount of an Annuity **In Exercises 73–76, you build an annuity by investing P dollars every month at interest rate r, compounded monthly. Find the amount A accrued after n months using the formula**

$$A = P\left[\frac{(1 + r/12)^n - 1}{r/12}\right], \text{ where } r \text{ is in decimal form.}$$

73. $P = \$25, r = 0.12, n = 48$ months

74. $P = \$100, r = 0.09, n = 60$ months

75. $P = \$200, r = 0.06, n = 72$ months

76. $P = \$75, r = 0.03, n = 24$ months

77. MODELING DATA

There are three options for investing \$500. The first earns 7% compounded annually, the second earns 7% compounded quarterly, and the third earns 7% compounded continuously.

(a) Find equations that model the growth of each investment and use a graphing utility to graph each model in the same viewing window over a 20-year period.

(b) Use the graph from part (a) to determine which investment yields the highest return after 20 years. What are the differences in earnings among the three investments?

78. Radioactive Decay Let Q represent a mass, in grams, of radioactive radium (^{226}Ra), whose half-life is 1600 years. The quantity of radium present after t years is given by $Q = 25\left(\frac{1}{2}\right)^{t/1600}$.

(a) Determine the initial quantity (when $t = 0$).

(b) Determine the quantity present after 1000 years.

(c) Use a graphing utility to graph the function over the interval $t = 0$ to $t = 5000$.

(d) When will the quantity of radium be 0 grams? Explain.

79. Radioactive Decay Let Q represent a mass, in grams, of carbon 14 (^{14}C), whose half-life is 5700 years. The quantity present after t years is given by $Q = 10\left(\frac{1}{2}\right)^{t/5700}$.

(a) Determine the initial quantity (when $t = 0$).

(b) Determine the quantity present after 2000 years.

(c) Sketch the graph of the function over the interval $t = 0$ to $t = 10{,}000$.

80. Algebraic-Graphical-Numerical Assume the annual rate of inflation is 4% for the next 10 years. The approximate cost C of goods or services during these years is $C(t) = P(1.04)^t$, where t is the time (in years) and P is the present cost. An oil change for your car presently costs \$26.88. Use the following methods to approximate the cost 10 years from now.

(a) Use a graphing utility to graph the function and then use the *value* feature.

(b) Use the *table* feature of the graphing utility to find a numerical approximation.

(c) Use a calculator to evaluate the cost function algebraically.

81. Population Growth The projected populations of California for the years 2020 through 2060 can be modeled by $P = 36.308e^{0.0065t}$, where P is the population (in millions) and t is the time (in years), with $t = 20$ corresponding to 2020. (Source: California Department of Finance)

(a) Use a graphing utility to graph the function for the years 2020 through 2060.

(b) Use the *table* feature of the graphing utility to create a table of values for the same time period as in part (a).

(c) According to the model, in what year will the population of California exceed 51 million?

82. *Why you should learn it* *(p. 324)* In early 2014, a new sedan had a manufacturer's suggested retail price of \$31,340. After t years, the sedan's value is given by

$$V(t) = 31{,}340\left(\tfrac{4}{5}\right)^t.$$

(a) Use a graphing utility to graph the function.

(b) Use the graphing utility to create a table of values that shows the value V for $t = 1$ to $t = 10$ years.

(c) According to the model, when will the sedan have no value?

Conclusions

True or False? **In Exercises 83 and 84, determine whether the statement is true or false. Justify your answer.**

83. $f(x) = 1^x$ is not an exponential function.

84. $e = \dfrac{271{,}801}{99{,}990}$

85. Library of Parent Functions Determine which equation(s) may be represented by the graph shown. (There may be more than one correct answer.)

(a) $y = e^x + 1$

(b) $y = -e^{-x} + 1$

(c) $y = e^{-x} - 1$

(d) $y = e^{-x} + 1$

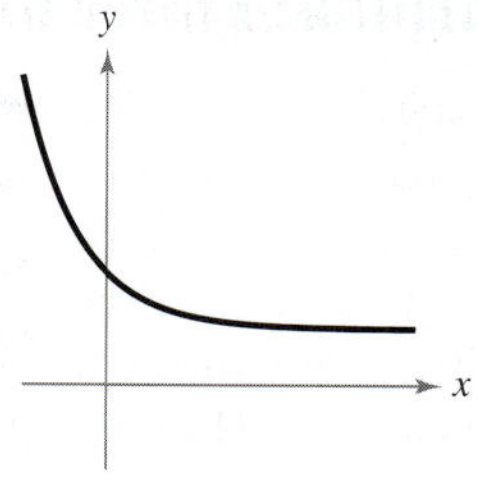

86. Exploration Use a graphing utility to graph $y_1 = e^x$ and each of the functions $y_2 = x^2$, $y_3 = x^3$, $y_4 = \sqrt{x}$, and $y_5 = |x|$ in the same viewing window.

(a) Which function increases at the fastest rate for "large" values of x?

(b) Use the result of part (a) to make a conjecture about the rates of growth of $y_1 = e^x$ and $y = x^n$, where n is a natural number and x is "large."

(c) Use the results of parts (a) and (b) to describe what is implied when it is stated that a quantity is growing exponentially.

87. Think About It Graph $y = 3^x$ and $y = 4^x$. Use the graph to solve the inequality $3^x < 4^x$.

88. HOW DO YOU SEE IT? The figure shows the graphs of $y = 2^x$, $y = e^x$, $y = 10^x$, $y = 2^{-x}$, $y = e^{-x}$, and $y = 10^{-x}$. Match each function with its graph. [The graphs are labeled (a) through (f).] Explain.

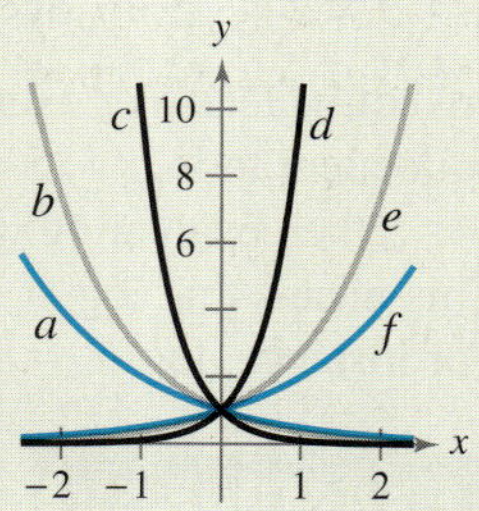

Think About It **In Exercises 89–92, place the correct symbol (< or >) between the two numbers.**

89. e^{π} ☐ π^e

90. 2^{10} ☐ 10^2

91. 5^{-3} ☐ 3^{-5}

92. $4^{1/2}$ ☐ $\left(\frac{1}{2}\right)^4$

Cumulative Mixed Review

Inverse Functions **In Exercises 93–96, determine whether the function has an inverse function. If it does, find f^{-1}.**

93. $f(x) = 5x - 7$

94. $f(x) = -\frac{2}{3}x + \frac{5}{2}$

95. $f(x) = \sqrt[3]{x + 8}$

96. $f(x) = \sqrt{x^2 + 6}$

---

97. *Make a Decision* To work an extended application analyzing the population per square mile in the United States, visit this textbook's website at *LarsonPrecalculus.com*. (Data Source: U.S. Census Bureau)

4.2 Logarithmic Functions and Their Graphs

Logarithmic Functions

In Section 1.7, you studied the concept of an inverse function. There, you learned that when a function is one-to-one—that is, when the function has the property that no horizontal line intersects its graph more than once—the function must have an inverse function. By looking back at the graphs of the exponential functions introduced in Section 4.1, you will see that every function of the form $f(x) = a^x$, where $a > 0$ and $a \neq 1$, passes the Horizontal Line Test and therefore must have an inverse function. This inverse function is called the **logarithmic function with base *a*.**

What you should learn

- Recognize and evaluate logarithmic functions with base a.
- Graph logarithmic functions with base a.
- Recognize, evaluate, and graph natural logarithmic functions.
- Use logarithmic functions to model and solve real-life problems.

Why you should learn it

Logarithmic functions are useful in modeling data that represent quantities that increase or decrease slowly. For instance, Exercise 114 on page 345 shows how to use a logarithmic function to model the minimum required ventilation rates in public school classrooms.

Definition of Logarithmic Function

For $x > 0$, $a > 0$, and $a \neq 1$,

$$y = \log_a x \quad \text{if and only if} \quad x = a^y.$$

The function given by

$$f(x) = \log_a x \qquad \text{Read as "log base } a \text{ of } x."$$

is called the **logarithmic function with base *a*.**

From the definition above, you can see that every logarithmic equation can be written in an equivalent exponential form and every exponential equation can be written in logarithmic form. So, the equations

$$y = \log_a x \qquad \text{and} \qquad x = a^y$$

are equivalent. For example, $2 = \log_3 9$ is equivalent to $9 = 3^2$, and $5^3 = 125$ is equivalent to $\log_5 125 = 3$.

When evaluating logarithms, remember that *a logarithm is an exponent.* This means that $\log_a x$ is the exponent to which a must be raised to obtain x. For instance, $\log_2 8 = 3$ because 2 must be raised to the third power to get 8.

EXAMPLE 1 Evaluating Logarithms

Evaluate each logarithm at the indicated value of x.

a. $f(x) = \log_2 x, \quad x = 32$

b. $f(x) = \log_3 x, \quad x = 1$

c. $f(x) = \log_4 x, \quad x = 2$

d. $f(x) = \log_{10} x, \quad x = \frac{1}{100}$

Solution

a. $f(32) = \log_2 32 = 5 \quad \text{because} \quad 2^5 = 32.$

b. $f(1) = \log_3 1 = 0 \quad \text{because} \quad 3^0 = 1.$

c. $f(2) = \log_4 2 = \frac{1}{2} \quad \text{because} \quad 4^{1/2} = \sqrt{4} = 2.$

d. $f\left(\frac{1}{100}\right) = \log_{10} \frac{1}{100} = -2 \quad \text{because} \quad 10^{-2} = \frac{1}{10^2} = \frac{1}{100}.$

Checkpoint Audio-video solution in English & Spanish at LarsonPrecalculus.com.

Evaluate each logarithm at the indicated value of x.

a. $f(x) = \log_6 x, x = 1$ **b.** $f(x) = \log_5 x, x = \frac{1}{125}$ **c.** $f(x) = \log_{10} x, x = 10{,}000$

Remark

In this text, the parentheses in $\log_a(u)$ are sometimes omitted when u is an expression involving exponents, radicals, products, or quotients. For instance, $\log_{10}(2x)$ can be written as $\log_{10} 2x$. To evaluate $\log_{10} 2x$, find the logarithm of the product $2x$.

The logarithmic function with base 10 is called the **common logarithmic function.** On most calculators, this function is denoted by [LOG]. Example 2 shows how to use a calculator to evaluate common logarithmic functions. You will learn how to use a calculator to calculate logarithms to any base in the next section.

EXAMPLE 2 Evaluating Common Logarithms on a Calculator

Use a calculator to evaluate the function $f(x) = \log_{10} x$ at each value of x.

a. $x = 10$ **b.** $x = \frac{1}{3}$ **c.** $x = 2.5$ **d.** $x = -2$

Solution

	Function Value	*Graphing Calculator Keystrokes*	*Display*
a.	$f(10) = \log_{10} 10$	[LOG] 10 [ENTER]	1
b.	$f(\frac{1}{3}) = \log_{10} \frac{1}{3}$	[LOG] [(] 1 [÷] 3 [)] [ENTER]	−0.4771213
c.	$f(2.5) = \log_{10} 2.5$	[LOG] 2.5 [ENTER]	0.3979400
d.	$f(-2) = \log_{10}(-2)$	[LOG] [(−)] 2 [ENTER]	ERROR

Note that the calculator displays an error message when you try to evaluate $\log_{10}(-2)$. In this case, there is no *real* power to which 10 can be raised to obtain -2.

✓ Checkpoint Audio-video solution in English & Spanish at LarsonPrecalculus.com.

Use a calculator to evaluate the function $f(x) = \log_{10} x$ at each value of x.

a. $x = 275$ **b.** $x = 0.275$ **c.** $x = -\frac{1}{2}$ **d.** $x = \frac{1}{2}$

Technology Tip

Some graphing utilities do not give an error message for $\log_{10}(-2)$. Instead, the graphing utility will display a complex number. For the purpose of this text, however, the domain of a logarithmic function is the set of positive *real* numbers.

The properties of logarithms listed below follow directly from the definition of the logarithmic function with base a.

Properties of Logarithms

1. $\log_a 1 = 0$ because $a^0 = 1$.

2. $\log_a a = 1$ because $a^1 = a$.

3. $\log_a a^x = x$ and $a^{\log_a x} = x$. Inverse Properties

4. If $\log_a x = \log_a y$, then $x = y$. One-to-One Property

EXAMPLE 3 Using Properties of Logarithms

a. Solve for x: $\log_2 x = \log_2 3$

b. Solve for x: $\log_4 4 = x$

c. Simplify: $\log_5 5^x$

d. Simplify: $7^{\log_7 14}$

Solution

a. Using the One-to-One Property (Property 4), you can conclude that $x = 3$.

b. Using Property 2, you can conclude that $x = 1$.

c. Using the Inverse Property (Property 3), it follows that $\log_5 5^x = x$.

d. Using the Inverse Property (Property 3), it follows that $7^{\log_7 14} = 14$.

✓ Checkpoint Audio-video solution in English & Spanish at LarsonPrecalculus.com.

a. Solve for x: $\log_{10} x = \log_{10} 2$ **b.** Simplify: $20^{\log_{20} 3}$

Graphs of Logarithmic Functions

To sketch the graph of $y = \log_a x$, you can use the fact that the graphs of inverse functions are reflections of each other in the line $y = x$.

EXAMPLE 4 Graphs of Exponential and Logarithmic Functions

In the same coordinate plane, sketch the graph of each function by hand.

a. $f(x) = 2^x$

b. $g(x) = \log_2 x$

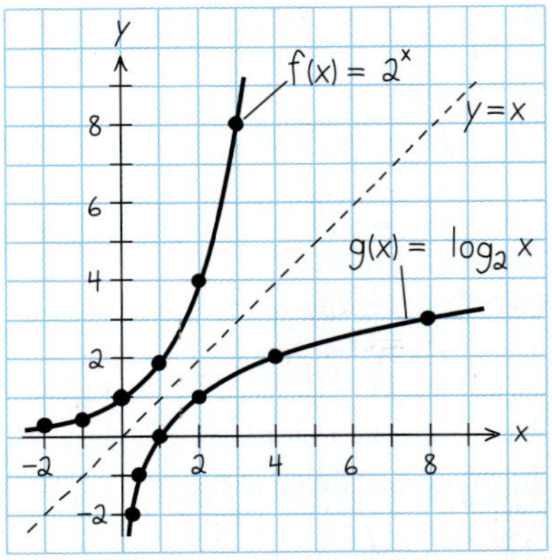

Figure 4.12

Solution

a. For $f(x) = 2^x$, construct a table of values. By plotting these points and connecting them with a smooth curve, you obtain the graph of f shown in Figure 4.12.

x	-2	-1	0	1	2	3
$f(x) = 2^x$	$\frac{1}{4}$	$\frac{1}{2}$	1	2	4	8

b. Because $g(x) = \log_2 x$ is the inverse function of $f(x) = 2^x$, the graph of g is obtained by plotting the points $(f(x), x)$ and connecting them with a smooth curve. The graph of g is a reflection of the graph of f in the line $y = x$, as shown in Figure 4.12.

✓ **Checkpoint** Audio-video solution in English & Spanish at LarsonPrecalculus.com.

In the same coordinate plane, sketch the graphs of (a) $f(x) = 8^x$ and (b) $g(x) = \log_8 x$.

Before you can confirm the result of Example 4 using a graphing utility, you need to know how to enter $\log_2 x$. You will learn how to do this using the *change-of-base formula* discussed in the next section.

EXAMPLE 5 Sketching the Graph of a Logarithmic Function

Sketch the graph of the common logarithmic function $f(x) = \log_{10} x$ by hand.

Solution

Begin by constructing a table of values. Note that some of the values can be obtained without a calculator by using the Inverse Property of Logarithms. Others require a calculator. Next, plot the points and connect them with a smooth curve, as shown in Figure 4.13. Note that $x = 0$ (the y-axis) is a vertical asymptote of the graph.

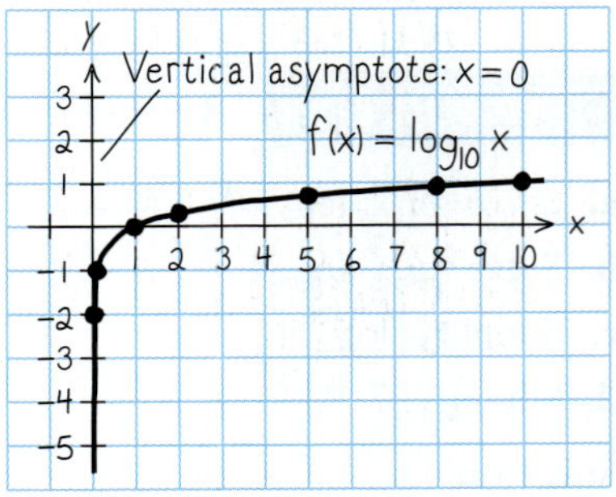

Figure 4.13

	Without calculator				With calculator		
x	$\frac{1}{100}$	$\frac{1}{10}$	1	10	2	5	8
$f(x) = \log_{10} x$	-2	-1	0	1	0.301	0.699	0.903

✓ **Checkpoint** Audio-video solution in English & Spanish at LarsonPrecalculus.com.

Without using a calculator, sketch the graph of $f(x) = \log_3 x$ by hand.

The nature of the graph in Figure 4.13 is typical of functions of the form $f(x) = \log_a x$, $a > 1$. They have one x-intercept and one vertical asymptote. Notice how slowly the graph rises for $x > 1$.

Library of Parent Functions: Logarithmic Function

The *parent logarithmic function* $f(x) = \log_a x$, $a > 0, a \neq 1$ is the inverse function of the exponential function. Its domain is the set of positive real numbers and its range is the set of all real numbers. This is the opposite of the exponential function. Moreover, the logarithmic function has the y-axis as a vertical asymptote, whereas the exponential function has the x-axis as a horizontal asymptote. Many real-life phenomena with slow rates of growth can be modeled by logarithmic functions. The basic characteristics of the logarithmic function are summarized below and on the inside cover of this text.

Graph of $f(x) = \log_a x$, $a > 1$

Domain: $(0, \infty)$

Range: $(-\infty, \infty)$

Intercept: $(1, 0)$

Increasing on $(0, \infty)$

y-axis is a vertical asymptote ($\log_a x \to -\infty$ as $x \to 0^+$)

Continuous

Reflection of graph of $f(x) = a^x$ in the line $y = x$

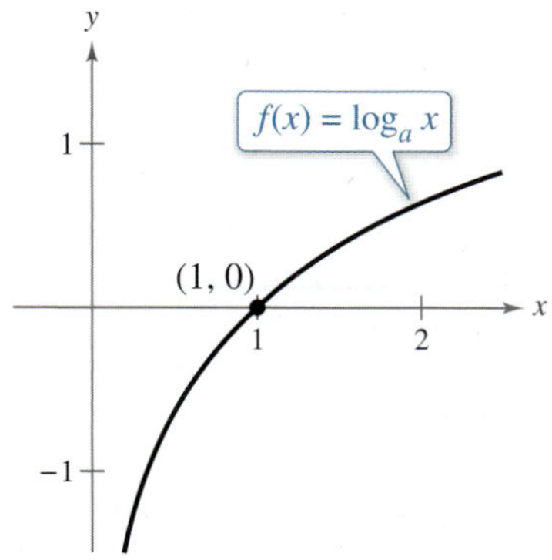

Explore the Concept

Use a graphing utility to graph $y = \log_{10} x$ and $y = 8$ in the same viewing window. Find a viewing window that shows the point of intersection. What is the point of intersection? Use the point of intersection to complete the equation $\log_{10} \square = 8$.

EXAMPLE 6 Library of Parent Functions $f(x) = \log_a x$

See LarsonPrecalculus.com for an interactive version of this type of example.

Each of the following functions is a transformation of the graph of $f(x) = \log_{10} x$.

a. Because $g(x) = \log_{10}(x - 1) = f(x - 1)$, the graph of g can be obtained by shifting the graph of f one unit to the *right*, as shown in Figure 4.14.

b. Because $h(x) = 2 + \log_{10} x = 2 + f(x)$, the graph of h can be obtained by shifting the graph of f two units *upward*, as shown in Figure 4.15.

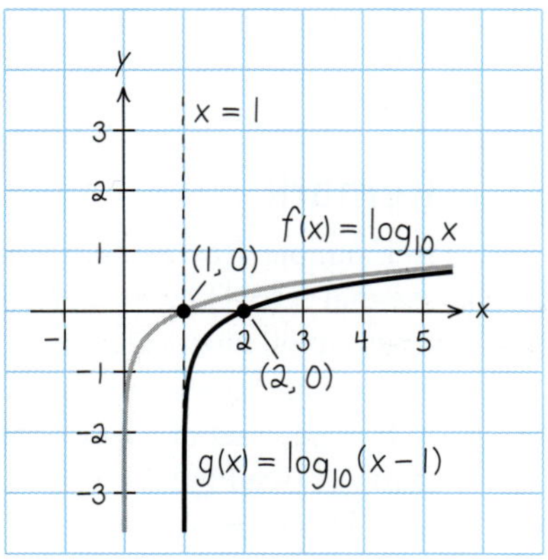

Figure 4.14

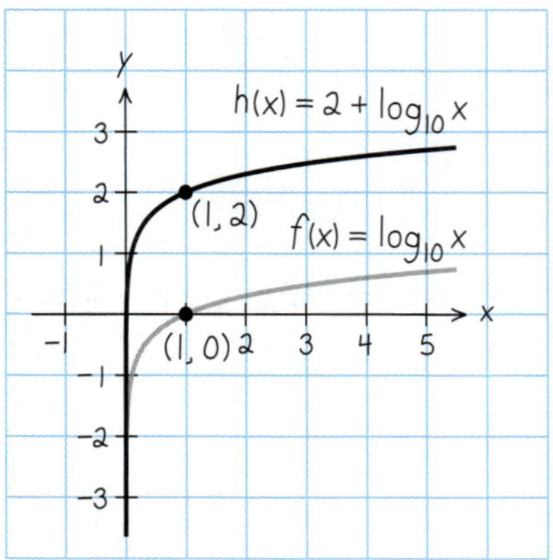

Figure 4.15

Notice that the transformation in Figure 4.15 keeps the y-axis as a vertical asymptote, but the transformation in Figure 4.14 yields the new vertical asymptote $x = 1$.

✓ **Checkpoint** Audio-video solution in English & Spanish at LarsonPrecalculus.com.

Use the graph of $f(x) = \log_{10} x$ to sketch the graph of each function by hand.

a. $g(x) = -1 + \log_{10} x$ **b.** $h(x) = \log_{10}(x + 3)$

The Natural Logarithmic Function

By looking back at the graph of the natural exponential function introduced in Section 4.1, you will see that $f(x) = e^x$ is one-to-one and so has an inverse function. This inverse function is called the **natural logarithmic function** and is denoted by the special symbol $\ln x$, read as "the natural log of x" or "el en of x."

The Natural Logarithmic Function

For $x > 0$,

$$y = \ln x \quad \text{if and only if} \quad x = e^y.$$

The function given by

$$f(x) = \log_e x = \ln x$$

is called the **natural logarithmic function.**

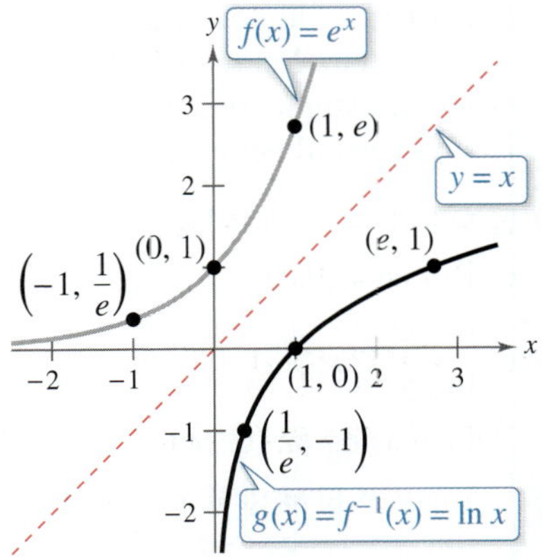

Reflection of graph of $f(x) = e^x$ in the line $y = x$

Figure 4.16

The equations $y = \ln x$ and $x = e^y$ are equivalent. Note that the natural logarithm $\ln x$ is written without a base. The base is understood to be e.

Because the functions $f(x) = e^x$ and $g(x) = \ln x$ are inverse functions of each other, their graphs are reflections of each other in the line $y = x$. This reflective property is illustrated in Figure 4.16.

EXAMPLE 7 Evaluating the Natural Logarithmic Function

Use a calculator to evaluate the function $f(x) = \ln x$ at each value of x.

a. $x = 2$

b. $x = 0.3$

c. $x = -1$

d. $x = 1 + \sqrt{2}$

Solution

	Function Value	*Graphing Calculator Keystrokes*	*Display*
a.	$f(2) = \ln 2$	LN 2 ENTER	0.6931472
b.	$f(0.3) = \ln 0.3$	LN .3 ENTER	−1.2039728
c.	$f(-1) = \ln(-1)$	LN (−) 1 ENTER	ERROR
d.	$f(1 + \sqrt{2}) = \ln(1 + \sqrt{2})$	LN (1 + √ 2) ENTER	0.8813736

Checkpoint *Audio-video solution in English & Spanish at LarsonPrecalculus.com.*

Use a calculator to evaluate the function $f(x) = \ln x$ at each value of x.

a. $x = 0.01$ **b.** $x = 4$ **c.** $x = \sqrt{3} + 2$ **d.** $x = \sqrt{3} - 2$

Technology Tip

On most calculators, the natural logarithm is denoted by LN, as illustrated in Example 7.

Remark

In Example 7(c), be sure you see that $\ln(-1)$ gives an error message on most calculators. This occurs because the domain of $\ln x$ is the set of *positive* real numbers (see Figure 4.16). So, $\ln(-1)$ is undefined.

The four properties of logarithms listed on page 337 are also valid for natural logarithms.

Properties of Natural Logarithms

1. $\ln 1 = 0$ because $e^0 = 1$.
2. $\ln e = 1$ because $e^1 = e$.
3. $\ln e^x = x$ and $e^{\ln x} = x$. Inverse Properties
4. If $\ln x = \ln y$, then $x = y$. One-to-One Property

EXAMPLE 8 Using Properties of Natural Logarithms

Use the properties of natural logarithms to rewrite each expression.

a. $\ln \frac{1}{e}$ **b.** $e^{\ln 5}$ **c.** $4 \ln 1$ **d.** $2 \ln e$

Solution

a. $\ln \frac{1}{e} = \ln e^{-1} = -1$ Inverse Property

b. $e^{\ln 5} = 5$ Inverse Property

c. $4 \ln 1 = 4(0) = 0$ Property 1

d. $2 \ln e = 2(1) = 2$ Property 2

✓ Checkpoint Audio-video solution in English & Spanish at LarsonPrecalculus.com.

Use the properties of natural logarithms to simplify each expression.

a. $\ln e^{1/3}$ **b.** $5 \ln 1$ **c.** $\frac{3}{4} \ln e$ **d.** $e^{\ln 7}$

EXAMPLE 9 Finding the Domains of Logarithmic Functions

Find the domain of each function.

a. $f(x) = \ln(x - 2)$ **b.** $g(x) = \ln(2 - x)$ **c.** $h(x) = \ln x^2$

Algebraic Solution

a. Because $\ln(x - 2)$ is defined only when

$$x - 2 > 0$$

it follows that the domain of f is $(2, \infty)$.

b. Because $\ln(2 - x)$ is defined only when

$$2 - x > 0$$

it follows that the domain of g is $(-\infty, 2)$.

c. Because $\ln x^2$ is defined only when

$$x^2 > 0$$

it follows that the domain of h is all real numbers except $x = 0$.

Graphical Solution

a.

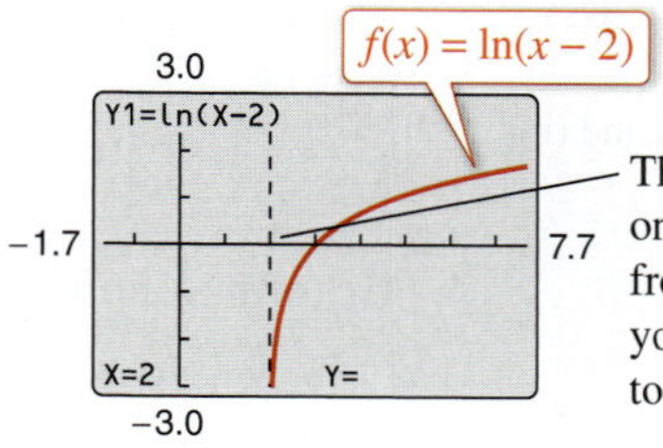

The x-coordinates of points on the graph appear to extend from the right of 2 to ∞. So, you can estimate the domain to be $(2, \infty)$.

b.

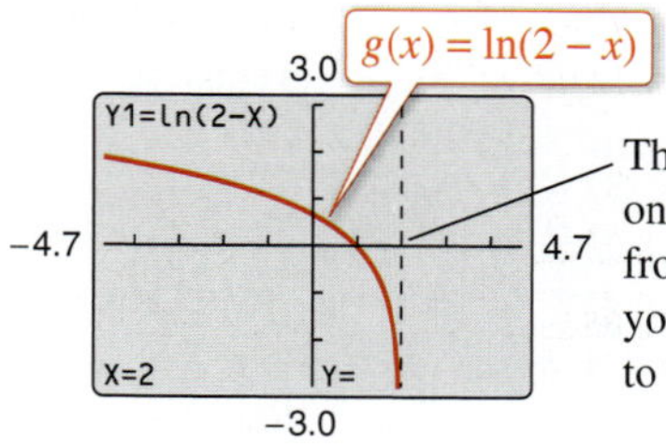

The x-coordinates of points on the graph appear to extend from $-\infty$ to the left of 2. So, you can estimate the domain to be $(-\infty, 2)$.

c.

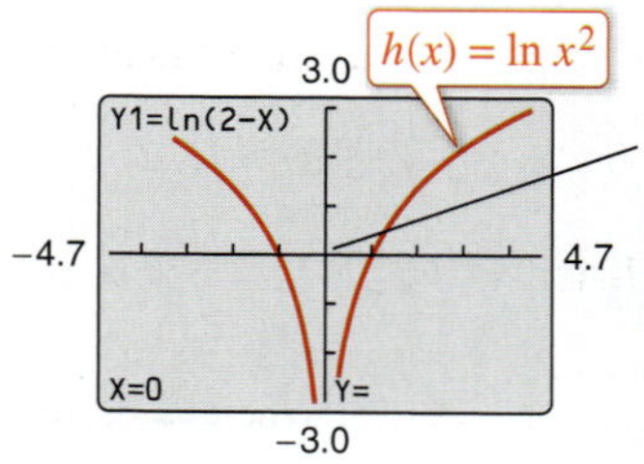

The x-coordinates of points on the graph appear to include all real numbers except 0. So, you can estimate the domain to be all real numbers except $x = 0$.

✓ Checkpoint Audio-video solution in English & Spanish at LarsonPrecalculus.com.

Find the domain of $f(x) = \ln(x + 3)$.

In Example 9, suppose you had been asked to analyze the function $h(x) = \ln|x - 2|$. How would the domain of this function compare with the domains of the functions given in parts (a) and (b) of the example?

Application

EXAMPLE 10 Psychology

Students participating in a psychology experiment attended several lectures on a subject and were given an exam. Every month for a year after the exam, the students were retested to see how much of the material they remembered. The average scores for the group are given by the *human memory model*

$$f(t) = 75 - 6 \ln(t + 1), \quad 0 \le t \le 12$$

where t is the time in months. The graph of f is shown in Figure 4.17.

a. What was the average score on the original exam ($t = 0$)?

b. What was the average score at the end of $t = 2$ months?

c. What was the average score at the end of $t = 6$ months?

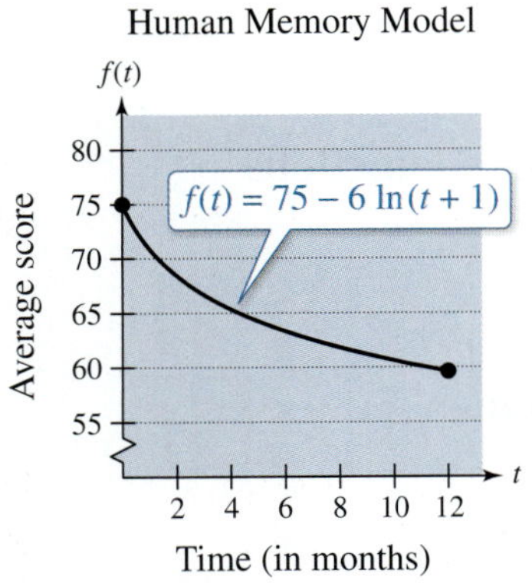

Figure 4.17

Psychologist

Algebraic Solution

a. The original average score was

$$\begin{aligned} f(0) &= 75 - 6 \ln(0 + 1) \\ &= 75 - 6 \ln 1 \\ &= 75 - 6(0) \\ &= 75. \end{aligned}$$

b. After 2 months, the average score was

$$\begin{aligned} f(2) &= 75 - 6 \ln(2 + 1) \\ &= 75 - 6 \ln 3 \\ &\approx 75 - 6(1.0986) \\ &\approx 68.41. \end{aligned}$$

c. After 6 months, the average score was

$$\begin{aligned} f(6) &= 75 - 6 \ln(6 + 1) \\ &= 75 - 6 \ln 7 \\ &\approx 75 - 6(1.9459) \\ &\approx 63.32. \end{aligned}$$

Graphical Solution

a.

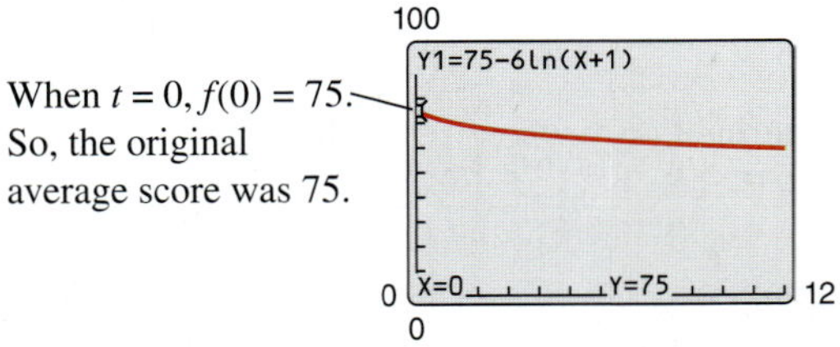

b.

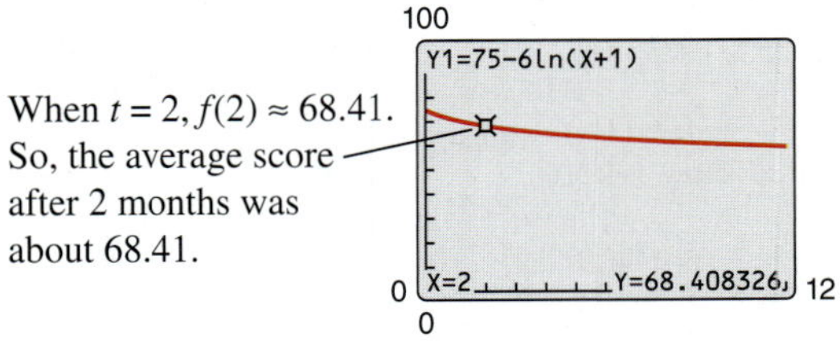

c.

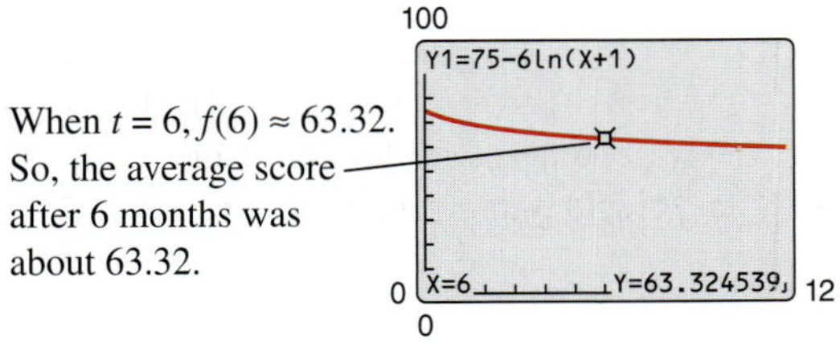

✓ Checkpoint Audio-video solution in English & Spanish at LarsonPrecalculus.com.

In Example 10, find the average score at the end of (a) $t = 1$ month, (b) $t = 9$ months, and (c) $t = 12$ months.

4.2 Exercises

See *CalcChat.com* for tutorial help and worked-out solutions to odd-numbered exercises. For instructions on how to use a graphing utility, see Appendix A.

Vocabulary and Concept Check

In Exercises 1–4, fill in the blank(s).

1. The inverse function of the exponential function $f(x) = a^x$ is called the _______ with base a.

2. The base of the _______ logarithmic function is 10, and the base of the _______ logarithmic function is e.

3. The inverse properties of logarithms are $\log_a a^x = x$ and _______ .

4. If $x = e^y$, then $y =$ _______ .

5. What exponential equation is equivalent to the logarithmic equation $\log_a b = c$?

6. For what value(s) of x is $\ln x = \ln 7$?

Procedures and Problem Solving

Rewriting Logarithmic Equations **In Exercises 7–14, write the logarithmic equation in exponential form. For example, the exponential form of $\log_5 25 = 2$ is $5^2 = 25$.**

7. $\log_4 64 = 3$ **8.** $\log_3 81 = 4$

9. $\log_7 \frac{1}{49} = -2$ **10.** $\log_{10} \frac{1}{1000} = -3$

11. $\log_{32} 4 = \frac{2}{5}$ **12.** $\log_{16} 8 = \frac{3}{4}$

13. $\log_2 \sqrt{2} = \frac{1}{2}$ **14.** $\log_5 \sqrt[3]{25} = \frac{2}{3}$

Rewriting Exponential Equations **In Exercises 15–22, write the exponential equation in logarithmic form. For example, the logarithmic form of $2^3 = 8$ is $\log_2 8 = 3$.**

15. $5^3 = 125$ **16.** $8^2 = 64$

17. $81^{1/4} = 3$ **18.** $9^{3/2} = 27$

19. $6^{-2} = \frac{1}{36}$ **20.** $10^{-3} = 0.001$

21. $g^a = 4$ **22.** $n^t = 10$

Evaluating Logarithms **In Exercises 23–26, use the definition of logarithmic function to evaluate the function at the indicated value of x without using a calculator.**

	Function	*Value*
23.	$f(x) = \log_2 x$	$x = 16$
24.	$f(x) = \log_{16} x$	$x = \frac{1}{4}$
25.	$g(x) = \log_{10} x$	$x = \frac{1}{1000}$
26.	$g(x) = \log_{10} x$	$x = 100{,}000$

Evaluating Common Logarithms on a Calculator **In Exercises 27–30, use a calculator to evaluate the function at the indicated value of x. Round your result to three decimal places.**

	Function	*Value*
27.	$f(x) = \log_{10} x$	$x = 345$
28.	$f(x) = \log_{10} x$	$x = \frac{4}{5}$
29.	$h(x) = 6 \log_{10} x$	$x = 14.8$
30.	$h(x) = 1.9 \log_{10} x$	$x = 4.3$

Using Properties of Logarithms **In Exercises 31–36, solve the equation for x.**

31. $\log_7 x = \log_7 9$ **32.** $\log_5 5 = \log_5 x$

33. $\log_4 4^2 = x$ **34.** $\log_3 3^{-5} = x$

35. $\log_8 x = \log_8 10^{-1}$ **36.** $\log_4 4^3 = \log_4 x$

Using Properties of Logarithms **In Exercises 37–40, use the properties of logarithms to simplify the expression.**

37. $\log_4 4^{3x}$ **38.** $6^{\log_6 36x}$

39. $3 \log_2 \frac{1}{2}$ **40.** $\frac{1}{4} \log_4 16$

Graphs of Exponential and Logarithmic Functions **In Exercises 41–44, sketch the graph of f. Then use the graph of f to sketch the graph of g.**

41. $f(x) = 6^x$, $g(x) = \log_6 x$ **42.** $f(x) = 5^x$, $g(x) = \log_5 x$

43. $f(x) = 15^x$, $g(x) = \log_{15} x$ **44.** $f(x) = 4^x$, $g(x) = \log_4 x$

Sketching the Graph of a Logarithmic Function **In Exercises 45–50, find the domain, vertical asymptote, and x-intercept of the logarithmic function, and sketch its graph by hand.**

45. $y = \log_{10}(x + 2)$ **46.** $y = \log_{10}(x - 1)$

47. $y = 1 + \log_{10} x$

48. $y = 2 - \log_{10} x$

49. $y = 1 + \log_{10}(x - 2)$

50. $y = 2 + \log_{10}(x + 1)$

Library of Parent Functions **In Exercises 51–54, use the graph of $y = \log_3 x$ to match the function with its graph. [The graphs are labeled (a), (b), (c), and (d).]**

(a)

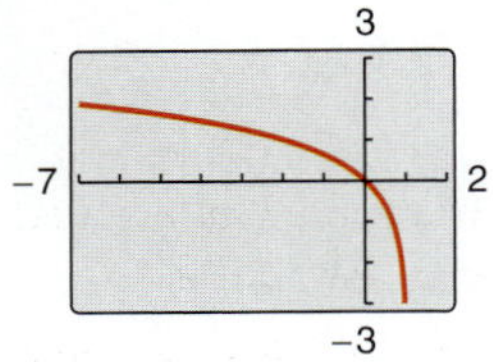

(b)

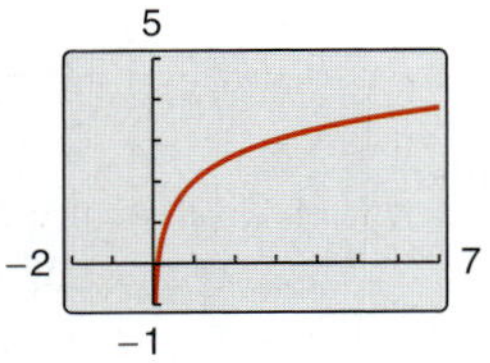

(c)

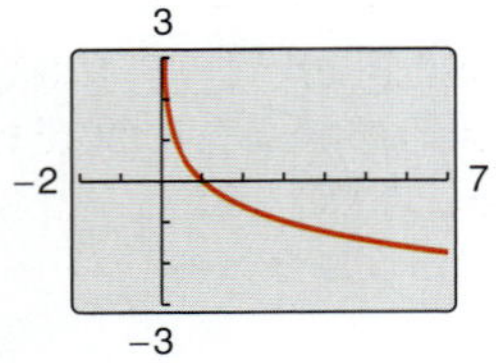

(d)

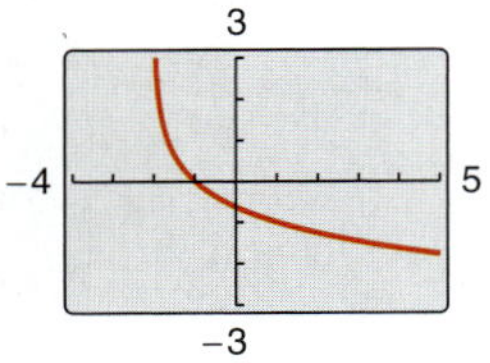

51. $f(x) = \log_3 x + 2$

52. $f(x) = -\log_3 x$

53. $f(x) = -\log_3(x + 2)$

54. $f(x) = \log_3(1 - x)$

Library of Parent Functions **In Exercises 55–60, describe the transformation of the graph of f that yields the graph of g.**

55. $f(x) = \log_{10} x, \quad g(x) = \log_{10}(-x)$

56. $f(x) = \log_{10} x, \quad g(x) = \log_{10}(x + 7)$

57. $f(x) = \log_2 x, \quad g(x) = 4 - \log_2 x$

58. $f(x) = \log_2 x, \quad g(x) = -3 + \log_2 x$

59. $f(x) = \log_8 x, \quad g(x) = -2 + \log_8(x - 3)$

60. $f(x) = \log_8 x, \quad g(x) = 4 + \log_8(x - 1)$

Rewriting Logarithmic Equations **In Exercises 61–68, write the logarithmic equation in exponential form. For example, the exponential form of $\ln 5 = 1.6094 \ldots$ is $e^{1.6094\ldots} = 5$.**

61. $\ln 6 = 1.7917 \ldots$ **62.** $\ln 4 = 1.3862 \ldots$

63. $\ln e = 1$ **64.** $\ln e^3 = 3$

65. $\ln \sqrt{e} = \dfrac{1}{2}$ **66.** $\ln \dfrac{1}{e^2} = -2$

67. $\ln 9 = 2.1972 \ldots$ **68.** $\ln \sqrt[3]{e} = \frac{1}{3}$

Rewriting Exponential Equations **In Exercises 69–76, write the exponential equation in logarithmic form. For example, the logarithmic form of $e^2 = 7.3890 \ldots$ is $\ln 7.3890 \ldots = 2$.**

69. $e^3 = 20.0855 \ldots$ **70.** $e^0 = 1$

71. $e^{1.3} = 3.6692 \ldots$ **72.** $e^{2.5} = 12.1824 \ldots$

73. $\sqrt[3]{e} = 1.3956 \ldots$ **74.** $\dfrac{1}{e^4} = 0.0183 \ldots$

75. $\sqrt{e^3} = 4.4816 \ldots$ **76.** $e^{3/4} = 2.1170 \ldots$

Evaluating the Natural Logarithmic Function **In Exercises 77–80, use a calculator to evaluate the function at the indicated value of x. Round your result to three decimal places.**

Function	*Value*
77. $f(x) = \ln x$	$x = 11$
78. $f(x) = \ln x$	$x = 18.31$
79. $f(x) = -\ln x$	$x = \frac{1}{2}$
80. $f(x) = 3 \ln x$	$x = \sqrt{0.65}$

Using Properties of Natural Logarithms **In Exercises 81–88, use the properties of natural logarithms to rewrite the expression.**

81. $\ln e^2$ **82.** $-\ln e$

83. $e^{\ln 1.8}$ **84.** $7 \ln e^0$

85. $e \ln 1$ **86.** $e^{\ln 22}$

87. $\ln e^{\ln e}$ **88.** $\ln \dfrac{1}{e^4}$

Library of Parent Functions **In Exercises 89–92, find the domain, vertical asymptote, and x-intercept of the logarithmic function, and sketch its graph by hand. Verify using a graphing utility.**

89. $f(x) = \ln(x - 1)$ **90.** $h(x) = \ln(x + 1)$

91. $g(x) = \ln(-x)$ **92.** $f(x) = \ln(3 - x)$

Library of Parent Functions **In Exercises 93–98, use the graph of $f(x) = \ln x$ to describe the transformation that yields the graph of g.**

93. $g(x) = \ln(x + 8)$ **94.** $g(x) = \ln(x - 4)$

95. $g(x) = \ln x - 5$ **96.** $g(x) = \ln x + 4$

97. $g(x) = \ln(x - 1) + 2$ **98.** $g(x) = \ln(x + 2) - 5$

Analyzing Graphs of Functions **In Exercises 99–108, (a) use a graphing utility to graph the function, (b) find the domain, (c) use the graph to find the open intervals on which the function is increasing and decreasing, and (d) approximate any relative maximum or minimum values of the function. Round your results to three decimal places.**

99. $f(x) = \dfrac{x}{2} - \ln \dfrac{x}{4}$ **100.** $g(x) = 6x \ln x$

101. $h(x) = \dfrac{14 \ln x}{x}$ **102.** $f(x) = \dfrac{x}{\ln x}$

103. $f(x) = \ln \dfrac{x + 2}{x - 1}$ **104.** $f(x) = \ln \dfrac{2x}{x + 2}$

105. $f(x) = \ln \dfrac{x^2}{10}$ **106.** $f(x) = \ln \dfrac{x}{x^2 + 1}$

107. $f(x) = \sqrt{\ln x}$ **108.** $f(x) = (\ln x)^2$

109. Psychology Students in a mathematics class were given an exam and then tested monthly with an equivalent exam. The average scores for the class are given by the human memory model

$$f(t) = 80 - 17\log_{10}(t+1), \quad 0 \le t \le 12$$

where t is the time in months.

(a) What was the average score on the original exam $(t = 0)$?

(b) What was the average score after 2 months?

(c) What was the average score after 11 months?

Verify your answers in parts (a), (b), and (c) using a graphing utility.

110. MODELING DATA

The table shows the temperatures T (in degrees Fahrenheit) at which water boils at selected pressures p (in pounds per square inch). (Source: Standard Handbook of Mechanical Engineers)

Spreadsheet at LarsonPrecalculus.com

Pressure, p	Temperature, T
5	162.24°
10	193.21°
14.696 (1 atm)	212.00°
20	227.96°
30	250.33°
40	267.25°
60	292.71°
80	312.03°
100	327.81°

A model that approximates the data is

$$T = 87.97 + 34.96 \ln p + 7.91\sqrt{p}.$$

(a) Use a graphing utility to plot the data and graph the model in the same viewing window. How well does the model fit the data?

(b) Use the graph to estimate the pressure at which the boiling point of water is 300°F.

(c) Calculate T when the pressure is 74 pounds per square inch. Verify your answer graphically.

111. Finance A principal P, invested at $3\frac{1}{2}\%$ and compounded continuously, increases to an amount K times the original principal after t years, where $T = (\ln K)/0.035$.

(a) Complete the table and interpret your results.

K	1	2	4	6	8	10	12
t							

(b) Use a graphing utility to graph the function.

112. Audiology The relationship between the number of decibels β and the intensity of a sound I in watts per square meter is given by

$$\beta = 10\log_{10}\left(\frac{I}{10^{-12}}\right).$$

(a) Determine the number of decibels of a sound with an intensity of 1 watt per square meter.

(b) Determine the number of decibels of a sound with an intensity of 10^{-2} watt per square meter.

(c) The intensity of the sound in part (a) is 100 times as great as that in part (b). Is the number of decibels 100 times as great? Explain.

113. Real Estate The model

$$t = 16.625 \ln \frac{x}{x - 750}, \quad x > 750$$

approximates the length of a home mortgage of \$150,000 at 6% in terms of the monthly payment. In the model, t is the length of the mortgage in years and x is the monthly payment in dollars.

(a) Use the model to approximate the lengths of a \$150,000 mortgage at 6% when the monthly payment is \$897.72 and when the monthly payment is \$1659.24.

(b) Approximate the total amounts paid over the term of the mortgage with a monthly payment of \$897.72 and with a monthly payment of \$1659.24. What amount of the total is interest costs for each payment?

114. *Why you should learn it* (p. 336) The rate of ventilation required in a public school classroom depends on the volume of air space per child. The model

$$y = 80.4 - 11\ln x, \quad 100 \le x \le 1500$$

approximates the minimum required rate of ventilation y (in cubic feet per minute per child) in a classroom with x cubic feet of air space per child.

(a) Use a graphing utility to graph the function and approximate the required rate of ventilation in a room with 300 cubic feet of air space per child.

(b) A classroom of 30 students has an air conditioning system that moves 450 cubic feet of air per minute. Determine the rate of ventilation per child.

(c) Use the graph in part (a) to estimate the minimum required air space per child for the classroom in part (b).

(d) The classroom in part (b) has 960 square feet of floor space and a ceiling that is 12 feet high. Is the rate of ventilation for this classroom adequate? Explain.

Conclusions

True or False? **In Exercises 115 and 116, determine whether the statement is true or false. Justify your answer.**

115. You can determine the graph of $f(x) = \log_6 x$ by graphing $g(x) = 6^x$ and reflecting it about the x-axis.

116. The graph of $f(x) = \log_3 x$ contains the point $(27, 3)$.

Think About It **In Exercises 117–120, find the value of the base b so that the graph of $f(x) = \log_b x$ contains the given point.**

117. $(32, 5)$ **118.** $(81, 4)$

119. $\left(\frac{1}{81}, 2\right)$ **120.** $\left(\frac{1}{64}, 3\right)$

Library of Parent Functions **In Exercises 121 and 122, determine which equation(s) may be represented by the graph shown. (There may be more than one correct answer.)**

121.

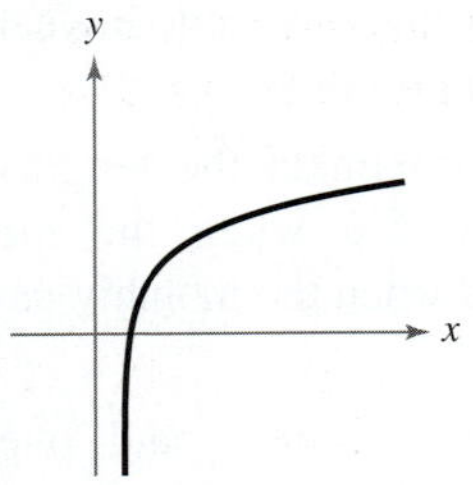

(a) $y = \log_2(x + 1) + 2$

(b) $y = \log_2(x - 1) + 2$

(c) $y = 2 - \log_2(x - 1)$

(d) $y = \log_2(x + 2) + 1$

122.

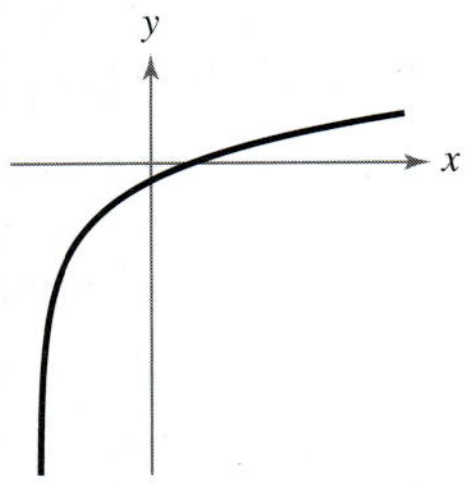

(a) $y = \ln(x - 1) + 2$

(b) $y = \ln(x + 2) - 1$

(c) $y = 2 - \ln(x - 1)$

(d) $y = \ln(x - 2) + 1$

123. Writing Explain why $\log_a x$ is defined only for $0 < a < 1$ and $a > 1$.

124. Exploration Let $f(x) = \ln x$ and $g(x) = x^{1/n}$.

(a) Use a graphing utility to graph g (for $n = 2$) and f in the same viewing window.

(b) Determine which function is increasing at a greater rate as x approaches infinity.

(c) Repeat parts (a) and (b) for $n = 3, 4,$ and 5. What do you notice?

125. Exploration

(a) Use a graphing utility to compare the graph of the function $y = \ln x$ with the graph of each function.

$y_1 = x - 1$, $y_2 = (x - 1) - \frac{1}{2}(x - 1)^2$,

$y_3 = (x - 1) - \frac{1}{2}(x - 1)^2 + \frac{1}{3}(x - 1)^3$

(b) Identify the pattern of successive polynomials given in part (a). Extend the pattern one more term and compare the graph of the resulting polynomial function with the graph of $y = \ln x$. What do you think the pattern implies?

126. HOW DO YOU SEE IT? The figure shows the graphs of $f(x) = 3^x$ and $g(x) = \log_3 x$. [The graphs are labeled m and n.]

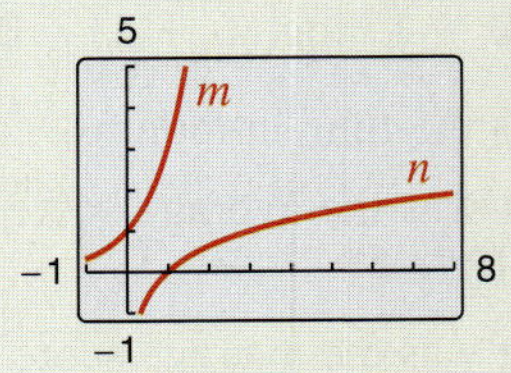

(a) Match each function with its graph.

(b) Given that $f(a) = b$, what is $g(b)$? Explain.

127. Exploration

(a) Use a graphing utility to complete the table for the function $f(x) = (\ln x)/x$.

x	1	5	10	10^2	10^4	10^6
$f(x)$						

(b) Use the table in part (a) to determine what value $f(x)$ approaches as x increases without bound. Use the graphing utility to confirm your result.

128. Writing Use a graphing utility to determine how many months it would take for the average score in Example 10 to decrease to 60. Explain your method of solving the problem. Describe another way that you can use the graphing utility to determine the answer. Also, based on the shape of the graph, does the rate at which a student forgets information *increase* or *decrease* with time? Explain.

Cumulative Mixed Review

Factoring a Polynomial **In Exercises 129–136, factor the polynomial.**

129. $x^2 + 2x - 3$ **130.** $x^2 + 4x - 5$

131. $12x^2 + 5x - 3$ **132.** $16x^2 - 54x - 7$

133. $16x^2 - 25$ **134.** $36x^2 - 49$

135. $2x^3 + x^2 - 45x$ **136.** $3x^3 - 5x^2 - 12x$

Evaluating an Arithmetic Combination of Functions **In Exercises 137 and 138, evaluate the function for $f(x) = 3x + 2$ and $g(x) = x^3 - 1$.**

137. $(f + g)(2)$ **138.** $(f - g)(-1)$

Using Graphs **In Exercises 139–142, solve the equation graphically.**

139. $5x - 7 = 7 + 5x$ **140.** $-2x + 3 = 8x$

141. $\sqrt{3x - 2} = 9$ **142.** $\sqrt{x - 11} = x + 2$

4.3 Properties of Logarithms

Change of Base

Most calculators have only two types of log keys, one for common logarithms (base 10) and one for natural logarithms (base e). Although common logs and natural logs are the most frequently used, you may occasionally need to evaluate logarithms to other bases. To do this, you can use the following **change-of-base formula.**

Change-of-Base Formula

Let a, b, and x be positive real numbers such that $a \neq 1$ and $b \neq 1$. Then $\log_a x$ can be converted to a different base using any of the following formulas.

Base b	*Base 10*	*Base e*
$\log_a x = \dfrac{\log_b x}{\log_b a}$	$\log_a x = \dfrac{\log_{10} x}{\log_{10} a}$	$\log_a x = \dfrac{\ln x}{\ln a}$

One way to look at the change-of-base formula is that logarithms to base a are simply *constant multiples* of logarithms to base b. The constant multiplier is $1/(\log_b a)$.

EXAMPLE 1 Changing Bases Using Common Logarithms

a. $\log_4 25 = \dfrac{\log_{10} 25}{\log_{10} 4}$ $\qquad \log_a x = \dfrac{\log_{10} x}{\log_{10} a}$

$\approx \dfrac{1.39794}{0.60206}$ $\qquad$ Use a calculator.

≈ 2.32 $\qquad$ Simplify.

b. $\log_3 17 = \dfrac{\log_{10} 17}{\log_{10} 3} \approx \dfrac{1.23045}{0.47712} \approx 2.58$

✓ **Checkpoint** *Audio-video solution in English & Spanish at LarsonPrecalculus.com.*

Evaluate $\log_2 12$ using the change-of-base formula and common logarithms.

EXAMPLE 2 Changing Bases Using Natural Logarithms

a. $\log_4 25 = \dfrac{\ln 25}{\ln 4}$ $\qquad \log_a x = \dfrac{\ln x}{\ln a}$

$\approx \dfrac{3.21888}{1.38629}$ $\qquad$ Use a calculator.

≈ 2.32 $\qquad$ Simplify.

b. $\log_3 17 = \dfrac{\ln 17}{\ln 3} \approx \dfrac{2.83321}{1.09861} \approx 2.58$

✓ **Checkpoint** *Audio-video solution in English & Spanish at LarsonPrecalculus.com.*

Evaluate $\log_2 12$ using the change-of-base formula and natural logarithms.

Notice in Examples 1 and 2 that the result is the same whether common logarithms or natural logarithms are used in the change-of-base formula.

What you should learn

- Rewrite logarithms with different bases.
- Use properties of logarithms to evaluate or rewrite logarithmic expressions.
- Use properties of logarithms to expand or condense logarithmic expressions.
- Use logarithmic functions to model and solve real-life problems.

Why you should learn it

Logarithmic functions can be used to model and solve real-life problems, such as the model for the number of decibels of a sound in Exercise 107 on page 352.

Properties of Logarithms

You know from the previous section that the logarithmic function with base a is the *inverse function* of the exponential function with base a. So, it makes sense that the properties of exponents (see Section 4.1) should have corresponding properties involving logarithms. For instance, the exponential property

$$a^0 = 1$$

has the corresponding logarithmic property

$$\log_a 1 = 0.$$

Properties of Logarithms (See the proof on page 395.)

Let a be a positive real number such that $a \neq 1$, and let n be a real number. If u and v are positive real numbers, then the following properties are true.

	Logarithm with Base a	*Natural Logarithm*
1. Product Property:	$\log_a(uv) = \log_a u + \log_a v$	$\ln(uv) = \ln u + \ln v$
2. Quotient Property:	$\log_a \frac{u}{v} = \log_a u - \log_a v$	$\ln \frac{u}{v} = \ln u - \ln v$
3. Power Property:	$\log_a u^n = n \log_a u$	$\ln u^n = n \ln u$

Remark

There is no general property that can be used to rewrite $\log_a(u \pm v)$. Specifically, $\log_a(x + y)$ is *not* equal to $\log_a x + \log_a y$.

EXAMPLE 3 Using Properties of Logarithms

Write each logarithm in terms of ln 2 and ln 3.

a. $\ln 6$ **b.** $\ln \frac{2}{27}$

Solution

a. $\ln 6 = \ln(2 \cdot 3)$ Rewrite 6 as $2 \cdot 3$.

$= \ln 2 + \ln 3$ Product Property

b. $\ln \frac{2}{27} = \ln 2 - \ln 27$ Quotient Property

$= \ln 2 - \ln 3^3$ Rewrite 27 as 3^3.

$= \ln 2 - 3 \ln 3$ Power Property

✓ **Checkpoint** *Audio-video solution in English & Spanish at LarsonPrecalculus.com.*

Write each logarithm in terms of $\log_{10} 3$ and $\log_{10} 5$.

a. $\log_{10} 75$ **b.** $\log_{10} \frac{9}{125}$

EXAMPLE 4 Using Properties of Logarithms

Use the properties of logarithms to verify that $-\log_{10} \frac{1}{100} = \log_{10} 100$.

Solution

$-\log_{10} \frac{1}{100} = -\log_{10}(100^{-1})$ Rewrite $\frac{1}{100}$ as 100^{-1}.

$= -(-1) \log_{10} 100$ Power Property

$= \log_{10} 100$ Simplify.

✓ **Checkpoint** *Audio-video solution in English & Spanish at LarsonPrecalculus.com.*

Use the properties of logarithms to verify that $\ln e^5 = 5$.

Rewriting Logarithmic Expressions

The properties of logarithms are useful for rewriting logarithmic expressions in forms that simplify the operations of algebra. This is true because they convert complicated products, quotients, and exponential forms into simpler sums, differences, and products, respectively.

EXAMPLE 5 Expanding Logarithmic Expressions

Use the properties of logarithms to expand each expression.

a. $\log_4 5x^3y$ **b.** $\ln \dfrac{\sqrt{3x-5}}{7}$

Solution

a. $\log_4 5x^3y = \log_4 5 + \log_4 x^3 + \log_4 y$ Product Property

$= \log_4 5 + 3\log_4 x + \log_4 y$ Power Property

b. $\ln \dfrac{\sqrt{3x-5}}{7} = \ln \dfrac{(3x-5)^{1/2}}{7}$ Rewrite radical using rational exponent.

$= \ln(3x-5)^{1/2} - \ln 7$ Quotient Property

$= \dfrac{1}{2}\ln(3x-5) - \ln 7$ Power Property

✓ **Checkpoint** Audio-video solution in English & Spanish at LarsonPrecalculus.com.

Use the properties of logarithms to expand the expression $\log_3 \dfrac{4x^2}{\sqrt{y}}$.

In Example 5, the properties of logarithms were used to *expand* logarithmic expressions. In Example 6, this procedure is reversed and the properties of logarithms are used to *condense* logarithmic expressions.

EXAMPLE 6 Condensing Logarithmic Expressions

See LarsonPrecalculus.com for an interactive version of this type of example.

Use the properties of logarithms to condense each expression.

a. $\frac{1}{2}\log_{10} x + 3\log_{10}(x+1)$

b. $2\ln(x+2) - \ln x$

c. $\frac{1}{3}[\log_2 x + \log_2(x-4)]$

Solution

a. $\frac{1}{2}\log_{10} x + 3\log_{10}(x+1) = \log_{10} x^{1/2} + \log_{10}(x+1)^3$ Power Property

$= \log_{10}\left[\sqrt{x}(x+1)^3\right]$ Product Property

b. $2\ln(x+2) - \ln x = \ln(x+2)^2 - \ln x$ Power Property

$= \ln \dfrac{(x+2)^2}{x}$ Quotient Property

c. $\frac{1}{3}[\log_2 x + \log_2(x-4)] = \frac{1}{3}\{\log_2[x(x-4)]\}$ Product Property

$= \log_2[x(x-4)]^{1/3}$ Power Property

$= \log_2 \sqrt[3]{x(x-4)}$ Rewrite with a radical.

✓ **Checkpoint** Audio-video solution in English & Spanish at LarsonPrecalculus.com.

Use the properties of logarithms to condense the expression

$2[\log_{10}(x+3) - 2\log_{10}(x-2)]$.

Explore the Concept

Use a graphing utility to graph the functions

$$y = \ln x - \ln(x-3)$$

and

$$y = \ln \frac{x}{x-3}$$

in the same viewing window. Does the graphing utility show the functions with the same domain? Should it? Explain your reasoning.

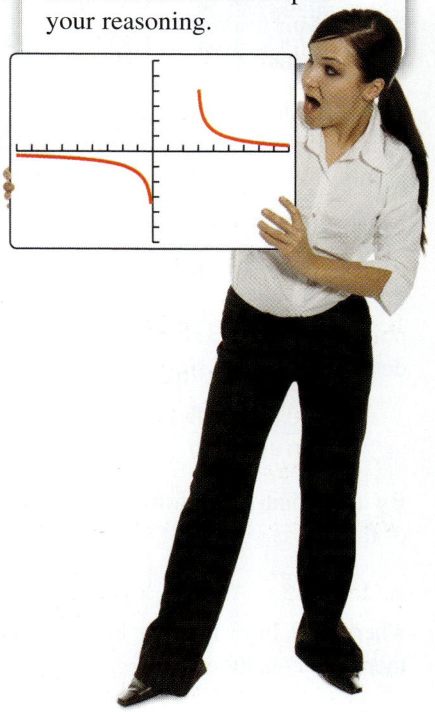

Application

EXAMPLE 7 Finding a Mathematical Model

The table shows the mean distance x from the sun and the period y (the time it takes a planet to orbit the sun) for each of the six planets that are closest to the sun. In the table, the mean distance is given in astronomical units (where the Earth's mean distance is defined as 1.0), and the period is given in years. The points in the table are plotted in Figure 4.18. Find an equation that relates y and x.

DATA

Planet	Mercury	Venus	Earth	Mars	Jupiter	Saturn
Mean distance, x	0.387	0.723	1.000	1.524	5.203	9.537
Period, y	0.241	0.615	1.000	1.881	11.863	29.447

Spreadsheet at LarsonPrecalculus.com

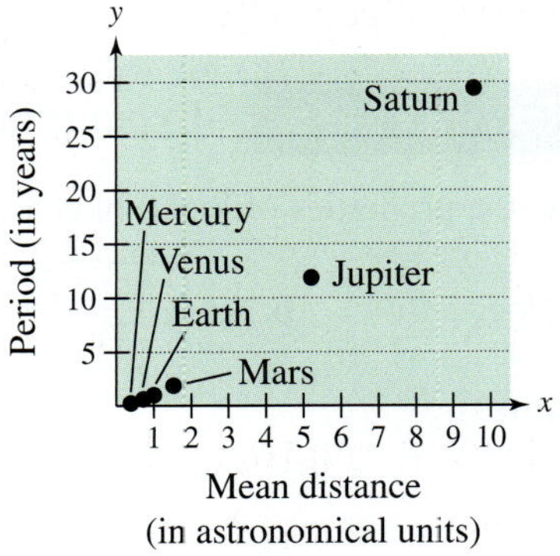

Figure 4.18

Solution

From Figure 4.18, it is not clear how to find an equation that relates y and x. To solve this problem, take the natural log of each of the x- and y-values in the table. This produces the following results.

Planet	Mercury	Venus	Earth	Mars	Jupiter	Saturn
$\ln x = X$	−0.949	−0.324	0.000	0.421	1.649	2.255
$\ln y = Y$	−1.423	−0.486	0.000	0.632	2.473	3.383

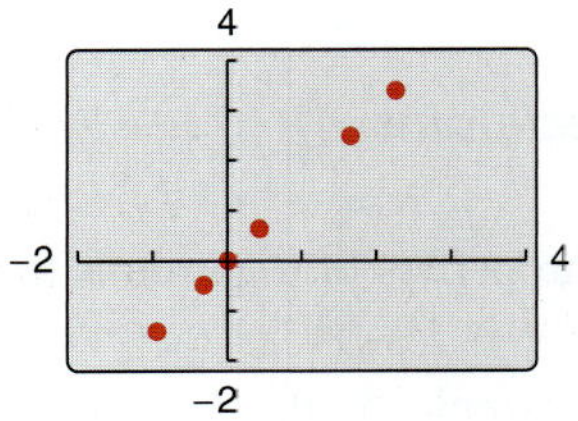

Figure 4.19

Now, by plotting the points in the table, you can see that all six of the points appear to lie in a line, as shown in Figure 4.19. To find an equation of the line through these points, you can use one of the following methods.

Method 1: Algebraic

Choose any two points to determine the slope of the line. Using the two points (0.421, 0.632) and (0, 0), you can determine that the slope of the line is

$$m = \frac{0.632 - 0}{0.421 - 0} \approx 1.5 = \frac{3}{2}.$$

By the point-slope form, the equation of the line is

$$Y = \tfrac{3}{2}X$$

where $Y = \ln y$ and $X = \ln x$. You can therefore conclude that

$$\ln y = \frac{3}{2} \ln x.$$

Method 2: Graphical

Using the *linear regression* feature of a graphing utility, you can find a linear model for the data, as shown in Figure 4.20. You can approximate this model to be $Y = 1.5X$, where $Y = \ln y$ and $X = \ln x$. From the model, you can see that the slope of the line is $\frac{3}{2}$. So, you can conclude that $\ln y = \frac{3}{2} \ln x$.

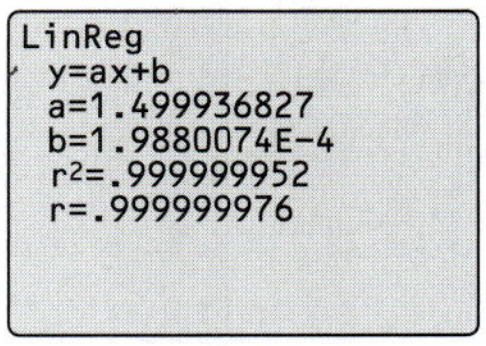

Figure 4.20

✓ Checkpoint *Audio-video solution in English & Spanish at LarsonPrecalculus.com.*

Find a logarithmic equation that relates y and x for the following ordered pairs.

(0.37, 0.51), (1.00, 1.00), (2.72, 1.95), (7.39, 3.79), (20.09, 7.39)

In Example 7, try to convert the final equation to $y = f(x)$ form. You will get a function of the form $y = ax^b$, which is called a *power model.*

4.3 Exercises

See CalcChat.com for tutorial help and worked-out solutions to odd-numbered exercises.
For instructions on how to use a graphing utility, see Appendix A.

Vocabulary and Concept Check

In Exercises 1 and 2, fill in the blank(s).

1. You can evaluate logarithms to any base using the _______ formula.

2. Two properties of logarithms are _______ $= n \log_a u$ and $\ln(uv) =$ _______ .

3. Is $\log_3 24 = \dfrac{\ln 3}{\ln 24}$ or $\log_3 24 = \dfrac{\ln 24}{\ln 3}$ correct?

4. Which property of logarithms can you use to condense the expression $\ln x - \ln 2$?

Procedures and Problem Solving

Changing the Base **In Exercises 5–12, rewrite the logarithm as a ratio of (a) common logarithms and (b) natural logarithms.**

5. $\log_5 x$ **6.** $\log_3 x$

7. $\log_{1/6} x$ **8.** $\log_{1/4} x$

9. $\log_a \frac{3}{10}$ **10.** $\log_a \frac{4}{5}$

11. $\log_{2.6} x$ **12.** $\log_{7.1} x$

Changing the Base **In Exercises 13–20, evaluate the logarithm using the change-of-base formula. Round your result to three decimal places.**

13. $\log_3 7$ **14.** $\log_9 4$

15. $\log_{1/2} 16$ **16.** $\log_{1/8} 64$

17. $\log_6 0.9$ **18.** $\log_4 0.045$

19. $\log_{15} 1460$ **20.** $\log_{20} 175$

Using Properties of Logarithms **In Exercises 21–24, rewrite the expression in terms of ln 4 and ln 5.**

21. $\ln 20$ **22.** $\ln 500$

23. $\ln \frac{25}{4}$ **24.** $\ln \frac{5}{2}$

Using Properties to Evaluate Logarithms **In Exercises 25–28, approximate the logarithm using the properties of logarithms, given the values $\log_b 2 \approx 0.3562$, $\log_b 3 \approx 0.5646$, and $\log_b 5 \approx 0.8271$. Round your result to four decimal places.**

25. $\log_b 8$ **26.** $\log_b 30$

27. $\log_b \frac{16}{25}$ **28.** $\log_b \sqrt{3}$

Graphing a Logarithm with Any Base **In Exercises 29–36, use the change-of-base formula $\log_a x = (\ln x)/(\ln a)$ and a graphing utility to graph the function.**

29. $f(x) = \log_3(x + 1)$ **30.** $f(x) = \log_2(x - 1)$

31. $f(x) = \log_{1/2}(x - 2)$ **32.** $f(x) = \log_{1/3}(x + 2)$

33. $f(x) = \log_{1/4} x^2$ **34.** $f(x) = \log_3 \sqrt{x}$

35. $f(x) = \log_5\left(\dfrac{x}{2}\right)$ **36.** $f(x) = \log_{1/3}\left(\dfrac{x}{3}\right)$

Simplifying a Logarithm **In Exercises 37–44, use the properties of logarithms to rewrite and simplify the logarithmic expression.**

37. $\log_4 8$ **38.** $\log_9 243$

39. $\log_2 4^2 \cdot 3^4$ **40.** $\log_3 9^2 \cdot 2^4$

41. $\ln 5e^6$ **42.** $\ln 8e^3$

43. $\ln \dfrac{6}{e^2}$ **44.** $\ln \dfrac{e^5}{7}$

Using Properties of Logarithms **In Exercises 45 and 46, use the properties of logarithms to verify the equation.**

45. $\log_5 \frac{1}{250} = -3 - \log_5 2$

46. $-\ln 24 = -(3 \ln 2 + \ln 3)$

Expanding Logarithmic Expressions **In Exercises 47–64, use the properties of logarithms to expand the expression as a sum, difference, and/or constant multiple of logarithms. (Assume all variables are positive.)**

47. $\log_{10} 10x$ **48.** $\log_{10} 100x$

49. $\log_{10} \dfrac{t}{8}$ **50.** $\log_{10} \dfrac{7}{z}$

51. $\log_8 x^4$ **52.** $\log_6 z^{-3}$

53. $\ln \sqrt{z}$ **54.** $\ln \sqrt[3]{t}$

55. $\ln xyz$ **56.** $\ln \dfrac{xy}{z}$

57. $\log_6 ab^3c^{-2}$ **58.** $\log_4 xy^6 z^4$

59. $\ln \sqrt[3]{\dfrac{x^4}{y^3}}$ **60.** $\ln \sqrt{\dfrac{x^2}{y^3}}$

61. $\ln \dfrac{x^2 - 1}{x^3}, \quad x > 1$ **62.** $\ln \dfrac{x}{\sqrt{x^2 + 1}}$

63. $\log_b \dfrac{x^4\sqrt{y}}{z^5}$ **64.** $\log_b \dfrac{\sqrt{xy^4}}{z^4}$

Algebraic-Graphical-Numerical In Exercises 65–68, (a) use a graphing utility to graph the two equations in the same viewing window and (b) use the *table* feature of the graphing utility to create a table of values for each equation. (c) What do the graphs and tables suggest? Verify your conclusion algebraically.

65. $y_1 = \ln[x^2(x-4)]$
$y_2 = 2\ln x + \ln(x-4)$

66. $y_1 = \ln 9x^3$
$y_2 = \ln 9 + 3\ln x$

67. $y_1 = \ln\left(\dfrac{x^4}{x-2}\right)$
$y_2 = 4\ln x - \ln(x-2)$

68. $y_1 = \ln\left(\dfrac{\sqrt{x}}{x+3}\right)$
$y_2 = \dfrac{1}{2}\ln x - \ln(x+3)$

Condensing Logarithmic Expressions In Exercises 69–84, use the properties of logarithms to condense the expression.

69. $\ln x + \ln 4$
70. $\ln y + \ln z$
71. $\log_4 z - \log_4 y$
72. $\log_5 8 - \log_5 t$
73. $4\log_3(x+2)$
74. $\frac{5}{2}\log_7(z-4)$
75. $\frac{1}{2}\ln(x^2+4) + \ln x$
76. $2\ln x + \ln(x+1)$
77. $\ln x - 3\ln(x+1)$
78. $\ln x - 2\ln(x+2)$
79. $\ln(x-2) + \ln 2 - 3\ln y$
80. $3\ln x + 2\ln y - 4\ln z$
81. $\ln x - 2[\ln(x+2) + \ln(x-2)]$
82. $4[\ln z + \ln(z+5)] - 2\ln(z-5)$
83. $\frac{1}{3}[2\ln(x+3) + \ln x - \ln(x^2-1)]$
84. $2[\ln x - \ln(x+1) - \ln(x-1)]$

Algebraic-Graphical-Numerical In Exercises 85–88, (a) use a graphing utility to graph the two equations in the same viewing window and (b) use the *table* feature of the graphing utility to create a table of values for each equation. (c) What do the graphs and tables suggest? Verify your conclusion algebraically.

85. $y_1 = 2[\ln 8 - \ln(x^2+1)]$, $y_2 = \ln\left[\dfrac{64}{(x^2+1)^2}\right]$

86. $y_1 = 2[\ln 6 + \ln(x^2+1)]$, $y_2 = \ln[36(x^2+1)^2]$

87. $y_1 = \ln x + \frac{1}{2}\ln(x+1)$, $y_2 = \ln\left(x\sqrt{x+1}\right)$

88. $y_1 = \frac{1}{2}\ln x - \ln(x+2)$, $y_2 = \ln\left(\dfrac{\sqrt{x}}{x+2}\right)$

Using Properties to Evaluate Logarithms In Exercises 89–102, find the exact value of the logarithm without using a calculator. If this is not possible, state the reason.

89. $\log_3 9$
90. $\log_6 6$
91. $\log_4 16^{3.4}$
92. $\log_5\left(\frac{1}{125}\right)$
93. $\log_2(-4)$
94. $\log_4(-16)$
95. $\log_5 375 - \log_5 3$
96. $\log_4 2 + \log_4 32$
97. $\ln e^3 - \ln e^7$
98. $\ln e^6 - 2\ln e^7$
99. $2\ln e^4$
100. $3\ln e^5$
101. $\ln \dfrac{1}{\sqrt{e}}$
102. $\ln \sqrt[5]{e^3}$

Algebraic-Graphical-Numerical In Exercises 103–106, (a) use a graphing utility to graph the two equations in the same viewing window and (b) use the *table* feature of the graphing utility to create a table of values for each equation. (c) Are the expressions equivalent? Explain. Verify your conclusion algebraically.

103. $y_1 = \ln x^2$, $y_2 = 2\ln x$
104. $y_1 = 2(\ln 2 + \ln x)$, $y_2 = \ln 4x^2$
105. $y_1 = \ln(x-2) + \ln(x+2)$, $y_2 = \ln(x^2-4)$
106. $y_1 = \frac{1}{4}\ln[x^4(x^2+1)]$, $y_2 = \ln x + \frac{1}{4}\ln(x^2+1)$

107. ***Why you should learn it*** *(p. 347)* The relationship between the number of decibels β and the intensity of a sound I in watts per square meter is given by

$$\beta = 10\log_{10}\left(\frac{I}{10^{-12}}\right).$$

(a) Use the properties of logarithms to write the formula in a simpler form.

(b) Use a graphing utility to complete the table. Verify your answers algebraically.

I	10^{-4}	10^{-6}	10^{-8}	10^{-10}	10^{-12}	10^{-14}
β						

108. Nail Length The table shows the approximate lengths and diameters (in inches) of common nails. Find a logarithmic equation that relates the diameter y of a common nail to its length x.

DATA

Length, x	Diameter, y
1	0.072
2	0.120
3	0.148
4	0.203
5	0.238
6	0.284

Spreadsheet at LarsonPrecalculus.com

109. MODELING DATA

A beaker of liquid at an initial temperature of 78°C is placed in a room at a constant temperature of 21°C. The temperature of the liquid is measured every 5 minutes during a half-hour period. The results are recorded as ordered pairs of the form (t, T), where t is the time (in minutes) and T is the temperature (in degrees Celsius). *(Spreadsheet at LarsonPrecalculus.com)*

DATA $(0, 78.0°)$, $(5, 66.0°)$, $(10, 57.5°)$, $(15, 51.2°)$, $(20, 46.3°)$, $(25, 42.5°)$, $(30, 39.6°)$

(a) The graph of the temperature of the room should be an asymptote of the graph of the model for the data. Subtract the room temperature from each of the temperatures in the ordered pairs. Use a graphing utility to plot the data points (t, T) and $(t, T - 21)$.

(b) An exponential model for the data $(t, T - 21)$ is given by

$$T - 21 = 54.4(0.964)^t.$$

Solve for T and graph the model. Compare the result with the plot of the original data.

(c) Take the natural logarithms of the revised temperatures. Use the graphing utility to plot the points $(t, \ln(T - 21))$ and observe that the points appear linear. Use the *regression* feature of the graphing utility to fit a line to the data. The resulting line has the form

$$\ln(T - 21) = at + b.$$

Use the properties of logarithms to solve for T. Verify that the result is equivalent to the model in part (b).

(d) Fit a rational model to the data. Take the reciprocals of the y-coordinates of the revised data points to generate the points

$$\left(t, \frac{1}{T - 21}\right).$$

Use the graphing utility to plot these points and observe that they appear linear. Use the *regression* feature of the graphing utility to fit a line to the data. The resulting line has the form

$$\frac{1}{T - 21} = at + b.$$

Solve for T and use the graphing utility to graph the rational function and the original data points.

110. Writing Write a short paragraph explaining why the transformations of the data in Exercise 109 were necessary to obtain the models. Why did taking the logarithms of the temperatures lead to a linear scatter plot? Why did taking the reciprocals of the temperatures lead to a linear scatter plot?

Conclusions

True or False? **In Exercises 111–116, determine whether the statement is true or false given that $f(x) = \ln x$, where $x > 0$. Justify your answer.**

111. $f(ax) = f(a) + f(x), \ a > 0$

112. $f(x - a) = f(x) - f(a), \ x > a$

113. $\sqrt{f(x)} = \frac{1}{2}f(x)$

114. $[f(x)]^n = nf(x)$

115. If $f(x) < 0$, then $0 < x < 1$.

116. If $f(x) > 0$, then $x > e$.

117. Error Analysis Describe the error.

$$\ln\left(\frac{x^2}{\sqrt{x^2+4}}\right) = \frac{\ln x^2}{\ln\sqrt{x^2+4}}$$ (crossed out)

118. Think About It Consider the functions below.

$$f(x) = \ln\frac{x}{2}, \quad g(x) = \frac{\ln x}{\ln 2}, \quad h(x) = \ln x - \ln 2$$

Which two functions have identical graphs? Verify your answer by using a graphing utility to graph all three functions in the same viewing window.

119. Exploration For how many integers between 1 and 20 can the natural logarithms be approximated given that $\ln 2 \approx 0.6931$, $\ln 3 \approx 1.0986$, and $\ln 5 \approx 1.6094$? Approximate these logarithms without a calculator.

120. HOW DO YOU SEE IT? The figure shows the graphs of $y = \ln x$, $y = \ln x^2$, $y = \ln 2x$, and $y = \ln 2$. Match each function with its graph. (The graphs are labeled A through D.) Explain your reasoning.

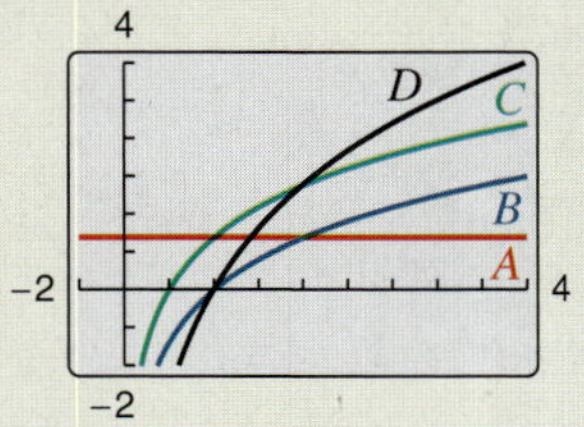

121. Think About It Does $y_1 = \ln[x(x - 2)]$ have the same domain as $y_2 = \ln x + \ln(x - 2)$? Explain.

122. Proof Prove that $\dfrac{\log_a x}{\log_{a/b} x} = 1 + \log_a \dfrac{1}{b}$.

Cumulative Mixed Review

Using Properties of Exponents **In Exercises 123–126, simplify the expression.**

123. $(64x^3y^4)^{-3}(8x^3y^2)^4$

124. $xy(x^{-1} + y^{-1})^{-1}$

125. $\dfrac{24xy^{-2}}{16x^{-3}y}$

126. $\left(\dfrac{2x^3}{3y}\right)^{-3}$

4.4 Solving Exponential and Logarithmic Equations

Introduction

So far in this chapter, you have studied the definitions, graphs, and properties of exponential and logarithmic functions. In this section, you will study procedures for *solving equations* involving exponential and logarithmic functions.

There are two basic strategies for solving exponential or logarithmic equations. The first is based on the One-to-One Properties and the second is based on the Inverse Properties. For $a > 0$ and $a \neq 1$, the following properties are true for all x and y for which $\log_a x$ and $\log_a y$ are defined.

One-to-One Properties

$a^x = a^y$ if and only if $x = y$.

$\log_a x = \log_a y$ if and only if $x = y$.

Inverse Properties

$a^{\log_a x} = x$

$\log_a a^x = x$

What you should learn

- Solve simple exponential and logarithmic equations.
- Solve more complicated exponential equations.
- Solve more complicated logarithmic equations.
- Use exponential and logarithmic equations to model and solve real-life problems.

Why you should learn it

Exponential and logarithmic equations can be used to model and solve real-life problems. For instance, Exercise 148 on page 363 shows how to use an exponential function to model the average heights of men and women.

EXAMPLE 1 Solving Simple Equations

	Original Equation	*Rewritten Equation*	*Solution*	*Property*
a.	$2^x = 32$	$2^x = 2^5$	$x = 5$	One-to-One
b.	$\log_4 x - \log_4 8 = 0$	$\log_4 x = \log_4 8$	$x = 8$	One-to-One
c.	$\ln x - \ln 3 = 0$	$\ln x = \ln 3$	$x = 3$	One-to-One
d.	$\left(\frac{1}{3}\right)^x = 9$	$3^{-x} = 3^2$	$x = -2$	One-to-One
e.	$e^x = 7$	$\ln e^x = \ln 7$	$x = \ln 7$	Inverse
f.	$\ln x = -3$	$e^{\ln x} = e^{-3}$	$x = e^{-3}$	Inverse
g.	$\log_{10} x = -1$	$10^{\log_{10} x} = 10^{-1}$	$x = 10^{-1} = \frac{1}{10}$	Inverse
h.	$\log_3 x = 4$	$3^{\log_3 x} = 3^4$	$x = 81$	Inverse

Checkpoint Audio-video solution in English & Spanish at LarsonPrecalculus.com.

Solve each equation for x.

a. $2^x = 512$ **b.** $\log_6 x = 3$ **c.** $5 - e^x = 0$ **d.** $9^x = \frac{1}{3}$

The strategies used in Example 1 are summarized as follows.

Strategies for Solving Exponential and Logarithmic Equations

1. Rewrite the original equation in a form that allows the use of the One-to-One Properties of exponential or logarithmic functions.
2. Rewrite an *exponential* equation in logarithmic form and apply the Inverse Property of logarithmic functions.
3. Rewrite a *logarithmic* equation in exponential form and apply the Inverse Property of exponential functions.

Solving Exponential Equations

EXAMPLE 2 Solving Exponential Equations

Solve each equation.

a. $e^x = 72$

b. $3(2^x) = 42$

Algebraic Solution

a.

$e^x = 72$	Write original equation.
$\ln e^x = \ln 72$	Take natural log of each side.
$x = \ln 72$	Inverse Property
$x \approx 4.28$	Use a calculator.

The solution is $x = \ln 72 \approx 4.28$. Check this in the original equation.

b.

$3(2^x) = 42$	Write original equation.
$2^x = 14$	Divide each side by 3.
$\log_2 2^x = \log_2 14$	Take log (base 2) of each side.
$x = \log_2 14$	Inverse Property
$x = \dfrac{\ln 14}{\ln 2}$	Change-of-base formula
$x \approx 3.81$	Use a calculator.

The solution is $x = \log_2 14 \approx 3.81$. Check this in the original equation.

Graphical Solution

To solve an equation using a graphing utility, you can graph the left- and right-hand sides of the equation and use the *intersect* feature.

a.

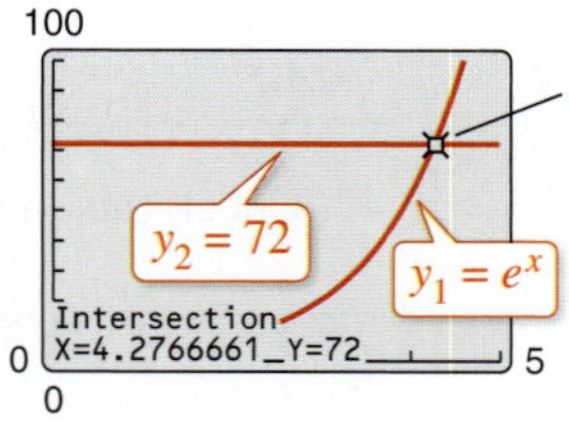

The intersection point is about (4.28, 72). So, the solution is $x \approx 4.28$.

b.

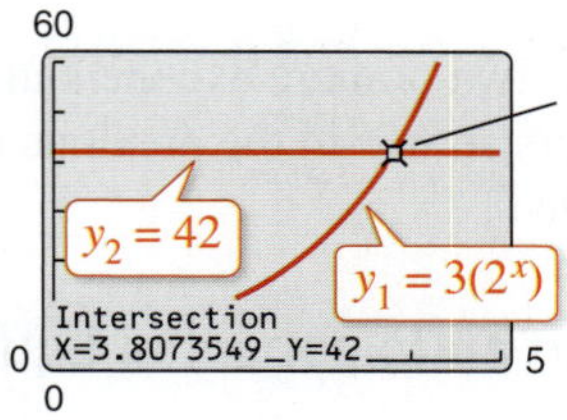

The intersection point is about (3.81, 42). So, the solution is $x \approx 3.81$.

✓ Checkpoint Audio-video solution in English & Spanish at LarsonPrecalculus.com.

Solve (a) $e^x = 10$ and (b) $2(5^x) = 32$.

EXAMPLE 3 Solving an Exponential Equation

Solve $4e^{2x} - 3 = 2$.

Algebraic Solution

$4e^{2x} - 3 = 2$	Write original equation.
$4e^{2x} = 5$	Add 3 to each side.
$e^{2x} = \frac{5}{4}$	Divide each side by 4.
$\ln e^{2x} = \ln \frac{5}{4}$	Take natural log of each side.
$2x = \ln \frac{5}{4}$	Inverse Property
$x = \frac{1}{2} \ln \frac{5}{4}$	Divide each side by 2.
$x \approx 0.11$	Use a calculator.

The solution is $x = \frac{1}{2} \ln \frac{5}{4} \approx 0.11$. Check this in the original equation.

Graphical Solution

Rather than using the procedure in Example 2, another way to solve the equation graphically is first to rewrite the equation as $4e^{2x} - 5 = 0$, and then use a graphing utility to graph $y = 4e^{2x} - 5$. Use the *zero* or *root* feature of the graphing utility to approximate the value of x for which $y = 0$, as shown in the figure.

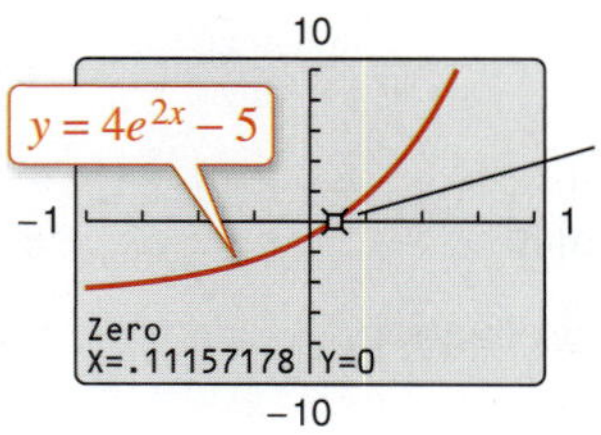

The zero occurs at $x \approx 0.11$. So, the solution of the original equation is $x \approx 0.11$.

✓ Checkpoint Audio-video solution in English & Spanish at LarsonPrecalculus.com.

Solve $2e^{3x} + 5 = 29$.

EXAMPLE 4 Solving an Exponential Equation

Solve $2(3^{2t-5}) - 4 = 11$.

Solution

$2(3^{2t-5}) - 4 = 11$	Write original equation.
$2(3^{2t-5}) = 15$	Add 4 to each side.
$3^{2t-5} = \frac{15}{2}$	Divide each side by 2.
$\log_3 3^{2t-5} = \log_3 \frac{15}{2}$	Take log (base 3) of each side.
$2t - 5 = \log_3 \frac{15}{2}$	Inverse Property
$2t = 5 + \log_3 7.5$	Add 5 to each side.
$t = \frac{5}{2} + \frac{1}{2}\log_3 7.5$	Divide each side by 2.
$t \approx 3.42$	Use a calculator.

The solution is $t = \frac{5}{2} + \frac{1}{2}\log_3 7.5 \approx 3.42$. Check this in the original equation.

Remark

Remember that to evaluate a logarithm such as $\log_3 7.5$, you need to use the change-of-base formula.

$$\log_3 7.5 = \frac{\ln 7.5}{\ln 3} \approx 1.834$$

✓ ***Checkpoint*** *Audio-video solution in English & Spanish at LarsonPrecalculus.com.*

Solve $6(2^{t+5}) + 4 = 11$.

When an equation involves two or more exponential expressions, you can still use a procedure similar to that demonstrated in the previous three examples. However, the algebra is a bit more complicated.

EXAMPLE 5 Solving an Exponential Equation in Quadratic Form

Solve $e^{2x} - 3e^x + 2 = 0$.

Algebraic Solution

$e^{2x} - 3e^x + 2 = 0$	Write original equation.
$(e^x)^2 - 3e^x + 2 = 0$	Write in quadratic form.
$(e^x - 2)(e^x - 1) = 0$	Factor.
$e^x - 2 = 0$	Set 1st factor equal to 0.
$e^x = 2$	Add 2 to each side.
$x = \ln 2$	Solution
$e^x - 1 = 0$	Set 2nd factor equal to 0.
$e^x = 1$	Add 1 to each side.
$x = \ln 1$	Inverse Property
$x = 0$	Solution

The solutions are

$$x = \ln 2 \approx 0.69 \quad \text{and} \quad x = 0.$$

Check these in the original equation.

Graphical Solution

Use a graphing utility to graph $y = e^{2x} - 3e^x + 2$ and then find the zeros.

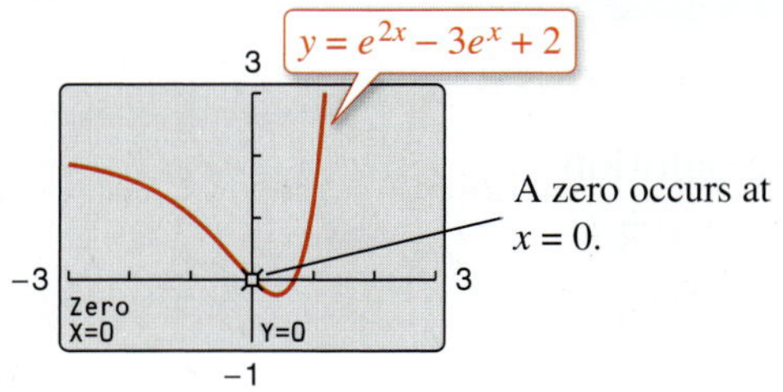

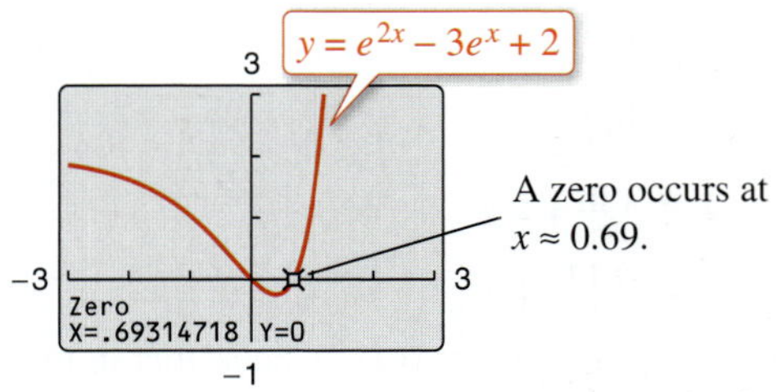

From the figures, you can conclude that the solutions are $x = 0$ and $x \approx 0.69$.

✓ ***Checkpoint*** *Audio-video solution in English & Spanish at LarsonPrecalculus.com.*

Solve $e^{2x} - 7e^x + 12 = 0$.

Solving Logarithmic Equations

To solve a logarithmic equation, you can write it in exponential form.

$\ln x = 3$	Logarithmic form
$e^{\ln x} = e^3$	Exponentiate each side.
$x = e^3$	Exponential form

This procedure is called *exponentiating* each side of an equation. It is applied after the logarithmic expression has been isolated.

EXAMPLE 6 Solving Logarithmic Equations

Solve each logarithmic equation.

a. $\ln 3x = 2$

b. $\log_3(5x - 1) = \log_3(x + 7)$

Solution

a.

$\ln 3x = 2$	Write original equation.
$e^{\ln 3x} = e^2$	Exponentiate each side.
$3x = e^2$	Inverse Property
$x = \frac{1}{3}e^2$	Multiply each side by $\frac{1}{3}$.
$x \approx 2.46$	Use a calculator.

The solution is $x = \frac{1}{3}e^2 \approx 2.46$. Check this in the original equation.

b.

$\log_3(5x - 1) = \log_3(x + 7)$	Write original equation.
$5x - 1 = x + 7$	One-to-One Property
$x = 2$	Solve for x.

The solution is $x = 2$. Check this in the original equation.

✓ Checkpoint Audio-video solution in English & Spanish at LarsonPrecalculus.com.

Solve each logarithmic equation.

a. $\ln x = \frac{2}{3}$ **b.** $\log_2(2x - 3) = \log_2(x + 4)$

EXAMPLE 7 Solving a Logarithmic Equation

Solve $5 + 2 \ln x = 4$.

Algebraic Solution

$5 + 2 \ln x = 4$	Write original equation.
$2 \ln x = -1$	Subtract 5 from each side.
$\ln x = -\frac{1}{2}$	Divide each side by 2.
$e^{\ln x} = e^{-1/2}$	Exponentiate each side.
$x = e^{-1/2}$	Inverse Property
$x \approx 0.61$	Use a calculator.

The solution is $x = e^{-1/2} \approx 0.61$. Check this in the original equation.

Graphical Solution

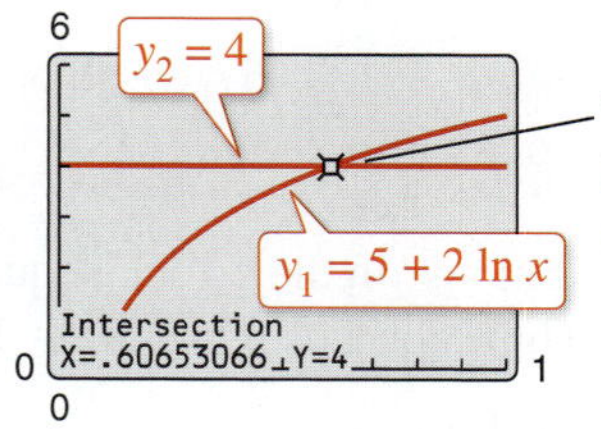

The intersection point is about (0.61, 4). So, the solution is $x \approx 0.61$.

✓ Checkpoint Audio-video solution in English & Spanish at LarsonPrecalculus.com.

Solve $7 + 3 \ln x = 5$.

EXAMPLE 8 Solving a Logarithmic Equation

Solve $2 \log_5 3x = 4$.

Solution

$2 \log_5 3x = 4$ Write original equation.

$\log_5 3x = 2$ Divide each side by 2.

$5^{\log_5 3x} = 5^2$ Exponentiate each side (base 5).

$3x = 25$ Inverse Property

$x = \frac{25}{3}$ Divide each side by 3.

The solution is $x = \frac{25}{3}$. Check this in the original equation. Or, perform a graphical check by graphing

$$y_1 = 2 \log_5 3x = 2\left(\frac{\log_{10} 3x}{\log_{10} 5}\right) \quad \text{and} \quad y_2 = 4$$

in the same viewing window. The two graphs should intersect at $x = \frac{25}{3} \approx 8.33$ and $y = 4$, as shown in Figure 4.21.

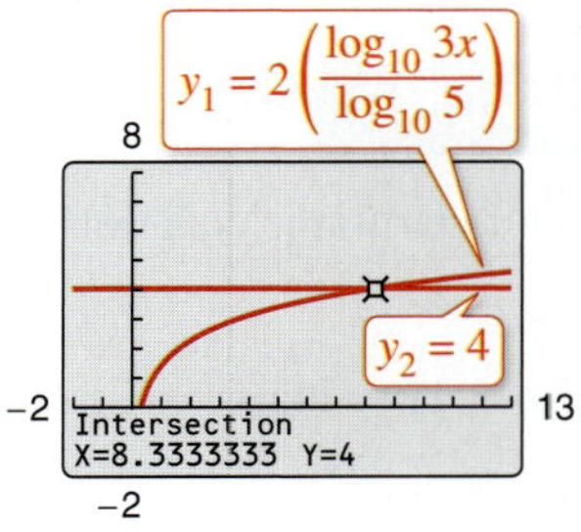

Figure 4.21

Checkpoint Audio-video solution in English & Spanish at LarsonPrecalculus.com.

Solve $3 \log_4 6x = 9$.

Because the domain of a logarithmic function generally does not include all real numbers, you should be sure to check for extraneous solutions of logarithmic equations, as shown in the next example.

EXAMPLE 9 Checking for Extraneous Solutions

Solve $\ln(x - 2) + \ln(2x - 3) = 2 \ln x$.

Algebraic Solution

$\ln(x - 2) + \ln(2x - 3) = 2 \ln x$ Write original equation.

$\ln[(x - 2)(2x - 3)] = \ln x^2$ Use properties of logarithms.

$\ln(2x^2 - 7x + 6) = \ln x^2$ Multiply binomials.

$2x^2 - 7x + 6 = x^2$ One-to-One Property

$x^2 - 7x + 6 = 0$ Write in general form.

$(x - 6)(x - 1) = 0$ Factor.

$x - 6 = 0 \Rightarrow x = 6$ Set 1st factor equal to 0.

$x - 1 = 0 \Rightarrow x = 1$ Set 2nd factor equal to 0.

Finally, by checking these two "solutions" in the original equation, you can conclude that $x = 1$ is not valid. This is because when $x = 1$,

$$\ln(x - 2) + \ln(2x - 3) = \ln(-1) + \ln(-1)$$

which is invalid because -1 is not in the domain of the natural logarithmic function. So, the only solution is $x = 6$.

Graphical Solution

First, rewrite the original equation as

$$\ln(x - 2) + \ln(2x - 3) - 2 \ln x = 0.$$

Then, use a graphing utility to graph the equation

$$y = \ln(x - 2) + \ln(2x - 3) - 2 \ln x$$

and find the zeros, as shown in the figure.

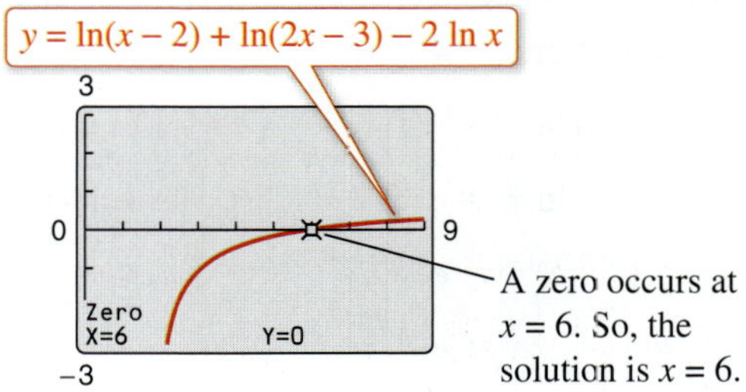

Checkpoint Audio-video solution in English & Spanish at LarsonPrecalculus.com.

Solve $\log_{10} x + \log_{10}(x - 9) = 1$.

EXAMPLE 10 The Change-of-Base Formula

Prove the change-of-base formula: $\log_a x = \dfrac{\log_b x}{\log_b a}$.

Solution

Begin by letting $y = \log_a x$ and writing the equivalent exponential form

$$a^y = x.$$

Now, taking the logarithms *with base b* of each side produces the following.

$$\log_b a^y = \log_b x$$

$$y \log_b a = \log_b x \qquad \text{Power Property}$$

$$y = \frac{\log_b x}{\log_b a} \qquad \text{Divide each side by } \log_b a.$$

$$\log_a x = \frac{\log_b x}{\log_b a} \qquad \text{Replace } y \text{ with } \log_a x.$$

✓ **Checkpoint** *Audio-video solution in English & Spanish at LarsonPrecalculus.com.*

Prove that $\log_a \dfrac{1}{x} = -\log_a x$.

Equations that involve combinations of algebraic functions, exponential functions, and logarithmic functions can be very difficult to solve by algebraic procedures. Here again, you can take advantage of a graphing utility.

EXAMPLE 11 Approximating the Solution of an Equation

Approximate (to three decimal places) the solution of $\ln x = x^2 - 2$.

Solution

First, rewrite the equation as

$$\ln x - x^2 + 2 = 0.$$

Then, use a graphing utility to graph

$$y = -x^2 + 2 + \ln x$$

as shown in Figure 4.22. From this graph, you can see that the equation has two solutions. Next, using the *zero* or *root* feature, you can approximate the two solutions to be $x \approx 0.138$ and $x \approx 1.564$.

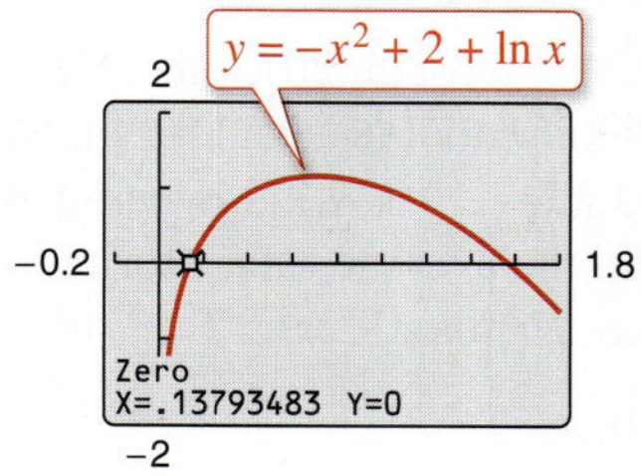

Figure 4.22

Check

$$\ln x = x^2 - 2 \qquad \text{Write original equation.}$$

$$\ln(0.138) \overset{?}{\approx} (0.138)^2 - 2 \qquad \text{Substitute 0.138 for } x.$$

$$-1.9805 \approx -1.9810 \qquad \text{Solution checks. ✓}$$

$$\ln(1.564) \overset{?}{\approx} (1.564)^2 - 2 \qquad \text{Substitute 1.564 for } x.$$

$$0.4472 \approx 0.4461 \qquad \text{Solution checks. ✓}$$

So, the two solutions $x \approx 0.138$ and $x \approx 1.564$ seem reasonable.

✓ **Checkpoint** *Audio-video solution in English & Spanish at LarsonPrecalculus.com.*

Approximate (to three decimal places) the solution of $5e^{0.3x} = 17$.

Applications

EXAMPLE 12 Doubling an Investment

See LarsonPrecalculus.com for an interactive version of this type of example.

You have deposited \$500 in an account that pays 6.75% interest, compounded continuously. How long will it take your money to double?

Solution

Using the formula for continuous compounding, you can find that the balance in the account is $A = Pe^{rt} = 500e^{0.0675t}$. To find the time required for the balance to double, let $A = 1000$ and solve the resulting equation for t.

$500e^{0.0675t} = 1000$ — Substitute 1000 for A.

$e^{0.0675t} = 2$ — Divide each side by 500.

$\ln e^{0.0675t} = \ln 2$ — Take natural log of each side.

$0.0675t = \ln 2$ — Inverse Property

$t = \dfrac{\ln 2}{0.0675}$ — Divide each side by 0.0675.

$t \approx 10.27$ — Use a calculator.

The balance in the account will double after approximately 10.27 years. This result is demonstrated graphically in Figure 4.23.

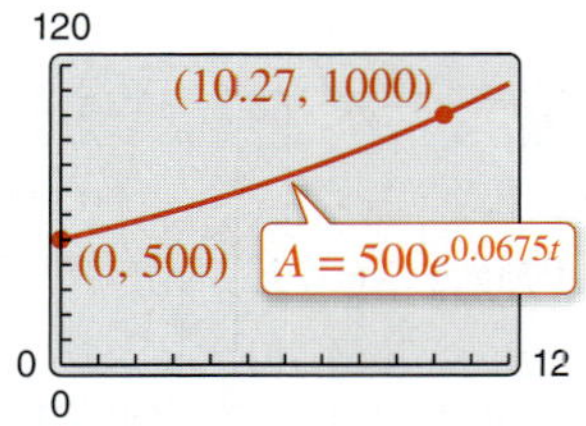

Figure 4.23

✓ ***Checkpoint*** *Audio-video solution in English & Spanish at LarsonPrecalculus.com.*

You invest \$500 at an annual interest rate of 5.25%, compounded continuously. How long will it take your money to double? Compare your result with that of Example 12.

EXAMPLE 13 Average Salary for Public School Teachers

From 2004 through 2013, the average salary y (in thousands of dollars) for public school teachers for the year t can be modeled by the equation $y = 33.374 + 9.162 \ln t$, where $t = 4$ represents 2004. During which year did the average salary for public school teachers reach \$50 thousand? (Source: National Center for Education Statistics)

Solution

$33.374 + 9.162 \ln t = y$ — Write original equation.

$33.374 + 9.162 \ln t = 50$ — Substitute 50 for y.

$9.162 \ln t = 16.626$ — Subtract 33.374 from each side.

$\ln t = \dfrac{16.626}{9.162}$ — Divide each side by 9.162.

$e^{\ln t} = e^{16.626/9.162}$ — Exponentiate each side.

$t = e^{16.626/9.162}$ — Inverse Property

$t \approx 6.14$ — Use a calculator.

The solution is $t \approx 6.14$ years. Because $t = 4$ represents 2004, it follows that the average salary for public school teachers reached \$50 thousand in 2006.

Teacher

✓ ***Checkpoint*** *Audio-video solution in English & Spanish at LarsonPrecalculus.com.*

In Example 13, during which year did the average salary reach \$52.5 thousand?

4.4 Exercises

See CalcChat.com for tutorial help and worked-out solutions to odd-numbered exercises.
For instructions on how to use a graphing utility, see Appendix A.

Vocabulary and Concept Check

In Exercises 1 and 2, fill in the blank.

1. To solve exponential and logarithmic equations, you can use the following One-to-One and Inverse Properties.
   (a) $a^x = a^y$ if and only if _______ .
   (b) $\log_a x = \log_a y$ if and only if _______ .
   (c) $a^{\log_a x} =$ _______
   (d) $\log_a a^x =$ _______
2. A(n) _______ solution does not satisfy the original equation.
3. What is the value of $\ln e^7$?
4. Can you solve $5^x = 125$ using a One-to-One Property?
5. What is the first step in solving the equation $3 + \ln x = 10$?
6. Do you solve $\log_4 x = 2$ by using a One-to-One Property or an Inverse Property?

Procedures and Problem Solving

Checking Solutions **In Exercises 7–14, determine whether each x-value is a solution of the equation.**

7. $4^{2x-7} = 64$
   (a) $x = 5$
   (b) $x = 2$
8. $2^{3x+1} = 128$
   (a) $x = -1$
   (b) $x = 2$
9. $3e^{x+2} = 75$
   (a) $x = -2 + e^{25}$
   (b) $x = -2 + \ln 25$
   (c) $x \approx 1.2189$
10. $4e^{x-1} = 60$
   (a) $x = 1 + \ln 15$
   (b) $x \approx 3.7081$
   (c) $x = \ln 16$
11. $\log_4(3x) = 3$
   (a) $x \approx 21.3560$
   (b) $x = -4$
   (c) $x = \frac{64}{3}$
12. $\log_6\left(\frac{5}{3}x\right) = 2$
   (a) $x \approx 20.2882$
   (b) $x = \frac{108}{5}$
   (c) $x = 21.6$
13. $\ln(x - 1) = 3.8$
   (a) $x = 1 + e^{3.8}$
   (b) $x \approx 45.7012$
   (c) $x = 1 + \ln 3.8$
14. $\ln(2 + x) = 2.5$
   (a) $x = e^{2.5} - 2$
   (b) $x \approx \frac{4073}{400}$
   (c) $x = \frac{1}{2}$

Solving Equations Graphically **In Exercises 15–22, use a graphing utility to graph f and g in the same viewing window. Approximate the point of intersection of the graphs of f and g. Then solve the equation $f(x) = g(x)$ algebraically.**

15. $f(x) = 64^x$
    $g(x) = 8$
16. $f(x) = 27^x$
    $g(x) = 9$
17. $f(x) = 5^{x-2} - 15$
    $g(x) = 10$
18. $f(x) = 2^{-x+1} - 3$
    $g(x) = 13$
19. $f(x) = 4\log_3 x$
    $g(x) = 20$
20. $f(x) = 3\log_5 x$
    $g(x) = 6$
21. $f(x) = \ln e^{x+1}$
    $g(x) = 2x + 5$
22. $f(x) = \ln e^{x-2}$
    $g(x) = 3x + 2$

Solving an Exponential Equation **In Exercises 23–36, solve the exponential equation.**

23. $4^x = 16$
24. $3^x = 243$
25. $5^x = \frac{1}{625}$
26. $7^x = \frac{1}{49}$
27. $\left(\frac{1}{8}\right)^x = 64$
28. $\left(\frac{1}{2}\right)^x = 32$
29. $\left(\frac{2}{3}\right)^x = \frac{81}{16}$
30. $\left(\frac{3}{4}\right)^x = \frac{64}{27}$
31. $e^x = 14$
32. $e^x = 66$
33. $6(10^x) = 216$
34. $5(8^x) = 325$
35. $2^{x+3} = 256$
36. $4^{x-1} = 256$

Solving a Logarithmic Equation **In Exercises 37–46, solve the logarithmic equation.**

37. $\ln x - \ln 5 = 0$
38. $\ln x - \ln 2 = 0$
39. $\ln x = -9$
40. $\ln x = -14$
41. $\log_x 625 = 4$
42. $\log_x 81 = 2$
43. $\log_{11} x = -1$
44. $\log_{10} x = -\frac{1}{4}$
45. $\ln(2x - 1) = 5$
46. $\ln(3x + 5) = 8$

Using Inverse Properties **In Exercises 47–54, simplify the expression.**

47. $\ln e^{x^2}$
48. $\ln e^{2x-1}$
49. $e^{\ln(x^2-3)}$
50. $e^{\ln x^2}$
51. $-1 + \ln e^{2x}$
52. $-4 + e^{\ln x^4}$
53. $5 - e^{\ln(x^2+1)}$
54. $3 - \ln(e^{x^2+2})$

Solving an Exponential Equation In Exercises 55–80, solve the exponential equation algebraically. Round your result to three decimal places. Use a graphing utility to verify your answer.

55. $8^{3x} = 360$ **56.** $6^{5x} = 3000$

57. $5^{-t/2} = 0.20$ **58.** $4^{-2t} = 0.0625$

59. $250e^{0.02x} = 10{,}000$ **60.** $100e^{0.005x} = 125{,}000$

61. $500e^{-x} = 300$ **62.** $1000e^{-4x} = 75$

63. $7 - 2e^x = 1$ **64.** $-14 + 3e^x = 11$

65. $5(2^{3-x}) - 13 = 100$ **66.** $6(8^{-2-x}) + 15 = 2601$

67. $\left(1 + \frac{0.10}{12}\right)^{12t} = 2$ **68.** $\left(16 + \frac{0.878}{26}\right)^{3t} = 30$

69. $5000\left[\frac{(1 + 0.005)^x}{0.005}\right] = 250{,}000$

70. $250\left[\frac{(1 + 0.01)^x}{0.01}\right] = 150{,}000$

71. $e^{2x} - 4e^x - 5 = 0$ **72.** $e^{2x} - e^x - 6 = 0$

73. $e^x = e^{x^2-2}$ **74.** $e^{2x} = e^{x^2-8}$

75. $e^{x^2-3x} = e^{x-2}$ **76.** $e^{-x^2} = e^{x^2-2x}$

77. $\frac{400}{1 + e^{-x}} = 350$ **78.** $\frac{525}{1 + e^{-x}} = 275$

79. $\frac{40}{1 - 5e^{-0.01x}} = 200$ **80.** $\frac{50}{1 - 2e^{-0.001x}} = 1000$

Algebraic-Graphical-Numerical In Exercises 81–84, (a) complete the table to find an interval containing the solution of the equation, (b) use a graphing utility to graph both sides of the equation to estimate the solution, and (c) solve the equation algebraically. Round your results to three decimal places.

81. $e^{3x} = 12$

x	0.6	0.7	0.8	0.9	1.0
e^{3x}					

82. $4e^{5x} = 24$

x	0.1	0.2	0.3	0.4	0.5
$4e^{5x}$					

83. $20(100 - e^{x/2}) = 500$

x	5	6	7	8	9
$20(100 - e^{x/2})$					

84. $11(77 - e^{x-4}) = 264$

x	5	6	7	8	9
$11(77 - e^{x-4})$					

Solving an Exponential Equation Graphically In Exercises 85–88, use the *zero* or *root* feature or the *zoom* and *trace* features of a graphing utility to approximate the solution of the exponential equation accurate to three decimal places.

85. $\left(1 + \frac{0.065}{365}\right)^{365t} = 4$ **86.** $\left(4 - \frac{2.471}{40}\right)^{9t} = 21$

87. $\frac{7000}{5 + e^{3x}} = 2$ **88.** $\frac{119}{e^{6x} - 14} = 7$

Finding the Zero of a Function In Exercises 89–92, use a graphing utility to graph the function and approximate its zero accurate to three decimal places.

89. $g(x) = 6e^{1-x} - 25$ **90.** $f(x) = 3e^{3x/2} - 962$

91. $g(t) = e^{0.09t} - 3$ **92.** $h(t) = e^{-0.125t} - 8$

Solving a Logarithmic Equation In Exercises 93–114, solve the logarithmic equation algebraically. Round the result to three decimal places. Verify your answer(s) using a graphing utility.

93. $\ln x = -3$ **94.** $\ln x = -4$

95. $\ln 4x = 2.1$ **96.** $\ln 2x = 1.5$

97. $\log_5(3x + 2) = \log_5(-x)$ **98.** $\log_9(4 + x) = \log_9 2x$

99. $-2 + 2 \ln 3x = 17$ **100.** $3 + 2 \ln x = 10$

101. $7 \log_4(0.6x) = 12$ **102.** $4 \log_{10}(x - 6) = 11$

103. $\log_{10}(z^2 + 19) = 2$ **104.** $\log_{12} x^2 = 6$

105. $\ln \sqrt{x + 2} = 1$ **106.** $\ln \sqrt{x - 8} = 5$

107. $\ln(x + 1)^2 = 2$ **108.** $\ln(x - 4)^2 = 5$

109. $\log_4 x - \log_4(x - 1) = \frac{1}{2}$

110. $\log_3 x + \log_3(x - 8) = 2$

111. $\ln(x + 5) = \ln(x - 1) + \ln(x + 1)$

112. $\ln(x + 1) - \ln(x - 2) = \ln x$

113. $\log_{10} 8x - \log_{10}(1 + \sqrt{x}) = 2$

114. $\log_{10} 4x - \log_{10}(12 + \sqrt{x}) = 2$

Algebraic-Graphical-Numerical In Exercises 115–118, (a) complete the table to find an interval containing the solution of the equation, (b) use a graphing utility to graph both sides of the equation to estimate the solution, and (c) solve the equation algebraically. Round your results to three decimal places.

115. $\ln 2x = 2.4$

x	2	3	4	5	6
$\ln 2x$					

116. $3 \ln 5x = 10$

x	4	5	6	7	8
$3 \ln 5x$					

117. $6\log_3(0.5x) = 11$

x	12	13	14	15	16
$6\log_3(0.5x)$					

118. $5\log_{10}(x-2) = 11$

x	150	155	160	165	170
$5\log_{10}(x-2)$					

Approximating the Solution of an Equation **In Exercises 119–124, use the *zero* or *root* feature of a graphing utility to approximate the solution of the logarithmic equation.**

119. $\log_{10} x = x^3 - 3$

120. $\log_{10} x = (x-3)^2$

121. $\log_{10} x + e^{0.5x} = 6$

122. $e^x \log_{10} x = 7$

123. $\ln(x+2) - 3^{x-2} + 10 = 5$

124. $\ln x^2 - e^x = -3 - \ln x^2$

Finding the Point of Intersection **In Exercises 125–130, use a graphing utility to approximate the point of intersection of the graphs. Round your result to three decimal places.**

125. $y_1 = 7$, $y_2 = 2^{x-1} - 5$

126. $y_1 = 4$, $y_2 = 3^{x+1} - 2$

127. $y_1 = 80$, $y_2 = 4e^{-0.2x}$

128. $y_1 = 500$, $y_2 = 1500e^{-x/2}$

129. $y_1 = 3.25$, $y_2 = \frac{1}{2}\ln(x+2)$

130. $y_1 = 1.05$, $y_2 = \ln\sqrt{x-2}$

Solving Exponential and Logarithmic Equations **In Exercises 131–138, solve the equation algebraically. Round the result to three decimal places. Verify your answer using a graphing utility.**

131. $2x^2e^{2x} + 2xe^{2x} = 0$

132. $-x^2e^{-x} + 2xe^{-x} = 0$

133. $-xe^{-x} + e^{-x} = 0$

134. $e^{-2x} - 2xe^{-2x} = 0$

135. $2x\ln x + x = 0$

136. $\dfrac{1-\ln x}{x^2} = 0$

137. $\dfrac{1+\ln x}{2} = 0$

138. $3x\ln\left(\dfrac{1}{x}\right) - x = 0$

Solving a Model for *x* **In Exercises 139–142, the equation represents the given type of model, which you will use in Section 4.5. Solve the equation for *x*.**

Model Type	*Equation*
139. Exponential growth	$y = ae^{bx}$
140. Exponential decay	$y = ae^{-bx}$
141. Gaussian	$y = ae^{-(x-b)^2/c}$
142. Logarithmic	$y = a + b\ln x$

Doubling and Tripling an Investment **In Exercises 143–146, find the time required for a \$1000 investment to (a) double at interest rate *r*, compounded continuously, and (b) triple at interest rate *r*, compounded continuously. Round your results to two decimal places.**

143. $r = 7\%$

144. $r = 6\%$

145. $r = 2.5\%$

146. $r = 3.75\%$

147. Economics The percent p (in decimal form) of the United States population who own a smartphone is given by

$$p = \frac{1}{1 + e^{-(t-93)/22.5}}$$

where t is the number of months after smartphones were available on the market. Find the number of months t when the percent of the population owning smartphones is (a) 50% and (b) 80%.

148. ***Why you should learn it*** *(p. 354)* The percent m of American males between the ages of 18 and 24 who are no more than x inches tall is modeled by

$$m(x) = \frac{100}{1 + e^{-0.6114(x-69.71)}}$$

and the percent f of American females between the ages of 18 and 24 who are no more than x inches tall is modeled by

$$f(x) = \frac{100}{1 + e^{-0.66607(x-64.51)}}.$$

(Source: U.S. National Center for Health Statistics)

(a) Use a graphing utility to graph the two functions in the same viewing window.

(b) Use the graphs in part (a) to determine the horizontal asymptotes of the functions. Interpret their meanings in the context of the problem.

(c) What is the average height for each sex?

149. Finance The numbers y of commercial banks in the United States from 2007 through 2013 can be modeled by

$$y = 11{,}912 - 2340.1\ln t, \quad 7 \le t \le 13$$

where t represents the year, with $t = 7$ corresponding to 2007. In what year were there about 6300 commercial banks? (Source: Federal Deposit Insurance Corp.)

150. Forestry The yield V (in millions of cubic feet per acre) for a forest at age t years is given by $V = 6.7e^{-48.1/t}$.

(a) Use a graphing utility to graph the function.

(b) Determine the horizontal asymptote of the function. Interpret its meaning in the context of the problem.

(c) Find the time necessary to obtain a yield of 1.3 million cubic feet.

151. Science An object at a temperature of 160°C was removed from a furnace and placed in a room at 20°C. The temperature T (in degrees Celsius) of the object was measured after each hour h and recorded in the table. A model for the data is given by $T = 20[1 + 7(2^{-h})]$.

DATA

Hour, h	Temperature, T
0	160°
1	90°
2	56°
3	38°
4	29°
5	24°

Spreadsheet at LarsonPrecalculus.com

(a) Use a graphing utility to plot the data and graph the model in the same viewing window.

(b) Identify the horizontal asymptote of the graph. Interpret its meaning in the context of the problem.

(c) Approximate the time when the temperature of the object is 100°C.

(d) How long does it take the object to cool to 80°C, one-half the initial temperature?

152. MODELING DATA

The table shows the numbers N of college-bound seniors intending to major in engineering who took the SAT exam from 2008 through 2013. The data can be modeled by the logarithmic function

$N = -152{,}656 + 111{,}959.9 \ln t$

where t represents the year, with $t = 8$ corresponding to 2008. (Source: The College Board)

DATA

Year	Number, N
2008	81,338
2009	88,719
2010	108,389
2011	116,746
2012	127,061
2013	132,275

Spreadsheet at LarsonPrecalculus.com

(a) According to the model, in what year would 150,537 seniors intending to major in engineering take the SAT exam?

(b) Use a graphing utility to graph the model with the data, and use the graph to verify your answer in part (a).

(c) Do you think this is a good model for predicting future values? Explain.

Conclusions

True or False? **In Exercises 153 and 154, determine whether the statement is true or false. Justify your answer.**

153. An exponential equation must have at least one solution.

154. A logarithmic equation can have at most one extraneous solution.

155. Error Analysis Describe the error.

$$2e^x = 10$$
$$\ln(2e^x) = \ln 10$$
$$2x = \ln 10$$
$$x = \tfrac{1}{2} \ln 10$$

156. HOW DO YOU SEE IT? Solving $\log_3 x + \log_3(x - 8) = 2$ algebraically, the solutions appear to be $x = 9$ and $x = -1$. Use the graph of $y = \log_3 x + \log_3(x - 8) - 2$ to determine whether each value is an actual solution of the equation. Explain your reasoning.

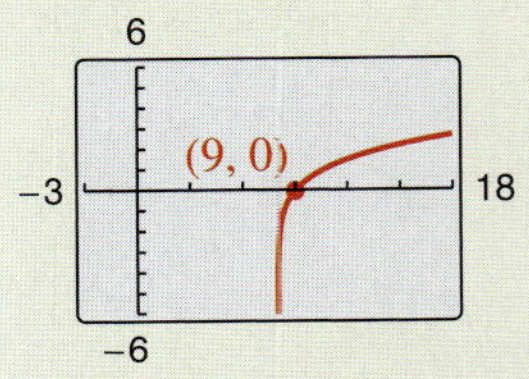

157. Exploration Let $f(x) = \log_a x$ and $g(x) = a^x$, where $a > 1$.

(a) Let $a = 1.2$ and use a graphing utility to graph the two functions in the same viewing window. What do you observe? Approximate any points of intersection of the two graphs.

(b) Determine the value(s) of a for which the two graphs have one point of intersection.

(c) Determine the value(s) of a for which the two graphs have two points of intersection.

158. Think About It Is the time required for a continuously compounded investment to quadruple twice as long as the time required for it to double? Give a reason for your answer and verify your answer algebraically.

Cumulative Mixed Review

Sketching Graphs **In Exercises 159–162, sketch the graph of the function.**

159. $f(x) = 3x^3 - 4$

160. $f(x) = |x - 2| - 8$

161. $f(x) = \begin{cases} 2x + 1, & x < 0 \\ -x^2, & x \ge 0 \end{cases}$

162. $f(x) = \begin{cases} x - 9, & x \le 1 \\ x^2 + 1, & x > 1 \end{cases}$

4.5 Exponential and Logarithmic Models

Introduction

There are many examples of exponential and logarithmic models in real life. In Section 4.1, you used the formula

$A = Pe^{rt}$ Exponential model

to find the balance in an account when the interest was compounded continuously. In Section 4.2, Example 10, you used the human memory model

$f(t) = 75 - 6\ln(t + 1).$ Logarithmic model

The five most common types of mathematical models involving exponential functions or logarithmic functions are listed below.

1. **Exponential growth model:** $y = ae^{bx}, \quad b > 0$
2. **Exponential decay model:** $y = ae^{-bx}, \quad b > 0$
3. **Gaussian model:** $y = ae^{-(x-b)^2/c}$
4. **Logistic growth model:** $y = \dfrac{a}{1 + be^{-rx}}$
5. **Logarithmic models:** $y = a + b\ln x, \quad y = a + b\log_{10} x$

The basic shapes of the graphs of these models are shown in Figure 4.24.

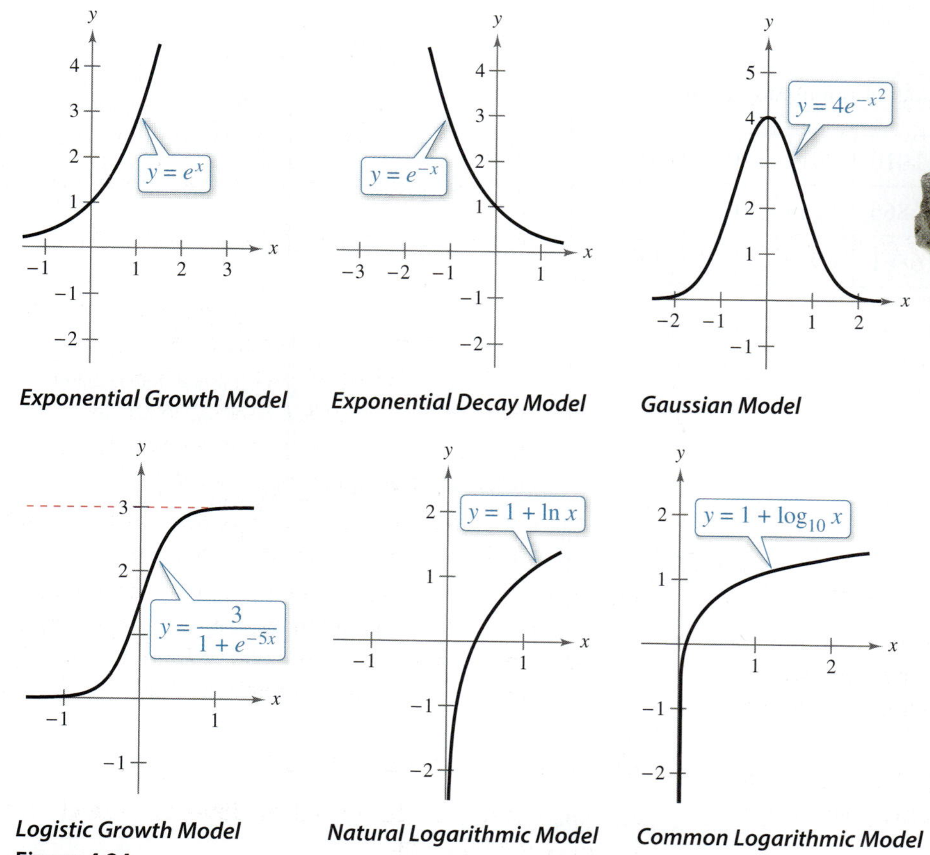

Figure 4.24

You can often gain quite a bit of insight into a situation modeled by an exponential or logarithmic function by identifying and interpreting the function's asymptotes.

What you should learn

- Recognize the five most common types of models involving exponential or logarithmic functions.
- Use exponential growth and decay functions to model and solve real-life problems.
- Use Gaussian functions to model and solve real-life problems.
- Use logistic growth functions to model and solve real-life problems.
- Use logarithmic functions to model and solve real-life problems.

Why you should learn it

Exponential decay models are used in carbon dating. For instance, in Exercise 39 on page 374, you will use an exponential decay model to estimate the age of a piece of ancient charcoal.

Exponential Growth and Decay

EXAMPLE 1 Demography

See LarsonPrecalculus.com for an interactive version of this type of example.

The table shows estimates of the world population (in millions) from 2007 through 2013. A scatter plot of the data is shown in the figure at the right. (Source: U.S. Census Bureau)

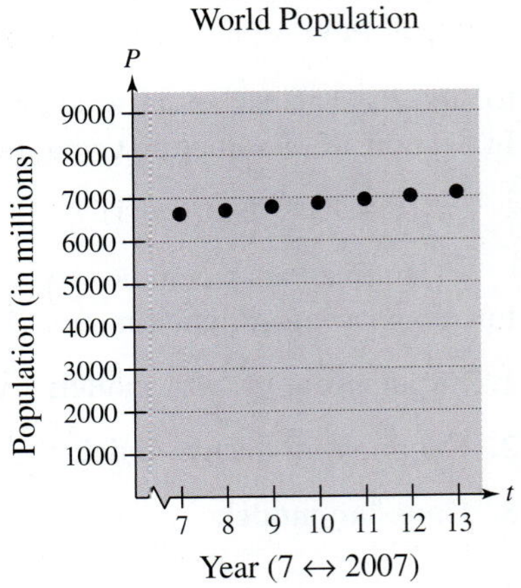

DATA

Year	Population, P
2007	6631
2008	6710
2009	6788
2010	6866
2011	6943
2012	7021
2013	7098

Spreadsheet at LarsonPrecalculus.com

An exponential growth model that approximates these data is given by

$$P = 6128e^{0.0113t}$$

where P is the population (in millions) and $t = 7$ represents 2007. Compare the values given by the model with the estimates shown in the table. According to this model, when will the world population reach 7.6 billion?

Algebraic Solution

The following table compares the two sets of population figures.

Year	2007	2008	2009	2010	2011	2012	2013
Population	6631	6710	6788	6866	6943	7021	7098
Model	6632	6708	6784	6861	6939	7018	7098

From the table, it appears that the model is a good fit for the data. To find when the world population will reach 7.6 billion, let $P = 7600$ in the model and solve for t.

$6128e^{0.0113t} = P$ — Write original equation.

$6128e^{0.0113t} = 7600$ — Substitute 7600 for P.

$e^{0.0113t} \approx 1.24021$ — Divide each side by 6128.

$\ln e^{0.0113t} \approx \ln 1.24021$ — Take natural log of each side.

$0.0113t \approx 0.21528$ — Inverse Property

$t \approx 19.1$ — Divide each side by 0.0113.

According to the model, the world population will reach 7.6 billion in 2019.

Graphical Solution

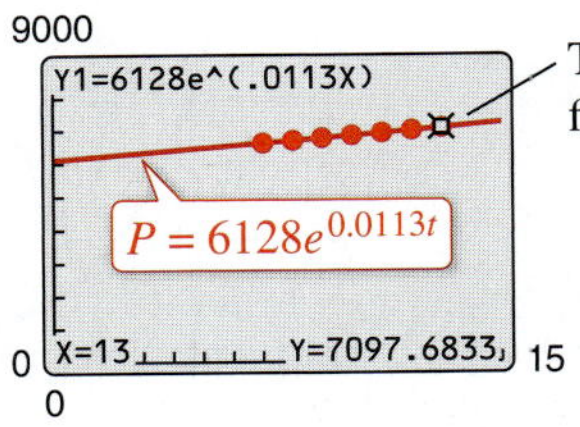

The model appears to fit the data closely.

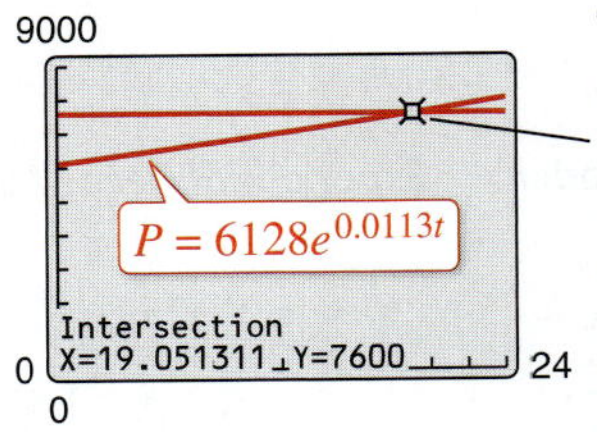

The intersection point of the model and the line $y = 7600$ is about (19.1, 7600). So, according to the model, the world population will reach 7.6 billion in 2019.

✓ **Checkpoint** *Audio-video solution in English & Spanish at LarsonPrecalculus.com.*

In Example 1, when will the world population reach 9 billion?

An exponential model increases (or decreases) by the same percent each year. What is the annual percent increase for the model in Example 1?

In Example 1, you were given the exponential growth model. Sometimes you must find such a model. One technique for doing this is shown in Example 2.

EXAMPLE 2 Modeling Population Growth

In a research experiment, a population of fruit flies is increasing according to the law of exponential growth. After 2 days there are 100 flies, and after 4 days there are 300 flies. How many flies will there be after 5 days?

Entomologist

Solution

Let y be the number of flies at time t (in days). From the given information, you know that $y = 100$ when $t = 2$ and $y = 300$ when $t = 4$. Substituting this information into the model $y = ae^{bt}$ produces

$$100 = ae^{2b} \quad \text{and} \quad 300 = ae^{4b}.$$

To solve for b, solve for a in the first equation.

$$100 = ae^{2b} \Rightarrow a = \frac{100}{e^{2b}}$$ Solve for a in the first equation.

Then substitute the result into the second equation.

$300 = ae^{4b}$ Write second equation.

$300 = \left(\frac{100}{e^{2b}}\right)e^{4b}$ Substitute $\frac{100}{e^{2b}}$ for a.

$300 = 100e^{2b}$ Simplify.

$3 = e^{2b}$ Divide each side by 100.

$\ln 3 = \ln e^{2b}$ Take natural log of each side.

$\ln 3 = 2b$ Inverse Property

$\frac{1}{2}\ln 3 = b$ Solve for b.

Using $b = \frac{1}{2}\ln 3$ and the equation you found for a, you can determine that

$a = \frac{100}{e^{2[(1/2)\ln 3]}}$ Substitute $\frac{1}{2}\ln 3$ for b.

$= \frac{100}{e^{\ln 3}}$ Simplify.

$= \frac{100}{3}$ Inverse Property

$\approx 33.33.$ Simplify.

So, with $a \approx 33.33$ and $b = \frac{1}{2}\ln 3 \approx 0.5493$, the exponential growth model is

$$y = 33.33e^{0.5493t}$$

as shown in Figure 4.25. After 5 days, the population will be

$$y = 33.33e^{0.5493(5)} \approx 520 \text{ flies.}$$

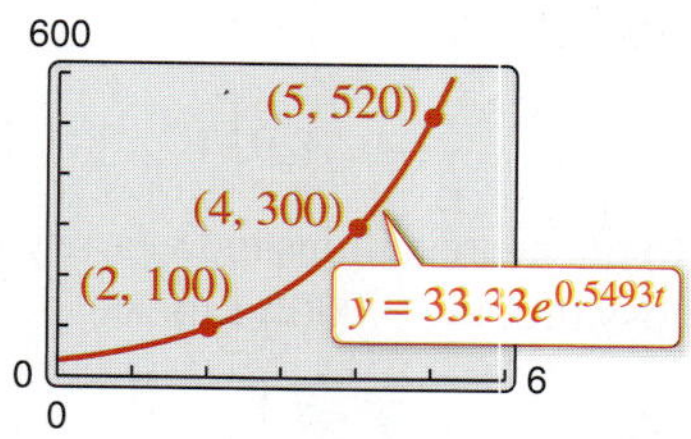

Figure 4.25

Checkpoint Audio-video solution in English & Spanish at LarsonPrecalculus.com.

The number of bacteria in a culture is increasing according to the law of exponential growth. After 1 hour there are 100 bacteria, and after 2 hours there are 200 bacteria. How many bacteria will there be after 3 hours?

In living organic material, the ratio of the content of radioactive carbon isotopes (carbon 14) to the content of nonradioactive carbon isotopes (carbon 12) is about 1 to 10^{12}. When organic material dies, its carbon 12 content remains fixed, whereas its radioactive carbon 14 begins to decay with a half-life of about 5700 years. To estimate the age of dead organic material, scientists use the following formula, which denotes the ratio of carbon 14 to carbon 12 present at any time t (in years).

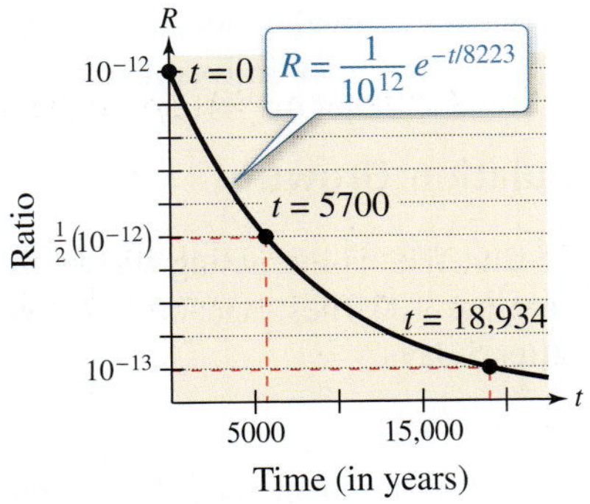

Figure 4.26

$$R = \frac{1}{10^{12}}e^{-t/8223} \qquad \text{Carbon dating model}$$

The graph of R is shown in Figure 4.26. Note that R decreases as t increases.

EXAMPLE 3 Carbon Dating

The ratio of carbon 14 to carbon 12 in a newly discovered fossil is

$$R = \frac{1}{10^{13}}.$$

Estimate the age of the fossil.

Algebraic Solution

In the carbon dating model, substitute the given value of R and solve for t.

$$\frac{1}{10^{12}}e^{-t/8223} = R \qquad \text{Write original model.}$$

$$\frac{e^{-t/8223}}{10^{12}} = \frac{1}{10^{13}} \qquad \text{Substitute } \frac{1}{10^{13}} \text{ for } R.$$

$$e^{-t/8223} = \frac{1}{10} \qquad \text{Multiply each side by } 10^{12}.$$

$$\ln e^{-t/8223} = \ln \frac{1}{10} \qquad \text{Take natural log of each side.}$$

$$-\frac{t}{8223} \approx -2.3026 \qquad \text{Inverse Property}$$

$$t \approx 18,934 \qquad \text{Multiply each side by } -8223.$$

So, to the nearest thousand years, you can estimate the age of the fossil to be 19,000 years.

Graphical Solution

Use a graphing utility to graph the formula for the ratio of carbon 14 to carbon 12 at any time t as

$$y_1 = \frac{1}{10^{12}}e^{-x/8223}.$$

In the same viewing window, graph $y_2 = 1/(10^{13})$, as shown in the figure.

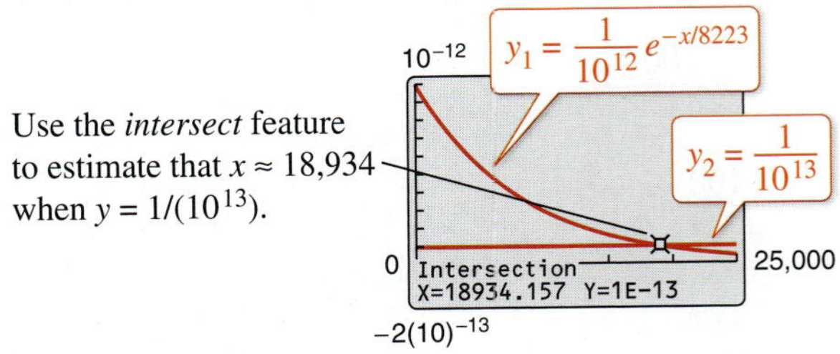

So, to the nearest thousand years, you can estimate the age of the fossil to be 19,000 years.

✓ **Checkpoint** *Audio-video solution in English & Spanish at LarsonPrecalculus.com.*

Estimate the age of a newly discovered fossil for which the ratio of carbon 14 to carbon 12 is $R = 1/10^{14}$.

The carbon dating model in Example 3 assumed that the carbon 14 to carbon 12 ratio was one part in 10,000,000,000,000. Suppose an error in measurement occurred and the actual ratio was only one part in 8,000,000,000,000. The fossil age corresponding to the actual ratio would then be approximately 17,000 years. Try checking this result.

Gaussian Models

As mentioned at the beginning of this section, Gaussian models are of the form

$$y = ae^{-(x-b)^2/c}.$$

This type of model is commonly used in probability and statistics to represent populations that are **normally distributed.** For *standard* normal distributions, the model takes the form

$$y = \frac{1}{\sqrt{2\pi}}e^{-x^2/2}.$$

The graph of a Gaussian model is called a **bell-shaped curve.** Try graphing the standard normal distribution curve with a graphing utility. Can you see why it is called a bell-shaped curve?

The average value for a population can be found from the bell-shaped curve by observing where the maximum y-value of the function occurs. The x-value corresponding to the maximum y-value of the function represents the average value of the independent variable—in this case, x.

EXAMPLE 4 SAT Scores

In 2013, the Scholastic Aptitude Test (SAT) mathematics scores for college-bound seniors roughly followed the normal distribution

$$y = 0.0034e^{-(x-514)^2/27{,}848}, \qquad 200 \le x \le 800$$

where x is the SAT score for mathematics. Use a graphing utility to graph this function and estimate the average SAT mathematics score. (Source: College Board)

Solution

The graph of the function is shown in Figure 4.27. On this bell-shaped curve, the maximum value of the curve represents the average score. Using the *maximum* feature of the graphing utility, you can see that the average mathematics score for college-bound seniors in 2013 was 514.

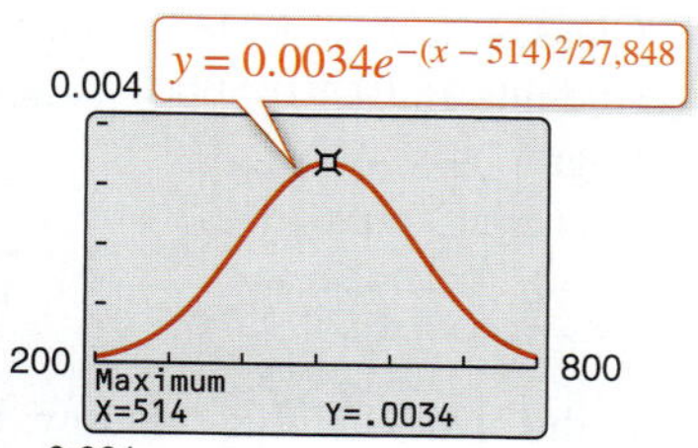

Figure 4.27

Checkpoint *Audio-video solution in English & Spanish at LarsonPrecalculus.com.*

In 2013, the Scholastic Aptitude Test (SAT) critical reading scores for college-bound seniors roughly followed the normal distribution

$$y = 0.0035e^{-(x-496)^2/26{,}450}$$

where x is the SAT score for critical reading. Use a graphing utility to graph this function and estimate the average SAT critical reading score. (Source: College Board)

In Example 4, note that 50% of the seniors who took the test received scores lower than 514 (see Figure 4.28).

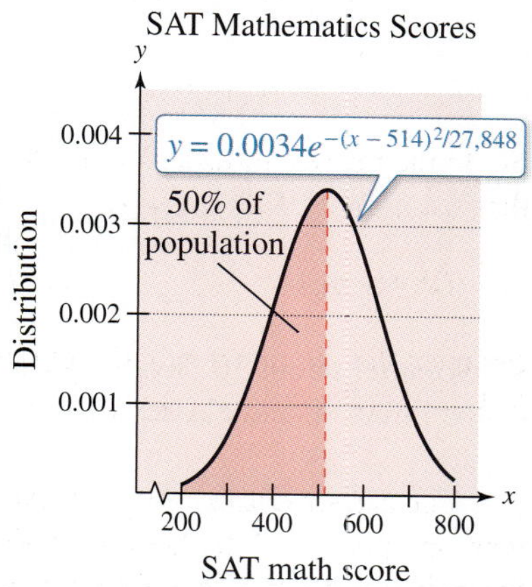

Figure 4.28

Logistic Growth Models

Some populations initially have rapid growth, followed by a declining rate of growth, as indicated by the graph in Figure 4.29. One model for describing this type of growth pattern is the **logistic curve** given by the function

$$y = \frac{a}{1 + be^{-rx}}$$

where y is the population size and x is the time. An example is a bacteria culture that is initially allowed to grow under ideal conditions, and then under less favorable conditions that inhibit growth. A logistic growth curve is also called a **sigmoidal curve.**

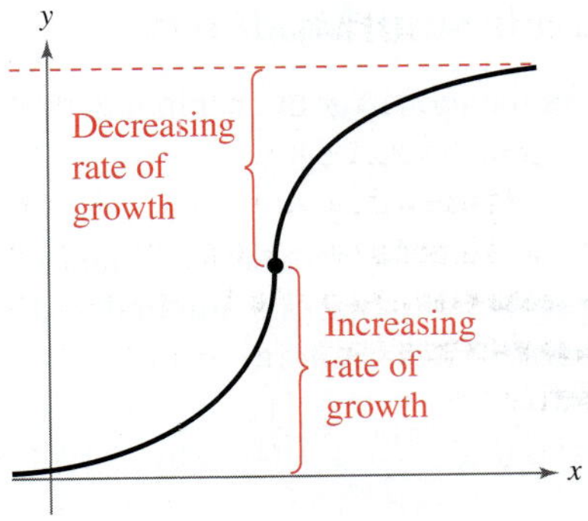

Figure 4.29 *Logistic Curve*

EXAMPLE 5 Spread of a Virus

On a college campus of 5000 students, one student returns from vacation with a contagious flu virus. The spread of the virus is modeled by

$$y = \frac{5000}{1 + 4999e^{-0.8t}}, \quad t \geq 0$$

where y is the total number of students infected after t days. The college will cancel classes when 40% or more of the students are infected.

a. How many students are infected after 5 days?

b. After how many days will the college cancel classes?

Algebraic Solution

a. After 5 days, the number of students infected is

$$y = \frac{5000}{1 + 4999e^{-0.8(5)}} = \frac{5000}{1 + 4999e^{-4}} \approx 54.$$

b. Classes are canceled when the number of infected students is $(0.40)(5000) = 2000$.

$$2000 = \frac{5000}{1 + 4999e^{-0.8t}}$$

$$1 + 4999e^{-0.8t} = 2.5$$

$$e^{-0.8t} = \frac{1.5}{4999}$$

$$\ln e^{-0.8t} = \ln \frac{1.5}{4999}$$

$$-0.8t = \ln \frac{1.5}{4999}$$

$$t = -\frac{1}{0.8} \ln \frac{1.5}{4999}$$

$$t \approx 10.14$$

So, after about 10 days, at least 40% of the students will be infected, and classes will be canceled.

Graphical Solution

a.

Use the *value* feature to estimate that $y \approx 54$ when $x = 5$. So, after 5 days, about 54 students will be infected.

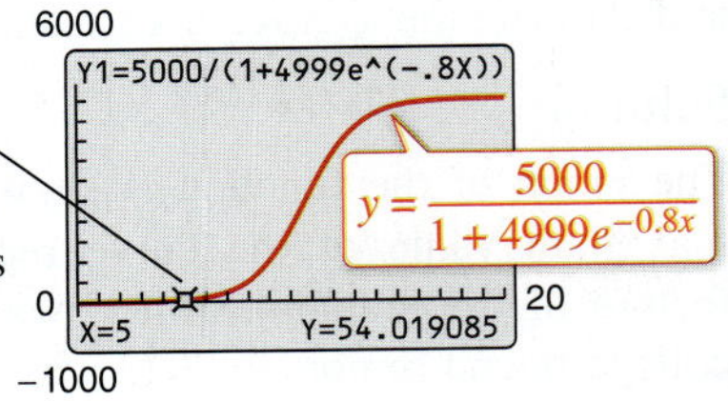

b. Classes are canceled when the number of infected students is $(0.40)(5000) = 2000$. Use a graphing utility to graph

$$y_1 = \frac{5000}{1 + 4999e^{-0.8x}} \quad \text{and} \quad y_2 = 2000$$

in the same viewing window. Use the *intersect* feature of the graphing utility to find the point of intersection of the graphs, as shown in the figure.

The intersection point occurs near $x \approx 10.14$. So, after about 10 days, at least 40% of the students will be infected, and classes will be canceled.

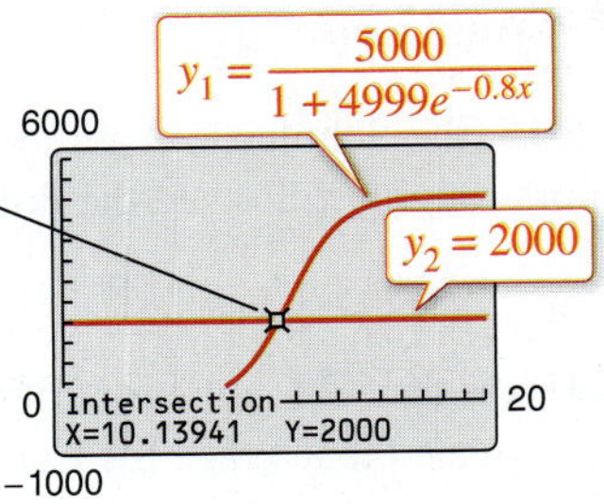

✓ Checkpoint *Audio-video solution in English & Spanish at LarsonPrecalculus.com.*

In Example 5, after how many days are 250 students infected?

Logarithmic Models

On the Richter scale, the magnitude R of an earthquake of intensity I is given by

$$R = \log_{10} \frac{I}{I_0}$$

where $I_0 = 1$ is the minimum intensity used for comparison. Intensity is a measure of the wave energy of an earthquake.

Seismologist

EXAMPLE 6 Magnitudes of Earthquakes

In 2014, Edgefield, South Carolina, experienced an earthquake that measured 4.1 on the Richter scale. Also in 2014, Nago, Japan, experienced an earthquake that measured 6.5 on the Richter scale. (Source: U.S. Geological Survey)

a. Find the intensity of each earthquake.

b. Compare the two intensities.

Solution

a. Because $I_0 = 1$ and $R = 4.1$, the intensity of the earthquake in South Carolina can be found as shown.

$R = \log_{10} \frac{I}{I_0}$ Write original model.

$4.1 = \log_{10} \frac{I}{1}$ Substitute 1 for I_0 and 4.1 for R.

$10^{4.1} = 10^{\log_{10} I}$ Exponentiate each side.

$10^{4.1} = I$ Inverse Property

So, the intensity was about $10^{4.1} \approx 12{,}589$. For the earthquake in Japan, $R = 6.5$. The intensity of this earthquake was

$R = \log_{10} \frac{I}{I_0}$ Write original model.

$6.5 = \log_{10} \frac{I}{1}$ Substitute 1 for I_0 and 6.5 for R.

$10^{6.5} = 10^{\log_{10} I}$ Exponentiate each side.

$10^{6.5} = I.$ Inverse Property

So, the intensity was about $10^{6.5} \approx 3{,}162{,}278$.

b. Note that an increase of 2.4 units on the Richter scale (from 4.1 to 6.5) represents an increase in intensity by a factor of

$$\frac{10^{6.5}}{10^{4.1}} = 10^{2.4}$$

$$\approx 251$$

In other words, the intensity of the earthquake in Japan was about 251 times as great as the intensity of the earthquake in South Carolina.

✓ **Checkpoint** *Audio-video solution in English & Spanish at LarsonPrecalculus.com.*

Find the intensities of earthquakes whose magnitudes are (a) $R = 6.0$ and (b) $R = 7.9$.

4.5 Exercises

See *CalcChat.com* for tutorial help and worked-out solutions to odd-numbered exercises.
For instructions on how to use a graphing utility, see Appendix A.

Vocabulary and Concept Check

1. Match each equation with its model.

(a) Exponential growth model (i) $y = ae^{-bx}, b > 0$
(b) Exponential decay model (ii) $y = a + b \ln x$
(c) Logistic growth model (iii) $y = \dfrac{a}{1 + be^{-rx}}$
(d) Gaussian model (iv) $y = ae^{bx}, b > 0$
(e) Natural logarithmic model (v) $y = a + b \log_{10} x$
(f) Common logarithmic model (vi) $y = ae^{-(x-b)^2/c}$

In Exercises 2 and 3, fill in the blank.

2. Gaussian models are commonly used in probability and statistics to represent populations that are _______ distributed.

3. Logistic growth curves are also called _______ curves.

4. Which model in Exercise 1 has a graph called a bell-shaped curve?

5. Does the model $y = 120e^{-0.25x}$ represent exponential growth or exponential decay?

6. Which model in Exercise 1 has a graph with two horizontal asymptotes?

Procedures and Problem Solving

Identifying Graphs of Models **In Exercises 7–12, match the function with its graph. [The graphs are labeled (a), (b), (c), (d), (e), and (f).]**

(a)

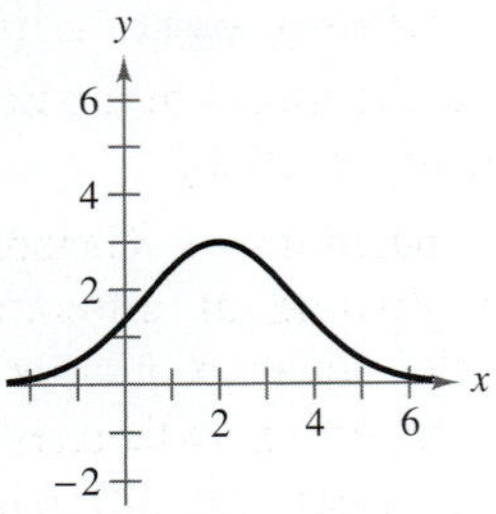

(b)

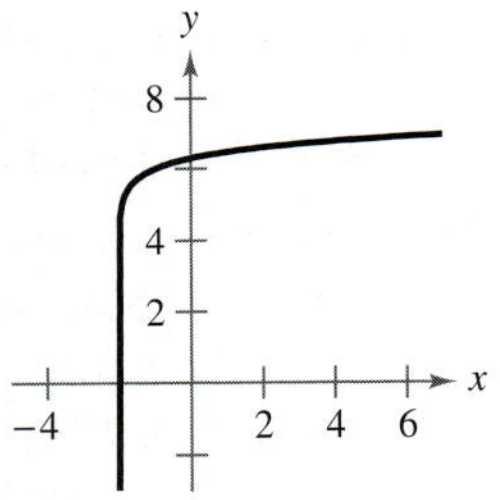

(c)

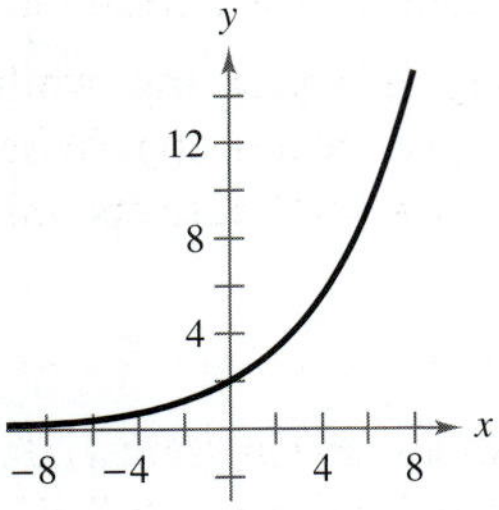

(d)

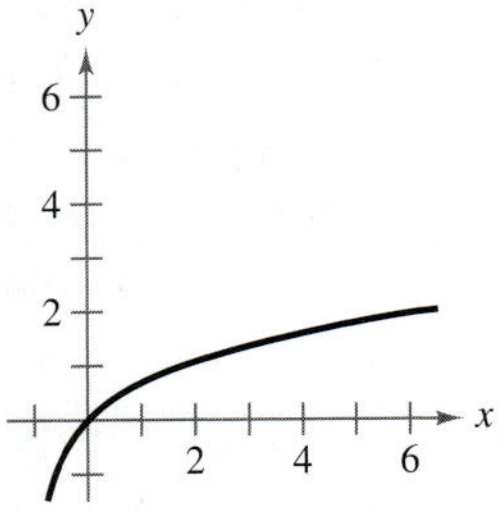

(e)

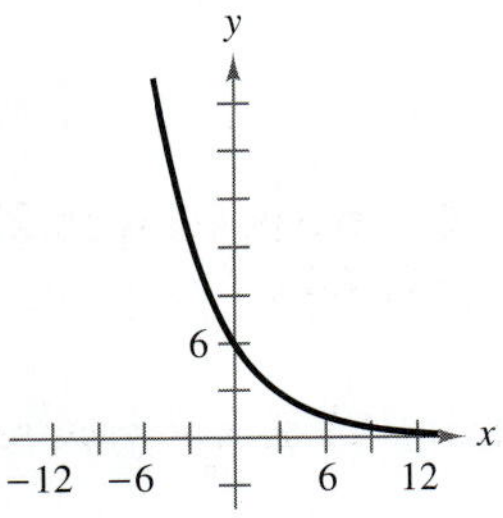

(f)

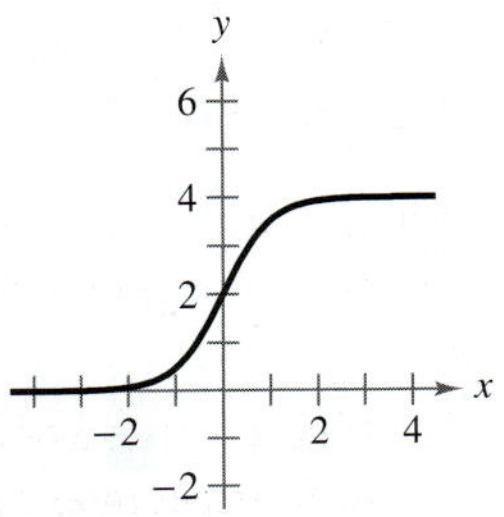

7. $y = 2e^{x/4}$ **8.** $y = 6e^{-x/4}$

9. $y = 6 + \log_{10}(x + 2)$ **10.** $y = 3e^{-(x-2)^2/5}$

11. $y = \ln(x + 1)$ **12.** $y = \dfrac{4}{1 + e^{-2x}}$

Using a Compound Interest Formula **In Exercises 13–20, complete the table for a savings account in which interest is compounded continuously.**

	Initial Investment	*Annual % Rate*	*Time to Double*	*Amount After 10 Years*
13.	\$10,000	4%		
14.	\$2000	2%		
15.	\$7500		21 years	
16.	\$1000		12 years	
17.	\$5000			\$5665.74
18.	\$300			\$385.21
19.		4.5%		\$100,000.00
20.		3.5%		\$2500.00

21. Tripling an Investment Complete the table for the time t (in years) necessary for P dollars to triple when interest is compounded continuously at rate r. Create a scatter plot of the data.

r	2%	4%	6%	8%	10%	12%
t						

22. Tripling an Investment Complete the table for the time t (in years) necessary for P dollars to triple when interest is compounded annually at rate r. Create a scatter plot of the data.

r	2%	4%	6%	8%	10%	12%
t						

23. Finance When \$1 is invested in an account over a 10-year period, the amount A in the account after t years is given by

$$A = 1 + 0.075t \quad \text{or} \quad A = e^{0.07t}$$

depending on whether the account pays simple interest at $7\frac{1}{2}\%$ or continuous compound interest at 7%. Use a graphing utility to graph each function in the same viewing window. Which grows at a greater rate?

24. Finance When \$1 is invested in an account over a 10-year period, the amount A in the account after t years is given by

$$A = 1 + 0.06t \quad \text{or} \quad A = \left(1 + \frac{0.055}{365}\right)^{365t}$$

depending on whether the account pays simple interest at 6% or compound interest at $5\frac{1}{2}\%$ compounded daily. Use a graphing utility to graph each function in the same viewing window. Which grows at a greater rate?

Radioactive Decay **In Exercises 25–30, complete the table for the radioactive isotope.**

	Isotope	*Half-Life (years)*	*Initial Quantity*	*Amount After 1000 Years*
25.	^{226}Ra	1600	10 g	
26.	^{14}C	5730	3 g	
27.	^{231}Pa	32,800		1.5 g
28.	^{239}Pu	24,100		0.4 g
29.	^{241}Am	432.2	26.4 g	
30.	^{238}Pu	87.74		0.1 g

Writing a Model **In Exercises 31–34, find the exponential model $y = ae^{bx}$ that fits the points shown in the graph or table.**

31.

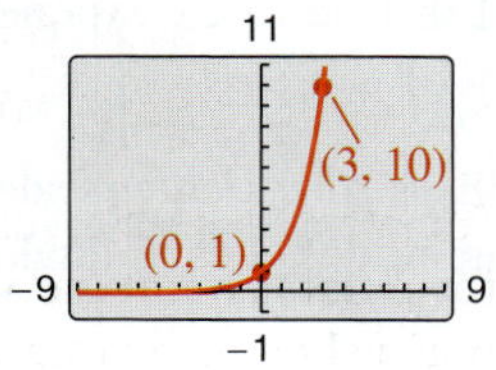

32.

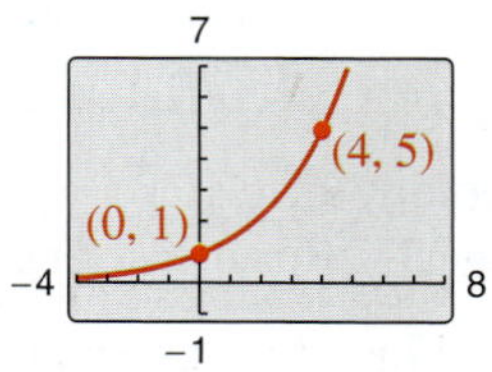

33.

x	0	5
y	4	1

34.

x	0	3
y	5	$\frac{1}{2}$

35. Demography The populations P (in thousands) of Antioch, California, from 2006 through 2012 can be modeled by $P = 90e^{0.013t}$, where t is the year, with $t = 6$ corresponding to 2006. (Source: U.S. Census Bureau)

(a) According to the model, was the population of Antioch increasing or decreasing from 2006 through 2012? Explain your reasoning.

(b) What were the populations of Antioch in 2006, 2009, and 2012?

(c) According to the model, when will the population of Antioch be approximately 116,000?

36. MODELING DATA

The table shows the populations (in millions) of five countries in 2013 and the projected populations (in millions) for 2025. (Source: U.S. Census Bureau)

Country	2013	2025
Australia	22.3	25.1
Canada	34.6	37.6
Hungary	9.9	9.6
Philippines	105.7	128.9
Turkey	80.7	90.5

(a) Find the exponential growth or decay model, $y = ae^{bt}$ or $y = ae^{-bt}$, for the population of each country, where t is the year, with $t = 13$ corresponding to 2013. Use the model to predict the population of each country in 2040.

(b) You can see that the populations of Canada and the Philippines are growing at different rates. What constant in the equation $y = ae^{bt}$ is determined by these different growth rates? Discuss the relationship between the different growth rates and the magnitude of the constant.

(c) The population of Turkey is increasing while the population of Hungary is decreasing. What constant in the equation $y = ae^{bt}$ reflects this difference? Explain.

37. Demography The populations P (in thousands) of Cameron County, Texas, from 2006 through 2012 can be modeled by

$$P = 339.2e^{kt}$$

where t is the year, with $t = 6$ corresponding to 2006. In 2011, the population was 412,600. (Source: U.S. Census Bureau)

(a) Find the value of k for the model. Round your result to four decimal places.

(b) Use your model to predict the population in 2018.

38. Demography The populations P (in thousands) of Pineville, North Carolina, from 2006 through 2012 can be modeled by $P = 5.4e^{kt}$, where t is the year, with $t = 6$ corresponding to 2006. In 2008, the population was 7000. (Source: U.S. Census Bureau)

(a) Find the value of k for the model. Round your result to four decimal places.

(b) Use your model to predict the population in 2018.

39. *Why you should learn it* *(p. 365)* Carbon 14 (^{14}C) dating assumes that the carbon dioxide on Earth today has the same radioactive content as it did centuries ago. If this is true, then the amount of ^{14}C absorbed by a tree that grew several centuries ago should be the same as the amount of ^{14}C absorbed by a tree growing today. A piece of ancient charcoal contains only 15% as much radioactive carbon as a piece of modern charcoal. How long ago was the tree burned to make the ancient charcoal, given that the half-life of ^{14}C is about 5700 years?

40. Radioactive Decay The half-life of radioactive radium (^{226}Ra) is 1600 years. What percent of a present amount of radioactive radium will remain after 100 years?

41. MODELING DATA

A new 2014 luxury sedan that sold for \$39,780 has a book value V of \$25,459 after 2 years.

(a) Find a linear model for the value V of the sedan.

(b) Find an exponential model for the value V of the sedan. Round the numbers in the model to four decimal places.

(c) Use a graphing utility to graph the two models in the same viewing window.

(d) Which model represents a greater depreciation rate in the first year?

(e) For what years is the value of the sedan greater using the linear model? the exponential model?

42. MODELING DATA

A new laptop computer that sold for \$1200 in 2014 has a book value V of \$650 after 2 years.

(a) Find a linear model for the value V of the laptop.

(b) Find an exponential model for the value V of the laptop. Round the numbers in the model to four decimal places.

(c) Use a graphing utility to graph the two models in the same viewing window.

(d) Which model represents a greater depreciation rate in the first year?

(e) For what years is the value of the laptop greater using the linear model? the exponential model?

43. Psychology The IQ scores for adults roughly follow the normal distribution $y = 0.0266e^{-(x-100)^2/450}$, $70 \le x \le 115$, where x is the IQ score.

(a) Use a graphing utility to graph the function.

(b) Use the graph in part (a) to estimate the average IQ score.

44. Marketing The sales S (in thousands of units) of a cleaning solution after x hundred dollars is spent on advertising are given by $S = 10(1 - e^{kx})$. When \$500 is spent on advertising, 2500 units are sold.

(a) Complete the model by solving for k.

(b) Estimate the number of units that will be sold when advertising expenditures are raised to \$700.

45. Forestry A conservation organization releases 100 animals of an endangered species into a game preserve. The organization believes that the preserve has a carrying capacity of 1000 animals and that the growth of the herd will follow the logistic curve

$$p(t) = \frac{1000}{1 + 9e^{-0.1656t}}$$

where t is measured in months.

(a) What is the population after 5 months?

(b) After how many months will the population reach 500?

(c) Use a graphing utility to graph the function. Use the graph to determine the values of p at which the horizontal asymptotes occur. Identify the asymptote that is most relevant in the context of the problem and interpret its meaning.

46. Biology The number Y of yeast organisms in a culture is given by the model

$$Y = \frac{663}{1 + 72e^{-0.547t}}$$

where t represents the time (in hours).

(a) Use a graphing utility to graph the model.

(b) Use the model to predict the populations for the 19th hour and the 30th hour.

(c) According to this model, what is the limiting value of the population?

(d) Why do you think this population of yeast follows a logistic growth model instead of an exponential growth model?

Geology In Exercises 47 and 48, use the Richter scale (see page 371) for measuring the magnitudes of earthquakes.

47. Find the intensities I of the following earthquakes measuring R on the Richter scale (let $I_0 = 1$). (Source: U.S. Geological Survey)

(a) Falkland Islands in 2013, $R = 7.0$

(b) Sea of Okhotsk in 2013, $R = 8.3$

(c) Kunisaki-shi, Japan in 2014, $R = 6.3$

48. Find the magnitudes R of the following earthquakes of intensity I (let $I_0 = 1$).

(a) $I = 39{,}811{,}000$

(b) $I = 12{,}589{,}000$

(c) $I = 251{,}200$

Audiology In Exercises 49–52, use the following information for determining sound intensity. The level of sound β (in decibels) with an intensity I is

$$\beta = 10 \log_{10} \frac{I}{I_0}$$

where I_0 is an intensity of 10^{-12} watt per square meter, corresponding roughly to the faintest sound that can be heard by the human ear. In Exercises 49 and 50, find the level of each sound β.

49. (a) $I = 10^{-10}$ watt per m^2 (quiet room)

(b) $I = 10^{-5}$ watt per m^2 (busy street corner)

(c) $I \approx 10^0$ watt per m^2 (threshold of pain)

50. (a) $I = 10^{-4}$ watt per m^2 (door slamming)

(b) $I = 10^{-3}$ watt per m^2 (loud car horn)

(c) $I = 10^{-2}$ watt per m^2 (siren at 30 meters)

51. As a result of the installation of a muffler, the noise level of an engine was reduced from 88 to 72 decibels. Find the percent decrease in the intensity level of the noise due to the installation of the muffler.

52. As a result of the installation of noise suppression materials, the noise level in an auditorium was reduced from 93 to 80 decibels. Find the percent decrease in the intensity level of the noise due to the installation of these materials.

Chemistry In Exercises 53–56, use the acidity model

$$\text{pH} = -\log[\text{H}^+]$$

where acidity (pH) is a measure of the hydrogen ion concentration $[\text{H}^+]$ (in moles of hydrogen per liter) of a solution.

53. Find the pH when $[\text{H}^+] = 2.3 \times 10^{-5}$.

54. Compute $[\text{H}^+]$ for a solution for which pH $= 5.8$.

55. A grape has a pH of 3.5, and baking soda has a pH of 8.0. The hydrogen ion concentration of the grape is how many times that of the baking soda?

56. The pH of a solution is decreased by one unit. The hydrogen ion concentration is increased by what factor?

57. Finance The total interest u paid on a home mortgage of P dollars at interest rate r for t years is given by

$$u = P\left[\frac{rt}{1 - \left(\dfrac{1}{1 + r/12}\right)^{12t}} - 1\right].$$

Consider a \$230,000 home mortgage at 3%.

(a) Use a graphing utility to graph the total interest function.

(b) Approximate the length of the mortgage when the total interest paid is the same as the amount of the mortgage. Is it possible that a person could pay twice as much in interest charges as the amount of the mortgage?

58. Finance A \$200,000 home mortgage for 30 years at 4.25% has a monthly payment of \$983.88. Part of the monthly payment goes toward the interest charge on the unpaid balance, and the remainder of the payment is used to reduce the principal. The amount that goes toward the interest is given by

$$u = M - \left(M - \frac{Pr}{12}\right)\left(1 + \frac{r}{12}\right)^{12t}$$

and the amount that goes toward reduction of the principal is given by

$$v = \left(M - \frac{Pr}{12}\right)\left(1 + \frac{r}{12}\right)^{12t}.$$

In these formulas, P is the amount of the mortgage, r is the interest rate, M is the monthly payment, and t is the time (in years).

(a) Use a graphing utility to graph each function in the same viewing window. (The viewing window should show all 30 years of mortgage payments.)

(b) In the early years of the mortgage, the larger part of the monthly payment goes for what purpose? Approximate the time when the monthly payment is evenly divided between interest and principal reduction.

(c) Repeat parts (a) and (b) for a repayment period of 20 years ($M = \$1238.47$). What can you conclude?

59. Forensics At 8:30 A.M., a coroner was called to the home of a person who had died during the night. In order to estimate the time of death, the coroner took the person's temperature twice. At 9:00 A.M. the temperature was 85.7°F, and at 11:00 A.M. the temperature was 82.8°F. From these two temperatures the coroner was able to determine that the time elapsed since death and the body temperature were related by the formula

$$t = -10 \ln \frac{T - 70}{98.6 - 70}$$

where t is the time (in hours elapsed since the person died) and T is the temperature (in degrees Fahrenheit) of the person's body. Assume that the person had a normal body temperature of 98.6°F at death and that the room temperature was a constant 70°F. Use the formula to estimate the time of death of the person. (This formula is derived from a general cooling principle called Newton's Law of Cooling.)

60. **Culinary Arts** You take a five-pound package of steaks out of a freezer at 11 A.M. and place it in a refrigerator. Will the steaks be thawed in time to be grilled at 6 P.M.? Assume that the refrigerator temperature is 40°F and the freezer temperature is 0°F. Use the formula (derived from Newton's Law of Cooling)

$$t = -5.05 \ln \frac{T - 40}{0 - 40}$$

where t is the time in hours (with $t = 0$ corresponding to 11 A.M.) and T is the temperature of the package of steaks (in degrees Fahrenheit).

Conclusions

True or False? **In Exercises 61 and 62, determine whether the statement is true or false. Justify your answer.**

61. The domain of a logistic growth function cannot be the set of real numbers.

62. The graph of a logistic growth function will always have an x-intercept.

63. **Think About It** Can the graph of a Gaussian model ever have an x-intercept? Explain.

64. **HOW DO YOU SEE IT?** For each graph, state whether an exponential, Gaussian, logarithmic, logistic, or quadratic model will fit the data best. Explain your reasoning. Then describe a real-life situation that could be represented by the data.

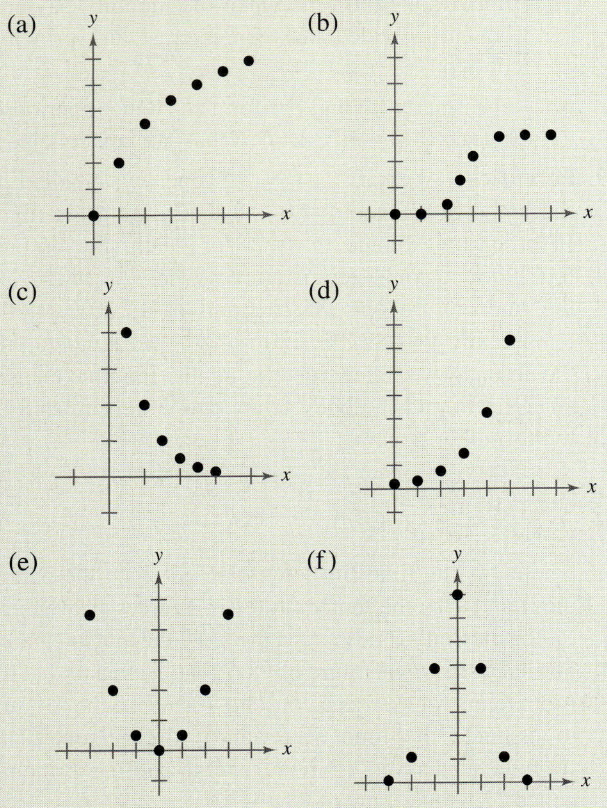

Identifying Models **In Exercises 65–68, identify the type of model you studied in this section that has the given characteristic.**

65. The maximum value of the function occurs at the average value of the independent variable.

66. A horizontal asymptote of its graph represents the limiting value of a population.

67. Its graph shows a steadily increasing rate of growth.

68. The only asymptote of its graph is a vertical asymptote.

Cumulative Mixed Review

Identifying Graphs of Linear Equations **In Exercises 69–72, match the equation with its graph and identify any intercepts. [The graphs are labeled (a), (b), (c), and (d).]**

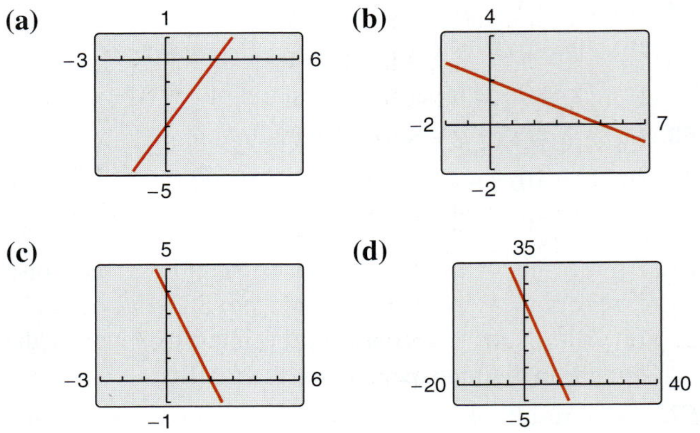

69. $4x - 3y - 9 = 0$

70. $2x + 5y - 10 = 0$

71. $y = 25 - 2.25x$

72. $\frac{x}{2} + \frac{y}{4} = 1$

Applying the Leading Coefficient Test **In Exercises 73–76, use the Leading Coefficient Test to determine the right-hand and left-hand behavior of the graph of the polynomial function.**

73. $f(x) = 2x^3 - 3x^2 + x - 1$

74. $f(x) = 5 - x^2 - 4x^4$

75. $g(x) = -1.6x^5 + 4x^2 - 2$

76. $g(x) = 7x^6 + 9.1x^5 - 3.2x^4 + 25x^3$

Using Synthetic Division **In Exercises 77 and 78, divide using synthetic division.**

77. $(2x^3 - 8x^2 + 3x - 9) \div (x - 4)$

78. $(x^4 - 3x + 1) \div (x + 5)$

79. ***Make a Decision*** To work an extended application analyzing the sales per share for Kohl's Corporation from 1995 through 2012, visit this textbook's website at *LarsonPrecalculus.com*. (Data Source: Kohl's Corp.)

4.6 Nonlinear Models

Classifying Scatter Plots

In Section 2.7, you saw how to fit linear models to data, and in Section 3.7, you saw how to fit quadratic models to data. In real life, many relationships between two variables are represented by different types of growth patterns. You can use a scatter plot to get an idea of which type of model will best fit a set of data.

EXAMPLE 1 Classifying Scatter Plots

See LarsonPrecalculus.com for an interactive version of this type of example.

Decide whether each set of data could best be modeled by a linear model, $y = ax + b$, an exponential model, $y = ab^x$, or a logarithmic model, $y = a + b \ln x$.

a. (2, 1), (2.5, 1.2), (3, 1.3), (3.5, 1.5), (4, 1.8), (4.5, 2), (5, 2.4), (5.5, 2.5), (6, 3.1), (6.5, 3.8), (7, 4.5), (7.5, 5), (8, 6.5), (8.5, 7.8), (9, 9), (9.5, 10)

b. (2, 2), (2.5, 3.1), (3, 3.8), (3.5, 4.3), (4, 4.6), (4.5, 5.3), (5, 5.6), (5.5, 5.9), (6, 6.2), (6.5, 6.4), (7, 6.9), (7.5, 7.2), (8, 7.6), (8.5, 7.9), (9, 8), (9.5, 8.2)

c. (2, 1.9), (2.5, 2.5), (3, 3.2), (3.5, 3.6), (4, 4.3), (4.5, 4.7), (5, 5.2), (5.5, 5.7), (6, 6.4), (6.5, 6.8), (7, 7.2), (7.5, 7.9), (8, 8.6), (8.5, 8.9), (9, 9.5), (9.5, 9.9)

Solution

a. From the figure, it appears that the data can best be modeled by an exponential function.

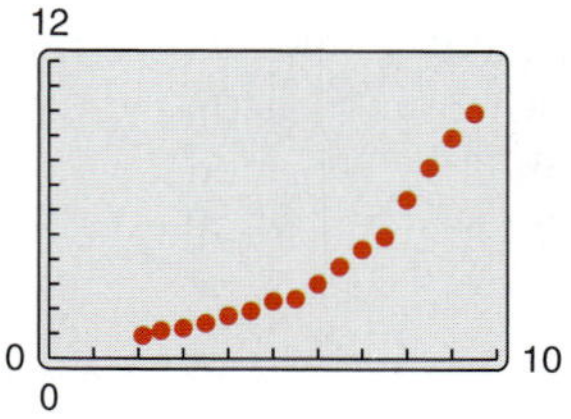

b. From the figure, it appears that the data can best be modeled by a logarithmic function.

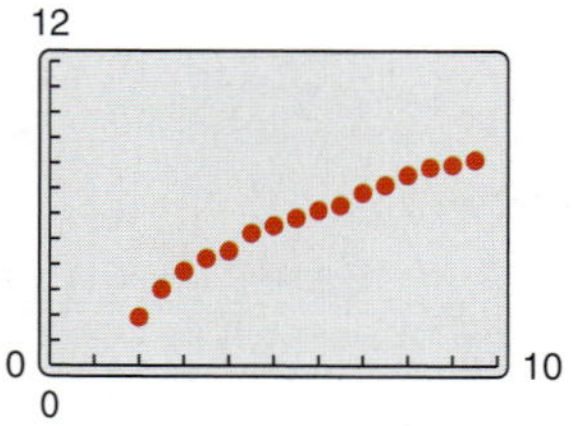

c. From the figure, it appears that the data can best be modeled by a linear function.

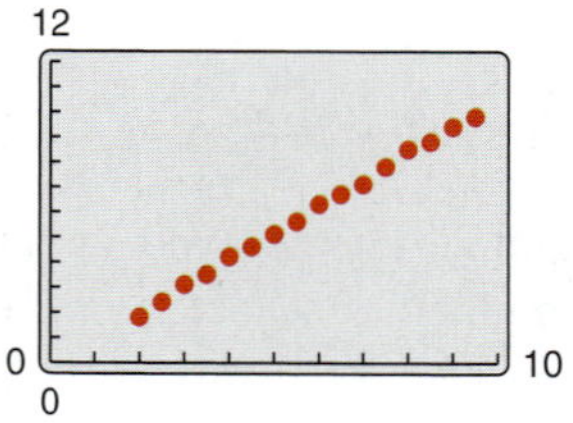

✓ Checkpoint *Audio-video solution in English & Spanish at LarsonPrecalculus.com.*

Decide whether the set of data could best be modeled by a linear model, $y = ax + b$, an exponential model, $y = ab^x$, or a logarithmic model, $y = a + b \ln x$.

(0.5, 12), (1, 8), (1.5, 5.7), (2, 4), (2.5, 3.1), (3, 2), (3.5, 1.4), (4, 1), (4.5, 0.8), (5, 0.6), (5.5, 0.4), (6, 0.25), (6.5, 0.18), (7, 0.13), (7.5, 0.09), (8, 0.06)

What you should learn

- Classify scatter plots.
- Use scatter plots and a graphing utility to find models for data and choose the model that best fits a set of data.
- Use a graphing utility to find exponential and logistic models for data.

Why you should learn it

Many real-life applications can be modeled by nonlinear equations. For instance, in Exercise 34 on page 384, you are asked to find a nonlinear model that relates air pressure to altitude.

Fitting Nonlinear Models to Data

Once you have used a scatter plot to determine the type of model that would best fit a set of data, there are several ways that you can actually find the model. Each method is best used with a computer or calculator, rather than with hand calculations.

EXAMPLE 2 Fitting a Model to Data

Fit the data from Example 1(a) to an exponential model and a power model. Identify the coefficient of determination and determine which model fits the data better.

(2, 1), (2.5, 1.2), (3, 1.3), (3.5, 1.5), (4, 1.8), (4.5, 2), (5, 2.4), (5.5, 2.5), (6, 3.1), (6.5, 3.8), (7, 4.5), (7.5, 5), (8, 6.5), (8.5, 7.8), (9, 9), (9.5, 10)

Solution

Begin by entering the data into a graphing utility. Then use the *regression* feature of the graphing utility to find exponential and power models for the data, as shown in Figure 4.30.

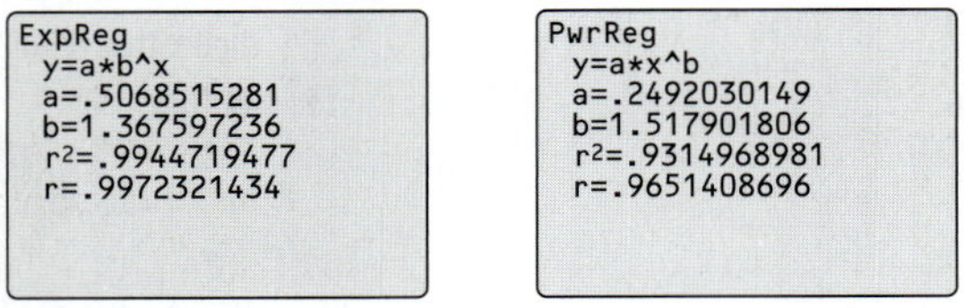

Exponential Model **Power Model**
Figure 4.30

So, an exponential model for the data is $y = 0.507(1.368)^x$, and a power model for the data is $y = 0.249x^{1.518}$. Plot the data and each model in the same viewing window, as shown in Figure 4.31. To determine which model fits the data better, compare the coefficients of determination for each model. The model whose r^2-value is closest to 1 is the model that better fits the data. In this case, the better-fitting model is the exponential model.

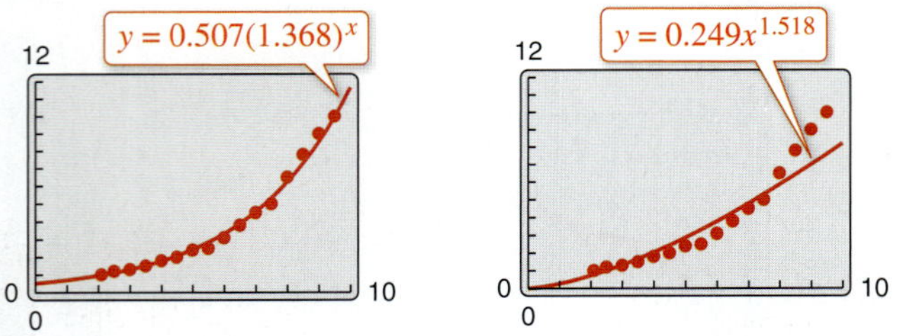

Exponential Model **Power Model**
Figure 4.31

✓ **Checkpoint** *Audio-video solution in English & Spanish at LarsonPrecalculus.com.*

Fit the data from the Checkpoint for Example 1 to an exponential model and a power model. Identify the coefficient of determination and determine which model fits the data better.

(0.5, 12), (1, 8), (1.5, 5.7), (2, 4), (2.5, 3.1), (3, 2), (3.5, 1.4), (4, 1), (4.5, 0.8), (5, 0.6), (5.5, 0.4), (6, 0.25), (6.5, 0.18), (7, 0.13), (7.5, 0.09), (8, 0.06)

Deciding which model best fits a set of data is a question that is studied in detail in statistics. Recall from Section 2.7 that the model that best fits a set of data is the one whose *sum of squared differences* is the least. In Example 2, the sums of squared differences are 0.90 for the exponential model and 14.30 for the power model.

EXAMPLE 3 Fitting a Model to Data

The table shows the yield y (in milligrams) of a chemical reaction after x minutes. Use a graphing utility to find a logarithmic model and a linear model for the data and identify the coefficient of determination for each model. Determine which model fits the data better.

DATA

Minutes, x	Yield, y
1	1.5
2	7.4
3	10.2
4	13.4
5	15.8
6	16.3
7	18.2
8	18.3

Spreadsheet at LarsonPrecalculus.com

Solution

Begin by entering the data into a graphing utility. Then use the *regression* feature of the graphing utility to find logarithmic and linear models for the data, as shown in Figure 4.32.

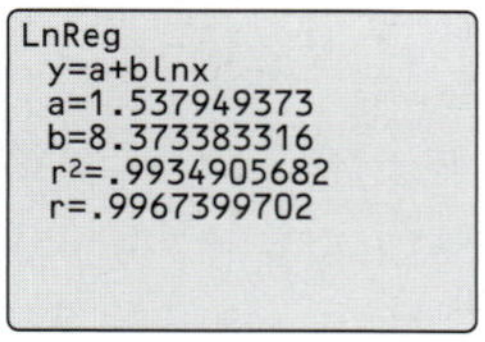

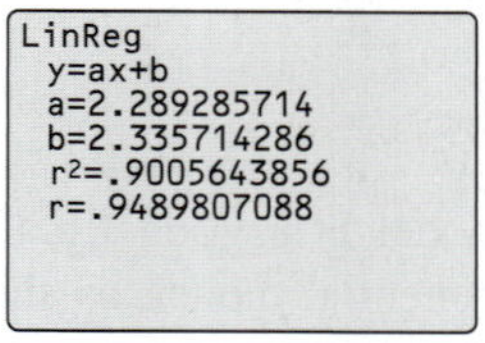

Logarithmic Model **Linear Model**
Figure 4.32

So, a logarithmic model for the data is $y = 1.538 + 8.373 \ln x$, and a linear model for the data is $y = 2.29x + 2.3$. Plot the data and each model in the same viewing window, as shown in Figure 4.33. To determine which model fits the data better, compare the coefficients of determination for each model. The model whose coefficient of determination is closer to 1 is the model that better fits the data. In this case, the better-fitting model is the logarithmic model.

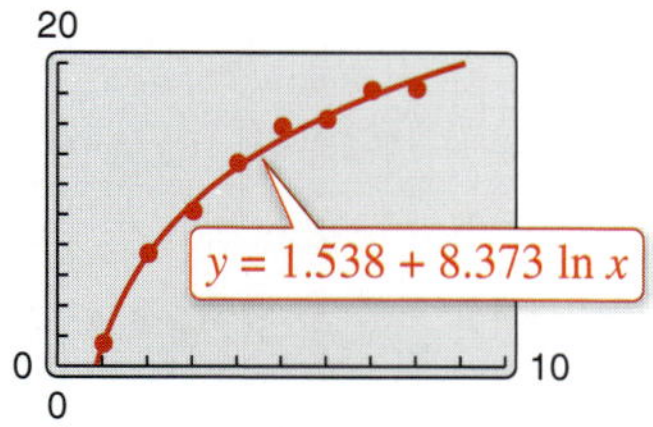

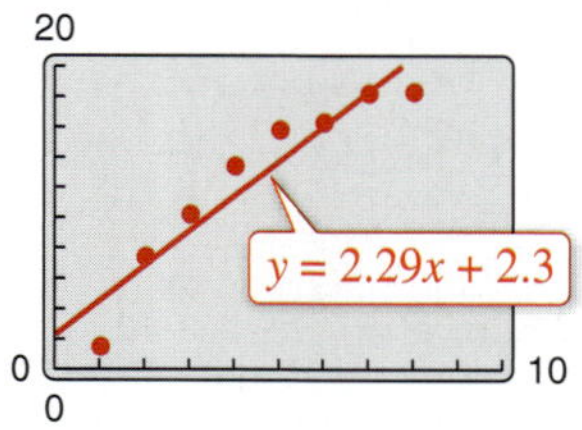

Logarithmic Model **Linear Model**
Figure 4.33

✓ *Checkpoint* *Audio-video solution in English & Spanish at LarsonPrecalculus.com.*

Use a graphing utility to find a logarithmic model and a linear model for the set of data. Identify the coefficient of determination for each model and determine which model fits the data better.

(0.5, 1.7), (1, 2.5), (1.5, 3), (2, 3.2), (2.5, 3.5), (3, 3.7),
(3.5, 3.8), (4, 3.9), (4.5, 3.9), (5, 4.1), (5.5, 4.2), (6, 4.3),
(6.5, 4.5), (7, 4.6), (7.5, 4.8), (8, 5)

In Example 3, the sum of the squared differences for the logarithmic model is 1.59 and the sum of the squared differences for the linear model is 24.31.

Explore the Concept

Use a graphing utility to find a quadratic model for the data in Example 3. Do you think this model fits the data better than the logarithmic model in Example 3? Explain your reasoning.

Modeling With Exponential and Logistic Functions

EXAMPLE 4 Fitting an Exponential Model to Data

The table shows the amount y (in grams) of a radioactive substance remaining after x days. Use a graphing utility to find a model for the data. How much of the substance remains after 15 days?

DATA

Day, x	Amount, y
0	10.0
1	8.9
2	7.8
3	6.7
4	6.0
5	5.3
6	4.7
7	4.0
8	3.5

Spreadsheet at LarsonPrecalculus.com

Solution

Begin by entering the data into a graphing utility and displaying the scatter plot, as shown in Figure 4.34.

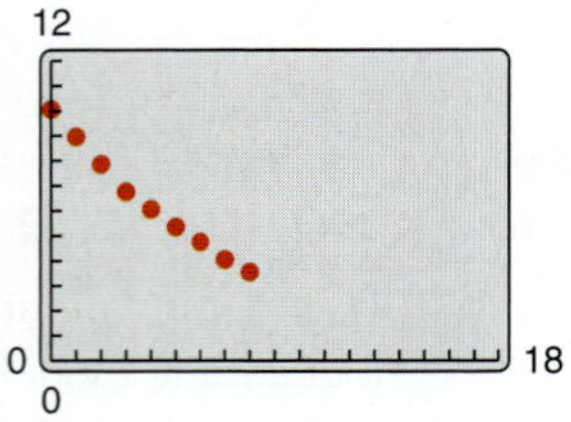

Figure 4.34

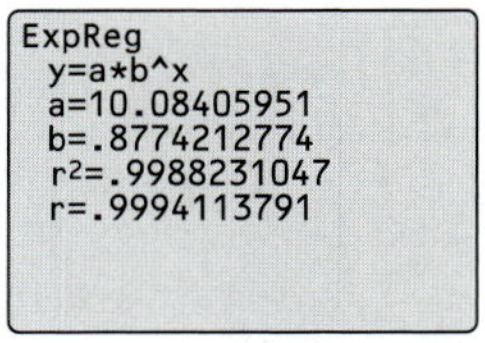

Figure 4.35

From the scatter plot, it appears that an exponential model is a good fit. Use the *regression* feature of the graphing utility to find the exponential model, as shown in Figure 4.35. Change the model to a natural exponential model, as follows.

$y = 10.08(0.877)^x$ Write original model.

$= 10.08e^{(\ln 0.877)x}$ $b = e^{\ln b}$

$\approx 10.08e^{-0.131x}$ Simplify.

Graph the data and the natural exponential model

$$y = 10.08e^{-0.131x}$$

in the same viewing window, as shown in Figure 4.36. From the model, you can see that the amount of the substance decreases by about 13% each day. Also, the amount of the substance remaining after 15 days is

$y = 10.08e^{-0.131x}$ Write natural exponential model.

$= 10.08e^{-0.131(15)}$ Substitute 15 for x.

≈ 1.4 grams. Use a calculator.

You can also use the *value* feature of the graphing utility to approximate the amount remaining after 15 days, as shown in Figure 4.37.

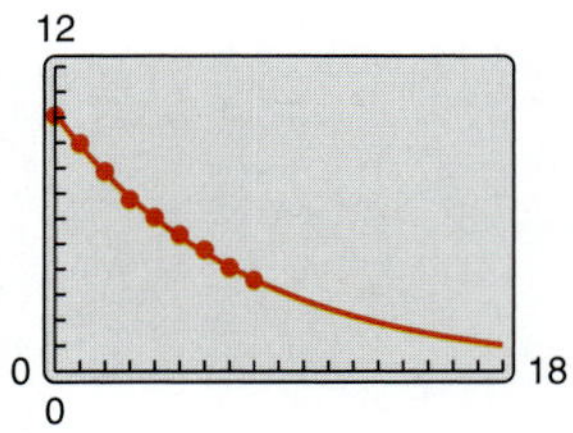

Figure 4.36

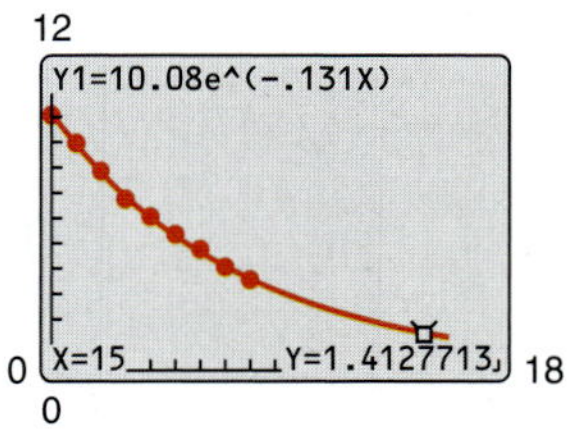

Figure 4.37

Remark

You can change an exponential model of the form

$$y = ab^x$$

to one of the form

$$y = ae^{cx}$$

by rewriting b in the form

$$b = e^{\ln b}.$$

For instance,

$$y = 3(2^x)$$

can be written as

$$y = 3(2^x)$$
$$= 3e^{(\ln 2)x}$$
$$\approx 3e^{0.693x}.$$

✓ ***Checkpoint*** *Audio-video solution in English & Spanish at LarsonPrecalculus.com.*

In Example 4, the initial amount of the substance is 10 grams. According to the natural exponential model, after how many days is the amount of the substance 5 grams? Interpret this value in the context of the problem.

The next example demonstrates how to use a graphing utility to fit a logistic model to data.

Forester

EXAMPLE 5 Fitting a Logistic Model to Data

To estimate the amount of defoliation caused by the gypsy moth during a given year, a forester counts the number x of egg masses on $\frac{1}{40}$ of an acre (circle of radius 18.6 feet) in the fall. The percent of defoliation y the next spring is shown in the table. (Source: USDA, Forest Service)

Egg masses, x	Percent of defoliation, y
0	12
25	44
50	81
75	96
100	99

a. Use the *regression* feature of a graphing utility to find a logistic model for the data.

b. How closely does the model represent the data?

Graphical Solution

a. Enter the data into a graphing utility. Using the *regression* feature of the graphing utility, you can find the logistic model, as shown in the figure.

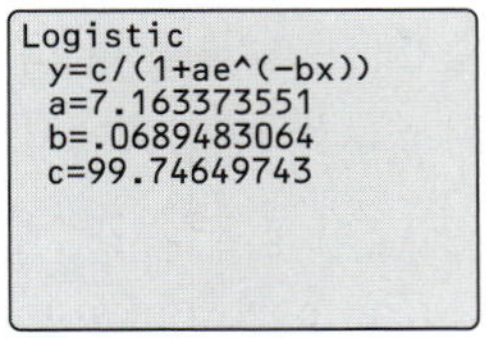

You can approximate this model to be

$$y = \frac{100}{1 + 7e^{-0.069x}}.$$

b. You can use the graphing utility to graph the actual data and the model in the same viewing window. In the figure, it appears that the model is a good fit for the actual data.

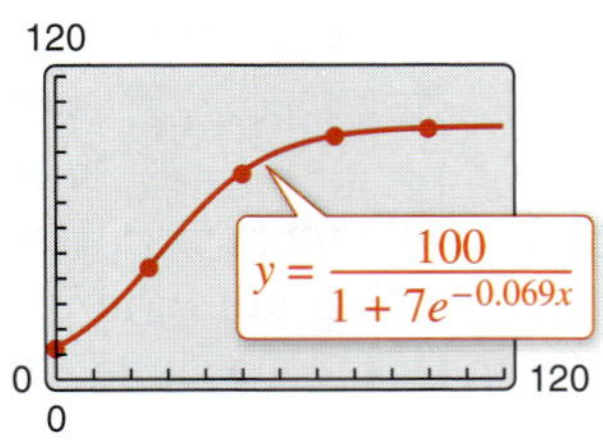

Numerical Solution

a. Enter the data into a graphing utility. Using the *regression* feature of the graphing utility, you can approximate the logistic model to be

$$y = \frac{100}{1 + 7e^{-0.069x}}.$$

b. You can see how well the model fits the data by comparing the actual values of y with the values of y given by the model, which are labeled y^* in the table below.

x	0	25	50	75	100
y	12	44	81	96	99
y^*	12.5	44.5	81.8	96.2	99.3

In the table, you can see that the model appears to be a good fit for the actual data.

✓ Checkpoint Audio-video solution in English & Spanish at LarsonPrecalculus.com.

Use a graphing utility to graph the logistic model in Example 5. Use the graph to determine the horizontal asymptotes. Interpret the asymptotes in the context of the real-life situation.

4.6 Exercises

See CalcChat.com for tutorial help and worked-out solutions to odd-numbered exercises.
For instructions on how to use a graphing utility, see Appendix A.

Vocabulary and Concept Check

In Exercises 1 and 2, fill in the blank.

1. A power model has the form _______ .
2. An exponential model of the form $y = ab^x$ can be rewritten as a natural exponential model of the form _______ .
3. What type of visual display can you create to get an idea of which type of model will best fit the data set?
4. A power model for a set of data has a coefficient of determination of $r^2 \approx 0.901$ and an exponential model for the data has a coefficient of determination of $r^2 \approx 0.967$. Which model fits the data better?

Procedures and Problem Solving

Classifying Scatter Plots **In Exercises 5–12, determine whether the scatter plot could best be modeled by a linear model, a quadratic model, an exponential model, a logarithmic model, or a logistic model.**

5. 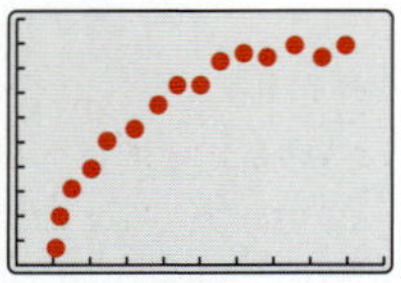

6. 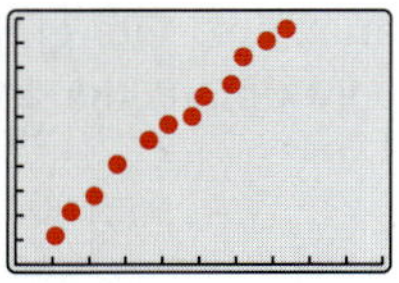

7. 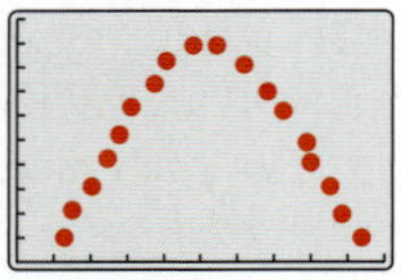

8. 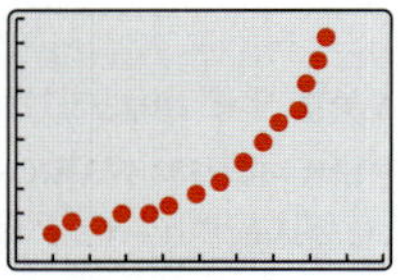

9. 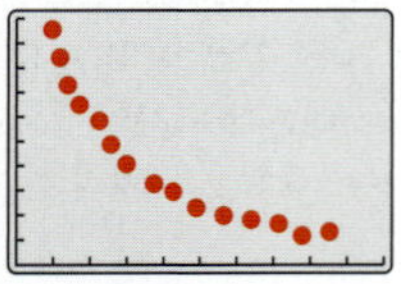

10. 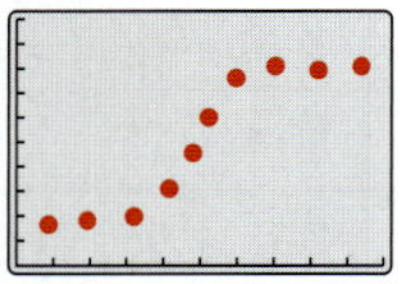

11. 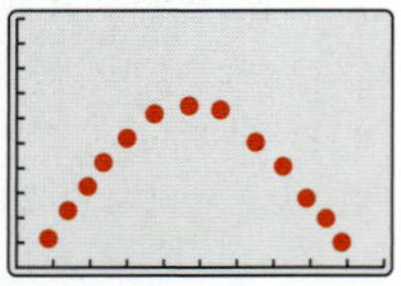

12. 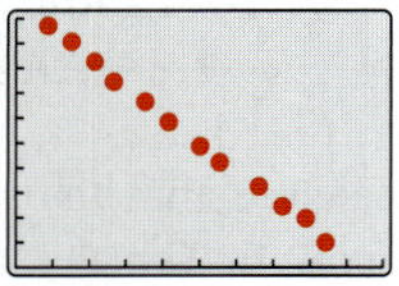

Classifying Scatter Plots **In Exercises 13–18, use a graphing utility to create a scatter plot of the data. Decide whether the data could best be modeled by a linear model, an exponential model, or a logarithmic model.**

13. (1, 2.0), (1.5, 3.5), (2, 4.0), (4, 5.8), (6, 7.0), (8, 7.8)
14. (1, 5.8), (1.5, 6.0), (2, 6.5), (4, 7.6), (6, 8.9), (8, 10.0)
15. (1, 4.4), (1.5, 4.7), (2, 5.5), (4, 9.9), (6, 18.1), (8, 33.0)
16. (1, 11.0), (1.5, 9.6), (2, 8.2), (4, 4.5), (6, 2.5), (8, 1.4)
17. (1, 7.5), (1.5, 7.0), (2, 6.8), (4, 5.0), (6, 3.5), (8, 2.0)
18. (1, 5.0), (1.5, 6.0), (2, 6.4), (4, 7.8), (6, 8.6), (8, 9.0)

Finding an Exponential Model **In Exercises 19–22, use the *regression* feature of a graphing utility to find an exponential model $y = ab^x$ for the data and identify the coefficient of determination. Use the graphing utility to plot the data and graph the model in the same viewing window.**

19. (0, 5), (1, 6), (2, 7), (3, 9), (4, 13)
20. (0, 4.0), (2, 6.9), (4, 18.0), (6, 32.3), (8, 59.1), (10, 118.5)
21. (0, 8.3), (1, 6.1), (2, 4.6), (3, 3.8), (4, 3.6)
22. (−3, 102.2), (0, 80.5), (3, 67.8), (6, 58.2), (10, 55.0)

Finding a Logarithmic Model **In Exercises 23–26, use the *regression* feature of a graphing utility to find a logarithmic model $y = a + b \ln x$ for the data and identify the coefficient of determination. Use the graphing utility to plot the data and graph the model in the same viewing window.**

23. (1, 2.0), (2, 3.0), (3, 3.5), (4, 4.0), (5, 4.1), (6, 4.2), (7, 4.5)
24. (1, 8.5), (2, 11.4), (4, 12.8), (6, 13.6), (8, 14.2), (10, 14.6)
25. (1, 11), (2, 6), (3, 5), (4, 4), (5, 3), (6, 2)
26. (3, 14.6), (6, 11.0), (9, 9.0), (12, 7.6), (15, 6.5)

Finding a Power Model **In Exercises 27–30, use the *regression* feature of a graphing utility to find a power model $y = ax^b$ for the data and identify the coefficient of determination. Use the graphing utility to plot the data and graph the model in the same viewing window.**

27. (1, 2.0), (2, 3.4), (5, 6.7), (6, 7.3), (10, 12.0)
28. (0.5, 1.0), (2, 12.5), (4, 33.2), (6, 65.7), (8, 98.5), (10, 150.0)
29. (1, 10.0), (2, 4.0), (3, 0.7), (4, 0.1)
30. (2, 450), (4, 385), (6, 345), (8, 332), (10, 312)

31. MODELING DATA

The table shows the yearly sales S (in millions of dollars) of Whole Foods Market for the years 2006 through 2013. (Source: Whole Foods Market)

DATA

Year	Sales, S
2006	5,607.4
2007	6,591.8
2008	7,953.9
2009	8,031.6
2010	9,005.8
2011	10,108.0
2012	11,699.0
2013	12,917.0

Spreadsheet at LarsonPrecalculus.com

(a) Use the *regression* feature of a graphing utility to find an exponential model and a power model for the data and identify the coefficient of determination for each model. Let t represent the year, with $t = 6$ corresponding to 2006.

(b) Use the graphing utility to graph each model with the data.

(c) Use the coefficients of determination to determine which model fits the data better.

32. MODELING DATA

The table shows the numbers of single beds B (in thousands) on North American cruise ships from 2007 through 2012. (Source: Cruise Lines International Association)

DATA

Year	Beds, B
2007	260.0
2008	270.7
2009	284.8
2010	307.7
2011	321.2
2012	333.7

Spreadsheet at LarsonPrecalculus.com

(a) Use the *regression* feature of a graphing utility to find a linear model, an exponential model, and a logarithmic model for the data and identify the coefficient of determination for each model. Let t represent the year, with $t = 7$ corresponding to 2007.

(b) Which model is the best fit for the data? Explain.

(c) Use the model you chose in part (b) to predict the number of beds in 2017. Is the number reasonable?

33. MODELING DATA

The populations P (in thousands) of Luxembourg for the years 1999 through 2013 are shown in the table, where t represents the year, with $t = 9$ corresponding to 1999. (Source: European Commission Eurostat)

DATA

Year	Population, P
1999	427.4
2000	433.6
2001	439.0
2002	444.1
2003	448.3
2004	455.0
2005	461.2
2006	469.1
2007	476.2
2008	483.8
2009	493.5
2010	502.1
2011	511.8
2012	524.9
2013	537.0

Spreadsheet at LarsonPrecalculus.com

(a) Use the *regression* feature of a graphing utility to find a linear model for the data and to identify the coefficient of determination. Plot the model and the data in the same viewing window.

(b) Use the *regression* feature of the graphing utility to find a power model for the data and to identify the coefficient of determination. Plot the model and the data in the same viewing window.

(c) Use the *regression* feature of the graphing utility to find an exponential model for the data and to identify the coefficient of determination. Plot the model and the data in the same viewing window.

(d) Use the *regression* feature of the graphing utility to find a logarithmic model for the data and to identify the coefficient of determination. Plot the model and the data in the same viewing window.

(e) Which model is the best fit for the data? Explain.

(f) Use each model to predict the populations of Luxembourg for the years 2014 through 2018.

(g) Which model is the best choice for predicting the future population of Luxembourg? Explain.

(h) Were your choices of models the same for parts (e) and (g)? If not, explain why your choices were different.

34. ***Why you should learn it*** *(p. 377)* The atmospheric pressure decreases with increasing altitude. At sea level, the average air pressure is approximately 1.03323 kilograms per square centimeter, and this pressure is called one atmosphere. Variations in weather conditions cause changes in the atmospheric pressure of up to ±5 percent. The ordered pairs (h, p) give the pressures p (in atmospheres) for various altitudes h (in kilometers). *(Spreadsheet at LarsonPrecalculus.com)*

DATA $(0, 1), (10, 0.25), (20, 0.06), (5, 0.55),$
$(15, 0.12), (25, 0.02)$

(a) Use the *regression* feature of a graphing utility to attempt to find the logarithmic model $p = a + b \ln h$ for the data. Explain why the result is an error message.

(b) Use the *regression* feature of the graphing utility to find the logarithmic model $h = a + b \ln p$ for the data.

(c) Use the graphing utility to plot the data and graph the logarithmic model in the same viewing window.

(d) Use the model to estimate the altitude at which the pressure is 0.75 atmosphere.

(e) Use the graph in part (c) to estimate the pressure at an altitude of 13 kilometers.

35. MODELING DATA

The table shows the annual sales S (in billions of dollars) of Starbucks for the years from 2009 through 2013. (Source: Starbucks Corp.)

Year	Sales, S
2009	9.77
2010	10.71
2011	11.70
2012	13.30
2013	14.89

(a) Use the *regression* feature of a graphing utility to find an exponential model for the data. Let t represent the year, with $t = 9$ corresponding to 2009.

(b) Rewrite the model from part (a) as a natural exponential model.

(c) Use the natural exponential model to predict the annual sales of Starbucks in 2018. Is the value reasonable?

36. MODELING DATA

A beaker of liquid at an initial temperature of 78°C is placed in a room at a constant temperature of 21°C. The temperature of the liquid is measured every 5 minutes for a period of $\frac{1}{2}$ hour. The results are recorded in the table, where t is the time (in minutes) and T is the temperature (in degrees Celsius).

DATA

Time, t	Temperature, T
0	78.0°
5	66.0°
10	57.5°
15	51.2°
20	46.3°
25	42.5°
30	39.6°

Spreadsheet at LarsonPrecalculus.com

(a) Use the *regression* feature of a graphing utility to find a linear model for the data. Use the graphing utility to plot the data and graph the model in the same viewing window. Do the data appear linear? Explain.

(b) Use the *regression* feature of the graphing utility to find a quadratic model for the data. Use the graphing utility to plot the data and graph the model in the same viewing window. Do the data appear quadratic? Even though the quadratic model appears to be a good fit, explain why it might not be a good model for predicting the temperature of the liquid when $t = 60$.

(c) The graph of the temperature of the room should be an asymptote of the graph of the model. Subtract the room temperature from each of the temperatures in the table. Use the *regression* feature of the graphing utility to find an exponential model for the revised data. Add the room temperature to this model. Use the graphing utility to plot the original data and graph the model in the same viewing window.

(d) Explain why the procedure in part (c) was necessary for finding the exponential model.

37. MODELING DATA

The table shows the percents P of women in different age groups (in years) who have been married at least once. (Source: U.S. Census Bureau)

Age group	Percent, P
18–24	14.6
25–29	49.0
30–34	70.3
35–39	79.9
40–44	85.0
45–49	87.0
50–54	89.5
55–59	91.1

Spreadsheet at LarsonPrecalculus.com

(a) Use the *regression* feature of a graphing utility to find a logistic model for the data. Let x represent the midpoint of the age group.

(b) Use the graphing utility to graph the model with the original data. How closely does the model represent the data?

38. MODELING DATA

The table shows the lengths y (in centimeters) of yellowtail snappers caught off the coast of Brazil for different ages (in years). (Source: Brazilian Journal of Oceanography)

Age, x	Length, y
1	11.21
2	20.77
3	28.94
4	35.92
5	41.87
6	46.96
7	51.30
8	55.01
9	58.17
10	60.87
11	63.18
12	65.15
13	66.84
14	68.27
15	69.50

Spreadsheet at LarsonPrecalculus.com

(a) Use the *regression* feature of a graphing utility to find a logistic model and a power model for the data.

(b) Use the graphing utility to graph each model from part (a) with the data. Use the graphs to determine which model better fits the data.

(c) Use the model from part (b) to predict the length of a 17-year-old yellowtail snapper.

Conclusions

True or False? **In Exercises 39 and 40, determine whether the statement is true or false. Justify your answer.**

39. The exponential model $y = ae^{bx}$ represents a growth model when $b > 0$.

40. To change an exponential model of the form $y = ab^x$ to one of the form $y = ae^{cx}$, rewrite b as $b = \ln e^b$.

41. Writing In your own words, explain how to fit a model to a set of data using a graphing utility.

42. HOW DO YOU SEE IT? Each graphing utility screen below shows a model that fits the set of data. The equations and r^2-values for the models are given. Determine the equation and coefficient of determination that represents each graph. Explain how you found your result.

(i)

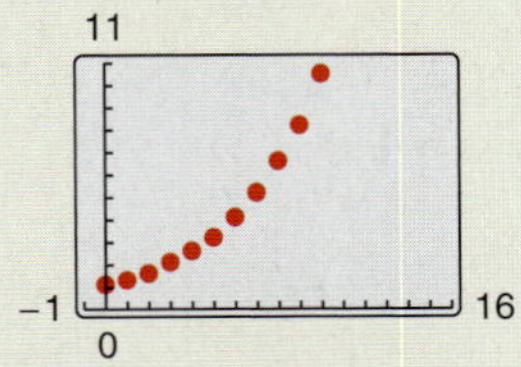

(ii)

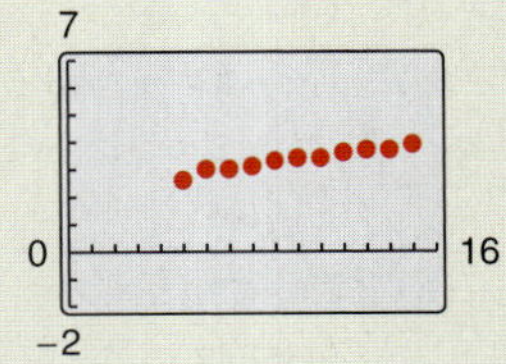

$y = 1.05e^{0.02288x}$ $\quad r^2 = 0.968$

$y = 1.06 \ln x + 0.96$ $\quad r^2 = 0.9991$

Cumulative Mixed Review

Using the Slope-Intercept Form **In Exercises 43–46, find the slope and y-intercept of the equation of the line. Then sketch the line by hand.**

43. $2x + 5y = 10$

44. $3x - 2y = 9$

45. $0.4x - 2.5y = 12.5$

46. $1.2x + 3.5y = 10.5$

Writing the Equation of a Parabola in Standard Form **In Exercises 47–50, write an equation of the parabola in standard form.**

47.

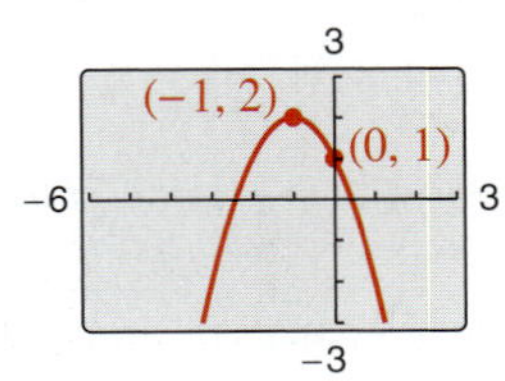

48.

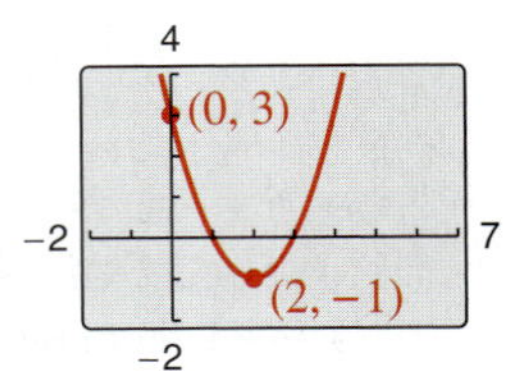

49.

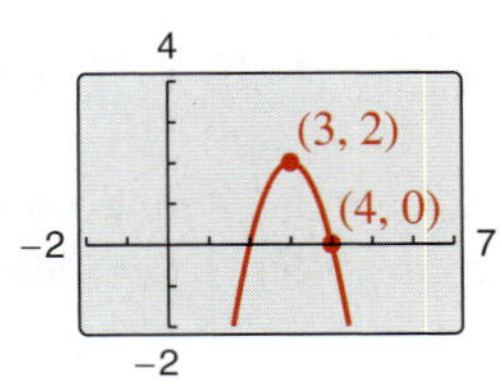

50.

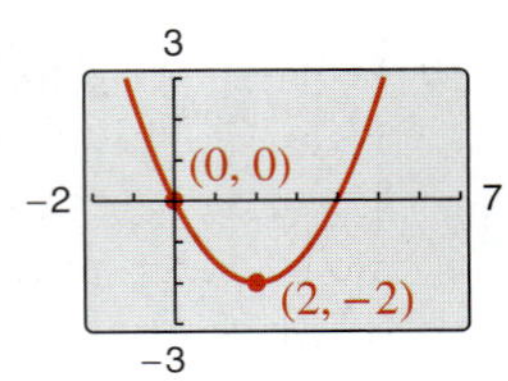

4 Chapter Summary

	What did you learn?	Explanation and Examples	Review Exercises
4.1	**Recognize and evaluate exponential functions with base *a* *(p. 324)*.**	The exponential function f with base a is denoted by $f(x) = a^x$, where $a > 0$, $a \neq 1$, and x is any real number.	1–4
	Graph exponential functions with base *a* *(p. 325)*.	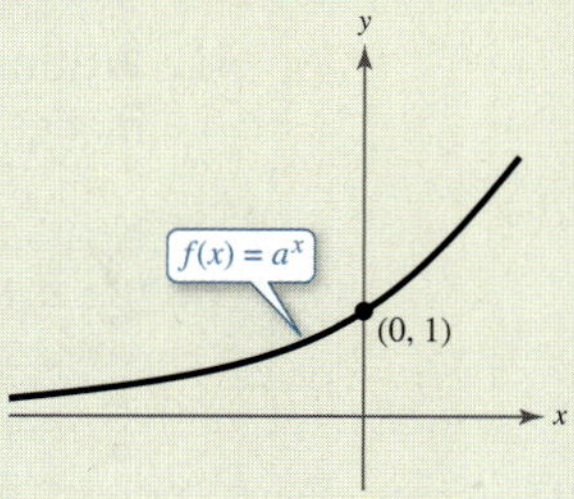 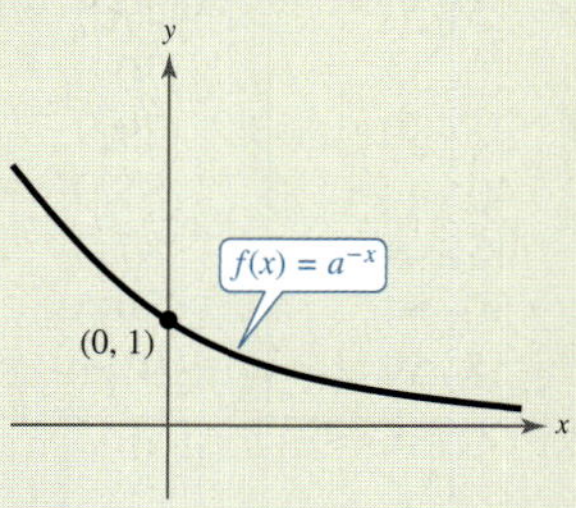	5–12
	Recognize, evaluate, and graph exponential functions with base *e* *(p. 328)*.	The function $f(x) = e^x$ is called the natural exponential function. 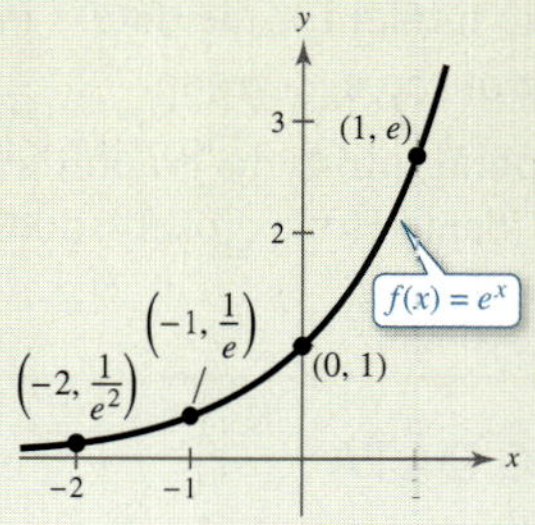	13–18
	Use exponential functions to model and solve real-life problems *(p. 330)*.	Exponential functions are used in compound interest formulas (see Example 8) and in radioactive decay models (see Example 10).	19–22
4.2	**Recognize and evaluate logarithmic functions with base *a* *(p. 336)*.**	For $x > 0$, $a > 0$, and $a \neq 1$, $y = \log_a x$ if and only if $x = a^y$. The function $f(x) = \log_a x$ is called the logarithmic function with base a.	23–36
	Graph logarithmic functions with base *a* *(p. 338)*, and recognize, evaluate, and graph natural logarithmic functions *(p. 340)*.	The graphs of $g(x) = \log_a x$ and $f(x) = a^x$ are reflections of each other in the line $y = x$. 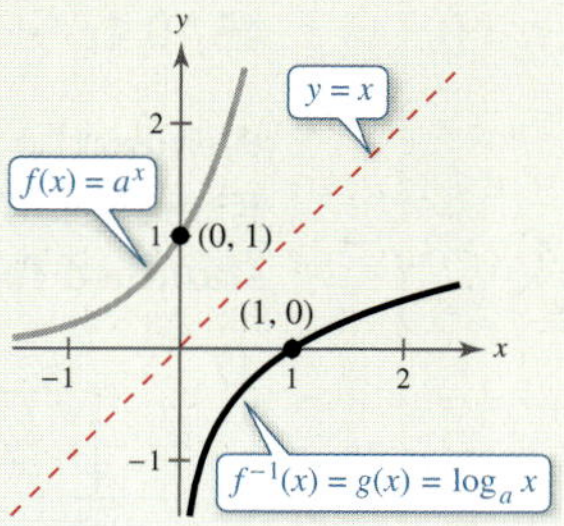 The function defined by $g(x) = \ln x$, $x > 0$, is called the natural logarithmic function. 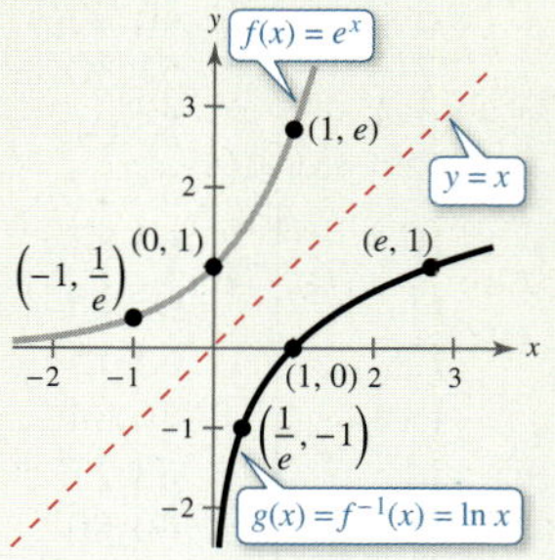	37–48
	Use logarithmic functions to model and solve real-life problems *(p. 342)*.	A logarithmic function is used in the human memory model. (See Example 10.)	49, 50
4.3	**Rewrite logarithms with different bases *(p. 347)*.**	Let a, b, and x be positive real numbers such that $a \neq 1$ and $b \neq 1$. Then $\log_a x$ can be converted to a different base as follows. *Base b*: $\log_a x = \dfrac{\log_b x}{\log_b a}$ *Base 10*: $\log_a x = \dfrac{\log_{10} x}{\log_{10} a}$ *Base e*: $\log_a x = \dfrac{\ln x}{\ln a}$	51–58

	What did you learn?	Explanation and Examples	Review Exercises
4.3	**Use properties of logarithms to evaluate, rewrite, expand, or condense logarithmic expressions *(p. 348).***	Let a be a real positive number ($a \neq 1$), n be a real number, and u and v be positive real numbers. **1. Product Property:** $\log_a(uv) = \log_a u + \log_a v$ $\ln(uv) = \ln u + \ln v$ **2. Quotient Property:** $\log_a(u/v) = \log_a u - \log_a v$ $\ln(u/v) = \ln u - \ln v$ **3. Power Property:** $\log_a u^n = n \log_a u, \quad \ln u^n = n \ln u$	59–78
	Use logarithmic functions to model and solve real-life problems *(p. 350).*	Logarithmic functions can be used to find an equation that relates the periods of several planets and their distances from the sun. (See Example 7.)	79, 80
4.4	**Solve simple exponential and logarithmic equations *(p. 354).***	Solve simple exponential or logarithmic equations using the One-to-One Properties and Inverse Properties of exponential and logarithmic functions.	81–94
	Solve more complicated exponential *(p. 355)* and logarithmic *(p. 357)* equations.	To solve more complicated equations, rewrite the equations so that the One-to-One Properties and Inverse Properties of exponential and logarithmic functions can be used. (See Examples 2–9.)	95–118
	Use exponential and logarithmic equations to model and solve real-life problems *(p. 360).*	Exponential and logarithmic equations can be used to find how long it will take to double an investment (see Example 12) and to find the year in which the average salary for public school teachers reached \$50,000 (see Example 13).	119, 120
4.5	**Recognize the five most common types of models involving exponential or logarithmic functions *(p. 365).***	**1. Exponential growth model:** $y = ae^{bx}, \quad b > 0$ **2. Exponential decay model:** $y = ae^{-bx}, \quad b > 0$ **3. Gaussian model:** $y = ae^{-(x-b)^2/c}$ **4. Logistic growth model:** $y = \dfrac{a}{1 + be^{-rx}}$ **5. Logarithmic models:** $y = a + b \ln x,$ $y = a + b \log_{10} x$	121–126
	Use exponential growth and decay functions to model and solve real-life problems *(p. 366).*	An exponential growth function can be used to model the world population (see Example 1) and an exponential decay function can be used to estimate the age of a fossil (see Example 3).	127
	Use Gaussian functions *(p. 369)*, logistic growth functions *(p. 370)*, and logarithmic functions *(p. 371)* to model and solve real-life problems.	A Gaussian function can be used to model SAT mathematics scores for college-bound seniors (see Example 4). A logistic growth function can be used to model the spread of a flu virus (see Example 5). A logarithmic function can be used to find the intensity of an earthquake using its magnitude (see Example 6).	128–130
4.6	**Classify scatter plots *(p. 377)*, and use scatter plots and a graphing utility to find models for data and choose the model that best fits a set of data *(p. 378).***	You can use a scatter plot and a graphing utility to choose a model that best fits a set of data that represents the yield of a chemical reaction. (See Example 3.)	131, 132
	Use a graphing utility to find exponential and logistic models for data *(p. 380).*	An exponential model can be used to estimate the amount of a radioactive substance remaining after a given day (see Example 4), and a logistic model can be used to estimate the percent of defoliation caused by the gypsy moth (see Example 5).	133, 134

4 Review Exercises

See *CalcChat.com* for tutorial help and worked-out solutions to odd-numbered exercises.
For instructions on how to use a graphing utility, see Appendix A.

4.1

Evaluating Exponential Functions **In Exercises 1–4, use a calculator to evaluate the function at the indicated value of x. Round your result to four decimal places.**

1. $f(x) = 1.45^x$, $x = 2\pi$ **2.** $f(x) = 7^x$, $x = \sqrt{11}$

3. $g(x) = 60^{2x}$, $x = -0.5$ **4.** $g(x) = 25^{-3x}$, $x = \frac{3}{2}$

Library of Parent Functions **In Exercises 5–8, use the graph of $y = 4^x$ to match the function with its graph. [The graphs are labeled (a), (b), (c), and (d).]**

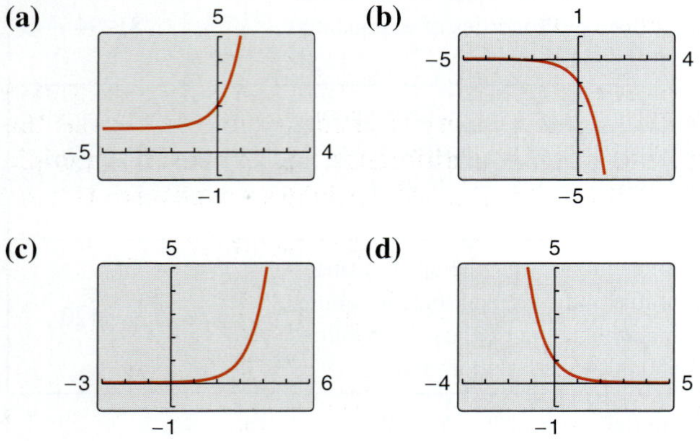

5. $f(x) = 4^{x-3}$ **6.** $f(x) = 4^{-x}$

7. $f(x) = -4^x$ **8.** $f(x) = 4^x + 1$

Graphs of $y = a^x$ and $y = a^{-x}$ **In Exercises 9–12, graph the exponential function by hand. Identify any asymptotes and intercepts and determine whether the graph of the function is increasing or decreasing.**

9. $f(x) = 8^x$ **10.** $f(x) = 0.3^x$

11. $g(x) = 8^{-x}$ **12.** $g(x) = 0.3^{-x}$

Graphing Natural Exponential Functions **In Exercises 13–18, use a graphing utility to construct a table of values for the function. Then sketch the graph of the function. Identify any asymptotes of the graph.**

13. $h(x) = e^{x-1}$ **14.** $f(x) = e^{x+2}$

15. $h(x) = -e^x + 3$ **16.** $f(x) = -2e^{-x}$

17. $f(x) = 4e^{-0.5x}$ **18.** $f(x) = 2 + e^{x-3}$

Finding the Balance for Compound Interest **In Exercises 19 and 20, complete the table to determine the balance A for \$10,000 invested at rate r for t years, compounded continuously.**

t	1	10	20	30	40	50
A						

19. $r = 8\%$ **20.** $r = 3\%$

21. Economics A new SUV costs \$30,795. The value V of the SUV after t years is modeled by $V(t) = 30{,}795\left(\frac{4}{5}\right)^t$.

(a) Use a graphing utility to graph the function.

(b) Find the value of the SUV after 2 years.

(c) According to the model, when does the SUV depreciate most rapidly? Is this realistic? Explain.

22. Radioactive Decay Let Q represent the mass, in grams, of a quantity of plutonium 241 (^{241}Pu), whose half-life is about 14 years. The quantity of plutonium present after t years is given by $Q = 50\left(\frac{1}{2}\right)^{t/14}$.

(a) Determine the initial quantity (when $t = 0$).

(b) Determine the quantity present after 10 years.

(c) Use a graphing utility to graph the function over the interval $t = 0$ to $t = 50$.

4.2

Rewriting Equations **In Exercises 23–32, write the logarithmic equation in exponential form, or write the exponential equation in logarithmic form.**

23. $\log_5 625 = 4$ **24.** $\log_9 81 = 2$

25. $\log_{64} 2 = \frac{1}{6}$ **26.** $\log_{10}\left(\frac{1}{100}\right) = -2$

27. $4^3 = 64$ **28.** $3^5 = 243$

29. $e^{3.2} = 24.532 \ldots$ **30.** $\ln 5 = 1.6094 \ldots$

31. $\left(\frac{1}{2}\right)^{-3} = 8$ **32.** $\left(\frac{2}{3}\right)^{-2} = \frac{9}{4}$

Evaluating Logarithms **In Exercises 33–36, use the definition of a logarithmic function to evaluate the function at the indicated value of x without using a calculator.**

	Function	*Value*
33.	$f(x) = \log_6 x$	$x = 216$
34.	$f(x) = \log_7 x$	$x = 1$
35.	$f(x) = \log_4 x$	$x = \frac{1}{4}$
36.	$f(x) = \log_{10} x$	$x = 0.00001$

Sketching the Graph of a Logarithmic Function **In Exercises 37–40, find the domain, vertical asymptote, and x-intercept of the logarithmic function, and sketch its graph by hand.**

37. $g(x) = -\log_{10} x + 5$ **38.** $g(x) = \log_{10}(x - 3)$

39. $f(x) = \log_{10}(x - 1) + 6$ **40.** $f(x) = \log_{10}(x + 2) - 3$

Evaluating the Natural Logarithmic Function **In Exercises 41–44, use a calculator to evaluate the function $f(x) = \ln x$ at the indicated value of x. Round your result to three decimal places.**

41. $x = 21.5$ **42.** $x = 0.46$

43. $x = \sqrt{6}$ **44.** $x = \frac{5}{6}$

Analyzing Graphs of Functions **In Exercises 45–48, use a graphing utility to graph the logarithmic function. Determine the domain and identify any vertical asymptote and x-intercept.**

45. $f(x) = \ln x + 3$
46. $f(x) = \ln(x - 3)$
47. $h(x) = \frac{1}{2}\ln x$
48. $f(x) = 4\ln x$

49. Aeronautics The time t (in minutes) for a small plane to climb to an altitude of h feet is given by

$$t = 50\log_{10}[18{,}000/(18{,}000 - h)]$$

where 18,000 feet is the plane's absolute ceiling.

(a) Determine the domain of the function appropriate for the context of the problem.

(b) Use a graphing utility to graph the function and identify any asymptotes.

(c) As the plane approaches its absolute ceiling, what can be said about the time required to further increase its altitude?

(d) Find the amount of time it will take for the plane to climb to an altitude of 4000 feet.

50. Real Estate The model

$$t = 12.542\ln[x/(x - 1000)], \quad x > 1000$$

approximates the length of a home mortgage of \$150,000 at 8% in terms of the monthly payment. In the model, t is the length of the mortgage in years and x is the monthly payment in dollars.

(a) Use the model to approximate the length of a \$150,000 mortgage at 8% when the monthly payment is \$1254.68.

(b) Approximate the total amount paid over the term of the mortgage with a monthly payment of \$1254.68. What amount of the total is interest costs?

4.3

Changing the Base **In Exercises 51–54, evaluate the logarithm using the change-of-base formula. Do each problem twice, once with common logarithms and once with natural logarithms. Round your results to three decimal places.**

51. $\log_4 9$
52. $\log_{1/2} 10$
53. $\log_{14} 15.6$
54. $\log_3 0.28$

Graphing a Logarithm with Any Base **In Exercises 55–58, use the change-of-base formula and a graphing utility to graph the function.**

55. $f(x) = \log_2(x - 2)$
56. $f(x) = 2 - \log_3 x$
57. $f(x) = -\log_{1/2}(x - 2)$
58. $f(x) = \log_{1/3}(x - 1) + 1$

Using Properties to Evaluate Logarithms **In Exercises 59–62, approximate the logarithm using the properties of logarithms, given the values $\log_b 2 \approx 0.3562$, $\log_b 3 \approx 0.5646$, and $\log_b 5 \approx 0.8271$.**

59. $\log_b 9$
60. $\log_b \frac{4}{9}$
61. $\log_b \sqrt{5}$
62. $\log_b 50$

Simplifying a Logarithm **In Exercises 63–66, use the properties of logarithms to rewrite and simplify the logarithmic expression.**

63. $\ln(5e^{-2})$
64. $\ln\sqrt{e^5}$
65. $\log_{10} 200$
66. $\log_{10} 0.002$

Expanding Logarithmic Expressions **In Exercises 67–72, use the properties of logarithms to expand the expression as a sum, difference, and/or constant multiple of logarithms. (Assume all variables are positive.)**

67. $\log_5 5x^2$
68. $\log_4 16xy^2$
69. $\log_{10}\dfrac{5\sqrt{y}}{x^2}$
70. $\ln\dfrac{\sqrt{x}}{4}$
71. $\ln\dfrac{x + 3}{xy}$
72. $\ln\dfrac{xy^5}{\sqrt{z}}$

Condensing Logarithmic Expressions **In Exercises 73–78, condense the expression to the logarithm of a single quantity.**

73. $\log_2 9 + \log_2 x$
74. $\log_6 y - 2\log_6 z$
75. $\frac{1}{2}\ln(2x - 1) - 2\ln(x + 1)$
76. $5\ln(x - 2) - \ln(x + 2) + 3\ln x$
77. $\ln 3 + \frac{1}{3}\ln(4 - x^2) - \ln x$
78. $3[\ln x - 2\ln(x^2 + 1)] + 2\ln 5$

79. Public Service The number of miles s of roads cleared of snow in 1 hour is approximated by the model

$$s = 25 - \frac{13\ln(h/12)}{\ln 3}, \quad 2 \le h \le 15$$

where h is the depth of the snow (in inches).

(a) Use a graphing utility to graph the function.

(b) Complete the table.

h	4	6	8	10	12	14
s						

(c) Using the graph of the function and the table, what conclusion can you make about the number of miles of roads cleared as the depth of the snow increases?

80. Psychology Students in a sociology class were given an exam and then retested monthly with an equivalent exam. The average scores for the class are given by the human memory model $f(t) = 85 - 17\log_{10}(t + 1)$, where t is the time in months and $0 \le t \le 10$. When will the average score decrease to 68?

4.4

Solving an Exponential or Logarithmic Equation **In Exercises 81–94, solve the equation without using a calculator.**

81. $10^x = 10{,}000$ **82.** $7^x = 343$

83. $6^x = \frac{1}{216}$ **84.** $6^{x-2} = \frac{1}{1296}$

85. $2^{x+1} = \frac{1}{16}$ **86.** $4^{x/2} = 64$

87. $\log_8 x = 4$ **88.** $\log_x 729 = 6$

89. $\log_2(x - 1) = 3$ **90.** $\log_5(2x + 27) = 2$

91. $\ln x = 4$ **92.** $\ln x = -3$

93. $\ln(x - 1) = 2$ **94.** $\ln(2x + 1) = -4$

Solving an Exponential Equation **In Exercises 95–104, solve the exponential equation algebraically. Round your result to three decimal places.**

95. $3e^{-5x} = 132$ **96.** $14e^{3x+2} = 560$

97. $2^x + 13 = 35$ **98.** $6^x - 28 = -8$

99. $-4(5^x) = -68$ **100.** $2(12^x) = 190$

101. $2e^{x-3} - 1 = 4$ **102.** $-e^{x/2} + 1 = \frac{1}{2}$

103. $e^{2x} - 7e^x + 10 = 0$ **104.** $e^{2x} - 6e^x + 8 = 0$

Solving a Logarithmic Equation **In Exercises 105–114, solve the logarithmic equation algebraically. Round your result to three decimal places.**

105. $\ln 3x = 6.4$ **106.** $\ln 5x = 4.5$

107. $\ln x - \ln 5 = 2$ **108.** $\ln x - \ln 3 = 4$

109. $\ln \sqrt{x + 1} = 2$

110. $\ln \sqrt{x + 40} = 3$

111. $\log_4(x + 5) = \log_4(13 - x) - \log_4(x - 3)$

112. $\log_5(x + 2) - \log_5 x = \log_5(x + 5)$

113. $\log_{10}(1 - x) = -1$

114. $\log_{10}(-x - 4) = 2$

Solving an Exponential or Logarithmic Equation **In Exercises 115–118, solve the equation algebraically. Round your result to three decimal places.**

115. $xe^x + e^x = 0$ **116.** $2xe^{2x} + e^{2x} = 0$

117. $x \ln x + x = 0$ **118.** $\dfrac{1 + \ln x}{x^2} = 0$

119. Finance You deposit $7550 in an account that pays 6.9% interest, compounded continuously. How long will it take for the money to double?

120. Economics The demand x for a 32-inch plasma television is modeled by

$$p = 5000\left(1 - \frac{4}{4 + e^{-0.0005x}}\right).$$

Find the demands x for prices of (a) $p = \$500$ and (b) $p = \$350$.

4.5

Identifying Graphs of Models **In Exercises 121–126, match the function with its graph. [The graphs are labeled (a), (b), (c), (d), (e), and (f).]**

(a)

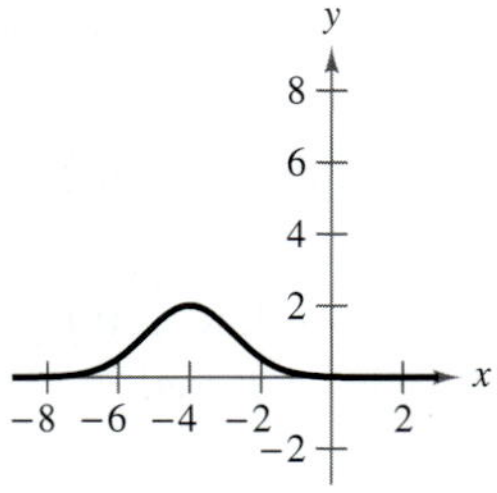

(b)

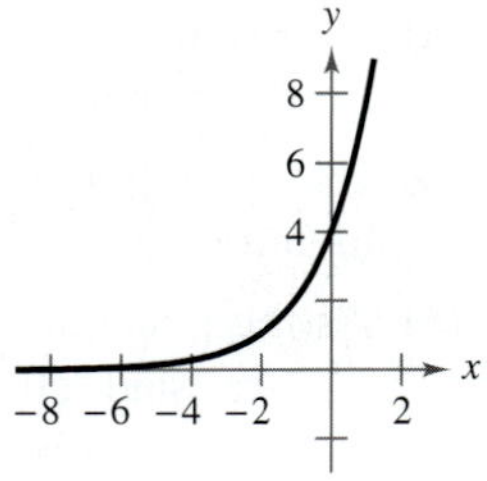

(c)

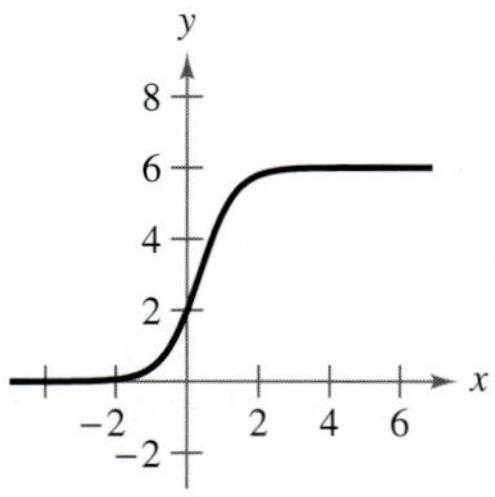

(d)

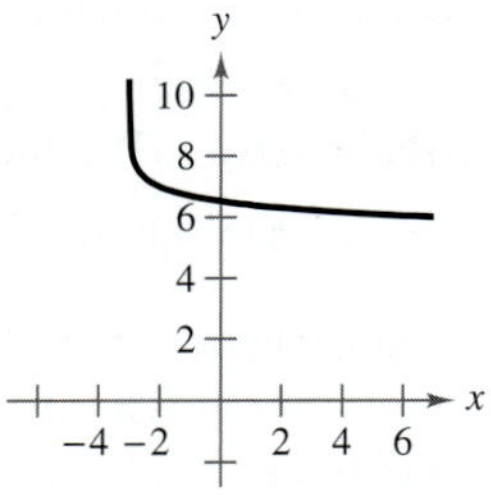

(e)

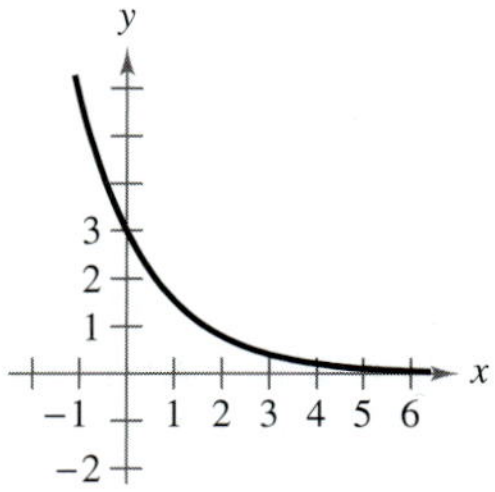

(f)

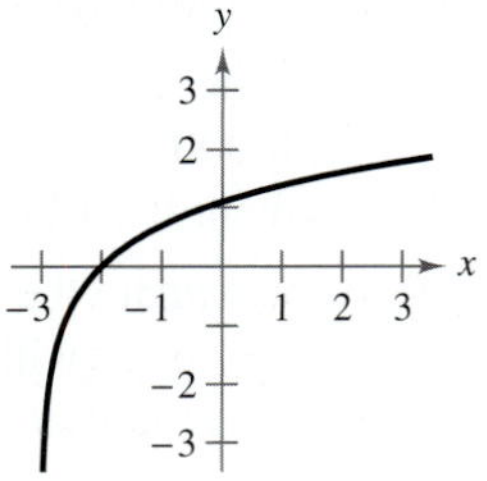

121. $y = 3e^{-2x/3}$ **122.** $y = 4e^{2x/3}$

123. $y = \ln(x + 3)$ **124.** $y = 7 - \log_{10}(x + 3)$

125. $y = 2e^{-(x+4)^2/3}$ **126.** $y = \dfrac{6}{1 + 2e^{-2x}}$

127. Demography The populations P (in millions) of North Carolina from 2000 through 2012 can be modeled by $P = 8.06e^{kt}$, where t is the year, with $t = 0$ corresponding to 2000. In 2001, the population was about 8,210,000. Find the value of k and use the result to predict the population in the year 2020. (Source: U.S. Census Bureau)

128. Education The scores for a biology test follow a normal distribution modeled by $y = 0.0499e^{-(x-74)^2/128}$, where x is the test score and $40 \le x \le 100$.

(a) Use a graphing utility to graph the function.

(b) Use the graph to estimate the average test score.

129. Education The average number N of words per minute that the students in a first grade class could read orally after t weeks of school is modeled by

$$N = \frac{62}{1 + 5.4e^{-0.24t}}.$$

Find the numbers of weeks it took the class to read at average rates of (a) 45 words per minute and (b) 61 words per minute.

130. Geology On the Richter scale, the magnitude R of an earthquake of intensity I is modeled by

$$R = \log_{10} \frac{I}{I_0}$$

where $I_0 = 1$ is the minimum intensity used for comparison. Find the intensities I of the following earthquakes measuring R on the Richter scale.

(a) $R = 7.1$ (b) $R = 9.2$ (c) $R = 3.8$

4.6

Classifying Scatter Plots In Exercises 131 and 132, determine whether the scatter plot could best be modeled by a linear model, an exponential model, a logarithmic model, or a logistic model.

131.

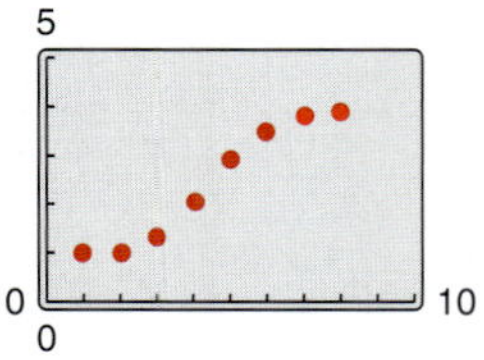

132.

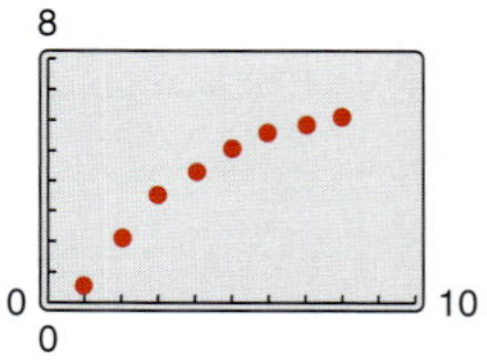

133. MODELING DATA

Each ordered pair (t, N) represents the year t and the number N (in thousands) of female participants in high school athletic programs during 13 school years, with $t = 1$ corresponding to the 2000–2001 school year. *(Spreadsheet at LarsonPrecalculus.com)* (Source: National Federation of State High School Associations)

DATA (1, 2784), (2, 2807), (3, 2856), (4, 2865), (5, 2908), (6, 2953), (7, 3022), (8, 3057), (9, 3114), (10, 3173), (11, 3174), (12, 3208), (13, 3223)

(a) Use the *regression* feature of a graphing utility to find a linear model, an exponential model, and a power model for the data and identify the coefficient of determination for each model.

(b) Use the graphing utility to graph each model with the original data.

(c) Determine which model best fits the data. Explain.

(d) Use the model you chose in part (c) to predict the school year in which about 3,697,000 girls will participate.

134. MODELING DATA

You plant a tree when it is 1 meter tall and check its height h (in meters) every 10 years, as shown in the table.

DATA — Spreadsheet at LarsonPrecalculus.com

Year	Height, h
0	1
10	3
20	7.5
30	14.5
40	19
50	20.5
60	21

(a) Use the *regression* feature of a graphing utility to find a logistic model for the data. Let x represent the year.

(b) Use the graphing utility to graph the model with the original data.

(c) How closely does the model represent the data?

(d) What is the limiting height of the tree?

Conclusions

True or False? In Exercises 135–138, determine whether the equation or statement is true or false. Justify your answer.

135. $e^{x-1} = \dfrac{e^x}{e}$

136. $\ln(x + y) = \ln(xy)$

137. The domain of the function $f(x) = \ln x$ is the set of all real numbers.

138. The logarithm of the quotient of two numbers is equal to the difference of the logarithms of the numbers.

139. Think About It Without using a calculator, explain why you know that $2^{\sqrt{2}}$ is greater than 2, but less than 4.

140. Exploration

(a) Use a graphing utility to compare the graph of the function $y = e^x$ with the graph of each function below. [$n!$ (read as "n factorial") is defined as $n! = 1 \cdot 2 \cdot 3 \cdots (n-1) \cdot n$.]

$$y_1 = 1 + \frac{x}{1!},\ y_2 = 1 + \frac{x}{1!} + \frac{x^2}{2!},$$

$$y_3 = 1 + \frac{x}{1!} + \frac{x^2}{2!} + \frac{x^3}{3!}$$

(b) Identify the pattern of successive polynomials given in part (a). Extend the pattern one more term and compare the graph of the resulting polynomial function with the graph of $y = e^x$. What do you think this pattern implies?

4 Chapter Test

See *CalcChat.com* for tutorial help and worked-out solutions to odd-numbered exercises.
For instructions on how to use a graphing utility, see Appendix A.

Take this test as you would take a test in class. After you are finished, check your work against the answers given in the back of the book.

In Exercises 1–3, use a graphing utility to construct a table of values for the function. Then sketch a graph of the function. Identify any asymptotes and intercepts.

1. $f(x) = 10^{-x}$ **2.** $f(x) = -6^{x-2}$ **3.** $f(x) = 1 - e^{2x}$

In Exercises 4–6, evaluate the expression.

4. $\log_7 7^{-0.89}$ **5.** $4.6 \ln e^2$ **6.** $5 - \log_{10} 1000$

In Exercises 7–9, find the domain, vertical asymptote, and *x*-intercept of the logarithmic function, and sketch its graph by hand.

7. $f(x) = -\log_{10} x - 6$ **8.** $f(x) = \ln(x + 4)$ **9.** $f(x) = 1 + \ln(x - 6)$

In Exercises 10–12, evaluate the logarithm using the change-of-base formula. Round your result to three decimal places.

10. $\log_7 44$ **11.** $\log_{2/5} 1.3$ **12.** $\log_{12} 64$

In Exercises 13–15, use the properties of logarithms to expand the expression as a sum, difference, and/or multiple of logarithms.

13. $\log_2 3a^4$ **14.** $\ln \dfrac{5\sqrt{x}}{6}$ **15.** $\ln \dfrac{x\sqrt{x+1}}{2e^4}$

In Exercises 16–18, condense the expression.

16. $\log_3 13 + \log_3 y$ **17.** $4 \ln x - 4 \ln y$

18. $\ln x - \ln(x + 2) + \ln(2x - 3)$

In Exercises 19–22, solve the equation.

19. $3^x = 81$ **20.** $5^{2x} = 2500$

21. $\log_7 x = 6$ **22.** $\log_{10}(x - 4) = 5$

In Exercises 23–26, solve the equation algebraically. Round your result to three decimal places.

23. $\dfrac{1025}{8 + e^{4x}} = 5$ **24.** $-xe^{-x} + e^{-x} = 0$

25. $\log_{10} x - \log_{10}(8 - 5x) = 2$ **26.** $2x \ln x - x = 0$

27. The half-life of radioactive actinium (^{227}Ac) is about 22 years. What percent of a present amount of radioactive actinium will remain after 15 years?

28. The table shows the annual revenues R (in billions of dollars) for Amazon.com from 2006 through 2013. (Source: Amazon.com)

(a) Use the *regression* feature of a graphing utility to find a logarithmic model, an exponential model, and a power model for the data. Let t represent the year, with $t = 6$ corresponding to 2006.

(b) Use the graphing utility to graph each model with the original data.

(c) Determine which model best fits the data. Use the model to predict the revenue of Amazon.com in 2019.

DATA

Year	Revenue, R
2006	10.711
2007	14.836
2008	19.166
2009	24.509
2010	34.204
2011	48.077
2012	61.093
2013	74.452

Spreadsheet at LarsonPrecalculus.com

Table for 28

Library of Parent Functions Review

In Exercises 1–9, fill in the blank to identify the parent function represented by the graph.

1. $f(x) =$ ___________

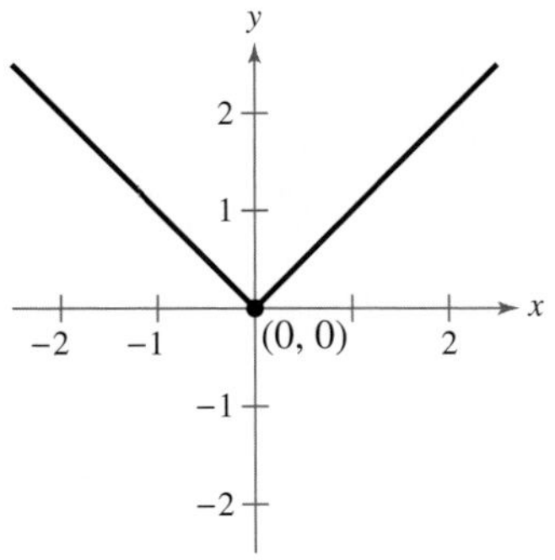

2. $f(x) =$ ___________

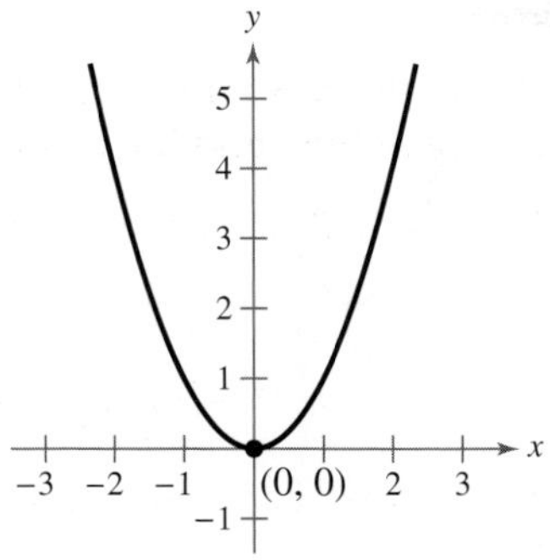

3. $f(x) =$ ___________

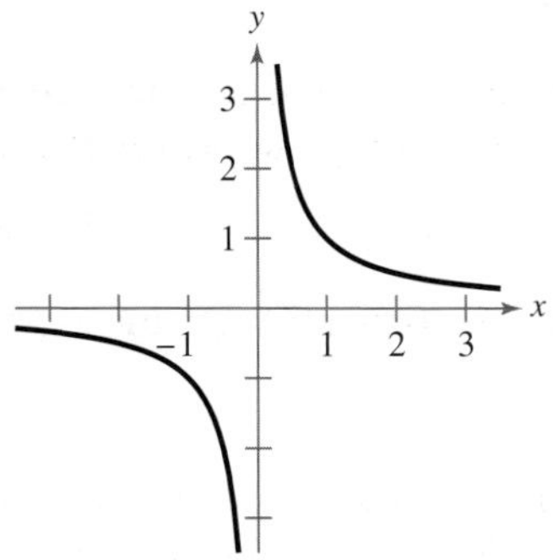

4. $f(x) =$ ___________

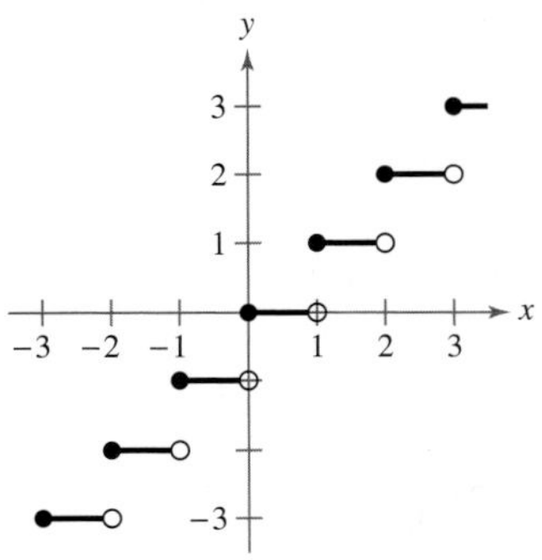

5. $f(x) =$ ___________

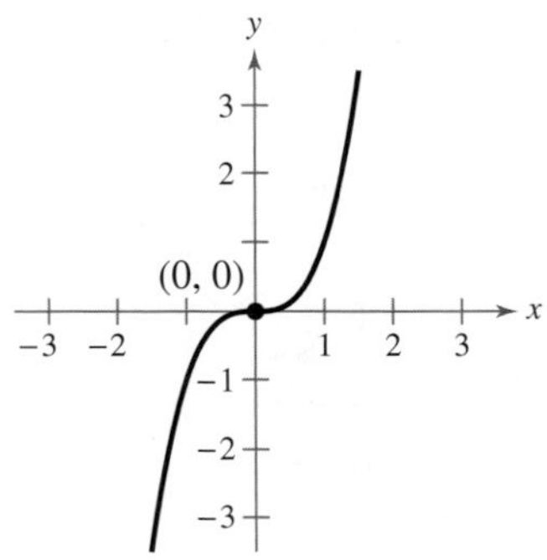

6. $f(x) =$ ___________

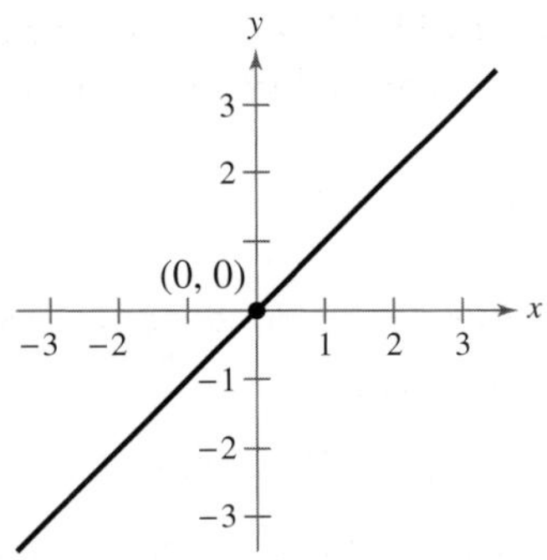

7. $f(x) =$ ___________

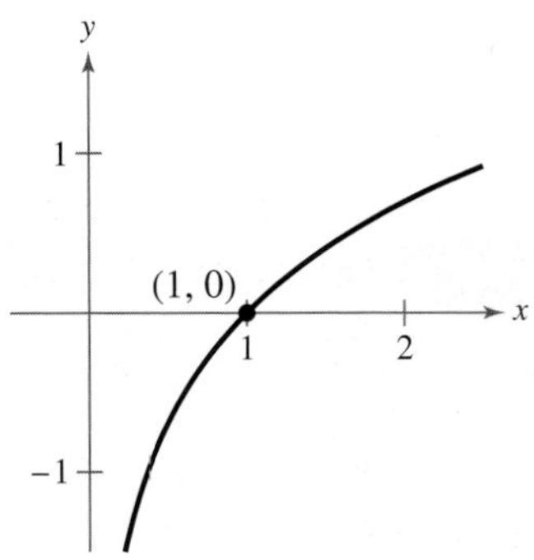

8. $f(x) =$ ___________

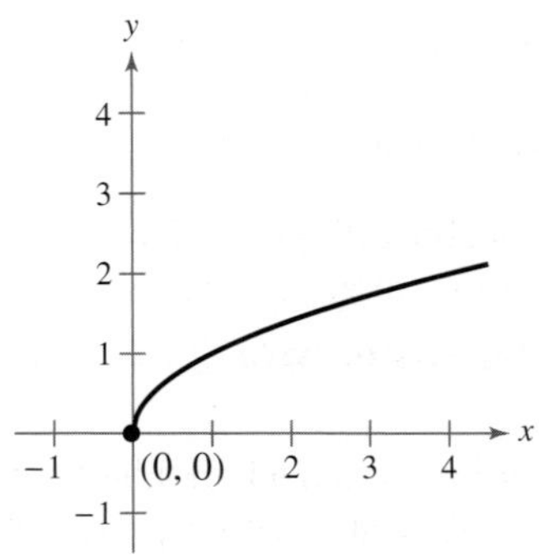

9. $f(x) =$ ___________

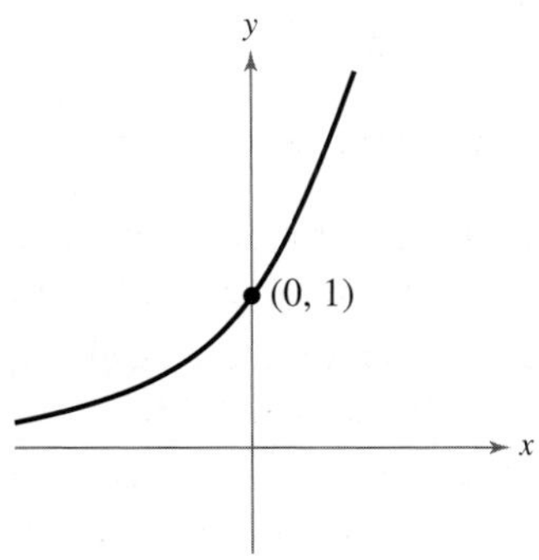

In Exercises 10–18, identify the parent function and describe the transformation shown in the graph. Write an equation for the graphed function. (There may be more than one correct answer.)

10.
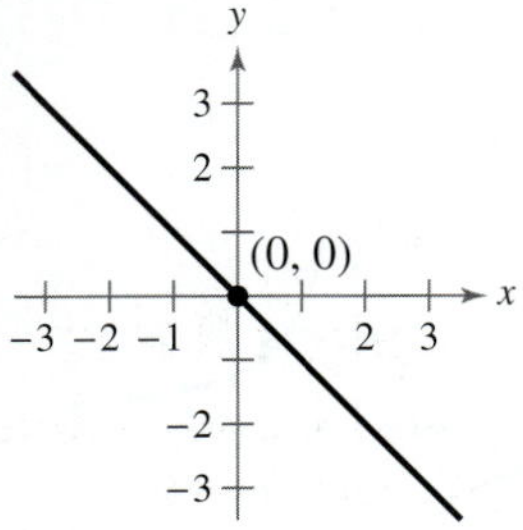

11.
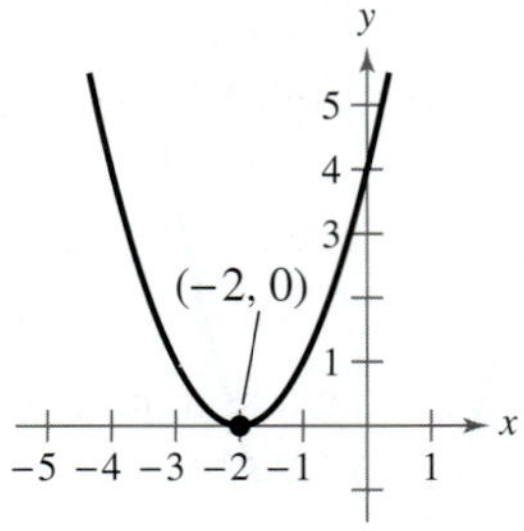

12.
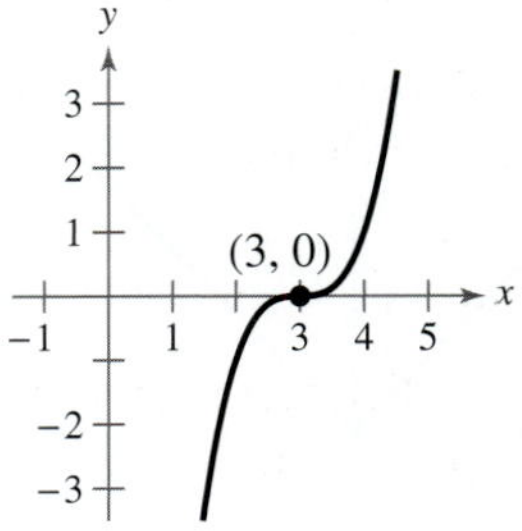

13.
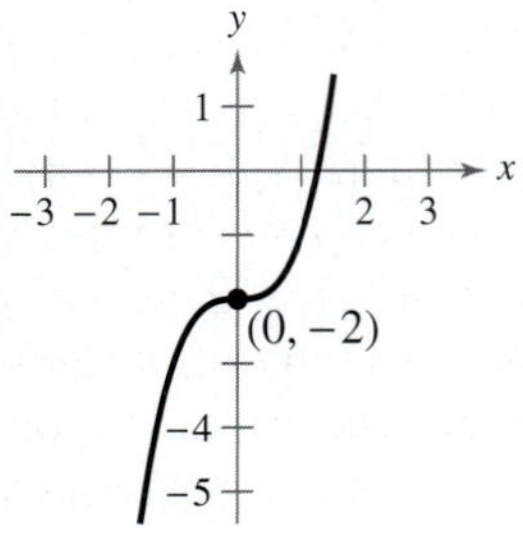

14.
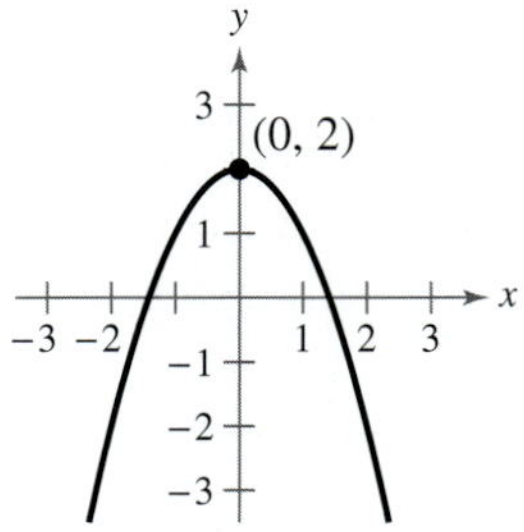

15.
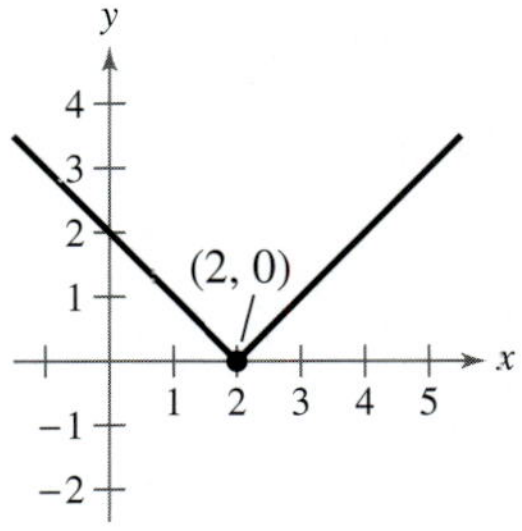

16.
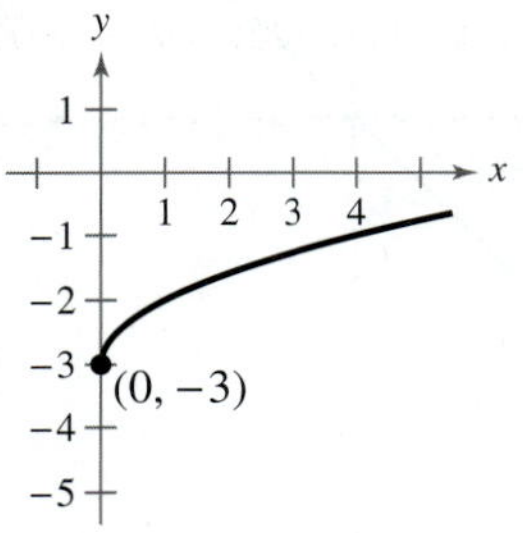

17.
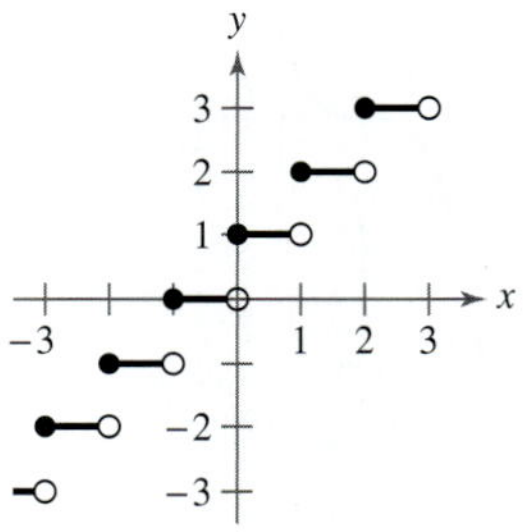

18.
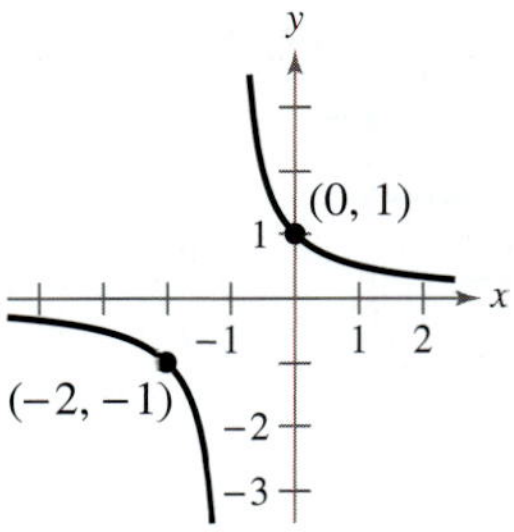

In Exercises 19 and 20, determine which function is represented by the graph. Do not use a calculator.

19. (a) $y = \log_{10}(x + 1) + 3$
(b) $y = \log_{10}(x + 1) - 3$
(c) $y = \log_{10}(x - 1) + 3$
(d) $y = \log_{10}(x + 3) + 1$

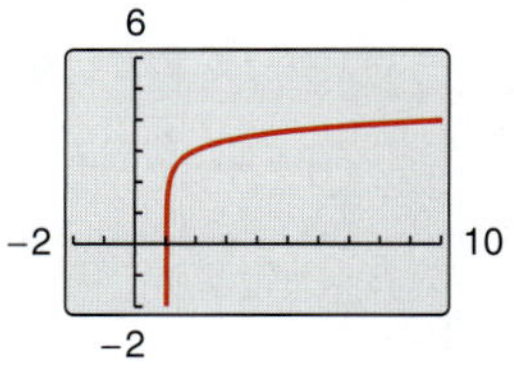

20. (a) $y = 2^{-x} + 1$
(b) $y = -2^{-x} + 1$
(c) $y = 2^{-x+1}$
(d) $y = 2^{-x-1}$

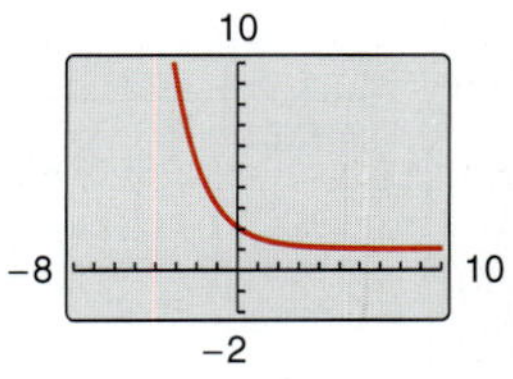

In Exercises 21–24, sketch the graph of the function by hand.

21. $y = (x - 1)^2 - 4$

22. $y = -(x - 4)^3$

23. $y = |x + 1| - 3$

24. $y = \sqrt{x - 1} + 2$

Proofs in Mathematics

Proofs in Mathematics

Each of the following three properties of logarithms can be proved by using properties of exponential functions.

Properties of Logarithms (p. 348)

Let a be a positive real number such that $a \neq 1$, and let n be a real number. If u and v are positive real numbers, then the following properties are true.

	Logarithm with Base a	*Natural Logarithm*
1. Product Property:	$\log_a(uv) = \log_a u + \log_a v$	$\ln(uv) = \ln u + \ln v$
2. Quotient Property:	$\log_a \frac{u}{v} = \log_a u - \log_a v$	$\ln\frac{u}{v} = \ln u - \ln v$
3. Power Property:	$\log_a u^n = n \log_a u$	$\ln u^n = n \ln u$

Slide Rules

The slide rule was invented by William Oughtred (1574–1660) in 1625. The slide rule is a computational device with a sliding portion and a fixed portion. A slide rule enables you to perform multiplication by using the Product Property of logarithms. There are other slide rules that allow for the calculation of roots and trigonometric functions. Slide rules were used by mathematicians and engineers until the invention of the handheld calculator in 1972.

Proof

Let

$$x = \log_a u \quad \text{and} \quad y = \log_a v.$$

The corresponding exponential forms of these two equations are

$$a^x = u \quad \text{and} \quad a^y = v.$$

To prove the Product Property, multiply u and v to obtain

$$uv = a^x a^y = a^{x+y}.$$

The corresponding logarithmic form of $uv = a^{x+y}$ is

$$\log_a(uv) = x + y.$$

So,

$$\log_a(uv) = \log_a u + \log_a v.$$

To prove the Quotient Property, divide u by v to obtain

$$\frac{u}{v} = \frac{a^x}{a^y} = a^{x-y}.$$

The corresponding logarithmic form of $u/v = a^{x-y}$ is

$$\log_a \frac{u}{v} = x - y.$$

So,

$$\log_a \frac{u}{v} = \log_a u - \log_a v.$$

To prove the Power Property, substitute a^x for u in the expression $\log_a u^n$, as follows.

$$\log_a u^n = \log_a(a^x)^n \qquad \text{Substitute } a^x \text{ for } u.$$

$$= \log_a a^{nx} \qquad \text{Property of exponents}$$

$$= nx \qquad \text{Inverse Property of logarithms}$$

$$= n \log_a u \qquad \text{Substitute } \log_a u \text{ for } x.$$

So, $\log_a u^n = n \log_a u$. ■

Progressive Summary (Chapters P–4)

This chart outlines the topics that have been covered so far in this text. Progressive Summary charts appear after Chapters 2, 4, and 7. In each Progressive Summary, new topics encountered for the first time appear in blue.

Algebraic Functions

Polynomial, Rational, Radical

Rewriting

Polynomial form ↔ Factored form
Operations with polynomials
Rationalize denominators
Simplify rational expressions
Exponent form ↔ Radical form
Operations with complex numbers

Solving

Equation	*Strategy*
Linear	Isolate variable
Quadratic	Factor, set to zero Extract square roots Complete the square Quadratic Formula
Polynomial	Factor, set to zero Rational Zero Test
Rational	Multiply by LCD
Radical	Isolate, raise to power
Absolute value	Isolate, form two equations

Analyzing

Graphically
Intercepts
Symmetry
Slope
Asymptotes
End behavior
Minimum values
Maximum values

Algebraically
Domain, Range
Transformations
Composition
Standard forms of equations
Leading Coefficient Test
Synthetic division
Descartes's Rule of Signs

Numerically
Table of values

Transcendental Functions

Exponential, Logarithmic

Rewriting

Exponential form ↔ Logarithmic form
Condense/expand logarithmic expressions

Solving

Equation	*Strategy*
Exponential	Take logarithm of each side
Logarithmic	Exponentiate each side

Analyzing

Graphically
Intercepts
Asymptotes

Algebraically
Domain, Range
Transformations
Composition
Inverse Properties

Numerically
Table of values

Other Topics

Rewriting

Solving

Analyzing

5 Linear Systems and Matrices

Section 5.5, Example 11
Softball Team Expenses

5.1 Solving Systems of Equations

The Methods of Substitution and Graphing

So far in this text, most problems have involved either a function of one variable or a single equation in two variables. However, many problems in science, business, and engineering involve two or more equations in two or more variables. To solve such problems, you need to find solutions of **systems of equations.** Here is an example of a system of two equations in two unknowns, x and y.

$$\begin{cases} 2x + y = 5 & \text{Equation 1} \\ 3x - 2y = 4 & \text{Equation 2} \end{cases}$$

A **solution** of this system is an ordered pair that satisfies each equation in the system. Finding the set of all such solutions is called **solving the system of equations.** For instance, the ordered pair (2, 1) is a solution of this system. To check this, you can substitute 2 for x and 1 for y in *each* equation.

Check (2, 1) *in Equation 1:*

$$2x + y = 5$$
$$2(2) + 1 \stackrel{?}{=} 5$$
$$5 = 5 \checkmark$$

Check (2, 1) *in Equation 2:*

$$3x - 2y = 4$$
$$3(2) - 2(1) \stackrel{?}{=} 4$$
$$4 = 4 \checkmark$$

In this section, you will study two ways to solve systems of equations, beginning with the **method of substitution.**

The Method of Substitution

To use the **method of substitution** to solve a system of two equations in x and y, perform the following steps.

1. Solve one of the equations for one variable in terms of the other.
2. Substitute the expression found in Step 1 into the other equation to obtain an equation in one variable.
3. Solve the equation obtained in Step 2.
4. Back-substitute the value(s) obtained in Step 3 into the expression obtained in Step 1 to find the value(s) of the other variable.
5. Check that each solution satisfies *both* of the original equations.

Another method is the **method of graphing.** When using this method, note that the solution of the system corresponds to the **point(s) of intersection** of the graphs.

The Method of Graphing

To use the **method of graphing** to solve a system of two equations in x and y, perform the following steps.

1. Solve both equations for y in terms of x.
2. Use a graphing utility to graph both equations in the same viewing window.
3. Use the *intersect* feature or the *zoom* and *trace* features of the graphing utility to approximate the point(s) of intersection of the graphs.
4. Check that each solution satisfies *both* of the original equations.

What you should learn

- Use the methods of substitution and graphing to solve systems of equations in two variables.
- Use systems of equations to model and solve real-life problems.

Why you should learn it

You can use systems of equations in situations in which the variables must satisfy two or more conditions. For instance, Exercise 88 on page 406 shows how to use a system of equations to compare two models for estimating the number of board feet in a 16-foot log.

Remark

When using the method of substitution, it does not matter which variable you choose to solve for first. Whether you solve for y first or x first, you will obtain the same solution. When making your choice, you should choose the variable and equation that are easier to work with.

EXAMPLE 1 Solving a System of Equations

Solve the system of equations.

$$\begin{cases} x + y = 4 \\ x - y = 2 \end{cases}$$ Equation 1; Equation 2

Algebraic Solution

Begin by solving for y in Equation 1.

$y = 4 - x$ Solve for y in Equation 1.

Next, substitute this expression for y into Equation 2 and solve the resulting single-variable equation for x.

$x - y = 2$ Write Equation 2.

$x - (4 - x) = 2$ Substitute $4 - x$ for y.

$x - 4 + x = 2$ Distributive Property

$2x - 4 = 2$ Combine like terms.

$2x = 6$ Add 4 to each side.

$x = 3$ Divide each side by 2.

Finally, you can solve for y by *back-substituting* $x = 3$ into the equation $y = 4 - x$ to obtain

$y = 4 - x$ Write revised Equation 1.

$y = 4 - 3$ Substitute 3 for x.

$y = 1.$ Solve for y.

The solution is the ordered pair (3, 1). Check this as follows.

Check (3, 1) *in Equation 1:*

$x + y = 4$ Write Equation 1.

$3 + 1 \stackrel{?}{=} 4$ Substitute for x and y.

$4 = 4$ Solution checks in Equation 1. ✓

Check (3, 1) *in Equation 2:*

$x - y = 2$ Write Equation 2.

$3 - 1 \stackrel{?}{=} 2$ Substitute for x and y.

$2 = 2$ Solution checks in Equation 2. ✓

Because (3, 1) satisfies both equations in the system, it is a solution of the system of equations.

Graphical Solution

Begin by solving both equations for y. Then use a graphing utility to graph the equations

$$y_1 = 4 - x$$

and

$$y_2 = x - 2$$

in the same viewing window. Use the *intersect* feature (see figure) to approximate the point of intersection of the graphs.

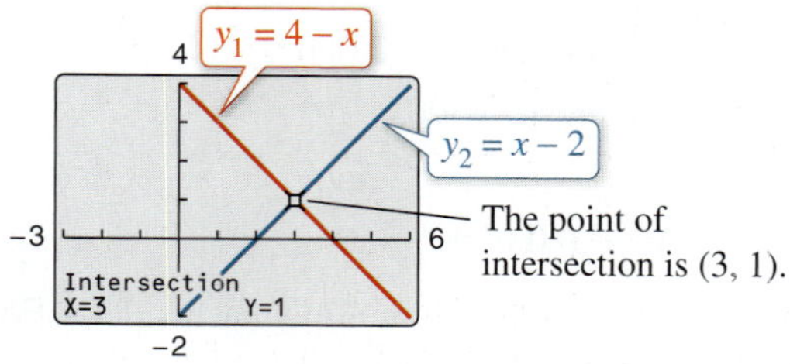

Check that (3, 1) is the exact solution as follows.

Check (3, 1) *in Equation 1:*

$x + y = 4$ Write Equation 1.

$3 + 1 \stackrel{?}{=} 4$ Substitute for x and y.

$4 = 4$ Solution checks in Equation 1. ✓

Check (3, 1) *in Equation 2:*

$x - y = 2$ Write Equation 2.

$3 - 1 \stackrel{?}{=} 2$ Substitute for x and y.

$2 = 2$ Solution checks in Equation 2. ✓

Because (3, 1) satisfies both equations in the system, it is a solution of the system of equations.

✓ Checkpoint *Audio-video solution in English & Spanish at LarsonPrecalculus.com.*

Solve the system of equations by the method of substitution.

$$\begin{cases} x - y = 0 \\ 5x - 3y = 6 \end{cases}$$

In the algebraic solution to Example 1, note that the term *back-substitution* implies that you work *backwards*. First you solve for one of the variables, and then you substitute that value *back* into one of the equations in the system to find the value of the other variable.

EXAMPLE 2 Solving a System by Substitution

A total of $12,000 is invested in two funds paying 5% and 3% simple interest. The yearly interest is $500. How much is invested at each rate?

Solution

Verbal Model: Amount in 5% fund + Amount in 3% fund = Total investment

Interest for 5% fund + Interest for 3% fund = Total interest

Labels: Amount in 5% fund $= x$ (dollars)

Interest for 5% fund $= 0.05x$ (dollars)

Amount in 3% fund $= y$ (dollars)

Interest for 3% fund $= 0.03y$ (dollars)

Total investment $= 12{,}000$ (dollars)

Total interest $= 500$ (dollars)

System: $\begin{cases} x + y = 12{,}000 & \text{Equation 1} \\ 0.05x + 0.03y = 500 & \text{Equation 2} \end{cases}$

To begin, it is convenient to multiply each side of Equation 2 by 100. This eliminates the need to work with decimals.

$$5x + 3y = 50{,}000 \qquad \text{Revised Equation 2}$$

To solve this system, you can solve for x in Equation 1.

$$x = 12{,}000 - y \qquad \text{Revised Equation 1}$$

Next, substitute this expression for x into revised Equation 2 and solve the resulting equation for y.

$$5x + 3y = 50{,}000 \qquad \text{Write revised Equation 2.}$$

$$5(12{,}000 - y) + 3y = 50{,}000 \qquad \text{Substitute } 12{,}000 - y \text{ for } x.$$

$$60{,}000 - 5y + 3y = 50{,}000 \qquad \text{Distributive Property}$$

$$-2y = -10{,}000 \qquad \text{Combine like terms.}$$

$$y = 5000 \qquad \text{Divide each side by } -2.$$

Finally, back-substitute the value $y = 5000$ to solve for x.

$$x = 12{,}000 - y \qquad \text{Write revised Equation 1.}$$

$$x = 12{,}000 - 5000 \qquad \text{Substitute 5000 for } y.$$

$$x = 7000 \qquad \text{Simplify.}$$

The solution is (7000, 5000). So, $7000 is invested at 5% and $5000 is invested at 3% to yield yearly interest of $500. Check this in the original system.

✓ ***Checkpoint*** *Audio-video solution in English & Spanish at LarsonPrecalculus.com.*

A total of $25,000 is invested in two funds paying 6.5% and 8.5% simple interest. The yearly interest is $2000. How much is invested at each rate?

The equations in Examples 1 and 2 are linear. Substitution and graphing can also be used to solve systems in which one or both of the equations are nonlinear.

Technology Tip

Remember that a good way to check the answers you obtain in this section is to use a graphing utility. For instance, enter the two equations in Example 2

$$y_1 = 12{,}000 - x$$

$$y_2 = \frac{500 - 0.05x}{0.03}$$

and find an appropriate viewing window that shows where the lines intersect. Then use the *intersect* feature or the *zoom* and *trace* features to find the point of intersection.

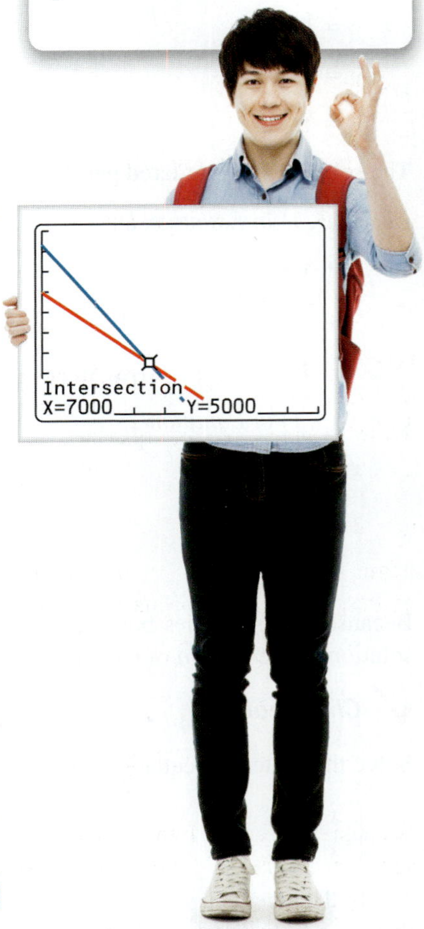

EXAMPLE 3 Substitution: Two-Solution Case

Solve the system of equations.

$$\begin{cases} x^2 + 4x - y = 7 & \text{Equation 1} \\ 2x - y = -1 & \text{Equation 2} \end{cases}$$

Algebraic Solution

Begin by solving for y in Equation 2 to obtain

$$y = 2x + 1. \qquad \text{Solve for } y \text{ in Equation 2.}$$

Next, substitute this expression for y into Equation 1 and solve for x.

$$x^2 + 4x - y = 7 \qquad \text{Write Equation 1.}$$

$$x^2 + 4x - (2x + 1) = 7 \qquad \text{Substitute } 2x + 1 \text{ for } y.$$

$$x^2 + 4x - 2x - 1 = 7 \qquad \text{Distributive Property}$$

$$x^2 + 2x - 8 = 0 \qquad \text{Write in general form.}$$

$$(x + 4)(x - 2) = 0 \qquad \text{Factor.}$$

$$x + 4 = 0 \;\Rightarrow\; x = -4 \qquad \text{Set 1st factor equal to 0.}$$

$$x - 2 = 0 \;\Rightarrow\; x = 2 \qquad \text{Set 2nd factor equal to 0.}$$

Back-substituting these values of x into revised Equation 2 produces

$$y = 2(-4) + 1 = -7 \quad \text{and} \quad y = 2(2) + 1 = 5.$$

So, the solutions are $(-4, -7)$ and $(2, 5)$. Check these in the original system.

Graphical Solution

Solve each equation for y and use a graphing utility to graph the equations in the same viewing window. From the figures below, the solutions are $(-4, -7)$ and $(2, 5)$. Check these in the original system.

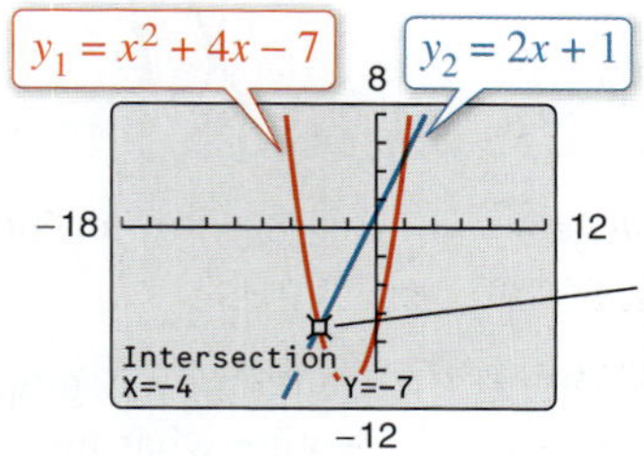

One of the points of intersection is $(-4, -7)$.

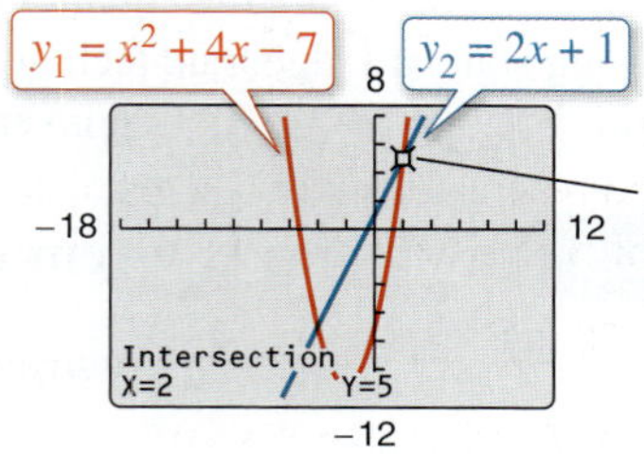

One of the points of intersection is $(2, 5)$.

✓ Checkpoint *Audio-video solution in English & Spanish at LarsonPrecalculus.com.*

Solve the system of equations.

$$\begin{cases} -2x + y = 5 \\ x^2 - y + 3x = 1 \end{cases}$$

EXAMPLE 4 Substitution: No-Solution Case

To solve the system of equations

$$\begin{cases} -x + y = 4 & \text{Equation 1} \\ x^2 + y = 3 & \text{Equation 2} \end{cases}$$

begin by solving for y in Equation 1 to obtain $y = x + 4$. Next, substitute this expression for y into Equation 2 and solve for x.

$$x^2 + (x + 4) = 3 \qquad \text{Substitute } x + 4 \text{ for } y \text{ in Equation 2.}$$

$$x^2 + x + 1 = 0 \qquad \text{Write in general form.}$$

$$x = \frac{-1 \pm \sqrt{3}i}{2} \qquad \text{Quadratic Formula}$$

Because this yields two imaginary values, the equation $x^2 + x + 1 = 0$ has no *real* solution. So, the original system of equations has no *real* solution.

✓ Checkpoint *Audio-video solution in English & Spanish at LarsonPrecalculus.com.*

Solve $\begin{cases} 2x - y = -3 \\ 2x^2 + 4x - y^2 = 0 \end{cases}$.

Explore the Concept

Graph the system of equations in Example 4. Do the graphs of the equations intersect? Why or why not?

From Examples 1, 3, and 4, you can see that a system of two equations in two unknowns can have exactly one solution, more than one solution, or no solution. For instance, in Figure 5.1, the two equations in Example 1 graph as two lines with a *single point* of intersection. The two equations in Example 3 graph as a parabola and a line with *two points* of intersection, as shown in Figure 5.2. The two equations in Example 4 graph as a line and a parabola that have *no points* of intersection, as shown in Figure 5.3.

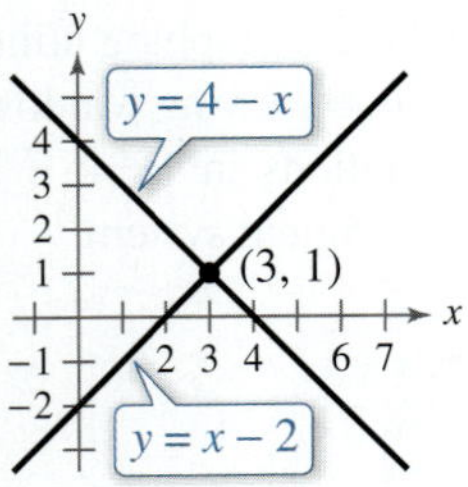

One Intersection Point
Figure 5.1

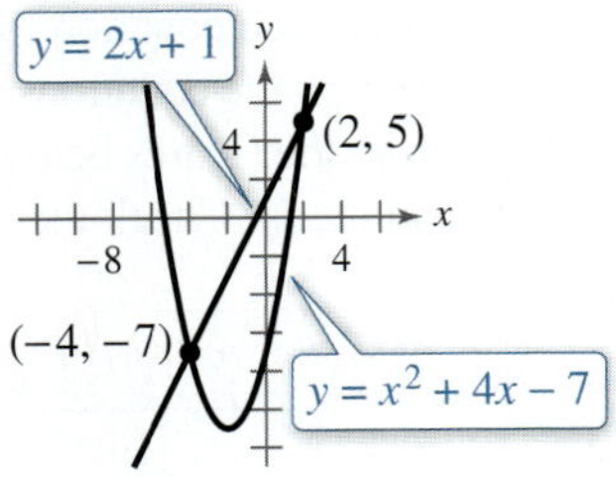

Two Intersection Points
Figure 5.2

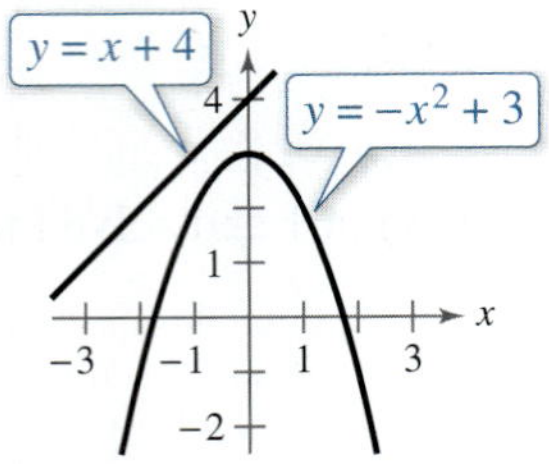

No Intersection Points
Figure 5.3

Example 5 shows the benefit of a graphical approach to solving systems of equations in two variables. Notice what happens when you try only the substitution method in Example 5. You obtain the equation $x + \ln x = 1$. It is difficult to solve this equation for x using standard algebraic techniques. In such cases, a graphical approach to solving systems of equations is more convenient.

EXAMPLE 5 Solving a System of Equations Graphically

See LarsonPrecalculus.com for an interactive version of this type of example.

Solve the system of equations.

$$\begin{cases} y = \ln x & \text{Equation 1} \\ x + y = 1 & \text{Equation 2} \end{cases}$$

Solution

From the graphs of these equations, it is clear that there is only one point of intersection. Use the *intersect* feature of a graphing utility to approximate the solution point as $(1, 0)$, as shown in the figure.

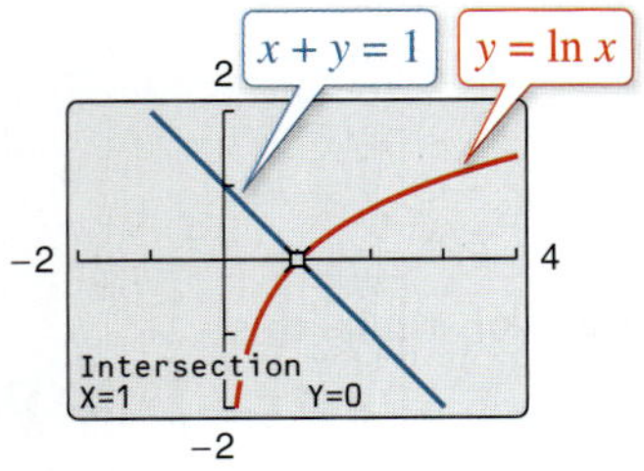

You can confirm this by substituting $(1, 0)$ into *both* equations.

Check $(1, 0)$ *in Equation 1:*

$y = \ln x$ Write Equation 1.

$0 = \ln 1$ Equation 1 checks. ✓

Check $(1, 0)$ *in Equation 2:*

$x + y = 1$ Write Equation 2.

$1 + 0 = 1$ Equation 2 checks. ✓

✓ ***Checkpoint*** *Audio-video solution in English & Spanish at LarsonPrecalculus.com.*

Solve the system of equations.

$$\begin{cases} y = 3 - \log_{10} x \\ -2x + y = 1 \end{cases}$$

Application

The total cost C of producing x units of a product typically has two components: the initial cost and the cost per unit. When enough units have been sold so that the total revenue R equals the total cost C, the sales are said to have reached the **break-even point.** You will find that the break-even point corresponds to the point of intersection of the cost and revenue curves.

EXAMPLE 6 Break-Even Analysis

A small business invests \$10,000 in equipment to produce a new soft drink. Each bottle of the soft drink costs \$0.65 to produce and is sold for \$1.20. How many bottles must be sold before the business breaks even?

Solution

The total cost of producing x bottles is

Total cost = Cost per bottle · Number of bottles + Initial cost

$$C = 0.65x + 10{,}000. \qquad \text{Equation 1}$$

The revenue obtained by selling x bottles is

Total revenue = Price per bottle · Number of bottles

$$R = 1.20x. \qquad \text{Equation 2}$$

Because the break-even point occurs when $R = C$, you have

$$C = 1.20x$$

and the system of equations to solve is

$$\begin{cases} C = 0.65x + 10{,}000 \\ C = 1.20x \end{cases}.$$

Now you can solve by substitution.

$$C = 0.65x + 10{,}000 \qquad \text{Write Equation 1.}$$

$$1.20x = 0.65x + 10{,}000 \qquad \text{Substitute } 1.20x \text{ for } C.$$

$$0.55x = 10{,}000 \qquad \text{Subtract } 0.65x \text{ from each side.}$$

$$x = \frac{10{,}0000}{0.55} \qquad \text{Divide each side by 0.55.}$$

$$x \approx 18{,}182 \text{ bottles} \qquad \text{Use a calculator.}$$

So, the business must sell about 18,182 bottles to break even. Note in Figure 5.4 that revenue less than the break-even point corresponds to an overall loss, whereas revenue greater than the break-even point corresponds to a profit. Verify the break-even point using the *intersect* feature or the *zoom* and *trace* features of a graphing utility.

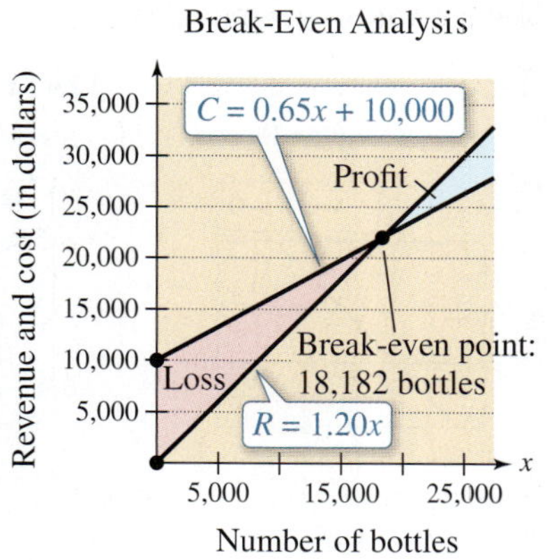

Figure 5.4

Checkpoint *Audio-video solution in English & Spanish at LarsonPrecalculus.com.*

In Example 6, each bottle of the soft drink costs \$0.50 to produce. How many bottles must be sold before the business breaks even?

Another way to view the solution in Example 6 is to consider the profit function $P = R - C$. The break-even point occurs when the profit is 0, which is the same as saying that $R = C$.

5.1 Exercises

See *CalcChat.com* for tutorial help and worked-out solutions to odd-numbered exercises.
For instructions on how to use a graphing utility, see Appendix A.

Vocabulary and Concept Check

In Exercises 1–4, fill in the blank(s).

1. A set of two or more equations in two or more unknowns is called a _______ of _______ .

2. A _______ of a system of equations is an ordered pair that satisfies each equation in the system.

3. The first step in solving a system of two equations in x and y by the method of _______ is to solve one of the equations for one variable in terms of the other.

4. A point of intersection of the graphs of the equations of a system is a _______ of the system.

5. What is the point of intersection of the graphs of the cost and revenue functions called?

6. The graphs of the equations of a system do not intersect. What can you conclude about the system?

Procedures and Problem Solving

Checking Solutions **In Exercises 7–10, determine whether each ordered pair is a solution of the system of equations.**

7. $\begin{cases} 4x - y = 1 \\ 6x + y = -6 \end{cases}$ (a) $(0, -3)$ (b) $(-1, -5)$ (c) $\left(-\frac{3}{2}, 3\right)$ (d) $\left(-\frac{1}{2}, -3\right)$

8. $\begin{cases} 4x^2 + y = 3 \\ -x - y = 11 \end{cases}$ (a) $(2, -13)$ (b) $(-2, -9)$ (c) $\left(-\frac{3}{2}, 6\right)$ (d) $\left(-\frac{7}{4}, -\frac{37}{4}\right)$

9. $\begin{cases} y = -2e^x \\ 3x - y = 2 \end{cases}$ (a) $(-2, 0)$ (b) $(0, -2)$ (c) $(0, -3)$ (d) $(-1, -5)$

10. $\begin{cases} -\log_{10} x + 3 = y \\ \frac{1}{9}x + y = \frac{28}{9} \end{cases}$ (a) $(100, 1)$ (b) $(10, 2)$ (c) $(1, 3)$ (d) $(1, 1)$

Solving a System by Substitution **In Exercises 11–18, solve the system by the method of substitution. Check your solution graphically.**

11. $\begin{cases} 2x + y = 6 \\ -x + y = 0 \end{cases}$

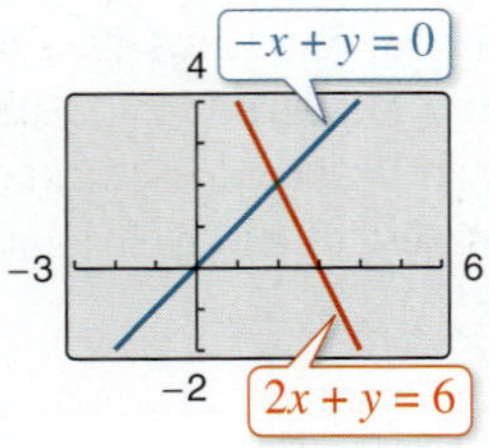

12. $\begin{cases} x - y = -4 \\ x + 2y = 5 \end{cases}$

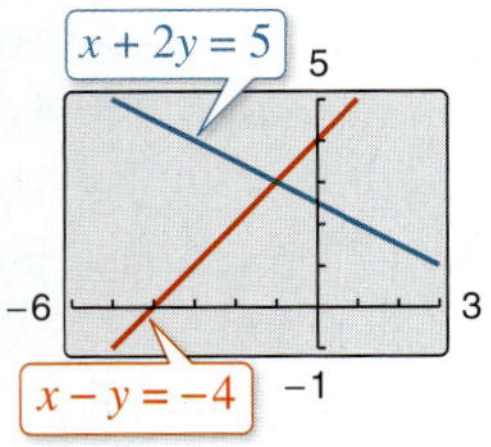

13. $\begin{cases} x - y = -4 \\ x^2 - y = -2 \end{cases}$

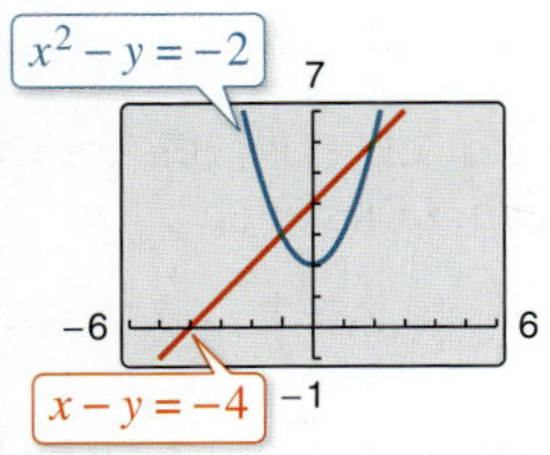

14. $\begin{cases} -2x + y = -5 \\ x^2 + y^2 = 25 \end{cases}$

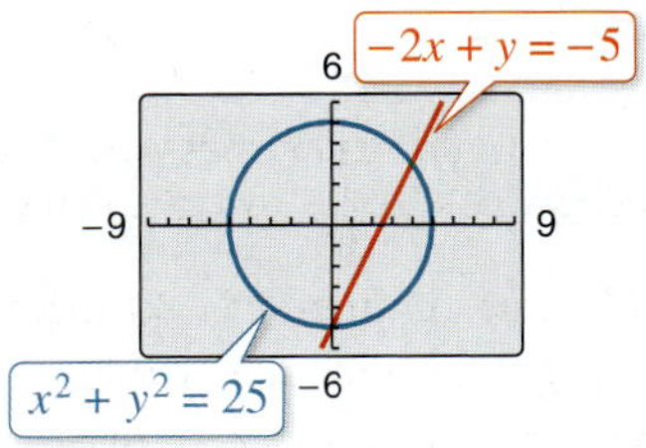

15. $\begin{cases} 3x + y = 2 \\ x^3 - 2 + y = 0 \end{cases}$

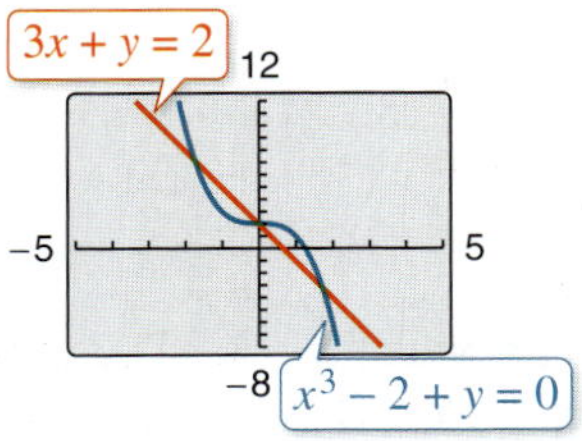

16. $\begin{cases} x + y = 0 \\ x^3 - 5x - y = 0 \end{cases}$

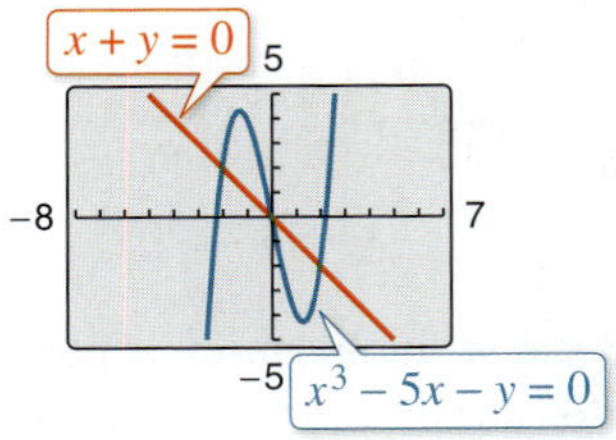

17. $\begin{cases} -\frac{7}{2}x - y = -18 \\ 8x^2 - 2y^3 = 0 \end{cases}$

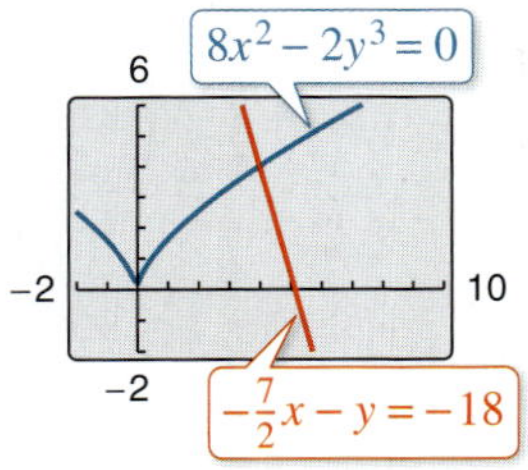

18. $\begin{cases} y = x^3 - 3x^2 + 4 \\ y = -2x + 4 \end{cases}$

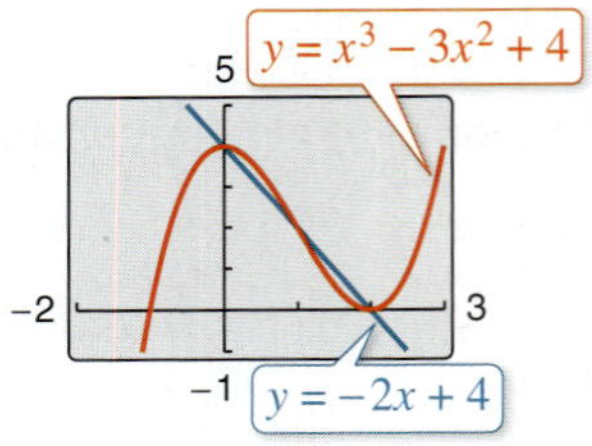

Solving a System of Equations **In Exercises 19–28, solve the system by the method of substitution. Use a graphing utility to verify your results.**

19. $\begin{cases} x + y = 0 \\ 4x + 3y = 10 \end{cases}$

20. $\begin{cases} x + 2y = 1 \\ 5x - 4y = -23 \end{cases}$

21. $\begin{cases} 2x - y + 2 = 0 \\ 4x + y - 5 = 0 \end{cases}$

22. $\begin{cases} 6x - 3y - 4 = 0 \\ x + 2y - 4 = 0 \end{cases}$

23. $\begin{cases} 1.5x + 0.8y = 2.3 \\ 0.3x - 0.2y = 0.1 \end{cases}$

24. $\begin{cases} -0.5x + 4y = 7.8 \\ 0.2x - 1.6y = -3.6 \end{cases}$

25. $\begin{cases} \frac{1}{5}x + \frac{1}{2}y = 8 \\ x + y = 20 \end{cases}$

26. $\begin{cases} \frac{1}{2}x + \frac{3}{4}y = 10 \\ \frac{3}{4}x - y = 4 \end{cases}$

27. $\begin{cases} -\frac{5}{3}x + y = 5 \\ -5x + 3y = 6 \end{cases}$

28. $\begin{cases} -\frac{2}{3}x + y = 2 \\ 3x - \frac{1}{2}y = 4 \end{cases}$

Solving a System by Substitution **In Exercises 29–32, you are given the yearly interest earned from a total of \$18,000 invested in two funds paying the given rates of simple interest. Write and solve a system of equations to find the amount invested at each rate.**

	Yearly Interest	*Rate 1*	*Rate 2*
29.	\$400	4%	2%
30.	\$840	6%	3%
31.	\$1182	5.6%	6.8%
32.	\$684	2.75%	4.25%

Solving a System with a Nonlinear Equation **In Exercises 33–38, solve the system by the method of substitution. Use a graphing utility to verify your results.**

33. $\begin{cases} x^2 - 2x + y = 8 \\ x - y = -2 \end{cases}$

34. $\begin{cases} 2x^2 - 2x - y = 14 \\ 2x - y = -2 \end{cases}$

35. $\begin{cases} 2x^2 - y = 1 \\ x - y = 2 \end{cases}$

36. $\begin{cases} 2x^2 + y = 3 \\ x + y = 4 \end{cases}$

37. $\begin{cases} x^3 - y = 0 \\ x - y = 0 \end{cases}$

38. $\begin{cases} y = -x \\ y = x^3 + 3x^2 + 2x \end{cases}$

Solving a System of Equations Graphically **In Exercises 39–46, solve the system graphically. Verify your solutions algebraically.**

39. $\begin{cases} -2x + y = 7 \\ x + 3y = 0 \end{cases}$

40. $\begin{cases} x + y = 8 \\ 4x + 4y = 0 \end{cases}$

41. $\begin{cases} x - 2y = -3 \\ 5x + 6y = 17 \end{cases}$

42. $\begin{cases} -5x + 2y = -2 \\ x - 2y = 6 \end{cases}$

43. $\begin{cases} x^2 + y = -1 \\ -x + 2y = 5 \end{cases}$

44. $\begin{cases} x^2 - y = 3 \\ x - y = 1 \end{cases}$

45. $\begin{cases} -x - y = 3 \\ x^2 + y^2 - 4x - 21 = 0 \end{cases}$

46. $\begin{cases} y^2 - x^2 + 9 = 0 \\ -\frac{1}{2}x + y = \frac{3}{2} \end{cases}$

Solving a System of Equations Graphically **In Exercises 47–60, use a graphing utility to approximate all points of intersection of the graphs of equations in the system. Round your results to three decimal places. Verify your solutions by checking them in the original system.**

47. $\begin{cases} 7x + 8y = 24 \\ x - 8y = 8 \end{cases}$

48. $\begin{cases} x - y = 0 \\ 5x - 2y = 6 \end{cases}$

49. $\begin{cases} x - y^2 = -1 \\ x - y = 5 \end{cases}$

50. $\begin{cases} x - y^2 = -2 \\ x - 2y = 6 \end{cases}$

51. $\begin{cases} x^2 + y^2 = 8 \\ y = x^2 \end{cases}$

52. $\begin{cases} x^2 + y^2 = 25 \\ (x - 8)^2 + y^2 = 41 \end{cases}$

53. $\begin{cases} y = e^x \\ x - y + 1 = 0 \end{cases}$

54. $\begin{cases} y = -4e^{-x} \\ y + 3x + 8 = 0 \end{cases}$

55. $\begin{cases} x + 2y = 8 \\ y = 2 + \ln x \end{cases}$

56. $\begin{cases} 3y + 2x = 9 \\ y = -2 + \ln(x - 1) \end{cases}$

57. $\begin{cases} y = \sqrt{x} + 4 \\ y = 2x + 1 \end{cases}$

58. $\begin{cases} x - y = 3 \\ \sqrt{x} - y = 1 \end{cases}$

59. $\begin{cases} x^2 + y^2 = 169 \\ x^2 - 8y = 104 \end{cases}$

60. $\begin{cases} x^2 + y^2 = 4 \\ 2x^2 - y = 2 \end{cases}$

Choosing a Solution Method **In Exercises 61–74, solve the system graphically or algebraically. Explain your choice of method.**

61. $\begin{cases} 2x - y = 0 \\ x^2 - y = -1 \end{cases}$

62. $\begin{cases} x + y = 4 \\ x^2 + y = 2 \end{cases}$

63. $\begin{cases} 3x - 7y = -6 \\ x^2 - y^2 = 4 \end{cases}$

64. $\begin{cases} x^2 + y^2 = 25 \\ 2x + y = 10 \end{cases}$

65. $\begin{cases} x^2 + y^2 = 1 \\ x + y = 4 \end{cases}$

66. $\begin{cases} x^2 + y^2 = 4 \\ x - y = 5 \end{cases}$

67. $\begin{cases} y = 2x + 1 \\ y = \sqrt{x + 2} \end{cases}$

68. $\begin{cases} y = 2x - 1 \\ y = \sqrt{x + 1} \end{cases}$

69. $\begin{cases} y - e^{-x} = 1 \\ y - \ln x = 3 \end{cases}$

70. $\begin{cases} 2 \ln x + y = 4 \\ e^x - y = 0 \end{cases}$

71. $\begin{cases} y = x^3 - 2x^2 + 1 \\ y = 1 - x^2 \end{cases}$

72. $\begin{cases} y = x^3 - 2x^2 + x - 1 \\ y = -x^2 + 3x - 1 \end{cases}$

73. $\begin{cases} xy - 1 = 0 \\ -5x - 2y + 1 = 0 \end{cases}$

74. $\begin{cases} xy - 2 = 0 \\ 3x - 2y + 4 = 0 \end{cases}$

Break-Even Analysis **In Exercises 75–78, use a graphing utility to graph the cost and revenue functions in the same viewing window. Find the sales x necessary to break even ($R = C$) and the corresponding revenue R obtained by selling x units. (Round to the nearest whole unit.)**

	Cost	*Revenue*
75.	$C = 8650x + 250{,}000$	$R = 9950x$
76.	$C = 2.65x + 350{,}000$	$R = 4.15x$
77.	$C = 5.5\sqrt{x} + 10{,}000$	$R = 3.29x$
78.	$C = 7.8\sqrt{x} + 18{,}500$	$R = 12.84x$

Geometry **In Exercises 79 and 80, find the dimensions of the rectangle meeting the specified conditions.**

79. The perimeter is 30 meters and the length is 3 meters greater than the width.

80. The perimeter is 280 centimeters and the width is 20 centimeters less than the length.

81. Marketing Research The daily DVD rentals of a newly released animated film and a newly released horror film from a movie rental store can be modeled by the equations

$$\begin{cases} N = 360 - 24x & \text{Animated film} \\ N = 24 + 18x & \text{Horror film} \end{cases}$$

where N is the number of DVDs rented and x represents the week, with $x = 1$ corresponding to the first week of release.

(a) Use the *table* feature of a graphing utility to find the numbers of rentals of each movie for each of the first 12 weeks of release.

(b) Use the results of part (a) to determine the solution to the system of equations.

(c) Solve the system of equations algebraically.

(d) Compare your results from parts (b) and (c).

(e) Interpret the results in the context of the situation.

82. Economics You want to buy either a wood pellet stove or an electric furnace. The pellet stove costs \$3650 and produces heat at a cost of \$19.15 per 1 million Btu (British thermal units). The electric furnace costs \$2780 and produces heat at a cost of \$33.25 per 1 million Btu.

(a) Write a function for the total cost y of buying the pellet stove and producing x million Btu of heat.

(b) Write a function for the total cost y of buying the electric furnace and producing x million Btu of heat.

(c) Use a graphing utility to graph and solve the system of equations formed by the two cost functions.

(d) Solve the system of equations algebraically.

(e) Interpret the results in the context of the situation.

83. Break-Even Analysis A small software company invests \$16,000 to produce a software package that will sell for \$55.95. Each unit can be produced for \$9.45.

(a) Write the cost and revenue functions for x units produced and sold.

(b) Use a graphing utility to graph the cost and revenue functions in the same viewing window. Use the graph to approximate the number of units that must be sold to break even and verify the result algebraically.

84. Professional Sales You are offered two jobs selling college textbooks. One company offers an annual salary of \$33,000 plus a year-end bonus of 1% of your total sales. The other company offers an annual salary of \$30,000 plus a year-end bonus of 2.5% of your total sales. How much would you have to sell in a year to make the second offer the better offer?

85. Geometry What are the dimensions of a rectangular tract of land with a perimeter of 40 miles and an area of 96 square miles?

86. Geometry What are the dimensions of an isosceles right triangle with a two-inch hypotenuse and an area of 1 square inch?

87. Finance You are deciding how to invest a total of \$20,000 in two funds paying 5.5% and 7.5% simple interest. You want to earn a total of \$1300 in interest from the investments each year.

(a) Write a system of equations in which one equation represents the total amount invested and the other equation represents the \$1300 yearly interest. Let x and y represent the amounts invested at 5.5% and 7.5%, respectively.

(b) Use a graphing utility to graph the two equations in the same viewing window.

(c) How much of the \$20,000 should you invest at 5.5% to earn \$1300 in interest per year? Explain your reasoning.

88. *Why you should learn it* *(p. 398)* You are offered two different rules for estimating the number of board feet in a 16-foot log. (A board foot is a unit of measure for lumber equal to a board 1 foot square and 1 inch thick.) One rule is the Doyle Log Rule modeled by

$$V = (D - 4)^2, \quad 5 \le D \le 40$$

where D is the diameter (in inches) of the log and V is its volume (in board feet). The other rule is the Scribner Log Rule modeled by

$$V = 0.79D^2 - 2D - 4, \quad 5 \le D \le 40.$$

(a) Use a graphing utility to graph the two log rules in the same viewing window.

(b) For what diameter do the two rules agree?

(c) You are selling large logs by the board foot. Which rule would you use? Explain your reasoning.

89. Algebraic-Graphical-Numerical The populations (in thousands) of Colorado C and Minnesota M from 2008 through 2012 can be modeled by the system

$$\begin{aligned} C &= 74.0t + 4303 && \text{Colorado} \\ M &= 33.0t + 4988 && \text{Minnesota} \end{aligned}$$

where t is the year, with $t = 8$ corresponding to 2008. (Source: U.S. Census Bureau)

(a) Record in a table the populations predicted by the models for the two states in the years 2013 through 2020.

(b) According to the table, in what year(s) will the population of Colorado be greater than that of Minnesota?

(c) Use a graphing utility to graph the models in the same viewing window. Estimate the point of intersection of the models.

(d) Find the point of intersection algebraically.

(e) Summarize your findings of parts (b) through (d).

90. MODELING DATA

The table shows the yearly revenues (in millions of dollars) of the online travel companies Expedia and Priceline.com from 2008 through 2013. (Sources: Expedia and Priceline.com)

DATA

Year	Expedia	Priceline.com
2008	2937	1885
2009	2955	2338
2010	3348	3085
2011	3449	4356
2012	4030	5261
2013	4771	6793

Spreadsheet at LarsonPrecalculus.com

(a) Use the *regression* feature of a graphing utility to find a quadratic model for the yearly revenue E of Expedia and a linear model for the yearly revenue P of Priceline.com. Let x represent the year, with $x = 8$ corresponding to 2008.

(b) Use the graphing utility to graph the models with the original data in the same viewing window.

(c) Use the graph in part (b) to approximate the first year when the revenues of Priceline.com will be greater than the revenues of Expedia.

(d) Algebraically approximate the first year when the revenues of Priceline.com will be greater than the revenues of Expedia.

(e) Compare your results from parts (c) and (d).

Conclusions

True or False? **In Exercises 91 and 92, determine whether the statement is true or false. Justify your answer.**

91. In order to solve a system of equations by substitution, you must always solve for y in one of the two equations and then back-substitute.

92. If a system consists of a parabola and a circle, then it can have at most two solutions.

93. Think About It When solving a system of equations by substitution, how do you recognize that the system has no solution?

94. Exploration Find equations of lines whose graphs intersect the graph of the parabola $y = x^2$ at (a) two points, (b) one point, and (c) no points. (There are many correct answers.)

95. Exploration Create systems of two linear equations in two variables that have (a) no solution, (b) one distinct solution, and (c) infinitely many solutions. (There are many correct answers.)

96. Exploration Create a system of linear equations in two variables that has the solution $(2, -1)$ as its only solution. (There are many correct answers.)

97. Exploration Consider the system of equations.

$$\begin{cases} y = b^x \\ y = x^b \end{cases}$$

(a) Use a graphing utility to graph the system of equations for $b = 2$ and $b = 4$.

(b) For a fixed value of $b > 1$, make a conjecture about the number of points of intersection of the graphs in part (a).

98. HOW DO YOU SEE IT? The cost C of producing x units and the revenue R obtained by selling x units are shown in the figure.

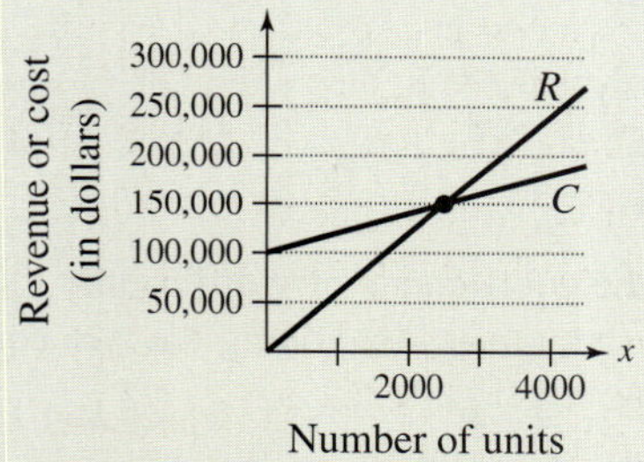

(a) Estimate the point of intersection. What does this point represent?

(b) Use the figure to identify the x-values that correspond to (i) an overall loss and (ii) a profit. Explain your reasoning.

Cumulative Mixed Review

Finding the Slope-Intercept Form **In Exercises 99–102, write an equation of the line passing through the two points. Use the slope-intercept form, if possible. If not possible, explain why.**

99. $(3, 4), (10, 6)$

100. $(6, 3), (10, 3)$

101. $(4, -2), (4, 5)$

102. $\left(\frac{3}{5}, 0\right), (4, 6)$

Finding the Domain and Asymptotes of a Function **In Exercises 103–108, find the domain of the function and identify any horizontal or vertical asymptotes.**

103. $f(x) = \dfrac{5}{x - 6}$

104. $f(x) = \dfrac{2x - 7}{3x + 2}$

105. $f(x) = \dfrac{x^2 + 2}{x^2 - 16}$

106. $f(x) = 3 - \dfrac{2}{x^2}$

107. $f(x) = \dfrac{x + 1}{x^2 + 1}$

108. $f(x) = \dfrac{x - 4}{x^2 + 16}$

5.2 Systems of Linear Equations in Two Variables

The Method of Elimination

In Section 5.1, you studied two methods for solving a system of equations: substitution and graphing. Now you will study the **method of elimination.** The key step in this method is to obtain, for one of the variables, coefficients that differ only in sign so that *adding* the equations eliminates the variable.

$$\begin{array}{rll} 3x + 5y = & 7 & \text{Equation 1} \\ -3x - 2y = & -1 & \text{Equation 2} \\ \hline 3y = & 6 & \text{Add equations.} \end{array}$$

Note that by adding the two equations, you eliminate the x-terms and obtain a single equation in y. Solving this equation for y produces

$$y = 2$$

which you can then back-substitute into one of the original equations to solve for x.

What you should learn

- Use the method of elimination to solve systems of linear equations in two variables.
- Graphically interpret the number of solutions of a system of linear equations in two variables.
- Use systems of linear equations in two variables to model and solve real-life problems.

Why you should learn it

You can use systems of linear equations to model many business applications. For instance, Exercise 84 on page 415 shows how to use a system of linear equations to determine the number of running shoes sold.

EXAMPLE 1 Solving a System by Elimination

Solve the system of linear equations.

$$\begin{cases} 3x + 2y = 4 & \text{Equation 1} \\ 5x - 2y = 8 & \text{Equation 2} \end{cases}$$

Solution

Because the coefficients of y differ only in sign, you can eliminate the y-terms by adding the two equations.

$$\begin{array}{rll} 3x + 2y = & 4 & \text{Write Equation 1.} \\ 5x - 2y = & 8 & \text{Write Equation 2.} \\ \hline 8x = & 12 & \text{Add equations.} \\ x = & \frac{3}{2} & \text{Solve for } x. \end{array}$$

So, $x = \frac{3}{2}$. By back-substituting into Equation 1, you can solve for y.

$$\begin{array}{rl} 3x + 2y = 4 & \text{Write Equation 1.} \\ 3\left(\frac{3}{2}\right) + 2y = 4 & \text{Substitute } \frac{3}{2} \text{ for } x. \\ y = -\frac{1}{4} & \text{Solve for } y. \end{array}$$

The solution is $\left(\frac{3}{2}, -\frac{1}{4}\right)$. You can check the solution *algebraically* by substituting into the original system, or graphically, as shown in Section 5.1.

Check

$$\begin{array}{rl} 3\left(\frac{3}{2}\right) + 2\left(-\frac{1}{4}\right) \stackrel{?}{=} 4 & \text{Substitute into Equation 1.} \\ \frac{9}{2} - \frac{1}{2} = 4 & \text{Equation 1 checks. ✓} \\ 5\left(\frac{3}{2}\right) - 2\left(-\frac{1}{4}\right) \stackrel{?}{=} 8 & \text{Substitute into Equation 2.} \\ \frac{15}{2} + \frac{1}{2} = 8 & \text{Equation 2 checks. ✓} \end{array}$$

✓ ***Checkpoint*** *Audio-video solution in English & Spanish at LarsonPrecalculus.com.*

Solve the system of linear equations.

$$\begin{cases} 2x + y = 4 \\ 2x - y = -1 \end{cases}$$

Explore the Concept

Use the method of substitution to solve the system given in Example 1. Which method is easier?

The Method of Elimination

To use the **method of elimination** to solve a system of two linear equations in x and y, perform the following steps.

1. Obtain coefficients for x (or y) that differ only in sign by multiplying all terms of one or both equations by suitably chosen constants.
2. Add the equations to eliminate one variable; solve the resulting equation.
3. Back-substitute the value obtained in Step 2 into either of the original equations and solve for the other variable.
4. Check your solution in both of the original equations.

EXAMPLE 2 Solving a System by Elimination

Solve the system of linear equations.

$$\begin{cases} 5x + 3y = 9 & \text{Equation 1} \\ 2x - 4y = 14 & \text{Equation 2} \end{cases}$$

Algebraic Solution

You can obtain coefficients of y that differ only in sign by multiplying Equation 1 by 4 and multiplying Equation 2 by 3.

$5x + 3y = 9$	⇒	$20x + 12y = 36$	Multiply Equation 1 by 4.
$2x - 4y = 14$	⇒	$6x - 12y = 42$	Multiply Equation 2 by 3.
		$26x = 78$	Add equations.

From this equation, you can see that $x = 3$. By back-substituting this value of x into Equation 2, you can solve for y.

$2x - 4y = 14$ Write Equation 2.

$2(3) - 4y = 14$ Substitute 3 for x.

$-4y = 8$ Combine like terms.

$y = -2$ Solve for y.

The solution is $(3, -2)$. You can check the solution algebraically by substituting into the original system.

Graphical Solution

Solve each equation for y and use a graphing utility to graph the equations in the same viewing window. From the figure, the solution is $(3, -2)$. Check this in the original system.

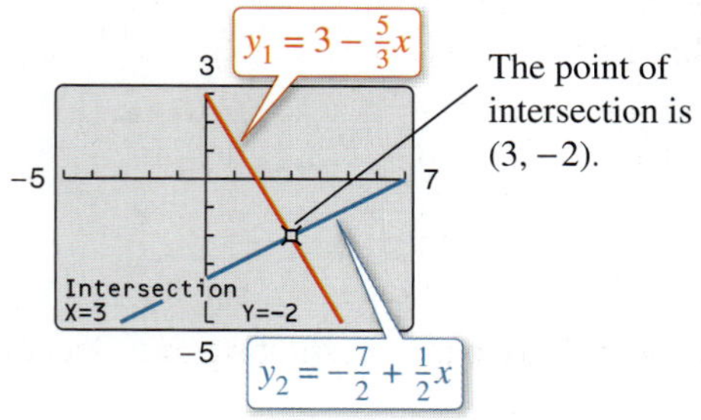

Checkpoint Audio-video solution in English & Spanish at LarsonPrecalculus.com.

Solve the system of linear equations.

$$\begin{cases} 3x + 2y = 7 \\ 2x + 5y = 1 \end{cases}$$

In Example 2, the two systems of linear equations (the original system and the system obtained by multiplying by constants)

$$\begin{cases} 5x + 3y = 9 \\ 2x - 4y = 14 \end{cases} \quad \text{and} \quad \begin{cases} 20x + 12y = 36 \\ 6x - 12y = 42 \end{cases}$$

are called **equivalent systems** because they have precisely the same solution set. The operations that can be performed on a system of linear equations to produce an equivalent system are (1) interchanging any two equations, (2) multiplying an equation by a nonzero constant, and (3) adding a multiple of one equation to any other equation in the system.

Graphical Interpretation of Two-Variable Systems

It is possible for any system of equations to have exactly one solution, two or more solutions, or no solution. If a system of *linear* equations has two different solutions, then it must have an *infinite* number of solutions. To see why this is true, consider the following graphical interpretations of a system of two linear equations in two variables.

Graphical Interpretations of Solutions

For a system of two linear equations in two variables, the number of solutions is one of the following.

Number of Solutions	*Graphical Interpretation*
1. Exactly one solution	The two lines intersect at one point.
2. Infinitely many solutions	The two lines are coincident (identical).
3. No solution	The two lines are parallel.

A system of linear equations is **consistent** when it has at least one solution. It is **inconsistent** when it has no solution.

EXAMPLE 3 Recognizing Graphs of Linear Systems

See LarsonPrecalculus.com for an interactive version of this type of example.

Match each system of linear equations (a, b, c) with its graph (i, ii, iii). Describe the number of solutions. Then state whether the system is consistent or inconsistent.

a. $\begin{cases} 2x - 3y = 3 \\ -4x + 6y = 6 \end{cases}$ **b.** $\begin{cases} 2x - 3y = 3 \\ x + 2y = 5 \end{cases}$ **c.** $\begin{cases} 2x - 3y = 3 \\ -4x + 6y = -6 \end{cases}$

i.

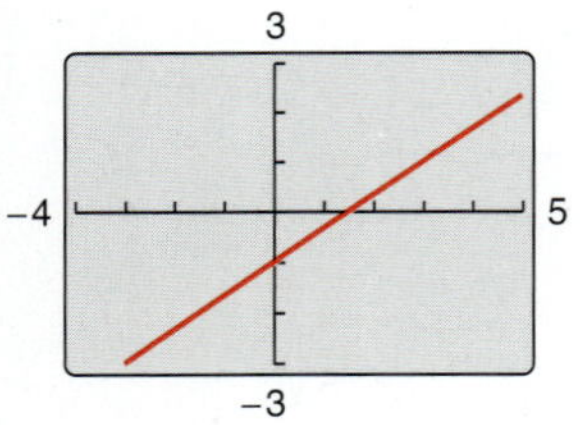

ii.

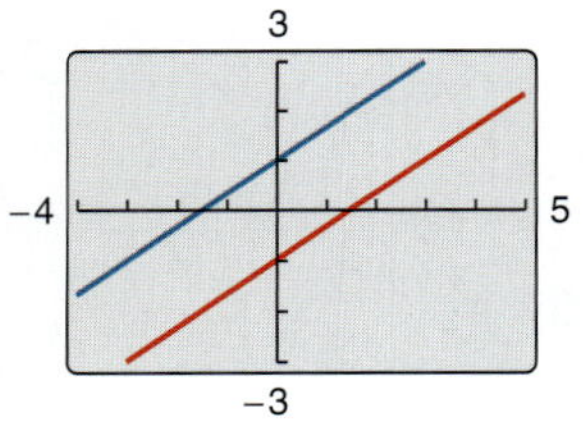

iii.

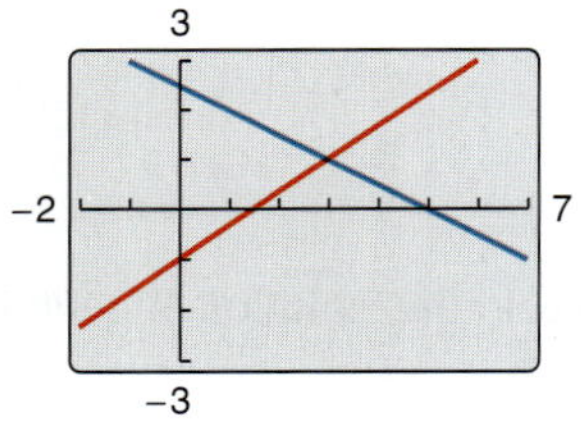

Solution

a. The graph of system (a) is a pair of parallel lines (ii). The lines have no point of intersection, so the system has no solution. The system is inconsistent.

b. The graph of system (b) is a pair of intersecting lines (iii). The lines have one point of intersection, so the system has exactly one solution. The system is consistent.

c. The graph of system (c) is a pair of lines that coincide (i). The lines have infinitely many points of intersection, so the system has infinitely many solutions. The system is consistent.

✓ ***Checkpoint*** *Audio-video solution in English & Spanish at LarsonPrecalculus.com.*

Sketch the graph of the system of linear equations. Then describe the number of solutions and state whether the system is consistent or inconsistent.

$$\begin{cases} -2x + 3y = 6 \\ 4x - 6y = -9 \end{cases}$$

Explore the Concept

Rewrite each system of equations in slope-intercept form and use a graphing utility to graph each system. What is the relationship between the slopes of the two lines and the number of points of intersection?

a. $\begin{cases} y = 5x + 1 \\ y - x = -5 \end{cases}$

b. $\begin{cases} 3y = 4x - 1 \\ -8x + 2 = -6y \end{cases}$

c. $\begin{cases} 2y = -x + 3 \\ -4 = y + \frac{1}{2}x \end{cases}$

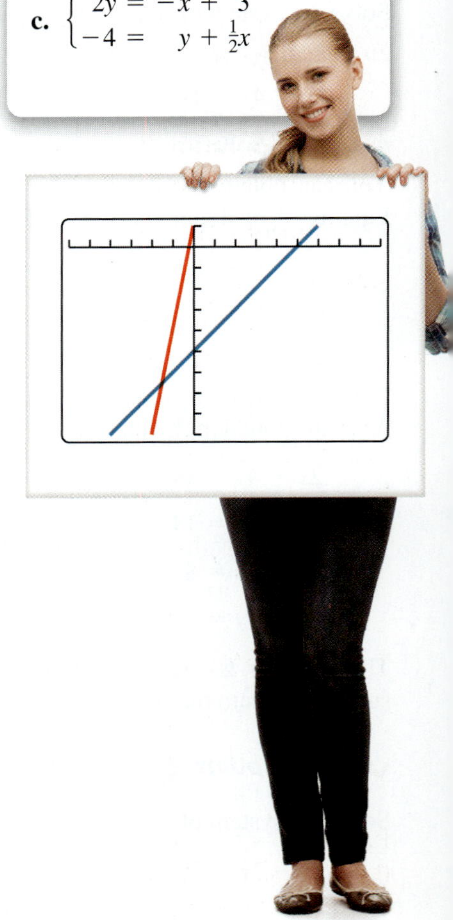

In Examples 4 and 5, note how you can use the method of elimination to determine that a system of linear equations has no solution or infinitely many solutions.

EXAMPLE 4 Method of Elimination: No Solution

Solve the system of linear equations.

$$\begin{cases} x - 2y = 3 & \text{Equation 1} \\ -2x + 4y = 1 & \text{Equation 2} \end{cases}$$

Algebraic Solution

To obtain coefficients that differ only in sign, multiply Equation 1 by 2.

$$\begin{array}{rcr} x - 2y = 3 & \Rightarrow & 2x - 4y = 6 \\ \underline{-2x + 4y = 1} & \Rightarrow & \underline{-2x + 4y = 1} \\ & & 0 = 7 \end{array}$$

By adding the equations, you obtain $0 = 7$. Because there are no values of x and y for which

$$0 = 7 \qquad \text{False statement}$$

this is a false statement. So, you can conclude that the system is inconsistent and has no solution.

Graphical Solution

Solve each equation for y and use a graphing utility to graph the equations in the same viewing window. From the figure, it appears that the system has no solution. When you use the *intersect* feature, the graphing utility cannot find a point of intersection and you will get an error message.

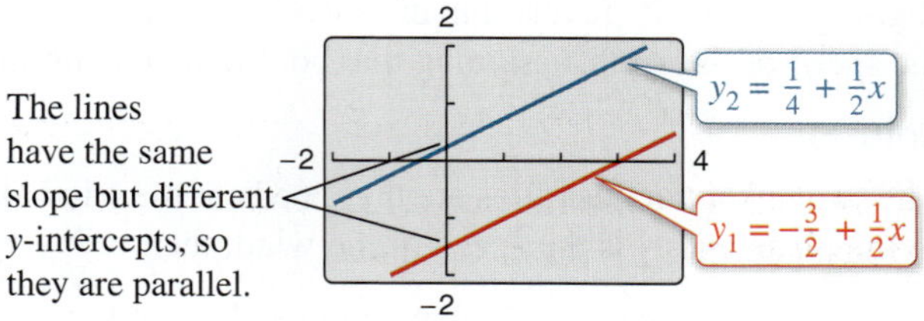

✓ **Checkpoint** *Audio-video solution in English & Spanish at LarsonPrecalculus.com.*

Solve the system of linear equations.

$$\begin{cases} 6x - 5y = 3 \\ -12x + 10y = 5 \end{cases}$$

EXAMPLE 5 Method of Elimination: Infinitely Many Solutions

Solve the system of linear equations.

$$\begin{cases} 2x - y = 1 & \text{Equation 1} \\ 4x - 2y = 2 & \text{Equation 2} \end{cases}$$

Solution

To obtain coefficients that differ only in sign, multiply Equation 1 by -2.

$$\begin{array}{rcrl} 2x - y = 1 & \Rightarrow & -4x + 2y = -2 & \text{Multiply Equation 1 by } -2. \\ \underline{4x - 2y = 2} & \Rightarrow & \underline{4x - 2y = 2} & \text{Write Equation 2.} \\ & & 0 = 0 & \text{Add equations.} \end{array}$$

Because $0 = 0$ for all values of x and y, the two equations turn out to be equivalent (have the same solution set). You can conclude that the system has infinitely many solutions. The solution set consists of all points (x, y) lying on the line $2x - y = 1$, as shown in Figure 5.5.

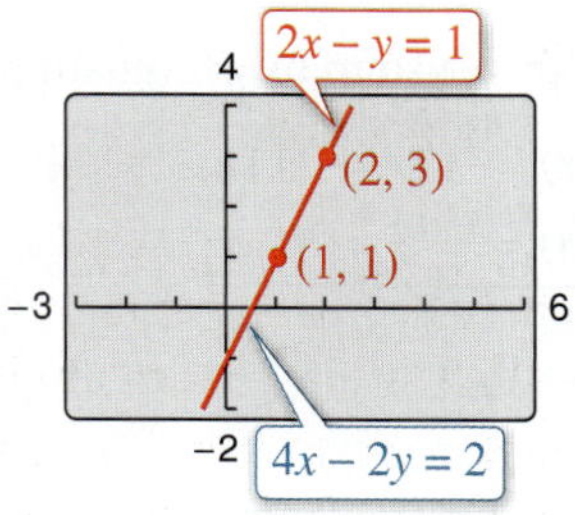

Figure 5.5

✓ **Checkpoint** *Audio-video solution in English & Spanish at LarsonPrecalculus.com.*

Solve the system of linear equations.

$$\begin{cases} \frac{1}{2}x - \frac{1}{8}y = -\frac{3}{8} \\ -4x + y = 3 \end{cases}$$

In Example 4, note that the occurrence of the false statement $0 = 7$ indicates that the system has no solution. In Example 5, note that the occurrence of a statement that is true for all values of the variables—in this case, $0 = 0$—indicates that the system has infinitely many solutions.

Application

At this point, you may be asking the question "How can I tell which application problems can be solved using a system of linear equations?" The answer comes from the following considerations.

1. Does the problem involve more than one unknown quantity?

2. Are there two (or more) equations or conditions to be satisfied?

When one or both of these conditions are met, the appropriate mathematical model for the problem may be a system of linear equations.

EXAMPLE 6 Aviation

An airplane flying into a headwind travels the 2000-mile flying distance between Cleveland, Ohio, and Fresno, California, in 4 hours and 24 minutes. On the return flight, the airplane travels this distance in 4 hours. Find the airspeed of the plane and the speed of the wind, assuming that both remain constant.

Solution

The two unknown quantities are the speeds of the wind and the plane. If r_1 is the speed of the plane and r_2 is the speed of the wind, then

$r_1 - r_2 =$ speed of the plane *against* the wind

$r_1 + r_2 =$ speed of the plane *with* the wind

as shown in Figure 5.6. Using the formula

distance = (rate)(time)

for these two speeds, you obtain the following equations.

$$2000 = (r_1 - r_2)\left(4 + \frac{24}{60}\right)$$

$$2000 = (r_1 + r_2)(4)$$

These two equations simplify as follows.

$$\begin{cases} 5000 = 11r_1 - 11r_2 & \text{Equation 1} \\ 500 = r_1 + r_2 & \text{Equation 2} \end{cases}$$

To solve this system by elimination, multiply Equation 2 by 11.

$$\begin{array}{lcl} 5000 = 11r_1 - 11r_2 & \Rightarrow & 5000 = 11r_1 - 11r_2 \quad \text{Write Equation 1.} \\ 500 = r_1 + r_2 & \Rightarrow & 5500 = 11r_1 + 11r_2 \quad \text{Multiply Equation 2 by 11.} \\ \hline & & 10{,}500 = 22r_1 \quad \text{Add equations.} \end{array}$$

So,

$$r_1 = \frac{10{,}500}{22} = \frac{5250}{11} \approx 477.27 \text{ miles per hour} \qquad \text{Speed of plane}$$

and

$$r_2 = 500 - \frac{5250}{11} = \frac{250}{11} \approx 22.73 \text{ miles per hour} \qquad \text{Speed of wind}$$

Check this solution in the original statement of the problem.

✓ **Checkpoint** *Audio-video solution in English & Spanish at LarsonPrecalculus.com.*

In Example 6, the return flight takes 4 hours and 6 minutes. Find the airspeed of the plane and the speed of the wind, assuming that both remain constant.

What's Wrong?

You use a graphing utility to graph the system

$$\begin{cases} 100y - x = 200 \\ 99y - x = -198 \end{cases}$$

as shown in the figure. You use the graph to conclude that the system has no solution. What's wrong?

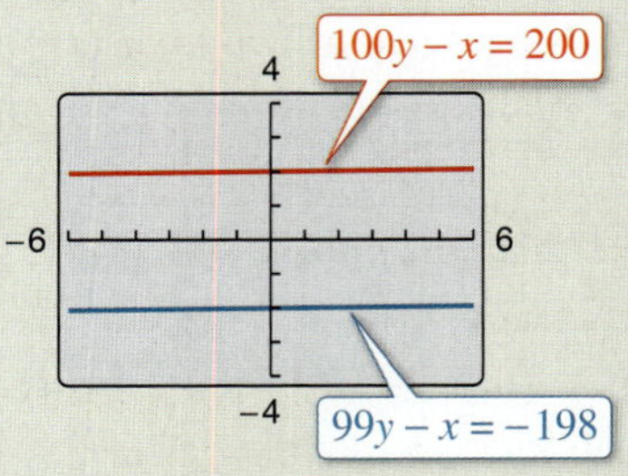

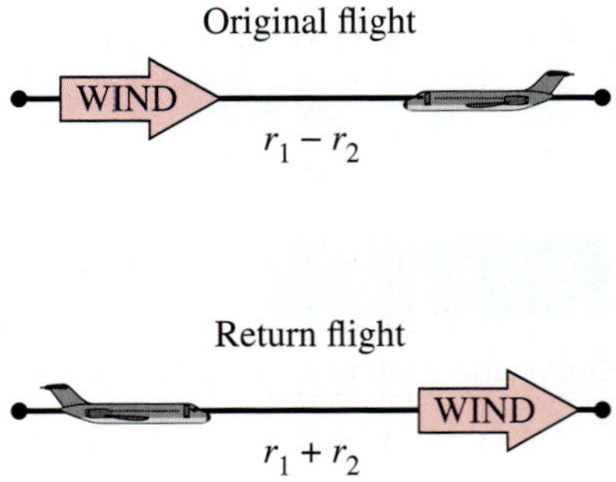

Figure 5.6

5.2 Exercises

See *CalcChat.com* for tutorial help and worked-out solutions to odd-numbered exercises.
For instructions on how to use a graphing utility, see Appendix A.

Vocabulary and Concept Check

In Exercises 1 and 2, fill in the blank(s).

1. The first step in solving a system of equations by the _______ of _______ is to obtain coefficients for x (or y) that differ only in sign.
2. Two systems of equations that have the same solution set are called _______ systems.

3. Is a system of linear equations with no solution consistent or inconsistent?
4. Is a system of linear equations with at least one solution consistent or inconsistent?
5. Is a system of two linear equations consistent when the lines are coincident?
6. When a system of linear equations has no solution, do the lines intersect?

Procedures and Problem Solving

Solving a System by Elimination **In Exercises 7–12, solve the system by the method of elimination. Label each line with its equation.**

7. $\begin{cases} 2x + y = 5 \\ x - y = 1 \end{cases}$

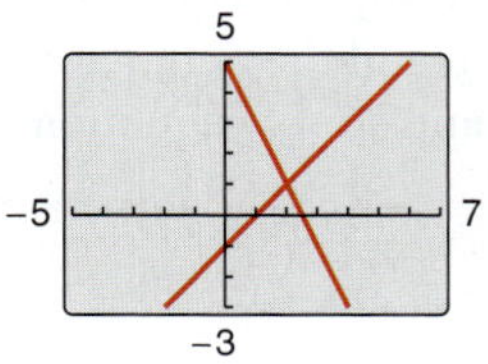

8. $\begin{cases} x + 3y = 1 \\ -x + 2y = 4 \end{cases}$

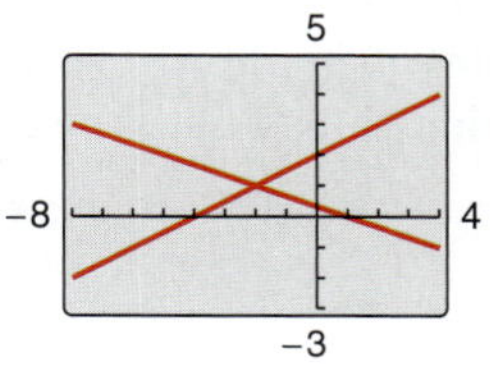

9. $\begin{cases} x + y = 0 \\ 3x + 2y = 1 \end{cases}$

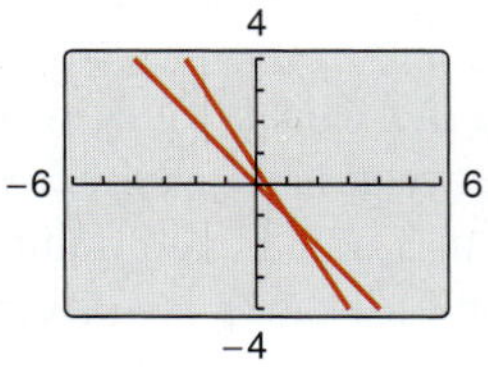

10. $\begin{cases} 2x - y = 3 \\ 4x + 3y = 21 \end{cases}$

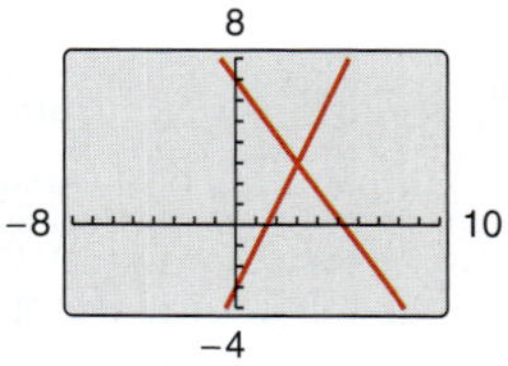

11. $\begin{cases} x - y = 2 \\ -2x + 2y = 5 \end{cases}$

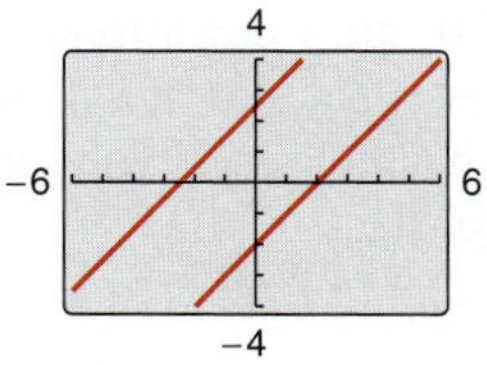

12. $\begin{cases} 3x - 2y = 5 \\ -6x + 4y = -10 \end{cases}$

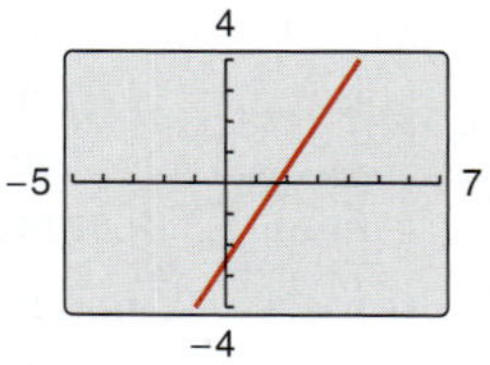

Solving a System by Elimination **In Exercises 13–26, solve the system by the method of elimination and check any solutions algebraically.**

13. $\begin{cases} x + 2y = 3 \\ x - 2y = 1 \end{cases}$

14. $\begin{cases} 3x - 5y = 2 \\ 2x + 5y = 13 \end{cases}$

15. $\begin{cases} 2x + 3y = 18 \\ 5x - y = 11 \end{cases}$

16. $\begin{cases} x + 7y = 12 \\ 3x - 5y = 10 \end{cases}$

17. $\begin{cases} 3r + 2s = -6 \\ 2r + 6s = 3 \end{cases}$

18. $\begin{cases} 8r + 16s = 20 \\ 16r + 50s = 55 \end{cases}$

19. $\begin{cases} 5u + 6v = 24 \\ 3u + 5v = 18 \end{cases}$

20. $\begin{cases} 3u + 11v = 4 \\ -2u - 5v = 9 \end{cases}$

21. $\begin{cases} -6x + 5y = -15 \\ 4x + 12y = 10 \end{cases}$

22. $\begin{cases} 9x + 3y = 18 \\ 2x - 7y = -19 \end{cases}$

23. $\begin{cases} 1.8x + 1.2y = 4 \\ 9x + 6y = 3 \end{cases}$

24. $\begin{cases} 3.1x - 2.9y = -10.2 \\ 31x - 12y = 34 \end{cases}$

25. $\begin{cases} 3x + \frac{1}{4}y = 1 \\ 2x - \frac{1}{3}y = 0 \end{cases}$

26. $\begin{cases} \frac{1}{2}x + \frac{2}{5}y = -2 \\ -\frac{5}{2}x + 2y = 10 \end{cases}$

Recognizing Graphs of Linear Systems **In Exercises 27–30, match the system of linear equations with its graph. State the number of solutions. Then state whether the system is consistent or inconsistent. [The graphs are labeled (a), (b), (c), and (d).]**

(a)

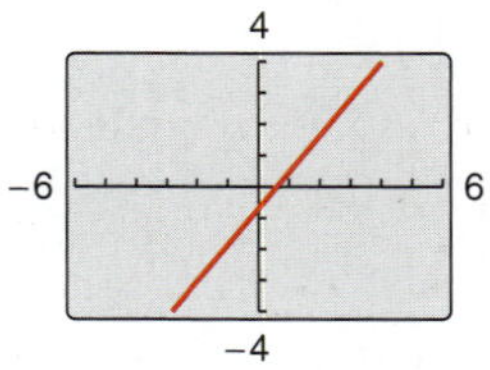

(b)

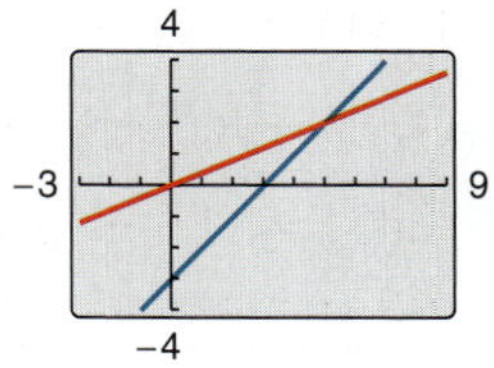

(c)

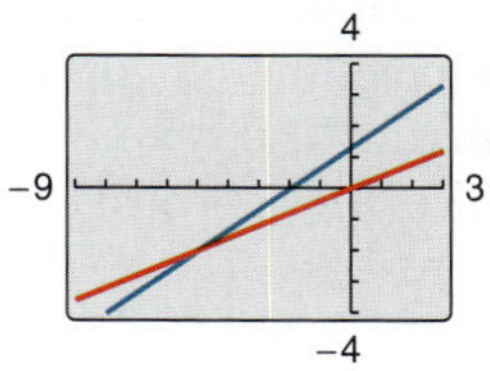

(d)

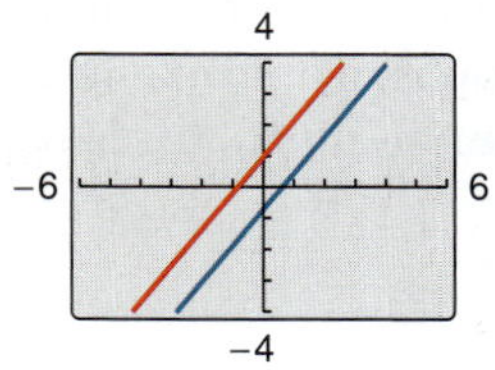

27. $\begin{cases} 2x - 5y = 0 \\ x - y = 3 \end{cases}$

28. $\begin{cases} -7x + 6y = -4 \\ 14x - 12y = 8 \end{cases}$

29. $\begin{cases} 7x - 6y = -6 \\ -7x + 6y = -4 \end{cases}$

30. $\begin{cases} 2x - 5y = 0 \\ 2x - 3y = -4 \end{cases}$

Solving a System by Elimination **In Exercises 31–46, solve the system by the method of elimination and check any solutions using a graphing utility.**

31. $\begin{cases} 5x + 3y = 6 \\ 3x - y = 5 \end{cases}$

32. $\begin{cases} x + 5y = 10 \\ 3x - 10y = -5 \end{cases}$

33. $\begin{cases} \frac{2}{5}x - \frac{3}{2}y = 4 \\ \frac{1}{5}x - \frac{3}{4}y = -2 \end{cases}$

34. $\begin{cases} \frac{2}{3}x + \frac{1}{6}y = \frac{2}{3} \\ 4x + y = 4 \end{cases}$

35. $\begin{cases} \frac{3}{4}x + y = \frac{1}{8} \\ \frac{9}{4}x + 3y = \frac{3}{8} \end{cases}$

36. $\begin{cases} \frac{1}{4}x + \frac{1}{6}y = 1 \\ -3x - 2y = 0 \end{cases}$

37. $\begin{cases} \dfrac{x + 2}{4} + \dfrac{y - 1}{4} = 1 \\ x - y = 4 \end{cases}$

38. $\begin{cases} \dfrac{2x + 5}{2} + \dfrac{y - 1}{3} = -1 \\ 2x - y = 12 \end{cases}$

39. $\begin{cases} -5x + 6y = -3 \\ 20x - 24y = 12 \end{cases}$

40. $\begin{cases} 7x + 8y = 16 \\ -14x - 16y = -4 \end{cases}$

41. $\begin{cases} 2.5x - 3y = 1.5 \\ x - 1.2y = -3.6 \end{cases}$

42. $\begin{cases} 6.3x + 7.2y = 5.4 \\ 5.6x + 6.4y = 4.8 \end{cases}$

43. $\begin{cases} 0.2x - 0.5y = -27.8 \\ 0.3x + 0.4y = 68.7 \end{cases}$

44. $\begin{cases} 0.2x + 0.6y = -1 \\ x - 0.5y = 2 \end{cases}$

45. $\begin{cases} \dfrac{1}{x} + \dfrac{3}{y} = 2 \\ \dfrac{4}{x} - \dfrac{1}{y} = -5 \end{cases}$

46. $\begin{cases} \dfrac{2}{x} - \dfrac{1}{y} = 5 \\ \dfrac{6}{x} + \dfrac{1}{y} = 11 \end{cases}$

Solving a System Graphically **In Exercises 47–52, use a graphing utility to graph the lines in the system. Use the graphs to determine whether the system is consistent or inconsistent. If the system is consistent, determine the solution. Verify your results algebraically.**

47. $\begin{cases} 3x + 5y = -2 \\ 4x - y = 5 \end{cases}$

48. $\begin{cases} -x + 3y = 0 \\ 3x - 9y = 14 \end{cases}$

49. $\begin{cases} 6x + 3y = -8 \\ -x - \frac{1}{2}y = \frac{4}{3} \end{cases}$

50. $\begin{cases} -\frac{1}{4}x - \frac{1}{2}y = 1 \\ 5x + y = 1 \end{cases}$

51. $\begin{cases} 3.2x - 16y = 7.5 \\ x - 5y = -9 \end{cases}$

52. $\begin{cases} -6x + 4y = -9 \\ 4.5x - 3y = 6.75 \end{cases}$

Solving a System Graphically **In Exercises 53–60, use a graphing utility to graph the two equations. Use the graphs to approximate the solution of the system. Round your results to three decimal places.**

53. $\begin{cases} 6y = 42 \\ 6x - y = 16 \end{cases}$

54. $\begin{cases} 4y = -8 \\ 7x - 2y = 25 \end{cases}$

55. $\begin{cases} \frac{3}{2}x - \frac{1}{5}y = 8 \\ -2x + 3y = 3 \end{cases}$

56. $\begin{cases} \frac{3}{4}x - \frac{5}{2}y = -9 \\ -x + 6y = 28 \end{cases}$

57. $\begin{cases} \frac{1}{3}x + y = -\frac{1}{3} \\ 5x - 3y = 7 \end{cases}$

58. $\begin{cases} 5x - y = -4 \\ 2x + \frac{3}{5}y = \frac{2}{5} \end{cases}$

59. $\begin{cases} 0.5x + 2.2y = 9 \\ 6x + 0.4y = -22 \end{cases}$

60. $\begin{cases} 2.4x + 3.8y = -17.6 \\ 4x - 0.2y = -3.2 \end{cases}$

Solving a System **In Exercises 61–68, use any method to solve the system.**

61. $\begin{cases} 3x - 5y = 7 \\ 2x + y = 9 \end{cases}$

62. $\begin{cases} -x + 3y = 17 \\ 4x + 3y = 7 \end{cases}$

63. $\begin{cases} y = 2x - 5 \\ y = 5x - 11 \end{cases}$

64. $\begin{cases} 7x + 3y = 16 \\ y = x + 2 \end{cases}$

65. $\begin{cases} x - 5y = 21 \\ 6x + 5y = 21 \end{cases}$

66. $\begin{cases} y = -2x - 17 \\ y = 2 - 3x \end{cases}$

67. $\begin{cases} -5x + 9y = 13 \\ y = x - 4 \end{cases}$

68. $\begin{cases} 4x - 3y = 6 \\ -5x + 7y = -1 \end{cases}$

Finding a System **In Exercises 69–72, find a system of linear equations that has the given solution. (There are many correct answers.)**

69. $(5, 0)$

70. $(-6, 1)$

71. $(2.5, -4)$

72. $\left(-\frac{3}{4}, 12\right)$

Economics **In Exercises 73–76, find the *point of equilibrium* of the demand and supply equations. The point of equilibrium is the price p and the number of units x that satisfy both the demand and supply equations.**

	Demand	*Supply*
73.	$p = 500 - 0.4x$	$p = 380 + 0.1x$
74.	$p = 100 - 0.05x$	$p = 25 + 0.1x$
75.	$p = 140 - 0.00002x$	$p = 80 + 0.00001x$
76.	$p = 400 - 0.0002x$	$p = 225 + 0.0005x$

77. **Aviation** An airplane flying into a headwind travels the 1800-mile flying distance between New York City and Albuquerque, New Mexico, in 3 hours and 36 minutes. On the return flight, the same distance is traveled in 3 hours. Find the airspeed of the plane and the speed of the wind, assuming that both remain constant.

78. **Navigation** A motorboat traveling with the current takes 40 minutes to travel 20 miles downstream. The return trip takes 60 minutes. Find the speed of the current and the speed of the boat relative to the current, assuming that both remain constant.

79. **Business** A minor league baseball team had a total attendance one evening of 1175. The tickets for adults and children sold for \$15 and \$12, respectively. The ticket revenue that night was \$16,275.

(a) Create a system of linear equations to find the numbers of adults A and children C at the game.

(b) Solve your system of equations by elimination or by substitution. Explain your choice.

(c) Use the *intersect* feature or the *zoom* and *trace* features of a graphing utility to solve your system.

80. Chemistry Thirty liters of a 40% acid solution is obtained by mixing a 25% solution with a 50% solution.

(a) Write a system of equations in which one equation represents the amount of final mixture required and the other represents the percent of acid in the final mixture. Let x and y represent the amounts of the 25% and 50% solutions, respectively.

(b) Use a graphing utility to graph the two equations in part (a) in the same viewing window. As the amount of the 25% solution increases, how does the amount of the 50% solution change?

(c) How much of each solution is required to obtain the specified concentration of the final mixture?

81. Business A grocer sells oranges for \$0.95 each and grapefruits for \$1.05 each. You purchased a mix of 16 oranges and grapefruits and paid \$15.90. How many of each type of fruit did you buy?

82. Nutrition Two cheeseburgers and one small order of french fries from a fast-food restaurant contain a total of 830 calories. Three cheeseburgers and two small orders of french fries contain a total of 1360 calories. Find the number of calories in each item.

83. Sales The projected sales S (in millions of dollars) of two clothing retailers from 2015 through 2020 can be modeled by

$$\begin{cases} S - 149.9t = 415.5 & \text{Retailer A} \\ S - 183.1t = 117.3 & \text{Retailer B} \end{cases}$$

where t is the year, with $t = 5$ corresponding to 2015.

(a) Solve the system of equations using the method of your choice. Explain why you chose that method.

(b) Interpret the meaning of the solution in the context of the problem.

(c) Interpret the meaning of the coefficient of the t-term in each model.

(d) Suppose the coefficients were equal and the models remained the same otherwise. How would this affect your answers in parts (a) and (b)?

84. *Why you should learn it* *(p. 408)* On a Saturday night, the manager of a shoe store evaluates the receipts of the previous week's sales. Two hundred fifty pairs of two different styles of running shoes were sold. One style sold for \$79.50 and the other sold for \$89.95. The receipts totaled \$20,711.00. The cash register that was supposed to record the number of each type of shoe sold malfunctioned. Can you recover the information? If so, how many shoes of each type were sold?

Fitting a Line to Data To find the least squares regression line $y = ax + b$ for a set of points

$$(x_1, y_1), (x_2, y_2), \ldots, (x_n, y_n)$$

you can solve the following system for a and b.

$$\begin{cases} nb + \left(\sum_{i=1}^{n} x_i\right)a = \left(\sum_{i=1}^{n} y_i\right) \\ \left(\sum_{i=1}^{n} x_i\right)b + \left(\sum_{i=1}^{n} x_i^2\right)a = \left(\sum_{i=1}^{n} x_i y_i\right) \end{cases}$$

In Exercises 85–88, the sums have been evaluated. Solve the given system for a and b to find the least squares regression line for the points. Use a graphing utility to confirm the results.

85. $\begin{cases} 5b + 10a = 20.2 \\ 10b + 30a = 50.1 \end{cases}$

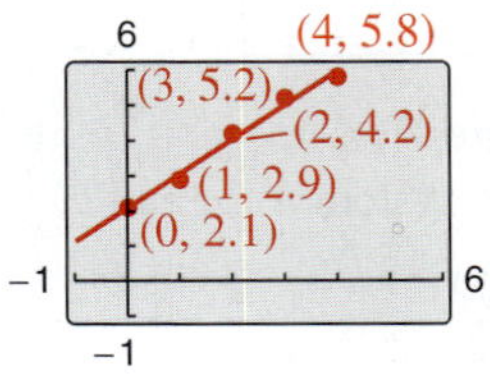

86. $\begin{cases} 5b + 10a = 11.7 \\ 10b + 30a = 25.6 \end{cases}$

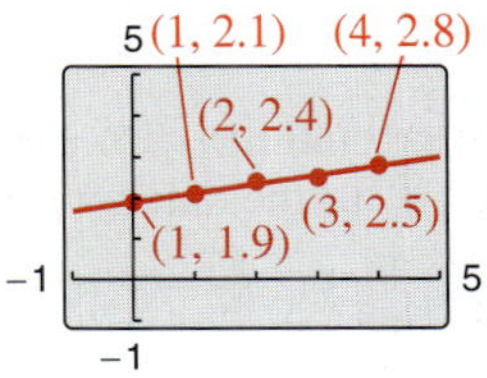

87. $\begin{cases} 6b + 15a = 23.6 \\ 15b + 55a = 48.8 \end{cases}$

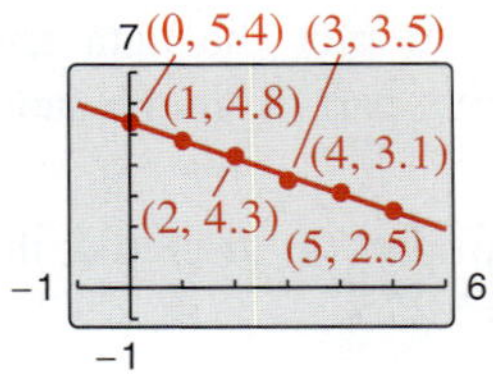

88. $\begin{cases} 7b + 21a = 13.1 \\ 21b + 91a = -2.8 \end{cases}$

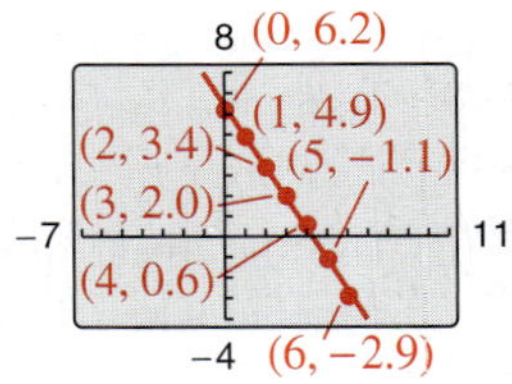

89. MODELING DATA

Four test plots were used to explore the relationship between wheat yield y (in bushels per acre) and the amount of fertilizer applied x (in hundreds of pounds per acre). The results are given by the ordered pairs of the form (x, y).

$$(1.0, 32), \quad (1.5, 41), \quad (2.0, 48), \quad (2.5, 53)$$

(a) Find the least squares regression line $y = ax + b$ for the data by solving the system for a and b.

$$\begin{cases} 4b + 7.0a = 174 \\ 7b + 13.5a = 322 \end{cases}$$

(b) Use the *regression* feature of a graphing utility to confirm the result in part (a).

(c) Use the graphing utility to plot the data and graph the linear model from part (a) in the same viewing window.

(d) Use the linear model from part (a) to predict the yield for a fertilizer application of 160 pounds per acre.

90. MODELING DATA

A candy store manager wants to know the demand for a candy bar as a function of the price. The daily sales for different prices of the product are shown in the table.

Price, x	Demand, y
\$1.00	45
\$1.20	37
\$1.50	23

(a) Find the least squares regression line $y = ax + b$ for the data by solving the system for a and b.

$$\begin{cases} 3.00b + 3.70a = 105.00 \\ 3.70b + 4.69a = 123.90 \end{cases}$$

(b) Use the *regression* feature of a graphing utility to confirm the result in part (a).

(c) Use the graphing utility to plot the data and graph the linear model from part (a) in the same viewing window.

(d) Use the linear model from part (a) to predict the demand when the price is \$1.75.

Conclusions

True or False? **In Exercises 91–93, determine whether the statement is true or false. Justify your answer.**

91. If a system of linear equations has two distinct solutions, then it has an infinite number of solutions.

92. If a system of linear equations has no solution, then the lines must be parallel.

93. Solving a system of equations graphically using a graphing utility always yields an exact solution.

94. **Writing** Briefly explain whether or not it is possible for a consistent system of linear equations to have exactly two solutions.

95. **Think About It** Find all value(s) of k for which the system of linear equations

$$\begin{cases} x + 3y = 9 \\ 2x + 6y = k \end{cases}$$

has (a) infinitely many solutions and (b) no solution.

96. **HOW DO YOU SEE IT?** Use the graphs of the two equations shown below.

(a) Describe the graphs of the two equations.

(b) Can you conclude that the system of equations shown in the graph is inconsistent? Explain.

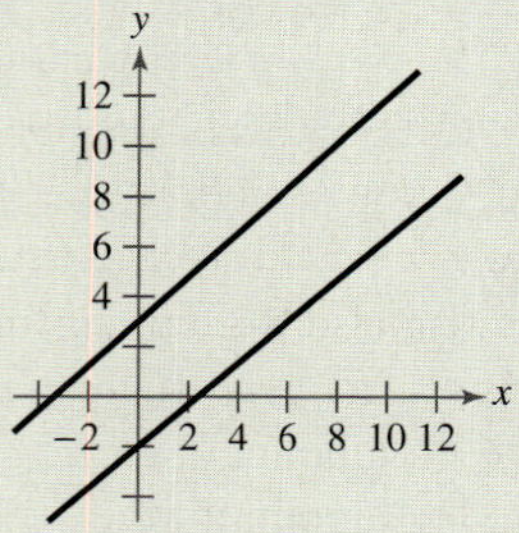

Solving a System **In Exercises 97 and 98, solve the system of equations for u and v. While solving for these variables, consider the transcendental functions as constants. (Systems of this type are found in a course in differential equations.)**

97. $\begin{cases} ue^x + vxe^x = 0 \\ ue^x + v(x + 1)e^x = e^x \ln x \end{cases}$

98. $\begin{cases} ue^{2x} + vxe^{2x} = 0 \\ u(2e^{2x}) + v(2x + 1)e^{2x} = \dfrac{e^{2x}}{x} \end{cases}$

Cumulative Mixed Review

Solving an Inequality **In Exercises 99–104, solve the inequality and graph the solution on a real number line.**

99. $-11 - 6x \geq 33$

100. $-6 \leq 3x - 10 < 6$

101. $|x - 8| < 10$

102. $|x + 10| \geq -3$

103. $2x^2 + 3x - 35 < 0$

104. $3x^2 + 12x > 0$

Rewriting a Logarithmic Expression **In Exercises 105–110, write the expression as the logarithm of a single quantity.**

105. $\ln x + \ln 6$

106. $\ln x - 5 \ln(x + 3)$

107. $\log_9 12 - \log_9 x$

108. $\frac{1}{4} \log_6 3 + \frac{1}{4} \log_6 x$

109. $2 \ln x - \ln(x + 2)$

110. $\frac{1}{2} \ln(x^2 + 4) - \ln x$

111. ***Make a Decision*** To work an extended application analyzing the average undergraduate tuition, room, and board charges at private degree-granting institutions in the United States from 1990 through 2012, visit this textbook's website at *LarsonPrecalculus.com.* (Data Source: U.S. Department of Education)

5.3 Multivariable Linear Systems

Row-Echelon Form and Back-Substitution

The method of elimination can be applied to a system of linear equations in more than two variables. When elimination is used to solve a system of linear equations, the goal is to rewrite the system in a form to which back-substitution can be applied. To see how this works, consider the following two systems of linear equations.

System of Three Linear Equations in Three Variables (See Example 2):

$$\begin{cases} x - 2y + 3z = 9 \\ -x + 3y + z = -2 \\ 2x - 5y + 5z = 17 \end{cases}$$

Equivalent System in Row-Echelon Form (See Example 1):

$$\begin{cases} x - 2y + 3z = 9 \\ y + 4z = 7 \\ z = 2 \end{cases}$$

The second system is said to be in **row-echelon form,** which means that it has a "stair-step" pattern with leading coefficients of 1. After comparing the two systems, it should be clear that it is easier to solve the system in row-echelon form, using back-substitution.

EXAMPLE 1 Using Back-Substitution in Row-Echelon Form

Solve the system of linear equations.

$$\begin{cases} x - 2y + 3z = 9 & \text{Equation 1} \\ y + 4z = 7 & \text{Equation 2} \\ z = 2 & \text{Equation 3} \end{cases}$$

Solution

From Equation 3, you know the value of z. To solve for y, substitute $z = 2$ into Equation 2 to obtain

$y + 4(2) = 7$ Substitute 2 for z.

$y = -1.$ Solve for y.

Next, substitute $y = -1$ and $z = 2$ into Equation 1 to obtain

$x - 2(-1) + 3(2) = 9$ Substitute -1 for y and 2 for z.

$x = 1.$ Solve for x.

The solution is

$x = 1,\ y = -1,$ and $z = 2$

which can be written as the **ordered triple**

$(1, -1, 2).$

Check this in the original system of equations.

✓ Checkpoint Audio-video solution in English & Spanish at LarsonPrecalculus.com.

Solve the system of linear equations.

$$\begin{cases} 2x - y + 5z = 22 \\ y + 3z = 6 \\ z = 3 \end{cases}$$

What you should learn

- Use back-substitution to solve linear systems in row-echelon form.
- Use Gaussian elimination to solve systems of linear equations.
- Solve nonsquare systems of linear equations.
- Graphically interpret three-variable linear systems.
- Use systems of linear equations to write partial fraction decompositions of rational expressions.
- Use systems of linear equations in three or more variables to model and solve real-life problems.

Why you should learn it

Systems of linear equations in three or more variables can be used to model and solve real-life problems. For instance, Exercise 94 on page 429 shows how to use a system of linear equations to analyze the numbers of par-3, par-4, and par-5 holes on a golf course.

Gaussian Elimination

Two systems of equations are *equivalent* when they have the same solution set. To solve a system that is not in row-echelon form, first convert it to an *equivalent* system that *is* in row-echelon form by using one or more of the elementary row operations shown below. This process is called **Gaussian elimination,** after the German mathematician Carl Friedrich Gauss (1777–1855).

Elementary Row Operations for Systems of Equations

1. Interchange two equations.
2. Multiply one of the equations by a nonzero constant.
3. Add a multiple of one equation to another equation.

EXAMPLE 2 Using Gaussian Elimination to Solve a System

Solve the system of linear equations.

$$\begin{cases} x - 2y + 3z = 9 & \text{Equation 1} \\ -x + 3y + z = -2 & \text{Equation 2} \\ 2x - 5y + 5z = 17 & \text{Equation 3} \end{cases}$$

Solution

Because the leading coefficient of the first equation is 1, you can begin by saving the x at the upper left and eliminating the other x-terms from the first column.

$$\begin{cases} x - 2y + 3z = 9 \\ y + 4z = 7 \\ 2x - 5y + 5z = 17 \end{cases}$$

Adding the first equation to the second equation produces a new second equation.

$$\begin{cases} x - 2y + 3z = 9 \\ y + 4z = 7 \\ -y - z = -1 \end{cases}$$

Adding -2 times the first equation to the third equation produces a new third equation.

Now that all but the first x have been eliminated from the first column, go to work on the second column. (You need to eliminate y from the third equation.)

$$\begin{cases} x - 2y + 3z = 9 \\ y + 4z = 7 \\ 3z = 6 \end{cases}$$

Adding the second equation to the third equation produces a new third equation.

Finally, you need a coefficient of 1 for z in the third equation.

$$\begin{cases} x - 2y + 3z = 9 \\ y + 4z = 7 \\ z = 2 \end{cases}$$

Multiplying the third equation by $\frac{1}{3}$ produces a new third equation.

This is the same system that was solved in Example 1. So, the solution is $x = 1$, $y = -1$, and $z = 2$.

Remark

Arithmetic errors are often made when elementary row operations are performed. You should note the operation performed in each step so that you can go back and check your work.

✓ **Checkpoint** *Audio-video solution in English & Spanish at LarsonPrecalculus.com.*

Solve the system of linear equations.

$$\begin{cases} x + y + z = 6 \\ 2x - y + z = 3 \\ 3x + y - z = 2 \end{cases}$$

The goal of Gaussian elimination is to use elementary row operations on a system in order to isolate one variable. You can then solve for the value of the variable and use back-substitution to find the values of the remaining variables.

The next example involves an inconsistent system—one that has no solution. The key to recognizing an inconsistent system is that at some stage in the elimination process, you obtain a false statement such as $0 = -2$.

EXAMPLE 3 An Inconsistent System

Solve the system of linear equations.

$$\begin{cases} x - 3y + z = 1 & \text{Equation 1} \\ 2x - y - 2z = 2 & \text{Equation 2} \\ x + 2y - 3z = -1 & \text{Equation 3} \end{cases}$$

Solution

$$\begin{cases} x - 3y + z = 1 \\ 5y - 4z = 0 \\ x + 2y - 3z = -1 \end{cases}$$

Adding -2 times the first equation to the second equation produces a new second equation.

$$\begin{cases} x - 3y + z = 1 \\ 5y - 4z = 0 \\ 5y - 4z = -2 \end{cases}$$

Adding -1 times the first equation to the third equation produces a new third equation.

$$\begin{cases} x - 3y + z = 1 \\ 5y - 4z = 0 \\ 0 = -2 \end{cases}$$

Adding -1 times the second equation to the third equation produces a new third equation.

Because $0 = -2$ is a false statement, you can conclude that this system is inconsistent and has no solution. Moreover, because this system is equivalent to the original system, you can conclude that the original system also has no solution.

✓ **Checkpoint** *Audio-video solution in English & Spanish at LarsonPrecalculus.com.*

Solve the system of linear equations.

$$\begin{cases} x + y - 2z = 3 \\ 3x - 2y + 4z = 1 \\ 2x - 3y + 6z = 8 \end{cases}$$

As with a system of linear equations in two variables, the number of solutions of a system of linear equations in more than two variables must fall into one of three categories.

The Number of Solutions of a Linear System

For a system of linear equations, exactly one of the following is true.

1. There is exactly one solution.
2. There are infinitely many solutions.
3. There is no solution.

A system of linear equations is called *consistent* when it has at least one solution. A consistent system with exactly one solution is **independent.** A consistent system with infinitely many solutions is **dependent.** A system of linear equations is called *inconsistent* when it has no solution.

EXAMPLE 4 A System with Infinitely Many Solutions

Solve the system of linear equations.

$$\begin{cases} x + y - 3z = -1 & \text{Equation 1} \\ y - z = 0 & \text{Equation 2} \\ -x + 2y = 1 & \text{Equation 3} \end{cases}$$

Solution

$$\begin{cases} x + y - 3z = -1 \\ y - z = 0 \\ 3y - 3z = 0 \end{cases}$$

Adding the first equation to the third equation produces a new third equation.

$$\begin{cases} x + y - 3z = -1 \\ y - z = 0 \\ 0 = 0 \end{cases}$$

Adding -3 times the second equation to the third equation produces a new third equation.

This result means that Equation 3 depends on Equations 1 and 2 in the sense that it gives no additional information about the variables. So, the original system is equivalent to

$$\begin{cases} x + y - 3z = -1 \\ y - z = 0 \end{cases}.$$

In the last equation, solve for y in terms of z to obtain $y = z$. Back-substituting for y in the previous equation produces

$$x = 2z - 1.$$

Finally, letting $z = a$, where a is a real number, the solutions of the original system are all of the form

$$x = 2a - 1, \quad y = a, \quad \text{and} \quad z = a.$$

So, every ordered triple of the form

$$(2a - 1, a, a) \qquad a \text{ is a real number.}$$

is a solution of the system.

Checkpoint *Audio-video solution in English & Spanish at LarsonPrecalculus.com.*

Solve the system of linear equations.

$$\begin{cases} x + 2y - 7z = -4 \\ 2x + 3y + z = 5 \\ 3x + 7y - 36z = -25 \end{cases}$$

In Example 4, there are other ways to write the same infinite set of solutions. For instance, the solutions could have been written as

$$\left(b, \tfrac{1}{2}(b + 1), \tfrac{1}{2}(b + 1)\right). \qquad b \text{ is a real number.}$$

This description produces the same set of solutions, as shown below.

Substitution	*Solution*	
$a = 0$	$(2(0) - 1, 0, 0) = (-1, 0, 0)$	Same solution
$b = -1$	$\left(-1, \tfrac{1}{2}(-1 + 1), \tfrac{1}{2}(-1 + 1)\right) = (-1, 0, 0)$	
$a = 1$	$(2(1) - 1, 1, 1) = (1, 1, 1)$	Same solution
$b = 1$	$\left(1, \tfrac{1}{2}(1 + 1), \tfrac{1}{2}(1 + 1)\right) = (1, 1, 1)$	
$a = 3$	$(2(3) - 1, 3, 3) = (5, 3, 3)$	Same solution
$b = 5$	$\left(5, \tfrac{1}{2}(5 + 1), \tfrac{1}{2}(5 + 1)\right) = (5, 3, 3)$	

Remark

There are an infinite number of solutions to Example 4, but they are all of a specific form. By selecting, for instance, a-values of 0, 1, and 3, you can verify that $(-1, 0, 0)$, $(1, 1, 1)$, and $(5, 3, 3)$ are specific solutions. It is incorrect to say simply that the solution to Example 4 is "infinite." You must also specify the form of the solutions.

Nonsquare Systems

So far, each system of linear equations you have looked at has been *square*, which means that the number of equations is equal to the number of variables. In a **nonsquare system of equations,** the number of equations differs from the number of variables. A system of linear equations cannot have a unique solution unless there are at least as many equations as there are variables in the system.

EXAMPLE 5 A System with Fewer Equations than Variables

See LarsonPrecalculus.com for an interactive version of this type of example.

Solve the system of linear equations.

$$\begin{cases} x - 2y + z = 2 & \text{Equation 1} \\ 2x - y - z = 1 & \text{Equation 2} \end{cases}$$

Solution

Begin by rewriting the system in row-echelon form.

$$\begin{cases} x - 2y + z = 2 \\ 3y - 3z = -3 \end{cases}$$

Adding −2 times the first equation to the second equation produces a new second equation.

$$\begin{cases} x - 2y + z = 2 \\ y - z = -1 \end{cases}$$

Multiplying the second equation by $\frac{1}{3}$ produces a new second equation.

Solve for y in terms of z to obtain

$$y = z - 1.$$

By back-substituting into Equation 1, you can solve for x as follows.

$$x - 2y + z = 2 \quad \text{Write Equation 1.}$$

$$x - 2(z - 1) + z = 2 \quad \text{Substitute } z - 1 \text{ for } y.$$

$$x - 2z + 2 + z = 2 \quad \text{Distributive Property}$$

$$x = z \quad \text{Solve for } x.$$

Finally, by letting $z = a$, where a is a real number, you have the solution

$$x = a, \quad y = a - 1, \quad \text{and} \quad z = a.$$

So, every ordered triple of the form

$$(a, a - 1, a) \quad a \text{ is a real number.}$$

is a solution of the system.

✓ **Checkpoint** *Audio-video solution in English & Spanish at LarsonPrecalculus.com.*

Solve the system of linear equations.

$$\begin{cases} x - y + 4z = 3 \\ 4x - z = 0 \end{cases}$$

In Example 5, try choosing some values of a to obtain different solutions of the system, such as

$$(1, 0, 1), \quad (2, 1, 2), \quad \text{and} \quad (3, 2, 3).$$

Then check each ordered triple in the original system to verify that it is a solution of the system.

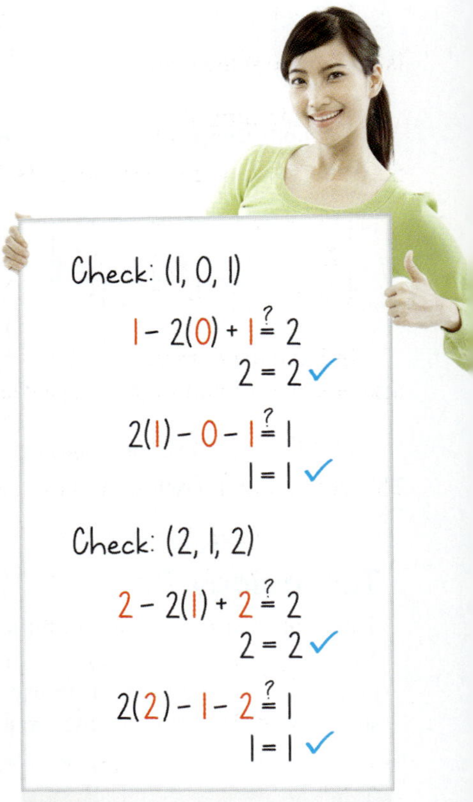

Graphical Interpretation of Three-Variable Systems

Solutions of equations in three variables can be represented graphically using a **three-dimensional coordinate system.** To construct such a system, begin with the xy-coordinate plane in a horizontal position. Then draw the z-axis as a vertical line through the origin.

Every ordered triple (x, y, z) corresponds to a point on the three-dimensional coordinate system. For instance, the points corresponding to $(-2, 5, 4)$, $(2, -5, 3)$, and $(3, 3, -2)$ are shown in Figure 5.7.

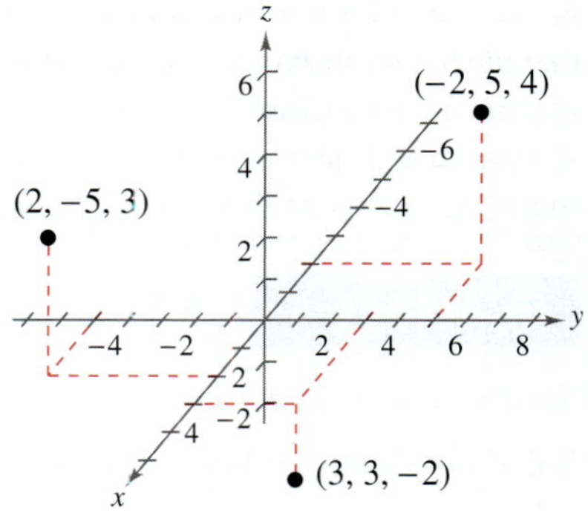

Figure 5.7

The **graph of an equation in three variables** consists of all points (x, y, z) that are solutions of the equation. The graph of a linear equation in three variables is a *plane*. Sketching graphs on a three-dimensional coordinate system is difficult because the sketch itself is only two-dimensional.

One technique for sketching a plane is to find the three points at which the plane intersects the axes. For instance, the plane

$$3x + 2y + 4z = 12$$

intersects the x-axis at the point $(4, 0, 0)$, the y-axis at the point $(0, 6, 0)$, and the z-axis at the point $(0, 0, 3)$. By plotting these three points, connecting them with line segments, and shading the resulting triangular region, you can sketch a portion of the graph, as shown in Figure 5.8.

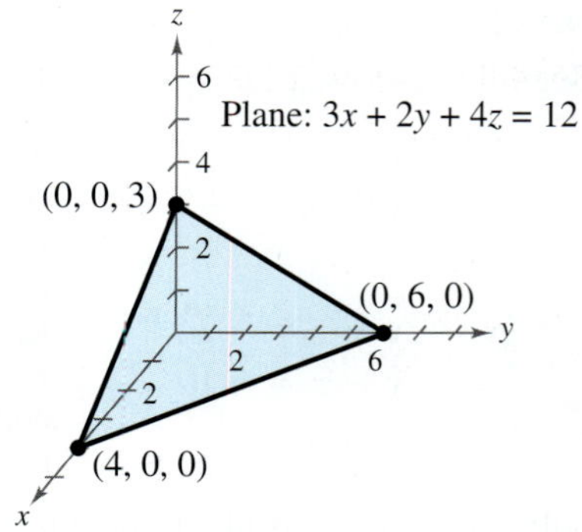

Figure 5.8

The graph of a system of three linear equations in three variables consists of *three* planes. When these planes intersect in a single point, the system has exactly one solution (see Figure 5.9). When the three planes have no point in common, the system has no solution (see Figures 5.10 and 5.11). When the three planes intersect in a line or a plane, the system has infinitely many solutions (see Figures 5.12 and 5.13).

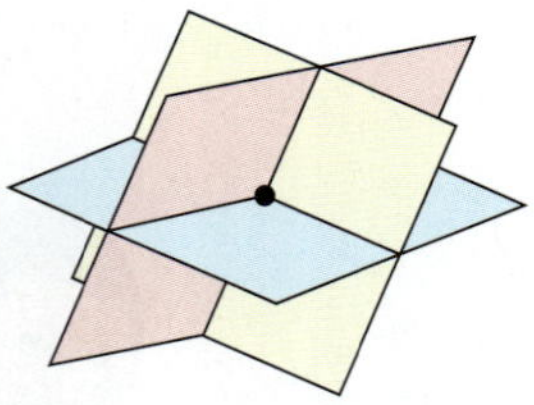

Solution: One point
Figure 5.9

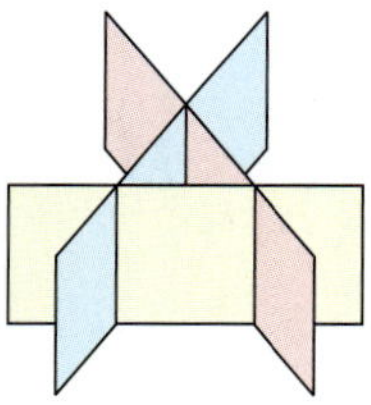

Solution: None
Figure 5.10

Solution: None
Figure 5.11

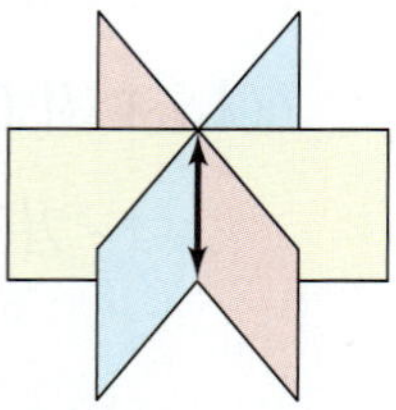

Solution: One line
Figure 5.12

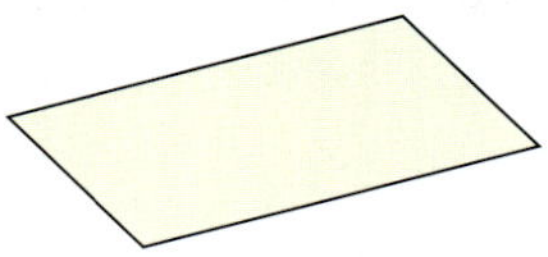

Solution: One plane
Figure 5.13

Technology Tip

Three-dimensional graphing utilities and computer algebra systems, such as *Maple* and *Mathematica*, are very efficient in producing three-dimensional graphs. They are good tools to use while studying calculus. If you have access to such a utility, try reproducing the plane shown in Figure 5.8.

Partial Fraction Decomposition

A rational expression can often be written as the sum of two or more simpler rational expressions. For example, the rational expression

$$\frac{x + 7}{x^2 - x - 6}$$

can be written as the sum of two fractions with linear denominators. That is,

$$\frac{x + 7}{x^2 - x - 6} = \underbrace{\overbrace{\frac{2}{x - 3}}^{}}_{\text{Partial fraction}} + \underbrace{\frac{-1}{x + 2}}_{\text{Partial fraction}}.$$

Partial fraction decomposition of $\frac{x + 7}{x^2 - x - 6}$

Each fraction on the right side of the equation is a **partial fraction,** and together they make up the **partial fraction decomposition** of the left side.

Decomposition of $N(x)/D(x)$ into Partial Fractions

1. *Divide when improper:* When $N(x)/D(x)$ is an improper fraction [degree of $N(x) \geq$ degree of $D(x)$], divide the denominator into the numerator to obtain

$$\frac{N(x)}{D(x)} = (\text{polynomial}) + \frac{N_1(x)}{D(x)}$$

and apply Steps 2, 3, and 4 (below) to the proper rational expression

$$\frac{N_1(x)}{D(x)}.$$

2. *Factor the denominator:* Completely factor the denominator into factors of the form

$$(px + q)^m \quad \text{and} \quad (ax^2 + bx + c)^n$$

where $(ax^2 + bx + c)$ is irreducible over the reals.

3. *Linear factors:* For *each* factor of the form

$$(px + q)^m$$

the partial fraction decomposition must include the following sum of m fractions.

$$\frac{A_1}{(px + q)} + \frac{A_2}{(px + q)^2} + \cdot \cdot \cdot + \frac{A_m}{(px + q)^m}$$

4. *Quadratic factors:* For *each* factor of the form

$$(ax^2 + bx + c)^n$$

the partial fraction decomposition must include the following sum of n fractions.

$$\frac{B_1x + C_1}{ax^2 + bx + c} + \frac{B_2x + C_2}{(ax^2 + bx + c)^2} + \cdot \cdot \cdot + \frac{B_nx + C_n}{(ax^2 + bx + c)^n}$$

One of the most important applications of partial fractions is in calculus. Partial fractions can be used in a calculus operation called antidifferentiation.

EXAMPLE 6 Distinct Linear Factors

Write the partial fraction decomposition of

$$\frac{x+7}{x^2-x-6}.$$

Solution

The expression is proper, so factor the denominator. Because

$$x^2 - x - 6 = (x-3)(x+2)$$

you should include one partial fraction with a constant numerator for each linear factor of the denominator and write

$$\frac{x+7}{x^2-x-6} = \frac{A}{x-3} + \frac{B}{x+2}.$$

Multiplying each side of this equation by $(x-3)(x+2)$, the least common denominator (LCD), leads to the basic equation

$$x + 7 = A(x+2) + B(x-3) \qquad \text{Basic equation}$$

$$= Ax + 2A + Bx - 3B \qquad \text{Distributive Property}$$

$$= (A+B)x + 2A - 3B. \qquad \text{Write in polynomial form.}$$

Because two polynomials are equal if and only if the coefficients of like terms are equal, you can equate the coefficients of like terms to opposite sides of the equation.

$$x + 7 = (A+B)x + (2A - 3B) \qquad \text{Equate coefficients of like terms.}$$

You can now write the following system of linear equations.

$$\begin{cases} A + B = 1 & \text{Equation 1} \\ 2A - 3B = 7 & \text{Equation 2} \end{cases}$$

You can solve the system of linear equations as follows.

$A + B = 1$	⇒	$3A + 3B = 3$	Multiply Equation 1 by 3.
$2A - 3B = 7$	⇒	$2A - 3B = 7$	Write Equation 2.
		$5A = 10$	Add equations.

From this equation, you can see that $A = 2$. By back-substituting this value of A into Equation 1, you can solve for B as follows.

$$A + B = 1 \qquad \text{Write Equation 1.}$$

$$2 + B = 1 \qquad \text{Substitute 2 for } A.$$

$$B = -1 \qquad \text{Solve for } B.$$

So, the partial fraction decomposition is

$$\frac{x+7}{x^2-x-6} = \frac{2}{x-3} - \frac{1}{x+2}.$$

Check this result by combining the two partial fractions on the right side of the equation, or by using a graphing utility.

✓ Checkpoint *Audio-video solution in English & Spanish at LarsonPrecalculus.com.*

Write the partial fraction decomposition of

$$\frac{x+5}{2x^2-x-1}.$$

Technology Tip

You can graphically check the decomposition found in Example 6. To do this, use a graphing utility to graph

$$y_1 = \frac{x+7}{x^2-x-6} \quad \text{and}$$

$$y_2 = \frac{2}{x-3} - \frac{1}{x+2}$$

in the same viewing window. The graphs should be identical.

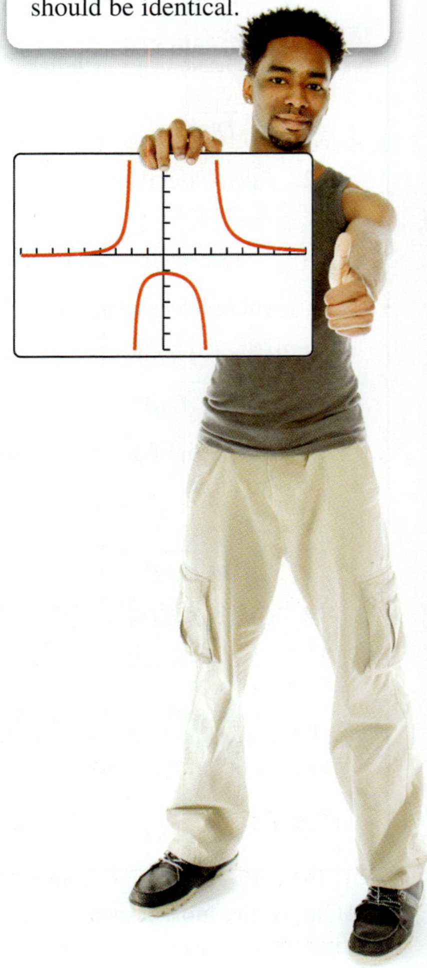

The next example shows how to find the partial fraction decomposition for a rational function whose denominator has a repeated linear factor.

EXAMPLE 7 Repeated Linear Factors

Write the partial fraction decomposition of

$$\frac{5x^2 + 20x + 6}{x^3 + 2x^2 + x}.$$

Solution

The expression is proper, so factor the denominator. Because the denominator factors as

$$x^3 + 2x^2 + x = x(x^2 + 2x + 1) = x(x + 1)^2$$

you should include one partial fraction with a constant numerator for each power of x and $(x + 1)$ and write

$$\frac{5x^2 + 20x + 6}{x^3 + 2x^2 + x} = \frac{A}{x} + \frac{B}{x + 1} + \frac{C}{(x + 1)^2}.$$

Multiplying by the LCD $x(x + 1)^2$ leads to the basic equation

$$5x^2 + 20x + 6 = A(x + 1)^2 + Bx(x + 1) + Cx \qquad \text{Basic equation}$$

$$= Ax^2 + 2Ax + A + Bx^2 + Bx + Cx \qquad \text{Expand.}$$

$$= (A + B)x^2 + (2A + B + C)x + A. \qquad \text{Polynomial form}$$

Equating coefficients of like terms on opposite sides of the equation

$$5x^2 + 20x + 6 = (A + B)x^2 + (2A + B + C)x + A$$

produces the following system of linear equations.

$$\begin{cases} A + B = 5 & \text{Equation 1} \\ 2A + B + C = 20 & \text{Equation 2} \\ A = 6 & \text{Equation 3} \end{cases}$$

Substituting 6 for A in Equation 1 yields

$$6 + B = 5$$

$$B = -1.$$

Substituting 6 for A and -1 for B in Equation 2 yields

$$2(6) + (-1) + C = 20$$

$$C = 9.$$

So, the partial fraction decomposition is

$$\frac{5x^2 + 20x + 6}{x^3 + 2x^2 + x} = \frac{6}{x} - \frac{1}{x + 1} + \frac{9}{(x + 1)^2}.$$

Check this result by combining the three partial fractions on the right side of the equation, or by using a graphing utility.

✓ **Checkpoint** *Audio-video solution in English & Spanish at LarsonPrecalculus.com.*

Write the partial fraction decomposition of

$$\frac{x + 4}{x^3 + x^2}.$$

Explore the Concept

Partial fraction decomposition is practical only for rational functions whose denominators factor "nicely." For example, the factorization of the expression $x^2 - x - 5$ is

$$\left(x - \frac{1 - \sqrt{21}}{2}\right)\left(x - \frac{1 + \sqrt{21}}{2}\right).$$

Write the basic equation and try to complete the decomposition for

$$\frac{x + 7}{x^2 - x - 5}.$$

What problems do you encounter?

Applications

EXAMPLE 8 Vertical Motion

The height at time t of an object that is moving in a (vertical) line with constant acceleration a is given by the *position equation* $s = \frac{1}{2}at^2 + v_0t + s_0$. The height s is measured in feet, the acceleration a is measured in feet per second squared, t is measured in seconds, v_0 is the initial velocity (in feet per second) at $t = 0$, and s_0 is the initial height (in feet). Find the values of a, v_0, and s_0 when

$$s = 52 \text{ at } t = 1, \quad s = 52 \text{ at } t = 2, \quad \text{and } s = 20 \text{ at } t = 3$$

and interpret the result. (See Figure 5.14.)

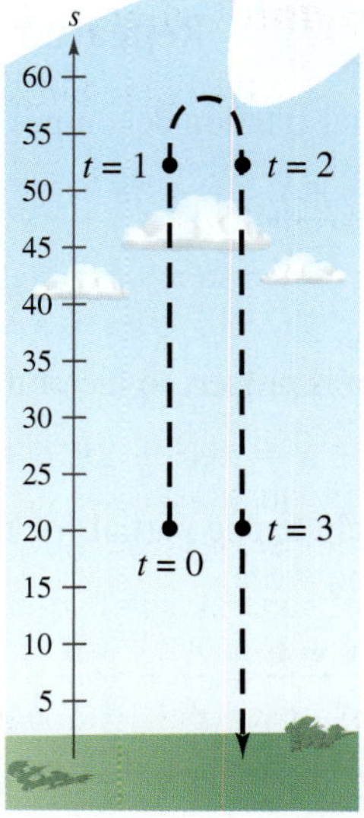

Figure 5.14

Solution

You can obtain three linear equations in a, v_0, and s_0 as follows.

When $t = 1$: $\frac{1}{2}a(1)^2 + v_0(1) + s_0 = 52$ ⟹ $a + 2v_0 + 2s_0 = 104$

When $t = 2$: $\frac{1}{2}a(2)^2 + v_0(2) + s_0 = 52$ ⟹ $2a + 2v_0 + s_0 = 52$

When $t = 3$: $\frac{1}{2}a(3)^2 + v_0(3) + s_0 = 20$ ⟹ $9a + 6v_0 + 2s_0 = 40$

Solving this system yields $a = -32$, $v_0 = 48$, and $s_0 = 20$. This solution results in a position equation of

$$s = -16t^2 + 48t + 20$$

and implies that the object was thrown upward at a velocity of 48 feet per second from a height of 20 feet.

✓ Checkpoint *Audio-video solution in English & Spanish at LarsonPrecalculus.com.*

For the position equation $s = \frac{1}{2}at^2 + v_0t + s_0$ given in Example 8, find the values of a, v_0, and s_0 when $s = 104$ at $t = 1$, $s = 76$ at $t = 2$, and $s = 16$ at $t = 3$, and interpret the result.

EXAMPLE 9 Data Analysis: Curve-Fitting

Find a quadratic equation $y = ax^2 + bx + c$ whose graph passes through the points $(-1, 3)$, $(1, 1)$, and $(2, 6)$.

Solution

Because the graph of $y = ax^2 + bx + c$ passes through the points $(-1, 3)$, $(1, 1)$, and $(2, 6)$, you can write the following.

When $x = -1$, $y = 3$: $a(-1)^2 + b(-1) + c = 3$

When $x = 1$, $y = 1$: $a(1)^2 + b(1) + c = 1$

When $x = 2$, $y = 6$: $a(2)^2 + b(2) + c = 6$

This produces the following system of linear equations.

$$\begin{cases} a - b + c = 3 & \text{Equation 1} \\ a + b + c = 1 & \text{Equation 2} \\ 4a + 2b + c = 6 & \text{Equation 3} \end{cases}$$

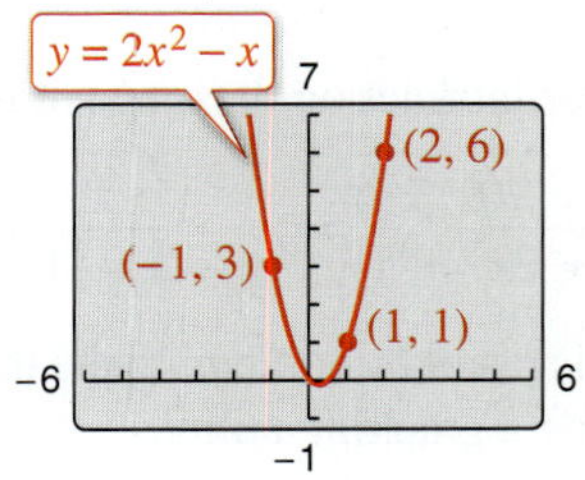

Figure 5.15

The solution of this system is $a = 2$, $b = -1$, and $c = 0$. So, the equation of the parabola is $y = 2x^2 - x$, and its graph is shown in Figure 5.15.

✓ Checkpoint *Audio-video solution in English & Spanish at LarsonPrecalculus.com.*

Find a quadratic equation $y = ax^2 + bx + c$ whose graph passes through the points $(0, 0)$, $(3, -3)$, and $(6, 0)$.

5.3 Exercises

See *CalcChat.com* for tutorial help and worked-out solutions to odd-numbered exercises.
For instructions on how to use a graphing utility, see Appendix A.

Vocabulary and Concept Check

In Exercises 1–6, fill in the blank.

1. A system of equations that is in _______ form has a "stair-step" pattern with leading coefficients of 1.

2. A solution of a system of three linear equations in three unknowns can be written as an _______ , which has the form (x, y, z).

3. The process used to write a system of equations in row-echelon form is called _______ elimination.

4. A system of equations is called _______ when the number of equations differs from the number of variables in the system.

5. Solutions of equations in three variables can be pictured using a _______ coordinate system.

6. The process of writing a rational expression as the sum of two or more simpler rational expressions is called _______ .

7. Is a consistent system with exactly one solution *independent* or *dependent*?

8. Is a consistent system with infinitely many solutions *independent* or *dependent*?

Procedures and Problem Solving

Checking Solutions **In Exercises 9–12, determine whether each ordered triple is a solution of the system of equations.**

9. $\begin{cases} 3x - y + z = 1 \\ 2x - 3z = -14 \\ 5y + 2z = 8 \end{cases}$

(a) $(3, 5, -3)$ (b) $(-1, 0, 4)$
(c) $(0, -1, 3)$ (d) $(1, 0, 4)$

10. $\begin{cases} 3x + 4y - z = 17 \\ 5x - y + 2z = -2 \\ 2x - 3y + 7z = -21 \end{cases}$

(a) $(1, 5, 6)$ (b) $(-2, -4, 2)$
(c) $(1, 3, -2)$ (d) $(0, 7, 0)$

11. $\begin{cases} 4x + y - z = 0 \\ -8x - 6y + z = -\frac{7}{4} \\ 3x - y = -\frac{9}{4} \end{cases}$

(a) $(0, 1, 1)$ (b) $\left(-\frac{3}{2}, \frac{5}{4}, -\frac{5}{4}\right)$
(c) $\left(-\frac{1}{2}, \frac{3}{4}, -\frac{5}{4}\right)$ (d) $\left(-\frac{1}{2}, \frac{1}{6}, -\frac{3}{4}\right)$

12. $\begin{cases} -4x - y - 8z = -6 \\ y + z = 0 \\ 4x - 7y = 6 \end{cases}$

(a) $(-2, -2, 2)$ (b) $\left(-\frac{33}{2}, -10, 10\right)$
(c) $\left(\frac{1}{8}, -\frac{1}{2}, \frac{1}{2}\right)$ (d) $\left(-\frac{11}{2}, -4, 4\right)$

Using Back-Substitution **In Exercises 13–18, use back-substitution to solve the system of linear equations.**

13. $\begin{cases} 2x - y + 5z = 16 \\ y + 2z = 2 \\ z = 2 \end{cases}$

14. $\begin{cases} 4x - 3y - 2z = 21 \\ 6y - 5z = -8 \\ z = -2 \end{cases}$

15. $\begin{cases} 2x + y - 3z = 10 \\ y + z = 12 \\ z = 2 \end{cases}$

16. $\begin{cases} x - y + 2z = 22 \\ 3y - 8z = -9 \\ z = -3 \end{cases}$

17. $\begin{cases} 4x - 2y + z = 8 \\ -y + z = 4 \\ z = 11 \end{cases}$

18. $\begin{cases} 5x - 8z = 22 \\ 3y - 5z = 10 \\ z = -4 \end{cases}$

Performing Row Operations **In Exercises 19 and 20, perform the row operation and write the equivalent system. What did the operation accomplish?**

19. Add Equation 1 to Equation 2.

$\begin{cases} x - 2y + 3z = 5 & \text{Equation 1} \\ -x + 3y - 5z = 4 & \text{Equation 2} \\ 2x - 3z = 0 & \text{Equation 3} \end{cases}$

20. Add -2 times Equation 1 to Equation 3.

$\begin{cases} x - 2y + 3z = 5 & \text{Equation 1} \\ -x + 3y - 5z = 4 & \text{Equation 2} \\ 2x - 3z = 0 & \text{Equation 3} \end{cases}$

Solving a System of Linear Equations In Exercises 21–42, solve the system of linear equations and check any solution algebraically.

21. $\begin{cases} x + y + z = 6 \\ 2x - y + z = 3 \\ 3x \quad\quad - z = 0 \end{cases}$

22. $\begin{cases} x + y + z = 5 \\ x - 2y + 4z = -1 \\ 3y + 4z = -1 \end{cases}$

23. $\begin{cases} 2x \quad\quad + 2z = 2 \\ 5x + 3y \quad\quad = 4 \\ 3y - 4z = 4 \end{cases}$

24. $\begin{cases} 2x + 4y + z = 2 \\ -2y - 3z = -8 \\ x \quad\quad - z = -1 \end{cases}$

25. $\begin{cases} 4x + y - 3z = 11 \\ 2x - 3y + 2z = 9 \\ x + y + z = -3 \end{cases}$

26. $\begin{cases} 5x - 3y + 2z = 3 \\ 2x + 4y - z = 7 \\ x - 11y + 4z = 3 \end{cases}$

27. $\begin{cases} 3x - 2y + 4z = 1 \\ x + y - 2z = 3 \\ 2x - 3y + 6z = 8 \end{cases}$

28. $\begin{cases} 2x + 4y + z = -4 \\ 2x - 4y + 6z = 13 \\ 4x - 2y + z = 6 \end{cases}$

29. $\begin{cases} 3x + 3y + 5z = 1 \\ 3x + 5y + 9z = 0 \\ 5x + 9y + 17z = 0 \end{cases}$

30. $\begin{cases} 2x + y + 3z = 1 \\ 2x + 6y + 8z = 3 \\ 6x + 8y + 18z = 5 \end{cases}$

31. $\begin{cases} 3x - 3y + 6z = 6 \\ x + 2y - z = 5 \\ 5x - 8y + 13z = 7 \end{cases}$

32. $\begin{cases} -x + 3y + z = 4 \\ 4x - 2y - 5z = -7 \\ 2x + 4y - 3z = 12 \end{cases}$

33. $\begin{cases} x - 2y + 3z = 4 \\ 3x - y + 2z = 0 \\ x + 3y - 4z = -2 \end{cases}$

34. $\begin{cases} x \quad\quad + 4z = 13 \\ 4x - 2y + z = 7 \\ 2x - 2y - 7z = -19 \end{cases}$

35. $\begin{cases} x + 2y + z = 1 \\ x - 2y + 3z = -3 \\ 2x + y + z = -1 \end{cases}$

36. $\begin{cases} 3x - 2y - 6z = -4 \\ -3x + 2y + 6z = 1 \\ x - y - 5z = -3 \end{cases}$

37. $\begin{cases} x \quad\quad + 4z = 1 \\ x + y + 10z = 10 \\ 2x - y + 2z = -5 \end{cases}$

38. $\begin{cases} x - 2y + z = 2 \\ 2x + 2y - 3z = -4 \\ 5x \quad\quad + z = 1 \end{cases}$

39. $\begin{cases} x - 2y + 5z = 2 \\ 4x \quad\quad - z = 0 \end{cases}$

40. $\begin{cases} 2x - 3y + z = -2 \\ -4x + 9y \quad\quad = 7 \end{cases}$

41. $\begin{cases} 12x + 5y + z = 0 \\ 23x + 4y - z = 0 \end{cases}$

42. $\begin{cases} 10x - 3y + 2z = 0 \\ 19x - 5y - z = 0 \end{cases}$

Finding a System In Exercises 43–48, find a system of linear equations that has the given solution. (There are many correct answers.)

43. $(3, -4, 2)$

44. $(-5, -2, 1)$

45. $\left(-6, -\frac{1}{2}, -\frac{7}{4}\right)$

46. $\left(-\frac{3}{2}, 4, -7\right)$

47. $(a, a + 4, a)$, where a is a real number

48. $(3a, a, a + 2)$, where a is a real number

Sketching a Plane In Exercises 49–52, sketch the plane represented by the linear equation. Then list four points that lie in the plane.

49. $x + y + z = 8$

50. $x + 2y + z = 4$

51. $3x + 2y + 2z = 12$

52. $5x + y + 3z = 15$

Writing the Partial Fraction Decomposition In Exercises 53–60, write the form of the partial fraction decomposition of the rational expression. Do not solve for the constants.

53. $\dfrac{7}{x^2 - 14x}$

54. $\dfrac{x - 2}{x^2 + 4x + 3}$

55. $\dfrac{12}{x^3 - 10x^2}$

56. $\dfrac{x^2 - 3x + 2}{4x^3 + 11x^2}$

57. $\dfrac{4x^2 + 3}{(x - 5)^3}$

58. $\dfrac{6x + 5}{(x + 2)^4}$

59. $\dfrac{x - 1}{x(x^2 + 1)^2}$

60. $\dfrac{x + 4}{x^2(3x - 1)^2}$

Partial Fraction Decomposition In Exercises 61–74, write the partial fraction decomposition for the rational expression. Check your result algebraically by combining fractions, and check your result graphically by using a graphing utility to graph the rational expression and the partial fractions in the same viewing window.

61. $\dfrac{1}{x^2 + x}$

62. $\dfrac{1}{4x^2 - 9}$

63. $\dfrac{5 - x}{2x^2 + x - 1}$

64. $\dfrac{x - 2}{x^2 + 4x + 3}$

65. $\dfrac{x^2 + 12x + 12}{x^3 - 4x}$

66. $\dfrac{x^2 + 12x - 9}{x^3 - 9x}$

67. $\dfrac{4x^2 + 2x - 1}{x^2(x + 1)}$

68. $\dfrac{2x - 3}{(x - 1)^2}$

69. $\dfrac{2x^3 - x^2 + x + 5}{x^2 + 3x + 2}$

70. $\dfrac{x^3 + 2x^2 - x + 1}{x^2 + 3x - 4}$

71. $\dfrac{x^4}{(x - 1)^3}$

72. $\dfrac{4x^4}{(2x - 1)^3}$

73. $\dfrac{x}{x^3 - x^2 - 2x + 2}$

74. $\dfrac{2x^2 + x + 8}{x^4 + 8x^2 + 16}$

Writing the Partial Fraction Decomposition In Exercises 75 and 76, write the partial fraction decomposition for the rational function. Identify the graph of the rational function and the graph of each term of its decomposition. State any relationship between the vertical asymptotes of the rational function and the vertical asymptotes of the terms of the decomposition.

75. $y = \dfrac{x - 12}{x(x - 4)}$

76. $y = \dfrac{2(4x - 3)}{x^2 - 9}$

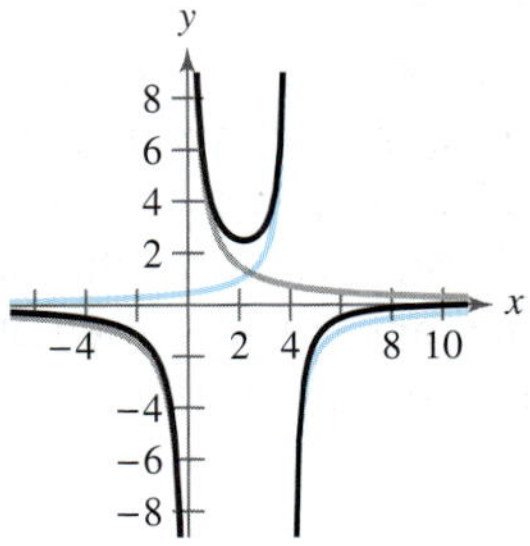

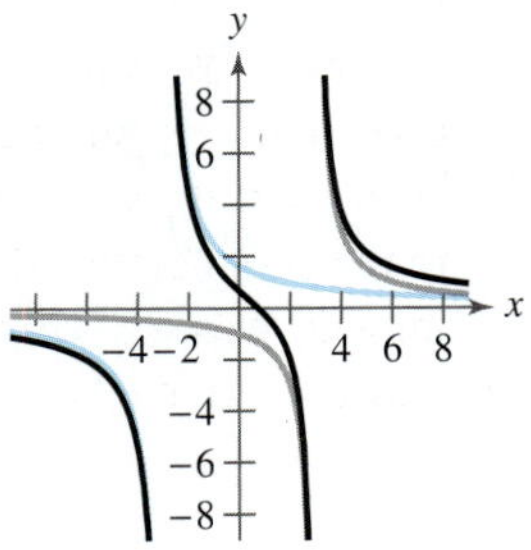

Vertical Motion **In Exercises 77–80, an object moving vertically is at the given heights at the specified times. Find the position equation $s = \frac{1}{2}at^2 + v_0 t + s_0$ for the object.**

77. At $t = 1$ second, $s = 128$ feet.
At $t = 2$ seconds, $s = 80$ feet.
At $t = 3$ seconds, $s = 0$ feet.

78. At $t = 1$ second, $s = 32$ feet.
At $t = 2$ seconds, $s = 32$ feet.
At $t = 3$ seconds, $s = 0$ feet.

79. At $t = 1$ second, $s = 352$ feet.
At $t = 2$ seconds, $s = 272$ feet.
At $t = 3$ seconds, $s = 160$ feet.

80. At $t = 1$ second, $s = 132$ feet.
At $t = 2$ seconds, $s = 100$ feet.
At $t = 3$ seconds, $s = 36$ feet.

Data Analysis: Curve-Fitting **In Exercises 81–84, find the equation of the parabola**

$$y = ax^2 + bx + c$$

that passes through the points. To verify your result, use a graphing utility to plot the points and graph the parabola.

81. $(0, 0), (3, 0), (4, 4)$ **82.** $(0, 5), (1, 6), (2, 5)$

83. $(-1, 1), (0, -4), (1, -13)$

84. $(-2, 9), (-1, 0), (1, 6)$

Finding the Equation of a Circle **In Exercises 85–88, find the equation of the circle**

$$x^2 + y^2 + Dx + Ey + F = 0$$

that passes through the points. To verify your result, use a graphing utility to plot the points and graph the circle.

85. $(0, 0), (5, 5), (10, 0)$ **86.** $(0, 0), (0, 6), (3, 3)$

87. $(-3, -1), (2, 4), (-6, 8)$

88. $(0, 0), (0, -2), (3, 0)$

89. **Finance** A college student borrowed \$30,000 to pay for tuition, room, and board. Some of the money was borrowed at 4%, some at 6%, and some at 8%. How much was borrowed at each rate, given that the annual interest was \$1550 and the amount borrowed at 8% was three times the amount borrowed at 6%?

90. **Finance** A small corporation borrowed \$775,000 to expand its software line. Some of the money was borrowed at 8%, some at 9%, and some at 10%. How much was borrowed at each rate, given that the annual interest was \$67,500 and the amount borrowed at 8% was four times the amount borrowed at 10%?

Investment Portfolio **In Exercises 91 and 92, consider an investor with a portfolio totaling \$500,000 that is invested in certificates of deposit, municipal bonds, blue-chip stocks, and growth or speculative stocks. How much is invested in each type of investment?**

91. The certificates of deposit pay 3% annually, and the municipal bonds pay 5% annually. Over a five-year period, the investor expects the blue-chip stocks to return 8% annually and the growth stocks to return 10% annually. The investor wants a combined annual return of 5% and also wants to have only one-fourth of the portfolio invested in stocks.

92. The certificates of deposit pay 2% annually, and the municipal bonds pay 4% annually. Over a five-year period, the investor expects the blue-chip stocks to return 10% annually and the growth stocks to return 14% annually. The investor wants a combined annual return of 6% and also wants to have only one-fourth of the portfolio invested in stocks.

93. **Physical Education** In the 2013 Women's NCAA Championship basketball game, the University of Connecticut defeated the University of Louisville by a score of 93 to 60. Connecticut won by scoring a combination of two-point baskets, three-point baskets, and one-point free throws. The number of two-point baskets was 12 more than the number of free throws. The number of free throws was three less than the number of three-point baskets. What combination of scoring accounted for Connecticut's 93 points? (Source: NCAA)

94. ***Why you should learn it*** *(p. 417)* The Augusta National Golf Club in Augusta, Georgia, is an 18-hole course that consists of par-3 holes, par-4 holes, and par-5 holes. A golfer who shoots par has a total of 72 strokes for the entire course. There are two more par-4 holes than twice the number of par-5 holes, and the number of par-3 holes is equal to the number of par-5 holes. Find the numbers of par-3, par-4, and par-5 holes on the course. (Source: Augusta National, Inc.)

95. **Electrical Engineering** When Kirchhoff's Laws are applied to the electrical network in the figure, the currents I_1, I_2, and I_3 (in amperes) are the solution of the system

$$\begin{cases} I_1 - I_2 + I_3 = 0 \\ 3I_1 + 2I_2 = 7. \\ 2I_2 + 4I_3 = 8 \end{cases}$$

Find the currents.

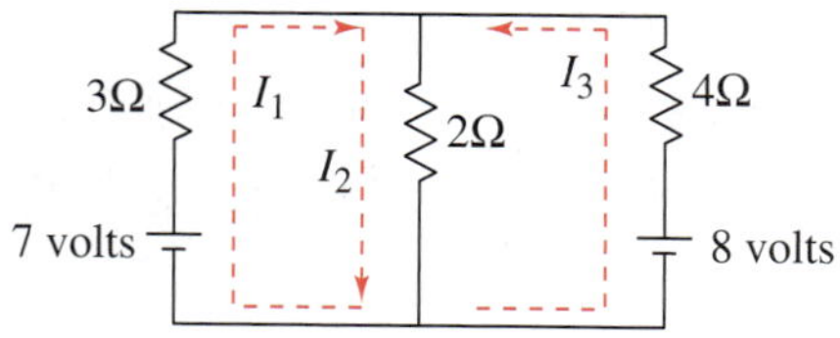

96. Physics A system of pulleys is loaded with 128-pound and 32-pound weights (see figure). The tensions t_1 and t_2 in the ropes and the acceleration a of the 32-pound weight are modeled by the following system, where t_1 and t_2 are measured in pounds and a is in feet per second squared. Solve the system.

$$\begin{cases} t_1 - 2t_2 = 0 \\ t_1 - 2a = 128 \\ t_2 + a = 32 \end{cases}$$

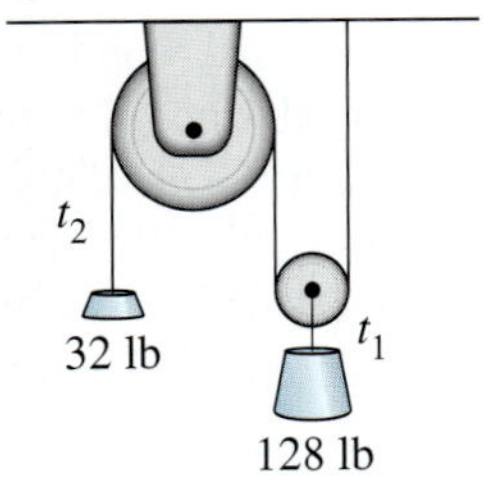

Fitting a Parabola To find the least squares regression parabola $y = ax^2 + bx + c$ for a set of points (x_1, y_1), (x_2, y_2), . . ., (x_n, y_n) you can solve the following system of linear equations for a, b, and c.

$$\begin{cases} nc + \left(\sum_{i=1}^{n} x_i\right)b + \left(\sum_{i=1}^{n} x_i^2\right)a = \sum_{i=1}^{n} y_i \\ \left(\sum_{i=1}^{n} x_i\right)c + \left(\sum_{i=1}^{n} x_i^2\right)b + \left(\sum_{i=1}^{n} x_i^3\right)a = \sum_{i=1}^{n} x_i y_i \\ \left(\sum_{i=1}^{n} x_i^2\right)c + \left(\sum_{i=1}^{n} x_i^3\right)b + \left(\sum_{i=1}^{n} x_i^4\right)a = \sum_{i=1}^{n} x_i^2 y_i \end{cases}$$

In Exercises 97–100, the sums have been evaluated. Solve the given system for a and b to find the least squares regression parabola for the points. Use a graphing utility to confirm the result.

97. $\begin{cases} 4c + 40a = 19 \\ 40b = -12 \\ 40c + 544a = 160 \end{cases}$

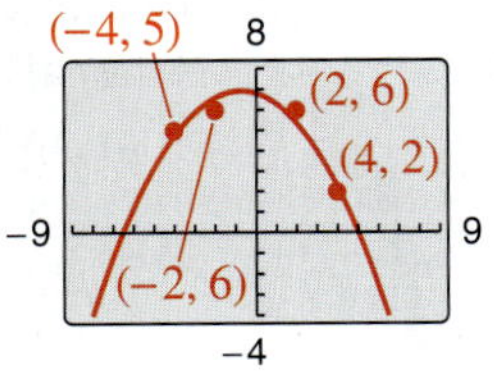

98. $\begin{cases} 5c + 10a = 8 \\ 10b = 12 \\ 10c + 34a = 22 \end{cases}$

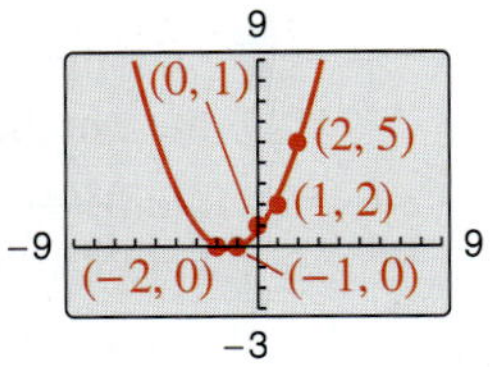

99. $\begin{cases} 4c + 9b + 29a = 20 \\ 9c + 29b + 99a = 70 \\ 29c + 99b + 353a = 254 \end{cases}$

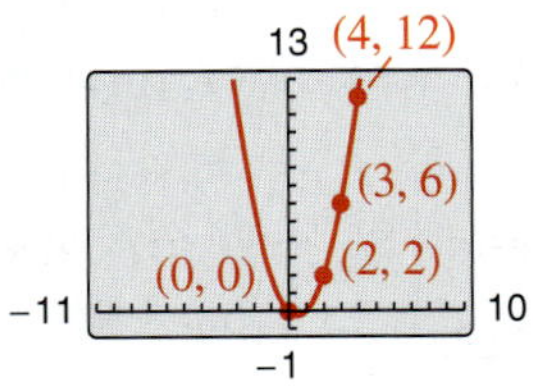

100. $\begin{cases} 4c + 6b + 14a = 25 \\ 6c + 14b + 36a = 21 \\ 14c + 36b + 98a = 33 \end{cases}$

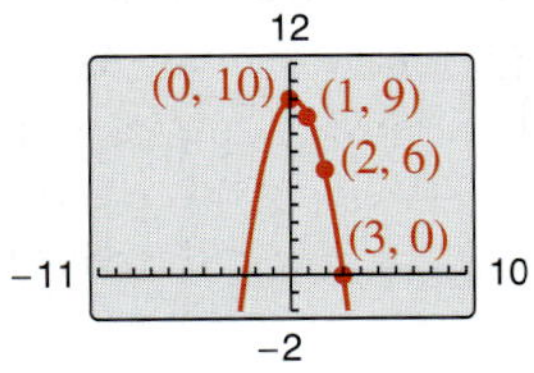

101. MODELING DATA

During the testing of a new automobile braking system, the speeds x (in miles per hour) and the stopping distances y (in feet) were recorded in the table.

Speed, x	Stopping distance, y
30	55
40	105
50	188

(a) Use the data to create a system of linear equations. Then find the least squares regression parabola for the data by solving the system.

(b) Use a graphing utility to graph the parabola and the data in the same viewing window.

(c) Use the model to estimate the stopping distance for a speed of 70 miles per hour.

102. MODELING DATA

A wildlife management team studied the reproduction rates of deer in three five-acre tracts of a wildlife preserve. In each tract, the number of females x and the percent of females y that had offspring the following year were recorded. The results are shown in the table.

Number, x	Percent, y
120	68
140	55
160	30

(a) Use the data to create a system of linear equations. Then find the least squares regression parabola for the data by solving the system.

(b) Use a graphing utility to graph the parabola and the data in the same viewing window.

(c) Use the model to predict the percent of females that had offspring when there were 170 females.

103. Environment The predicted cost C (in thousands of dollars) for a company to remove $p\%$ of a chemical from its wastewater is given by the model

$$C = \frac{120p}{10{,}000 - p^2}, \quad 0 \le p < 100.$$

Write the partial fraction decomposition of the rational function. Verify your result by using the *table* feature of a graphing utility to create a table comparing the original function with the partial fractions.

104. Thermodynamics The magnitude of the range R of exhaust temperatures (in degrees Fahrenheit) in an experimental diesel engine is approximated by the model

$$R = \frac{2000(4 - 3x)}{(11 - 7x)(7 - 4x)}, \quad 0 \le x \le 1$$

where x is the relative load (in foot-pounds).

(a) Write the partial fraction decomposition of the rational function.

(b) The decomposition in part (a) is the difference of two fractions. The absolute values of the terms give the expected maximum and minimum temperatures of the exhaust gases. Use a graphing utility to graph each term.

Conclusions

True or False? **In Exercises 105 and 106, determine whether the statement is true or false. Justify your answer.**

105. The system

$$\begin{cases} x + 4y - 5z = 8 \\ \quad\;\; 2y + \;\, z = 5 \\ \qquad\qquad\;\; z = 1 \end{cases}$$

is in row-echelon form.

106. If a system of three linear equations is inconsistent, then its graph has no points common to all three equations.

107. Error Analysis You are tutoring a student in algebra. In trying to find a partial fraction decomposition, your student writes the following.

$$\frac{x^2 + 1}{x(x - 1)} = \frac{A}{x} + \frac{B}{x - 1}$$

$$x^2 + 1 = A(x - 1) + Bx \qquad \text{Basic equation}$$

$$x^2 + 1 = (A + B)x - A$$

Your student then forms the following system of linear equations.

$$\begin{cases} A + B = 0 \\ -A \quad\;\; = 1 \end{cases}$$

Solve the system and check the partial fraction decomposition it yields. Has your student worked the problem correctly? If not, what went wrong?

108. Think About It Find values of a, b, and c (if possible) such that the system of linear equations has (a) a unique solution, (b) no solution, and (c) an infinite number of solutions.

$$\begin{cases} x + y \qquad\; = 2 \\ \qquad y + z = 2 \\ x \qquad + z = 2 \\ ax + by + cz = 0 \end{cases}$$

109. Think About It Are the two systems of equations equivalent? Give reasons for your answer.

$$\begin{cases} x + 3y - \;z = 6 \\ 2x - \;y + 2z = 1 \\ 3x + 2y - \;z = 2 \end{cases} \qquad \begin{cases} x + 3y - \;z = 6 \\ \quad - 7y + 4z = 1 \\ \quad - 7y - 4z = -16 \end{cases}$$

110. HOW DO YOU SEE IT? The number of sides x and the combined number of sides and diagonals y for each of three regular polygons are shown below. Write a system of linear equations to find an equation of the form $y = ax^2 + bx + c$ that represents the relationship between x and y for the three polygons.

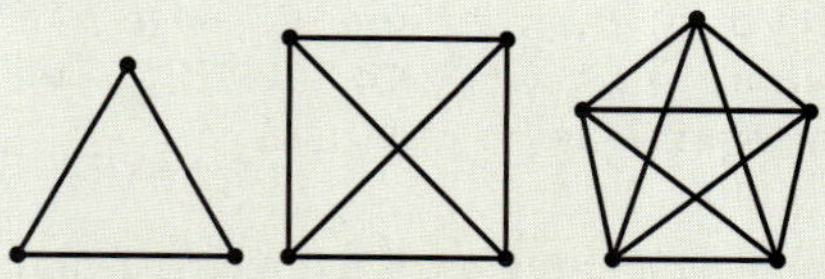

$x = 3, y = 3$ $\quad$ $x = 4, y = 6$ $\quad$ $x = 5, y = 10$

111. Exploration Find a system of equations in three variables that has exactly two equations and no solution.

112. Writing When using Gaussian elimination to solve a system of linear equations, explain how you can recognize that the system has no solution. Give an example that illustrates your answer.

Lagrange Multiplier **In Exercises 113 and 114, find values of x, y, and λ that satisfy the system. These systems arise in certain optimization problems in calculus. (λ is called a *Lagrange multiplier*.)**

113. $\begin{cases} \quad\;\; y + \lambda = 0 \\ x \quad\;\; + \lambda = 0 \\ x + y - 10 = 0 \end{cases}$ **114.** $\begin{cases} 2x \qquad\; + \lambda = 0 \\ \qquad 2y + \lambda = 0 \\ x + \;\; y - 4 = 0 \end{cases}$

Cumulative Mixed Review

Finding Zeros **In Exercises 115–118, (a) determine the real zeros of f and (b) sketch the graph of f.**

115. $f(x) = x^3 + x^2 - 12x$ **116.** $f(x) = -8x^4 + 32x^2$

117. $f(x) = 2x^3 + 5x^2 - 21x - 36$

118. $f(x) = 6x^3 - 29x^2 - 6x + 5$

Solving an Equation Incolving Radicals **In Exercises 119 and 120, solve the equation.**

119. $5\sqrt{x} - 15 = 0$ **120.** $\sqrt{x} + \sqrt{x - 4} = 4$

121. *Make a Decision* To work an extended application analyzing the earnings per share for Wal-Mart Stores, Inc. from 1990 through 2012, visit this textbook's website at *LarsonPrecalculus.com*. (Data Source: Wal-Mart Stores, Inc.)

5.4 Matrices and Systems of Equations

Matrices

In this section, you will study a streamlined technique for solving systems of linear equations. This technique involves the use of a rectangular array of real numbers called a **matrix.** The plural of matrix is *matrices*.

Definition of Matrix

If m and n are positive integers, then an $m \times n$ (read "m by n") matrix is a rectangular array

$$\begin{array}{c} \text{Row 1} \\ \text{Row 2} \\ \text{Row 3} \\ \vdots \\ \text{Row } m \end{array} \overset{\begin{array}{ccccc} \text{Column 1} & \text{Column 2} & \text{Column 3} & \cdots & \text{Column } n \end{array}}{\begin{bmatrix} a_{11} & a_{12} & a_{13} & \cdots & a_{1n} \\ a_{21} & a_{22} & a_{23} & \cdots & a_{2n} \\ a_{31} & a_{32} & a_{33} & \cdots & a_{3n} \\ \vdots & \vdots & \vdots & & \vdots \\ a_{m1} & a_{m2} & a_{m3} & \cdots & a_{mn} \end{bmatrix}}$$

in which each **entry** a_{ij} of the matrix is a real number. An $m \times n$ matrix has m rows and n columns.

The entry in the ith row and jth column of a matrix is denoted by the *double subscript* notation a_{ij}. For instance, the entry a_{23} is the entry in the second row and third column. A matrix having m rows and n columns is said to be of **dimension** $m \times n$. If $m = n$, then the matrix is **square** of dimension $m \times m$ (or $n \times n$). For a square matrix, the entries

$$a_{11}, a_{22}, a_{33}, \ldots$$

are the **main diagonal** entries.

EXAMPLE 1 Dimension of a Matrix

Determine the dimension of each matrix.

a. $[2]$ **b.** $\begin{bmatrix} 1 & -3 & 0 & \frac{1}{2} \end{bmatrix}$ **c.** $\begin{bmatrix} 0 & 0 \\ 0 & 0 \end{bmatrix}$ **d.** $\begin{bmatrix} 5 & 0 \\ 2 & -2 \\ -7 & 4 \end{bmatrix}$ **e.** $\begin{bmatrix} -2 \\ 0 \\ 1 \end{bmatrix}$

Solution

a. This matrix has *one* row and *one* column. The dimension of the matrix is 1×1.

b. This matrix has *one* row and *four* columns. The dimension of the matrix is 1×4.

c. This matrix has *two* rows and *two* columns. The dimension of the matrix is 2×2.

d. This matrix has *three* rows and *two* columns. The dimension of the matrix is 3×2.

e. This matrix has *three* rows and *one* column. The dimension of the matrix is 3×1.

✓ **Checkpoint** *Audio-video solution in English & Spanish at LarsonPrecalculus.com.*

Determine the dimension of the matrix $\begin{bmatrix} 14 & 7 & 10 \\ -2 & -3 & -8 \end{bmatrix}$.

A matrix that has only one row [such as the matrix in Example 1(b)] is called a **row matrix,** and a matrix that has only one column [such as the matrix in Example 1(e)] is called a **column matrix.**

What you should learn

- Write matrices and identify their dimensions.
- Perform elementary row operations on matrices.
- Use matrices and Gaussian elimination to solve systems of linear equations.
- Use matrices and Gauss-Jordan elimination to solve systems of linear equations.

Why you should learn it

Matrices can be used to solve systems of linear equations in two or more variables. For instance, Exercise 95 on page 444 shows how a matrix can be used to help model an equation for the average annual consumer costs for health insurance.

Physician

A matrix derived from a system of linear equations (each written in standard form with the constant term on the right) is the **augmented matrix** of the system. Moreover, the matrix derived from the coefficients of the system (but not including the constant terms) is the **coefficient matrix** of the system.

System:
$$\begin{cases} x - 4y + 3z = 5 \\ -x + 3y - z = -3 \\ 2x \quad\quad - 4z = 6 \end{cases}$$

Augmented Matrix:
$$\left[\begin{array}{rrr:r} 1 & -4 & 3 & 5 \\ -1 & 3 & -1 & -3 \\ 2 & 0 & -4 & 6 \end{array}\right]$$

Coefficient Matrix:
$$\begin{bmatrix} 1 & -4 & 3 \\ -1 & 3 & -1 \\ 2 & 0 & -4 \end{bmatrix}$$

Note the use of 0 for the missing coefficient of the y-variable in the third equation, and also note the fourth column (of constant terms) in the augmented matrix.

When forming either the coefficient matrix or the augmented matrix of a system, you should begin by vertically aligning the variables in the equations and using 0's for any missing coefficients of variables.

Remark

The optional dotted line in the augmented matrix helps to separate the coefficients of the linear system from the constant terms.

EXAMPLE 2 Writing an Augmented Matrix

Write the augmented matrix for the system of linear equations. What is the dimension of the augmented matrix?

$$\begin{cases} x + 3y = 9 \\ -y + 4z = -2 \\ x - 5z = 0 \end{cases}$$

Solution

Begin by writing the linear system and aligning the variables.

$$\begin{cases} x + 3y \quad\quad\quad = 9 \\ \quad\;\; -y + 4z = -2 \\ x \quad\quad\;\; - 5z = 0 \end{cases}$$

Next, use the coefficients and constant terms as the matrix entries. Include zeros for the coefficients of the missing variables.

$$\begin{array}{c} R_1 \\ R_2 \\ R_3 \end{array}\left[\begin{array}{rrr:r} 1 & 3 & 0 & 9 \\ 0 & -1 & 4 & -2 \\ 1 & 0 & -5 & 0 \end{array}\right]$$

The augmented matrix has three rows and four columns, so it is a 3×4 matrix. The notation R_n is used to designate each row in the matrix. For instance, Row 1 is represented by R_1.

✓ Checkpoint *Audio-video solution in English & Spanish at LarsonPrecalculus.com.*

Write the augmented matrix for the system of linear equations. What is the dimension of the augmented matrix?

$$\begin{cases} x + y + z = 2 \\ 2x - y + 3z = -1 \\ -x + 2y - z = 4 \end{cases}$$

Elementary Row Operations

In Section 5.3, you studied three operations that can be used on a system of linear equations to produce an equivalent system. Recall that the operations are (1) interchanging any two equations, (2) multiplying one of the equations by a nonzero constant, and (3) adding a multiple of one equation to another equation.

In matrix terminology, these three operations correspond to **elementary row operations.** An elementary row operation on an augmented matrix of a given system of linear equations produces a new augmented matrix corresponding to a new (but equivalent) system of linear equations. Two matrices are **row-equivalent** when one can be obtained from the other by a sequence of elementary row operations.

> **Elementary Row Operations for Matrices**
>
> **1.** Interchange two rows.
>
> **2.** Multiply one of the rows by a nonzero constant.
>
> **3.** Add a multiple of one row to another row.

Although elementary row operations are simple to perform, they involve a lot of arithmetic. Because it is easy to make a mistake, you should get in the habit of noting the elementary row operations performed in each step so that you can go back and check your work. Example 3 demonstrates the elementary row operations described above.

EXAMPLE 3 Elementary Row Operations

a. Interchange the first and second rows of the original matrix.

Original Matrix *New Row-Equivalent Matrix*

$$\begin{bmatrix} 0 & 1 & 3 & 4 \\ -1 & 2 & 0 & 3 \\ 2 & -3 & 4 & 1 \end{bmatrix} \qquad \begin{matrix} R_2 \\ R_1 \\ \end{matrix} \begin{bmatrix} -1 & 2 & 0 & 3 \\ 0 & 1 & 3 & 4 \\ 2 & -3 & 4 & 1 \end{bmatrix}$$

b. Multiply the first row of the original matrix by $\frac{1}{2}$.

Original Matrix *New Row-Equivalent Matrix*

$$\begin{bmatrix} 2 & -4 & 6 & -2 \\ 1 & 3 & -3 & 0 \\ 5 & -2 & 1 & 2 \end{bmatrix} \qquad \begin{matrix} \frac{1}{2}R_1 \rightarrow \\ \\ \\ \end{matrix} \begin{bmatrix} 1 & -2 & 3 & -1 \\ 1 & 3 & -3 & 0 \\ 5 & -2 & 1 & 2 \end{bmatrix}$$

c. Add -2 times the first row of the original matrix to the third row.

Original Matrix *New Row-Equivalent Matrix*

$$\begin{bmatrix} 1 & 2 & -4 & 3 \\ 0 & 3 & -2 & -1 \\ 2 & 1 & 5 & -2 \end{bmatrix} \qquad \begin{matrix} \\ \\ -2R_1 + R_3 \rightarrow \end{matrix} \begin{bmatrix} 1 & 2 & -4 & 3 \\ 0 & 3 & -2 & -1 \\ 0 & -3 & 13 & -8 \end{bmatrix}$$

Note that the elementary row operation is written beside the row that is *changed*.

✓ ***Checkpoint*** *Audio-video solution in English & Spanish at LarsonPrecalculus.com.*

Fill in the blank using an elementary row operation to form a row-equivalent matrix.

$$\begin{bmatrix} 1 & 2 & 3 \\ 3 & 12 & 5 \end{bmatrix} \qquad \begin{bmatrix} 1 & 2 & 3 \\ 0 & \square & -4 \end{bmatrix}$$

■

> ## Technology Tip
> Most graphing utilities can perform elementary row operations on matrices. For instructions on how to use the *matrix* feature and the elementary row operations features of a graphing utility, see Appendix A; for specific keystrokes, go to this textbook's *Companion Website*.

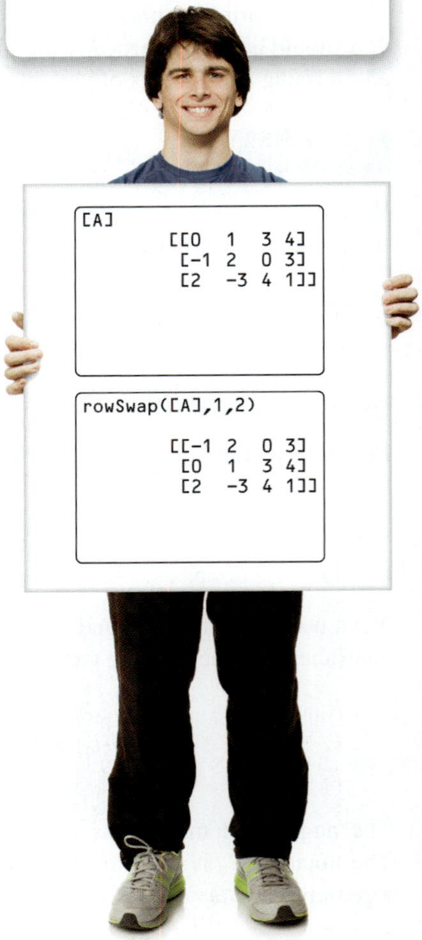

Gaussian Elimination with Back-Substitution

In Example 2 of Section 5.3, you used Gaussian elimination with back-substitution to solve a system of linear equations. The next example demonstrates the matrix version of Gaussian elimination. The basic difference between the two methods is that with matrices you do not need to keep writing the variables.

EXAMPLE 4 Comparing Linear Systems and Matrix Operations

Linear System

$$\begin{cases} x - 2y + 3z = 9 \\ -x + 3y + z = -2 \\ 2x - 5y + 5z = 17 \end{cases}$$

Associated Augmented Matrix

$$\begin{bmatrix} 1 & -2 & 3 & \vdots & 9 \\ -1 & 3 & 1 & \vdots & -2 \\ 2 & -5 & 5 & \vdots & 17 \end{bmatrix}$$

Add the first equation to the second equation.

$$\begin{cases} x - 2y + 3z = 9 \\ y + 4z = 7 \\ 2x - 5y + 5z = 17 \end{cases}$$

Add the first row to the second row: $R_1 + R_2$.

$$R_1 + R_2 \rightarrow \begin{bmatrix} 1 & -2 & 3 & \vdots & 9 \\ 0 & 1 & 4 & \vdots & 7 \\ 2 & -5 & 5 & \vdots & 17 \end{bmatrix}$$

Add -2 times the first equation to the third equation.

$$\begin{cases} x - 2y + 3z = 9 \\ y + 4z = 7 \\ -y - z = -1 \end{cases}$$

Add -2 times the first row to the third row: $-2R_1 + R_3$.

$$-2R_1 + R_3 \rightarrow \begin{bmatrix} 1 & -2 & 3 & \vdots & 9 \\ 0 & 1 & 4 & \vdots & 7 \\ 0 & -1 & -1 & \vdots & -1 \end{bmatrix}$$

Add the second equation to the third equation.

$$\begin{cases} x - 2y + 3z = 9 \\ y + 4z = 7 \\ 3z = 6 \end{cases}$$

Add the second row to the third row: $R_2 + R_3$.

$$R_2 + R_3 \rightarrow \begin{bmatrix} 1 & -2 & 3 & \vdots & 9 \\ 0 & 1 & 4 & \vdots & 7 \\ 0 & 0 & 3 & \vdots & 6 \end{bmatrix}$$

Multiply the third equation by $\frac{1}{3}$.

$$\begin{cases} x - 2y + 3z = 9 \\ y + 4z = 7 \\ z = 2 \end{cases}$$

Multiply the third row by $\frac{1}{3}$: $\frac{1}{3}R_3$.

$$\tfrac{1}{3}R_3 \rightarrow \begin{bmatrix} 1 & -2 & 3 & \vdots & 9 \\ 0 & 1 & 4 & \vdots & 7 \\ 0 & 0 & 1 & \vdots & 2 \end{bmatrix}$$

At this point, you can use back-substitution to find x and y.

$y + 4(2) = 7$ Substitute 2 for z.

$y = -1$ Solve for y.

$x - 2(-1) + 3(2) = 9$ Substitute -1 for y and 2 for z.

$x = 1$ Solve for x.

The solution is $x = 1$, $y = -1$, and $z = 2$.

Remark

Remember that you should check a solution by substituting the values of x, y, and z into each equation of the original system. For instance, you can check the solution to Example 4 as shown.

Equation 1:
$1 - 2(-1) + 3(2) = 9$ ✓

Equation 2:
$-1 + 3(-1) + 2 = -2$ ✓

Equation 3:
$2(1) - 5(-1) + 5(2) = 17$ ✓

Checkpoint *Audio-video solution in English & Spanish at LarsonPrecalculus.com.*

Compare solving the linear system to solving it using its associated augmented matrix.

$$\begin{cases} 2x + y - z = -3 \\ 4x - 2y + 2z = -2 \\ -6x + 5y + 4z = 10 \end{cases}$$

The last matrix in Example 4 is in **row-echelon form.** The term *echelon* refers to the stair-step pattern formed by the nonzero elements of the matrix. To be in this form, a matrix must have the following properties.

Row-Echelon Form and Reduced Row-Echelon Form

A matrix in **row-echelon form** has the following properties.

1. Any rows consisting entirely of zeros occur at the bottom of the matrix.
2. For each row that does not consist entirely of zeros, the first nonzero entry is 1 (called a **leading 1**).
3. For two successive (nonzero) rows, the leading 1 in the higher row is farther to the left than the leading 1 in the lower row.

A matrix in *row-echelon form* is in **reduced row-echelon form** when every column that has a leading 1 has zeros in every position above and below its leading 1.

It is worth mentioning that the row-echelon form of a matrix is not unique. That is, two different sequences of elementary row operations may yield different row-echelon forms. The *reduced* row-echelon form of a given matrix, however, is unique.

Technology Tip

Some graphing utilities can automatically transform a matrix to row-echelon form and reduced row-echelon form. For instructions on how to use the *row-echelon form* feature and the *reduced row-echelon form* feature of a graphing utility, see Appendix A; for specific keystrokes, go to this textbook's *Companion Website*.

EXAMPLE 5 Row-Echelon Form

Determine whether each matrix is in row-echelon form. If it is, determine whether the matrix is in reduced row-echelon form.

a. $\begin{bmatrix} 1 & 2 & -1 & 4 \\ 0 & 1 & 0 & 3 \\ 0 & 0 & 1 & -2 \end{bmatrix}$ **b.** $\begin{bmatrix} 1 & 2 & -1 & 2 \\ 0 & 0 & 0 & 0 \\ 0 & 1 & 2 & -4 \end{bmatrix}$

c. $\begin{bmatrix} 1 & -5 & 2 & -1 & 3 \\ 0 & 0 & 1 & 3 & -2 \\ 0 & 0 & 0 & 1 & 4 \\ 0 & 0 & 0 & 0 & 1 \end{bmatrix}$ **d.** $\begin{bmatrix} 1 & 0 & 0 & -1 \\ 0 & 1 & 0 & 2 \\ 0 & 0 & 1 & 3 \\ 0 & 0 & 0 & 0 \end{bmatrix}$

e. $\begin{bmatrix} 1 & 2 & -3 & 4 \\ 0 & 2 & 1 & -1 \\ 0 & 0 & 1 & -3 \end{bmatrix}$ **f.** $\begin{bmatrix} 0 & 1 & 0 & 5 \\ 0 & 0 & 1 & 3 \\ 0 & 0 & 0 & 0 \end{bmatrix}$

Solution

The matrices in (a), (c), (d), and (f) are in row-echelon form. The matrices in (d) and (f) are in *reduced* row-echelon form because every column that has a leading 1 has zeros in every position above and below its leading 1. The matrix in (b) is not in row-echelon form because the row of all zeros does not occur at the bottom of the matrix. The matrix in (e) is not in row-echelon form because the first nonzero entry in Row 2 is not a leading 1.

✔ ***Checkpoint*** *Audio-video solution in English & Spanish at LarsonPrecalculus.com.*

Determine whether the matrix is in row-echelon form. If it is, determine whether it is in reduced row-echelon form.

$$\begin{bmatrix} 1 & 0 & -2 & 4 \\ 0 & 1 & 11 & 3 \\ 0 & 0 & 0 & 0 \end{bmatrix}$$

Every matrix is row-equivalent to a matrix in row-echelon form. For instance, in Example 5, you can change the matrix in part (e) to row-echelon form by multiplying its second row by $\frac{1}{2}$.

Gaussian elimination with back-substitution works well for solving systems of linear equations by hand or with a computer. For this algorithm, the order in which the elementary row operations are performed is important. You should operate *from left to right by columns*, using elementary row operations to obtain zeros in all entries directly below the leading 1's.

EXAMPLE 6 Gaussian Elimination with Back-Substitution

Solve the system of equations.

$$\begin{cases} y + z - 2w = -3 \\ x + 2y - z = 2 \\ 2x + 4y + z - 3w = -2 \\ x - 4y - 7z - w = -19 \end{cases}$$

Solution

$$\begin{bmatrix} 0 & 1 & 1 & -2 & \vdots & -3 \\ 1 & 2 & -1 & 0 & \vdots & 2 \\ 2 & 4 & 1 & -3 & \vdots & -2 \\ 1 & -4 & -7 & -1 & \vdots & -19 \end{bmatrix}$$ Write augmented matrix.

$$\begin{matrix} R_2 \\ R_1 \\ \\ \\ \end{matrix} \begin{bmatrix} 1 & 2 & -1 & 0 & \vdots & 2 \\ 0 & 1 & 1 & -2 & \vdots & -3 \\ 2 & 4 & 1 & -3 & \vdots & -2 \\ 1 & -4 & -7 & -1 & \vdots & -19 \end{bmatrix}$$ Interchange R_1 and R_2 so first column has leading 1 in upper left corner.

$$\begin{matrix} \\ \\ -2R_1 + R_3 \rightarrow \\ -R_1 + R_4 \rightarrow \end{matrix} \begin{bmatrix} 1 & 2 & -1 & 0 & \vdots & 2 \\ 0 & 1 & 1 & -2 & \vdots & -3 \\ 0 & 0 & 3 & -3 & \vdots & -6 \\ 0 & -6 & -6 & -1 & \vdots & -21 \end{bmatrix}$$ Perform operations on R_3 and R_4 so first column has zeros below its leading 1.

$$\begin{matrix} \\ \\ \\ 6R_2 + R_4 \rightarrow \end{matrix} \begin{bmatrix} 1 & 2 & -1 & 0 & \vdots & 2 \\ 0 & 1 & 1 & -2 & \vdots & -3 \\ 0 & 0 & 3 & -3 & \vdots & -6 \\ 0 & 0 & 0 & -13 & \vdots & -39 \end{bmatrix}$$ Perform operations on R_4 so second column has zeros below its leading 1.

$$\begin{matrix} \\ \\ \frac{1}{3}R_3 \rightarrow \\ -\frac{1}{13}R_4 \rightarrow \end{matrix} \begin{bmatrix} 1 & 2 & -1 & 0 & \vdots & 2 \\ 0 & 1 & 1 & -2 & \vdots & -3 \\ 0 & 0 & 1 & -1 & \vdots & -2 \\ 0 & 0 & 0 & 1 & \vdots & 3 \end{bmatrix}$$ Perform operations on R_3 and R_4 so third and fourth columns have leading 1's.

The matrix is now in row-echelon form, and the corresponding system is

$$\begin{cases} x + 2y - z = 2 \\ y + z - 2w = -3 \\ z - w = -2 \\ w = 3 \end{cases}$$

Using back-substitution, you can determine that the solution is

$x = -1, \quad y = 2, \quad z = 1, \quad$ and $\quad w = 3.$

Checkpoint *Audio-video solution in English & Spanish at LarsonPrecalculus.com.*

Solve the system of equations.

$$\begin{cases} -3x + 5y + 3z = -19 \\ 3x + 4y + 4z = 8 \\ 4x - 8y - 6z = 26 \end{cases}$$

The following steps summarize the procedure used in Example 6.

Gaussian Elimination with Back-Substitution

1. Write the augmented matrix of the system of linear equations.
2. Use elementary row operations to rewrite the augmented matrix in row-echelon form.
3. Write the system of linear equations corresponding to the matrix in row-echelon form and use back-substitution to find the solution.

Remember that it is possible for a system to have no solution. If, in the elimination process, you obtain a row with zeros except for the last entry, then you can conclude that the system is inconsistent.

EXAMPLE 7 A System with No Solution

Solve the system of equations.

$$\begin{cases} x - y + 2z = 4 \\ x + z = 6 \\ 2x - 3y + 5z = 4 \\ 3x + 2y - z = 1 \end{cases}$$

Solution

$$\begin{bmatrix} 1 & -1 & 2 & \vdots & 4 \\ 1 & 0 & 1 & \vdots & 6 \\ 2 & -3 & 5 & \vdots & 4 \\ 3 & 2 & -1 & \vdots & 1 \end{bmatrix}$$ Write augmented matrix.

$$\begin{array}{r} \\ -R_1 + R_2 \rightarrow \\ -2R_1 + R_3 \rightarrow \\ -3R_1 + R_4 \rightarrow \end{array} \begin{bmatrix} 1 & -1 & 2 & \vdots & 4 \\ 0 & 1 & -1 & \vdots & 2 \\ 0 & -1 & 1 & \vdots & -4 \\ 0 & 5 & -7 & \vdots & -11 \end{bmatrix}$$ Perform row operations.

$$\begin{array}{r} \\ \\ R_2 + R_3 \rightarrow \\ \\ \end{array} \begin{bmatrix} 1 & -1 & 2 & \vdots & 4 \\ 0 & 1 & -1 & \vdots & 2 \\ 0 & 0 & 0 & \vdots & -2 \\ 0 & 5 & -7 & \vdots & -11 \end{bmatrix}$$ Perform row operations.

Note that the third row of this matrix consists of zeros except for the last entry. This means that the original system of linear equations is *inconsistent*. You can see why this is true by converting back to a system of linear equations. Because the third equation is not possible, the system has no solution.

$$\begin{cases} x - y + 2z = 4 \\ y - z = 2 \\ 0 = -2 \\ 5y - 7z = -11 \end{cases}$$

✓ Checkpoint *Audio-video solution in English & Spanish at LarsonPrecalculus.com.*

Solve the system of equations.

$$\begin{cases} x + y + z = 1 \\ x + 2y + 2z = 2 \\ x - y - z = 1 \end{cases}$$

Gauss–Jordan Elimination

With Gaussian elimination, elementary row operations are applied to a matrix to obtain a (row-equivalent) row-echelon form of the matrix. A second method of elimination, called **Gauss-Jordan elimination** after Carl Friedrich Gauss (1777–1855) and Wilhelm Jordan (1842–1899), continues the reduction process until a *reduced* row-echelon form is obtained. This procedure is demonstrated in Example 8.

EXAMPLE 8 Gauss–Jordan Elimination

See LarsonPrecalculus.com for an interactive version of this type of example.

Use Gauss-Jordan elimination to solve the system.

$$\begin{cases} x - 2y + 3z = 9 \\ -x + 3y + z = -2 \\ 2x - 5y + 5z = 17 \end{cases}$$

Solution

In Example 4, Gaussian elimination was used to obtain the row-echelon form

$$\begin{bmatrix} 1 & -2 & 3 & \vdots & 9 \\ 0 & 1 & 4 & \vdots & 7 \\ 0 & 0 & 1 & \vdots & 2 \end{bmatrix}.$$

Now, rather than using back-substitution, apply additional elementary row operations until you obtain a matrix in *reduced* row-echelon form. To do this, you must produce zeros above each of the leading 1's, as follows.

$$2R_2 + R_1 \rightarrow \begin{bmatrix} 1 & 0 & 11 & \vdots & 23 \\ 0 & 1 & 4 & \vdots & 7 \\ 0 & 0 & 1 & \vdots & 2 \end{bmatrix}$$

Perform operations on R_1 so second column has a zero above its leading 1.

$$\begin{matrix} -11R_3 + R_1 \rightarrow \\ -4R_3 + R_2 \rightarrow \\ \end{matrix} \begin{bmatrix} 1 & 0 & 0 & \vdots & 1 \\ 0 & 1 & 0 & \vdots & -1 \\ 0 & 0 & 1 & \vdots & 2 \end{bmatrix}$$

Perform operations on R_1 and R_2 so third column has zeros above its leading 1.

The matrix is now in reduced row-echelon form. Converting back to a system of linear equations, you have

$$\begin{cases} x = 1 \\ y = -1. \\ z = 2 \end{cases}$$

Now you can simply read the solution, $x = 1$, $y = -1$, and $z = 2$, which can be written as the ordered triple $(1, -1, 2)$. You can check this result using the *reduced row-echelon form* feature of a graphing utility, as shown in Figure 5.16.

✓ Checkpoint *Audio-video solution in English & Spanish at LarsonPrecalculus.com.*

Use Gauss-Jordan elimination to solve the system.

$$\begin{cases} -3x + 7y + 2z = 1 \\ -5x + 3y - 5z = -8 \\ 2x - 2y - 3z = 15 \end{cases}$$

In Example 8, note that the solution is the same as the one obtained using Gaussian elimination in Example 4. The advantage of Gauss-Jordan elimination is that, from the reduced row-echelon form, you can simply read the solution without the need for back-substitution.

Technology Tip

For a demonstration of a graphical approach to Gauss-Jordan elimination on a 2×3 matrix, see the Visualizing Row Operations Program, available for several models of graphing calculators at this textbook's *Companion Website*.

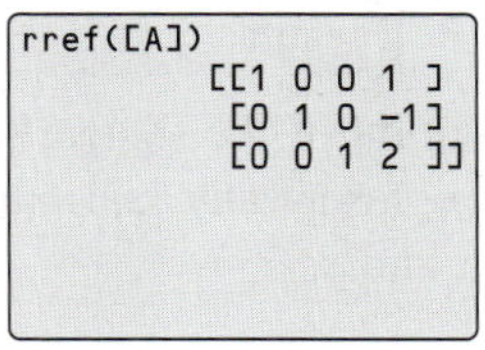

Figure 5.16

The elimination procedures described in this section employ an algorithmic approach that is easily adapted to computer programs. However, the procedure makes no effort to avoid fractional coefficients. For instance, in the elimination procedure for the system

$$\begin{cases} 2x - 5y + 5z = 17 \\ 3x - 2y + 3z = 11 \\ -3x + 3y \qquad = -6 \end{cases}$$

you may be inclined to multiply the first row by $\frac{1}{2}$ to produce a leading 1, which will result in working with fractional coefficients. For hand computations, you can sometimes avoid fractions by judiciously choosing the order in which you apply elementary row operations.

EXAMPLE 9 A System with an Infinite Number of Solutions

Solve the system.

$$\begin{cases} 2x + 4y - 2z = 0 \\ 3x + 5y \qquad = 1 \end{cases}$$

Solution

$$\begin{bmatrix} 2 & 4 & -2 & \vdots & 0 \\ 3 & 5 & 0 & \vdots & 1 \end{bmatrix}$$

$$\tfrac{1}{2}R_1 \rightarrow \begin{bmatrix} 1 & 2 & -1 & \vdots & 0 \\ 3 & 5 & 0 & \vdots & 1 \end{bmatrix}$$

$$-3R_1 + R_2 \rightarrow \begin{bmatrix} 1 & 2 & -1 & \vdots & 0 \\ 0 & -1 & 3 & \vdots & 1 \end{bmatrix}$$

$$-R_2 \rightarrow \begin{bmatrix} 1 & 2 & -1 & \vdots & 0 \\ 0 & 1 & -3 & \vdots & -1 \end{bmatrix}$$

$$-2R_2 + R_1 \rightarrow \begin{bmatrix} 1 & 0 & 5 & \vdots & 2 \\ 0 & 1 & -3 & \vdots & -1 \end{bmatrix}$$

The corresponding system of equations is

$$\begin{cases} x + 5z = 2 \\ y - 3z = -1 \end{cases}.$$

Solving for x and y in terms of z, you have

$$x = -5z + 2 \quad \text{and} \quad y = 3z - 1.$$

To write a solution of the system that does not use any of the three variables of the system, let a represent any real number and let $z = a$. Now substitute a for z in the equations for x and y.

$$x = -5z + 2 = -5a + 2 \quad \text{and} \quad y = 3z - 1 = 3a - 1$$

So, the solution set has the form $(-5a + 2, 3a - 1, a)$. Recall from Section 5.3 that a solution set of this form represents an infinite number of solutions. Try substituting values for a to obtain a few solutions. Then check each solution in the original system of equations.

✓ Checkpoint *Audio-video solution in English & Spanish at LarsonPrecalculus.com.*

Solve the system.

$$\begin{cases} 2x - 6y + 6z = 46 \\ 2x - 3y \qquad = 31 \end{cases}$$

5.4 Exercises

See *CalcChat.com* for tutorial help and worked-out solutions to odd-numbered exercises.
For instructions on how to use a graphing utility, see Appendix A.

Vocabulary and Concept Check

In Exercises 1–3, fill in the blank.

1. A rectangular array of real numbers that can be used to solve a system of linear equations is called a _______ .

2. A matrix in row-echelon form is in _______ when every column that has a leading 1 has zeros in every position above and below its leading 1.

3. The process of using row operations to write a matrix in reduced row-echelon form is called _______ .

In Exercises 4–6, refer to the system of linear equations $\begin{cases} -2x + 3y = 5 \\ 6x + 7y = 4 \end{cases}$.

4. Is the coefficient matrix for the system a *square* matrix?

5. Is the augmented matrix for the system of dimension 3×2?

6. Is the augmented matrix row-equivalent to its reduced row-echelon form?

Procedures and Problem Solving

Dimension of a Matrix **In Exercises 7–12, determine the dimension of the matrix.**

7. $\begin{bmatrix} 7 & 0 \end{bmatrix}$

8. $\begin{bmatrix} 3 & -1 & 2 & 6 \end{bmatrix}$

9. $\begin{bmatrix} 4 \\ 32 \\ 3 \end{bmatrix}$

10. $\begin{bmatrix} 5 & 4 & 2 \\ 3 & -5 & 1 \\ 7 & -2 & 9 \end{bmatrix}$

11. $\begin{bmatrix} 33 & 45 \\ -9 & 20 \end{bmatrix}$

12. $\begin{bmatrix} 3 & -1 & 6 & 4 \\ -2 & 5 & 7 & 7 \end{bmatrix}$

Writing an Augmented Matrix **In Exercises 13–18, write the augmented matrix for the system of linear equations. What is the dimension of the augmented matrix?**

13. $\begin{cases} 4x - 3y = -5 \\ -x + 3y = 12 \end{cases}$

14. $\begin{cases} 7x + 4y = 22 \\ 5x - 9y = 15 \end{cases}$

15. $\begin{cases} x + 10y - 2z = 2 \\ 5x - 3y + 4z = 0 \\ 2x + y = 6 \end{cases}$

16. $\begin{cases} x - 3y + z = 1 \\ 4y = 0 \\ 7z = -5 \end{cases}$

17. $\begin{cases} 7x - 5y + z = 13 \\ 19x - 8z = 10 \end{cases}$

18. $\begin{cases} 9x + 2y - 3z = 20 \\ -25y + 11z = -5 \end{cases}$

Writing a System of Equations **In Exercises 19–22, write the system of linear equations represented by the augmented matrix. (Use the variables *x*, *y*, *z*, and *w*, if applicable.)**

19. $\left[\begin{array}{cc|c} 3 & 4 & 0 \\ 1 & -1 & 7 \end{array}\right]$

20. $\left[\begin{array}{cc|c} 7 & -5 & 2 \\ 8 & 0 & -2 \end{array}\right]$

21. $\left[\begin{array}{ccc|c} 0 & 12 & 3 & 0 \\ -2 & 18 & 0 & 10 \\ 1 & 7 & -8 & 43 \end{array}\right]$

22. $\left[\begin{array}{cccc|c} 6 & 2 & -1 & -5 & -25 \\ -1 & 0 & 7 & 3 & 7 \\ 4 & -1 & -10 & 6 & 23 \\ 0 & 8 & 1 & -11 & -21 \end{array}\right]$

Identifying an Elementary Row Operation **In Exercises 23–26, identify the elementary row operation performed to obtain the new row-equivalent matrix.**

	Original Matrix	*New Row-Equivalent Matrix*
23.	$\begin{bmatrix} -4 & 8 & -20 \\ -2 & 6 & 7 \end{bmatrix}$	$\begin{bmatrix} 1 & -2 & 5 \\ -2 & 6 & 7 \end{bmatrix}$
24.	$\begin{bmatrix} -3 & 6 & 0 \\ 5 & 2 & -2 \end{bmatrix}$	$\begin{bmatrix} -18 & 0 & 6 \\ 5 & 2 & -2 \end{bmatrix}$
25.	$\begin{bmatrix} -1 & -2 & 3 & -2 \\ 2 & -5 & 1 & -7 \\ 5 & 4 & -7 & 6 \end{bmatrix}$	$\begin{bmatrix} -1 & -2 & 3 & -2 \\ 2 & -5 & 1 & -7 \\ 0 & -6 & 8 & -4 \end{bmatrix}$
26.	$\begin{bmatrix} 0 & -1 & -5 & 5 \\ -1 & 3 & -7 & 6 \\ 4 & -5 & 1 & 3 \end{bmatrix}$	$\begin{bmatrix} -1 & 3 & -7 & 6 \\ 0 & -1 & -5 & 5 \\ 4 & -5 & 1 & 3 \end{bmatrix}$

Elementary Row Operations **In Exercises 27–30, fill in the blank(s) using elementary row operations to form a row-equivalent matrix.**

27. $\begin{bmatrix} 1 & 4 & 3 \\ 2 & 10 & 5 \end{bmatrix}$

$\begin{bmatrix} 1 & 4 & 3 \\ 0 & \square & -1 \end{bmatrix}$

28. $\begin{bmatrix} 3 & 6 & 8 \\ 4 & -3 & 6 \end{bmatrix}$

$\begin{bmatrix} 1 & \square & \frac{8}{3} \\ 4 & -3 & 6 \end{bmatrix}$

29. $\begin{bmatrix} 1 & 1 & 4 & -1 \\ 3 & 8 & 10 & 3 \\ -2 & 1 & 12 & 6 \end{bmatrix}$

$\begin{bmatrix} 1 & 1 & 4 & -1 \\ 0 & 5 & \square & \square \\ 0 & 3 & \square & \square \end{bmatrix}$

$\begin{bmatrix} 1 & 1 & 4 & -1 \\ 0 & 1 & \square & \square \\ 0 & 3 & \square & \square \end{bmatrix}$

30. $\begin{bmatrix} 2 & 4 & 8 & 3 \\ 1 & -1 & -3 & 2 \\ 2 & 6 & 4 & 9 \end{bmatrix}$

$\begin{bmatrix} 1 & -1 & -3 & 2 \\ 2 & \square & \square & \square \\ 2 & 6 & 4 & 9 \end{bmatrix}$

$\begin{bmatrix} 1 & -1 & -3 & 2 \\ 0 & 6 & \square & \square \\ 0 & 8 & \square & \square \end{bmatrix}$

Comparing Linear Systems and Matrix Operations **In Exercises 31 and 32, (a) perform the row operations to solve the augmented matrix, (b) write and solve the system of linear equations represented by the augmented matrix, and (c) compare the two solution methods. Which do you prefer?**

31. $\left[\begin{array}{cc|c} -3 & 4 & 22 \\ 6 & -4 & -28 \end{array}\right]$

(i) Add R_2 to R_1.

(ii) Add -2 times R_1 to R_2.

(iii) Multiply R_2 by $-\frac{1}{4}$.

(iv) Multiply R_1 by $\frac{1}{3}$.

32. $\left[\begin{array}{ccc|c} 7 & 13 & 1 & -4 \\ -3 & -5 & -1 & -4 \\ 3 & 6 & 1 & -2 \end{array}\right]$

(i) Add R_2 to R_1.

(ii) Multiply R_1 by $\frac{1}{4}$.

(iii) Add R_3 to R_2.

(iv) Add -3 times R_1 to R_3.

(v) Add -2 times R_2 to R_1.

33. Repeat steps (i) through (iv) for the matrix in Exercise 31 using a graphing utility.

34. Repeat steps (i) through (v) for the matrix in Exercise 32 using a graphing utility.

Row-Echelon Form **In Exercises 35–40, determine whether the matrix is in row-echelon form. If it is, determine if it is also in reduced row-echelon form.**

35. $\begin{bmatrix} 1 & 0 & 0 & 0 \\ 0 & 1 & 1 & 5 \\ 0 & 0 & 0 & 0 \end{bmatrix}$ **36.** $\begin{bmatrix} 1 & 0 & 1 & 0 \\ 0 & 1 & 0 & 2 \\ 0 & 0 & 1 & 0 \end{bmatrix}$

37. $\begin{bmatrix} 3 & 0 & 3 & 7 \\ 0 & -2 & 0 & 4 \\ 0 & 0 & 1 & 5 \end{bmatrix}$ **38.** $\begin{bmatrix} 1 & 3 & 0 & 0 \\ 0 & 0 & 1 & 8 \\ 0 & 0 & 0 & 0 \end{bmatrix}$

39. $\begin{bmatrix} 1 & 0 & 2 & 1 \\ 0 & 1 & -3 & 10 \\ 0 & 0 & 1 & 0 \end{bmatrix}$ **40.** $\begin{bmatrix} 1 & 0 & 0 & 1 \\ 0 & 1 & 0 & -1 \\ 0 & 0 & 0 & 2 \end{bmatrix}$

Using Gaussian Elimination **In Exercises 41–44, write the matrix in row-echelon form. Remember that the row-echelon form of a matrix is not unique.**

41. $\begin{bmatrix} 1 & -4 & 5 \\ -2 & 6 & -6 \end{bmatrix}$ **42.** $\begin{bmatrix} 1 & -3 & 2 \\ 5 & 0 & 7 \end{bmatrix}$

43. $\begin{bmatrix} 1 & 2 & -1 & 3 \\ 3 & 7 & -5 & 14 \\ -2 & -1 & -3 & 8 \end{bmatrix}$

44. $\begin{bmatrix} 1 & -3 & 0 & -7 \\ -3 & 10 & 1 & 23 \\ 4 & -10 & 2 & -24 \end{bmatrix}$

Using a Graphing Utility **In Exercises 45–48, use the matrix capabilities of a graphing utility to write the matrix in reduced row-echelon form.**

45. $\begin{bmatrix} 3 & 3 & 3 \\ -1 & 0 & -4 \\ 2 & 4 & -2 \end{bmatrix}$ **46.** $\begin{bmatrix} 1 & 3 & 2 \\ 5 & 15 & 9 \\ 2 & 6 & 10 \end{bmatrix}$

47. $\begin{bmatrix} -4 & 1 & 0 & 6 \\ 1 & -2 & 3 & -4 \end{bmatrix}$

48. $\begin{bmatrix} 5 & 1 & 2 & 4 \\ -1 & 5 & 10 & -32 \end{bmatrix}$

Using Back-Substitution **In Exercises 49–52, write the system of linear equations represented by the augmented matrix. Then use back-substitution to find the solution. (Use the variables x, y, and z, if applicable.)**

49. $\left[\begin{array}{cc|c} 1 & -2 & 4 \\ 0 & 1 & -3 \end{array}\right]$ **50.** $\left[\begin{array}{cc|c} 1 & 8 & 12 \\ 0 & 1 & 3 \end{array}\right]$

51. $\left[\begin{array}{ccc|c} 1 & -1 & 4 & 0 \\ 0 & 1 & -1 & 2 \\ 0 & 0 & 1 & -2 \end{array}\right]$

52. $\left[\begin{array}{ccc|c} 1 & 0 & -2 & -7 \\ 0 & 1 & 1 & 9 \\ 0 & 0 & 1 & -3 \end{array}\right]$

Interpreting Reduced Row-Echelon Form **In Exercises 53–56, an augmented matrix that represents a system of linear equations (in the variables x and y or x, y, and z) has been reduced using Gauss-Jordan elimination. Write the solution represented by the augmented matrix.**

53. $\left[\begin{array}{cc|c} 1 & 0 & 7 \\ 0 & 1 & -5 \end{array}\right]$ **54.** $\left[\begin{array}{cc|c} 1 & 0 & -2 \\ 0 & 1 & 4 \end{array}\right]$

55. $\left[\begin{array}{ccc|c} 1 & 0 & 0 & -4 \\ 0 & 1 & 0 & -8 \\ 0 & 0 & 1 & 2 \end{array}\right]$

56. $\left[\begin{array}{ccc|c} 1 & 0 & 0 & 3 \\ 0 & 1 & 0 & -1 \\ 0 & 0 & 1 & 0 \end{array}\right]$

Gaussian Elimination with Back-Substitution In Exercises 57–64, use matrices to solve the system of equations, if possible. Use Gaussian elimination with back-substitution.

57. $\begin{cases} x + 2y = 7 \\ 2x + y = 8 \end{cases}$

58. $\begin{cases} 2x + 6y = 16 \\ 2x + 3y = 7 \end{cases}$

59. $\begin{cases} -x + y = -22 \\ 3x + 4y = 4 \\ 4x - 8y = 32 \end{cases}$

60. $\begin{cases} x + 2y = 0 \\ x + y = 6 \\ 3x - 2y = 8 \end{cases}$

61. $\begin{cases} x + 2y - 3z = -28 \\ 4y + 2z = 0 \\ -x + y - z = -5 \end{cases}$

62. $\begin{cases} -x + y - 2z = -4 \\ 4x + 2y - 3z = 8 \\ 2x + 4y - 7z = 1 \end{cases}$

63. $\begin{cases} 3x + 2y - z + w = 0 \\ x - y + 4z + 2w = 25 \\ -2x + y + 2z - w = 2 \\ x + y + z + w = 6 \end{cases}$

64. $\begin{cases} x - 4y + 3z - 2w = 9 \\ 3x - 2y + z - 4w = -13 \\ -4x + 3y - 2z + w = -4 \\ -2x + y - 4z + 3w = -10 \end{cases}$

Gauss-Jordan Elimination In Exercises 65–72, use matrices to solve the system of equations, if possible. Use Gauss-Jordan elimination.

65. $\begin{cases} x - 3z = -2 \\ 3x + y - 2z = 5 \\ 2x + 2y + z = 4 \end{cases}$

66. $\begin{cases} 2x - y + 3z = 24 \\ 2y - z = 14 \\ 7x - 5y = 6 \end{cases}$

67. $\begin{cases} -x - y - 3z = -12 \\ 2x - y - 4z = 6 \\ -2x + 4y + 14z = 19 \end{cases}$

68. $\begin{cases} 2x + 3z = 3 \\ 4x - 3y + 7z = 5 \\ 8x - 9y + 15z = 9 \end{cases}$

69. $\begin{cases} x + y - 5z = 3 \\ x - 2z = 1 \\ 2x - y - z = 0 \end{cases}$

70. $\begin{cases} 2x + 2y - z = 2 \\ x - 3y + z = 28 \\ -x + y = 14 \end{cases}$

71. $\begin{cases} x + 2y + z = 8 \\ 3x + 7y + 6z = 26 \end{cases}$

72. $\begin{cases} x + y + 4z = 5 \\ 2x + y - z = 9 \end{cases}$

Using a Graphing Utility In Exercises 73–76, use the matrix capabilities of a graphing utility to reduce the augmented matrix corresponding to the system of equations, and solve the system.

73. $\begin{cases} x + y + 4z = 2 \\ 2x + 5y + 20z = 10 \\ -x + 3y + 8z = -2 \end{cases}$

74. $\begin{cases} x + y + z = 0 \\ 2x + 3y + z = 0 \\ 3x + 5y + z = 0 \end{cases}$

75. $\begin{cases} 2x + 10y + 2z = 6 \\ x + 5y + 2z = 6 \\ x + 5y + z = 3 \\ -3x - 15y - 3z = -9 \end{cases}$

76. $\begin{cases} 2x + y - z + 2w = -6 \\ 3x + 4y + w = 1 \\ x + 5y + 2z + 6w = -3 \\ 5x + 2y - z - w = 3 \end{cases}$

Comparing Solutions of Two Systems In Exercises 77–80, determine whether the two systems of linear equations yield the same solution. If so, find the solution.

77. (a) $\begin{cases} x - 2y + z = -6 \\ y - 5z = 16 \\ z = -3 \end{cases}$ (b) $\begin{cases} x + y - 2z = 6 \\ y + 3z = -8 \\ z = -3 \end{cases}$

78. (a) $\begin{cases} x - 3y + 4z = -11 \\ y - z = -4 \\ z = 2 \end{cases}$ (b) $\begin{cases} x + 4y = -11 \\ y + 3z = 4 \\ z = 2 \end{cases}$

79. (a) $\begin{cases} x - 4y + 5z = 27 \\ y - 7z = -54 \\ z = 8 \end{cases}$ (b) $\begin{cases} x - 6y + z = 15 \\ y + 5z = 42 \\ z = 8 \end{cases}$

80. (a) $\begin{cases} x + 3y - z = 19 \\ y + 6z = -18 \\ z = -4 \end{cases}$ (b) $\begin{cases} x - y + 3z = -21 \\ y - 2z = 14 \\ z = -4 \end{cases}$

Data Analysis: Curve Fitting In Exercises 81–84, use a system of equations to find the equation of the parabola $y = ax^2 + bx + c$ that passes through the points. Solve the system using matrices. Use a graphing utility to verify your result.

81.

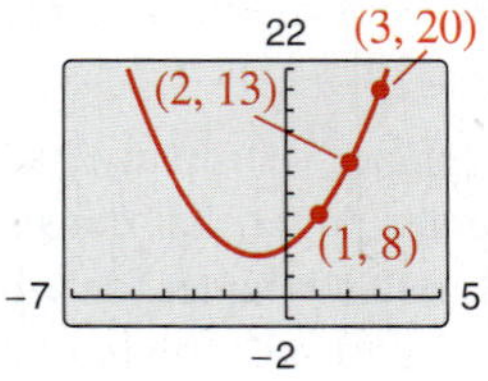

82.

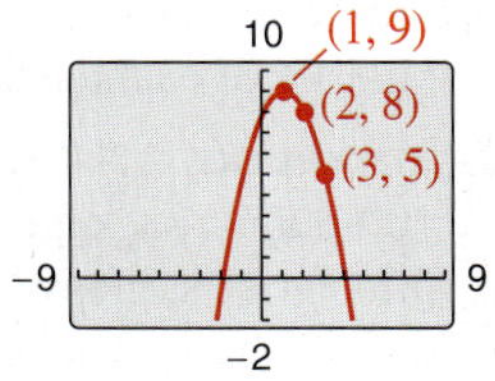

83.

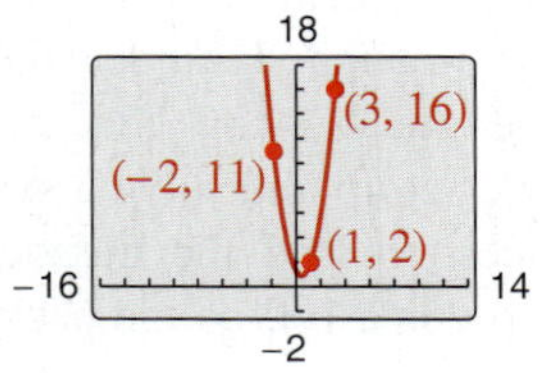

84.

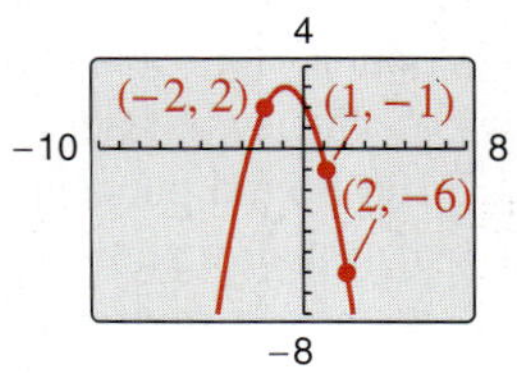

Curve Fitting In Exercises 85 and 86, use a system of equations to find the quadratic function $f(x) = ax^2 + bx + c$ that satisfies the equations. Solve the system using matrices.

85. $f(-2) = -15$
$f(-1) = 7$
$f(1) = -3$

86. $f(-2) = -3$
$f(1) = -3$
$f(2) = -11$

Curve Fitting In Exercises 87 and 88, use a system of equations to find the cubic function $f(x) = ax^3 + bx^2 + cx + d$ that satisfies the equations. Solve the system using matrices.

87. $f(-2) = -7$
$f(-1) = 2$
$f(1) = -4$
$f(2) = -7$

88. $f(-2) = -17$
$f(-1) = -5$
$f(1) = 1$
$f(2) = 7$

89. Electrical Engineering The currents in an electrical network are given by the solution of the system

$$\begin{cases} I_1 - I_2 + I_3 = 0 \\ 2I_1 + 2I_2 = 7 \\ 2I_2 + 4I_3 = 8 \end{cases}$$

where I_1, I_2, and I_3 are measured in amperes. Solve the system of equations using matrices.

90. Finance A corporation borrowed \$1,500,000 to expand its line of shoes. Some of the money was borrowed at 3%, some at 4%, and some at 6%. Use a system of equations to determine how much was borrowed at each rate, given that the annual interest was \$74,000 and the amount borrowed at 6% was four times the amount borrowed at 3%. Solve the system using matrices.

91. Using Matrices A food server examines the amount of money earned in tips after working an 8-hour shift. The server has a total of \$95 in denominations of \$1, \$5, \$10, and \$20 bills. The total number of paper bills is 26. The number of \$5 bills is 4 times the number of \$10 bills, and the number of \$1 bills is 1 less than twice the number of \$5 bills. Write a system of linear equations to represent the situation. Then use matrices to find the number of each denomination.

92. Marketing A wholesale paper company sells a 100-pound package of computer paper that consists of three grades, glossy, semi-gloss, and matte, for printing photographs. Glossy costs \$5.50 per pound, semi-gloss costs \$4.25 per pound, and matte costs \$3.75 per pound. One half of the 100-pound package consists of the two less expensive grades. The cost of the 100-pound package is \$480. Set up and solve a system of equations, using matrices, to find the number of pounds of each grade of paper in a 100-pound package.

93. Partial Fractions Use a system of equations to write the partial fraction decomposition of the rational expression. Solve the system using matrices.

$$\frac{8x^2}{(x-1)^2(x+1)} = \frac{A}{x+1} + \frac{B}{x-1} + \frac{C}{(x-1)^2}$$

94. MODELING DATA

A video of the path of a ball thrown by a baseball player was analyzed with a grid covering the TV screen. The video was paused three times, and the position of the ball was measured each time. The coordinates obtained are shown in the table (x and y are measured in feet).

Horizontal distance, x	Height, y
0	5.0
15	9.6
30	12.4

(a) Use a system of equations to find the equation of the parabola $y = ax^2 + bx + c$ that passes through the points. Solve the system using matrices.

(b) Use a graphing utility to graph the parabola.

(c) Graphically approximate the maximum height of the ball and the point at which the ball strikes the ground.

(d) Algebraically approximate the maximum height of the ball and the point at which the ball strikes the ground.

95. ***Why you should learn it*** *(p. 432)* The table shows the average annual consumer costs y (in dollars) for health insurance from 2010 to 2012. (Source: U.S. Bureau of Labor Statistics)

Year	Cost, y
2010	1831
2011	1922
2012	2061

(a) Use a system of equations to find the equation of the parabola $y = at^2 + bt + c$ that passes through the points. Let t represent the year, with $t = 0$ corresponding to 2010. Solve the system using matrices.

(b) Use a graphing utility to graph the parabola and plot the data points.

(c) Use the equation in part (a) to estimate the average consumer costs in 2015, 2020, and 2025.

(d) Are your estimates in part (c) reasonable? Explain.

96. MODELING DATA

The table shows the average annual salaries y (in thousands of dollars) for public school classroom teachers in the United States from 2011 through 2013. (Source: National Education Association)

Year	Annual salary, y
2011	55.5
2012	55.4
2013	56.4

(a) Use a system of equations to find the equation of the parabola $y = at^2 + bt + c$ that passes through the points. Let t represent the year, with $t = 1$ corresponding to 2011. Solve the system using matrices.

(b) Use a graphing utility to graph the parabola and plot the data points.

(c) Use the equation in part (a) to estimate the average annual salaries in 2015, 2020, and 2025.

(d) Are your estimates in part (c) reasonable? Explain.

97. Network Analysis Water flowing through a network of pipes (in thousands of cubic meters per hour) is shown below.

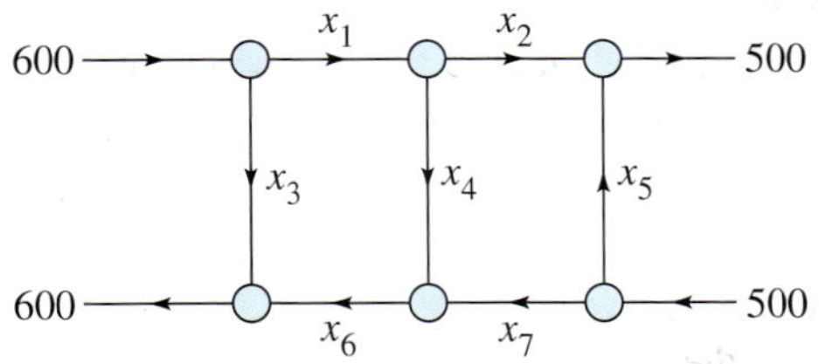

(a) Use matrices to solve this system for the water flow represented by x_i, $i = 1, 2, 3, 4, 5, 6$, and 7.

(b) Find the network flow pattern when $x_6 = 0$ and $x_7 = 0$.

(c) Find the network flow pattern when $x_5 = 400$ and $x_6 = 500$.

98. Network Analysis The flow of water (in thousands of cubic meters per hour) into and out of the right side of the network of pipes in Exercise 97 is increased from 500 to 700.

(a) Draw a diagram of the new network.

(b) Use matrices to solve this system for the water flow represented by x_i, $i = 1, 2, 3, 4, 5, 6$, and 7.

(c) Find the network flow pattern when $x_1 = 600$ and $x_7 = 200$.

(d) Find the network flow pattern when $x_4 = 150$ and $x_5 = 350$.

Conclusions

True or False? **In Exercises 99 and 100, determine whether the statement is true or false. Justify your answer.**

99. When using Gaussian elimination to solve a system of linear equations, you may conclude that the system is inconsistent before you complete the process of rewriting the augmented matrix in row-echelon form.

100. You cannot write an augmented matrix for a dependent system of linear equations in reduced row-echelon form.

101. Think About It The augmented matrix represents a system of linear equations (in the variables x, y, and z) that has been reduced using Gauss-Jordan elimination. Write a system of three equations in three variables with *nonzero* coefficients that is represented by the reduced matrix. (There are many correct answers.)

$$\begin{bmatrix} 1 & 0 & 3 & \vdots & -2 \\ 0 & 1 & 4 & \vdots & 1 \\ 0 & 0 & 0 & \vdots & 0 \end{bmatrix}$$

102. HOW DO YOU SEE IT? Determine whether the matrix below is in row-echelon form, reduced row-echelon form, or neither when it satisfies the given conditions.

$$\begin{bmatrix} 1 & b \\ c & 1 \end{bmatrix}$$

(a) $b = 0, c = 0$ (b) $b \neq 0, c = 0$

(c) $b = 0, c \neq 0$ (d) $b \neq 0, c \neq 0$

103. Think About It Can a 2×4 augmented matrix whose entries are all nonzero real numbers represent an independent system of linear equations? Explain.

104. Think About It Determine all values of a and b for which the augmented matrix has each given number of solutions.

$$\begin{bmatrix} 1 & 2 & \vdots & -4 \\ 0 & a & \vdots & b \end{bmatrix}$$

(a) Exactly one solution

(b) Infinitely many solutions

(c) No solution

Cumulative Mixed Review

Graphing a Rational Function **In Exercises 105–108, sketch the graph of the function. Identify any asymptotes.**

105. $f(x) = \dfrac{7}{-x - 1}$

106. $f(x) = \dfrac{4x}{5x^2 + 2}$

107. $f(x) = \dfrac{x^2 - 2x - 3}{x - 4}$

108. $f(x) = \dfrac{x^2 - 36}{x + 1}$

5.5 Operations with Matrices

Equality of Matrices

In Section 5.4, you used matrices to solve systems of linear equations. There is a rich mathematical theory of matrices, and its applications are numerous. This section and the next two introduce some fundamentals of matrix theory. It is standard mathematical convention to represent matrices in any of the following three ways.

Representation of Matrices

1. A matrix can be denoted by an uppercase letter such as A, B, or C.
2. A matrix can be denoted by a representative element enclosed in brackets, such as

   $[a_{ij}]$, $[b_{ij}]$, or $[c_{ij}]$.
3. A matrix can be denoted by a rectangular array of numbers such as

$$A = [a_{ij}] = \begin{bmatrix} a_{11} & a_{12} & a_{13} & \cdots & a_{1n} \\ a_{21} & a_{22} & a_{23} & \cdots & a_{2n} \\ a_{31} & a_{32} & a_{33} & \cdots & a_{3n} \\ \vdots & \vdots & \vdots & & \vdots \\ a_{m1} & a_{m2} & a_{m3} & \cdots & a_{mn} \end{bmatrix}.$$

What you should learn

- Decide whether two matrices are equal.
- Add and subtract matrices and multiply matrices by scalars.
- Multiply two matrices.
- Use matrix operations to model and solve real-life problems.

Why you should learn it

Matrix algebra provides a systematic way of performing mathematical operations on large arrays of numbers. In Exercise 89 on page 458, you will use matrix multiplication to help analyze the labor and wage requirements for a boat manufacturer.

Two matrices

$A = [a_{ij}]$ and $B = [b_{ij}]$

are **equal** when they have the same dimension $(m \times n)$ and all of their corresponding entries are equal.

EXAMPLE 1 Equality of Matrices

Solve for a_{11}, a_{12}, a_{21}, and a_{22} in the following matrix equation.

$$\begin{bmatrix} a_{11} & a_{12} \\ a_{21} & a_{22} \end{bmatrix} = \begin{bmatrix} 2 & -1 \\ -3 & 0 \end{bmatrix}$$

Solution

Because two matrices are equal only when their corresponding entries are equal, you can conclude that

$$a_{11} = 2, \quad a_{12} = -1, \quad a_{21} = -3, \quad \text{and} \quad a_{22} = 0.$$

Checkpoint *Audio-video solution in English & Spanish at LarsonPrecalculus.com.*

Solve for a_{11}, a_{12}, a_{21}, and a_{22} in the following matrix equation.

$$\begin{bmatrix} a_{11} & a_{12} \\ a_{21} & a_{22} \end{bmatrix} = \begin{bmatrix} 6 & 3 \\ -2 & 4 \end{bmatrix}$$

Be sure you see that for two matrices to be equal, they must have the same dimension *and* their corresponding entries must be equal. For instance,

$$\begin{bmatrix} 2 & -1 \\ \sqrt{4} & \frac{1}{2} \end{bmatrix} = \begin{bmatrix} 2 & -1 \\ 2 & 0.5 \end{bmatrix} \quad \text{but} \quad \begin{bmatrix} 2 & -1 \\ 3 & 4 \\ 0 & 0 \end{bmatrix} \neq \begin{bmatrix} 2 & -1 \\ 3 & 4 \end{bmatrix}.$$

Matrix Addition and Scalar Multiplication

Definition of Matrix Addition

If $A = [a_{ij}]$ and $B = [b_{ij}]$ are matrices of dimension $m \times n$, then their sum is the $m \times n$ matrix given by $A + B = [a_{ij} + b_{ij}]$. The sum of two matrices of different dimensions is undefined.

Remark

With matrix addition, you can add two matrices (of the same dimension) by adding their corresponding entries.

EXAMPLE 2 Addition of Matrices

a. $\begin{bmatrix} -1 & 2 \\ 0 & 1 \end{bmatrix} + \begin{bmatrix} 1 & 3 \\ -1 & 2 \end{bmatrix} = \begin{bmatrix} -1 + 1 & 2 + 3 \\ 0 + (-1) & 1 + 2 \end{bmatrix} = \begin{bmatrix} 0 & 5 \\ -1 & 3 \end{bmatrix}$

b. $\begin{bmatrix} 1 \\ -3 \\ -2 \end{bmatrix} + \begin{bmatrix} -1 \\ 3 \\ 2 \end{bmatrix} = \begin{bmatrix} 0 \\ 0 \\ 0 \end{bmatrix}$

c. The sum of

$$A = \begin{bmatrix} 2 & 1 & 0 \\ 4 & 0 & -1 \end{bmatrix} \quad \text{and} \quad B = \begin{bmatrix} 0 & 1 \\ -1 & 3 \end{bmatrix}$$

is undefined because A is of dimension 2×3 and B is of dimension 2×2.

Checkpoint *Audio-video solution in English & Spanish at LarsonPrecalculus.com.*

Evaluate the expression.

$$\begin{bmatrix} 4 & -1 \\ 2 & -3 \end{bmatrix} + \begin{bmatrix} 2 & -1 \\ 0 & 6 \end{bmatrix}$$

Technology Tip

Try using a graphing utility to find the sum of the two matrices in Example 2(c). Your graphing utility should display an error message similar to the one shown below.

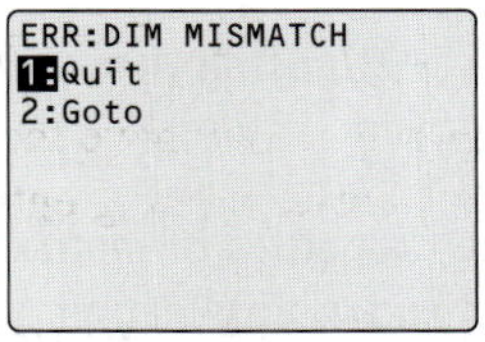

In operations with matrices, numbers are usually referred to as **scalars.** In this text, scalars will always be real numbers. You can multiply a matrix A by a scalar c by multiplying each entry in A by c.

Definition of Scalar Multiplication

If $A = [a_{ij}]$ is an $m \times n$ matrix and c is a scalar, then the **scalar multiple** of A by c is the $m \times n$ matrix given by $cA = [ca_{ij}]$.

EXAMPLE 3 Scalar Multiplication

Find $3A$ using $A = \begin{bmatrix} 2 & 2 & 4 \\ -3 & 0 & -1 \\ 2 & 1 & 2 \end{bmatrix}$.

Solution

$$3A = 3\begin{bmatrix} 2 & 2 & 4 \\ -3 & 0 & -1 \\ 2 & 1 & 2 \end{bmatrix} = \begin{bmatrix} 3(2) & 3(2) & 3(4) \\ 3(-3) & 3(0) & 3(-1) \\ 3(2) & 3(1) & 3(2) \end{bmatrix} = \begin{bmatrix} 6 & 6 & 12 \\ -9 & 0 & -3 \\ 6 & 3 & 6 \end{bmatrix}$$

Checkpoint *Audio-video solution in English & Spanish at LarsonPrecalculus.com.*

Find $3A$ using $A = \begin{bmatrix} 4 & -1 \\ 0 & 4 \\ -3 & 8 \end{bmatrix}$.

The symbol $-A$ represents the negation of A, which is the scalar product $(-1)A$. Moreover, if A and B are of the same dimension, then $A - B$ represents the sum of A and $(-1)B$. That is,

$$A - B = A + (-1)B.$$

The order of operations for matrix expressions is similar to that for real numbers. In particular, you perform scalar multiplication before matrix addition and subtraction, as shown in Example 4.

Explore the Concept

What do you observe about the relationship between the corresponding entries of A and B below? Use a graphing utility to find $A + B$. What conclusion can you make about the entries of A and B and the sum $A + B$?

$$A = \begin{bmatrix} -1 & 5 \\ 2 & -6 \end{bmatrix}$$

$$B = \begin{bmatrix} 1 & -5 \\ -2 & 6 \end{bmatrix}$$

EXAMPLE 4 Scalar Multiplication and Matrix Subtraction

Find $3A - B$ using $A = \begin{bmatrix} 2 & 2 & 4 \\ -3 & 0 & -1 \\ 2 & 1 & 2 \end{bmatrix}$ and $B = \begin{bmatrix} 2 & 0 & 0 \\ 1 & -4 & 3 \\ -1 & 3 & 2 \end{bmatrix}$.

Solution

Note that A is the same matrix from Example 3, where you found $3A$.

$$3A - B = \begin{bmatrix} 6 & 6 & 12 \\ -9 & 0 & -3 \\ 6 & 3 & 6 \end{bmatrix} - \begin{bmatrix} 2 & 0 & 0 \\ 1 & -4 & 3 \\ -1 & 3 & 2 \end{bmatrix}$$ Perform scalar multiplication first.

$$= \begin{bmatrix} 4 & 6 & 12 \\ -10 & 4 & -6 \\ 7 & 0 & 4 \end{bmatrix}$$ Subtract corresponding entries.

✓ Checkpoint *Audio-video solution in English & Spanish at LarsonPrecalculus.com.*

Find $3A - 2B$ using $A = \begin{bmatrix} 4 & -1 \\ 0 & 4 \\ -3 & 8 \end{bmatrix}$ and $B = \begin{bmatrix} 0 & 4 \\ -1 & 3 \\ 1 & 7 \end{bmatrix}$. ■

The properties of matrix addition and scalar multiplication are similar to those of addition and multiplication of real numbers. One important property of addition of real numbers is that the number 0 is the additive identity. That is, $c + 0 = c$ for any real number c. For matrices, a similar property holds. That is, if A is an $m \times n$ matrix and O is the $m \times n$ **zero matrix** consisting entirely of zeros, then $A + O = A$. In other words, O is the **additive identity** for the set of all $m \times n$ matrices. For example, the following matrices are the additive identities for the sets of all 2×3 and 2×2 matrices.

$$O = \underbrace{\begin{bmatrix} 0 & 0 & 0 \\ 0 & 0 & 0 \end{bmatrix}}_{2 \times 3 \text{ zero matrix}} \quad \text{and} \quad O = \underbrace{\begin{bmatrix} 0 & 0 \\ 0 & 0 \end{bmatrix}}_{2 \times 2 \text{ zero matrix}}$$

Properties of Matrix Addition and Scalar Multiplication

Let A, B, and C be $m \times n$ matrices and let c and d be scalars.

1. $A + B = B + A$ — Commutative Property of Matrix Addition
2. $A + (B + C) = (A + B) + C$ — Associative Property of Matrix Addition
3. $(cd)A = c(dA)$ — Associative Property of Scalar Multiplication
4. $1A = A$ — Scalar Identity
5. $A + O = A$ — Additive Identity
6. $c(A + B) = cA + cB$ — Distributive Property
7. $(c + d)A = cA + dA$ — Distributive Property

Remark

Note that the Associative Property of Matrix Addition allows you to write expressions such as $A + B + C$ without ambiguity because the same sum occurs no matter how the matrices are grouped. This same reasoning applies to sums of four or more matrices.

EXAMPLE 5 Using the Distributive Property

$$3\left(\begin{bmatrix}-2 & 0\\ 4 & 1\end{bmatrix}+\begin{bmatrix}4 & -2\\ 3 & 7\end{bmatrix}\right)=3\begin{bmatrix}-2 & 0\\ 4 & 1\end{bmatrix}+3\begin{bmatrix}4 & -2\\ 3 & 7\end{bmatrix}$$
$$=\begin{bmatrix}-6 & 0\\ 12 & 3\end{bmatrix}+\begin{bmatrix}12 & -6\\ 9 & 21\end{bmatrix}$$
$$=\begin{bmatrix}6 & -6\\ 21 & 24\end{bmatrix}$$

Remark

In Example 5, you could add the two matrices first and then multiply the resulting matrix by 3. The result would be the same.

✓ ***Checkpoint*** *Audio-video solution in English & Spanish at LarsonPrecalculus.com.*

Evaluate the expression using the Distributive Property.

$$2\left(\begin{bmatrix}1 & 3\\ -2 & 2\end{bmatrix}+\begin{bmatrix}-4 & 0\\ -3 & 1\end{bmatrix}\right)$$

The algebra of real numbers and the algebra of matrices have many similarities. For example, compare the following solutions.

Real Numbers (Solve for x.)	*$m \times n$ Matrices (Solve for X.)*
$x + a = b$	$X + A = B$
$x + a + (-a) = b + (-a)$	$X + A + (-A) = B + (-A)$
$x + 0 = b - a$	$X + O = B - A$
$x = b - a$	$X = B - A$

The algebra of real numbers and the algebra of matrices also have important differences, which will be discussed later.

EXAMPLE 6 Solving a Matrix Equation

Solve for X in the equation $3X + A = B$, where

$$A=\begin{bmatrix}1 & -2\\ 0 & 3\end{bmatrix} \quad \text{and} \quad B=\begin{bmatrix}-3 & 4\\ 2 & 1\end{bmatrix}.$$

Solution

Begin by solving the matrix equation for X.

$$3X + A = B \;\Rightarrow\; 3X = B - A \;\Rightarrow\; X = \tfrac{1}{3}(B - A)$$

Now, using the matrices A and B, you have

$$X=\frac{1}{3}\left(\begin{bmatrix}-3 & 4\\ 2 & 1\end{bmatrix}-\begin{bmatrix}1 & -2\\ 0 & 3\end{bmatrix}\right)$$ Substitute the matrices.

$$=\frac{1}{3}\begin{bmatrix}-4 & 6\\ 2 & -2\end{bmatrix}$$ Subtract matrix A from matrix B.

$$=\begin{bmatrix}-\frac{4}{3} & 2\\ \frac{2}{3} & -\frac{2}{3}\end{bmatrix}.$$ Multiply the resulting matrix by $\frac{1}{3}$.

✓ ***Checkpoint*** *Audio-video solution in English & Spanish at LarsonPrecalculus.com.*

Solve for X in the equation $2X - A = B$, where

$$A=\begin{bmatrix}6 & 1\\ 0 & 3\end{bmatrix} \quad \text{and} \quad B=\begin{bmatrix}4 & -1\\ -2 & 5\end{bmatrix}.$$

Matrix Multiplication

Another basic matrix operation is **matrix multiplication.**

Definition of Matrix Multiplication

If $A = [a_{ij}]$ is an $m \times n$ matrix and $B = [b_{ij}]$ is an $n \times p$ matrix, then the product AB is an $m \times p$ matrix given by

$$AB = [c_{ij}]$$

where

$$c_{ij} = a_{i1}b_{1j} + a_{i2}b_{2j} + a_{i3}b_{3j} + \cdot \cdot \cdot + a_{in}b_{nj}.$$

Remark

At first glance, this definition may seem unusual. You will see later, however, that the definition of the product of two matrices has many practical applications.

The definition of matrix multiplication indicates a *row-by-column* multiplication, where the entry in the ith row and jth column of the product AB is obtained by multiplying the entries in the ith row of A by the corresponding entries in the jth column of B and then adding the results. The general pattern for matrix multiplication is as follows.

$$\begin{bmatrix} a_{11} & a_{12} & a_{13} & \cdots & a_{1n} \\ a_{21} & a_{22} & a_{23} & \cdots & a_{2n} \\ a_{31} & a_{32} & a_{33} & \cdots & a_{3n} \\ \vdots & \vdots & \vdots & & \vdots \\ a_{i1} & a_{i2} & a_{i3} & \cdots & a_{in} \\ \vdots & \vdots & \vdots & & \vdots \\ a_{m1} & a_{m2} & a_{m3} & \cdots & a_{mn} \end{bmatrix} \begin{bmatrix} b_{11} & b_{12} & \cdots & b_{1j} & \cdots & b_{1p} \\ b_{21} & b_{22} & \cdots & b_{2j} & \cdots & b_{2p} \\ b_{31} & b_{32} & \cdots & b_{3j} & \cdots & b_{3p} \\ \vdots & \vdots & & \vdots & & \vdots \\ b_{n1} & b_{n2} & \cdots & b_{nj} & \cdots & b_{np} \end{bmatrix} = \begin{bmatrix} c_{11} & c_{12} & \cdots & c_{1j} & \cdots & c_{1p} \\ c_{21} & c_{22} & \cdots & c_{2j} & \cdots & c_{2p} \\ \vdots & \vdots & & \vdots & & \vdots \\ c_{i1} & c_{i2} & \cdots & c_{ij} & \cdots & c_{ip} \\ \vdots & \vdots & & \vdots & & \vdots \\ c_{m1} & c_{m2} & \cdots & c_{mj} & \cdots & c_{mp} \end{bmatrix}$$

$$a_{i1}b_{1j} + a_{i2}b_{2j} + a_{i3}b_{3j} + \cdot \cdot \cdot + a_{in}b_{nj} = c_{ij}$$

EXAMPLE 7 Finding the Product of Two Matrices

Find the product AB using $A = \begin{bmatrix} -1 & 3 \\ 4 & -2 \\ 5 & 0 \end{bmatrix}$ and $B = \begin{bmatrix} -3 & 2 \\ -4 & 1 \end{bmatrix}$.

Solution

First, note that the product AB is defined because the number of columns of A is equal to the number of rows of B. Moreover, the product AB has dimension 3×2. To find the entries of the product, multiply each row of A by each column of B.

$$\begin{aligned} AB &= \begin{bmatrix} -1 & 3 \\ 4 & -2 \\ 5 & 0 \end{bmatrix} \begin{bmatrix} -3 & 2 \\ -4 & 1 \end{bmatrix} \\ &= \begin{bmatrix} (-1)(-3) + (3)(-4) & (-1)(2) + (3)(1) \\ (4)(-3) + (-2)(-4) & (4)(2) + (-2)(1) \\ (5)(-3) + (0)(-4) & (5)(2) + (0)(1) \end{bmatrix} \\ &= \begin{bmatrix} -9 & 1 \\ -4 & 6 \\ -15 & 10 \end{bmatrix} \end{aligned}$$

✓ Checkpoint Audio-video solution in English & Spanish at LarsonPrecalculus.com.

Find the product AB using $A = \begin{bmatrix} -1 & 4 \\ 2 & 0 \\ 1 & 2 \end{bmatrix}$ and $B = \begin{bmatrix} 1 & -2 \\ 0 & 7 \end{bmatrix}$.

Be sure you understand that for the product of two matrices to be defined, the number of *columns* of the first matrix must equal the number of *rows* of the second matrix. That is, the middle two indices must be the same. The outside two indices give the dimension of the product, as shown in the following diagram.

$$\underset{m \times n}{A} \quad \times \quad \underset{n \times p}{B} \quad = \quad \underset{m \times p}{AB}$$

Equal (the middle indices n and n)

Dimension of AB (the outside indices m and p)

EXAMPLE 8 Matrix Multiplication

See LarsonPrecalculus.com for an interactive version of this type of example.

a. $\begin{bmatrix} 1 & 0 & 3 \\ 2 & -1 & -2 \end{bmatrix}\begin{bmatrix} -2 & 4 & 2 \\ 1 & 0 & 0 \\ -1 & 1 & -1 \end{bmatrix} = \begin{bmatrix} -5 & 7 & -1 \\ -3 & 6 & 6 \end{bmatrix}$

$2 \times 3 \qquad 3 \times 3 \qquad 2 \times 3$

b. $\begin{bmatrix} 3 & 4 \\ -2 & 5 \end{bmatrix}\begin{bmatrix} 1 & 0 \\ 0 & 1 \end{bmatrix} = \begin{bmatrix} 3 & 4 \\ -2 & 5 \end{bmatrix}$

$2 \times 2 \qquad 2 \times 2 \qquad 2 \times 2$

c. $\begin{bmatrix} 1 & 2 \\ 1 & 1 \end{bmatrix}\begin{bmatrix} -1 & 2 \\ 1 & -1 \end{bmatrix} = \begin{bmatrix} 1 & 0 \\ 0 & 1 \end{bmatrix}$

$2 \times 2 \qquad 2 \times 2 \qquad 2 \times 2$

d. $\begin{bmatrix} 1 & -2 & -3 \end{bmatrix}\begin{bmatrix} 2 \\ -1 \\ 1 \end{bmatrix} = \begin{bmatrix} 1 \end{bmatrix}$

$1 \times 3 \qquad 3 \times 1 \qquad 1 \times 1$

e. $\begin{bmatrix} 2 \\ -1 \\ 1 \end{bmatrix}\begin{bmatrix} 1 & -2 & -3 \end{bmatrix} = \begin{bmatrix} 2 & -4 & -6 \\ -1 & 2 & 3 \\ 1 & -2 & -3 \end{bmatrix}$

$3 \times 1 \qquad 1 \times 3 \qquad 3 \times 3$

f. The product AB for the following matrices is not defined.

$$A = \begin{bmatrix} -2 & 1 \\ 1 & -3 \\ 1 & 4 \end{bmatrix} \quad \text{and} \quad B = \begin{bmatrix} -2 & 3 & 1 & 4 \\ 0 & 1 & -1 & 2 \\ 2 & -1 & 0 & 1 \end{bmatrix}$$

$3 \times 2 \qquad 3 \times 4$

✓ Checkpoint *Audio-video solution in English & Spanish at LarsonPrecalculus.com.*

Find the product AB using $A = \begin{bmatrix} 0 & 4 & -3 \\ 2 & 1 & 7 \\ 3 & -2 & 1 \end{bmatrix}$ and $B = \begin{bmatrix} -2 & 0 \\ 0 & -4 \\ 1 & 2 \end{bmatrix}$. ■

In parts (d) and (e) of Example 8, note that the two products are different. Even when both AB and BA are defined, matrix multiplication is not, in general, commutative. That is, for most matrices,

$$AB \neq BA.$$

This is one way in which the algebra of real numbers and the algebra of matrices differ.

Explore the Concept

Use the following matrices to find AB, BA, $(AB)C$, and $A(BC)$. What do your results tell you about matrix multiplication and commutativity and associativity?

$$A = \begin{bmatrix} 1 & 2 \\ 3 & 4 \end{bmatrix}$$

$$B = \begin{bmatrix} 0 & 1 \\ 2 & 3 \end{bmatrix}$$

$$C = \begin{bmatrix} 3 & 0 \\ 0 & 1 \end{bmatrix}$$

EXAMPLE 9 Matrix Multiplication

Use a graphing utility to find the product AB using

$$A = \begin{bmatrix} 1 & 2 & 3 \\ 2 & -5 & 1 \end{bmatrix} \quad \text{and} \quad B = \begin{bmatrix} -3 & 2 & 1 \\ 4 & -2 & 0 \\ 1 & 2 & 3 \end{bmatrix}.$$

Solution

Note that the dimension of A is 2×3 and the dimension of B is 3×3. So, the product will have dimension 2×3. Use the *matrix editor* to enter A and B into the graphing utility. Then, find the product, as shown in Figure 5.17.

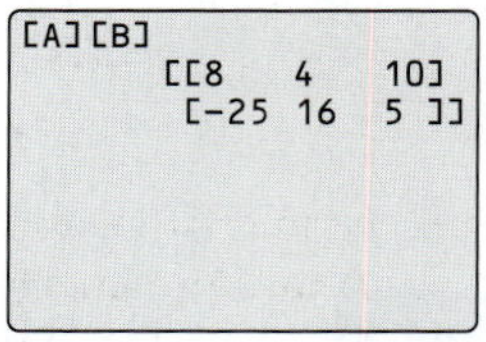

Figure 5.17

✓ ***Checkpoint*** *Audio-video solution in English & Spanish at LarsonPrecalculus.com.*

Find the product AB using $A = \begin{bmatrix} 7 & 5 & -4 \\ -2 & 5 & 1 \\ 10 & -4 & -7 \end{bmatrix}$ and $B = \begin{bmatrix} 2 & -2 & 3 \\ 8 & 1 & 4 \\ -4 & 2 & -8 \end{bmatrix}$.

Properties of Matrix Multiplication

Let A, B, and C be matrices and let c be a scalar.

1. $A(BC) = (AB)C$ — Associative Property of Matrix Multiplication
2. $A(B + C) = AB + AC$ — Left Distributive Property
3. $(A + B)C = AC + BC$ — Right Distributive Property
4. $c(AB) = (cA)B = A(cB)$ — Associative Property of Scalar Multiplication

Definition of Identity Matrix

The $n \times n$ matrix that consists of 1's on its main diagonal and 0's elsewhere is called the **identity matrix of dimension** $\boldsymbol{n \times n}$ and is denoted by

$$I_n = \begin{bmatrix} 1 & 0 & 0 & \cdots & 0 \\ 0 & 1 & 0 & \cdots & 0 \\ 0 & 0 & 1 & \cdots & 0 \\ \vdots & \vdots & \vdots & & \vdots \\ 0 & 0 & 0 & \cdots & 1 \end{bmatrix}. \qquad \text{Identity matrix}$$

Note that an identity matrix must be *square*. When the dimension is understood to be $n \times n$, you can denote I_n simply by I.

If A is an $n \times n$ matrix, then the identity matrix has the property that $AI_n = A$ and $I_nA = A$. For example,

$$\begin{bmatrix} 3 & -2 & 5 \\ 1 & 0 & 4 \\ -1 & 2 & -3 \end{bmatrix}\begin{bmatrix} 1 & 0 & 0 \\ 0 & 1 & 0 \\ 0 & 0 & 1 \end{bmatrix} = \begin{bmatrix} 3 & -2 & 5 \\ 1 & 0 & 4 \\ -1 & 2 & -3 \end{bmatrix} \qquad AI = A$$

and

$$\begin{bmatrix} 1 & 0 & 0 \\ 0 & 1 & 0 \\ 0 & 0 & 1 \end{bmatrix}\begin{bmatrix} 3 & -2 & 5 \\ 1 & 0 & 4 \\ -1 & 2 & -3 \end{bmatrix} = \begin{bmatrix} 3 & -2 & 5 \\ 1 & 0 & 4 \\ -1 & 2 & -3 \end{bmatrix}. \qquad IA = A$$

Applications

Matrix multiplication can be used to represent a system of linear equations. Note how the system

$$\begin{cases} a_{11}x_1 + a_{12}x_2 + a_{13}x_3 = b_1 \\ a_{21}x_1 + a_{22}x_2 + a_{23}x_3 = b_2 \\ a_{31}x_1 + a_{32}x_2 + a_{33}x_3 = b_3 \end{cases}$$

can be written as the matrix equation $AX = B$, where A is the *coefficient matrix* of the system, and X and B are column matrices. The column matrix B is also called a *constant matrix*. Its entries are the constant terms in the system of equations.

$$\underset{A}{\begin{bmatrix} a_{11} & a_{12} & a_{13} \\ a_{21} & a_{22} & a_{23} \\ a_{31} & a_{32} & a_{33} \end{bmatrix}} \times \underset{X}{\begin{bmatrix} x_1 \\ x_2 \\ x_3 \end{bmatrix}} = \underset{B}{\begin{bmatrix} b_1 \\ b_2 \\ b_3 \end{bmatrix}}$$

EXAMPLE 10 Solving a System of Linear Equations

For the system of linear equations, (a) write the system as a matrix equation $AX = B$ and (b) use Gauss-Jordan elimination on $[A \vdots B]$ to solve for the matrix X.

$$\begin{cases} x_1 - 2x_2 + x_3 = -4 \\ x_2 + 2x_3 = 4 \\ 2x_1 + 3x_2 - 2x_3 = 2 \end{cases}$$

Solution

a. In matrix form $AX = B$, the system can be written as follows.

$$\begin{bmatrix} 1 & -2 & 1 \\ 0 & 1 & 2 \\ 2 & 3 & -2 \end{bmatrix} \begin{bmatrix} x_1 \\ x_2 \\ x_3 \end{bmatrix} = \begin{bmatrix} -4 \\ 4 \\ 2 \end{bmatrix}$$

b. The augmented matrix is

$$[A \vdots B] = \left[\begin{array}{ccc|c} 1 & -2 & 1 & -4 \\ 0 & 1 & 2 & 4 \\ 2 & 3 & -2 & 2 \end{array}\right].$$

Using Gauss-Jordan elimination, you can rewrite this equation as

$$[I \vdots X] = \left[\begin{array}{ccc|c} 1 & 0 & 0 & -1 \\ 0 & 1 & 0 & 2 \\ 0 & 0 & 1 & 1 \end{array}\right].$$

So, the solution of the system of linear equations is $x_1 = -1$, $x_2 = 2$, and $x_3 = 1$. The solution of the matrix equation is

$$X = \begin{bmatrix} x_1 \\ x_2 \\ x_3 \end{bmatrix} = \begin{bmatrix} -1 \\ 2 \\ 1 \end{bmatrix}.$$

Technology Tip

Most graphing utilities can be used to obtain the reduced row-echelon form of a matrix. The screen below shows how one graphing utility displays the reduced row-echelon form of the augmented matrix in Example 10.

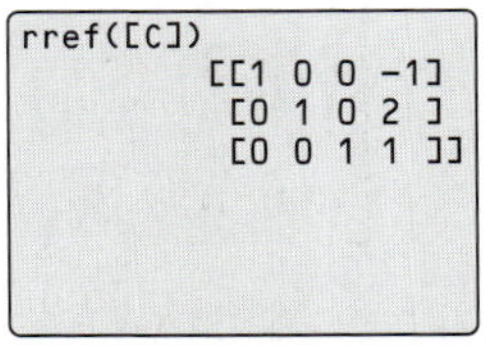

✓ ***Checkpoint*** *Audio-video solution in English & Spanish at LarsonPrecalculus.com.*

For the system of linear equations, (a) write the system as a matrix equation $AX = B$ and (b) use Gauss-Jordan elimination on $[A \vdots B]$ to solve for the matrix X.

$$\begin{cases} -2x_1 - 3x_2 = -4 \\ 6x_1 + x_2 = -36 \end{cases}$$

EXAMPLE 11 Softball Team Expenses

Two softball teams submit equipment lists to their sponsors, as shown in the table.

Equipment	Women's team	Men's team
Bats	12	15
Balls	45	38
Gloves	15	17

The equipment costs are as follows.

Bats: \$90 per bat

Balls: \$6 per ball

Gloves: \$60 per glove

Use matrices to find the total cost of equipment for each team.

Solution

The equipment list E can be written in matrix form as

$$E = \begin{bmatrix} 12 & 15 \\ 45 & 38 \\ 15 & 17 \end{bmatrix}.$$

The costs per item C can be written in matrix form as

$$C = [90 \quad 6 \quad 60].$$

You can find the total cost of the equipment for each team using the product CE because the number of columns of C (3 columns) equals the number of rows of E (3 rows). Therefore, the total cost of equipment for each team is given by

$$\begin{aligned} CE &= [90 \quad 6 \quad 60]\begin{bmatrix} 12 & 15 \\ 45 & 38 \\ 15 & 17 \end{bmatrix} \\ &= [90(12) + 6(45) + 60(15) \quad 90(15) + 6(38) + 60(17)] \\ &= [2250 \quad 2598]. \end{aligned}$$

Remark

Notice in Example 11 that you cannot find the total cost using the product EC because EC is not defined. That is, the number of columns of E (2 columns) does not equal the number of rows of C (1 row).

So, the total cost of equipment for the women's team is \$2250, and the total cost of equipment for the men's team is \$2598. You can use a graphing utility to check this result, as shown in Figure 5.18.

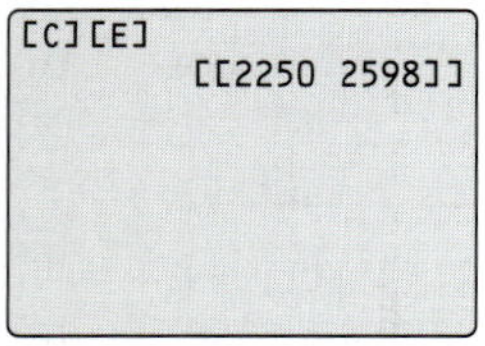

Figure 5.18

✓ Checkpoint Audio-video solution in English & Spanish at LarsonPrecalculus.com.

Repeat Example 11 when each bat costs \$100, each ball costs \$7, and each glove costs \$65.

5.5 Exercises

See *CalcChat.com* for tutorial help and worked-out solutions to odd-numbered exercises.
For instructions on how to use a graphing utility, see Appendix A.

Vocabulary and Concept Check

In Exercises 1–4, fill in the blank(s).

1. Two matrices are _______ when they have the same dimension and all of their corresponding entries are equal.

2. When working with matrices, real numbers are often referred to as _______ .

3. A matrix consisting entirely of zeros is called a _______ matrix and is denoted by _______ .

4. The $n \times n$ matrix consisting of 1's on its main diagonal and 0's elsewhere is called the _______ matrix of dimension n.

In Exercises 5 and 6, match the matrix property with the correct form. *A*, *B*, and *C* are matrices, and *c* and *d* are scalars.

5. (a) $(cd)A = c(dA)$ (i) Commutative Property of Matrix Addition
(b) $A + B = B + A$ (ii) Associative Property of Matrix Addition
(c) $1A = A$ (iii) Associative Property of Scalar Multiplication
(d) $c(A + B) = cA + cB$ (iv) Scalar Identity
(e) $A + (B + C) = (A + B) + C$ (v) Distributive Property

6. (a) $A(B + C) = AB + AC$ (i) Associative Property of Matrix Multiplication
(b) $c(AB) = (cA)B = A(cB)$ (ii) Left Distributive Property
(c) $A(BC) = (AB)C$ (iii) Right Distributive Property
(d) $(A + B)C = AC + BC$ (iv) Associative Property of Scalar Multiplication

7. In general, when multiplying matrices A and B, does $AB = BA$?

8. What is the dimension of AB when A is a 2×3 matrix and B is a 3×4 matrix?

Procedures and Problem Solving

Equality of Matrices **In Exercises 9–12, find *x* and *y* or *x*, *y*, and *z*.**

9. $\begin{bmatrix} x & -7 \\ 9 & y \end{bmatrix} = \begin{bmatrix} 5 & -7 \\ 9 & -8 \end{bmatrix}$

10. $\begin{bmatrix} -5 & x \\ y & 8 \end{bmatrix} = \begin{bmatrix} -5 & 13 \\ 12 & 8 \end{bmatrix}$

11. $\begin{bmatrix} 4 & 5 & 4 \\ 13 & 15 & 3y \\ 2 & 2z-6 & 0 \end{bmatrix} = \begin{bmatrix} 4 & 2x+7 & 4 \\ 13 & 15 & 12 \\ 2 & 3z-14 & 0 \end{bmatrix}$

12. $\begin{bmatrix} x+4 & 8 & -3 \\ 1 & 22 & 2y \\ 7 & -2 & z+2 \end{bmatrix} = \begin{bmatrix} 2x+9 & 8 & -3 \\ 1 & 22 & -8 \\ 7 & -2 & 11 \end{bmatrix}$

Operations with Matrices **In Exercises 13–20, find, if possible, (a) *A* + *B*, (b) *A* − *B*, (c) 3*A*, and (d) 3*A* − 2*B*. Use the matrix capabilities of a graphing utility to verify your results.**

13. $A = \begin{bmatrix} 5 & -2 \\ 3 & 1 \end{bmatrix}$, $B = \begin{bmatrix} 3 & 1 \\ -2 & 6 \end{bmatrix}$

14. $A = \begin{bmatrix} 1 & 2 \\ 2 & 1 \end{bmatrix}$, $B = \begin{bmatrix} -3 & -2 \\ 4 & 2 \end{bmatrix}$

15. $A = \begin{bmatrix} 8 & -1 \\ 2 & 3 \\ -4 & 5 \end{bmatrix}$, $B = \begin{bmatrix} 1 & 6 \\ -1 & -5 \\ 1 & 10 \end{bmatrix}$

16. $A = \begin{bmatrix} 1 & -1 & 3 \\ 0 & 6 & 9 \end{bmatrix}$, $B = \begin{bmatrix} -2 & 0 & -5 \\ -3 & 4 & -7 \end{bmatrix}$

17. $A = \begin{bmatrix} 4 & 5 & -1 & 3 & 4 \\ 1 & 2 & -2 & -1 & 0 \end{bmatrix}$,
$B = \begin{bmatrix} 1 & 0 & -1 & 1 & 0 \\ -6 & 8 & 2 & -3 & -7 \end{bmatrix}$

18. $A = \begin{bmatrix} -1 & 4 & 0 \\ 3 & -2 & 2 \\ 5 & 4 & -1 \\ 0 & 8 & -6 \\ -4 & -1 & 0 \end{bmatrix}$, $B = \begin{bmatrix} -3 & 5 & 1 \\ 2 & -4 & -7 \\ 10 & -9 & -1 \\ 3 & 2 & -4 \\ 0 & 1 & -2 \end{bmatrix}$

19. $A = \begin{bmatrix} 6 & 0 & 3 \\ -1 & -4 & 0 \end{bmatrix}$, $B = \begin{bmatrix} 8 & -1 \\ 4 & -3 \end{bmatrix}$

20. $A = \begin{bmatrix} 3 \\ 2 \\ -1 \end{bmatrix}$, $B = \begin{bmatrix} -4 & 6 & 2 \end{bmatrix}$

Evaluating an Expression In Exercises 21–24, evaluate the expression.

21. $\begin{bmatrix} -5 & 0 \\ 3 & -6 \end{bmatrix} + \begin{bmatrix} 7 & 1 \\ -2 & -1 \end{bmatrix} + \begin{bmatrix} -10 & -8 \\ 14 & 6 \end{bmatrix}$

22. $\begin{bmatrix} 6 & 9 \\ -1 & 0 \\ 7 & 1 \end{bmatrix} + \begin{bmatrix} 0 & 5 \\ -2 & -1 \\ 3 & -6 \end{bmatrix} + \begin{bmatrix} -13 & -7 \\ 4 & -1 \\ -6 & 0 \end{bmatrix}$

23. $\frac{1}{3}\left(\begin{bmatrix} -4 & 0 & 1 \\ 0 & 2 & -12 \end{bmatrix} - \begin{bmatrix} 5 & 1 & -2 \\ 12 & -6 & 3 \end{bmatrix}\right)$

24. $\frac{1}{2}([3 \quad -2 \quad 4 \quad 0] - [10 \quad -6 \quad -18 \quad 9])$

Operations with Matrices In Exercises 25–28, use the matrix capabilities of a graphing utility to evaluate the expression. Round your results to the nearest thousandths, if necessary.

25. $\frac{3}{7}\begin{bmatrix} 2 & 5 \\ -1 & -4 \end{bmatrix} + 6\begin{bmatrix} -3 & 0 \\ 2 & 2 \end{bmatrix}$

26. $\frac{4}{5}\begin{bmatrix} 14 & -11 \\ -22 & 19 \end{bmatrix} + 7\begin{bmatrix} -22 & 20 \\ 13 & 6 \end{bmatrix}$

27. $-5\begin{bmatrix} 3.211 & 6.829 \\ -1.004 & 4.914 \\ 0.055 & -3.889 \end{bmatrix} - \frac{1}{4}\begin{bmatrix} 1.630 & -3.090 \\ 5.256 & 8.335 \\ -9.768 & 4.251 \end{bmatrix}$

28. $-3\begin{bmatrix} 10 & 15 \\ -20 & 10 \\ 12 & 4 \end{bmatrix} - \frac{1}{8}\left(\begin{bmatrix} 12 & 11 \\ 7 & 0 \\ 6 & 9 \end{bmatrix} + \begin{bmatrix} -3 & 13 \\ -3 & 8 \\ -14 & 15 \end{bmatrix}\right)$

Solving a Matrix Equation In Exercises 29–32, solve for X when

$$A = \begin{bmatrix} -2 & -1 \\ 1 & 0 \\ 3 & -4 \end{bmatrix} \text{ and } B = \begin{bmatrix} 0 & 3 \\ 2 & 0 \\ -4 & -1 \end{bmatrix}.$$

29. $3X = A + 3B$

30. $2X = 2A + B$

31. $2A + 4B = -2X$

32. $-3X - 3A = 9B$

Finding the Product of Two Matrices In Exercises 33–40, find AB, if possible.

33. $A = \begin{bmatrix} 3 & -1 \\ 4 & -5 \\ 2 & 6 \end{bmatrix}, \quad B = \begin{bmatrix} 6 & 0 \\ 7 & -1 \end{bmatrix}$

34. $A = \begin{bmatrix} -1 & 6 \\ -4 & 5 \\ 0 & 3 \end{bmatrix}, \quad B = \begin{bmatrix} 2 & 3 \\ 0 & 9 \end{bmatrix}$

35. $A = \begin{bmatrix} 2 & 1 \\ -3 & 4 \\ -1 & 6 \end{bmatrix}, \quad B = \begin{bmatrix} 0 & -3 & 0 \\ 4 & 0 & 2 \\ 8 & -2 & 7 \end{bmatrix}$

36. $A = \begin{bmatrix} 1 & 0 & 3 & -2 \\ 6 & 13 & 8 & -17 \end{bmatrix}, \quad B = \begin{bmatrix} 1 & 6 \\ 4 & 2 \end{bmatrix}$

37. $A = \begin{bmatrix} 6 & 0 & 0 \\ 0 & 4 & 0 \\ 0 & 0 & -2 \end{bmatrix}, \quad B = \begin{bmatrix} \frac{1}{3} & 0 & 0 \\ 0 & -\frac{1}{4} & 0 \\ 0 & 0 & \frac{1}{6} \end{bmatrix}$

38. $A = \begin{bmatrix} 5 & 0 & 0 \\ 0 & -8 & 0 \\ 0 & 0 & 7 \end{bmatrix}, \quad B = \begin{bmatrix} \frac{1}{5} & 0 & 0 \\ 0 & -\frac{1}{8} & 0 \\ 0 & 0 & \frac{1}{2} \end{bmatrix}$

39. $A = \begin{bmatrix} 5 \\ -3 \\ 4 \end{bmatrix}, \quad B = [2 \quad -8 \quad 4]$

40. $A = \begin{bmatrix} 5 \\ 6 \end{bmatrix}, \quad B = [-3 \quad -1 \quad -5 \quad -9]$

Operations with Matrices In Exercises 41–46, find, if possible, (a) AB, (b) BA, and (c) A^2. (*Note:* $A^2 = AA$.) Use the matrix capabilities of a graphing utility to verify your results.

41. $A = \begin{bmatrix} 1 & 2 \\ 4 & 2 \end{bmatrix}, \quad B = \begin{bmatrix} 2 & -1 \\ -1 & 8 \end{bmatrix}$

42. $A = \begin{bmatrix} 6 & 3 \\ -2 & -4 \end{bmatrix}, \quad B = \begin{bmatrix} -2 & 0 \\ 2 & 4 \end{bmatrix}$

43. $A = \begin{bmatrix} 3 & -1 \\ 1 & 3 \end{bmatrix}, \quad B = \begin{bmatrix} 1 & -3 \\ 3 & 1 \end{bmatrix}$

44. $A = \begin{bmatrix} 1 & -1 \\ 1 & 1 \end{bmatrix}, \quad B = \begin{bmatrix} 1 & 3 \\ -3 & 1 \end{bmatrix}$

45. $A = \begin{bmatrix} 7 \\ 8 \\ -1 \end{bmatrix}, \quad B = [1 \quad 1 \quad 2]$

46. $A = [3 \quad 2 \quad 1], \quad B = \begin{bmatrix} 2 \\ 3 \\ 0 \end{bmatrix}$

Matrix Multiplication In Exercises 47–50, use the matrix capabilities of a graphing utility to find AB, if possible.

47. $A = \begin{bmatrix} 1 & -12 & 4 \\ 14 & 10 & 12 \\ 6 & -15 & 3 \end{bmatrix}, \quad B = \begin{bmatrix} 12 & 10 \\ -6 & 12 \\ 10 & 16 \end{bmatrix}$

48. $A = \begin{bmatrix} -3 & 8 & -6 & 8 \\ -12 & 15 & 9 & 6 \\ 5 & -1 & 1 & 5 \end{bmatrix}, \quad B = \begin{bmatrix} 3 & 1 & 6 \\ 24 & 15 & 14 \\ 16 & 10 & 21 \\ 8 & -4 & 10 \end{bmatrix}$

49. $A = \begin{bmatrix} 7 & 6 & 9 & -4 \\ 3 & -4 & 11 & -2 \\ -5 & -8 & 1 & 12 \end{bmatrix}, \quad B = \begin{bmatrix} 15 & 8 \\ 23 & -17 \\ 9 & 10 \end{bmatrix}$

50. $A = \begin{bmatrix} -2 & 6 & 12 \\ 21 & -5 & 6 \\ 13 & -2 & 9 \end{bmatrix}, \quad B = \begin{bmatrix} 3 & 0 \\ -7 & 18 \\ 34 & 14 \\ 0.5 & 1.4 \end{bmatrix}$

Operations with Matrices **In Exercises 51–54, use the matrix capabilities of a graphing utility to evaluate the expression.**

51. $\begin{bmatrix} 3 & 1 \\ 0 & -2 \end{bmatrix}\begin{bmatrix} 1 & 0 \\ -2 & 2 \end{bmatrix}\begin{bmatrix} 1 & 0 \\ 2 & 4 \end{bmatrix}$

52. $\begin{bmatrix} 6 & 5 & -1 \\ 1 & -2 & 0 \end{bmatrix}\begin{bmatrix} 0 & 3 \\ -1 & -3 \\ 4 & 1 \end{bmatrix}\begin{bmatrix} -2 & 2 \\ 0 & -1 \end{bmatrix}$

53. $\begin{bmatrix} 0 & 2 & -2 \\ 4 & 1 & 2 \end{bmatrix}\left(\begin{bmatrix} 4 & 0 \\ 0 & -1 \\ -1 & 2 \end{bmatrix} + \begin{bmatrix} -2 & 3 \\ -3 & 5 \\ 0 & -3 \end{bmatrix}\right)$

54. $\begin{bmatrix} 3 \\ -1 \\ 5 \\ 7 \end{bmatrix}([5 \quad -6] + [7 \quad -1] + [-8 \quad 9])$

Matrix Multiplication **In Exercises 55–58, use matrix multiplication to determine whether each matrix is a solution of the system of equations. Use a graphing utility to verify your results.**

55. $\begin{cases} x + 2y = 4 \\ 3x + 2y = 0 \end{cases}$

(a) $\begin{bmatrix} 2 \\ 1 \end{bmatrix}$ (b) $\begin{bmatrix} -2 \\ 3 \end{bmatrix}$ (c) $\begin{bmatrix} -4 \\ 4 \end{bmatrix}$ (d) $\begin{bmatrix} 2 \\ -3 \end{bmatrix}$

56. $\begin{cases} 6x + 2y = 0 \\ -x + 5y = 16 \end{cases}$

(a) $\begin{bmatrix} -1 \\ 3 \end{bmatrix}$ (b) $\begin{bmatrix} 2 \\ -6 \end{bmatrix}$ (c) $\begin{bmatrix} 3 \\ -9 \end{bmatrix}$ (d) $\begin{bmatrix} -3 \\ 9 \end{bmatrix}$

57. $\begin{cases} -2x - 3y = -6 \\ 4x + 2y = 20 \end{cases}$

(a) $\begin{bmatrix} 3 \\ 0 \end{bmatrix}$ (b) $\begin{bmatrix} 4 \\ 2 \end{bmatrix}$ (c) $\begin{bmatrix} -6 \\ 6 \end{bmatrix}$ (d) $\begin{bmatrix} 6 \\ -2 \end{bmatrix}$

58. $\begin{cases} 5x - 7y = -15 \\ 3x + y = 17 \end{cases}$

(a) $\begin{bmatrix} -4 \\ -5 \end{bmatrix}$ (b) $\begin{bmatrix} 5 \\ 2 \end{bmatrix}$ (c) $\begin{bmatrix} 4 \\ 5 \end{bmatrix}$ (d) $\begin{bmatrix} 2 \\ 11 \end{bmatrix}$

Solving a System of Linear Equations **In Exercises 59–66, (a) write the system of equations as a matrix equation $AX = B$ and (b) use Gauss-Jordan elimination on the augmented matrix $[A \vdots B]$ to solve for the matrix X. Use a graphing utility to check your solution.**

59. $\begin{cases} -x_1 + x_2 = 4 \\ -2x_1 + x_2 = 0 \end{cases}$

60. $\begin{cases} 2x_1 + 3x_2 = 5 \\ x_1 + 4x_2 = 10 \end{cases}$

61. $\begin{cases} -2x_1 - 3x_2 = -4 \\ 6x_1 + x_2 = -36 \end{cases}$

62. $\begin{cases} -4x_1 + 9x_2 = -13 \\ x_1 - 3x_2 = 12 \end{cases}$

63. $\begin{cases} x_1 - 2x_2 + 3x_3 = 9 \\ -x_1 + 3x_2 - x_3 = -6 \\ 2x_1 - 5x_2 + 5x_3 = 17 \end{cases}$

64. $\begin{cases} x_1 + x_2 - 3x_3 = 9 \\ -x_1 + 2x_2 = 6 \\ x_1 - x_2 + x_3 = -5 \end{cases}$

65. $\begin{cases} x_1 - 5x_2 + 2x_3 = -20 \\ -3x_1 + x_2 - x_3 = 8 \\ -2x_2 + 5x_3 = -16 \end{cases}$

66. $\begin{cases} x_1 - x_2 + 4x_3 = 17 \\ x_1 + 3x_2 = -11 \\ -6x_2 + 5x_3 = 40 \end{cases}$

Operations with Matrices **In Exercises 67–72, use a graphing utility to perform the operations for the matrices A, B, and C and the scalar c. Write a brief statement comparing the results of parts (a) and (b).**

$$A = \begin{bmatrix} 1 & 2 & -1 \\ 0 & -2 & 3 \\ 4 & -3 & 2 \end{bmatrix}, \quad B = \begin{bmatrix} 2 & 3 & 0 \\ 4 & 1 & -2 \\ -1 & 2 & 0 \end{bmatrix},$$

$$C = \begin{bmatrix} 3 & -2 & 1 \\ -4 & 0 & 3 \\ -1 & 3 & -2 \end{bmatrix}, \quad \text{and} \quad c = 3$$

67. (a) $A(B + C)$ (b) $AB + AC$

68. (a) $(B + C)A$ (b) $BA + CA$

69. (a) $A(BC)$ (b) $(AB)C$

70. (a) $c(AB)$ (b) $(cA)B$

71. (a) $(A + B)^2$ (b) $A^2 + AB + BA + B^2$

72. (a) $(A - B)^2$ (b) $A^2 - AB - BA + B^2$

Operations with Matrices **In Exercises 73–80, perform the operations (a) using a graphing utility and (b) by hand algebraically. If it is not possible to perform the operation(s), state the reason.**

$$A = \begin{bmatrix} 1 & 2 & -2 \\ -1 & 1 & 0 \end{bmatrix}, \quad B = \begin{bmatrix} -1 & 4 & -1 \\ -2 & -1 & 0 \end{bmatrix},$$

$$C = \begin{bmatrix} 1 & 2 \\ -2 & 3 \\ 1 & 0 \end{bmatrix}, \quad c = -2, \quad \text{and} \quad d = -3$$

73. $A + cB$

74. $B + dA$

75. $c(AB)$

76. $A(B + C)$

77. $CA - BC$

78. dAB^2

79. $cdA + B$

80. $cA + dB$

Operations with Matrices **In Exercises 81 and 82, use a graphing utility to perform the indicated operations.**

$$A = \begin{bmatrix} 2 & 0 \\ 4 & 5 \end{bmatrix} \quad \text{and} \quad B = \begin{bmatrix} 5 & 3 \\ 1 & 4 \end{bmatrix}$$

81. $A^2 - 5A + 2I_2$

82. $B^2 - 7B + 6I_2$

83. Manufacturing A corporation that makes sunglasses has four factories, each of which manufactures standard and designer sunglasses. The production levels are represented by A.

Factory: 1, 2, 3, 4 (columns); Sunglass type: Standard, Designer (rows)

$$A = \begin{bmatrix} 100 & 120 & 60 & 40 \\ 140 & 160 & 200 & 80 \end{bmatrix}$$

Find the production levels when production is decreased by 15%.

84. Manufacturing A corporation has four factories, each of which manufactures sport utility vehicles and pickup trucks. The production levels are represented by A.

Factory: 1, 2, 3, 4 (columns); Vehicle type: SUV, Pickup (rows)

$$A = \begin{bmatrix} 100 & 90 & 70 & 30 \\ 40 & 20 & 60 & 60 \end{bmatrix}$$

Find the production levels when production is decreased by 10%.

85. Manufacturing A corporation has three factories, each of which manufactures acoustic guitars and electric guitars. The production levels are represented by A.

Factory: A, B, C (columns); Guitar type: Acoustic, Electric (rows)

$$A = \begin{bmatrix} 70 & 50 & 25 \\ 35 & 100 & 70 \end{bmatrix}$$

Find the production levels when production is increased by 20%.

86. Tourism A vacation service has identified four resort hotels with a special all-inclusive package (room and meals included) at a popular travel destination. The quoted room rates are for double and family (maximum of four people) occupancy for 5 days and 4 nights. The current rates for the two types of rooms at the four hotels are represented by matrix A.

Hotel: w, x, y, z (columns); Occupancy: Double, Family (rows)

$$A = \begin{bmatrix} 668 & 2666 & 973 & 918 \\ 1214 & 4637 & 1656 & 1699 \end{bmatrix}$$

Room rates are guaranteed not to increase by more than 12%. What is the maximum rate per package per hotel during the next year?

87. Agriculture A fruit grower raises two crops, apples and peaches. Each of these crops is shipped to three different outlets. The shipment levels are represented by A.

Outlet: 1, 2, 3 (columns); Crop: Apples, Peaches (rows)

$$A = \begin{bmatrix} 125 & 100 & 75 \\ 100 & 175 & 125 \end{bmatrix}$$

(a) The grower earns a \$3.50 profit per unit of apples and a \$6.00 profit per unit of peaches. Organize the profit per unit of each crop in a matrix B.

(b) Find the product BA and interpret the result.

88. Physical Education The numbers of calories burned per hour by individuals of different weights performing different types of aerobic exercises are shown in the matrix.

130-lb person, 155-lb person (columns); Bicycling, Jogging, Swimming (rows)

$$B = \begin{bmatrix} 325 & 387 \\ 236 & 281 \\ 354 & 422 \end{bmatrix}$$

(a) A 130-pound person and a 155-pound person bicycle for 30 minutes, jog for 15 minutes, and swim for 60 minutes. Organize the time spent exercising in a matrix A.

(b) Find the product AB and interpret the result.

89. *Why you should learn it* *(p. 446)* A company manufactures boats. Its labor-hour and wage requirements are represented by L and W, respectively. Compute LW and interpret the result.

Department: Cutting, Assembly, Packaging (columns); Boat size: Small, Medium, Large (rows)

$$L = \begin{bmatrix} 1.0\text{ h} & 0.5\text{ h} & 0.2\text{ h} \\ 1.6\text{ h} & 1.0\text{ h} & 0.2\text{ h} \\ 2.5\text{ h} & 2.0\text{ h} & 1.4\text{ h} \end{bmatrix}$$

Plant: A, B (columns); Department: Cutting, Assembly, Packaging (rows)

$$W = \begin{bmatrix} \$15 & \$13 \\ \$12 & \$11 \\ \$11 & \$10 \end{bmatrix}$$

90. Inventory Control A company sells five models of computers through three retail outlets. The inventories are represented by S. The wholesale and retail prices are represented by T. Compute ST and interpret the result.

Model: A, B, C, D, E (columns); Outlet: 1, 2, 3 (rows)

$$S = \begin{bmatrix} 3 & 2 & 2 & 3 & 0 \\ 0 & 2 & 3 & 4 & 3 \\ 4 & 2 & 1 & 3 & 2 \end{bmatrix}$$

Price: Wholesale, Retail (columns); Model: A, B, C, D, E (rows)

$$T = \begin{bmatrix} \$840 & \$1100 \\ \$1200 & \$1350 \\ \$1450 & \$1650 \\ \$2650 & \$3000 \\ \$3050 & \$3200 \end{bmatrix}$$

91. Politics The matrix

$$P = \begin{array}{c} \text{From} \\ \begin{array}{ccc} R & D & I \end{array} \\ \begin{bmatrix} 0.6 & 0.1 & 0.1 \\ 0.2 & 0.7 & 0.1 \\ 0.2 & 0.2 & 0.8 \end{bmatrix} \end{array} \begin{array}{c} 1 \\ 2 \\ 3 \end{array} \Big\} \text{To}$$

is called a *stochastic matrix*. Each entry $p_{ij} (i \neq j)$ represents the proportion of the voting population that changes from party i to party j, and p_{ii} represents the proportion that remains loyal to the party from one election to the next. Compute and interpret P^2.

92. Politics Use a graphing utility to find P^3, P^4, P^5, P^6, P^7, and P^8 for the matrix given in Exercise 91. Can you detect a pattern as P is raised to higher powers?

Conclusions

True or False? In Exercises 93 and 94, determine whether the statement is true or false. Justify your answer.

93. Two matrices can be added only when they have the same dimension.

94. $\begin{bmatrix} -6 & -2 \\ 2 & -6 \end{bmatrix}\begin{bmatrix} 4 & 0 \\ 0 & -1 \end{bmatrix} = \begin{bmatrix} 4 & 0 \\ 0 & -1 \end{bmatrix}\begin{bmatrix} -6 & -2 \\ 2 & -6 \end{bmatrix}$

Think About It In Exercises 95–102, let matrices A, B, C, and D be of dimensions 2×3, 2×3, 3×2, and 2×2, respectively. Determine whether the matrices are of proper dimension to perform the operation(s). If so, give the dimension of the answer.

95. $A + 2C$

96. $B - 3C$

97. CD

98. BC

99. $CA - D$

100. $CB - D$

101. $D(A - 3B)$

102. $C(A + 2B)$

Think About It In Exercises 103–106, use the matrices

$$A = \begin{bmatrix} 2 & -1 \\ 1 & 3 \end{bmatrix} \text{ and } B = \begin{bmatrix} -1 & 1 \\ 0 & -2 \end{bmatrix}.$$

103. Show that $(A + B)^2 \neq A^2 + 2AB + B^2$.

104. Show that $(A - B)^2 \neq A^2 - 2AB + B^2$.

105. Show that $(A + B)(A - B) \neq A^2 - B^2$.

106. Show that $(A + B)^2 = A^2 + AB + BA + B^2$.

107. Think About It If a, b, and c are real numbers such that $c \neq 0$ and $ac = bc$, then $a = b$. However, if A, B, and C are nonzero matrices such that $AC = BC$, then A is *not necessarily* equal to B. Illustrate this using the following matrices.

$$A = \begin{bmatrix} 0 & 1 \\ 0 & 1 \end{bmatrix}, \quad B = \begin{bmatrix} 1 & 0 \\ 1 & 0 \end{bmatrix}, \quad C = \begin{bmatrix} 2 & 3 \\ 2 & 3 \end{bmatrix}$$

108. Think About It If a and b are real numbers such that $ab = 0$, then $a = 0$ or $b = 0$. However, if A and B are matrices such that $AB = O$, it is *not necessarily* true that $A = O$ or $B = O$. Illustrate this using the following matrices.

$$A = \begin{bmatrix} 3 & 3 \\ 4 & 4 \end{bmatrix}, \quad B = \begin{bmatrix} 1 & -1 \\ -1 & 1 \end{bmatrix}$$

109. Exploration Let $i = \sqrt{-1}$ and let

$$A = \begin{bmatrix} i & 0 \\ 0 & i \end{bmatrix} \text{ and } B = \begin{bmatrix} 0 & -i \\ i & 0 \end{bmatrix}.$$

(a) Find A^2, A^3, and A^4. Identify any similarities with i^2, i^3, and i^4.

(b) Find and identify B^2.

110. Exploration Let A and B be unequal diagonal matrices of the same dimension. (A **diagonal matrix** is a square matrix in which each entry not on the main diagonal is zero.) Determine the products AB for several pairs of such matrices. Make a conjecture about a quick rule for such products.

111. Think About It Let matrices A and B be of dimensions 3×2 and 2×2, respectively. Answer the following questions and explain your reasoning.

(a) Is it possible that $A = B$?

(b) Is $A + B$ defined?

(c) Is AB defined?

112. HOW DO YOU SEE IT? An electronics manufacturer produces two models of LCD televisions, which are shipped to three warehouses. The shipment levels are represented by A.

$$A = \begin{array}{c} \text{Warehouse} \\ \begin{array}{ccc} 1 & 2 & 3 \end{array} \\ \begin{bmatrix} 5{,}000 & 4{,}000 & 8{,}000 \\ 6{,}000 & 10{,}000 & 5{,}000 \end{bmatrix} \end{array} \begin{array}{c} A \\ B \end{array} \Big\} \text{Model}$$

(a) Interpret the value of a_{21}.

(b) How could you find the shipment levels when the shipments are increased by 15%?

(c) Each model A television sells for \$700 and each model B television sells for \$900. How could you use matrices to find the cost of the two models of LCD televisions at each warehouse?

Cumulative Mixed Review

Condensing a Logarithmic Expression In Exercises 113 and 114, condense the expression to the logarithm of a single quantity.

113. $3 \ln 4 - \frac{1}{3} \ln(x^2 + 3)$

114. $2 \ln 7t^4 - \frac{3}{5} \ln t^5$

5.6 The Inverse of a Square Matrix

The Inverse of a Matrix

This section further develops the algebra of matrices. To begin, consider the real number equation $ax = b$. To solve this equation for x, multiply each side of the equation by a^{-1} (provided that $a \neq 0$).

$$ax = b$$
$$(a^{-1}a)x = a^{-1}b$$
$$(1)x = a^{-1}b$$
$$x = a^{-1}b$$

The number a^{-1} is called the *multiplicative inverse* of a because $a^{-1}a = 1$. The definition of the multiplicative **inverse of a matrix** is similar.

Definition of the Inverse of a Square Matrix

Let A be an $n \times n$ matrix and let I_n be the $n \times n$ identity matrix. If there exists a matrix A^{-1} such that

$$AA^{-1} = I_n = A^{-1}A$$

then A^{-1} is called the **inverse** of A. The symbol A^{-1} is read "A inverse."

What you should learn

- Verify that two matrices are inverses of each other.
- Use Gauss-Jordan elimination to find inverses of matrices.
- Use a formula to find inverses of 2×2 matrices.
- Use inverse matrices to solve systems of linear equations.

Why you should learn it

A system of equations can be solved using the inverse of the coefficient matrix. This method is particularly useful when the coefficients are the same for several systems, but the constants are different. Exercise 66 on page 468 shows how to use an inverse matrix to find a model for the number of international travelers to the United States from South America.

EXAMPLE 1 The Inverse of a Matrix

Show that B is the inverse of A, where

$$A = \begin{bmatrix} -1 & 2 \\ -1 & 1 \end{bmatrix} \quad \text{and} \quad B = \begin{bmatrix} 1 & -2 \\ 1 & -1 \end{bmatrix}.$$

Solution

To show that B is the inverse of A, show that $AB = I = BA$, as follows.

$$AB = \begin{bmatrix} -1 & 2 \\ -1 & 1 \end{bmatrix}\begin{bmatrix} 1 & -2 \\ 1 & -1 \end{bmatrix} = \begin{bmatrix} -1+2 & 2-2 \\ -1+1 & 2-1 \end{bmatrix} = \begin{bmatrix} 1 & 0 \\ 0 & 1 \end{bmatrix}$$

$$BA = \begin{bmatrix} 1 & -2 \\ 1 & -1 \end{bmatrix}\begin{bmatrix} -1 & 2 \\ -1 & 1 \end{bmatrix} = \begin{bmatrix} -1+2 & 2-2 \\ -1+1 & 2-1 \end{bmatrix} = \begin{bmatrix} 1 & 0 \\ 0 & 1 \end{bmatrix}$$

As you can see,

$$AB = I = BA.$$

This is an example of a square matrix that has an inverse. Note that not all square matrices have inverses.

✓ Checkpoint Audio-video solution in English & Spanish at LarsonPrecalculus.com.

Show that B is the inverse of A, where

$$A = \begin{bmatrix} 2 & -1 \\ -3 & 1 \end{bmatrix} \quad \text{and} \quad B = \begin{bmatrix} -1 & -1 \\ -3 & -2 \end{bmatrix}.$$

Recall that it is not always true that $AB = BA$, even when both products are defined. However, if A and B are both square matrices and $AB = I_n$, then it can be shown that $BA = I_n$. So, in Example 1, you need only check that $AB = I_2$.

Finding Inverse Matrices

When a matrix A has an inverse, A is called **invertible** (or **nonsingular**); otherwise, A is called **singular.** A nonsquare matrix cannot have an inverse. To see this, note that if A is of dimension $m \times n$ and B is of dimension $n \times m$ (where $m \neq n$), then the products AB and BA are of different dimensions and so cannot be equal to each other. Not all square matrices have inverses, as you will see later in this section. When a matrix does have an inverse, however, that inverse is unique. Example 2 shows how to use systems of equations to find the inverse of a matrix.

EXAMPLE 2 Finding the Inverse of a Matrix

See LarsonPrecalculus.com for an interactive version of this type of example.

Find the inverse of

$$A = \begin{bmatrix} 1 & 4 \\ -1 & -3 \end{bmatrix}.$$

Solution

To find the inverse of A, try to solve the matrix equation $AX = I$ for X.

$$\overset{A}{\begin{bmatrix} 1 & 4 \\ -1 & -3 \end{bmatrix}} \overset{X}{\begin{bmatrix} x_{11} & x_{12} \\ x_{21} & x_{22} \end{bmatrix}} = \overset{I}{\begin{bmatrix} 1 & 0 \\ 0 & 1 \end{bmatrix}}$$

$$\begin{bmatrix} x_{11} + 4x_{21} & x_{12} + 4x_{22} \\ -x_{11} - 3x_{21} & -x_{12} - 3x_{22} \end{bmatrix} = \begin{bmatrix} 1 & 0 \\ 0 & 1 \end{bmatrix}$$

Equating corresponding entries, you obtain the following two systems of linear equations.

$$\begin{cases} x_{11} + 4x_{21} = 1 \\ -x_{11} - 3x_{21} = 0 \end{cases}$$ Linear system with two variables, x_{11} and x_{21}.

$$\begin{cases} x_{12} + 4x_{22} = 0 \\ -x_{12} - 3x_{22} = 1 \end{cases}$$ Linear system with two variables, x_{12} and x_{22}.

Solve the first system using elementary row operations to determine that $x_{11} = -3$ and $x_{21} = 1$. From the second system you can determine that $x_{12} = -4$ and $x_{22} = 1$. Therefore, the inverse of A is

$$A^{-1} = X$$
$$= \begin{bmatrix} -3 & -4 \\ 1 & 1 \end{bmatrix}.$$

You can use matrix multiplication to check this result.

Check

$$AA^{-1} = \begin{bmatrix} 1 & 4 \\ -1 & -3 \end{bmatrix}\begin{bmatrix} -3 & -4 \\ 1 & 1 \end{bmatrix} = \begin{bmatrix} 1 & 0 \\ 0 & 1 \end{bmatrix} \checkmark$$

$$A^{-1}A = \begin{bmatrix} -3 & -4 \\ 1 & 1 \end{bmatrix}\begin{bmatrix} 1 & 4 \\ -1 & -3 \end{bmatrix} = \begin{bmatrix} 1 & 0 \\ 0 & 1 \end{bmatrix} \checkmark$$

✓ ***Checkpoint*** *Audio-video solution in English & Spanish at LarsonPrecalculus.com.*

Find the inverse of

$$A = \begin{bmatrix} 1 & -2 \\ -1 & 3 \end{bmatrix}.$$

Explore the Concept

Most graphing utilities are capable of finding the inverse of a square matrix. Try using a graphing utility to find the inverse of the matrix

$$A = \begin{bmatrix} 2 & -3 & 1 \\ -1 & 2 & -1 \\ -2 & 0 & 1 \end{bmatrix}.$$

After you find A^{-1}, store it as $[B]$ and use the graphing utility to find $[A] \times [B]$ and $[B] \times [A]$. What can you conclude?

In Example 2, note that the two systems of linear equations have the *same coefficient matrix* A. Rather than solve the two systems represented by

$$\left[\begin{array}{rr|r} 1 & 4 & 1 \\ -1 & -3 & 0 \end{array}\right]$$

and

$$\left[\begin{array}{rr|r} 1 & 4 & 0 \\ -1 & -3 & 1 \end{array}\right]$$

separately, you can solve them *simultaneously* by *adjoining* the identity matrix to the coefficient matrix to obtain

$$\overbrace{\left[\begin{array}{rr}1 & 4 \\ -1 & -3\end{array}\right.}^{A}\ \vdots\ \overbrace{\left.\begin{array}{rr}1 & 0 \\ 0 & 1\end{array}\right]}^{I}.$$

This "doubly augmented" matrix can be represented as

$[A \ \vdots\ I]$.

By applying Gauss-Jordan elimination to this matrix, you can solve *both* systems with a single elimination process.

$$\left[\begin{array}{rr|rr} 1 & 4 & 1 & 0 \\ -1 & -3 & 0 & 1 \end{array}\right]$$

$$R_1 + R_2 \rightarrow \left[\begin{array}{rr|rr} 1 & 4 & 1 & 0 \\ 0 & 1 & 1 & 1 \end{array}\right]$$

$$-4R_2 + R_1 \rightarrow \left[\begin{array}{rr|rr} 1 & 0 & -3 & -4 \\ 0 & 1 & 1 & 1 \end{array}\right]$$

So, from the "doubly augmented" matrix $[A \ \vdots\ I]$, you obtained the matrix $[I \ \vdots\ A^{-1}]$.

$$\overbrace{\left[\begin{array}{rr}1 & 4 \\ -1 & -3\end{array}\right.}^{A}\ \vdots\ \overbrace{\left.\begin{array}{rr}1 & 0 \\ 0 & 1\end{array}\right]}^{I} \Rightarrow \overbrace{\left[\begin{array}{rr}1 & 0 \\ 0 & 1\end{array}\right.}^{I}\ \vdots\ \overbrace{\left.\begin{array}{rr}-3 & -4 \\ 1 & 1\end{array}\right]}^{A^{-1}}$$

This procedure (or algorithm) works for any square matrix that has an inverse.

Finding an Inverse Matrix

Let A be a square matrix of dimension $n \times n$.

1. Write the $n \times 2n$ matrix that consists of the given matrix A on the left and the $n \times n$ identity matrix I on the right to obtain

   $[A \ \vdots\ I]$.

2. If possible, row reduce A to I using elementary row operations on the *entire* matrix

   $[A \ \vdots\ I]$.

   The result will be the matrix

   $[I \ \vdots\ A^{-1}]$.

   If this is not possible, then A is not invertible.

3. Check your work by multiplying to see that

   $AA^{-1} = I = A^{-1}A$.

Explore the Concept

Select two 2×2 matrices A and B that have inverses. Enter them into your graphing utility and calculate $(AB)^{-1}$. Then calculate $B^{-1}A^{-1}$ and $A^{-1}B^{-1}$. Make a conjecture about the inverse of the product of two invertible matrices.

EXAMPLE 3 Finding the Inverse of a Matrix

Find the inverse of $A = \begin{bmatrix} 1 & -1 & 0 \\ 1 & 0 & -1 \\ 6 & -2 & -3 \end{bmatrix}$.

Solution

Begin by adjoining the identity matrix to A to form the matrix

$$[A \ \vdots \ I] = \left[\begin{array}{rrr|rrr} 1 & -1 & 0 & 1 & 0 & 0 \\ 1 & 0 & -1 & 0 & 1 & 0 \\ 6 & -2 & -3 & 0 & 0 & 1 \end{array}\right].$$

Use elementary row operations to obtain the form $[I \ \vdots \ A^{-1}]$, as follows.

$$\left[\begin{array}{rrr|rrr} 1 & 0 & 0 & -2 & -3 & 1 \\ 0 & 1 & 0 & -3 & -3 & 1 \\ 0 & 0 & 1 & -2 & -4 & 1 \end{array}\right]$$

Therefore, the matrix A is invertible and its inverse is

$$A^{-1} = \begin{bmatrix} -2 & -3 & 1 \\ -3 & -3 & 1 \\ -2 & -4 & 1 \end{bmatrix}.$$

Confirm this result by multiplying A by A^{-1} to obtain I as follows.

Check

$$AA^{-1} = \begin{bmatrix} 1 & -1 & 0 \\ 1 & 0 & -1 \\ 6 & -2 & -3 \end{bmatrix}\begin{bmatrix} -2 & -3 & 1 \\ -3 & -3 & 1 \\ -2 & -4 & 1 \end{bmatrix} = \begin{bmatrix} 1 & 0 & 0 \\ 0 & 1 & 0 \\ 0 & 0 & 1 \end{bmatrix} = I$$

✓ **Checkpoint** *Audio-video solution in English & Spanish at LarsonPrecalculus.com.*

Find the inverse of $A = \begin{bmatrix} 1 & -2 & -1 \\ 0 & -1 & 2 \\ 1 & -2 & 0 \end{bmatrix}$.

The algorithm shown in Example 3 applies to any $n \times n$ matrix A. When using this algorithm, if the matrix A does not reduce to the identity matrix, then A does not have an inverse. For instance, the following matrix has no inverse.

$$A = \begin{bmatrix} 1 & 2 & 0 \\ 3 & -1 & 2 \\ -2 & 3 & -2 \end{bmatrix}$$

To see why matrix A above has no inverse, begin by adjoining the identity matrix to A to form

$$[A \ \vdots \ I] = \left[\begin{array}{rrr|rrr} 1 & 2 & 0 & 1 & 0 & 0 \\ 3 & -1 & 2 & 0 & 1 & 0 \\ -2 & 3 & -2 & 0 & 0 & 1 \end{array}\right].$$

Then, use elementary row operations to obtain

$$\left[\begin{array}{rrr|rrr} 1 & 2 & 0 & 1 & 0 & 0 \\ 0 & -7 & 2 & -3 & 1 & 0 \\ 0 & 0 & 0 & -1 & 1 & 1 \end{array}\right].$$

At this point in the elimination process, you can see that it is impossible to obtain the identity matrix I on the left. Therefore, A is not invertible.

Technology Tip

Most graphing utilities can find the inverse of a matrix by using the inverse key $\boxed{x^{-1}}$. For instructions on how to use the inverse key to find the inverse of a matrix, see Appendix A; for specific keystrokes, go to this textbook's *Companion Website*.

The Inverse of a 2 × 2 Matrix

Using Gauss-Jordan elimination to find the inverse of a matrix works well (even as a computer technique) for matrices of dimension 3×3 or greater. For 2×2 matrices, however, many people prefer to use a formula for the inverse rather than Gauss-Jordan elimination. This simple formula, which works *only* for 2×2 matrices, is explained as follows. If A is the 2×2 matrix given by

$$A = \begin{bmatrix} a & b \\ c & d \end{bmatrix}$$

then A is invertible if and only if

$$ad - bc \neq 0.$$

If $ad - bc \neq 0$, then the inverse is given by

$$A^{-1} = \frac{1}{ad - bc}\begin{bmatrix} d & -b \\ -c & a \end{bmatrix}.$$ Formula for the inverse of a 2×2 matrix

The denominator

$$ad - bc$$

is called the *determinant* of the 2×2 matrix A. You will study determinants in the next section.

EXAMPLE 4 Finding the Inverse of a 2 × 2 Matrix

If possible, find the inverse of each matrix.

a. $A = \begin{bmatrix} 3 & -1 \\ -2 & 2 \end{bmatrix}$

b. $B = \begin{bmatrix} 3 & -1 \\ -6 & 2 \end{bmatrix}$

Solution

a. For the matrix A, apply the formula for the inverse of a 2×2 matrix to obtain

$$\begin{aligned} ad - bc &= (3)(2) - (-1)(-2) \\ &= 4. \end{aligned}$$

Because this quantity is not zero, the inverse is formed by interchanging the entries on the main diagonal, changing the signs of the other two entries, and multiplying by the scalar $\frac{1}{4}$, as follows.

$$A^{-1} = \frac{1}{4}\begin{bmatrix} 2 & 1 \\ 2 & 3 \end{bmatrix}$$ Substitute for a, b, c, d, and the determinant.

$$= \begin{bmatrix} \frac{1}{2} & \frac{1}{4} \\ \frac{1}{2} & \frac{3}{4} \end{bmatrix}$$ Multiply by the scalar $\frac{1}{4}$.

b. For the matrix B, you have

$$\begin{aligned} ad - bc &= (3)(2) - (-1)(-6) \\ &= 0 \end{aligned}$$

which means that B is not invertible.

✓ **Checkpoint** *Audio-video solution in English & Spanish at LarsonPrecalculus.com.*

If possible, find the inverse of $A = \begin{bmatrix} 5 & -1 \\ 3 & 4 \end{bmatrix}$.

Explore the Concept

Use a graphing utility to find the inverse of the matrix

$$A = \begin{bmatrix} 1 & -3 \\ -2 & 6 \end{bmatrix}.$$

What message appears on the screen? Why does the graphing utility display this message?

Systems of Linear Equations

You know that a system of linear equations can have exactly one solution, infinitely many solutions, or no solution. If the coefficient matrix A of a *square* system (a system that has the same number of equations as variables) is invertible, then the system has a unique solution, which is defined as follows.

A System of Equations with a Unique Solution

If A is an invertible matrix, then the system of linear equations represented by $AX = B$ has a unique solution given by

$$X = A^{-1}B.$$

The formula $X = A^{-1}B$ is used on most graphing utilities to solve linear systems that have invertible coefficient matrices. That is, you enter the $n \times n$ coefficient matrix $[A]$ and the $n \times 1$ column matrix $[B]$. The solution X is given by $[A]^{-1}[B]$.

EXAMPLE 5 Solving a System of Equations Using an Inverse

Use an inverse matrix to solve the system.

$$\begin{cases} 2x + 3y + z = -1 \\ 3x + 3y + z = 1 \\ 2x + 4y + z = -2 \end{cases}$$

Solution

Begin by writing the system as $AX = B$.

$$\begin{bmatrix} 2 & 3 & 1 \\ 3 & 3 & 1 \\ 2 & 4 & 1 \end{bmatrix}\begin{bmatrix} x \\ y \\ z \end{bmatrix} = \begin{bmatrix} -1 \\ 1 \\ -2 \end{bmatrix}$$

Then, use Gauss-Jordan elimination to find A^{-1}.

$$A^{-1} = \begin{bmatrix} -1 & 1 & 0 \\ -1 & 0 & 1 \\ 6 & -2 & -3 \end{bmatrix}$$

Finally, multiply B by A^{-1} on the left to obtain the solution.

$$\begin{aligned} X &= A^{-1}B \\ &= \begin{bmatrix} -1 & 1 & 0 \\ -1 & 0 & 1 \\ 6 & -2 & -3 \end{bmatrix}\begin{bmatrix} -1 \\ 1 \\ -2 \end{bmatrix} \\ &= \begin{bmatrix} 2 \\ -1 \\ -2 \end{bmatrix} \end{aligned}$$

So, the solution is $x = 2$, $y = -1$, and $z = -2$. Use a graphing utility to verify A^{-1} for the system of equations.

Remark

Remember that matrix multiplication is not commutative. So, you must multiply matrices in the correct order. For instance, in Example 5, you must multiply B by A^{-1} on the left.

Checkpoint *Audio-video solution in English & Spanish at LarsonPrecalculus.com.*

Use an inverse matrix to solve the system.

$$\begin{cases} x + y + z = 6 \\ 3x + 5y + 4z = 27 \\ 3x + 6y + 5z = 35 \end{cases}$$

5.6 Exercises

See *CalcChat.com* for tutorial help and worked-out solutions to odd-numbered exercises.
For instructions on how to use a graphing utility, see Appendix A.

Vocabulary and Concept Check

In Exercises 1 and 2, fill in the blank(s).

1. If there exists an $n \times n$ matrix A^{-1} such that $AA^{-1} = I_n = A^{-1}A$, then A^{-1} is called the _______ of A.
2. If a matrix A has an inverse, then it is called invertible or _______; if it does not have an inverse, then it is called _______.
3. Do all square matrices have inverses?
4. Given that A and B are square matrices and $AB = I_n$, does $BA = I_n$?

Procedures and Problem Solving

The Inverse of a Matrix **In Exercises 5–10, show that B is the inverse of A.**

5. $A = \begin{bmatrix} 1 & 3 \\ -1 & -2 \end{bmatrix}$, $B = \begin{bmatrix} -2 & -3 \\ 1 & 1 \end{bmatrix}$

6. $A = \begin{bmatrix} -4 & 3 \\ 3 & -2 \end{bmatrix}$, $B = \begin{bmatrix} 2 & 3 \\ 3 & 4 \end{bmatrix}$

7. $A = \begin{bmatrix} 5 & 4 \\ 3 & 2 \end{bmatrix}$, $B = \begin{bmatrix} -1 & 2 \\ \frac{3}{2} & -\frac{5}{2} \end{bmatrix}$

8. $A = \begin{bmatrix} -\frac{1}{2} & -\frac{5}{4} \\ 1 & 2 \end{bmatrix}$, $B = \begin{bmatrix} 8 & 5 \\ -4 & -2 \end{bmatrix}$

9. $A = \begin{bmatrix} 2 & -17 & 11 \\ -1 & 11 & -7 \\ 0 & 3 & -2 \end{bmatrix}$, $B = \begin{bmatrix} 1 & 1 & 2 \\ 2 & 4 & -3 \\ 3 & 6 & -5 \end{bmatrix}$

10. $A = \begin{bmatrix} -4 & 1 & 5 \\ -1 & 2 & 4 \\ 0 & -1 & -1 \end{bmatrix}$, $B = \frac{1}{4}\begin{bmatrix} -2 & 4 & 6 \\ 1 & -4 & -11 \\ -1 & 4 & 7 \end{bmatrix}$

Finding the Inverse of a Matrix **In Exercises 11–20, find the inverse of the matrix (if it exists).**

11. $\begin{bmatrix} 2 & 0 \\ 0 & 3 \end{bmatrix}$

12. $\begin{bmatrix} 1 & 2 \\ 3 & 7 \end{bmatrix}$

13. $\begin{bmatrix} 1 & -2 \\ 2 & -3 \end{bmatrix}$

14. $\begin{bmatrix} -2 & 5 \\ 6 & -15 \\ 0 & 1 \end{bmatrix}$

15. $\begin{bmatrix} 2 & 7 & 1 \\ -3 & -9 & 2 \end{bmatrix}$

16. $\begin{bmatrix} -7 & 33 \\ 4 & -19 \end{bmatrix}$

17. $\begin{bmatrix} 1 & 0 & 0 \\ 3 & 5 & 0 \\ 2 & 5 & 0 \end{bmatrix}$

18. $\begin{bmatrix} 1 & 2 & 2 \\ 3 & 7 & 9 \\ -1 & -4 & -7 \end{bmatrix}$

19. $\begin{bmatrix} 1 & 1 & 1 \\ 3 & 5 & 4 \\ 3 & 6 & 5 \end{bmatrix}$

20. $\begin{bmatrix} 1 & 6 & 10 \\ 3 & 4 & 0 \\ 2 & 5 & 5 \end{bmatrix}$

Finding the Inverse of a Matrix **In Exercises 21–28, use the matrix capabilities of a graphing utility to find the inverse of the matrix (if it exists).**

21. $\begin{bmatrix} 1 & 1 & 2 \\ 3 & 1 & 0 \\ -2 & 0 & 3 \end{bmatrix}$

22. $\begin{bmatrix} 1 & 2 & -1 \\ 3 & 7 & -10 \\ -5 & -7 & -15 \end{bmatrix}$

23. $\begin{bmatrix} -\frac{1}{2} & \frac{3}{4} & \frac{1}{4} \\ 1 & 0 & -\frac{3}{2} \\ 0 & -1 & \frac{1}{2} \end{bmatrix}$

24. $\begin{bmatrix} -\frac{5}{6} & \frac{1}{3} & \frac{11}{6} \\ 0 & \frac{2}{3} & 2 \\ 1 & -\frac{1}{2} & -\frac{5}{2} \end{bmatrix}$

25. $\begin{bmatrix} 0.1 & 0.2 & 0.3 \\ -0.3 & 0.2 & 0.2 \\ 0.5 & 0.4 & 0.4 \end{bmatrix}$

26. $\begin{bmatrix} 0.6 & 0 & -0.3 \\ 0.7 & -1 & 0.2 \\ 1 & 0 & -0.9 \end{bmatrix}$

27. $\begin{bmatrix} -1 & 0 & 1 & 0 \\ 0 & 2 & 0 & -2 \\ 2 & 0 & -1 & 0 \\ 0 & -1 & 0 & 1 \end{bmatrix}$

28. $\begin{bmatrix} 1 & -2 & -1 & -2 \\ 3 & -5 & -2 & -3 \\ 2 & -5 & -2 & -5 \\ -1 & 4 & 4 & 11 \end{bmatrix}$

Finding the Inverse of a 2×2 Matrix **In Exercises 29–34, use the formula on page 464 to find the inverse of the 2×2 matrix (if it exists).**

29. $\begin{bmatrix} 4 & -3 \\ 8 & -6 \end{bmatrix}$

30. $\begin{bmatrix} 1 & -2 \\ -3 & 2 \end{bmatrix}$

31. $\begin{bmatrix} \frac{7}{2} & -\frac{3}{4} \\ \frac{1}{5} & \frac{4}{5} \end{bmatrix}$

32. $\begin{bmatrix} -\frac{1}{4} & -\frac{2}{3} \\ \frac{1}{3} & \frac{8}{9} \end{bmatrix}$

33. $\begin{bmatrix} 2 & 3 \\ -1 & 5 \end{bmatrix}$

34. $\begin{bmatrix} 7 & 12 \\ -8 & -5 \end{bmatrix}$

Finding a Matrix Entry **In Exercises 35 and 36, find the value of the constant k such that $B = A^{-1}$.**

35. $A = \begin{bmatrix} 1 & 2 \\ -2 & 0 \end{bmatrix}$, $B = \begin{bmatrix} k & -\frac{1}{2} \\ \frac{1}{2} & \frac{1}{4} \end{bmatrix}$

36. $A = \begin{bmatrix} -1 & 1 \\ 2 & 1 \end{bmatrix}$, $B = \begin{bmatrix} -\frac{1}{3} & \frac{1}{3} \\ k & \frac{1}{3} \end{bmatrix}$

Solving a System of Linear Equations **In Exercises 37–40, use the inverse matrix found in Exercise 13 to solve the system of linear equations.**

37. $\begin{cases} x - 2y = 5 \\ 2x - 3y = 10 \end{cases}$

38. $\begin{cases} x - 2y = 0 \\ 2x - 3y = 3 \end{cases}$

39. $\begin{cases} x - 2y = 4 \\ 2x - 3y = 2 \end{cases}$

40. $\begin{cases} x - 2y = 1 \\ 2x - 3y = -2 \end{cases}$

Solving a System of Linear Equations **In Exercises 41 and 42, use the inverse matrix found in Exercise 19 to solve the system of linear equations.**

41. $\begin{cases} x + y + z = 0 \\ 3x + 5y + 4z = 5 \\ 3x + 6y + 5z = 2 \end{cases}$

42. $\begin{cases} x + y + z = -1 \\ 3x + 5y + 4z = 2 \\ 3x + 6y + 5z = 0 \end{cases}$

Solving a System of Linear Equations **In Exercises 43 and 44, use the inverse matrix found in Exercise 28 and the matrix capabilities of a graphing utility to solve the system of linear equations.**

43. $\begin{cases} x_1 - 2x_2 - x_3 - 2x_4 = 0 \\ 3x_1 - 5x_2 - 2x_3 - 3x_4 = 1 \\ 2x_1 - 5x_2 - 2x_3 - 5x_4 = -1 \\ -x_1 + 4x_2 + 4x_3 + 11x_4 = 2 \end{cases}$

44. $\begin{cases} x_1 - 2x_2 - x_3 - 2x_4 = 1 \\ 3x_1 - 5x_2 - 2x_3 - 3x_4 = -2 \\ 2x_1 - 5x_2 - 2x_3 - 5x_4 = 0 \\ -x_1 + 4x_2 + 4x_3 + 11x_4 = -3 \end{cases}$

Solving a System of Equations Using an Inverse **In Exercises 45–52, use an inverse matrix to solve (if possible) the system of linear equations.**

45. $\begin{cases} 3x + 4y = -2 \\ 5x + 3y = 4 \end{cases}$

46. $\begin{cases} 18x + 12y = 13 \\ 30x + 24y = 23 \end{cases}$

47. $\begin{cases} -0.4x + 0.8y = 1.6 \\ 2x - 4y = 5 \end{cases}$

48. $\begin{cases} 0.2x - 0.6y = 2.4 \\ -x + 1.4y = -8.8 \end{cases}$

49. $\begin{cases} -\frac{1}{4}x + \frac{3}{8}y = -2 \\ \frac{3}{2}x + \frac{3}{4}y = -12 \end{cases}$

50. $\begin{cases} \frac{5}{6}x - y = -10 \\ -\frac{5}{4}x + \frac{3}{2}y = -2 \end{cases}$

51. $\begin{cases} 4x - y + z = -5 \\ 2x + 2y + 3z = 10 \\ 5x - 2y + 6z = 1 \end{cases}$

52. $\begin{cases} 4x - 2y + 3z = -2 \\ 2x + 2y + 5z = 16 \\ 8x - 5y - 2z = 4 \end{cases}$

Solving a System of Linear Equations **In Exercises 53–56, use the matrix capabilities of a graphing utility to solve (if possible) the system of linear equations.**

53. $\begin{cases} 5x - 3y + 2z = 2 \\ 2x + 2y - 3z = 3 \\ -x + 7y - 8z = 4 \end{cases}$

54. $\begin{cases} 2x + 3y + 5z = 4 \\ 3x + 5y - 9z = 7 \\ 5x + 9y + 17z = 13 \end{cases}$

55. $\begin{cases} 7x - 3y + 2w = 41 \\ -2x + y - w = -13 \\ 4x + z - 2w = 12 \\ -x + y - w = -8 \end{cases}$

56. $\begin{cases} 2x + 5y + w = 11 \\ x + 4y + 2z - 2w = -7 \\ 2x - 2y + 5z + w = 3 \\ x - 3w = -1 \end{cases}$

Investment Portfolio **In Exercises 57–60, consider a person who invests in AAA-rated bonds, A-rated bonds, and B-rated bonds. The average yields are 4.5% on AAA bonds, 5% on A bonds, and 7% on B bonds. The person invests twice as much in B bonds as in A bonds. Let x, y, and z represent the amounts (in dollars) invested in AAA, A, and B bonds, respectively.**

$$\begin{cases} x + y + z = \text{(total investment)} \\ 0.045x + 0.05y + 0.07z = \text{(annual return)} \\ 2y - z = 0 \end{cases}$$

Use the inverse of the coefficient matrix of this system to find the amount invested in each type of bond.

	Total Investment	*Annual Return*
57.	\$10,000	\$560
58.	\$20,000	\$955
59.	\$300,000	\$18,450
60.	\$500,000	\$28,000

Electrical Engineering **In Exercises 61 and 62, consider the circuit in the figure. The currents I_1, I_2, and I_3, in amperes, are given by the solution of the system of linear equations**

$$\begin{cases} 2I_1 + 4I_3 = E_1 \\ I_2 + 4I_3 = E_2 \\ I_1 + I_2 - I_3 = 0 \end{cases}$$

where E_1 and E_2 are voltages. Use the inverse of the coefficient matrix of this system to find the unknown currents for the given voltages.

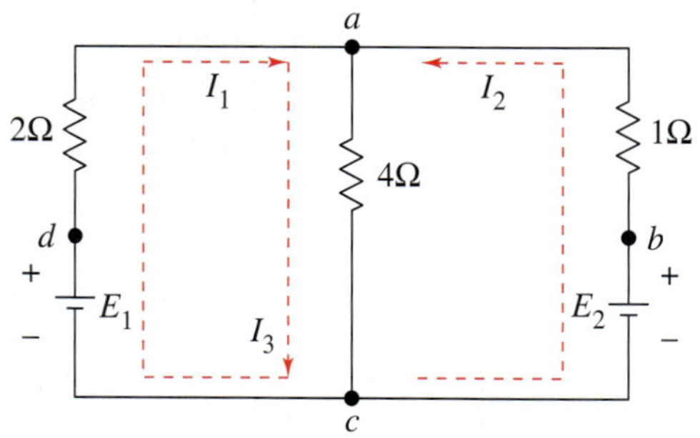

61. $E_1 = 15$ volts, $E_2 = 17$ volts

62. $E_1 = 10$ volts, $E_2 = 10$ volts

Horticulture **In Exercises 63 and 64, consider a company that specializes in potting soil. Each bag of potting soil for seedlings requires 2 units of sand, 1 unit of loam, and 1 unit of peat moss. Each bag of potting soil for general potting requires 1 unit of sand, 2 units of loam, and 1 unit of peat moss. Each bag of potting soil for hardwood plants requires 2 units of sand, 2 units of loam, and 2 units of peat moss. Find the numbers of bags of the three types of potting soil that the company can produce with the given amounts of raw materials.**

63. 500 units of sand
500 units of loam
400 units of peat moss

64. 500 units of sand
750 units of loam
450 units of peat moss

65. **Floral Design** A florist is creating 10 centerpieces for the tables at a wedding reception. Roses cost \$2.50 each, lilies cost \$4 each, and irises cost \$2 each. The customer has a budget of \$300 for the centerpieces and wants each centerpiece to contain 12 flowers, with twice as many roses as the number of irises and lilies combined.

(a) Write a linear system that represents the situation.

(b) Write a matrix equation that corresponds to your system.

(c) Solve your linear system using an inverse matrix. Find the number of flowers of each type that the florist can use to create the 10 centerpieces.

66. ***Why you should learn it*** *(p. 460)* The table shows the numbers of international travelers y (in thousands) to the United States from South America from 2010 through 2012. (Source: U.S. Department of Commerce)

Year	Travelers, y (in thousands)
2010	3250
2011	3757
2012	4416

(a) The data can be approximated by a parabola. Create a system of linear equations for the data. Let t represent the year, with $t = 0$ corresponding to 2010.

(b) Use the matrix capabilities of a graphing utility to find an inverse matrix to solve the system in part (a) and find the least squares regression parabola $y = at^2 + bt + c$.

(c) Use the graphing utility to graph the parabola with the data points.

(d) Use the result of part (b) to estimate the numbers of international travelers to the United States from South America in 2013, 2014, and 2015.

(e) Are your estimates from part (d) reasonable? Explain.

Conclusions

True or False? **In Exercises 67–69, determine whether the statement is true or false. Justify your answer.**

67. Multiplication of an invertible matrix and its inverse is commutative.

68. When the product of two square matrices is the identity matrix, the matrices are inverses of one another.

69. A nonsingular matrix can have dimension 2×3.

70. **Writing** Explain how to determine whether the inverse of a 2×2 matrix exists. If so, explain how to find the inverse.

71. **Writing** Explain in your own words how to write a system of three linear equations in three variables as a matrix equation $AX = B$, as well as how to solve the system using an inverse matrix.

72. **The Inverse of a 2 × 2 Matrix** If A is a 2×2 matrix given by $A = \begin{bmatrix} a & b \\ c & d \end{bmatrix}$, then A is invertible if and only if $ad - bc \neq 0$. If $ad - bc \neq 0$, verify that the inverse is $A^{-1} = \dfrac{1}{ad - bc}\begin{bmatrix} d & -b \\ -c & a \end{bmatrix}$.

73. **Exploration** Consider matrices of the form

$$A = \begin{bmatrix} a_{11} & 0 & 0 & 0 & \cdots & 0 \\ 0 & a_{22} & 0 & 0 & \cdots & 0 \\ 0 & 0 & a_{33} & 0 & \cdots & 0 \\ \vdots & \vdots & \vdots & \vdots & \cdots & \vdots \\ 0 & 0 & 0 & 0 & \cdots & a_{nn} \end{bmatrix}.$$

(a) Write a 2×2 matrix and a 3×3 matrix in the form of A. Find the inverse of each.

(b) Use the result of part (a) to make a conjecture about the inverse of a matrix in the form of A.

74. **HOW DO YOU SEE IT?** Let A be a 2×2 matrix given by

$$A = \begin{bmatrix} x & 0 \\ 0 & y \end{bmatrix}.$$

Use the determinant of A to determine the conditions under which A^{-1} exists.

Cumulative Mixed Review

Solving an Equation **In Exercises 75–78, solve the equation algebraically. Round your result to three decimal places.**

75. $e^{2x} + 2e^x - 15 = 0$

76. $e^{2x} - 10e^x + 24 = 0$

77. $7 \ln 3x = 12$

78. $\ln(x + 9) = 2$

5.7 The Determinant of a Square Matrix

The Determinant of a 2×2 Matrix

Every *square* matrix can be associated with a real number called its **determinant.** Determinants have many uses, and several will be discussed in this and the next section. Historically, the use of determinants arose from special number patterns that occur when systems of linear equations are solved. For instance, the system

$$\begin{cases} a_1x + b_1y = c_1 \\ a_2x + b_2y = c_2 \end{cases}$$

has a solution

$$x = \frac{c_1b_2 - c_2b_1}{a_1b_2 - a_2b_1}$$

and

$$y = \frac{a_1c_2 - a_2c_1}{a_1b_2 - a_2b_1}$$

provided that $a_1b_2 - a_2b_1 \neq 0$. Note that each fraction has the same denominator. This denominator is called the *determinant* of the coefficient matrix of the system.

Coefficient Matrix *Determinant*

$$A = \begin{bmatrix} a_1 & b_1 \\ a_2 & b_2 \end{bmatrix} \qquad \det(A) = a_1b_2 - a_2b_1$$

The determinant of the matrix A can also be denoted by vertical bars on both sides of the matrix, as indicated in the following definition.

Definition of the Determinant of a 2×2 Matrix

The **determinant** of the matrix

$$A = \begin{bmatrix} a_1 & b_1 \\ a_2 & b_2 \end{bmatrix}$$

is given by

$$\begin{aligned} \det(A) &= |A| \\ &= \begin{vmatrix} a_1 & b_1 \\ a_2 & b_2 \end{vmatrix} \\ &= a_1b_2 - a_2b_1. \end{aligned}$$

In this text, $\det(A)$ and $|A|$ are used interchangeably to represent the determinant of A. Although vertical bars are also used to denote the absolute value of a real number, the context will show which use is intended.

A convenient method for remembering the formula for the determinant of a 2×2 matrix is shown in the following diagram.

$$\det(A) = \begin{vmatrix} a_1 & b_1 \\ a_2 & b_2 \end{vmatrix} = a_1b_2 - a_2b_1$$

Note that the determinant is the difference of the products of the two diagonals of the matrix.

What you should learn

- Find the determinants of 2×2 matrices.
- Find minors and cofactors of square matrices.
- Find the determinants of square matrices.

Why you should learn it

Determinants and Cramer's Rule can be used to find the least squares regression parabola that models retail sales of family clothing stores, as shown in Exercise 27 of Section 5.8 on page 484.

EXAMPLE 1 The Determinant of a 2 × 2 Matrix

Find the determinant of each matrix.

a. $A = \begin{bmatrix} 2 & -3 \\ 1 & 2 \end{bmatrix}$

b. $B = \begin{bmatrix} 2 & 1 \\ 4 & 2 \end{bmatrix}$

c. $C = \begin{bmatrix} 0 & \frac{3}{2} \\ 2 & 4 \end{bmatrix}$

Solution

a. $\det(A) = \begin{vmatrix} 2 & -3 \\ 1 & 2 \end{vmatrix} = 2(2) - 1(-3) = 4 + 3 = 7$

b. $\det(B) = \begin{vmatrix} 2 & 1 \\ 4 & 2 \end{vmatrix} = 2(2) - 4(1) = 4 - 4 = 0$

c. $\det(C) = \begin{vmatrix} 0 & \frac{3}{2} \\ 2 & 4 \end{vmatrix} = 0(4) - 2\left(\frac{3}{2}\right) = 0 - 3 = -3$

✓ *Checkpoint* *Audio-video solution in English & Spanish at LarsonPrecalculus.com.*

Find the determinant of each matrix.

a. $A = \begin{bmatrix} 1 & 2 \\ 3 & -1 \end{bmatrix}$ **b.** $B = \begin{bmatrix} 5 & 0 \\ -4 & 2 \end{bmatrix}$ **c.** $C = \begin{bmatrix} 3 & 6 \\ 2 & 4 \end{bmatrix}$

Notice in Example 1 that the determinant of a matrix can be positive, zero, or negative.

The determinant of a matrix of dimension 1 × 1 is defined simply as the entry of the matrix. For instance, if $A = [-2]$, then $\det(A) = -2$.

Explore the Concept

Try using a graphing utility to find the determinant of

$$A = \begin{bmatrix} 3 & -1 & 1 \\ 0 & 2 & 1 \end{bmatrix}.$$

What message appears on the screen? Why does the graphing utility display this message?

EXAMPLE 2 Using a Graphing Utility

Use the matrix capabilities of a graphing utility to find the determinant of the matrix.

$$A = \begin{bmatrix} 1.4 & 0.7 \\ -0.3 & -2.5 \end{bmatrix}$$

Solution

Use the *matrix editor* to enter the matrix as $[A]$, as shown in Figure 5.19. Then choose the *determinant* feature. From Figure 5.20, $\det(A) = -3.29$.

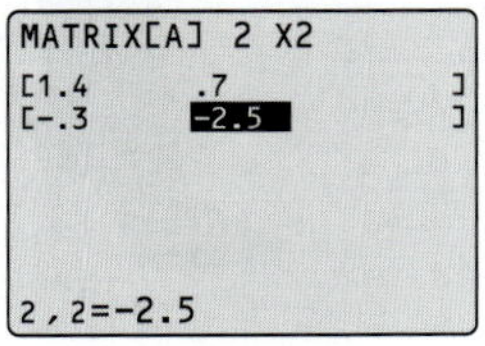

Figure 5.19

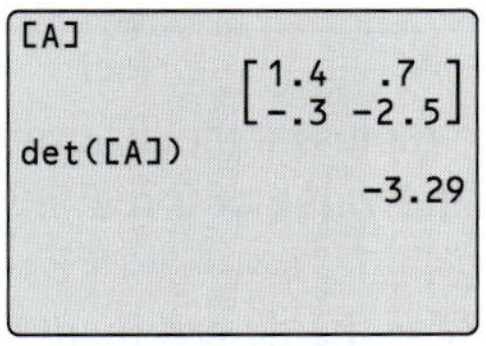

Figure 5.20

✓ *Checkpoint* *Audio-video solution in English & Spanish at LarsonPrecalculus.com.*

Use the matrix capabilities of a graphing utility to find the determinant of the matrix.

$$A = \begin{bmatrix} 0.1 & -0.7 \\ 0.8 & 0.8 \end{bmatrix}$$

Minors and Cofactors

To define the determinant of a square matrix of dimension 3×3 or greater, it is helpful to introduce the concepts of **minors** and **cofactors.**

> **Minors and Cofactors of a Square Matrix**
>
> If A is a square matrix, then the **minor** M_{ij} of the entry a_{ij} is the determinant of the matrix obtained by deleting the ith row and jth column of A. The **cofactor** C_{ij} of the entry a_{ij} is given by $C_{ij} = (-1)^{i+j}M_{ij}$.

Sign Patterns for Cofactors

$$\begin{bmatrix} + & - & + \\ - & + & - \\ + & - & + \end{bmatrix}$$

3×3 matrix

$$\begin{bmatrix} + & - & + & - \\ - & + & - & + \\ + & - & + & - \\ - & + & - & + \end{bmatrix}$$

4×4 matrix

$$\begin{bmatrix} + & - & + & - & + & \cdots \\ - & + & - & + & - & \cdots \\ + & - & + & - & + & \cdots \\ - & + & - & + & - & \cdots \\ + & - & + & - & + & \cdots \\ \vdots & \vdots & \vdots & \vdots & \vdots & \end{bmatrix}$$

$n \times n$ matrix

In the sign patterns for cofactors at the right, notice that *odd* positions (where $i + j$ is odd) have negative signs and *even* positions (where $i + j$ is even) have positive signs.

EXAMPLE 3 Finding the Minors and Cofactors of a Matrix

Find all the minors and cofactors of

$$A = \begin{bmatrix} 0 & 2 & 1 \\ 3 & -1 & 2 \\ 4 & 0 & 1 \end{bmatrix}.$$

Solution

To find the minor M_{11}, delete the first row and first column of A and evaluate the determinant of the resulting matrix.

$$\begin{bmatrix} 0 & 2 & 1 \\ 3 & -1 & 2 \\ 4 & 0 & 1 \end{bmatrix}, \quad M_{11} = \begin{vmatrix} -1 & 2 \\ 0 & 1 \end{vmatrix} = -1(1) - 0(2) = -1$$

Similarly, to find the minor M_{12}, delete the first row and second column.

$$\begin{bmatrix} 0 & 2 & 1 \\ 3 & -1 & 2 \\ 4 & 0 & 1 \end{bmatrix}, \quad M_{12} = \begin{vmatrix} 3 & 2 \\ 4 & 1 \end{vmatrix} = 3(1) - 4(2) = -5$$

Continuing this pattern, you obtain all the minors.

$$\begin{array}{lll} M_{11} = -1 & M_{12} = -5 & M_{13} = 4 \\ M_{21} = 2 & M_{22} = -4 & M_{23} = -8 \\ M_{31} = 5 & M_{32} = -3 & M_{33} = -6 \end{array}$$

Now, to find the cofactors, combine these minors with the checkerboard pattern of signs for a 3×3 matrix shown at the upper right.

$$\begin{array}{lll} C_{11} = -1 & C_{12} = 5 & C_{13} = 4 \\ C_{21} = -2 & C_{22} = -4 & C_{23} = 8 \\ C_{31} = 5 & C_{32} = 3 & C_{33} = -6 \end{array}$$

✔ **Checkpoint** *Audio-video solution in English & Spanish at LarsonPrecalculus.com.*

Find all the minors and cofactors of

$$A = \begin{bmatrix} 1 & 2 & 3 \\ 0 & -1 & 5 \\ 2 & 1 & 4 \end{bmatrix}.$$

The Determinant of a Square Matrix

The following definition is called *inductive* because it uses determinants of matrices of dimension $(n - 1) \times (n - 1)$ to define determinants of matrices of dimension $n \times n$.

> **Determinant of a Square Matrix**
>
> If A is a square matrix (of dimension 2×2 or greater), then the determinant of A is the sum of the entries in any row (or column) of A multiplied by their respective cofactors. For instance, expanding along the first row yields
>
> $$|A| = a_{11}C_{11} + a_{12}C_{12} + \cdot \cdot \cdot + a_{1n}C_{1n}.$$
>
> Applying this definition to find a determinant is called **expanding by cofactors.**

Try checking that for a 2×2 matrix

$$A = \begin{bmatrix} a_1 & b_1 \\ a_2 & b_2 \end{bmatrix}$$

this definition of the determinant yields $|A| = a_1b_2 - a_2b_1$, as previously defined.

EXAMPLE 4 The Determinant of a 3×3 Matrix

See LarsonPrecalculus.com for an interactive version of this type of example.

Find the determinant of $A = \begin{bmatrix} 0 & 2 & 1 \\ 3 & -1 & 2 \\ 4 & 0 & 1 \end{bmatrix}$.

Solution

Note that this is the same matrix that was in Example 3. There you found the cofactors of the entries in the first row to be

$$C_{11} = -1, \quad C_{12} = 5, \quad \text{and} \quad C_{13} = 4.$$

So, by the definition of the determinant of a square matrix, you have

$$\begin{aligned} |A| &= a_{11}C_{11} + a_{12}C_{12} + a_{13}C_{13} && \text{First-row expansion} \\ &= 0(-1) + 2(5) + 1(4) \\ &= 14. \end{aligned}$$

✓ **Checkpoint** *Audio-video solution in English & Spanish at LarsonPrecalculus.com.*

Find the determinant of $A = \begin{bmatrix} 3 & 4 & -2 \\ 3 & 5 & 0 \\ -1 & 4 & 1 \end{bmatrix}$.

In Example 4, the determinant was found by expanding by the cofactors in the first row. You could have used any row or column. For instance, you could have expanded along the second row to obtain

$$\begin{aligned} |A| &= a_{21}C_{21} + a_{22}C_{22} + a_{23}C_{23} && \text{Second-row expansion} \\ &= 3(-2) + (-1)(-4) + 2(8) \\ &= 14. \end{aligned}$$

When expanding by cofactors, you do not need to find cofactors of zero entries, because when $a_{ij} = 0$, you have $a_{ij}C_{ij} = (0)C_{ij} = 0$. So, the row (or column) containing the most zeros is usually the best choice for expansion by cofactors.

5.7 Exercises

See *CalcChat.com* for tutorial help and worked-out solutions to odd-numbered exercises. For instructions on how to use a graphing utility, see Appendix A.

Vocabulary and Concept Check

In Exercises 1 and 2, fill in the blank.

1. Both det(A) and $|A|$ represent the _______ of the matrix A.
2. The determinant of the matrix obtained by deleting the ith row and jth column of a square matrix A is called the _______ of the entry a_{ij}.
3. For a square matrix B, the minor $M_{23} = 5$. What is the cofactor C_{23} of matrix B?
4. To find the determinant of a matrix using expanding by cofactors, do you need to find all the cofactors?

Procedures and Problem Solving

The Determinant of a Matrix **In Exercises 5–12, find the determinant of the matrix.**

5. $[4]$
6. $[-12]$
7. $\begin{bmatrix} 8 & 4 \\ -2 & 3 \end{bmatrix}$
8. $\begin{bmatrix} -5 & 2 \\ 6 & 3 \end{bmatrix}$
9. $\begin{bmatrix} -7 & 6 \\ \frac{1}{2} & 3 \end{bmatrix}$
10. $\begin{bmatrix} 4 & -3 \\ 0 & 0 \end{bmatrix}$
11. $\begin{bmatrix} \sqrt{3} & 3 \\ 4 & \sqrt{3} \end{bmatrix}$
12. $\begin{bmatrix} 9 & \sqrt{5} \\ \sqrt{5} & 4 \end{bmatrix}$

Using a Graphing Utility **In Exercises 13–16, use the matrix capabilities of a graphing utility to find the determinant of the matrix.**

13. $\begin{bmatrix} 1.9 & -0.3 \\ 5.6 & 3.2 \end{bmatrix}$
14. $\begin{bmatrix} 7.2 & 0.8 \\ 3.6 & -0.4 \end{bmatrix}$
15. $\begin{bmatrix} 1.3 & 0.2 & 3.2 \\ 0.2 & 6.2 & 0.2 \\ -0.4 & 4.4 & 0.3 \end{bmatrix}$
16. $\begin{bmatrix} 5.1 & 0.2 & 7.3 \\ -6.3 & 0.2 & 0.2 \\ 0.5 & 3.4 & 0.4 \end{bmatrix}$

Finding the Minors and Cofactors of a Matrix **In Exercises 17–20, find all (a) minors and (b) cofactors of the matrix.**

17. $\begin{bmatrix} 3 & 4 \\ 2 & -5 \end{bmatrix}$
18. $\begin{bmatrix} 11 & 6 \\ -3 & 2 \end{bmatrix}$
19. $\begin{bmatrix} -4 & 6 & 3 \\ 7 & -2 & 8 \\ 1 & 0 & -5 \end{bmatrix}$
20. $\begin{bmatrix} -2 & 9 & 4 \\ 7 & -6 & 0 \\ 6 & 7 & -6 \end{bmatrix}$

Finding a Determinant **In Exercises 21–24, find the determinant of the matrix. Expand by cofactors on each indicated row or column.**

21. $\begin{bmatrix} -3 & 2 & 1 \\ 4 & 5 & 6 \\ 2 & -3 & 1 \end{bmatrix}$
    (a) Row 1
    (b) Column 2
22. $\begin{bmatrix} -3 & 4 & 2 \\ 6 & 3 & 1 \\ 4 & -7 & -8 \end{bmatrix}$
    (a) Row 2
    (b) Column 3
23. $\begin{bmatrix} 6 & 0 & -3 & 5 \\ 4 & 13 & 6 & -8 \\ -1 & 0 & 7 & 4 \\ 8 & 6 & 0 & 2 \end{bmatrix}$
    (a) Row 2
    (b) Column 2
24. $\begin{bmatrix} 10 & 8 & 3 & -7 \\ 4 & 0 & 5 & -6 \\ 0 & 3 & 2 & 7 \\ 1 & 0 & -3 & 2 \end{bmatrix}$
    (a) Row 3
    (b) Column 1

Finding a Determinant **In Exercises 25–34, find the determinant of the matrix. Expand by cofactors on the row or column that appears to make the computations easiest.**

25. $\begin{bmatrix} 1 & 4 & -2 \\ 3 & 2 & 0 \\ -1 & 4 & 3 \end{bmatrix}$
26. $\begin{bmatrix} -3 & 1 & 0 \\ 7 & 11 & 5 \\ 1 & 2 & 2 \end{bmatrix}$
27. $\begin{bmatrix} 6 & 3 & -7 \\ 0 & 0 & 0 \\ 4 & -6 & 3 \end{bmatrix}$
28. $\begin{bmatrix} 1 & 1 & 2 \\ 3 & -5 & 9 \\ 0 & 0 & 0 \end{bmatrix}$
29. $\begin{bmatrix} -1 & 2 & -5 \\ 0 & 3 & -4 \\ 0 & 0 & 3 \end{bmatrix}$
30. $\begin{bmatrix} 1 & 0 & 0 \\ -1 & -1 & 0 \\ 4 & 1 & 5 \end{bmatrix}$
31. $\begin{bmatrix} 2 & 6 & 6 & 2 \\ 2 & 7 & 3 & 6 \\ 1 & 5 & 0 & 1 \\ 3 & 7 & 0 & 7 \end{bmatrix}$
32. $\begin{bmatrix} 3 & 6 & -5 & 4 \\ -2 & 2 & 6 & 0 \\ 1 & 1 & 2 & 0 \\ 0 & 3 & -1 & -1 \end{bmatrix}$
33. $\begin{bmatrix} 3 & 2 & 4 & -1 & 5 \\ -2 & 0 & 1 & 3 & 2 \\ 1 & 0 & 0 & 4 & 0 \\ 6 & 0 & 2 & -1 & 0 \\ 3 & 0 & 5 & 1 & 0 \end{bmatrix}$
34. $\begin{bmatrix} 5 & 2 & 0 & 0 & -2 \\ 0 & 1 & 4 & 3 & 2 \\ 0 & 0 & 2 & 6 & 3 \\ 0 & 0 & 3 & 4 & 1 \\ 0 & 0 & 0 & 0 & 2 \end{bmatrix}$

Using a Graphing Utility In Exercises 35–38, use the matrix capabilities of a graphing utility to evaluate the determinant.

35. $\begin{vmatrix} 1 & -1 & 8 & 4 \\ 2 & 6 & 0 & -4 \\ 2 & 0 & 2 & 6 \\ 0 & 2 & 8 & 0 \end{vmatrix}$ **36.** $\begin{vmatrix} 0 & -3 & 8 & 2 \\ 8 & 1 & -1 & 6 \\ -4 & 6 & 0 & 9 \\ -7 & 0 & 0 & 14 \end{vmatrix}$

37. $\begin{vmatrix} 3 & -2 & 4 & 3 & 1 & 3 \\ -1 & 0 & 2 & 1 & 0 & 0 \\ 5 & -1 & 0 & 3 & 2 & 1 \\ 4 & 7 & -8 & 0 & 0 & -2 \\ 1 & 2 & 3 & 0 & 2 & 4 \\ 3 & -5 & 1 & 2 & 3 & 1 \end{vmatrix}$

38. $\begin{vmatrix} -2 & 0 & 1 & 4 & 3 & -2 \\ -3 & 3 & 0 & -2 & 1 & -1 \\ 4 & 5 & -1 & 0 & 7 & 3 \\ 2 & 4 & 3 & 2 & 0 & 1 \\ 1 & 3 & 4 & 2 & -4 & 0 \\ -5 & 2 & -1 & 3 & 2 & -3 \end{vmatrix}$

The Determinant of a Matrix Product In Exercises 39–42, find (a) $|A|$, (b) $|B|$, (c) AB, and (d) $|AB|$. What do you notice about $|AB|$?

39. $A = \begin{bmatrix} -2 & 1 \\ 4 & -2 \end{bmatrix}$, $B = \begin{bmatrix} 1 & 2 \\ 0 & -1 \end{bmatrix}$

40. $A = \begin{bmatrix} 4 & 0 \\ 3 & -2 \end{bmatrix}$, $B = \begin{bmatrix} -1 & 1 \\ -2 & 2 \end{bmatrix}$

41. $A = \begin{bmatrix} 3 & 2 & 0 \\ -1 & -3 & 4 \\ -2 & 0 & 1 \end{bmatrix}$, $B = \begin{bmatrix} -3 & 0 & 1 \\ 0 & 2 & -1 \\ -2 & -1 & 1 \end{bmatrix}$

42. $A = \begin{bmatrix} 2 & 0 & 1 \\ 1 & -4 & 2 \\ 3 & 1 & 0 \end{bmatrix}$, $B = \begin{bmatrix} 2 & -1 & 4 \\ 0 & 1 & 0 \\ 3 & -2 & 1 \end{bmatrix}$

Using a Graphing Utility In Exercises 43 and 44, use the matrix capabilities of a graphing utility to find (a) $|A|$, (b) $|B|$, (c) AB, and (d) $|AB|$. What do you notice about $|AB|$?

43. $A = \begin{bmatrix} 6 & 4 & 0 & 1 \\ 2 & -3 & -2 & -4 \\ 0 & 1 & 5 & 0 \\ -1 & 0 & -1 & 1 \end{bmatrix}$, $B = \begin{bmatrix} 0 & -5 & 0 & -2 \\ -2 & 4 & -1 & -4 \\ 3 & 0 & 1 & 0 \\ 1 & -2 & 3 & 0 \end{bmatrix}$

44. $A = \begin{bmatrix} -1 & 5 & 2 & 0 \\ 0 & 0 & 1 & 1 \\ 3 & -3 & -1 & 0 \\ 4 & 2 & 4 & -1 \end{bmatrix}$, $B = \begin{bmatrix} 1 & 5 & 0 & 0 \\ 10 & -1 & 2 & 4 \\ 2 & 0 & 0 & 1 \\ -3 & 2 & 5 & 0 \end{bmatrix}$

Verifying an Equation In Exercises 45–50, evaluate the determinants to verify the equation.

45. $\begin{vmatrix} w & x \\ y & z \end{vmatrix} = -\begin{vmatrix} y & z \\ w & x \end{vmatrix}$

46. $\begin{vmatrix} w & cx \\ y & cz \end{vmatrix} = c\begin{vmatrix} w & x \\ y & z \end{vmatrix}$

47. $\begin{vmatrix} w & x \\ y & z \end{vmatrix} = \begin{vmatrix} w & x + cw \\ y & z + cy \end{vmatrix}$

48. $\begin{vmatrix} w & x \\ cw & cx \end{vmatrix} = 0$

49. $\begin{vmatrix} 1 & x & x^2 \\ 1 & y & y^2 \\ 1 & z & z^2 \end{vmatrix} = (y - x)(z - x)(z - y)$

50. $\begin{vmatrix} a + b & a & a \\ a & a + b & a \\ a & a & a + b \end{vmatrix} = b^2(3a + b)$

Solving an Equation In Exercises 51–62, solve for x.

51. $\begin{vmatrix} x & 2 \\ 1 & x \end{vmatrix} = 2$

52. $\begin{vmatrix} x & 4 \\ -1 & x \end{vmatrix} = 20$

53. $\begin{vmatrix} 2x & -3 \\ -2 & 2x \end{vmatrix} = 3$

54. $\begin{vmatrix} x & 2 \\ 4 & 9x \end{vmatrix} = 8$

55. $\begin{vmatrix} x & 2 \\ 2 & x - 2 \end{vmatrix} = -1$

56. $\begin{vmatrix} x + 1 & 2 \\ -1 & x \end{vmatrix} = 4$

57. $\begin{vmatrix} x + 3 & 2 \\ 1 & x + 2 \end{vmatrix} = 0$

58. $\begin{vmatrix} x - 1 & 2 \\ 3 & x - 2 \end{vmatrix} = 0$

59. $\begin{vmatrix} 2x & 1 \\ -1 & x - 1 \end{vmatrix} = x$

60. $\begin{vmatrix} x - 1 & x \\ x + 1 & 2 \end{vmatrix} = -8$

61. $\begin{vmatrix} 1 & 2 & x \\ -1 & 3 & 2 \\ 3 & -2 & 1 \end{vmatrix} = 0$

62. $\begin{vmatrix} 1 & x & -2 \\ 1 & 3 & 3 \\ 0 & 2 & -2 \end{vmatrix} = 0$

Entries Involving Expressions In Exercises 63–68, evaluate the determinant, in which the entries are functions. Determinants of this type occur when changes of variables are made in calculus.

63. $\begin{vmatrix} 4u & -1 \\ -1 & 2v \end{vmatrix}$

64. $\begin{vmatrix} 3x^2 & -3y^2 \\ 1 & 1 \end{vmatrix}$

65. $\begin{vmatrix} e^{2x} & e^{3x} \\ 2e^{2x} & 3e^{3x} \end{vmatrix}$

66. $\begin{vmatrix} e^{-x} & xe^{-x} \\ -e^{-x} & (1 - x)e^{-x} \end{vmatrix}$

67. $\begin{vmatrix} x & \ln x \\ 1 & 1/x \end{vmatrix}$

68. $\begin{vmatrix} x & x \ln x \\ 1 & 1 + \ln x \end{vmatrix}$

Conclusions

True or False? **In Exercises 69 and 70, determine whether the statement is true or false. Justify your answer.**

69. If a square matrix has an entire row of zeros, then the determinant will always be zero.

70. If two columns of a square matrix are the same, then the determinant of the matrix will be zero.

71. Exploration Find a pair of 3×3 matrices A and B to demonstrate that $|A + B| \neq |A| + |B|$.

72. Think About It Let A be a 3×3 matrix such that $|A| = 5$. Can you use this information to find $|2A|$? Explain.

Exploration **In Exercises 73–76, (a) find the determinant of A, (b) find A^{-1}, (c) find $\det(A^{-1})$, and (d) compare your results from parts (a) and (c). Make a conjecture based on your results.**

73. $A = \begin{bmatrix} 1 & 2 \\ -2 & 2 \end{bmatrix}$ **74.** $A = \begin{bmatrix} 5 & -1 \\ 2 & -1 \end{bmatrix}$

75. $A = \begin{bmatrix} 1 & -3 & -2 \\ -1 & 3 & 1 \\ 0 & 2 & -2 \end{bmatrix}$ **76.** $A = \begin{bmatrix} -1 & 3 & 2 \\ 1 & 3 & -1 \\ 1 & 1 & -2 \end{bmatrix}$

Properties of Determinants **In Exercises 77–79, a property of determinants is given (A and B are square matrices). State how the property has been applied to the given determinants and use a graphing utility to verify the results.**

77. If B is obtained from A by interchanging two rows of A or by interchanging two columns of A, then $|B| = -|A|$.

(a) $\begin{vmatrix} 1 & 3 & 4 \\ -7 & 2 & -5 \\ 6 & 1 & 2 \end{vmatrix} = -\begin{vmatrix} 1 & 4 & 3 \\ -7 & -5 & 2 \\ 6 & 2 & 1 \end{vmatrix}$

(b) $\begin{vmatrix} 1 & 3 & 4 \\ -2 & 2 & 0 \\ 1 & 6 & 2 \end{vmatrix} = -\begin{vmatrix} 1 & 6 & 2 \\ -2 & 2 & 0 \\ 1 & 3 & 4 \end{vmatrix}$

78. If B is obtained from A by adding a multiple of a row of A to another row of A or by adding a multiple of a column of A to another column of A, then $|B| = |A|$.

(a) $\begin{vmatrix} 1 & -3 \\ 5 & 2 \end{vmatrix} = \begin{vmatrix} 1 & -3 \\ 0 & 17 \end{vmatrix}$

(b) $\begin{vmatrix} 5 & 4 & 2 \\ 2 & -3 & 4 \\ 7 & 6 & 3 \end{vmatrix} = \begin{vmatrix} 1 & 10 & -6 \\ 2 & -3 & 4 \\ 7 & 6 & 3 \end{vmatrix}$

79. If B is obtained from A by multiplying a row of A by a nonzero constant c or by multiplying a column of A by a nonzero constant c, then $|B| = c|A|$.

(a) $\begin{vmatrix} 1 & 5 \\ 6 & 9 \end{vmatrix} = 3\begin{vmatrix} 1 & 5 \\ 2 & 3 \end{vmatrix}$ (b) $\begin{vmatrix} 2 & 8 \\ 6 & 8 \end{vmatrix} = 8\begin{vmatrix} 1 & 2 \\ 3 & 2 \end{vmatrix}$

80. Exploration A **diagonal matrix** is a square matrix with all zero entries above and below its main diagonal. Evaluate the determinant of each diagonal matrix. Make a conjecture based on your results.

(a) $\begin{bmatrix} 7 & 0 \\ 0 & 4 \end{bmatrix}$ (b) $\begin{bmatrix} -1 & 0 & 0 \\ 0 & 5 & 0 \\ 0 & 0 & 2 \end{bmatrix}$ (c) $\begin{bmatrix} 2 & 0 & 0 & 0 \\ 0 & -2 & 0 & 0 \\ 0 & 0 & 1 & 0 \\ 0 & 0 & 0 & 3 \end{bmatrix}$

81. Exploration A **triangular matrix** is a square matrix with all zero entries either above or below its main diagonal. Such a matrix is **upper triangular** when it has all zeros below the main diagonal and **lower triangular** when it has all zeros above the main diagonal. A diagonal matrix is both upper and lower triangular. Evaluate the determinant of each triangular matrix. Make a conjecture based on your results.

(a) $\begin{bmatrix} 3 & -2 \\ 0 & 5 \end{bmatrix}$ (b) $\begin{bmatrix} 3 & -7 & 1 \\ 0 & -5 & -9 \\ 0 & 0 & 5 \end{bmatrix}$ (c) $\begin{bmatrix} 4 & 0 & 0 & 0 \\ 3 & -3 & 0 & 0 \\ 3 & 6 & 5 & 0 \\ 2 & -2 & 1 & 2 \end{bmatrix}$

82. HOW DO YOU SEE IT? Explain why the determinant of the matrix is equal to zero.

$$\begin{bmatrix} 2 & -4 & 5 \\ 1 & -2 & 3 \\ 0 & 0 & 0 \end{bmatrix}$$

83. Writing Describe the different methods you have learned for finding the determinant of a square matrix.

84. Exploration Consider square matrices in which the entries are consecutive integers. An example of such a matrix is shown below. Use a graphing utility to evaluate four determinants of this type. Make a conjecture based on the results. Then verify your conjecture.

$$\begin{bmatrix} 4 & 5 & 6 \\ 7 & 8 & 9 \\ 10 & 11 & 12 \end{bmatrix}$$

Cumulative Mixed Review

Factoring a Quadratic Expression **In Exercises 85 and 86, factor the expression.**

85. $4y^2 - 12y + 9$ **86.** $4y^2 - 28y + 49$

Solving a System of Equations **In Exercises 87 and 88, solve the system of equations using the method of substitution or the method of elimination.**

87. $\begin{cases} 3x - 10y = 46 \\ x + y = -2 \end{cases}$ **88.** $\begin{cases} 5x + 7y = 23 \\ -4x - 2y = -4 \end{cases}$

5.8 Applications of Matrices and Determinants

Area of a Triangle

In this section, you will study some additional applications of matrices and determinants. The first involves a formula for finding the area of a triangle whose vertices are given by three points on a rectangular coordinate system.

Area of a Triangle

The area of a triangle with vertices (x_1, y_1), (x_2, y_2), and (x_3, y_3) is

$$\text{Area} = \pm\frac{1}{2}\begin{vmatrix} x_1 & y_1 & 1 \\ x_2 & y_2 & 1 \\ x_3 & y_3 & 1 \end{vmatrix}$$

where the symbol $(\pm)$ indicates that the appropriate sign should be chosen to yield a positive area.

What you should learn

- Use determinants to find areas of triangles.
- Use determinants to determine whether points are collinear.
- Use Cramer's Rule to solve systems of linear equations.
- Use matrices to encode and decode messages.

Why you should learn it

Matrices can be used to decode a message, as shown in Exercise 36 on page 485.

Nfsphot | Dreamstime.com

EXAMPLE 1 Finding the Area of a Triangle

See LarsonPrecalculus.com for an interactive version of this type of example.

Find the area of the triangle whose vertices are $(1, 0)$, $(2, 2)$, and $(4, 3)$, as shown in the figure.

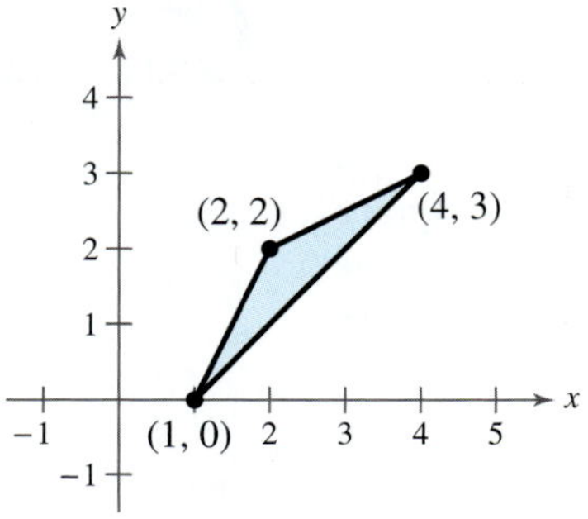

Solution

Begin by letting $(x_1, y_1) = (1, 0)$, $(x_2, y_2) = (2, 2)$, and $(x_3, y_3) = (4, 3)$. Then, to find the area of the triangle, evaluate the determinant by expanding along row 1.

$$\begin{vmatrix} x_1 & y_1 & 1 \\ x_2 & y_2 & 1 \\ x_3 & y_3 & 1 \end{vmatrix} = \begin{vmatrix} 1 & 0 & 1 \\ 2 & 2 & 1 \\ 4 & 3 & 1 \end{vmatrix}$$

$$= 1(-1)^2\begin{vmatrix} 2 & 1 \\ 3 & 1 \end{vmatrix} + 0(-1)^3\begin{vmatrix} 2 & 1 \\ 4 & 1 \end{vmatrix} + 1(-1)^4\begin{vmatrix} 2 & 2 \\ 4 & 3 \end{vmatrix}$$

$$= 1(-1) + 0 + 1(-2)$$

$$= -3$$

Using this value, you can conclude that the area of the triangle is

$$\text{Area} = -\frac{1}{2}\begin{vmatrix} 1 & 0 & 1 \\ 2 & 2 & 1 \\ 4 & 3 & 1 \end{vmatrix} = -\frac{1}{2}(-3) = \frac{3}{2} \text{ square units.}$$

✓ **Checkpoint** *Audio-video solution in English & Spanish at LarsonPrecalculus.com.*

Find the area of a triangle whose vertices are $(0, 0)$, $(4, 1)$, and $(2, 5)$.

Collinear Points

What if the three points in Example 1 had been on the same line? What would have happened had the area formula been applied to three such points? The answer is that the determinant would have been zero. Consider, for instance, the three collinear points (0, 1), (2, 2), and (4, 3), as shown in Figure 5.21. The area of the "triangle" that has these three points as vertices is

$$\frac{1}{2}\begin{vmatrix} 0 & 1 & 1 \\ 2 & 2 & 1 \\ 4 & 3 & 1 \end{vmatrix} = \frac{1}{2}\left[0(-1)^2\begin{vmatrix} 2 & 1 \\ 3 & 1 \end{vmatrix} + 1(-1)^3\begin{vmatrix} 2 & 1 \\ 4 & 1 \end{vmatrix} + 1(-1)^4\begin{vmatrix} 2 & 2 \\ 4 & 3 \end{vmatrix}\right]$$

$$= \frac{1}{2}[0 - 1(-2) + 1(-2)]$$

$$= 0$$

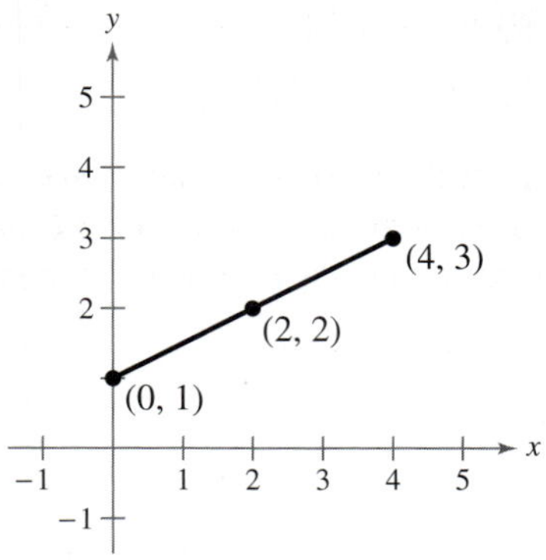

Figure 5.21

This result is generalized as follows.

Test for Collinear Points

Three points (x_1, y_1), (x_2, y_2), and (x_3, y_3) are **collinear** (lie on the same line) if and only if

$$\begin{vmatrix} x_1 & y_1 & 1 \\ x_2 & y_2 & 1 \\ x_3 & y_3 & 1 \end{vmatrix} = 0.$$

EXAMPLE 2 Testing for Collinear Points

Determine whether the points

$(-2, -2)$, $(1, 1)$, and $(7, 5)$

are collinear. (See Figure 5.22.)

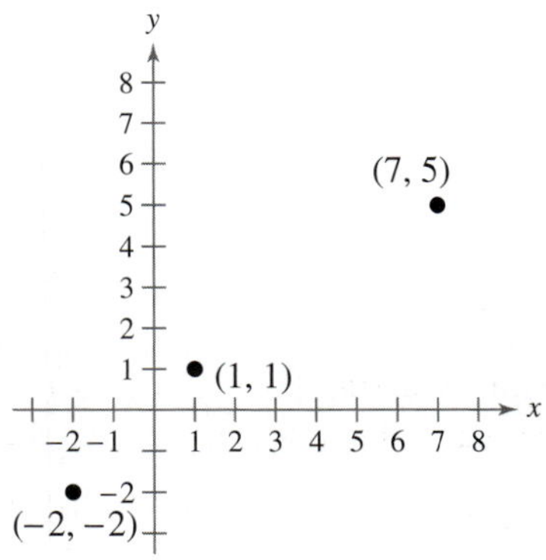

Figure 5.22

Solution

Begin by letting $(x_1, y_1) = (-2, -2)$, $(x_2, y_2) = (1, 1)$, and $(x_3, y_3) = (7, 5)$. Then by expanding along row 1, you have

$$\begin{vmatrix} x_1 & y_1 & 1 \\ x_2 & y_2 & 1 \\ x_3 & y_3 & 1 \end{vmatrix} = \begin{vmatrix} -2 & -2 & 1 \\ 1 & 1 & 1 \\ 7 & 5 & 1 \end{vmatrix}$$

$$= -2(-1)^2\begin{vmatrix} 1 & 1 \\ 5 & 1 \end{vmatrix} + (-2)(-1)^3\begin{vmatrix} 1 & 1 \\ 7 & 1 \end{vmatrix} + 1(-1)^4\begin{vmatrix} 1 & 1 \\ 7 & 5 \end{vmatrix}$$

$$= -2(-4) + 2(-6) + 1(-2)$$

$$= -6.$$

Because the value of this determinant is *not* zero, you can conclude that the three points are not collinear.

Checkpoint *Audio-video solution in English & Spanish at LarsonPrecalculus.com.*

Determine whether the points

$(-2, 4)$, $(3, -1)$, and $(6, -4)$

are collinear.

Cramer's Rule

So far, you have studied three methods for solving a system of linear equations: substitution, elimination with equations, and elimination with matrices. You will now study one more method, **Cramer's Rule,** named after Gabriel Cramer (1704–1752). This rule uses determinants to write the solution of a system of linear equations. To see how Cramer's Rule works, take another look at the solution described at the beginning of Section 5.7. There, it was pointed out that the system

$$\begin{cases} a_1x + b_1y = c_1 \\ a_2x + b_2y = c_2 \end{cases}$$

has a solution

$$x = \frac{c_1b_2 - c_2b_1}{a_1b_2 - a_2b_1}$$

and

$$y = \frac{a_1c_2 - a_2c_1}{a_1b_2 - a_2b_1}$$

provided that

$$a_1b_2 - a_2b_1 \neq 0.$$

Each numerator and denominator in this solution can be expressed as a determinant, as follows.

$$x = \frac{c_1b_2 - c_2b_1}{a_1b_2 - a_2b_1} = \frac{\begin{vmatrix} c_1 & b_1 \\ c_2 & b_2 \end{vmatrix}}{\begin{vmatrix} a_1 & b_1 \\ a_2 & b_2 \end{vmatrix}}$$

$$y = \frac{a_1c_2 - a_2c_1}{a_1b_2 - a_2b_1} = \frac{\begin{vmatrix} a_1 & c_1 \\ a_2 & c_2 \end{vmatrix}}{\begin{vmatrix} a_1 & b_1 \\ a_2 & b_2 \end{vmatrix}}$$

Relative to the original system, the denominators of x and y are simply the determinant of the *coefficient* matrix of the system. This determinant is denoted by D. The numerators of x and y are denoted by D_x and D_y, respectively. They are formed by using the column of constants as replacements for the coefficients of x and y, as follows.

Coefficient Matrix	D	D_x	D_y
$\begin{bmatrix} a_1 & b_1 \\ a_2 & b_2 \end{bmatrix}$	$\begin{vmatrix} a_1 & b_1 \\ a_2 & b_2 \end{vmatrix}$	$\begin{vmatrix} c_1 & b_1 \\ c_2 & b_2 \end{vmatrix}$	$\begin{vmatrix} a_1 & c_1 \\ a_2 & c_2 \end{vmatrix}$

For example, given the system

$$\begin{cases} 2x - 5y = 3 \\ -4x + 3y = 8 \end{cases}$$

the coefficient matrix, D, D_x, and D_y are as follows.

Coefficient Matrix	D	D_x	D_y
$\begin{bmatrix} 2 & -5 \\ -4 & 3 \end{bmatrix}$	$\begin{vmatrix} 2 & -5 \\ -4 & 3 \end{vmatrix}$	$\begin{vmatrix} 3 & -5 \\ 8 & 3 \end{vmatrix}$	$\begin{vmatrix} 2 & 3 \\ -4 & 8 \end{vmatrix}$

Cramer's Rule generalizes easily to systems of n equations in n variables. The value of each variable is given as the quotient of two determinants. The denominator is the determinant of the coefficient matrix, and the numerator is the determinant of the matrix formed by replacing the column corresponding to the variable being solved for with the column representing the constants. For instance, the solution for x_3 in the following system is shown.

$$\begin{cases} a_{11}x_1 + a_{12}x_2 + a_{13}x_3 = b_1 \\ a_{21}x_1 + a_{22}x_2 + a_{23}x_3 = b_2 \\ a_{31}x_1 + a_{32}x_2 + a_{33}x_3 = b_3 \end{cases} \qquad x_3 = \frac{|A_3|}{|A|} = \frac{\begin{vmatrix} a_{11} & a_{12} & b_1 \\ a_{21} & a_{22} & b_2 \\ a_{31} & a_{32} & b_3 \end{vmatrix}}{\begin{vmatrix} a_{11} & a_{12} & a_{13} \\ a_{21} & a_{22} & a_{23} \\ a_{31} & a_{32} & a_{33} \end{vmatrix}}$$

Cramer's Rule

If a system of n linear equations in n variables has a coefficient matrix A with a *nonzero* determinant $|A|$, then the solution of the system is

$$x_1 = \frac{|A_1|}{|A|}, \quad x_2 = \frac{|A_2|}{|A|}, \ldots, \quad x_n = \frac{|A_n|}{|A|}$$

where the ith column of A_i is the column of constants in the system of equations. If the determinant of the coefficient matrix is zero, then the system has either no solution or infinitely many solutions.

EXAMPLE 3 Using Cramer's Rule for a 2 × 2 System

Use Cramer's Rule to solve the system.

$$\begin{cases} 4x - 2y = 10 \\ 3x - 5y = 11 \end{cases}$$

Solution

To begin, find the determinant of the coefficient matrix.

$$D = \begin{vmatrix} 4 & -2 \\ 3 & -5 \end{vmatrix} = -20 - (-6) = -14$$

Because this determinant is not zero, apply Cramer's Rule.

$$x = \frac{D_x}{D} = \frac{\begin{vmatrix} 10 & -2 \\ 11 & -5 \end{vmatrix}}{-14} = \frac{(-50) - (-22)}{-14} = \frac{-28}{-14} = 2$$

$$y = \frac{D_y}{D} = \frac{\begin{vmatrix} 4 & 10 \\ 3 & 11 \end{vmatrix}}{-14} = \frac{44 - 30}{-14} = \frac{14}{-14} = -1$$

So, the solution is $x = 2$ and $y = -1$. Check this in the original system.

Checkpoint *Audio-video solution in English & Spanish at LarsonPrecalculus.com.*

Use Cramer's Rule to solve the system.

$$\begin{cases} 3x + 4y = 1 \\ 5x + 3y = 9 \end{cases}$$

EXAMPLE 4 Using Cramer's Rule for a 3 × 3 System

Use Cramer's Rule, if possible, to solve the system of linear equations.

Coefficient Matrix

$$\begin{cases} -x + 2y - 3z = 1 \\ 2x \qquad\; + z = 0 \\ 3x - 4y + 4z = 2 \end{cases} \Rightarrow \begin{bmatrix} -1 & 2 & -3 \\ 2 & 0 & 1 \\ 3 & -4 & 4 \end{bmatrix}$$

Solution

The coefficient matrix above can be expanded along the second row, as follows.

$$D = 2(-1)^3\begin{vmatrix} 2 & -3 \\ -4 & 4 \end{vmatrix} + 0(-1)^4\begin{vmatrix} -1 & -3 \\ 3 & 4 \end{vmatrix} + 1(-1)^5\begin{vmatrix} -1 & 2 \\ 3 & -4 \end{vmatrix}$$

$$= -2(-4) + 0 - 1(-2)$$

$$= 10$$

Because this determinant is not zero, you can apply Cramer's Rule.

$$x = \frac{D_x}{D} = \frac{\begin{vmatrix} 1 & 2 & -3 \\ 0 & 0 & 1 \\ 2 & -4 & 4 \end{vmatrix}}{10} = \frac{8}{10} = \frac{4}{5}$$

$$y = \frac{D_y}{D} = \frac{\begin{vmatrix} -1 & 1 & -3 \\ 2 & 0 & 1 \\ 3 & 2 & 4 \end{vmatrix}}{10} = \frac{-15}{10} = -\frac{3}{2}$$

$$z = \frac{D_z}{D} = \frac{\begin{vmatrix} -1 & 2 & 1 \\ 2 & 0 & 0 \\ 3 & -4 & 2 \end{vmatrix}}{10} = \frac{-16}{10} = -\frac{8}{5}$$

The solution is

$$\left(\tfrac{4}{5}, -\tfrac{3}{2}, -\tfrac{8}{5}\right).$$

Check this in the original system.

Checkpoint *Audio-video solution in English & Spanish at LarsonPrecalculus.com.*

Use Cramer's Rule to solve the system.

$$\begin{cases} 4x - \;\, y + \;\, z = 12 \\ 2x + 2y + 3z = \;\, 1 \\ 5x - 2y + 6z = 22 \end{cases}$$

Remember that Cramer's Rule does not apply when the determinant of the coefficient matrix is zero. This would create division by zero, which is undefined. For example, consider this system of linear equations.

$$\begin{cases} -x \qquad\;\; + \;\, z = \;\; 4 \\ 2x - y + \;\, z = -3 \\ \qquad\;\; y - 3z = \;\; 1 \end{cases}$$

Using a graphing utility to evaluate the determinant of the coefficient matrix A, you find that Cramer's Rule cannot be applied because $|A| = 0$.

Technology Tip

Try using a graphing utility to evaluate D_x/D for the system of linear equations shown at the lower left. You should obtain an error message similar to the one shown below.

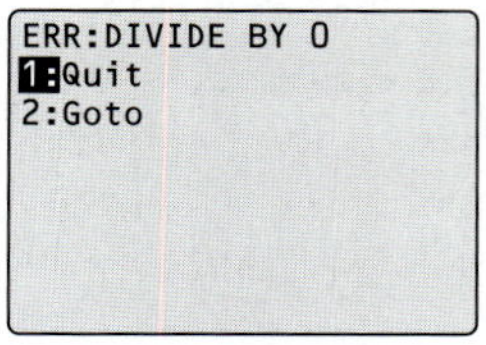

Cryptography

A **cryptogram** is a message written according to a secret code. (The Greek word *kryptos* means "hidden.") Matrix multiplication can be used to encode and decode messages. To begin, you need to assign a number to each letter in the alphabet (with 0 assigned to a blank space), as follows.

0 = _	9 = I	18 = R
1 = A	10 = J	19 = S
2 = B	11 = K	20 = T
3 = C	12 = L	21 = U
4 = D	13 = M	22 = V
5 = E	14 = N	23 = W
6 = F	15 = O	24 = X
7 = G	16 = P	25 = Y
8 = H	17 = Q	26 = Z

Then the message is converted to numbers and partitioned into **uncoded row matrices,** each having n entries, as demonstrated in Example 5.

Applied Cryptography Researcher

EXAMPLE 5 Forming Uncoded Row Matrices

Write the uncoded row matrices of dimension 1×3 for the message

MEET ME MONDAY.

Solution

Partitioning the message (including blank spaces, but ignoring punctuation) into groups of three produces the following uncoded row matrices.

[13 5 5] [20 0 13] [5 0 13] [15 14 4] [1 25 0]

M E E T _ M E _ M O N D A Y _

Note that a blank space is used to fill out the last uncoded row matrix.

✓ Checkpoint *Audio-video solution in English & Spanish at LarsonPrecalculus.com.*

Write the uncoded row matrices of order 1×3 for the message

OWLS ARE NOCTURNAL.

To encode a message, use the techniques demonstrated in Section 5.6 to choose an $n \times n$ invertible matrix such as

$$A = \begin{bmatrix} 1 & -2 & 2 \\ -1 & 1 & 3 \\ 1 & -1 & -4 \end{bmatrix}$$

and multiply the uncoded row matrices by A (on the right) to obtain **coded row matrices.** Here is an example.

Uncoded Matrix *Encoding Matrix A* *Coded Matrix*

$$\begin{bmatrix} 13 & 5 & 5 \end{bmatrix} \begin{bmatrix} 1 & -2 & 2 \\ -1 & 1 & 3 \\ 1 & -1 & -4 \end{bmatrix} = \begin{bmatrix} 13 & -26 & 21 \end{bmatrix}$$

This technique is further illustrated in Example 6.

Tom Wang | Dreamstime.com

EXAMPLE 6 Encoding a Message

Use the matrix A to encode the message MEET ME MONDAY.

$$A = \begin{bmatrix} 1 & -2 & 2 \\ -1 & 1 & 3 \\ 1 & -1 & -4 \end{bmatrix}$$

Solution

The coded row matrices are obtained by multiplying each of the uncoded row matrices found in Example 5 by the matrix A, as follows.

Uncoded Matrix *Encoding Matrix A* *Coded Matrix*

$$[13 \quad 5 \quad 5]\begin{bmatrix} 1 & -2 & 2 \\ -1 & 1 & 3 \\ 1 & -1 & -4 \end{bmatrix} = [13 \quad -26 \quad 21]$$

$$[20 \quad 0 \quad 13]\begin{bmatrix} 1 & -2 & 2 \\ -1 & 1 & 3 \\ 1 & -1 & -4 \end{bmatrix} = [33 \quad -53 \quad -12]$$

$$[5 \quad 0 \quad 13]\begin{bmatrix} 1 & -2 & 2 \\ -1 & 1 & 3 \\ 1 & -1 & -4 \end{bmatrix} = [18 \quad -23 \quad -42]$$

$$[15 \quad 14 \quad 4]\begin{bmatrix} 1 & -2 & 2 \\ -1 & 1 & 3 \\ 1 & -1 & -4 \end{bmatrix} = [5 \quad -20 \quad 56]$$

$$[1 \quad 25 \quad 0]\begin{bmatrix} 1 & -2 & 2 \\ -1 & 1 & 3 \\ 1 & -1 & -4 \end{bmatrix} = [-24 \quad 23 \quad 77]$$

So, the sequence of coded row matrices is

$$[13 \quad -26 \quad 21][33 \quad -53 \quad -12][18 \quad -23 \quad -42][5 \quad -20 \quad 56][-24 \quad 23 \quad 77].$$

Finally, removing the matrix notation produces the following cryptogram.

13 −26 21 33 −53 −12 18 −23 −42 5 −20 56 −24 23 77

✓ **Checkpoint** *Audio-video solution in English & Spanish at LarsonPrecalculus.com.*

Use the following invertible matrix to encode the message OWLS ARE NOCTURNAL.

$$A = \begin{bmatrix} 1 & -1 & 0 \\ 1 & 0 & -1 \\ 6 & -2 & -3 \end{bmatrix}$$

For those who do not know the encoding matrix A, decoding the cryptogram found in Example 6 is difficult. But for an authorized receiver who knows the encoding matrix A, decoding is simple. The receiver need only multiply the coded row matrices by A^{-1} (on the right) to retrieve the uncoded row matrices. Here is an example.

$$\underbrace{[13 \quad -26 \quad 21]}_{\text{Coded}}\underbrace{\begin{bmatrix} -1 & -10 & -8 \\ -1 & -6 & -5 \\ 0 & -1 & -1 \end{bmatrix}}_{A^{-1}} = \underbrace{[13 \quad 5 \quad 5]}_{\text{Uncoded}}$$

This technique is further illustrated in Example 7.

Technology Tip

An efficient method for encoding the message at the left with your graphing utility is to enter A as a 3×3 matrix. Let B be the 5×3 matrix whose rows are the uncoded row matrices

$$B = \begin{bmatrix} 13 & 5 & 5 \\ 20 & 0 & 13 \\ 5 & 0 & 13 \\ 15 & 14 & 4 \\ 1 & 25 & 0 \end{bmatrix}.$$

The product BA gives the coded row matrices.

EXAMPLE 7 Decoding a Message

Use the inverse of the matrix A in Example 6 to decode the cryptogram.

13 −26 21 33 −53 −12 18 −23 −42 5 −20 56 −24 23 77

Solution

First, find the decoding matrix A^{-1} by using the techniques demonstrated in Section 5.6. Next partition the message into groups of three to form the coded row matrices. Then multiply each coded row matrix by A^{-1} (on the right).

Coded Matrix *Decoding Matrix* A^{-1} *Decoded Matrix*

$$\begin{bmatrix} 13 & -26 & 21 \end{bmatrix}\begin{bmatrix} -1 & -10 & -8 \\ -1 & -6 & -5 \\ 0 & -1 & -1 \end{bmatrix} = \begin{bmatrix} 13 & 5 & 5 \end{bmatrix}$$

$$\begin{bmatrix} 33 & -53 & -12 \end{bmatrix}\begin{bmatrix} -1 & -10 & -8 \\ -1 & -6 & -5 \\ 0 & -1 & -1 \end{bmatrix} = \begin{bmatrix} 20 & 0 & 13 \end{bmatrix}$$

$$\begin{bmatrix} 18 & -23 & -42 \end{bmatrix}\begin{bmatrix} -1 & -10 & -8 \\ -1 & -6 & -5 \\ 0 & -1 & -1 \end{bmatrix} = \begin{bmatrix} 5 & 0 & 13 \end{bmatrix}$$

$$\begin{bmatrix} 5 & -20 & 56 \end{bmatrix}\begin{bmatrix} -1 & -10 & -8 \\ -1 & -6 & -5 \\ 0 & -1 & -1 \end{bmatrix} = \begin{bmatrix} 15 & 14 & 4 \end{bmatrix}$$

$$\begin{bmatrix} -24 & 23 & 77 \end{bmatrix}\begin{bmatrix} -1 & -10 & -8 \\ -1 & -6 & -5 \\ 0 & -1 & -1 \end{bmatrix} = \begin{bmatrix} 1 & 25 & 0 \end{bmatrix}$$

So, the message is as follows.

[13	5	5]	[20	0	13]	[5	0	13]	[15	14	4]	[1	25	0]
M	E	E	T		M	E		M	O	N	D	A	Y	

✓ **Checkpoint** *Audio-video solution in English & Spanish at LarsonPrecalculus.com.*

Use the inverse of the matrix A in the Checkpoint with Example 6 to decode the cryptogram.

110 −39 −59 25 −21 −3 23 −18 −5 47 −20 −24
149 −56 −75 87 −38 −37.

Technology Tip

An efficient method for decoding the cryptogram in Example 7 with your graphing utility is to enter A as a 3×3 matrix and then find A^{-1}. Let B be the 5×3 matrix whose rows are the coded row matrices, as shown below. The product BA^{-1} gives the decoded row matrices.

$$B = \begin{bmatrix} 13 & -26 & 21 \\ 33 & -53 & -12 \\ 18 & -23 & -42 \\ 5 & -20 & 56 \\ -24 & 23 & 77 \end{bmatrix}$$

5.8 Exercises

See *CalcChat.com* for tutorial help and worked-out solutions to odd-numbered exercises.
For instructions on how to use a graphing utility, see Appendix A.

Vocabulary and Concept Check

In Exercises 1 and 2, fill in the blank.

1. _______ is a method for using determinants to solve a system of linear equations.

2. A message written according to a secret code is called a _______ .

In Exercises 3 and 4, consider three points (x_1, y_1), (x_2, y_2), and (x_3, y_3), and the determinant shown at the right.

$$\begin{vmatrix} x_1 & y_1 & 1 \\ x_2 & y_2 & 1 \\ x_3 & y_3 & 1 \end{vmatrix}$$

3. Suppose the three points are vertices of a triangle and the value of the determinant is -6. What number do you multiply -6 by to find the area of the triangle?

4. Suppose the value of the determinant is 0. What can you conclude?

Procedures and Problem Solving

Finding an Area **In Exercises 5–10, use a determinant to find the area of the figure with the given vertices.**

5. $(-2, 4), (2, 3), (-1, 5)$ **6.** $(-3, 5), (2, 6), (3, -5)$

7. $\left(0, \frac{1}{2}\right), \left(\frac{5}{2}, 0\right), (4, 3)$ **8.** $\left(\frac{9}{2}, 0\right), (2, 6), \left(0, -\frac{3}{2}\right)$

9.

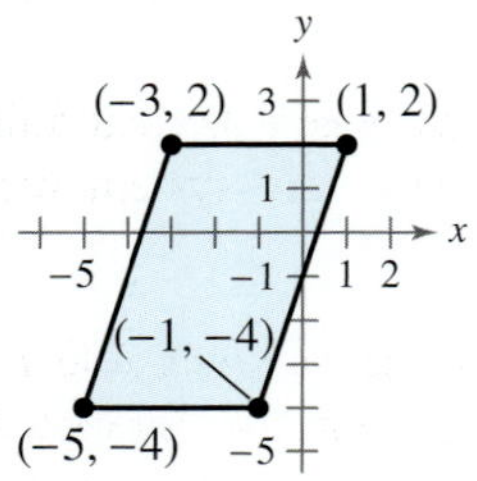

10.

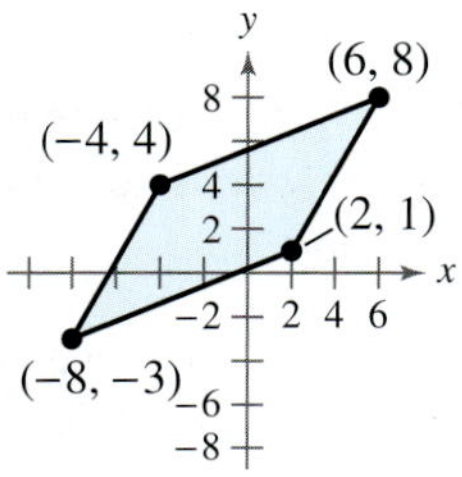

Finding a Coordinate **In Exercises 11 and 12, find x or y such that the triangle has an area of 4 square units.**

11. $(-1, 5), (-2, 0), (x, 2)$ **12.** $(-4, 2), (-3, 5), (-1, y)$

Testing for Collinear Points **In Exercises 13–16, use a determinant to determine whether the points are collinear.**

13. $(3, -1), (0, -3), (12, 5)$ **14.** $(3, -5), (6, 1), (4, 2)$

15. $\left(2, -\frac{1}{2}\right), (-4, 4), (6, -3)$ **16.** $\left(0, \frac{1}{2}\right), (2, -1), \left(-4, \frac{7}{2}\right)$

Finding a Coordinate **In Exercises 17 and 18, find x or y such that the points are collinear.**

17. $(1, -2), (x, 2), (5, 6)$ **18.** $(-6, 2), (-5, y), (-3, 5)$

Using Cramer's Rule **In Exercises 19–24, use Cramer's Rule to solve (if possible) the system of equations.**

19. $\begin{cases} -7x + 11y = -1 \\ 3x - 9y = 9 \end{cases}$ **20.** $\begin{cases} 3x + 2y = -2 \\ 6x + 4y = 4 \end{cases}$

21. $\begin{cases} 4x - 3y = -10 \\ 6x + 9y = 12 \end{cases}$ **22.** $\begin{cases} 6x - 5y = 17 \\ -13x + 3y = -76 \end{cases}$

23. $\begin{cases} 4x - y + z = -5 \\ 2x + 2y + 3z = 10 \\ 6x + y + 4z = -5 \end{cases}$ **24.** $\begin{cases} 4x - 2y + 3z = -2 \\ 2x + 2y + 5z = 16 \\ 8x - 5y - 2z = 4 \end{cases}$

Comparing Solution Methods **In Exercises 25 and 26, solve the system of equations using (a) Gaussian elimination and (b) Cramer's Rule. Which method do you prefer, and why?**

25. $\begin{cases} 3x + 3y + 5z = 1 \\ 3x + 5y + 9z = 2 \\ 5x + 9y + 17z = 4 \end{cases}$ **26.** $\begin{cases} 2x + 3y - 5z = 1 \\ 3x + 5y + 9z = -16 \\ 5x + 9y + 17z = -30 \end{cases}$

27. ***Why you should learn it*** *(p. 469)* The retail sales of family clothing stores in the United States from 2009 through 2013 are shown in the table. (Source: U.S. Census Bureau)

Year	Sales (in billions of dollars)
2009	81.4
2010	85.3
2011	89.9
2012	94.7
2013	99.5

The coefficients of the least squares regression parabola $y = at^2 + bt + c$, where y represents the retail sales (in billions of dollars) and t represents the year, with $t = 9$ corresponding to 2009, can be found by solving the system

$$\begin{cases} 80{,}499a + 6985b + 615c = 56{,}453.6 \\ 6985a + 615b + 55c = 5004.4 \\ 615a + 55b + 5c = 450.8 \end{cases}$$

(a) Use Cramer's Rule to solve the system and write the least squares regression parabola for the data.

(b) Use a graphing utility to graph the parabola with the data. How well does the model fit the data?

(c) Use the model to predict the retail sales of family clothing stores in the U.S. in the year 2015.

28. MODELING DATA

The retail sales y (in billions of dollars) of stores selling auto parts, accessories, and tires in the United States from 2009 through 2013 are given by the ordered pairs of the form $(t, y(t))$, where $t = 9$ represents 2009. (Source: U.S. Census Bureau)

(9, 74.1) (10, 77.7) (11, 82.7)
(12, 83.9) (13, 82.8)

The coefficients of the least squares regression parabola $y = at^2 + bt + c$ can be found by solving the system

$$\begin{cases} 80{,}499a + 6985b + 615c = 49{,}853.8 \\ 6985a + 615b + 55c = 4436.8. \\ 615a + 55b + 5c = 401.2 \end{cases}$$

(a) Use Cramer's Rule to solve the system and write the least squares regression parabola for the data.

(b) Use a graphing utility to graph the parabola with the data. How well does the model fit the data?

(c) Is this a good model for predicting retail sales in future years? Explain.

Encoding a Message **In Exercises 29 and 30, (a) write the uncoded 1 × 2 row matrices for the message, and then (b) encode the message using the encoding matrix.**

Message	*Encoding Matrix*
29. COME HOME SOON	$\begin{bmatrix} 1 & 2 \\ 3 & 5 \end{bmatrix}$
30. HELP IS ON THE WAY	$\begin{bmatrix} -2 & 3 \\ -1 & 1 \end{bmatrix}$

Encoding a Message **In Exercises 31 and 32, (a) write the uncoded 1 × 3 row matrices for the message, and then (b) encode the message using the encoding matrix.**

Message	*Encoding Matrix*
31. CALL ME TOMORROW	$\begin{bmatrix} 1 & -1 & 0 \\ 1 & 0 & -1 \\ -6 & 2 & 3 \end{bmatrix}$
32. HAPPY BIRTHDAY	$\begin{bmatrix} 1 & 2 & 2 \\ 3 & 7 & 9 \\ -1 & -4 & -7 \end{bmatrix}$

Decoding a Message **In Exercises 33 and 34, use A^{-1} to decode the cryptogram.**

33. $A = \begin{bmatrix} 1 & 2 \\ 3 & 5 \end{bmatrix}$ 11 21 64 112 25 50 29 53 23 46 40 75 55 92

34. $A = \begin{bmatrix} 3 & -4 & 2 \\ 0 & 2 & 1 \\ 4 & -5 & 3 \end{bmatrix}$ 112 −140 83 19 −25 13 72 −76 61 95 −118 71 20 21 38 35 −23 36 42 −48 32

Nfsphot | Dreamstime.com

35. Decoding a Message The following cryptogram was encoded with a 2 × 2 matrix.

5 2 25 11 −2 −7 −15 −15 32 14 −8 −13 38 19 −19 −19 37 16

The last word of the message is _SUE. What is the message?

36. ***Why you should learn it*** *(p. 476)* The following cryptogram was encoded with a 2 × 2 matrix.

8 21 −15 −10 −13 −13 5 10 5 25 5 19 −1 6 20 40 −18 −18 1 16

The last word of the message is _RON. What is the message?

Conclusions

True or False? **In Exercises 37 and 38, determine whether the statement is true or false. Justify your answer.**

37. Cramer's Rule cannot be used to solve a system of linear equations when the determinant of the coefficient matrix is zero.

38. In a system of linear equations, when the determinant of the coefficient matrix is zero, the system has no solution.

39. Writing Describe a way to use an invertible $n \times n$ matrix to encode a message that is converted to numbers and partitioned into uncoded column matrices.

40. HOW DO YOU SEE IT? At this point in the text, you have learned several methods for finding an equation of a line that passes through two given points. Briefly describe these methods and discuss the advantages and disadvantages of each.

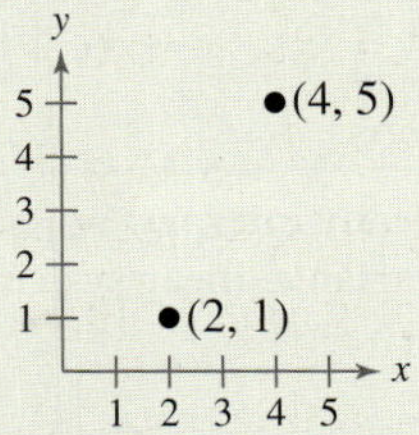

Cumulative Mixed Review

Equation of a Line **In Exercises 41–44, find the general form of the equation of the line that passes through the two points.**

41. $(-1, 5), (7, 3)$ **42.** $(0, -6), (-2, 10)$

43. $(3, -3), (10, -1)$ **44.** $(-4, 12), (4, 2)$

5 Chapter Summary

	What did you learn?	Explanation and Examples	Review Exercises
5.1	**Use the methods of substitution and graphing to solve systems of equations in two variables (*p. 398*), and use systems of equations to model and solve real-life problems (*p. 403*).**	**Substitution:** (1) Solve one of the equations for one variable in terms of the other. (2) Substitute the expression found in Step 1 into the other equation to obtain an equation in one variable and (3) solve the equation. (4) Back-substitute the value(s) obtained in Step 3 into the expression obtained in Step 1 to find the value(s) of the other variable. (5) Check the solution(s). **Graphing:** (1) Solve both equations for y in terms of x. (2) Use a graphing utility to graph both equations. (3) Use the *intersect* feature or the *zoom* and *trace* features of the graphing utility to approximate the point(s) of intersection of the graphs. (4) Check the solution(s).	1–20
5.2	**Use the method of elimination to solve systems of linear equations in two variables (*p. 408*), and graphically interpret the number of solutions of a system of linear equations in two variables (*p. 410*).**	**Elimination:** (1) Obtain coefficients for x (or y) that differ only in sign by multiplying all terms of one or both equations by suitably chosen constants. (2) Add the equations to eliminate one variable and solve the resulting equation. (3) Back-substitute the value obtained in Step 2 into either of the original equations and solve for the other variable. (4) Check the solution(s).	21–36
	Use systems of linear equations in two variables to model and solve real-life problems (*p. 412*).	A system of linear equations in two variables can be used to find the airspeed of an airplane and the speed of the wind. (See Example 6.)	37–40
5.3	**Use back-substitution to solve linear systems in row-echelon form (*p. 417*).**	$\begin{cases} x - 2y + 3z = 9 \\ -x + 3y = -4 \\ 2x - 5y + 5z = 17 \end{cases} \rightarrow \begin{cases} x - 2y + 3z = 9 \\ y + 3z = 5 \\ z = 2 \end{cases}$	41, 42
	Use Gaussian elimination to solve systems of linear equations (*p. 418*).	Use elementary row operations to convert a system of linear equations to row-echelon form. (1) Interchange two equations. (2) Multiply one of the equations by a nonzero constant. (3) Add a multiple of one equation to another equation.	43–48
	Solve nonsquare systems of linear equations (*p. 421*).	In a nonsquare system, the number of equations differs from the number of variables. A system of linear equations cannot have a unique solution unless there are at least as many equations as there are variables.	49, 50
	Graphically interpret three-variable linear systems (*p. 422*).	The graph of a system of three linear equations in three variables consists of three planes. When they intersect in a single point, the system has exactly one solution. When they have no point in common, the system has no solution. When they intersect in a line or a plane, the system has infinitely many solutions (see Figures 5.9–5.13).	51, 52
	Use systems of linear equations to write partial fraction decompositions of rational expressions (*p. 423*).	$\dfrac{9}{x^3 - 6x^2} = \dfrac{9}{x^2(x-6)} = \dfrac{A}{x} + \dfrac{B}{x^2} + \dfrac{C}{x-6}$	53–58
	Use systems of linear equations in three or more variables to model and solve real-life problems (*p. 426*).	A system of linear equations in three variables can be used to find the position equation of an object that is moving in a (vertical) line with constant acceleration. (See Example 8.)	59–62

	What did you learn?	Explanation and Examples	Review Exercises
5.4	**Write matrices and identify their dimensions (*p. 432*), and perform elementary row operations on matrices (*p. 434*).**	**Elementary Row Operations** **1.** Interchange two rows. **2.** Multiply a row by a nonzero constant. **3.** Add a multiple of a row to another row.	63–80
	Use matrices and Gaussian elimination to solve systems of linear equations (*p. 435*).	Write the augmented matrix of the system. Use elementary row operations to rewrite the augmented matrix in row-echelon form. Write the system of linear equations corresponding to the matrix in row-echelon form and use back-substitution to find the solution.	81–88
	Use matrices and Gauss-Jordan elimination to solve systems of linear equations (*p. 439*).	Gauss-Jordan elimination continues the reduction process on a matrix in row-echelon form until a reduced row-echelon form is obtained. (See Example 8.)	89–100
5.5	**Decide whether two matrices are equal (*p. 446*).**	Two matrices are equal when they have the same dimension and all of their corresponding entries are equal.	101–104
	Add and subtract matrices and multiply matrices by scalars (*p. 447*), multiply two matrices (*p. 450*), and use matrix operations to model and solve real-life problems (*p. 453*).	**1.** Let $A = [a_{ij}]$ and $B = [b_{ij}]$ be matrices of dimension $m \times n$ and let c be a scalar. $A + B = [a_{ij} + b_{ij}]$ $\quad cA = [ca_{ij}]$ **2.** Let $A = [a_{ij}]$ be an $m \times n$ matrix and let $B = [b_{ij}]$ be an $n \times p$ matrix. The product AB is an $m \times p$ matrix given by $AB = [c_{ij}]$, where $c_{ij} = a_{i1}b_{1j} + a_{i2}b_{2j} + a_{i3}b_{3j} + \cdot\cdot\cdot + a_{in}b_{nj}$.	105–128
5.6	**Verify that two matrices are inverses of each other (*p. 460*), and use Gauss-Jordan elimination to find inverses of matrices (*p. 461*).**	Write the $n \times 2n$ matrix $[A \ \vdots \ I]$. Row reduce A to I using elementary row operations. The result will be the matrix $[I \ \vdots \ A^{-1}]$.	129–140
	Use a formula to find inverses of 2 × 2 matrices (*p. 464*).	If $A = \begin{bmatrix} a & b \\ c & d \end{bmatrix}$ and $ad - bc \neq 0$, then $A^{-1} = \dfrac{1}{ad - bc}\begin{bmatrix} d & -b \\ -c & a \end{bmatrix}$.	141–146
	Use inverse matrices to solve systems of linear equations (*p. 465*).	If A is an invertible matrix, then the system of linear equations represented by $AX = B$ has a unique solution given by $X = A^{-1}B$.	147–156
5.7	**Find the determinants of 2 × 2 matrices (*p. 469*).**	$\det(A) = \|A\| = \begin{vmatrix} a_1 & b_1 \\ a_2 & b_2 \end{vmatrix} = a_1b_2 - a_2b_1$	157–162
	Find minors and cofactors of square matrices (*p. 471*), and find the determinants of square matrices (*p. 472*).	The determinant of a square matrix A (of dimension 2 × 2 or greater) is the sum of the entries in any row (or column) of A multiplied by their respective cofactors.	163–174
5.8	**Use determinants to find areas of triangles (*p. 476*) and to determine whether points are collinear (*p. 477*), use Cramer's Rule to solve systems of linear equations (*p. 478*), and use matrices to encode and decode messages (*p. 481*).**	If a system of n linear equations in n variables has a coefficient matrix A with a *nonzero* determinant $\|A\|$, then the solution of the system is $x_1 = \dfrac{\|A_1\|}{\|A\|}, x_2 = \dfrac{\|A_2\|}{\|A\|}, \ldots, x_n = \dfrac{\|A_n\|}{\|A\|}$ where the ith column of A_i is the column of constants in the system of equations.	175–201

5 Review Exercises

See CalcChat.com for tutorial help and worked-out solutions to odd-numbered exercises. For instructions on how to use a graphing utility, see Appendix A.

5.1

Solving a System by Substitution **In Exercises 1–10, solve the system by the method of substitution.**

1. $\begin{cases} x + y = 2 \\ x - y = 0 \end{cases}$

2. $\begin{cases} 2x - 3y = 3 \\ x - y = 0 \end{cases}$

3. $\begin{cases} 4x - y = 1 \\ -8x + 2y = 5 \end{cases}$

4. $\begin{cases} 10x + 6y = -14 \\ x + 9y = -7 \end{cases}$

5. $\begin{cases} 0.5x + y = 0.75 \\ 1.25x - 4.5y = -2.5 \end{cases}$

6. $\begin{cases} -x + \frac{2}{5}y = \frac{3}{5} \\ -x + \frac{1}{5}y = -\frac{4}{5} \end{cases}$

7. $\begin{cases} x^2 - y^2 = 9 \\ x - y = 1 \end{cases}$

8. $\begin{cases} x^2 + y^2 = 169 \\ 3x + 2y = 39 \end{cases}$

9. $\begin{cases} y = 2x^2 \\ y = x^4 - 2x^2 \end{cases}$

10. $\begin{cases} x = y + 3 \\ x = y^2 + 7 \end{cases}$

Solving a System of Equations Graphically **In Exercises 11–16, use a graphing utility to approximate all points of intersection of the graphs of the equations in the system. Verify your solutions by checking them in the original system.**

11. $\begin{cases} 5x + 6y = 7 \\ -x - 4y = 0 \end{cases}$

12. $\begin{cases} 8x - 3y = -3 \\ 2x + 5y = 28 \end{cases}$

13. $\begin{cases} y^2 - 4x = 0 \\ x + y = 0 \end{cases}$

14. $\begin{cases} y = 2x^2 - 4x + 1 \\ y = x^2 - 4x + 3 \end{cases}$

15. $\begin{cases} y = 2(6 - x) \\ y = 2^{x-2} \end{cases}$

16. $\begin{cases} y = \ln(x + 2) + 1 \\ x + y = 0 \end{cases}$

17. **Finance** You invest \$5000 in a greenhouse. The planter, potting soil, and seed for each plant cost \$6.43, and the selling price of each plant is \$12.68. How many plants must you sell to break even?

18. **Finance** You are offered two sales jobs. One company offers an annual salary of \$55,000 plus a year-end bonus of 1.5% of your total sales. The other company offers an annual salary of \$52,000 plus a year-end bonus of 2% of your total sales. How much would you have to sell in a year to make the second offer the better offer?

19. **Geometry** The perimeter of a rectangle is 480 meters and its length is 1.5 times its width. Find the dimensions of the rectangle.

20. **Geometry** The perimeter of a rectangle is 68 feet and its width is $\frac{8}{9}$ times its length. Find the dimensions of the rectangle.

5.2

Solving a System by Elimination **In Exercises 21–30, solve the system by the method of elimination.**

21. $\begin{cases} x + 3y = -5 \\ -x - 8y = 0 \end{cases}$

22. $\begin{cases} 7x + 2y = 12 \\ x - 2y = -4 \end{cases}$

23. $\begin{cases} 2x - y = 2 \\ 6x + 8y = 39 \end{cases}$

24. $\begin{cases} 7x + 12y = 63 \\ 2x + 3y = 15 \end{cases}$

25. $\begin{cases} \frac{1}{5}x + \frac{3}{10}y = 7 \\ \frac{2}{5}x + \frac{3}{5}y = 14 \end{cases}$

26. $\begin{cases} 0.2x + 0.3y = 0.14 \\ 0.4x + 0.5y = 0.20 \end{cases}$

27. $\begin{cases} 3x - 2y = 0 \\ 3x + 2(y + 5) = 10 \end{cases}$

28. $\begin{cases} 6x - 4(y - 1) = 19 \\ -3x + 2y = 3 \end{cases}$

29. $\begin{cases} 1.5x + 2.5y = 8.5 \\ 6x + 10y = 24 \end{cases}$

30. $\begin{cases} 1.25x - 2y = 3.5 \\ 5x - 8y = 14 \end{cases}$

Solving a System Graphically **In Exercises 31–36, use a graphing utility to graph the lines in the system. Use the graphs to determine whether the system is consistent or inconsistent. If the system is consistent, determine the solution. Verify your results algebraically.**

31. $\begin{cases} 3x + 2y = 0 \\ x - y = 4 \end{cases}$

32. $\begin{cases} x + y = 6 \\ -2x - 2y = -12 \end{cases}$

33. $\begin{cases} \frac{1}{4}x - \frac{1}{5}y = 2 \\ -5x + 4y = 8 \end{cases}$

34. $\begin{cases} \frac{7}{2}x - 7y = -1 \\ -x + 2y = 4 \end{cases}$

35. $\begin{cases} 8x - 2y = 10 \\ -6x + 1.5y = -7.5 \end{cases}$

36. $\begin{cases} -x + 3.2y = 10.4 \\ -2x - 9.6y = 6.4 \end{cases}$

Supply and Demand **In Exercises 37 and 38, find the point of equilibrium of the demand and supply equations.**

	Demand	*Supply*
37.	$p = 37 - 0.0002x$	$p = 22 + 0.00001x$
38.	$p = 120 - 0.0001x$	$p = 45 + 0.0002x$

39. **Aviation** Two planes leave Pittsburgh and Philadelphia at the same time, each going to the other city. One plane flies 25 miles per hour faster than the other. Find the airspeed of each plane given that the cities are 275 miles apart and the planes pass each other after 40 minutes of flying time.

40. **Economics** A total of \$46,000 is invested in two corporate bonds that pay 6.75% and 7.25% simple interest. The investor wants an annual interest income of \$3245 from the investments. What is the most that can be invested in the 6.75% bond?

5.3

Using Back-Substitution **In Exercises 41 and 42, use back-substitution to solve the system of linear equations.**

41. $\begin{cases} x - 4y + 3z = 3 \\ -y + z = -1 \\ z = -5 \end{cases}$

42. $\begin{cases} x - 7y + 8z = -14 \\ y - 9z = 26 \\ z = -3 \end{cases}$

Solving a System of Linear Equations **In Exercises 43–50, solve the system of linear equations and check any solution algebraically.**

43. $\begin{cases} x - 2y + 5z = -12 \\ 2x + 6z = -28 \\ x + y + 18z = -4 \end{cases}$

44. $\begin{cases} x + y + z = 2 \\ -x + 3y + 2z = 8 \\ 4x + y = 4 \end{cases}$

45. $\begin{cases} x - 2y + z = -6 \\ 2x - 3y = -7 \\ -x + 3y - 3z = 11 \end{cases}$

46. $\begin{cases} 2x + 6z = -9 \\ 3x - 2y + 11z = -16 \\ 3x - y + 7z = -11 \end{cases}$

47. $\begin{cases} x - 2y + 3z = -5 \\ 2x + 4y + 5z = 1 \\ x + 2y + z = 0 \end{cases}$

48. $\begin{cases} x - 2y + z = 5 \\ 2x + 3y + z = 5 \\ x + y + 2z = 3 \end{cases}$

49. $\begin{cases} 5x - 12y + 7z = 16 \\ 3x - 7y + 4z = 9 \end{cases}$

50. $\begin{cases} 2x + 5y - 19z = 34 \\ 3x + 8y - 31z = 54 \end{cases}$

Sketching a Plane **In Exercises 51 and 52, sketch the plane represented by the linear equation. Then list four points that lie in the plane.**

51. $x + 5y + z = 10$

52. $2x + 3y + 4z = 12$

Partial Fraction Decomposition **In Exercises 53–58, write the partial fraction decomposition for the rational expression. Check your result algebraically by combining the fractions, and check your result graphically by using a graphing utility to graph the rational expression and the partial fractions in the same viewing window.**

53. $\dfrac{4 - x}{x^2 + 6x + 8}$

54. $\dfrac{-x}{x^2 + 3x + 2}$

55. $\dfrac{9}{x^2 - 9}$

56. $\dfrac{x^2}{x^2 + 2x - 15}$

57. $\dfrac{x^2 + 2x}{x^3 - x^2 + x - 1}$

58. $\dfrac{2x^2 - x + 7}{x^4 + 8x^2 + 16}$

Data Analysis: Curve-Fitting **In Exercises 59 and 60, find the equation of the parabola $y = ax^2 + bx + c$ that passes through the points. To verify your result, use a graphing utility to plot the points and graph the parabola.**

59. $(-1, -4), (1, -2), (2, 5)$

60. $(-1, 0)\ (1, 4), (2, 3)$

61. Physical Education Pebble Beach Golf Links in Pebble Beach, California, is an 18-hole course that consists of par-3 holes, par-4 holes, and par-5 holes. There are two more par-4 holes than twice the number of par-5 holes, and the number of par-3 holes is equal to the number of par-5 holes. Find the number of par-3, par-4, and par-5 holes on the course. (Source: Pebble Beach Resorts)

62. Economics An inheritance of \$40,000 is divided among three investments yielding \$3500 in interest per year. The interest rates for the three investments are 7%, 9%, and 11%. Find the amount of each investment when the second and third are \$3000 and \$5000 less than the first, respectively.

5.4

Dimension of a Matrix **In Exercises 63–66, determine the dimension of the matrix.**

63. $\begin{bmatrix} -3 \\ 1 \\ 10 \end{bmatrix}$

64. $\begin{bmatrix} 3 & -1 & 0 & 6 \\ -2 & 7 & 1 & 4 \end{bmatrix}$

65. $[3]$

66. $[6 \quad 7 \quad -5 \quad 0 \quad -8]$

Writing an Augmented Matrix **In Exercises 67–70, write the augmented matrix for the system of linear equations.**

67. $\begin{cases} 6x - 7y = 11 \\ -2x + 5y = -1 \end{cases}$

68. $\begin{cases} -x + y = 12 \\ 10x = -90 \end{cases}$

69. $\begin{cases} 3x - 5y + z = 25 \\ -4x - 2z = -14 \\ 6x + y = 15 \end{cases}$

70. $\begin{cases} 8x - 7y + 4z = 12 \\ 3x - 5y + 2z = 20 \\ 5x + 3y - 3z = 26 \end{cases}$

Writing a System of Equations **In Exercises 71 and 72, write the system of linear equations represented by the augmented matrix. (Use the variables *x*, *y*, *z*, and *w*, if applicable.)**

71. $\begin{bmatrix} 5 & 1 & 7 & \vdots & -9 \\ 4 & 2 & 0 & \vdots & 10 \\ 9 & 4 & 2 & \vdots & 3 \end{bmatrix}$

72. $\begin{bmatrix} 13 & 16 & 7 & 3 & \vdots & 0 \\ 0 & 21 & 0 & 5 & \vdots & 12 \\ 4 & 10 & -4 & 3 & \vdots & -1 \end{bmatrix}$

Using Gaussian Elimination **In Exercises 73–76, write the matrix in row-echelon form. Remember that the row-echelon form of a matrix is not unique.**

73. $\begin{bmatrix} -1 & 4 & 0 \\ 3 & -5 & 9 \end{bmatrix}$

74. $\begin{bmatrix} 0 & -2 & 7 \\ 2 & 6 & -14 \end{bmatrix}$

75. $\begin{bmatrix} 0 & 1 & 1 \\ 1 & 2 & 3 \\ 2 & 2 & 2 \end{bmatrix}$

76. $\begin{bmatrix} 4 & 8 & 16 & -4 \\ 3 & -1 & 2 & 0 \\ -2 & 10 & 12 & 3 \end{bmatrix}$

Using a Graphing Utility **In Exercises 77–80, use the matrix capabilities of a graphing utility to write the matrix in reduced row-echelon form.**

77. $\begin{bmatrix} 3 & -2 \\ 4 & -3 \end{bmatrix}$

78. $\begin{bmatrix} 1 & 1 & 2 \\ -1 & 0 & 3 \\ 1 & 2 & 8 \end{bmatrix}$

79. $\begin{bmatrix} 1.5 & 3.6 & 4.2 \\ 0.2 & 1.4 & 1.8 \\ 2.0 & 4.4 & 6.4 \end{bmatrix}$

80. $\begin{bmatrix} 4.1 & 8.3 & 1.6 \\ 3.2 & -1.7 & 2.4 \\ -2.3 & 1.0 & 1.2 \end{bmatrix}$

Gaussian Elimination with Back-Substitution **In Exercises 81–88, use matrices to solve the system of equations, if possible. Use Gaussian elimination with back-substitution.**

81. $\begin{cases} 5x + 4y = 2 \\ -x + y = -22 \end{cases}$

82. $\begin{cases} 2x - 5y = 2 \\ 3x - 7y = 1 \end{cases}$

83. $\begin{cases} 0.6x - 0.9y = -0.3 \\ 0.4x - 0.6y = 0.3 \end{cases}$

84. $\begin{cases} 0.2x - 0.1y = 0.07 \\ 0.4x - 0.5y = -0.01 \end{cases}$

85. $\begin{cases} 2x + 3y + 3z = 3 \\ 6x + 6y + 12z = 13 \\ 12x + 9y - z = 2 \end{cases}$

86. $\begin{cases} x + 2y + 6z = 1 \\ 2x + 5y + 15z = 4 \\ 3x + y + 3z = -6 \end{cases}$

87. $\begin{cases} x + 2y - z = 1 \\ y + z = 0 \end{cases}$

88. $\begin{cases} x - y + 4z - w = 4 \\ x + 3y - 2z + w = -4 \\ y - z + w = -3 \\ 2x + z + w = 0 \end{cases}$

Gauss-Jordan Elimination **In Exercises 89–96, use matrices to solve the system of equations, if possible. Use Gauss-Jordan elimination.**

89. $\begin{cases} x + y + z = -3 \\ 4x + y - 2z = -12 \end{cases}$

90. $\begin{cases} 2x + 2y = 4 \\ -3x - y = -16 \end{cases}$

91. $\begin{cases} -x + y + 2z = 1 \\ 2x + 3y + z = -2 \\ 5x + 4y + 2z = 4 \end{cases}$

92. $\begin{cases} 4x + 4y + 4z = 5 \\ 4x - 2y - 8z = 1 \\ 5x + 3y + 8z = 6 \end{cases}$

93. $\begin{cases} x + y + 2z = 4 \\ x - y + 4z = 1 \\ x + 3z = -5 \end{cases}$

94. $\begin{cases} x + y + 4z = 0 \\ 2x + y + 2z = 0 \\ -x + y - 2z = -1 \end{cases}$

95. $\begin{cases} x + 2y - z = 3 \\ x - y - z = -3 \\ 2x + y + 3z = 10 \end{cases}$

96. $\begin{cases} x - 3y + z = 2 \\ 3x - y - z = -6 \\ -x - 5y + 3z = 10 \end{cases}$

Using a Graphing Utility **In Exercises 97–100, use the matrix capabilities of a graphing utility to reduce the augmented matrix corresponding to the system of equations and solve the system.**

97. $\begin{cases} x + 2y - z = 7 \\ -y - z = 4 \\ 4x - z = 16 \end{cases}$

98. $\begin{cases} 3x + 6z = 0 \\ -2x + y = 5 \\ y + 2z = 3 \end{cases}$

99. $\begin{cases} 4x + 12y + 2z = 20 \\ x + 6y + 4z = 12 \\ 5x + 6y - 8z = -4 \end{cases}$

100. $\begin{cases} 3x - y + 5z - 2w = -44 \\ x + 6y + 4z - w = 1 \\ 5x - y + z + 3w = -15 \\ 4y - z - 8w = 58 \end{cases}$

5.5

Equality of Matrices **In Exercises 101–104, find x and y.**

101. $\begin{bmatrix} -1 & x \\ y & 9 \end{bmatrix} = \begin{bmatrix} -1 & 12 \\ -7 & 9 \end{bmatrix}$

102. $\begin{bmatrix} -1 & 0 \\ x & 5 \\ -4 & y \end{bmatrix} = \begin{bmatrix} -1 & 0 \\ 8 & 5 \\ -4 & 0 \end{bmatrix}$

103. $\begin{bmatrix} x+3 & 4 & -4y \\ 0 & -3 & 2 \\ -2 & y+5 & 6x \end{bmatrix} = \begin{bmatrix} 5x-1 & 4 & -44 \\ 0 & -3 & 2 \\ -2 & 16 & 6 \end{bmatrix}$

104. $\begin{bmatrix} -9 & 4 & 2x & -5 \\ 0 & -3 & 7 & -4 \\ -5 & -2 & 1 & 0 \end{bmatrix} = \begin{bmatrix} -9 & 4 & x-10 & -5 \\ 0 & -3 & 7 & 2y \\ \frac{1}{2}x & y & 1 & 0 \end{bmatrix}$

Operations with Matrices **In Exercises 105–108, find, if possible, (a) $A + B$, (b) $A - B$, (c) $2A$, and (d) $A + 3B$.**

105. $A = \begin{bmatrix} 7 & 3 \\ -1 & 5 \end{bmatrix}, \quad B = \begin{bmatrix} 10 & -20 \\ 14 & -3 \end{bmatrix}$

106. $A = \begin{bmatrix} 6 & 0 & 7 \\ 5 & -1 & 2 \\ 3 & 2 & 3 \end{bmatrix}, \quad B = \begin{bmatrix} 0 & 5 & 1 \\ -4 & 8 & 6 \\ 2 & -1 & 1 \end{bmatrix}$

107. $A = \begin{bmatrix} -11 & 16 & 19 \\ -7 & -2 & 1 \end{bmatrix}, \quad B = \begin{bmatrix} 6 & 0 \\ 8 & -4 \\ -2 & 10 \end{bmatrix}$

108. $A = \begin{bmatrix} 6 & 5 & 8 \\ -1 & 2 & 4 \\ 3 & 9 & -2 \end{bmatrix}, \quad B = \begin{bmatrix} -8 \\ 3 \\ 7 \end{bmatrix}$

Evaluating an Expression **In Exercises 109–112, evaluate the expression.**

109. $\begin{bmatrix} 2 & 1 & 0 \\ 0 & 5 & -4 \end{bmatrix} - 3\begin{bmatrix} 5 & 3 & -6 \\ 0 & -2 & 5 \end{bmatrix}$

110. $-4\begin{bmatrix} 1 & 2 \\ 5 & -4 \\ 6 & 0 \end{bmatrix} + 8\begin{bmatrix} 7 & 1 \\ 1 & 2 \\ 1 & 4 \end{bmatrix}$

111. $-1\begin{bmatrix} 8 & -1 \\ -2 & 4 \end{bmatrix} + 5\left(\begin{bmatrix} -2 & 0 \\ 3 & -1 \end{bmatrix} + \begin{bmatrix} 7 & -8 \\ 4 & 3 \end{bmatrix}\right)$

112. $6\left(\begin{bmatrix} -4 & -1 & -3 & 4 \\ 2 & -5 & 7 & -10 \end{bmatrix} - \begin{bmatrix} -1 & 1 & 13 & -7 \\ 14 & -3 & 8 & -1 \end{bmatrix}\right)$

Operations with Matrices **In Exercises 113 and 114, use the matrix capabilities of a graphing utility to evaluate the expression.**

113. $-\frac{3}{8}\begin{bmatrix} 8 & -2 & 5 \\ 1 & 3 & -1 \end{bmatrix} + 6\begin{bmatrix} 4 & -2 & -3 \\ 2 & 7 & 6 \end{bmatrix}$

114. $-5\begin{bmatrix} 2.7 & 0.2 \\ 7.3 & -2.9 \\ 8.6 & 2.1 \end{bmatrix} + \frac{7}{4}\begin{bmatrix} 4.4 & -2.3 \\ 6.6 & 11.6 \\ -1.5 & 3.9 \end{bmatrix}$

Solving a Matrix Equation **In Exercises 115–118, solve for X when**

$$A = \begin{bmatrix} -4 & 0 \\ 1 & -5 \\ -3 & 2 \end{bmatrix} \text{ and } B = \begin{bmatrix} -2 & 2 \\ -2 & 1 \\ 4 & 4 \end{bmatrix}.$$

115. $2X = 3A - 2B$

116. $6X = 6A + 3B$

117. $3X + 2A = B$

118. $2A = 3X + 5B$

Finding the Product of Two Matrices **In Exercises 119–122, find AB, if possible.**

119. $A = \begin{bmatrix} 1 & 2 \\ 5 & -4 \\ 6 & 0 \end{bmatrix}$, $B = \begin{bmatrix} 6 & -2 & 8 \\ 4 & 0 & 0 \end{bmatrix}$

120. $A = \begin{bmatrix} 6 & -5 & 7 \end{bmatrix}$, $B = \begin{bmatrix} -1 & 2 & 1 \\ 4 & -3 & 0 \\ 8 & 1 & 1 \end{bmatrix}$

121. $A = \begin{bmatrix} 1 & 5 & 6 \\ 2 & -4 & 0 \end{bmatrix}$, $B = \begin{bmatrix} 7 & 5 & 2 \\ 0 & 1 & 0 \end{bmatrix}$

122. $A = \begin{bmatrix} 3 & 2 \\ 2 & -4 \\ -1 & 3 \end{bmatrix}$, $B = \begin{bmatrix} 4 & -3 & 2 \\ 0 & 3 & -1 \\ 0 & 6 & 2 \end{bmatrix}$

Operations with Matrices **In Exercises 123–126, use the matrix capabilities of a graphing utility to evaluate the expression.**

123. $\begin{bmatrix} 4 & 1 \\ 11 & -7 \\ 12 & 3 \end{bmatrix}\begin{bmatrix} 3 & -5 & 6 \\ 2 & -2 & -2 \end{bmatrix}$

124. $\begin{bmatrix} -2 & 3 & 10 \\ 4 & -2 & 2 \end{bmatrix}\begin{bmatrix} 1 & 1 \\ -5 & 2 \\ 3 & 2 \end{bmatrix}$

125. $\begin{bmatrix} 2 & 1 \\ 6 & 0 \end{bmatrix}\left(\begin{bmatrix} 4 & 2 \\ -3 & 1 \end{bmatrix} + \begin{bmatrix} -2 & 4 \\ 0 & 4 \end{bmatrix}\right)$

126. $\begin{bmatrix} 1 & -1 \\ 4 & 2 \end{bmatrix}\begin{bmatrix} 0 & 3 \\ 1 & 2 \end{bmatrix}\begin{bmatrix} 1 & 0 \\ 5 & -3 \end{bmatrix}$

127. Manufacturing A tire corporation has three factories, each of which manufactures two models of tires. The production levels are represented by A.

$$A = \overset{\text{Factory}}{\overset{\begin{matrix} 1 & \quad 2 & \quad 3 \end{matrix}}{\begin{bmatrix} 80 & 120 & 140 \\ 40 & 100 & 80 \end{bmatrix}}} \begin{matrix} \text{A} \\ \text{B} \end{matrix} \Big\} \text{Model}$$

(a) The price per unit for a model A tire is \$99 and the price per unit for a model B tire is \$112. Organize the price per unit of each tire in a matrix B.

(b) Find the product BA and interpret the result.

128. Manufacturing An electronics manufacturing company produces three different models of headphones that are shipped to two warehouses. The shipment levels are represented by A.

$$A = \overset{\text{Warehouse}}{\overset{\begin{matrix} 1 & \quad 2 \end{matrix}}{\begin{bmatrix} 8200 & 7400 \\ 6500 & 9800 \\ 5400 & 4800 \end{bmatrix}}} \begin{matrix} \text{A} \\ \text{B} \\ \text{C} \end{matrix} \Bigg\} \text{Model}$$

(a) The price per unit is \$79.99 for model A headphones, \$109.95 for model B headphones, and \$189.99 for model C headphones. Organize the price per unit of the headphones in a matrix B.

(b) Find the product BA and interpret the result.

5.6

The Inverse of a Matrix **In Exercises 129 and 130, show that B is the inverse of A.**

129. $A = \begin{bmatrix} -4 & -1 \\ 7 & 2 \end{bmatrix}$, $B = \begin{bmatrix} -2 & -1 \\ 7 & 4 \end{bmatrix}$

130. $A = \begin{bmatrix} 1 & 1 & 0 \\ 1 & 0 & 1 \\ 6 & 2 & 3 \end{bmatrix}$, $B = \begin{bmatrix} -2 & -3 & 1 \\ 3 & 3 & -1 \\ 2 & 4 & -1 \end{bmatrix}$

Finding the Inverse of a Matrix **In Exercises 131–134, find the inverse of the matrix (if it exists).**

131. $\begin{bmatrix} -6 & 5 \\ -5 & 4 \end{bmatrix}$

132. $\begin{bmatrix} -3 & -9 \\ 4 & 12 \end{bmatrix}$

133. $\begin{bmatrix} 2 & 0 & 3 \\ -1 & 1 & 1 \\ 2 & -2 & 1 \end{bmatrix}$

134. $\begin{bmatrix} 0 & -2 & 1 \\ -5 & -2 & -3 \\ 7 & 3 & 4 \end{bmatrix}$

Finding the Inverse of a Matrix **In Exercises 135–140, use the matrix capabilities of a graphing utility to find the inverse of the matrix (if it exists).**

135. $\begin{bmatrix} 2 & 6 \\ 3 & -6 \end{bmatrix}$

136. $\begin{bmatrix} 3 & -10 \\ 4 & 2 \end{bmatrix}$

137. $\begin{bmatrix} 1 & -1 & -2 \\ 0 & 3 & -2 \\ 1 & 2 & -4 \end{bmatrix}$

138. $\begin{bmatrix} 1 & 2 & 0 \\ -1 & 1 & 1 \\ 0 & -1 & 0 \end{bmatrix}$

139. $\begin{bmatrix} 1 & 3 & 1 & 6 \\ 4 & 4 & 2 & 6 \\ 3 & 4 & 1 & 2 \\ -1 & 2 & -1 & -2 \end{bmatrix}$

140. $\begin{bmatrix} 8 & 0 & 2 & 8 \\ 4 & -2 & 0 & -2 \\ 1 & 2 & 1 & 4 \\ -1 & 4 & 1 & 1 \end{bmatrix}$

Finding the Inverse of a 2 × 2 Matrix In Exercises 141–146, use the formula on page 464 to find the inverse of the 2 × 2 matrix (if it exists).

141. $\begin{bmatrix} -7 & 2 \\ -8 & 2 \end{bmatrix}$ **142.** $\begin{bmatrix} 10 & 4 \\ 7 & 3 \end{bmatrix}$

143. $\begin{bmatrix} -1 & 10 \\ 2 & -20 \end{bmatrix}$ **144.** $\begin{bmatrix} -6 & -5 \\ 3 & 3 \end{bmatrix}$

145. $\begin{bmatrix} -\frac{1}{2} & 20 \\ \frac{3}{10} & -6 \end{bmatrix}$ **146.** $\begin{bmatrix} -\frac{3}{4} & \frac{5}{2} \\ -\frac{4}{5} & -\frac{8}{3} \end{bmatrix}$

Solving a System of Equations Using an Inverse In Exercises 147–152, use an inverse matrix to solve (if possible) the system of linear equations.

147. $\begin{cases} -x + 4y = 8 \\ 2x - 7y = -5 \end{cases}$ **148.** $\begin{cases} 2x + 3y = -10 \\ 4x - y = 1 \end{cases}$

149. $\begin{cases} 3x + 2y - z = 6 \\ x - y + 2z = -1 \\ 5x + y + z = 7 \end{cases}$

150. $\begin{cases} -x + 4y - 2z = 12 \\ 2x - 9y + 5z = -25 \\ -x + 5y - 4z = 10 \end{cases}$

151. $\begin{cases} x + 2y + z - w = -2 \\ 2x + y + z + w = 1 \\ x - y + 2w = -4 \\ z + w = 1 \end{cases}$

152. $\begin{cases} x + y + z + w = 1 \\ x - y + 2z + w = -3 \\ y + w = 2 \\ x + w = 2 \end{cases}$

Solving a System of Linear Equations In Exercises 153–156, use the matrix capabilities of a graphing utility to solve (if possible) the system of linear equations.

153. $\begin{cases} x + 2y = -1 \\ 3x + 4y = -5 \end{cases}$ **154.** $\begin{cases} x + 3y = 23 \\ -6x + 2y = -18 \end{cases}$

155. $\begin{cases} -3x - 3y - 4z = 2 \\ y + z = -1 \\ 4x + 3y + 4z = -1 \end{cases}$

156. $\begin{cases} 2x + 3y - 4z = 1 \\ x - y + 2z = -4 \\ 3x + 7y - 10z = 0 \end{cases}$

5.7

The Determinant of a Matrix In Exercises 157–162, find the determinant of the matrix.

157. $[-23]$ **158.** $[0]$

159. $\begin{bmatrix} 8 & 5 \\ 2 & -4 \end{bmatrix}$ **160.** $\begin{bmatrix} -9 & 11 \\ 7 & -4 \end{bmatrix}$

161. $\begin{bmatrix} 50 & -30 \\ 10 & 5 \end{bmatrix}$ **162.** $\begin{bmatrix} 14 & -24 \\ 12 & -15 \end{bmatrix}$

Finding the Minors and Cofactors of a Matrix In Exercises 163–166, find all (a) minors and (b) cofactors of the matrix.

163. $\begin{bmatrix} 2 & -1 \\ 7 & 4 \end{bmatrix}$ **164.** $\begin{bmatrix} 3 & 6 \\ 5 & -4 \end{bmatrix}$

165. $\begin{bmatrix} 3 & 2 & -1 \\ -2 & 5 & 0 \\ 1 & 8 & 6 \end{bmatrix}$ **166.** $\begin{bmatrix} 8 & 3 & 4 \\ 6 & 5 & -9 \\ -4 & 1 & 2 \end{bmatrix}$

Finding a Determinant In Exercises 167–174, find the determinant of the matrix. Expand by cofactors on the row or column that appears to make the computations easiest.

167. $\begin{bmatrix} -2 & 4 & 1 \\ -6 & 0 & 2 \\ 5 & 3 & 4 \end{bmatrix}$ **168.** $\begin{bmatrix} 4 & 7 & -1 \\ 2 & -3 & 4 \\ -5 & 0 & -1 \end{bmatrix}$

169. $\begin{bmatrix} -2 & 5 & 0 \\ 2 & -1 & 0 \\ -1 & 1 & 3 \end{bmatrix}$ **170.** $\begin{bmatrix} 0 & 1 & -2 \\ 0 & 1 & 2 \\ -1 & -1 & 3 \end{bmatrix}$

171. $\begin{bmatrix} 1 & 4 & -2 \\ 3 & 1 & 0 \\ -2 & 2 & 1 \end{bmatrix}$ **172.** $\begin{bmatrix} 0 & 3 & 1 \\ 5 & -2 & 1 \\ 1 & 6 & 1 \end{bmatrix}$

173. $\begin{bmatrix} 3 & 0 & -4 & 0 \\ 0 & 8 & 1 & 2 \\ 6 & 1 & 8 & 2 \\ 0 & 3 & -4 & 1 \end{bmatrix}$ **174.** $\begin{bmatrix} -5 & 6 & 0 & 0 \\ 0 & 1 & -1 & 2 \\ -3 & 4 & -5 & 1 \\ 1 & 6 & 0 & 3 \end{bmatrix}$

5.8

Finding an Area In Exercises 175–180, use a determinant to find the area of the figure with the given vertices.

175. $(1, 0), (5, 0), (5, 8)$ **176.** $(-4, 0), (4, 0), (0, 6)$

177. $\left(\frac{1}{2}, 1\right), \left(2, -\frac{5}{2}\right) \left(\frac{3}{2}, 1\right)$ **178.** $\left(\frac{3}{2}, 1\right), \left(4, -\frac{1}{2}\right), (4, 2)$

179. $(-2, -1), (4, 9), (-2, -9), (4, 1)$

180. $(-4, 8), (4, 0), (-4, 0), (4, -8)$

Testing for Collinear Points In Exercises 181–184, use a determinant to determine whether the points are collinear.

181. $(2, 4), (5, 6), (4, 1)$ **182.** $(0, -5), (2, 1), (4, 7)$

183. $(-1, 7), (3, -9), (-3, 15)$

184. $(-3, 2), (2, -3), (-4, -4)$

Using Cramer's Rule In Exercises 185–192, use Cramer's Rule to solve (if possible) the system of equations.

185. $\begin{cases} x + 2y = 5 \\ -x + y = 1 \end{cases}$ **186.** $\begin{cases} 2x - y = -10 \\ 3x + 2y = -1 \end{cases}$

187. $\begin{cases} 8x - 4y = 3 \\ -12x + 6y = -15 \end{cases}$ **188.** $\begin{cases} 3x + 8y = -7 \\ 9x - 5y = 37 \end{cases}$

189. $\begin{cases} -2x + 3y - 5z = -11 \\ 4x - y + z = -3 \\ -x - 4y + 6z = 15 \end{cases}$

190. $\begin{cases} 5x - 2y + z = 15 \\ -8x + 3y - 9z = -39 \\ 2x - y - 7z = -3 \end{cases}$

191. $\begin{cases} x - 3y + 2z = 2 \\ 2x + 2y - 3z = 3 \\ x - 7y + 8z = -4 \end{cases}$ **192.** $\begin{cases} 14x - 21y - 7z = 10 \\ -4x + 2y - 2z = 4 \\ 56x - 21y + 7z = 5 \end{cases}$

Comparing Solution Methods **In Exercises 193 and 194, solve the system of equations using (a) Gaussian elimination and (b) Cramer's Rule. Which method do you prefer, and why?**

193. $\begin{cases} x - 3y + 2z = 5 \\ 2x + y - 4z = -1 \\ 2x + 4y + 2z = 3 \end{cases}$ **194.** $\begin{cases} x + 2y - z = -3 \\ 2x - y + z = -1 \\ 4x - 2y - z = 5 \end{cases}$

Encoding a Message **In Exercises 195 and 196, (a) write the uncoded 1 × 3 row matrices for the message, and then (b) encode the message using the encoding matrix.**

Message	*Encoding Matrix*
195. LOOK OUT BELOW	$\begin{bmatrix} 2 & -2 & 0 \\ 3 & 0 & -3 \\ -6 & 2 & 3 \end{bmatrix}$
196. JUST DO IT	$\begin{bmatrix} 2 & 1 & 0 \\ -6 & -6 & -2 \\ 3 & 2 & 1 \end{bmatrix}$

Decoding a Message **In Exercises 197–199, use A^{-1} to decode the cryptogram.**

197. $A = \begin{bmatrix} 1 & 0 & -1 \\ -1 & -2 & 0 \\ 1 & -2 & 2 \end{bmatrix}$

32 −46 37 9 −48 15 3 −14
10 −1 −6 2 −8 −22 −3

198. $A = \begin{bmatrix} 1 & 2 & 0 \\ -1 & 1 & 2 \\ 2 & -1 & 2 \end{bmatrix}$

30 −7 30 5 10 80 37 34 16 40 −7
38 −3 8 36 16 −1 58 23 46 0

199. $A = \begin{bmatrix} 1 & -1 & 0 \\ 0 & 1 & 2 \\ 1 & 1 & -2 \end{bmatrix}$

21 −11 14 29 −11 −18 32 −6 −26 31 −19
−12 10 6 26 13 −11 −2 37 28 −8 5 13 36

200. MODELING DATA

The populations (in millions) of Florida for the years 2010 through 2013 are shown in the table. (Source: U.S. Census Bureau)

Year	Population (in millions)
2010	18.8
2011	18.9
2012	19.0
2013	19.2

The coefficients of the least squares regression line $y = at + b$, where y is the population (in millions) and t is the year, with $t = 10$ corresponding to 2010, can be found by solving the system

$$\begin{cases} 4b + 46a = 75.9 \\ 46b + 534a = 873.5 \end{cases}.$$

(a) Use Cramer's Rule to solve the system and find the least squares regression line.

(b) Use a graphing utility to graph the line from part (a).

(c) Use the graph from part (b) to estimate when the population of Florida will exceed 20 million.

(d) Use your regression equation to find algebraically when the population will exceed 20 million.

Conclusions

True or False? **In Exercises 201 and 202, determine whether the statement is true or false. Justify your answer.**

201. Solving a system of equations graphically will always give an exact solution.

202. $\begin{vmatrix} a_{11} & a_{12} & a_{13} \\ a_{21} & a_{22} & a_{23} \\ a_{31} + c_1 & a_{32} + c_2 & a_{33} + c_3 \end{vmatrix} =$

$\begin{vmatrix} a_{11} & a_{12} & a_{13} \\ a_{21} & a_{22} & a_{23} \\ a_{31} & a_{32} & a_{33} \end{vmatrix} + \begin{vmatrix} a_{11} & a_{12} & a_{13} \\ a_{21} & a_{22} & a_{23} \\ c_1 & c_2 & c_3 \end{vmatrix}$

203. Writing What is the relationship between the three elementary row operations performed on an augmented matrix and the operations that lead to equivalent systems of equations?

204. Think About It Under what conditions does a matrix have an inverse?

5 Chapter Test

See CalcChat.com for tutorial help and worked-out solutions to odd-numbered exercises.
For instructions on how to use a graphing utility, see Appendix A.

Take this test as you would take a test in class. After you are finished, check your work against the answers given in the back of the book.

In Exercises 1–3, solve the system by the method of substitution. Check your solution graphically.

1. $\begin{cases} x - y = 6 \\ 3x + 5y = 2 \end{cases}$ **2.** $\begin{cases} y = x - 1 \\ y = (x - 1)^3 \end{cases}$ **3.** $\begin{cases} 4x - y^2 = 7 \\ x - y = -3 \end{cases}$

In Exercises 4–6, solve the system by the method of elimination.

4. $\begin{cases} 3x + 5y = -11 \\ 5x - y = 19 \end{cases}$ **5.** $\begin{cases} 1.5x + 3y = -9 \\ 2x - 1.2y = 5 \end{cases}$ **6.** $\begin{cases} x - 4y - z = 3 \\ 2x - 5y + z = 0 \\ 3x - 3y + 2z = -1 \end{cases}$

7. Find the equation of the parabola $y = ax^2 + bx + c$ that passes through the points $(0, 6)$, $(-2, 2)$, and $\left(3, \frac{9}{2}\right)$.

In Exercises 8 and 9, write the partial fraction decomposition for the rational expression.

8. $\dfrac{5x - 2}{(x - 1)^2}$ **9.** $\dfrac{x^3 + x^2 + x + 2}{x^4 + x^2}$

In Exercises 10 and 11, use matrices to solve the system of equations, if possible.

10. $\begin{cases} 2x + y + 2z = 4 \\ 2x + 2y = 5 \\ 2x - y + 6z = 2 \end{cases}$ **11.** $\begin{cases} 2x + 3y + z = 10 \\ 2x - 3y - 3z = 22 \\ 4x - 2y + 3z = -2 \end{cases}$

12. If possible, find (a) $A - B$, (b) $3A$, (c) $3A - 2B$, and (d) AB.

$$A = \begin{bmatrix} 5 & 4 & 4 \\ -4 & -4 & 0 \\ 1 & 2 & 0 \end{bmatrix}, \quad B = \begin{bmatrix} 4 & 4 & 0 \\ 3 & 2 & 1 \\ 1 & -2 & 0 \end{bmatrix}$$

13. Use the inverse of $A = \begin{bmatrix} -2 & 2 & 3 \\ 1 & -1 & 0 \\ 0 & 1 & 4 \end{bmatrix}$ to solve the system shown below.

$$\begin{cases} -2x + 2y + 3z = 7 \\ x - y = -5 \\ y + 4z = -1 \end{cases}$$

In Exercises 14 and 15, find the determinant of the matrix.

14. $\begin{bmatrix} -25 & 18 \\ 6 & -7 \end{bmatrix}$ **15.** $\begin{bmatrix} 4 & 0 & 3 \\ 1 & -8 & 2 \\ 3 & 2 & 2 \end{bmatrix}$

16. Use determinants to find the area of the parallelogram shown at the right.

17. Use Cramer's Rule to solve (if possible) $\begin{cases} 2x - 2y = 3 \\ x + 4y = -1 \end{cases}$.

18. The flow of traffic (in vehicles per hour) through a network of streets is shown at the right. Solve the system for the traffic flow represented by x_i, $i = 1, 2, 3, 4,$ and 5.

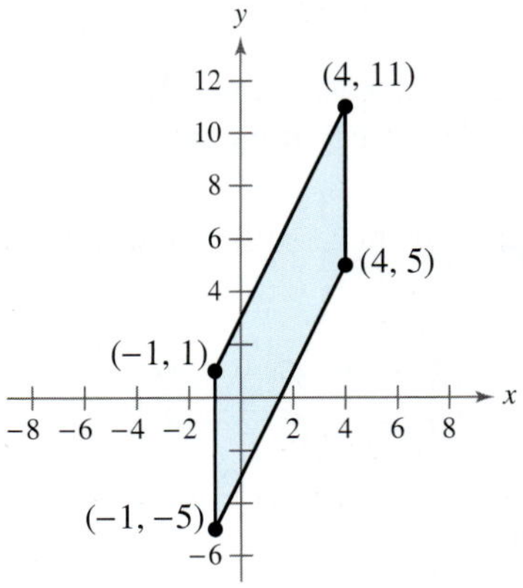

Figure for 16

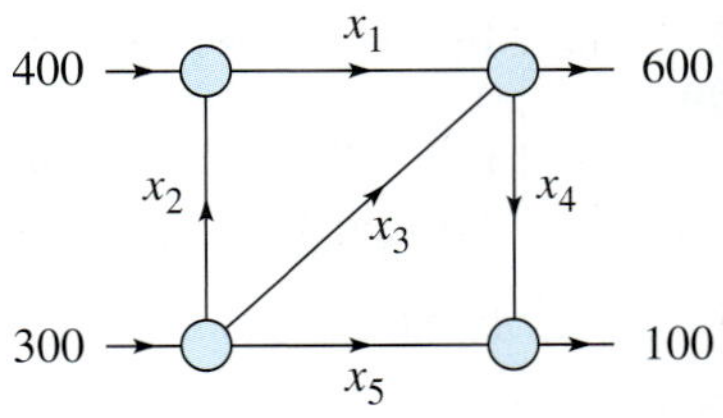

Figure for 18

3–5 Cumulative Test

See *CalcChat.com* for tutorial help and worked-out solutions to odd-numbered exercises.
For instructions on how to use a graphing utility, see Appendix A.

Take this test to review the material in Chapters 3–5. After you are finished, check your work against the answers given in the back of the book.

In Exercises 1 and 2, sketch the graph of the function. Use a graphing utility to verify the graph.

1. $f(x) = -(x - 2)^2 + 5$ **2.** $f(x) = \frac{1}{4}x(x - 2)^2$

3. Write $f(x) = x^3 + 2x^2 + 4x + 8$ as the product of linear factors, and list all the zeros of f.

4. Use a graphing utility to approximate any real zeros of $g(x) = x^3 + 5x^2 + 4$ accurate to three decimal places.

5. Divide $(2x^3 - 5x^2 + 6x - 20)$ by $(x - 6)$.

6. Find a polynomial function with real coefficients that has the zeros 0, -3, and $1 + \sqrt{5}i$.

In Exercises 7–9, sketch the graph of the rational function. Identify any asymptotes. Use a graphing utility to verify your graph.

7. $f(x) = \dfrac{2x}{x - 3}$ **8.** $f(x) = \dfrac{5x}{x^2 + x - 6}$ **9.** $f(x) = \dfrac{x^2 - 3x + 8}{x - 2}$

In Exercises 10 and 11, use the graph of *f* to describe the transformation that yields the graph of *g*.

10. $f(x) = 3^x$
$g(x) = 3^{x+1} - 5$

11. $f(x) = \log_{10} x$
$g(x) = -\log_{10}(x + 3)$

In Exercises 12–15, (a) determine the domain of the function, (b) find all intercepts, if any, of the graph of the function, and (c) identify any asymptotes of the graph of the function.

12. $f(x) = 8^{-x+1} + 2$ **13.** $f(x) = 2e^{0.01x}$

14. $f(x) = \dfrac{\ln x}{x}$ **15.** $f(x) = \log_6(2x - 1) + 1$

In Exercises 16–18, evaluate the expression without using a calculator.

16. $\log_2\left(\dfrac{1}{16}\right)$ **17.** $\ln e^{10}$ **18.** $\ln\dfrac{1}{e^3}$

In Exercises 19–21, evaluate the logarithm using the change-of-base formula. Round your result to three decimal places.

19. $\log_5 16$ **20.** $\log_9 7.3$ **21.** $\log_2\left(\frac{3}{2}\right)$

22. Use the properties of logarithms to expand $\ln\left(\dfrac{x^2 - 4}{x^2 + 1}\right)$.

23. Write $2 \ln x - \ln(x - 1) + \ln(x + 1)$ as a logarithm of a single quantity.

In Exercises 24–29, solve the equation algebraically. Round your result to three decimal places and verify your result graphically.

24. $3^x + 5 = 32$ **25.** $2(6^{x+1} - 32) = 100$

26. $3 \log_8(x + 1) = 2$ **27.** $2500e^{0.05x} = 500{,}000$

28. $2x^2e^{2x} - 2xe^{2x} = 0$ **29.** $\ln(2x - 5) - \ln x = 1$

In Exercises 30–35, use any method to solve the system of equations.

30. $\begin{cases} 4x - 5y = 29 \\ -x + 9y = 16 \end{cases}$ **31.** $\begin{cases} 2x - y^2 = 0 \\ x - y = 4 \end{cases}$ **32.** $\begin{cases} 4x - 3y = 0 \\ 2x - 9y = 5 \end{cases}$

33. $\begin{cases} x - y + 3z = -1 \\ 2x + 4y + z = 2 \end{cases}$ **34.** $\begin{cases} x + y - 5z = 3 \\ 4x + y - 11z = 6 \\ -x - 2z = -1 \end{cases}$ **35.** $\begin{cases} 2x + 5y - z = 4 \\ x - y - 3z = 6 \\ x + 2y + 2z = -6 \end{cases}$

In Exercises 36–39, perform the matrix operations given

$$A = \begin{bmatrix} -3 & 0 & -4 \\ 2 & 4 & 5 \\ -4 & 8 & 1 \end{bmatrix} \quad \text{and} \quad B = \begin{bmatrix} -1 & 5 & 2 \\ 6 & -3 & 3 \\ 0 & 4 & -2 \end{bmatrix}.$$

36. $3A - 2B$ **37.** $5A + 3B$ **38.** AB **39.** BA

In Exercises 40–42, find the determinant of the matrix.

40. $\begin{bmatrix} 4 & -5 \\ 6 & -2 \end{bmatrix}$ **41.** $\begin{bmatrix} 3 & 4 & \frac{1}{2} \\ 4 & -3 & 8 \\ -2 & 1 & 5 \end{bmatrix}$ **42.** $\begin{bmatrix} 1 & 3 & 6 \\ 0 & -2 & 8 \\ 0 & 0 & 7 \end{bmatrix}$

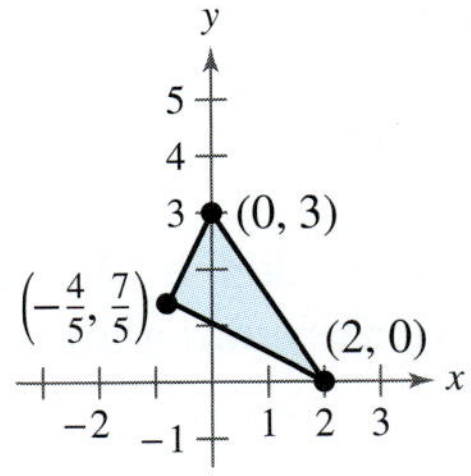

Figure for 43

43. Use a determinant to find the area of the triangle shown at the right.

44. Find the inverse of the matrix in Exercise 42 (if it exists).

45. A deposit of \$200 is made in an account that earns 6% interest. Find the balance after 30 years when the interest is compounded (a) monthly and (b) continuously. Recall that

$$A = P\left(1 + \frac{r}{n}\right)^{nt} \text{ and } A = Pe^{rt}.$$

46. The populations P (in thousands) of Oregon from 2000 through 2013 can be modeled by $P = 3434e^{kt}$, where t is the year, with $t = 0$ corresponding to 2000. In 2012, the population was 3,900,000. Find the value of k and use this result to predict the population in 2017. (Source: U.S. Census Bureau)

47. The table shows the annual sales S (in billions of dollars) for Wal-Mart from 2003 through 2012. (Source: Wal-Mart Stores, Inc.)

(a) Use the *regression* feature of a graphing utility to find a quadratic model, an exponential model, and a power model for the data and identify the coefficient of determination for each model. Let t represent the year, with $t = 3$ corresponding to 2003.

(b) Use the graphing utility to graph each model with the original data.

(c) Determine which model best fits the data. Explain.

(d) Use the model you chose in part (c) to predict the annual sales of Wal-Mart in 2020. Is your answer reasonable? Explain.

DATA

Year	Sales, S (in billions of dollars)
2003	258.681
2004	287.989
2005	315.654
2006	348.650
2007	378.799
2008	405.607
2009	408.214
2010	421.849
2011	446.950
2012	469.162

Spreadsheet at LarsonPrecalculus.com

Table for 47

48. An electronics company invests \$150,000 to produce an e-reader that will sell for \$200. Each unit can be produced for \$59.95.

(a) Write the cost and revenue functions for x units produced and sold.

(b) Use a graphing utility to graph the cost and revenue functions in the same viewing window. Use the graph to approximate the number of units that must be sold to break even, and verify the result algebraically.

49. What are the dimensions of a rectangle with a perimeter of 76 meters and an area of 352 square meters?

Proofs in Mathematics

An **indirect proof** can be useful in proving statements of the form "p implies q." Recall that the conditional statement $p \rightarrow q$ is false only when p is true and q is false. To prove a conditional statement indirectly, assume that p is true and q is false. If this assumption leads to an impossibility, then you have proved that the conditional statement is true. An indirect proof is also called a **proof by contradiction.**

You can use an indirect proof to prove the conditional statement

"If a is a positive integer and a^2 is divisible by 2, then a is divisible by 2"

as follows. First, assume that p, "a is a positive integer and a^2 is divisible by 2," is true and q, "a is divisible by 2," is false. This means that a is not divisible by 2. If so, then a is odd and can be written as

$$a = 2n + 1$$

where n is an integer.

$$a = 2n + 1 \quad \text{Definition of an odd integer}$$

$$a^2 = 4n^2 + 4n + 1 \quad \text{Square each side.}$$

$$a^2 = 2(2n^2 + 2n) + 1 \quad \text{Distributive Property}$$

So, by the definition of an odd integer, a^2 is odd. This contradicts the assumption, and you can conclude that a is divisible by 2.

EXAMPLE Using an Indirect Proof

Use an indirect proof to prove that $\sqrt{2}$ is an irrational number.

Solution

Begin by assuming that $\sqrt{2}$ is *not* an irrational number. Then $\sqrt{2}$ can be written as the quotient of two integers a and b ($b \neq 0$) that have no common factors.

$$\sqrt{2} = \frac{a}{b} \quad \text{Assume that } \sqrt{2} \text{ is a rational number.}$$

$$2 = \frac{a^2}{b^2} \quad \text{Square each side.}$$

$$2b^2 = a^2 \quad \text{Multiply each side by } b^2.$$

This implies that 2 is a factor of a^2. So, 2 is also a factor of a, and a can be written as $2c$, where c is an integer.

$$2b^2 = (2c)^2 \quad \text{Substitute } 2c \text{ for } a.$$

$$2b^2 = 4c^2 \quad \text{Simplify.}$$

$$b^2 = 2c^2 \quad \text{Divide each side by 2.}$$

This implies that 2 is a factor of b^2 and also a factor of b. So, 2 is a factor of both a and b. This contradicts the assumption that a and b have no common factors. So, you can conclude that $\sqrt{2}$ is an irrational number. ■

Proofs without words are pictures or diagrams that give a visual understanding of why a theorem or statement is true. They can also provide a starting point for writing a formal proof. The following proof shows that a 2×2 determinant is the area of a parallelogram.

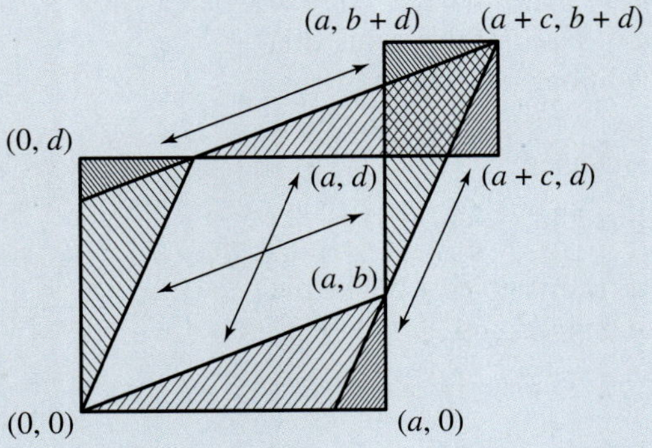

$\begin{vmatrix} a & b \\ c & d \end{vmatrix} = ad - bc =$ ‖▭‖ − ‖□‖ = ‖▱‖

The following is a color-coded version of the proof along with a brief explanation of why this proof works.

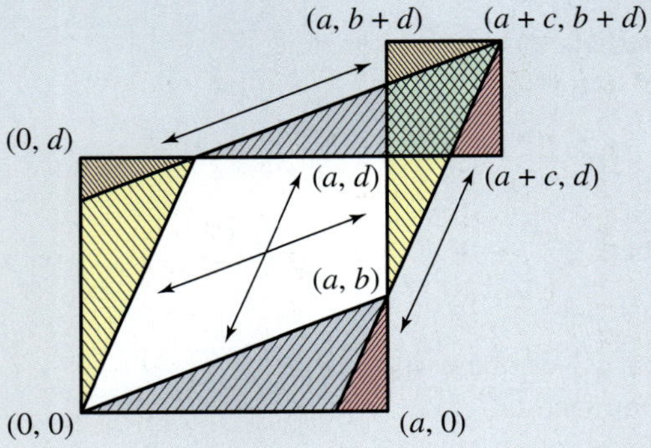

$\begin{vmatrix} a & b \\ c & d \end{vmatrix} = ad - bc =$ ‖▭‖ − ‖□‖ = ‖▱‖

Area of ▭ = Area of orange △ + Area of yellow △ + Area of blue △ + Area of pink △ + Area of white quadrilateral

Area of □ = Area of orange △ + Area of pink △ + Area of green quadrilateral

Area of ▱ = Area of white quadrilateral + Area of blue △ + Area of yellow △ − Area of green quadrilateral

= Area of ▭ − Area of □

From "Proof Without Words" by Solomon W. Golomb, *Mathematics Magazine*, March 1985. Vol. 58, No. 2, pg. 107.

6 Sequences, Series, and Probability

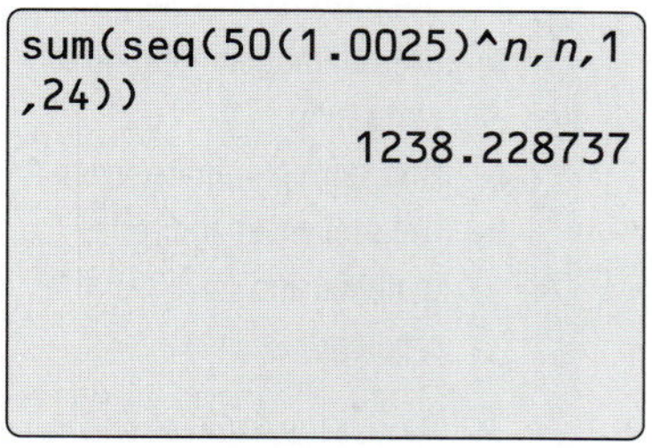

Section 6.3, Example 8
Increasing Annuity

6.1 Sequences and Series

What you should learn

- Use sequence notation to write the terms of sequences.
- Use factorial notation.
- Use summation notation to write sums.
- Find sums of infinite series.
- Use sequences and series to model and solve real-life problems.

Why you should learn it

Sequences and series are useful in modeling sets of values in order to identify patterns. For instance, Exercise 123 on page 509 shows how a sequence can be used to model the population of trout in a lake.

Sequences

In mathematics, the word *sequence* is used in much the same way as in ordinary English. Saying that a collection is listed in *sequence* means that it is ordered so that it has a first member, a second member, a third member, and so on.

Mathematically, you can think of a sequence as a *function* whose domain is the set of positive integers. Instead of using function notation, sequences are usually written using subscript notation, as shown in the following definition.

Definition of Sequence

An **infinite sequence** is a function whose domain is the set of positive integers. The function values

$$a_1, a_2, a_3, a_4, \ldots, a_n, \ldots$$

are the **terms** of the sequence. If the domain of a function consists of the first n positive integers only, then the sequence is a **finite sequence.**

On occasion, it is convenient to begin subscripting a sequence with 0 instead of 1 so that the terms of the sequence become

$$a_0, a_1, a_2, a_3, \ldots.$$

The domain of the function is the set of nonnegative integers.

EXAMPLE 1 Writing the Terms of a Sequence

Write the first four terms of each sequence.

a. $a_n = 3n - 2$

b. $a_n = 3 + (-1)^n$

Solution

a. The first four terms of the sequence given by $a_n = 3n - 2$ are

$a_1 = 3(1) - 2 = 1$ 1st term

$a_2 = 3(2) - 2 = 4$ 2nd term

$a_3 = 3(3) - 2 = 7$ 3rd term

$a_4 = 3(4) - 2 = 10.$ 4th term

b. The first four terms of the sequence given by $a_n = 3 + (-1)^n$ are

$a_1 = 3 + (-1)^1 = 3 - 1 = 2$ 1st term

$a_2 = 3 + (-1)^2 = 3 + 1 = 4$ 2nd term

$a_3 = 3 + (-1)^3 = 3 - 1 = 2$ 3rd term

$a_4 = 3 + (-1)^4 = 3 + 1 = 4.$ 4th term

✓ ***Checkpoint*** *Audio-video solution in English & Spanish at LarsonPrecalculus.com.*

Write the first four terms of the sequence given by $a_n = 2n + 1$.

EXAMPLE 2 Writing the Terms of a Sequence

Write the first five terms of the sequence given by $a_n = \dfrac{(-1)^n}{2n - 1}$.

Algebraic Solution

The first five terms of the sequence are as follows.

$$a_1 = \frac{(-1)^1}{2(1) - 1} = \frac{-1}{2 - 1} = -1 \qquad \text{1st term}$$

$$a_2 = \frac{(-1)^2}{2(2) - 1} = \frac{1}{4 - 1} = \frac{1}{3} \qquad \text{2nd term}$$

$$a_3 = \frac{(-1)^3}{2(3) - 1} = \frac{-1}{6 - 1} = -\frac{1}{5} \qquad \text{3rd term}$$

$$a_4 = \frac{(-1)^4}{2(4) - 1} = \frac{1}{8 - 1} = \frac{1}{7} \qquad \text{4th term}$$

$$a_5 = \frac{(-1)^5}{2(5) - 1} = \frac{-1}{10 - 1} = -\frac{1}{9} \qquad \text{5th term}$$

Numerical Solution

Set your graphing utility to *sequence* mode and enter the sequence. Use the *table* feature to create a table showing the terms of the sequence, as shown in the figure.

Use the *table* feature in *ask* mode.

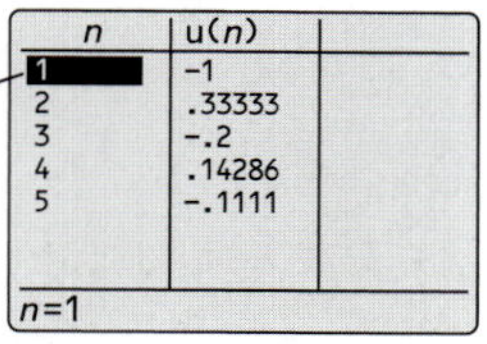

So, you can estimate the first five terms of the sequence as follows.

$u_1 = -1$, $u_2 = 0.33333 \approx \frac{1}{3}$, $u_3 = -0.2 = -\frac{1}{5}$,

$u_4 = 0.14286 \approx \frac{1}{7}$, and $u_5 = -0.1111 \approx -\frac{1}{9}$

Checkpoint Audio-video solution in English & Spanish at LarsonPrecalculus.com.

Write the first four terms of the sequence given by $a_n = \dfrac{2 + (-1)^n}{n}$.

Simply listing the first few terms is not sufficient to define a unique sequence—the nth term *must be given*. To see this, consider the following sequences, both of which have the same first three terms.

$$\frac{1}{2}, \frac{1}{4}, \frac{1}{8}, \frac{1}{16}, \ldots, \frac{1}{2^n}, \ldots$$

$$\frac{1}{2}, \frac{1}{4}, \frac{1}{8}, \frac{1}{15}, \ldots, \frac{6}{(n + 1)(n^2 - n + 6)}, \ldots$$

EXAMPLE 3 Finding the nth Term of a Sequence

Write an expression for the apparent nth term a_n of each sequence.

a. 1, 3, 5, 7, . . . **b.** 2, 5, 10, 17, . . .

Solution

a. n: 1 2 3 4 . . . n

Terms: 1 3 5 7 . . . a_n

Apparent Pattern: Each term is 1 less than twice n. So, the apparent nth term is $a_n = 2n - 1$.

b. n: 1 2 3 4 . . . n

Terms: 2 5 10 17 . . . a_n

Apparent Pattern: Each term is 1 more than the square of n. So, the apparent nth term is $a_n = n^2 + 1$.

Checkpoint Audio-video solution in English & Spanish at LarsonPrecalculus.com.

Write an expression for the apparent nth term a_n of each sequence.

a. 1, 5, 9, 13, 17, . . . **b.** 3, 5, 7, 9, 11, . . .

Some sequences are defined **recursively.** To define a sequence recursively, you need to be given one or more of the first few terms. All other terms of the sequence are then defined using previous terms.

EXAMPLE 4 A Recursive Sequence

Write the first five terms of the sequence defined recursively as

$a_1 = 3, a_k = 2a_{k-1} + 1, \quad \text{where} \quad k \geq 2.$

Solution

$a_1 = 3$ — 1st term is given.

$a_2 = 2a_{2-1} + 1 = 2a_1 + 1 = 2(3) + 1 = 7$ — Use recursion formula.

$a_3 = 2a_{3-1} + 1 = 2a_2 + 1 = 2(7) + 1 = 15$ — Use recursion formula.

$a_4 = 2a_{4-1} + 1 = 2a_3 + 1 = 2(15) + 1 = 31$ — Use recursion formula.

$a_5 = 2a_{5-1} + 1 = 2a_4 + 1 = 2(31) + 1 = 63$ — Use recursion formula.

✓ Checkpoint *Audio-video solution in English & Spanish at LarsonPrecalculus.com.*

Write the first five terms of the sequence defined recursively as

$a_1 = 6, \quad a_{k+1} = a_k + 1, \quad \text{where} \quad k \geq 2.$

In the next example you will study a well-known recursive sequence, the Fibonacci sequence.

EXAMPLE 5 The Fibonacci Sequence: A Recursive Sequence

The Fibonacci sequence is defined recursively as follows.

$a_0 = 1, a_1 = 1, a_k = a_{k-2} + a_{k-1}, \quad \text{where } k \geq 2$

Write the first six terms of this sequence.

Solution

$a_0 = 1$ — 0th term is given.

$a_1 = 1$ — 1st term is given.

$a_2 = a_{2-2} + a_{2-1} = a_0 + a_1 = 1 + 1 = 2$ — Use recursion formula.

$a_3 = a_{3-2} + a_{3-1} = a_1 + a_2 = 1 + 2 = 3$ — Use recursion formula.

$a_4 = a_{4-2} + a_{4-1} = a_2 + a_3 = 2 + 3 = 5$ — Use recursion formula.

$a_5 = a_{5-2} + a_{5-1} = a_3 + a_4 = 3 + 5 = 8$ — Use recursion formula.

You can check this result using the *table* feature of a graphing utility, as shown below.

n	u(n)	
0	1	
1	1	
2	2	
3	3	
4	5	
5	8	
n=		

✓ Checkpoint *Audio-video solution in English & Spanish at LarsonPrecalculus.com.*

Write the first five terms of the sequence defined recursively as

$a_0 = 1, \quad a_1 = 3, \quad a_k = a_{k-2} + a_{k-1}, \quad \text{where} \quad k \geq 2.$

Technology Tip

To graph a sequence using a graphing utility, set the mode to *dot* and *sequence* and enter the sequence. Try graphing the sequence in Example 4 and using the *trace* feature to identify the terms. For instructions on how to use the sequence mode, see Appendix A; for specific keystrokes, go to this textbook's *Companion Website.*

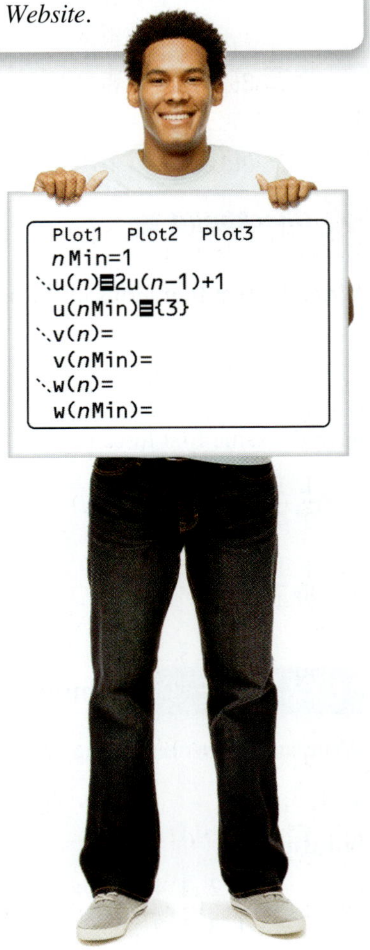

Factorial Notation

Some very important sequences in mathematics involve terms that are defined with special types of products called **factorials.**

> **Definition of Factorial**
>
> If n is a positive integer, then ***n* factorial** is defined as
>
> $$n! = 1 \cdot 2 \cdot 3 \cdot 4 \cdot \cdot \cdot (n-1) \cdot n.$$
>
> As a special case, zero factorial is defined as
>
> $$0! = 1.$$

Notice that $0! = 1$ and $1! = 1$. Here are some other values of $n!$.

$$2! = 1 \cdot 2 = 2 \quad 3! = 1 \cdot 2 \cdot 3 = 6 \quad 4! = 1 \cdot 2 \cdot 3 \cdot 4 = 24$$

Factorials follow the same conventions for order of operations as do exponents. For instance,

$$2n! = 2(n!) = 2(1 \cdot 2 \cdot 3 \cdot 4 \cdot \cdot \cdot n), \quad \text{whereas} \quad (2n)! = 1 \cdot 2 \cdot 3 \cdot 4 \cdot \cdot \cdot 2n.$$

Explore the Concept

Most graphing utilities have the capability to compute $n!$. Use your graphing utility to compare $3 \cdot 5!$ and $(3 \cdot 5)!$. How do they differ? Notice that the value of n does not have to be very large before the value of $n!$ becomes huge. How large a value of $n!$ will your graphing utility allow you to compute?

EXAMPLE 6 Writing the Terms of a Sequence Involving Factorials

Write the first five terms of the sequence given by $a_n = \dfrac{2^n}{n!}$. Begin with $n = 0$.

Algebraic Solution

$a_0 = \dfrac{2^0}{0!} = \dfrac{1}{1} = 1$ 0th term

$a_1 = \dfrac{2^1}{1!} = \dfrac{2}{1} = 2$ 1st term

$a_2 = \dfrac{2^2}{2!} = \dfrac{4}{2} = 2$ 2nd term

$a_3 = \dfrac{2^3}{3!} = \dfrac{8}{6} = \dfrac{4}{3}$ 3rd term

$a_4 = \dfrac{2^4}{4!} = \dfrac{16}{24} = \dfrac{2}{3}$ 4th term

Graphical Solution

Using a graphing utility set to *dot* and *sequence* modes, enter the sequence. Next, graph the sequence, as shown in the figure.

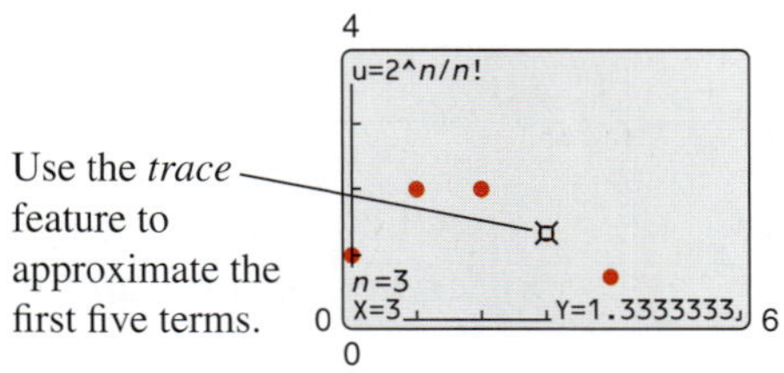

The first five terms are $u_0 = 1$, $u_1 = 2$, $u_2 = 2$, $u_3 \approx 1.333 \approx \frac{4}{3}$, and $u_4 \approx 0.667 \approx \frac{2}{3}$.

✔ **Checkpoint** Audio-video solution in English & Spanish at LarsonPrecalculus.com.

Write the first five terms of the sequence given by $a_n = \dfrac{3^n + 1}{n!}$. Begin with $n = 0$.

EXAMPLE 7 Simplifying Factorial Expressions

a. $\dfrac{8!}{2! \cdot 6!} = \dfrac{1 \cdot 2 \cdot \cancel{3 \cdot 4 \cdot 5 \cdot 6} \cdot 7 \cdot 8}{1 \cdot 2 \cdot \cancel{1 \cdot 2 \cdot 3 \cdot 4 \cdot 5 \cdot 6}} = \dfrac{7 \cdot 8}{2} = 28$

b. $\dfrac{n!}{(n-1)!} = \dfrac{\cancel{1 \cdot 2 \cdot 3 \cdot \cdot \cdot (n-1)} \cdot n}{\cancel{1 \cdot 2 \cdot 3 \cdot \cdot \cdot (n-1)}} = n$

✔ **Checkpoint** Audio-video solution in English & Spanish at LarsonPrecalculus.com.

Simplify the factorial expression $\dfrac{4!(n+1)!}{3!n!}$.

Remark

In Example 7(a), you can also simplify the computation as

$$\frac{8!}{2! \cdot 6!} = \frac{8 \cdot 7 \cdot \cancel{6!}}{2! \cdot \cancel{6!}} = 28.$$

Summation Notation

There is a convenient notation for the sum of the terms of a finite sequence. It is called **summation notation** or **sigma notation** because it involves the use of the uppercase Greek letter sigma, written as Σ.

Definition of Summation Notation

The sum of the first n terms of a sequence is represented by

$$\sum_{i=1}^{n} a_i = a_1 + a_2 + a_3 + a_4 + \cdot \cdot \cdot + a_n$$

where i is called the **index of summation,** n is the **upper limit of summation,** and 1 is the **lower limit of summation.**

Remark

Summation notation is an instruction to add the terms of a sequence. From the definition at the left, the upper limit of summation tells you where to end the sum. Summation notation helps you generate the appropriate terms of the sequence prior to finding the actual sum.

EXAMPLE 8 Sigma Notation for Sums

a. $\displaystyle\sum_{i=1}^{5} 3i = 3(1) + 3(2) + 3(3) + 3(4) + 3(5)$

$= 3(1 + 2 + 3 + 4 + 5)$

$= 3(15)$

$= 45$

b. $\displaystyle\sum_{k=3}^{6} (1 + k^2) = (1 + 3^2) + (1 + 4^2) + (1 + 5^2) + (1 + 6^2)$

$= 10 + 17 + 26 + 37$

$= 90$

c. $\displaystyle\sum_{n=0}^{8} \frac{1}{n!} = \frac{1}{0!} + \frac{1}{1!} + \frac{1}{2!} + \frac{1}{3!} + \frac{1}{4!} + \frac{1}{5!} + \frac{1}{6!} + \frac{1}{7!} + \frac{1}{8!}$

$= 1 + 1 + \frac{1}{2} + \frac{1}{6} + \frac{1}{24} + \frac{1}{120} + \frac{1}{720} + \frac{1}{5040} + \frac{1}{40,320}$

≈ 2.71828

For the summation in part (c), note that the sum is very close to the irrational number $e \approx 2.718281828$. It can be shown that as more terms of the sequence whose nth term is $1/n!$ are added, the sum becomes closer and closer to e.

✓ Checkpoint *Audio-video solution in English & Spanish at LarsonPrecalculus.com.*

Find the sum $\displaystyle\sum_{i=1}^{4} (4i + 1)$.

In Example 8, note that the lower limit of a summation does not have to be 1. Also note that the index of summation does not have to be the letter i. For instance, in part (b), the letter k is the index of summation.

Properties of Sums (See the proofs on page 562.)

1. $\displaystyle\sum_{i=1}^{n} c = cn$, c is a constant.

2. $\displaystyle\sum_{i=1}^{n} ca_i = c\sum_{i=1}^{n} a_i$, c is a constant.

3. $\displaystyle\sum_{i=1}^{n} (a_i + b_i) = \sum_{i=1}^{n} a_i + \sum_{i=1}^{n} b_i$

4. $\displaystyle\sum_{i=1}^{n} (a_i - b_i) = \sum_{i=1}^{n} a_i - \sum_{i=1}^{n} b_i$

Series

Many applications involve the sum of the terms of a finite or an infinite sequence. Such a sum is called a **series.**

Definition of a Series

Consider the infinite sequence $a_1, a_2, a_3, \ldots, a_i, \ldots$.

1. The sum of the first n terms of the sequence is called a **finite series** or the ***n*th partial sum** of the sequence and is denoted by

$$a_1 + a_2 + a_3 + \cdots + a_n = \sum_{i=1}^{n} a_i.$$

2. The sum of all the terms of the infinite sequence is called an **infinite series** and is denoted by

$$a_1 + a_2 + a_3 + \cdots + a_i + \cdots = \sum_{i=1}^{\infty} a_i.$$

EXAMPLE 9 Finding the Sum of a Series

See LarsonPrecalculus.com for an interactive version of this type of example.

For the series

$$\sum_{i=1}^{\infty} \frac{3}{10^i}$$

find (a) the third partial sum and (b) the sum.

Solution

a. The third partial sum is

$$\sum_{i=1}^{3} \frac{3}{10^i} = \frac{3}{10^1} + \frac{3}{10^2} + \frac{3}{10^3}$$
$$= 0.3 + 0.03 + 0.003$$
$$= 0.333.$$

b. The sum of the series is

$$\sum_{i=1}^{\infty} \frac{3}{10^i} = \frac{3}{10^1} + \frac{3}{10^2} + \frac{3}{10^3} + \frac{3}{10^4} + \frac{3}{10^5} + \cdots$$
$$= 0.3 + 0.03 + 0.003 + 0.0003 + 0.00003 + \cdots$$
$$= 0.33333\ldots$$
$$= \frac{1}{3}.$$

Checkpoint *Audio-video solution in English & Spanish at LarsonPrecalculus.com.*

For the series

$$\sum_{i=1}^{\infty} \frac{5}{10^i}$$

find (a) the fourth partial sum and (b) the sum.

Notice in Example 9(b) that the sum of an infinite series can be a finite number.

Technology Tip

Most graphing utilities are able to sum the first n terms of a sequence. Try using a graphing utility to confirm the results in Example 8 and Example 9(a).

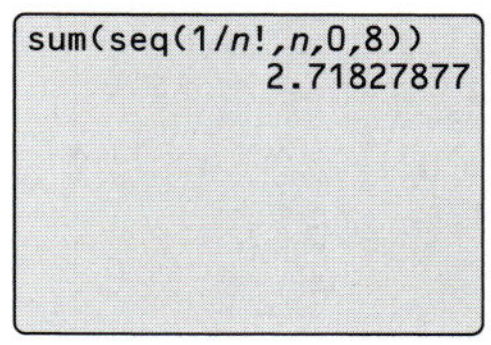

Application

EXAMPLE 10 Compound Interest

An investor deposits \$5000 in an account that earns 3% interest compounded quarterly. The balance in the account after n quarters is given by

$$A_n = 5000\left(1 + \frac{0.03}{4}\right)^n, \quad n = 0, 1, 2, \ldots .$$

Use this information to (a) write the first three terms of the sequence and (b) find the balance in the account after 10 years by computing the 40th term of the sequence.

Solution

a. The first three terms of the sequence are as follows.

$$A_0 = 5000\left(1 + \frac{0.03}{4}\right)^0 = \$5000.00 \qquad \text{Original deposit}$$

$$A_1 = 5000\left(1 + \frac{0.03}{4}\right)^1 = \$5037.50 \qquad \text{First-quarter balance}$$

$$A_2 = 5000\left(1 + \frac{0.03}{4}\right)^2 \approx \$5075.28 \qquad \text{Second-quarter balance}$$

b. The 40th term of the sequence is

$$A_{40} = 5000\left(1 + \frac{0.03}{4}\right)^{40} \approx \$6741.74. \qquad \text{Ten-year balance}$$

Checkpoint *Audio-video solution in English & Spanish at LarsonPrecalculus.com.*

An investor deposits \$1000 in an account that earns 3% interest compounded monthly. The balance in the account after n months is given by

$$A_n = 1000\left(1 + \frac{0.03}{12}\right)^n, \quad n = 0, 1, 2, \ldots .$$

Use this information to (a) write the first three terms of the sequence and (b) find the balance in the account after 4 years by computing the 48th term of the sequence.

Explore the Concept

A $3 \times 3 \times 3$ cube is created using 27 unit cubes (a unit cube has a length, width, and height of 1 unit), and only the faces of the cubes that are visible are painted blue (see the figure at the right). Complete the table below to determine how many unit cubes of the $3 \times 3 \times 3$ cube have no blue faces, one blue face, two blue faces, and three blue faces. Do the same for a $4 \times 4 \times 4$ cube, a $5 \times 5 \times 5$ cube, and a $6 \times 6 \times 6$ cube, and record your results in the table below. What type of pattern do you observe in the table? Write a formula you could use to determine the column values for an $n \times n \times n$ cube.

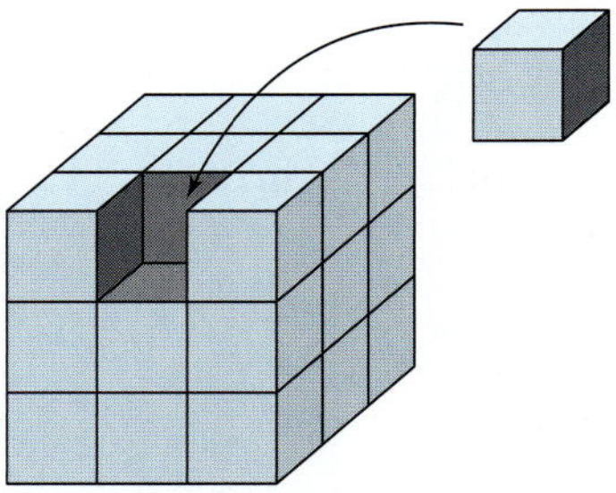

Cube \ Number of blue faces	0	1	2	3
$3 \times 3 \times 3$				

6.1 Exercises

See *CalcChat.com* for tutorial help and worked-out solutions to odd-numbered exercises. For instructions on how to use a graphing utility, see Appendix A.

Vocabulary and Concept Check

In Exercises 1–4, fill in the blank(s).

1. The function values $a_1, a_2, a_3, a_4, \ldots, a_n, \ldots$ are called the _______ of a sequence.

2. If you are given one or more of the first few terms of a sequence, and all other terms of the sequence are defined using previous terms, then the sequence is defined _______ .

3. For the sum $\sum_{i=1}^{n} a_i$, i is called the _______ of summation, n is the _______ of summation, and 1 is the _______ of summation.

4. The sum of the terms of a finite or an infinite sequence is called a _______ .

5. Which describes an infinite sequence? a finite sequence?

(a) The domain consists of the first n positive integers.

(b) The domain consists of the set of positive integers.

6. Write $1 \cdot 2 \cdot 3 \cdot 4 \cdot 5 \cdot 6$ in factorial notation.

Procedures and Problem Solving

Writing the Terms of a Sequence **In Exercises 7–16, write the first five terms of the sequence. (Assume n begins with 1.)**

7. $a_n = 2n - 5$ **8.** $a_n = 4n - 7$

9. $a_n = 3^n$ **10.** $a_n = \left(\frac{1}{2}\right)^n$

11. $a_n = \left(-\frac{1}{2}\right)^n$ **12.** $a_n = (-2)^n$

13. $a_n = \dfrac{n+1}{n}$ **14.** $a_n = \dfrac{n}{n+1}$

15. $a_n = \dfrac{n}{n^2+1}$ **16.** $a_n = \dfrac{2n}{n+1}$

Writing the Terms of a Sequence **In Exercises 17–26, write the first five terms of the sequence (a) using the *table* feature of a graphing utility and (b) algebraically. (Assume n begins with 1.)**

17. $a_n = \dfrac{1 + (-1)^n}{n}$ **18.** $a_n = \dfrac{1 + (-1)^n}{2n}$

19. $a_n = 1 - \dfrac{1}{2^n}$ **20.** $a_n = \dfrac{3^n}{4^n}$

21. $a_n = \dfrac{1}{n^{3/2}}$ **22.** $a_n = \dfrac{1}{\sqrt{n}}$

23. $a_n = \dfrac{(-1)^n}{n^2}$ **24.** $a_n = (-1)^n\left(\dfrac{n}{n+1}\right)$

25. $a_n = (2n - 1)(2n + 1)$ **26.** $a_n = n(n - 1)(n - 2)$

Using a Graphing Utility **In Exercises 27–32, use the *table* feature of a graphing utility to find the first 10 terms of the sequences. (Assume n begins with 1.)**

27. $a_n = 2(3n - 1) + 5$ **28.** $a_n = 2n(n + 1) - 7$

29. $a_n = 1 + \dfrac{n+1}{n}$ **30.** $a_n = \dfrac{4n^2}{n+2}$

31. $a_n = (-1)^n + 1$ **32.** $a_n = (-1)^{n+1} + 8$

Finding a Term of a Sequence **In Exercises 33–38, find the indicated term of the sequence.**

33. $a_n = \dfrac{n^2}{n^2+1}$, $a_{10} = \square$ **34.** $a_n = \dfrac{n^2}{2n+1}$, $a_5 = \square$

35. $a_n = (-1)^n(3n - 2)$, $a_{25} = \square$ **36.** $a_n = (-1)^{n-1}[n(n - 1)]$, $a_{16} = \square$

37. $a_n = \dfrac{2^{n+1}}{2^n+1}$, $a_7 = \square$ **38.** $a_n = \dfrac{3^n}{3^n+1}$, $a_6 = \square$

Finding the *n*th Term of a Sequence **In Exercises 39–52, write an expression for the apparent nth term of the sequence. (Assume n begins with 1.)**

39. 3, 8, 13, 18, 23, . . . **40.** 3, 7, 11, 15, 19, . . .

41. 7, 13, 19, 25, 31, . . . **42.** 9, 11, 13, 15, 17, . . .

43. 0, 3, 8, 15, 24, . . . **44.** 4, 7, 12, 19, 28, . . .

45. $\frac{2}{3}, \frac{3}{4}, \frac{4}{5}, \frac{5}{6}, \frac{6}{7}, \ldots$ **46.** $\frac{2}{1}, \frac{3}{3}, \frac{4}{5}, \frac{5}{7}, \frac{6}{9}, \ldots$

47. $\frac{1}{2}, \frac{-1}{4}, \frac{1}{8}, \frac{-1}{16}, \ldots$ **48.** $\frac{1}{3}, -\frac{2}{9}, \frac{4}{27}, -\frac{8}{81}, \ldots$

49. $1 + \frac{1}{1}, 1 + \frac{1}{2}, 1 + \frac{1}{3}, 1 + \frac{1}{4}, 1 + \frac{1}{5}, \ldots$

50. $1 + \frac{1}{3}, 1 + \frac{1}{6}, 1 + \frac{1}{11}, 1 + \frac{1}{18}, 1 + \frac{1}{27}, \ldots$

51. 1, 3, 1, 3, 1, . . .

52. 1, −1, 1, −1, 1, . . .

A Recursive Sequence **In Exercises 53–58, write the first five terms of the sequence defined recursively.**

53. $a_1 = 28,\ a_k = a_{k-1} - 4$

54. $a_1 = 15,\ a_k = a_{k-1} + 3$

55. $a_1 = 3,\ a_{k+1} = 2(a_k - 1)$

56. $a_1 = 32,\ a_{k+1} = \frac{1}{2}a_k$

57. $a_0 = 1,\ a_1 = 3,\ a_k = a_{k-2} + a_{k-1}$

58. $a_0 = -1,\ a_1 = 5,\ a_k = a_{k-2} + a_{k-1}$

Finding the *n*th Term of a Recursive Sequence **In Exercises 59–62, write the first five terms of the sequence defined recursively. Use the pattern to write the *n*th term of the sequence as a function of *n*. (Assume *n* begins with 1.)**

59. $a_1 = 6,\ a_{k+1} = a_k + 2$

60. $a_1 = 25,\ a_{k+1} = a_k - 5$

61. $a_1 = 81,\ a_{k+1} = \frac{1}{3}a_k$

62. $a_1 = 14,\ a_{k+1} = -2a_k$

Writing the Terms of a Sequence Involving Factorials **In Exercises 63–68, write the first five terms of the sequence (a) using the *table* feature of a graphing utility and (b) algebraically. (Assume *n* begins with 0.)**

63. $a_n = \dfrac{1}{n!}$

64. $a_n = \dfrac{1}{(n+1)!}$

65. $a_n = \dfrac{n^2}{(n+1)!}$

66. $a_n = \dfrac{n^3}{(n+2)!}$

67. $a_n = \dfrac{(-1)^{2n}}{(2n)!}$

68. $a_n = \dfrac{(-1)^{2n+1}}{(2n+1)!}$

Simplifying Factorial Expressions **In Exercises 69–76, simplify the factorial expression.**

69. $\dfrac{2!}{4!}$

70. $\dfrac{5!}{7!}$

71. $\dfrac{12!}{4! \cdot 8!}$

72. $\dfrac{10!}{5! \cdot 3!}$

73. $\dfrac{(n+3)!}{n!}$

74. $\dfrac{(n+2)!}{n!}$

75. $\dfrac{(2n-1)!}{(2n+1)!}$

76. $\dfrac{(2n-2)!}{(2n)!}$

Identifying a Graph of a Sequence **In Exercises 77–80, match the sequence with its graph. [The graphs are labeled (a), (b), (c), and (d).]**

(a)

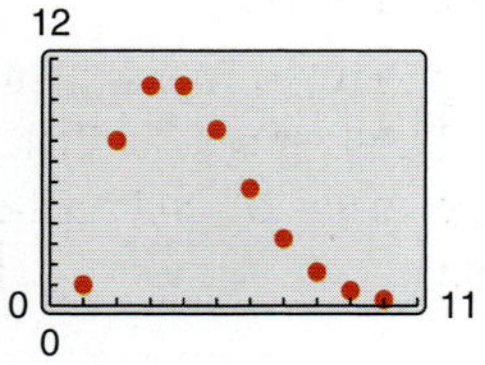

(b)

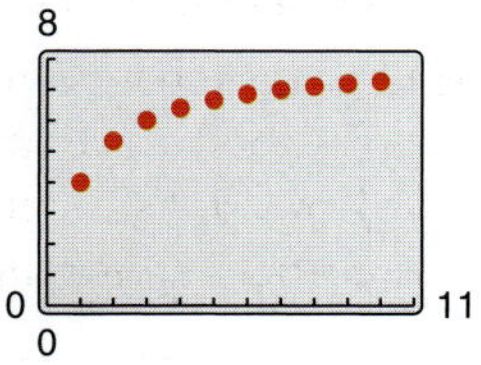

(c)

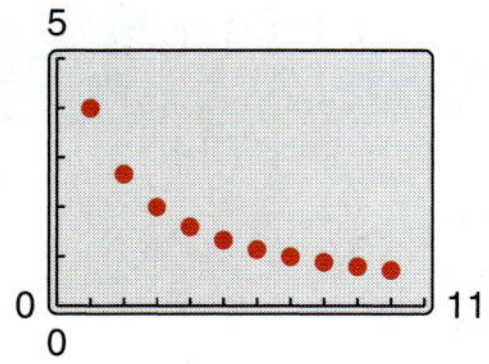

(d)

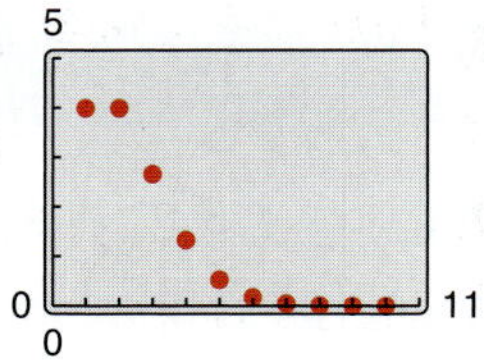

77. $a_n = \dfrac{8}{n+1}$

78. $a_n = \dfrac{8n}{n+1}$

79. $a_n = \dfrac{2^{n+2}}{2n!}$

80. $a_n = \dfrac{4^n}{n!}$

Graphing the Terms of a Sequence **In Exercises 81–86, use a graphing utility to graph the first 10 terms of the sequence. (Assume *n* begins with 1.)**

81. $a_n = \frac{2}{3}n$

82. $a_n = \frac{1}{2}n + 3$

83. $a_n = 16(-0.5)^{n-1}$

84. $a_n = 8(-0.75)^{n-1}$

85. $a_n = \dfrac{2n}{n+1}$

86. $a_n = \dfrac{3n^2}{n^2+1}$

Sigma Notation for Sums **In Exercises 87–98, find the sum.**

87. $\displaystyle\sum_{i=1}^{5}(2i+1)$

88. $\displaystyle\sum_{i=1}^{6}(3i-1)$

89. $\displaystyle\sum_{i=0}^{6}4i^2$

90. $\displaystyle\sum_{i=0}^{5}3i^2$

91. $\displaystyle\sum_{j=3}^{5}\frac{1}{j^2-3}$

92. $\displaystyle\sum_{j=3}^{5}\frac{1}{j+1}$

93. $\displaystyle\sum_{k=1}^{4}10$

94. $\displaystyle\sum_{k=1}^{5}4$

95. $\displaystyle\sum_{i=2}^{5}[(i-1)^3 + (i+1)^2]$

96. $\displaystyle\sum_{k=2}^{7}[(k+1) + (k-3)^2]$

97. $\displaystyle\sum_{i=0}^{4}2^i$

98. $\displaystyle\sum_{j=0}^{4}(-2)^j$

Using a Graphing Utility **In Exercises 99–102, use a graphing utility to find the sum.**

99. $\displaystyle\sum_{j=1}^{6}(24-3j)$

100. $\displaystyle\sum_{j=1}^{10}\frac{6}{3j+1}$

101. $\displaystyle\sum_{k=0}^{4}\frac{(-1)^k}{(k+1)!}$

102. $\displaystyle\sum_{k=0}^{4}\frac{(-1)^k}{k!}$

Writing a Sum Using Sigma Notation **In Exercises 103–112, use sigma notation to write the sum. Then use a graphing utility to find the sum.**

103. $\dfrac{1}{3(1)} + \dfrac{1}{3(2)} + \dfrac{1}{3(3)} + \cdots + \dfrac{1}{3(9)}$

104. $\dfrac{5}{1+1} + \dfrac{5}{1+2} + \dfrac{5}{1+3} + \cdots + \dfrac{5}{1+15}$

105. $\left[2\left(\frac{1}{8}\right)+3\right]+\left[2\left(\frac{2}{8}\right)+3\right]+\cdots+\left[2\left(\frac{8}{8}\right)+3\right]$

106. $\left[1-\left(\frac{1}{6}\right)^2\right]+\left[1-\left(\frac{2}{6}\right)^2\right]+\cdots+\left[1-\left(\frac{6}{6}\right)^2\right]$

107. $-3+9-27+81-243+729$

108. $1-\frac{1}{2}+\frac{1}{4}-\frac{1}{8}+\cdots-\frac{1}{128}$

109. $\frac{1}{1^2}-\frac{1}{2^2}+\frac{1}{3^2}-\frac{1}{4^2}+\cdots-\frac{1}{20^2}$

110. $\frac{1}{1\cdot 3}-\frac{1}{2\cdot 4}+\frac{1}{3\cdot 5}-\frac{1}{4\cdot 6}+\cdots-\frac{1}{10\cdot 12}$

111. $\frac{1}{4}+\frac{3}{8}+\frac{7}{16}+\frac{15}{32}+\frac{31}{64}$

112. $\frac{1}{2}+\frac{2}{4}+\frac{6}{8}+\frac{24}{16}+\frac{120}{32}+\frac{720}{64}$

Finding a Partial Sum **In Exercises 113–116, find the indicated partial sum of the series.**

113. $\sum_{i=1}^{\infty}\frac{7}{5^i}$

Fourth partial sum

114. $\sum_{i=1}^{\infty}\frac{2}{3^i}$

Fifth partial sum

115. $\sum_{n=1}^{\infty}4\left(-\frac{1}{2}\right)^n$

Third partial sum

116. $\sum_{n=1}^{\infty}8\left(-\frac{1}{4}\right)^n$

Fourth partial sum

Finding the Sum of a Series **In Exercises 117–120, find (a) the fourth partial sum and (b) the sum of the infinite series.**

117. $\sum_{i=1}^{\infty}\frac{6}{10^i}$

118. $\sum_{k=1}^{\infty}\frac{4}{10^k}$

119. $\sum_{k=1}^{\infty}7\left(\frac{1}{10}\right)^k$

120. $\sum_{i=1}^{\infty}2\left(\frac{1}{10}\right)^i$

121. Compound Interest A deposit of $5000 is made in an account that earns 3% interest compounded quarterly. The balance in the account after n quarters is given by

$$A_n = 5000\left(1+\frac{0.03}{4}\right)^n, \quad n = 1, 2, 3, \ldots.$$

(a) Compute the first eight terms of this sequence.

(b) Find the balance in this account after 10 years by computing the 40th term of the sequence.

122. Compound Interest An investor deposits $10,000 in an account that earns 3.5% interest compounded monthly. The balance in the account after n months is given by

$$A_n = 10{,}000\left(1+\frac{0.035}{12}\right)^n, \quad n = 1, 2, 3, \ldots.$$

(a) Write the first eight terms of the sequence.

(b) Find the balance in the account after 5 years by computing the 60th term of the sequence.

(c) Is the balance after 10 years twice the balance after 5 years? Explain.

123. ***Why you should learn it*** *(p. 500)* A landlocked lake has been selected to be stocked in the year 2015 with 5500 trout, and to be restocked each year thereafter with 500 trout. Each year the fish population declines 25% due to harvesting and other natural causes.

(a) Write a recursive sequence that gives the population p_n of trout in the lake in terms of the year n, with $n = 0$ corresponding to 2015.

(b) Use the recursion formula from part (a) to find the numbers of trout in the lake for $n = 1, 2, 3,$ and 4. Interpret these values in the context of the situation.

(c) Use a graphing utility to find the number of trout in the lake as time passes infinitely. Explain your result.

124. MODELING DATA

The revenues R_n (in millions of dollars) of Netflix from 2008 through 2013 are shown in the table. (Source: Netflix, Inc.)

Year	Revenue, R_n (in millions of dollars)
2008	1364.7
2009	1670.3
2010	2162.6
2011	3204.6
2012	3609.3
2013	4374.6

Spreadsheet at LarsonPrecalculus.com

(a) Use a graphing utility to graph the data. Let n represent the year, with $n = 8$ corresponding to 2008.

(b) Use the *regression* feature of the graphing utility to find a linear sequence and a quadratic sequence that model the data. Identify the coefficient of determination for each model.

(c) Graph each model with the data. Decide which model is a better fit for the data. Explain.

(d) Use the model you chose in part (c) to predict the revenue of Netflix in 2017. Does your answer seem reasonable? Explain.

(e) Use your model from part (c) to find when the revenue will reach 15 billion dollars.

(f) Use the model you chose in part (c) to approximate the total revenue from 2008 through 2013. Compare this sum with the result of adding the revenues shown in the table.

Conclusions

True or False? **In Exercises 125 and 126, determine whether the statement is true or false. Justify your answer.**

125. $\sum_{i=1}^{4}(i^2 + 2i) = \sum_{i=1}^{4} i^2 + 2\sum_{i=1}^{4} i$

126. $\sum_{j=1}^{4} 2^j = \sum_{j=3}^{6} 2^{j-2}$

Fibonacci Sequence **In Exercises 127 and 128, use the Fibonacci sequence. (See Example 5.)**

127. Write the first 12 terms of the Fibonacci sequence a_n and the first 10 terms of the sequence given by

$$b_n = \frac{a_{n+1}}{a_n}, \quad n > 0.$$

128. Using the definition of b_n given in Exercise 127, show that b_n can be defined recursively by

$$b_n = 1 + \frac{1}{b_{n-1}}.$$

Exploration **In Exercises 129–132, let**

$$a_n = \frac{(1 + \sqrt{5})^n - (1 - \sqrt{5})^n}{2^n\sqrt{5}}$$

be a sequence with *n*th term a_n.

129. Use the *table* feature of a graphing utility to find the first five terms of the sequence.

130. Do you recognize the terms of the sequence in Exercise 129? What sequence is it?

131. Find expressions for a_{n+1} and a_{n+2} in terms of n.

132. Use the result from Exercise 131 to show that $a_{n+2} = a_{n+1} + a_n$. Is this result the same as your answer to Exercise 129? Explain.

A Sequence Involving *x* **In Exercises 133–142, write the first five terms of the sequence.**

133. $a_n = \dfrac{x^n}{n!}$

134. $a_n = \dfrac{x^2}{n^2}$

135. $a_n = \dfrac{(-1)^n x^{2n+1}}{2n + 1}$

136. $a_n = \dfrac{(-1)^n x^{n+1}}{n + 1}$

137. $a_n = \dfrac{(-1)^n x^{2n}}{(2n)!}$

138. $a_n = \dfrac{(-1)^n x^{2n+1}}{(2n + 1)!}$

139. $a_n = \dfrac{(-1)^n x^n}{n!}$

140. $a_n = \dfrac{(-1)^n x^{n+1}}{(n + 1)!}$

141. $a_n = \dfrac{(-1)^{n+1}(x + 1)^n}{n!}$

142. $a_n = \dfrac{(-1)^n(x - 1)^n}{(n + 1)!}$

Writing Partial Sums **In Exercises 143–146, write the first five terms of the sequence. Then find an expression for the *n*th partial sum.**

143. $a_n = \dfrac{1}{2n} - \dfrac{1}{2n + 2}$

144. $a_n = \dfrac{1}{n} - \dfrac{1}{n + 1}$

145. $a_n = \dfrac{1}{n + 1} - \dfrac{1}{n + 2}$

146. $a_n = \dfrac{1}{n} - \dfrac{1}{n + 2}$

147. Think About It Does every finite series whose terms are integers have a finite sum? Explain.

148. HOW DO YOU SEE IT? The graph represents the first six terms of a sequence.

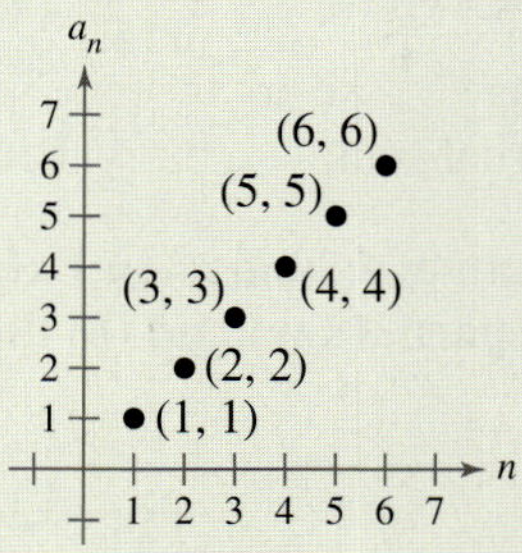

(a) Write the first six terms of the sequence.

(b) Write an expression for the apparent *n*th term a_n of the sequence.

(c) Use sigma notation to represent the partial sum of the first 50 terms of the sequence.

Cumulative Mixed Review

Operations with Matrices **In Exercises 149–152, find, if possible, (a) $A - B$, (b) $2B - 3A$, (c) AB, and (d) BA.**

149. $A = \begin{bmatrix} 6 \\ 3 \end{bmatrix}, \quad B = \begin{bmatrix} 4 \\ -3 \end{bmatrix}$

150. $A = \begin{bmatrix} 10 & 7 \\ -4 & 6 \end{bmatrix}, \quad B = \begin{bmatrix} 0 & -12 \\ 8 & 11 \end{bmatrix}$

151. $A = \begin{bmatrix} -2 & -3 & 6 \\ 4 & 5 & 7 \\ 1 & 7 & 4 \end{bmatrix}, \quad B = \begin{bmatrix} 1 & 4 & 2 \\ 0 & 1 & 6 \\ 0 & 3 & 1 \end{bmatrix}$

152. $A = \begin{bmatrix} -1 & 4 \\ 5 & 1 \\ 0 & -1 \end{bmatrix}, \quad B = \begin{bmatrix} 0 & 4 & 0 \\ 3 & 1 & -2 \end{bmatrix}$

6.2 Arithmetic Sequences and Partial Sums

Arithmetic Sequences

A sequence whose consecutive terms have a common difference is called an **arithmetic sequence.**

Definition of Arithmetic Sequence

A sequence is **arithmetic** when the differences between consecutive terms are the same. So, the sequence

$$a_1, a_2, a_3, a_4, \ldots, a_n, \ldots$$

is arithmetic when there is a number d such that

$$a_2 - a_1 = a_3 - a_2 = a_4 - a_3 = \cdots = d.$$

The number d is the **common difference** of the sequence.

What you should learn

- Recognize, write, and find the nth terms of arithmetic sequences.
- Find nth partial sums of arithmetic sequences.
- Use arithmetic sequences to model and solve real-life problems.

Why you should learn it

Arithmetic sequences can reduce the amount of time it takes to find the sum of a sequence of numbers with a common difference. In Exercise 83 on page 517, you will use an arithmetic sequence to find the number of bricks needed to lay a brick patio.

EXAMPLE 1 Examples of Arithmetic Sequences

a. The sequence whose nth term is

$$4n + 3$$

is arithmetic. The common difference between consecutive terms is 4.

$$\underbrace{7, 11}_{11 - 7 = 4}, 15, 19, \ldots, 4n + 3, \ldots$$ Begin with $n = 1$.

b. The sequence whose nth term is

$$7 - 5n$$

is arithmetic. The common difference between consecutive terms is -5.

$$\underbrace{2, -3}_{-3 - 2 = -5}, -8, -13, \ldots, 7 - 5n, \ldots$$ Begin with $n = 1$.

c. The sequence whose nth term is

$$\tfrac{1}{4}(n + 3)$$

is arithmetic. The common difference between consecutive terms is $\frac{1}{4}$.

$$\underbrace{1, \frac{5}{4}}_{\frac{5}{4} - 1 = \frac{1}{4}}, \frac{3}{2}, \frac{7}{4}, \ldots, \frac{n + 3}{4}, \ldots$$ Begin with $n = 1$.

✔ ***Checkpoint*** *Audio-video solution in English & Spanish at LarsonPrecalculus.com.*

Write the first four terms of the arithmetic sequence whose nth term is $3n - 1$. Then find the common difference between consecutive terms.

The sequence 1, 4, 9, 16, . . . , whose nth term is n^2, is *not* arithmetic. The difference between the first two terms is

$$a_2 - a_1 = 4 - 1 = 3$$

but the difference between the second and third terms is

$$a_3 - a_2 = 9 - 4 = 5.$$

The nth term of an arithmetic sequence can be derived from the pattern below.

$a_1 = a_1$ 1st term

$a_2 = a_1 + d$ 2nd term

$a_3 = a_1 + 2d$ 3rd term

$a_4 = a_1 + 3d$ 4th term

$a_5 = a_1 + 4d$ 5th term

1 less

⋮

$a_n = a_1 + (n - 1)d$ nth term

1 less

The result is summarized in the following definition.

The nth Term of an Arithmetic Sequence

The nth term of an arithmetic sequence has the form

$$a_n = a_1 + (n - 1)d$$

where d is the common difference between consecutive terms of the sequence and a_1 is the first term of the sequence.

EXAMPLE 2 Finding the nth Term of an Arithmetic Sequence

Find a formula for the nth term of the arithmetic sequence whose common difference is 3 and whose first term is 2.

Solution

You know that the formula for the nth term is of the form $a_n = a_1 + (n - 1)d$. Moreover, because the common difference is $d = 3$ and the first term is $a_1 = 2$, the formula must have the form

$$a_n = 2 + 3(n - 1) \quad \text{Substitute 2 for } a_1 \text{ and 3 for } d.$$

or $a_n = 3n - 1$. The sequence therefore has the following form.

$$2, 5, 8, 11, 14, \ldots, 3n - 1, \ldots$$

The figure below shows a graph of the first 15 terms of the sequence. Notice that the points lie on a line. This makes sense because a_n is a linear function of n. In other words, the terms "arithmetic" and "linear" are closely connected.

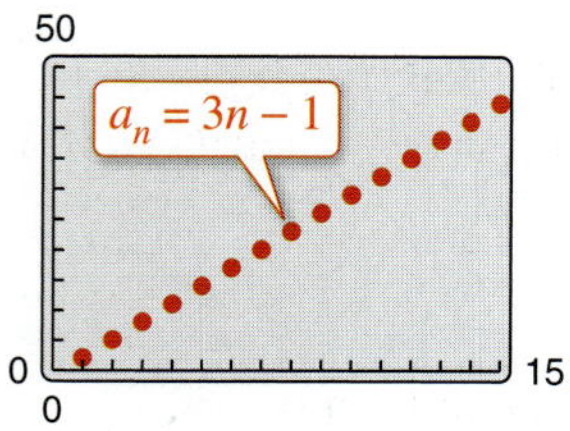

Checkpoint *Audio-video solution in English & Spanish at LarsonPrecalculus.com.*

Find a formula for the nth term of the arithmetic sequence whose common difference is 5 and whose first term is -1.

Explore the Concept

Consider the following sequences.

$1, 4, 7, 10, \ldots, 3n - 2, \ldots$

$-5, 1, 7, 13, \ldots, 6n - 11, \ldots$

$\frac{5}{2}, \frac{3}{2}, \frac{1}{2}, -\frac{1}{2}, \ldots, \frac{7}{2} - n, \ldots$

What relationship do you observe between successive terms of these sequences?

Technology Tip

You can use a graphing utility to generate the arithmetic sequence in Example 2 by using the following steps.

2 ENTER

3 + ANS

Now press the *enter* key repeatedly to generate the terms of the sequence.

EXAMPLE 3 Writing the Terms of an Arithmetic Sequence

The fourth term of an arithmetic sequence is 20, and the 13th term is 65. Write the first 13 terms of this sequence.

Solution

You know that $a_4 = 20$ and $a_{13} = 65$. So, you must add the common difference d nine times to the fourth term to obtain the 13th term. Therefore, the fourth and 13th terms of the sequence are related by

$$a_{13} = a_4 + 9d. \qquad a_4 \text{ and } a_{13} \text{ are nine terms apart.}$$

Using $a_4 = 20$ and $a_{13} = 65$, you have

$$65 = 20 + 9d.$$

So, you can conclude that $d = 5$, which implies that the sequence is as follows.

$$\begin{array}{cccccccccccccc} a_1, & a_2 & a_3 & a_4 & a_5 & a_6 & a_7 & a_8 & a_9 & a_{10} & a_{11} & a_{12} & a_{13} & \ldots \\ 5, & 10, & 15, & 20, & 25, & 30, & 35, & 40, & 45, & 50, & 55, & 60, & 65, & \ldots \end{array}$$

✓ Checkpoint *Audio-video solution in English & Spanish at LarsonPrecalculus.com.*

The eighth term of an arithmetic sequence is 25, and the 12th term is 41. Write the first 11 terms of this sequence.

When you know the nth term of an arithmetic sequence *and* you know the common difference of the sequence, you can find the $(n + 1)$th term by using the *recursion formula*

$$a_{n+1} = a_n + d. \qquad \text{Recursion formula}$$

With this formula, you can find any term of an arithmetic sequence, *provided* that you know the preceding term. For instance, when you know the first term, you can find the second term. Then, knowing the second term, you can find the third term, and so on.

EXAMPLE 4 Using a Recursion Formula

Find the ninth term of the arithmetic sequence whose first two terms are 2 and 9.

Solution

You know that the sequence is arithmetic. Also, $a_1 = 2$ and $a_2 = 9$. So, the common difference for this sequence is

$$d = 9 - 2 = 7.$$

There are two ways to find the ninth term. One way is simply to write out the first nine terms (by repeatedly adding 7).

$$2,\ 9,\ 16,\ 23,\ 30,\ 37,\ 44,\ 51,\ 58$$

Another way to find the ninth term is to first find a formula for the nth term. Because the common difference is $d = 7$ and the first term is $a_1 = 2$, the formula must have the form

$$a_n = 2 + 7(n - 1). \qquad \text{Substitute 2 for } a_1 \text{ and 7 for } d.$$

So, a formula for the nth term is $a_n = 7n - 5$, which implies that the ninth term is

$$a_9 = 7(9) - 5 = 58.$$

✓ Checkpoint *Audio-video solution in English & Spanish at LarsonPrecalculus.com.*

Find the tenth term of the arithmetic sequence that begins with 7 and 15.

Technology Tip

Most graphing utilities have a built-in function that will display the terms of an arithmetic sequence. For instructions on how to use the *sequence* feature, see Appendix A; for specific keystrokes, go to this textbook's *Companion Website.*

The Sum of a Finite Arithmetic Sequence

There is a simple formula for the *sum* of a finite arithmetic sequence.

The Sum of a Finite Arithmetic Sequence (See the proof on page 563.)

The sum of a finite arithmetic sequence with n terms is given by $S_n = \frac{n}{2}(a_1 + a_n)$.

> **Remark**
> Note that this formula works only for *arithmetic* sequences.

EXAMPLE 5 Sum of a Finite Arithmetic Sequence

Find the sum: $1 + 3 + 5 + 7 + 9 + 11 + 13 + 15 + 17 + 19$.

Solution

To begin, notice that the sequence is arithmetic (with a common difference of 2). Moreover, the sequence has 10 terms. So, the sum of the sequence is

$$S_n = 1 + 3 + 5 + 7 + 9 + 11 + 13 + 15 + 17 + 19$$

$$= \frac{n}{2}(a_1 + a_n) \quad \text{Formula for sum of an arithmetic sequence}$$

$$= \frac{10}{2}(1 + 19) \quad \text{Substitute 10 for } n\text{, 1 for } a_1\text{, and 19 for } a_n.$$

$$= 5(20) \quad \text{Simplify.}$$

$$= 100. \quad \text{Sum of the sequence}$$

✓ **Checkpoint** Audio-video solution in English & Spanish at LarsonPrecalculus.com.

Find the sum: $40 + 37 + 34 + 31 + 28 + 25 + 22$.

The sum of the first n terms of an infinite sequence is called the **nth partial sum.** The nth partial sum of an arithmetic sequence can be found by using the formula for the sum of a finite arithmetic sequence.

> **Remark**
> If you go on to take a course in calculus, you will study sequences and series in detail. You will learn that sequences and series play a major role in the study of calculus.

EXAMPLE 6 Partial Sum of an Arithmetic Sequence

Find the 150th partial sum of the arithmetic sequence 5, 16, 27, 38, 49, ...

Solution

For this arithmetic sequence, you have $a_1 = 5$ and $d = 16 - 5 = 11$. So,

$$a_n = 5 + 11(n - 1)$$

and the nth term is $a_n = 11n - 6$. Therefore, $a_{150} = 11(150) - 6 = 1644$, and the sum of the first 150 terms is

$$S_n = \frac{n}{2}(a_1 + a_n) \quad n\text{th partial sum formula}$$

$$= \frac{150}{2}(5 + 1644) \quad \text{Substitute 150 for } n\text{, 5 for } a_1\text{, and 1644 for } a_n.$$

$$= 75(1649) \quad \text{Simplify.}$$

$$= 123{,}675. \quad n\text{th partial sum}$$

✓ **Checkpoint** Audio-video solution in English & Spanish at LarsonPrecalculus.com.

Find the 120th partial sum of the arithmetic sequence

6, 12, 18, 24, 30, ...

Application

EXAMPLE 7 Total Sales

See LarsonPrecalculus.com for an interactive version of this type of example.

A small business sells $20,000 worth of sports memorabilia during its first year. The owner of the business has set a goal of increasing annual sales by $15,000 each year for 19 years. Assuming that this goal is met, find the total sales during the first 20 years this business is in operation.

Algebraic Solution

The annual sales form an arithmetic sequence in which $a_1 = 20{,}000$ and $d = 15{,}000$. So,

$$a_n = 20{,}000 + 15{,}000(n - 1)$$

and the nth term of the sequence is

$$a_n = 15{,}000n + 5000.$$

This implies that the 20th term of the sequence is

$$a_{20} = 15{,}000(20) + 5000$$
$$= 300{,}000 + 5000$$
$$= 305{,}000.$$

The sum of the first 20 terms of the sequence is

$$S_n = \frac{n}{2}(a_1 + a_n) \qquad n\text{th partial sum formula}$$
$$= \frac{20}{2}(20{,}000 + 305{,}000) \qquad \text{Substitute 20 for } n, \text{ 20,000 for } a_1, \text{ and 305,000 for } a_n.$$
$$= 10(325{,}000) \qquad \text{Simplify.}$$
$$= 3{,}250{,}000. \qquad \text{Simplify.}$$

So, the total sales for the first 20 years are $3,250,000.

Numerical Solution

The annual sales form an arithmetic sequence in which $a_1 = 20{,}000$ and $d = 15{,}000$. So,

$$a_n = 20{,}000 + 15{,}000(n - 1)$$

and the nth term of the sequence is

$$a_n = 15{,}000n + 5000.$$

Use the *list editor* of a graphing utility to create a table that shows the sales for each of the first 20 years and the total sales for the first 20 years, as shown in the figure.

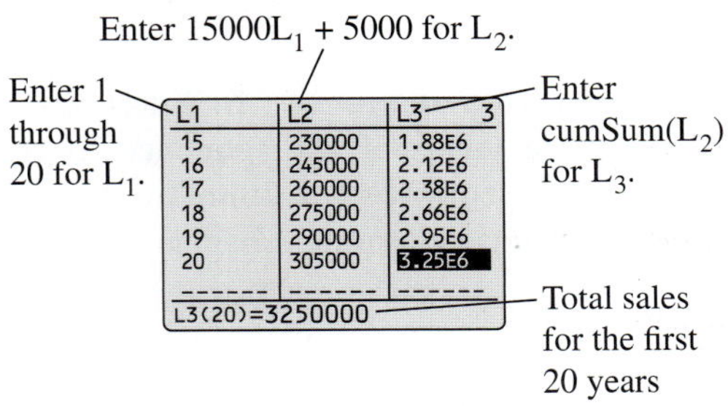

So, the total sales for the first 20 years are $3,250,000.

✓ Checkpoint Audio-video solution in English & Spanish at LarsonPrecalculus.com.

A company sells $160,000 worth of printing paper during its first year. The sales manager has set a goal of increasing annual sales of printing paper by $20,000 each year for 9 years. Assuming that this goal is met, find the total sales of printing paper during the first 10 years this company is in operation.

The figure below shows the annual sales for the business in Example 7. Notice that the annual sales for the business follow a *linear growth* pattern. In other words, saying that a quantity increases arithmetically is the same as saying that it increases linearly.

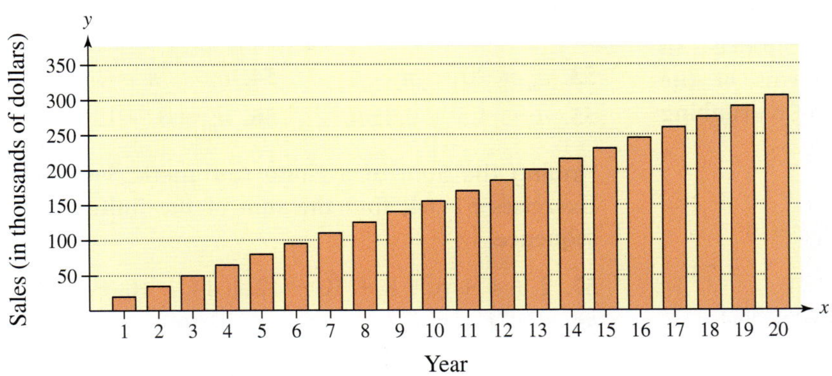

6.2 Exercises

See *CalcChat.com* for tutorial help and worked-out solutions to odd-numbered exercises.
For instructions on how to use a graphing utility, see Appendix A.

Vocabulary and Concept Check

In Exercises 1 and 2, fill in the blank.

1. The nth term of an arithmetic sequence has the form _______ .
2. The formula $S_n = \frac{n}{2}(a_1 + a_n)$ can be used to find the sum of the first n terms of an arithmetic sequence, called the _______ .
3. How do you know when a sequence is arithmetic?
4. Is 4 or 1 the common difference of the arithmetic sequence $a_n = 4n + 1$?

Procedures and Problem Solving

Identifying an Arithmetic Sequence **In Exercises 5–10, determine whether or not the sequence is arithmetic. If it is, find the common difference.**

5. $10, 12, 14, 16, 18, \ldots$
6. $4, 9, 14, 19, 24, \ldots$
7. $3, \frac{5}{2}, 2, \frac{3}{2}, 1, \ldots$
8. $3.7, 3.1, 2.5, 1.9, 1.3, \ldots$
9. $1^2, 2^2, 3^2, 4^2, 5^2, \ldots$
10. $\frac{1}{3}, \frac{2}{3}, \frac{4}{3}, \frac{8}{3}, \frac{16}{3}, \ldots$

Writing the Terms of a Sequence **In Exercises 11–20, write the first five terms of the sequence. Determine whether or not the sequence is arithmetic. If it is, find the common difference. (Assume n begins with 1.)**

11. $a_n = 8 + 13n$
12. $a_n = 150 - 7n$
13. $a_n = 53 - 4(n + 6)$
14. $a_n = 1 + (n - 1)4$
15. $a_n = 2^n + n$
16. $a_n = 2^{n-1}$
17. $a_n = 3 + 2(-1)^n$
18. $a_n = (-1)^n$
19. $a_n = \dfrac{(-1)^{2n}}{4}$
20. $a_n = (-1)^{2n+1}$

Finding the nth Term of an Arithmetic Sequence **In Exercises 21–30, find a formula for a_n for the arithmetic sequence.**

21. $a_1 = 1, d = 6$
22. $a_1 = 15, d = 4$
23. $a_1 = 43, d = -7$
24. $a_1 = 100, d = -8$
25. $3, \frac{9}{4}, \frac{3}{2}, \frac{3}{4}, 0, \ldots$
26. $4, \frac{3}{2}, -1, -\frac{7}{2}, \ldots$
27. $a_1 = -5, a_4 = 22$
28. $a_1 = -4, a_5 = 16$
29. $a_3 = 94, a_6 = 85$
30. $a_5 = 190, a_{10} = 115$

Writing the Terms of an Arithmetic Sequence **In Exercises 31–38, write the first five terms of the arithmetic sequence. Use the *table* feature of a graphing utility to verify your results.**

31. $a_1 = 5, d = 6$
32. $a_1 = 5, d = -\frac{3}{4}$
33. $a_1 = -2.6, d = 0.2$
34. $a_1 = -10, d = 9$
35. $a_8 = 26, a_{12} = 42$
36. $a_6 = -38, a_{11} = -73$
37. $a_3 = 19, a_{15} = -1.7$
38. $a_5 = 16, a_{14} = 38.5$

Writing the Terms of an Arithmetic Sequence **In Exercises 39–42, write the first five terms of the arithmetic sequence. Find the common difference and write the nth term of the sequence as a function of n.**

39. $a_1 = 15, a_{k+1} = a_k + 4$
40. $a_1 = 6, a_{k+1} = a_k + 5$
41. $a_1 = \frac{3}{5}, a_{k+1} = -\frac{1}{10} + a_k$
42. $a_1 = 1.5, a_{k+1} = a_k - 2.5$

Using a Recursion Formula **In Exercises 43–46, the first two terms of the arithmetic sequence are given. Find the missing term. Use the *table* feature of a graphing utility to verify your results.**

43. $a_1 = 5, \ a_2 = 11, \ a_{10} = \square$
44. $a_1 = 3, \ a_2 = 13, \ a_9 = \square$
45. $a_1 = 4.2, \ a_2 = 1.8, \ a_7 = \square$
46. $a_1 = -0.7, \ a_2 = -13.8, \ a_8 = \square$

Graphing the Terms of a Sequence **In Exercises 47–50, use a graphing utility to graph the first 10 terms of the sequence. (Assume n begins with 1.)**

47. $a_n = 15 - \frac{3}{2}n$
48. $a_n = -5 + 2n$
49. $a_n = 0.4n - 2$
50. $a_n = -1.3n + 7$

Finding the Terms of a Sequence **In Exercises 51–56, use the *table* feature of a graphing utility to find the first 10 terms of the sequence. (Assume n begins with 1.)**

51. $a_n = 4n - 5$
52. $a_n = 17 + 3n$
53. $a_n = 20 - \frac{3}{4}n$
54. $a_n = \frac{4}{5}n - 3$
55. $a_n = 1.5 + 0.05n$
56. $a_n = 8 - 12.5n$

Finding the Sum of a Finite Arithmetic Sequence **In Exercises 57–64, find the sum of the finite arithmetic sequence.**

57. $2 + 4 + 6 + 8 + 10 + 12 + 14 + 16 + 18 + 20$
58. $1 + 4 + 7 + 10 + 13 + 16 + 19$

59. $-1 + (-3) + (-5) + (-7) + (-9)$

60. $-2 + (-5) + (-8) + (-11) + (-14) + (-17)$

61. Sum of the first 100 positive integers

62. Sum of the first 50 negative integers

63. Sum of the integers from -100 to 30

64. Sum of the integers from -10 to 50

Finding a Partial Sum of an Arithmetic Sequence **In Exercises 65–70, find the indicated *n*th partial sum of the arithmetic sequence.**

65. $8, 20, 32, 44, \ldots; \ n = 10$

66. $2, 8, 14, 20, \ldots; \ n = 25$

67. $7.2, 6.4, 5.6, 4.8, \ldots; \ n = 10$

68. $4.2, 3.7, 3.2, 2.7, \ldots; \ n = 12$

69. $a_1 = 100, a_{25} = 220, \ n = 25$

70. $a_1 = 15, a_{100} = 307, \ n = 100$

Finding a Partial Sum of an Arithmetic Sequence **In Exercises 71–76, find the partial sum without using a graphing utility.**

71. $\sum_{n=1}^{50} n$

72. $\sum_{n=1}^{100} 2n$

73. $\sum_{n=11}^{30} n - \sum_{n=1}^{10} n$

74. $\sum_{n=51}^{100} n - \sum_{n=1}^{50} n$

75. $\sum_{n=1}^{500} (n + 8)$

76. $\sum_{n=1}^{250} (1000 - n)$

Finding a Partial Sum Using a Graphing Utility **In Exercises 77–82, use a graphing utility to find the partial sum.**

77. $\sum_{n=1}^{20} (2n + 1)$

78. $\sum_{n=1}^{50} (40 - 2n)$

79. $\sum_{n=0}^{100} \frac{n + 5}{2}$

80. $\sum_{n=0}^{100} \frac{4 - n}{4}$

81. $\sum_{i=1}^{60} \left(250 - \frac{2}{5}i\right)$

82. $\sum_{j=1}^{200} (10.5 + 0.025j)$

83. ***Why you should learn it*** *(p. 511)* A brick patio has the approximate shape of a trapezoid, as shown in the figure. The patio has 18 rows of bricks. The first row has 14 bricks and the 18th row has 31 bricks. How many bricks are in the patio?

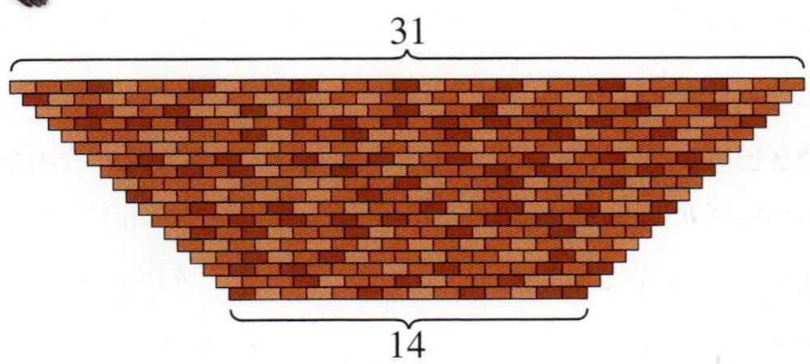

84. Performing Arts An auditorium has 20 rows of seats. There are 20 seats in the first row, 21 seats in the second row, 22 seats in the third row, and so on (see figure). How many seats are there in all 20 rows?

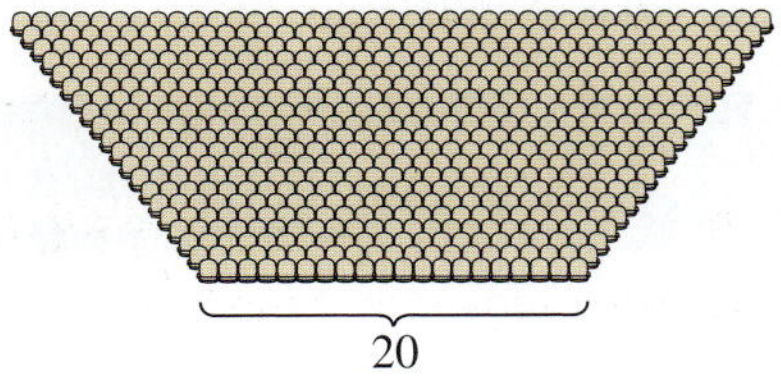

85. Business A hardware store makes a profit of \$30,000 during its first year. The store owner sets a goal of increasing profits by \$5000 each year for 4 years. Assuming that this goal is met, find the total profit during the first 5 years of business.

86. Physics An object with negligible air resistance is dropped from a plane. During the first second of fall, the object falls 16 feet; during the second second, it falls 48 feet; during the third second, it falls 80 feet; and during the fourth second, it falls 112 feet. Assume this pattern continues. How many feet will the object fall in 8 seconds?

87. MODELING DATA

The table shows the sales a_n (in billions of dollars) for PetSmart from 2006 through 2013. (Source: PetSmart, Inc.)

DATA

Year	Sales, a_n (in billions of dollars)
2006	4.23
2007	4.67
2008	5.07
2009	5.34
2010	5.69
2011	6.11
2012	6.76
2013	6.92

Spreadsheet at LarsonPrecalculus.com

(a) Use the *regression* feature of a graphing utility to find an arithmetic sequence for the data. Let n represent the year, with $n = 6$ corresponding to 2006.

(b) Use the sequence from part (a) to approximate the annual sales for PetSmart for the years 2006 through 2013. How well does the model fit the data?

(c) Use the sequence to find the total sales for PetSmart during the period from 2006 through 2013.

(d) Use the sequence to predict the total sales during the period from 2014 through 2021. Is your total reasonable? Explain.

88. MODELING DATA

The table shows the numbers a_n (in thousands) of master's degrees conferred in the United States from 2005 through 2012. (Source: U.S. National Center for Education Statistics)

DATA

Year	Master's degrees conferred, a_n (in thousands)
2005	580
2006	600
2007	611
2008	631
2009	662
2010	693
2011	731
2012	754

Spreadsheet at LarsonPrecalculus.com

(a) Use the *regression* feature of a graphing utility to find an arithmetic sequence for the data. Let n represent the year, with $n = 5$ corresponding to 2005.

(b) Use the sequence from part (a) to approximate the numbers of master's degrees conferred for the years 2005 through 2012. How well does the model fit the data?

(c) Use the sequence to find the total number of master's degrees conferred during the period from 2005 through 2012.

(d) Use the sequence to predict the total number of master's degrees conferred during the period from 2013 through 2023. Is your total reasonable? Explain.

Conclusions

True or False? **In Exercises 89 and 90, determine whether the statement is true or false. Justify your answer.**

89. Given the nth term and the common difference of an arithmetic sequence, it is possible to find the $(n + 1)$th term.

90. If the only known information about a finite arithmetic sequence is its first term and its last term, then it is possible to find the sum of the sequence.

Finding the Terms of a Sequence **In Exercises 91 and 92, find the first 10 terms of the sequence.**

91. $a_1 = x, d = 2x$ **92.** $a_1 = -y, d = 5y$

93. Writing Explain how to use the first two terms of an arithmetic sequence to find the nth term.

94. Think About It The sum of the first 20 terms of an arithmetic sequence with a common difference of 3 is 650. Find the first term.

95. Think About It The sum of the first n terms of an arithmetic sequence with first term a_1 and common difference d is S_n. Determine the sum when each term is increased by 5. Explain.

96. Think About It In each sequence, decide whether it is possible to fill in the blanks to form an arithmetic sequence. If so, find a recursion formula for the sequence. Explain how you found your answers.

(a) -7, ___, ___, ___, ___, ___, 11

(b) 2, 6, ___, ___, 162

(c) 4, 7.5, ___, ___, ___, ___, ___, 28.5

97. Writing Carl Friedrich Gauss, a famous nineteenth century mathematician, was a child prodigy. It was said that when Gauss was 10, he was asked by his teacher to add the numbers from 1 to 100. Almost immediately, Gauss found the answer by mentally finding the summation. Write an explanation of how he arrived at his conclusion, and then find the formula for the sum of the first n natural numbers.

98. HOW DO YOU SEE IT? The graph of a sequence is shown below. Determine which of the statements about the sequence is false. Justify your answer.

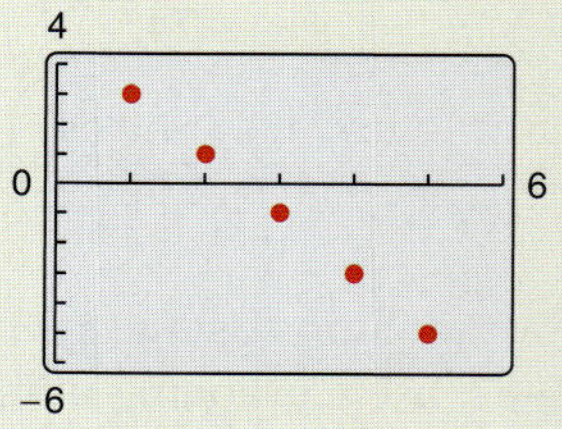

(a) The sequence is arithmetic.

(b) The common difference is 2.

(c) $a_1 = 3$.

Cumulative Mixed Review

Gauss-Jordan Elimination **In Exercises 99 and 100, use Gauss-Jordan elimination to solve the system of equations.**

99. $\begin{cases} 2x - y + 7z = -10 \\ 3x + 2y - 4z = 17 \\ 6x - 5y + z = -20 \end{cases}$

100. $\begin{cases} -x + 4y + 10z = 4 \\ 5x - 3y + z = 31 \\ 8x + 2y - 3z = -5 \end{cases}$

101. *Make a Decision* To work an extended application analyzing the net sales of Dollar Tree from 2001 through 2013, visit this textbook's website at *LarsonPrecalculus.com.* (Data Source: Dollar Tree, Inc.)

6.3 Geometric Sequences and Series

Geometric Sequences

In Section 6.2, you learned that a sequence whose consecutive terms have a common *difference* is an arithmetic sequence. In this section, you will study another important type of sequence called a **geometric sequence.** Consecutive terms of a geometric sequence have a common *ratio.*

Definition of Geometric Sequence

A sequence is **geometric** when the ratios of consecutive terms are the same. So, the sequence $a_1, a_2, a_3, a_4, \ldots, a_n, \ldots$ is geometric when there is a number r such that

$$\frac{a_2}{a_1} = \frac{a_3}{a_2} = \frac{a_4}{a_3} = \cdots = r, \quad r \neq 0.$$

The number r is the **common ratio** of the sequence.

What you should learn

- Recognize, write, and find the *n*th terms of geometric sequences.
- Find *n*th partial sums of geometric sequences.
- Find sums of infinite geometric series.
- Use geometric sequences to model and solve real-life problems.

Why you should learn it

Geometric sequences can reduce the amount of time it takes to find the sum of a sequence of numbers with a common ratio. For instance, Exercise 111 on page 528 shows how to use a geometric sequence to find the total vertical distance traveled by a bouncing ball.

EXAMPLE 1 Examples of Geometric Sequences

a. The sequence whose nth term is 2^n is geometric. For this sequence, the common ratio between consecutive terms is 2.

$$2, 4, 8, 16, \ldots, 2^n, \ldots \qquad \text{Begin with } n = 1.$$

$$\tfrac{4}{2} = 2$$

b. The sequence whose nth term is $4(3^n)$ is geometric. For this sequence, the common ratio between consecutive terms is 3.

$$12, 36, 108, 324, \ldots, 4(3^n), \ldots \qquad \text{Begin with } n = 1.$$

$$\tfrac{36}{12} = 3$$

c. The sequence whose nth term is $\left(-\frac{1}{3}\right)^n$ is geometric. For this sequence, the common ratio between consecutive terms is $-\frac{1}{3}$.

$$-\frac{1}{3}, \frac{1}{9}, -\frac{1}{27}, \frac{1}{81}, \ldots, \left(-\frac{1}{3}\right)^n, \ldots \qquad \text{Begin with } n = 1.$$

$$\frac{1/9}{-1/3} = -\frac{1}{3}$$

In parts (a), (b), and (c), notice that each of the geometric sequences has an nth term of the form ar^n, where r is the common ratio of the sequence.

Checkpoint Audio-video solution in English & Spanish at LarsonPrecalculus.com.

Write the first four terms of the geometric sequence whose nth term is $6(-2)^n$. Then find the common ratio of the consecutive terms.

The sequence $1, 4, 9, 16, \ldots,$ whose nth term is n^2, is *not* geometric. The ratio of the second term to the first term is

$$\frac{a_2}{a_1} = \frac{4}{1} = 4$$

but the ratio of the third term to the second term is

$$\frac{a_3}{a_2} = \frac{9}{4}.$$

The nth Term of a Geometric Sequence

The nth term of a geometric sequence has the form

$$a_n = a_1 r^{n-1}$$

where r is the common ratio of consecutive terms of the sequence. So, every geometric sequence can be written in the following form.

$$\begin{array}{ccccccccc} a_1, & a_2, & a_3, & a_4, & a_5, & \ldots, & a_n, & \ldots \\ \downarrow & \downarrow & \downarrow & \downarrow & \downarrow & & \downarrow & \\ a_1, & a_1r, & a_1r^2, & a_1r^3, & a_1r^4, & \ldots, & a_1r^{n-1}, & \ldots \end{array}$$

When you know the nth term of a geometric sequence, you can find the $(n + 1)$th term by multiplying by r. That is, $a_{n+1} = a_n r$.

EXAMPLE 2 Writing the Terms of a Geometric Sequence

Write the first five terms of the geometric sequence whose first term is $a_1 = 3$ and whose common ratio is $r = 2$.

Solution

Starting with 3, repeatedly multiply by 2 to obtain the following.

$a_1 = 3$ — 1st term

$a_2 = 3(2^1) = 6$ — 2nd term

$a_3 = 3(2^2) = 12$ — 3rd term

$a_4 = 3(2^3) = 24$ — 4th term

$a_5 = 3(2^4) = 48$ — 5th term

✓ Checkpoint *Audio-video solution in English & Spanish at LarsonPrecalculus.com.*

Write the first five terms of the geometric sequence whose first term is $a_1 = 2$ and whose common ratio is $r = 4$.

Technology Tip

You can use a graphing utility to generate the geometric sequence in Example 2 by using the following steps.

3 [ENTER]

2 [×] [ANS]

Now press the *enter* key repeatedly to generate the terms of the sequence.

```
3
                3
2*Ans
                6
2*Ans
               12
```

EXAMPLE 3 Finding a Term of a Geometric Sequence

Find the 15th term of the geometric sequence whose first term is 20 and whose common ratio is 1.05.

Algebraic Solution

$a_n = a_1 r^{n-1}$ — Formula for a geometric sequence

$a_{15} = 20(1.05)^{15-1}$ — Substitute 20 for a_1, 1.05 for r, and 15 for n.

≈ 39.60 — Use a calculator.

Numerical Solution

For this sequence, $r = 1.05$ and $a_1 = 20$. So, $a_n = 20(1.05)^{n-1}$. Use the *table* feature of a graphing utility to create a table that shows the terms of the sequence, as shown in the figure.

n	u(n)
9	29.549
10	31.027
11	32.578
12	34.207
13	35.917
14	37.713
15	39.599

u(n)=39.59863199

The number in the 15th row is the 15th term of the sequence.

So, $a_{15} \approx 39.60$.

✓ Checkpoint *Audio-video solution in English & Spanish at LarsonPrecalculus.com.*

Find the 12th term of the geometric sequence whose first term is 14 and whose common ratio is 1.2.

EXAMPLE 4 Finding a Term of a Geometric Sequence

Find a formula for the nth term of the geometric sequence 5, 15, 45, ... What is the ninth term of the sequence?

Solution

The common ratio of this sequence is $r = 15/3 = 3$. Because the first term is $a_1 = 5$, the formula must have the form

$$a_n = a_1 r^{n-1} = 5(3)^{n-1}.$$

You can determine the ninth term $(n = 9)$ to be

$$a_9 = 5(3)^{9-1} \quad \text{Substitute 9 for } n.$$
$$= 5(6561) \quad \text{Use a calculator.}$$
$$= 32{,}805. \quad \text{Simplify.}$$

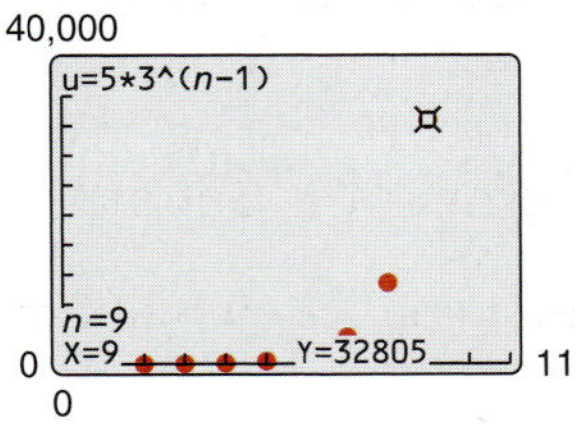

A graph of the first nine terms of the sequence is shown in the figure. Notice that the points lie on an exponential curve. This makes sense because a_n is an exponential function of n.

✓ Checkpoint *Audio-video solution in English & Spanish at LarsonPrecalculus.com.*

Find a formula for the nth term of the geometric sequence 4, 20, 100, ... What is the 12th term of the sequence?

When you know *any* two terms of a geometric sequence, you can use that information to find *any other* term of the sequence.

Technology Tip

Most graphing utilities have a built-in function that will display the terms of a geometric sequence. For instructions on how to use the *sequence* feature, see Appendix A; for specific keystrokes, go to this textbook's *Companion Website*.

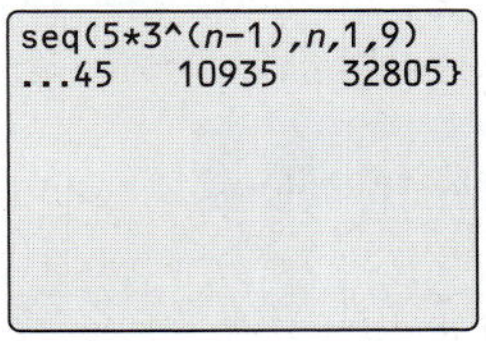

EXAMPLE 5 Finding a Term of a Geometric Sequence

The fourth term of a geometric sequence is 125, and the 10th term is 125/64. Find the 14th term. (Assume that the terms of the sequence are positive.)

Solution

The 10th term is related to the fourth term by the equation

$$a_{10} = a_4 r^6. \quad \text{Multiply 4th term by } r^{10-4}.$$

Because $a_{10} = 125/64$ and $a_4 = 125$, you can solve for r as follows.

$$\frac{125}{64} = 125r^6 \quad \text{Substitute } \tfrac{125}{64} \text{ for } a_{10} \text{ and 125 for } a_4.$$

$$\frac{1}{64} = r^6 \quad \text{Divide each side by 125.}$$

$$\frac{1}{2} = r \quad \text{Take the sixth root of each side.}$$

You can obtain the 14th term by multiplying the 10th term by $r^{14-10} = r^4$.

$$a_{14} = a_{10}r^4 = \frac{125}{64}\left(\frac{1}{2}\right)^4 = \frac{125}{1024}$$

✓ Checkpoint *Audio-video solution in English & Spanish at LarsonPrecalculus.com.*

The second term of a geometric sequence is 6, and the fifth term is 81/4. Find the eighth term. (Assume that the terms of the sequence are positive.)

Remark

Remember that r is the common ratio of consecutive terms of a geometric sequence. So, in Example 5,

$$a_{10} = a_1 r^9$$
$$= a_1 \cdot r \cdot r \cdot r \cdot r^6$$
$$= a_1 \cdot \frac{a_2}{a_1} \cdot \frac{a_3}{a_2} \cdot \frac{a_4}{a_3} \cdot r^6$$
$$= a_4 r^6.$$

The Sum of a Finite Geometric Sequence

The formula for the sum of a *finite* geometric sequence is as follows.

The Sum of a Finite Geometric Sequence (See the proof on page 563.)

The sum of the finite geometric sequence

$$a_1, a_1r, a_1r^2, a_1r^3, a_1r^4, \ldots, a_1r^{n-1}$$

with common ratio $r \neq 1$ is given by

$$S_n = \sum_{i=1}^{n} a_1 r^{i-1} = a_1\left(\frac{1-r^n}{1-r}\right).$$

EXAMPLE 6 Sum of a Finite Geometric Sequence

Find the following sum.

$$\sum_{n=1}^{12} 4(0.3)^n$$

Solution

By writing out a few terms, you have

$$\sum_{n=1}^{12} 4(0.3)^n = 4(0.3)^1 + 4(0.3)^2 + 4(0.3)^3 + \cdots + 4(0.3)^{12}.$$

Now, because

$$a_1 = 4(0.3), \qquad r = 0.3, \qquad \text{and} \qquad n = 12$$

you can apply the formula for the sum of a finite geometric sequence to obtain

$$\sum_{n=1}^{12} 4(0.3)^n = a_1\left(\frac{1-r^n}{1-r}\right)$$ Formula for sum of a finite geometric sequence

$$= 4(0.3)\left[\frac{1-(0.3)^{12}}{1-0.3}\right]$$ Substitute 4(0.3) for a_1, 0.3 for r, and 12 for n.

$$\approx 1.71.$$ Use a calculator.

Checkpoint *Audio-video solution in English & Spanish at LarsonPrecalculus.com.*

Find the sum $\sum_{i=1}^{10} 2(0.25)^{i-1}$.

When using the formula for the sum of a geometric sequence, be careful to check that the index begins at $i = 1$. For an index that begins at $i = 0$, you must adjust the formula for the nth partial sum. For instance, if the index in Example 6 had begun with $n = 0$, then the sum would have been

$$\sum_{n=0}^{12} 4(0.3)^n = 4(0.3)^0 + \sum_{n=1}^{12} 4(0.3)^n$$

$$= 4 + \sum_{n=1}^{12} 4(0.3)^n$$

$$\approx 4 + 1.71$$

$$= 5.71.$$

Technology Tip

Using the *sum sequence* feature of a graphing utility, you can calculate the sum of the sequence in Example 6 to be about 1.7142848, as shown below.

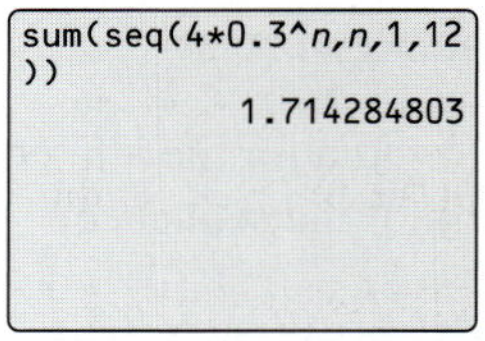

Calculate the sum beginning at $n = 0$. You should obtain a sum of about 5.7142848.

What's Wrong?

You use a graphing utility to find the sum

$$\sum_{n=1}^{6} 7(0.5)^n$$

as shown in the figure. What's wrong?

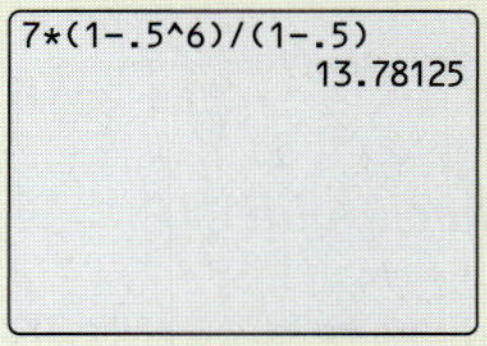

Geometric Series

The sum of the terms of an infinite geometric sequence is called an **infinite geometric series** or simply a **geometric series.**

The formula for the sum of a *finite geometric sequence* can, depending on the value of r, be extended to produce a formula for the sum of an *infinite geometric series.* Specifically, if the common ratio r has the property that $|r| < 1$, then it can be shown that r^n becomes arbitrarily close to zero as n increases without bound. Consequently,

$$a_1\left(\frac{1 - r^n}{1 - r}\right) \longrightarrow a_1\left(\frac{1 - 0}{1 - r}\right) \quad \text{as} \quad n \longrightarrow \infty.$$

This result is summarized as follows.

The Sum of an Infinite Geometric Sequence

If $|r| < 1$, then the infinite geometric series

$$a_1 + a_1r + a_1r^2 + a_1r^3 + \cdot \cdot \cdot + a_1r^{n-1} + \cdot \cdot \cdot$$

has the sum

$$S = \sum_{i=0}^{\infty} a_1r^i = \frac{a_1}{1 - r}.$$

Explore the Concept

Notice that the formula for the sum of an infinite geometric series requires that $|r| < 1$. What happens when $r = 1$ or $r = -1$? Give examples of infinite geometric series for which $|r| > 1$ and convince yourself that they do not have finite sums.

Note that when $|r| \geq 1$, the series does not have a sum.

EXAMPLE 7 Finding the Sum of an Infinite Geometric Series

Find each sum.

a. $\sum_{n=0}^{\infty} 4(0.6)^n$

b. $3 + 0.3 + 0.03 + 0.003 + \cdot \cdot \cdot$

Solution

a. $\sum_{n=0}^{\infty} 4(0.6)^n = 4 + 4(0.6) + 4(0.6)^2 + 4(0.6)^3 + \cdot \cdot \cdot + 4(0.6)^n + \cdot \cdot \cdot$

$= \dfrac{4}{1 - 0.6}$ $\qquad \dfrac{a_1}{a - r}$

$= 10$

b. $3 + 0.3 + 0.03 + 0.003 + \cdot \cdot \cdot = 3 + 3(0.1) + 3(0.1)^2 + 3(0.1)^3 + \cdot \cdot \cdot$

$= \dfrac{3}{1 - 0.1}$ $\qquad \dfrac{a_1}{a - r}$

$= \dfrac{10}{3}$

≈ 3.33

Checkpoint *Audio-video solution in English & Spanish at LarsonPrecalculus.com.*

Find each sum.

a. $\sum_{n=0}^{\infty} 5(0.5)^n$

b. $5 + 1 + 0.2 + 0.04 + \cdot \cdot \cdot$

Application

EXAMPLE 8 Increasing Annuity

See LarsonPrecalculus.com for an interactive version of this type of example.

A deposit of \$50 is made on the first day of each month in an account that pays 3% compounded monthly. What is the balance at the end of 2 years? (This type of savings plan is called an **increasing annuity.**)

Personal Financial Advisor

Solution

Recall from Section 4.1 that the compound interest formula is

$$A = P\left(1 + \frac{r}{n}\right)^{nt}. \qquad \text{Formula for compound interest}$$

To find the balance in the account after 24 months, consider each of the 24 deposits separately. The first deposit will gain interest for 24 months, and its balance will be

$$A_{24} = 50\left(1 + \frac{0.03}{12}\right)^{24}$$
$$= 50(1.0025)^{24}.$$

The second deposit will gain interest for 23 months, and its balance will be

$$A_{23} = 50\left(1 + \frac{0.03}{12}\right)^{23}$$
$$= 50(1.0025)^{23}.$$

The last deposit will gain interest for only 1 month, and its balance will be

$$A_1 = 50\left(1 + \frac{0.03}{12}\right)^{1}$$
$$= 50(1.0025).$$

The total balance in the annuity will be the sum of the balances of the 24 deposits. Using the formula for the sum of a finite geometric sequence, with $a_1 = 50(1.0025)$, $r = 1.0025$, and $n = 24$, you have

$$S_n = a_1\left(\frac{1 - r^n}{1 - r}\right) \qquad \text{Formula for sum of a finite geometric sequence}$$

$$S_{24} = 50(1.0025)\left[\frac{1 - (1.0025)^{24}}{1 - 1.0025}\right] \qquad \text{Substitute } 50(1.0025) \text{ for } a_1, 1.0025 \text{ for } r, \text{ and 24 for } n.$$

$$\approx \$1238.23. \qquad \text{Use a calculator.}$$

You can use the *sum sequence* feature of a graphing utility to verify this result, as shown in the figure.

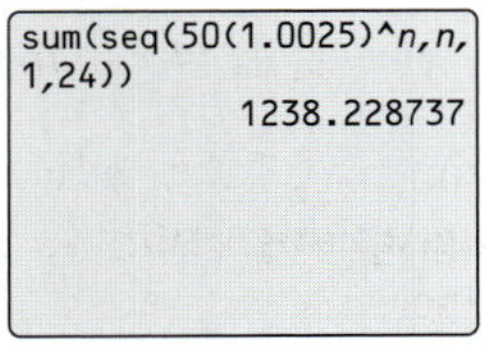

✓ Checkpoint *Audio-video solution in English & Spanish at LarsonPrecalculus.com.*

An investor deposits \$70 on the first day of each month in an account that pays 2% interest, compounded monthly. What is the balance at the end of 4 years?

6.3 Exercises

See *CalcChat.com* for tutorial help and worked-out solutions to odd-numbered exercises. For instructions on how to use a graphing utility, see Appendix A.

Vocabulary and Concept Check

In Exercises 1–3, fill in the blank(s).

1. A sequence is called a _______ sequence if the ratios of consecutive terms are the same. This ratio is called the _______ ratio.

2. The nth term of a geometric sequence has the form $a_n =$ _______ .

3. The sum of the terms of an infinite geometric sequence is called a _______ .

4. Can a geometric sequence have a common ratio of 0?

5. For what values of the common ratio r is it possible to find the sum of an infinite geometric series?

6. Which formula represents the sum of a *finite geometric sequence*? an *infinite geometric series*?

(a) $S = \dfrac{a_1}{1 - r}, \ |r| < 1$ (b) $S_n = a_1\left(\dfrac{1 - r^n}{1 - r}\right)$

Procedures and Problem Solving

Identifying a Geometric Sequence **In Exercises 7–16, determine whether or not the sequence is geometric. If it is, find the common ratio.**

7. $5, 15, 45, 135, \ldots$

8. $3, 12, 48, 192, \ldots$

9. $6, 18, 30, 42, \ldots$

10. $4, 19, 34, 49, \ldots$

11. $1, -\frac{1}{2}, \frac{1}{4}, -\frac{1}{8}, \ldots$

12. $9, -6, 4, -\frac{8}{3}, \ldots$

13. $20, 2, 0.2, 0.02, \ldots$

14. $5, 1, 0.2, 0.04, \ldots$

15. $1, \frac{1}{2}, \frac{1}{3}, \frac{1}{4}, \ldots$

16. $\frac{1}{5}, \frac{2}{7}, \frac{3}{9}, \frac{4}{11}, \ldots$

Writing the Terms of a Geometric Sequence **In Exercises 17–24, write the first five terms of the geometric sequence.**

17. $a_1 = 6, \ r = 3$

18. $a_1 = 4, \ r = 2$

19. $a_1 = 1, \ r = \frac{1}{2}$

20. $a_1 = 2, \ r = \frac{1}{3}$

21. $a_1 = 5, \ r = -\frac{1}{10}$

22. $a_1 = 6, \ r = -\frac{1}{4}$

23. $a_1 = 9, \ r = e$

24. $a_1 = 7, \ r = \sqrt{5}$

Finding the *n*th Term of a Geometric Sequence **In Exercises 25–30, write the first five terms of the geometric sequence. Find the common ratio and write the *n*th term of the sequence as a function of *n*.**

25. $a_1 = 64, a_{k+1} = \frac{1}{2}a_k$

26. $a_1 = 81, a_{k+1} = \frac{1}{3}a_k$

27. $a_1 = 9, a_{k+1} = 2a_k$

28. $a_1 = 5, a_{k+1} = 3a_k$

29. $a_1 = 6, a_{k+1} = -\frac{3}{2}a_k$

30. $a_1 = 30, a_{k+1} = -\frac{2}{3}a_k$

Finding a Term of a Geometric Sequence **In Exercises 31–38, find the indicated term of the geometric sequence (a) using the *table* feature of a graphing utility and (b) algebraically.**

31. $a_1 = 11, r = 1.03$, 12th term

32. $a_1 = 24, r = 2.6$, 8th term

33. $a_1 = 8, r = -\frac{4}{3}$, 7th term

34. $a_1 = 8, r = -\frac{3}{4}$, 9th term

35. $a_1 = -\frac{1}{4}, r = 8$, 6th term

36. $a_1 = -\frac{1}{128}, r = 2$, 12th term

37. $a_1 = 7, r = \sqrt{2}$, 14th term

38. $a_1 = 2, r = \sqrt{3}$, 11th term

Finding a Term of a Geometric Sequence **In Exercises 39–42, find a formula for the *n*th term of the geometric sequence. Then find the indicated term of the geometric sequence.**

39. 9th term: $7, 21, 63, \ldots$

40. 7th term: $3, 36, 432, \ldots$

41. 10th term: $5, 30, 180, \ldots$

42. 22nd term: $4, 8, 16, \ldots$

Finding a Term of a Geometric Sequence **In Exercises 43–46, find the indicated term of the geometric sequence.**

43. $a_1 = 4, a_4 = \frac{1}{2}$, 10th term

44. $a_1 = 5, a_3 = \frac{45}{4}$, 8th term

45. $a_2 = -18, a_5 = \frac{2}{3}$, 6th term

46. $a_2 = -8, a_5 = \frac{64}{27}$, 6th term

Graphing the Terms of a Sequence **In Exercises 47–50, use a graphing utility to graph the first 10 terms of the sequence.**

47. $a_n = 12(-0.75)^{n-1}$

48. $a_n = 20(0.85)^{n-1}$

49. $a_n = 2(1.3)^{n-1}$

50. $a_n = 10(-1.2)^{n-1}$

Finding a Sequence of Partial Sums **In Exercises 51 and 52, find the sequence of the first five partial sums S_1, S_2, S_3, S_4, and S_5 of the geometric sequence by adding terms.**

51. $8, -4, 2, -1, \frac{1}{2}, \ldots$

52. $8, 12, 18, 27, \frac{81}{2}, \ldots$

Finding a Sequence of Partial Sums **In Exercises 53 and 54, use a graphing utility to create a table showing the sequence of the first 10 partial sums $S_1, S_2, S_3, \ldots,$ and S_{10} for the series.**

53. $\sum_{n=1}^{\infty} 16\left(\frac{1}{2}\right)^{n-1}$

54. $\sum_{n=1}^{\infty} 4(0.2)^{n-1}$

Finding the Sum of a Finite Geometric Sequence **In Exercises 55–64, find the sum. Use a graphing utility to verify your result.**

55. $\sum_{n=1}^{9} 2^{n-1}$

56. $\sum_{n=1}^{9} (-2)^{n-1}$

57. $\sum_{i=1}^{7} 64\left(-\frac{1}{2}\right)^{i-1}$

58. $\sum_{i=1}^{6} 32\left(\frac{1}{4}\right)^{i-1}$

59. $\sum_{n=0}^{20} 12\left(\frac{6}{5}\right)^{n}$

60. $\sum_{n=0}^{15} 10\left(\frac{7}{6}\right)^{n}$

61. $\sum_{i=1}^{10} 8\left(-\frac{1}{4}\right)^{i-1}$

62. $\sum_{i=1}^{10} 5\left(-\frac{1}{3}\right)^{i-1}$

63. $\sum_{n=0}^{5} 300(1.06)^{n}$

64. $\sum_{n=0}^{6} 500(1.04)^{n}$

Using Summation Notation **In Exercises 65–68, use summation notation to write the sum.**

65. $5 + 15 + 45 + \cdots + 3645$

66. $7 + 14 + 28 + \cdots + 896$

67. $2 - \frac{1}{2} + \frac{1}{8} - \cdots + \frac{1}{2048}$

68. $15 - 3 + \frac{3}{5} - \cdots - \frac{3}{625}$

Finding the Sum of an Infinite Geometric Series **In Exercises 69–84, find the sum of the infinite geometric series, if possible. If not possible, explain why.**

69. $\sum_{n=0}^{\infty} 10\left(\frac{4}{5}\right)^{n}$

70. $\sum_{n=0}^{\infty} 6\left(\frac{2}{3}\right)^{n}$

71. $\sum_{n=0}^{\infty} 5\left(-\frac{1}{2}\right)^{n}$

72. $\sum_{n=0}^{\infty} 5\left(-\frac{1}{4}\right)^{n}$

73. $\sum_{n=1}^{\infty} 2\left(\frac{7}{3}\right)^{n-1}$

74. $\sum_{n=1}^{\infty} 8\left(\frac{5}{3}\right)^{n-1}$

75. $\sum_{n=0}^{\infty} 10(0.11)^{n}$

76. $\sum_{n=0}^{\infty} 5(0.45)^{n}$

77. $\sum_{n=0}^{\infty} [-3(-0.9)^{n}]$

78. $\sum_{n=0}^{\infty} [-10(-0.2)^{n}]$

79. $9 + 6 + 4 + \frac{8}{3} + \cdots$

80. $8 + 6 + \frac{9}{2} + \frac{27}{8} + \cdots$

81. $3 + \frac{15}{2} + \frac{75}{4} + \frac{375}{8} + \cdots$

82. $2 + \frac{7}{3} + \frac{49}{18} + \frac{343}{108} + \cdots$

83. $-7 + 2 - \frac{4}{7} + \frac{8}{49} - \cdots$

84. $-6 + 5 - \frac{25}{6} + \frac{125}{36} - \cdots$

Writing a Repeating Decimal as a Rational Number **In Exercises 85–88, find the rational number representation of the repeating decimal.**

85. $0.\overline{36}$

86. $0.\overline{297}$

87. $1.2\overline{5}$

88. $1.3\overline{8}$

Identifying a Sequence **In Exercises 89–96, determine whether the sequence associated with the series is arithmetic or geometric. Find the common difference or ratio and find the sum of the first 15 terms.**

89. $8 + 16 + 32 + 64 + \cdots$

90. $17 + 14 + 11 + 8 + \cdots$

91. $90 + 30 + 10 + \frac{10}{3} + \cdots$

92. $\frac{5}{4} + \frac{7}{4} + \frac{9}{4} + \frac{11}{4} + \cdots$

93. $\sum_{n=1}^{\infty} 6n$

94. $\sum_{n=1}^{\infty} 3^{n-1}$

95. $\sum_{n=0}^{\infty} 6(0.8)^{n}$

96. $\sum_{n=0}^{\infty} \frac{n+8}{8}$

97. Annuity A deposit of \$100 is made at the beginning of each month in an account that pays 3% interest, compounded monthly. The balance A in the account at the end of 5 years is given by

$$A = 100\left(1 + \frac{0.03}{12}\right)^{1} + \cdots + 100\left(1 + \frac{0.03}{12}\right)^{60}.$$

Find A.

98. Annuity A deposit of \$50 is made at the beginning of each month in an account that pays 2% interest, compounded monthly. The balance A in the account at the end of 6 years is given by

$$A = 50\left(1 + \frac{0.02}{12}\right)^{1} + \cdots + 50\left(1 + \frac{0.02}{12}\right)^{72}.$$

Find A.

99. Annuity A deposit of P dollars is made at the beginning of each month in an account earning an annual interest rate r, compounded monthly. The balance A after t years is given by

$$A = P\left(1 + \frac{r}{12}\right) + P\left(1 + \frac{r}{12}\right)^{2} + \cdots + P\left(1 + \frac{r}{12}\right)^{12t}.$$

Show that the balance is given by

$$A = P\left[\left(1 + \frac{r}{12}\right)^{12t} - 1\right]\left(1 + \frac{12}{r}\right).$$

100. **Annuity** A deposit of P dollars is made at the beginning of each month in an account earning an annual interest rate r, compounded continuously. The balance A after t years is given by $A = Pe^{r/12} + Pe^{2r/12} + \cdots + Pe^{12tr/12}$. Show that the balance is given by

$$A = \frac{Pe^{r/12}(e^{rt} - 1)}{e^{r/12} - 1}.$$

Annuities In Exercises 101–104, consider making monthly deposits of P dollars in a savings account earning an annual interest rate r. Use the results of Exercises 99 and 100 to find the balances A after t years when the interest is compounded (a) monthly and (b) continuously.

101. $P = \$50,\ r = 7\%,\ t = 20$ years

102. $P = \$75,\ r = 4\%,\ t = 25$ years

103. $P = \$100,\ r = 5\%,\ t = 40$ years

104. $P = \$20,\ r = 6\%,\ t = 50$ years

105. **Geometry** The sides of a square are 16 inches in length. A new square is formed by connecting the midpoints of the sides of the original square, and two of the resulting triangles are shaded (see figure). After this process is repeated five more times, determine the total area of the shaded region.

106. **Geometry** The sides of a square are 27 inches in length. New squares are formed by dividing the original square into nine squares. The center square is then shaded (see figure). After this process is repeated three more times, determine the total area of the shaded region.

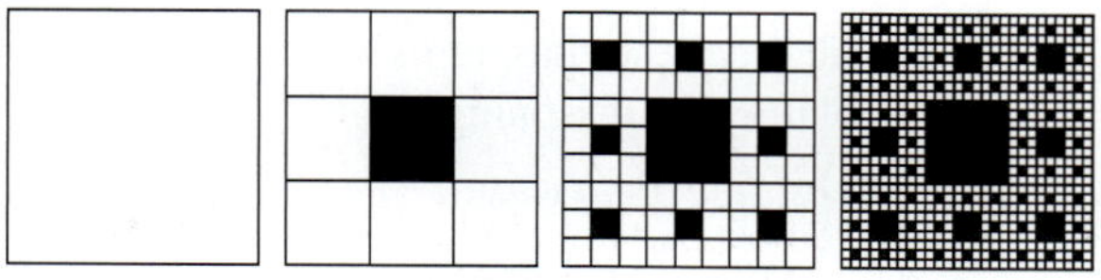

107. **Physics** The temperature of water in an ice cube tray is 70°F when it is placed in a freezer. Its temperature n hours after being placed in the freezer is 20% less than 1 hour earlier.

(a) Find a formula for the nth term of the geometric sequence that gives the temperature of the water n hours after it is placed in the freezer.

(b) Find the temperatures of the water 6 hours and 12 hours after it is placed in the freezer.

(c) Use a graphing utility to graph the sequence to approximate the time required for the water to freeze.

108. MODELING DATA

The mid-year populations a_n of China (in millions) from 2007 through 2013 are given by ordered pairs of the form (n, a_n), with $n = 7$ corresponding to 2007. *(Spreadsheet at LarsonPrecalculus.com)* (Source: U.S. Census Bureau)

DATA

(7, 1310.6) (8, 1317.1) (9, 1323.6)
(10, 1330.1) (11, 1336.7) (12, 1343.2)
(13, 1349.6)

(a) Use the *exponential regression* feature of a graphing utility to find a geometric sequence that models the data.

(b) Use the sequence from part (a) to describe the rate at which the population of China is growing.

(c) Use the sequence from part (a) to predict the population of China in 2018. The U.S. Census Bureau predicts that the population of China will be 1376.7 million in 2018. How does this value compare with your prediction?

(d) Use the sequence from part (a) to determine when the population of China will reach 1.40 billion.

109. **Fractals** In a *fractal,* a geometric figure is repeated at smaller and smaller scales. The sphereflake shown is a computer-generated fractal that was created by Eric Haines. The radius of the large sphere is 1. Attached to the large sphere are nine spheres of radius $\frac{1}{3}$. Attached to each of the smaller spheres are nine spheres of radius $\frac{1}{9}$. This process is continued infinitely.

(a) Write a formula in series notation that gives the surface area of the sphereflake.

(b) Write a formula in series notation that gives the volume of the sphereflake.

(c) Is the surface area of the sphereflake finite or infinite? Is the volume finite or infinite? If either is finite, find the value.

Eric Haines

110. Manufacturing The manufacturer of a new food processor plans to produce and sell 8000 units per year. Each year, 10% of all units sold become inoperative. So, 8000 units will be in use after 1 year, [8000 + 0.9(8000)] units will be in use after 2 years, and so on.

(a) Write a formula in series notation for the number of units that will be operative after n years.

(b) Find the numbers of units that will be operative after 10 years, 15 years, and 20 years.

(c) If this trend continues indefinitely, will the number of units that will be operative be finite? If so, how many? If not, explain your reasoning.

111. ***Why you should learn it*** *(p. 519)* A ball is dropped from a height of 6 feet and begins bouncing, as shown in the figure. The height of each bounce is three-fourths the height of the previous bounce. Find the total vertical distance the ball travels before coming to rest.

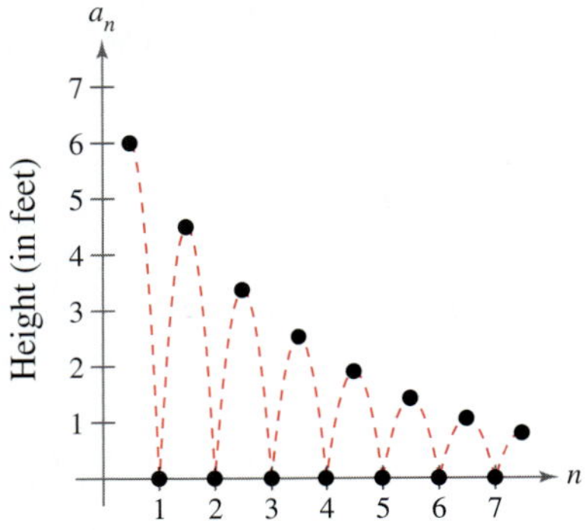

Conclusions

True or False? **In Exercises 112–114, determine whether the statement is true or false. Justify your answer.**

112. For a geometric sequence in which the quotient of term a_7 and term a_5 is 16, the common ratio is 8.

113. The first n terms of a geometric sequence with a common ratio of 1 are the same as the first n terms of an arithmetic sequence with a common difference of 0 if both sequences have the same first term.

114. You can find the nth term of a geometric sequence by multiplying its common ratio by the first term of the sequence raised to the $(n - 1)$th power.

Writing the Terms of a Geometric Sequence **In Exercises 115 and 116, write the first five terms of the geometric sequence.**

115. $a_1 = 3, r = \dfrac{x}{2}$

116. $a_1 = \frac{1}{4}, r = 7x$

Finding a Term of a Geometric Sequence **In Exercises 117 and 118, find the indicated term of the geometric sequence.**

117. $a_1 = 100, r = e^x$, 9th term

118. $a_1 = 4, r = (4x)/3$, 6th term

119. Exploration Use a graphing utility to graph each function. Identify the horizontal asymptote of the graph and determine its relationship to the sum.

(a) $f(x) = 6\left[\dfrac{1 - (0.5)^x}{1 - (0.5)}\right], \quad \displaystyle\sum_{n=0}^{\infty} 6\left(\frac{1}{2}\right)^n$

(b) $f(x) = 2\left[\dfrac{1 - (0.8)^x}{1 - (0.8)}\right], \quad \displaystyle\sum_{n=0}^{\infty} 2\left(\frac{4}{5}\right)^n$

120. Writing Write a brief paragraph explaining why the terms of a geometric sequence decrease in magnitude when $-1 < r < 1$.

121. Writing Write a brief paragraph explaining how to use the first two terms of a geometric sequence to find the nth term.

122. **HOW DO YOU SEE IT?** Use the figures shown below.

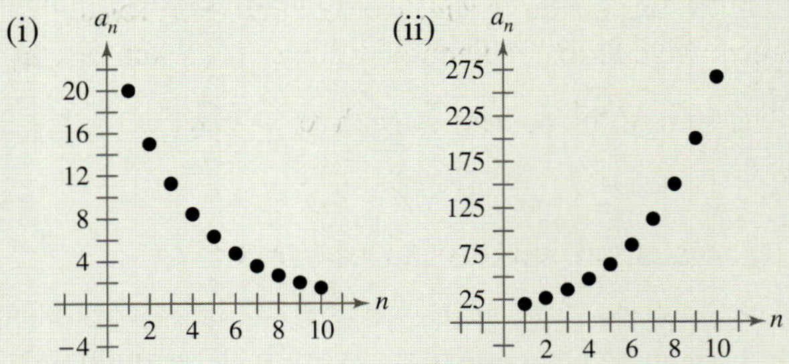

(a) Without performing any calculations, determine which figure shows the terms of the sequence $a_n = 20\left(\frac{4}{3}\right)^{n-1}$, and which shows the terms of $a_n = 20\left(\frac{3}{4}\right)^{n-1}$. Explain your reasoning,

(b) Which sequence has terms that can be summed? Explain your reasoning.

Cumulative Mixed Review

Finding a Determinant **In Exercises 123 and 124, find the determinant of the matrix.**

123. $\begin{bmatrix} -1 & 3 & 4 \\ -2 & 8 & 0 \\ 0 & 5 & -1 \end{bmatrix}$

124. $\begin{bmatrix} -1 & 0 & 4 \\ -4 & 3 & 5 \\ 0 & 2 & -3 \end{bmatrix}$

125. ***Make a Decision*** To work an extended application analyzing the monthly profits of a clothing manufacturer during a period of 36 months, visit the textbook's website at *LarsonPrecalculus.com.*

6.4 The Binomial Theorem

Binomial Coefficients

Recall that a *binomial* is a polynomial that has two terms. In this section, you will study a formula that provides a quick method of raising a binomial to a power. To begin, look at the expansion of

$$(x + y)^n$$

for several values of n.

$$(x + y)^0 = 1$$

$$(x + y)^1 = x + y$$

$$(x + y)^2 = x^2 + 2xy + y^2$$

$$(x + y)^3 = x^3 + 3x^2y + 3xy^2 + y^3$$

$$(x + y)^4 = x^4 + 4x^3y + 6x^2y^2 + 4xy^3 + y^4$$

$$(x + y)^5 = x^5 + 5x^4y + 10x^3y^2 + 10x^2y^3 + 5xy^4 + y^5$$

$$(x + y)^6 = x^6 + 6x^5y + 15x^4y^2 + 20x^3y^3 + 15x^2y^4 + 6xy^5 + y^6$$

There are several observations you can make about these expansions.

1. In each expansion, there are $n + 1$ terms.
2. In each expansion, x and y have symmetric roles. The powers of x decrease by 1 in successive terms, whereas the powers of y increase by 1.
3. The sum of the powers of each term is n. For instance, in the expansion of

   $$(x + y)^5$$

   the sum of the powers of each term is 5.

   $$(x + y)^5 = x^5 + \overbrace{5x^4y^1}^{4+1=5} + \overbrace{10x^3y^2}^{3+2=5} + 10x^2y^3 + 5x^1y^4 + y^5$$
4. The coefficients increase and then decrease in a symmetric pattern.

The coefficients of a binomial expansion are called **binomial coefficients.** To find them, you can use the **Binomial Theorem.**

The Binomial Theorem (See the proof on page 564.)

In the expansion of $(x + y)^n$

$$(x + y)^n = x^n + nx^{n-1}y + \cdot \cdot \cdot + {}_nC_rx^{n-r}y^r + \cdot \cdot \cdot + nxy^{n-1} + y^n$$

the coefficient of $x^{n-r}y^r$ is

$${}_nC_r = \frac{n!}{(n - r)!r!}.$$

The symbol

$$\binom{n}{r}$$

is sometimes used in place of ${}_nC_r$ to denote binomial coefficients.

What you should learn

- Use the Binomial Theorem to calculate binomial coefficients.
- Use binomial coefficients to write binomial expansions.
- Use Pascal's Triangle to calculate binomial coefficients.

Why you should learn it

You can use binomial coefficients to model and solve real-life problems. For instance, in Exercise 118 on page 536, you will use binomial coefficients to write the expansion of a model that represents the amount of child support collected in the United States.

EXAMPLE 1 Finding Binomial Coefficients

Find each binomial coefficient.

a. ${}_8C_2$ **b.** $\binom{10}{3}$ **c.** ${}_7C_0$ **d.** $\binom{8}{8}$

Solution

a. $${}_8C_2 = \frac{8!}{6! \cdot 2!} = \frac{(8 \cdot 7) \cdot \cancel{6!}}{\cancel{6!} \cdot 2!} = \frac{8 \cdot 7}{2 \cdot 1} = 28$$

b. $$\binom{10}{3} = \frac{10!}{7! \cdot 3!} = \frac{(10 \cdot 9 \cdot 8) \cdot \cancel{7!}}{\cancel{7!} \cdot 3!} = \frac{10 \cdot 9 \cdot 8}{3 \cdot 2 \cdot 1} = 120$$

c. $${}_7C_0 = \frac{\cancel{7!}}{\cancel{7!} \cdot 0!} = 1$$

d. $$\binom{8}{8} = \frac{\cancel{8!}}{0! \cdot \cancel{8!}} = 1$$

✓ **Checkpoint** Audio-video solution in English & Spanish at LarsonPrecalculus.com.

Find each binomial coefficient.

a. $\binom{11}{5}$ **b.** ${}_9C_2$ **c.** $\binom{5}{0}$ **d.** ${}_{15}C_{15}$

When $r \neq 0$ and $r \neq n$, as in parts (a) and (b) of Example 1, there is a simple pattern for evaluating binomial coefficients that works because there will always be factorial terms that divide out from the expression.

$${}_8C_2 = \frac{\overbrace{8 \cdot 7}^{\text{2 factors}}}{\underbrace{2 \cdot 1}_{\text{2 factorial}}} \quad \text{and} \quad \binom{10}{3} = \frac{\overbrace{10 \cdot 9 \cdot 8}^{\text{3 factors}}}{\underbrace{3 \cdot 2 \cdot 1}_{\text{3 factorial}}}$$

EXAMPLE 2 Finding Binomial Coefficients

Notice how the pattern shown above is used in parts (a) and (b).

a. $${}_7C_3 = \frac{7 \cdot \cancel{6} \cdot 5}{\cancel{3} \cdot \cancel{2} \cdot 1} = 35$$

b. $${}_7C_4 = \frac{7 \cdot \cancel{6} \cdot 5 \cdot \cancel{4}}{\cancel{4} \cdot \cancel{3} \cdot \cancel{2} \cdot 1} = 35$$

c. $${}_{12}C_1 = \frac{12}{1} = 12$$

d. $${}_{12}C_{11} = \frac{12!}{1! \cdot 11!} = \frac{(12) \cdot \cancel{11!}}{1! \cdot \cancel{11!}} = \frac{12}{1} = 12$$

✓ **Checkpoint** Audio-video solution in English & Spanish at LarsonPrecalculus.com.

Find each binomial coefficient.

a. ${}_7C_5$ **b.** $\binom{7}{2}$ **c.** ${}_{14}C_{13}$ **d.** $\binom{14}{1}$

It is not a coincidence that the results in parts (a) and (b) of Example 2 are the same and that the results in parts (c) and (d) are the same. In general, it is true that ${}_nC_r = {}_nC_{n-r}$.

Technology Tip

Most graphing utilities are programmed to evaluate ${}_nC_r$. For instructions on how to use the ${}_nC_r$ feature, see Appendix A; for specific keystrokes, go to this textbook's *Companion Website*.

Explore the Concept

Find each pair of binomial coefficients.

a. ${}_7C_0$, ${}_7C_7$ **d.** ${}_7C_1$, ${}_7C_6$

b. ${}_8C_0$, ${}_8C_8$ **e.** ${}_8C_1$, ${}_8C_7$

c. ${}_{10}C_0$, ${}_{10}C_{10}$ **f.** ${}_{10}C_1$, ${}_{10}C_9$

What do you observe about the pairs in (a), (b), and (c)? What do you observe about the pairs in (d), (e), and (f)? Write two conjectures from your observations. Develop a convincing argument for your two conjectures.

Binomial Expansions

As mentioned at the beginning of this section, when you write out the coefficients for a binomial that is raised to a power, you are **expanding a binomial.** The formula for binomial coefficients gives you a way to expand binomials.

EXAMPLE 3 Expanding a Binomial

Write the expansion of the expression $(x + 1)^3$.

Solution

The binomial coefficients are ${}_3C_0 = 1$, ${}_3C_1 = 3$, ${}_3C_2 = 3$, and ${}_3C_3 = 1$. So, the expansion is as follows.

$$(x+1)^3 = (1)x^3 + (3)x^2(1) + (3)x(1^2) + (1)(1^3)$$
$$= x^3 + 3x^2 + 3x + 1$$

✓ Checkpoint *Audio-video solution in English & Spanish at LarsonPrecalculus.com.*

Write the expansion of the expression $(x + 1)^4$.

To expand binomials representing *differences*, rather than sums, you alternate signs, as shown in the next example.

EXAMPLE 4 Expanding a Binomial

Write the expansion of the expression $(x - 1)^3$.

Solution

Expand using the binomial coefficients from Example 3.

$$(x-1)^3 = [x + (-1)]^3$$
$$= (1)x^3 + (3)x^2(-1) + (3)x(-1)^2 + (1)(-1)^3$$
$$= x^3 - 3x^2 + 3x - 1$$

✓ Checkpoint *Audio-video solution in English & Spanish at LarsonPrecalculus.com.*

Write the expansion of the expression $(x - 1)^4$.

EXAMPLE 5 Expanding Binomial Expressions

See LarsonPrecalculus.com for an interactive version of this type of example.

Write the expansion of (a) $(2x - 3)^4$ and (b) $(x - 2y)^4$.

Solution

The binomial coefficients are ${}_4C_0 = 1$, ${}_4C_1 = 4$, ${}_4C_2 = 6$, ${}_4C_3 = 4$, and ${}_4C_4 = 1$. So, the expansions are as follows.

a. $$(2x-3)^4 = (1)(2x)^4 - (4)(2x)^3(3) + (6)(2x)^2(3^2) - (4)(2x)(3^3) + (1)(3^4)$$
$$= 16x^4 - 96x^3 + 216x^2 - 216x + 81$$

b. $$(x-2y)^4 = (1)x^4 - (4)x^3(2y) + (6)x^2(2y)^2 - (4)x(2y)^3 + (1)(2y)^4$$
$$= x^4 - 8x^3y + 24x^2y^2 - 32xy^3 + 16y^4$$

✓ Checkpoint *Audio-video solution in English & Spanish at LarsonPrecalculus.com.*

Write the expansion of (a) $(3y - 1)^4$ and (b) $(2x - y)^5$.

Technology Tip

You can use a graphing utility to check the expansion in Example 5(a) by graphing the original binomial expression and the expansion in the same viewing window. The graphs should coincide, as shown below.

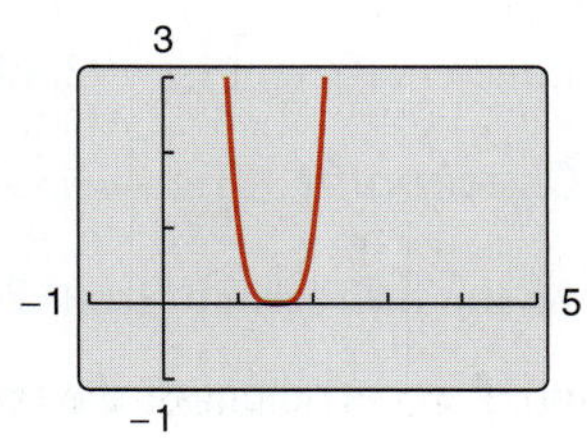

EXAMPLE 6 Expanding a Binomial

Write the expansion of the expression

$(x^2 + 4)^3$.

Solution

Expand using the binomial coefficients from Example 3.

$$(x^2 + 4)^3 = (1)(x^2)^3 + (3)(x^2)^2(4) + (3)x^2(4^2) + (1)(4^3)$$
$$= x^6 + 12x^4 + 48x^2 + 64$$

✓ **Checkpoint** Audio-video solution in English & Spanish at LarsonPrecalculus.com.

Write the expansion of the expression $(5 + y^2)^3$.

Sometimes you will need to find a specific term in a binomial expansion. Instead of writing out the entire expansion, you can use the fact that, from the Binomial Theorem, the $(r + 1)$th term is

$${}_nC_r x^{n-r} y^r.$$

For instance, to find the third term of the expression in Example 6, you could use the formula above with $n = 3$ and $r = 2$ to obtain

$${}_3C_2(x^2)^{3-2} \cdot 4^2 = 3(x^2) \cdot 16 = 48x^2.$$

EXAMPLE 7 Finding a Term or Coefficient

a. Find the sixth term of $(a + 2b)^8$.

b. Find the coefficient of the term a^6b^5 in the expansion of $(2a - 5b)^{11}$.

Solution

a. Because the formula is for the $(r + 1)$th term, r is one less than the number of the term you need. So, to find the sixth term in this binomial expansion, use $r = 5$, $n = 8$, $x = a$, and $y = 2b$.

$$\begin{aligned} {}_nC_r x^{n-r} y^r &= {}_8C_5 a^{8-5}(2b)^5 \\ &= 56 \cdot a^3 \cdot (2b)^5 \\ &= 56(2^5)a^3b^5 \\ &= 1792a^3b^5 \end{aligned}$$

b. Note that

$$(2a - 5b)^{11} = [2a + (-5b)]^{11}.$$

So, $n = 11$, $r = 5$, $x = 2a$, and $y = -5b$. Substitute these values to obtain

$$\begin{aligned} {}_nC_r x^{n-r} y^r &= {}_{11}C_5(2a)^6(-5b)^5 \\ &= (462)(64a^6)(-3125b^5) \\ &= -92{,}400{,}000a^6b^5. \end{aligned}$$

So, the coefficient is $-92{,}400{,}000$.

✓ **Checkpoint** Audio-video solution in English & Spanish at LarsonPrecalculus.com.

a. Find the fifth term of $(a + 2b)^8$.

b. Find the coefficient of the term a^4b^7 in the expansion of $(2a - 5b)^{11}$.

Pascal's Triangle

There is a convenient way to remember the pattern for binomial coefficients. By arranging the coefficients in a triangular pattern, you obtain the following array, which is called **Pascal's Triangle.** This triangle is named after the famous French mathematician Blaise Pascal (1623–1662).

1
1 1
1 2 1
1 3 3 1
1 4 6 4 1
1 5 10 10 5 1 $\quad 4 + 6 = 10$
1 6 15 20 15 6 1
1 7 21 35 35 21 7 1 $\quad 15 + 6 = 21$

The first and last number in each row of Pascal's Triangle is 1. Every other number in each row is formed by adding the two numbers immediately above the number. Pascal noticed that the numbers in this triangle are precisely the same numbers as the coefficients of binomial expansions, as follows.

$$(x+y)^0 = 1 \qquad \text{0th row}$$
$$(x+y)^1 = 1x + 1y \qquad \text{1st row}$$
$$(x+y)^2 = 1x^2 + 2xy + 1y^2 \qquad \text{2nd row}$$
$$(x+y)^3 = 1x^3 + 3x^2y + 3xy^2 + 1y^3 \qquad \text{3rd row}$$
$$(x+y)^4 = 1x^4 + 4x^3y + 6x^2y^2 + 4xy^3 + 1y^4 \qquad \vdots$$
$$(x+y)^5 = 1x^5 + 5x^4y + 10x^3y^2 + 10x^2y^3 + 5xy^4 + 1y^5$$
$$(x+y)^6 = 1x^6 + 6x^5y + 15x^4y^2 + 20x^3y^3 + 15x^2y^4 + 6xy^5 + 1y^6$$
$$(x+y)^7 = 1x^7 + 7x^6y + 21x^5y^2 + 35x^4y^3 + 35x^3y^4 + 21x^2y^5 + 7xy^6 + 1y^7$$

The top row of Pascal's Triangle is called the *zeroth row* because it corresponds to the binomial expansion $(x + y)^0 = 1$. Similarly, the next row is called the *first row* because it corresponds to the binomial expansion

$$(x + y)^1 = 1(x) + 1(y).$$

In general, the *nth row* of Pascal's Triangle gives the coefficients of $(x + y)^n$.

EXAMPLE 8 Using Pascal's Triangle

Use the seventh row of Pascal's Triangle to find the binomial coefficients.

$${}_8C_0 \quad {}_8C_1 \quad {}_8C_2 \quad {}_8C_3 \quad {}_8C_4 \quad {}_8C_5 \quad {}_8C_6 \quad {}_8C_7 \quad {}_8C_8$$

Solution

1 7 21 35 35 21 7 1 ← Seventh row
1 8 28 56 70 56 28 8 1

$${}_8C_0 \quad {}_8C_1 \quad {}_8C_2 \quad {}_8C_3 \quad {}_8C_4 \quad {}_8C_5 \quad {}_8C_6 \quad {}_8C_7 \quad {}_8C_8$$

✓ ***Checkpoint*** *Audio-video solution in English & Spanish at LarsonPrecalculus.com.*

Use the eighth row of Pascal's Triangle (see Example 8) to find the binomial coefficients.

$${}_9C_0 \quad {}_9C_1 \quad {}_9C_2 \quad {}_9C_3 \quad {}_9C_4 \quad {}_9C_5 \quad {}_9C_6 \quad {}_9C_7 \quad {}_9C_8 \quad {}_9C_9$$

Explore the Concept

Complete the table and describe the result.

n	r	${}_nC_r$	${}_nC_{n-r}$
9	5		
7	1		
12	4		
6	0		
10	7		

What characteristics of Pascal's Triangle are illustrated by this table?

6.4 Exercises

See *CalcChat.com* for tutorial help and worked-out solutions to odd-numbered exercises.
For instructions on how to use a graphing utility, see Appendix A.

Vocabulary and Concept Check

In Exercises 1 and 2, fill in the blanks.

1. The notation used to denote a binomial coefficient is _______ or _______ .
2. When you write out the coefficients for a binomial that is raised to a power, you are _______ a _______ .
3. List two ways to find binomial coefficients.
4. In the expression of $(x + y)^3$, what is the sum of the powers of the third term?

Procedures and Problem Solving

Finding Binomial Coefficients **In Exercises 5–16, find the binomial coefficient.**

5. ${}_7C_5$
6. ${}_8C_6$
7. ${}_{20}C_{15}$
8. ${}_{19}C_{12}$
9. ${}_{14}C_1$
10. ${}_{18}C_2$
11. $\binom{12}{0}$
12. $\binom{20}{20}$
13. $\binom{10}{4}$
14. $\binom{10}{6}$
15. $\binom{100}{98}$
16. $\binom{100}{2}$

Finding Binomial Coefficients **In Exercises 17–22, use a graphing utility to find ${}_nC_r$.**

17. ${}_{41}C_{36}$
18. ${}_{34}C_4$
19. ${}_{50}C_{46}$
20. ${}_{500}C_{498}$
21. ${}_{250}C_2$
22. ${}_{1000}C_2$

Expanding a Binomial **In Exercises 23–54, use the Binomial Theorem to expand and simplify the expression.**

23. $(x + 1)^5$
24. $(x + 1)^6$
25. $(a + 3)^3$
26. $(a + 2)^4$
27. $(y - 4)^3$
28. $(y - 5)^4$
29. $(x + y)^5$
30. $(x + y)^6$
31. $(2x - y)^5$
32. $(5x - y)^4$
33. $(4y - 3)^3$
34. $(2y - 5)^3$
35. $(2r - 3s)^6$
36. $(4x - 3y)^4$
37. $(x^2 + 2)^4$
38. $(y^2 + 2)^6$
39. $(5 - x^2)^5$
40. $(3 - y^2)^3$
41. $(x^2 + y^2)^4$
42. $(x^2 + y^2)^6$
43. $(3x^3 - y)^6$
44. $(2x^3 - y)^5$
45. $\left(\frac{1}{x} + y\right)^5$
46. $\left(\frac{1}{x} + y\right)^6$
47. $\left(\frac{2}{x} - 2y\right)^4$
48. $\left(\frac{2}{x} - 3y\right)^5$
49. $(4x - 1)^3 - 2(4x - 1)^4$
50. $(x + 3)^5 - 4(x + 3)^4$
51. $3(x + 1)^5 + 4(x + 1)^3$
52. $2(x - 3)^4 + 5(x - 3)^2$
53. $-3(x - 2)^3 - 4(x + 1)^6$
54. $-5(x + 2)^5 - 2(x - 1)^2$

Finding a Term in a Binomial Expansion **In Exercises 55–62, find the specified *n*th term in the expansion of the binomial.**

55. $(x + 8)^{10}$, $n = 4$
56. $(x + 6)^6$, $n = 7$
57. $(x - 6y)^5$, $n = 3$
58. $(x - 10z)^7$, $n = 4$
59. $(4x + 3y)^9$, $n = 8$
60. $(5a + 6b)^5$, $n = 5$
61. $(10x - 3y)^{12}$, $n = 10$
62. $(7x - 2y)^{15}$, $n = 7$

Finding a Coefficient in a Binomial Expansion **In Exercises 63–70, find the coefficient *a* of the given term in the expansion of the binomial.**

	Binomial	*Term*
63.	$(x + 3)^{12}$	ax^5
64.	$(x + 4)^{12}$	ax^4
65.	$(x - 2y)^{10}$	ax^8y^2
66.	$(4x - y)^{10}$	ax^2y^8
67.	$(3x - 2y)^9$	ax^6y^3
68.	$(2x - 3y)^8$	ax^4y^4
69.	$(x^2 + y)^{10}$	ax^8y^6
70.	$(z^2 + y)^{12}$	$az^{14}y^5$

Using Pascal's Triangle **In Exercises 71–74, use Pascal's Triangle to find the binomial coefficient.**

71. ${}_7C_4$
72. ${}_6C_3$
73. ${}_6C_5$
74. ${}_5C_2$

Using Pascal's Triangle **In Exercises 75–80, expand the binomial by using Pascal's Triangle to determine the coefficients.**

75. $(5y + 2)^5$ **76.** $(2v + 3)^6$

77. $(2x + 3y)^5$ **78.** $(3x + 4y)^5$

79. $(3t - 2v)^4$ **80.** $(5v - 2z)^4$

Using the Binomial Theorem **In Exercises 81–84, use the Binomial Theorem to expand and simplify the expression.**

81. $(3\sqrt{x} + 5)^3$ **82.** $(2\sqrt{t} - 7)^3$

83. $(x^{2/3} - y^{1/3})^3$ **84.** $(u^{3/5} + v^{1/5})^5$

Expanding an Expression **In Exercises 85–88, expand the expression in the difference quotient and simplify.**

$$\frac{f(x+h) - f(x)}{h}, \quad h \neq 0$$

85. $f(x) = x^3$ **86.** $f(x) = x^4$

87. $f(x) = x^6$ **88.** $f(x) = x^8$

Expanding a Complex Number **In Exercises 89–102, use the Binomial Theorem to expand the complex number. Simplify your result.** $\left(\textbf{Remember that } i = \sqrt{-1}.\right)$

89. $(1 + i)^4$ **90.** $(1 - i)^6$

91. $(4 + i)^4$ **92.** $(2 + i)^5$

93. $(2 - 3i)^6$ **94.** $(3 - 2i)^6$

95. $(5 + \sqrt{-16})^3$ **96.** $(5 + \sqrt{-9})^3$

97. $(4 + \sqrt{3}i)^4$ **98.** $(5 - \sqrt{3}i)^4$

99. $\left(-\frac{1}{2} + \frac{\sqrt{3}}{2}i\right)^3$ **100.** $\left(-\frac{1}{3} + \frac{\sqrt{3}}{3}i\right)^3$

101. $\left(\frac{1}{4} - \frac{\sqrt{3}}{4}i\right)^3$ **102.** $\left(\frac{1}{2} - \frac{\sqrt{3}}{2}i\right)^3$

Using the Binomial Theorem to Approximate **In Exercises 103–106, use the Binomial Theorem to approximate the quantity accurate to three decimal places. For example, in Exercise 103, use the expansion**

$$(1.02)^8 = (1 + 0.02)^8$$
$$= 1 + 8(0.02) + 28(0.02)^2 + \cdots.$$

103. $(1.02)^8$ **104.** $(2.005)^{10}$

105. $(2.99)^{12}$ **106.** $(1.98)^9$

Using the Binomial Theorem **In Exercises 107–110, use a graphing utility to graph f and g in the same viewing window. What is the relationship between the two graphs? Use the Binomial Theorem to write the polynomial function g in standard form.**

107. $f(x) = x^4 - 5x^2$
$g(x) = f(x - 2)$

108. $f(x) = x^3 - 4x$
$g(x) = f(x + 4)$

109. $f(x) = -x^3 + 3x^2 - 4$
$g(x) = f(x + 5)$

110. $f(x) = -x^4 + 4x^2 - 1$
$g(x) = f(x - 3)$

Using a Graphing Utility **In Exercises 111 and 112, use a graphing utility to graph the functions in the given order and in the same viewing window. Compare the graphs. Which two functions have identical graphs, and why?**

111. (a) $f(x) = (1 - x)^3$
(b) $g(x) = 1 - 3x$
(c) $h(x) = 1 - 3x + 3x^2$
(d) $p(x) = 1 - 3x + 3x^2 - x^3$

112. (a) $f(x) = \left(1 - \frac{1}{2}x\right)^4$
(b) $g(x) = 1 - 2x + \frac{3}{2}x^2$
(c) $h(x) = 1 - 2x + \frac{3}{2}x^2 - \frac{1}{2}x^3$
(d) $p(x) = 1 - 2x + \frac{3}{2}x^2 - \frac{1}{2}x^3 + \frac{1}{16}x^4$

Finding a Probability **In Exercises 113–116, consider n independent trials of an experiment in which each trial has two possible outcomes, success or failure. The probability of a success on each trial is p and the probability of a failure is $q = 1 - p$. In this context, the term ${}_nC_k p^k q^{n-k}$ in the expansion of $(p + q)^n$ gives the probability of k successes in the n trials of the experiment.**

113. A fair coin is tossed seven times. To find the probability of obtaining four heads, evaluate the term

$${}_7C_4\left(\tfrac{1}{2}\right)^4\left(\tfrac{1}{2}\right)^3$$

in the expansion of $\left(\frac{1}{2} + \frac{1}{2}\right)^7$.

114. The probability of a baseball player getting a hit during any given time at bat is $\frac{1}{4}$. To find the probability that the player gets three hits during the next 10 times at bat, evaluate the term

$${}_{10}C_3\left(\tfrac{1}{4}\right)^3\left(\tfrac{3}{4}\right)^7$$

in the expansion of $\left(\frac{1}{4} + \frac{3}{4}\right)^{10}$.

115. The probability of a sales representative making a sale to any one customer is $\frac{1}{3}$. The sales representative makes eight contacts a day. To find the probability of making four sales, evaluate the term

$${}_8C_4\left(\tfrac{1}{3}\right)^4\left(\tfrac{2}{3}\right)^4$$

in the expansion of $\left(\frac{1}{3} + \frac{2}{3}\right)^8$.

116. To find the probability that the sales representative in Exercise 115 makes four sales when the probability of a sale to any one customer is $\frac{1}{2}$, evaluate the term

$${}_8C_4\left(\tfrac{1}{2}\right)^4\left(\tfrac{1}{2}\right)^4$$

in the expansion of $\left(\frac{1}{2} + \frac{1}{2}\right)^8$.

117. MODELING DATA

The per capita consumptions of milk f (in gallons) in the United States from 1990 through 2012 can be approximated by the model

$$f(t) = 0.005t^2 - 0.37t + 25.7, \quad 0 \le t \le 22$$

where t represents the year, with $t = 0$ corresponding to 1990 (see figure). (Source: Economic Research Service, U.S. Department of Agriculture)

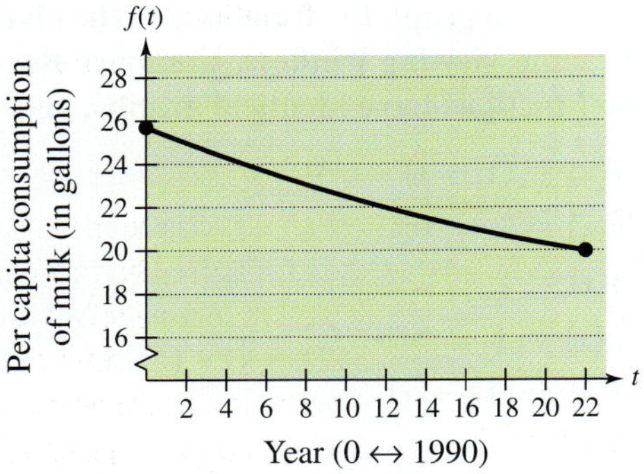

(a) Adjust the model so that $t = 0$ corresponds to 2000 rather than 1990. To do this, shift the graph of f 10 units to the left and obtain $g(t) = f(t + 10)$. Write $g(t)$ in standard form.

(b) Use a graphing utility to graph f and g in the same viewing window.

118. ***Why you should learn it*** *(p. 529)* The amounts f (in billions of dollars) of child support collected in the United States from 2000 through 2013 can be approximated by the model

$$f(t) = -0.039t^2 + 1.30t + 17.7,$$
$$0 \le t \le 13$$

where t represents the year, with $t = 0$ corresponding to 2000 (see figure). (Source: U.S. Department of Health and Human Services)

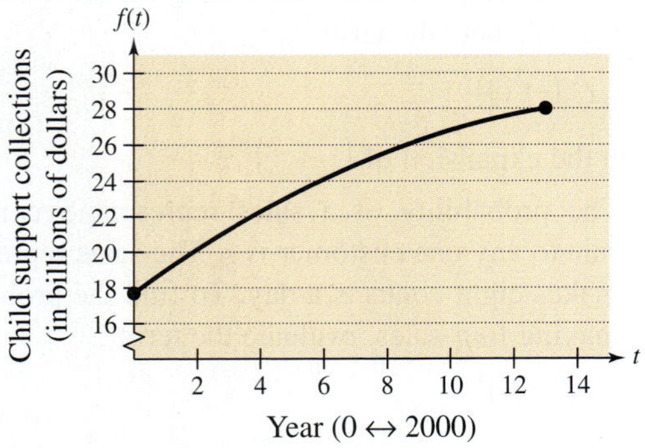

(a) Adjust the model so that $t = 0$ corresponds to 2005 rather than 2000. To do this, shift the graph of f five units to the left and obtain $g(t) = f(t + 5)$. Write $g(t)$ in standard form.

(b) Use a graphing utility to graph f and g in the same viewing window.

Conclusions

True or False? **In Exercises 119 and 120, determine whether the statement is true or false. Justify your answer.**

119. The Binomial Theorem can be used to produce each row of Pascal's Triangle.

120. A binomial that represents a difference cannot always be accurately expanded using the Binomial Theorem.

121. **Writing** In your own words, explain how to form the rows of Pascal's Triangle. Then form rows 8–10 of Pascal's Triangle.

122. **HOW DO YOU SEE IT?** The expansions of $(x + y)^4$, $(x + y)^5$, and $(x + y)^6$ are as follows.

$$(x + y)^4 = 1x^4 + 4x^3y + 6x^2y^2 + 4xy^3 + 1y^4$$
$$(x + y)^5 = 1x^5 + 5x^4y + 10x^3y^2 + 10x^2y^3 + 5xy^4 + 1y^5$$
$$(x + y)^6 = 1x^6 + 6x^5y + 15x^4y^2 + 20x^3y^3 + 15x^2y^4 + 6xy^5 + 1y^6$$

(a) Explain how the exponent of a binomial is related to the number of terms in its expansion.

(b) How many terms are in the expansion of $(x + y)^n$?

Proof **In Exercises 123–126, prove the property for all integers r and n, where $0 \le r \le n$.**

123. ${}_nC_r = {}_nC_{n-r}$

124. ${}_nC_0 - {}_nC_1 + {}_nC_2 - \cdots \pm {}_nC_n = 0$

125. ${}_{n+1}C_r = {}_nC_r + {}_nC_{r-1}$

126. The sum of the numbers in the nth row of Pascal's Triangle is 2^n.

127. **Error Analysis** You are a math instructor and receive the following solution from one of your students on a quiz. Find the error(s) in the solution and write a short paragraph discussing ways that your student could avoid the error(s) in the future. Find the second term in the expansion of $(2x - 3y)^5$.

~~$5(2x)^4(3y)^2 = 720x^4y^2$~~

Cumulative Mixed Review

Finding the Inverse of a Matrix **In Exercises 128 and 129, find the inverse of the matrix.**

128. $\begin{bmatrix} -1 & -4 \\ 1 & 2 \end{bmatrix}$ **129.** $\begin{bmatrix} 11 & -12 \\ 2 & -2 \end{bmatrix}$

6.5 Counting Principles

Simple Counting Problems

The last two sections of this chapter present a brief introduction to some of the basic counting principles and their application to probability. In the next section, you will see that much of probability has to do with counting the number of ways an event can occur.

What you should learn

- Solve simple counting problems.
- Use the Fundamental Counting Principle to solve more complicated counting problems.
- Use permutations to solve counting problems.
- Use combinations to solve counting problems.

Why you should learn it

You can use counting principles to solve counting problems that occur in real life. For instance, in Exercise 67 on page 545, you are asked to use counting principles to determine in how many ways a player can select six numbers in a Powerball lottery.

EXAMPLE 1 Selecting Pairs of Numbers at Random

You place eight pieces of paper, numbered from 1 to 8, in a box. You draw one piece of paper at random from the box, record its number, and *replace* the paper in the box. Then, you draw a second piece of paper at random from the box and record its number. Finally, you add the two numbers. How many different ways can you obtain a sum of 12?

Solution

To solve this problem, count the number of different ways that a sum of 12 can be obtained using two numbers from 1 to 8. As shown below, a sum of 12 can occur in five different ways.

First number	4	5	6	7	8
Second number	8	7	6	5	4

✓ Checkpoint *Audio-video solution in English & Spanish at LarsonPrecalculus.com.*

In Example 1, how many different ways can you obtain a sum of 14?

EXAMPLE 2 Selecting Pairs of Numbers at Random

You place eight pieces of paper, numbered from 1 to 8, in a box. You draw one piece of paper at random from the box, record its number, and *do not* replace the paper in the box. Then, you draw a second piece of paper at random from the box and record its number. Finally, you add the two numbers. How many different ways can you obtain a sum of 12?

Solution

To solve this problem, count the number of different ways that a sum of 12 can be obtained using two *different* numbers from 1 to 8. As shown below, a sum of 12 can occur in four different ways.

First number	4	5	7	8
Second number	8	7	5	4

✓ Checkpoint *Audio-video solution in English & Spanish at LarsonPrecalculus.com.*

Repeat Example 2 for drawing *three* pieces of paper.

The difference between the counting problems in Examples 1 and 2 can be described by saying that the random selection in Example 1 occurs **with replacement,** whereas the random selection in Example 2 occurs **without replacement,** which eliminates the possibility of choosing two 6's.

The Fundamental Counting Principle

Examples 1 and 2 describe simple counting problems in which you can *list* each possible way that an event can occur. When it is possible, this is always the best way to solve a counting problem. However, some events can occur in so many different ways that it is not feasible to write out the entire list. In such cases, you must rely on formulas and counting principles. The most important of these is the **Fundamental Counting Principle.**

> **Fundamental Counting Principle**
>
> Let E_1 and E_2 be two events. The first event E_1 can occur in m_1 different ways. After E_1 has occurred, E_2 can occur in m_2 different ways. The number of ways that the two events can occur is $m_1 \cdot m_2$.

The Fundamental Counting Principle can be extended to three or more events. For instance, the number of ways that three events E_1, E_2, and E_3 can occur is $m_1 \cdot m_2 \cdot m_3$.

EXAMPLE 3 Using the Fundamental Counting Principle

How many different pairs of letters from the English alphabet are possible?

Solution

There are two events in this situation. The first event is the choice of the first letter, and the second event is the choice of the second letter. Because the English alphabet contains 26 letters, it follows that the number of two-letter pairs is

$$26 \cdot 26 = 676.$$

✓ Checkpoint *Audio-video solution in English & Spanish at LarsonPrecalculus.com.*

A combination lock will open when you select the right choice of three numbers (from 1 to 30, inclusive). How many different lock combinations are possible?

EXAMPLE 4 Using the Fundamental Counting Principle

Telephone numbers in the United States currently have 10 digits. The first three are the *area code* and the next seven are the *local telephone number*. How many different telephone numbers are possible within each area code? (Note that at this time, a local telephone number cannot begin with 0 or 1.)

Solution

Because the first digit of a local number cannot be 0 or 1, there are only eight choices for the first digit. For each of the other six digits, there are 10 choices.

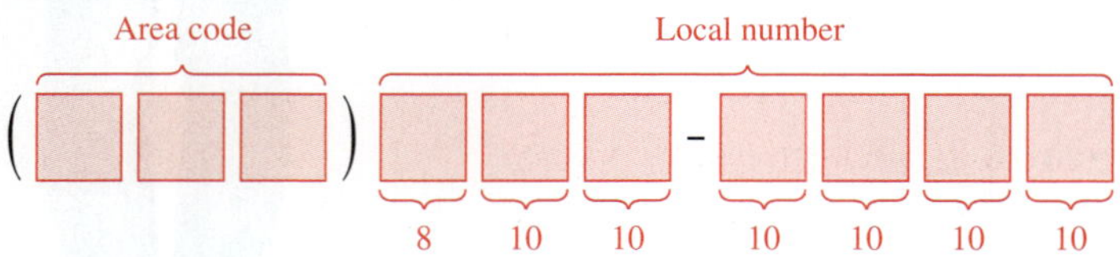

So, the number of local telephone numbers that are possible within each area code is

$$8 \cdot 10 \cdot 10 \cdot 10 \cdot 10 \cdot 10 \cdot 10 = 8{,}000{,}000.$$

✓ Checkpoint *Audio-video solution in English & Spanish at LarsonPrecalculus.com.*

A product's catalog number is made up of one letter from the English alphabet followed by a five-digit number. How many different catalog numbers are possible?

Permutations

One important application of the Fundamental Counting Principle is in determining the number of ways that n elements can be arranged (in order). An ordering of n elements is called a **permutation** of the elements.

Definition of Permutation

A **permutation** of n different elements is an ordering of the elements such that one element is first, one is second, one is third, and so on.

EXAMPLE 5 Finding the Number of Permutations

How many permutations of the letters

A, B, C, D, E, and F

are possible?

Solution

Consider the following reasoning.

First position:	Any of the *six* letters
Second position:	Any of the remaining *five* letters
Third position:	Any of the remaining *four* letters
Fourth position:	Any of the remaining *three* letters
Fifth position:	Either of the remaining *two* letters
Sixth position:	The *one* remaining letter

So, the numbers of choices for the six positions are as follows.

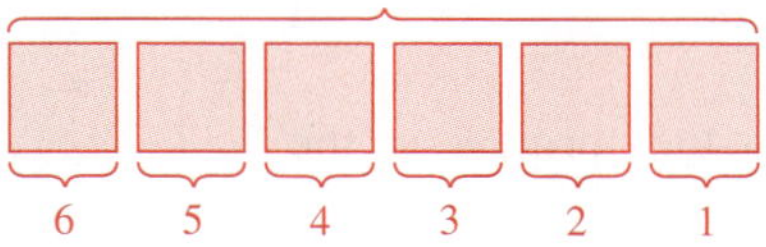

The total number of permutations of the six letters is

$$6! = 6 \cdot 5 \cdot 4 \cdot 3 \cdot 2 \cdot 1 = 720.$$

Checkpoint *Audio-video solution in English & Spanish at LarsonPrecalculus.com.*

How many permutations of the letters

W, X, Y, and Z

are possible?

The result obtained in Example 5 can be generalized to conclude that the number of permutations of n different elements is $n!$.

Number of Permutations of n Elements

The number of permutations of n elements is given by

$$n \cdot (n - 1) \cdot \cdot \cdot 4 \cdot 3 \cdot 2 \cdot 1 = n!.$$

In other words, there are $n!$ different ways that n elements can be ordered.

It is useful, on occasion, to order a *subset* of a collection of elements rather than the entire collection. For instance, you might want to order r elements out of a collection of n elements. Such an ordering is called a **permutation of n elements taken r at a time.** This ordering is demonstrated in the next example.

EXAMPLE 6 Counting Horse Race Finishes

Eight horses are running in a race. In how many different ways can these horses come in first, second, and third? (Assume that there are no ties.)

Solution

Here are the different possibilities.

Win (first position):	*Eight* choices
Place (second position):	*Seven* choices
Show (third position):	*Six* choices

The numbers of choices for the three positions are as follows.

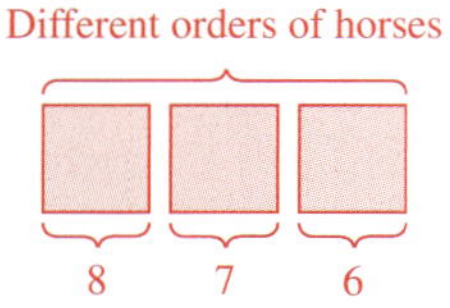

So, using the Fundamental Counting Principle, you can determine that there are

$$8 \cdot 7 \cdot 6 = 336$$

different ways in which the eight horses can come in first, second, and third.

Checkpoint *Audio-video solution in English & Spanish at LarsonPrecalculus.com.*

A coin club has five members. In how many ways can there be a president and a vice-president?

The result obtained in Example 6 can be generalized to the formula in the following definition.

Permutations of n Elements Taken r at a Time

The number of **permutations of n elements taken r at a time** is given by

$$_nP_r = \frac{n!}{(n-r)!} = n(n-1)(n-2) \cdot \cdot \cdot (n-r+1).$$

Using this formula, you can rework Example 6 to find that the number of permutations of eight horses taken three at a time is

$$\begin{aligned} _8P_3 &= \frac{8!}{(8-3)!} \\ &= \frac{8!}{5!} \\ &= \frac{8 \cdot 7 \cdot 6 \cdot \cancel{5!}}{\cancel{5!}} \\ &= 336 \end{aligned}$$

which is the same answer obtained in the example.

Technology Tip

Most graphing utilities are programmed to evaluate $_nP_r$. The figure below shows how one graphing utility evaluates the permutation $_8P_3$. For instructions on how to use the $_nP_r$ feature, see Appendix A; for specific keystrokes, go to this textbook's *Companion Website.*

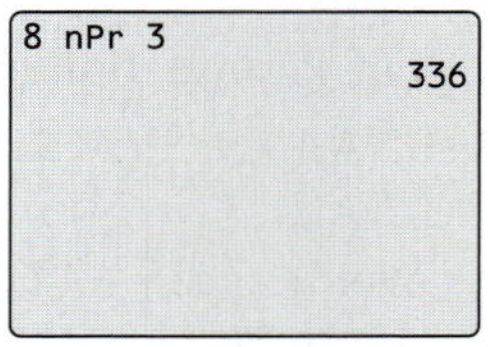

Remember that for permutations, order is important. So, to find the possible permutations of the letters A, B, C, and D taken three at a time, you count the permutations (A, B, D) and (B, A, D) as different because the *order* of the elements is different.

Consider, however, the possible permutations of the letters A, A, B, and C. The total number of permutations of the four letters would be

$${}_4P_4 = 4!.$$

However, not all of these arrangements would be *distinguishable* because there are two A's in the list. To find the number of distinguishable permutations, you can use the following formula.

Distinguishable Permutations

Suppose a set of n objects has n_1 of one kind of object, n_2 of a second kind, n_3 of a third kind, and so on, with

$$n = n_1 + n_2 + n_3 + \cdot \cdot \cdot + n_k.$$

The number of **distinguishable permutations** of the n objects is given by

$$\frac{n!}{n_1! \cdot n_2! \cdot n_3! \cdot \cdot \cdot \cdot \cdot n_k!}.$$

EXAMPLE 7 Distinguishable Permutations

See LarsonPrecalculus.com for an interactive version of this type of example.

In how many distinguishable ways can the letters in BANANA be written?

Solution

This word has six letters, of which three are A's, two are N's, and one is a B. So, the number of distinguishable ways in which the letters can be written is

$$\frac{6!}{3! \cdot 2! \cdot 1!} = \frac{6 \cdot 5 \cdot 4 \cdot \cancel{3!}}{\cancel{3!} \cdot 2!} = 60.$$

The 60 different distinguishable permutations are as follows.

AAABNN	AAANBN	AAANNB	AABANN
AABNAN	AABNNA	AANABN	AANANB
AANBAN	AANBNA	AANNAB	AANNBA
ABAANN	ABANAN	ABANNA	ABNAAN
ABNANA	ABNNAA	ANAABN	ANAANB
ANABAN	ANABNA	ANANAB	ANANBA
ANBAAN	ANBANA	ANBNAA	ANNAAB
ANNABA	ANNBAA	BAAANN	BAANAN
BAANNA	BANAAN	BANANA	BANNAA
BNAAAN	BNAANA	BNANAA	BNNAAA
NAAABN	NAAANB	NAABAN	NAABNA
NAANAB	NAANBA	NABAAN	NABANA
NABNAA	NANAAB	NANABA	NANBAA
NBAAAN	NBAANA	NBANAA	NBNAAA
NNAAAB	NNAABA	NNABAA	NNBAAA

✓ **Checkpoint** *Audio-video solution in English & Spanish at LarsonPrecalculus.com.*

In how many distinguishable ways can the letters in MITOSIS be written?

Combinations

When you count the number of possible permutations of a set of elements, order is important. As a final topic in this section, you will look at a method for selecting subsets of a larger set in which order *is not* important. Such subsets are called **combinations of *n* elements taken *r* at a time.** For instance, the combinations {A, B, C} and {B, A, C} are equivalent because both sets contain the same three elements, and the order in which the elements are listed is not important. So, you would count only one of the two sets. A common example of a combination is a card game in which the player is free to reorder the cards after they have been dealt.

Combinations of *n* Elements Taken *r* at a Time

The number of **combinations of *n* elements taken *r* at a time** is given by

$${}_nC_r = \frac{n!}{(n-r)!r!}.$$

Note that the formula for ${}_nC_r$ is the same one given for binomial coefficients.

EXAMPLE 8 Combinations of *n* Elements Taken *r* at a Time

a. In how many different ways can three letters be chosen from the letters A, B, C, D, and E? (The order of the three letters is not important.)

b. A standard poker hand consists of five cards dealt from a deck of 52. How many different poker hands are possible? (After the cards are dealt, the player may reorder them, so order is not important.)

Solution

a. You can find the number of different ways in which the letters can be chosen by using the formula for the number of combinations of five elements taken three at a time, as follows.

$${}_5C_3 = \frac{5!}{2!3!} = \frac{5 \cdot \overset{2}{\cancel{4}} \cdot \cancel{3!}}{\cancel{2} \cdot 1 \cdot \cancel{3!}} = 10$$

b. You can find the number of different poker hands by using the formula for the number of combinations of 52 elements taken five at a time, as follows.

$${}_{52}C_5 = \frac{52!}{47!5!} = \frac{52 \cdot 51 \cdot 50 \cdot 49 \cdot 48 \cdot \cancel{47!}}{5 \cdot 4 \cdot 3 \cdot 2 \cdot 1 \cdot \cancel{47!}} = 2{,}598{,}960$$

Remark

In Example 8(a), you could also make a list of the different combinations of three letters chosen from five letters.

$\{A, B, C\}$	$\{A, B, D\}$
$\{A, B, E\}$	$\{A, C, D\}$
$\{A, C, E\}$	$\{A, D, E\}$
$\{B, C, D\}$	$\{B, C, E\}$
$\{B, D, E\}$	$\{C, D, E\}$

From this list, you can conclude that there are 10 different ways in which three letters can be chosen from five letters, which is the same answer obtained in Example 8(a).

✔ **Checkpoint** *Audio-video solution in English & Spanish at LarsonPrecalculus.com.*

In how many different ways can two letters be chosen from the letters A, B, C, D, E, F, and G? (The order of the two letters is not important.)

When solving problems involving counting principles, you need to be able to distinguish among the various counting principles in order to determine which is necessary to solve the problem correctly. To do this, ask yourself the following questions.

1. Is the order of the elements important? *Permutation*
2. Are the chosen elements a subset of a larger set in which order is not important? *Combination*
3. Are there two or more separate events? *Fundamental Counting Principle*

6.5 Exercises

See *CalcChat.com* for tutorial help and worked-out solutions to odd-numbered exercises.
For instructions on how to use a graphing utility, see Appendix A.

Vocabulary and Concept Check

In Exercises 1 and 2, fill in the blank.

1. The _______ states that if there are m_1 ways for one event to occur and m_2 ways for a second event to occur, then there are $m_1 \cdot m_2$ ways for both events to occur.

2. The number of _______ of n objects is given by $\dfrac{n!}{n_1! \cdot n_2! \cdot n_3! \cdot \cdots \cdot n_k!}$.

3. Is the ordering of n elements called a *permutation* or a *combination* of the elements?

4. What do n and r represent in the formula ${}_nP_r = \dfrac{n!}{(n-r)!}$?

Procedures and Problem Solving

Selecting Numbers at Random **In Exercises 5–12, determine the number of ways in which a computer can randomly generate one or more such integers, or pairs of integers, from 1 through 15.**

5. An odd integer
6. An even integer
7. A prime integer
8. An integer that is greater than 11
9. A pair of integers whose sum is 20
10. A pair of integers whose sum is 22
11. A pair of distinct integers whose sum is 8
12. A pair of distinct integers whose sum is 24

13. **Audio Engineering** A customer can choose one of four amplifiers, one of ten stereo receivers, and one of five speaker models for an entertainment system. Determine the number of possible system configurations.

14. **Education** A college student is preparing a course schedule for the next semester. The student must select one of two mathematics courses, one of three science courses, and one of five courses from the social sciences and humanities. How many schedules are possible?

15. **Education** In how many ways can a 10-question true-false exam be answered? (Assume that no questions are omitted.)

16. **Physiology** In a physiology class, a student must dissect three different specimens. The student can select one of nine earthworms, one of four frogs, and one of seven fetal pigs. In how many ways can the student select the specimens?

17. **Fundamental Counting Principle** How many three-digit numbers can be formed under each condition?
    (a) The leading digit cannot be 0.
    (b) The leading digit cannot be 0 and no repetition of digits is allowed.
    (c) The leading digit cannot be 0 and the number must be a multiple of 5.

18. **Fundamental Counting Principle** How many four-digit numbers can be formed under each condition?
    (a) The leading digit cannot be 0 and the number must be less than 5000.
    (b) The leading digit cannot be 0 and the number must be even.

19. **Telecommunication** In 2014, the state of Nevada added a third area code. Using the information about telephone numbers given in Example 4, how many telephone numbers can Nevada's phone system accommodate?

20. **Transportation** In Pennsylvania, each standard automobile license plate number consists of three letters followed by a four-digit number. How many distinct Pennsylvania license plate numbers can be formed?

21. **Operations Research** In 1963, the United States Postal Service launched the Zoning Improvement Plan (ZIP) Code to streamline the mail-delivery system. A ZIP code consists of a five-digit sequence of numbers.
    (a) Find the number of ZIP codes consisting of five digits.
    (b) Find the number of ZIP codes consisting of five digits when the first digit is 1 or 2.

22. **Operations Research** In 1983, in order to identify small geographic segments within a delivery code, the post office began to use an expanded ZIP code called ZIP+4, which is composed of the original five-digit code plus a four-digit add-on code.
    (a) Find the number of ZIP codes consisting of five digits followed by the four additional digits.
    (b) Find the number of ZIP codes consisting of five digits followed by the four additional digits when the first number of the five-digit code is 1 or 2.

23. **Banking** ATM personal identification number (PIN) codes typically consist of four-digit sequences of numbers.
    (a) Find the total number of ATM codes possible.
    (b) Find the total number of ATM codes possible when the first digit is not 0.

24. **Radio Broadcasting** Typically, radio stations are identified by four "call letters." Radio stations east of the Mississippi River have call letters that start with the letter W, and radio stations west of the Mississippi River have call letters that start with the letter K.

(a) Find the number of different sets of radio station call letters that are possible in the United States.

(b) Find the number of different sets of radio station call letters that are possible when the call letters must include a Q.

25. **Finding Permutations** Three couples have reserved seats in a row for a concert. In how many different ways can they be seated when (a) there are no seating restrictions? (b) each couple sits together?

26. **Finding Permutations** In how many orders can five girls and three boys walk through a doorway single file when (a) there are no restrictions? (b) the girls walk through before the boys?

27. **Photography** In how many ways can five children posing for a photograph line up in a row?

28. **Transportation Design** In how many ways can eight people sit in an eight-passenger vehicle?

29. **Politics** The nine justices of the U.S. Supreme Court pose for a photograph while standing in a straight line, as opposed to the typical pose of two rows. How many different orders of the justices are possible for this photograph?

30. **Manufacturing** Four processes are involved in assembling a product, and they can be performed in any order. The management wants to test each order to determine which is the least time-consuming. How many different orders will have to be tested?

Evaluating a Permutation **In Exercises 31–36, evaluate $_nP_r$ using the formula from this section.**

31. $_4P_4$

32. $_5P_5$

33. $_8P_3$

34. $_{20}P_2$

35. $_6P_5$

36. $_7P_4$

Evaluating a Permutation Using a Graphing Utility **In Exercises 37–40, evaluate $_nP_r$ using a graphing utility.**

37. $_{30}P_6$

38. $_{10}P_8$

39. $_{120}P_4$

40. $_{100}P_5$

41. **Politics** From a pool of 12 candidates, the offices of president, vice-president, secretary, and treasurer will be filled. In how many ways can the offices be filled?

42. **Physical Education** How many different batting orders can a baseball coach create from a team of 15 players when there are nine positions to fill?

43. **Education** The graphic design department is holding a contest in which it will award scholarships of different values to those who finish in first place, second place, and third place. The department receives 104 entries. How many different orders of the top three places are possible?

44. **Athletics** Eight sprinters have qualified for the finals in the 100-meter dash at the NCAA national track meet. How many different orders of the top three finishes are possible? (Assume there are no ties.)

Writing Permutations **In Exercises 45 and 46, use the letters A, B, C, and D.**

45. Write all permutations of the letters.

46. Write all permutations of the letters when the letters B and C must remain between the letters A and D.

Distinguishable Permutations **In Exercises 47–50, find the number of distinguishable permutations of the group of letters.**

47. A, A, G, E, E, E, M

48. B, B, B, T, T, T, T, T

49. A, L, G, E, B, R, A

50. M, I, S, S, I, S, S, I, P, P, I

Evaluating a Combination **In Exercises 51–56, evaluate $_nC_r$ using the formula from this section.**

51. $_5C_2$

52. $_6C_3$

53. $_4C_1$

54. $_7C_1$

55. $_{25}C_0$

56. $_{20}C_0$

Evaluating a Combination Using a Graphing Utility **In Exercises 57–60, evaluate $_nC_r$ using a graphing utility.**

57. $_{33}C_4$

58. $_{10}C_7$

59. $_{42}C_5$

60. $_{50}C_6$

Writing Combinations **In Exercises 61 and 62, use the letters A, B, C, D, E, and F.**

61. Write all possible selections of two letters that can be formed from the letters. (The order of the two letters is not important.)

62. Write all possible selections of three letters that can be formed from the letters. (The order of the three letters is not important.)

63. **Politics** As of May 2014, the U.S. Senate Committee on Indian Affairs had 14 members. Party affiliation is not a factor in selection. How many different committees are possible from the 100 U.S. senators?

64. **Education** You can answer any 18 questions from a total of 20 questions on an exam. In how many different ways can you select the questions?

65. Jury Selection In how many different ways can a jury of 12 people be randomly selected from a group of 40 people?

66. Geometry Three points that are not collinear determine three lines. How many lines are determined by nine points, of which no three are collinear?

67. *Why you should learn it* (*p. 537*) Powerball is played with 59 white balls, numbered 1 through 59, and 35 red balls, numbered 1 through 35. Five white balls and one red ball, the Powerball, are drawn. In how many ways can a player select the six numbers?

68. Law A law office interviews paralegals for 10 openings. There are 13 paralegals with two years of experience and 20 paralegals with one year of experience. How many combinations of seven paralegals with two years of experience and three paralegals with one year of experience are possible?

69. Forming a Committee A six-member research committee is to be formed having one administrator, three faculty members, and two students. There are seven administrators, 12 faculty members, and 25 students in contention for the committee. How many six-member committees are possible?

70. Sociology The number of possible interpersonal relationships increases dramatically as the size of a group increases. Determine the numbers of different two-person relationships that are possible in groups of people of sizes (a) 3, (b) 8, (c) 12, and (d) 20.

71. Game Theory You are dealt five cards from an ordinary deck of 52 playing cards. In how many ways can you get a full house? (A full house consists of three of one kind and two of another. For example, 8-8-8-5-5 and K-K-K-10-10 are full houses.)

72. Quality Engineering A shipment of 30 flat screen televisions contains three defective units. In how many ways can a vending company purchase four of these units and receive (a) all good units, (b) two good units, and (c) at least two good units?

Geometry **In Exercises 73–76, find the number of diagonals of the polygon. (A line segment connecting any two nonadjacent vertices is called a *diagonal* of a polygon.)**

73. Pentagon
74. Hexagon
75. Octagon
76. Decagon (10 sides)

Solving an Equation **In Exercises 77–84, solve for n.**

77. $14 \cdot {}_nP_3 = {}_{n+2}P_4$
78. ${}_nP_5 = 18 \cdot {}_{n-2}P_4$
79. ${}_nP_4 = 10 \cdot {}_{n-1}P_3$
80. ${}_nP_6 = 12 \cdot {}_{n-1}P_5$
81. ${}_{n+1}P_3 = 4 \cdot {}_nP_2$
82. ${}_{n+2}P_3 = 6 \cdot {}_{n+2}P_1$
83. $4 \cdot {}_{n+1}P_2 = {}_{n+2}P_3$
84. $5 \cdot {}_{n-1}P_1 = {}_nP_2$

Conclusions

True or False? **In Exercises 85 and 86, determine whether the statement is true or false. Justify your answer.**

85. The number of pairs of letters that can be formed from any of the first 13 letters in the alphabet (A–M), where repetitions are allowed, is an example of a permutation.

86. The number of permutations of n elements can be derived by using the Fundamental Counting Principle.

87. Think About It Can your calculator evaluate ${}_{100}P_{80}$? If not, explain why.

88. Think About It Decide whether each scenario should be counted using permutations or combinations. Explain your reasoning. (Do not calculate.)

(a) Number of ways 10 people can line up in a row for concert tickets

(b) Number of different arrangements of three types of flowers from an array of 20 types

89. Writing Explain in your own words the meaning of ${}_nP_r$.

90. HOW DO YOU SEE IT? Without calculating, determine whether the value of ${}_nP_r$ is greater than the value of ${}_nC_r$ for the values of n and r given in the table. Complete the table using yes (Y) or no (N). Is the value of ${}_nP_r$ always greater than the value of ${}_nC_r$? Explain.

n \ r	0	1	2	3	4	5
1						
2						
3						
4						
5						

Proof **In Exercises 91–94, prove the identity.**

91. ${}_nP_{n-1} = {}_nP_n$
92. ${}_nC_n = {}_nC_0$
93. ${}_nC_{n-1} = {}_nC_1$
94. ${}_nC_r = \dfrac{{}_nP_r}{r!}$

Cumulative Mixed Review

Using Cramer's Rule **In Exercises 95 and 96, use Cramer's Rule to solve the system of equations.**

95. $\begin{cases} -5x + 3y = -14 \\ 7x - 2y = 2 \end{cases}$
96. $\begin{cases} -3x - 4y = -1 \\ 9x + 5y = -4 \end{cases}$

Denys Prokofyev | Dreamstime.com

6.6 Probability

The Probability of an Event

Any happening whose result is uncertain is called an **experiment.** The possible results of the experiment are **outcomes,** the set of all possible outcomes of the experiment is the **sample space** of the experiment, and any subcollection of a sample space is an **event.**

For instance, when a six-sided die is tossed, the sample space can be represented by the numbers 1 through 6. For the experiment to be fair, each of the outcomes must be *equally likely*.

To describe a sample space in such a way that each outcome is equally likely, you must sometimes distinguish between or among various outcomes in ways that appear artificial. Example 1 illustrates such a situation.

What you should learn

- Find probabilities of events.
- Find probabilities of mutually exclusive events.
- Find probabilities of independent events.

Why you should learn it

You can use probability to solve a variety of problems that occur in real life. For instance, in Exercise 58 on page 553, you are asked to use probability to help analyze the age distribution of unemployed workers.

EXAMPLE 1 Finding the Sample Space

Find the sample space for each experiment.

a. One coin is tossed.

b. Two coins are tossed.

c. Three coins are tossed.

Solution

a. Because the coin will land either heads up (denoted by H) or tails up (denoted by T), the sample space is $S = \{H, T\}$.

b. Because either coin can land heads up or tails up, the possible outcomes are as follows.

HH = heads up on both coins

HT = heads up on first coin and tails up on second coin

TH = tails up on first coin and heads up on second coin

TT = tails up on both coins

So, the sample space is $S = \{HH, HT, TH, TT\}$.

Note that this list distinguishes between the two cases

HT and TH

even though these two outcomes appear to be similar.

c. Following the notation in part (b), the sample space is

$S = \{HHH, HHT, HTH, HTT, THH, THT, TTH, TTT\}$.

Note that this list distinguishes among the cases

HHT, HTH, and THH

and among the cases

HTT, THT, and TTH.

✓ **Checkpoint** Audio-video solution in English & Spanish at LarsonPrecalculus.com.

Find the sample space for the following experiment.

You toss a coin twice and a six-sided die once.

To calculate the probability of an event, count the number of outcomes in the event and in the sample space. The *number of equally likely outcomes* in event E is denoted by $n(E)$, and the number of equally likely outcomes in the sample space S is denoted by $n(S)$. The probability that event E will occur is given by $n(E)/n(S)$.

Explore the Concept

Toss two coins 40 times and write down the number of heads that occur on each toss (0, 1, or 2). How many times did two heads occur? Without performing the experiment, how many times would you expect two heads to occur when two coins are tossed 1000 times?

The Probability of an Event

If an event E has $n(E)$ equally likely outcomes and its sample space S has $n(S)$ equally likely outcomes, then the **probability** of event E is given by

$$P(E) = \frac{n(E)}{n(S)}.$$

Because the number of outcomes in an event must be less than or equal to the number of outcomes in the sample space, the probability of an event must be a number from 0 to 1, inclusive. That is,

$$0 \le P(E) \le 1$$

as shown in the figure. If $P(E) = 0$, then event E *cannot occur,* and E is called an **impossible event.** If $P(E) = 1$, then event E *must occur,* and E is called a **certain event.**

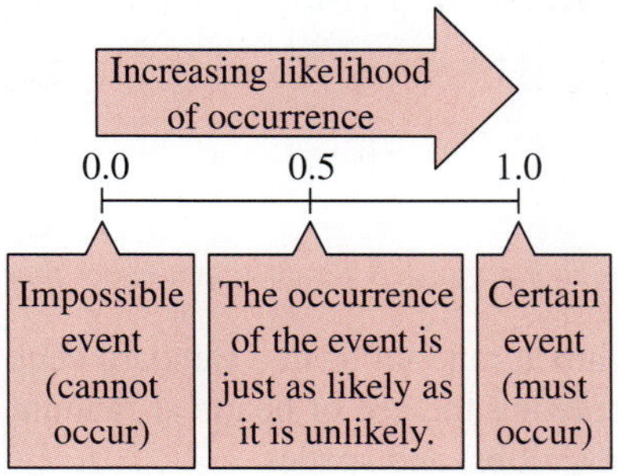

EXAMPLE 2 Finding the Probability of an Event

See LarsonPrecalculus.com for an interactive version of this type of example.

a. Two coins are tossed. What is the probability that both land heads up?

b. A card is drawn at random from a standard deck of playing cards. What is the probability that it is an ace?

Solution

a. Following the procedure in Example 1(b), let

$E = \{HH\}$ and $S = \{HH, HT, TH, TT\}$.

The probability of getting two heads is

$$P(E) = \frac{n(E)}{n(S)} = \frac{1}{4}.$$

b. Because there are 52 cards in a standard deck of playing cards and there are four aces (one of each suit), the probability of drawing an ace is

$$P(E) = \frac{n(E)}{n(S)} = \frac{4}{52} = \frac{1}{13}.$$

Remark

Note that a probability can be written as a fraction, a decimal, or a percent. For instance, in Example 2(a), the probability of getting two heads can be written as $\frac{1}{4}$, 0.25, or 25%.

✓ Checkpoint *Audio-video solution in English & Spanish at LarsonPrecalculus.com.*

a. You toss three coins. What is the probability that all three land tails up?

b. You draw one card at random from a standard deck of playing cards. What is the probability that it is a diamond?

You could have written out the sample space in Example 2(b) and simply counted the outcomes in the desired event. For larger sample spaces, however, using the counting principles discussed in Section 6.5 should save you time.

EXAMPLE 3 Finding the Probability of an Event

Twelve-sided dice, as shown in the figure at the right, can be constructed (in the shape of regular dodecahedrons) such that each of the numbers from 1 to 6 appears twice on each die. Show that these dice can be used in any game requiring ordinary six-sided dice without changing the probabilities of the various events.

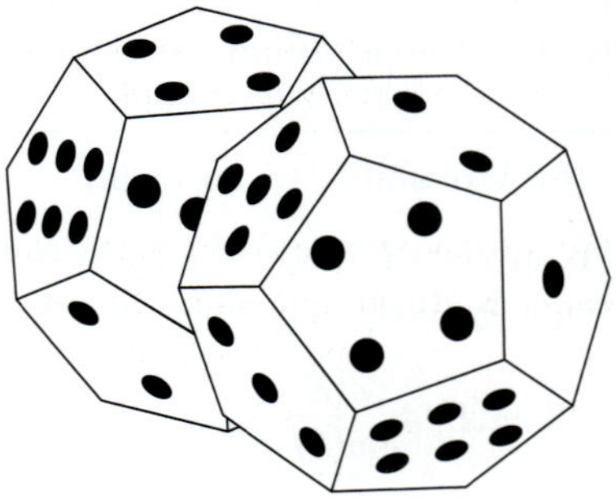

Solution

For an ordinary six-sided die, each of the numbers 1, 2, 3, 4, 5, and 6 occurs only once, so the probability of rolling any particular number is

$$P(E) = \frac{n(E)}{n(S)} = \frac{1}{6}.$$

For a 12-sided die, each number occurs twice, so the probability of rolling any particular number is

$$P(E) = \frac{n(E)}{n(S)} = \frac{2}{12} = \frac{1}{6}.$$

✓ **Checkpoint** Audio-video solution in English & Spanish at LarsonPrecalculus.com.

Show that the probability of drawing a club at random from a standard deck of playing cards is the same as the probability of drawing the ace of hearts at random from a set consisting of the aces of hearts, diamonds, clubs, and spades.

EXAMPLE 4 Random Selection

The numbers of colleges and universities in various regions of the United States in 2012 are shown in the figure. One institution is selected at random. What is the probability that the institution is in one of the three southern regions? (Source: U.S. National Center for Education Statistics)

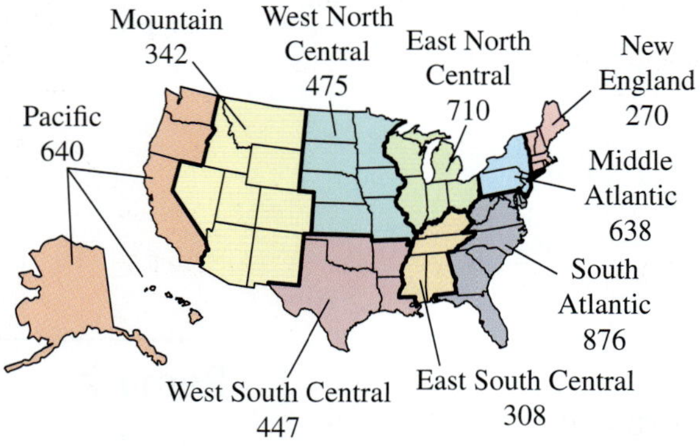

Solution

From the figure, the total number of colleges and universities is 4706. Because there are

$$447 + 308 + 876 = 1631$$

colleges and universities in the three southern regions, the probability that the institution is in one of these regions is

$$P(E) = \frac{n(E)}{n(S)} = \frac{1631}{4706} \approx 0.347.$$

✓ **Checkpoint** Audio-video solution in English & Spanish at LarsonPrecalculus.com.

In Example 4, what is the probability that the randomly selected institution is in the Pacific region?

Mutually Exclusive Events

Two events A and B (from the same sample space) are **mutually exclusive** when A and B have no outcomes in common. In the terminology of sets, the intersection of A and B is the empty set, which implies that

$$P(A \cap B) = 0.$$

For instance, when two dice are tossed, the event A of rolling a sum of 6 and the event B of rolling a sum of 9 are mutually exclusive. To find the probability that one or the other of two mutually exclusive events will occur, you can *add* their individual probabilities.

Probability of the Union of Two Events

If A and B are events in the same sample space, then the probability of A or B occurring is given by

$$P(A \cup B) = P(A) + P(B) - P(A \cap B).$$

If A and B are mutually exclusive, then

$$P(A \cup B) = P(A) + P(B).$$

EXAMPLE 5 The Probability of a Union

One card is selected at random from a standard deck of 52 playing cards. What is the probability that the card is either a heart or a face card?

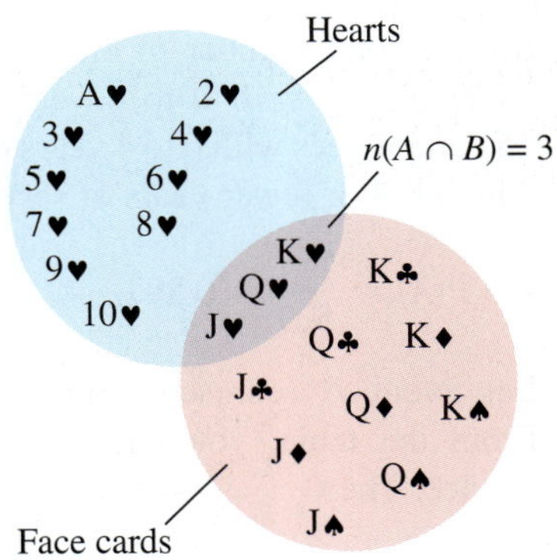

Figure 6.1

Solution

Because the deck has 13 hearts, the probability of selecting a heart (event A) is

$$P(A) = \frac{13}{52}.$$

Similarly, because the deck has 12 face cards, the probability of selecting a face card (event B) is

$$P(B) = \frac{12}{52}.$$

Because three of the cards are hearts and face cards (see Figure 6.1), it follows that

$$P(A \cap B) = \frac{3}{52}.$$

Finally, applying the formula for the probability of the union of two events, you can conclude that the probability of selecting either a heart or a face card is

$$\begin{aligned} P(A \cup B) &= P(A) + P(B) - P(A \cap B) \\ &= \frac{13}{52} + \frac{12}{52} - \frac{3}{52} \\ &= \frac{22}{52} \\ &\approx 0.423. \end{aligned}$$

✓ **Checkpoint** *Audio-video solution in English & Spanish at LarsonPrecalculus.com.*

You draw one card at random from a standard deck of 52 playing cards. What is the probability that the card is either an ace or a spade?

EXAMPLE 6 Probability of Mutually Exclusive Events

The human resources department of a company has compiled data on the numbers of employees who have been with the company for various periods of time. The results are shown in the table.

DATA

Years of service	Number of employees
0–4	157
5–9	89
10–14	74
15–19	63
20–24	42
25–29	38
30–34	37
35–39	21
40–44	8

Spreadsheet at LarsonPrecalculus.com

Human Resources Manager

An employee is chosen at random. What is the probability that the employee has

a. 4 or fewer years of service?

b. 9 or fewer years of service?

Solution

a. To begin, add the numbers of employees.

$$157 + 89 + 74 + 63 + 42 + 38 + 37 + 21 + 8 = 529$$

So, the total number of employees is 529. Next, let event A represent choosing an employee with 0 to 4 years of service. From the table, you know there are 157 employees with 0 to 4 years of service. The probability of choosing an employee who has 4 or fewer years of service is

$$P(A) = \frac{157}{529} \approx 0.297.$$

b. Let event B represent choosing an employee with 5 to 9 years of service. From the table, you know that there are 89 employees with 5 to 9 years of service. Then

$$P(B) = \frac{89}{529}.$$

Because event A from part (a) and event B have no outcomes in common, you can conclude that these two events are mutually exclusive and that

$$\begin{aligned} P(A \cup B) &= P(A) + P(B) \\ &= \frac{157}{529} + \frac{89}{529} \\ &= \frac{246}{529} \\ &\approx 0.465. \end{aligned}$$

So, the probability of choosing an employee who has 9 or fewer years of service is about 0.465.

✓ **Checkpoint** *Audio-video solution in English & Spanish at LarsonPrecalculus.com.*

In Example 6, what is the probability that an employee chosen at random has 30 or more years of service?

Independent Events

Two events are **independent** when the occurrence of one has no effect on the occurrence of the other. For instance, rolling a total of 12 with two six-sided dice has no effect on the outcome of future rolls of the dice. To find the probability that two independent events will occur, *multiply* the probabilities of each.

> **Probability of Independent Events**
>
> If A and B are **independent events,** then the probability that both A and B will occur is given by
>
> $$P(A \text{ and } B) = P(A) \cdot P(B).$$
>
> This rule can be extended to any number of independent events.

EXAMPLE 7 Probability of Independent Events

A random number generator on a computer selects three integers from 1 to 20. What is the probability that all three numbers are less than or equal to 5?

Solution

The probability of selecting a number from 1 to 5 is

$$P(A) = \frac{5}{20} = \frac{1}{4}.$$

So, the probability that all three numbers are less than or equal to 5 is

$$P(A) \cdot P(A) \cdot P(A) = \left(\frac{1}{4}\right)\left(\frac{1}{4}\right)\left(\frac{1}{4}\right) = \frac{1}{64}.$$

✓ **Checkpoint** *Audio-video solution in English & Spanish at LarsonPrecalculus.com.*

A random number generator on a computer selects two integers from 1 to 30. What is the probability that both numbers are less than 12?

EXAMPLE 8 Probability of Independent Events

In 2014, approximately 44% of the adult population in the United States said their cell phone would be very hard to give up. In a survey, 10 people were chosen at random from the adult population. What is the probability that all 10 say their cell phone would be very hard to give up? (Source: Pew Research Center)

Solution

Let A represent choosing an adult who says his or her cell phone would be very hard to give up. The probability of choosing an adult who says his or her cell phone would be very hard to give up is 0.44, the probability of choosing a second adult who says his or her cell phone would be very hard to give up is 0.44, and so on. Because these events are independent, the probability that all 10 people say their cell phone would be very hard to give up is

$$[P(A)]^{10} = (0.44)^{10} \approx 0.000272.$$

✓ **Checkpoint** *Audio-video solution in English & Spanish at LarsonPrecalculus.com.*

In 2014, approximately 35% of the adult population in the United States said their television would be very hard to give up. In a survey, eight people were chosen at random from the adult population. What is the probability that all eight say their television would be very hard to give up? (Source: Pew Research Center)

6.6 Exercises

See *CalcChat.com* for tutorial help and worked-out solutions to odd-numbered exercises.
For instructions on how to use a graphing utility, see Appendix A.

Vocabulary and Concept Check

In Exercises 1–4, fill in the blank(s).

1. The set of all possible outcomes of an experiment is called the _______ .

2. To determine the probability of an event, use the formula $P(E) = \frac{n(E)}{n(S)}$, where $n(E)$ is _______ and $n(S)$ is _______ .

3. If two events from the same sample space have no outcomes in common, then the two events are _______ .

4. If the occurrence of one event has no effect on the occurrence of a second event, then the events are _______ .

5. Write an inequality that represents the possible values of the probability $P(E)$ of an event.

6. What is the probability of an impossible event?

7. What is the probability of a certain event?

8. Match the probability formula with the correct probability name.

(a) Probability of the union of two events (i) $P(A \cup B) = P(A) + P(B)$

(b) Probability of mutually exclusive events (ii) $P(A \cup B) = P(A) + P(B) - P(A \cap B)$

(c) Probability of independent events (iii) $P(A \text{ and } B) = P(A) \cdot P(B)$

Procedures and Problem Solving

Finding the Sample Space **In Exercises 9–12, determine the sample space for the experiment.**

9. A coin and a six-sided die are tossed.

10. A six-sided die is tossed twice and the results of roll 1 and roll 2 are recorded.

11. A taste tester has to rank three varieties of orange juice, A, B, and C, according to preference.

12. Two county supervisors are selected from five supervisors, A, B, C, D, and E, to study a recycling plan.

Finding the Probability of an Event **In Exercises 13–16, find the probability for the experiment of tossing a coin three times. Use the sample space**

$S = \{HHH, HHT, HTH, HTT, THH, THT, TTH, TTT\}$.

13. The probability of getting exactly two tails

14. The probability of getting a head on the first toss

15. The probability of getting at least one head

16. The probability of getting at least two heads

Finding the Probability of an Event **In Exercises 17–20, find the probability for the experiment of selecting one card at random from a standard deck of 52 playing cards.**

17. The card is a face card. **18.** The card is a black card.

19. The card is a red face card.

20. The card is a numbered card (2–10).

Finding the Probability of an Event **In Exercises 21–24, use the sample space from Exercise 10 to find the probability for the experiment of tossing a six-sided die twice.**

21. The sum is 6. **22.** The sum is at least 8.

23. The sum is less than 5.

24. The sum is odd or prime.

Random Selection **In Exercises 25–28, one of a team's 2200 season ticket holders is selected at random to win a prize. The circle graph shows the ages of the season ticket holders. Find the probability of the event.**

Ages of Season Ticket Holders

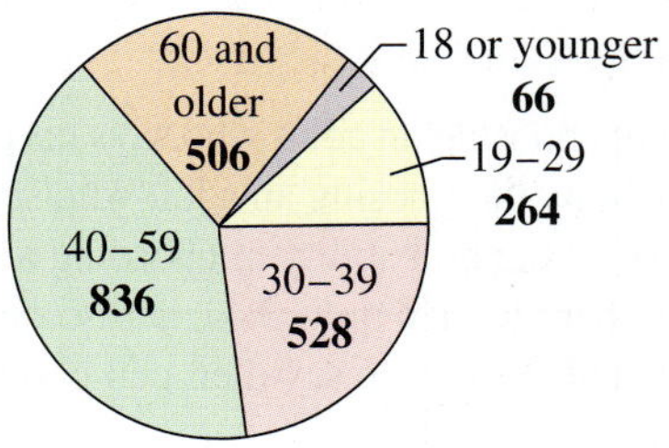

25. The winner is younger than 19 years old.

26. The winner is older than 39 years old.

27. The winner is 19 to 39 years old.

28. The winner is younger than 19 years old or older than 59 years old.

Using Combinations **In Exercises 29–32, find the probability for the experiment of drawing two marbles at random (without replacement) from a bag containing one green, two yellow, and three red marbles. (*Hint:* Use combinations to find the numbers of outcomes for the given event and sample space.)**

29. Both marbles are red.

30. Both marbles are yellow.

31. Neither marble is yellow.

32. The marbles are of different colors.

Finding the Probability of a Complement **The *complement of an event A* is the collection of all outcomes in the sample space that are *not* in A. If the probability of A is P(A), then the probability of the complement A′ is given by $P(A') = 1 - P(A)$. In Exercises 33–36, you are given the probability that an event *will* happen. Find the probability that the event *will not* happen.**

33. $P(E) = 0.75$ **34.** $P(E) = 0.36$

35. $P(E) = \frac{2}{3}$ **36.** $P(E) = \frac{7}{8}$

Using the Probability of a Complement **In Exercises 37–40, you are given the probability that an event *will not* happen. Find the probability that the event *will* happen.**

37. $P(E') = 0.12$ **38.** $P(E') = 0.84$

39. $P(E') = \frac{13}{20}$ **40.** $P(E') = \frac{61}{100}$

The Probability of a Union **In Exercises 41–44, one card is selected at random from a standard deck of 52 playing cards. Use a formula to find the probability of the union of the two events.**

41. The card is a club or a king.

42. The card is a face card or a black card.

43. The card is a 5 or a 2.

44. The card is a heart or a spade.

Probability of Mutually Exclusive Events **In Exercises 45–50, use the table, which shows the age groups of students in a college sociology class.**

Age	Number of students
18–19	11
20–21	18
22–30	2
31–40	1

A student from the class is randomly chosen for a project. Find the probability that the student is the given age.

45. 18 or 19 years old

46. 22 to 30 years old

47. 20 to 30 years old

48. 18 to 21 years old

49. Older than 21 years old

50. Younger than 31 years old

Probability of Independent Events **In Exercises 51–56, a random number generator selects three numbers from 1 through 10. Find the probability of the event.**

51. All three numbers are even.

52. All three numbers are a factor of 8.

53. All three numbers are less than or equal to 3.

54. All three numbers are greater than or equal to 9.

55. Two numbers are less than 5 and the other number is 10.

56. One number is 2, 4, or 6, and the other two numbers are odd.

57. Political Science Taylor, Moore, and Perez are candidates for public office. It is estimated that Moore and Perez have about the same probability of winning, and Taylor is believed to be twice as likely to win as either of the others. Find the probability of each candidate's winning the election.

58. ***Why you should learn it*** *(p. 546)* In 2013, there were approximately 11.46 million unemployed workers in the United States. The circle graph shows the age profile of these unemployed workers. (Source: U.S. Bureau of Labor Statistics)

Ages of Unemployed Workers

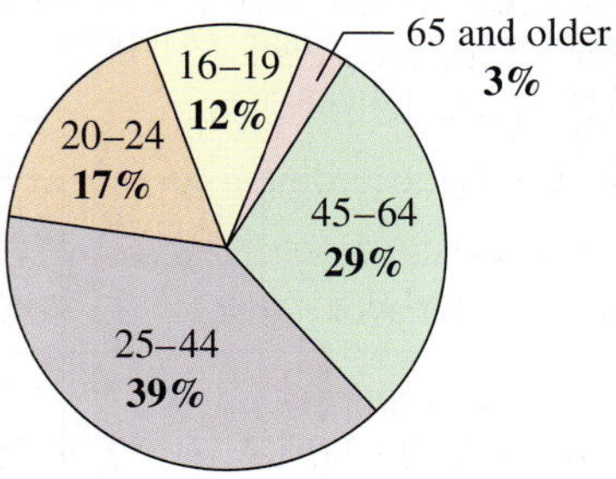

(a) Estimate the number of unemployed workers in the 16–19 age group.

(b) Estimate the number of unemployed workers who are 45 years or older.

(c) What is the probability that a person selected at random from the population of unemployed workers is in the 25–44 age group?

(d) What is the probability that a person selected at random from the population of unemployed workers is 25 to 64 years old?

(e) What is the probability that a person selected at random from the population of unemployed workers is 45 or older?

59. Education The levels of educational attainment of the United States population age 25 years or older in 2013 are shown in the circle graph. Use the fact that the population of people 25 years or older was 206.9 million in 2013. (Source: U.S. Census Bureau)

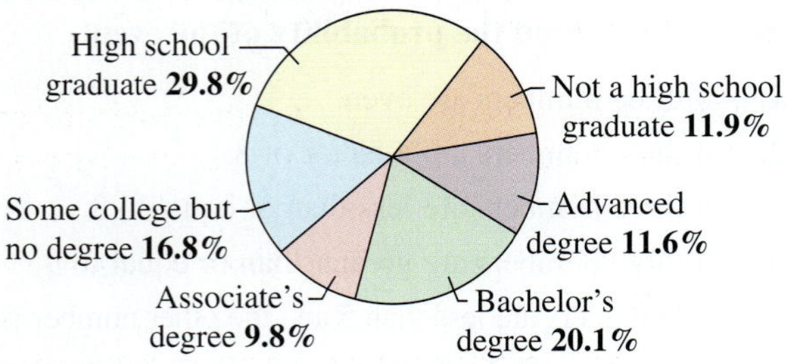

(a) Estimate the number of people 25 years old or older who have advanced degrees.

(b) Find the probability that a person 25 years old or older selected at random has earned a Bachelor's degree or higher.

(c) Find the probability that a person 25 years old or older selected at random has earned a high school diploma or gone on to post-secondary education.

(d) Find the probability that a person 25 years old or older selected at random has earned a degree.

60. Political Science One hundred college students were interviewed to determine their political party affiliations and whether they favored a balanced budget amendment to the U.S. Constitution. The results of the study are listed in the table, where D represents Democrat and R represents Republican.

	D	R	Total
Favor	23	32	55
Oppose	25	9	34
Unsure	7	4	11
Total	55	45	100

A person is selected at random from the sample. Find the probability that the person selected is (a) a person who does not favor the amendment, (b) a Republican, and (c) a Democrat who favors the amendment.

In Exercises 61–64, the sample spaces are large and you should use the counting principles discussed in Section 6.5.

61. Using Counting Principles On a game show, you are given five digits to arrange in the proper order to form the price of a car. If you are correct, you win the car. What is the probability of winning when you (a) randomly guess the position of each digit and (b) know the first digit and randomly guess the others?

62. Using Counting Principles The deck for a card game is made up of 108 cards. Twenty-five each are red, yellow, blue, and green, and eight are wild cards. Each player is randomly dealt a seven-card hand. What is the probability that a hand will contain (a) exactly two wild cards, and (b) two wild cards, two red cards, and three blue cards?

63. Using Counting Principles A shipment of 12 microwave ovens contains three defective units. A vending company has ordered four of these units, and because all are packaged identically, the selection will be random. What is the probability that (a) all four units are good, (b) exactly two units are good, and (c) at least two units are good?

64. Using Counting Principles Two integers from 1 through 40 are chosen by a random number generator. What is the probability that (a) the numbers are both even, (b) one number is even and one is odd, (c) both numbers are less than 30, and (d) the same number is chosen twice?

65. Marketing The circle graph shows the methods used by shoppers to pay for merchandise. Two shoppers are chosen at random. What is the probability that (a) both shoppers paid for their purchases only in cash, (b) at least one shopper paid only in cash, and (c) neither shopper paid only in cash?

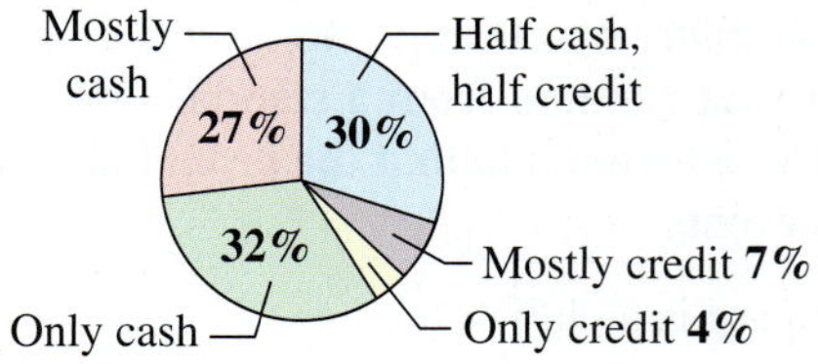

66. Aerospace Engineering A space vehicle has an independent backup system for one of its communication networks. The probability that either system will function satisfactorily for the duration of a flight is 0.985. What is the probability that during a given flight (a) both systems function satisfactorily, (b) at least one system functions satisfactorily, and (c) both systems fail?

67. Marketing A sales representative makes sales on approximately one-fifth of all calls. On a given day, the representative contacts six potential clients. What is the probability that a sale will be made with (a) all six contacts, (b) none of the contacts, and (c) at least one contact?

68. Genetics Assume that the probability of the birth of a child of a particular sex is 50%. In a family with four children, what is the probability that (a) all the children are boys, (b) all the children are the same sex, and (c) there is at least one boy?

69. Estimating π A coin of diameter d is randomly dropped onto a paper that contains a grid of squares d units on a side. (See figure.)

(a) Find the probability that the coin covers a vertex of one of the squares on the grid.

(b) Perform the experiment 100 times and use the results to approximate π.

70. Geometry You and a friend agree to meet at your favorite fast food restaurant between 5:00 and 6:00 P.M. The one who arrives first will wait 15 minutes for the other, after which the first person will leave (see figure). What is the probability that the two of you will actually meet, assuming that your arrival times are random within the hour?

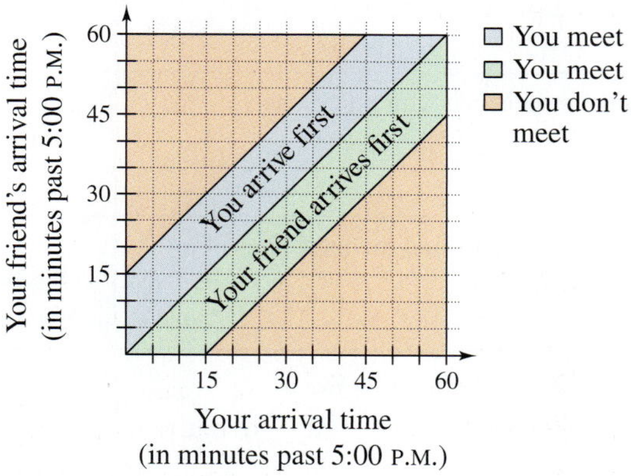

Conclusions

True or False? **In Exercises 71 and 72, determine whether the statement is true or false. Justify your answer.**

71. If the probability of an outcome in a sample space is 1, then the probability of the other outcomes in the sample space is 0.

72. If A and B are independent events with nonzero probabilities, then A can occur when B occurs.

73. Exploration Consider a group of n people.

(a) Explain why the following pattern gives the probability that the n people have distinct birthdays.

$$n = 2: \quad \frac{365}{365} \cdot \frac{364}{365} = \frac{365 \cdot 364}{365^2}$$

$$n = 3: \quad \frac{365}{365} \cdot \frac{364}{365} \cdot \frac{363}{365} = \frac{365 \cdot 364 \cdot 363}{365^3}$$

(b) Use the pattern in part (a) to write an expression for the probability that four people ($n = 4$) have distinct birthdays.

(c) Let P_n be the probability that the n people have distinct birthdays. Verify that this probability can be obtained recursively by

$$P_1 = 1 \quad \text{and} \quad P_n = \frac{365 - (n - 1)}{365} P_{n-1}.$$

(d) Explain why $Q_n = 1 - P_n$ gives the probability that at least two people in a group of n people have the same birthday.

(e) Use the results of parts (c) and (d) to complete the table.

n	10	15	20	23	30	40	50
P_n							
Q_n							

(f) How many people must be in a group so that the probability of at least two of them having the same birthday is greater than $\frac{1}{2}$? Explain.

74. Writing Write a paragraph describing in your own words the difference between mutually exclusive events and independent events.

75. Think About It Let A and B be two events from the same sample space such that $P(A) = 0.76$ and $P(B) = 0.58$.

(a) Is it possible that A and B are mutually exclusive? Explain. Draw a diagram to support your answer.

(b) Is it possible that A' and B' are mutually exclusive? Explain. Draw a diagram to support your answer.

(c) Determine the possible range of $P(A \cup B)$.

76. HOW DO YOU SEE IT? Use the figure to determine whether events A and B are mutually exclusive. Explain your reasoning.

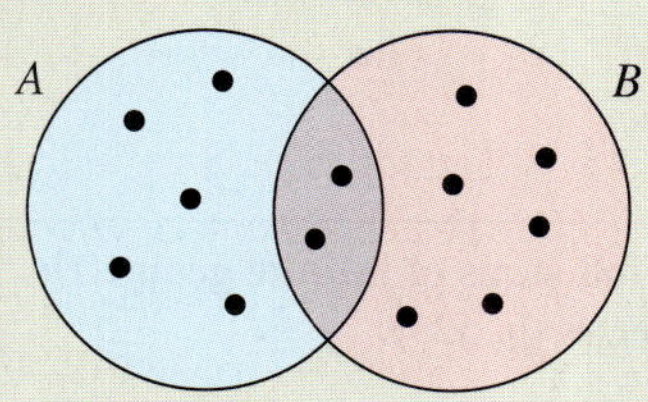

Cumulative Mixed Review

Evaluating a Combination **In Exercises 77–80, evaluate $_nC_r$. Verify your result using a graphing utility.**

77. $_8C_4$

78. $_9C_5$

79. $_{11}C_8$

80. $_{16}C_{13}$

6 Chapter Summary

	What did you learn?	Explanation and Examples	Review Exercises
6.1	**Use sequence notation to write the terms of sequences *(p. 500)*.**	$a_n = 7n - 4$; $a_1 = 7(1) - 4 = 3$, $a_2 = 7(2) - 4 = 10$, $a_3 = 7(3) - 4 = 17$, $a_4 = 7(4) - 4 = 24$	1–8
	Use factorial notation *(p. 503)*.	If n is a positive integer, then $n! = 1 \cdot 2 \cdot 3 \cdot 4 \cdot \cdot \cdot (n - 1) \cdot n$.	9–12
	Use summation notation to write sums *(p. 504)*.	The sum of the first n terms of a sequence is represented by $\sum_{i=1}^{n} a_i = a_1 + a_2 + a_3 + a_4 + \cdot \cdot \cdot + a_n$.	13–20
	Find sums of infinite series *(p. 505)*.	$\sum_{i=1}^{\infty} \frac{5}{10^i} = \frac{5}{10^1} + \frac{5}{10^2} + \frac{5}{10^3} + \frac{5}{10^4} + \frac{5}{10^5} + \cdot \cdot \cdot$ $= 0.5 + 0.05 + 0.005 + 0.0005 + 0.00005 + \cdot \cdot \cdot$ $= 0.55555 \ldots$ $= \frac{5}{9}$	21–24
	Use sequences and series to model and solve real-life problems *(p. 506)*.	A sequence can be used to model the balance in an account that earns compound interest. (See Example 10.)	25, 26
6.2	**Recognize, write, and find the *n*th terms of arithmetic sequences *(p. 511)*.**	$a_n = 9n + 5$; $a_1 = 9(1) + 5 = 14$, $a_2 = 9(2) + 5 = 23$, $a_3 = 9(3) + 5 = 32$, $a_4 = 9(4) + 5 = 41$	27–38
	Find *n*th partial sums of arithmetic sequences *(p. 514)*.	The sum of a finite arithmetic sequence with n terms is $S_n = \frac{n}{2}(a_1 + a_n)$.	39–42
	Use arithmetic sequences to model and solve real-life problems *(p. 515)*.	An arithmetic sequence can be used to find the total sales of a small business. (See Example 7.)	43, 44
6.3	**Recognize, write, and find the *n*th terms of geometric sequences *(p. 519)*.**	$a_n = 3(4^n)$; $a_1 = 3(4^1) = 12$, $a_2 = 3(4^2) = 48$, $a_3 = 3(4^3) = 192$, $a_4 = 3(4^4) = 768$	45–56
	Find *n*th partial sums of geometric sequences *(p. 522)*.	The sum of the finite geometric sequence $a_1, a_1r, a_1r^2, \ldots, a_1r^{n-1}$ with common ratio $r \neq 1$ is given by $S_n = \sum_{i=1}^{n} a_1r^{i-1} = a_1\left(\frac{1 - r^n}{1 - r}\right)$.	57–60
	Find sums of infinite geometric series *(p. 523)*.	If $\lvert r \rvert < 1$, then the infinite geometric series $a_1 + a_1r + a_1r^2 + \cdot \cdot \cdot + a_1r^{n-1} + \cdot \cdot \cdot$ has the sum $S = \sum_{i=0}^{\infty} a_1r^i = \frac{a_1}{1 - r}$.	61–64
	Use geometric sequences to model and solve real-life problems *(p. 524)*.	A finite geometric sequence can be used to find the balance in an annuity at the end of 2 years. (See Example 8.)	65, 66

	What did you learn?	Explanation and Examples	Review Exercises
6.4	**Use the Binomial Theorem to calculate binomial coefficients *(p. 529).***	**The Binomial Theorem:** In the expansion of $(x + y)^n = x^n + nx^{n-1}y + \cdot \cdot \cdot + {}_nC_r x^{n-r}y^r + \cdot \cdot \cdot + nxy^{n-1} + y^n$, the coefficient of $x^{n-r}y^r$ is $${}_nC_r = \frac{n!}{(n-r)!r!}.$$	67–70
	Use binomial coefficients to write binomial expansions *(p. 531).*	$(x + 1)^3 = x^3 + 3x^2 + 3x + 1$ $(x - 1)^4 = x^4 - 4x^3 + 6x^2 - 4x + 1$	71–74
	Use Pascal's Triangle to calculate binomial coefficients *(p. 533).*	First several rows of Pascal's triangle: 1 1 1 1 2 1 1 3 3 1 1 4 6 4 1	75–78
6.5	**Solve simple counting problems *(p. 537).***	A computer randomly generates an integer from 1 through 15. The computer can generate an integer that is divisible by 3 in five ways (3, 6, 9, 12, and 15).	79, 80
	Use the Fundamental Counting Principle to solve more complicated counting problems *(p. 538).*	**Fundamental Counting Principle:** Let E_1 and E_2 be two events. The first event E_1 can occur in m_1 different ways. After E_1 has occurred, E_2 can occur in m_2 different ways. The number of ways that the two events can occur is $m_1 \cdot m_2$.	81, 82
	Use permutations to solve counting problems *(p. 539).*	A permutation of n different elements is an ordering of the elements such that one element is first, one is second, one is third, and so on. The number of permutations of n elements is given by $$n \cdot (n-1) \cdot \cdot \cdot 4 \cdot 3 \cdot 2 \cdot 1 = n!.$$ In other words, there are $n!$ different ways that n elements can be ordered. The number of permutations of n elements taken r at a time is $${}_nP_r = \frac{n!}{(n-r)!}.$$	83–88, 91
	Use combinations to solve counting problems *(p. 542).*	The number of combinations of n elements taken r at a time is $${}_nC_r = \frac{n!}{(n-r)!r!}.$$	89, 90, 92
6.6	**Find probabilities of events *(p. 546).***	If an event E has $n(E)$ equally likely outcomes and its sample space S has $n(S)$ equally likely outcomes, then the probability of event E is $$P(E) = \frac{n(E)}{n(S)}.$$	93, 94
	Find probabilities of mutually exclusive events *(p. 549).*	If A and B are events in the same sample space, then the probability of A *or* B occurring is $$P(A \cup B) = P(A) + P(B) - P(A \cap B).$$ If A and B are mutually exclusive, then $$P(A \cup B) = P(A) + P(B).$$	95
	Find probabilities of independent events *(p. 551).*	If A and B are independent events, then the probability that both A and B will occur is $P(A \text{ and } B) = P(A) \cdot P(B)$.	96

6 Review Exercises

See CalcChat.com for tutorial help and worked-out solutions to odd-numbered exercises.
For instructions on how to use a graphing utility, see Appendix A.

6.1

Writing the Terms of a Sequence In Exercises 1 and 2, write the first five terms of the sequence. (Assume n begins with 1.)

1. $a_n = \dfrac{2^n}{2^n + 1}$ **2.** $a_n = \dfrac{(-1)^n 5n}{2n - 1}$

Finding the nth Term of a Sequence In Exercises 3–6, write an expression for the apparent nth term of the sequence. (Assume n begins with 1.)

3. 7, 12, 17, 22, 27, . . . **4.** 50, 48, 46, 44, 42, . . .

5. $2, \frac{2}{3}, \frac{2}{5}, \frac{2}{7}, \frac{2}{9}, \ldots$ **6.** $\frac{3}{2}, \frac{4}{3}, \frac{5}{4}, \frac{6}{5}, \frac{7}{6}, \ldots$

A Recursive Sequence In Exercises 7 and 8, write the first five terms of the sequence defined recursively.

7. $a_1 = 9,\ a_{k+1} = \frac{1}{3}a_k$ **8.** $a_1 = 49,\ a_{k+1} = a_k + 6$

Simplifying Factorial Expressions In Exercises 9–12, simplify the factorial expression.

9. $\dfrac{18!}{20!}$ **10.** $\dfrac{10!}{8!}$

11. $\dfrac{(n+1)!}{(n-1)!}$ **12.** $\dfrac{2n!}{(n+1)!}$

Sigma Notation for Sums In Exercises 13–16, find the sum.

13. $\displaystyle\sum_{i=1}^{6} 5$ **14.** $\displaystyle\sum_{k=2}^{5} 4k$

15. $\displaystyle\sum_{j=1}^{4} \frac{6}{j^2}$ **16.** $\displaystyle\sum_{i=1}^{8} \frac{i}{i+1}$

Writing a Sum Using Sigma Notation In Exercises 17–20, use sigma notation to write the sum. Then use a graphing utility to find the sum.

17. $\dfrac{1}{2(1)} + \dfrac{1}{2(2)} + \dfrac{1}{2(3)} + \cdots + \dfrac{1}{2(20)}$

18. $3(1^2) + 3(2^2) + 3(3^2) + \cdots + 3(9^2)$

19. $\dfrac{1}{2} + \dfrac{2}{3} + \dfrac{3}{4} + \cdots + \dfrac{9}{10}$

20. $1 - \dfrac{1}{3} + \dfrac{1}{9} - \dfrac{1}{27} + \cdots$

Finding the Sum of a Series In Exercises 21–24, find (a) the fourth partial sum and (b) the sum of the infinite series.

21. $\displaystyle\sum_{j=1}^{\infty} \frac{8}{10^j}$ **22.** $\displaystyle\sum_{j=1}^{\infty} \frac{3}{100^j}$

23. $\displaystyle\sum_{k=1}^{\infty} 4(0.01)^k$ **24.** $\displaystyle\sum_{k=1}^{\infty} 2(0.1)^k$

25. Compound Interest A deposit of \$3000 is made in an account that earns 2% interest compounded quarterly. The balance in the account after n quarters is given by

$$a_n = 3000\left(1 + \frac{0.02}{4}\right)^n; \quad n = 1, 2, 3, \ldots.$$

(a) Compute the first eight terms of this sequence.

(b) Find the balance in this account after 10 years by computing the 40th term of the sequence.

26. Education The average salaries a_n (in thousands of dollars) of college faculty members in the United States from 2001 through 2012 can be approximated by the model

$$a_n = -0.028n^2 + 2.31n + 56.7; \quad n = 1, 2, 3, \ldots, 12$$

where n is the year, with $n = 1$ corresponding to 2001. (Source: American Association of University Professors)

(a) Find the terms of this finite sequence for the given values of n.

(b) Use a graphing utility to graph the sequence for the given values of n.

(c) Use the sequence to predict the average salaries of college faculty members for the years 2013 through 2019. Do your results seem reasonable? Explain.

6.2

Identifying an Arithmetic Sequence In Exercises 27–30, determine whether or not the sequence is arithmetic. If it is, find the common difference.

27. 5, 3, 1, −1, −3, . . . **28.** 0, 1, 3, 6, 10, . . .

29. $\frac{1}{4}, \frac{1}{8}, \frac{1}{16}, \frac{1}{32}, \ldots$ **30.** $\frac{9}{9}, \frac{8}{9}, \frac{7}{9}, \frac{6}{9}, \frac{5}{9}, \ldots$

Writing the Terms of an Arithmetic Sequence In Exercises 31–34, write the first five terms of the arithmetic sequence. Use the *table* feature of a graphing utility to verify your results.

31. $a_1 = 3, \quad d = 4$ **32.** $a_1 = 8, \quad d = -2$

33. $a_4 = 10, \quad a_{10} = 28$ **34.** $a_2 = 14, \quad a_6 = 22$

Writing the Terms of an Arithmetic Sequence In Exercises 35 and 36, write the first five terms of the arithmetic sequence. Find the common difference and write the nth term of the sequence as a function of n.

35. $a_1 = 35, \quad a_{k+1} = a_k - 3$

36. $a_1 = 15, \quad a_{k+1} = a_k + \frac{5}{2}$

Finding the nth Term of an Arithmetic Sequence In Exercises 37 and 38, find a formula for a_n for the arithmetic sequence and find the sum of the first 25 terms of the sequence.

37. $a_1 = 100,\quad d = -3$

38. $a_1 = 10,\quad a_3 = 28$

Finding the Sum of a Finite Arithmetic Sequence In Exercises 39–42, find the sum of the finite arithmetic sequence. Use a graphing utility to verify your result.

39. $\sum_{j=1}^{11} (2.5j - 5)$

40. $\sum_{k=1}^{11} \left(\frac{2}{3}k + 4\right)$

41. Sum of the first 50 positive multiples of 5

42. Sum of the integers from 10 to 70

43. Accounting The starting salary for an accountant is \$53,300 with a guaranteed salary increase of \$1066 per year. Determine (a) the salary during the fifth year and (b) the total compensation through 5 full years of employment.

44. Agriculture In his first trip around a field, a farmer makes 123 bales. In his second trip, he makes 11 fewer bales. Because each trip is shorter than the preceding trip, the farmer estimates that the same pattern will continue. Estimate the total number of bales made when there are another five trips around the field.

6.3

Identifying a Geometric Sequence In Exercises 45–48, determine whether or not the sequence is geometric. If it is, find the common ratio.

45. $5, 10, 20, 40, \ldots$

46. $54, -18, 6, -2, \ldots$

47. $\frac{1}{2}, \frac{2}{3}, \frac{3}{4}, \frac{4}{5}, \ldots$

48. $\frac{1}{3}, \frac{2}{3}, 1, \frac{4}{3}, \ldots$

Writing the Terms of a Geometric Sequence In Exercises 49–52, write the first five terms of the geometric sequence.

49. $a_1 = 4,\quad r = -\frac{1}{4}$

50. $a_1 = 2,\quad r = \frac{3}{2}$

51. $a_1 = 9,\quad a_2 = 6$

52. $a_1 = 2,\quad a_2 = 12$

Finding the nth Term of a Geometric Sequence In Exercises 53 and 54, write the first five terms of the geometric sequence. Find the common ratio and write the nth term of the sequence as a function of n.

53. $a_1 = 120, a_{k+1} = \frac{1}{3}a_k$

54. $a_1 = 200, a_{k+1} = 0.1a_k$

Finding a Term of a Geometric Sequence In Exercises 55 and 56, find the indicated term of the geometric sequence (a) algebraically and (b) using the *table* feature of a graphing utility.

55. $a_1 = 16,\quad a_2 = -8,\quad$ 6th term

56. $a_3 = 6,\quad a_4 = 1,\quad$ 9th term

Finding the Sum of a Finite Geometric Sequence In Exercises 57–60, find the sum. Use a graphing utility to verify your result.

57. $\sum_{i=1}^{7} 5^{i-1}$

58. $\sum_{i=1}^{5} 3^{i-1}$

59. $\sum_{n=1}^{7} 3(-4)^{n-1}$

60. $\sum_{n=1}^{4} 12\left(-\frac{1}{2}\right)^{n-1}$

Finding the Sum of an Infinite Geometric Series In Exercises 61–64, find the sum of the infinite geometric series, if possible. If not possible, explain why.

61. $\sum_{i=1}^{\infty} 4\left(\frac{7}{8}\right)^{i-1}$

62. $\sum_{i=1}^{\infty} 6\left(\frac{1}{3}\right)^{i-1}$

63. $\sum_{k=1}^{\infty} 4\left(\frac{3}{2}\right)^{k-1}$

64. $\sum_{k=1}^{\infty} 1.3\left(\frac{1}{10}\right)^{k-1}$

65. Finance A company buys a fleet of six vans for \$130,000. During the next 5 years, the fleet will depreciate at a rate of 30% per year. (That is, at the end of each year, the depreciated value will be 70% of the value at the beginning of the year.)

(a) Find the formula for the nth term of a geometric sequence that gives the value of the fleet t full years after it was purchased.

(b) Find the depreciated value of the fleet at the end of 5 full years.

66. Finance A deposit of \$80 is made at the beginning of each month in an account that pays 4% interest, compounded monthly. The balance A in the account at the end of 4 years is given by

$$A = 80\left(1 + \frac{0.04}{12}\right)^1 + \cdots + 80\left(1 + \frac{0.04}{12}\right)^{48}.$$

Find A.

6.4

Finding Binomial Coefficients In Exercises 67–70, find the binomial coefficient. Use a graphing utility to verify your result.

67. ${}_{10}C_8$

68. ${}_{12}C_5$

69. ${}_{9}C_4$

70. ${}_{14}C_{12}$

Expanding a Binomial In Exercises 71–74, use the Binomial Theorem to expand and simplify the expression.

71. $(x + 5)^4$

72. $(y - 3)^3$

73. $(a - 4b)^5$

74. $(3x + y)^7$

Using Pascal's Triangle In Exercises 75–78, use Pascal's Triangle to find the binomial coefficient.

75. ${}_{7}C_2$

76. ${}_{9}C_7$

77. ${}_{8}C_4$

78. ${}_{10}C_5$

6.5

79. Probability Slips of paper numbered 1 through 15 are placed in a hat. In how many ways can two numbers be drawn so that the sum of the numbers is 12? Assume the random selection is without replacement.

80. Fundamental Counting Principle A customer can choose one of six speaker systems, one of five DVD players, and one of six flat screen televisions to design a home entertainment system. Determine the number of possible system configurations.

81. Education A college student is preparing a course schedule of four classes for the next semester. The student can choose from the open sections shown in the table.

Course	Sections
Math 100	001–004
Economics 110	001–003
English 105	001–006
Humanities 101	001–003

(a) Find the number of possible schedules that the student can create from the offerings.

(b) Find the number of possible schedules that the student can create from the offerings when two of the Math 100 sections are closed.

(c) Find the number of possible schedules that the student can create from the offerings when two of the Math 100 sections and four of the English 105 sections are closed.

82. Marketing A telemarketing firm is making calls to prospective customers by randomly dialing seven-digit phone numbers within an area code.

(a) Find the number of possible calls that the telemarketer can make.

(b) The telemarketing firm is calling only within exchanges that begin with a "7" or a "6." How many different calls are possible?

Evaluating a Permutation **In Exercises 83 and 84, evaluate $_nP_r$ using the formula from Section 6.5. Use a graphing utility to verify your result.**

83. $_{12}P_5$ **84.** $_6P_4$

Distinguishable Permutations **In Exercises 85 and 86, find the number of distinguishable permutations of the group of letters.**

85. C, A, L, C, U, L, U, S **86.** I, N, T, E, G, R, A, T, E

Solving an Equation **In Exercises 87 and 88, solve for n.**

87. $_{n+1}P_2 = 4 \cdot {_nP_1}$ **88.** $8 \cdot {_nP_2} = {_{n+1}P_3}$

Evaluating a Combination **In Exercises 89 and 90, evaluate $_nC_r$ using the formula from Section 6.5. Use a graphing utility to verify your result.**

89. $_8C_6$ **90.** $_{50}C_{48}$

91. Athletics There are 12 bicyclists entered in a race. In how many different orders could the 12 bicyclists finish? (Assume there are no ties.)

92. Athletics From a pool of seven juniors and eleven seniors, four co-captains will be chosen for the football team. How many different combinations are possible if two juniors and two seniors are to be chosen?

6.6

93. Random Selection A man has five pairs of socks (no two pairs are the same color). He randomly selects two socks from a drawer. What is the probability that he gets a matched pair?

94. Library Science A child returns a five-volume set of books to a bookshelf. The child is not able to read and so cannot distinguish one volume from another. What is the probability that the books are shelved in the correct order?

95. Education A sample of college students, faculty members, and administrators were asked whether they favored a proposed increase in the annual activity fee to enhance student life on campus. The results of the study are shown in the table.

	Favor	Oppose	Total
Students	237	163	400
Faculty	37	38	75
Admin.	18	7	25
Total	292	208	500

A person is selected at random from the sample. Find the probability that the person selected is (a) a person who does not favor the proposal, (b) a student, and (c) a faculty member who favors the proposal.

96. Probability A six-sided die is rolled six times. What is the probability that each side appears exactly once?

Conclusions

True or False? **In Exercises 97 and 98, determine whether the statement is true or false. Justify your answer.**

97. $\dfrac{(n+2)!}{n!} = (n+2)(n+1)$ **98.** $\sum_{j=1}^{6} 3^j = \sum_{j=3}^{8} 3^{j-2}$

99. Writing In your own words, explain what makes a sequence (a) arithmetic and (b) geometric.

6 Chapter Test

See *CalcChat.com* for tutorial help and worked-out solutions to odd-numbered exercises.
For instructions on how to use a graphing utility, see Appendix A.

Take this test as you would take a test in class. After you are finished, check your work against the answers given in the back of the book.

In Exercises 1–4, write the first five terms of the sequence.

1. $a_n = 3\left(\frac{2}{3}\right)^{n-1}$ (Begin with $n = 1$.)

2. $a_1 = 12$ and $a_{k+1} = a_k + 4$

3. $b_n = \dfrac{(-1)^n x^n}{n}$ (Begin with $n = 1$.)

4. $b_n = \dfrac{(-1)^{2n+1} x^{2n+1}}{n!}$ (Begin with $n = 1$.)

In Exercises 5–7, simplify the factorial expression.

5. $\dfrac{11! \cdot 4!}{4! \cdot 7!}$

6. $\dfrac{n!}{(n+1)!}$

7. $\dfrac{2n!}{(n-1)!}$

8. Write an expression for the apparent nth term of the sequence

2, 5, 10, 17, 26, ... (Assume n begins with 1.)

In Exercises 9 and 10, find a formula for the *n*th term of the sequence.

9. Arithmetic: $a_1 = 5000$, $d = -100$

10. Geometric: $a_1 = 4$, $a_{k+1} = \frac{1}{2}a_k$

11. Use sigma notation to write $\dfrac{2}{3(1)+1} + \dfrac{2}{3(2)+1} + \cdot \cdot \cdot + \dfrac{2}{3(12)+1}$.

12. Use sigma notation to write $2 + \frac{1}{2} + \frac{1}{8} + \frac{1}{32} + \frac{1}{128} + \cdot \cdot \cdot$.

In Exercises 13–15, find the sum.

13. $\displaystyle\sum_{n=1}^{8} 49\left(\frac{1}{7}\right)^{n-1}$

14. $\displaystyle\sum_{n=1}^{7} (8n - 5)$

15. $5 - 2 + \frac{4}{5} - \frac{8}{25} + \frac{16}{125} - \cdot \cdot \cdot$

16. Use the Binomial Theorem to expand and simplify $(3a - 5b)^4$.

In Exercises 17–20, evaluate the expression.

17. ${}_9C_3$

18. ${}_{20}C_3$

19. ${}_{70}P_3$

20. ${}_9P_2$

21. Solve for n in $4 \cdot {}_nP_3 = {}_{n+1}P_4$.

22. How many distinct license plates can be issued consisting of one letter followed by a three-digit number?

23. Four students are randomly selected from a class of 25 to answer questions from a reading assignment. In how many ways can the four be selected?

24. A card is drawn from a standard deck of 52 playing cards. Find the probability that it is a face card in the suit of hearts.

25. In 2014, six of the twelve men's basketball teams in the Big Ten Conference participated in the NCAA Men's Basketball Championship Tournament. What is the probability that a random selection of six teams from the Big Ten Conference will be the same as the six teams that participated in the tournament?

26. Two integers from 1 to 60 are chosen by a random number generator. What is the probability that (a) both numbers are odd, (b) both numbers are less than 12, and (c) the same number is chosen twice?

Proofs in Mathematics

Properties of Sums (p. 504)

1. $\sum_{i=1}^{n} c = cn,$ c is a constant.
2. $\sum_{i=1}^{n} ca_i = c\sum_{i=1}^{n} a_i,$ c is a constant.
3. $\sum_{i=1}^{n} (a_i + b_i) = \sum_{i=1}^{n} a_i + \sum_{i=1}^{n} b_i$
4. $\sum_{i=1}^{n} (a_i - b_i) = \sum_{i=1}^{n} a_i - \sum_{i=1}^{n} b_i$

Proof

Each of these properties follows directly from the properties of real numbers.

1. $\sum_{i=1}^{n} c = c + c + c + \cdot \cdot \cdot + c = cn$ n terms

The Distributive Property is used in the proof of Property 2.

2. $$\begin{aligned}\sum_{i=1}^{n} ca_i &= ca_1 + ca_2 + ca_3 + \cdot \cdot \cdot + ca_n\\ &= c(a_1 + a_2 + a_3 + \cdot \cdot \cdot + a_n)\\ &= c\sum_{i=1}^{n} a_i\end{aligned}$$

The proof of Property 3 uses the Commutative and Associative Properties of Addition.

3. $$\begin{aligned}\sum_{i=1}^{n} (a_i + b_i) &= (a_1 + b_1) + (a_2 + b_2) + (a_3 + b_3) + \cdot \cdot \cdot + (a_n + b_n)\\ &= (a_1 + a_2 + a_3 + \cdot \cdot \cdot + a_n) + (b_1 + b_2 + b_3 + \cdot \cdot \cdot + b_n)\\ &= \sum_{i=1}^{n} a_i + \sum_{i=1}^{n} b_i\end{aligned}$$

The proof of Property 4 uses the Commutative and Associative Properties of Addition and the Distributive Property.

4. $$\begin{aligned}\sum_{i=1}^{n} (a_i - b_i) &= (a_1 - b_1) + (a_2 - b_2) + (a_3 - b_3) + \cdot \cdot \cdot + (a_n - b_n)\\ &= (a_1 + a_2 + a_3 + \cdot \cdot \cdot + a_n) + (-b_1 - b_2 - b_3 - \cdot \cdot \cdot - b_n)\\ &= (a_1 + a_2 + a_3 + \cdot \cdot \cdot + a_n) - (b_1 + b_2 + b_3 + \cdot \cdot \cdot + b_n)\\ &= \sum_{i=1}^{n} a_i - \sum_{i=1}^{n} b_i\end{aligned}$$

■

Infinite Series

The study of infinite series was considered a novelty in the fourteenth century. Logician Richard Suiseth, whose nickname was Calculator, solved this problem.

> *If throughout the first half of a given time interval a variation continues at a certain intensity; throughout the next quarter of the interval at double the intensity; throughout the following eighth at triple the intensity and so ad infinitum; The average intensity for the whole interval will be the intensity of the variation during the second subinterval (or double the intensity).*

This is the same as saying that the sum of the infinite series

$$\frac{1}{2} + \frac{2}{4} + \frac{3}{8} + \cdot \cdot \cdot + \frac{n}{2^n} + \cdot \cdot \cdot$$

is 2.

The Sum of a Finite Arithmetic Sequence (p. 514)

The sum of a finite arithmetic sequence with n terms is given by

$$S_n = \frac{n}{2}(a_1 + a_n).$$

Proof

Begin by generating the terms of the arithmetic sequence in two ways. In the first way, repeatedly add d to the first term to obtain

$$\begin{aligned} S_n &= a_1 + a_2 + a_3 + \cdot \cdot \cdot + a_{n-2} + a_{n-1} + a_n \\ &= a_1 + [a_1 + d] + [a_1 + 2d] + \cdot \cdot \cdot + [a_1 + (n-1)d]. \end{aligned}$$

In the second way, repeatedly subtract d from the nth term to obtain

$$\begin{aligned} S_n &= a_n + a_{n-1} + a_{n-2} + \cdot \cdot \cdot + a_3 + a_2 + a_1 \\ &= a_n + [a_n - d] + [a_n - 2d] + \cdot \cdot \cdot + [a_n - (n-1)d]. \end{aligned}$$

When you add these two versions of S_n, the multiples of d subtract out and you obtain

$$2S_n = (a_1 + a_n) + (a_1 + a_n) + (a_1 + a_n) + \cdot \cdot \cdot + (a_1 + a_n) \qquad n \text{ terms}$$

$$2S_n = n(a_1 + a_n)$$

$$S_n = \frac{n}{2}(a_1 + a_n).$$

The Sum of a Finite Geometric Sequence (p. 522)

The sum of the finite geometric sequence

$$a_1, \quad a_1r, \quad a_1r^2, \quad a_1r^3, \quad a_1r^4, \quad . . . , \quad a_1r^{n-1}$$

with common ratio $r \neq 1$ is given by

$$S_n = \sum_{i=1}^{n} a_1r^{i-1} = a_1\left(\frac{1 - r^n}{1 - r}\right).$$

Proof

$$S_n = a_1 + a_1r + a_1r^2 + \cdot \cdot \cdot + a_1r^{n-2} + a_1r^{n-1}$$

$$rS_n = a_1r + a_1r^2 + a_1r^3 + \cdot \cdot \cdot + a_1r^{n-1} + a_1r^n \qquad \text{Multiply by } r.$$

Subtracting the second equation from the first yields

$$S_n - rS_n = a_1 - a_1r^n.$$

So, $S_n(1 - r) = a_1(1 - r^n)$, and, because $r \neq 1$, you have

$$S_n = a_1\left(\frac{1 - r^n}{1 - r}\right).$$

The Binomial Theorem (p. 529)

In the expansion of $(x + y)^n$

$$(x + y)^n = x^n + nx^{n-1}y + \cdots + {}_nC_r x^{n-r}y^r + \cdots + nxy^{n-1} + y^n$$

the coefficient of $x^{n-r}y^r$ is

$$ {}_nC_r = \frac{n!}{(n-r)!r!}. $$

Proof

The Binomial Theorem can be proved quite nicely using **mathematical induction.** (For information about mathematical induction, see Appendix F at this textbook's *Companion Website*.) The steps are straightforward but look a little messy, so only an outline of the proof is presented.

1. For $n = 1$, you have

   $$(x + y)^1 = x^1 + y^1 = {}_1C_0x + {}_1C_1y$$

   and the formula is valid.

2. Assuming that the formula is true for $n = k$, the coefficient of $x^{k-r}y^r$ is

   $$ {}_kC_r = \frac{k!}{(k-r)!r!} = \frac{k(k-1)(k-2)\cdots(k-r+1)}{r!}. $$

   To show that the formula is true for $n = k + 1$, look at the coefficient of $x^{k+1-r}y^r$ in the expansion of

   $$(x + y)^{k+1} = (x + y)^k(x + y).$$

   From the right-hand side, you can determine that the term involving $x^{k+1-r}y^r$ is the sum of two products.

   $$\begin{aligned}
   &({}_kC_r x^{k-r}y^r)(x) + ({}_kC_{r-1}x^{k+1-r}y^{r-1})(y)\\
   &= \left[\frac{k!}{(k-r)!r!} + \frac{k!}{(k+1-r)!(r-1)!}\right]x^{k+1-r}y^r\\
   &= \left[\frac{(k+1-r)k!}{(k+1-r)!r!} + \frac{k!r}{(k+1-r)!r!}\right]x^{k+1-r}y^r\\
   &= \left[\frac{k!(k+1-r+r)}{(k+1-r)!r!}\right]x^{k+1-r}y^r\\
   &= \left[\frac{(k+1)!}{(k+1-r)!r!}\right]x^{k+1-r}y^r\\
   &= {}_{k+1}C_r x^{k+1-r}y^r
   \end{aligned}$$

So, by mathematical induction, the Binomial Theorem is valid for all positive integers n. ■

The Principle of Mathematical Induction

Let P_n be a statement involving the positive integer n. If

1. P_1 is true, and
2. the truth of P_k implies the truth of P_{k+1} for every positive integer k,

then P_n must be true for all positive integers n.

Appendix A: Technology Support

Introduction

Graphing utilities, such as graphing calculators and computers with graphing software, are very valuable tools for visualizing mathematical principles, verifying solutions of equations, exploring mathematical ideas, and developing mathematical models. Although graphing utilities are extremely helpful in learning mathematics, their use does not mean that learning algebra is any less important. In fact, the combination of knowledge of mathematics and the use of graphing utilities enables you to explore mathematics more easily and to a greater depth. If you are using a graphing utility in this course, it is up to you to learn its capabilities and to practice using this tool to enhance your mathematical learning.

In this text, there are many opportunities to use a graphing utility, some of which are described below.

Uses of a Graphing Utility

1. Check or validate answers to problems obtained using algebraic methods.
2. Discover and explore algebraic properties, rules, and concepts.
3. Graph functions, and approximate solutions of equations involving functions.
4. Efficiently perform complicated mathematical procedures, such as those found in many real-life applications.
5. Find mathematical models for sets of data.

In this appendix, the features of graphing utilities are discussed from a generic perspective and are listed in alphabetical order. To learn how to use the features of a specific graphing utility, consult your user's manual or go to this textbook's *Companion Website*. Additional keystroke guides are available for most graphing utilities, and your college library may have a resource on how to use your graphing utility.

Many graphing utilities are designed to act as "function graphers." In this course, functions and their graphs are studied in detail. You may recall from previous courses that a function can be thought of as a rule that describes the relationship between two variables. These rules are frequently written in terms of x and y. For instance, the equation

$$y = 3x + 5$$

represents y as a function of x.

Many graphing utilities have an *equation editor* feature that requires that an equation be written in "$y =$ " form in order to be entered, as shown in Figure A.1. (You should note that your *equation editor* screen may not look like the screen shown in Figure A.1.)

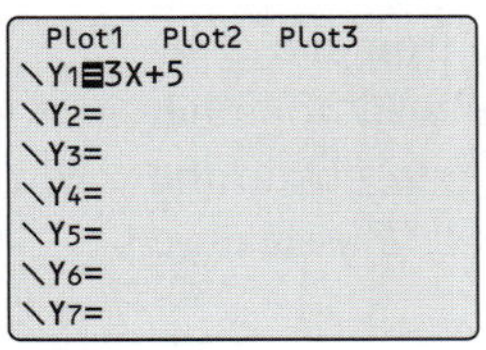

Figure A.1

Cumulative Sum Feature

The *cumulative sum* feature finds partial sums of a series. For instance, to find the first four partial sums of the series

$$\sum_{k=1}^{4} 2(0.1)^k$$

choose the *cumulative sum* feature, which is found in the *operations* menu of the *list* feature (see Figure A.2). To use this feature, you will also have to use the *sequence* feature (see Figure A.2 and page A15). You must enter an expression for the sequence, a variable, the lower limit of summation, and the upper limit of summation, as shown in Figure A.3. After pressing ENTER, you can see that the first four partial sums are 0.2, 0.22, 0.222, and 0.2222. You may have to scroll to the right in order to see all the partial sums.

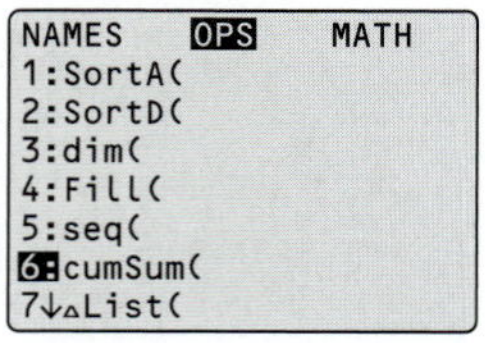

Figure A.2

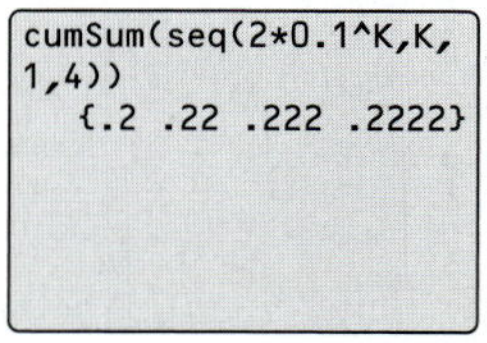

Figure A.3

Technology Tip

As you use your graphing utility, be aware of how parentheses are inserted in an expression. Some graphing utilities automatically insert the left parenthesis when certain calculator buttons are pressed. The placement of parentheses can make a difference between a correct answer and an incorrect answer.

Determinant Feature

The *determinant* feature evaluates the determinant of a square matrix. For instance, to evaluate the determinant of

$$A = \begin{bmatrix} 7 & -1 & 0 \\ 2 & 2 & 3 \\ -6 & 4 & 1 \end{bmatrix}$$

enter the 3×3 matrix in the graphing utility using the *matrix editor*, as shown in Figure A.4. Then choose the *determinant* feature from the *math* menu of the *matrix* feature, as shown in Figure A.5. Once you choose the matrix name, A, press ENTER and you should obtain a determinant of -50, as shown in Figure A.6.

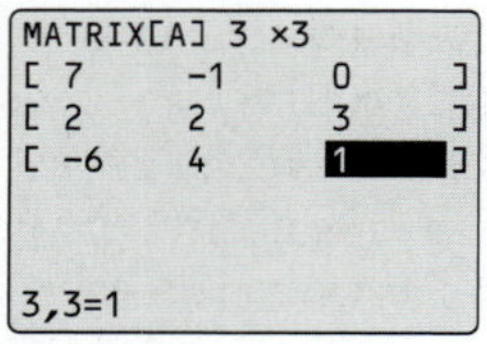

Figure A.4

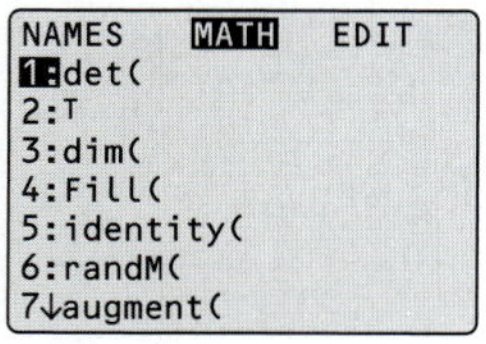

Figure A.5

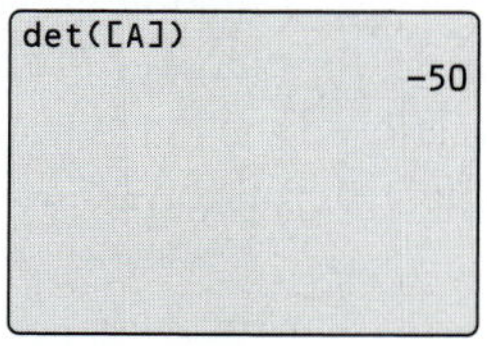

Figure A.6

Draw Inverse Feature

The *draw inverse* feature graphs the inverse function of a *one-to-one* function. For instance, to graph the inverse function of $f(x) = x^3 + 4$, first enter the function in the *equation editor* (see Figure A.7) and graph the function (using a square viewing window), as shown in Figure A.8. Then choose the *draw inverse* feature from the *draw* feature menu, as shown in Figure A.9. You must enter the function you want to graph the inverse function of, as shown in Figure A.10. Finally, press ENTER to obtain the graph of the inverse function of $f(x) = x^3 + 4$, as shown in Figure A.11. This feature can be used only when the graphing utility is in *function* mode.

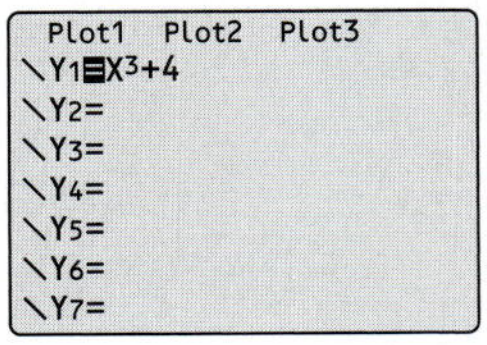

Figure A.7

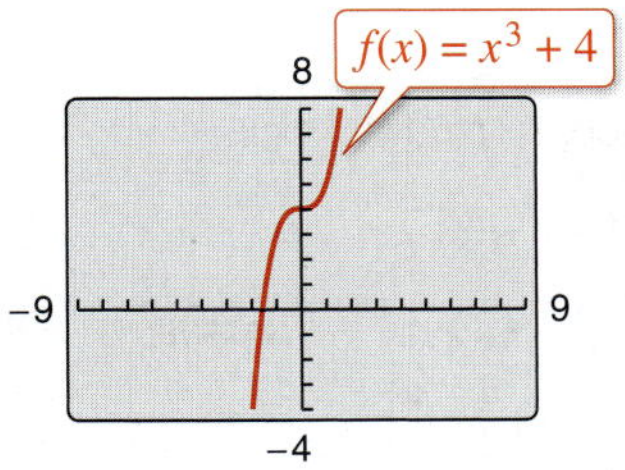

Figure A.8

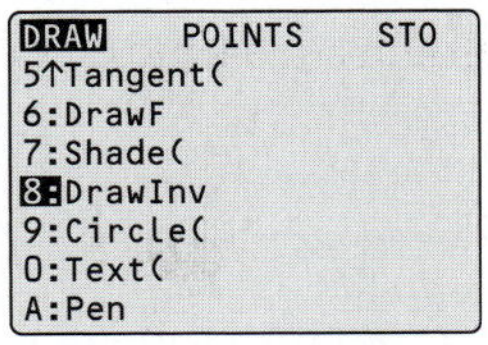

Figure A.9

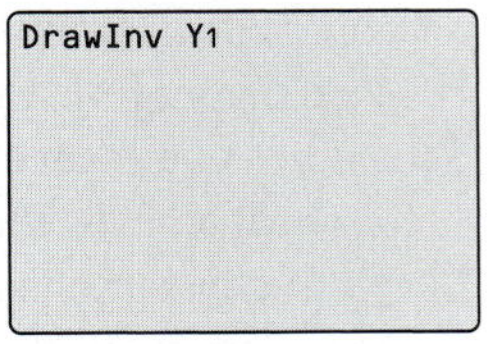

Figure A.10

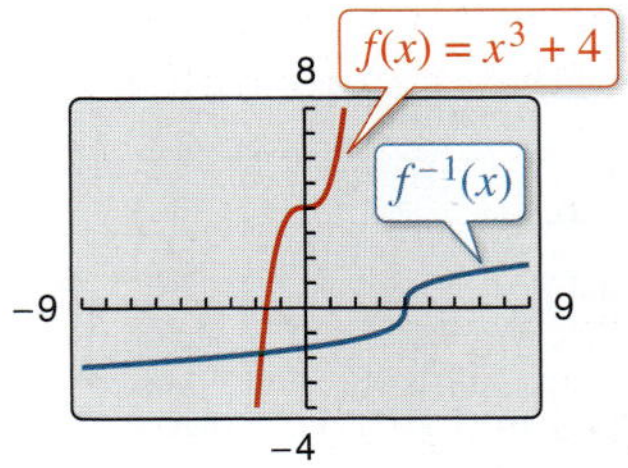

Figure A.11

Elementary Row Operations Features

Most graphing utilities can perform elementary row operations on matrices.

Row Swap Feature

The *row swap* feature interchanges two rows of a matrix. To interchange rows 1 and 3 of the matrix

$$A = \begin{bmatrix} -1 & -2 & 1 & 2 \\ 2 & -4 & 6 & -2 \\ 1 & 3 & -3 & 0 \end{bmatrix}$$

first enter the matrix in the graphing utility using the *matrix editor*, as shown in Figure A.12. Then choose the *row swap* feature from the *math* menu of the *matrix* feature, as shown in Figure A.13. When using this feature, you must enter the name of the matrix and the two rows that are to be interchanged. After pressing ENTER, you should obtain the matrix shown in Figure A.14. Because the resulting matrix will be used to demonstrate the other elementary row operation features, use the *store* feature to copy the resulting matrix to [A], as shown in Figure A.15.

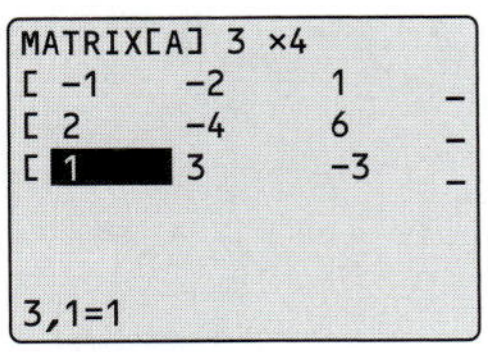

Figure A.12

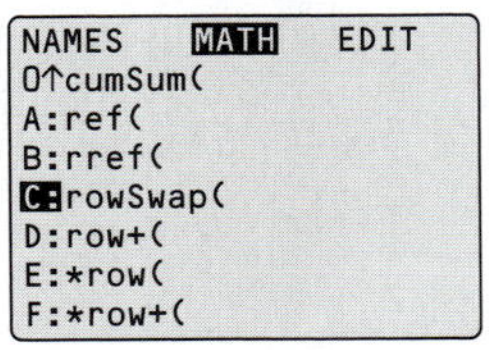

Figure A.13

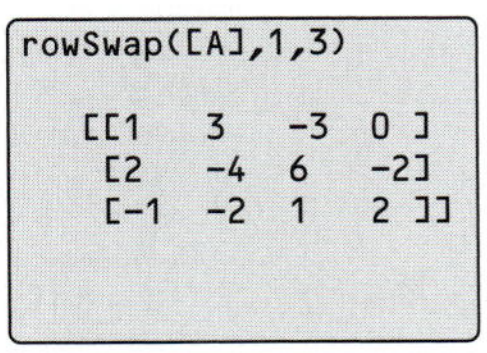

Figure A.14

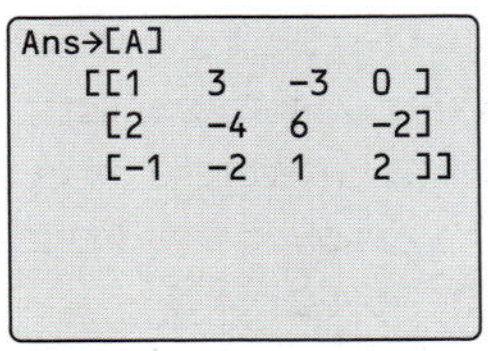

Figure A.15

Technology Tip

The *store* feature of a graphing utility is used to store a value in a variable or to copy one matrix to another matrix. For instance, as shown at the left, after performing a row operation on a matrix, you can copy the answer to another matrix (see Figure A.15). You can then perform another row operation on the copied matrix. To continue performing row operations to obtain a matrix in row-echelon form or reduced row-echelon form, you must copy the resulting matrix to a new matrix before each operation.

Row Addition and Row Multiplication and Addition Features

The *row addition* and *row multiplication and addition* features add a row or a multiple of a row of a matrix to another row of the same matrix. To add row 1 to row 3 of the matrix stored in [A], choose the *row addition* feature from the *math* menu of the *matrix* feature, as shown in Figure A.16. When using this feature, you must enter the name of the matrix and the two rows that are to be added. After pressing ENTER, you should obtain the matrix shown in Figure A.17. Copy the resulting matrix to [A].

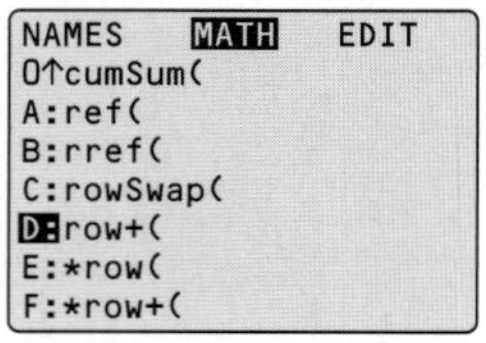

Figure A.16

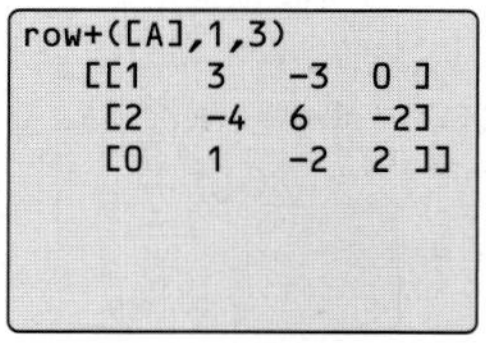

Figure A.17

To add −2 times row 1 to row 2 of the matrix stored in [A], choose the *row multiplication and addition* feature from the *math* menu of the *matrix* feature, as shown in Figure A.18. When using this feature, you must enter the constant, the name of the matrix, the row the constant is multiplied by, and the row to be added to. After pressing ENTER, you should obtain the matrix shown in Figure A.19. Copy the resulting matrix to [A].

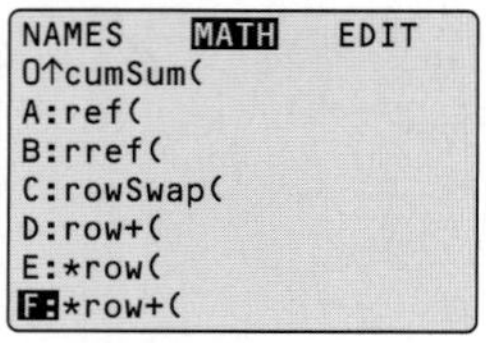

Figure A.18

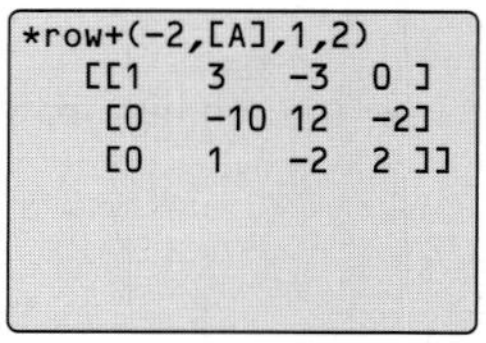

Figure A.19

Row Multiplication Feature

The *row multiplication* feature multiplies a row of a matrix by a nonzero constant. To multiply row 2 of the matrix stored in [A] by

$$-\frac{1}{10}$$

choose the *row multiplication* feature from the *math* menu of the *matrix* feature, as shown in Figure A.20. When using this feature, you must enter the constant, the name of the matrix, and the row to be multiplied. After pressing ENTER, you should obtain the matrix shown in Figure A.21.

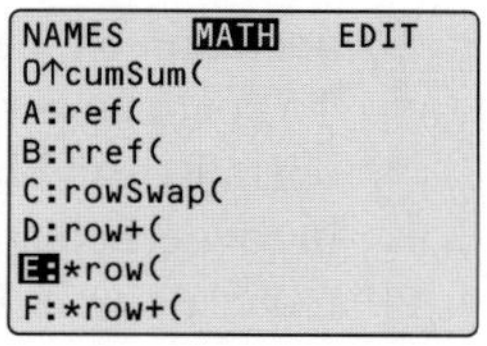

Figure A.20

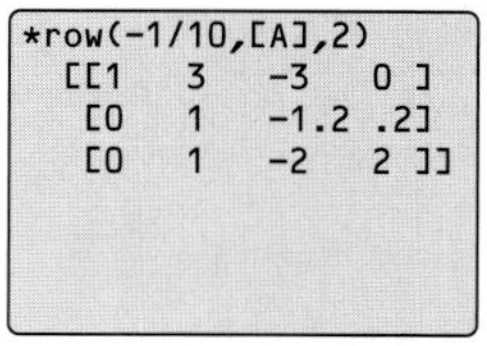

Figure A.21

Intersect Feature

The *intersect* feature finds the point(s) of intersection of two graphs. The *intersect* feature is found in the *calculate* menu (see Figure A.22).

To find the point(s) of intersection of the graphs of $y_1 = -x + 2$ and $y_2 = x + 4$, first enter the equations in the *equation editor*, as shown in Figure A.23. Then graph the equations, as shown in Figure A.24. Next, use the *intersect* feature to find the point of intersection. Trace the cursor along the graph of y_1 near the intersection and press ENTER (see Figure A.25). Then trace the cursor along the graph of y^2 near the intersection and press ENTER (see Figure A.26). Marks are then placed on the graph at these points (see Figure A.27). Finally, move the cursor near the point of intersection and press ENTER. In Figure A.28, you can see that the coordinates of the point of intersection are displayed at the bottom of the window. So, the point of intersection is $(-1, 3)$.

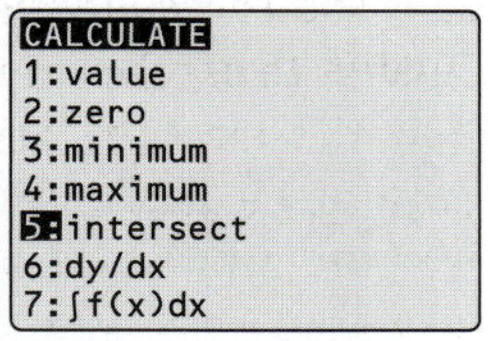

Figure A.22

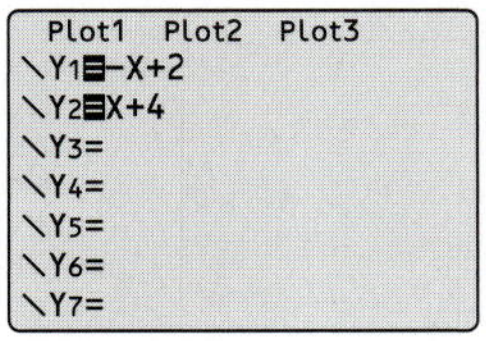

Figure A.23

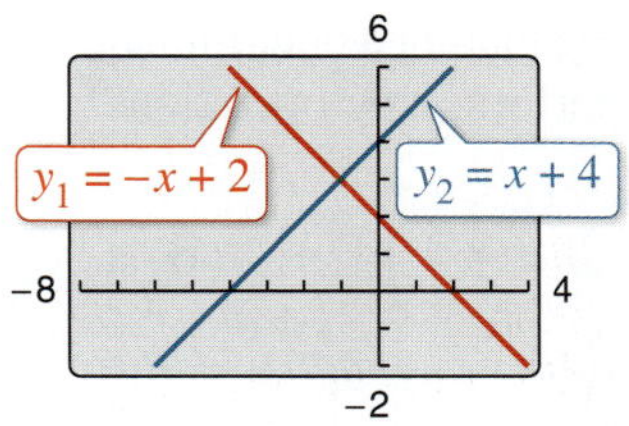

Figure A.24

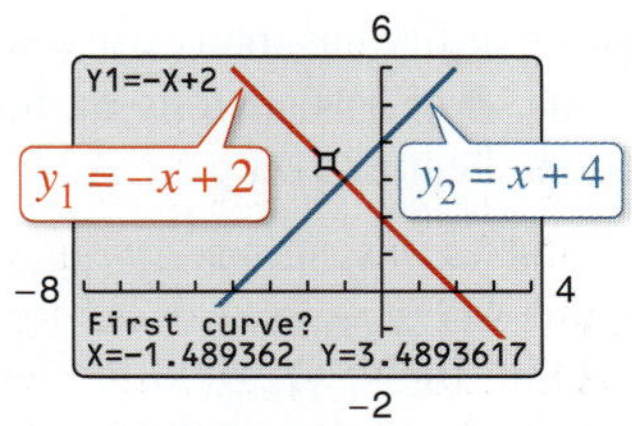

Figure A.25

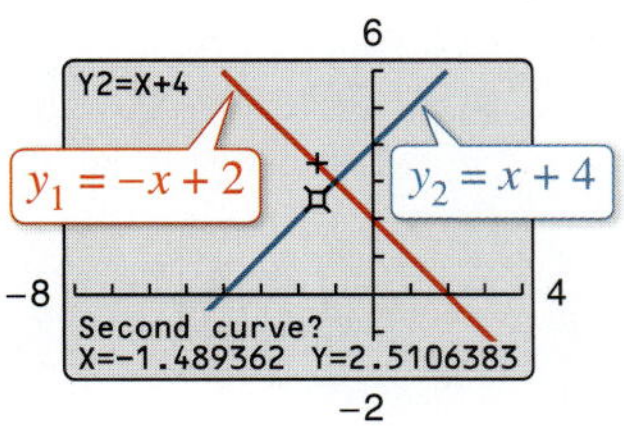

Figure A.26

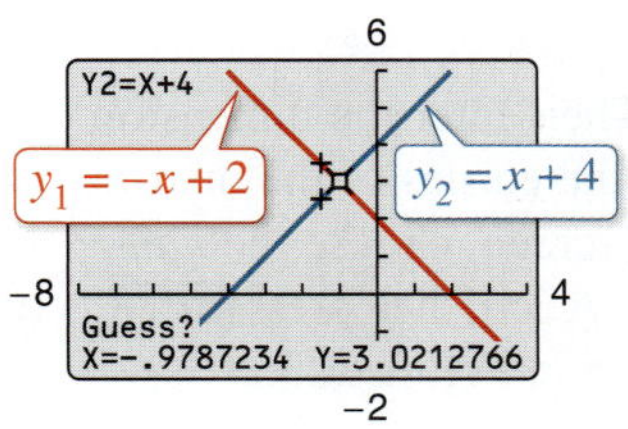

Figure A.27

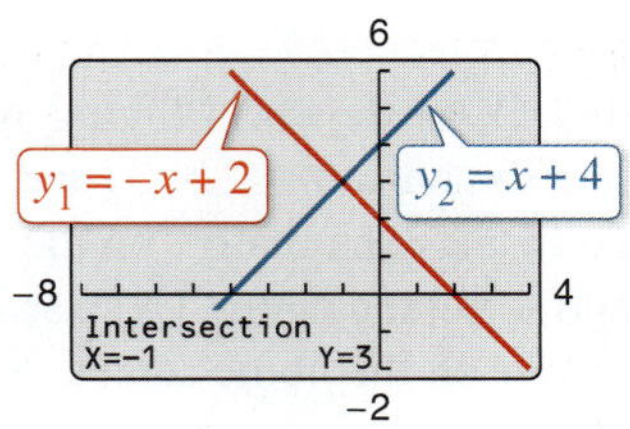

Figure A.28

List Editor

Most graphing utilities can store data in lists. The *list editor* can be used to create tables and to store statistical data. The *list editor* can be found in the *edit* menu of the *statistics* feature, as shown in Figure A.29. To enter the numbers 1 through 10 in a list, first choose a list (L_1) and then begin entering the data into each row, as shown in Figure A.30.

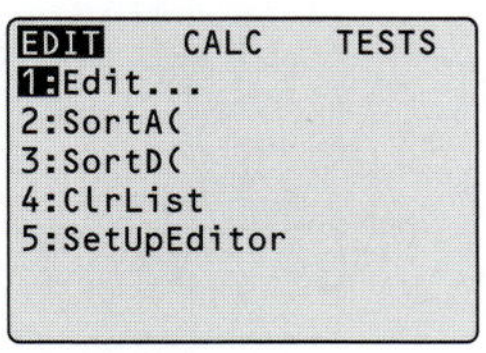

Figure A.29

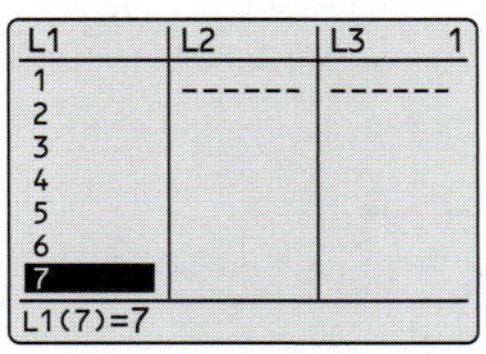

Figure A.30

You can also attach a formula to a list. For instance, you can multiply each of the data values in L_1 by 3. First, display the *list editor* and move the cursor to the top line.

Then move the cursor onto the list to which you want to attach the formula (L_2). Finally, enter the formula $3*L_1$ (see Figure A.31) and then press ENTER. You should obtain the list shown in Figure A.32.

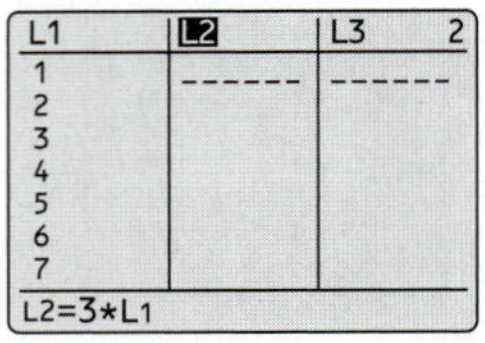

Figure A.31

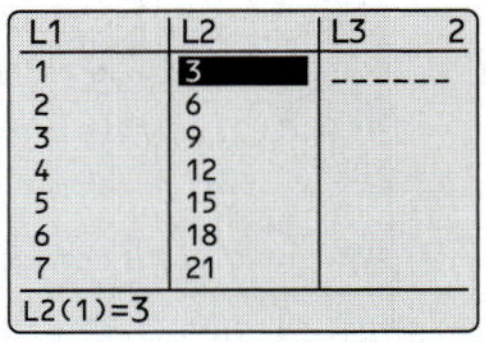

Figure A.32

Matrix Feature

The *matrix* feature of a graphing utility has many uses, such as evaluating a determinant and performing row operations. (For instructions on how to perform row operations, see Elementary Row Operations Features on page A3.)

Matrix Editor

You can define, display, and edit matrices using the *matrix editor*. The *matrix editor* can be found in the *edit* menu of the *matrix* feature. For instance, to enter the matrix

$$A = \begin{bmatrix} 6 & -3 & 4 \\ 9 & 0 & -1 \end{bmatrix}$$

first choose the matrix name [A], as shown in Figure A.33. Then enter the dimension of the matrix (in this case, the dimension is 2×3) and enter the entries of the matrix, as shown in Figure A.34. To display the matrix on the home screen, choose the *name* menu of the *matrix* feature and select the matrix [A] (see Figure A.35), then press ENTER. The matrix A should now appear on the home screen, as shown in Figure A.36.

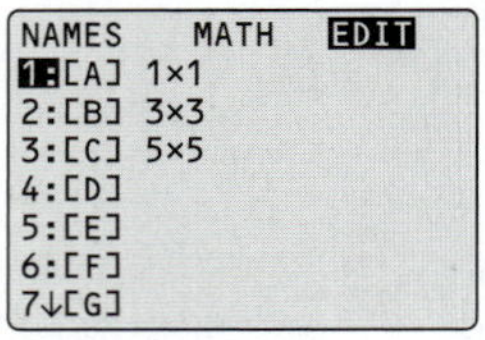

Figure A.33

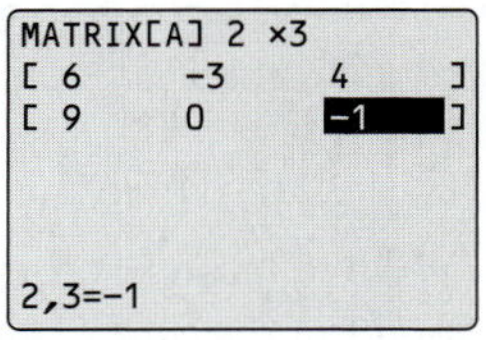

Figure A.34

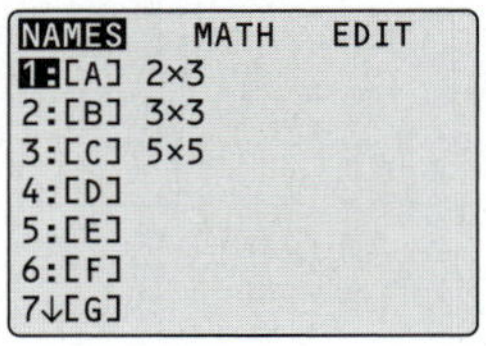

Figure A.35

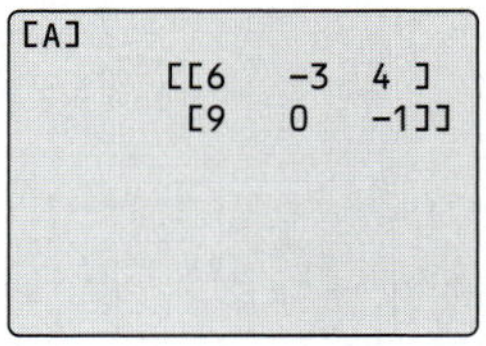

Figure A.36

Matrix Operations

Most graphing utilities can perform matrix operations. To find the sum $A + B$ of the matrices shown at the right, first enter the matrices in the *matrix editor* as [A] and [B]. Then find the sum, as shown in Figure A.37. Scalar multiplication can be performed in a similar manner. For instance, you can evaluate $7A$, where A is the matrix at the right, as shown in Figure A.38. To find the product AB of the matrices A and B at the right, first be sure that the product is defined. Because the number of columns of A (2 columns) equals the number of rows of B (2 rows), you can find the product AB, as shown in Figure A.39.

$$A = \begin{bmatrix} -3 & 5 \\ 0 & 4 \end{bmatrix}$$

$$B = \begin{bmatrix} 7 & -2 \\ -1 & 2 \end{bmatrix}$$

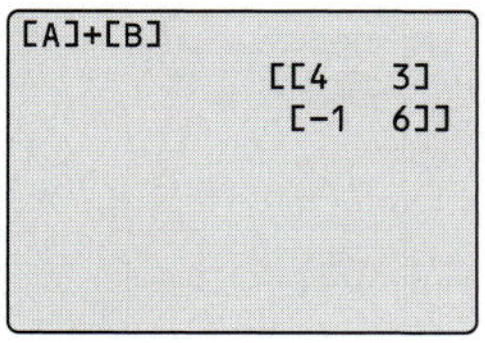

Figure A.37

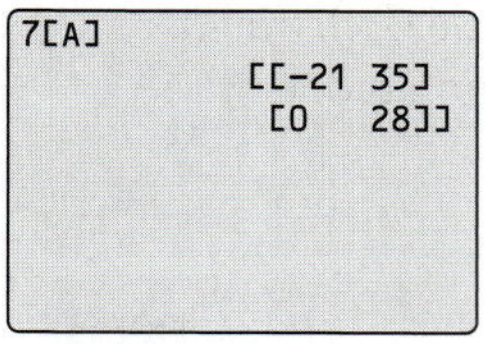

Figure A.38

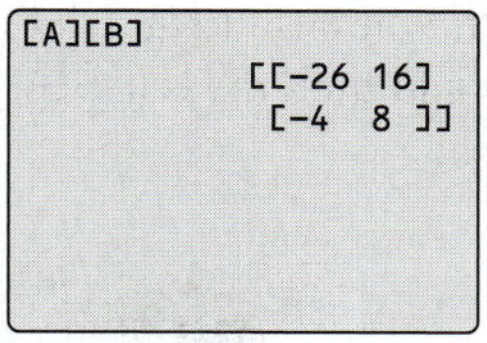

Figure A.39

Inverse Matrix

Some graphing utilities may not have an *inverse matrix* feature. However, you can find the inverse of a square matrix by using the inverse key $\boxed{x^{-1}}$. To find the inverse of the matrix

$$A = \begin{bmatrix} 1 & -2 & 1 \\ -1 & 3 & 0 \\ 2 & 4 & 5 \end{bmatrix}$$

enter the matrix in the *matrix editor* as [A]. Then find the inverse, as shown in Figure A.40.

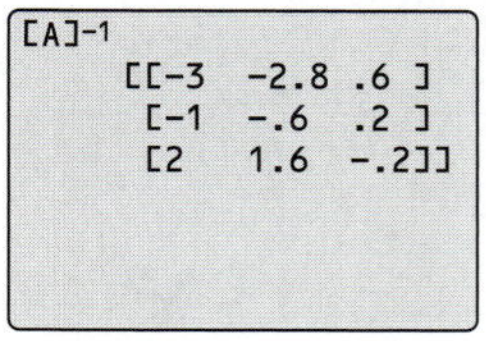

Figure A.40

Maximum and Minimum Features

The *maximum* and *minimum* features find relative extrema of a function. For instance, the graph of

$$y = x^3 - 3x$$

is shown in Figure A.41. In the figure, the graph appears to have a relative maximum at $x = -1$ and a relative minimum at $x = 1$. To find the exact values of the relative extrema, you can use the *maximum* and *minimum* features found in the *calculate* menu (see Figure A.42). First, to find the relative maximum, choose the *maximum* feature and trace the cursor along the graph to a point left of the maximum and press ENTER (see Figure A.43). Then trace the cursor along the graph to a point right of the maximum and press ENTER (see Figure A.44). Note the two arrows near the top of the display marking the left and right bounds, as shown in Figure A.45. Next, trace the cursor along the graph between the two bounds and as close to the maximum as you can (see Figure A.45) and press ENTER. In Figure A.46, you can see that the coordinates of the maximum point are displayed at the bottom of the window. So, the relative maximum occurs at the point $(-1, 2)$.

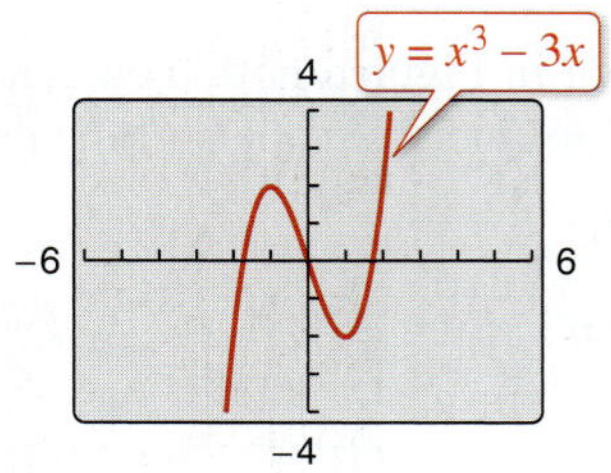

Figure A.41

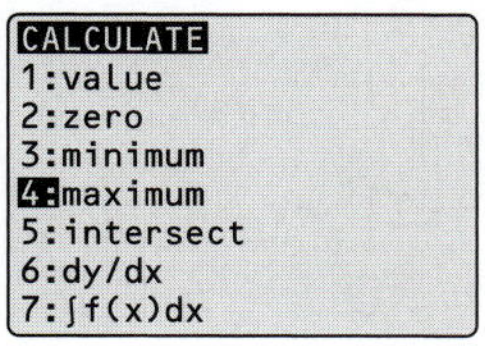

Figure A.42

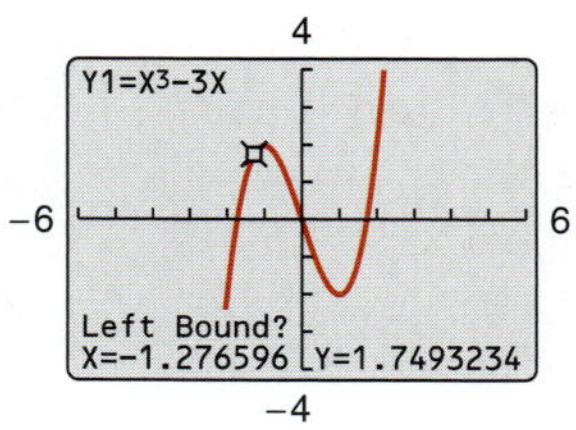

Figure A.43

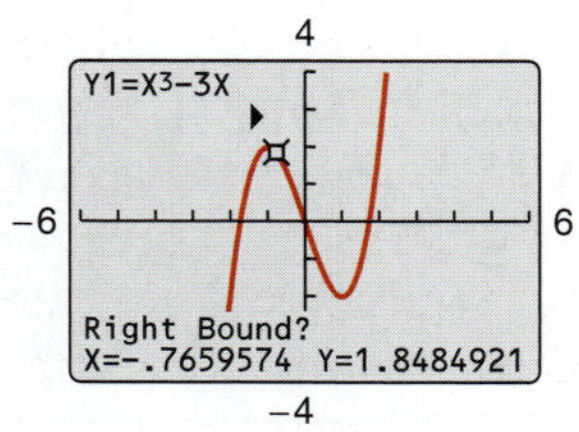

Figure A.44

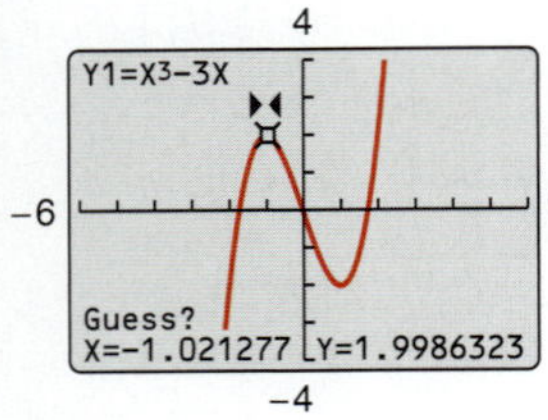

Figure A.45

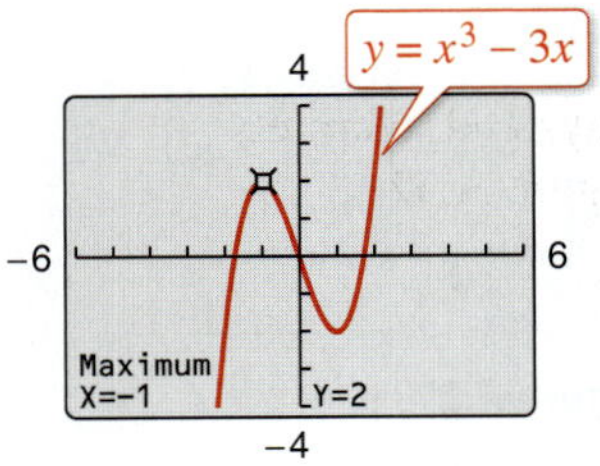

Figure A.46

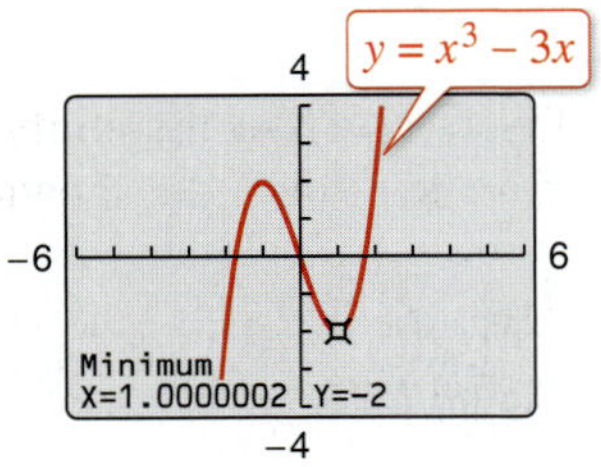

Figure A.47

You can find the relative minimum in a similar manner. In Figure A.47, you can see that the relative minimum occurs at the point $(1, -2)$.

Mean and Median Features

In real-life applications, you often encounter large data sets and want to calculate statistical values. The *mean* and *median* features calculate the mean and median of a data set. For instance, in a survey, 100 people were asked how much money (in dollars) per week they withdraw from an automatic teller machine (ATM). The results are shown in the table below. The frequency represents the number of responses.

Amount	10	20	30	40	50	60	70	80	90	100
Frequency	3	8	10	19	24	13	13	7	2	1

To find the mean and median of the data set, first enter the data in the *list editor*, as shown in Figure A.48. Enter the amount in L_1 and the frequency in L_2. Then choose the *mean* feature from the *math* menu of the *list* feature, as shown in Figure A.49. When using this feature, you must enter a list and a frequency list (if applicable). In this case, the list is L_1 and the frequency list is L_2. After pressing ENTER, you should obtain a mean of

$49.80

as shown in Figure A.50. You can follow the same steps (except choose the *median* feature) to find the median of the data. You should obtain a median of

$50

as shown in Figure A.51.

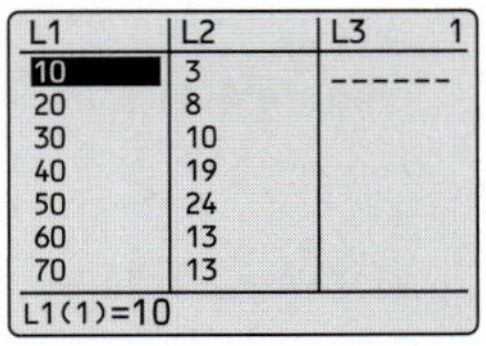

Figure A.48

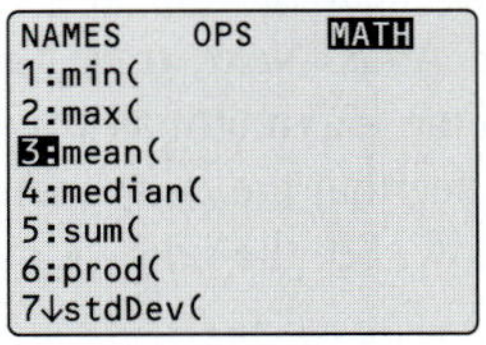

Figure A.49

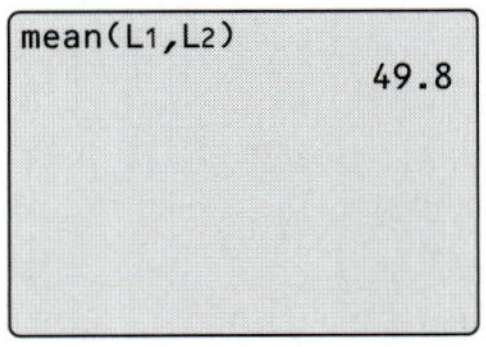

Figure A.50

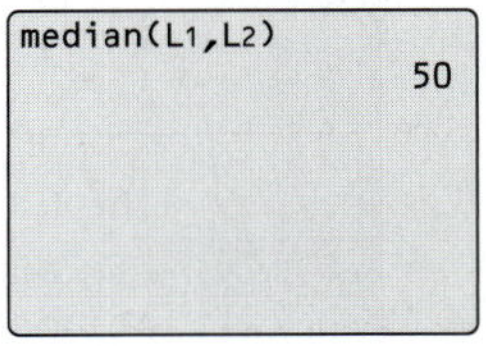

Figure A.51

Mode Settings

Mode settings of a graphing utility control how the utility displays and interprets numbers and graphs. The default mode settings are shown in Figure A.52.

Figure A.52

Radian and Degree Modes

The trigonometric functions can be applied to angles measured in either radians or degrees. When your graphing utility is in *radian* mode, it interprets angle values as radians and displays answers in radians. When your graphing utility is in *degree* mode, it interprets angle values as degrees and displays answers in degrees. For instance, to calculate $\sin(\pi/6)$, make sure the calculator is in *radian* mode. You should obtain an answer of 0.5, as shown in Figure A.53. To calculate $\sin 45°$, make sure the calculator is in *degree* mode, as shown in Figure A.54. You should obtain an approximate answer of 0.7071, as shown in Figure A.55. Without changing the mode of the calculator before evaluating $\sin 45°$, you would obtain an answer of approximately 0.8509, which is the sine of 45 radians.

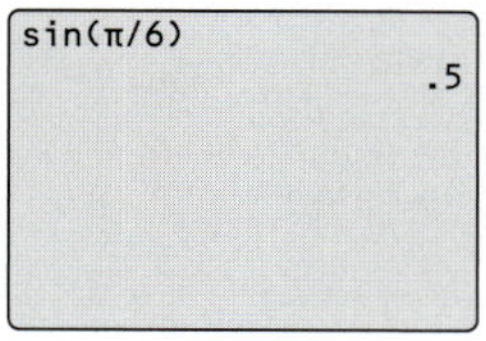

Figure A.53

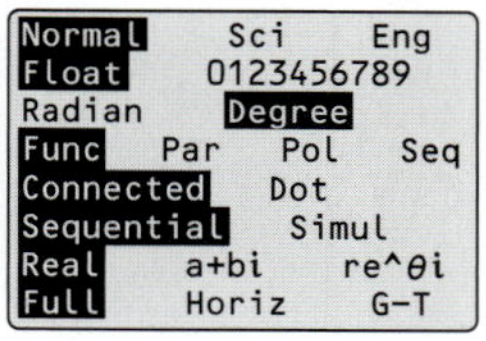

Figure A.54

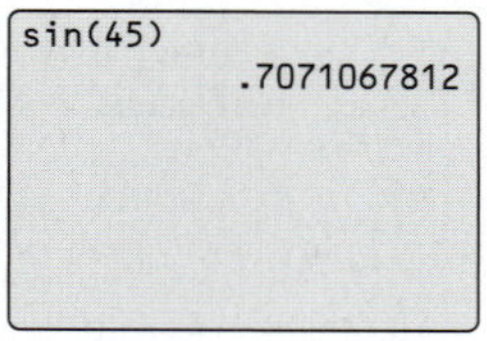

Figure A.55

Function, Parametric, Polar, and Sequence Modes

Most graphing utilities can graph using four different modes.

Function Mode The *function* mode is used to graph standard algebraic and trigonometric functions. For instance, to graph $y = 2x^2$, use the *function* mode, as shown in Figure A.52. Then enter the equation in the *equation editor*, as shown in Figure A.56. Using a standard viewing window (see Figure A.57), you obtain the graph shown in Figure A.58.

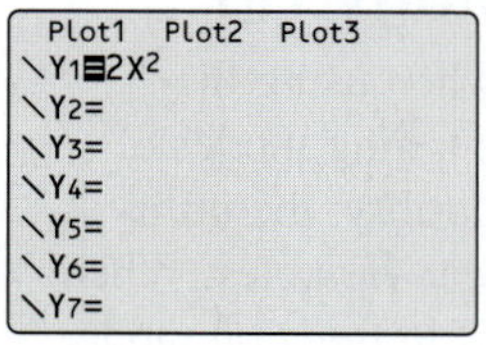

Figure A.56

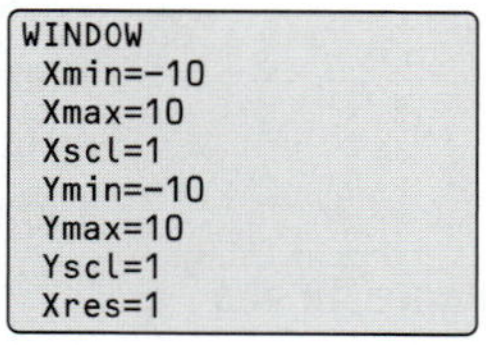

Figure A.57

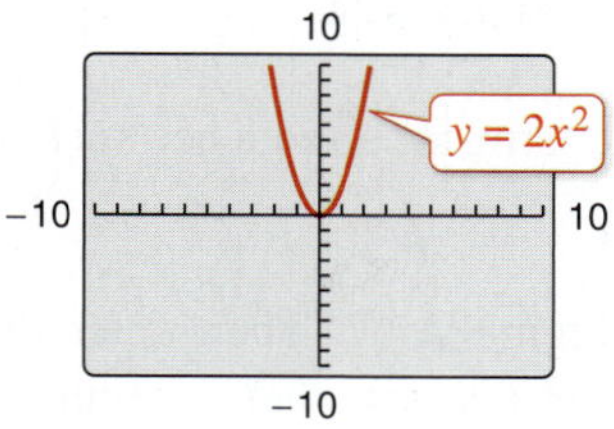

Figure A.58

Parametric Mode To graph parametric equations such as

$$x = t + 1 \quad \text{and} \quad y = t^2$$

use the *parametric* mode, as shown in Figure A.59. Then enter the equations in the *equation editor*, as shown in Figure A.60. Using the viewing window shown in Figure A.61, you obtain the graph shown in Figure A.62.

Figure A.59

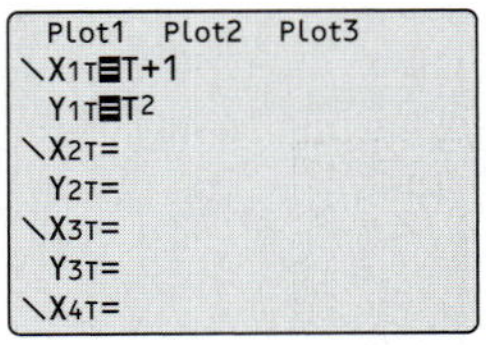

Figure A.60

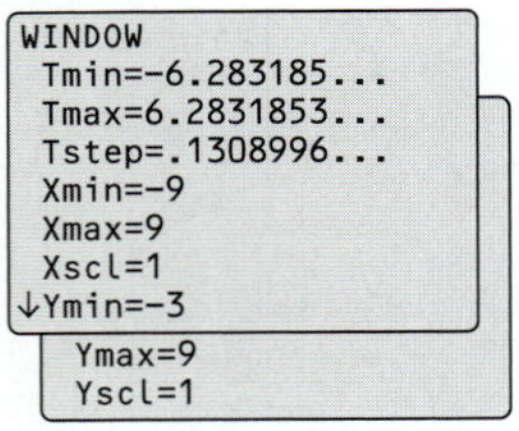

Figure A.61

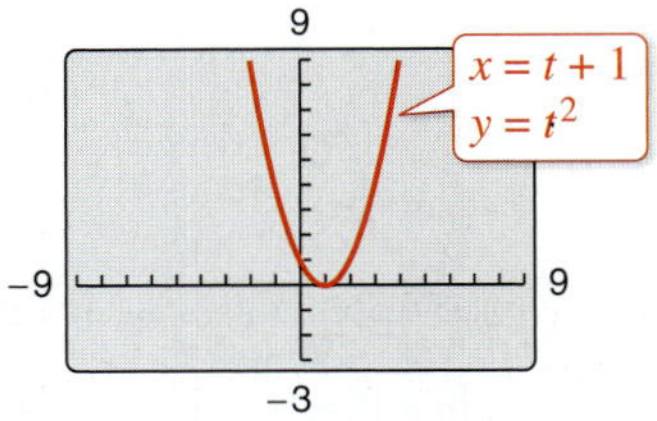

Figure A.62

Polar Mode To graph polar equations of the form $r = f(\theta)$, you can use the *polar* mode of a graphing utility. For instance, to graph the polar equation

$$r = 2\cos\theta$$

use the *polar* mode (and *radian* mode), as shown in Figure A.63. Then enter the equation in the *equation editor*, as shown in Figure A.64. Using the viewing window shown in Figure A.65, you obtain the graph shown in Figure A.66.

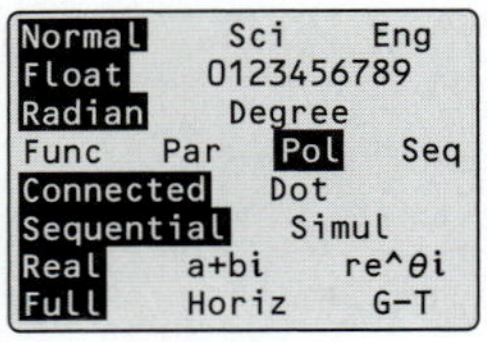

Figure A.63

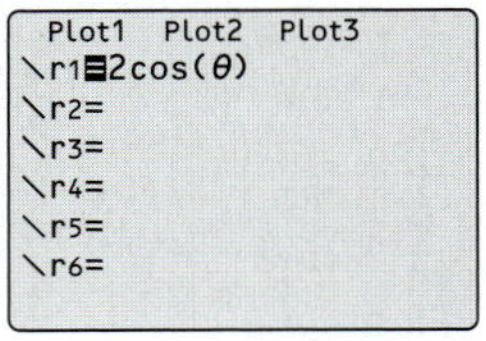

Figure A.64

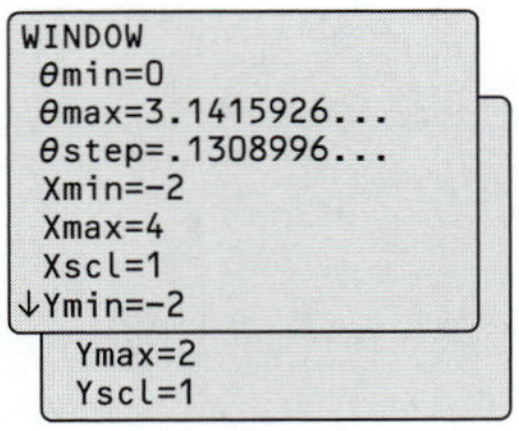

Figure A.65

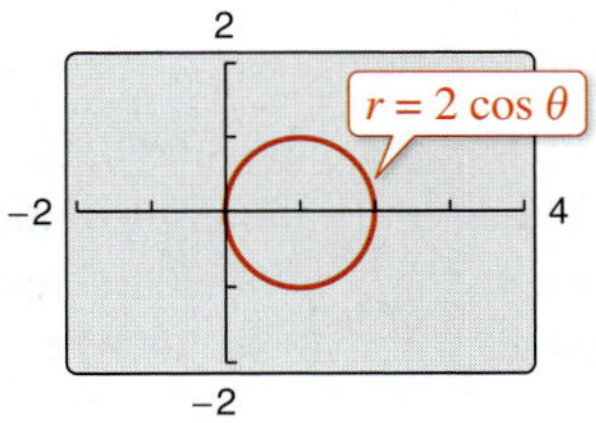

Figure A.66

Sequence Mode To graph the first five terms of a sequence such as

$$a_n = 4n - 5$$

use the *sequence* mode, as shown in Figure A.67. Then enter the sequence in the *equation editor*, as shown in Figure A.68 (assume that n begins with 1). Using the viewing window shown in Figure A.69, you obtain the graph shown in Figure A.70.

Figure A.67

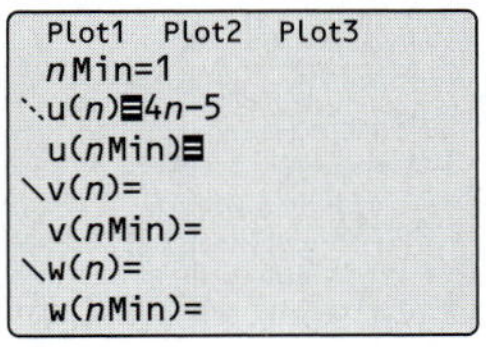

Figure A.68

Technology Tip

Note that when using the different graphing modes of a graphing utility, the utility uses different variables. When the utility is in *function* mode, it uses the variables x and y. In *parametric* mode, the utility uses the variables x, y, and t. In *polar* mode, the utility uses the variables r and θ. In *sequence* mode, the utility uses the variables u (instead of a) and n.

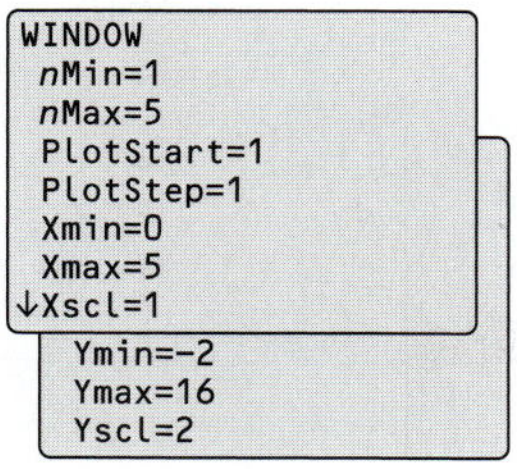

Figure A.69

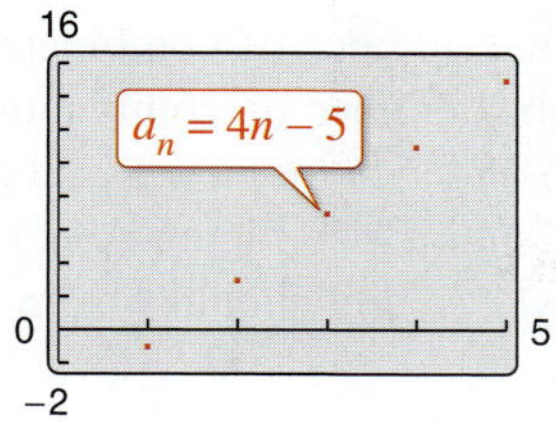

Figure A.70

Connected and Dot Modes

Graphing utilities use the point-plotting method to graph functions. When a graphing utility is in *connected* mode, the utility connects the points that are plotted. When the utility is in *dot* mode, it does not connect the points that are plotted. For instance, the graph of

$$y = x^3$$

in *connected* mode is shown in Figure A.71. To graph this function using *dot* mode, first change the mode to *dot* mode (see Figure A.72) and then graph the equation, as shown in Figure A.73. As you can see in Figure A.73, the graph is a collection of dots.

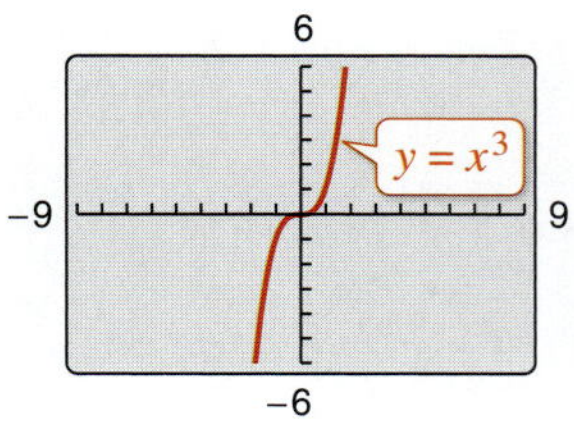

Figure A.71

Figure A.72

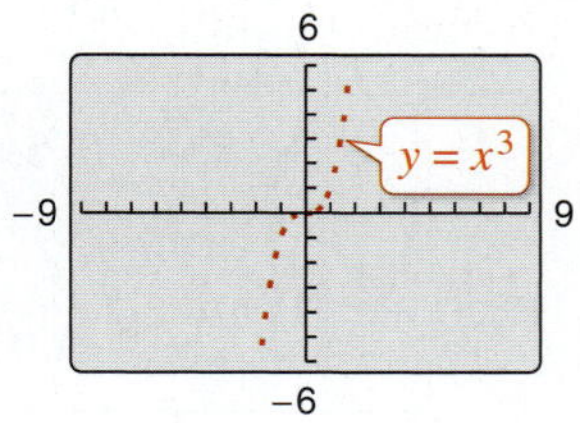

Figure A.73

A problem arises in some graphing utilities when the *connected* mode is used. Graphs with vertical asymptotes, such as rational functions and tangent functions, appear to be connected. For instance, the graph of

$$y = \frac{1}{x + 3}$$

is shown in Figure A.74. Notice how the two portions of the graph appear to be connected with a vertical line at $x = -3$. From your study of rational functions, you know that the graph has a vertical asymptote at $x = -3$ and therefore is undefined when $x = -3$. If your graphing utility draws extraneous vertical lines when graphing functions with vertical asymptotes, you can use the *dot* mode to eliminate the vertical lines, as shown in Figure A.75.

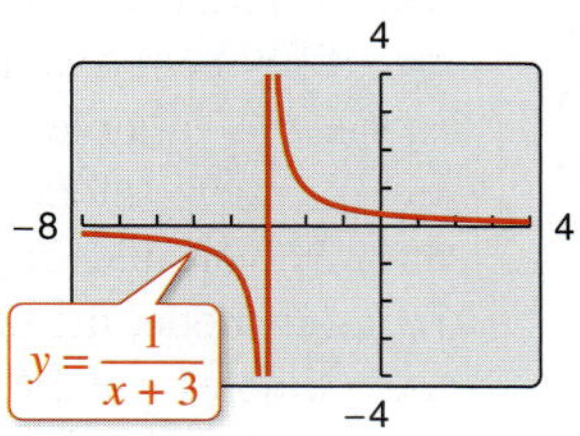

Figure A.74

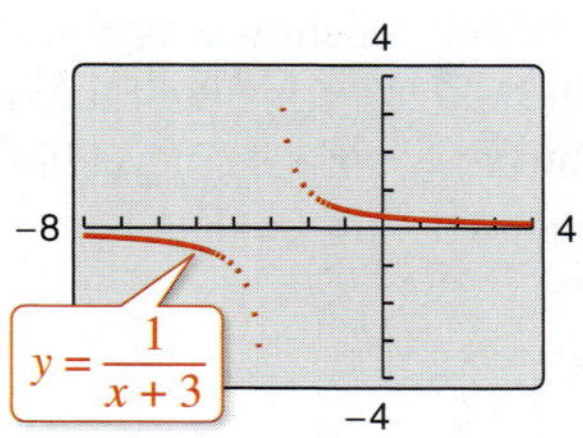

Figure A.75

${}_nC_r$ Feature

The ${}_nC_r$ feature calculates binomial coefficients and the number of combinations of n elements taken r at a time. For instance, to find the number of combinations of eight elements taken five at a time, enter 8 (the n-value) on the home screen and choose the ${}_nC_r$ feature from the *probability* menu of the *math* feature (see Figure A.76). Next, enter 5 (the r-value) on the home screen and press ENTER. You should obtain 56, as shown in Figure A.77.

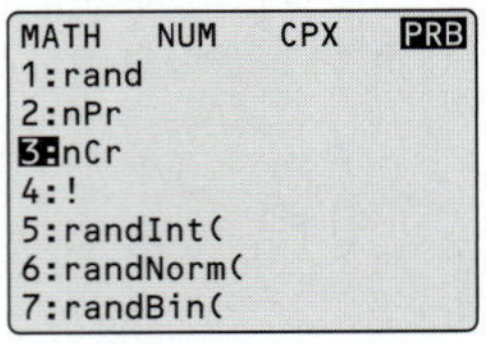

Figure A.76

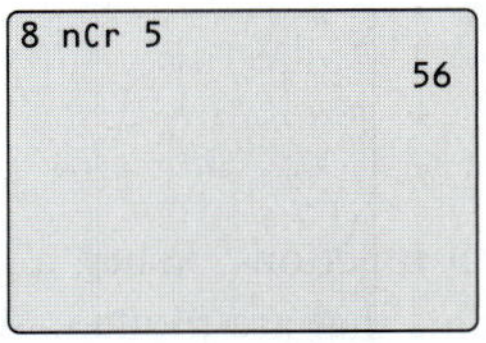

Figure A.77

${}_nP_r$ Feature

The ${}_nP_r$ feature calculates the number of permutations of n elements taken r at a time. For instance, to find the number of permutations of six elements taken four at a time, enter 6 (the n-value) on the home screen and choose the ${}_nP_r$ feature from the *probability* menu of the *math* feature (see Figure A.78). Next, enter 4 (the r-value) on the home screen and press ENTER. You should obtain 360, as shown in Figure A.79.

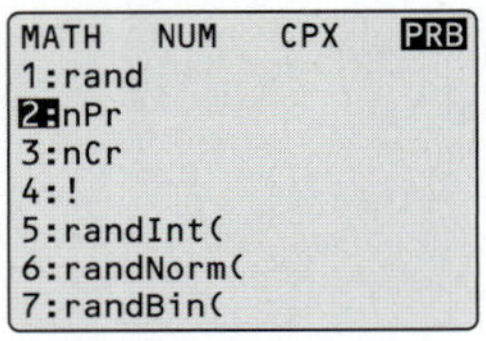

Figure A.78

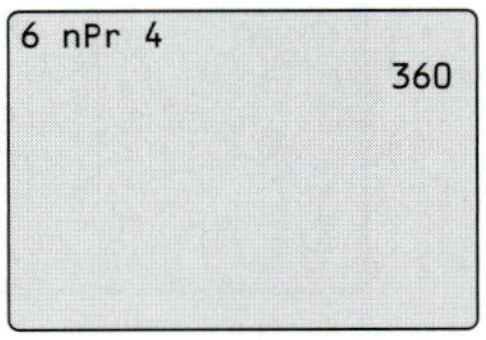

Figure A.79

One-Variable Statistics Feature

Graphing utilities are useful in calculating statistical values for a set of data. The *one-variable statistics* feature analyzes data with one measured variable. This feature outputs the mean of the data, the sum of the data, the sum of the data squared, the sample standard deviation of the data, the population standard deviation of the data, the number of data points, the minimum data value, the maximum data value, the first quartile of the data, the median of the data, and the third quartile of the data. Consider the following data, which show the hourly earnings (in dollars) of 12 retail sales associates.

8.05, 10.25, 8.45, 9.15, 8.90, 8.20, 9.25, 10.30, 8.60, 9.60, 10.05, 11.35

You can use the *one-variable statistics* feature to determine the mean and standard deviation of the data. First, enter the data in the *list editor*, as shown in Figure A.80. Then choose the *one-variable statistics* feature from the *calculate* menu of the *statistics* feature, as shown in Figure A.81. When using this feature, you must enter a list. In this case, the list is L_1. In Figure A.82, you can see that the mean of the data is

$$\bar{x} \approx 9.35$$

and the standard deviation of the data is

$$\sigma x \approx 0.95.$$

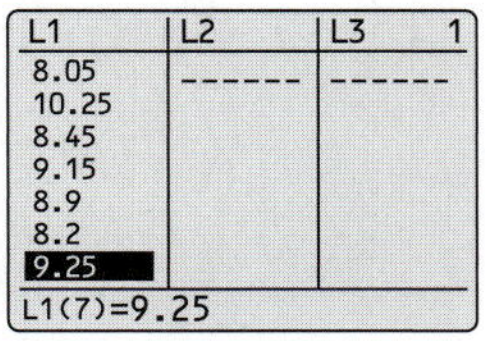

Figure A.80

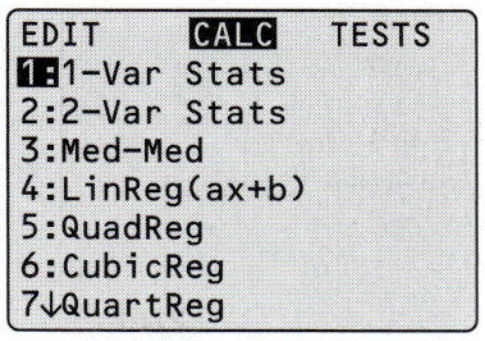

Figure A.81

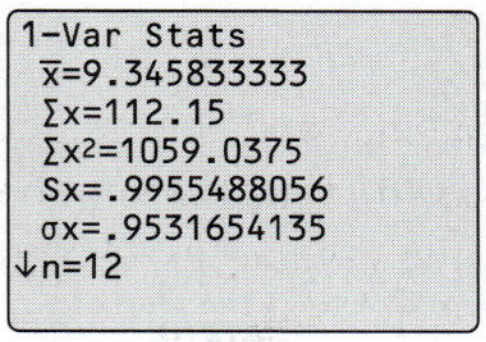

Figure A.82

Regression Feature

Throughout the text, you are asked to use the *regression* feature of a graphing utility to find models for sets of data. Most graphing utilities have built-in regression programs for the following.

Regression	*Form of Model*
Linear	$y = ax + b$ or $y = a + bx$
Quadratic	$y = ax^2 + bx + c$
Cubic	$y = ax^3 + bx^2 + cx + d$
Quartic	$y = ax^4 + bx^3 + cx^2 + dx + e$
Logarithmic	$y = a + b \ln x$
Exponential	$y = ab^x$
Power	$y = ax^b$
Logistic	$y = \dfrac{c}{1 + ae^{-bx}}$
Sine	$y = a \sin(bx + c) + d$

For instance, you can find a linear model for the data in the table.

x	y
6	222.8
7	228.7
8	235.0
9	240.3
10	245.0
11	248.2
12	254.4
13	260.2
14	268.3
15	287.0

Enter the data in the *list editor*, as shown in Figure A.83. Note that L_1 contains the x-values, and L_2 contains the y-values. Now choose the *linear regression* feature from the *calculate* menu of the *statistics* feature, as shown in Figure A.84. In Figure A.85, you can see that a linear model for the data is given by

$$y = 6.22x + 183.7.$$

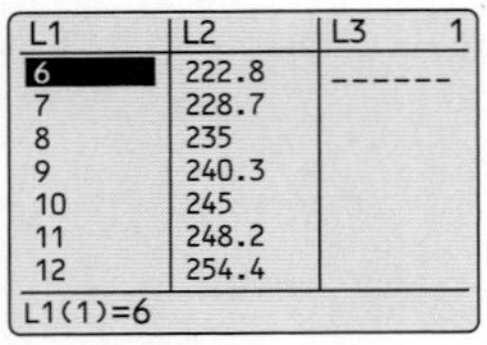

Figure A.83

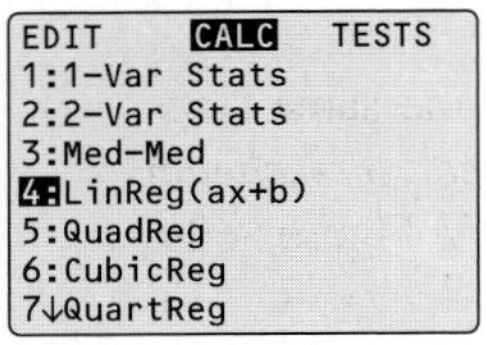

Figure A.84

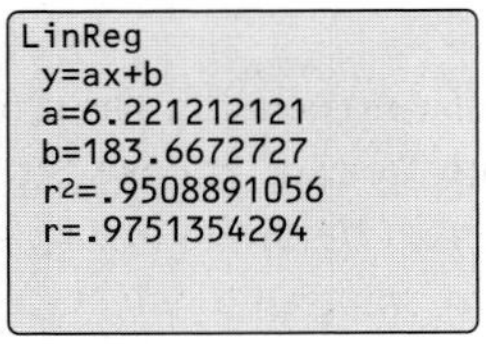

Figure A.85

When you use the *regression* feature of a graphing utility, you will notice that the program may also output an "r-value." (For some calculators, make sure you select the *diagnostics on* feature before you use the *regression* feature. Otherwise, the calculator will not output an r-value.) The r-value, or *correlation coefficient,* measures how well the linear model fits the data. The closer the value of $|r|$ is to 1, the better the fit. For the data above,

$$r \approx 0.97514$$

which implies that the model is a good fit for the data.

Row-Echelon and Reduced Row-Echelon Features

Some graphing utilities have features that can automatically transform a matrix to row-echelon form and reduced row-echelon form. These features can be used to check your solutions to systems of equations.

Row-Echelon Feature

Consider the system of equations and the corresponding augmented matrix shown below.

Linear System

$$\begin{cases} 2x + 5y - 3z = 4 \\ 4x + y = 2 \\ -x + 3y - 2z = -1 \end{cases}$$

Augmented Matrix

$$\begin{bmatrix} 2 & 5 & -3 & \vdots & 4 \\ 4 & 1 & 0 & \vdots & 2 \\ -1 & 3 & -2 & \vdots & -1 \end{bmatrix}$$

You can use the *row-echelon* feature of a graphing utility to write the augmented matrix in row-echelon form. First, enter the matrix in the graphing utility using the *matrix editor*, as shown in Figure A.86. Next, choose the *row-echelon* feature from the *math* menu of the *matrix* feature, as shown in Figure A.87. When using this feature, you must enter the name of the matrix. In this case, the name of the matrix is [A]. You should obtain the matrix shown in Figure A.88. You may have to scroll to the right in order to see all the entries of the matrix.

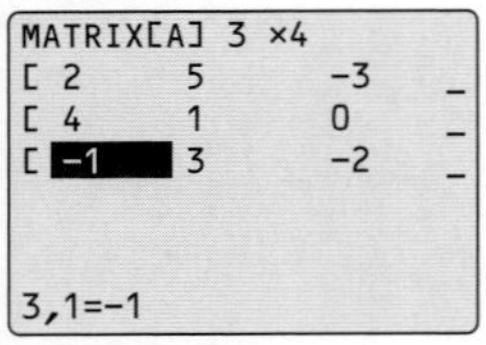

Figure A.86

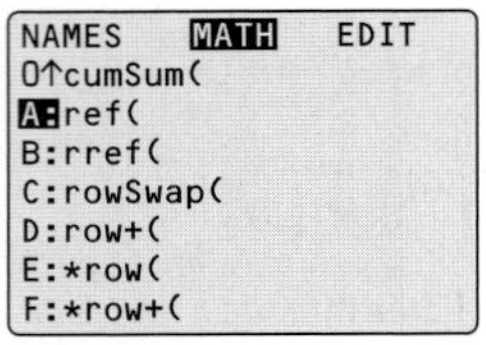

Figure A.87

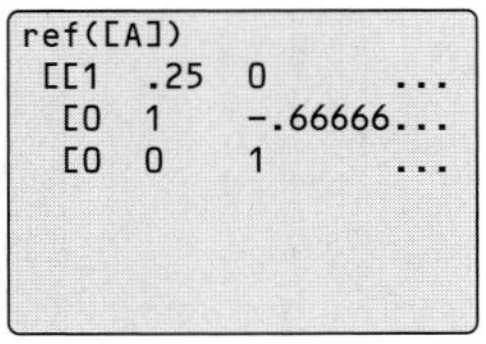

Figure A.88

Reduced Row-Echelon Feature

To write the augmented matrix in reduced row-echelon form, follow the same steps used to write a matrix in row-echelon form except choose the *reduced row-echelon* feature, as shown in Figure A.89. You should obtain the matrix shown in Figure A.90. From Figure A.90, you can conclude that the solution to the system is $x = 3$, $y = -10$, and $z = -16$.

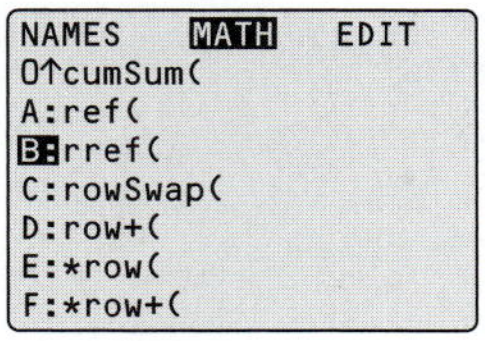

Figure A.89

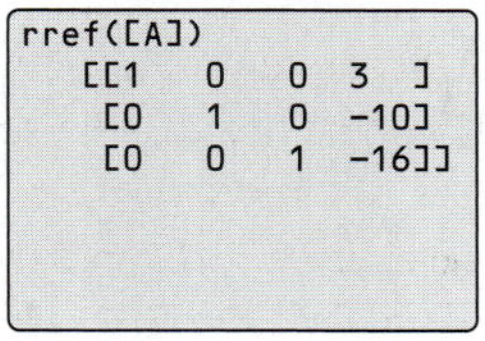

Figure A.90

Sequence Feature

The *sequence* feature is used to display the terms of sequences. For instance, to determine the first five terms of the arithmetic sequence

$$a_n = 3n + 5 \qquad \text{Assume } n \text{ begins with 1.}$$

set the graphing utility to *sequence* mode. Then choose the *sequence* feature from the *operations* menu of the *list* feature, as shown in Figure A.91. When using this feature, you must enter the sequence, the variable (in this case n), the beginning value (in this case 1), and the end value (in this case 5). The first five terms of the sequence are 8, 11, 14, 17, and 20, as shown in Figure A.92. You may have to scroll to the right in order to see all the terms of the sequence.

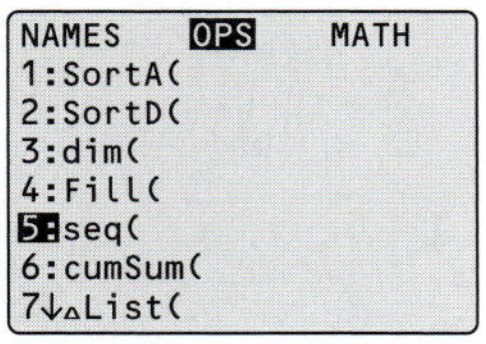

Figure A.91

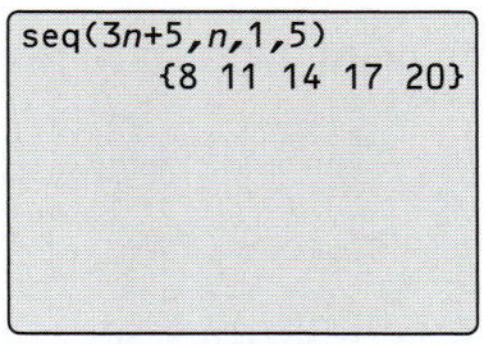

Figure A.92

Shade Feature

Most graphing utilities have a *shade* feature that can be used to graph inequalities. For instance, to graph the inequality

$$y \leq 2x - 3$$

first enter the equation $y = 2x - 3$ in the *equation editor*, as shown in Figure A.93. Next, using a standard viewing window (see Figure A.94), graph the equation, as shown in Figure A.95.

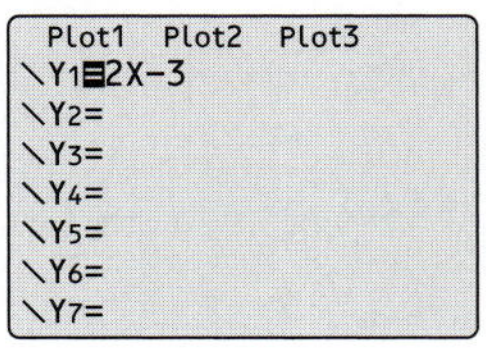

Figure A.93

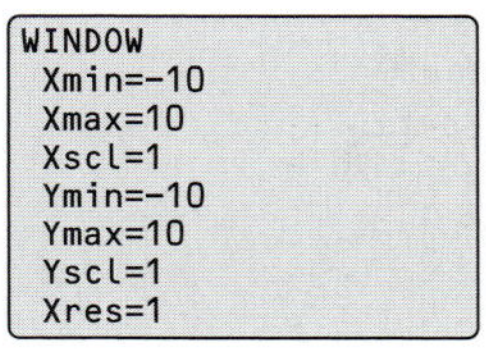

Figure A.94

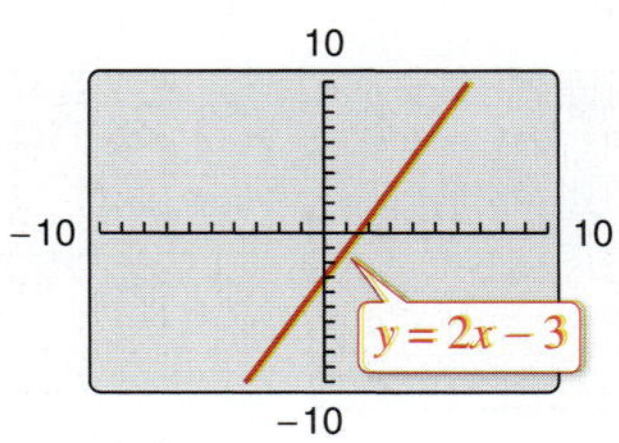

Figure A.95

Because the inequality sign is $\leq$, you want to shade the region below the line $y = 2x - 3$. Choose the *shade* feature from the *draw* feature menu, as shown in Figure A.96. You must enter a lower function and an upper function. In this case, the lower function is -10 (this is the least y-value in the viewing window) and the upper function is Y_1 $(y = 2x - 3)$, as shown in Figure A.97. Then press ENTER to obtain the graph shown in Figure A.98.

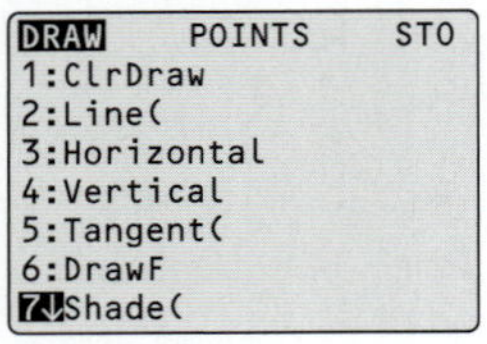

Figure A.96

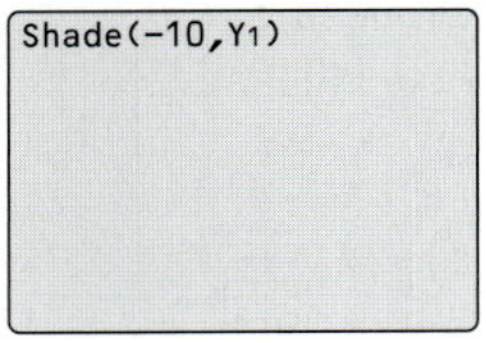

Figure A.97

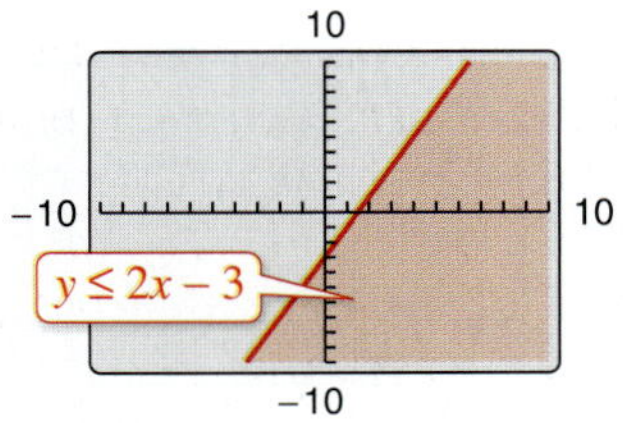

Figure A.98

To graph the inequality

$$y \geq 2x - 3$$

(using a standard viewing window), you would enter the lower function as Y_1 $(y = 2x - 3)$ and the upper function as 10 (the greatest y-value in the viewing window).

Statistical Plotting Feature

The *statistical plotting* feature plots data that are stored in lists. Most graphing utilities can display the following types of plots.

Plot Type	*Variables*
Scatter plot	x-list, y-list
xy line graph	x-list, y-list
Histogram	x-list, frequency
Box-and-whisker plot	x-list, frequency
Normal probability plot	data list, data axis

For instance, use a box-and-whisker plot to represent the following set of data. Then use the graphing utility plot to find the least number, the lower quartile, the median, the upper quartile, and the greatest number.

17, 19, 21, 27, 29, 30, 37, 27, 15, 23, 19, 16

First, use the *list editor* to enter the values in a list, as shown in Figure A.99. Then go to the *statistical plotting editor*. In this editor, you will turn the plot on, select the box-and-whisker plot, select the list you entered in the *list editor*, and enter the frequency of each item in the list, as shown in Figure A.100. Now use the *zoom* feature and choose the *zoom stat* option to set the viewing window and plot the graph, as shown in Figure A.101. Use the *trace* feature to find that the least number is 15, the lower quartile is 18, the median is 22, the upper quartile is 28, and the greatest number is 37.

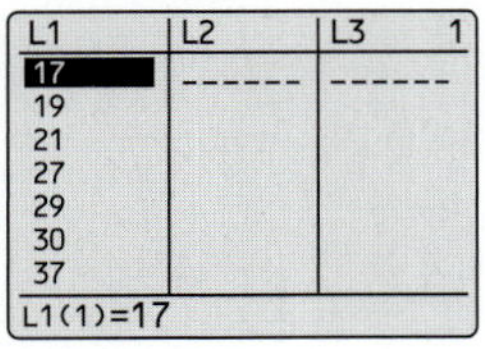

Figure A.99

Figure A.100

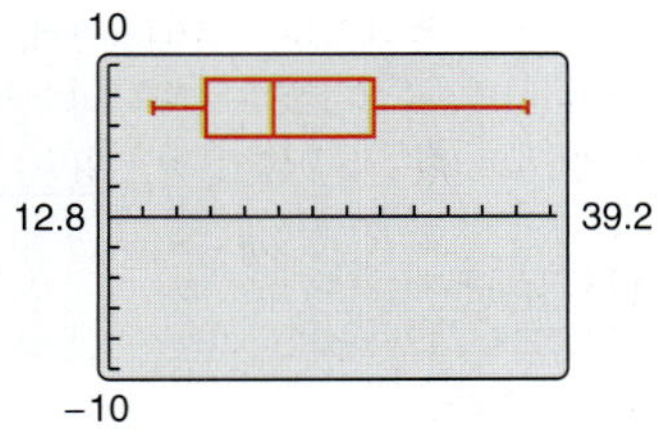

Figure A.101

Sum Feature

The *sum* feature finds the sum of a list of data. For instance, the data below represent a student's quiz scores on 10 quizzes throughout an algebra course.

22, 23, 19, 24, 20, 15, 25, 21, 18, 24

To find the total quiz points the student earned, enter the data in the *list editor*, as shown in Figure A.102. To find the sum, choose the *sum* feature from the *math* menu of the *list* feature, as shown in Figure A.103. You must enter a list. In this case, the list is L_1. You should obtain a sum of 211, as shown in Figure A.104.

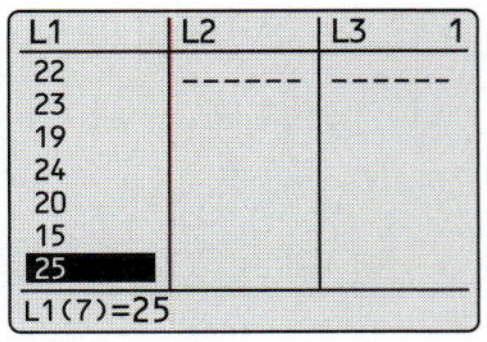

Figure A.102

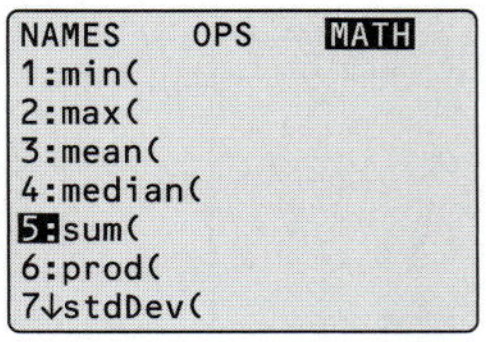

Figure A.103

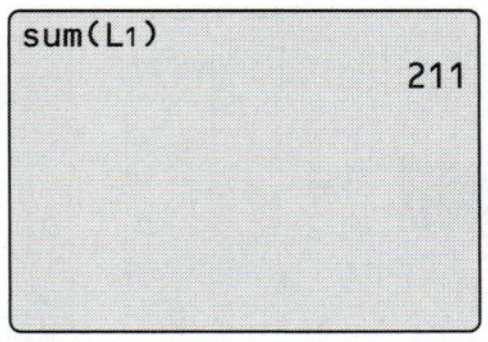

Figure A.104

Sum Sequence Feature

The *sum* feature and the *sequence* feature can be combined to find the sum of a sequence or series. For instance, to find the sum

$$\sum_{k=0}^{10} 5^{k+1}$$

first choose the *sum* feature from the *math* menu of the *list* feature, as shown in Figure A.105. Then choose the *sequence* feature from the *operations* menu of the *list* feature, as shown in Figure A.106. You must enter an expression for the sequence, a variable, the lower limit of summation, and the upper limit of summation. After pressing ENTER, you should obtain the sum 61,035,155, as shown in Figure A.107.

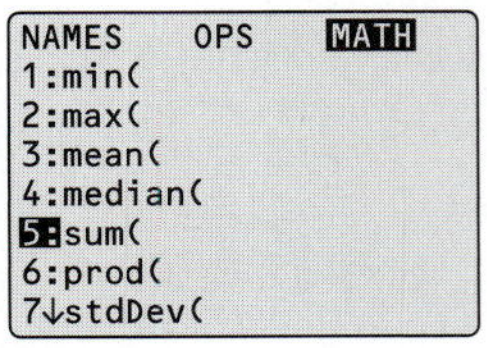

Figure A.105

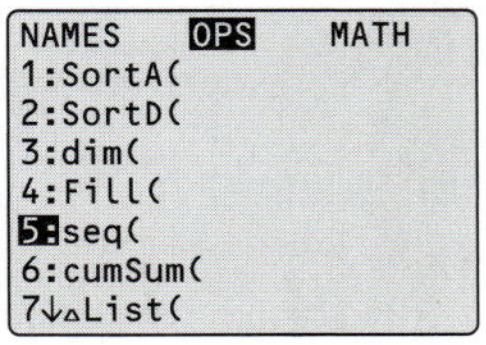

Figure A.106

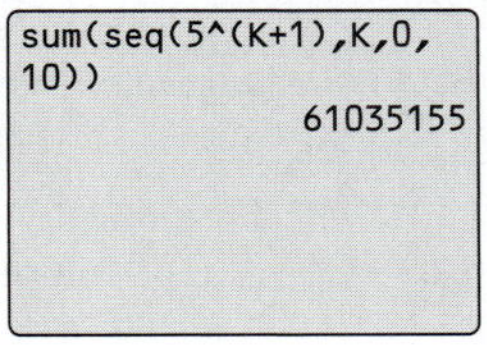

Figure A.107

Table Feature

Most graphing utilities are capable of displaying a table of values with x-values and one or more corresponding y-values. These tables can be used to check solutions of an equation and to generate ordered pairs to assist in graphing an equation by hand.

To use the *table* feature, enter an equation in the *equation editor*. The table may have a setup screen, which allows you to select the starting x-value and the table step or x-increment. You may then have the option of automatically generating values for x and y or building your own table using the *ask* mode (see Figure A.108).

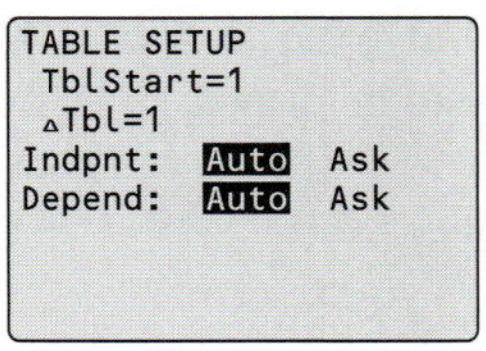

Figure A.108

For instance, enter the equation

$$y = \frac{3x}{x + 2}$$

in the *equation editor*, as shown in Figure A.109. In the table setup screen, set the table to start at $x = -4$ and set the table step to 1, as shown in Figure A.110. When you view the table, notice that the first x-value is -4 and that each successive value increases by 1. Also, notice that the Y_1 column gives the resulting y-value for each x-value, as shown in Figure A.111. The table shows that the y-value for $x = -2$ is ERROR. This means that the equation is undefined when $x = -2$.

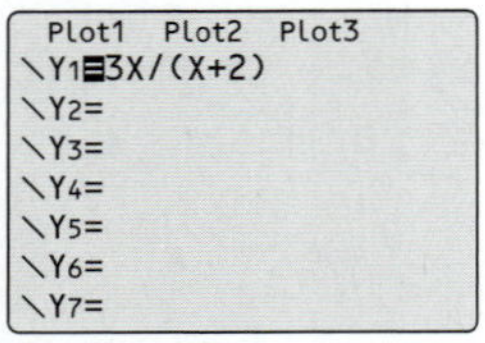

Figure A.109

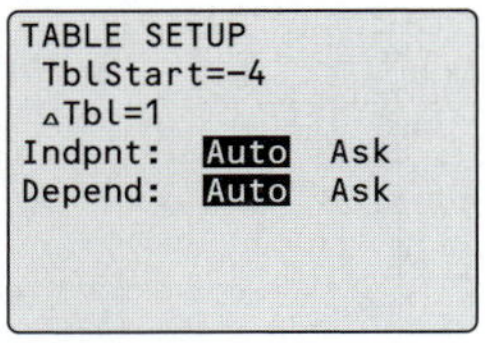

Figure A.110

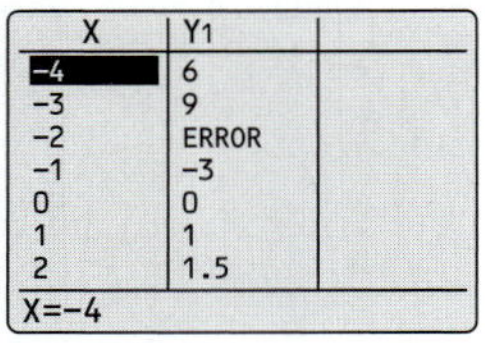

Figure A.111

With the same equation in the *equation editor*, set the independent variable in the table to *ask* mode, as shown in Figure A.112. In this mode, you do not need to set the starting x-value or the table step because you are entering any value you choose for x. You may enter any real value for x—integers, fractions, decimals, irrational numbers, and so forth. When you enter

$$x = 1 + \sqrt{3}$$

the graphing utility may rewrite the number as a decimal approximation, as shown in Figure A.113. You can continue to build your own table by entering additional x-values in order to generate y-values, as shown in Figure A.114.

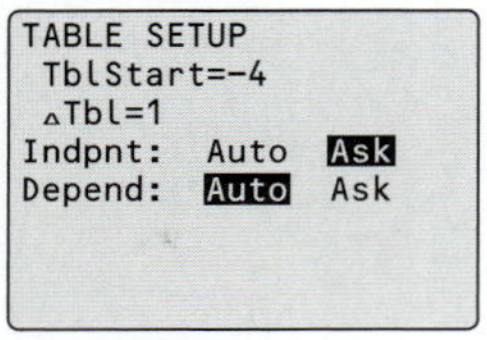

Figure A.112

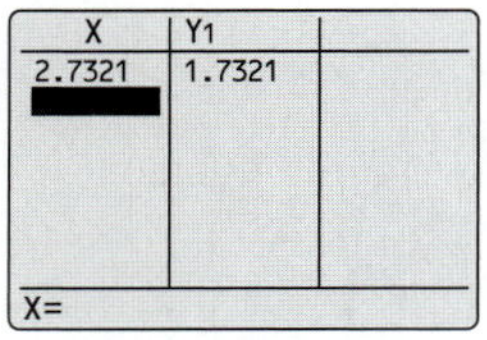

Figure A.113

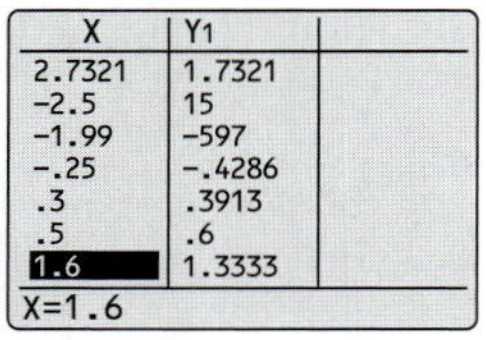

Figure A.114

When you have several equations in the *equation editor*, the table may generate y-values for each equation.

Tangent Feature

Some graphing utilities have the capability of drawing a tangent line to a graph at a given point. For instance, consider the equation

$$y = -x^3 + x + 2.$$

To draw the line tangent to the point $(1, 2)$ on the graph of y, enter the equation in the *equation editor*, as shown in Figure A.115. Using the viewing window shown in Figure A.116, graph the equation, as shown in Figure A.117. Next, choose the *tangent* feature from the *draw* feature menu, as shown in Figure A.118. You can either move the cursor to select a point or enter the x-value at which you want the tangent line to be drawn.

Because you want the tangent line to the point

$$(1, 2)$$

enter 1 (see Figure A.119) and then press ENTER. The x-value you entered and the equation of the tangent line are displayed at the bottom of the window, as shown in Figure A.120.

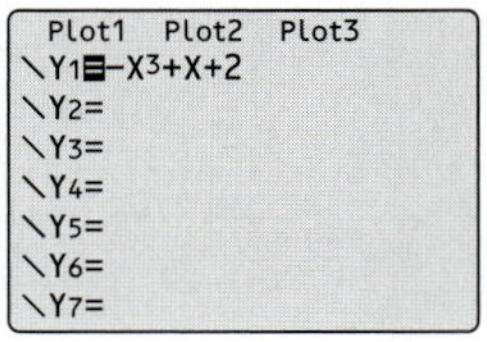

Figure A.115

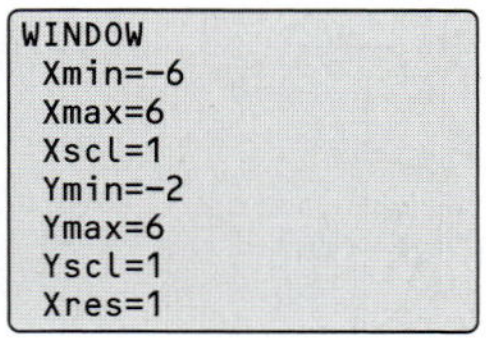

Figure A.116

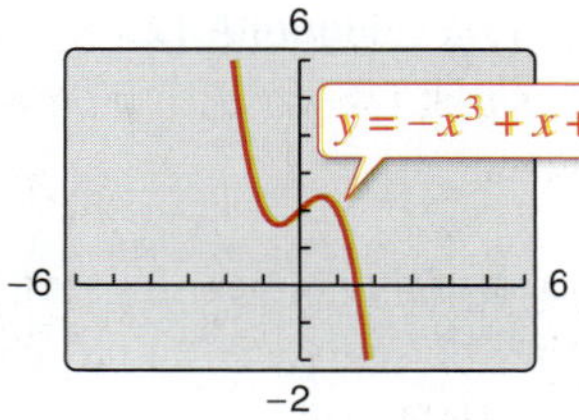

Figure A.117

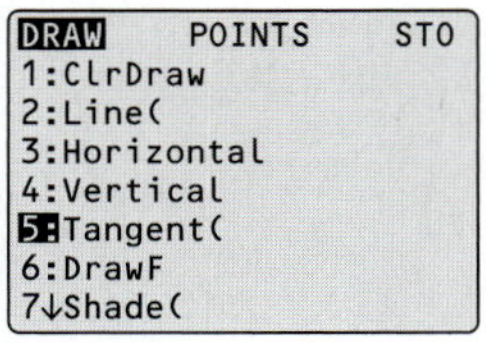

Figure A.118

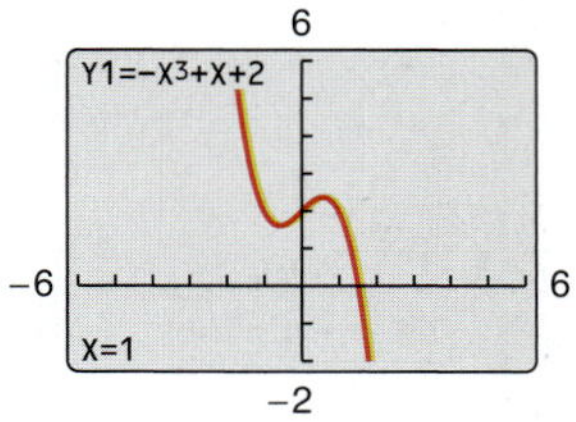

Figure A.119

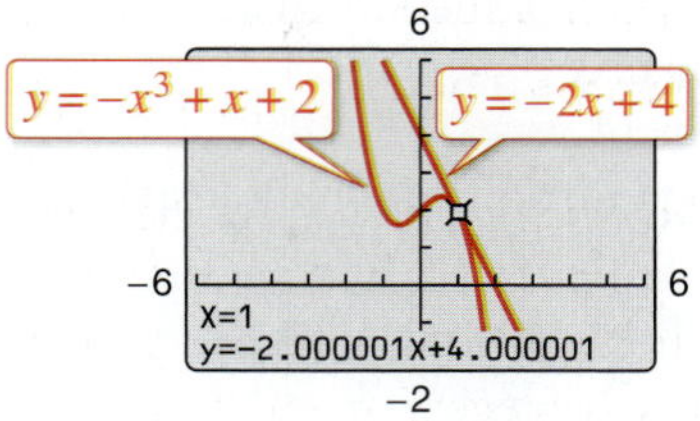

Figure A.120

Trace Feature

For instructions on how to use the *trace* feature, see Zoom and Trace Features on page A23.

Value Feature

The *value* feature finds the value of a function y for a given x-value. To find the value of a function such as

$$f(x) = 0.5x^2 - 1.5x$$

at $x = 1.8$, first enter the function in the *equation editor* (see Figure A.121) and then graph the function (using a standard viewing window), as shown in Figure A.122. Next, choose the *value* feature from the *calculate* menu, as shown in Figure A.123. You will see "X = " displayed at the bottom of the window. Enter the x-value, in this case $x = 1.8$, as shown in Figure A.124. When entering an x-value, be sure it is between the Xmin and Xmax values you entered for the viewing window. Then press ENTER. In Figure A.125, you can see that when $x = 1.8$,

$$y = -1.08.$$

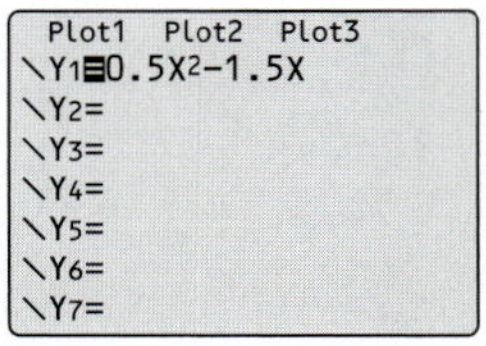

Figure A.121

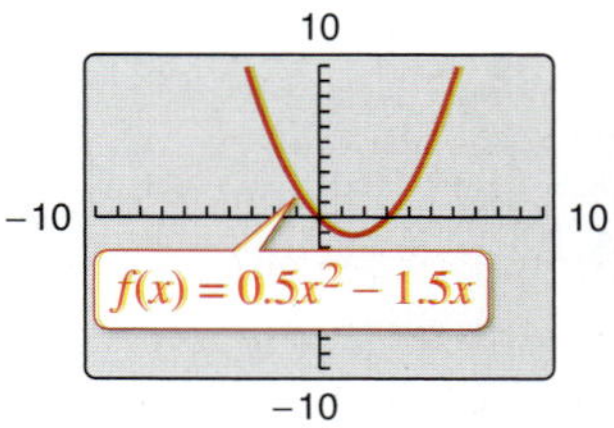

Figure A.122

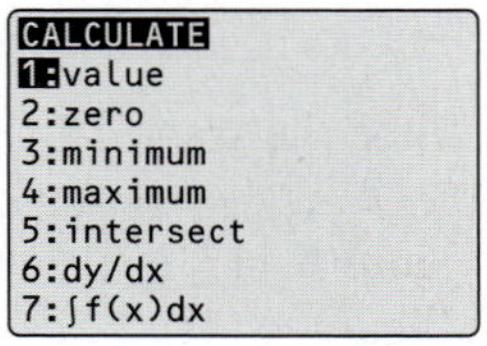

Figure A.123

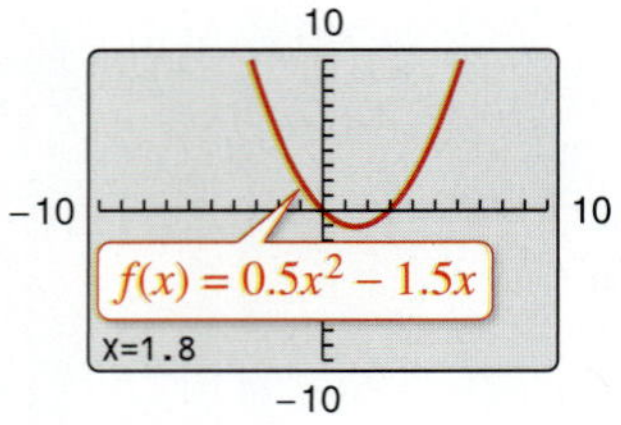

Figure A.124

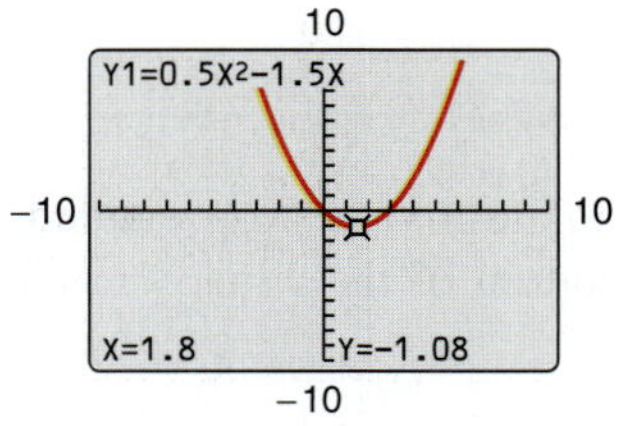

Figure A.125

Viewing Window

A viewing window for a graph is a rectangular portion of the coordinate plane. A viewing window is determined by the following six values (see Figure A.126).

Xmin = the minimum value of x
Xmax = the maximum value of x
Xscl = the number of units per tick mark on the x-axis
Ymin = the minimum value of y
Ymax = the maximum value of y
Yscl = the number of units per tick mark on the y-axis

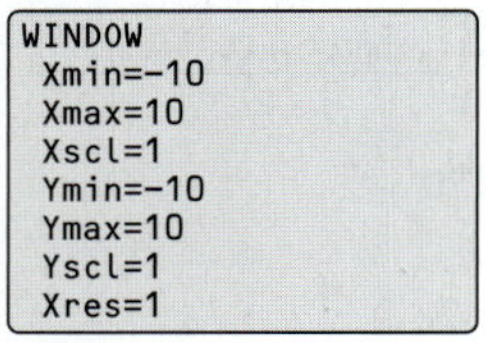

Figure A.126

When you enter these six values in a graphing utility, you are setting the viewing window. On some graphing utilities, there is a seventh value for the viewing window labeled Xres. This sets the pixel resolution (1 through 8). For instance, when

Xres = 1

functions are evaluated and graphed at each pixel on the x-axis. Some graphing utilities have a standard viewing window, as shown in Figure A.127. To initialize the standard viewing window quickly, choose the *standard viewing window* feature from the *zoom* feature menu (see page A23), as shown in Figure A.128.

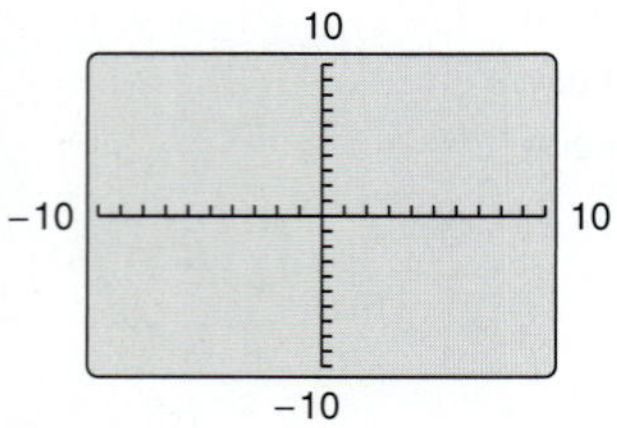

Figure A.127

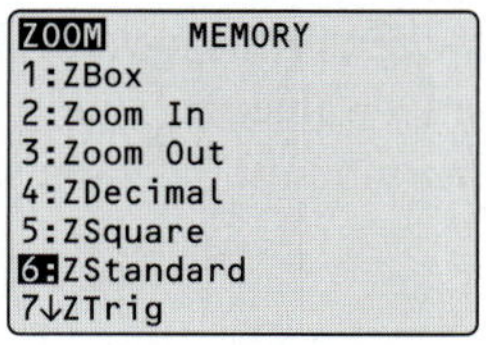

Figure A.128

By choosing different viewing windows for a graph, it is possible to obtain different impressions of the graph's shape. For instance, Figure A.129 shows four different viewing windows for the graph of

$$y = 0.1x^4 - x^3 + 2x^2.$$

Of these viewing windows, the one shown in part (a) is the most complete.

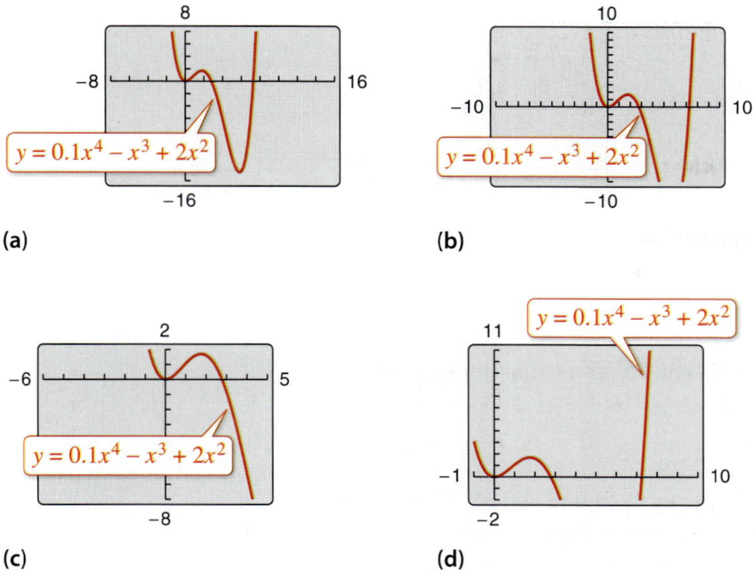

Figure A.129

On most graphing utilities, the display screen is two-thirds as high as it is wide. On such screens, you can obtain a graph with a true geometric perspective by using a square setting—one in which

$$\frac{\text{Ymax} - \text{Ymin}}{\text{Xmax} - \text{Xmin}} = \frac{2}{3}.$$

One such setting is shown in Figure A.130. Notice that the x and y tick marks are equally spaced on a square setting but not on a standard setting (see Figure A.127). To initialize the square viewing window quickly, choose the *square viewing window* feature from the *zoom* feature menu (see page A23), as shown in Figure A.131.

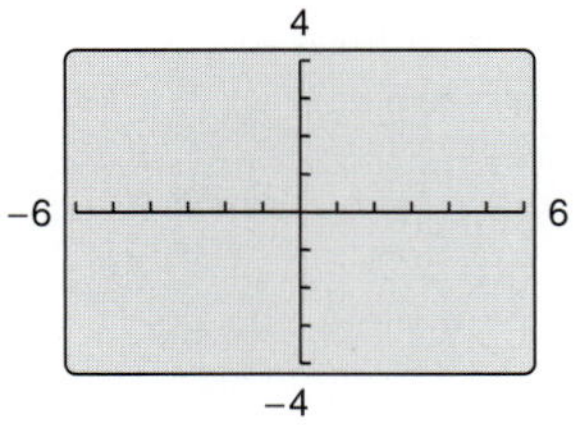

Figure A.130

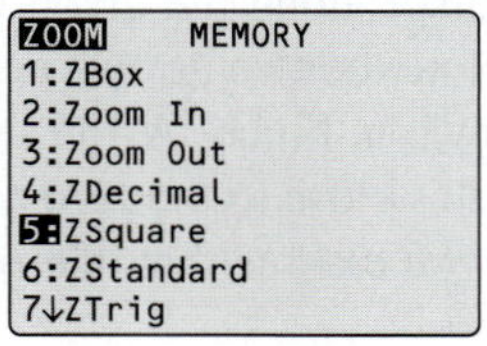

Figure A.131

To see how the viewing window affects the geometric perspective, graph the semicircles

$$y_1 = \sqrt{9 - x^2}$$

and

$$y_2 = -\sqrt{9 - x^2}$$

using a standard viewing window, as shown in Figure A.132. Notice how the circle appears elliptical rather than circular. Now graph y_1 and y_2 using a square viewing window, as shown in Figure A.133. Notice how the circle appears circular. (Note that when you graph the two semicircles, your graphing utility may not connect them. This is because some graphing utilities are limited in their resolution.)

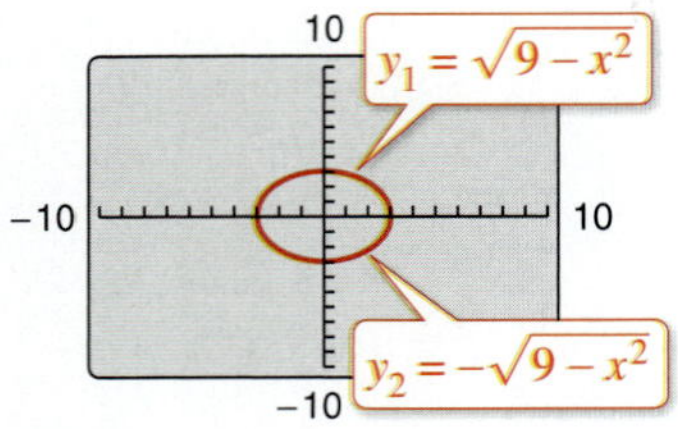

Figure A.132

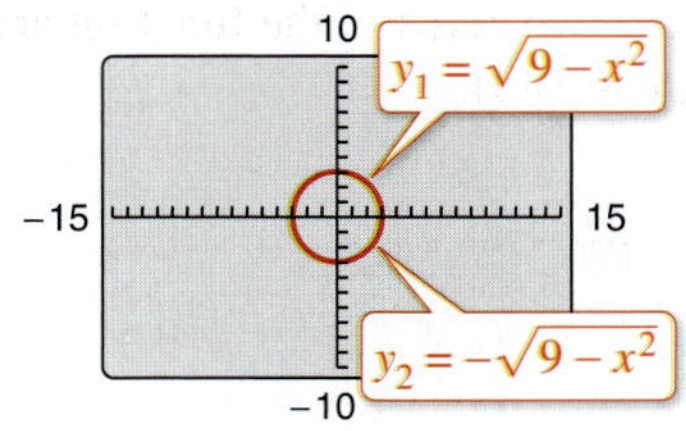

Figure A.133

Zero or Root Feature

The *zero* or *root* feature finds the real zeros of the various types of functions studied in this text. To find the zeros of a function such as

$$f(x) = 2x^3 - 4x$$

first enter the function in the *equation editor*, as shown in Figure A.134. Now graph the equation (using a standard viewing window), as shown in Figure A.135. From the figure, you can see that the graph of the function crosses the x-axis three times, so the function has three real zeros.

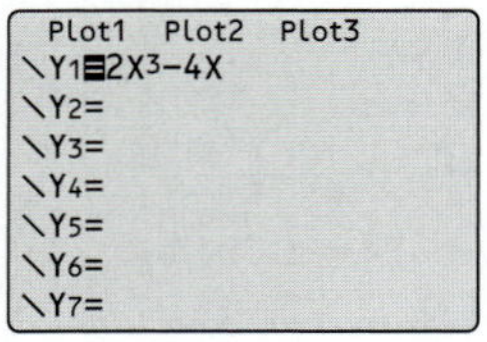

Figure A.134

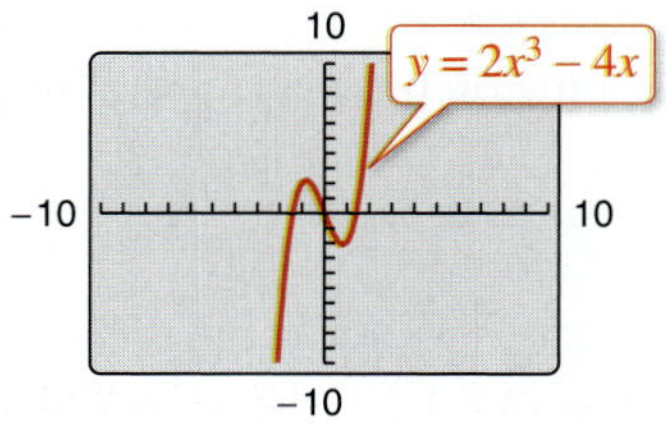

Figure A.135

To find these zeros, choose the *zero* feature found in the *calculate* menu (see Figure A.136). Next, trace the cursor along the graph to a point left of one of the zeros and press ENTER (see Figure A.137). Then trace the cursor along the graph to a point right of the zero and press ENTER (see Figure A.138). Note the two arrows near the top of the display marking the left and right bounds, as shown in Figure A.139. Now trace the cursor along the graph between the two bounds and as close to the zero as you can (see Figure A.140) and press ENTER. In Figure A.141, you can see that one zero of the function is $x \approx -1.414214$.

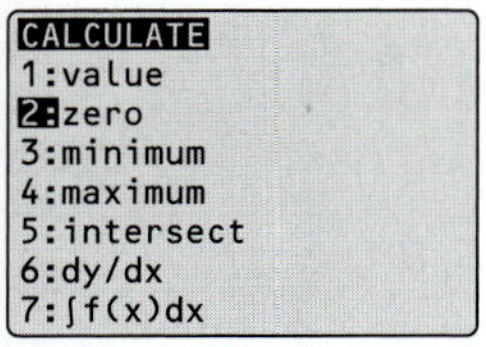

Figure A.136

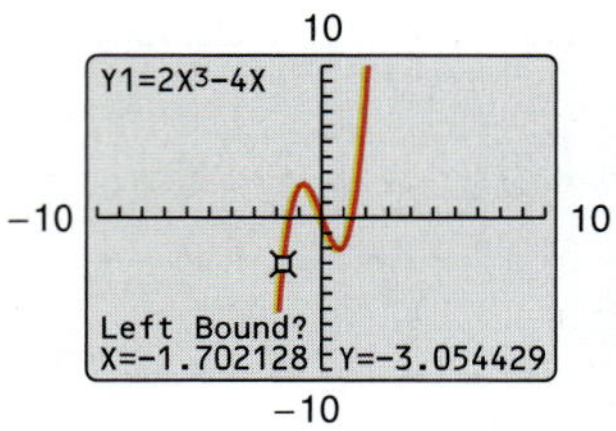

Figure A.137

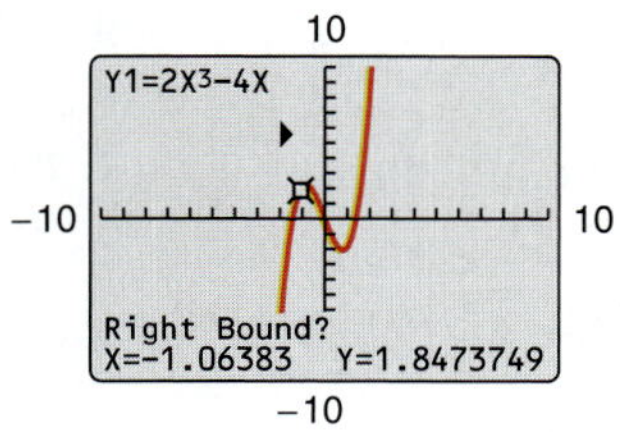

Figure A.138

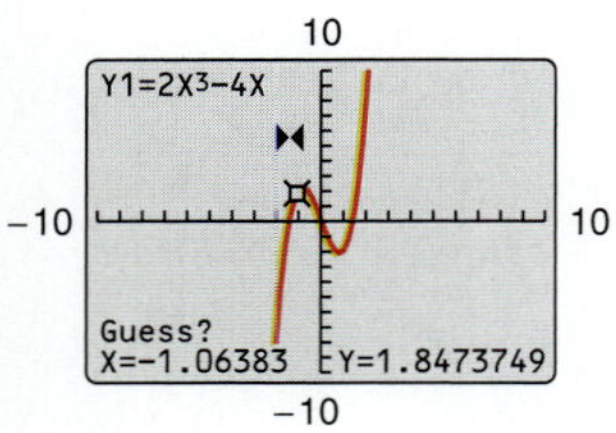

Figure A.139

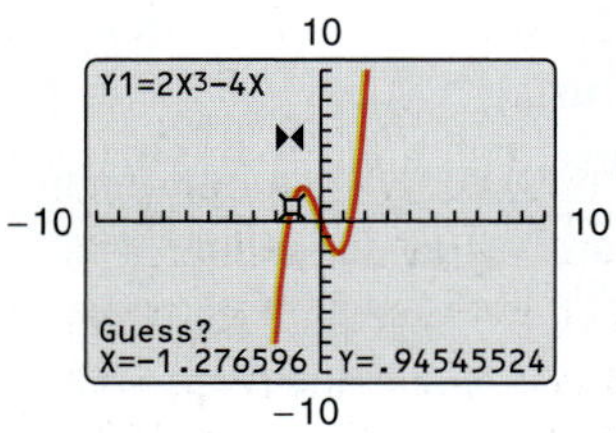

Figure A.140

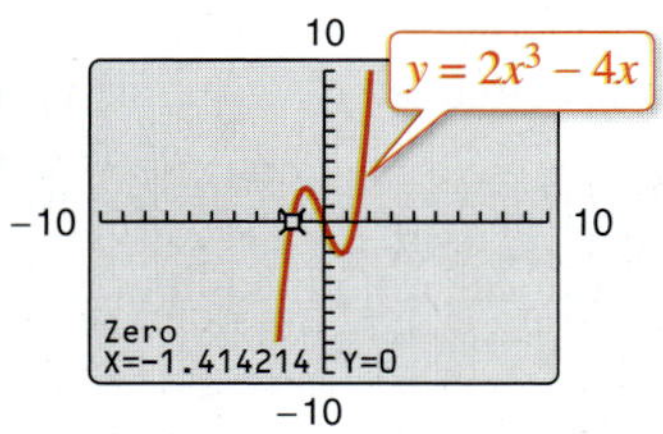

Figure A.141

Repeat this process to determine that the other two zeros of the function are $x = 0$ (see Figure A.142) and $x \approx 1.414214$ (see Figure A.143).

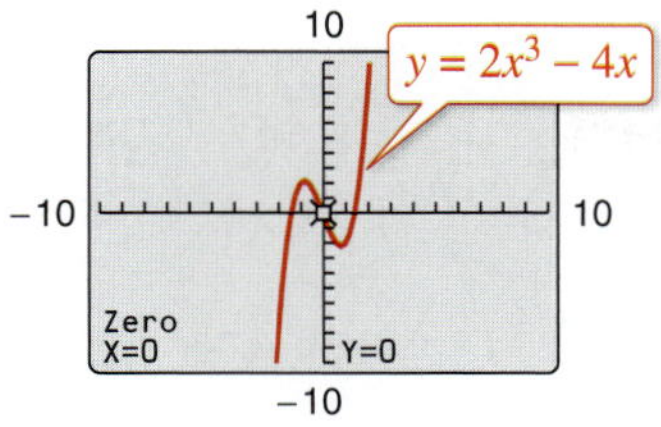

Figure A.142

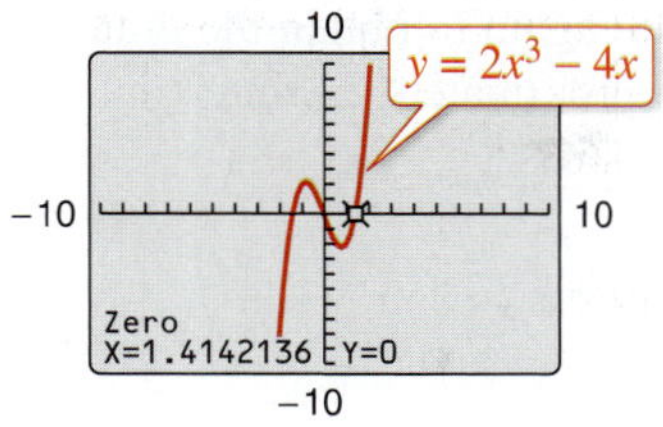

Figure A.143

Zoom and Trace Features

The *zoom* feature enables you to adjust the viewing window of a graph quickly (see Figure A.144). For example, the *zoom box* feature allows you to create a new viewing window by drawing a box around any part of the graph.

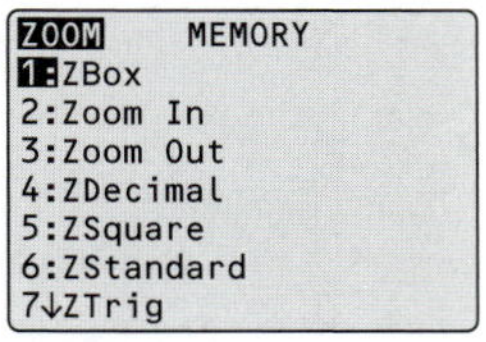

Figure A.144

The *trace* feature moves from point to point along a graph. For instance, enter the equation

$$y = 2x^3 - 3x + 2$$

in the *equation editor* (see Figure A.145) and graph the equation, as shown in Figure A.146. To activate the *trace* feature, press TRACE; then use the arrow keys to move the cursor along the graph. As you trace the graph, the coordinates of each point are displayed, as shown in Figure A.147.

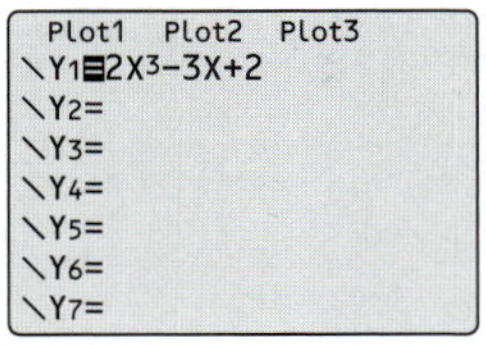

Figure A.145

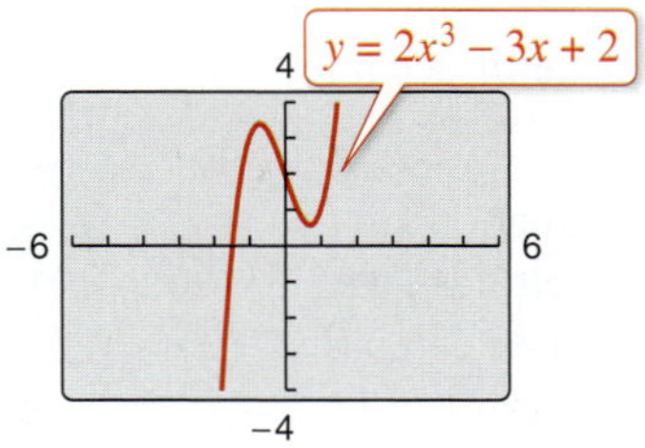

Figure A.146

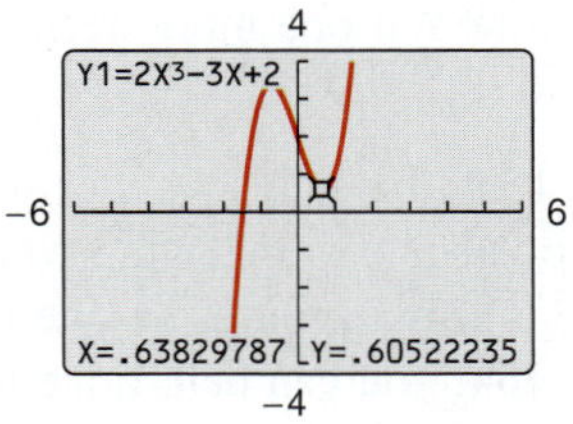

Figure A.147

The *trace* feature combined with the *zoom* feature enables you to obtain better and better approximations of desired points on a graph. For instance, you can use the *zoom* feature to approximate the x-intercept of the graph of

$$y = 2x^3 - 3x + 2.$$

From the viewing window shown in Figure A.146, the graph appears to have only one x-intercept. This intercept lies between -2 and -1. To zoom in on the x-intercept, choose the *zoom-in* feature from the *zoom* feature menu, as shown in Figure A.148. Next, trace the cursor to the point you want to zoom in on, in this case the x-intercept (see Figure A.149). Then press ENTER. You should obtain the graph shown in Figure A.150.

Now, using the *trace* feature, you can approximate the x-intercept to be

$$x \approx -1.468085$$

as shown in Figure A.151. Use the *zoom-in* feature again to obtain the graph shown in Figure A.152. Using the *trace* feature, you can approximate the x-intercept to be

$$x \approx -1.476064$$

as shown in Figure A.153.

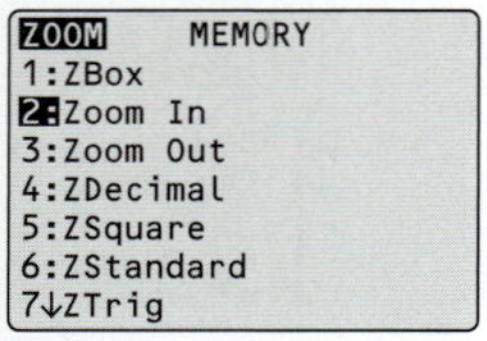

Figure A.148

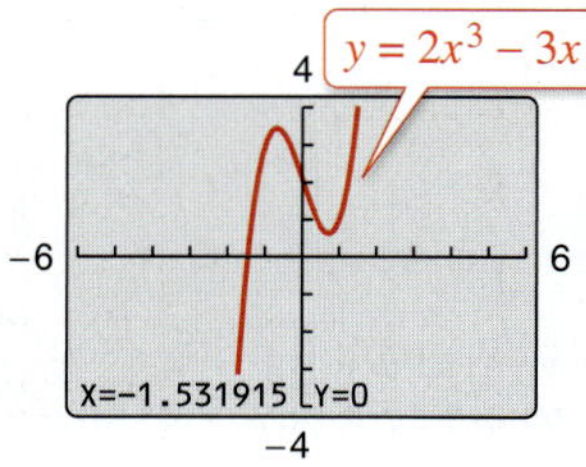

Figure A.149

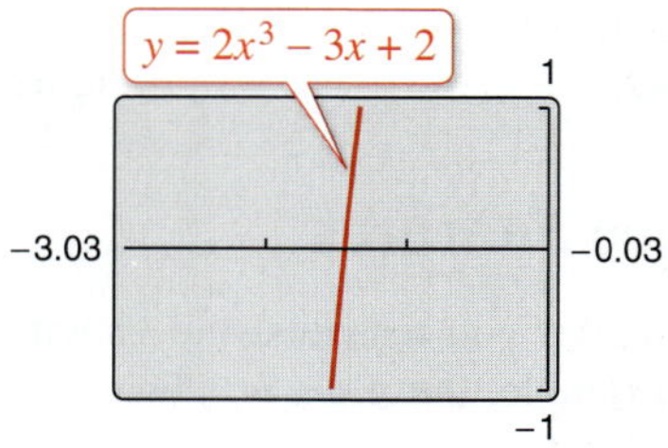

Figure A.150

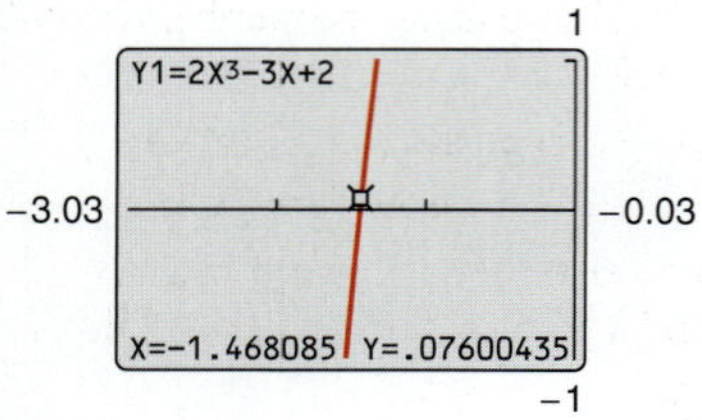

Figure A.151

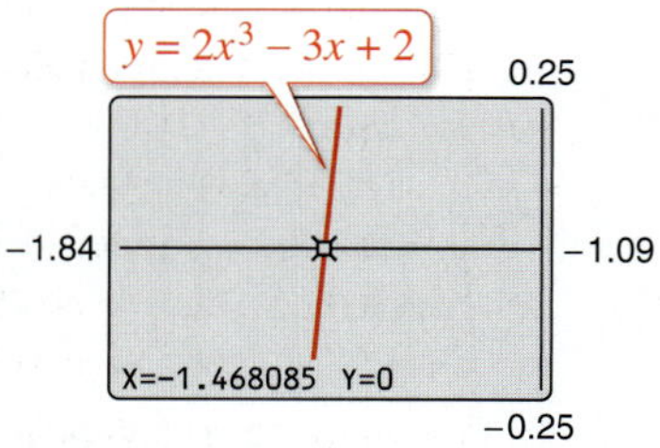

Figure A.152

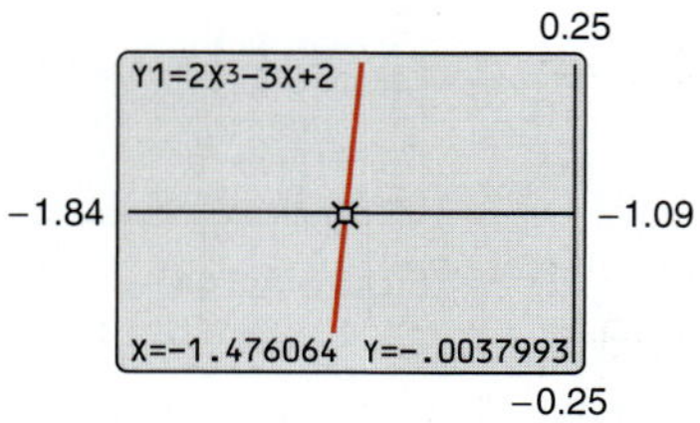

Figure A.153

Here are some suggestions for using the *zoom* feature.

1. With each successive zoom-in, adjust the x-scale so that the viewing window shows at least one tick mark on each side of the x-intercept.
2. The error in your approximation will be less than the distance between two scale marks.
3. The *trace* feature can usually be used to add one more decimal place of accuracy without changing the viewing window.

You can adjust the scale in Figure A.153 to obtain a better approximation of the x-intercept. Using the suggestions above, change the viewing window settings so that the viewing window shows at least one tick mark on each side of the x-intercept, as shown in Figure A.154. From Figure A.154, you can determine that the error in your approximation will be less than 0.001 (the Xscl value). Then, using the *trace* feature, you can improve the approximation, as shown in Figure A.155. To three decimal places, the x-intercept is

$$x \approx -1.476.$$

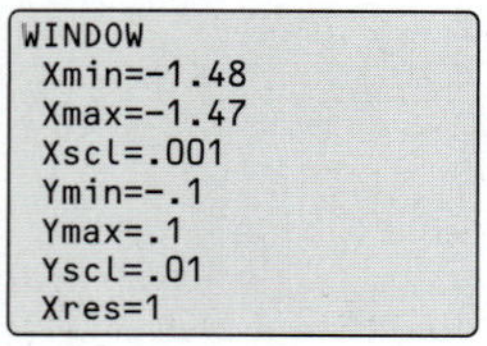

Figure A.154

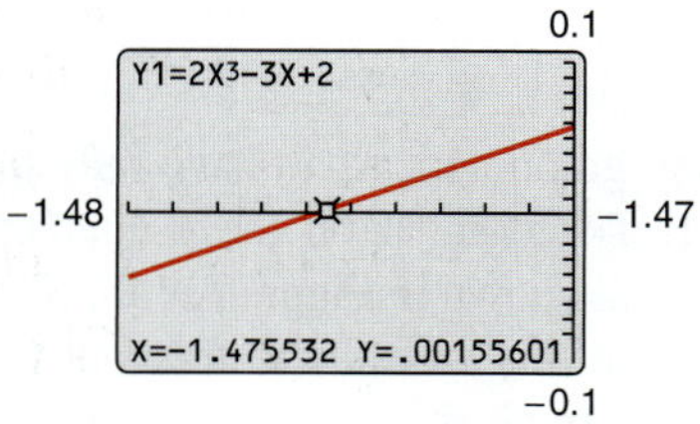

Figure A.155

Answers to Odd-Numbered Exercises and Tests

Chapter P

Section P.1 (page 9)

1. rational **3.** prime **5.** terms **7.** c
8. d **9.** e **10.** b **11.** f **12.** a
13. (a) 5, 1 (b) 5, 0, 1 (c) $-9, 5, 0, 1, -4, -1$
(d) $-9, -\frac{7}{2}, 5, \frac{2}{3}, 0, 1, -4, -1$ (e) $\sqrt{2}$
15. (a) 1, 20 (b) 1, 20 (c) $-13, 1, -10, 20$
(d) $2.01, 0.666\ldots, -13, 1, -10, 20$
(e) $0.010110111\ldots$
17. (a) $\frac{6}{3}, 3$ (b) $\frac{6}{3}, 3$ (c) $\frac{6}{3}, -2, 3, -3$
(d) $-\frac{1}{3}, \frac{6}{3}, -7.5, -2, 3, -3$ (e) $-\pi, \frac{1}{2}\sqrt{2}$
19. 0.3125 **21.** $0.\overline{123}$ **23.** $-9.\overline{09}$ **25.** $\frac{32}{5}$
27. $-\frac{123}{10}$ **29.** $-1 < 2.5$
31.
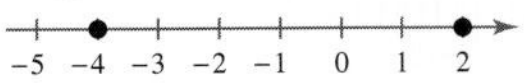
$-4 < 2$
33.
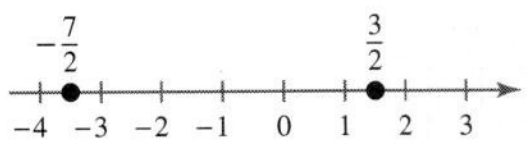
$\frac{3}{2} > -\frac{7}{2}$
35.
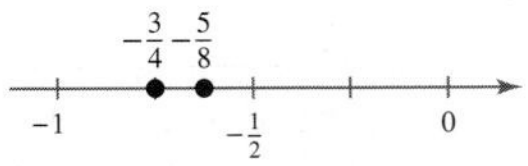
$-\frac{3}{4} < -\frac{5}{8}$
37. (a) $x \le 5$ is the set of all real numbers less than or equal to 5.
(b)
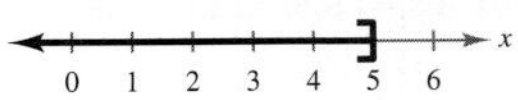
(c) Unbounded
39. (a) $x < 0$ is the set of all negative real numbers.
(b)
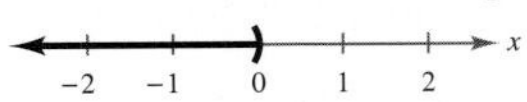
(c) Unbounded
41. (a) $-2 < x < 2$ is the set of all real numbers greater than -2 and less than 2.
(b)
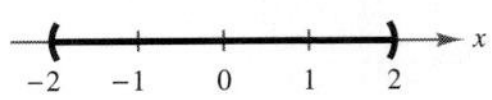
(c) Bounded
43. (a) $-1 \le x < 0$ is the set of all negative real numbers greater than or equal to -1.
(b)
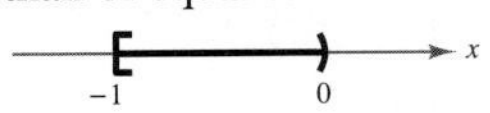
(c) Bounded
45. $x < 0$; $(-\infty, 0)$ **47.** $z \ge 10$; $[10, \infty)$
49. $9 \le t \le 24$; $[9, 24]$ **51.** $0 < m \le 5$; $(0, 5]$
53. $[-3, 8)$ **55.** $[-a, a+4]$
57. The set of all real numbers greater than -6
59. The set of all real numbers less than or equal to 2
61. 10 **63.** -6 **65.** -1 **67.** (a) 1 (b) -1
69. 11 **71.** $\frac{7}{2}$ **73.** $|-3| > -|-3|$ **75.** $-5 = -|5|$
77. $-|-1| < -(-1)$ **79.** 51 **81.** 7
83. $\frac{128}{75}$ **85.** $|x-5| \le 3$ **87.** $|y-0| \ge 6$
89. \$1091.2 billion; \$290.3 billion
91. \$2025.2 billion; \$236.2 billion
93. \$2524.0 billion; \$458.5 billion
95. $|\$113{,}356 - \$112{,}700| = \$656 > \$500$
$0.05(\$112{,}700) = \$5635$
Because the actual expenses differ from the budget by more than \$500, there is failure to meet the "budget variance test."
97. $|\$37{,}335 - \$37{,}600| = \$265 < \$500$
$0.05(\$37{,}600) = \$1880$
Because the difference between the actual expenses and the budget is less than \$500 and less than 5% of the budgeted amount, there is compliance with the "budget variance test."
99. Terms: $7x, 4$; coefficient: 7
101. Terms: $\sqrt{3}x^2, -8x, -11$; coefficients: $\sqrt{3}, -8$
103. Terms: $4x^3, \frac{x}{2}, -5$; coefficients: $4, \frac{1}{2}$
105. (a) 3 (b) -6 **107.** (a) 0 (b) 0
109. Distributive Property
111. Commutative Property of Addition
113. Associative Property of Addition
115. $\frac{1}{2}$ **117.** $\frac{3}{8}$ **119.** $\frac{x}{2}$ **121.** $\frac{96}{x}$ **123.** $-\frac{7}{5}$
125. 179 mi
127. False. 0 is nonnegative but not positive.
129. False. A contradiction can be shown using the numbers $a = 2$ and $b = 1$. $\left(2 > 1, \text{ but } \frac{1}{2} \not> \frac{1}{1}.\right)$
131. (a) $-A$ is negative. (b) $-C$ is positive.
(c) $B - A$ is negative. (d) $A - C$ is positive.
133. When u and v have the same sign, $|u+v| = |u| + |v|$.
When u and v have different signs, $|u+v| < |u| + |v|$.

Section P.2 (page 20)

1. exponent, base **3.** principal nth root
5. rationalizing **7.** $2 - 3\sqrt{5}$
9. No **11.** (a) 81 (b) $\frac{1}{9}$
13. (a) 1 (b) 1296 **15.** (a) 1728 (b) -729
17. (a) $\frac{5}{6}$ (b) 27 **19.** -54 **21.** 5 **23.** $\frac{7}{4}$
25. (a) x^5 (b) $3x^9$ **27.** (a) $9x^2$ (b) 1
29. (a) $\frac{7}{x}$ (b) $\frac{4}{3}(x+y)^2, x+y \ne 0$
31. (a) $\frac{x^2}{y^2}, x \ne 0$ (b) $\frac{b^5}{a^5}, b \ne 0$
33. -1600 **35.** 2.125 **37.** 5103 **39.** 9.735×10^2
41. 1.0252484×10^4 **43.** -1.11025×10^3
45. 2.485×10^{-4} **47.** -2.5×10^{-6}
49. 5.73×10^7 **51.** 8.99×10^{-5} **53.** 10,800
55. -0.000000765 **57.** 514 **59.** 0.00009
61. 60,000 **63.** 175,000,000 **65.** 5×10^4 or 50,000
67. (a) 4.907×10^{17} (b) 1.479
69. (a) 67,082.039 (b) 9.390
71. 11 **73.** 4 **75.** Not a real number. **77.** $-\frac{1}{3}$
79. 12.651 **81.** 0.005 **83.** -281.088 **85.** 0.149
87. 2.709 **89.** 1,010,000.128 **91.** 20 **93.** 6
95. (a) $3\sqrt{5}$ (b) $2\sqrt[3]{4a^2/b^2}$
97. (a) $2x\sqrt[3]{2x^2}$ (b) $\frac{5|x|\sqrt{3}}{y^2}$
99. (a) $34\sqrt{2}$ (b) $22\sqrt{2}$
101. (a) $13\sqrt{x+1}$ (b) $18\sqrt{5x}$ **103.** $\sqrt{\frac{3}{11}} = \frac{\sqrt{3}}{\sqrt{11}}$

105. $5 > \sqrt{3^2 + 2^2}$ **107.** $\frac{\sqrt{3}}{3}$ **109.** $\frac{\sqrt{14} + 2}{2}$

111. $\frac{3}{\sqrt{3}}$ **113.** $\frac{2}{3(\sqrt{5} - \sqrt{3})}$

115. (a) $\sqrt{3}$ (b) $\sqrt[3]{(x + 1)^2}$ **117.** $64^{1/3}$ **119.** $\sqrt[5]{\frac{1}{32}}$

121. $\sqrt[5]{-243}$ **123.** $81^{3/4}$ **125.** $\frac{2}{|x|}$ **127.** $\frac{1}{x^3}, x > 0$

129. $\frac{1}{8}$ **131.** -3 **133.** $\sqrt[4]{35}$ **135.** $3\sqrt[4]{3(x + 1)}$

137. Brazil: 1.09×10^4 Iran: 5.16×10^3
Canada: 5.29×10^4 Ireland: 4.62×10^4
Germany: 4.43×10^4 Mexico: 1.12×10^4
India: 1.44×10^3

139. $t \approx 13.29$ sec

141. False. $(a^n)^k = a^{nk}$. In general, $nk \neq n^k$.

143. True. $x^{-1} + y^{-1} = \frac{1}{x} + \frac{1}{y} = \frac{y}{xy} + \frac{x}{xy} = \frac{x + y}{xy}$

145. $1 = \frac{a^n}{a^n} = a^{n-n} = a^0$

147. No. Rationalizing the denominator produces a number equivalent to the original fraction; squaring does not.

Section P.3 (page 31)

1. n, a_n **3.** First, Outer, Inner, Last

5. When each of its factors is prime.

7. d **8.** e **9.** b **10.** a **11.** f **12.** c

13. Answers will vary, but first term is $-2x^3$.

15. Answers will vary, but first term has the form $-ax^4, a > 0$.

17. $4x^2 + 3x + 2$
Degree: 2; leading coefficient: 4

19. $x^7 - 8$
Degree: 7; leading coefficient: 1

21. $-2x^5 + 6x^4 - x + 1$
Degree: 5; leading coefficient: -2

23. Polynomial: $-8y^2 + 2y$ **25.** Not a polynomial

27. $3x + 10$ **29.** $2x + 17$ **31.** $x^2 - 20$

33. $8.1x^3 + 29.7x^2 + 11$ **35.** $5z^2 - 40z$ **37.** $6y^2 - 20y$

39. $3x^3 - 6x^2 + 3x$ **41.** $x^2 + 7x + 12$

43. $6x^2 - 7x - 5$ **45.** $16y^2 + 56y + 49$

47. $25x^2 - 20x + 4$ **49.** $x^2 - 81$ **51.** $x^2 - 4y^2$

53. $x^3 + 3x^2 + 3x + 1$ **55.** $8x^3 - 12x^2y + 6xy^2 - y^3$

57. $\frac{1}{16}x^2 - 9$ **59.** $\frac{25}{4}x^2 + 15x + 9$

61. $-3x^4 - x^3 - 12x^2 - 19x - 5$ **63.** $x^2 + 2xz + z^2 - 25$

65. $x^2 + 2xy + y^2 - 6y - 6x + 9$ **67.** $5(x - 8)$

69. $2x(x^2 - 3)$ **71.** $(x - 5)(3x + 8)$

73. $(5x - 4)(5x - 3)$ **75.** $(x + 6)(x - 6)$

77. $3(4y + 3)(4y - 3)$ **79.** $\left(2x - \frac{1}{3}\right)\left(2x + \frac{1}{3}\right)$

81. $[(x - 1) - 2][(x - 1) + 2] = (x - 3)(x + 1)$

83. $(x - 2)^2$ **85.** $\left(x + \frac{1}{2}\right)^2$ **87.** $(2x - 3)^2$

89. $\left(2x - \frac{1}{3}\right)^2$ **91.** $(x - 2)(x^2 + 2x + 4)$

93. $(z + 1)(z^2 - z + 1)$ **95.** $\left(x + \frac{1}{3}\right)\left(x^2 - \frac{1}{3}x + \frac{1}{9}\right)$

97. $(5v - 1)(25v^2 + 5v + 1)$

99. $(-x + y + 1)(x^2 + xy + x + y^2 + 2y + 1)$

101. $(x - 1)(x + 2)$ **103.** $(s - 2)(s - 3)$

105. $-(y - 4)(y + 5)$ **107.** $(3x - 2)(x + 5)$

109. $(5x + 1)(x + 5)$ **111.** $-(5u - 2)(u + 3)$

113. $\frac{1}{96}(3x + 2)(4x - 3)$ **115.** $(x - 1)(x^2 + 2)$

117. $(x^2 + 1)(x - 5)$ **119.** $(x - 4)(x + 5)$

121. $(3x + 2)(2x - 1)$ **123.** $10(x - 2)(x + 2)$

125. $y(y + 1)(y - 1)$ **127.** $(x - 1)^2$ **129.** $(2x - 1)^2$

131. $-2x(x^2 - x - 3)$ **133.** $(9x + 1)(x + 1)$

135. $(3x + 1)(x^2 + 5)$ **137.** $(u + 2)(3 - u^2)$

139. $(x - 2)(x + 2)(2x + 1)$ **141.** $(x + 1)^2(x - 1)^2$

143. $3(t + 2)(t^2 - 2t + 4)$ **145.** $2(2x - 1)(4x - 1)$

147. $-(x + 1)(x - 3)(x + 9)$

149. (a) $1000r^2 + 2000r + 1000$

(b)

r	1%	$1\frac{1}{2}$%	2%
$1000(1 + r)^2$	1020.10	1030.23	1040.40

r	$2\frac{1}{2}$%	3%
$1000(1 + r)^2$	1050.63	1060.9

(c) The amount increases with increasing r.

151. (a) $T(x) = 0.0475x^2 + 1.099x + 0.23$

(b)

x (mi/h)	30	40	55
T (ft)	75.95	120.19	204.36

(c) Stopping distance increases as speed increases.

153. a

155.

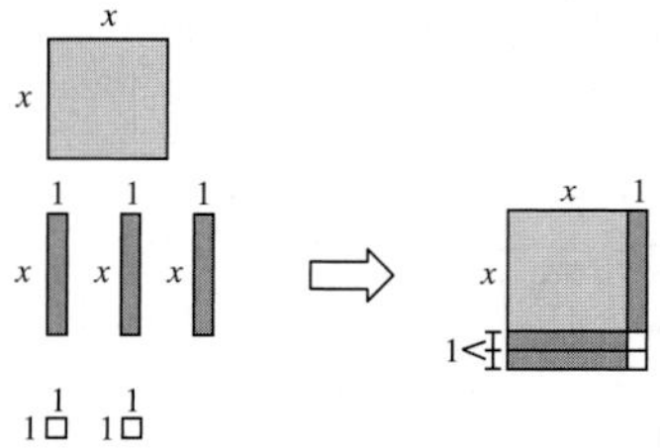

157.

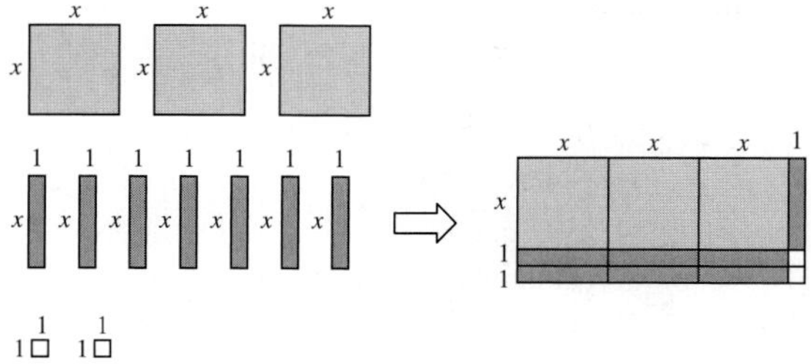

159. $4\pi(r + 1)$ **161.** $(4x^3)(2x + 1)^3(2x^2 + 2x + 1)$

163. $(2x - 5)^3(5x - 4)^2(70x - 107)$

165. $-\frac{8(x - 2)}{(2x + 3)^3}$ **167.** $-14, 14, -2, 2$

169. $-51, 51, -15, 15, -27, 27$

171. Sample answer: $-2, -6$ **173.** Sample answer: $2, -3$

175. (a) $V = \pi h(R - r)(R + r)$

(b) $V = 2\pi\left[\left(\frac{R + r}{2}\right)(R - r)\right]h$

177. False. $(x^2 - 1)(x^2 + 1)$ becomes a fourth-degree polynomial.

179. False. If the leading coefficients are opposites, the sum will be a first-degree polynomial.

181. False. $(3x - 6)(x + 1) = 3(x - 2)(x + 1)$

183. Degree: $m + n$

185. Answers will vary. Sample answer: To cube a binomial difference, cube the first term. Next subtract 3 times the square of the first term multiplied by the second term. Then add 3 times the first term multiplied by the square of the second term. Lastly, subtract the cube of the second term $(x - y)^3 = x^3 - 3x^2y + 3xy^2 - y^3$.

187. A 3 was not factored out of the second binomial.

189. $(x^n + y^n)(x^n - y^n)$

Section P.4 (page 42)

1. domain **3.** complex fractions

5. When its numerator and denominator have no factors in common aside from ± 1

7. All real numbers x **9.** All positive real numbers x

11. All real numbers x except $x = -2$

13. All real numbers x except $x = -7$

15. All real numbers x except $x = \pm\frac{1}{3}$

17. All real numbers x such that $x \geq -10$

19. All real numbers x such that $x \leq 4$

21. All real numbers x such that $x > -1$

23. $3x, x \neq 0$ **25.** $(x + 1), x \neq -1$ **27.** $x + 2, x \neq -2$

29. $\frac{3x}{2}, x \neq 0$ **31.** $\frac{3y}{xy + x}$ **33.** $-\frac{4y}{5}, y \neq \frac{1}{2}$

35. $-\frac{1}{2}, x \neq 5$ **37.** $y - 4, y \neq -4$

39. $\frac{x(x + 3)}{x - 2}, x \neq -2$ **41.** $\frac{y - 4}{y + 6}, y \neq 3$

43. $-(x^2 + 1), x \neq 2$ **45.** $z - 2$

47.

x	-4	-3	-2	-1	0	1	2
$\frac{x^2 + 2x - 3}{x - 1}$	-1	0	1	2	3	Undef.	5
$x + 3$	-1	0	1	2	3	4	5

The expressions are equivalent except at $x = 1$.

49. $\frac{1}{5(x - 2)}, x \neq 1$ **51.** $-\frac{r + 1}{r}, r \neq 1, -1$

53. $\frac{t - 3}{(t + 3)(t - 2)}, t \neq -2$ **55.** $\frac{3}{2}, x \neq -y$

57. $\frac{\pi}{4}$ **59.** $\frac{x + 5}{x - 1}$ **61.** $-\frac{2x^2 - 5x - 18}{(2x + 1)(x + 3)}$

63. $-\frac{2}{x - 2}$ **65.** $-\frac{x^2 + 3}{(x + 1)(x - 2)(x - 3)}$

67. $\frac{4x + 1}{(x - 1)(x + 1)}$ **69.** $\frac{5x^2 - 5x - 6}{x^2(x + 1)}$ **71.** $\frac{1}{2}, x \neq 2$

73. $x(x + 1), x \neq -1, 0$ **75.** $-\frac{2x + h}{x^2(x + h)^2}, h \neq 0$

77. $\frac{2x - 1}{2x}, x > 0$ **79.** $x^{-2}(x^7 - 2) = \frac{x^7 - 2}{x^2}$

81. $-\frac{1}{(x^2 + 1)^5}$ **83.** $\frac{2x^3 - 2x^2 - 5}{(x - 1)^{1/2}}$ **85.** $\frac{2x^2 - 1}{x^{5/2}}$

87. $-\frac{(x - 1)(x^2 + x + 2)}{x^2(x^2 + 1)^{3/2}}$ **89.** $\frac{-2(3x^2 + 3x - 5)}{\sqrt{4x + 3}(x^2 + 5)^2}$

91. $\frac{1}{\sqrt{x + 4} + \sqrt{x}}$ **93.** $\frac{1}{\sqrt{x + 2} + \sqrt{2}}, x \neq 0$

95. $-\frac{1}{\sqrt{1 - x} + 1}, x \neq 0$ **97.** $\frac{x}{2(2x + 1)}$

99. (a) 6.39% (b) $\frac{288(MN - P)}{N(MN + 12P)}$; 6.39%

101. (a) $\frac{1}{50}$ min (b) $\frac{x}{50}$ min (c) 2.4 min

103. (a)

Year	2005	2006	2007	2008
Births (in millions)	4.152	4.266	4.341	4.265
Population (in millions)	295.9	298.5	301.2	303.8

Year	2009	2010	2011	2012
Births (in millions)	4.125	4.027	3.971	3.939
Population (in millions)	306.5	309.1	311.7	314.4

(b) The models are close to the actual data.

(c) $\frac{0.06815t^2 - 0.9865t + 3.948}{(0.01753t^2 - 0.2530t + 1)(2.64t + 282.7)}$

(d)

Year	2005	2006	2007	2008
Ratio	0.0140	0.0143	0.0144	0.0140

Year	2009	2010	2011	2012
Ratio	0.0135	0.0130	0.0127	0.0125

The ratio has remained fairly constant over time.

105. $2x + h, h \neq 0$

107. $\frac{-2x - h}{x^2(x + h)^2}, h \neq 0$ **109.** $\frac{4n^2 + 6n + 26}{3}, n \neq 0$

111. False. The domain of the left-hand side is $x^n \neq 1$.

113. The expression will still have a radical in the denominator.

115. Only common factors of the numerator and denominator can be canceled. In this case, factors of terms were incorrectly canceled.

117. Answers will vary. Counterexample: Let $x = y = 1$.

$$\frac{1}{1 + 1} = \frac{1}{2} \neq \frac{1}{1} + \frac{1}{1}$$

119. $\frac{2x^2 - 1}{x^4}$; Answers will vary.

Section P.5 (page 53)

1. Cartesian **3.** Midpoint Formula

5. c **6.** f **7.** a **8.** d **9.** e **10.** b

11. A: (2, 6); B: (−6, −2); C: (4, −4); D: (−3, 2)

13.

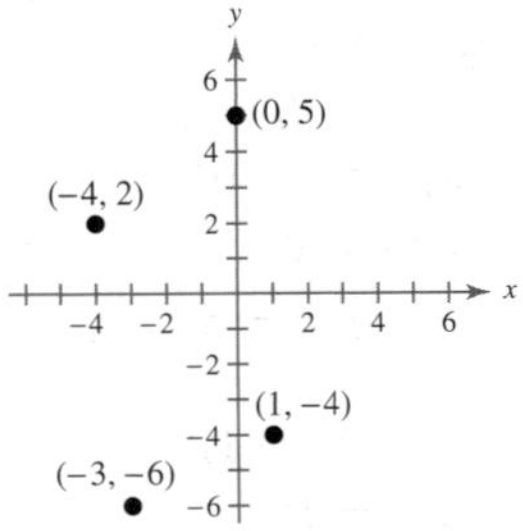

15.

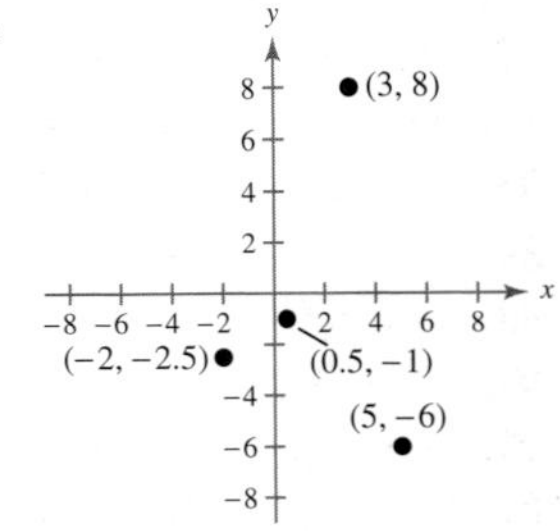

17. $(-5, 4)$ **19.** $(0, -6)$ **21.** Quadrant IV
23. Quadrant II **25.** Quadrant III or IV
27. Quadrant III **29.** Quadrant I or III
31.

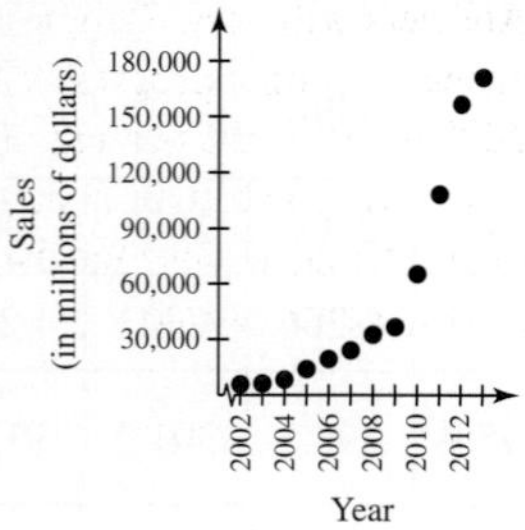

33. 8 **35.** 13 **37.** $5\sqrt{2}$ **39.** $\dfrac{\sqrt{277}}{6}$ **41.** $\sqrt{71.78}$

43. (a) 3, 4, 5 (b) $3^2 + 4^2 = 5^2$

45. (a) 10, 3, $\sqrt{109}$ (b) $10^2 + 3^2 = (\sqrt{109})^2$

47. $(\sqrt{5})^2 + (\sqrt{45})^2 = (\sqrt{50})^2$

49. Two equal sides of length $\sqrt{29}$

51. Opposite sides have equal lengths of $2\sqrt{5}$ and $\sqrt{85}$.

53. The diagonals are of equal length $(\sqrt{58})$. The slope of the line between $(-5, 6)$ and $(0, 8)$ is $\frac{2}{5}$. The slope of the line between $(-5, 6)$ and $(-3, 1)$ is $-\frac{5}{2}$. The slopes are negative reciprocals, indicating perpendicular lines, which form a right angle.

55. $5\sqrt{74} \approx 43$ yd **57.** 192.1 km

59. (a)

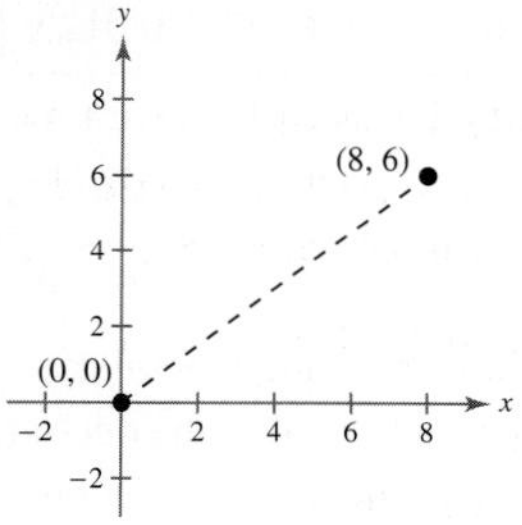

(b) $(4, 3)$

61. (a)

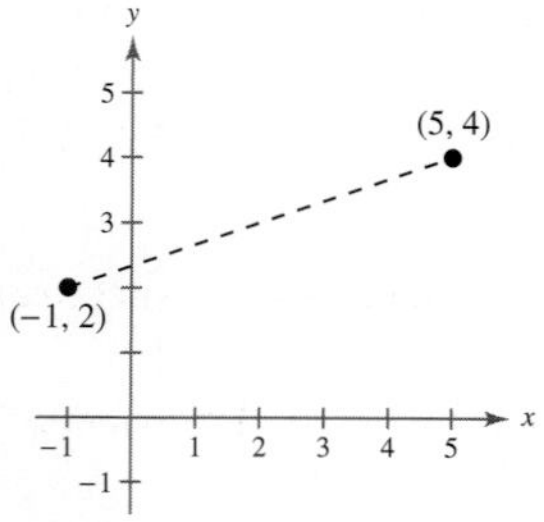

(b) $(2, 3)$

63. (a)

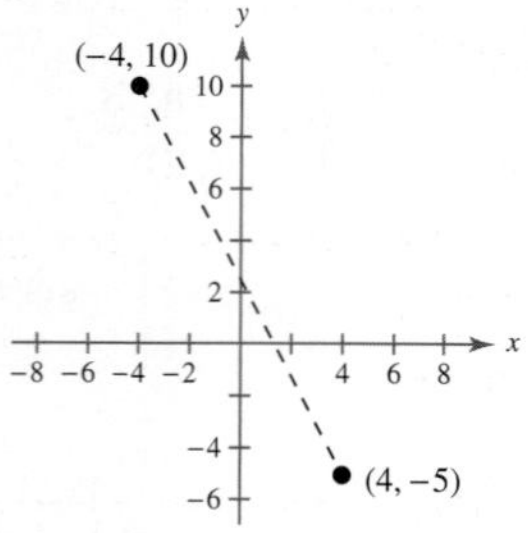

(b) $\left(0, \frac{5}{2}\right)$

65. (a)

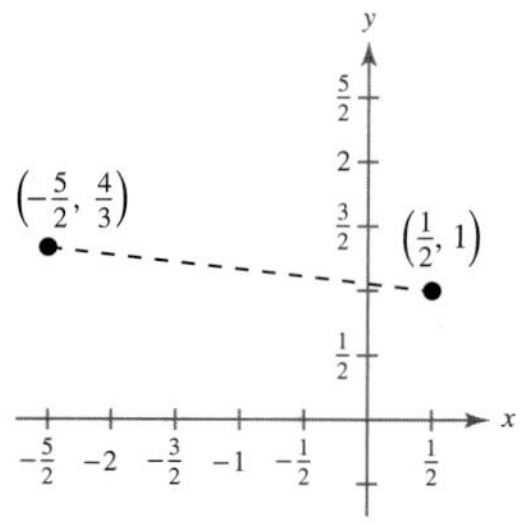

(b) $\left(-1, \dfrac{7}{6}\right)$

67. (a)

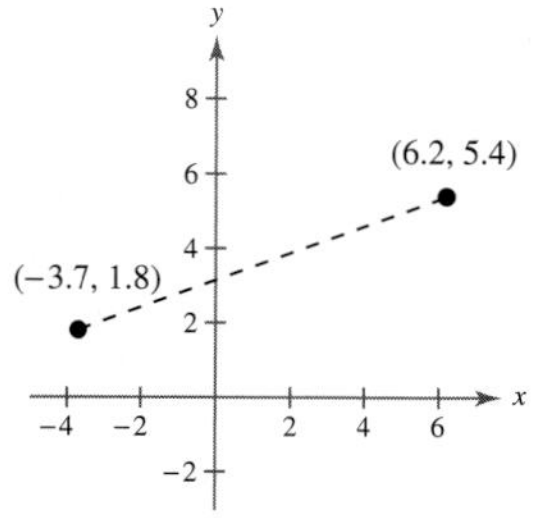

(b) $(1.25, 3.6)$

69. \$1182.5 million **71.** $x^2 + y^2 = 25$

73. $(x - 2)^2 + (y + 1)^2 = 16$ **75.** $(x + 1)^2 + (y - 2)^2 = 5$

77. $x^2 + y^2 = 17$ **79.** $(x + 2)^2 + (y - 1)^2 = 1$

81. $(x - 3)^2 + (y + 6)^2 = 16$

83. $(x - 2)^2 + (y + 1)^2 = 16$

85. Center: $(0, 0)$
Radius: 5

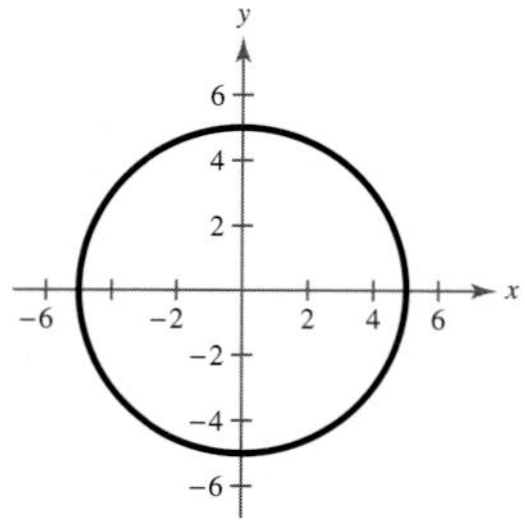

87. Center: $(6, 0)$
Radius: 3

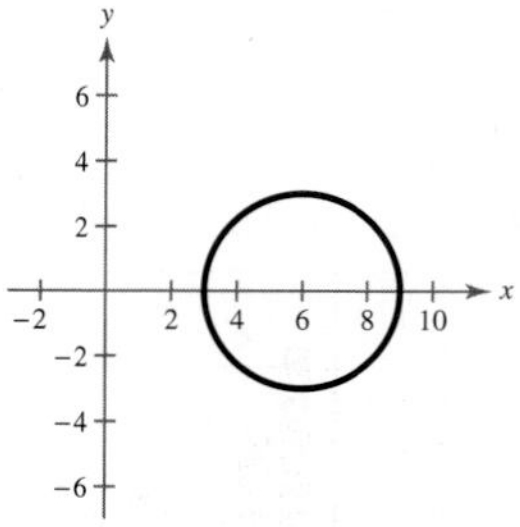

89. Center: $\left(\frac{1}{3}, -\frac{2}{3}\right)$
Radius: $\frac{4}{3}$

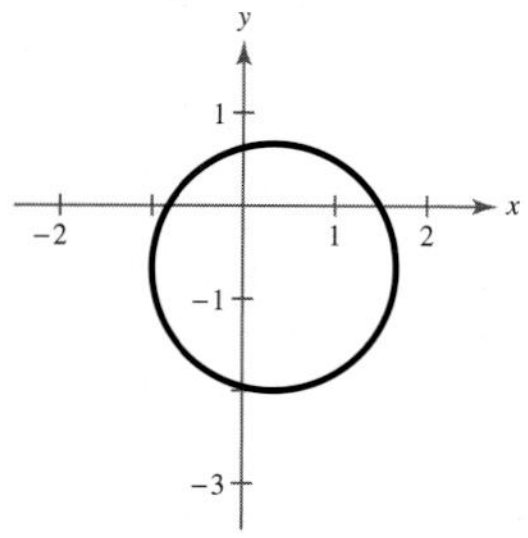

91. $(0, 1), (4, 2), (1, 4)$

93. $(-1, 5), (-4, 8), (-6, 5), (-3, 2)$

95. (a) 65
(b) No. There are many variables that will affect the final exam score.

97. True. The lengths of the sides from $(-8, 4)$ to $(2, 11)$ and from $(2, 11)$ to $(-5, 1)$ are both $\sqrt{149}$.

99. 0; 0

101. $\left(\frac{3x_1 + x_2}{4}, \frac{3y_1 + y_2}{4}\right), \left(\frac{x_1 + x_2}{2}, \frac{y_1 + y_2}{2}\right),$
$\left(\frac{x_1 + 3x_2}{4}, \frac{y_1 + 3y_2}{4}\right)$

(a) $\left(\frac{7}{4}, -\frac{7}{4}\right), \left(\frac{5}{2}, -\frac{3}{2}\right), \left(\frac{13}{4}, -\frac{5}{4}\right)$

(b) $\left(-\frac{3}{2}, -\frac{9}{4}\right), \left(-1, -\frac{3}{2}\right), \left(-\frac{1}{2}, -\frac{3}{4}\right)$

103. Proof

Section P.6 (page 62)

1. Line plots **3.** c **4.** d **5.** a **6.** b

7. (a) \$3.52 (b) \$0.56

9.

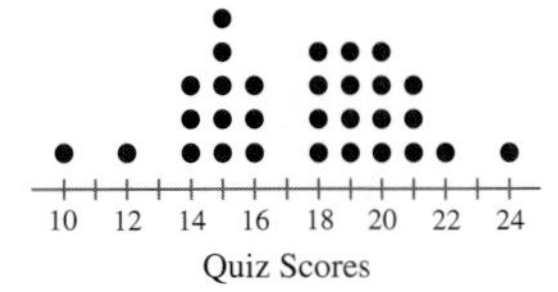

15

11.

Interval	Tally
[7, 10)	||
[10, 13)	𝍸 𝍸 𝍸
[13, 16)	𝍸 𝍸 𝍸 |||
[16, 19)	𝍸 𝍸 |
[19, 22)	||||

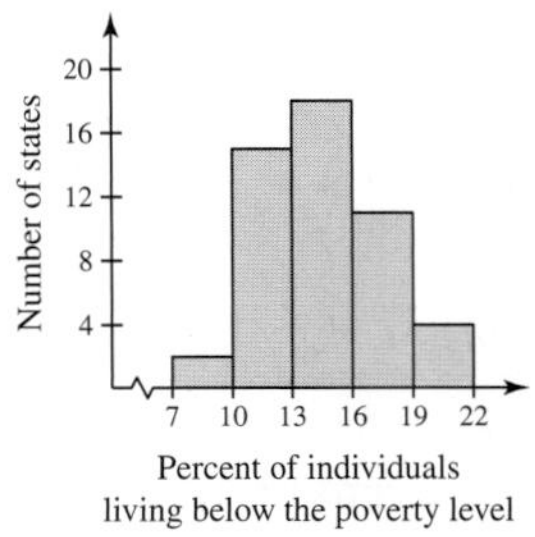

13.

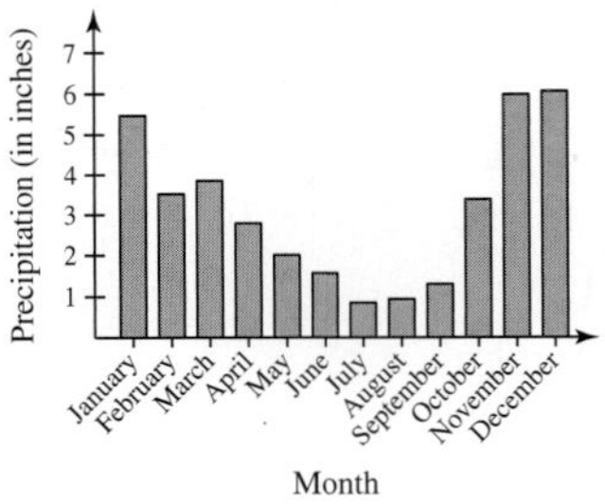

Answers will vary. Sample answer: The amount of precipitation decreases at a fairly constant rate from January to July, and then it starts to increase at a fairly constant rate until December.

15.

Year	2008	2009	2010
Differences in tuition charges (in dollars)	16,681	17,058	16,693

Year	2011	2012	2013
Differences in tuition charges (in dollars)	17,110	17,291	18,044

17.

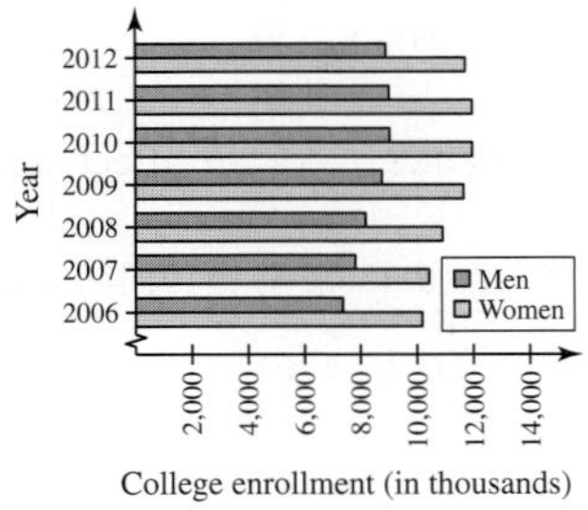

19. The price increased at a fairly constant rate from 2005 to 2008.

21. About 13% **23.** \$2.03; December

25. From November to December

27.

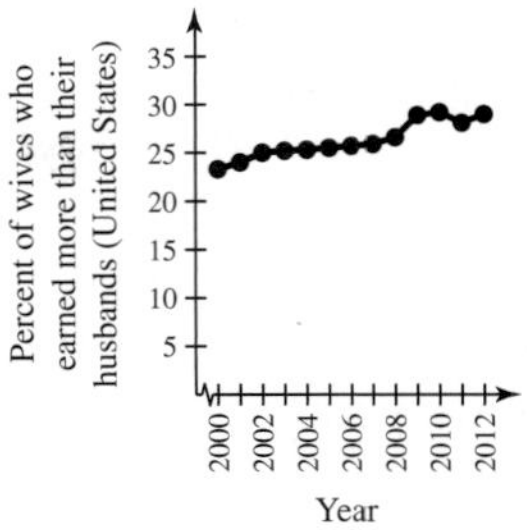

Answers will vary. Sample answer: As time progresses from 2000 to 2012, the percent of wives who earned more income than their husbands increases at a fairly constant rate.

29.

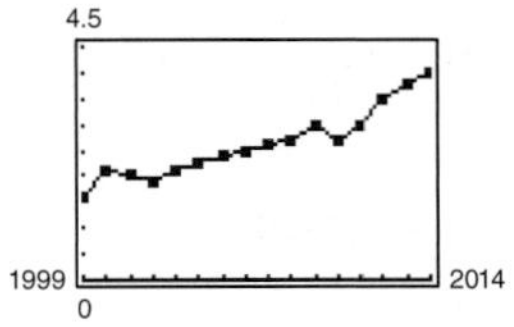

31.

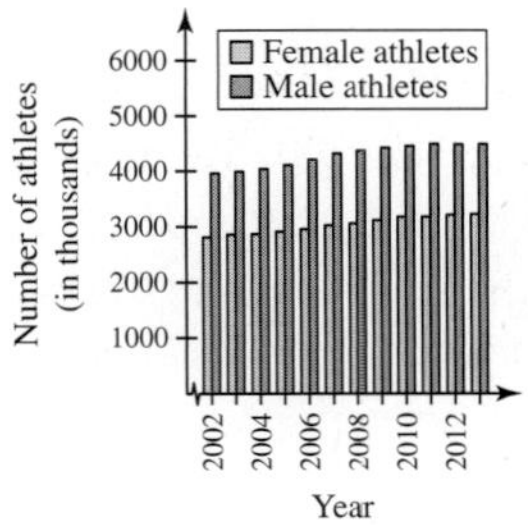

Answers will vary. Sample answer: A double bar graph is best because there are two different sets of data within the same time interval that do not deal primarily with increasing or decreasing behavior.

33. Answers will vary.

35. Line plots are useful for ordering small sets.
Histograms or bar graphs can be used to organize larger sets.
Line graphs are used to show trends over periods of time.

Review Exercises (page 68)

1. (a) 11 (b) 11 (c) $11, -14$
(d) $11, -14, -\frac{8}{9}, \frac{5}{2}, 0.4$ (e) $\sqrt{6}$

3. (a) $0.8\overline{3}$ (b) 0.875

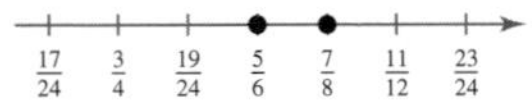

$\frac{5}{6} < \frac{7}{8}$

5. (a) The set consists of all real numbers x greater than or equal to -6.
(b)

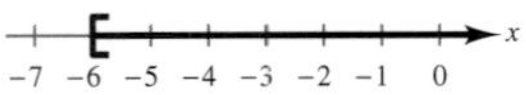

(c) Unbounded

7. (a) The set contains all real numbers x greater than or equal to -4 and less than or equal to 0.
(b)

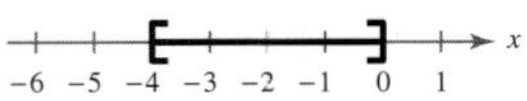

(c) Bounded

9. 122 **11.** $|x - 7| \geq 6$ **13.** (a) -11 (b) 25

15. (a) Division by zero is undefined. (b) $-\frac{3}{2}$

17. Associative Property of Addition

19. Additive Identity Property

21. Multiplicative Inverse Property

23. (a) $-8z^3$ (b) $3a^3b^2$ **25.** (a) $\frac{3u^2}{v^4}$ (b) m^{-2}

27. 2.585×10^9 **29.** -1.25×10^{-7} **31.** 128,000

33. 0.000018 **35.** 78 **37.** 2 **39.** $5a\sqrt{a}$ **41.** $\frac{3}{4}$

43. $\dfrac{5|x|\sqrt{3}}{y^2}$ **45.** $\sqrt{3}$ **47.** $3\sqrt{3x}$ **49.** $\sqrt{2x}(2x + 1)$

51. $\dfrac{3 + \sqrt{5}}{4}$ **53.** $\dfrac{5}{2\sqrt{5}}$ **55.** 32,768 **57.** $6x^{1/3}, x \neq 0$

59. $-2x^5 - x^4 + 3x^3 + 15x^2 + 5$; degree: 5; leading coefficient: -2

61. $3x^2 + 7x - 1$ **63.** $2x^3 - x^2 + 3x - 9$

65. $-2a^3 - 2a^2 + 6a$ **67.** $x^2 + 13x + 36$ **69.** $x^2 - 64$

71. $x^3 - 12x^2 + 48x - 64$ **73.** $m^2 - 14m - n^2 + 49$

75. $(x + 3)(x + 5) = 5(x + 3) + x(x + 3)$; Distributive Property

77. $7(x + 5)$ **79.** $2x(x^2 + 9x - 2)$ **81.** $(x - 3)(x + 4)$

83. $(x - 13)(x + 13)$ **85.** $(x + 3)^2$

87. $(x + 6)(x^2 - 6x + 36)$ **89.** $(x - 9)(x + 3)$

91. $(2x + 1)(x + 10)$ **93.** $(x - 4)(x^2 - 3)$

95. $(x - 3)(2x + 5)$ **97.** All real numbers x

99. All real numbers x except $x = \frac{3}{2}$

101. $\dfrac{x}{x^2 + 7}, x \neq 0$ **103.** $\dfrac{x - 6}{x - 5}, x \neq -5$

105. $\dfrac{1}{x^2}, x \neq \pm 2$ **107.** $\dfrac{1}{5}x(5x - 6), x \neq 0, -\dfrac{3}{2}$

109. $\dfrac{x^3 - x + 3}{(x - 1)(x + 2)}$ **111.** $\dfrac{x + 1}{x(x^2 + 1)}$

113. $-\dfrac{1}{xy(x + y)}, x \neq y$ **115.** $\dfrac{x(3x^2 + 1)}{(2x^2 + 1)^4}$

117.

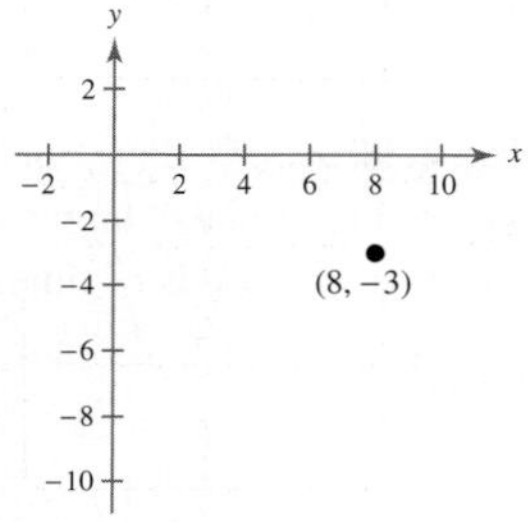

Quadrant IV

119.

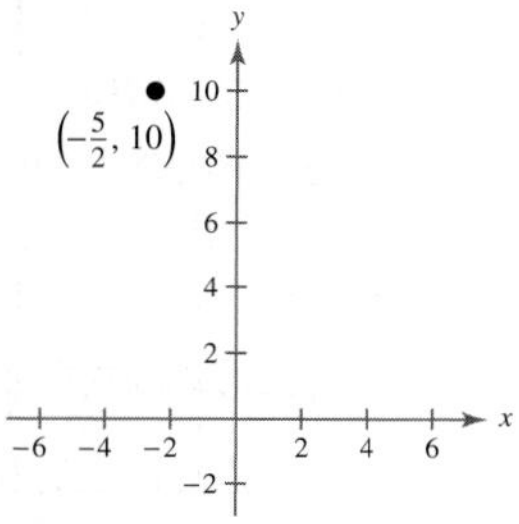

Quadrant II

121.

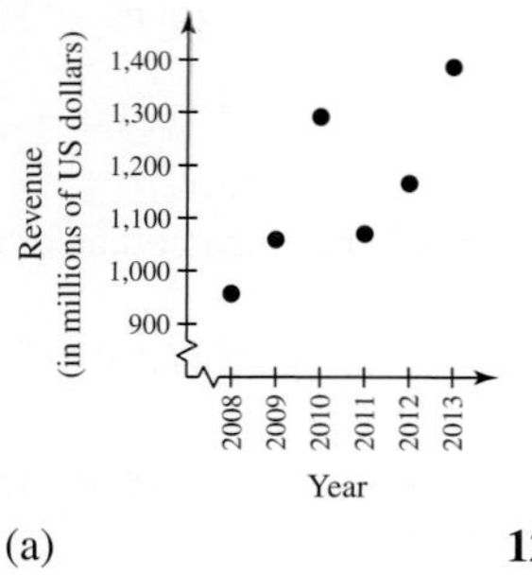

123. (a)

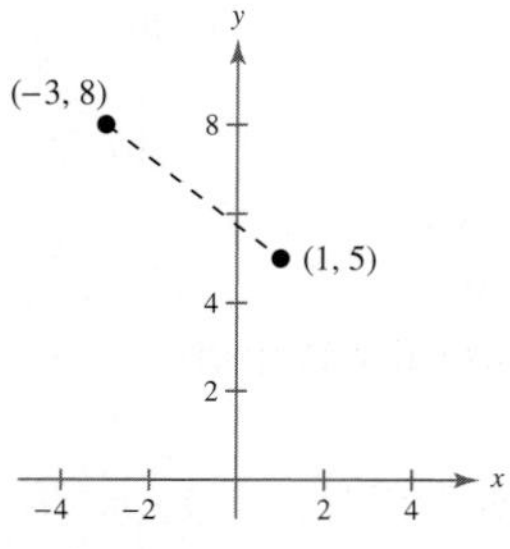

(b) Distance: 5

(c) Midpoint: $(-1, 6.5)$

125. (a)

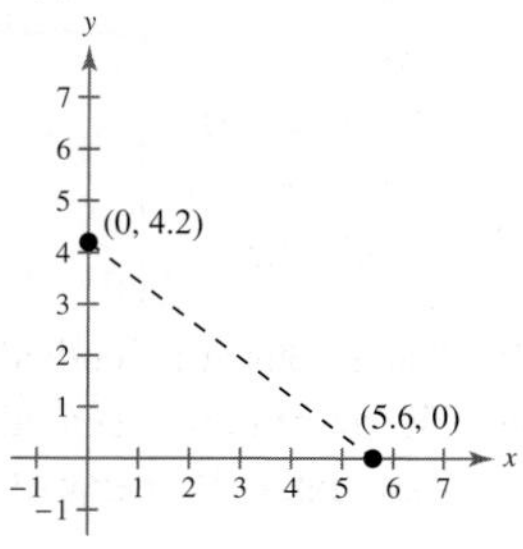

(b) Distance: 7

(c) Midpoint: $(2.8, 2.1)$

127. $(x - 3)^2 + (y + 1)^2 = 68$

129. $(2, 5), (4, 5), (2, 0), (4, 0)$

131.

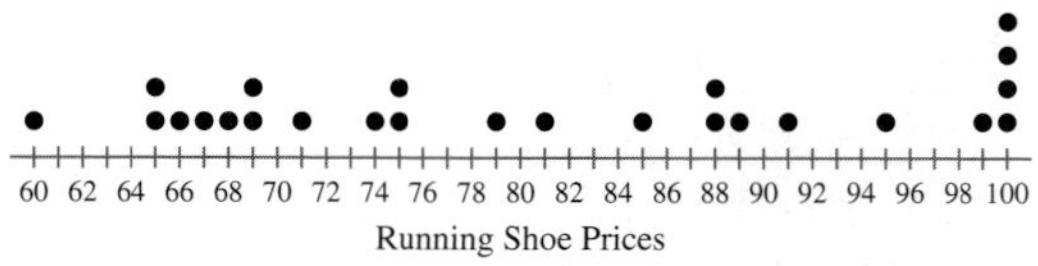

133.

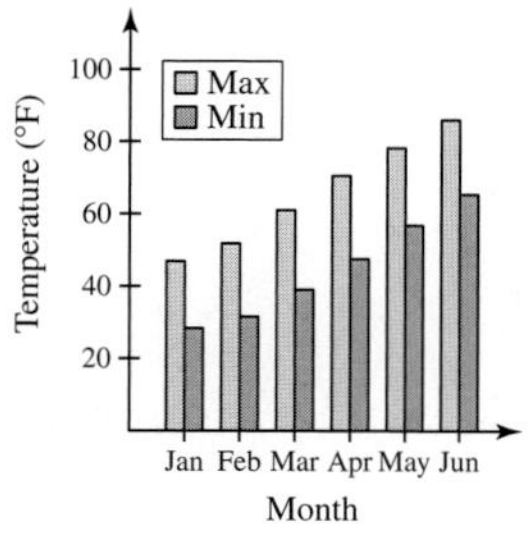

Chapter Test (page 71)

1. $-\frac{10}{3} > -|-4|$ **2.** 54

3. Commutative Property of Multiplication

4. (a) $\frac{8}{729}$ (b) 25 (c) 1.8×10^5 (d) 2.7×10^{13}

5. (a) $12z^8$ (b) $\dfrac{1}{(u - 2)^7}$ (c) $\dfrac{3x^2}{y^2}$

6. (a) $15z\sqrt{2z}$ (b) $-10\sqrt{y}$ (c) $\dfrac{2}{v}\sqrt[3]{\dfrac{2}{v^2}}$

7. $-2x^5 - x^4 + 3x^3 + 3$

Degree: 5; leading coefficient: -2

8. $2x^2 - 3x - 5$ **9.** $8x^3 - 20x^2 + 12x - 30$

10. $8, x \neq 3$ **11.** $\dfrac{x - 1}{2x}, x \neq \pm 1$ **12.** $x^2 - 5$

13. $x^3 - 6x^2 + 12x - 8$ **14.** $x^2 + 2xy + y^2 - z^2$

15. $x^2(2x + 1)(x - 2)$ **16.** $(x - 2)(x + 2)^2$

17. $8(x - 2)(x^2 + 2x + 4)$

18. (a) All real numbers x except $x = \pm 4$

(b) All real numbers x such that $x \leq 7$

19. (a) $\dfrac{\sqrt[3]{16^5}}{16} = 4\sqrt[3]{4}$ (b) $-3 - 3\sqrt{3}$

(c) $\dfrac{\sqrt{x + 2} + \sqrt{2}}{x}$

20. $\frac{5}{6}\sqrt{3}x^2$

21.

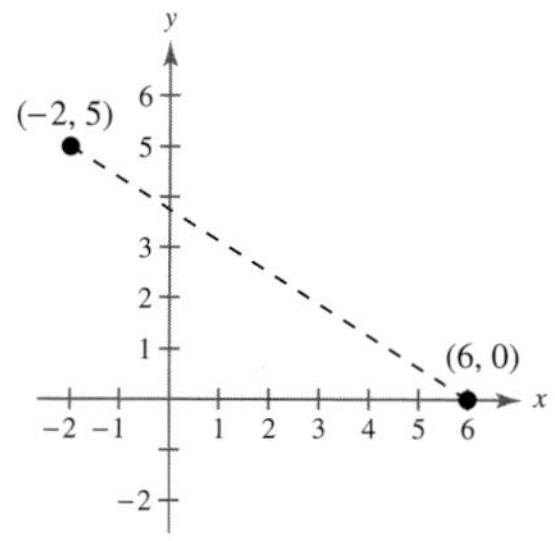

Midpoint: $\left(2, \frac{5}{2}\right)$

Distance: $\sqrt{89}$

22. (a) $(-8, 4)$ and $(-4, -8)$ (b) $(x + 6)^2 + (y + 2)^2 = 40$

23.

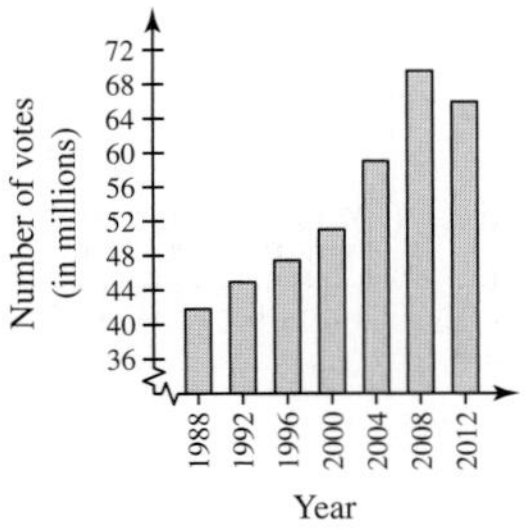

Chapter 1

Section 1.1 (page 81)

1. solution point **3.** Algebraic, graphical, numerical
5. (a) Yes (b) Yes **7.** (a) No (b) Yes
9. (a) No (b) Yes **11.** (a) Yes (b) No

13.

x	-2	0	$\frac{2}{3}$	1	2
y	-4	-1	0	$\frac{1}{2}$	2
Solution point	$(-2, -4)$	$(0, -1)$	$\left(\frac{2}{3}, 0\right)$	$\left(1, \frac{1}{2}\right)$	$(2, 2)$

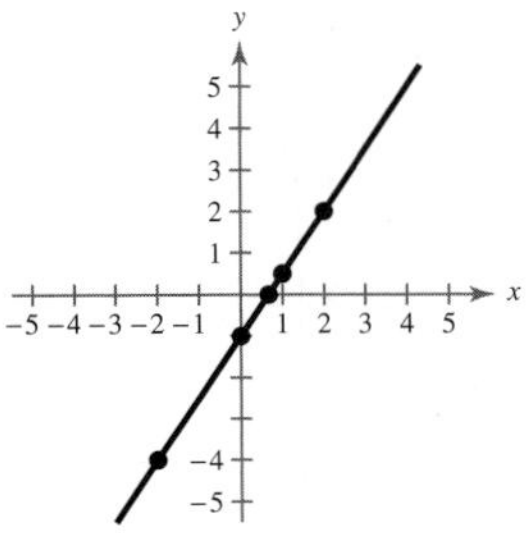

15.

x	-1	0	1	2	3
y	3	0	-1	0	3
Solution point	$(-1, 3)$	$(0, 0)$	$(1, -1)$	$(2, 0)$	$(3, 3)$

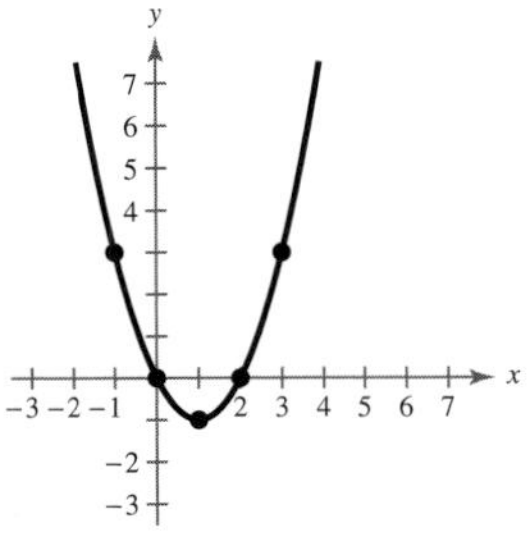

17. b **18.** d **19.** c **20.** a **21.** e **22.** f

23.

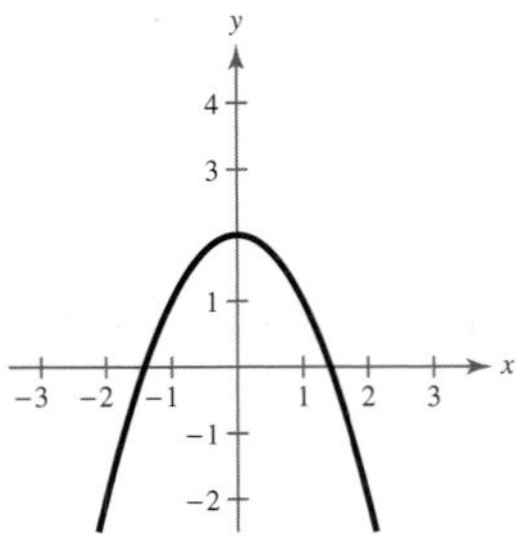

25.

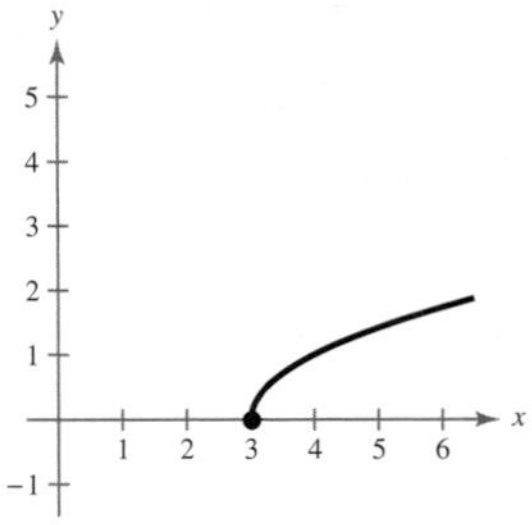

27.

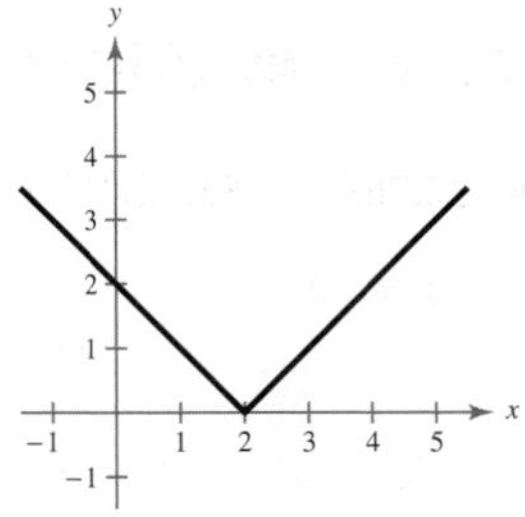

29.

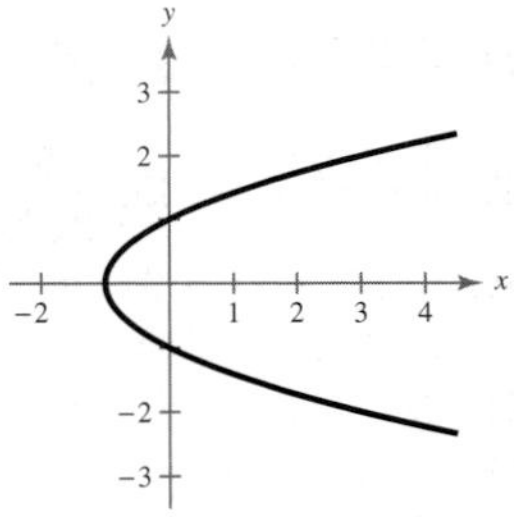

31.

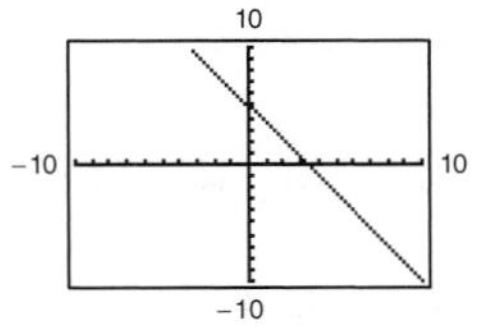

Intercepts: $\left(\frac{10}{3}, 0\right)$, $(0, 5)$

33.

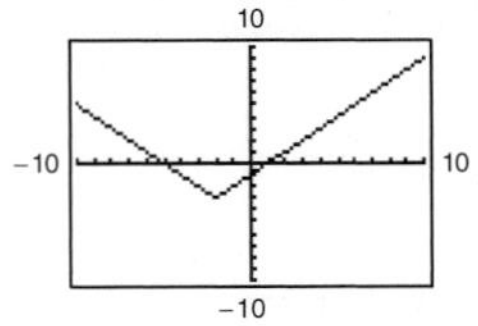

Intercepts: $(-5, 0)$, $(1, 0)$, $(0, -1)$

35.

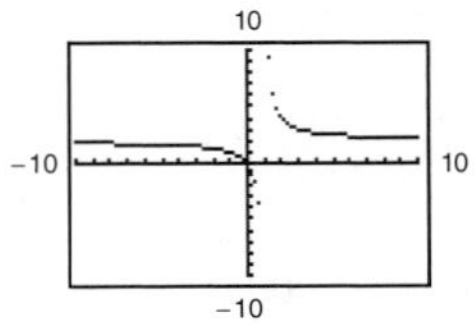

Intercept: $(0, 0)$

37.

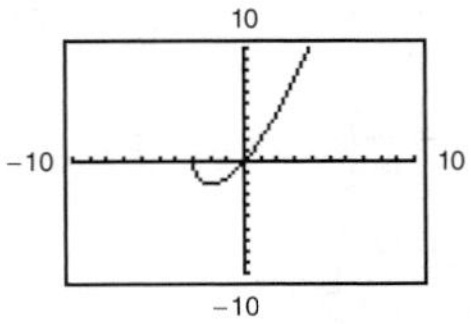

Intercepts: $(-3, 0)$, $(0, 0)$

39.

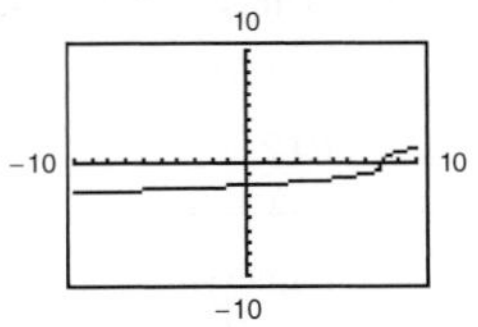

Intercepts: $(0, -2)$, $(8, 0)$

41.

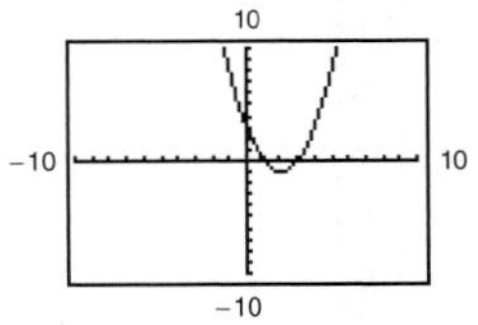

Intercepts: $(0, 3)$, $(1, 0)$, $(3, 0)$

43.

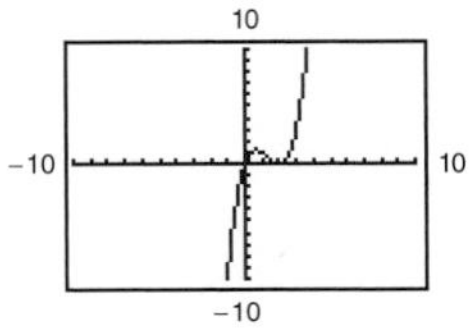

Intercepts: $(0, 0)$, $(2, 0)$

45.

Xmin = -10
Xmax = 10
Xscl = 2
Ymin = -50
Ymax = 100
Yscl = 25

47. The graphs are identical. Distributive Property

49. The graphs are identical.
Associative Property of Multiplication

51.

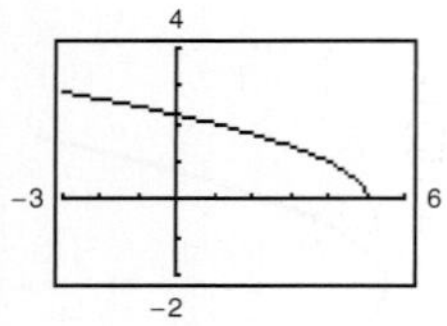

(a) $(3, 1.41)$
(b) $(-4, 3)$

53.

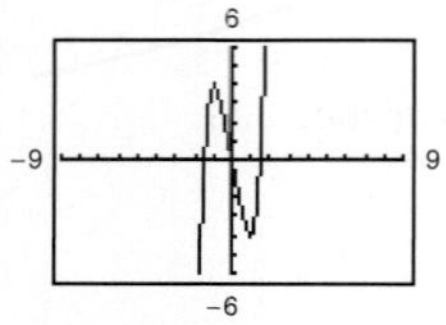

(a) $(-0.5, 2.47)$
(b) $(-1.58, -2), (0.40, -2), (1.37, -2)$

55. $y_1 = \sqrt{16 - x^2}$
$y_2 = -\sqrt{16 - x^2}$

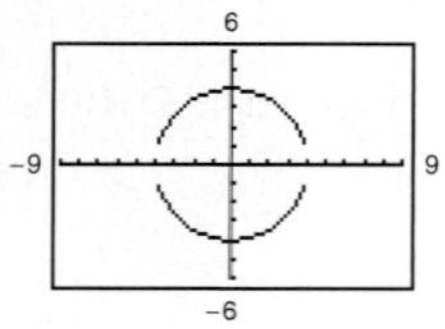

57. $y_1 = 2 + \sqrt{49 - (x - 1)^2}$
$y_2 = 2 - \sqrt{49 - (x - 1)^2}$

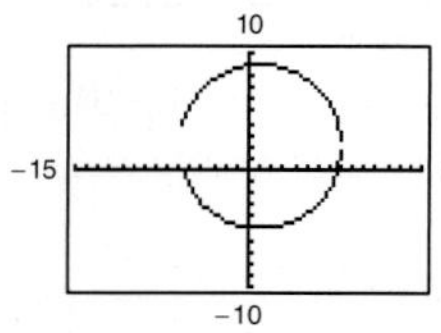

59. b, c, d

61. (a)

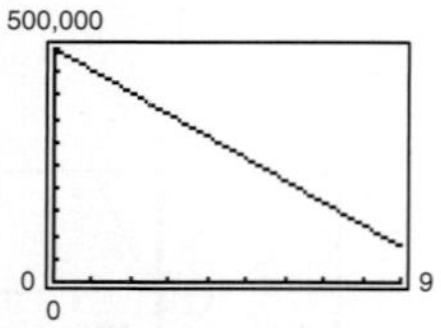

(b) \$227,400
(c) 7.3 yr

63. (a)

Year	2006	2007	2008	2009
New houses (in thousands)	410.5	290.9	198.1	132.1

Year	2010	2011	2012	2013
New houses (in thousands)	93.0	80.7	95.3	136.7

The model fits the data well.

(b)

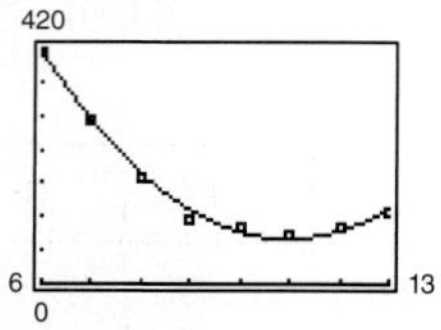

The model fits the data well.

(c) 2015: 300,000; 2017: 570,680; Yes. Answers will vary.
(d) 2009, 2012

65. False. $y = x^2 - 1$ has two x-intercepts.

67. Use the equations $y = 3400 + 0.05x$ and $y = 3000 + 0.07x$ to model the situation. When the sales level x equals \$20,000, both equations yield \$4400.

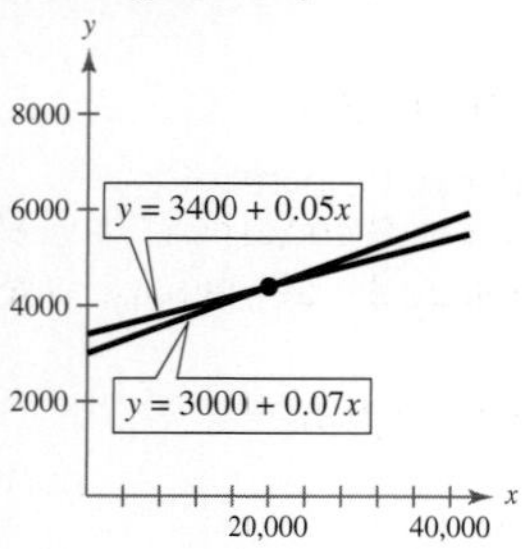

69. $2x^2 + 8x + 11$

Section 1.2 (page 92)

1. (a) iii (b) i (c) v (d) ii (e) iv **3.** parallel
5. 0 **7.** (a) L_2 (b) L_3 (c) L_1 **9.** $\frac{3}{2}$

11.

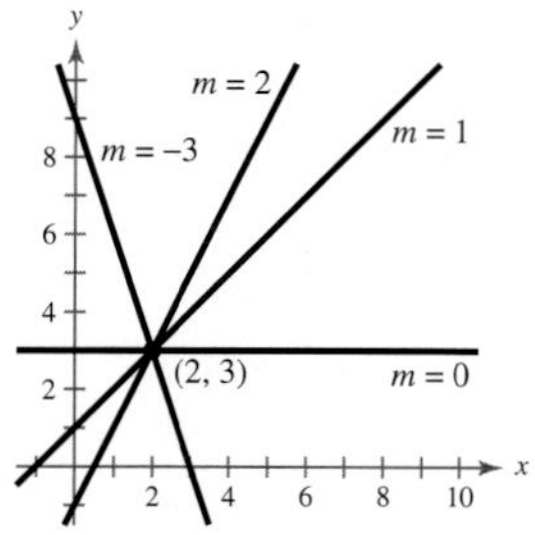

13. $m = -\frac{5}{2}$

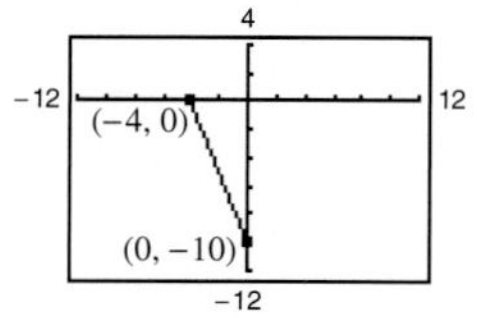

15. m is undefined.

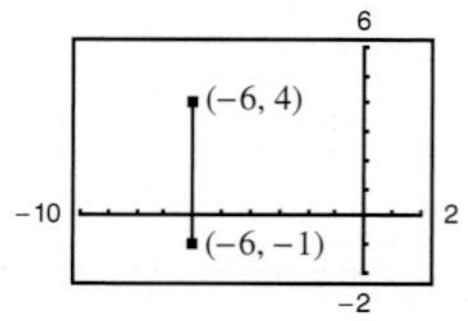

17. $(0, 1), (3, 1), (-1, 1)$ **19.** $(1, 4), (1, 6), (1, 9)$
21. $(-1, -7), (-2, -5), (-5, 1)$
23. $(3, -4), (5, -3), (9, -1)$

25. $y = 3x - 2$

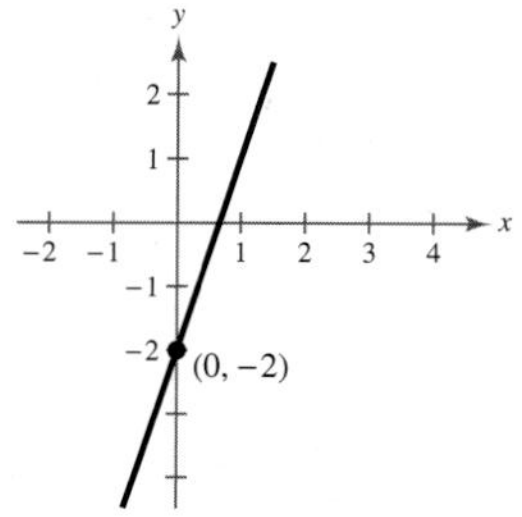

27. $y = -\frac{1}{2}x - 2$

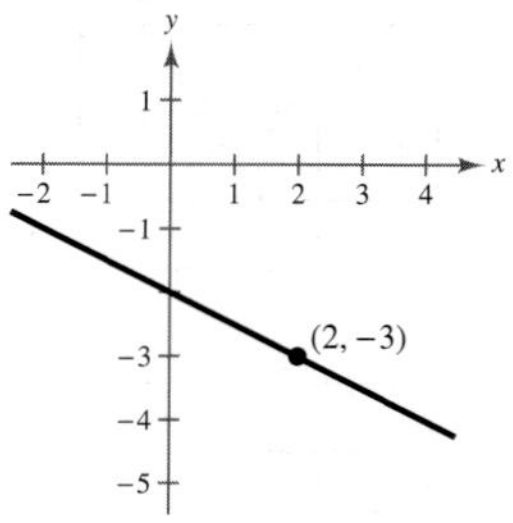

29. $x = 6$

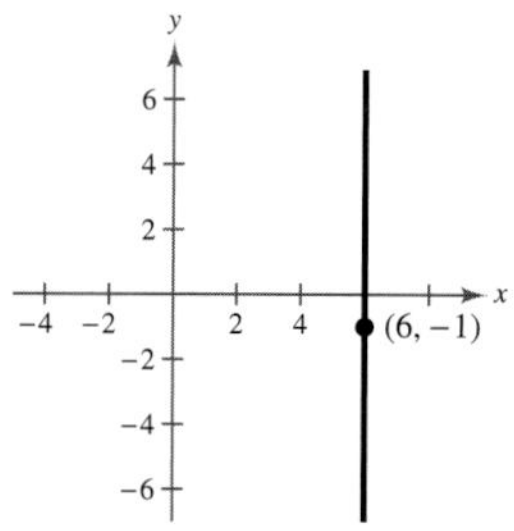

31. $y = \frac{3}{2}$

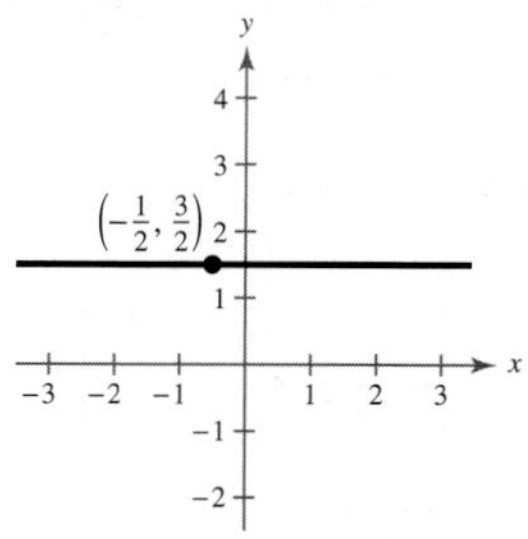

33. $y = \frac{1}{30}x + \frac{19}{15}$; about \$1.9 million
35. $m = \frac{2}{3}$; y-intercept: $(0, -3)$; a line that rises from left to right
37. $m = \frac{2}{5}$; y-intercept: $(0, 2)$; a line that rises from left to right
39. Slope is undefined; no y-intercept; a vertical line at $x = -6$
41. $m = 0$; y-intercept: $\left(0, -\frac{2}{3}\right)$; a horizontal line at $y = -\frac{2}{3}$

43. (a) $m = 5$; y-intercept: $(0, 3)$
(b)

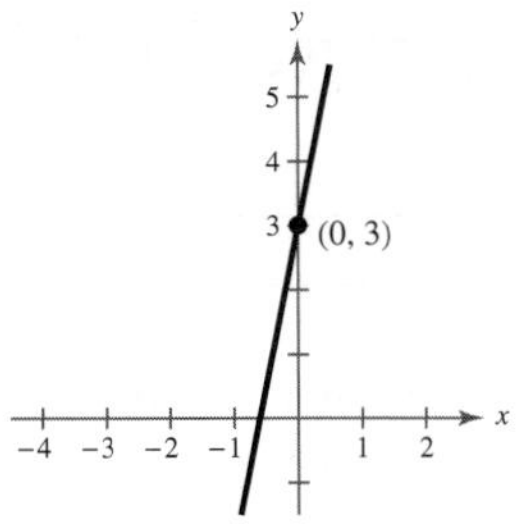

45. (a) Slope is undefined; there is no y-intercept.
(b)

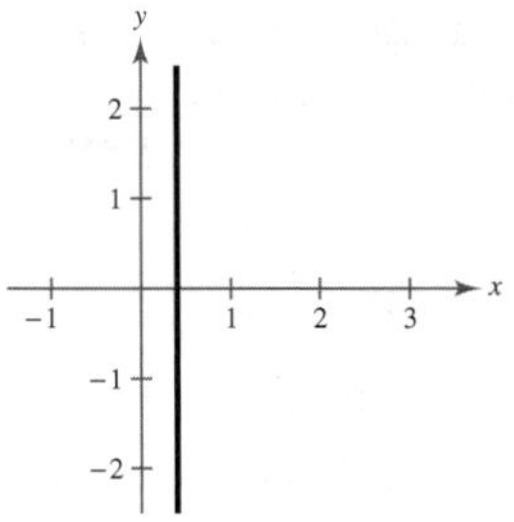

47. (a) $m = 0$; y-intercept: $\left(0, -\frac{5}{3}\right)$
(b)

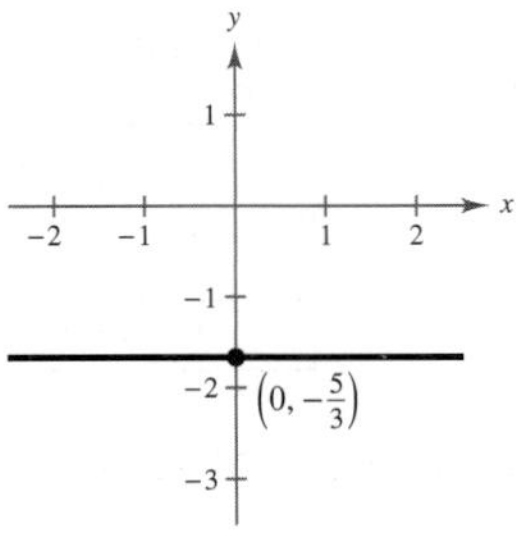

49. $y = 2x - 5$

51. $y = -\frac{3}{5}x + 2$

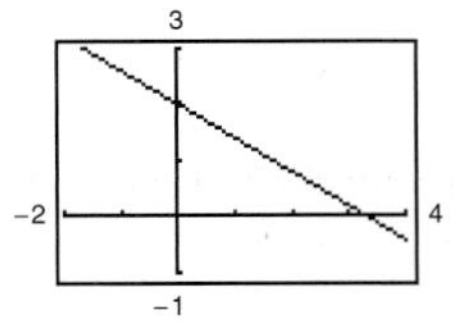

53. $x + 8 = 0$

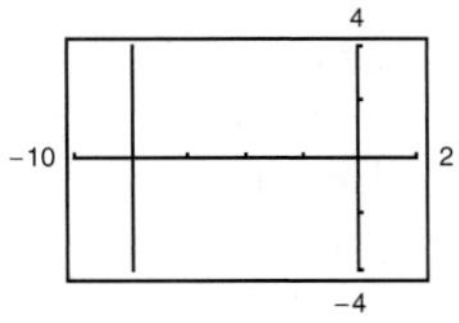

55. $y = -\frac{1}{2}x + \frac{3}{2}$

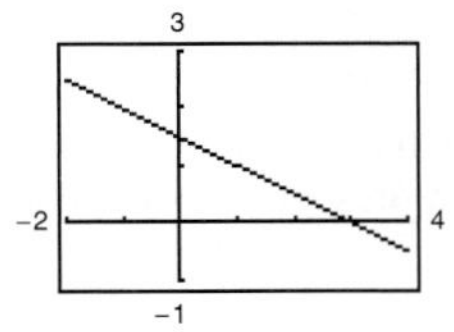

57. $y = -\frac{6}{5}x - \frac{18}{25}$

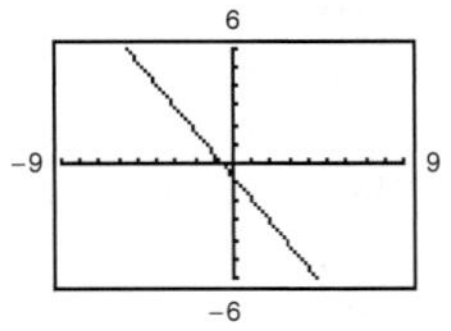

59. $y = \frac{2}{5}x + \frac{1}{5}$

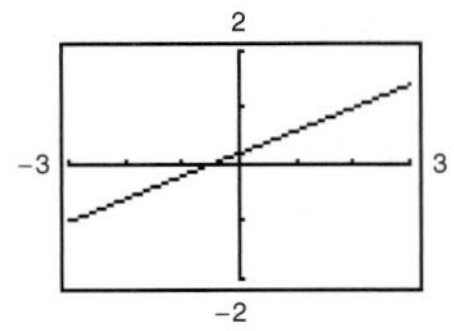

61.

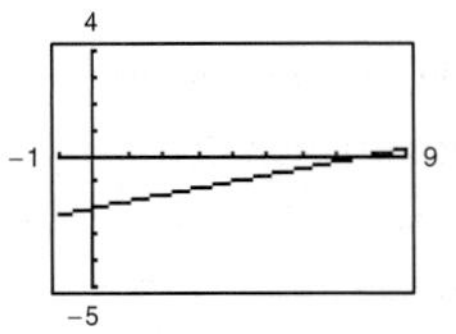

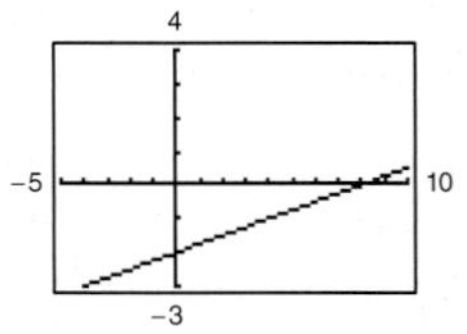

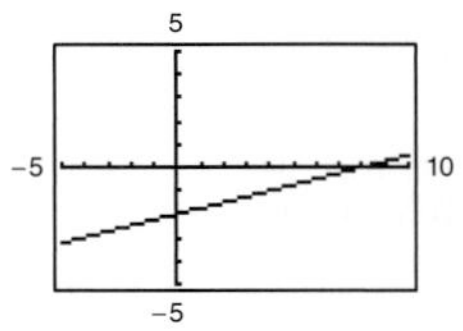

The first and second graphs do not give an accurate view of the slope. The third graph is best because it uses a square setting.

63. Perpendicular **65.** Parallel

67. (a) $y = 2x - 3$ (b) $y = -\frac{1}{2}x + 2$

69. (a) $y = -\frac{3}{4}x + \frac{3}{8}$ (b) $y = \frac{4}{3}x + \frac{127}{72}$

71. (a) $y = -\frac{6}{5}x - 6.08$ (b) $y = \frac{5}{6}x + 1.85$

73. (a) $x = 3$ (b) $y = -2$

75. (a) $y = 1$ (b) $x = -5$

77. $y = 2x + 1$ **79.** $y = -\frac{1}{2}x + 1$

81. The lines $y = \frac{1}{4}x$ and $y = -4x$ are perpendicular.

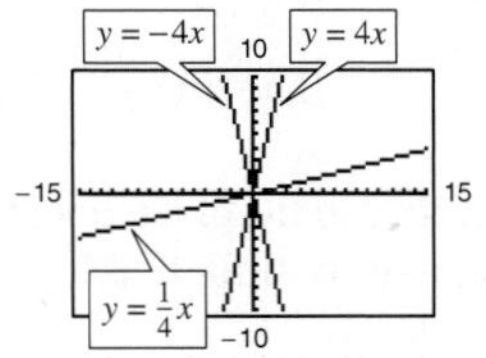

83. The lines $y = -\frac{1}{2}x$ and $y = -\frac{1}{2}x + 3$ are parallel. Both are perpendicular to $y = 2x - 4$.

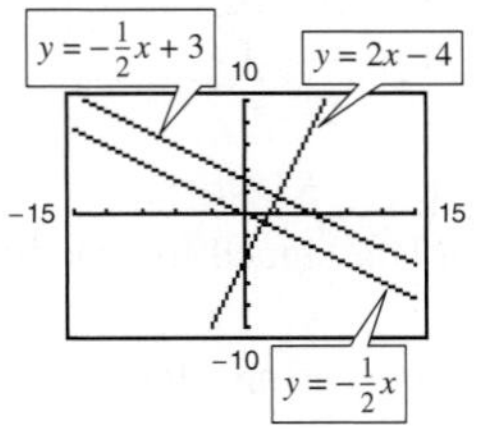

85. 12 ft

87. (a) The greatest increase was from 2010 to 2011, and the greatest decrease was from 2008 to 2009.
(b) $y = 3.559x + 5.309$
(c) There is an increase of about \$3.559 billion per year.
(d) \$65.812 billion; Answers will vary.

89. $V = 125t + 665$ **91.** $V = -2000t + 50{,}400$

93. (a) $V = 25{,}000 - 2300t$
(b)

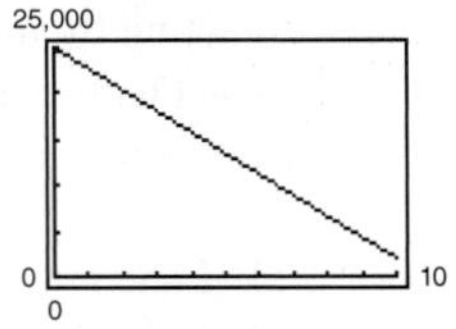

t	0	1	2	3	4
V	25,000	22,700	20,400	18,100	15,800

t	5	6	7	8	9	10
V	13,500	11,200	8900	6600	4300	2000

95. (a) $C = 30.75t + 36{,}500$
(b) $R = 80t$
(c) $P = 49.25t - 36{,}500$
(d) $t \approx 741.1$ h

97. (a) Increase of about 1295 students per year
(b) 76,090; 89,040; 95,515
(c) $y = 1295x + 73{,}500$; where $x = 0$ corresponds to 1994; $m = 1295$; The slope determines the average increase in enrollment.

99. False. The slopes $\left(\frac{2}{7}\text{ and } -\frac{11}{7}\right)$ are not equal.

101.

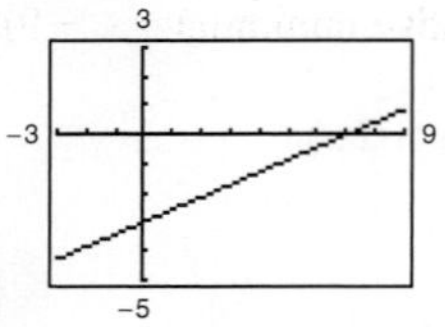

a and *b* represent the *x*- and *y*-intercepts.

103.

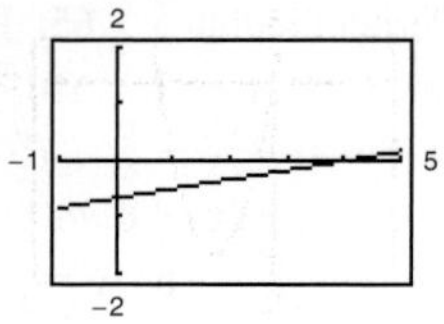

a and *b* represent the *x*- and *y*-intercepts.

105. $9x + 2y - 18 = 0$ **107.** $12x + 3y + 2 = 0$

109. a **111.** c

113. No. Answers will vary. Sample answer: The line $y = 2$ does not have an x-intercept.

115. Yes. Once a parallel line is established to the given line, there are an infinite number of distances away from that line, and thus an infinite number of parallel lines.

117. Yes; $x + 20$ **119.** No **121.** No

123. $(x - 9)(x + 3)$ **125.** $(2x - 5)(x + 8)$

127. Answers will vary.

Section 1.3 (page 105)

1. domain, range, function **3.** No **5.** No

7. Yes. Each element of the domain is assigned to exactly one element of the range.

9. No. The National Football Conference, an element in the domain, is assigned to three elements in the range, the Giants, the Saints, and the Seahawks; The American Football Conference, an element in the domain, is also assigned to three elements in the range, the Patriots, the Ravens, and the Steelers.

11. Yes. Each input value is matched with one output value.

13. No. The inputs 1, 2, and 3 are each matched with two outputs.

15. (a) Function
(b) Not a function because the element 1 in A corresponds to two elements, -2 and 1, in B.
(c) Not a function from A to B. It is a function from B to A.

17. Yes; yes; Each input value (year) is matched with exactly one output value (average price) for both name brand and generic drug prescriptions.

19. Not a function **21.** Function **23.** Function

25. Not a function **27.** Function **29.** Not a function

31. (a) 7 (b) -11 (c) $3t + 7$

33. (a) 0 (b) -0.75 (c) $x^2 - 10x + 24$

35. (a) 1 (b) 2.5 (c) $3 - 2|x|$

37. (a) Undefined (b) $-\frac{1}{5}$ (c) $\frac{1}{y^2 + 6y}$

39. (a) 1 (b) -1 (c) $\frac{|t|}{t}$

41. (a) -1 (b) 2 (c) 6

43. (a) 6 (b) 3 (c) 10

45. (a) 0 (b) 4 (c) 5

47. $\{(-2, 9), (-1, 4), (0, 1), (1, 0), (2, 1)\}$

49. $\{(-2, 4), (-1, 3), (0, 2), (1, 3), (2, 4)\}$

51.

t	-5	-4	-3	-2	-1
$h(t)$	1	$\frac{1}{2}$	0	$\frac{1}{2}$	1

53. 5 **55.** $\frac{4}{9}$ **57.** All real numbers x

59. All real numbers t except $t = 0$

61. All real numbers x

63. All real numbers x except $x = 0, -2$

65. All real numbers y such that $y > 10$

67.

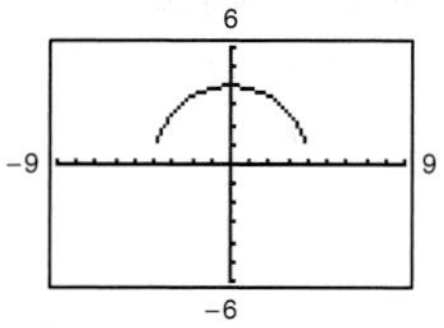

Domain: $[-4, 4]$
Range: $[0, 4]$

69.

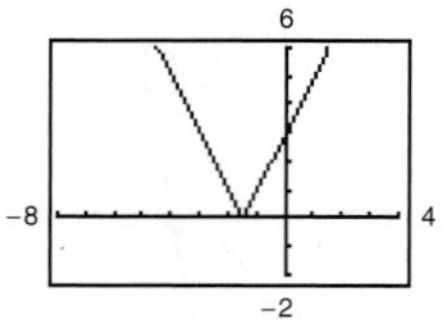

Domain: $(-\infty, \infty)$
Range: $[0, \infty)$

71. $A = \frac{C^2}{4\pi}$

73. (a) 1024 cm^3
(b)

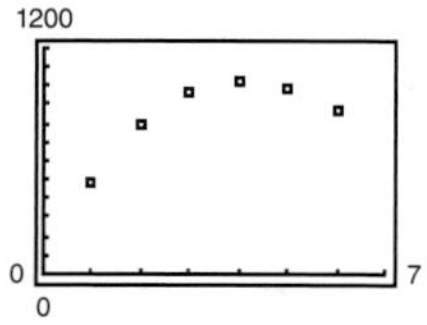

Yes, it is a function.
(c) $V = x(24 - 2x)^2, 0 < x < 12$
(d)

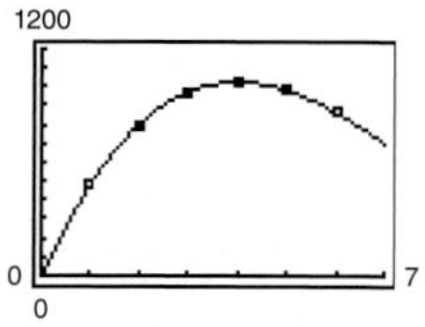

The function fits the data points; Answers will vary.

75. $A = 2xy = 2x\sqrt{36 - x^2}, 0 < x < 6$

77. (a) $C = 68.75x + 248{,}000$ (b) $R = 99.99x$
(c) $P = 31.24x - 248{,}000$
(d) 376,800; 376,800 profit for 20,000 units
(e) $-248{,}000$; \$248,000 loss for 0 units

79. (a) The independent variable is t and represents the year. The dependent variable is n and represents the numbers of miles traveled.
(b)

t	0	1	2	3	4	5
$n(t)$	3.95	3.96	3.98	3.99	4.00	4.02

t	6	7	8	9	10	11
$n(t)$	4.03	4.04	4.05	4.07	4.08	4.09

(c) The model fits the data well.
(d) Sample answer: No. The function may not accurately model other years.

81. Yes; The ball is 6 feet high when it reaches the other child.

83. $2, c \neq 0$ **85.** $3 + h, h \neq 0$

87. False. The range is $[-1, \infty)$.

89. $f(x) = \sqrt{x} + 2$
Domain: $x \geq 0$
Range: $[2, \infty)$

91. No, f is not the independent variable because the value of f depends on the value of x.

93. $\frac{12x + 20}{x + 2}$ **95.** $\frac{(x + 6)(x + 10)}{5(x + 3)}, x \neq 0, \frac{1}{2}$

Section 1.4 (page 118)

1. decreasing **3.** $[1, 4]$ **5.** Relative maximum

7. Domain: $(-\infty, \infty)$
Range: $(-\infty, 1]$
$f(0) = 1$

9. Domain: $[-4, 4]$
Range: $[0, 4]$
$f(0) = 4$

11.
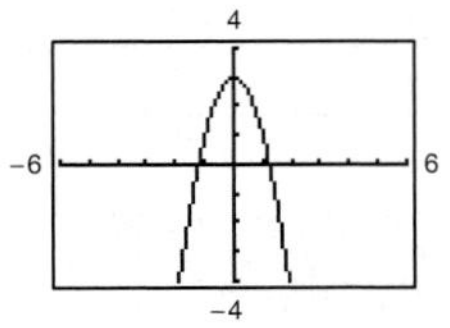

Domain: $(-\infty, \infty)$
Range: $(-\infty, 3]$

13.
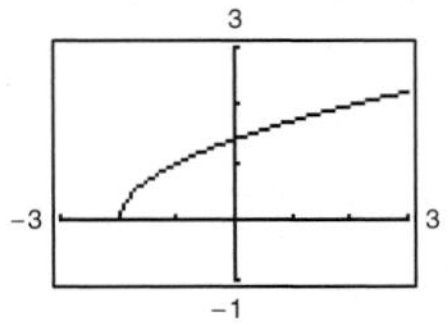

Domain: $[-2, \infty)$
Range: $[0, \infty)$

15.
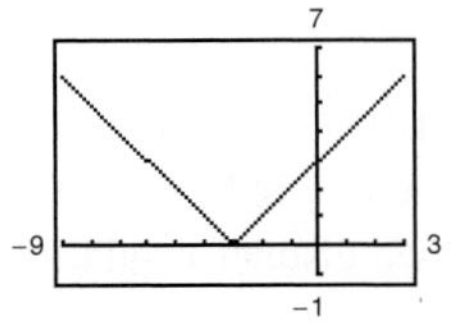

Domain: $(-\infty, \infty)$
Range: $[0, \infty)$

17. (a) $(-\infty, \infty)$ (b) $[-2, \infty)$ (c) $-1, 3$
(d) x-intercepts (e) -1 (f) y-intercept
(g) -2; $(1, -2)$ (h) $0, (-1, 0)$ (i) 2

19. Function. Graph the given function over the window shown in the figure.

21. Not a function. Solve for y and graph the two resulting functions.

23. Increasing on $(-\infty, \infty)$

25. Increasing on $(-\infty, 0), (2, \infty)$
Decreasing on $(0, 2)$

27. (a)
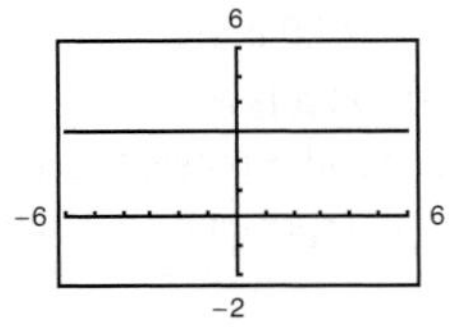

(b) Constant: $(-\infty, \infty)$

29. (a)
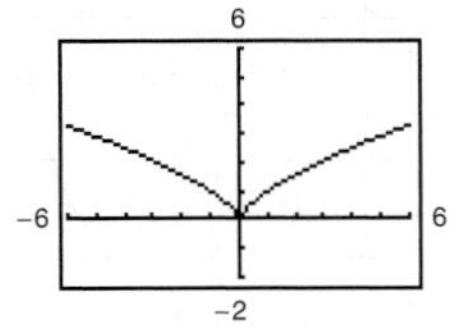

(b) Decreasing on $(-\infty, 0)$
Increasing on $(0, \infty)$

31. (a)
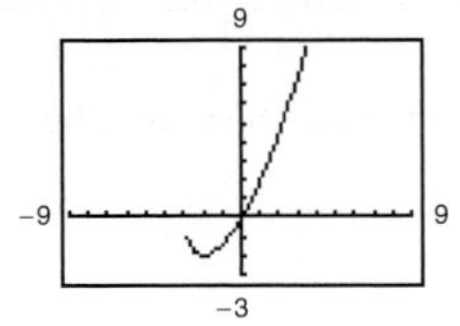

(b) Increasing on $(-2, \infty)$
Decreasing on $(-3, -2)$

33. (a)
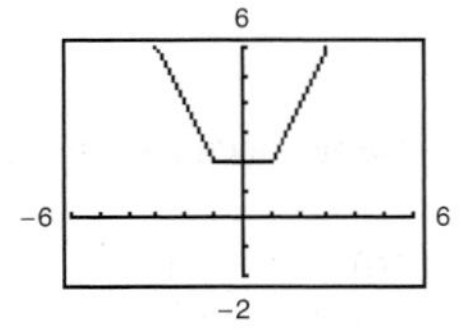

(b) Decreasing on $(-\infty, -1)$
Constant on $(-1, 1)$
Increasing on $(1, \infty)$

35.
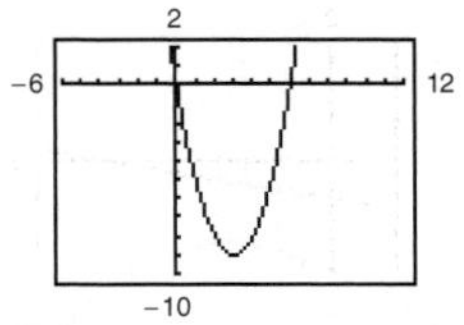

Relative minimum: $(3, -9)$

37.
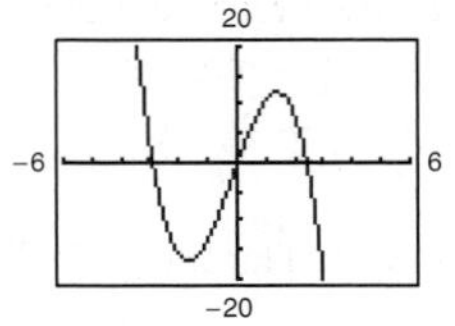

Relative minimum: $(-1.70, -16.86)$
Relative maximum: $(1.37, 12.16)$

39.
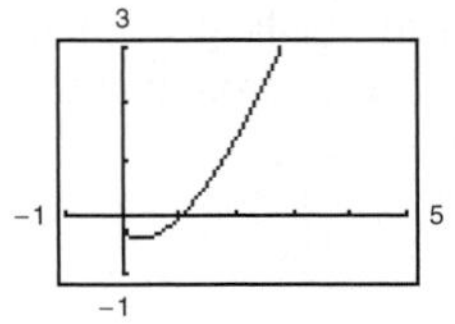

Relative minimum: $(0.33, -0.38)$

41.
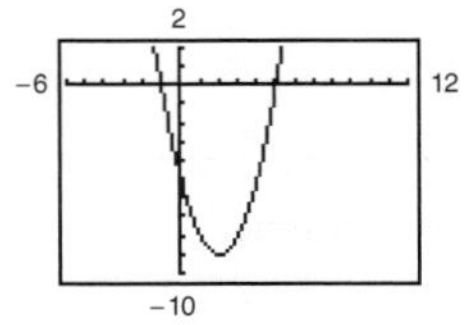

Relative minimum: $(2, -9)$

43.
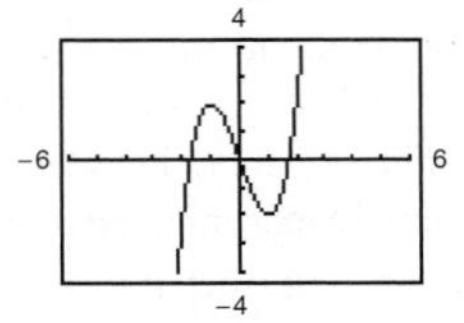

Relative minimum: $(1, -2)$
Relative maximum: $(-1, 2)$

45.
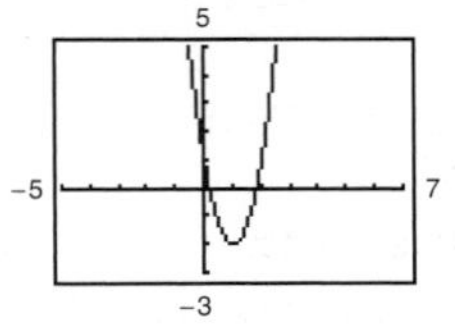

Relative minimum: $(1, -2)$

47.
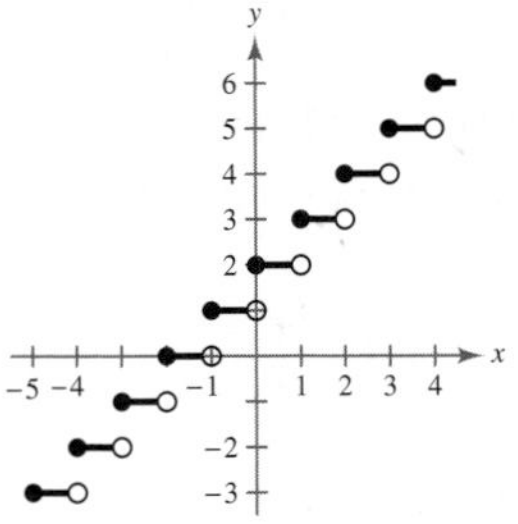

49.
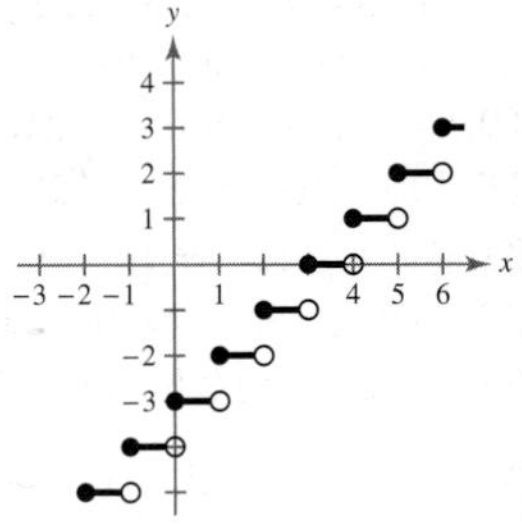

51.
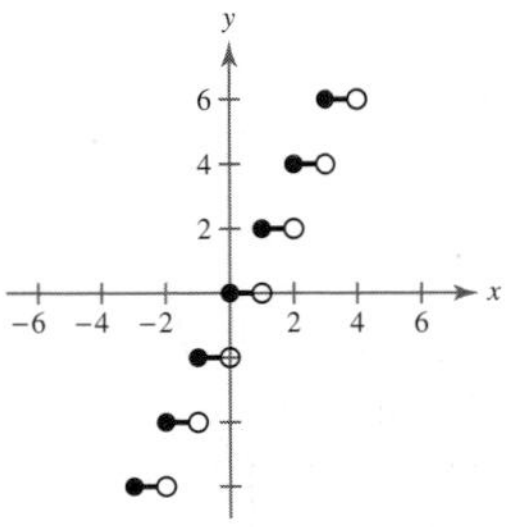

53.
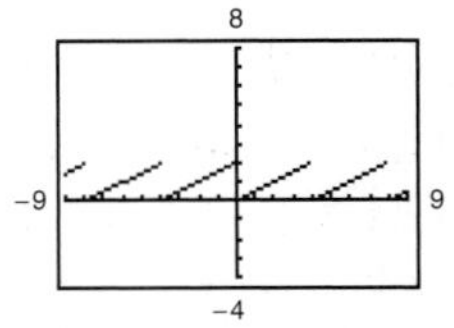

Domain: $(-\infty, \infty)$
Range: $[0, 2)$
Sawtooth pattern

55.

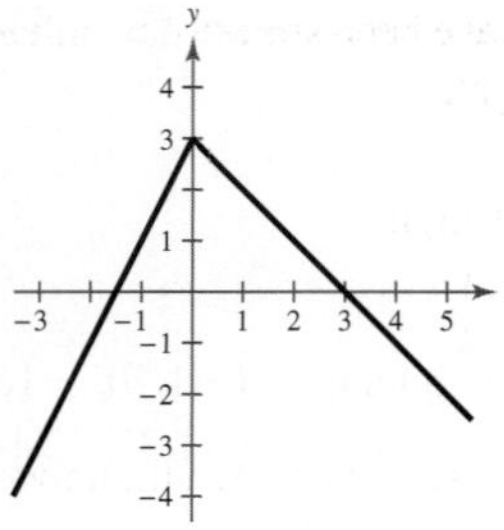

57.

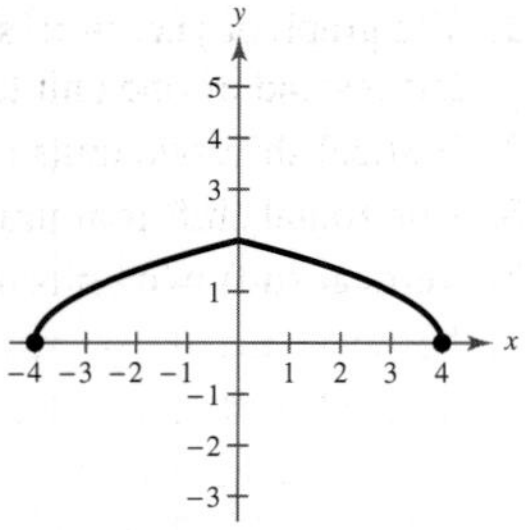

59.

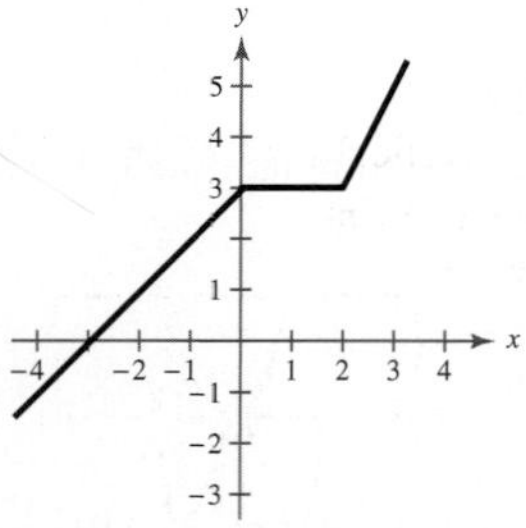

61.

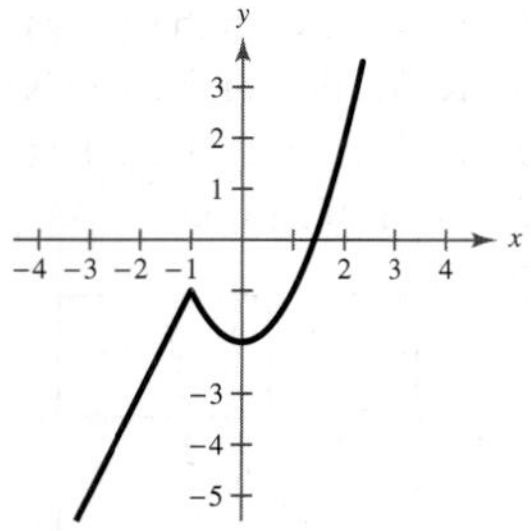

63. Even

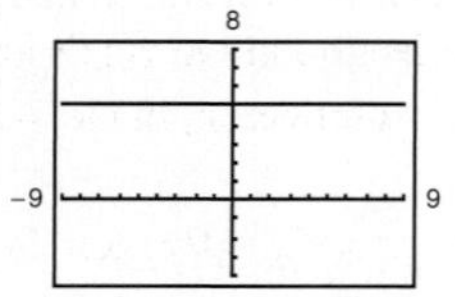

65. Neither

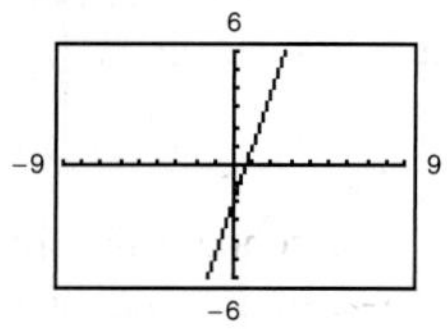

67. Even

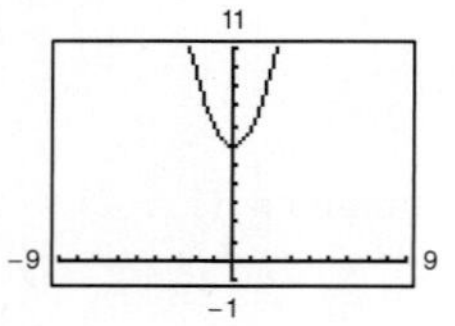

69. Neither

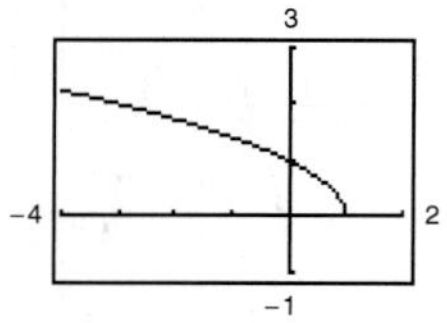

71. Neither

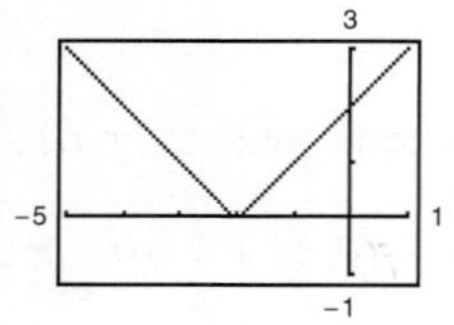

73. (a) $\left(-\frac{3}{2}, 4\right)$ (b) $\left(-\frac{3}{2}, -4\right)$

75. (a) $(2, -9)$ (b) $(2, 9)$

77. (a) $(-x, -y)$ (b) $(-x, y)$

79. (a)–(c) Neither **81.** (a)–(c) Odd

83. (a)–(c) Odd **85.** (a)–(c) Even

87.

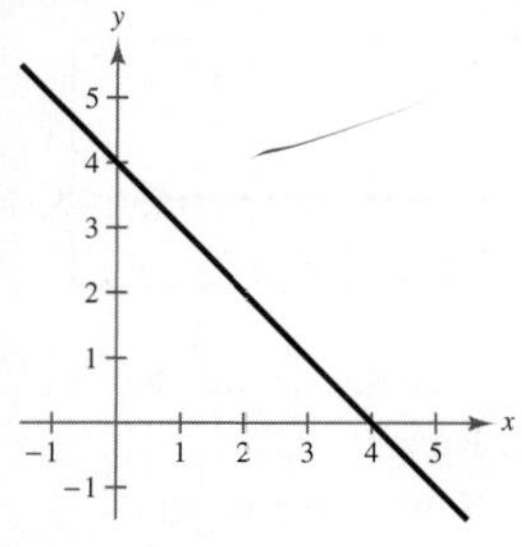

$(-\infty, 4]$

89.

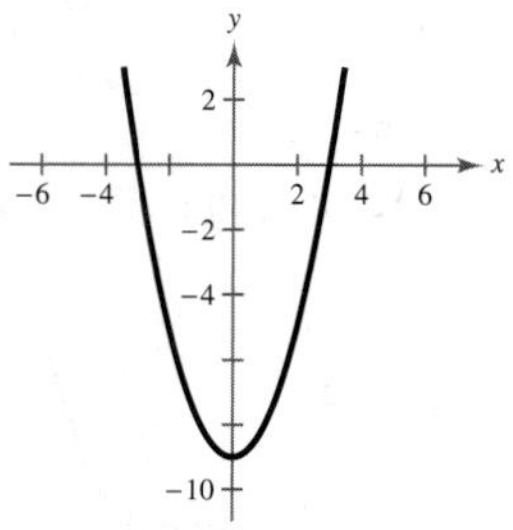

$(-\infty, -3], [3, \infty)$

91. (a) C_2 is the appropriate model. The cost of the first hour is \$1.00 and the cost increases \$0.50 when the next hour begins, and so on.

(b)

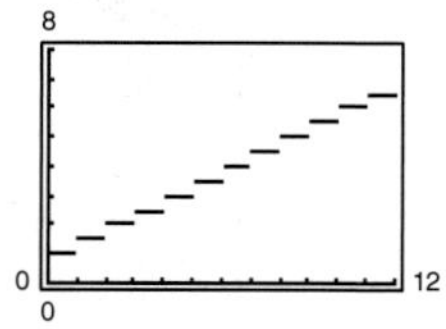

\$4.50

93. $h = -x^2 + 4x - 3, 1 \le x \le 3$

95. (a)

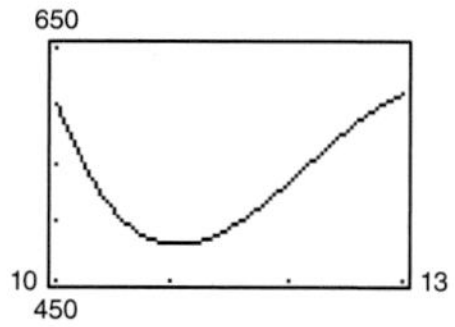

(b) Decreasing from 2010 to 2011; increasing from 2011 to 2013

(c) About 480

97. False. Counterexample: $f(x) = \sqrt{1 + x^2}$

99. c **100.** d **101.** b **102.** e **103.** a **104.** f

105. No. Each y-value corresponds to two distinct x-values when $y > 0$.

107. Yes

109. (a) Even. g is a reflection in the x-axis.

(b) Even. g is a reflection in the y-axis.

(c) Even. g is a vertical shift downward.

(d) Neither. g is shifted to the left and reflected in the x-axis.

111. Proof

113. Terms: $-2x^2, 11x, 3$; coefficients: $-2, 11$

115. Terms: $\frac{x}{3}, -5x^2, x^3$; coefficients: $\frac{1}{3}, -5, 1$

117. (a) -17 (b) -17 (c) $-x^2 + 3x + 1$

119. $h + 4, h \neq 0$

Section 1.5 (page 128)

1. Horizontal shifts, vertical shifts, reflections

3. $-f(x), f(-x)$

5.

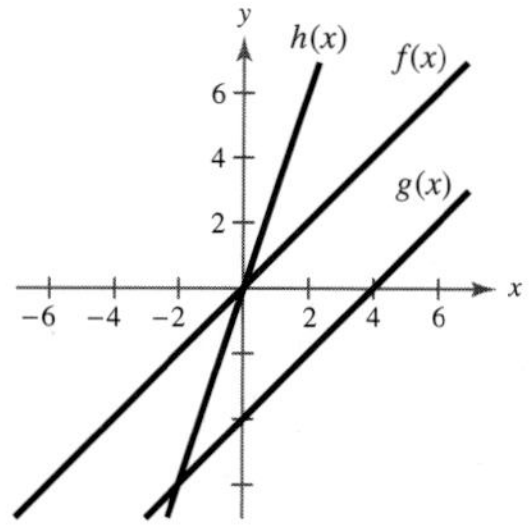

7.

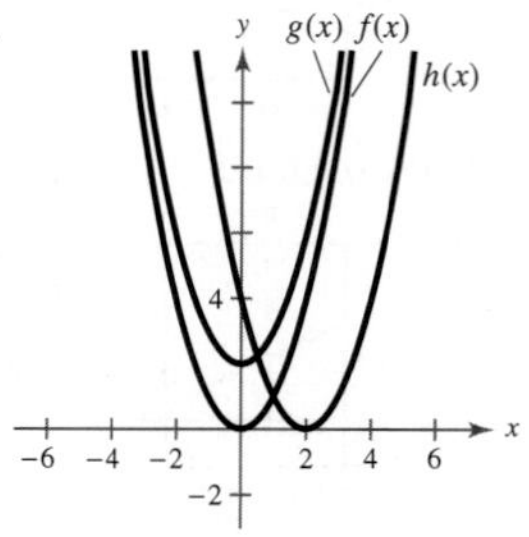

9.

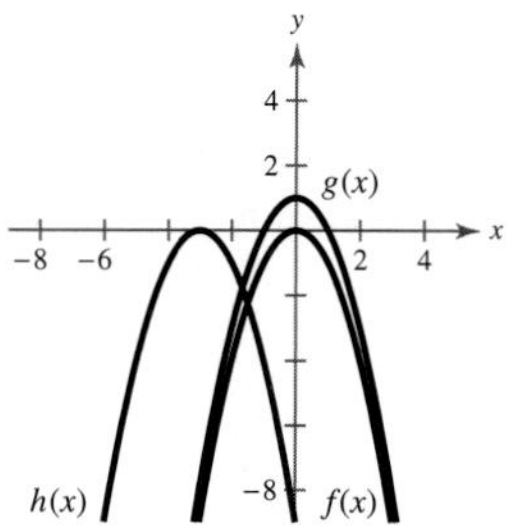

11.

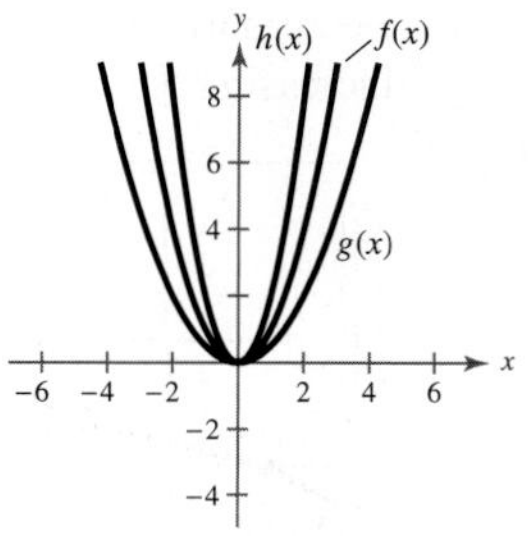

13.

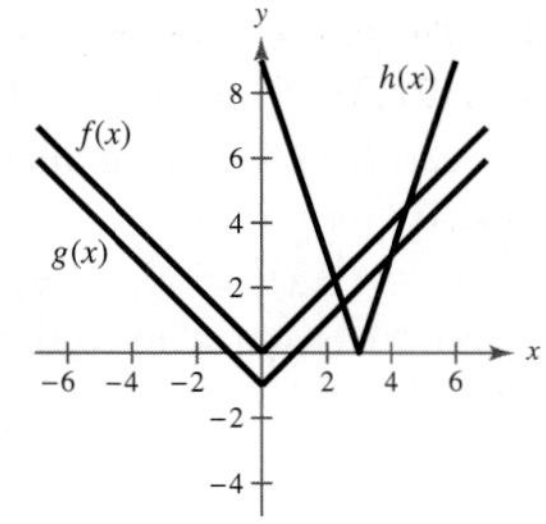

15.

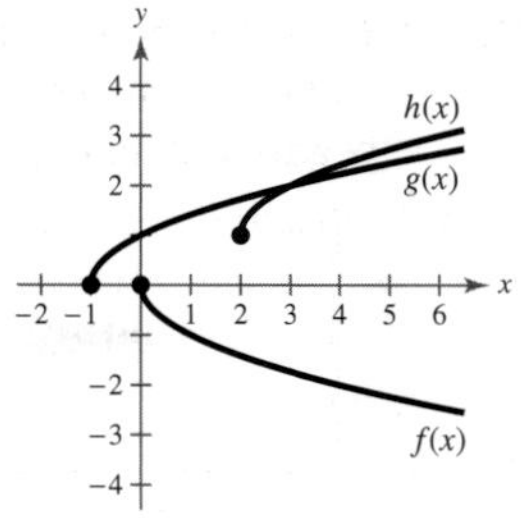

17.

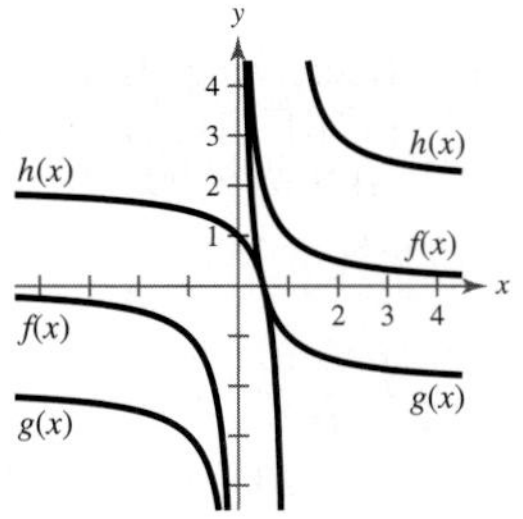

19. (a)

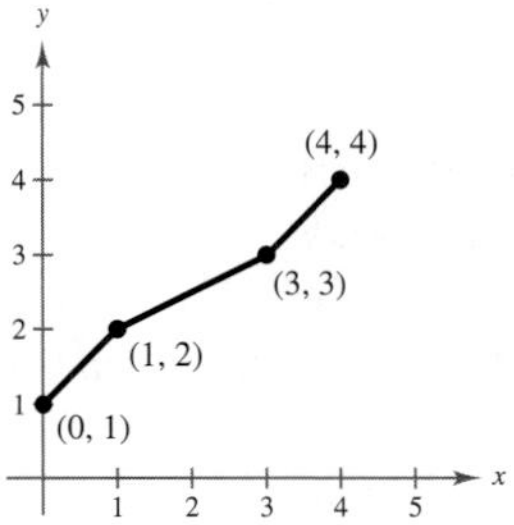

(b)

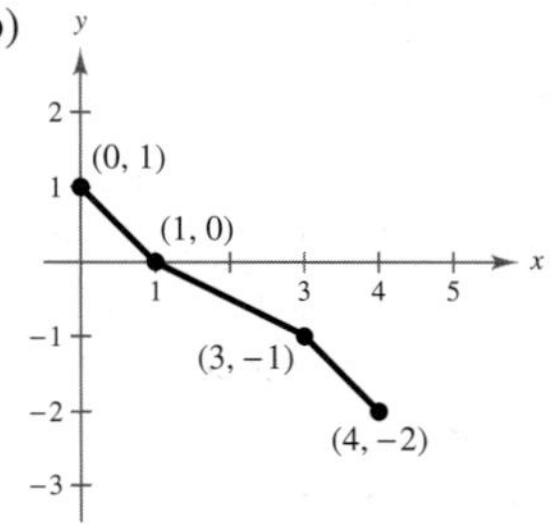

(c)

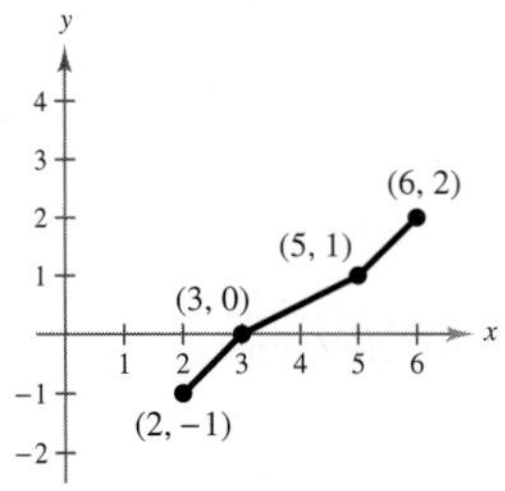

(d)

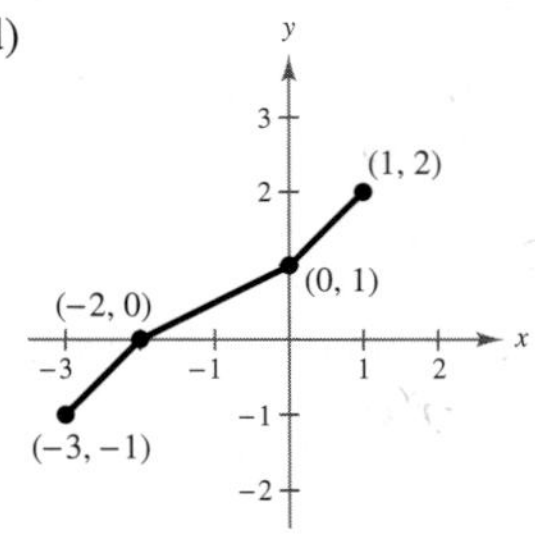

(e)

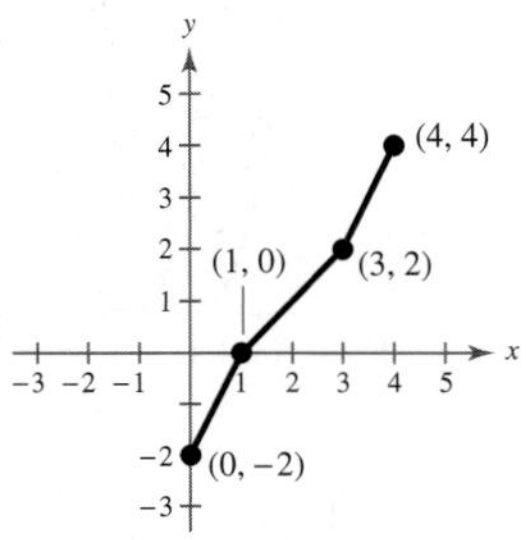

(f)

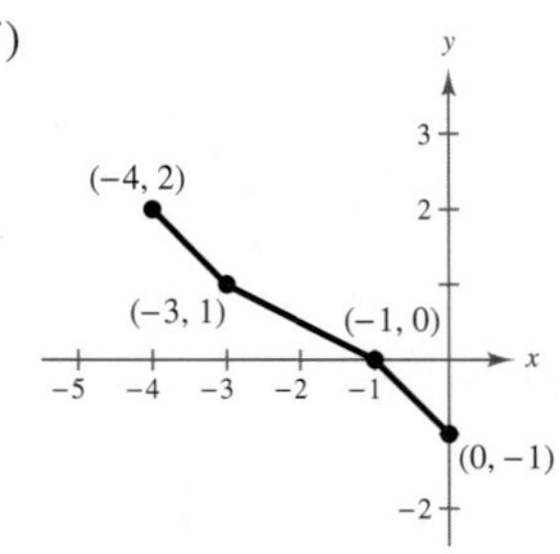

(g)

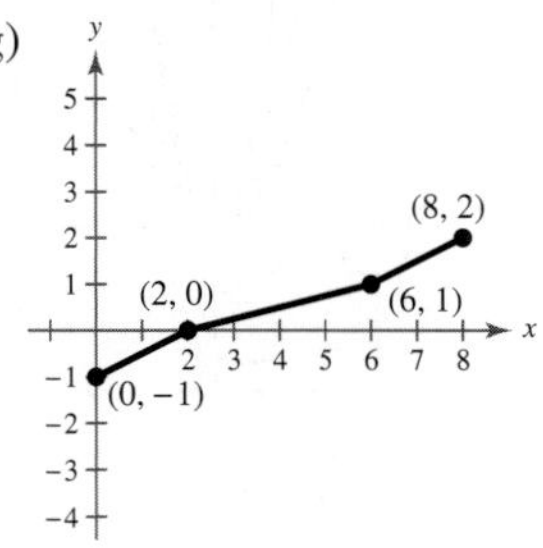

21. The graph of $f(x) = x^2$ should have been shifted one unit to the left instead of one unit to the right.

23. Vertical shift two units upward

25. Horizontal shift four units to the right

27. Vertical shift two units downward

29. Vertical shift of $y = x$; $y = x + 3$

31. Vertical shift of $y = x^2$; $y = x^2 - 3$

33. Reflection in the x-axis and a vertical shift one unit upward of $y = \sqrt{x}$; $y = 1 - \sqrt{x}$

35. Reflection in the x-axis

37. Reflection in the y-axis (identical)

39. Reflection in the x-axis **41.** Vertical stretch

43. Vertical shrink **45.** Horizontal shrink

47.

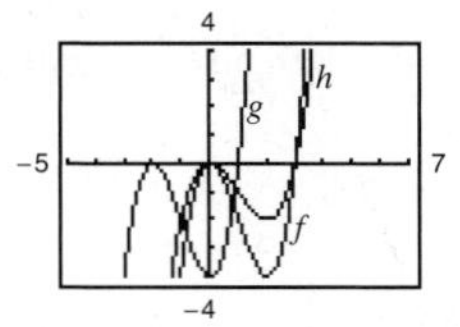

g is a horizontal shift and h is a vertical shrink.

49.

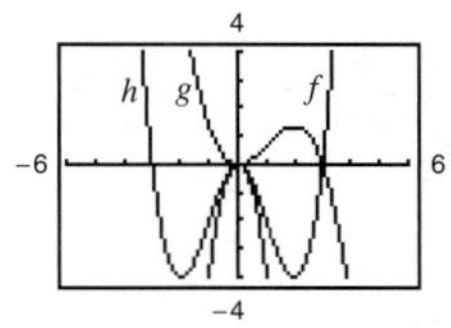

g is a vertical shrink and a reflection in the x-axis and h is a reflection in the y-axis.

51. (a) $f(x) = x^2$

(b) Horizontal shift five units to the left, reflection in the x-axis, and vertical shift two units upward

(c)

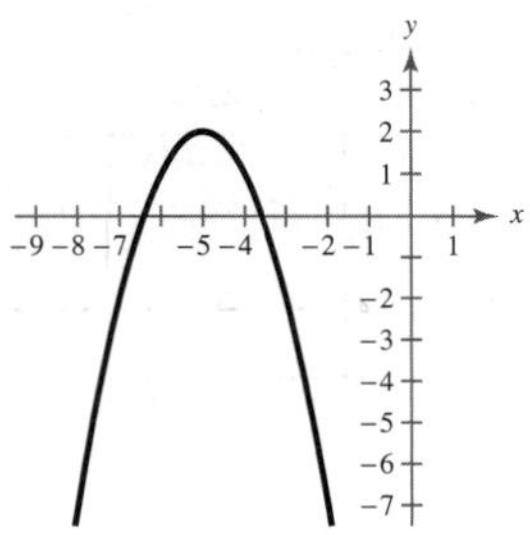

(d) $g(x) = 2 - f(x + 5)$

53. (a) $f(x) = x^2$

(b) Horizontal shift four units to the right, vertical stretch, and vertical shift three units upward

(c)

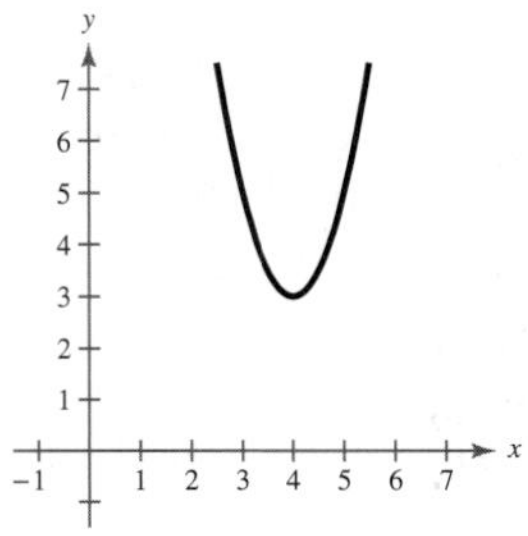

(d) $f(x) = 3 + 2f(x - 4)$

55. (a) $f(x) = x^3$

(b) Horizontal shift two units to the right and vertical shrink

(c)

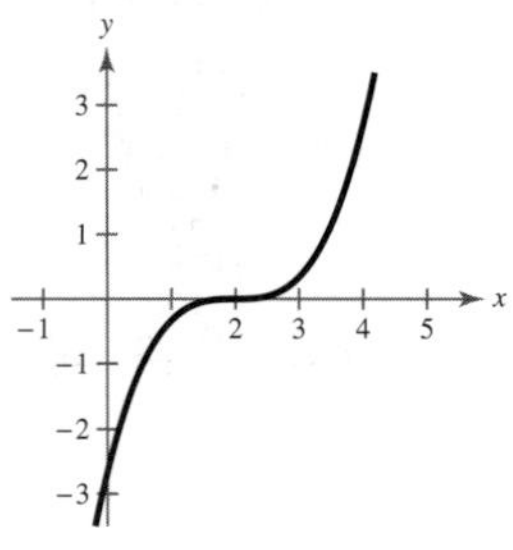

(d) $g(x) = \frac{1}{3}f(x - 2)$

57. (a) $f(x) = x^3$
(b) Horizontal shift one unit to the right and vertical shift two units upward
(c)

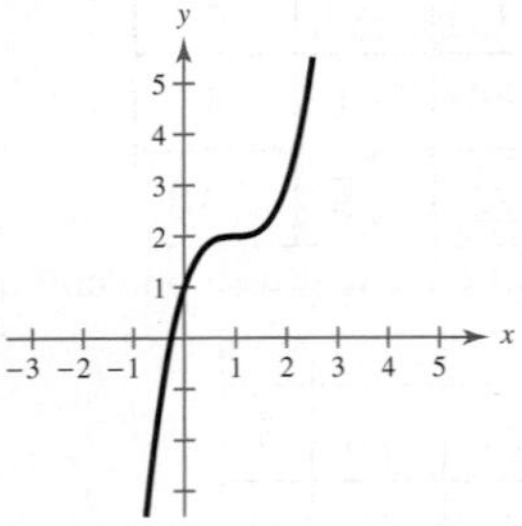

(d) $g(x) = f(x - 1) + 2$

59. (a) $f(x) = \frac{1}{x}$
(b) Horizontal shift eight units to the left and vertical shift nine units downward
(c)

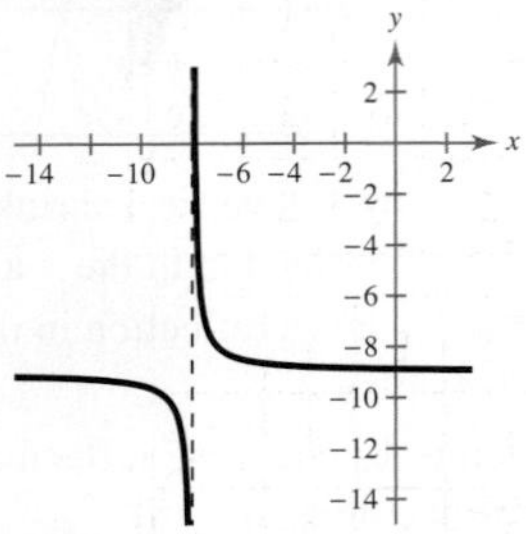

(d) $g(x) = f(x + 8) - 9$

61. (a) $f(x) = |x|$
(b) Horizontal shift one unit to the right, reflection in the x-axis, vertical stretch, and vertical shift four units downward
(c)

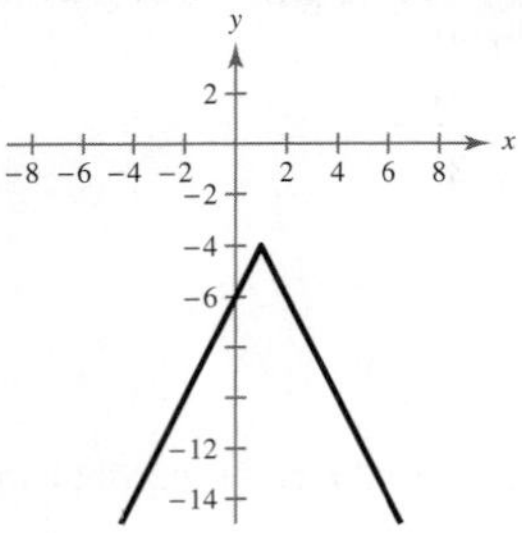

(d) $g(x) = -2f(x - 1) - 4$

63. (a) $f(x) = \sqrt{x}$
(b) Horizontal shift three units to the left, reflection in the x-axis, vertical shrink, and vertical shift one unit downward
(c)

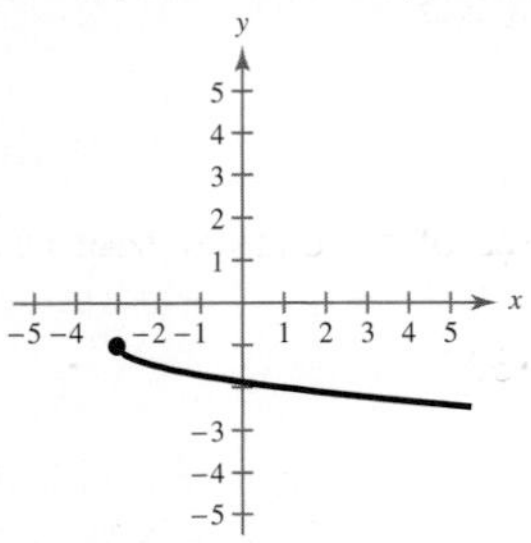

(d) $g(x) = -\frac{1}{2}f(x + 3) - 1$

65. (a) Horizontal shift 26.17 units to the right, reflection in the x-axis, vertical shrink, and vertical shift 126.5 units upward
(b)

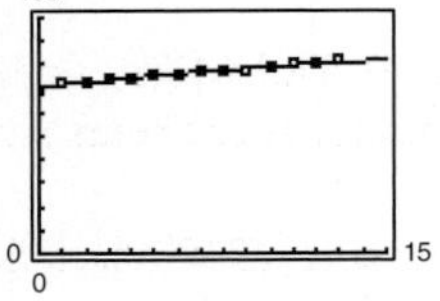

(c) $N(t) = -0.03[(t + 9) - 26.17]^2 + 126.5$
$= -0.03(t - 17.17)^2 + 126.5$
To make a horizontal shift 9 years backward (9 units left), add 9 to t.

67. False. When $f(x) = x^2$, $f(-x) = (-x)^2 = x^2$. Because $f(x) = f(-x)$ in this case, $y = f(-x)$ is not a reflection of $y = f(x)$ across the x-axis in all cases.

69. $x = -2$ and $x = 3$

71. Cannot be determined because it is a vertical shift

73. c **75.** c **77.** Answers will vary.

79. (a) The graph of g is a vertical shrink of the graph of f.
(b) The graph of g is a vertical stretch of the graph of f.

81. Neither

83. All real numbers x except $x = 9$

85. All real numbers x such that $-10 \le x \le 10$

Section 1.6 (page 137)

1. addition, subtraction, multiplication, division

3. $g(x)$ **5.** $2x$

7.

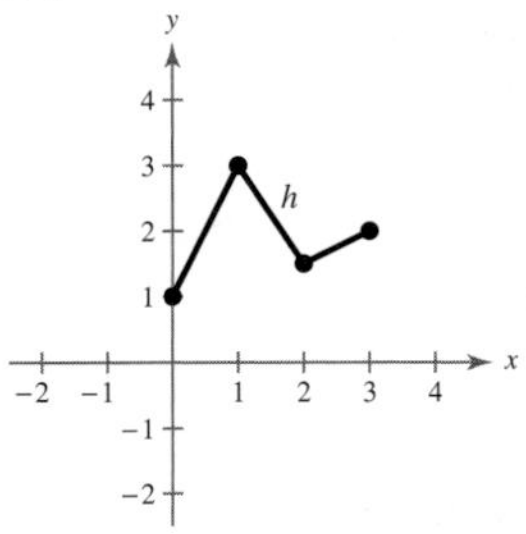

9.

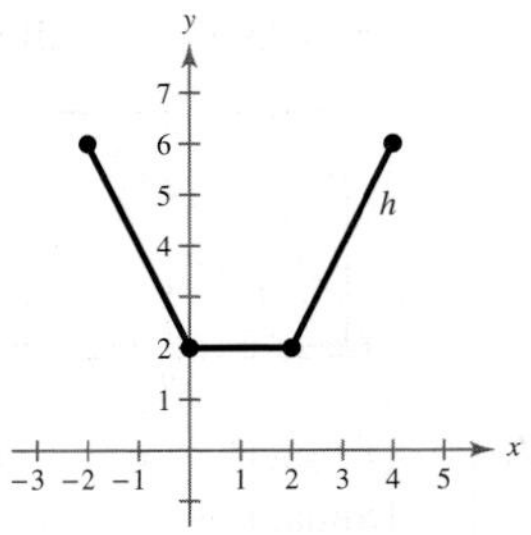

11. (a) $2x$
(b) 6
(c) $x^2 - 9$
(d) $\frac{x + 3}{x - 3}$
All real numbers x, except $x = 3$

13. (a) $3x^2 - 5x + 6$
(b) $3x^2 + 5x - 6$
(c) $-15x^3 + 18x^2$
(d) $\frac{3x^2}{6 - 5x}$
All real numbers x, except $x = \frac{6}{5}$

15. (a) $x^2 + 5 + \sqrt{1 - x}$ (b) $x^2 + 5 - \sqrt{1 - x}$
(c) $(x^2 + 5)\sqrt{1 - x}$ (d) $\frac{x^2 + 5}{\sqrt{1 - x}}, x < 1$

17. (a) $\frac{x + 1}{x^2}$ (b) $\frac{x - 1}{x^2}$ (c) $\frac{1}{x^3}$ (d) $x, x \neq 0$

19. 9 **21.** 1 **23.** -280 **25.** $-\frac{24}{7}$ **27.** $t^2 + t + 1$

29. $-125t^3 - 50^2 + 5t + 2$ **31.** $\frac{t^2 - 8t + 15}{t - 6}$

33.

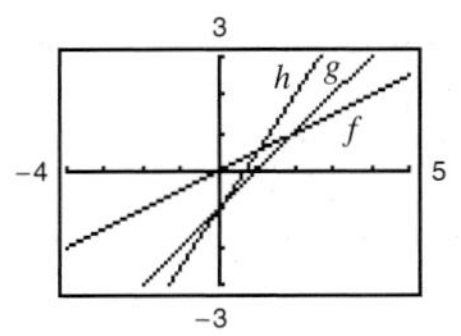

35.

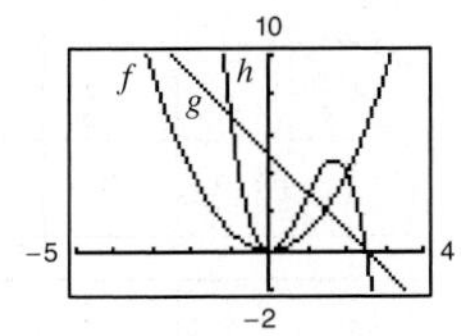

37.

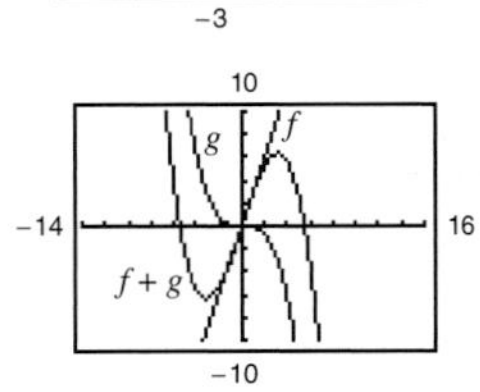

$f(x), 0 \le x \le 2;$
$g(x), x > 6$

39.

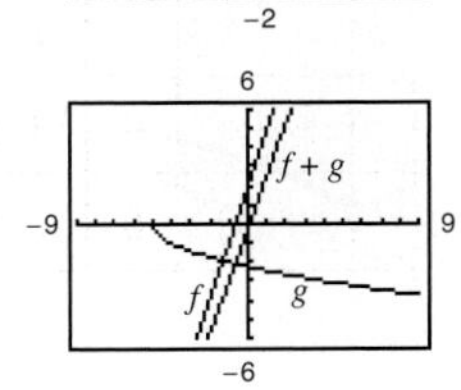

$f(x), 0 \le x \le 2;$
$f(x), x > 6$

41. (a) $2(x + 4)^2$ (b) $2x^2 + 4$ (c) 32

43. (a) $20 - 3x$ (b) $-3x$ (c) 20

45. (a) All real numbers x such that $x \geq 7$
(b) All real numbers x
(c) All real numbers x such that $x \leq -\frac{\sqrt{7}}{2}$ or $x \geq \frac{\sqrt{7}}{2}$

47. (a) All real numbers x
(b) All real numbers x such that $x \geq 0$
(c) All real numbers x such that $x \geq 0$

49. (a) All real numbers x except $x = 0$
(b) All real numbers x except $x = -3$
(c) All real numbers x except $x = -3$

51. (a) All real numbers x (b) All real numbers x
(c) All real numbers x

53. (a) All real numbers x
(b) All real numbers x except $x = \pm 2$
(c) All real numbers x except $x = \pm 2$

55. (a) $(f \circ g)(x) = \sqrt{x^2 + 4}$
$(g \circ f)(x) = x + 4, x \geq -4$;
Domain of $f \circ g$: all real numbers x
(b) 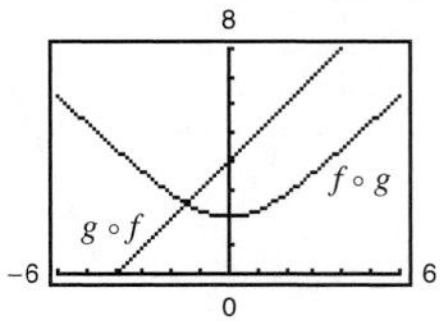

$f \circ g \neq g \circ f$

57. (a) $(f \circ g)(x) = x$; $(g \circ f)(x) = x$;
Domain of $f \circ g$: all real numbers x
(b) 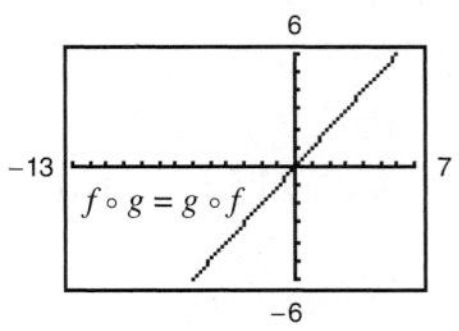

$f \circ g = g \circ f$

59. (a) $(f \circ g)(x) = x^4$; $(g \circ f)(x) = x^4$;
Domain of $f \circ g$: all real numbers x
(b) 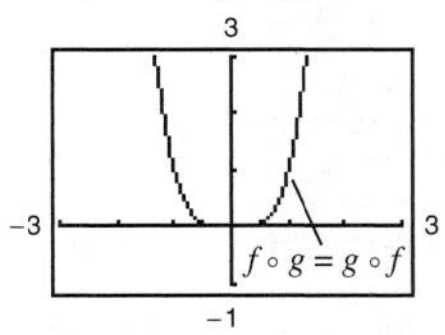

$f \circ g = g \circ f$

61. (a) $(f \circ g)(x) = x$; $(g \circ f)(x) = x$ (b) $x = x$
(c)

x	0	1	2	3
$g(x)$	$-\frac{4}{5}$	$-\frac{3}{5}$	$-\frac{2}{5}$	$-\frac{1}{5}$
$(f \circ g)(x)$	0	1	2	3

x	0	1	2	3
$f(x)$	4	9	14	19
$(g \circ f)(x)$	0	1	2	3

63. (a) $(f \circ g)(x) = \sqrt{x^2 + 1}$; $(g \circ f)(x) = x + 1, x \geq -6$
(b) $x + 1 \neq \sqrt{x^2 + 1}$
(c)

x	0	1	2	3
$g(x)$	-5	-4	-1	4
$(f \circ g)(x)$	1	$\sqrt{2}$	$\sqrt{5}$	$\sqrt{10}$

x	0	1	2	3
$f(x)$	$\sqrt{6}$	$\sqrt{7}$	$\sqrt{8}$	3
$(g \circ f)(x)$	1	2	3	4

65. (a) $(f \circ g)(x) = |2x^3|$; $(g \circ f)(x) = 2|x|^3$
(b) $|2x^3| = 2|x|^3$
(c)

x	-2	-1	0	1	2
$g(x)$	-16	-2	0	2	16
$(f \circ g)(x)$	16	2	0	2	16

x	-2	-1	0	1	2
$f(x)$	2	1	0	1	2
$(g \circ f)(x)$	16	2	0	2	16

67. (a) 3 (b) 0 **69.** (a) 2 (b) 4

71. $f(x) = x^2$, $g(x) = 2x + 1$ **73.** $f(x) = \sqrt[3]{x}$, $g(x) = x^2 - 4$

75. $f(x) = \frac{1}{x}$, $g(x) = x + 2$

77. $f(x) = x^2 + 2x$, $g(x) = x + 4$

79. (a) $T = \frac{3}{4}x + \frac{1}{15}x^2$
(b) 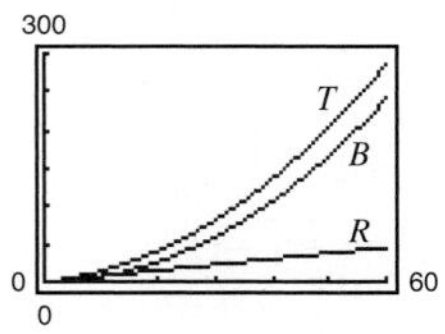

(c) B. For example, $B(60) = 240$, whereas $R(60)$ is only 45.

81. (a) $r(x) = \frac{x}{2}$ (b) $A(r) = \pi r^2$
(c) $(A \circ r)(x) = \pi\left(\frac{x}{2}\right)^2$
$(A \circ r)(x)$ represents the area of the circular base of the tank with radius $x/2$.

83. (a) $T = 1.3t^2 + 72.4t + 1322$
(b) 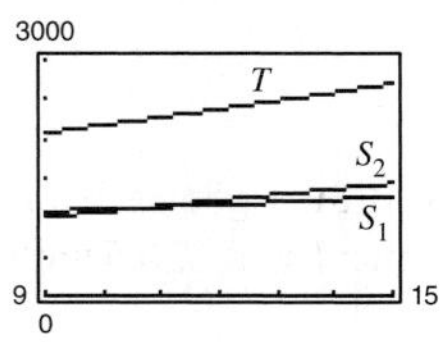

85. (a) $N(T(t))$ or $(N \circ T)(t) = 40t^2 + 590$; $N(T(t))$ or $(N \circ T)(t)$ represents the number of bacteria after t hours outside the refrigerator.
(b) $(N \circ T)(12) = 6350$; There are 6350 bacteria in a refrigerated food product after 12 hours outside the refrigerator.
(c) About 3.9 h

87. $s(t) = \sqrt{(150 - 450t)^2 + (200 - 450t)^2}$
$= 50\sqrt{162t^2 - 126t + 25}$

89. False. $g(x) = x - 3$

91. a, c **93.** Odd (Proofs will vary.)

95. (a) $Y = O - 12$; Answers will vary.
(b) Middle child is 7 years old, oldest child is 14 years old.

97. $(0, -5), (1, -5), (2, -7)$ **99.** $(0, 7), (1, 4\sqrt{3}), (2, \sqrt{45})$

Section 1.7 (page 148)

1. inverse, f^{-1} **3.** $y = x$ **5.** At most once

7. $f^{-1}(x) = \dfrac{x}{6}$ **9.** $f^{-1}(x) = x - 11$

11. $f^{-1}(x) = 2x + 1$ **13.** $f^{-1}(x) = x^3$

15. c **16.** b **17.** a **18.** d

19. $f(g(x)) = f(\sqrt[3]{x}) = (\sqrt[3]{x})^3 = x$
$g(f(x)) = g(x^3) = \sqrt[3]{x^3} = x$

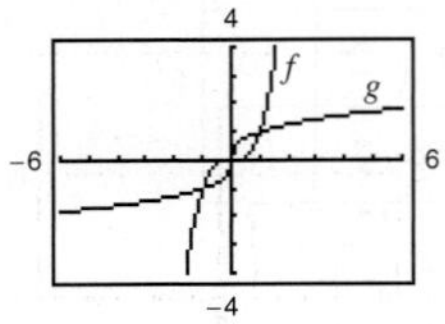

Reflections in the line $y = x$

21. $f(g(x)) = f(x^2 + 4), x \geq 0$
$= \sqrt{(x^2 + 4) - 4} = x$
$g(f(x)) = g(\sqrt{x - 4})$
$= (\sqrt{x - 4})^2 + 4 = x$

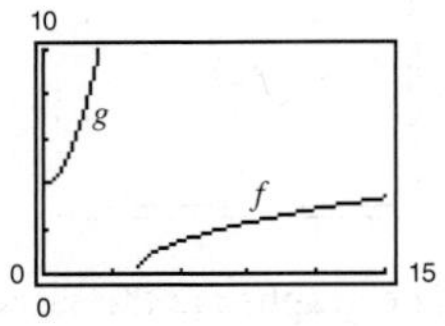

Reflections in the line $y = x$

23. $f(g(x)) = f(\sqrt[3]{1 - x}) = 1 - (\sqrt[3]{1 - x})^3 = x$
$g(f(x)) = g(1 - x^3) = \sqrt[3]{1 - (1 - x^3)} = x$

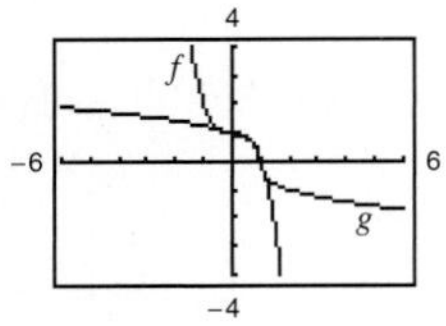

Reflections in the line $y = x$

25. (a) $f(g(x)) = f\left(-\dfrac{2x + 6}{7}\right)$
$= \dfrac{-7}{2}\left(-\dfrac{2x + 6}{7}\right) - 3 = x$
$g(f(x)) = g\left(-\dfrac{7}{2}x - 3\right)$
$= -\dfrac{2\left(-\dfrac{7}{2}x - 3\right) + 6}{7} = x$

(b)

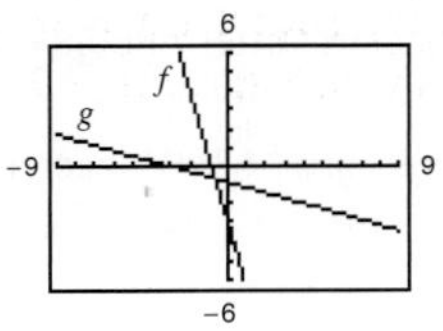

(c)

x	0	2	-2	6
$f(x)$	-3	-10	4	-24

x	-3	-10	4	-24
$g(x)$	0	2	-6	6

27. (a) $f(g(x)) = f(\sqrt[3]{x - 5}) = (\sqrt[3]{x - 5})^3 + 5 = x$
$g(f(x)) = g(x^3 + 5) = \sqrt[3]{(x^3 + 5) - 5} = x$

(b)

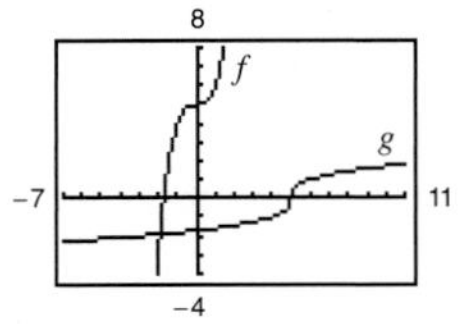

(c)

x	0	1	-1	-2	4
$f(x)$	5	6	4	-3	69

x	5	6	4	-3	69
$g(x)$	0	1	-1	-2	4

29. (a) $f(g(x)) = f(8 + x^2)$
$= -\sqrt{(8 + x^2) - 8}$
$= -\sqrt{x^2} = -(-x) = x, x \leq 0$
$g(f(x)) = g(-\sqrt{x - 8})$
$= 8 + (-\sqrt{x - 8})^2$
$= 8 + (x - 8) = x, x \geq 8$

(b)

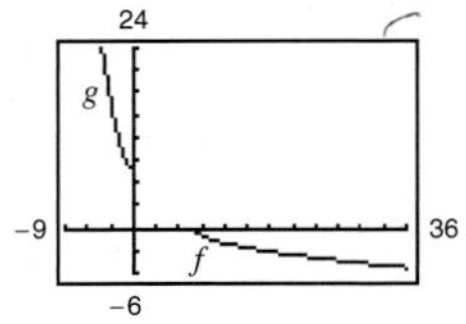

(c)

x	8	9	12	15
$f(x)$	0	-1	-2	$-\sqrt{7}$

x	0	-1	-2	$-\sqrt{7}$
$g(x)$	8	9	12	15

31. (a) $f(g(x)) = f\left(\dfrac{x}{2}\right)$
$= 2\left(\dfrac{x}{2}\right) = x$
$g(f(x)) = g(2x)$
$= \dfrac{2x}{2} = x$

(b)

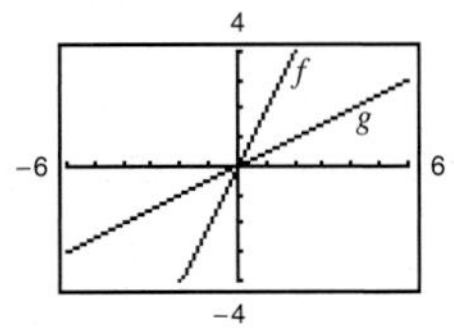

(c)

x	-4	-2	0	2	4
$f(x)$	-8	-4	0	4	8

x	-8	-4	0	4	8
$g(x)$	-4	-2	0	2	4

33. (a) $f(g(x)) = f\left(-\frac{5x+1}{x-1}\right)$

$$= \frac{\left(-\frac{5x+1}{x-1}\right) - 1}{\left(-\frac{5x+1}{x-1}\right) + 5} = \frac{\frac{6x}{x-1}}{\frac{6}{x-1}} = x, x \neq 1$$

$g(f(x)) = g\left(\frac{x-1}{x+5}\right)$

$$= -\frac{5\left(\frac{x-1}{x+5}\right) + 1}{\left(\frac{x-1}{x+5}\right) - 1} = \frac{\frac{6x}{x+5}}{\frac{6}{x+5}} = x, x \neq -5$$

(b)

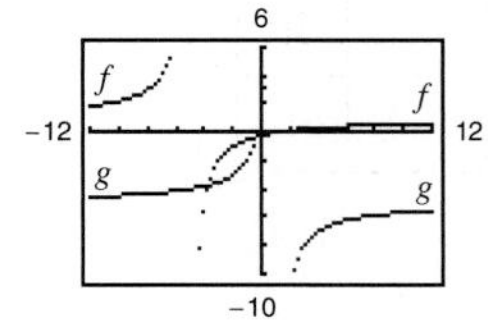

(c)

x	-3	-2	-1	0	2	3	4
$f(x)$	-2	-1	$-\frac{1}{2}$	$-\frac{1}{5}$	$\frac{1}{7}$	$\frac{1}{4}$	$\frac{1}{3}$

x	-2	-1	$-\frac{1}{2}$	$-\frac{1}{5}$	$\frac{1}{7}$	$\frac{1}{4}$	$\frac{1}{3}$
$g(x)$	-3	-2	-1	0	2	3	4

35. Yes. No two elements in the domain of f correspond to the same element in the range of f.

37. No. -3 and 0 both correspond to 6, so f is not one-to-one.

39. Not a function **41.** Function; one-to-one

43. Function; one-to-one

45.

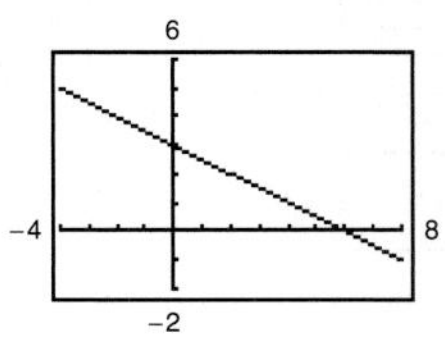

One-to-one

47.

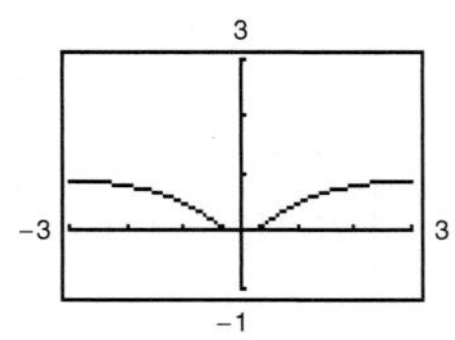

Not one-to-one

49.

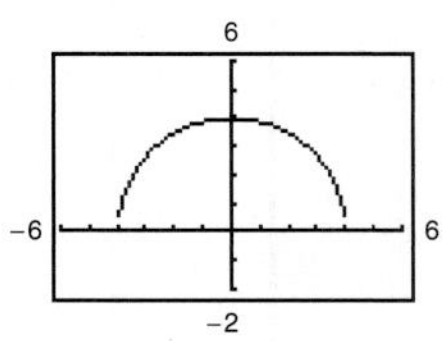

Not one-to-one

51.

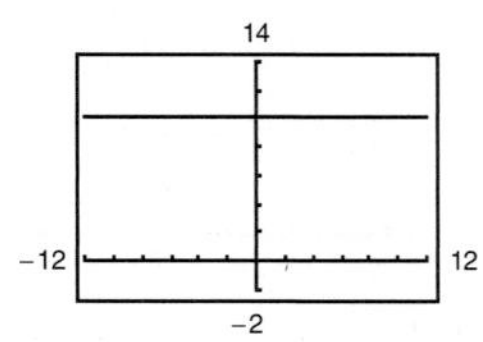

Not one-to-one

53.

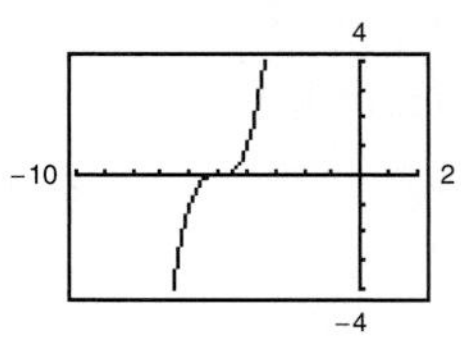

One-to-one

55.

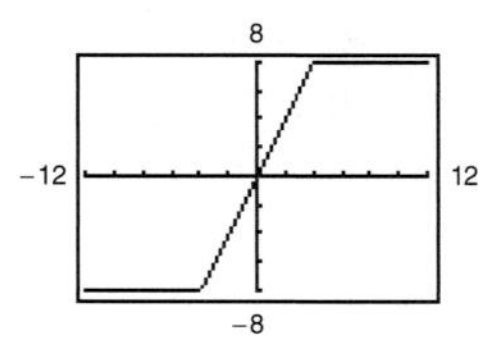

Not one-to-one

57.

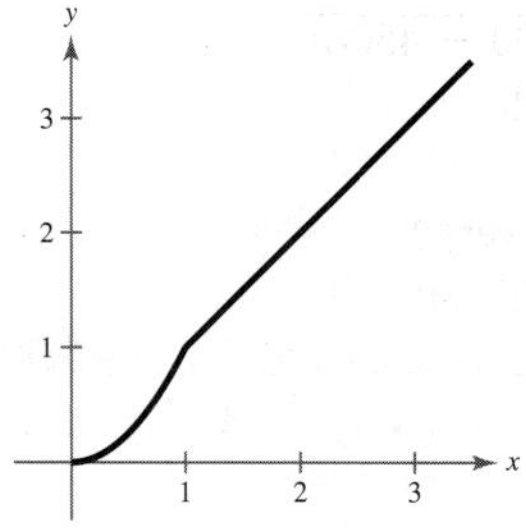

An inverse function exists.

59. Not one-to-one **61.** $f^{-1}(x) = \frac{5x-4}{3}$

63. Not one-to-one **65.** $f^{-1}(x) = \sqrt{x} - 3$

67. $f^{-1}(x) = \frac{x^2-3}{2}, x \geq 0$ **69.** $f^{-1}(x) = 2 - x, x \geq 0$

71. $f^{-1}(x) = \frac{x+9}{4}$

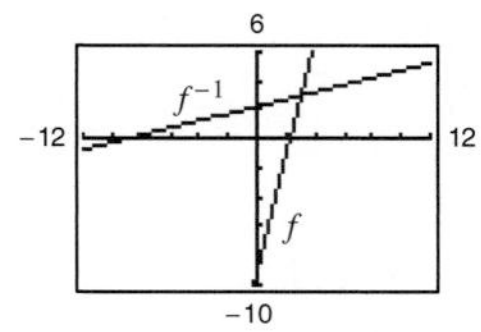

Reflections in the line $y = x$

73. $f^{-1}(x) = \sqrt[5]{x}$

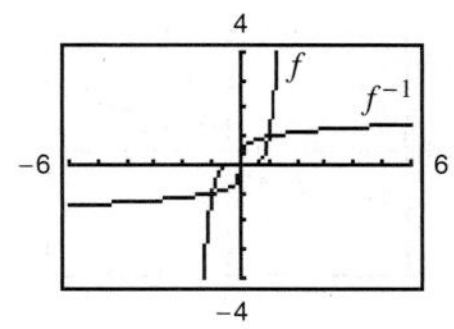

Reflections in the line $y = x$

75. $f^{-1}(x) = -\sqrt[4]{x}$

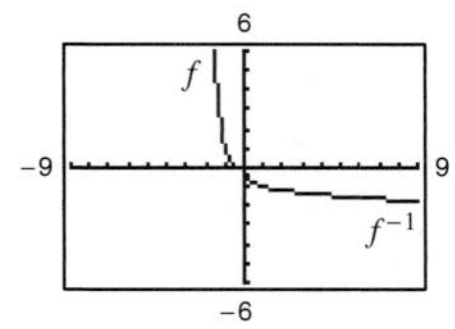

Reflections in the line $y = x$

77. $f^{-1}(x) = \sqrt{4 - x^2}$, $0 \leq x \leq 2$

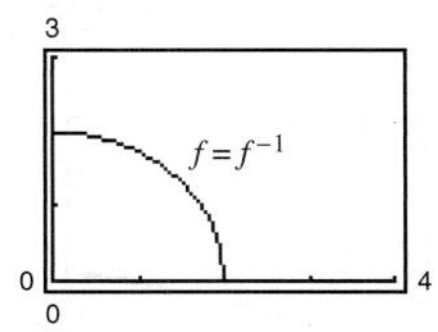

The graphs are the same.

79. $f^{-1}(x) = \sqrt[3]{\frac{4}{x}}$

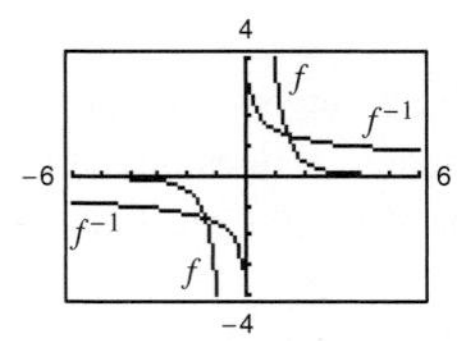

Reflections in the line $y = x$

81. $f^{-1}(x) = \sqrt{x} + 2$

Domain of f: all real numbers x such that $x \geq 2$

Range of f: all real numbers y such that $y \geq 0$

Domain of f^{-1}: all real numbers x such that $x \geq 0$

Range of f^{-1}: all real numbers y such that $y \geq 2$

83. $f^{-1}(x) = x - 2$

Domain of f: all real numbers x such that $x \geq -2$

Range of f: all real numbers y such that $y \geq 0$

Domain of f^{-1}: all real numbers x such that $x \geq 0$

Range of f^{-1}: all real numbers y such that $y \geq -2$

85. $f^{-1}(x) = \sqrt{x} - 3$
Domain of f: all real numbers x such that $x \geq -3$
Range of f: all real numbers y such that $y \geq 0$
Domain of f^{-1}: all real numbers x such that $x \geq 0$
Range of f^{-1}: all real numbers y such that $y \geq -3$

87. $f^{-1}(x) = \dfrac{\sqrt{-2(x+5)}}{2}$
Domain of f: all real numbers x such that $x \geq 0$
Range of f: all real numbers y such that $y \leq -5$
Domain of f^{-1}: all real numbers x such that $x \leq -5$
Range of f^{-1}: all real numbers y such that $y \geq 0$

89. $f^{-1}(x) = x + 3$
Domain of f: all real numbers x such that $x \geq 4$
Range of f: all real numbers y such that $y \geq 1$
Domain of f^{-1}: all real numbers x such that $x \geq 1$
Range of f^{-1}: all real numbers y such that $y \geq 4$

91.

x	-4	-2	2	3
$f^{-1}(x)$	-2	-1	1	3

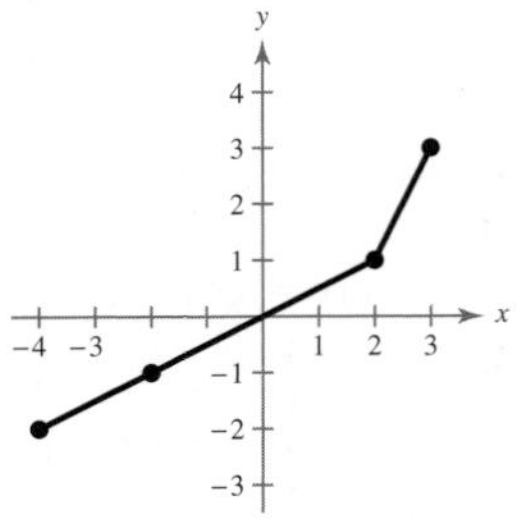

93. 4 **95.** -2 **97.** 0 **99.** 2

101. (a) and (b)

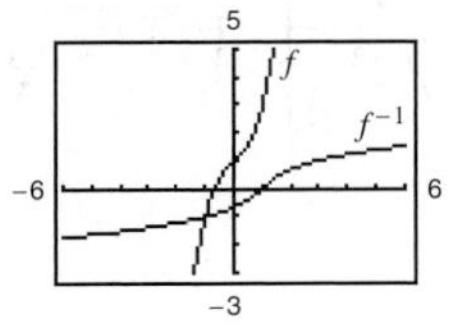

(c) Inverse function because it satisfies the Vertical Line Test

103. (a) and (b)

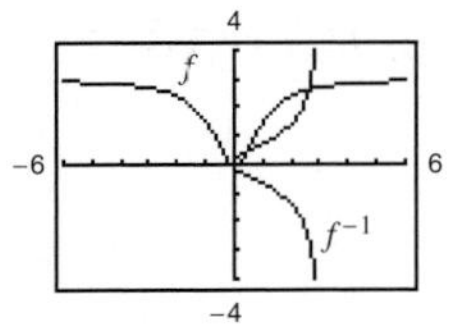

(c) Not an inverse function because it does not satisfy the Vertical Line Test

105. 32 **107.** -168 **109.** $8\sqrt[3]{x} + 24$

111. $\dfrac{x+1}{2}$ **113.** $\dfrac{x+1}{2}$

115. (a) f is one-to-one because no two elements in the domain (men's U.S. shoe sizes) correspond to the same element in the range (men's European shoe sizes).
(b) 44 (c) 10 (d) 41 (e) 12

117. (a) 19 9 64 64 4 69 29 4 64 9 104 29 94
(b) $f^{-1}(x) = \dfrac{x-4}{5}$; What time

119. False. For example, $y = x^2$ is even, but does not have an inverse.

121. This situation could be represented by a one-to-one function. The inverse function would represent the number of miles completed in terms of time in hours.

123. This function could not be represented by a one-to-one function because it oscillates.

125. The graph of f^{-1} is a reflection of the graph of f in the line $y = x$.

127. (a) The function will be one-to-one because no two values of x will produce the same value for $f(x)$.
(b) $f^{-1}(50)$ represents the value of 50 degrees Fahrenheit in degrees Celsius.

129. Constant function **131.** Proof **133.** $9x, x \neq 0$
135. $-(x+6), x \neq 6$ **137.** Not a function
139. Function

Review Exercises (page 154)

1.

x	-2	0	2	3	4
y	3	2	1	$\frac{1}{2}$	0
Solution point	$(-2, 3)$	$(0, 2)$	$(2, 1)$	$\left(3, \frac{1}{2}\right)$	$(4, 0)$

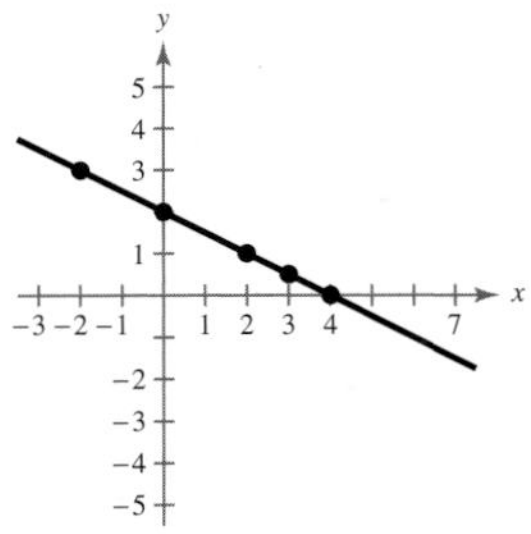

3.

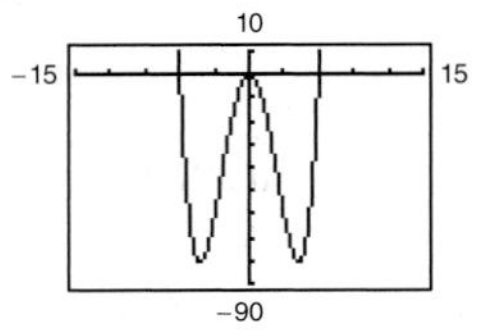

Intercepts: $(0, 0), (\pm 6, 0)$

5.

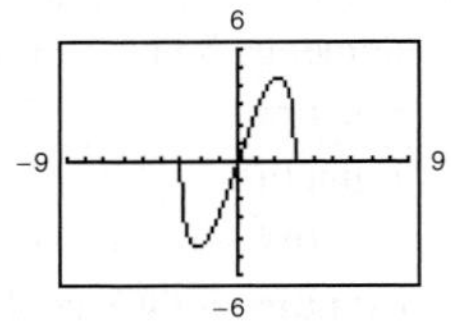

Intercepts: $(0, 0), (\pm 3, 0)$

7. (a)
Xmin = 0
Xmax = 6
Xscl = 1
Ymin = 9000
Ymax = 18,000
Yscl = 1000

(b)

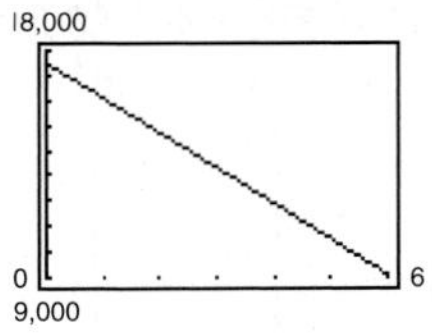

(c) 4

9.

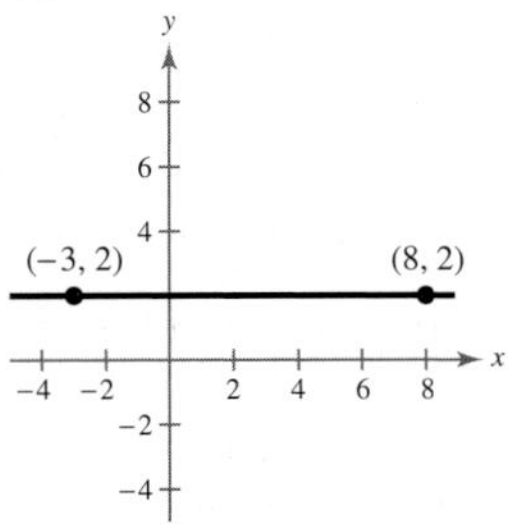

$m = 0$

11.

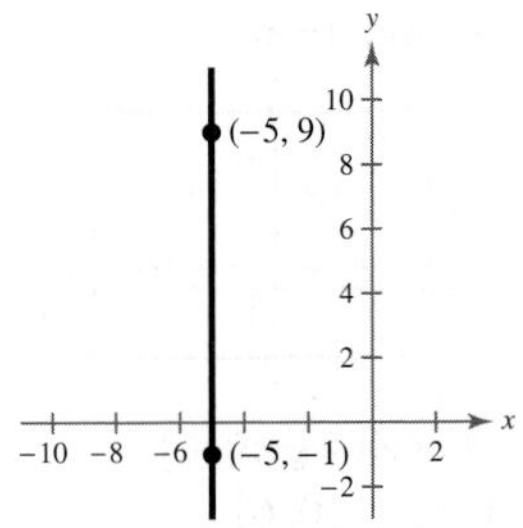

m is undefined.

13.

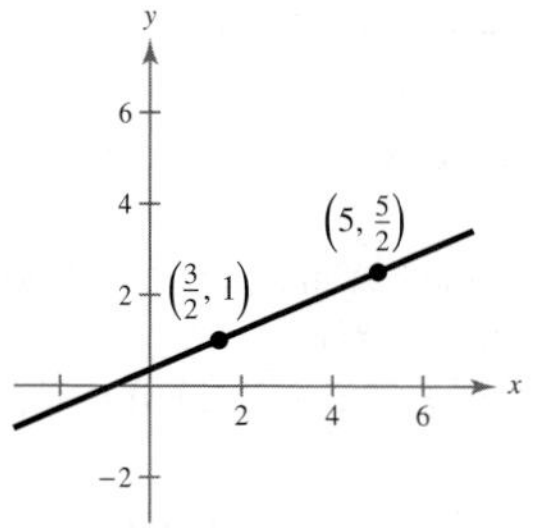

$m = \frac{3}{7}$

15.

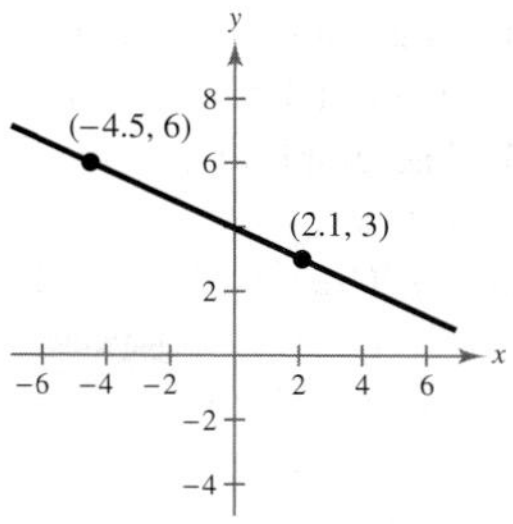

$m = -\frac{5}{11}$

17. $x - 4y - 6 = 0$; $(6, 0)$, $(10, 1)$, $(-2, -2)$

19. $3x + 2y + 10 = 0$; $(4, -11)$, $(2, -8)$, $(-2, -2)$

21. $y - 6 = 0$; $(0, 6)$, $(1, 6)$, $(-1, 6)$

23. $y = -1$

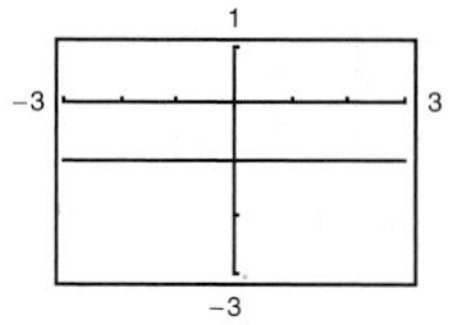

25. $x = \frac{5}{6}$
The line cannot be graphed because the slope is undefined.

27. $y = \frac{2}{7}x + \frac{2}{7}$

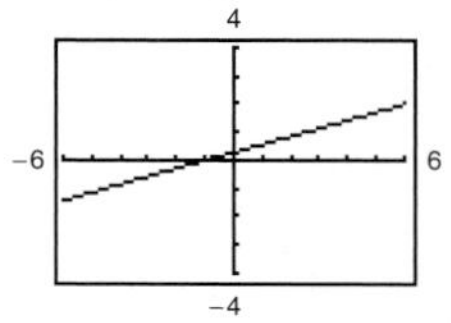

29. $y = -\frac{1}{2}x + \frac{1}{2}$

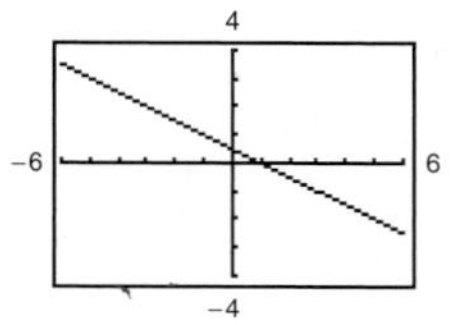

31. (a) $y = \frac{5}{4}x - \frac{23}{4}$ (b) $y = -\frac{4}{5}x + \frac{2}{5}$

33. (a) Not a function because element 20 in A corresponds to two elements, 4 and 6, in B.
(b) Function

35. Not a function **37.** Function **39.** Function

41. (a) 2 (b) 10 (c) $b^6 + 1$ (d) $x^2 - 2x + 2$

43. All real numbers x except $x = -2$

45. All real numbers x such that $-5 \le x \le 5$

47. (a) $C = 5.25x + 17{,}500$ (b) $P = 3.18x - 17{,}500$

49. $2h + 4x + 3$, $h \ne 0$

51.

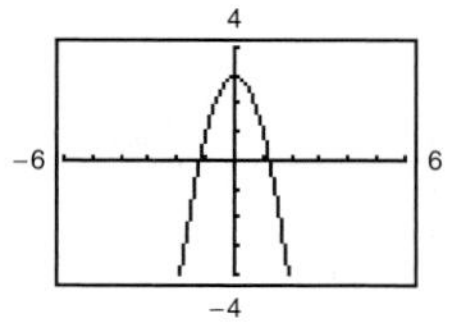

Domain: all real numbers x
Range: $(-\infty, 3]$

53.

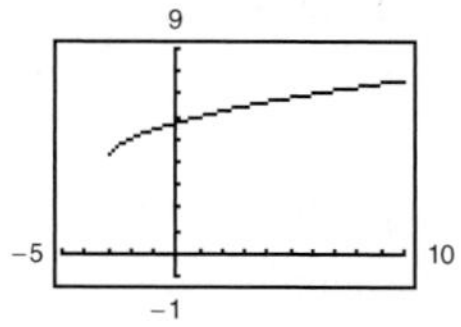

Domain: $[-3, \infty)$
Range: $[4, \infty)$

55.

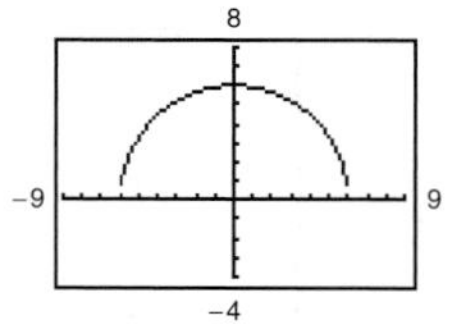

Domain: $[-6, 6]$
Range: $[0, 6]$

57.

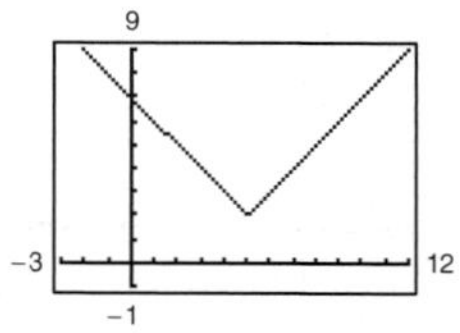

Domain: all real numbers x
Range: $[2, \infty)$

59. Function. Solve for y and graph the resulting function.

61. (a)

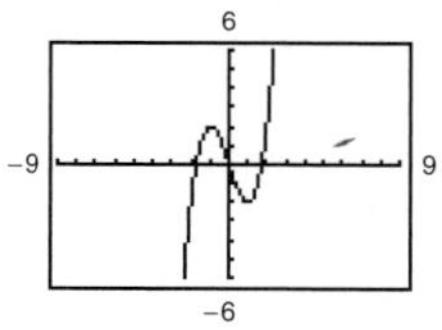

(b) Increasing on $(-\infty, -1)$, $(1, \infty)$
Decreasing on $(-1, 1)$

63. (a)

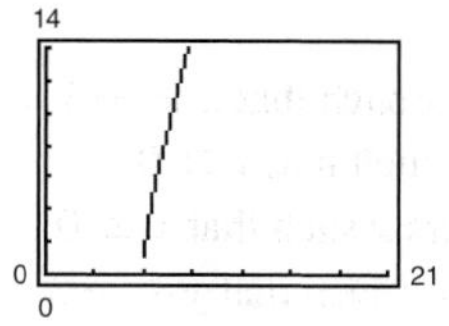

(b) Increasing on $(6, \infty)$

65. Relative minima: $(-2, 0)$, $(2, 0)$
Relative maximum: $(0, 16)$

67.

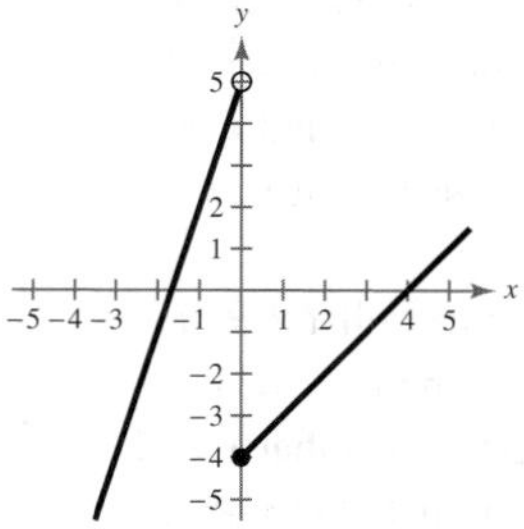

69.

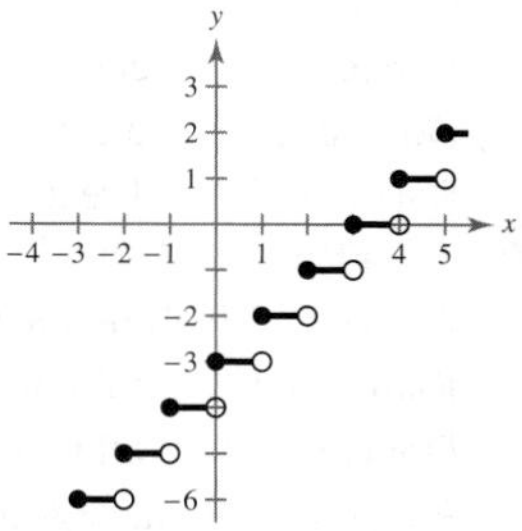

71. Even **73.** Even **75.** Neither **77.** Odd

79. Rational function $f(x) = \dfrac{1}{x}$
Horizontal shift three units right
$g(x) = \dfrac{1}{x - 3}$

81. Absolute value function $f(x) = |x|$
Vertical shift three units upward
$g(x) = |x| + 3$

83.

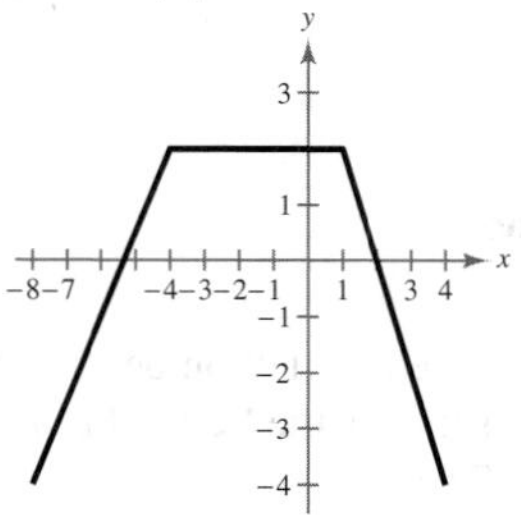

85.

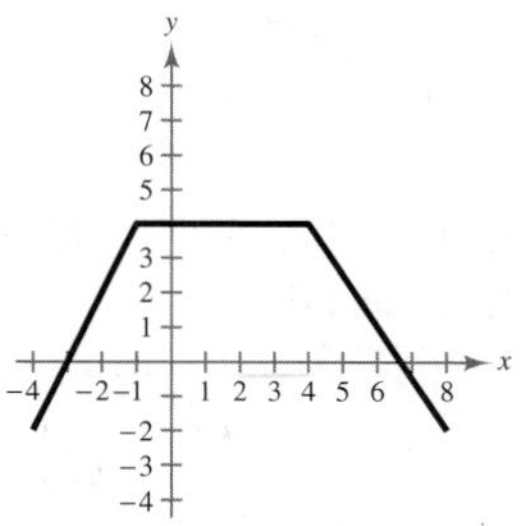

87. (a) Rational function $f(x) = \dfrac{1}{x}$
(b) Vertical shift six units downward
(c)

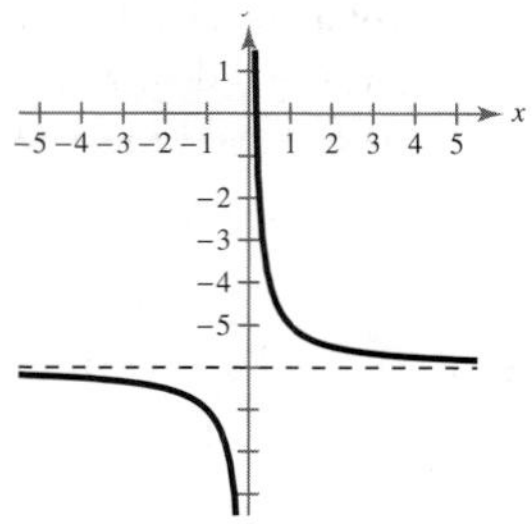

(d) $h(x) = f(x) - 6$

89. (a) Cubic function $f(x) = x^3$
(b) Horizontal shift two units right, vertical shift five units upward
(c)

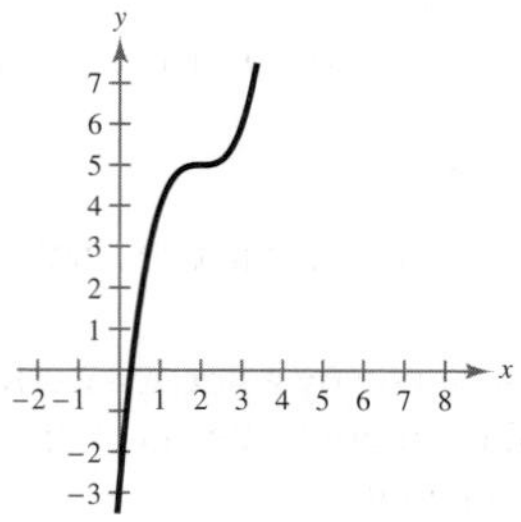

(d) $h(x) = f(x - 2) + 5$

91. (a) Square root function $f(x) = \sqrt{x}$
(b) Reflection in the x-axis, vertical shift six units downward
(c)
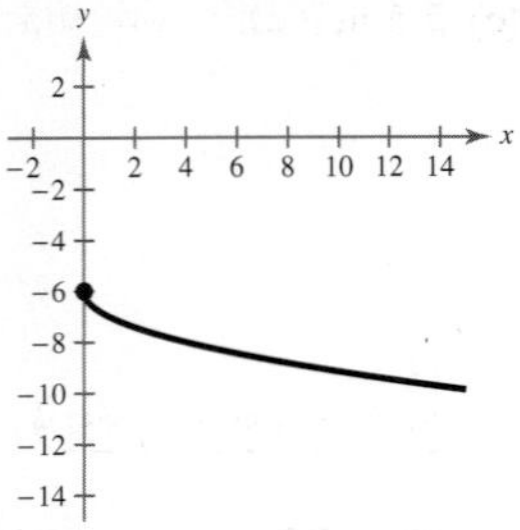
(d) $h(x) = -f(x) - 6$

93. (a) Absolute value function $f(x) = |x|$
(b) Horizontal shrink by a factor of $\frac{1}{3}$, vertical shift nine units upward
(c)
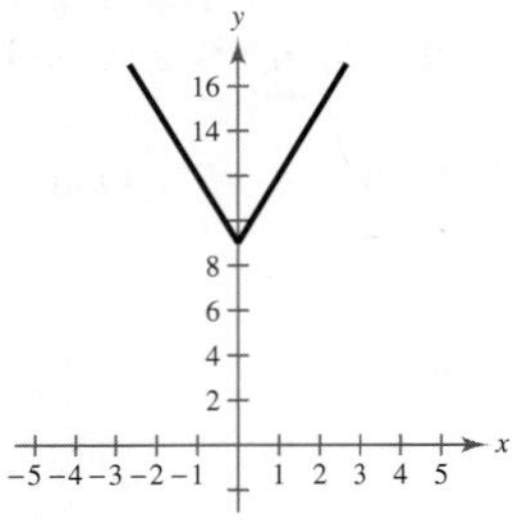
(d) $h(x) = f(3x) + 9$

95. -7 **97.** -42 **99.** 5 **101.** 17

103. $f(x) = x^2, g(x) = x + 3$ **105.** $f(x) = \sqrt{x}, g(x) = 4x + 2$

107. $f(x) = \dfrac{4}{x}, g(x) = x + 2$

109.
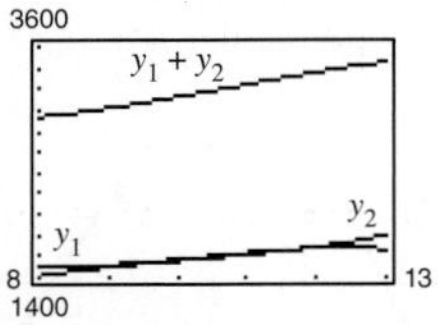

111. $f^{-1}(x) = \dfrac{x}{6}$ **113.** $f^{-1}(x) = 2x - 6$

115. Answers will vary.

117.
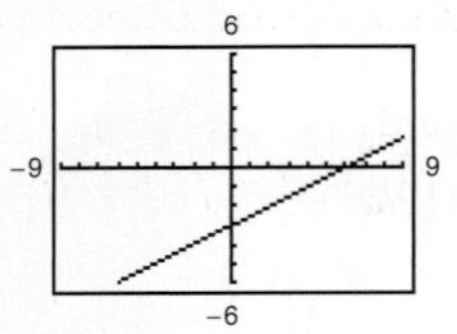

One-to-one

119.
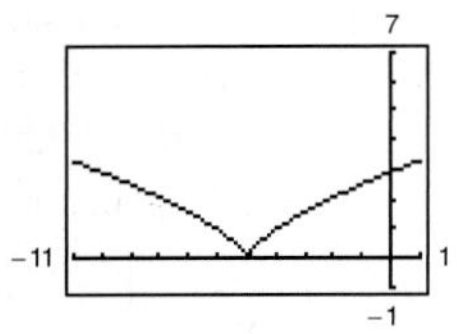

Not one-to-one

121. $f^{-1}(x) = 2x + 10$ **123.** $f^{-1}(x) = \sqrt[3]{-\dfrac{x + 3}{5}}$

125. $f^{-1}(x) = x^2 - 10, x \ge 0$ **127.** $f^{-1}(x) = \sqrt{4x - 4}$

Chapter Test (page 157)

1.
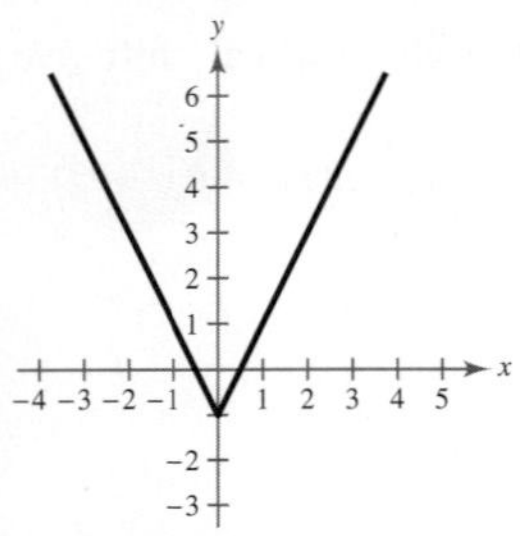
Intercepts: $(0, 1), \left(-\frac{1}{2}, 0\right), \left(\frac{1}{2}, 0\right)$

2.
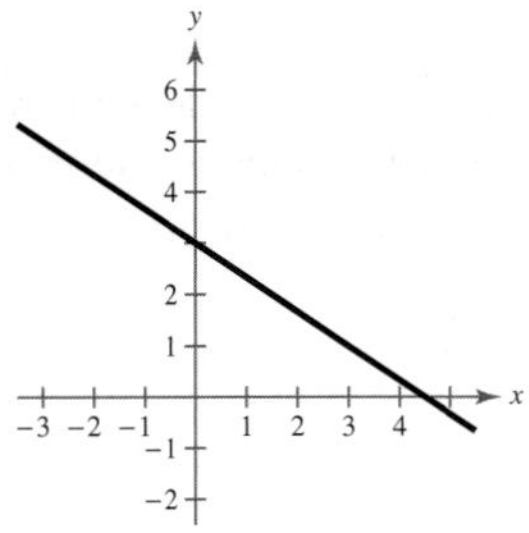
Intercepts: $(0, 3), \left(\frac{9}{2}, 0\right)$

3.
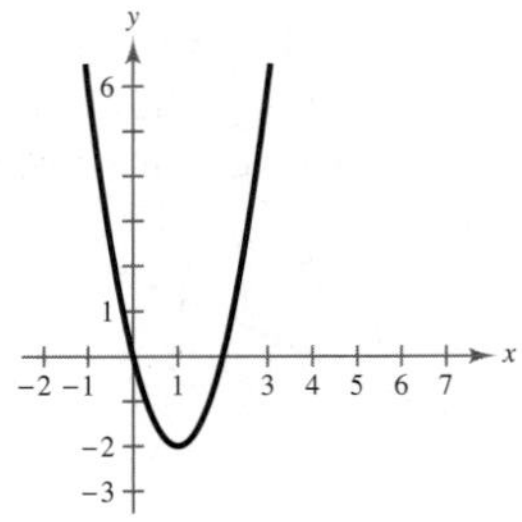
Intercepts: $(0, 0), (2, 0)$

4.
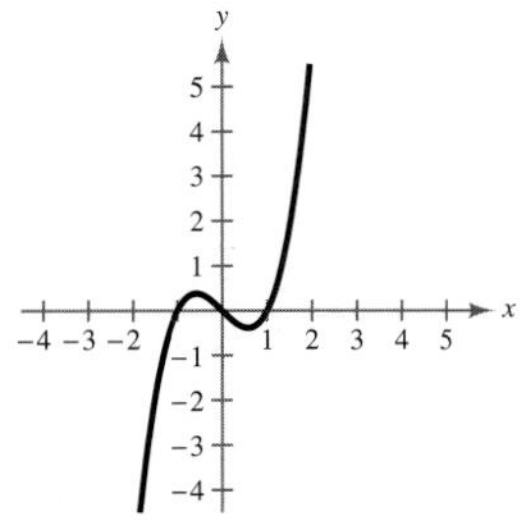
Intercepts: $(-1, 0), (0, 0), (1, 0)$

5.
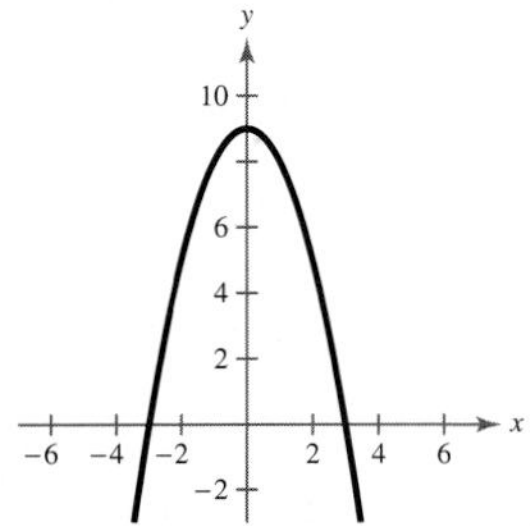
Intercepts: $(\pm 3, 0), (0, 9)$

6.
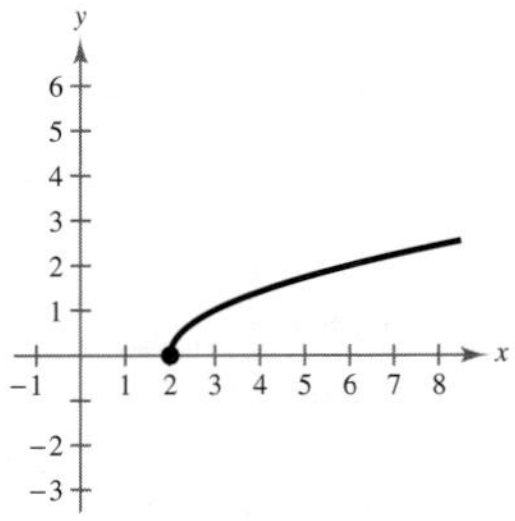
Intercept: $(2, 0)$

7. (a) $5x + 2y - 8 = 0$ (b) $2x - 5y + 20 = 0$

8. $y = -x + 1$

9. No. For some values of x there correspond more than one value of y.

10. (a) -9 (b) 1 (c) $|t - 4| - 15$ **11.** $(-\infty, 3]$

12. $C = 24.60x + 25{,}000$
$P = 76.9x - 25{,}000$

13. Odd **14.** Even

15. Increasing: $(-2, 0), (2, \infty)$
Decreasing: $(-\infty, -2), (0, 2)$

16. Increasing: $(-2, 2)$
Constant: $(-\infty, -2), (2, \infty)$

17.
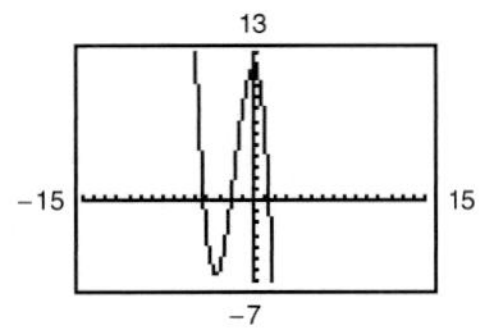

Relative minimum: $(-3.33, -6.52)$
Relative maximum: $(0, 12)$

18.
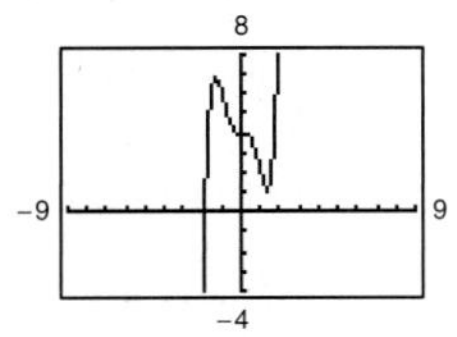

Relative minimum: $(1.34, 1.10)$
Relative maximum: $(-1.34, 6.90)$

19. (a) $f(x) = x^3$
(b) Horizontal shift five units to the right, reflection in the x-axis, vertical stretch, and vertical shift three units upward
(c)
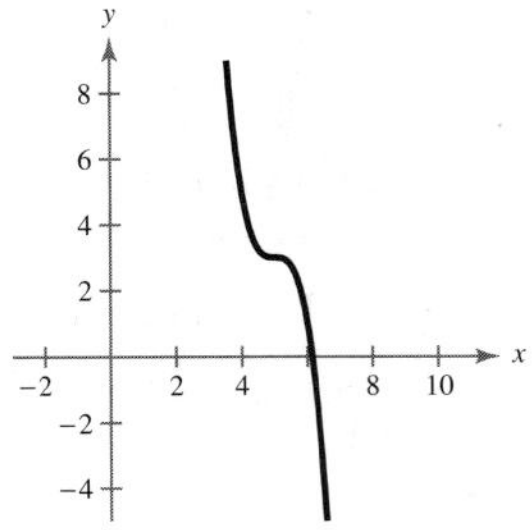

20. (a) $f(x) = \sqrt{x}$
(b) Reflection in the y-axis, horizontal shift two units to the left, horizontal shrink by a factor of $\frac{1}{7}$
(c)
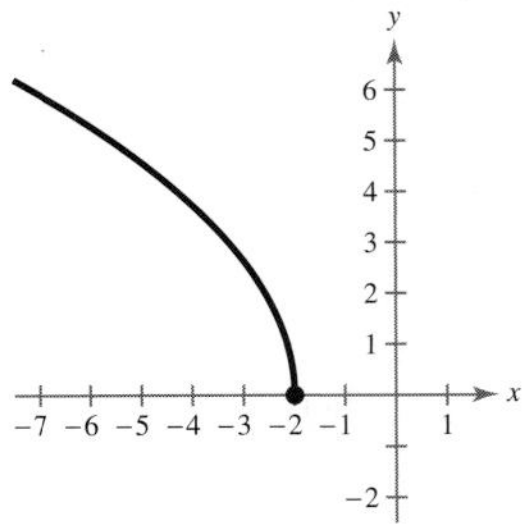

21. (a) $f(x) = |x|$
(b) Reflection in the y-axis (no effect), vertical stretch, and vertical shift seven units downward
(c)
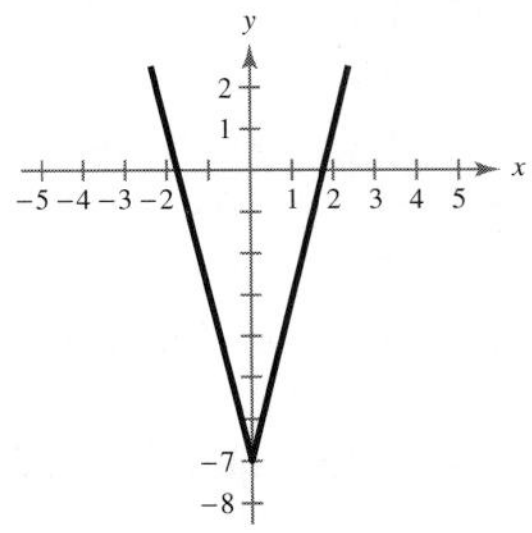

22. (a) $x^2 - \sqrt{2 - x}, (-\infty, 2]$
(b) $\dfrac{x^2}{\sqrt{2 - x}}, (-\infty, 2)$
(c) $2 - x, (-\infty, 2]$
(d) $\sqrt{2 - x^2}, \left[-\sqrt{2}, \sqrt{2}\right]$

23. $f^{-1}(x) = \sqrt[3]{x - 8}$ **24.** No inverse

25. $f^{-1}(x) = \left(\frac{8}{3}x\right)^{2/3}, x \geq 0$

Chapter 2

Section 2.1 (page 166)

1. equation **3.** extraneous **5.** Conditional equation
7. (a) Yes (b) No (c) No (d) No
9. (a) No (b) No (c) No (d) Yes
11. Identity **13.** Contradiction **15.** Conditional equation
17. Conditional equation **19.** $-\frac{96}{23}$ **21.** $\frac{20}{81}$ **23.** 12
25. -9 **27.** -10 **29.** 4 **31.** $-\frac{16}{33}$ **33.** $\frac{1}{2}$
35. 5 **37.** $\frac{5}{3}$ **39.** No solution **41.** No solution
43. $h = \dfrac{2A}{b}$ **45.** $P = A\left(1 + \dfrac{r}{n}\right)^{-nt}$ **47.** $h = \dfrac{V}{\pi r^2}$
49. 161.14 cm
51. (a)
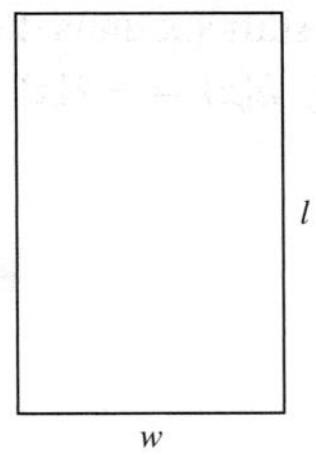

(b) $l = 1.5w$; $P = 5w$
(c) 7.5 m long × 5 m wide

53. (a) Test average $= \dfrac{\text{test 1} + \text{test 2} + \text{test 3} + \text{test 4}}{4}$ (b) 92

55. 2.5 h **57.** 46.3 mi/h **59.** 30 ft **61.** 1.25%
63. 50 lb of each kind
65. \$8000 in notebook computers; \$32,000 in tablet computers
67. $h = 27$ ft
69. (a)
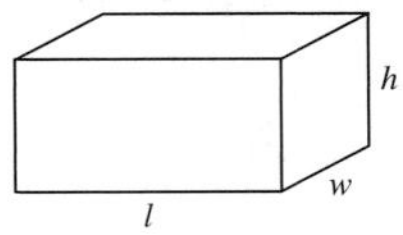

(b) 24 in. × 12 in. × 8 in.

71.
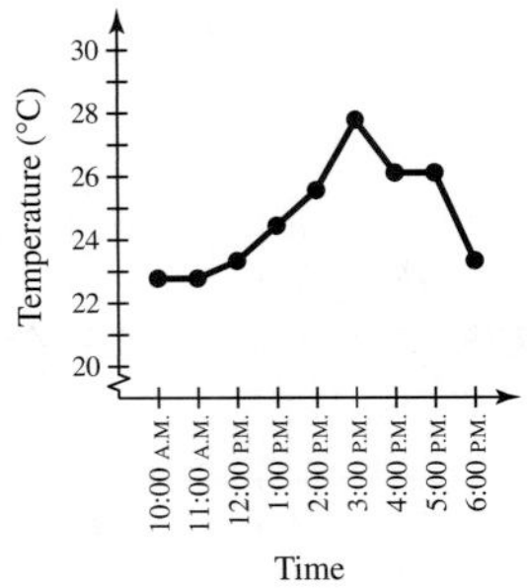

73. $\frac{200}{3}$ km/h **75.** $x = 6$ ft
77. False. It is quadratic; $x(3 - x) = 10 \implies 3x - x^2 = 10$.
79. $9x + 27 = 0$ **81.** $2x + \frac{7}{2} = 4$
83. There is no solution because $x = 1$ is an extraneous solution.
85. Find the least common denominator (LCD) of all terms in the equation and multiply every term by this LCD.
It is possible to introduce an extraneous solution because you are multiplying by a variable.
You can substitute the answer for the variable in the original equation, or you can graph the original equation.

87.
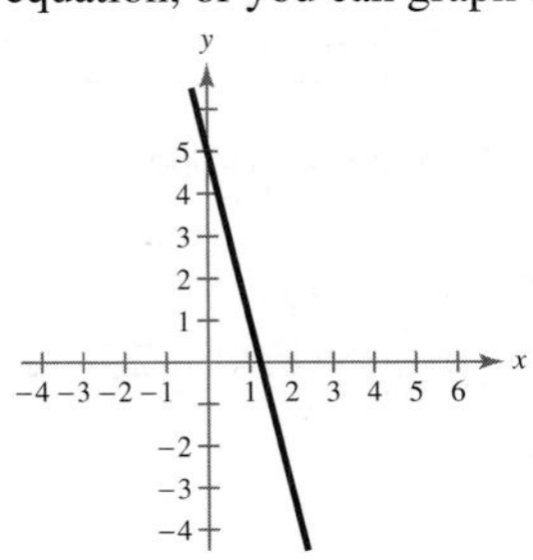

89.
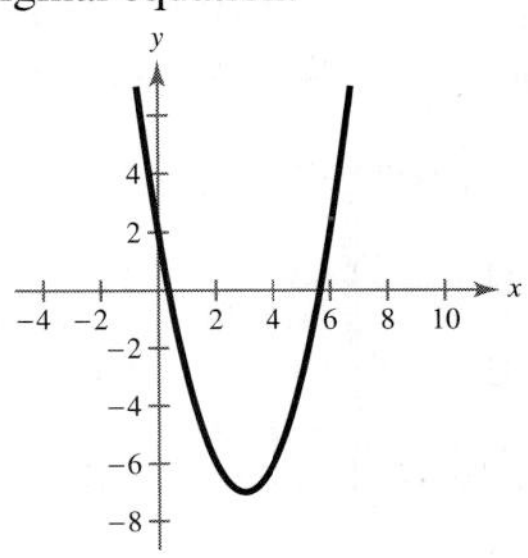

91.
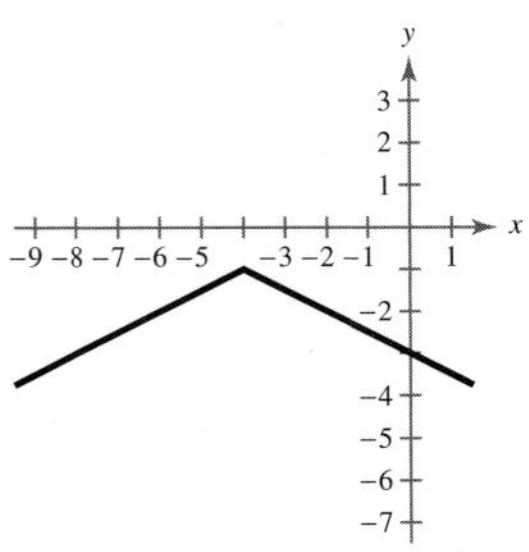

CHAPTER 2

Section 2.2 (page 176)

1. x-intercept, y-intercept **3.** $(1, 0), (-1, 0)$ **5.** $-1, 1$
7. $(5, 0), (0, -5)$ **9.** $(0, 2)$ **11.** $(-2, 0), (0, 0)$
13. $(2, 0)$ **15.** $(1, 0), \left(0, \frac{1}{2}\right)$

17.

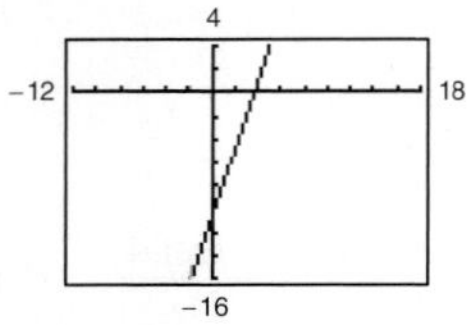

$\left(\frac{11}{3}, 0\right), (0, -11)$

19.

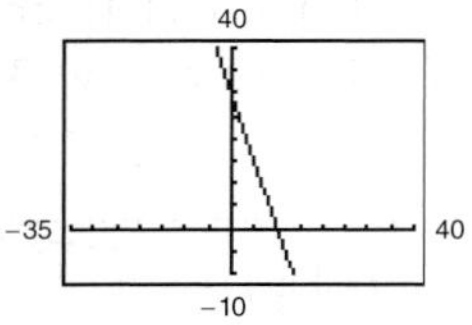

$(10, 0), (0, 30)$

21–25. Answers will vary.
27. $\frac{12}{23}$; $f(x) = 2.3x - 1.2 = 0$ **29.** 29; $f(x) = 3x - 87 = 0$
31. $\frac{38}{7}$; $f(x) = 7x - 38 = 0$
33. -194; $f(x) = 0.20x + 38.8 = 0$
35. $\frac{9}{7}$; $f(x) = 9 - 7x = 0$ **37.** 15; $f(x) = 3x - 45 = 0$
39. $-\frac{3}{10}$; $f(x) = 10x + 3 = 0$ **41.** -1.379
43. 2.172, 7.828 **45.** 0.5, -3, 3 **47.** -1.861
49. -1.333 **51.** ± 3.162 **53.** -4, 2 **55.** -1, 2.333
57. 11 **59.** -4
61. (a)

x	-1	0	1	2	3	4
$3.2x - 5.8$	-9	-5.8	-2.6	0.6	3.8	7

$1 < x < 2$; Answers will vary.

(b)

x	1.5	1.6	1.7	1.8	1.9	2
$3.2x - 5.8$	-1	-0.68	-0.36	-0.04	0.28	0.60

$1.8 < x < 1.9$; Answers will vary. Sample answer: Use smaller intervals of x to increase accuracy.

63. $(2, 4)$ **65.** $(2, 2)$ **67.** $(8, -2)$ **69.** $(-1, 3)$
71. $(4, 1)$ **73.** $(0, 0), (1, 1)$ **75.** $(1.67, 1.66)$
77. (a) 6.46
(b) $\dfrac{1.73}{0.27} \approx 6.41$
Yes, the more rounding performed, the less accurate the result.
79. (a) $t(x) = \dfrac{x}{63} + \dfrac{280 - x}{54}$
(b)

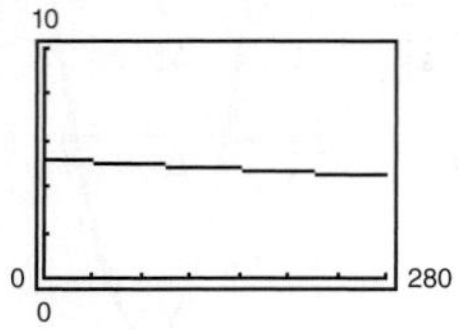

$0 \le x \le 280$

(c) 164.5 mi

81. (a) $A(x) = 12x$
(b)

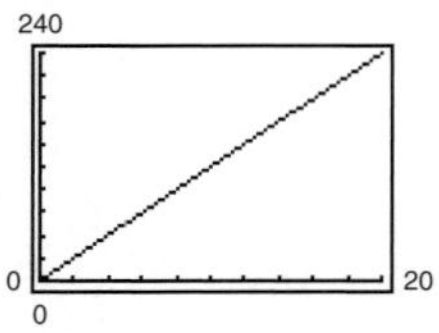

(c) 15 units

83. (a) $T = 10{,}000 + \frac{1}{2}x$ (b) \$6800 (c) \$7600 (d) \$7500

85. (a)

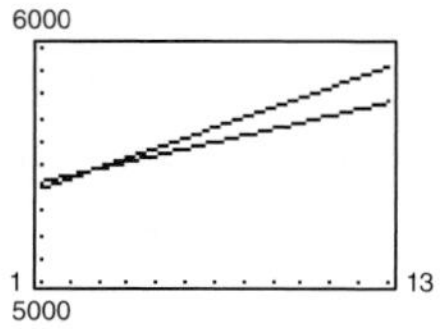

$(2.9, 5479.4)$; In 2002, both states had the same population.
(b) $(2.9, 5479.4)$; In 2002, both states had the same population.
(c) Change in population per year; Maryland's population is growing faster.
(d) Maryland: 6,049,400; Wisconsin: 5,852,400
Answers will vary.
87. True. A point lies on the y-axis when $x = 0$.
89. Answers will vary.
91. (a) $(0, 2), (-1, 0)$ (b) Answers will vary.
(c) $(-1, 0), (3, 8)$
93. $2 + \sqrt{14}$ **95.** $\dfrac{14\left(3\sqrt{10} + 1\right)}{89}$
97. $12x^2 + 31x - 91$ **99.** $16x^2 + 8x + 1$

Section 2.3 (Page 184)

1. (a) ii (b) iii (c) i **3.** complex, $a + bi$
5. $-2 + 4i$ **7.** $a = -9, b = 4$ **9.** $a = 6, b = 5$
11. $4 + 3i$ **13.** 12 **15.** $1 - 8i$ **17.** -11 **19.** $0.3i$
21. $-3 + 3i$ **23.** $-14 + 20i$ **25.** $\frac{19}{6} + \frac{37}{6}i$
27. $-4.2 + 7.5i$ **29.** $17 - \sqrt{3}i$ **31.** $-2\sqrt{3}$
33. -10 **35.** $12 + 20i$ **37.** $5 + i$ **39.** $-20 + 32i$
41. 24 **43.** $-13 + 84i$ **45.** $80i$ **47.** $6 + 2i$; 40
49. $-1 - \sqrt{7}i$; 8 **51.** $-\sqrt{29}i$; 29 **53.** $9 + \sqrt{6}i$; 87
55. $-6i$ **57.** $\frac{8}{41} + \frac{10}{41}i$ **59.** $\frac{4}{5} - \frac{3}{5}i$ **61.** $-\frac{40}{1681} - \frac{9}{1681}i$
63. $-\frac{1}{2} - \frac{5}{2}i$ **65.** $\frac{62}{949} + \frac{297}{949}i$ **67.** $-1 + 6i$
69. $-375\sqrt{3}i$ **71.** i
73. (a) 1 (b) i (c) $-i$ (d) -1
75. False. Any real number is equal to its conjugate.
77. False. Example: $(1 + i) + (1 - i) = 2$, which is not an imaginary number.
79. True. Answers will vary.
81. $\sqrt{-6}\sqrt{-6} = \left(\sqrt{6}i\right)\left(\sqrt{6}i\right) = 6i^2 = -6$
83. $16x^2 - 25$ **85.** $3x^2 + \frac{23}{2}x - 2$

Section 2.4 (page 196)

1. quadratic equation
3. factoring, extracting square roots, completing the square, Quadratic Formula
5. $2x^2 + 5x - 3 = 0$ **7.** $3x^2 - 60x - 10 = 0$
9. $0, -\frac{1}{3}$ **11.** 3, 7 **13.** 4 **15.** $-2, \frac{4}{3}$ **17.** $-7, -4$
19. $-a - b, -a + b$ **21.** ± 7 **23.** $\pm 3\sqrt{3}$; ± 5.20
25. ± 10 **27.** 16, 8 **29.** $\dfrac{1 \pm \sqrt{6}i}{3}$; $0.33 \pm 0.82i$
31. 2 **33.** -8, 4 **35.** $3 \pm \sqrt{7}$ **37.** $2 \pm 3i$
39. $-4 \pm 4i$ **41.** $1 \pm \sqrt{5}i$ **43.** $1 \pm \dfrac{\sqrt{6}}{3}$
45. $-\dfrac{5}{4} \pm \dfrac{\sqrt{89}}{4}$

47. (a)

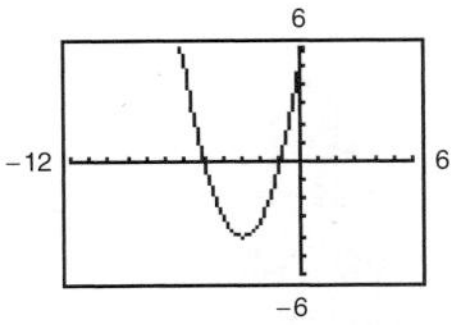

(b) and (c) $(-1, 0), (-5, 0)$

49. (a)

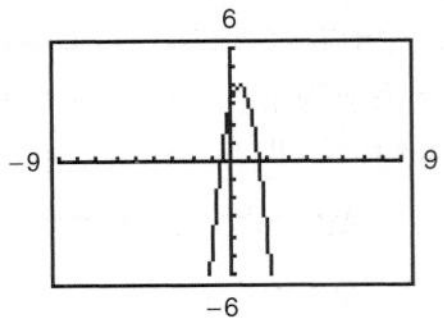

(b) and (c) $(-0.5, 0), (1.5, 0)$

51. One real solution **53.** One real solution

55. No real solutions

57. $\frac{9 \pm \sqrt{5}}{2}$ **59.** $-\frac{3}{2} \pm \frac{\sqrt{23}}{2}i$ **61.** $-2 \pm 2\sqrt{3}$

63. $\frac{1}{2}$ **65.** $-\frac{3}{4}$ **67.** $-2 \pm \frac{1}{2}i$ **69.** $-1, 4$

71. $6, -12$ **73.** $1 \pm \frac{3}{2}i$ **75.** $\frac{3 \pm \sqrt{29}}{10}$

77. $\frac{-7 \pm \sqrt{73}}{4}$ **79.** $\frac{3}{2}$

81. $x^2 + x - 30 = 0$; $2x^2 + 2x - 60 = 0$

83. $21x^2 + 31x - 42 = 0$; $3x^2 + \frac{31}{7}x - 6 = 0$

85. $x^2 - 75 = 0$; $\frac{x^2}{5} - 15 = 0$

87. $x^2 - 2x - 11 = 0$; $5x^2 - 10x - 55 = 0$

89. $x^2 - 4x + 5 = 0$; $-x^2 + 4x - 5 = 0$

91. (a)

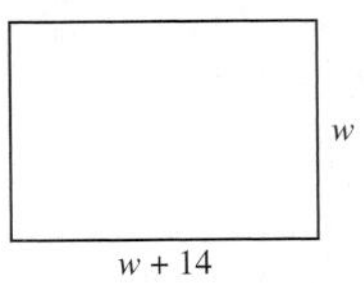

(b) $1632 = w^2 + 14w$

(c) Width: 34 ft; length: 48 ft

93. (a) $A(x) = -\frac{8}{3}x^2 + \frac{200}{3}x, 0 < x < 100$

(b)

x	y	Area
2	$\frac{92}{3}$	$\frac{368}{3} \approx 123$
4	28	224
6	$\frac{76}{3}$	304
8	$\frac{68}{3}$	$\frac{1088}{3} \approx 363$
10	20	400
12	$\frac{52}{3}$	416
14	$\frac{44}{3}$	$\frac{1232}{3} \approx 411$

Approximate dimensions for maximum area: 24 m × $\frac{52}{3}$ m

(c)

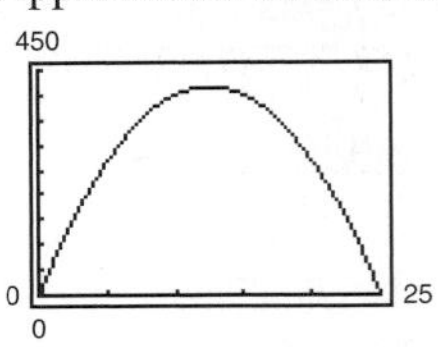

Approximate dimensions for maximum area: 25 m × $\frac{50}{3}$ m

(d) and (e) $15 \times 23\frac{1}{3}$ or 35×10 m

95. (a) $s = -16t^2 + 2080$

(b)

t	0	2	4	6	8	10	12
s	2080	2016	1824	1504	1056	480	−224

(c) (10, 12); 11.40 sec

97. (a) 22.36 sec (b) 3.73 mi

99. (a) 2006 (b) Answers will vary.

(c)

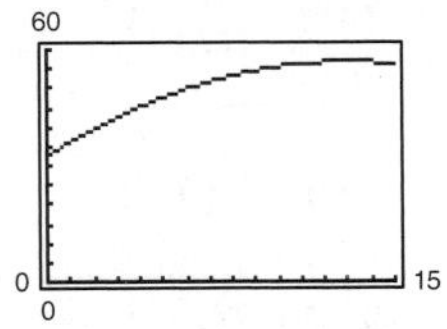

(d) 2010 (e) Answers will vary.

101. (a)

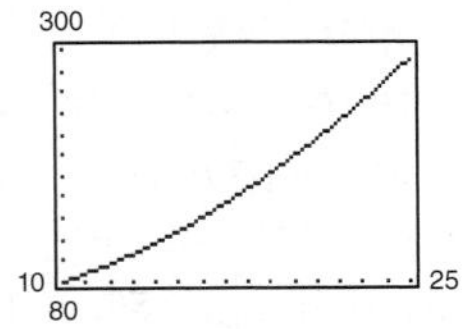

(b) 16.8°C (c) 2.5

103. (a) $x^2 + 225 = l^2$ (b) About 73.5 ft

105. False. Both solutions are complex.

107. False. Imaginary solutions are always complex conjugates of each other.

109. Proof

111. (a) Quadratic Formula (iv)

(b) Completing the square (iii)

(c) Factoring (i)

(d) Extracting the square roots (ii)

Answers will vary.

113. $x^2(x - 3)(x^2 + 3x + 9)$

115. $(x + 5)(x - \sqrt{2})(x + \sqrt{2})$ **117.** Answers will vary.

Section 2.5 (page 207)

1. polynomial **3.** Square both sides of the equation.

5. $0, \pm 2$ **7.** $0, \pm 3i$ **9.** $-3, 0$ **11.** $\pm 1, 5$

13. $\pm 1, \pm\sqrt{3}$ **15.** $\pm\frac{\sqrt{7}}{6}, \pm i$ **17.** $\frac{100}{9}$ **19.** 26

21. $-\frac{243}{2}$ **23.** -256.5 **25.** 6, 7 **27.** 0 **29.** 0

31. $\frac{1}{4}$ **33.** 9 **35.** No solution **37.** $1, -\frac{125}{8}$

39. -5 **41.** $-116, 134$ **43.** 2, 3 **45.** 1

47. $2, -\frac{3}{2}$ **49.** $4, -5$ **51.** $\frac{-3 \pm \sqrt{21}}{6}$ **53.** $-\frac{1}{5}, -\frac{1}{3}$

55. $\frac{1 \pm \sqrt{17}}{2}$ **57.** $2, -\frac{3}{5}$ **59.** $8, -3$ **61.** $2\sqrt{6}, -6$

63. $3, \frac{-1 - \sqrt{17}}{2}$

65. (a)

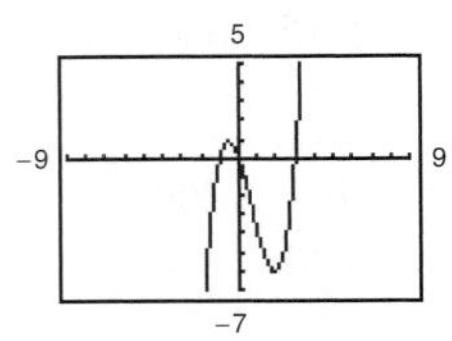

(b) and (c) $(-1, 0), (0, 0), (3, 0)$

(d) They are the same.

CHAPTER 2

67. (a)

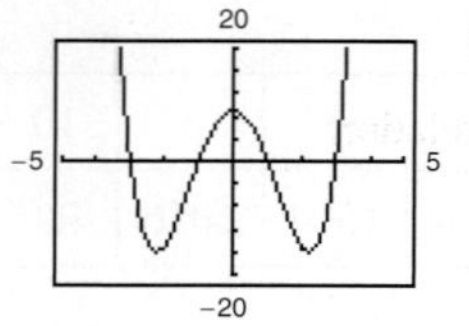

(b) and (c) $(-3, 0), (-1, 0), (1, 0), (3, 0)$

(d) They are the same.

69. (a)

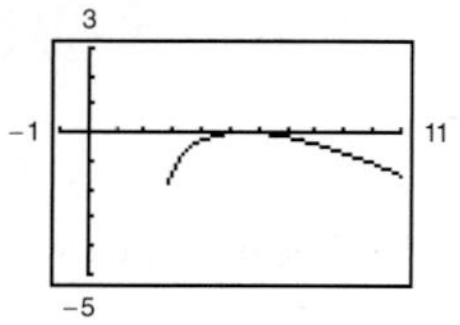

(b) and (c) $x = 5, 6$

(d) They are the same.

71. (a)

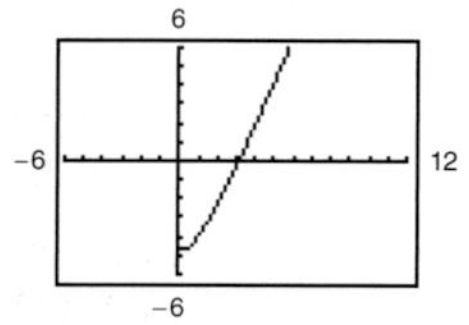

(b) and (c) $x \approx 3.09164$

(d) They are the same.

73. (a)

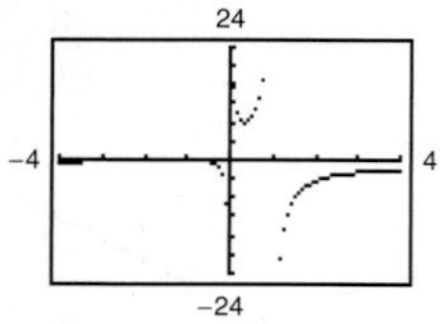

(b) and (c) $x = -1$

(d) They are the same.

75. (a)

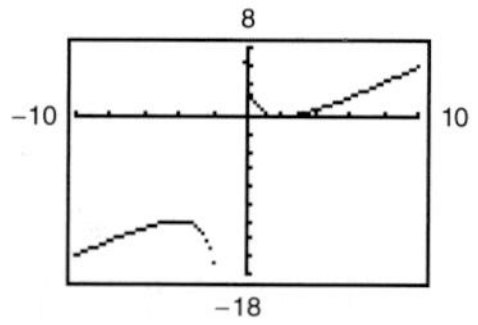

(b) and (c) $x = 2$

(d) They are the same.

77. (a)

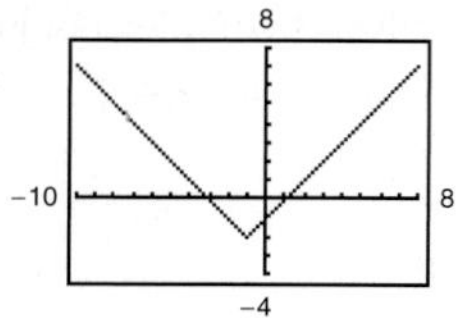

(b) and (c) $x = 1, -3$ (d) They are the same.

79. 34 students **81.** About 4.5% **83.** 14.806 lb/in.^{2}

85. (a) $C = 126{,}720(8 - x) + 39{,}600\sqrt{16x^2 + 9}$

(b) \$1,123,424.95 (c) 2 mi or 0.382 mi

(d)

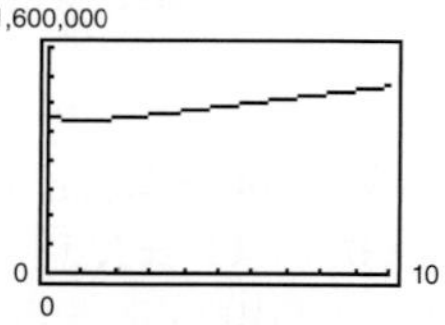

(e) $x = 1$

87. (a)

Year	2008	2009	2010	2011	2012
Crimes (in millions)	11.17	10.73	10.41	10.23	10.20

(b) 2009 (c) $t \approx 8.36$

(d)

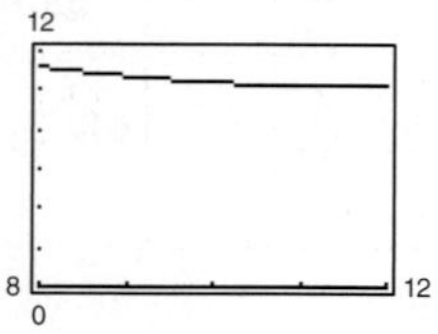

2008

89. False. See Example 7.

91. (a) $x = 6, x = -4$ (b) $x = 4, x = -20$

93. Dividing by x loses the solution $x = 0$.

95. $\dfrac{25}{6x}$ **97.** $\dfrac{-3z^2 - 2z + 4}{z(z + 2)}$ **99.** 11

Section 2.6 (page 219)

1. double **3.** $x \le -a, x \ge a$ **5.** No

7. d **8.** a **9.** f **10.** b **11.** e **12.** c

13. (a) Yes (b) No (c) Yes (d) No

15. (a) No (b) Yes (c) Yes (d) No

17. $x > 7$

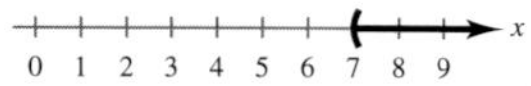

19. $x < -2$

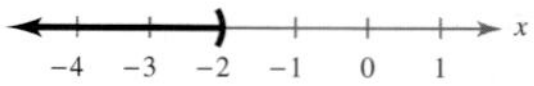

21. $x > -1$

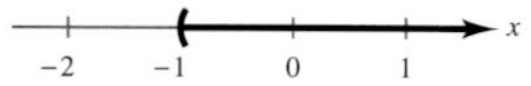

23. $x \ge 4$

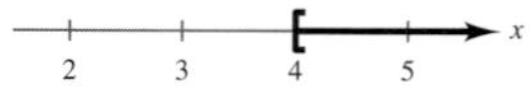

25. $-1 \le x \le 3$

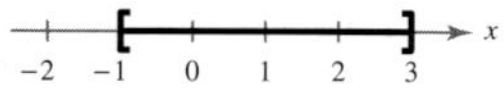

27. $-2 < x \le 5$

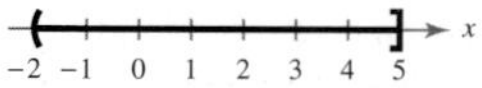

29. $-\frac{9}{2} < x < \frac{15}{2}$

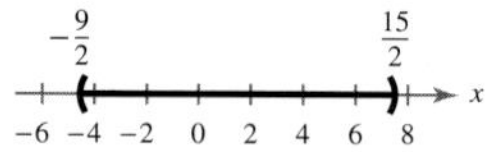

31.

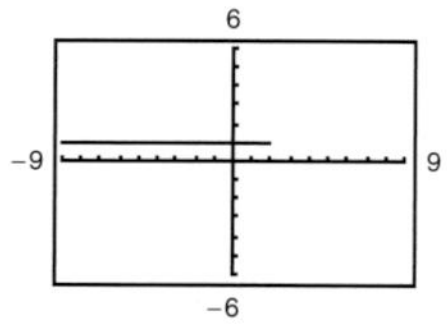

$x \le 2$

33.

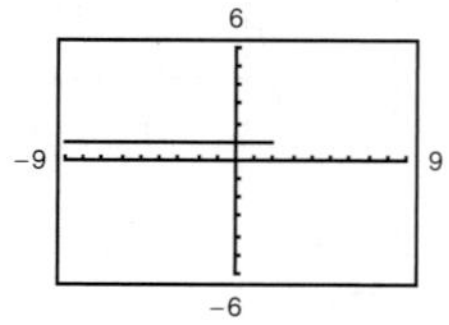

$x < 2$

35.

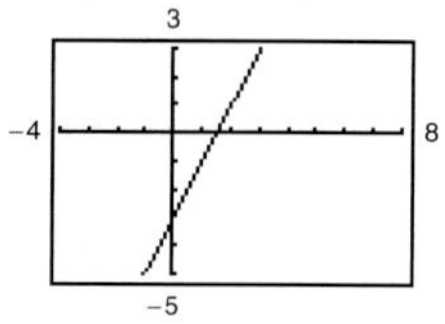

(a) $x \ge 2$ (b) $x \le \frac{3}{2}$

37.

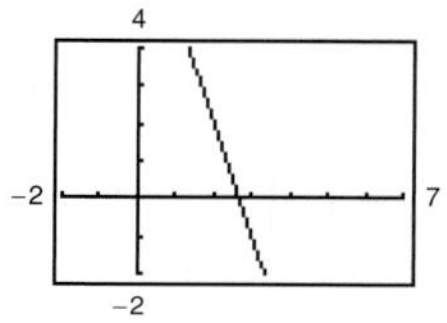

(a) $\frac{5}{3} \le x \le 3$ (b) $x \ge \frac{8}{3}$

39. $x < -2, x > 2$

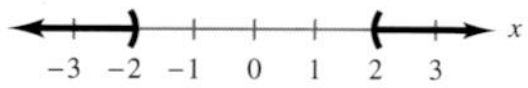

41. $1 \le x \le 13$

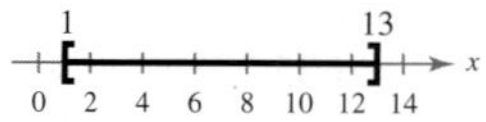

43. $x \le -7, x \ge 13$

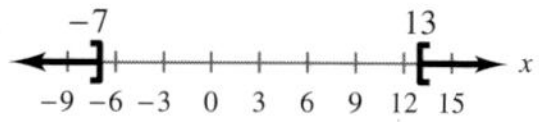

45. $\frac{1}{2} < x < \frac{3}{2}$

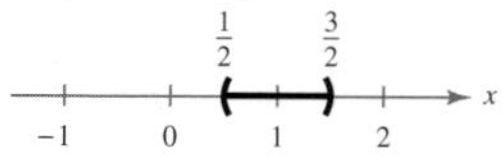

47.

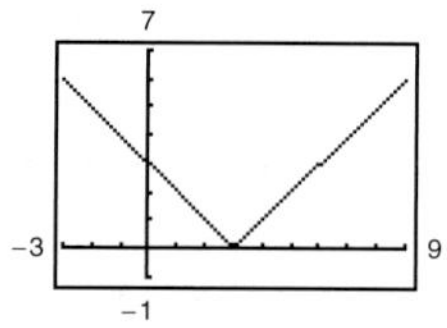

(a) $1 \le x \le 5$ (b) $x \le -1, x \ge 7$

49. $|x| \le 3$ **51.** $|x + 1| \ge 4$ **53.** $|x - 6| < 10$

55. $|x + 1| > 3$

57. Positive on: $(-\infty, -1) \cup (5, \infty)$

Negative on: $(-1, 5)$

59. Positive on: $\left(-\infty, \dfrac{2 - \sqrt{10}}{2}\right) \cup \left(\dfrac{2 + \sqrt{10}}{2}, \infty\right)$

Negative on: $\left(\dfrac{2 - \sqrt{10}}{2}, \dfrac{2 + \sqrt{10}}{2}\right)$

61. Negative on: $(-\infty, \infty)$

63. $(-\infty, -5], [1, \infty)$

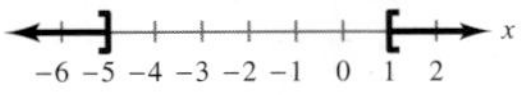

65. $(-7, 3)$

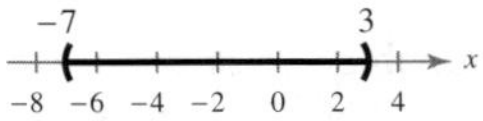

67. $0, [4, \infty)$

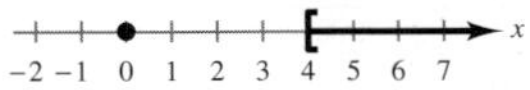

69. $(-3, -1), \left(\frac{3}{2}, \infty\right)$

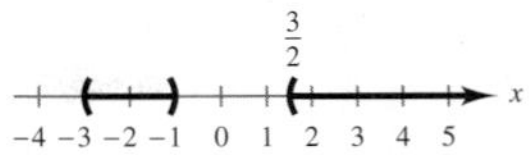

71. $(-1, 1), (3, \infty)$

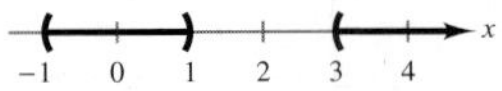

73. No solution

75. $\left(-\infty, \frac{1}{2}\right), \left(\frac{1}{2}, \infty\right)$

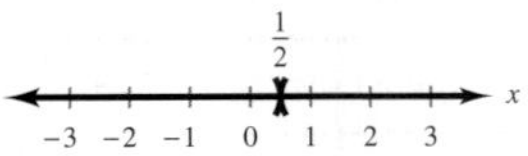

77. (a) $x = 1$ (b) $x \geq 1$ (c) $x > 1$

79.

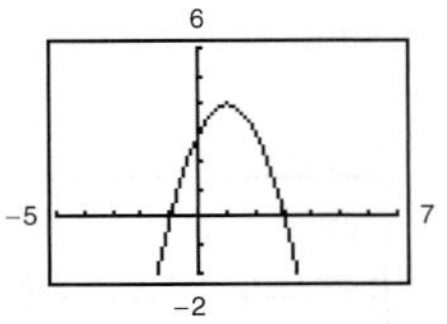

(a) $x \leq -1, x \geq 3$ (b) $0 \leq x \leq 2$

81. $(-\infty, -1), (0, 1)$

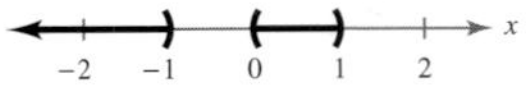

83. $(-\infty, -1), [4, \infty)$

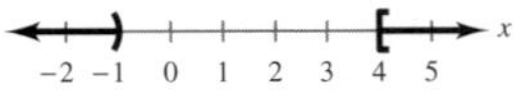

85.

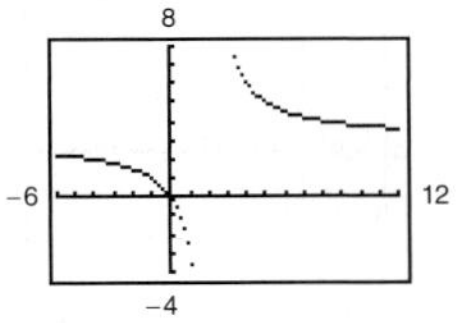

(a) $0 \leq x < 2$ (b) $2 < x \leq 4$

87. $[5, \infty)$ **89.** $[-3, 4]$ **91.** $\left(-\infty, -\frac{1}{3}\right], [7, \infty)$

93. (a) 2005 (b) (2003, 2005); (2005, 2012)

95. (a) 10 sec (b) (4, 6)

97. (a)

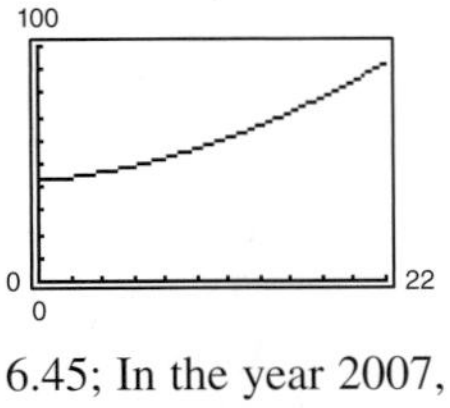

(b) (1996, 2001)
(c) $6.605 < x < 11.646$

99. $t \geq 6.45$; In the year 2007, there were at least 900 Bed Bath & Beyond stores.

101. $t \approx 1.04$; In 2001, there were the same number of Bed Bath & Beyond stores as Williams-Sonoma stores.

103. $333\frac{1}{3}$ vibrations/sec **105.** $1.2 < t < 2.4$

107. False. $10 \geq -x$

109. False. Cube roots have no restrictions on the domain.

111. (a) iv (b) ii (c) iii (d) i

113. $y^{-1} = \dfrac{x}{12}$ **115.** $y^{-1} = \sqrt[3]{x - 7}$

117. Answers will vary.

Section 2.7 (page 228)

1. positive **3.** Negative correlation

5. (a)

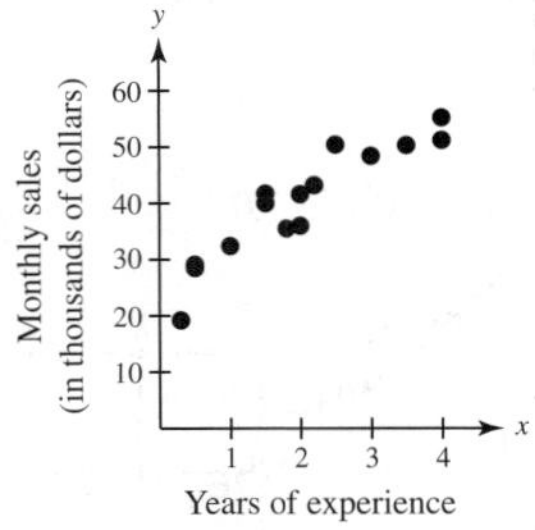

(b) Yes, the data appear somewhat linear. More experience, x, corresponds to higher sales, y.

7. Negative correlation **9.** No correlation

11. (a) and (b)

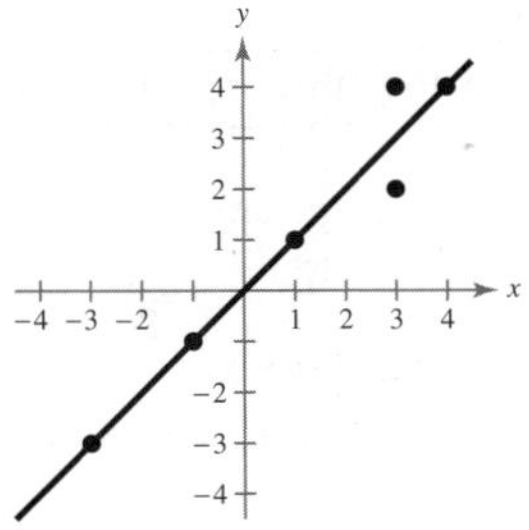

(c) $y = x$

13. (a) and (b)

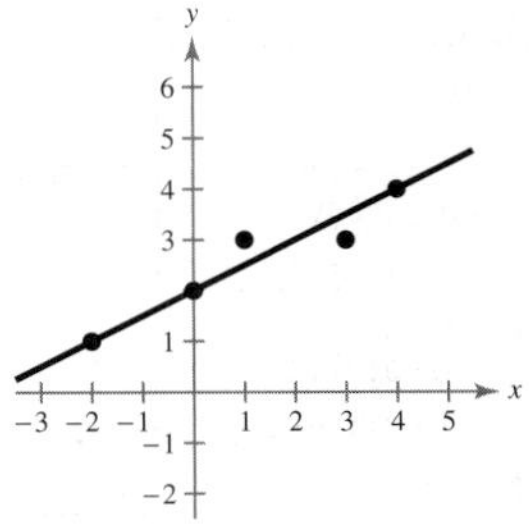

(c) $y = \frac{1}{2}x + 2$

15. (a) and (b)

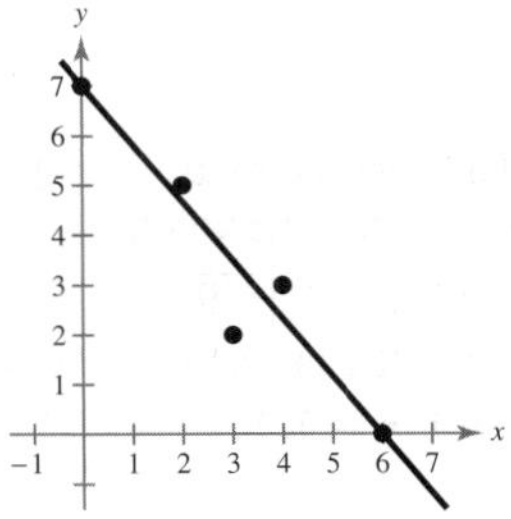

(c) $y = -\frac{7}{6}x + 7$

17. $y = 0.46x + 1.6$

(a)

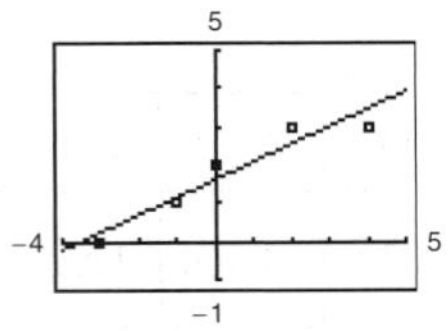

(b)

x	-3	-1	0	2	4
Linear equation	0.22	1.14	1.6	2.52	3.44
Given data	0	1	2	3	3

The model fits the data well.

19. (a)

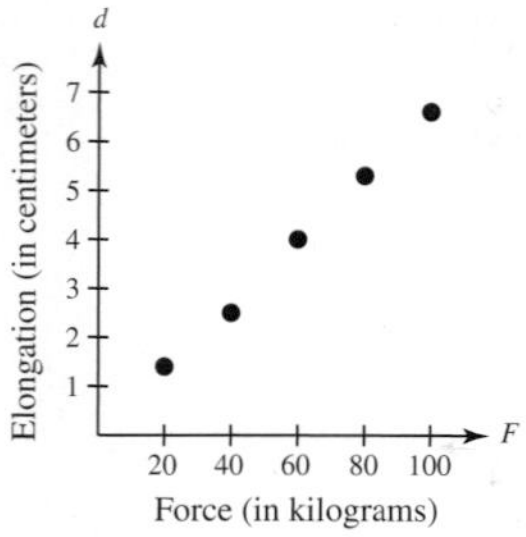

(b) $d = 0.07F - 0.3$
(c) $d = 0.066F$
(d) 3.63 cm

21. (a) and (c) (b) $V = 22.8t + 13$

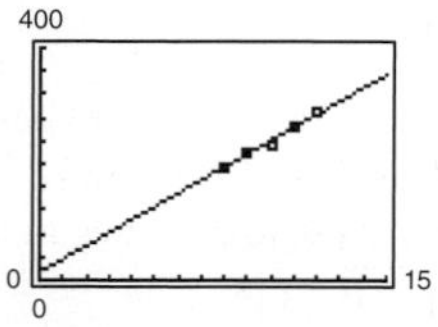

The model fits the data well.
(d) 2015: $355 million; 2020: $469 million; Answers will vary.
(e) 22.8; The slope represents the average annual increase in enterprise values (in millions of dollars).

23. (a) and (c) (b) $P = 37.3t + 8416$

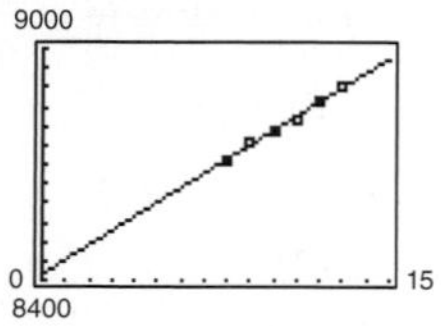

(d)

Year	2008	2009	2010	2011	2012	2013
Actual	8711	8756	8792	8821	8868	8899
Model	8714.4	8751.7	8789	8826.3	8863.6	8900.9

The model fits the data well.
(e) 10,280,800 people; Answers will vary.

25. (a) $45.70x + 108.0$
(b)

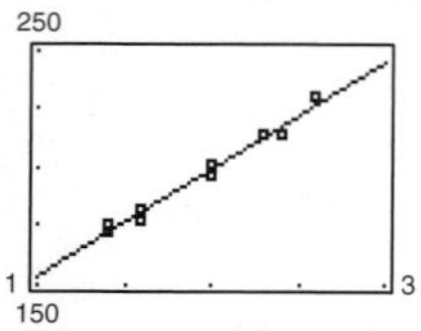

(c) The slope represents the increase in sales due to increased advertising.
(d) $176,550

27. (a) $T = -0.015t + 4.79$
$r \approx -0.866$
(b) The negative slope means that the winning times are generally decreasing over time.
(c)

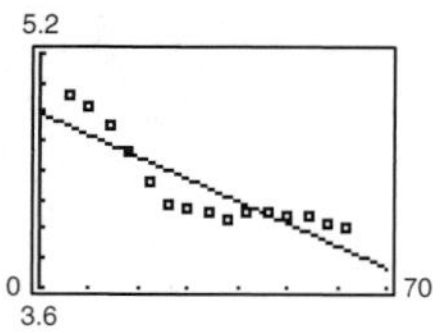

(d)

Year	1956	1960	1964	1968	1972
Actual	4.91	4.84	4.72	4.53	4.32
Model	4.70	4.64	4.58	4.52	4.46

Year	1976	1980	1984	1988	1992
Actual	4.16	4.15	4.12	4.06	4.12
Model	4.40	4.34	4.28	4.22	4.16

Year	1996	2000	2004	2008	2012
Actual	4.12	4.10	4.09	4.05	4.02
Model	4.10	4.04	3.98	3.92	3.86

The model does not fit the data well.
(e) The closer $|r|$ is to 1, the better the model fits the data.
(f) No; The winning times have leveled off in recent years, but the model values continue to decrease to unrealistic times.

29. True. To have positive correlation, the y-values tend to increase as x increases.
31. Answers will vary.
33. (a) 10 (b) $2w^2 + 5w + 7$ **35.** $-\frac{3}{5}$ **37.** $-\frac{1}{4}, \frac{3}{2}$

Review Exercises (page 234)

1. (a) No (b) No (c) No (d) Yes **3.** $x = 45$
5. $x = \frac{11}{3}$ **7.** $x = \frac{1}{2}$ **9.** No solution **11.** $x = \frac{7}{3}$
13. September: $325,000; October: $364,000
15. (a) (b) $h = \frac{64}{3}$ m

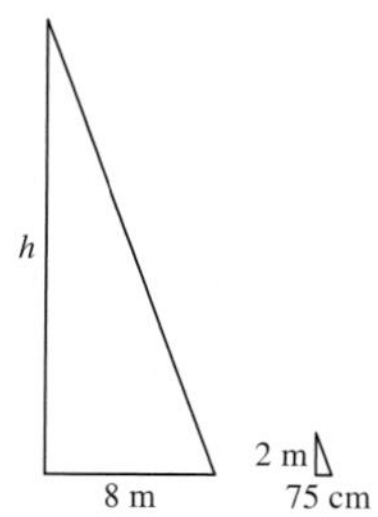

17. $-2.06°C$ **19.** $(-3, 0), (0, 3)$ **21.** $(1, 0), (8, 0), (0, 8)$
23. $x = 2.2$ **25.** No real solution **27.** $x = -0.402, 2.676$
29. $(1, -2)$ **31.** $(-1, 5), (2, -1)$ **33.** $3 - 4i$
35. $-4 - i$ **37.** $8 + 8i$ **39.** $40 + 65i$ **41.** $-26 + 7i$
43. $3 + 9i$ **45.** $-4 - 46i$ **47.** -80 **49.** $1 - 6i$
51. $\frac{17}{26} + \frac{7}{26}i$ **53.** $\pm\frac{7}{3}$ **55.** 0, 2 **57.** $-1, 8$
59. $-4 \pm 3\sqrt{2}$ **61.** $\frac{1}{3} \pm \frac{2}{3}i$ **63.** $-2 \pm \sqrt{13}$
65. $6 \pm \sqrt{6}$ **67.** $-4, 7$ **69.** $5 \pm \sqrt{34}$ **71.** $\frac{1}{2}, -5$
73. $\dfrac{-1 \pm \sqrt{61}}{2}$ **75.** $-2 \pm \sqrt{6}i$ **77.** $\dfrac{3 \pm \sqrt{33}i}{2}$
79. (a)

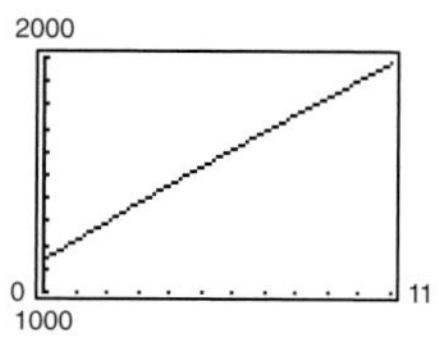

(b) and (c) 2008
(d) Answers will vary.

81. $0, \frac{2}{3}, 8$ **83.** $0, \frac{12}{5}$ **85.** $\pm 2, \pm\sqrt{3}i$ **87.** $\pm 2, \pm\sqrt{7}$

89. 5 **91.** $1, \frac{9}{4}$ **93.** No solution **95.** $-124, 126$

97. $-2 \pm \frac{\sqrt{95}}{5}, -4$ **99.** $-4, 1$ **101.** $\pm\sqrt{10}$

103. $-\frac{8}{3}, 2$ **105.** $-5, 15$ **107.** $1, 3$

109. 4 investors **111.** 191.5 mi/h **113.** 3%

115. (a)

Year	2000	2001	2002	2003	2004
Enrollment (in millions)	3.09	3.17	3.25	3.33	3.40

Year	2005	2006	2007	2008	2009
Enrollment (in millions)	3.48	3.55	3.62	3.69	3.76

Year	2010	2011	2012
Enrollment (in millions)	3.83	3.89	3.96

(b)

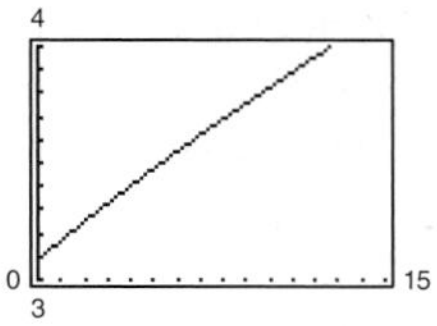

(c) and (d) 2005
(e) 2051; Answers will vary.
(f) Answers will vary.

117. $(-\infty, 9)$

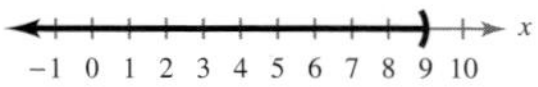

119. $\left(-\frac{5}{3}, \infty\right)$

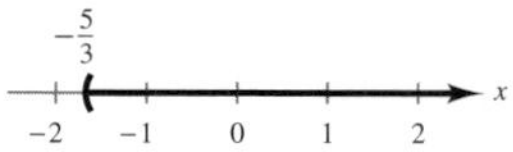

121. $[-3, 9)$

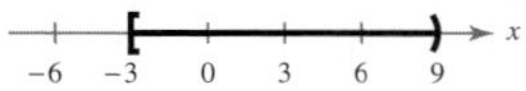

123. $[-4, 4]$

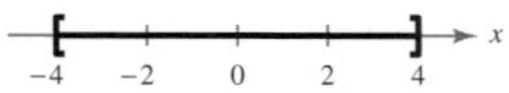

125. $(-\infty, 0), (3, \infty)$

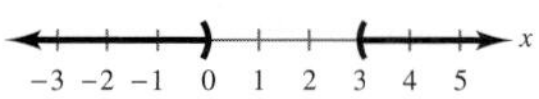

127. $\left[-\frac{1}{2}, \frac{7}{2}\right]$

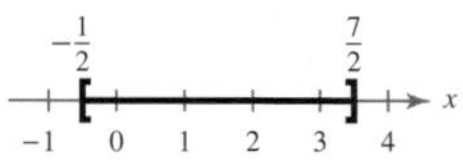

129. $(-\infty, -1), (3, \infty)$

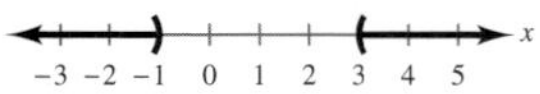

131. $\left[-\frac{1}{4}, 6\right]$

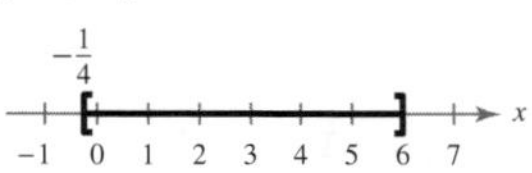

133. $[-4, 0], [4, \infty)$

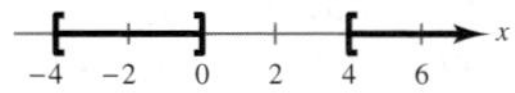

135. $\left(-\infty, -\frac{1}{3}\right), (0, \infty)$

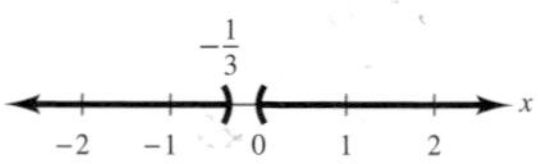

137. $(-\infty, 3), [20, \infty)$

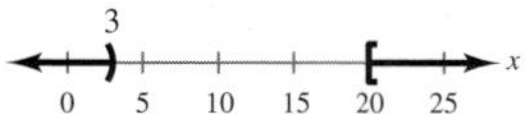

139. $[-4, 4]$ **141.** \$0.37 per gallon

143. (a)

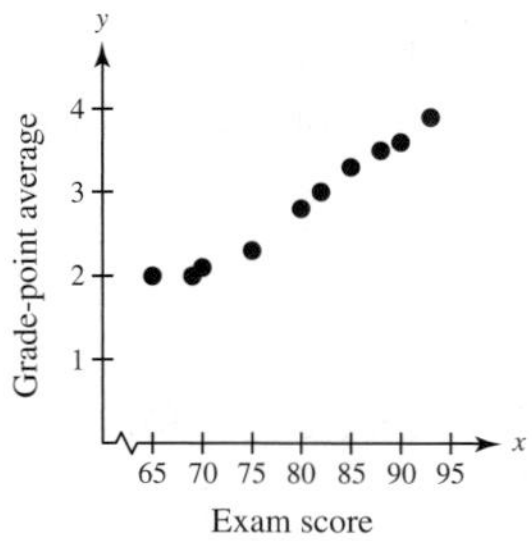

(b) Yes. Answers will vary.

145. (a)

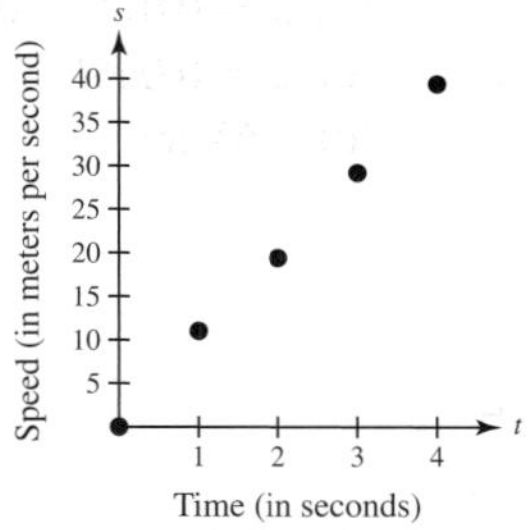

(b) Answers will vary. Sample answer: $S = 10t - 0.4$
(c) $s = 9.70t + 0.4$; $r \approx 0.999$
(d) 24.7 m/sec

147. False. A graph with two distinct y-intercepts is not a function.

149. False. A regression line can have a positive or negative slope.

151. Answers will vary. **153.** $\sqrt{-8}\sqrt{-8} = 8i^2 = -8$

155. (a) 1 (b) i (c) -1 (d) $-i$

Chapter Test (page 238)

1. $x = 3$ **2.** $x = \frac{2}{15}$ **3.** $-9 - 18i$

4. $6 + \left(2\sqrt{5} + \sqrt{14}\right)i$ **5.** $13 + 4i$ **6.** $-17 + 14i$

7. $5 - 8i$ **8.** $\frac{4}{13} + \frac{7}{13}i$ **9.** ± 1.414 **10.** ± 0.5

11. 0 **12.** $\pm 1, 0$ **13.** 7, 8 **14.** $-6 \pm \sqrt{38}$ **15.** $\pm\frac{9}{2}$

16. $-\frac{7}{10} \pm \frac{\sqrt{71}}{10}i$ **17.** $\pm 2, \frac{4}{3}$ **18.** 2

19. $\pm\sqrt{58}$ **20.** $-\frac{5}{2}, \frac{11}{4}$

21. $(-3, \infty)$

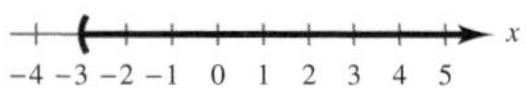

22. $(3, 13)$

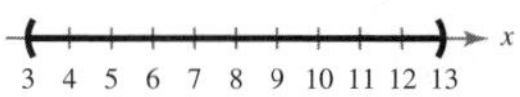

23. $\left(-\infty, -\frac{1}{2}\right], \left[-\frac{1}{3}, \infty\right)$

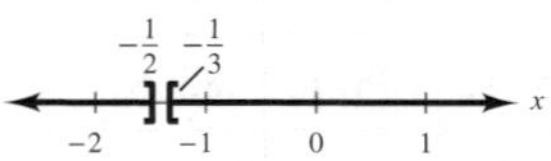

24. $\left[-12, \frac{2}{3}\right)$

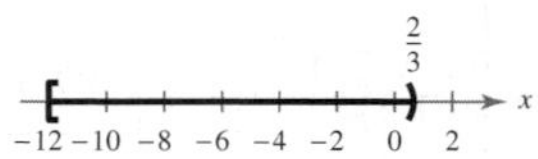

25. $(1.8, 12.2)$ **26.** $C = 2.485t + 30.36$; 2021

Cumulative Test for Chapters P–2 (page 239)

1. $\frac{7x^3}{16y^5}, x \neq 0$ **2.** $9\sqrt{15}$ **3.** $2x^2y\sqrt{7y}$

4. $7x - 10$ **5.** $x^3 - x^2 - 5x + 6$ **6.** $\frac{x - 1}{(x + 1)(x + 3)}$

7. $-(x - 10)(x + 2)$ **8.** $x(1 + x)(1 - 6x)$

9. $2(3 - 2x)(9 + 6x + 4x^2)$

10. Midpoint: $\left(-\frac{1}{2}, -2\right)$; $d = 6\sqrt{5} \approx 13.42$

11. $\left(x + \frac{1}{2}\right)^2 + (y + 8)^2 = 16$

12.

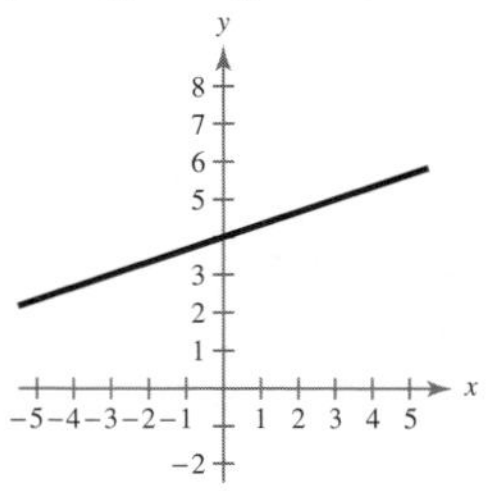

13.

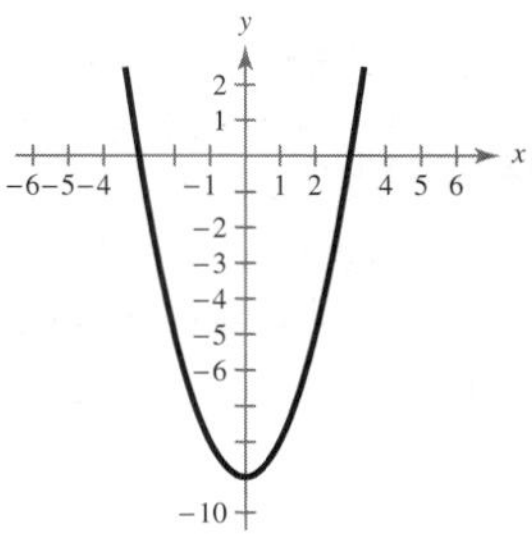

CHAPTER 2

14.

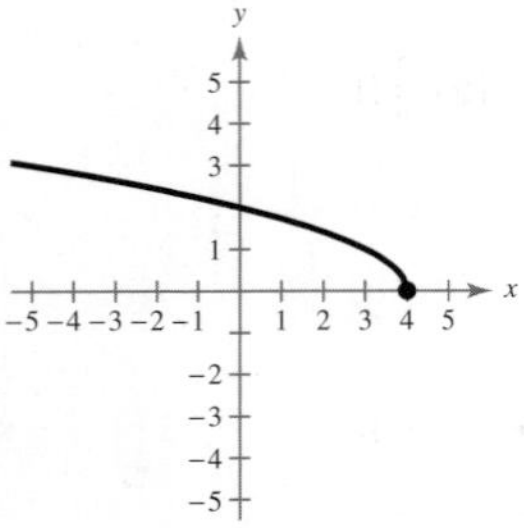

15. (a) $x + y = 3$
(b) Answers will vary. Sample answer: $(-1, 4), (0, 3), (1, 2)$

16. (a) $2x + y = 0$
(b) Answers will vary. Sample answer: $(0, 0), (1, -2), (2, -4)$

17. (a) $x = -\frac{3}{7}$
(b) Answers will vary. Sample answer:
$\left(-\frac{3}{7}, 0\right), \left(-\frac{3}{7}, 1\right), \left(-\frac{3}{7}, -3\right)$

18. (a) $y - 6x = -9$ (b) $y + \frac{1}{6}x = \frac{10}{3}$

19. (a) $\frac{5}{3}$ (b) Undefined (c) $\frac{5 + 4s}{3 + 4s}$

20. (a) -32 (b) 4 (c) 20 **21.** $(-\infty, \infty)$

22. $\left[-\frac{5}{7}, \infty\right)$ **23.** $[-3, 3]$ **24.** $\left(-\infty, -\frac{2}{5}\right) \cup \left(-\frac{2}{5}, \infty\right)$

25. Odd **26.** No. It does not pass the Vertical Line Test.

27.

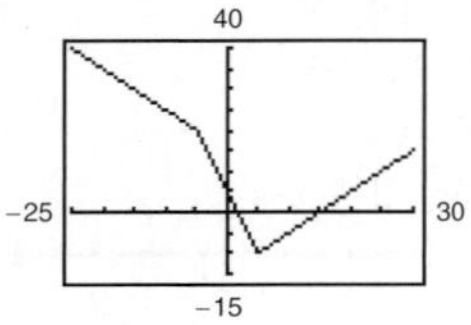

Decreasing on $(-\infty, 5)$
Increasing on $(5, \infty)$

28. (a) Vertical shrink (b) Vertical shift two units up
(c) Horizontal shift two units left, reflection in x-axis

29. $x^2 + 4x - 2$ **30.** $x^2 - 2x + 2$ **31.** $3x^2 + 3x - 2$

32. $3x^3 + x^2 - 2x$ **33.** Yes; $f^{-1}(x) = -\frac{x - 4}{5}$

34. No **35.** Yes; $f^{-1}(x) = (x + 2)^3$ **36.** 0, 1, 2

37. -3 **38.** $-0.667, -2$ **39.** 4.444

40. $\left(-\infty, \frac{120}{7}\right]$

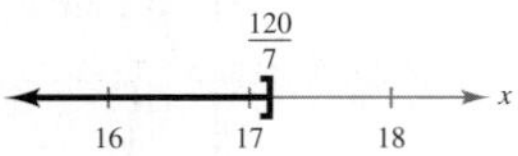

41. $(-\infty, -3], \left[\frac{5}{2}, \infty\right)$

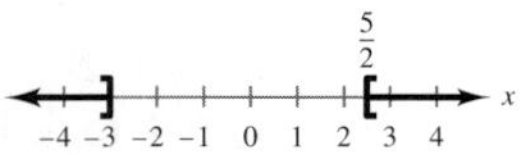

42. $\left(-\infty, -\frac{3}{2}\right), \left(-\frac{1}{4}, \infty\right)$

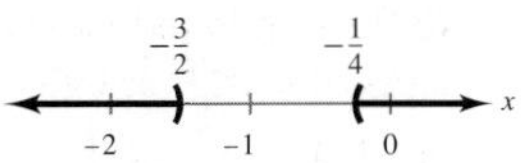

43. $(-1, 2]$

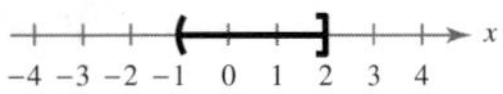

44. 4.456 in.

45. (a) $A = x(273 - x)$
(b)

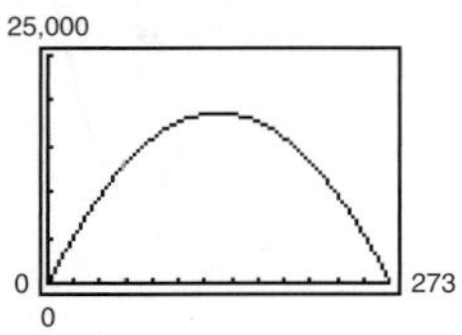

$0 < x < 273$

(c) 76.2 ft × 196.8 ft

46. (a)

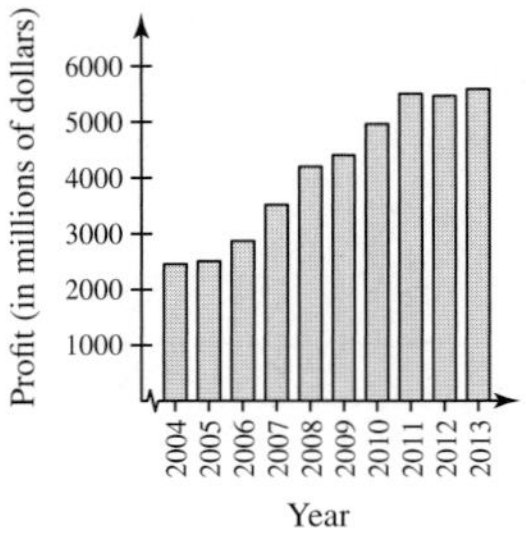

McDonald's profits appear to be increasing at a fairly constant rate.

(b) $P = 403.05t + 722.69$; $r \approx 0.9811$

(c)

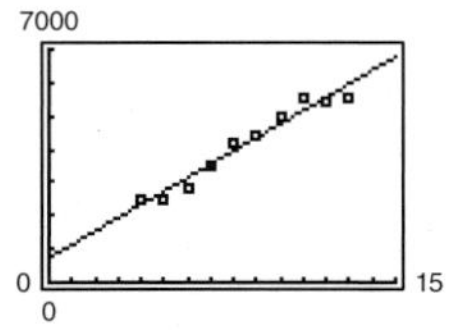

(d) \$6,768,440,000; \$7,977,590,000

(e) Answers will vary.

Chapter 3

Section 3.1 (page 250)

1. nonnegative integer, real **3.** Yes; (2, 3)

5. c **6.** d **7.** b **8.** a

9.

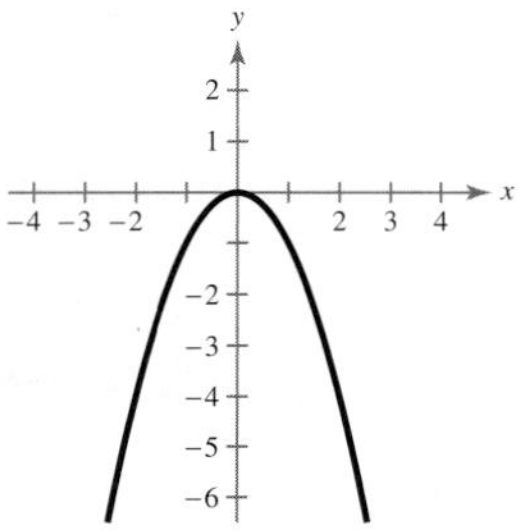

Reflection in the x-axis

11.

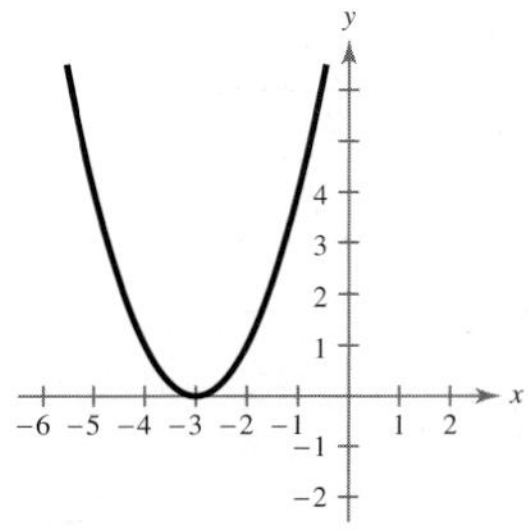

Horizontal shift three units to the left

13.

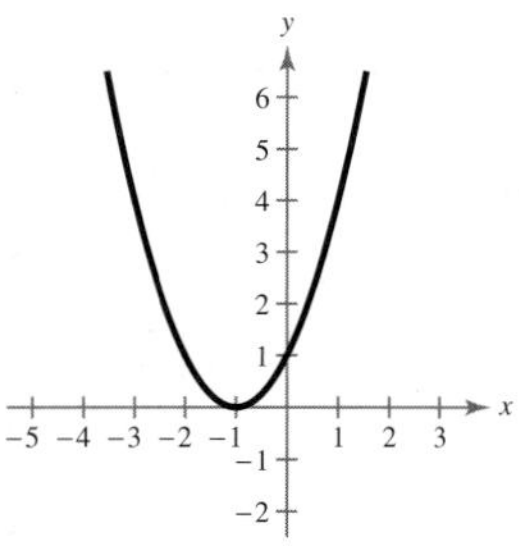

Horizontal shift one unit to the left

15.

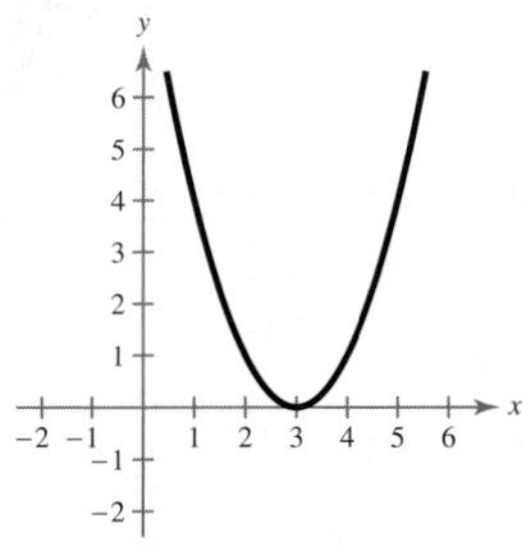

Horizontal shift three units to the right

17. Parabola opening downward
Vertex: (0, 20)

19. Parabola opening upward
Vertex: $(0, -5)$

21. Parabola opening upward
Vertex: $(-3, -4)$

23. Parabola opening upward
Vertex: (1, 0)

25. Parabola opening upward
Vertex: $\left(\frac{1}{2}, 1\right)$

27. Parabola opening downward
Vertex: $(1, 6)$

29. Parabola opening upward
Vertex: $\left(\frac{1}{2}, 20\right)$

31. Parabola opening upward
Vertex: $(-4, -5)$
x-intercepts: $\left(-4 \pm \sqrt{5}, 0\right)$

33. Parabola opening downward
Vertex: $(1, 16)$
x-intercepts: $(-3, 0), (5, 0)$

35. Parabola opening downward
Vertex: $(4, 1)$
x-intercepts: $\left(4 \pm \frac{1}{2}\sqrt{2}, 0\right)$

37. $y = -(x + 1)^2 + 4$ **39.** $f(x) = (x + 2)^2 + 5$

41. $y = 4(x - 1)^2 - 2$ **43.** $y = -\frac{104}{125}\left(x - \frac{1}{2}\right)^2 + 1$

45. $(5, 0), (-1, 0)$ **47.** $(-4, 0)$

49.

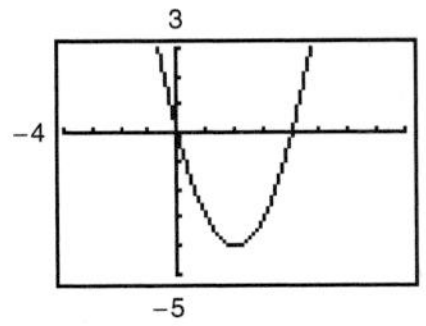

$(0, 0), (4, 0)$

51.

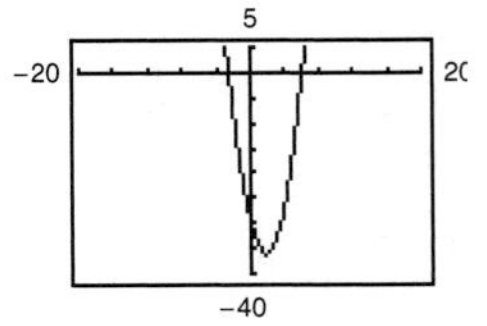

$\left(-\frac{5}{2}, 0\right), (6, 0)$

53.

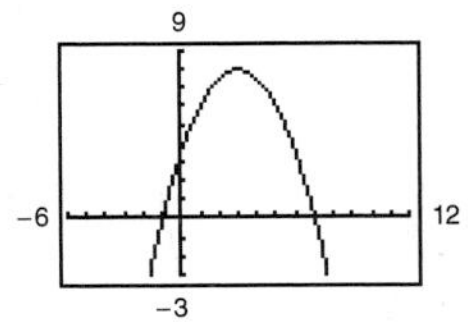

$(7, 0), (-1, 0)$

55. $f(x) = x^2 - 2x - 3$
$g(x) = -x^2 + 2x + 3$

57. $f(x) = 2x^2 + 7x + 3$
$g(x) = -2x^2 - 7x - 3$

59. 55, 55 **61.** 12, 6

63. (a)

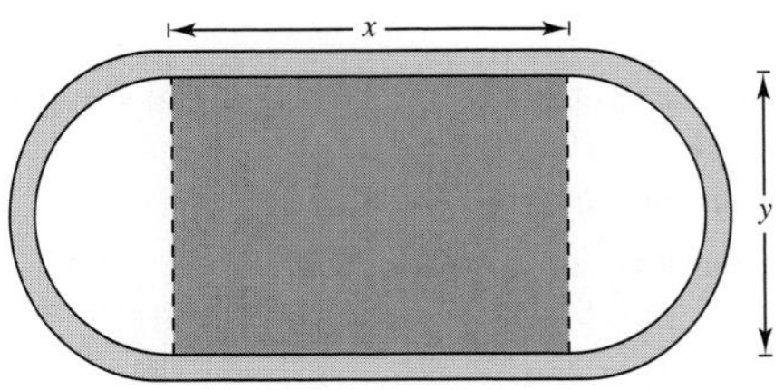

(b) $r = \frac{1}{2}y$; $d = y\pi$ (c) $y = \dfrac{200 - 2x}{\pi}$

(d) $A = x\left(\dfrac{200 - 2x}{\pi}\right)$

(e)

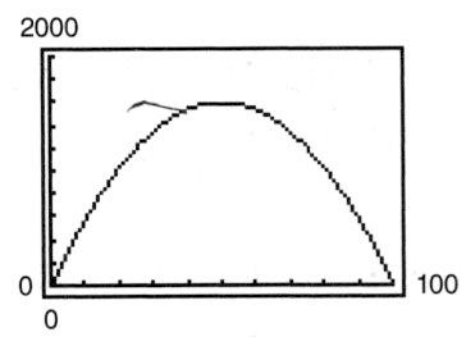

$x = 50$ m, $y = \dfrac{100}{\pi}$ m

65. (a)

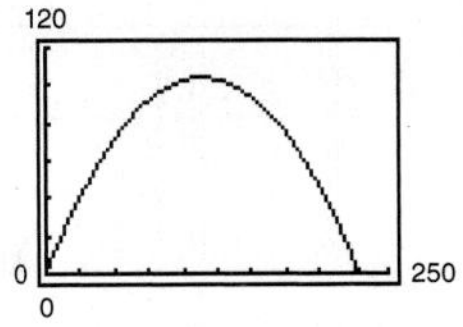

(b) $\frac{3}{2}$ ft

(c) About 104 ft (d) About 228.6 ft

67. (a) $A = -2x^2 + 112x - 600$ (b) $x = 28$ in.

69. (a)

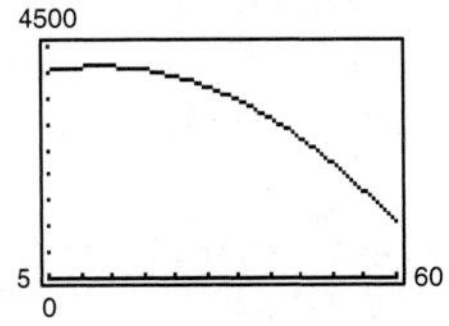

(b) 1963; 4110 cigarettes
(c) Yes; Consumption began decreasing.
(d) 1057 cigarettes per year; about 3 cigarettes per day

71. True. The vertex is $(0, -1)$ and the parabola opens down.

73. False. The vertex of $f(x)$ is $(-1, 4)$ and the vertex of $g(x)$ is $(-1, -4)$.

75. c, d **77.** Horizontal shift z units to the right

79. Vertical stretch $(z > 1)$ or shrink $(0 < z < 1)$ and horizontal shift three units to the right

81. $b = \pm 20$ **83.** $b = \pm 8$ **85.** Proofs

87. $y = -x^2 + 5x - 4$; Answers will vary.

89. Model (a). The profits are positive and rising.

91. $(4, 2)$ **93.** $(2, 11)$

Section 3.2 (page 262)

1. continuous **3.** (a) solution (b) $(x - a)$ (c) $(a, 0)$

5. No **7.** $f(x_2) > 0$

9. f **10.** h **11.** c **12.** a **13.** e

14. d **15.** g **16.** b

17.

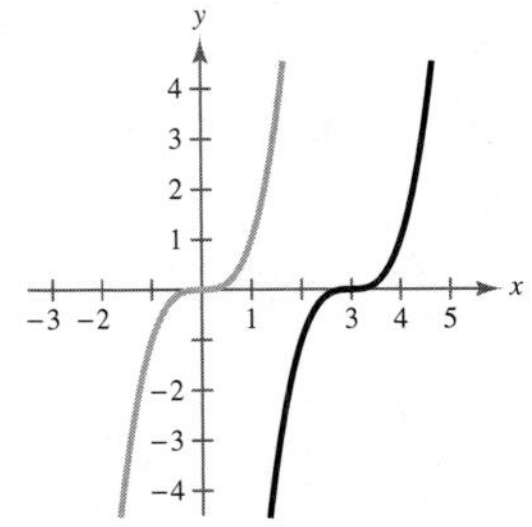

Horizontal shift three units to the right

19.

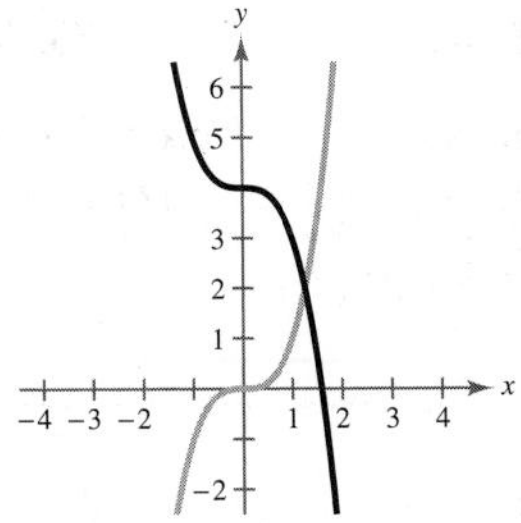

Reflection in the x-axis and vertical shift four units upward

21.

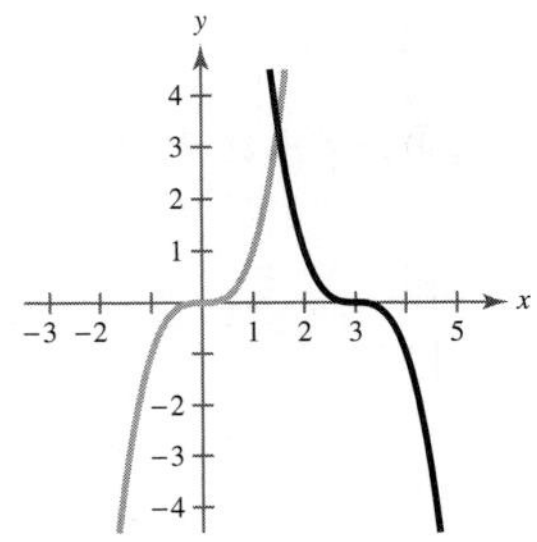

Reflection in the x-axis and horizontal shift three units to the right

23.

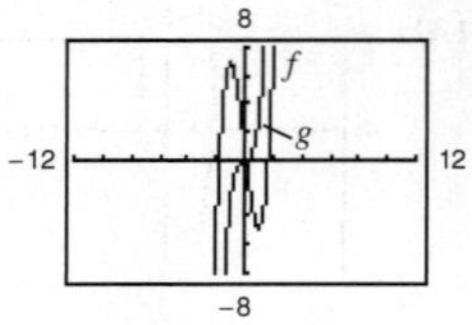

Yes, because both graphs have the same leading coefficient.

25.

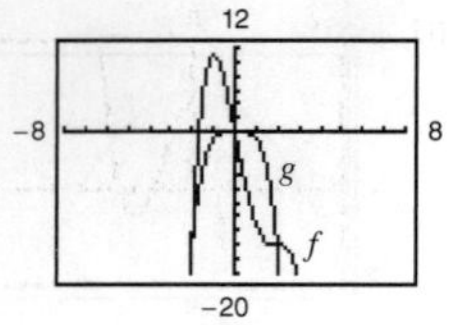

Yes, because both graphs have the same leading coefficient.

27.

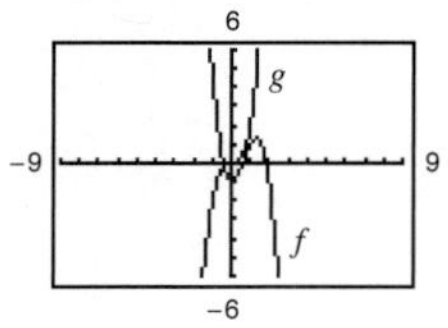

No, because the graphs have different leading coefficients.

29. Rises to the left, rises to the right

31. Falls to the left, falls to the right

33. Falls to the left, rises to the right

35. Falls to the left, falls to the right

37. (a) $\left(2 \pm \sqrt{3}, 0\right)$

(b)

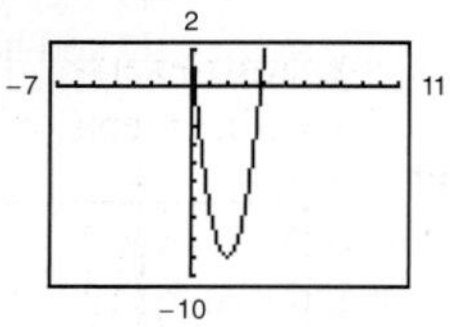

(c) (0.27, 0), (3.73, 0); Answers are approximately the same.

39. (a) $(-1, 0), (1, 0)$

(b)

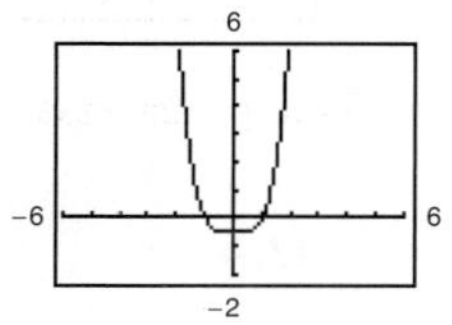

(c) $(-1, 0), (1, 0)$; Answers are the same.

41. (a) $(0, 0), \left(\pm\sqrt{2}, 0\right)$

(b)

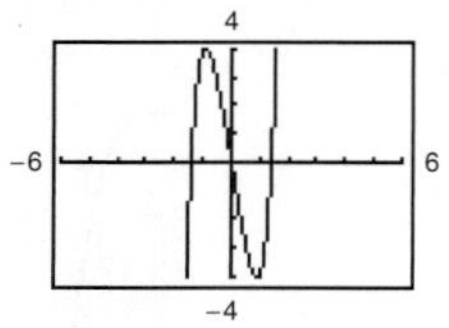

(c) $(-1.41, 0), (0, 0), (1.41, 0)$; Answers are approximately the same.

43. (a) $\left(\pm\sqrt{5}, 0\right)$

(b)

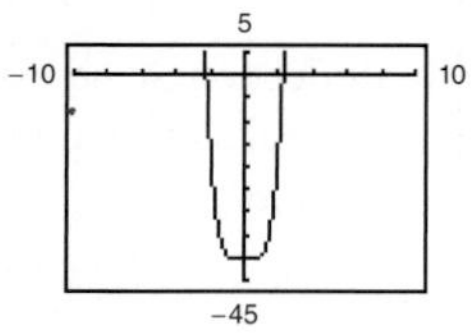

(c) $(-2.236, 0), (2.236, 0)$; Answers are approximately the same.

45. (a) $(4, 0), (\pm 5, 0)$

(b)

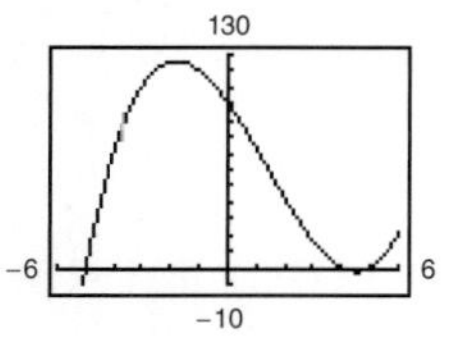

(c) $(-5, 0), (4, 0), (5, 0)$; Answers are the same.

47. (a) $(0, 0), \left(\frac{5}{2}, 0\right)$

(b)

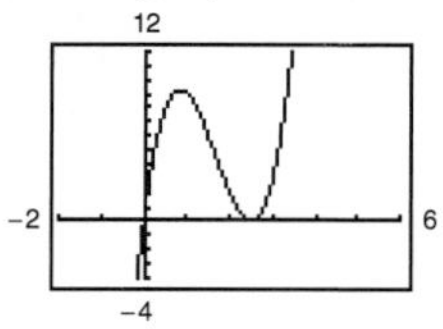

(c) (0, 0), (2.5, 0) Answers are the same.

49. ± 5 (multiplicity 1)

51. 3 (multiplicity 2)

53. $1, -2$ (multiplicity 1)

55. 2 (multiplicity 2), 0 (multiplicity 1)

57. $\dfrac{-5 \pm \sqrt{37}}{2}$ (multiplicity 1)

59.

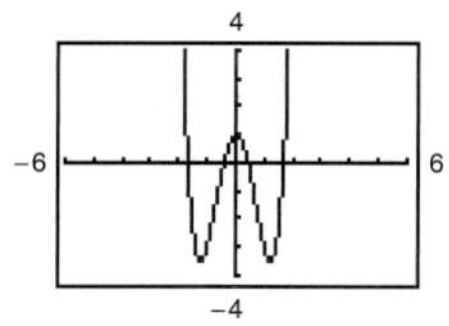

Zeros: $\pm 1.680, \pm 0.421$

Relative minima: $(\pm 1.225, -3.500)$

Relative maximum: (0, 1)

61.

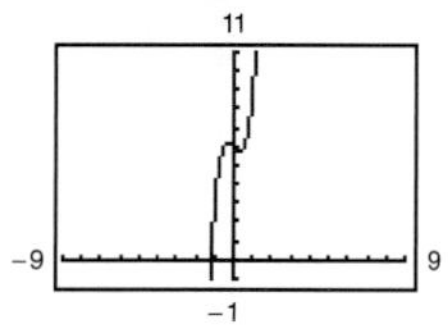

Zero: -1.178

Relative minimum: (0.324, 5.782)

Relative maximum: $(-0.324, 6.218)$

63.

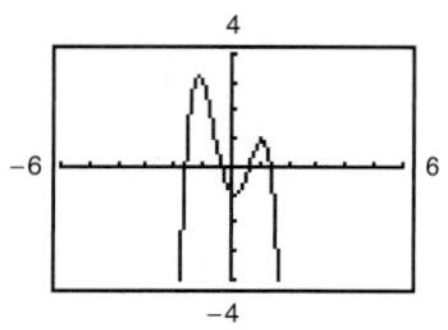

Zeros: $-1.618, -0.366, 0.618, 1.366$

Relative minimum: $(0.101, -1.050)$

Relative maxima: $(-1.165, 3.267), (1.064, 1.033)$

65. $f(x) = x^2 - 7x$

67. $f(x) = x^3 + 6x^2 + 8x$

69. $f(x) = x^4 - 4x^3 - 9x^2 + 36x$

71. $f(x) = x^2 - 2x - 1$

73. $f(x) = x^3 - 6x^2 + 7x + 2$

75. $f(x) = x^3 + 5x^2 + 8x + 4$

77. $f(x) = x^4 + 2x^3 - 23x^2 - 24x + 144$

79. $f(x) = -x^3 - 4x^2 - 5x - 2$

81.

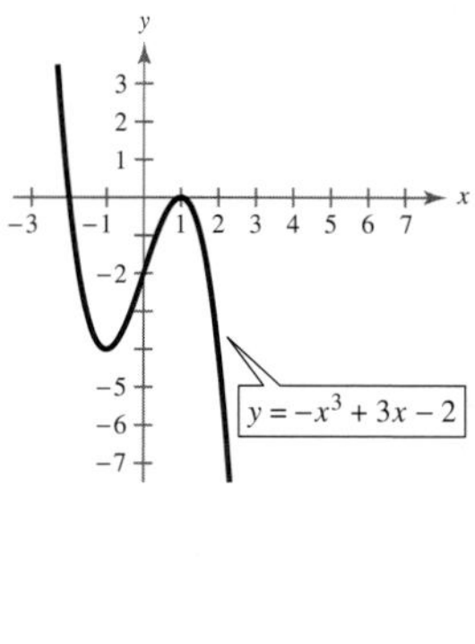

83.

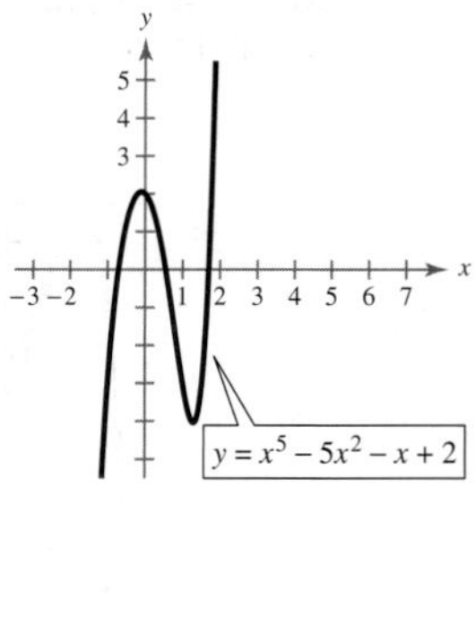

85. (a) Falls to the left, rises to the right
(b) $(0, 0), (3, 0), (-3, 0)$
(c) and (d)

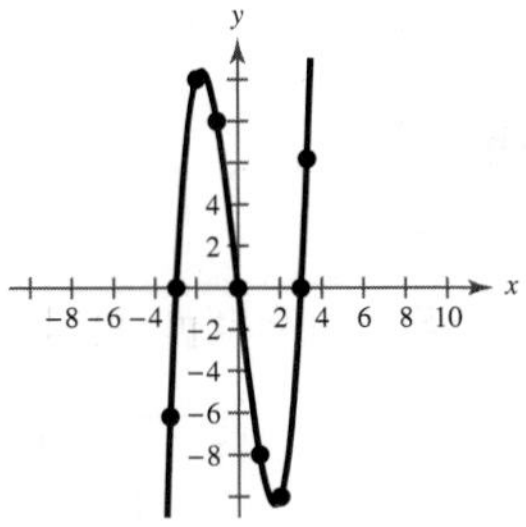

87. (a) Falls to the left, rises to the right
(b) $(0, 0), (3, 0)$
(c) and (d)

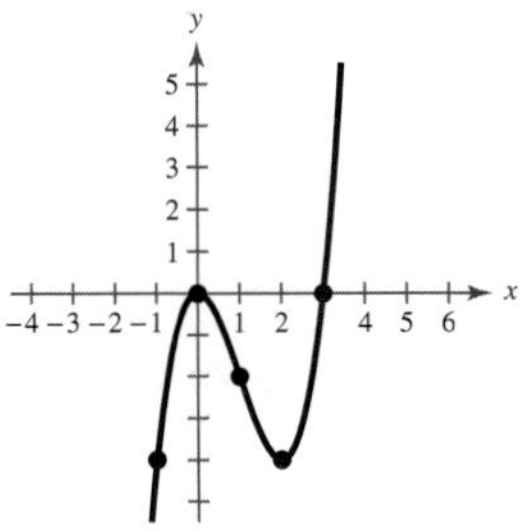

89. (a) Falls to the left, falls to the right
(b) $(\pm 2, 0), (\pm\sqrt{5}, 0)$
(c) and (d)

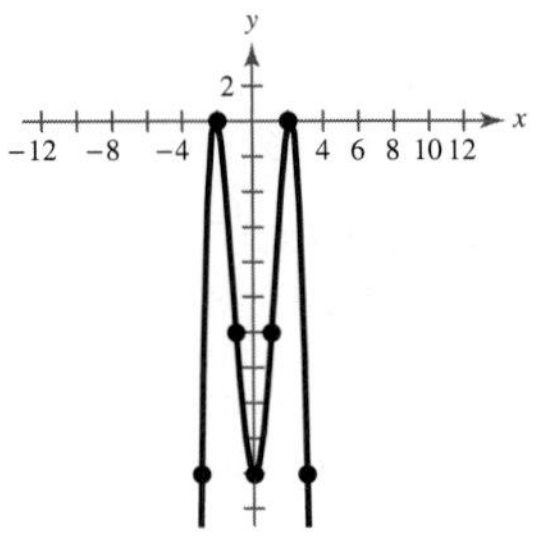

91. (a) Falls to the left, rises to the right
(b) $(\pm 3, 0)$
(c) and (d)

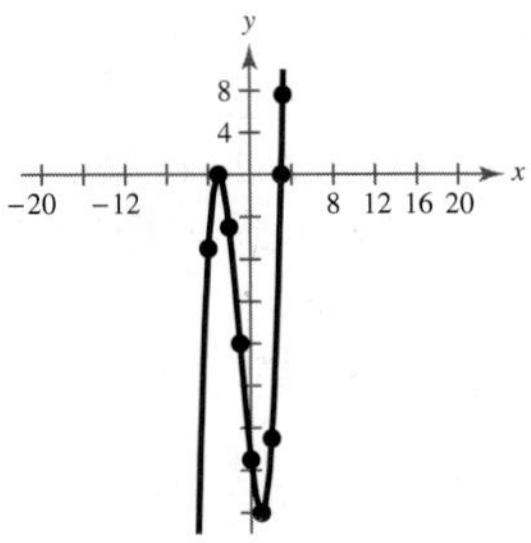

93. (a) Falls to the left, falls to the right
(b) $(-2, 0), (2, 0)$
(c) and (d)

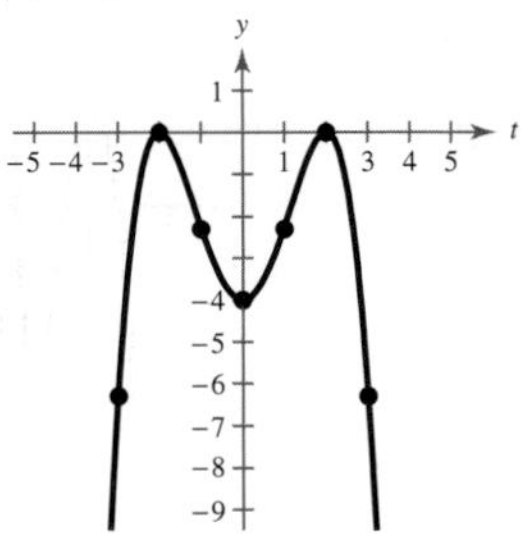

95. (a)

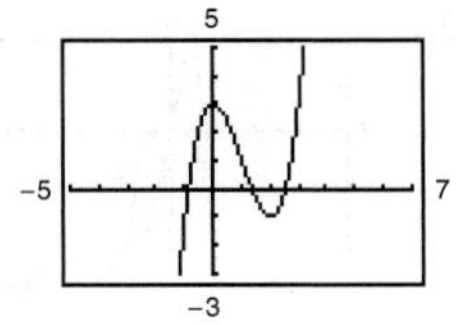

$(-1, 0), (1, 2), (2, 3)$
(b) $-0.879, 1.347, 2.532$

97. (a)

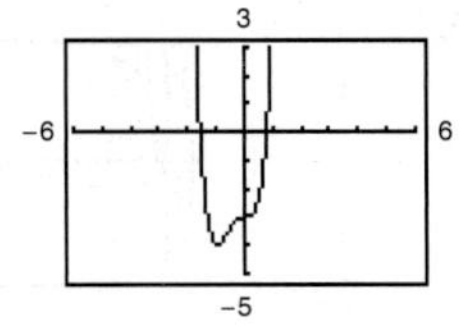

$(-2, -1), (0, 1)$
(b) $-1.585, 0.779$

99. (a)

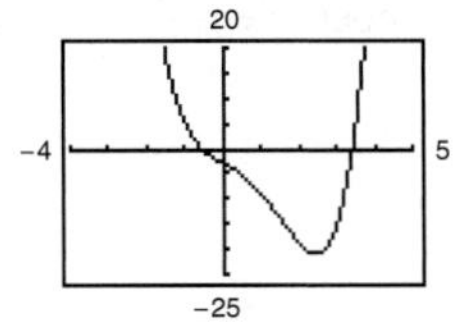

$(-1, 0), (3, 4)$
(b) $-0.578, 3.418$

101.

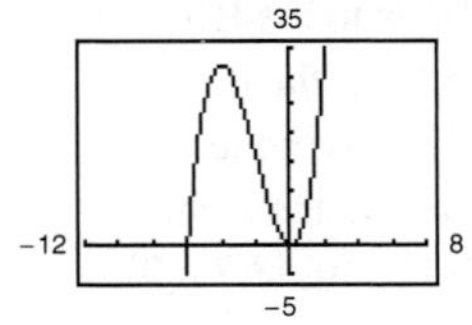

Two x-intercepts

103.

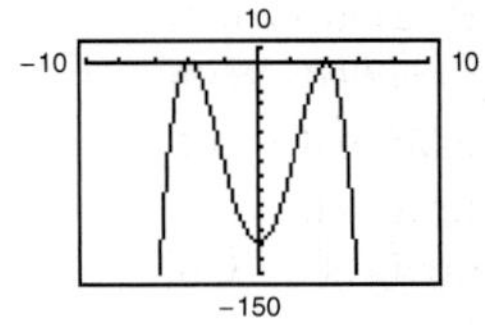

y-axis symmetry
Two x-intercepts

105.

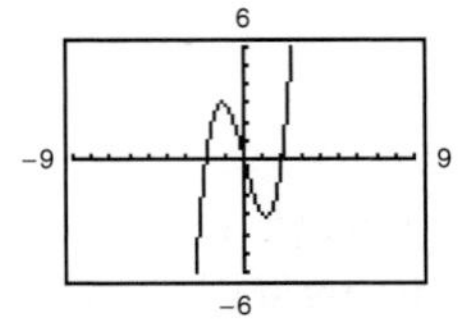

Origin symmetry
Three x-intercepts

107.

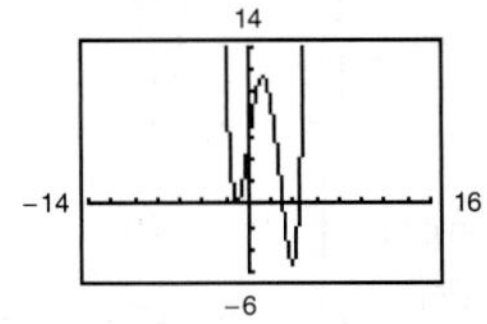

Three x-intercepts

109. (a) Answers will vary.
(b) Domain: $0 < x < 18$
(c)

Height, x	Volume, V
1	1156
2	2048
3	2700
4	3136
5	3380
6	3456
7	3388

$5 < x < 7$

(d)

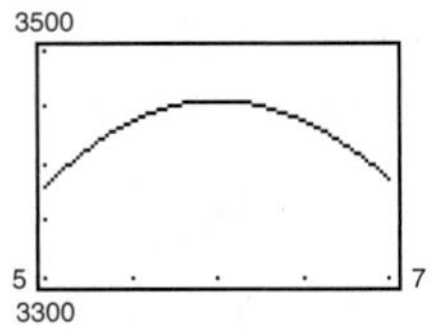

$x = 6$

111. $(200, 160)$

113. (a)

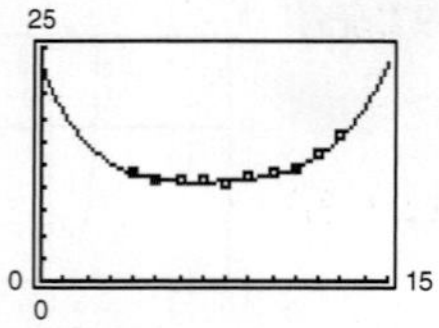

The model fits the data well.

(b)

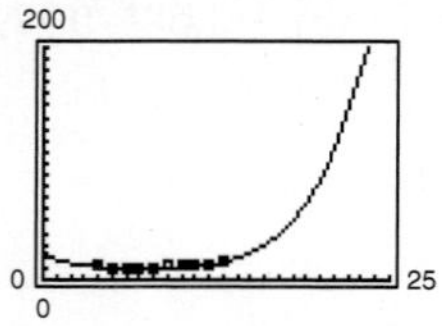

Answers will vary. Sample answer: You could use the model to estimate production in 2015 because the result is somewhat reasonable, but you would not use the model to estimate the 2020 production because the result is unreasonably high.

(c)

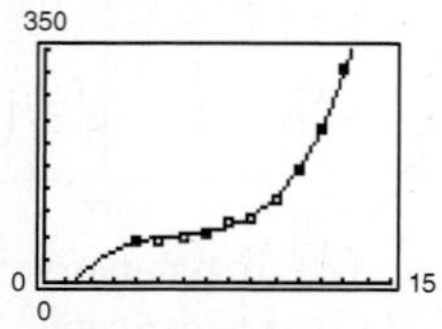

The model fits the data well.

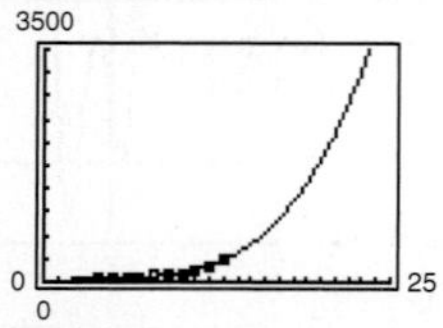

Answers will vary. Sample answer: You could use the model to estimate production in 2015 because the result is somewhat reasonable, but you would not use the model to estimate the 2020 production because the result is unreasonably high.

115. False. The graph will always cross the x-axis.

117. True. The degree is odd and the leading coefficient is -1.

119. False. The graph crosses the x-axis at $x = -3$ and $x = 0$.

121.

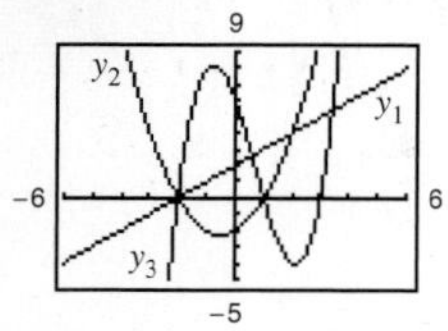

123. 69 **125.** $-\frac{1408}{49} \approx -28.73$ **127.** 109

129. $x > -8$

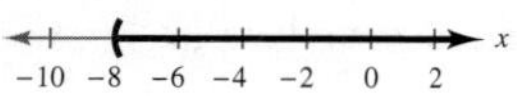

131. $-26 \le x < 7$

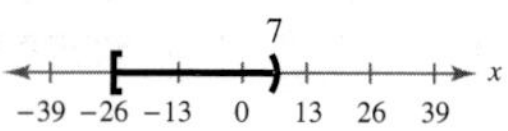

Section 3.3 (page 277)

1. $f(x)$ is the dividend, $d(x)$ is the divisor, $q(x)$ is the quotient, and $r(x)$ is the remainder.

3. constant term, leading coefficient **5.** upper, lower

7. 7 **9.** $(x + 3)(x + 2)$ **11.** $(x + 2)(x + 6)(x - 3)$

13. $x^2 - x - 20 - \dfrac{54}{x - 3}$ **15.** $7x^2 - 14x + 28 - \dfrac{53}{x + 2}$

17. $5x - 1$ **19.** $x - \dfrac{x + 9}{x^2 + 1}$ **21.** $2x - \dfrac{17x - 5}{x^2 - 2x + 1}$

23. $3x^2 - 2x + 5, x \neq 5$ **25.** $6x^2 + 25x + 74 + \dfrac{248}{x - 3}$

27. $9x^2 - 16, x \neq 2$ **29.** $x^2 - 8x + 64, x \neq -8$

31. $4x^2 + 14x - 30, x \neq -\frac{1}{2}$

33.

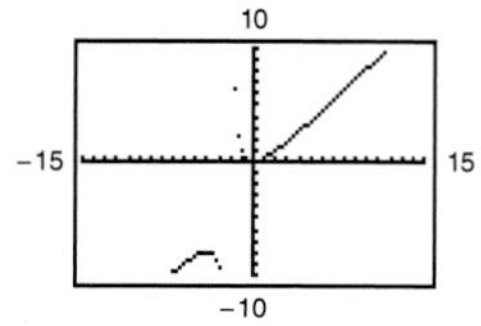

35.

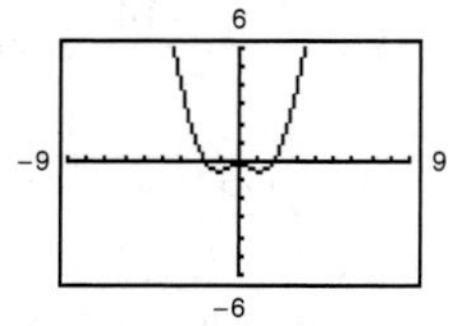

37. $f(x) = (x - 4)(x^2 + 3x - 2) + 3, f(4) = 3$

39. $f(x) = \left(x - \sqrt{2}\right)\left[x^2 + \left(3 + \sqrt{2}\right)x + 3\sqrt{2}\right] - 8,$
$f\left(\sqrt{2}\right) = -8$

41. $f(x) = \left(x - 1 + \sqrt{3}\right)\left[4x^2 - \left(2 + 4\sqrt{3}\right)x - \left(2 + 2\sqrt{3}\right)\right],$
$f\left(1 - \sqrt{3}\right) = 0$

43. (a) -2 (b) 1 (c) $-\frac{1}{4}$ (d) 5

45. (a) -35 (b) -22 (c) -10 (d) -211

47. $(x + 1)(x + 3)(x - 4)$
Zeros: $-1, -3, 4$

49. $(2x + 3)(x - 3)(x - 7)$
Zeros: $-\frac{3}{2}, 3, 7$

51. $\left(x - \sqrt{3}\right)\left(x + \sqrt{3}\right)(x + 2)$
Zeros: $\pm\sqrt{3}, -2$

53. (a) Answers will vary. (b) $(2x - 1), (x - 1)$
(c) $(x + 2)(x - 1)(2x - 1)$ (d) $-2, 1, \frac{1}{2}$

55. (a) Answers will vary. (b) $(x - 1), (x - 2)$
(c) $(x - 5)(x + 4)(x - 1)(x - 2)$ (d) $-4, 1, 2, 5$

57. (a) Answers will vary. (b) $(x + 7), (3x - 2)$
(c) $(2x + 1)(3x - 2)(x + 7)$ (d) $-7, -\frac{1}{2}, \frac{2}{3}$

59. $\pm1, \pm3; \pm1, -3$

61. $\pm1, \pm3, \pm5, \pm9, \pm15, \pm45, \pm\frac{1}{2}, \pm\frac{3}{2}, \pm\frac{5}{2}, \pm\frac{9}{2}, \pm\frac{15}{2}, \pm\frac{45}{2}; -1, \frac{3}{2}, 3, 5$

63. 4, 2, or 0 positive real zeros, no negative real zeros

65. 2 or 0 positive real zeros, 1 negative real zero

67. (a) 1 positive real zero, 2 or 0 negative real zeros
(b) $\pm1, \pm2, \pm4$
(c)

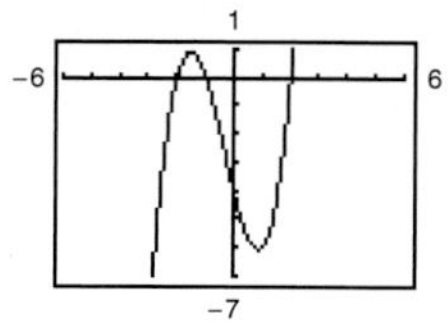

(d) $-2, -1, 2$

69. (a) 3 or 1 positive real zero(s), 1 negative real zero
(b) $\pm1, \pm2, \pm4, \pm8, \pm\frac{1}{2}$
(c)

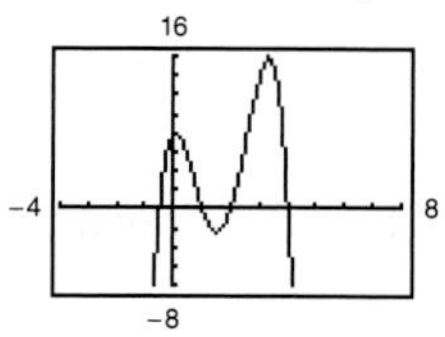

(d) $-\frac{1}{2}, 1, 2, 4$

71. (a) 2 or 0 positive real zeros, 1 negative real zero
(b) $\pm1, \pm3, \pm\frac{1}{2}, \pm\frac{3}{2}, \pm\frac{1}{4}, \pm\frac{3}{4}, \pm\frac{1}{8}, \pm\frac{3}{8}, \pm\frac{1}{16}, \pm\frac{3}{16}, \pm\frac{1}{32}, \pm\frac{3}{32}$
(c)

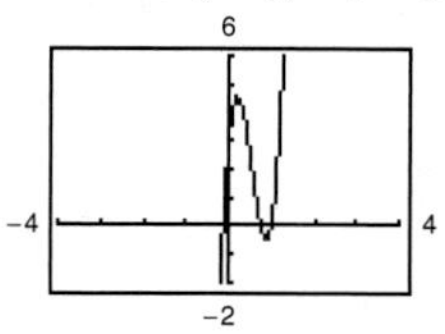

(d) $-\frac{1}{8}, \frac{3}{4}, 1$

73. Answers will vary; 1.937, 3.705

75. Answers will vary; ± 2 **77.** $\pm 2, \pm\frac{3}{2}$

79. $\pm 1, \frac{1}{4}$ **81.** d **82.** a **83.** b **84.** c

85. $-\frac{1}{2}, 2 \pm \sqrt{3}, 1$ **87.** $-1, \frac{3}{2}, 4 \pm \sqrt{17}$

89. $-3, 0, \dfrac{3 \pm \sqrt{14}}{5}$ **91.** $-1, 2$ **93.** $-6, \dfrac{1}{2}, 1$

95. $-3, -\dfrac{3}{2}, \dfrac{1}{2}, 4$ **97.** $-3, 0, \dfrac{-1 \pm \sqrt{7}}{4}$

99. $-\frac{3}{2}, -\frac{5}{4}, -1, 1, 2$

101. (a) $-2, 0.268, 3.732$ (b) -2
(c) $h(t) = (t + 2)\left(t - 2 + \sqrt{3}\right)\left(t - 2 - \sqrt{3}\right)$

103. (a) $0, 3, 4, -1.414, 1.414$ (b) $0, 3, 4$
(c) $h(x) = x(x - 3)(x - 4)\left(x + \sqrt{2}\right)\left(x - \sqrt{2}\right)$

105. (a)

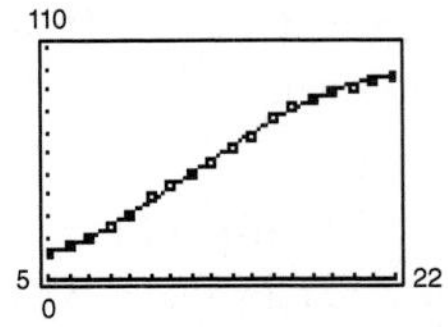

(b) The model fits the data well.
(c) 45.6 subscriptions; Sample answer: No, the number of subscriptions is unlikely to decrease.

107. (a) Answers will vary.
(b)

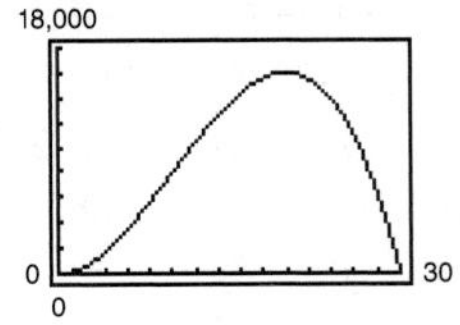

$20 \times 20 \times 40$
(c) $15, \dfrac{15 \pm 15\sqrt{5}}{2}$;
$\dfrac{15 - 15\sqrt{5}}{2}$ represents a negative volume.

109. False. If $(7x + 4)$ is a factor of f, then $-\frac{4}{7}$ is a zero of f.

111. $-2(x - 1)^2(x + 2)$

113. $-(x - 2)(x + 2)(x + 1)(x - 1)$

115. (a) $x + 1,\ x \neq 1$ (b) $x^2 + x + 1,\ x \neq 1$
(c) $x^3 + x^2 + x + 1,\ x \neq 1$
$$\frac{x^n - 1}{x - 1} = x^{n-1} + x^{n-2} + \cdots + x^2 + x + 1, x \neq 1$$

117. $\pm\dfrac{\sqrt{17}}{2}$ **119.** $-\dfrac{4}{3}, 5$

Section 3.4 (page 286)

1. Fundamental Theorem, Algebra **3.** n linear factors

5. c **6.** d **7.** b **8.** a

9–11. Answers will vary.

13. Zeros: $4, -i, i$. One real zero; they are the same.

15. Zeros: $\sqrt{2}i, \sqrt{2}i, -\sqrt{2}i, -\sqrt{2}i$. No real zeros; they are the same.

17. $2 \pm \sqrt{3}$
$\left(x - 2 - \sqrt{3}\right)\left(x - 2 + \sqrt{3}\right)$

19. $6 \pm \sqrt{10}$
$\left(x - 6 - \sqrt{10}\right)\left(x - 6 + \sqrt{10}\right)$

21. $\pm 5i$
$(x + 5i)(x - 5i)$

23. $\pm\frac{3}{2}, \pm\frac{3}{2}i$
$(2x - 3)(2x + 3)(2x - 3i)(2x + 3i)$

25. $\dfrac{1 \pm \sqrt{223}i}{2}$
$\left(z - \dfrac{1 - \sqrt{223}i}{2}\right)\left(z - \dfrac{1 + \sqrt{223}i}{2}\right)$

27. $\pm i, \pm 3i$
$(x + i)(x - i)(x + 3i)(x - 3i)$

29. $\frac{5}{3}, \pm 4i$
$(3x - 5)(x - 4i)(x + 4i)$

31. $-5, 4 \pm 3i$
$(t + 5)(t - 4 + 3i)(t - 4 - 3i)$

33. $1 \pm \sqrt{5}i, -\frac{1}{5}$
$(5x + 1)\left(x - 1 + \sqrt{5}i\right)\left(x - 1 - \sqrt{5}i\right)$

35. $2, 2, \pm 2i$
$(x - 2)^2(x + 2i)(x - 2i)$

37. (a) $7 \pm \sqrt{3}$ (b) $\left(x - 7 - \sqrt{3}\right)\left(x - 7 + \sqrt{3}\right)$
(c) $\left(7 \pm \sqrt{3}, 0\right)$

39. (a) $\frac{3}{2}, \pm 2i$ (b) $(2x - 3)(x - 2i)(x + 2i)$ (c) $\left(\frac{3}{2}, 0\right)$

41. (a) $-6, 3 \pm 4i$ (b) $(x + 6)(x - 3 - 4i)(x - 3 + 4i)$
(c) $(-6, 0)$

43. (a) $\pm 4i, \pm 3i$ (b) $(x + 4i)(x - 4i)(x + 3i)(x - 3i)$
(c) None

45. $f(x) = x^3 - 5x^2 + x - 5$

47. $f(x) = x^4 - 2x^3 + 5x^2 - 8x + 4$

49. $f(x) = x^4 + 3x^3 - 7x^2 + 15x$

51. (a) $-2(x + 1)(x - 2)(x + i)(x - i)$
(b) $f(x) = -2x^4 + 2x^3 + 2x^2 + 2x + 4$

53. (a) $3(x + 1)\left(x - 2 + \sqrt{5}i\right)\left(x - 2 - \sqrt{5}i\right)$
(b) $3x^3 - 9x^2 + 15x + 27$

55. (a) $(x^2 + 1)(x^2 - 7)$ (b) $(x^2 + 1)\left(x + \sqrt{7}\right)\left(x - \sqrt{7}\right)$
(c) $(x + i)(x - i)\left(x + \sqrt{7}\right)\left(x - \sqrt{7}\right)$

57. (a) $(x^2 - 6)(x^2 - 2x + 3)$
(b) $\left(x + \sqrt{6}\right)\left(x - \sqrt{6}\right)(x^2 - 2x + 3)$
(c) $\left(x + \sqrt{6}\right)\left(x - \sqrt{6}\right)\left(x - 1 - \sqrt{2}i\right)\left(x - 1 + \sqrt{2}i\right)$

59. $-\frac{3}{2}, \pm 5i$ **61.** $-3, 5 \pm 2i$ **63.** $-\frac{2}{3}, 1 \pm \sqrt{3}i$

65. (a) 1.000, 2.000 (b) $-3 \pm \sqrt{2}i$

67. (a) 0.750 (b) $\dfrac{1}{2} \pm \dfrac{\sqrt{5}}{2}i$

69. No. Setting $h = 50$ and solving the resulting equation yields imaginary roots.

71. False. A third-degree polynomial must have at least one real zero.

73. Answers will vary.

75. Parabola opening upward
Vertex: $\left(\frac{7}{2}, -\frac{81}{4}\right)$

77. Parabola opening upward
Vertex: $\left(-\frac{5}{12}, -\frac{169}{24}\right)$

Section 3.5 (page 293)

1. rational functions **3.** vertical asymptote

5. (a) Domain: all real numbers x except $x = 1$

(b)

x	$f(x)$
0.5	-2
0.9	-10
0.99	-100
0.999	-1000

x	$f(x)$
1.5	2
1.1	10
1.01	100
1.001	1000

x	$f(x)$
5	0.25
10	$0.\overline{1}$
100	$0.\overline{01}$
1000	$0.\overline{001}$

x	$f(x)$
-5	$-0.1\overline{6}$
-10	$-0.\overline{09}$
-100	$-0.\overline{0099}$
-1000	$-0.\overline{000999}$

(c) f approaches $-\infty$ from the left and ∞ from the right of $x = 1$.

7. (a) Domain: all real numbers x except $x = 1$

(b)

x	$f(x)$
0.5	3
0.9	27
0.99	297
0.999	2997

x	$f(x)$
1.5	9
1.1	33
1.01	303
1.001	3003

x	$f(x)$
5	3.75
10	$3.\overline{33}$
100	$3.\overline{03}$
1000	$3.\overline{003}$

x	$f(x)$
-5	-2.5
-10	-2.727
-100	-2.97
-1000	-2.997

(c) f approaches ∞ from both the left and the right of $x = 1$.

9. (a) Domain: all real numbers x except $x = \pm 1$

(b)

x	$f(x)$
0.5	-1
0.9	-12.79
0.99	-147.8
0.999	-1498

x	$f(x)$
1.5	5.4
1.1	17.29
1.01	152.3
1.001	1502.3

x	$f(x)$
5	3.125
10	$3.\overline{03}$
100	$3.\overline{0003}$
1000	3.000003

x	$f(x)$
-5	3.125
-10	$3.\overline{03}$
-100	$3.\overline{0003}$
-1000	3.000003

(c) f approaches ∞ from the left and $-\infty$ from the right of $x = -1$. f approaches $-\infty$ from the left and ∞ from the right of $x = 1$.

11. a **12.** d **13.** c **14.** e **15.** b **16.** f

17. Vertical asymptote: $x = 0$
Horizontal asymptote: $y = 0$

19. Vertical asymptotes: $x = -3, 2$
Horizontal asymptote: $y = 2$

21. Vertical asymptote: $x = 2$
Horizontal asymptote: $y = -1$
Hole at $x = 0$

23. Vertical asymptotes: $x = 0, x = -8$
Horizontal asymptote: $y = 1$
No hole

25. (a) Domain: all real numbers x
(b) Continuous
(c) Horizontal asymptote: $y = 5$

27. (a) Domain: all real numbers x except $x = 3$
(b) Not continuous
(c) Vertical asymptote: $x = 3$
Horizontal asymptote: $y = 0$

29. (a) Domain of f: all real numbers x except $x = 4$
Domain of g: all real numbers x
(b) Vertical asymptote: none; Hole in f at $x = 4$
(c)

x	1	2	3	4	5	6	7
$f(x)$	5	6	7	Undef.	9	10	11
$g(x)$	5	6	7	8	9	10	11

(d)

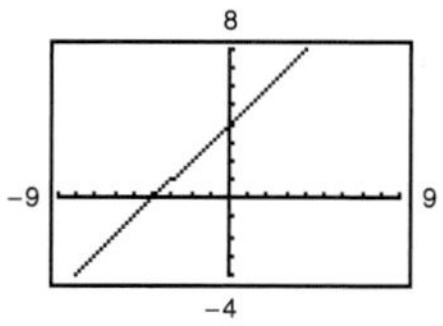

(e) Graphing utilities are limited in their resolution and therefore may not show a hole in a graph.

31. (a) Domain of f: all real numbers x except $x = -1, 3$
Domain of g: all real numbers x except $x = 3$
(b) Vertical asymptote: $x = 3$; Hole in f at $x = -1$
(c)

x	-2	-1	0	1	2	3	4
$f(x)$	$\frac{3}{5}$	Undef.	$\frac{1}{3}$	0	-1	Undef.	3
$g(x)$	$\frac{3}{5}$	$\frac{1}{2}$	$\frac{1}{3}$	0	-1	Undef.	3

(d)

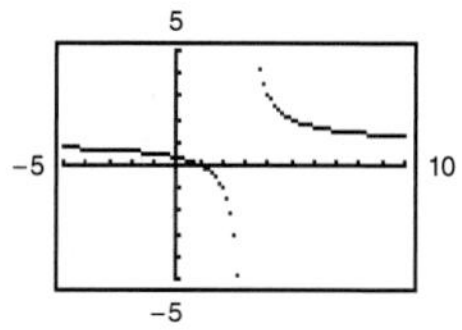

(e) Graphing utilities are limited in their resolution and therefore may not show a hole in a graph.

33. 4; less than; greater than **35.** 2; greater than; less than

37. ± 3 **39.** -3 **41.** $-4, 5$ **43.** -7

45. (a)

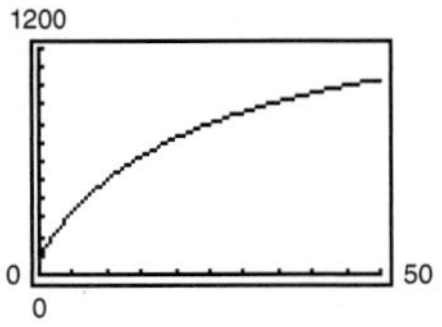

(b) 333 deer, 500 deer, 800 deer

(c) 1500. Because the degrees of the numerator and the denominator are equal, the limiting size is the ratio of the leading coefficients, $60/0.04 = 1500$.

47. (a)

M	200	400	600	800	1000
t	0.472	0.596	0.710	0.817	0.916

M	1200	1400	1600	1800	2000
t	1.009	1.096	1.178	1.255	1.328

The greater the mass, the more time required per oscillation.

(b) $M \approx 1306$ g

49. (a)

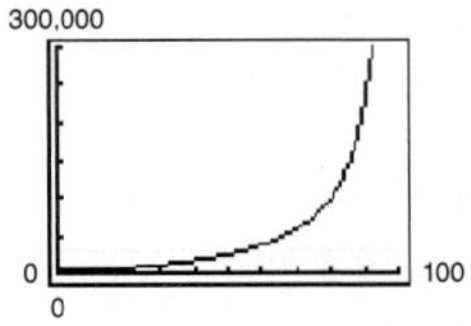

(b) \$4411.76; \$25,000; \$225,000

(c) No. The function is undefined at $p = 100$.

51. False. $\dfrac{1}{x^2 + 1}$ has no vertical asymptote.

53. No. If $x = c$ is also a zero in the denominator, then f is undefined at $x = c$.

55.

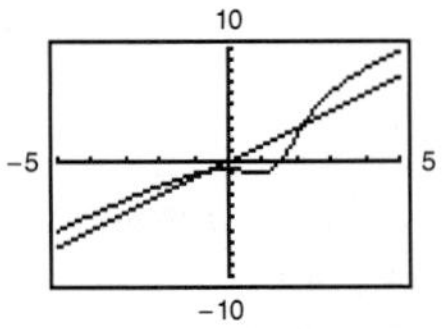

Both graphs have the same slope.

57. $x - y - 1 = 0$ **59.** $3x - y + 1 = 0$

61. $x + 9 + \dfrac{42}{x - 4}$ **63.** $2x^2 - 9 + \dfrac{34}{x^2 + 5}$

Section 3.6 (page 303)

1. slant, asymptote **3.** Yes

5.

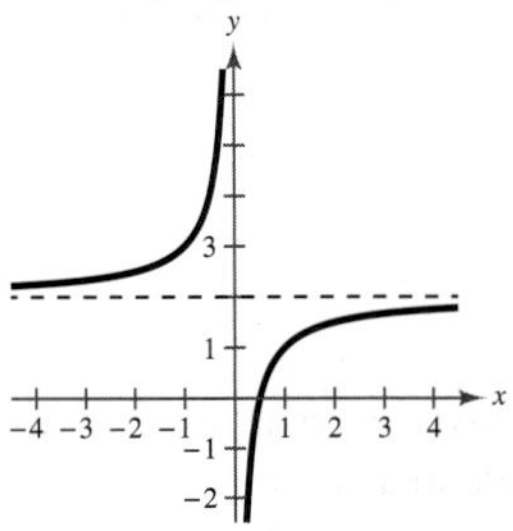

Reflection in the x-axis, vertical shift two units upward

7.

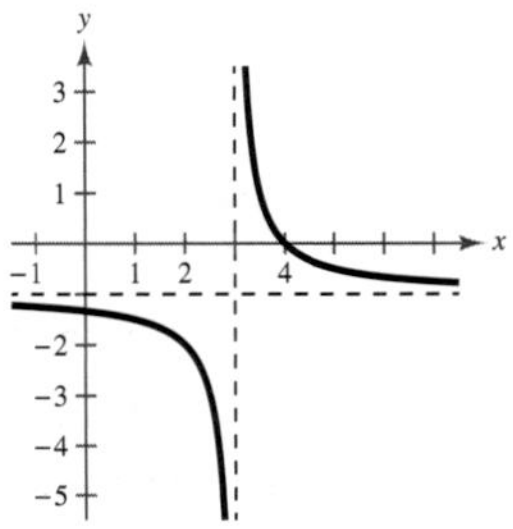

Horizontal shift three units to the right, vertical shift one unit downward

9.

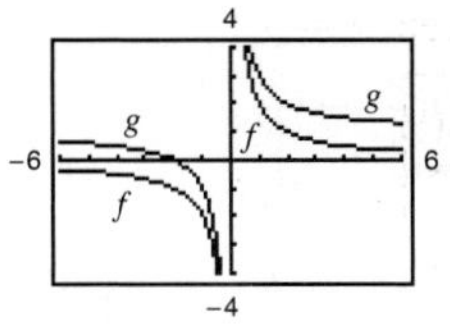

Vertical shift

11.

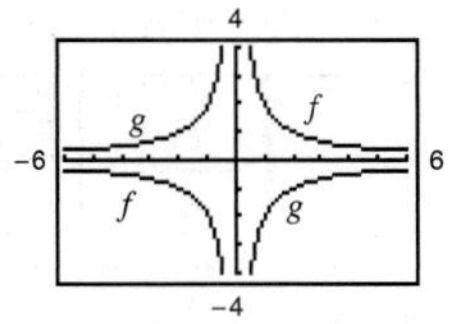

Reflection in the x-axis

13.

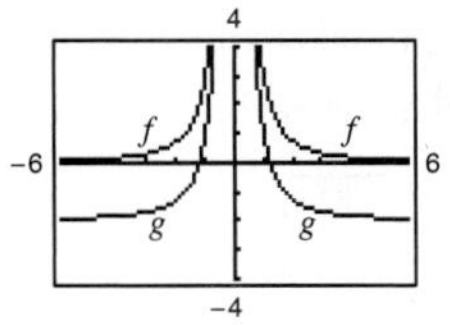

Vertical shift

15.

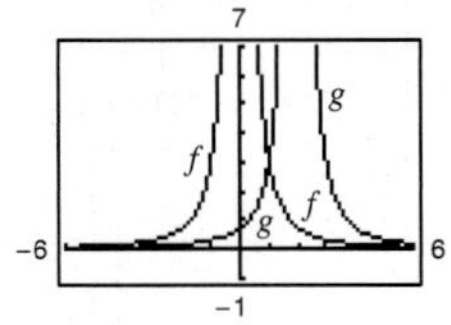

Horizontal shift

17.

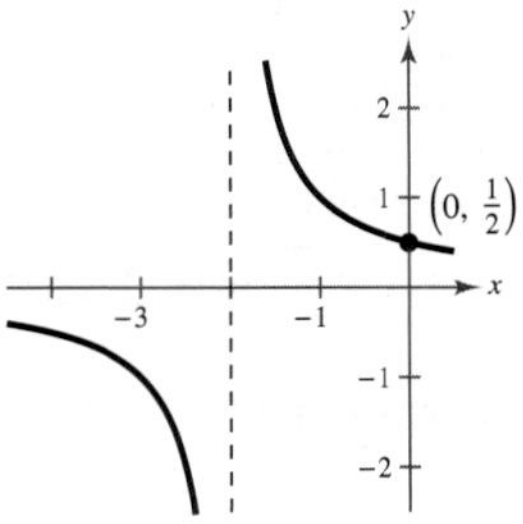

19.

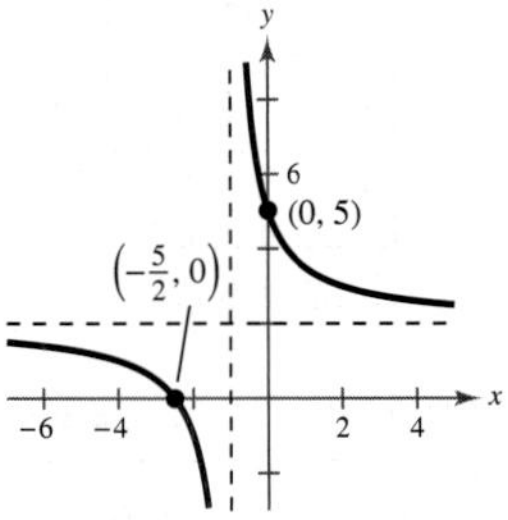

21.

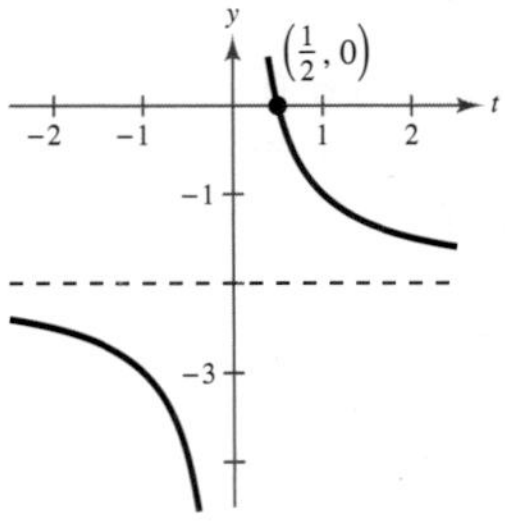

23.

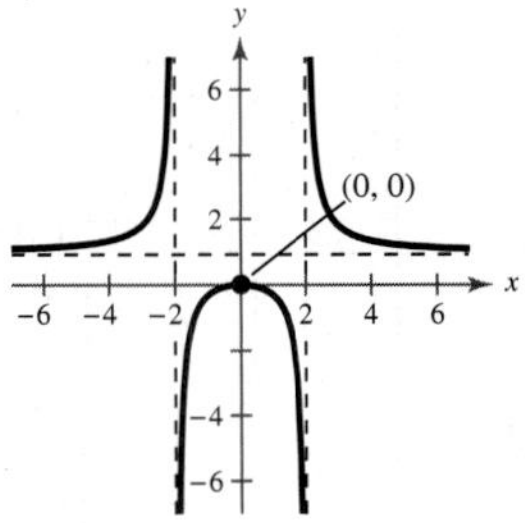

25.

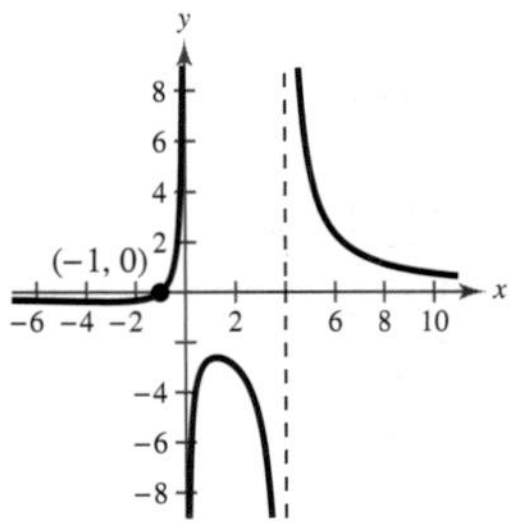

27.

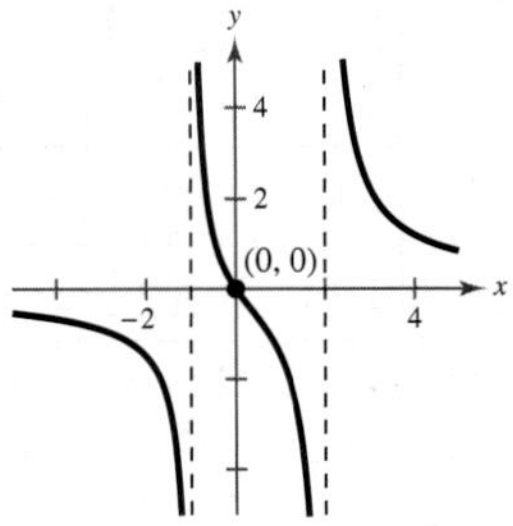

29.

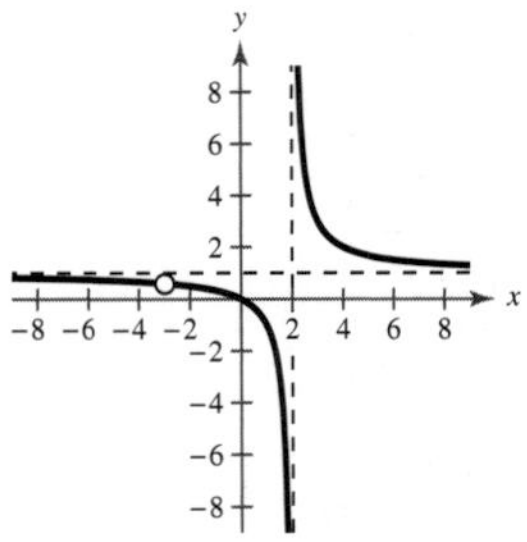

There is a hole at $x = -3$.

31.

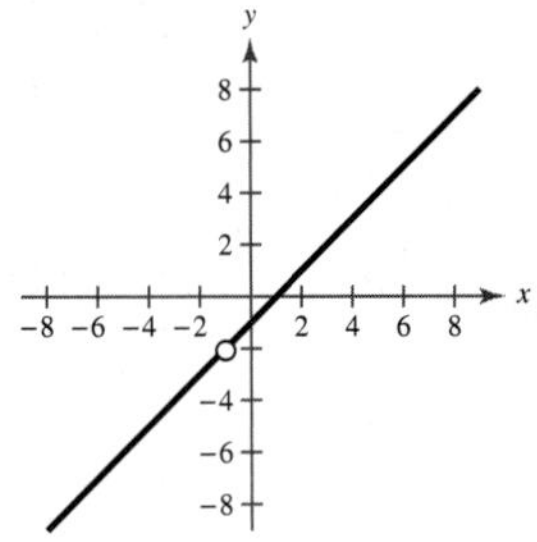

There is a hole at $x = -1$.

33.

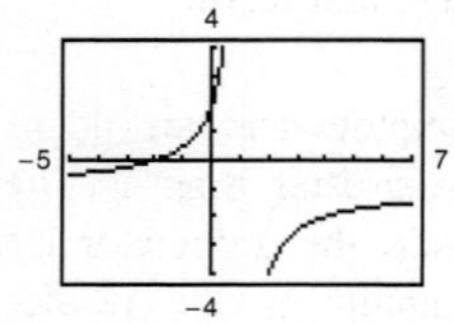

Domain: $(-\infty, 1), (1, \infty)$
Vertical asymptote: $x = 1$
Horizontal asymptote: $y = -1$

35.

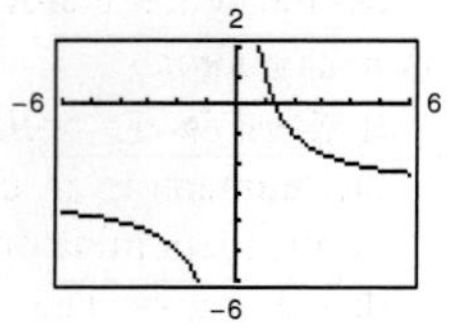

Domain: $(-\infty, 0), (0, \infty)$
Vertical asymptote: $x = 0$
Horizontal asymptote: $y = -3$

37.

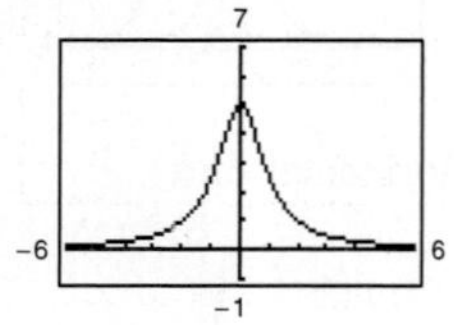

Domain: $(-\infty, \infty)$
Horizontal asymptote: $y = 0$

39.

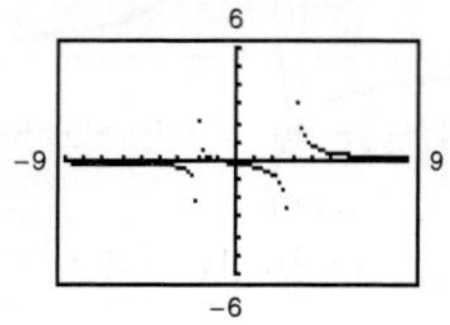

Domain: $(-\infty, -2), (-2, 3), (3, \infty)$
Vertical asymptotes: $x = -2, x = 3$
Horizontal asymptote: $y = 0$

41.

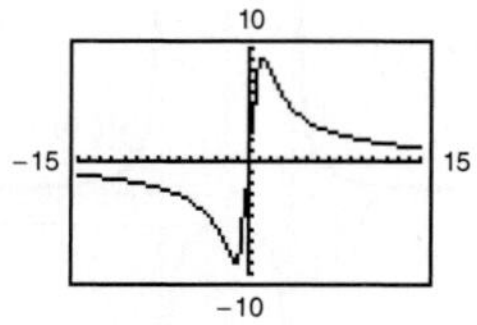

Domain: $(-\infty, 0), (0, \infty)$
Vertical asymptote: $x = 0$
Horizontal asymptote: $y = 0$

43.

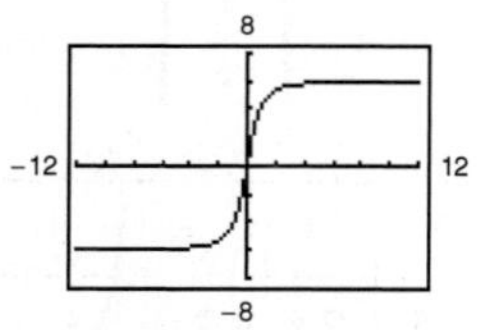

There are two horizontal asymptotes, $y = \pm 6$.

45.

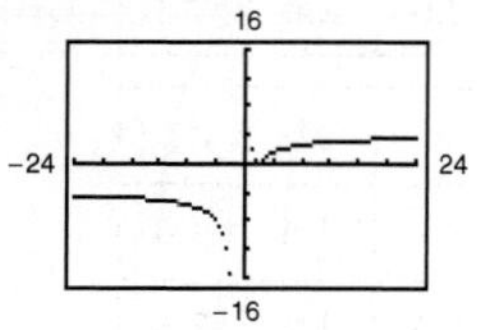

There are two horizontal asymptotes, $y = \pm 4$, and one vertical asymptote, $x = -1$.

47.

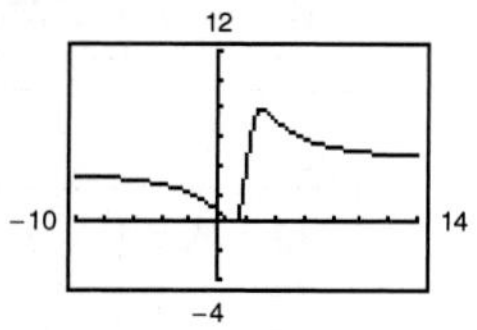

The graph crosses the horizontal asymptote, $y = 4$.

49.

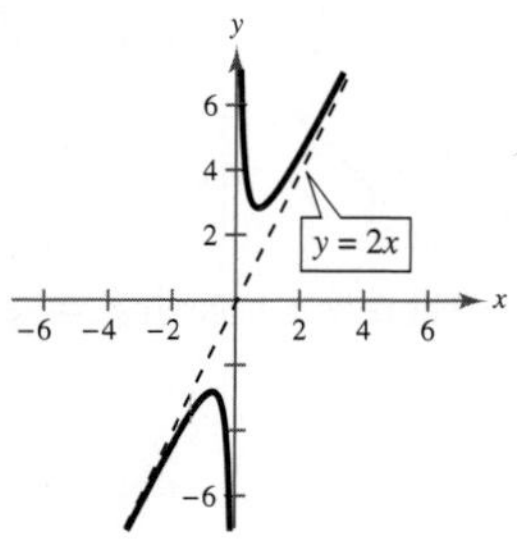

51.

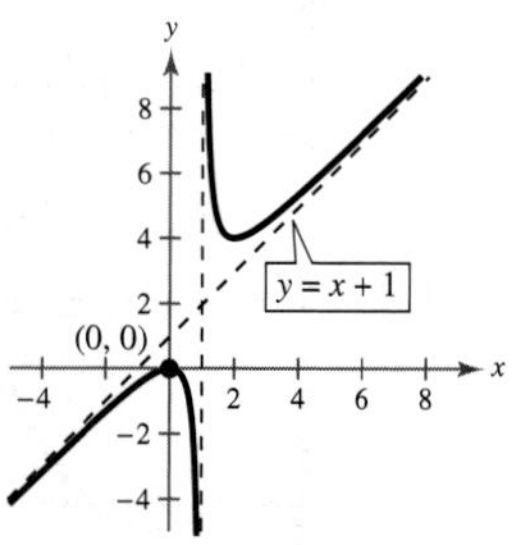

53.

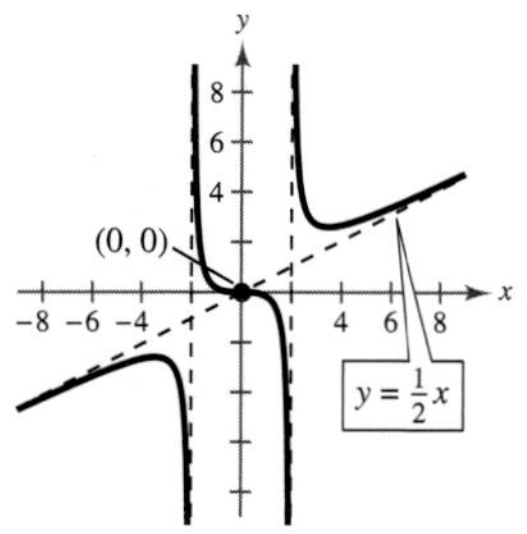

55.

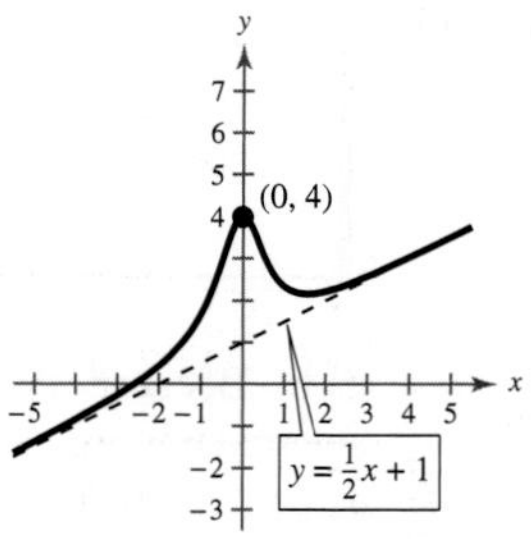

57. $(-1, 0)$ **59.** $(1, 0), (-1, 0)$

61.

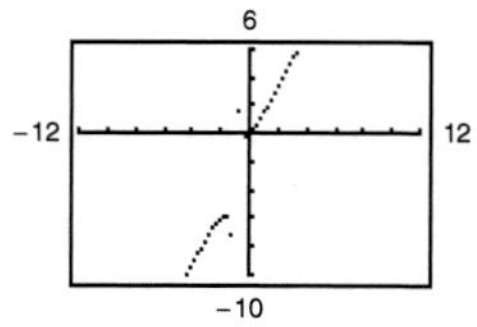

Domain: $(-\infty, -1), (-1, \infty)$
Vertical asymptote: $x = -1$
Slant asymptote: $y = 2x - 1$

63.

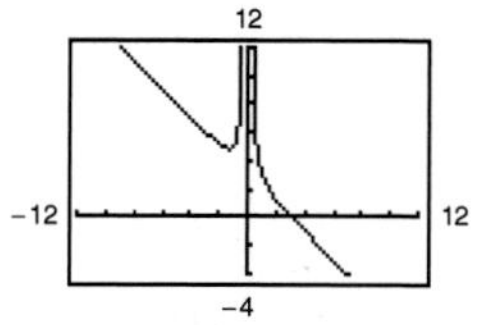

Domain: $(-\infty, 0), (0, \infty)$
Vertical asymptote: $x = 0$
Slant asymptote: $y = -x + 3$

65. Vertical asymptotes: $x = \pm 2$; horizontal asymptote: $y = 1$; slant asymptote: none; holes: none

67. Vertical asymptote: $x = -\frac{3}{2}$; horizontal asymptote: $y = 1$; slant asymptote: none; hole at $x = 2$

69. Vertical asymptote: $x = -2$; horizontal asymptote: none; slant asymptote: $y = 2x - 7$; hole at $x = -1$

71.

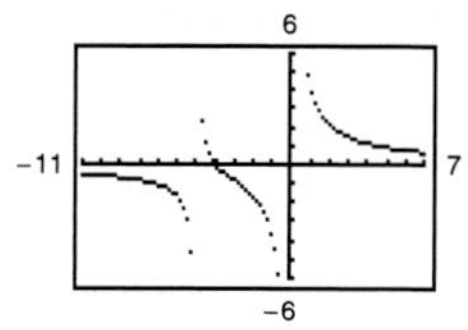

$(-4, 0)$

73.

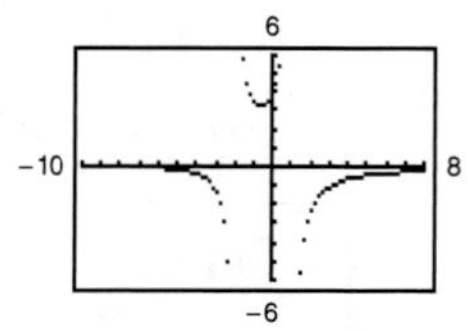

$(-8, 0)$

75.

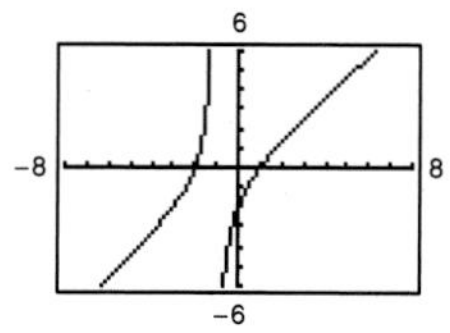

$(-2, 0), (1, 0)$

77.

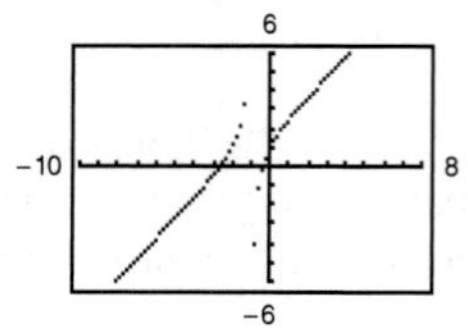

$\left(\frac{-3 \pm \sqrt{5}}{2}, 0\right)$

79.

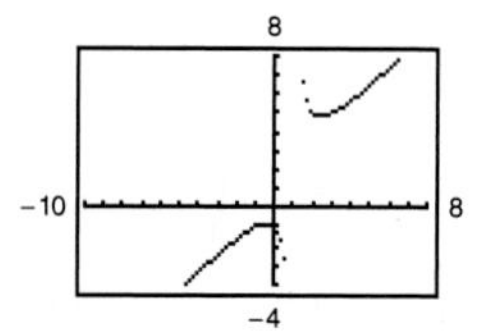

None

81.

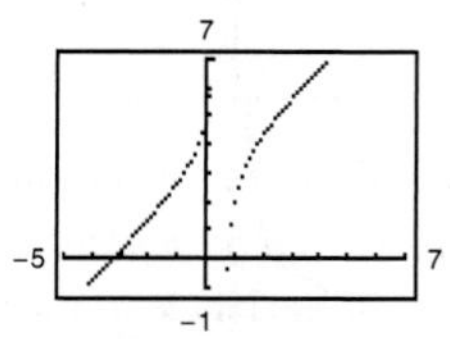

$\left(\frac{-5 \pm \sqrt{65}}{4}, 0\right)$

83. (a) Answers will vary. (b) $[0, 950]$

(c)

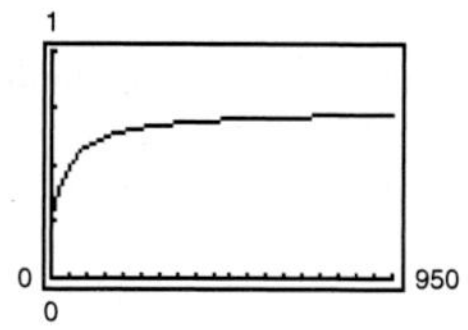

The concentration increases more slowly; 75%

85. (a) Answers will vary. (b) $(2, \infty)$

(c)

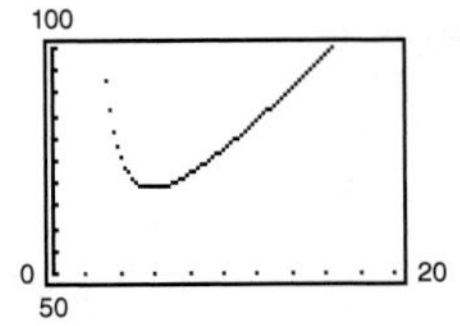

5.9 in $\times$ 11.8 in.

87.

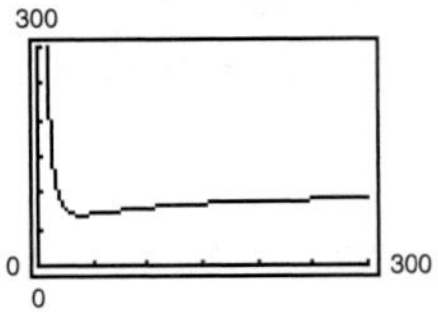

$x \approx 40$

89. (a) $C = 0$. The chemical will eventually dissipate.

(b)

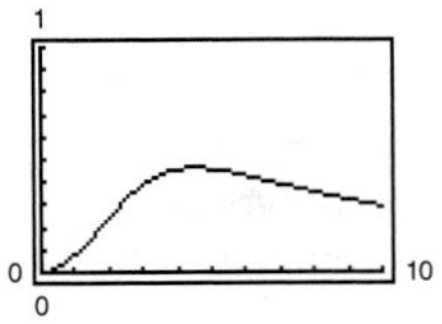

$t \approx 4.5$ h

(c) Before about 2.6 hours and after about 8.3 hours

91. (a) $A = -0.14t + 3.1$

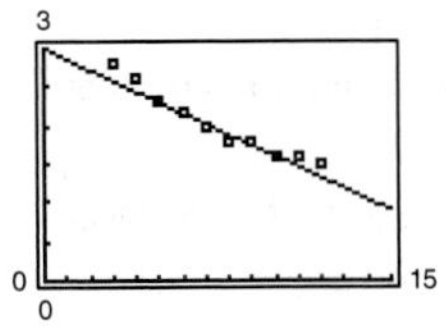

(b) $A = \dfrac{1}{0.035t + 0.253}$

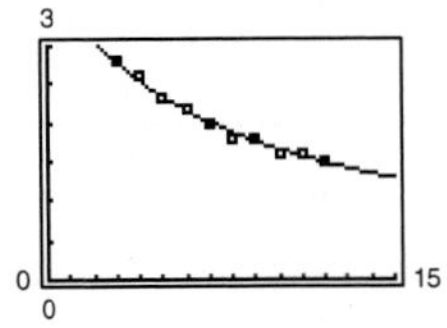

(c) Answers will vary.

93. False. The graph of a rational function is continuous when the polynomial in the denominator has no real zeros.

95.

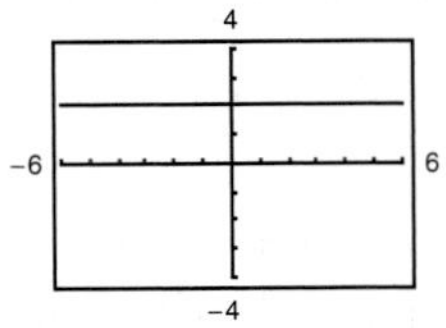

The denominator is a factor of the numerator.

97. *Horizontal asymptotes:*

If the degree of the numerator is greater than the degree of the denominator, then there is no horizontal asymptote.

If the degree of the numerator is less than the degree of the denominator, then there is a horizontal asymptote at $y = 0$.

If the degree of the numerator is equal to the degree of the denominator, then there is a horizontal asymptote at the line given by the ratio of the leading coefficients.

Vertical asymptotes:

Set the denominator equal to zero and solve.

Slant asymptotes:

If there is no horizontal asymptote and the degree of the numerator is exactly one greater than the degree of the denominator, then divide the numerator by the denominator. The slant asymptote is the result, not including the remainder.

99. $\dfrac{512}{x^3}$ **101.** 3

103.

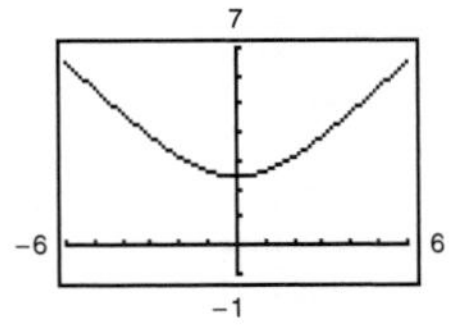

Domain: $(-\infty, \infty)$
Range: $[\sqrt{6}, \infty)$

105.

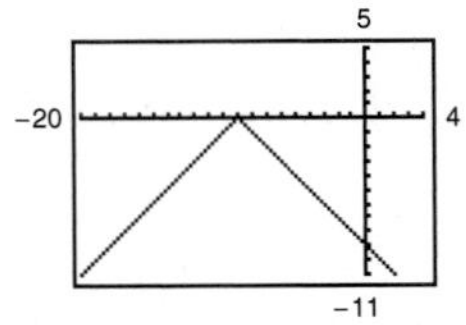

Domain: $(-\infty, \infty)$
Range: $(-\infty, 0]$

107. Answers will vary.

Section 3.7 (page 311)

1. Quadratic **3.** Quadratic **5.** Linear **7.** Neither

9. (a)

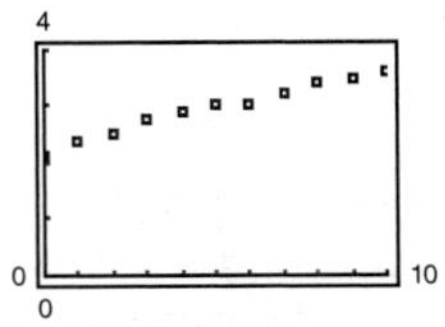

(b) Linear

(c) $y = 0.14x + 2.2$

(d)

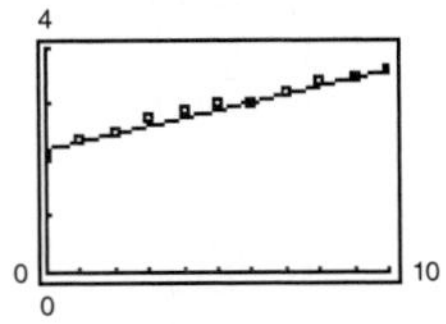

(e)

x	0	1	2	3	4	5
Actual, y	2.1	2.4	2.5	2.8	2.9	3.0
Model, y	2.2	2.3	2.5	2.6	2.8	2.9

x	6	7	8	9	10
Actual, y	3.0	3.2	3.4	3.5	3.6
Model, y	3.0	3.2	3.3	3.5	3.6

11. (a)

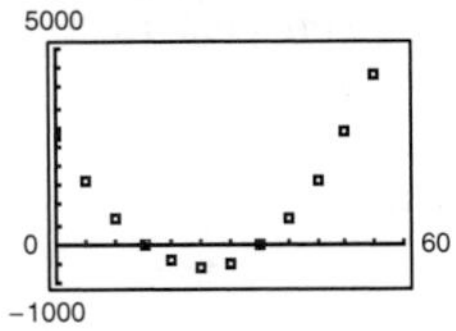

(b) Quadratic

(c) $y = 5.50x^2 - 274.6x + 2822$

(d)

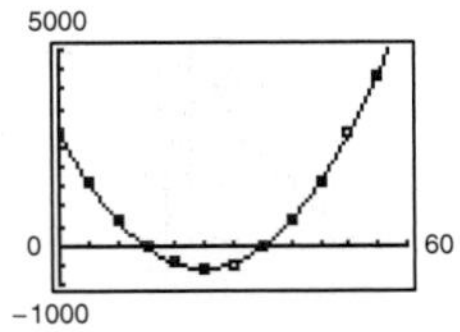

(e)

x	0	5	10	15	20
Actual, y	2795	1590	650	−30	−450
Model, y	2822	1586.5	626	−59.5	−470

x	25	30	35	40
Actual, y	−615	−520	−55	625
Model, y	−605.5	−466	−51.5	638

x	45	50	55
Actual, y	1630	2845	4350
Model, y	1602.5	2842	4356.5

13. (a)

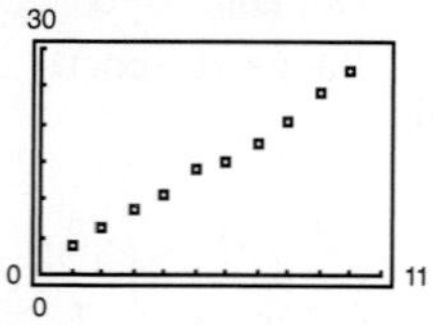

(b) Linear

(c) $y = 2.48x + 1.1$

(d)

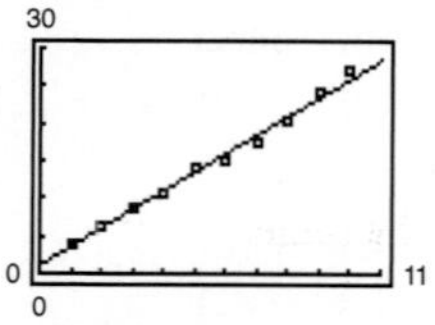

(e)

x	1	2	3	4	5
Actual, y	4.0	6.5	8.8	10.6	13.9
Model, y	3.6	6.1	8.5	11.0	13.5

x	6	7	8	9	10
Actual, y	15.0	17.5	20.1	24.0	27.1
Model, y	16.0	18.5	20.9	23.4	25.9

15. (a)

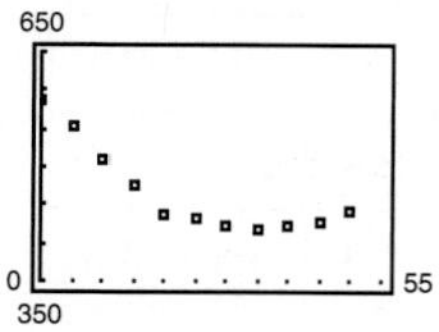

(b) Quadratic (c) $y = 0.14x^2 - 9.9x + 591$

(d)

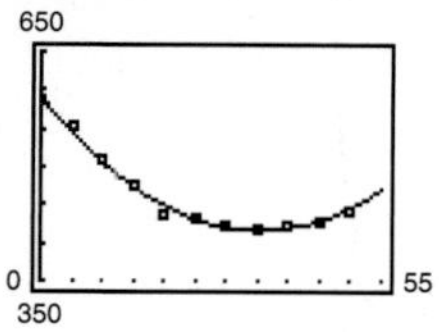

(e)

x	0	5	10	15	20	25
Actual, y	587	551	512	478	436	430
Model, y	591	545	506	474	449	431

x	30	35	40	45	50
Actual, y	424	420	423	429	444
Model, y	420	416	419	429	446

17. (a) and (c)

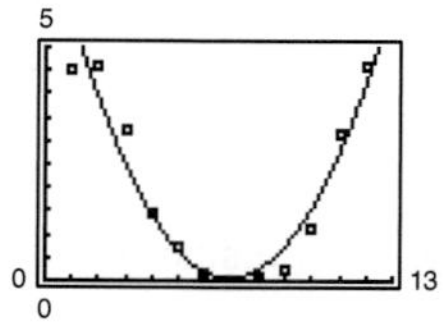

(b) $P = 0.1601t^2 - 2.206t + 7.65$; $r^2 \approx 0.9185$

(d) July (e) July; They are the same.

19. (a) and (c)

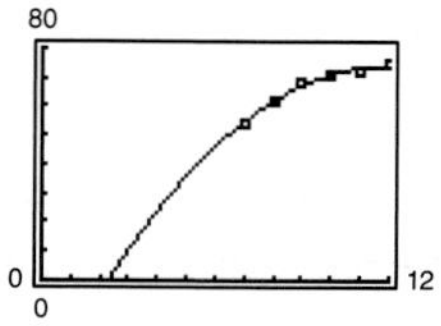

(b) $P = -0.805t^2 + 19.10t - 39.3$

(d) 2017; No, by 2022 the model gives negative values.

21. (a)

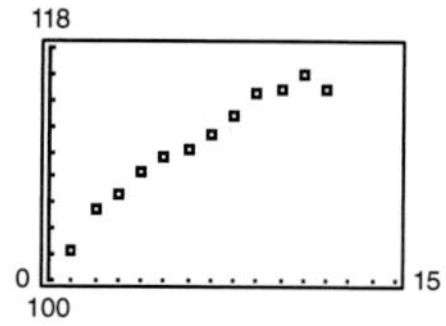

(b) $T = 1.23t + 102.4$
$r^2 \approx 0.9468$

(c)

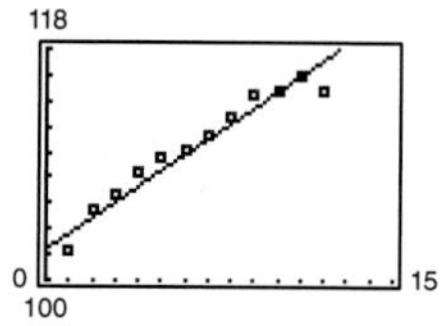

(d) $T = -0.076t^2 + 2.14t + 100.8$
$r^2 \approx 0.9870$

(e)

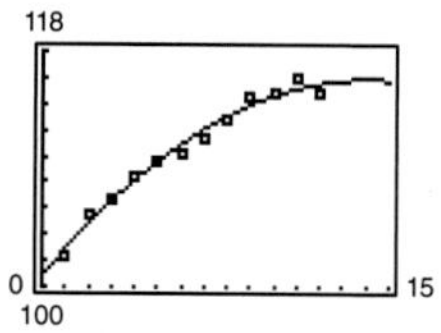

(f) Quadratic. Its coefficient of determination is closer to 1.

(g) Linear: 2015
Quadratic: Not possible

23. True. See "Basic Characteristics of Quadratic Functions" on page 245.

25. The model is consistently above the data points.

27. (a) $(f \circ g)(x) = 2x^2 + 5$ (b) $(g \circ f)(x) = 4(x^2 - x + 1)$

29. (a) $(f \circ g)(x) = x$ (b) $(g \circ f)(x) = x$

31. $f^{-1}(x) = \dfrac{x-5}{2}$ **33.** $f^{-1}(x) = \sqrt{x-5}$

35. $1 + 3i$; 10 **37.** $5i$; 25

Review Exercises (page 316)

1.

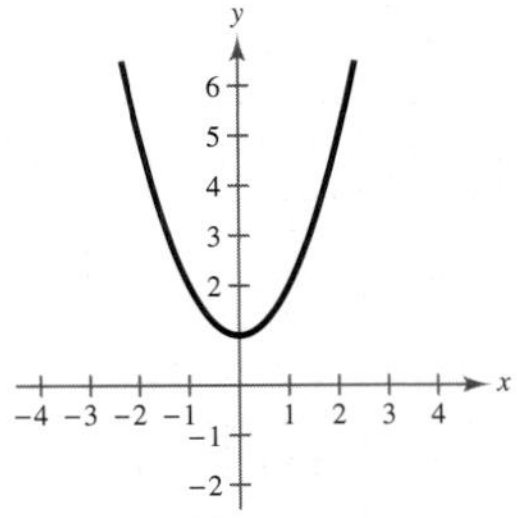

Vertical shift one unit upward

3.

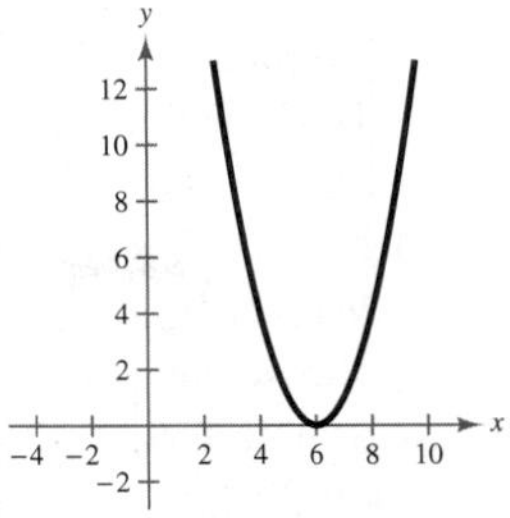

Horizontal shift six units to the right

5.

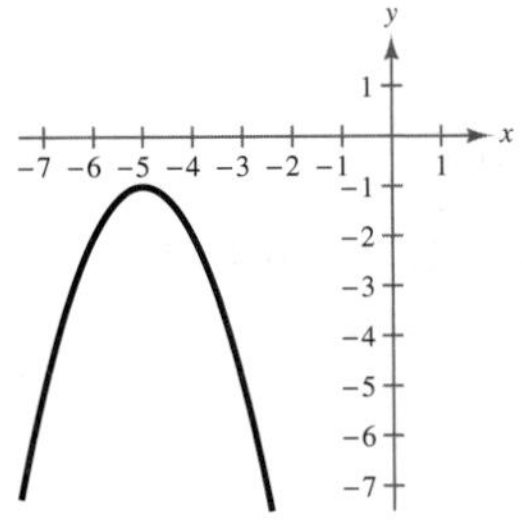

Reflection in the x-axis, horizontal shift five units to the left, vertical shift one unit downward

7. Parabola opening upward
Vertex: $\left(-\frac{3}{2}, 1\right)$

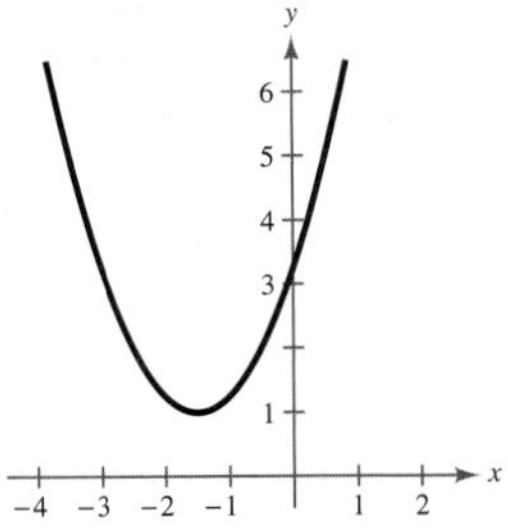

x-intercept: none

9. Parabola opening upward
Vertex: $\left(-\frac{5}{2}, -\frac{41}{12}\right)$

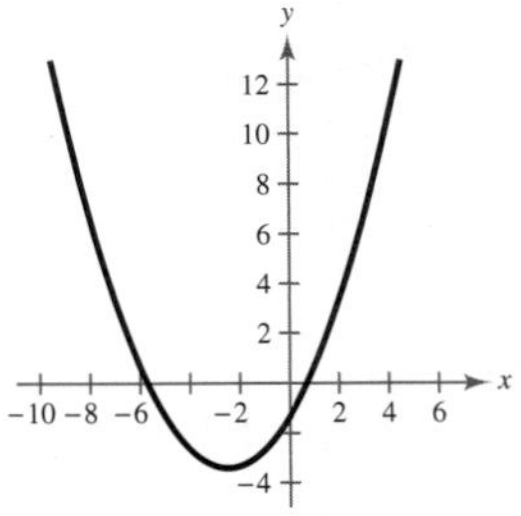

x-intercepts:
$$\left(\frac{-5 \pm \sqrt{41}}{2}, 0\right)$$

11. $f(x) = -(x + 3)^2 + 4$

13. (a) $A = x\left(\frac{8 - x}{2}\right), 0 < x < 8$

(b)

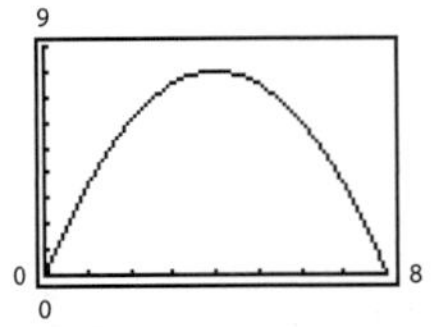

$x = 4, y = 2$

(c) $A = -\frac{1}{2}(x - 4)^2 + 8$; $x = 4$, $y = 2$; They are the same.

15.

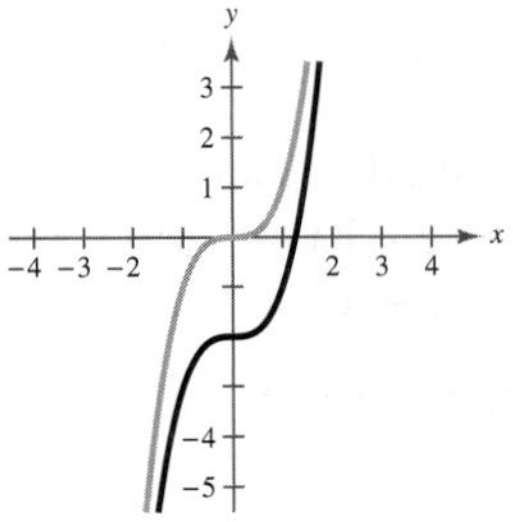

Vertical shift two units downward

17.

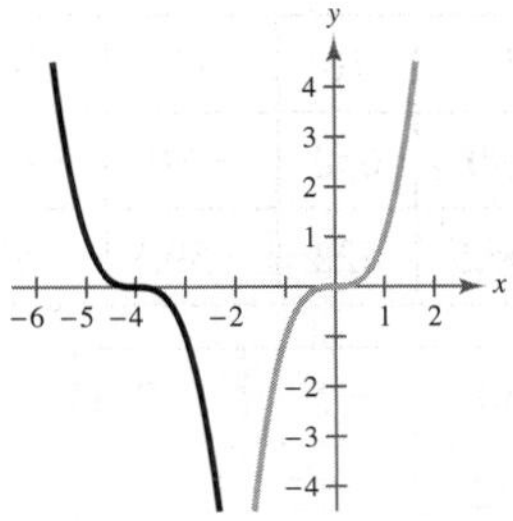

Reflection in the x-axis, horizontal shift four units to the left

19.

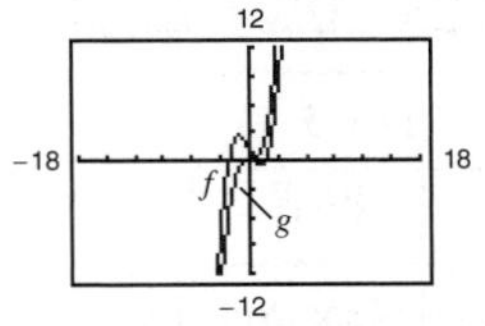

Yes. Both functions are of the same degree and have positive leading coefficients.

21. Rises to the left, rises to the right

23. (a) $x = -1, 0, 0, 2$

(b)

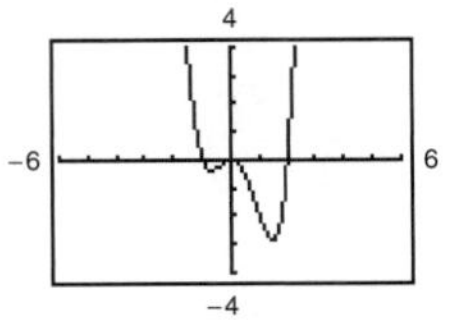

(c) $x = -1, 0, 0, 2$; They are the same.

25. (a) $t = 0, \pm\sqrt{3}$

(b)

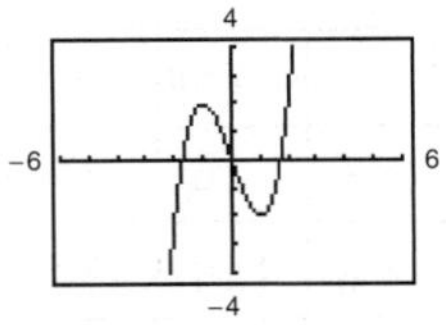

(c) $t = 0, \pm 1.73$; They are the same.

27. (a) $x = -3, -3, 0$

(b)

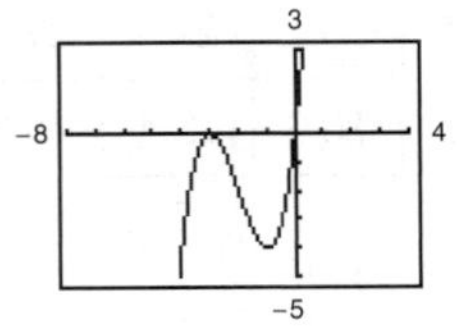

(c) $x = -3, -3, 0$; They are the same.

29. $f(x) = x^3 - 12x - 16$

31. $f(x) = x^3 - 7x^2 + 13x - 3$

33. (a) Rises to the left, rises to the right

(b) $x = -3, -1, 3, 3$

(c) and (d)

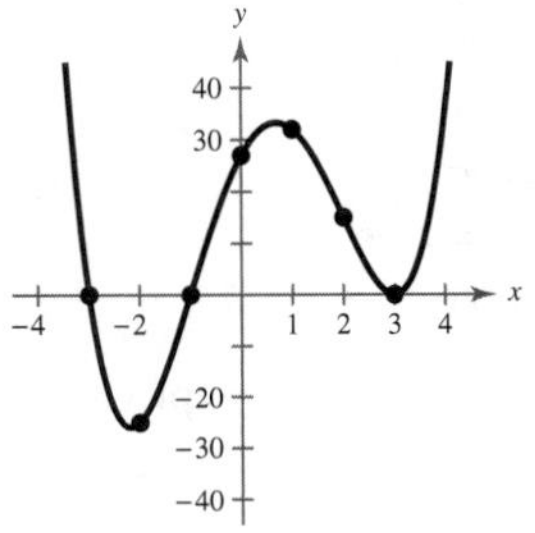

35. (a) $(-3, -2), (-1, 0), (0, 1)$
(b) $x = -2.25, -0.56, 0.80$

37. (a) $(-3, -2), (-1, 0), (3, 4)$ (b) $x = -2.98, -0.55, 3.53$

39. $8x + 5 + \dfrac{2}{3x - 2}$ **41.** $x^2 - 2,\ x \neq \pm 1$

43. $5x + 2,\ x \neq \dfrac{3 \pm \sqrt{5}}{2}$ **45.** $3x^2 + 2x + 20 + \dfrac{58}{x - 4}$

47. $0.25x^3 - 4.5x^2 + 9x - 18 + \dfrac{36}{x + 2}$

49. $6x^3 - 27x,\ x \neq \dfrac{2}{3}$ **51.** (a) -421 (b) -156

53. (a) Answers will vary. (b) $(x + 1)(x + 7)$
(c) $f(x) = (x - 4)(x + 1)(x + 7)$ (d) $x = 4, -1, -7$

55. $\pm 1, \pm 3, \pm\frac{3}{2}, \pm\frac{3}{4}, \pm\frac{1}{2}, \pm\frac{1}{4}$

57. 2 or 0 positive real zeros
1 negative real zero

59. Answers will vary.
$x = \frac{3}{4}$

61. $x = -1, 2, 3$ **63.** $x = -1, \frac{3}{2}, 3, \frac{2}{3}$

65 and 67. Answers will vary. **69.** $x = 0, -3$

71. $x = 4, \dfrac{3 \pm \sqrt{15}i}{2}$;
$(x - 4)\left(x - \dfrac{3 + \sqrt{15}i}{2}\right)\left(x - \dfrac{3 - \sqrt{15}i}{2}\right)$

73. $x = 2, -\frac{3}{2}, 1 \pm i$;
$(x - 2)(2x + 3)(x - 1 + i)(x - 1 - i)$

75. $x = 0, -1, \pm\sqrt{5}i$; $x^2(x + 1)(x + \sqrt{5}i)(x - \sqrt{5}i)$

77. (a) $x = 2, 1 \pm i$
(b) $(x - 2)(x - 1 - i)(x - 1 + i)$ (c) $(2, 0)$

79. (a) $x = -6, -1, \frac{2}{3}$ (b) $-(x + 1)(x + 6)(3x - 2)$
(c) $(-6, 0), (-1, 0), (\frac{2}{3}, 0)$

81. (a) $\pm 3i, \pm 5i$
(b) $(x - 3i)(x + 3i)(x - 5i)(x + 5i)$ (c) None

83. $f(x) = x^3 - 5x^2 + 9x - 45$

85. $f(x) = x^3 + 6x^2 + 28x + 40$

87. (a) $(x^2 + 9)(x^2 - 2x - 1)$
(b) $(x^2 + 9)(x - 1 + \sqrt{2})(x - 1 - \sqrt{2})$
(c) $(x + 3i)(x - 3i)(x - 1 + \sqrt{2})(x - 1 - \sqrt{2})$

89. $x = -3, \pm 2i$

91. (a) Domain: all real numbers x except $x = -3$
(b) Not continuous
(c) Vertical asymptote: $x = -3$
Horizontal asymptote: $y = -1$

93. (a) Domain: all real numbers x except $x = 6, -3$
(b) Not continuous
(c) Vertical asymptotes: $x = 6, x = -3$
Horizontal asymptote: $y = 0$

95. (a) Domain: all real numbers x except $x = 7$
(b) Not continuous
(c) Vertical asymptote: $x = 7$
Horizontal asymptote: $y = -1$

97. (a) Domain: all real numbers x except $x = \pm\dfrac{\sqrt{6}}{2}$
(b) Not continuous
(c) Vertical asymptotes: $x = \pm\dfrac{\sqrt{6}}{2}$
Horizontal asymptote: $y = 2$

99. (a) Domain: all real numbers x except $x = 5, -3$
(b) Not continuous
(c) Vertical asymptote: $x = -3$
Horizontal asymptote: $y = 0$

101. (a) Domain: all real numbers x except $x = -3, 0, 8$
(b) Not continuous
(c) Vertical asymptotes: $x = -3, x = 0, x = 8$
Horizontal asymptote: $y = 0$

103. (a) \$176 million; \$528 million; \$1584 million
(b)
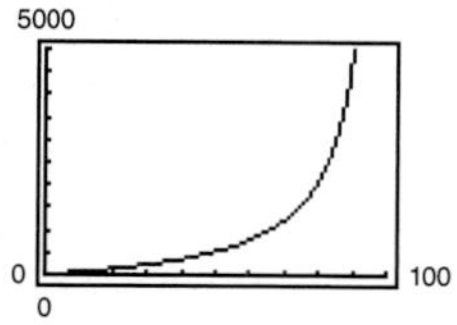

Answers will vary.
(c) No. As $p \to 100$, the cost approaches ∞.

105. Vertical asymptote: $x = -1$
Horizontal asymptote: $y = 1$
Hole at $x = 1$

107. Vertical asymptote: $x = -1$
Slant asymptote: $y = 3x + 2$

109.
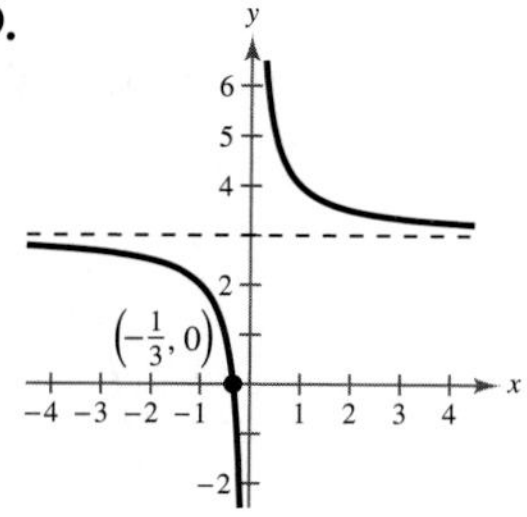

111.
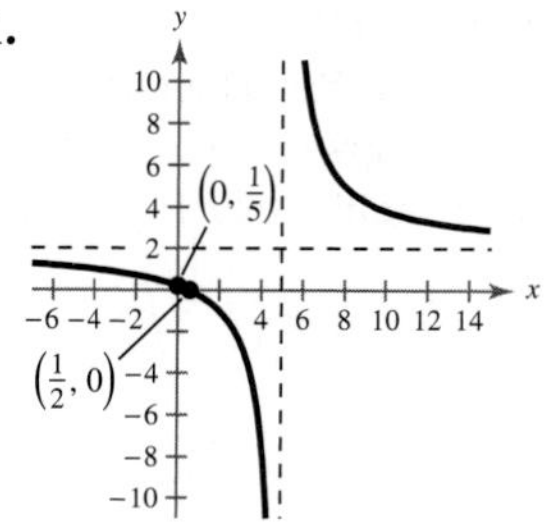

113.
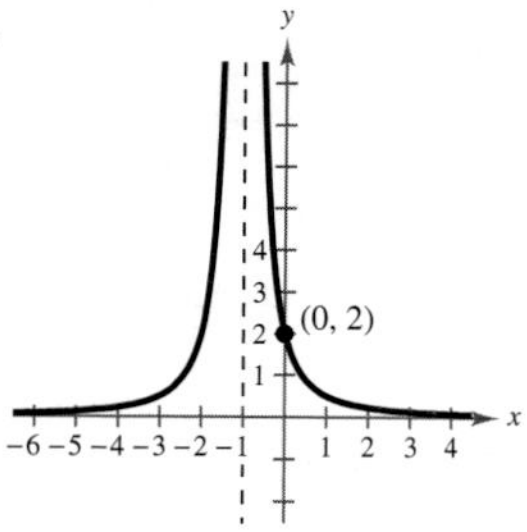

115.
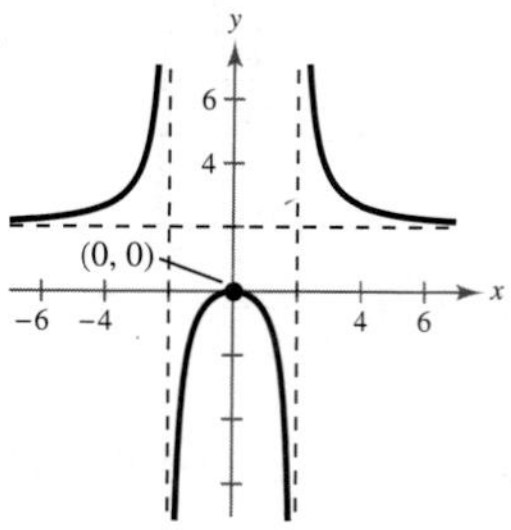

117.
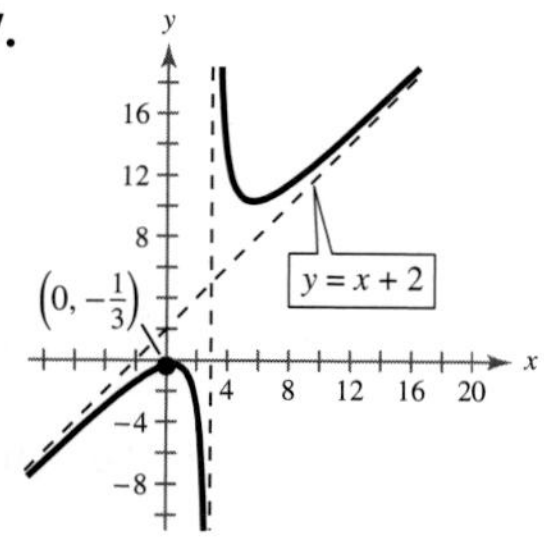

119. (a)
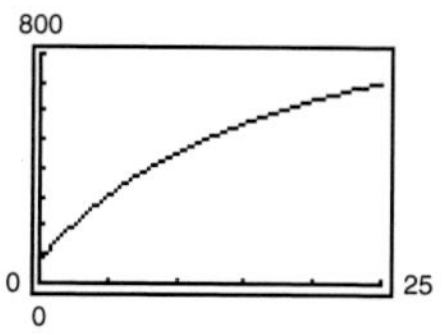

304,000; 453,333; 702,222
(b) 1,200,000, because N has a horizontal asymptote at $y = 1200$.

121. Quadratic

123. (a)

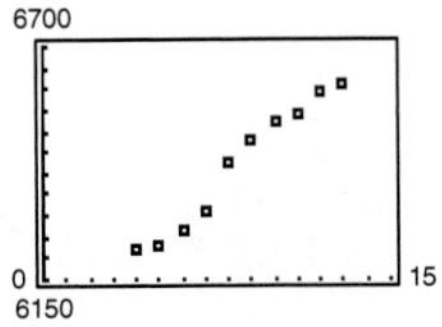

(b)

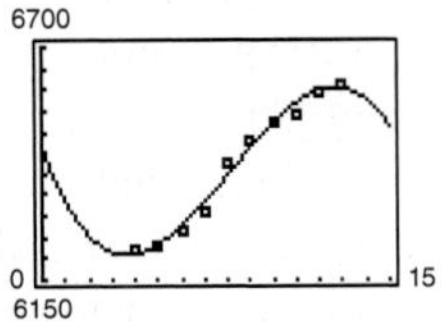

(c) $S = -1.13t^2 + 68.9t + 5926$

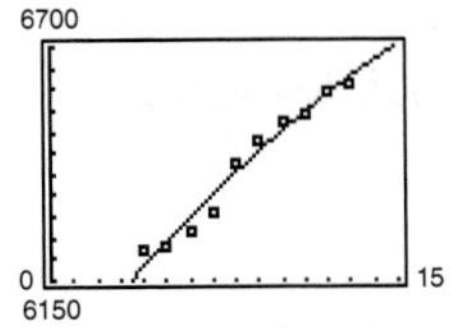

(d) Cubic; The cubic model more closely follows the pattern of the data.

(e) 6511 stations

125. False. Example: $(1 + 2i) + (1 - 2i) = 2$.

Chapter Test (page 320)

1. Vertex: $\left(-\frac{5}{2}, -\frac{1}{4}\right)$
Intercepts: $(0, 6)$, $(-3, 0)$, $(-2, 0)$

2. $y = (x - 3)^2 - 6$

3. (a) 50 ft
(b) 5; Yes, changing this value would act as a vertical shift upward or downward.

4. 0, multiplicity 1; $\frac{3}{2}$, multiplicity 2

5.

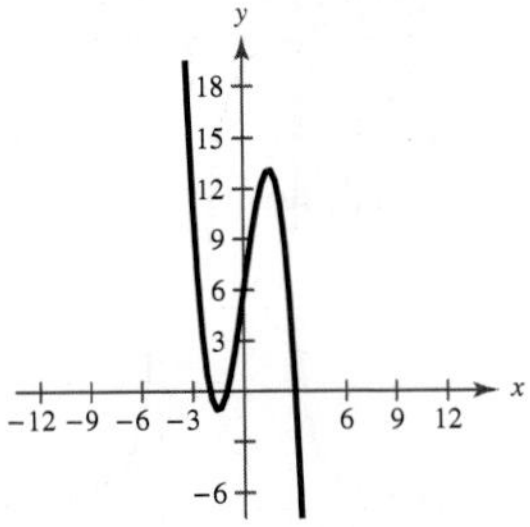

6. $2x + \dfrac{x - 3}{x^2 + 2}$ **7.** $2x^3 + 4x^2 + 3x + 6 + \dfrac{9}{x - 2}$

8. 13

9. $\pm1, \pm2, \pm3, \pm4, \pm6, \pm8, \pm12, \pm24, \pm\frac{1}{2}, \pm\frac{3}{2}$

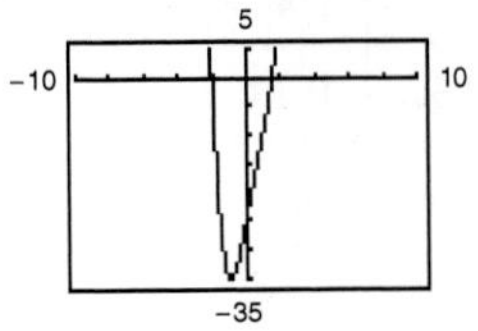

$t = -2, \frac{3}{2}$

10. $\pm1, \pm2, \pm\frac{1}{3}, \pm\frac{2}{3}$

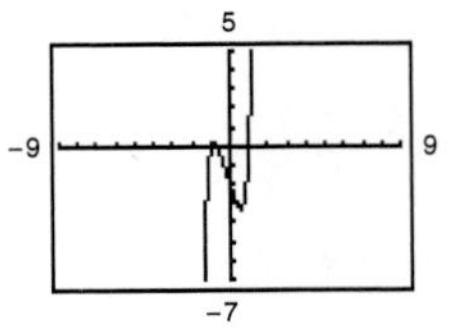

$\pm1, -\frac{2}{3}$

11. $x = -1, 4 \pm \sqrt{3}i$
$(x + 1)(x - 4 + \sqrt{3}i)(x - 4 - \sqrt{3}i)$

12. $x^4 - 6x^3 + 13x^2 - 10x$

13. $x^4 - 6x^3 + 16x^2 - 24x + 16$ **14.** $x^3 - 2x^2 + 2x$

15.

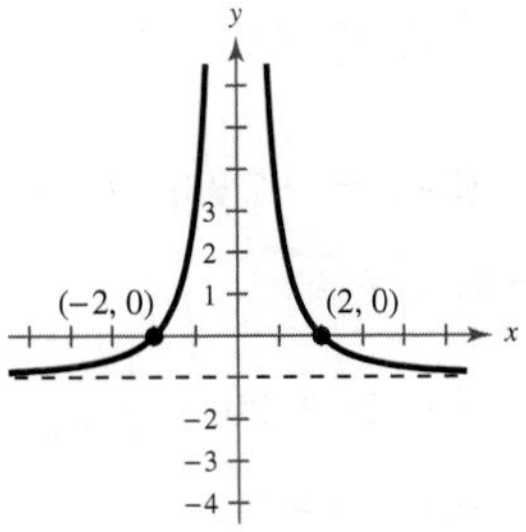

16.

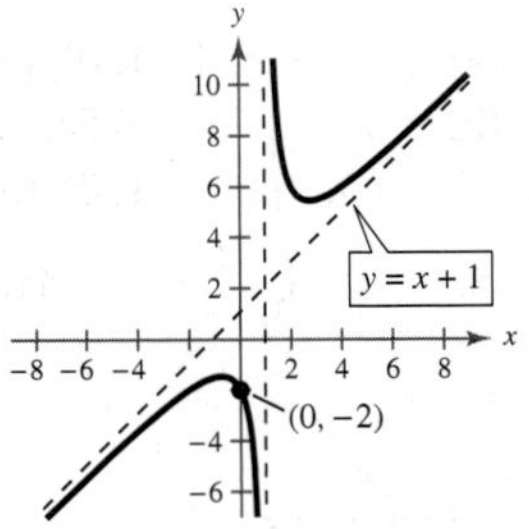

17.

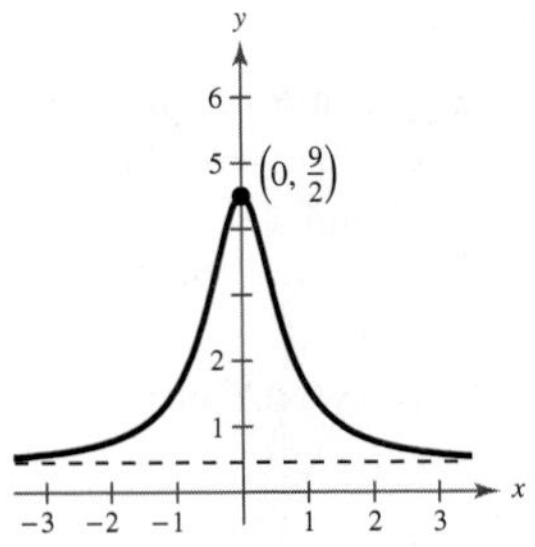

18. (a) and (c)

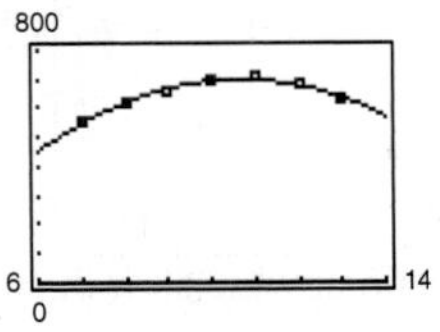

The model fits the data well.

(b) $A = -11.313t^2 + 241.07t - 585.7$

(d) \$484.9 billion, −\$289.5 billion

(e) No. By 2019, the model gives negative values.

Chapter 4

Section 4.1 (page 333)

1. transcendental **3.** Horizontal shift one unit to the left

5. 4112.033 **7.** 0.006

9.

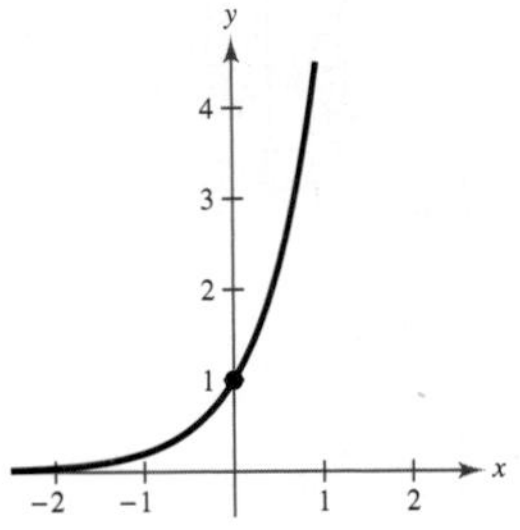

$y = 0$, $(0, 1)$, increasing

11.

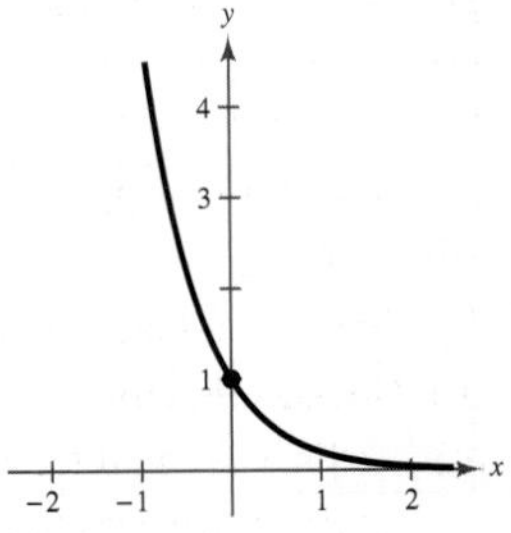

$y = 0$, $(0, 1)$, decreasing

13.

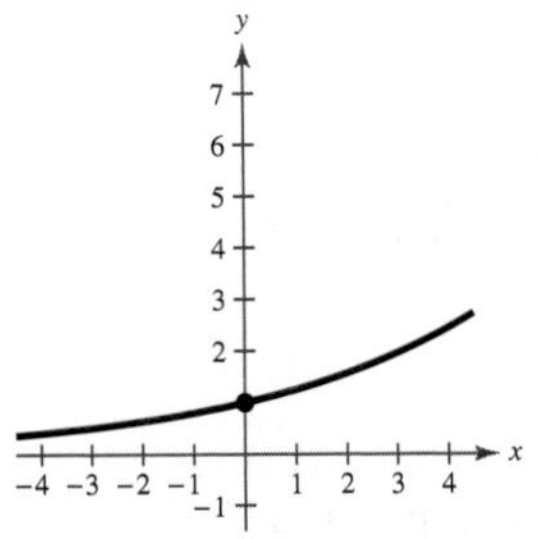

$y = 0$, $(0, 1)$, increasing

15.

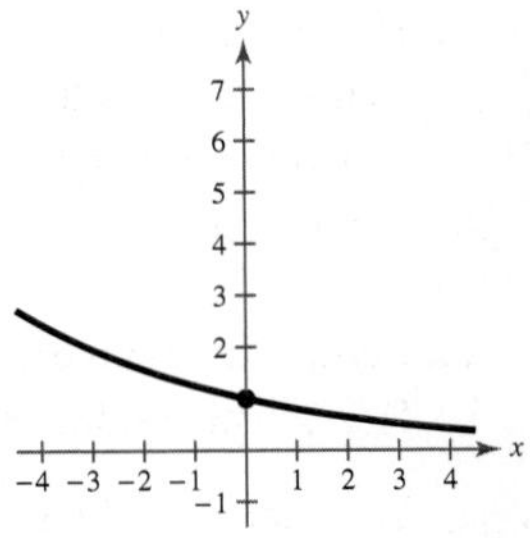

$y = 0$, $(0, 1)$, decreasing

17. d **18.** a **19.** c **20.** b

21. Right shift of five units

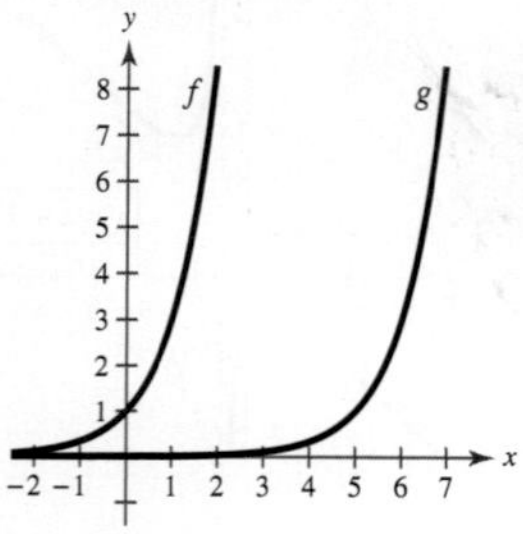

23. Upward shift of five units

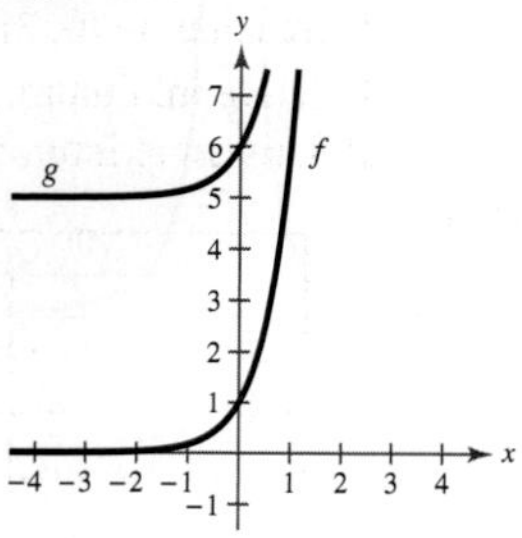

25. Left shift of four units and reflection in the x-axis

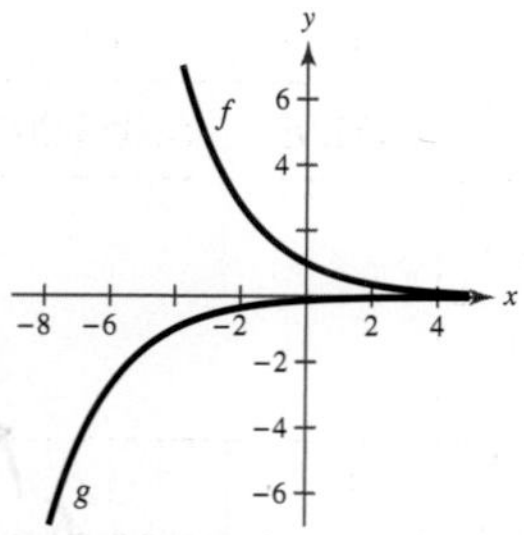

27. Right shift of two units and downward shift of three units

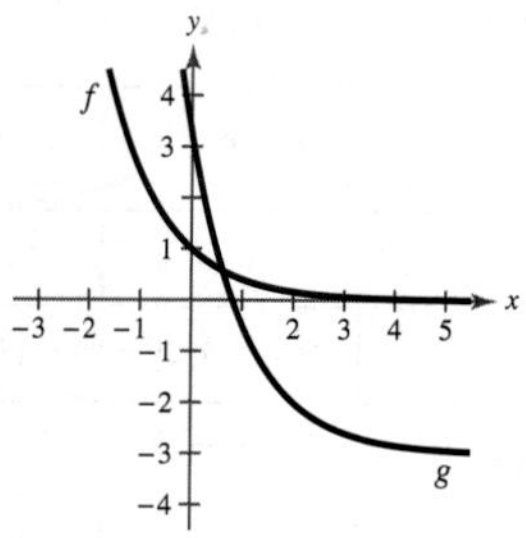

29. Reflection in the y-axis and upward shift of two units

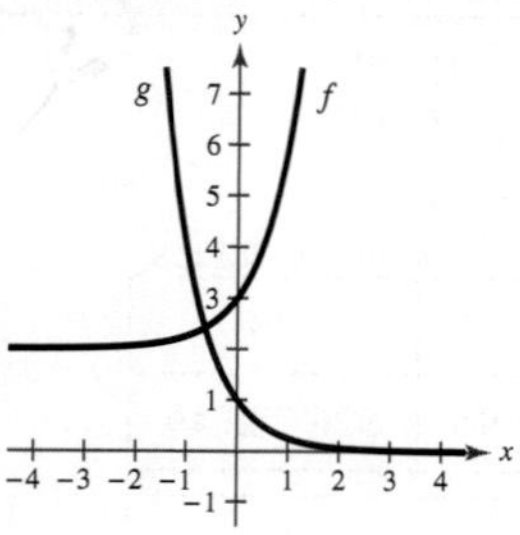

31. (a)

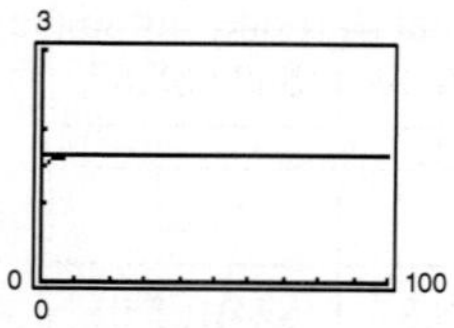

(b) $e^{0.5} \approx 1.6487$

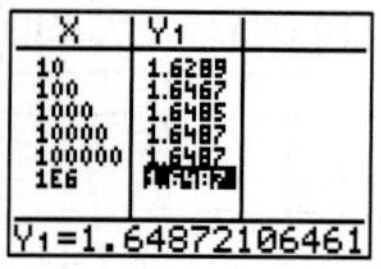

33. 9897.129 **35.** 54.164

37.

x	-2	-1	0	1	2
$f(x)$	0.16	0.4	1	2.5	6.25

Asymptote: $y = 0$

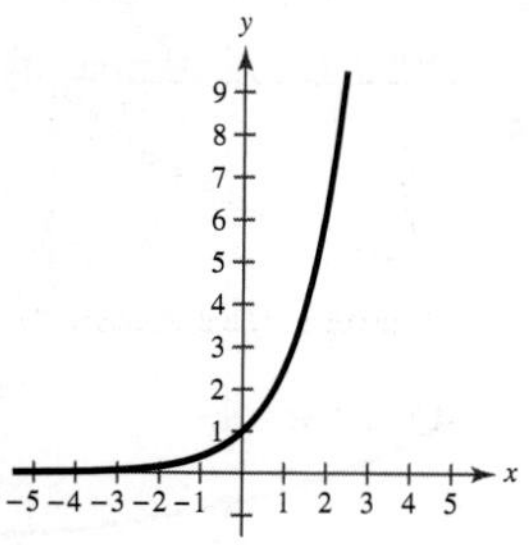

39.

x	-2	-1	0	1	2
$f(x)$	0.03	0.17	1	6	36

Asymptote: $y = 0$

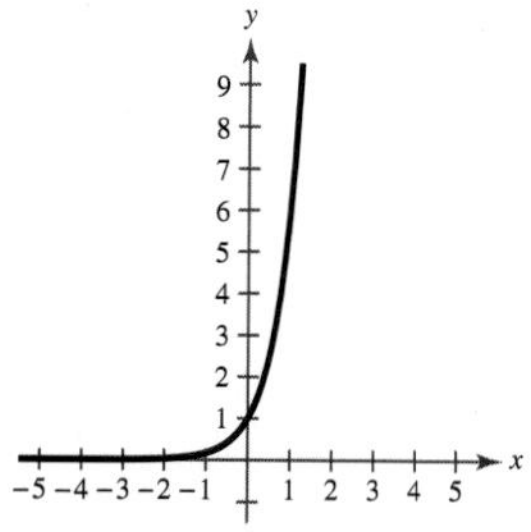

41.

x	-3	-2	-1	0	1
$f(x)$	0.33	1	3	9	27

Asymptote: $y = 0$

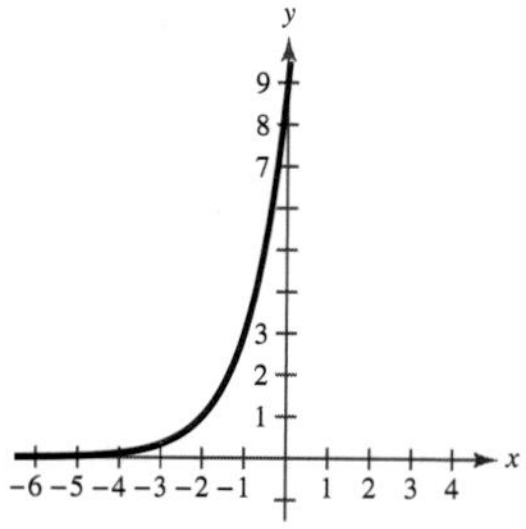

43.

x	-1	0	1	2	3	4
y	1.04	1.11	1.33	2	4	10

Asymptote: $y = 1$

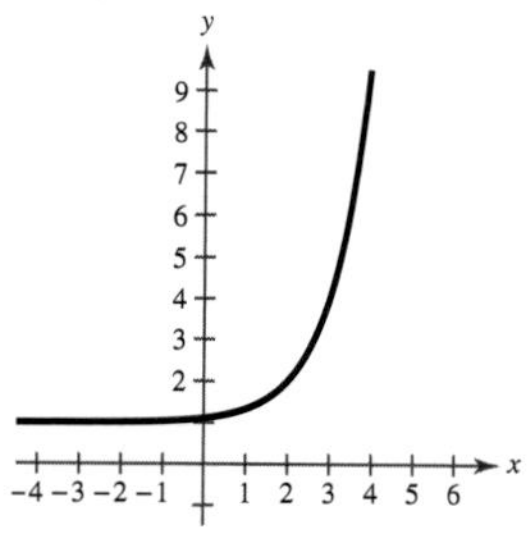

45.

x	-2	-1	0	1	2
$f(x)$	7.39	2.72	1	0.37	0.14

Asymptote: $y = 0$

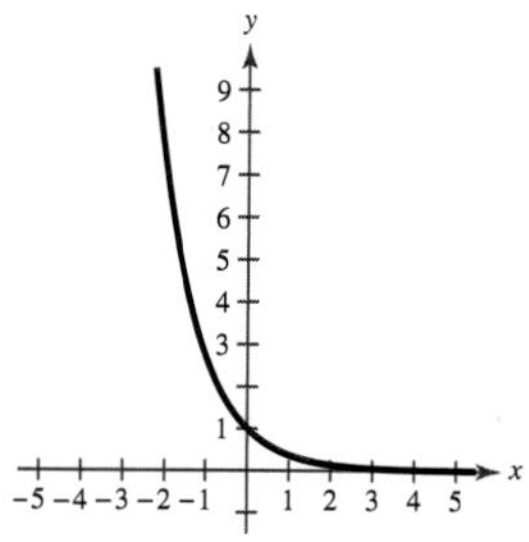

47.

x	-6	-5	-4	-3	-2
$f(x)$	0.41	1.10	3	8.15	22.17

Asymptote: $y = 0$

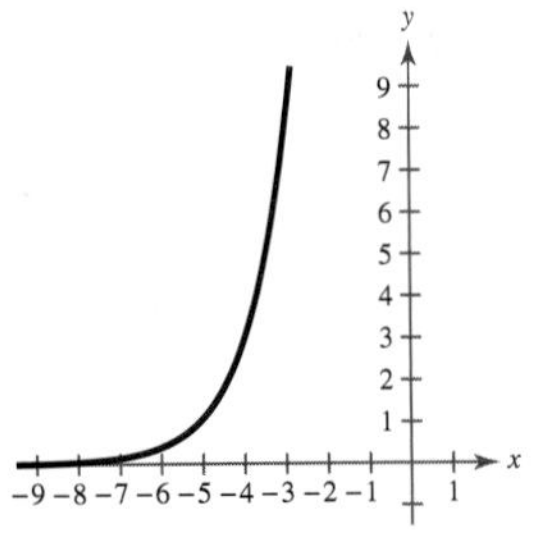

49.

x	3	4	5	6	7
$f(x)$	2.14	2.37	3	4.72	9.39

Asymptote: $y = 2$

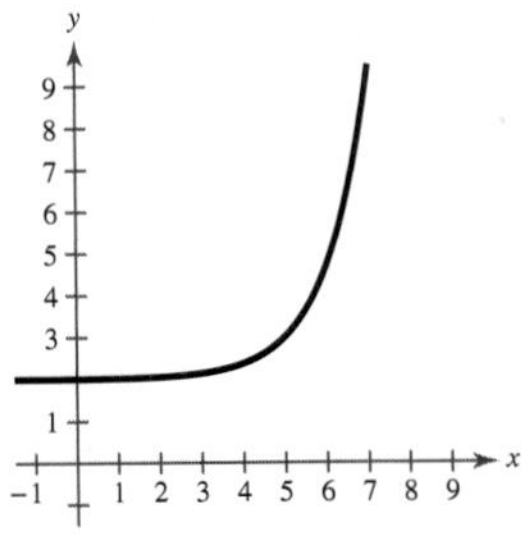

51.

t	-2	-1	0	1	2
$s(t)$	2.54	2.26	2	1.77	1.57

Asymptote: $y = 0$

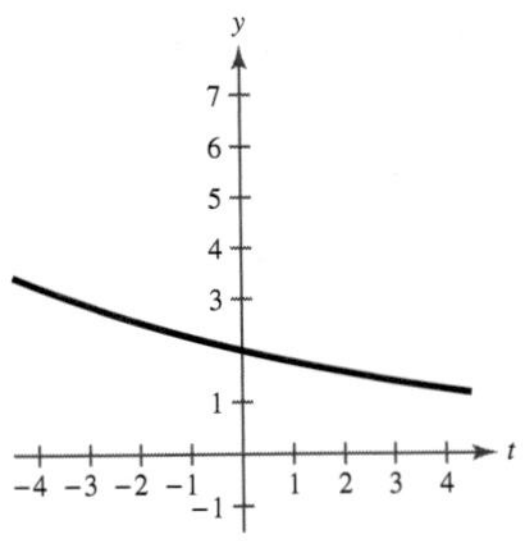

53. (a)

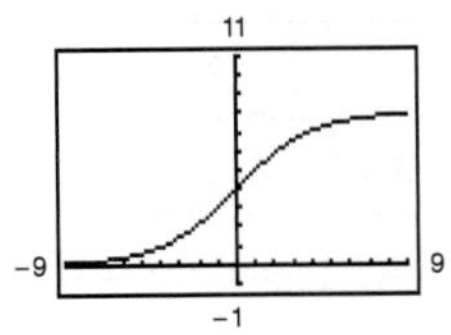

(b) $y = 0, y = 8$

55. (a)

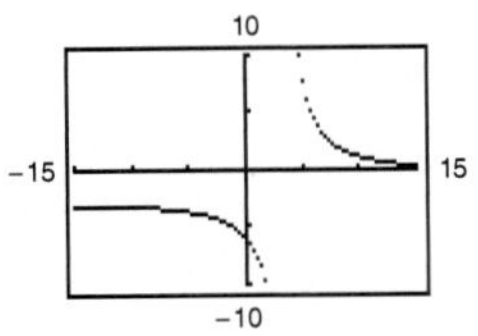

(b) $y = -3, y = 0, x \approx 3.47$

57. (86.350, 1500) **59.** $(-8.419, 0.2)$

61. (a)

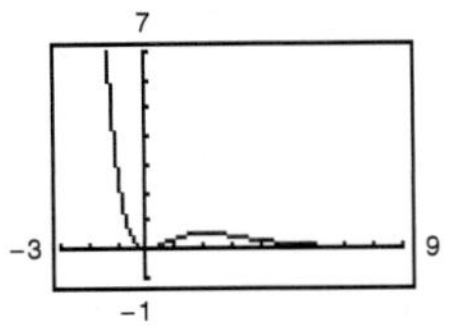

(b) Decreasing on $(-\infty, 0), (2, \infty)$
Increasing on $(0, 2)$

(c) Relative minimum: $(0, 0)$
Relative maximum: $(2, 0.54)$

63. (a)

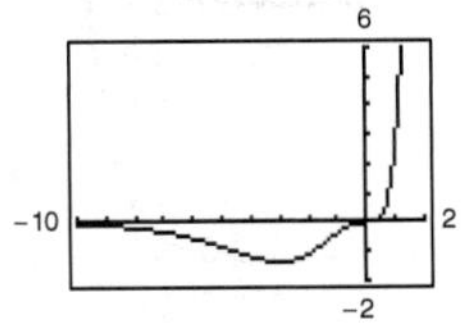

(b) Decreasing on $(-\infty, -3)$
Increasing on $(-3, \infty)$

(c) Relative minimum: $(-3, -1.344)$

65.

n	1	2	4	12
A	\$3047.49	\$3050.48	\$3051.99	\$3053.00

n	365	Continuous
A	\$3053.49	\$3053.51

67.

n	1	2	4	12
A	\$5477.81	\$5520.10	\$5541.79	\$5556.46

n	365	Continuous
A	\$5563.61	\$5563.85

69.

t	1	10	20
A	\$12,489.73	\$17,901.90	\$26,706.49

t	30	40	50
A	\$39,841.40	\$59,436.39	\$88,668.67

71.

t	1	10	20
A	\$12,427.44	\$17,028.81	\$24,165.03

t	30	40	50
A	\$34,291.81	\$48,662.40	\$69,055.23

73. \$1530.57 **75.** \$17,281.77

77. (a) $y_1 = 500(1 + 0.07)^t$

$y_2 = 500\left(1 + \frac{0.07}{4}\right)^{4t}$

$y_3 = 500e^{0.07t}$

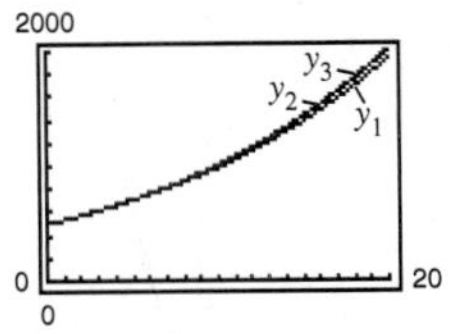

(b) y_3 yields the highest return after 20 years.
$y_2 - y_1 = \$68.36$
$y_3 - y_2 = \$24.40$
$y_3 - y_1 = \$92.76$

79. (a) 10 g (b) 7.84 g

(c)

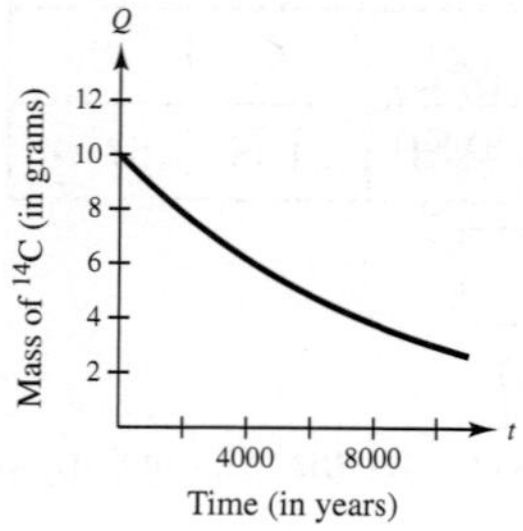

81. (a)

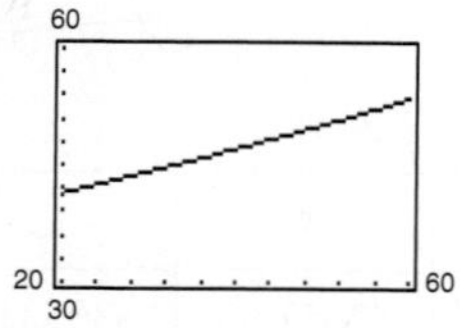

(b)

t	20	21	22	23	24	25
P	41.35	41.62	41.89	42.16	42.44	42.71

t	26	27	28	29	30	31
P	42.99	43.27	43.56	43.84	44.13	44.41

t	32	33	34	35	36	37
P	44.70	44.99	45.29	45.58	45.88	46.18

t	38	39	40	41	42	43
P	46.48	46.78	47.09	47.40	47.71	48.02

t	44	45	46	47	48	49
P	48.33	48.64	48.96	49.28	49.60	49.93

t	50	51	52	53	54	55
P	50.25	50.58	50.91	51.24	51.58	51.91

t	56	57	58	59	60
P	52.25	52.59	52.93	53.28	53.63

(c) 2053

83. True. The definition of an exponential function is $f(x) = a^x$, $a > 0, a \neq 1$.

85. d

87.

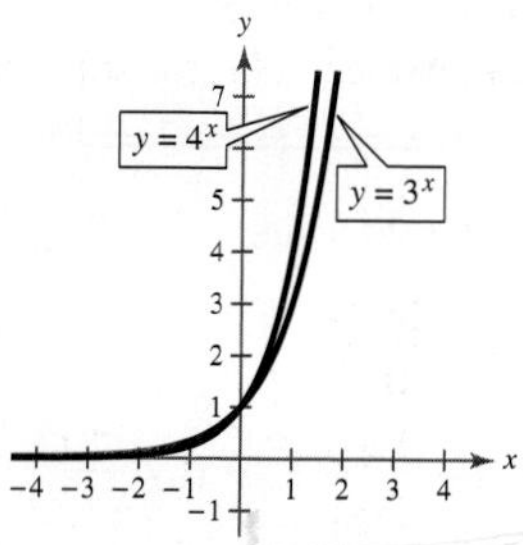

$3^x < 4^x$ when $x > 0$

89. $>$ **91.** $>$ **93.** $f^{-1}(x) = \dfrac{x + 7}{5}$

95. $f^{-1}(x) = x^3 - 8$ **97.** Answers will vary.

Section 4.2 (page 343)

1. logarithmic function **3.** $a^{\log_a x} = x$ **5.** $b = a^c$

7. $4^3 = 64$ **9.** $7^{-2} = \frac{1}{49}$ **11.** $32^{2/5} = 4$

13. $2^{1/2} = \sqrt{2}$ **15.** $\log_5 125 = 3$ **17.** $\log_{81} 3 = \frac{1}{4}$

19. $\log_6 \frac{1}{36} = -2$ **21.** $\log_g 4 = a$ **23.** 4 **25.** -3

27. 2.538 **29.** 7.022 **31.** 9 **33.** 2 **35.** $\frac{1}{10}$

37. $3x$ **39.** -3

41.

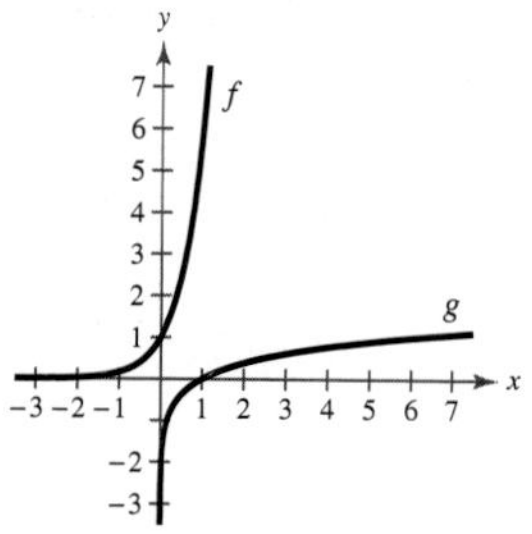

43.

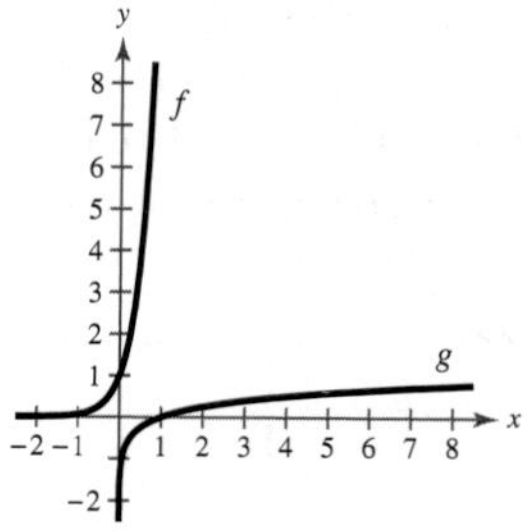

45.

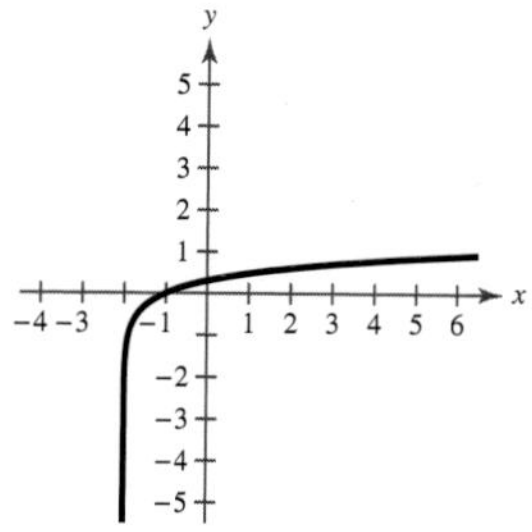

Domain: $(-2, \infty)$
Vertical asymptote: $x = -2$
x-intercept: $(-1, 0)$

47.

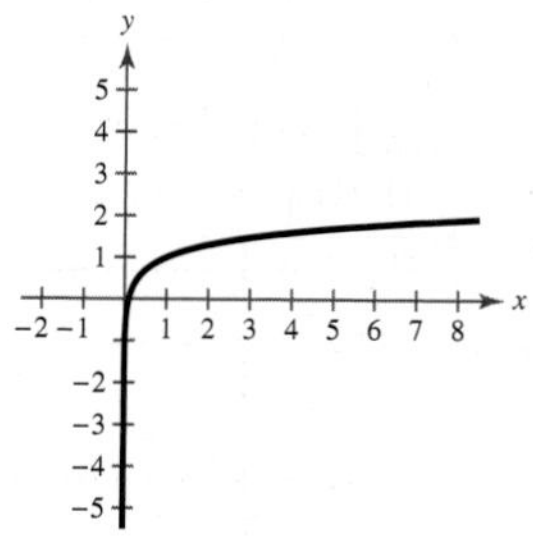

Domain: $(0, \infty)$
Vertical asymptote: $x = 0$
x-intercept: $\left(\frac{1}{10}, 0\right)$

49.

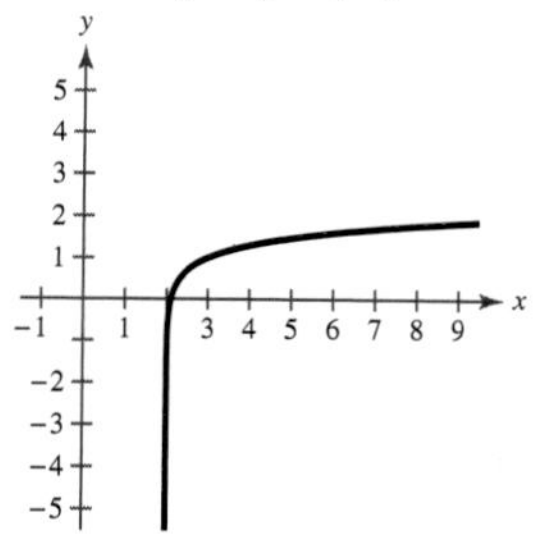

Domain: $(2, \infty)$
Vertical asymptote: $x = 2$
x-intercept: $\left(\frac{21}{10}, 0\right)$

51. b **52.** c **53.** d **54.** a

55. Reflection in the y-axis

57. Reflection in the x-axis, vertical shift four units upward

59. Horizontal shift three units to the right and vertical shift two units downward

61. $e^{1.7917\ldots} = 6$ **63.** $e^1 = e$ **65.** $e^{1/2} = \sqrt{e}$

67. $e^{2.1972\ldots} = 9$ **69.** $\ln 20.0855\ldots = 3$

71. $\ln 3.6692\ldots = 1.3$ **73.** $\ln 1.3956\ldots = \frac{1}{3}$

75. $\ln 4.4816\ldots = \frac{3}{2}$ **77.** 2.398 **79.** 0.693

81. 2 **83.** 1.8 **85.** 0 **87.** 1

89. 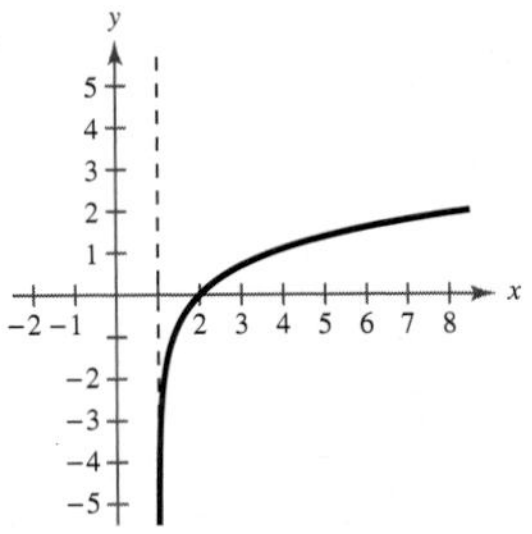

Domain: $(1, \infty)$
Vertical asymptote: $x = 1$
x-intercept: $(2, 0)$

91. 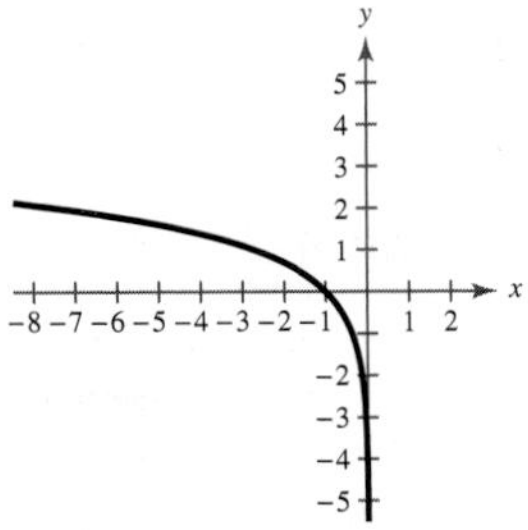

Domain: $(-\infty, 0)$
Vertical asymptote: $x = 0$
x-intercept: $(-1, 0)$

93. Horizontal shift eight units to the left

95. Vertical shift five units downward

97. Horizontal shift one unit to the right and vertical shift two units upward

99. (a) 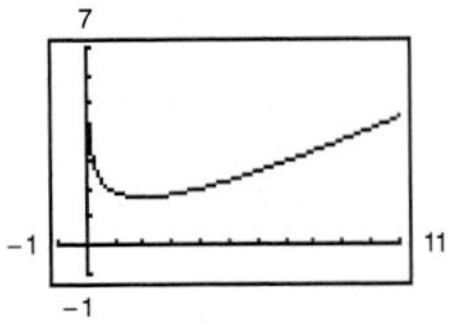
(b) Domain: $(0, \infty)$
(c) Decreasing on $(0, 2)$; increasing on $(2, \infty)$
(d) Relative minimum: $(2, 1.693)$

101. (a) 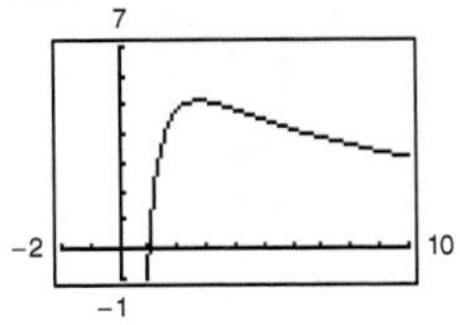
(b) Domain: $(0, \infty)$
(c) Decreasing on $(2.718, \infty)$; increasing on $(0, 2.718)$
(d) Relative maximum: $(2.718, 5.150)$

103. (a) 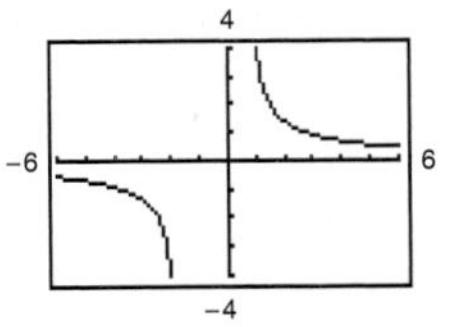
(b) Domain: $(-\infty, -2), (1, \infty)$
(c) Decreasing on $(-\infty, -2)$; $(1, \infty)$
(d) No relative maxima or minima

105. (a) 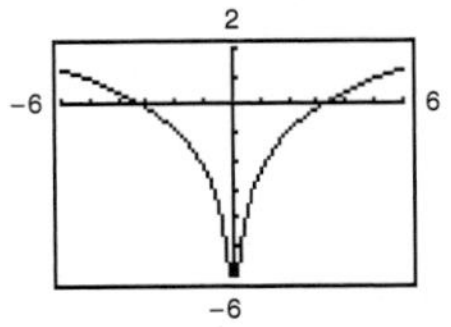
(b) Domain: $(-\infty, 0), (0, \infty)$
(c) Decreasing on $(-\infty, 0)$; increasing on $(0, \infty)$
(d) No relative maxima or minima

107. (a) 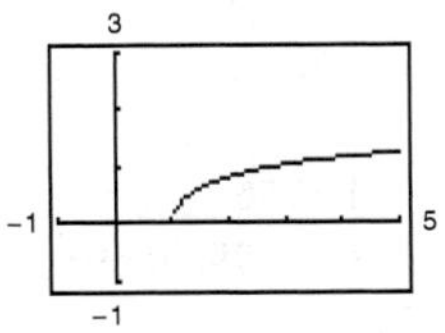
(b) Domain: $[1, \infty)$
(c) Increasing on $(1, \infty)$
(d) Relative minimum: $(1, 0)$

109. (a) 80 (b) about 71.89 (c) about 61.65

111. (a)

K	1	2	4	6	8
t	0	19.80	39.61	51.19	59.41

K	10	12
t	65.79	71.00

It takes about 19.80 years for the principal to double.

(b) 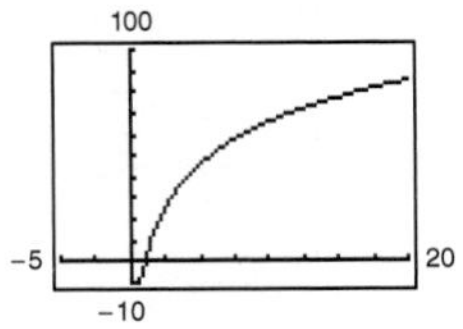

113. (a) 30 yr, 10 yr
(b) Total amount: \$323,179.20; interest: \$173,179.20
Total amount: \$199,108.80; interest: \$49,108.80

115. False. Reflect $g(x)$ about the line $y = x$.

117. 2 **119.** $\frac{1}{9}$ **121.** b

123. $\log_a x$ is the inverse of a^x only if $0 < a < 1$ and $a > 1$, so $\log_a x$ is defined only for $0 < a < 1$ and $a > 1$.

125. (a) 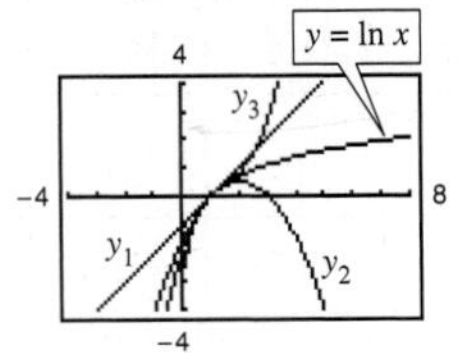

(b) 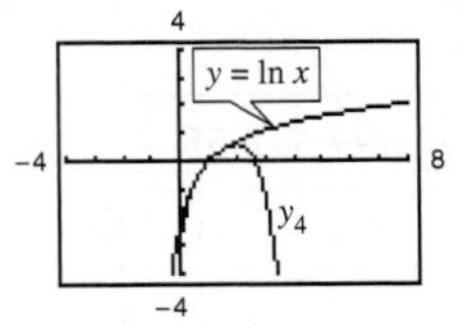

$y_4 = (x - 1) - \frac{1}{2}(x - 1)^2 + \frac{1}{3}(x - 1)^3 - \frac{1}{4}(x - 1)^4$
Answers will vary.

127. (a)

x	1	5	10	10^2
$f(x)$	0	0.32	0.23	0.046

x	10^4	10^6
$f(x)$	0.00092	0.0000138

(b) 0

129. $(x + 3)(x - 1)$ **131.** $(4x + 3)(3x - 1)$

133. $(4x + 5)(4x - 5)$ **135.** $x(2x - 9)(x + 5)$

137. 15 **139.** The graphs do not intersect. No real solution.

141. 27.67

Section 4.3 (page 351)

1. change-of-base **3.** $\log_3 24 = \dfrac{\ln 24}{\ln 3}$

5. (a) $\dfrac{\log_{10} x}{\log_{10} 5}$ (b) $\dfrac{\ln x}{\ln 5}$ **7.** (a) $\dfrac{\log_{10} x}{\log_{10} \frac{1}{6}}$ (b) $\dfrac{\ln x}{\ln \frac{1}{6}}$

9. (a) $\dfrac{\log_{10} \frac{3}{10}}{\log_{10} a}$ (b) $\dfrac{\ln \frac{3}{10}}{\ln a}$ **11.** (a) $\dfrac{\log_{10} x}{\log_{10} 2.6}$ (b) $\dfrac{\ln x}{\ln 2.6}$

13. 1.771 **15.** -4 **17.** -0.059 **19.** 2.691
21. $\ln 5 + \ln 4$ **23.** $2 \ln 5 - \ln 4$
25. 1.0686 **27.** -0.2294

29.

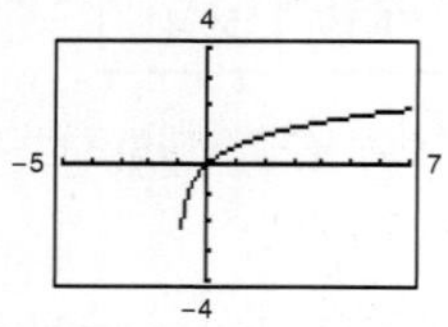

31.

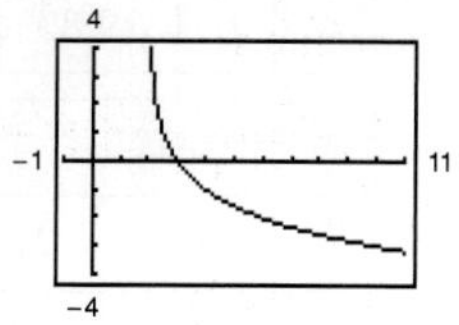

33.

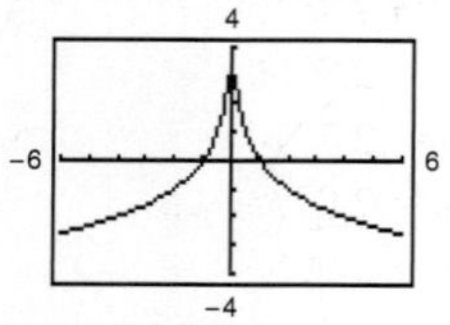

35.

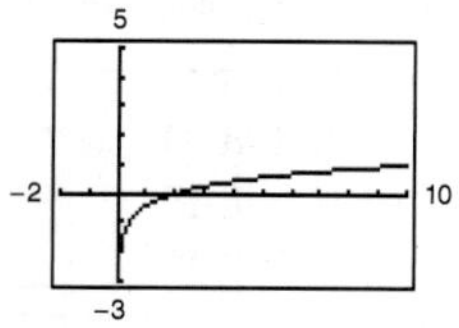

37. $\frac{3}{2}$ **39.** $4 + 4 \log_2 3$ **41.** $6 + \ln 5$ **43.** $\ln 6 - 2$
45. Answers will vary. **47.** $1 + \log_{10} x$
49. $\log_{10} t - \log_{10} 8$ **51.** $4 \log_8 x$ **53.** $\frac{1}{2} \ln z$
55. $\ln x + \ln y + \ln z$ **57.** $\log_6 a + 3 \log_6 b - 2 \log_6 c$
59. $\frac{4}{3} \ln x - \frac{1}{3} \ln y$
61. $\ln(x + 1) + \ln(x - 1) - 3 \ln x, x > 1$
63. $4 \log_b x + \frac{1}{2} \log_b y - 5 \log_b z$
65. (a)

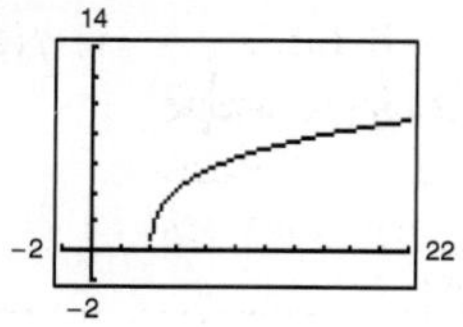

(b)

x	4	5	6	7	8	9	10
y_1	Error	3.22	4.28	4.99	5.55	6.00	6.40
y_2	Error	3.22	4.28	4.99	5.55	6.00	6.40

(c) $y_1 = y_2$

67. (a)

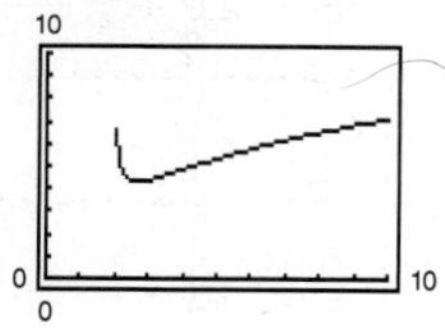

(b)

x	0	3	4	5	6	7	8
y_1	Error	4.39	4.85	5.34	5.78	6.17	6.53
y_2	Error	4.39	4.85	5.34	5.78	6.17	6.53

(c) $y_1 = y_2$

69. $\ln 4x$ **71.** $\log_4 \frac{z}{y}$ **73.** $\log_3(x + 2)^4$

75. $\ln\left(x\sqrt{x^2 + 4}\right)$ **77.** $\ln \frac{x}{(x + 1)^3}$ **79.** $\ln \frac{2(x - 2)}{y^3}$

81. $\ln \frac{x}{(x^2 - 4)^2}$ **83.** $\ln \sqrt[3]{\frac{x(x + 3)^2}{x^2 - 1}}$

85. (a)

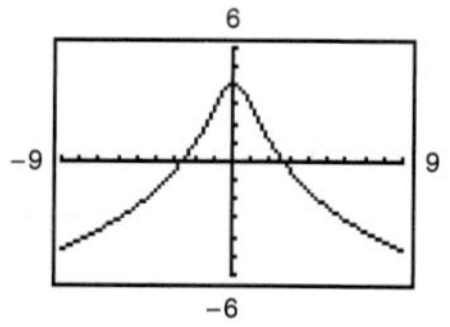

(b)

x	-5	-4	-3	-2	-1
y_1	-2.36	-1.51	-0.45	0.94	2.77
y_2	-2.36	-1.51	-0.45	0.94	2.77

x	0	1	2	3	4	5
y_1	4.16	2.77	0.94	-0.45	-1.51	-2.36
y_2	4.16	2.77	0.94	-0.45	-1.51	-2.36

(c) $y_1 = y_2$

87. (a)

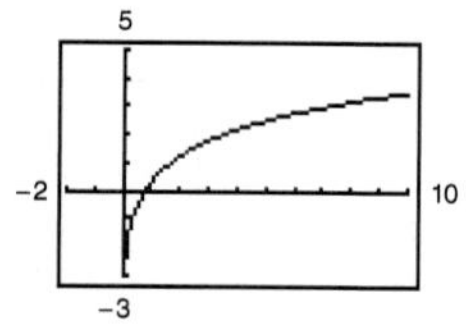

(b)

x	-1	0	1	2
y_1	Error	Error	0.35	1.24
y_2	Error	Error	0.35	1.24

x	3	4	5
y_1	1.79	2.19	2.51
y_2	1.79	2.19	2.51

(c) $y_1 = y_2$

89. 2 **91.** 6.8
93. Not possible; -4 is not in the domain of $\log_2 x$.
95. 3 **97.** -4 **99.** 8 **101.** $-\frac{1}{2}$
103. (a)

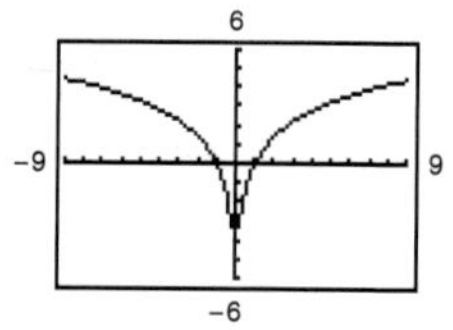

(b)

x	-5	-4	-3	-2	-1
y_1	3.22	2.77	2.20	1.39	0
y_2	Error	Error	Error	Error	Error

x	0	1	2	3	4	5
y_1	Error	0	1.39	2.20	2.77	3.22
y_2	Error	0	1.39	2.20	2.77	3.22

(c) No. The domains differ.

105. (a)

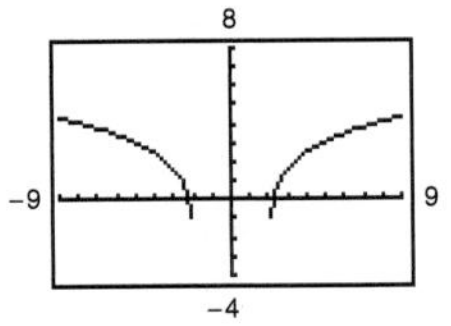

(b)

x	-4	-3	0	3
y_1	Error	Error	Error	1.61
y_2	2.48	1.61	Error	1.61

x	4	5	6
y_1	2.48	3.04	3.47
y_2	2.48	3.04	3.47

(c) No. The domains differ.

107. (a) $\beta = 120 + 10 \log_{10} I$

(b)

I	10^{-4}	10^{-6}	10^{-8}	10^{-10}	10^{-12}	10^{-14}
β	80	60	40	20	0	-20

109. (a)

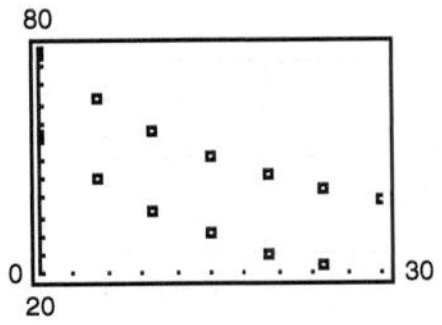

(b) $T = 54.4(0.964)^t + 21$

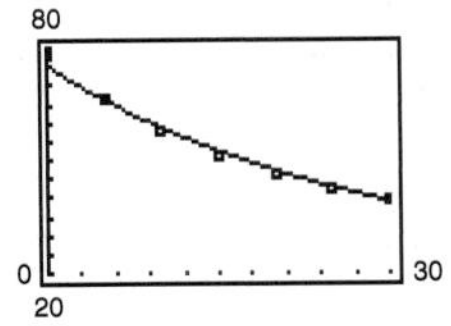

The model fits the data.

(c)

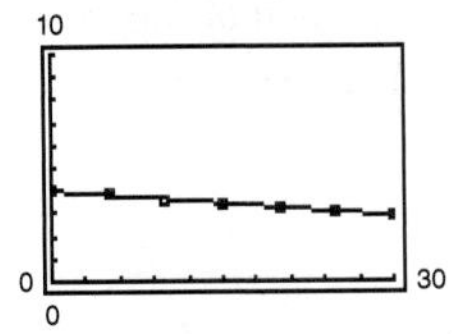

$\ln(T - 21) = -0.037t + 3.997$

$T = e^{(-0.037t + 3.997)} + 21$

(d)

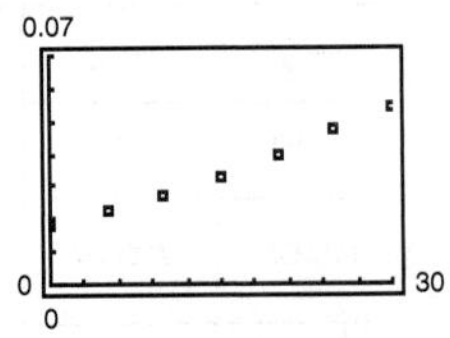

$$\frac{1}{T - 21} = 0.0012t + 0.0162$$

$$T = \frac{1}{0.0012t + 0.0162} + 21$$

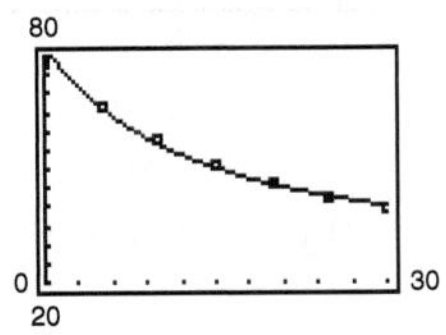

111. True **113.** False. $f(\sqrt{x}) = \frac{1}{2}f(x)$

115. True. $\ln x < 0$ when $0 < x < 1$.

117. The error is an improper use of the Quotient Property of logarithms.

$$\ln \frac{x^2}{\sqrt{x^2 + 4}} = \ln x^2 - \ln \sqrt{x^2 + 4} = 2 \ln x - \frac{1}{2} \ln(x^2 + 4)$$

119. $\ln 1 = 0$ $\quad \ln 9 \approx 2.1972$

$\ln 2 \approx 0.6931$ $\quad \ln 10 \approx 2.3025$

$\ln 3 \approx 1.0986$ $\quad \ln 12 \approx 2.4848$

$\ln 4 \approx 1.3862$ $\quad \ln 15 \approx 2.7080$

$\ln 5 \approx 1.6094$ $\quad \ln 16 \approx 2.7724$

$\ln 6 \approx 1.7917$ $\quad \ln 18 \approx 2.8903$

$\ln 8 \approx 2.0793$ $\quad \ln 20 \approx 2.9956$

121. No.

Domain of y_1: $(-\infty, 0), (2, \infty)$

Domain of y_2: $(2, \infty)$

123. $\frac{x^3}{64y^4}, x \neq 0$ **125.** $\frac{3x^4}{2y^3}, x \neq 0$

Section 4.4 (page 361)

1. (a) $x = y$ (b) $x = y$ (c) x (d) x **3.** 7

5. Subtract 3 from both sides. **7.** (a) Yes (b) No

9. (a) No (b) Yes (c) Yes, approximate

11. (a) Yes, approximate (b) No (c) Yes

13. (a) Yes (b) Yes, approximate (c) No

15.

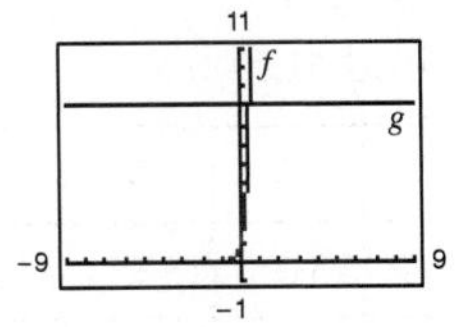

$\left(\frac{1}{2}, 8\right); \frac{1}{2}$

17.

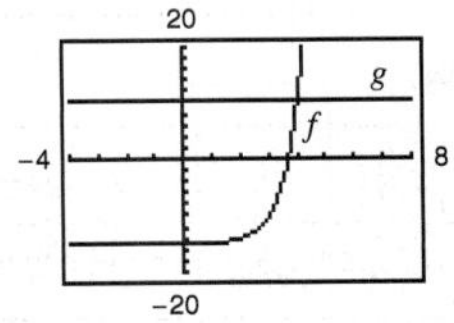

$(4, 10); 4$

19.

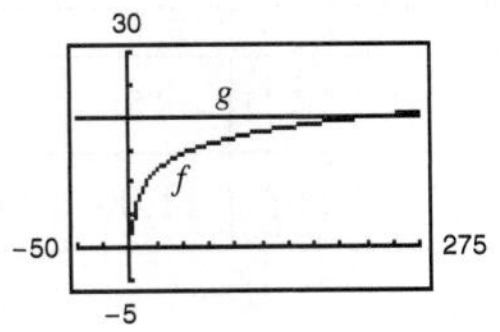

$(243, 20); 243$

21.

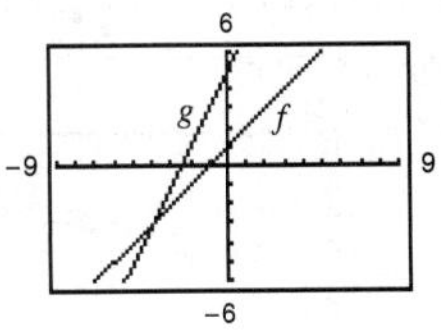

$(-4, -3); -4$

23. 2 **25.** -4 **27.** -2 **29.** -4 **31.** $\ln 14$

33. $\log_{10} 36$ **35.** 5 **37.** 5 **39.** e^{-9} **41.** 5

43. $\frac{1}{11}$ **45.** $\frac{e^5 + 1}{2}$ **47.** x^2 **49.** $x^2 - 3$

51. $2x - 1$ **53.** $-x^2 + 4$ **55.** 0.944 **57.** 2

59. 184.444 **61.** 0.511 **63.** 1.099 **65.** -1.498

67. 6.960 **69.** -277.951 **71.** 1.609 **73.** $-1, 2$

75. 0.586, 3.414 **77.** 1.946 **79.** 183.258

81. (a)

x	0.6	0.7	0.8	0.9	1.0
e^{3x}	6.05	8.17	11.02	14.88	20.09

$(0.8, 0.9)$

(b)

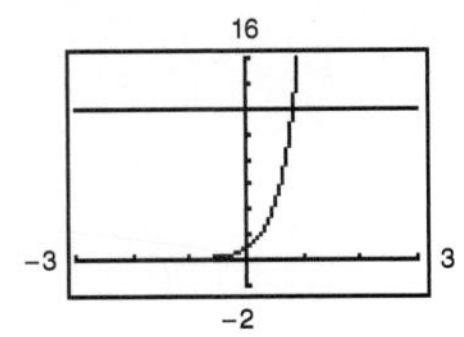

(c) 0.828

CHAPTER 4

83. (a)

x	5	6	7	8	9
$20(100 - e^{x/2})$	1756	1598	1338	908	200

(8, 9)

(b)

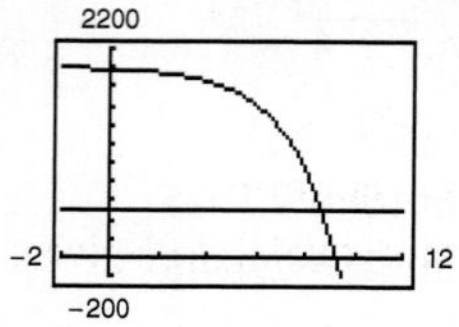

(c) 8.635

85. 21.330 **87.** 2.720

89.

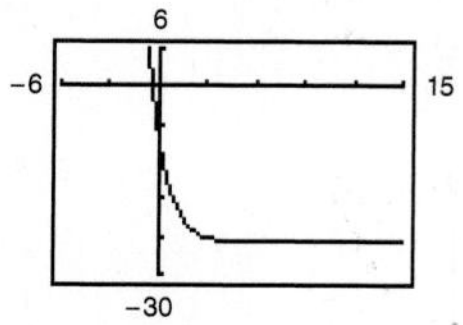

$x = -0.427$

91.

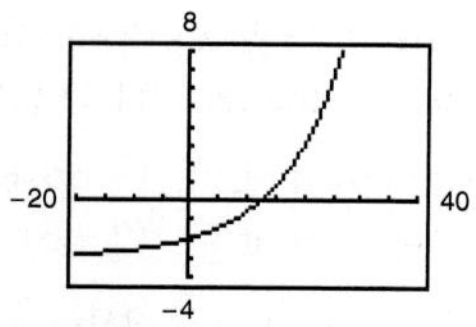

$x = 12.207$

93. 0.050 **95.** 2.042 **97.** $-\frac{1}{2}$ **99.** 4453.242

101. 17.945 **103.** ± 9 **105.** 5.389

107. 1.718, -3.718 **109.** 2 **111.** 3 **113.** 180.384

115. (a)

x	2	3	4	5	6
$\ln 2x$	1.39	1.79	2.08	2.30	2.48

(5, 6)

(b)

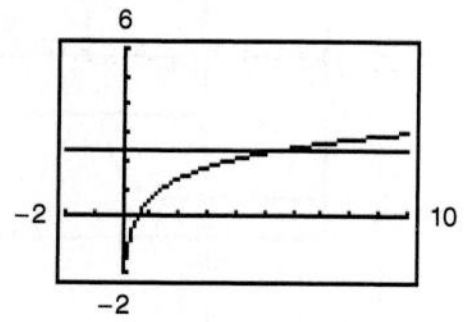

5.512

(c) 5.512

117. (a)

x	12	13	14	15	16
$6 \log_3(0.5x)$	9.79	10.22	10.63	11.00	11.36

(14, 15)

(b)

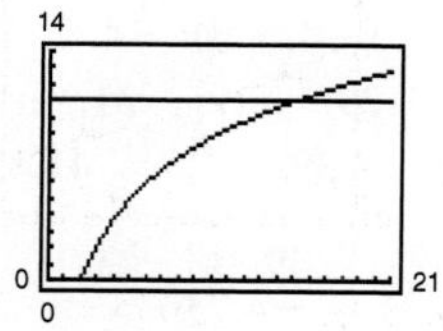

14.988

(c) 14.988

119. 1.469, 0.001 **121.** 3.398 **123.** -1.993, 3.738

125.

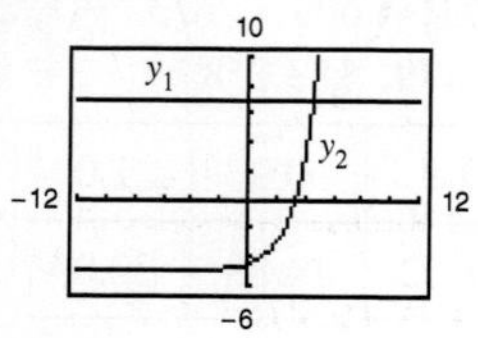

(4.585, 7)

127.

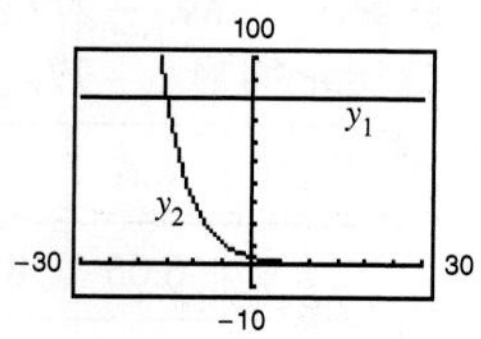

(-14.979, 80)

129.

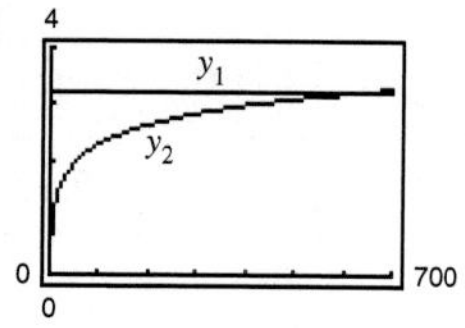

(663.142, 3.25)

131. $-1, 0$ **133.** 1 **135.** $e^{-1/2} \approx 0.607$

137. $e^{-1} \approx 0.368$ **139.** $\dfrac{\ln y - \ln a}{b}$

141. $b \pm \sqrt{c(\ln y - \ln a)}$ **143.** (a) 9.90 yr (b) 15.69 yr

145. (a) 27.73 yr (b) 43.94 yr

147. (a) 93 months (b) 124 months **149.** 2011

151. (a)

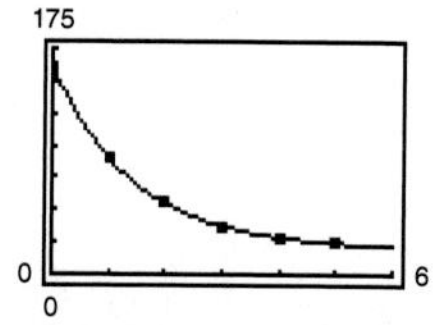

(b) $y = 20$. The object's temperature cannot cool below the room's temperature.

(c) 0.81 h (d) About 1.22 h

153. False; $e^x = 0$ has no solutions.

155. The error is that both sides of the equation should be divided by 2 before taking the natural log of both sides.

$$2e^x = 10$$
$$e^x = 5$$
$$\ln e^x = \ln 5$$
$$x = \ln 5$$

157. (a)

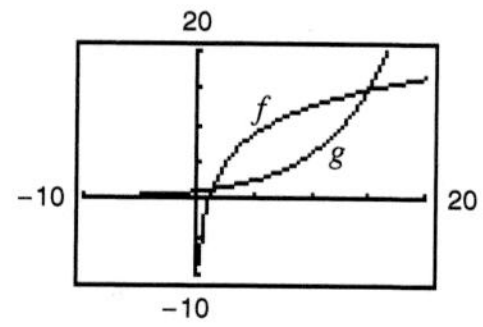

(1.258, 1.258), (14.767, 14.767)

(b) $a = e^{(1/e)} \approx 1.445$ (c) $1 < a < e^{(1/e)}$

159.

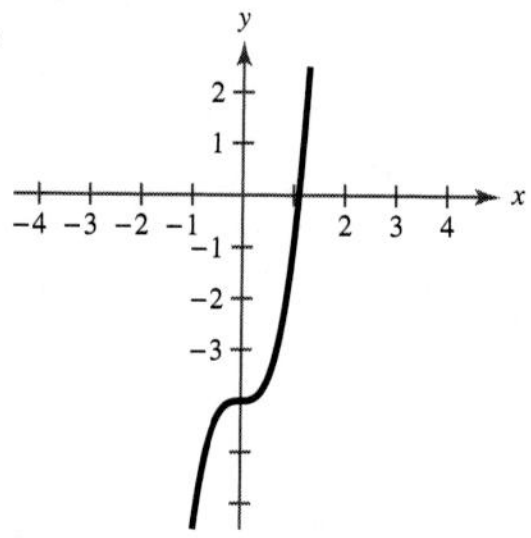

161.

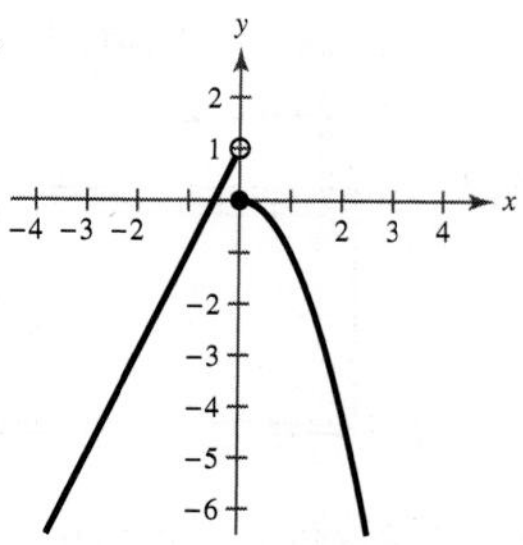

Section 4.5 (page 372)

1. (a) iv (b) i (c) iii (d) vi (e) ii (f) v

3. sigmoidal **5.** Exponential decay

7. c **8.** e **9.** b **10.** a **11.** d **12.** f

	Initial Investment	*Annual % Rate*	*Time to Double*	*Amount After 10 Years*
13.	$10,000	4%	17.3 yr	$14,918.25
15.	$7500	3.30%	21 yr	$10,432.26
17.	$5000	1.25%	55.45 yr	$5665.74
19.	$63,762.82	4.5%	15.40 yr	$100,000.00

21.

r	2%	4%	6%	8%	10%	12%
t	54.93	27.47	18.31	13.73	10.99	9.16

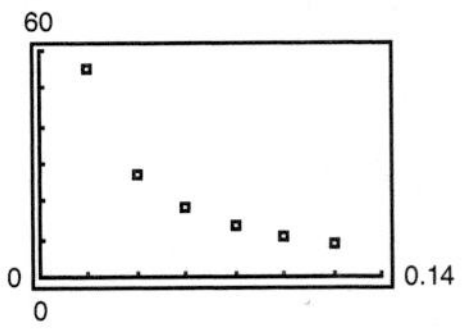

23. 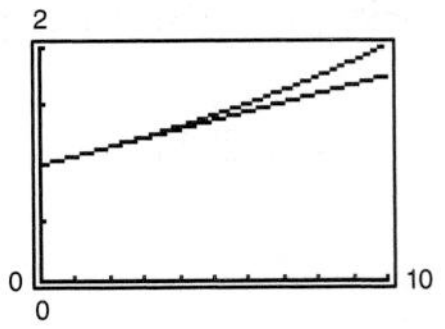

Continuous compounding

	Isotope	Half-Life (years)	Initial Quantity	Amount After 1000 Years
25.	^{226}Ra	1600	10 g	6.48 g
27.	^{231}Pa	32,800	1.53 g	1.5 g
29.	^{241}Am	432.2	26.4 g	5.31 g

31. $y = e^{0.768x}$ **33.** $y = 4e^{-0.2773x}$

35. (a) Increasing. The positive exponent indicates that the model is increasing.
(b) 97,301 people; 101,171 people; 105,194 people
(c) 2019

37. (a) 0.0178 (b) About 467,309 people

39. About 15,601 yr ago

41. (a) $V = -7160.5t + 39{,}780$
(b) $V = 39{,}780e^{-0.2231t}$
(c) 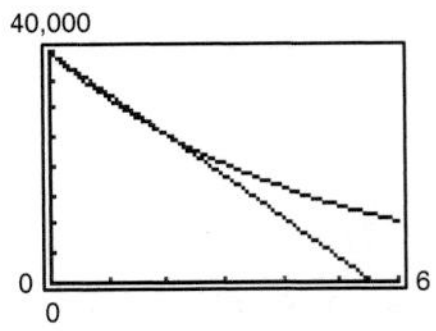

(d) Exponential model
(e) $0 < t < 2$; $t \geq 2$

43. (a) 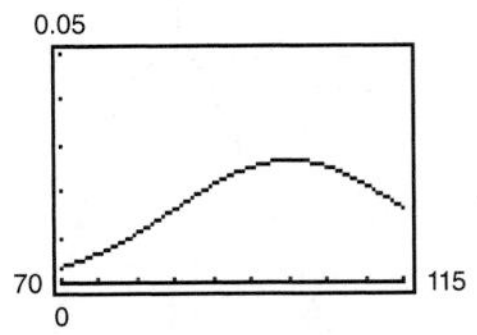

(b) 100

45. (a) About 203 animals (b) About 13 mo
(c) 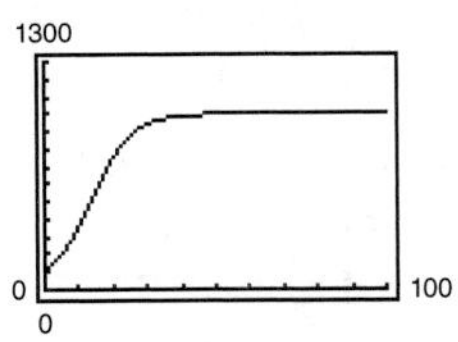

The horizontal asymptotes occur at $p = 1000$ and $p = 0$. The asymptote at $p = 1000$ means there will not be more than 1000 animals in the preserve.

47. (a) 10,000,000 (b) 199,526,232 (c) 1,995,262

49. (a) 20 dB (b) 70 dB (c) 120 dB

51. 97.49% **53.** 4.64 **55.** About 31,623 times

57. (a) 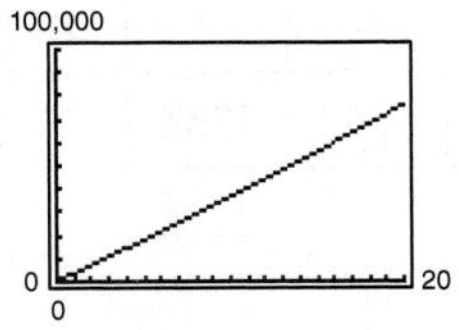

(b) 53.08 yr; yes

59. 3:00 A.M.

61. False. The domain can be all real numbers.

63. No. Any x-value in the Gaussian model will give a positive y-value.

65. Gaussian model **67.** Exponential growth model

69. a; $(0, -3)$, $\left(\frac{9}{4}, 0\right)$ **71.** d; $(0, 25)$, $\left(\frac{100}{9}, 0\right)$

73. Falls to the left, rises to the right

75. Rises to the left, falls to the right

77. $2x^2 + 3 + \dfrac{3}{x - 4}$ **79.** Answers will vary.

Section 4.6 (page 382)

1. $y = ax^b$ **3.** Scatter plot **5.** Logarithmic model

7. Quadratic model **9.** Exponential model

11. Linear model

13. 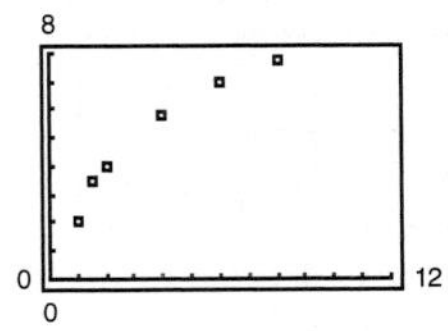

Logarithmic model

15. 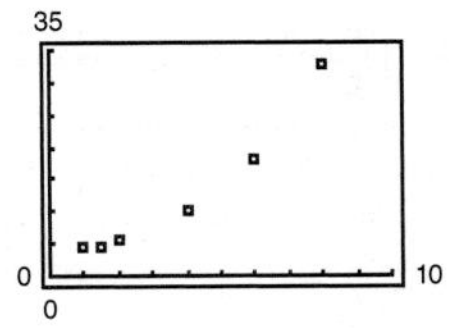

Exponential model

17. 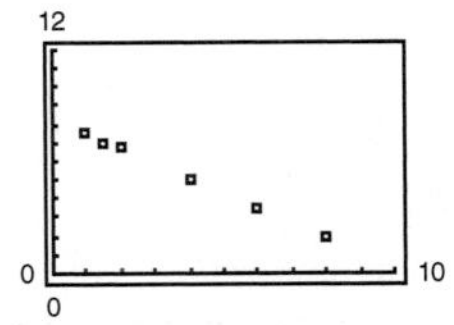

Linear model

19. $y = 4.752(1.2607)^x$; 0.96773
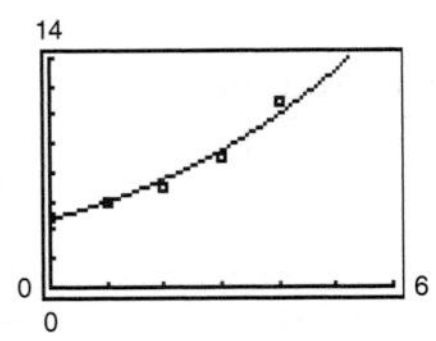

21. $y = 7.707(0.8070)^x$; 0.9448
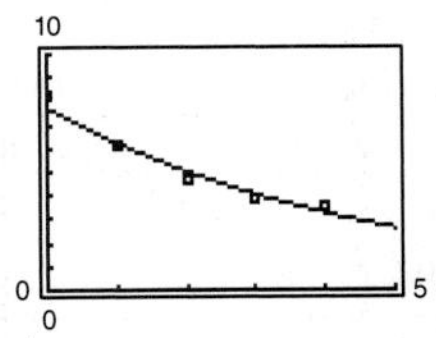

23. $y = 2.083 + 1.257 \ln x$; 0.98672
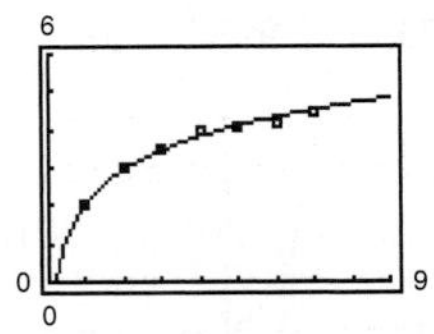

25. $y = -4.730 \ln x + 10.353$; 0.9661
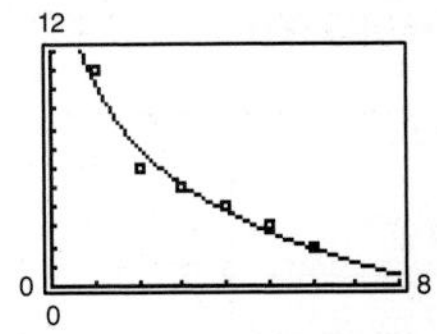

27. $y = 1.985x^{0.760}$; 0.99686
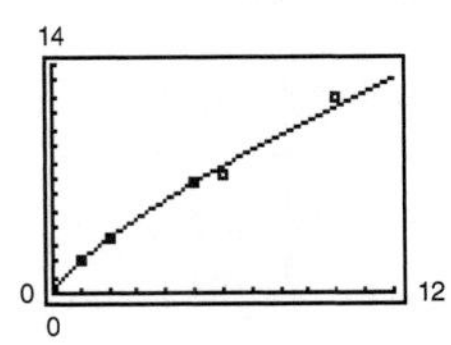

29. $y = 16.103x^{-3.174}$; 0.88161
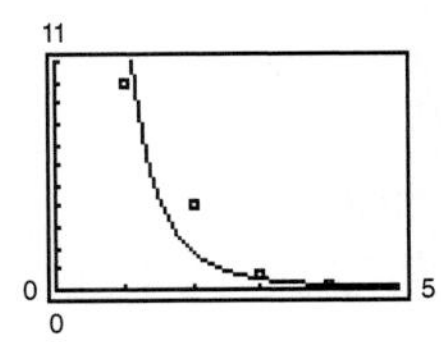

31. (a) Exponential model: $S = 2952.790(1.1203)^t$; 0.9821
Power model: $S = 877.314t^{1.0326}$; 0.9782

(b) Exponential model: Power model:

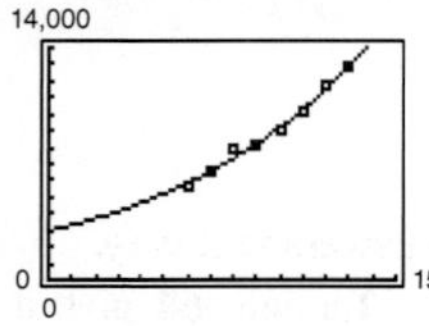

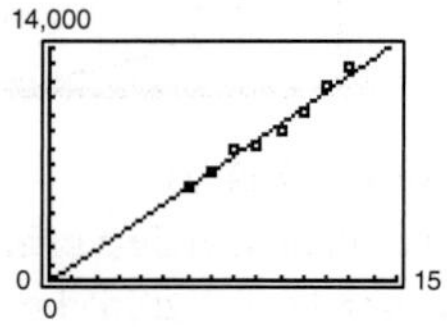

(c) Exponential model

33. (a) Linear model: $P = 7.57t + 352.7$; $r^2 \approx 0.9786$

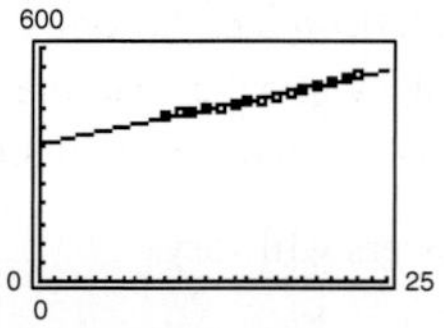

(b) Power model: $P = 248.55t^{0.2351}$; $r^2 \approx 0.9438$

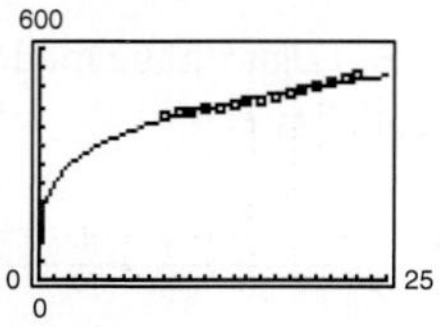

(c) Exponential model: $P = 366.630(1.0160)^t$; $r^2 \approx 0.9866$

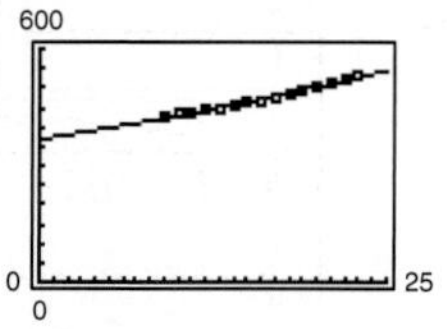

(d) Logarithmic model: $P = 111.66 \ln t + 168.6$;
$r^2 \approx 0.9289$

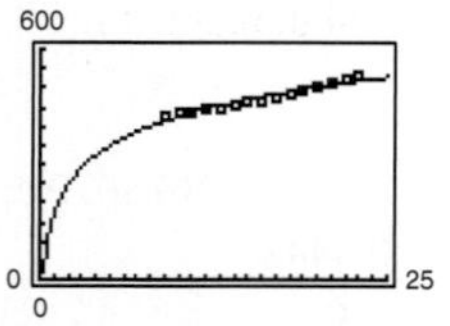

(e) The exponential model is the best fit because its coefficient of determination is closest to 1.

(f)

Linear:

Year	2014	2015	2016	2017	2018
Population (in thousands)	534.4	542.0	549.5	557.1	564.7

Power:

Year	2014	2015	2016	2017	2018
Population (in thousands)	524.7	529.7	534.7	539.4	544.1

Exponential:

Year	2014	2015	2016	2017	2018
Population (in thousands)	536.6	545.2	553.9	562.8	571.8

Logarithmic:

Year	2014	2015	2016	2017	2018
Population (in thousands)	523.5	528.0	532.4	536.6	540.7

(g) and (h) Answers will vary.

35. (a) $S = 3.723(1.1117)^t$
(b) $S = 3.723(e^{0.1059t})$
(c) \$25.05 billion; Answers will vary.

37. (a) $P = \dfrac{88.2676}{1 + 901.9894e^{-0.2560x}}$

(b)

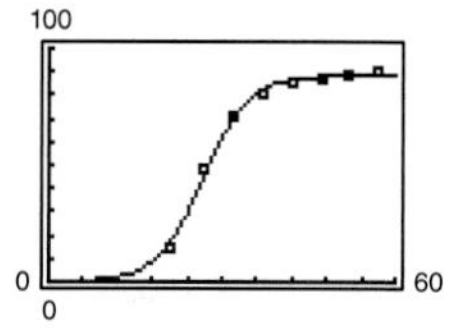

The model fits the data well.

39. True. See page 365. **41.** Answers will vary.

43. Slope: $-\frac{2}{5}$; y-intercept: $(0, 2)$

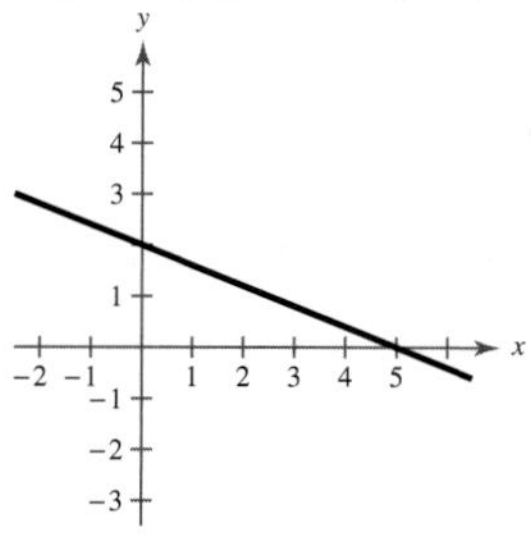

45. Slope: 0.16; y-intercept: $(0, -5)$

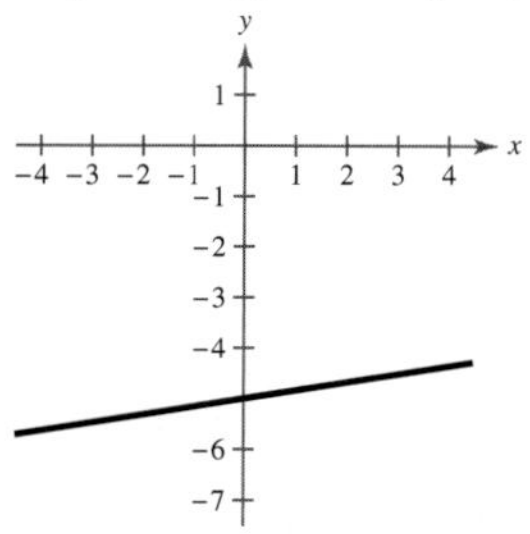

47. $y = -(x + 1)^2 + 2$ **49.** $y = -2(x - 3)^2 + 2$

Review Exercises (page 388)

1. 10.3254 **3.** 0.0167 **5.** c **6.** d **7.** b **8.** a

9.

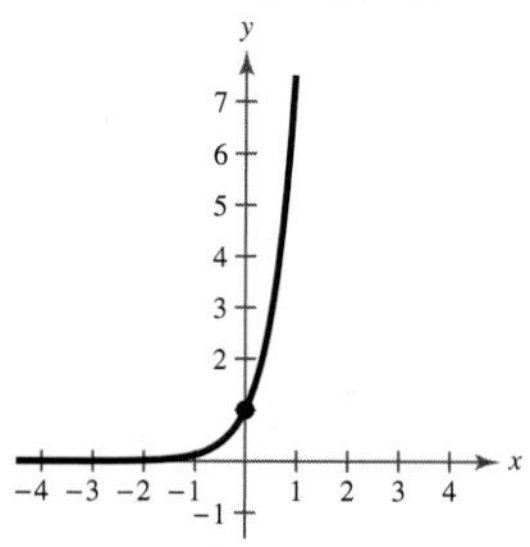

Horizontal asymptote: $y = 0$
y-intercept: $(0, 1)$
Increasing on $(-\infty, \infty)$

11.

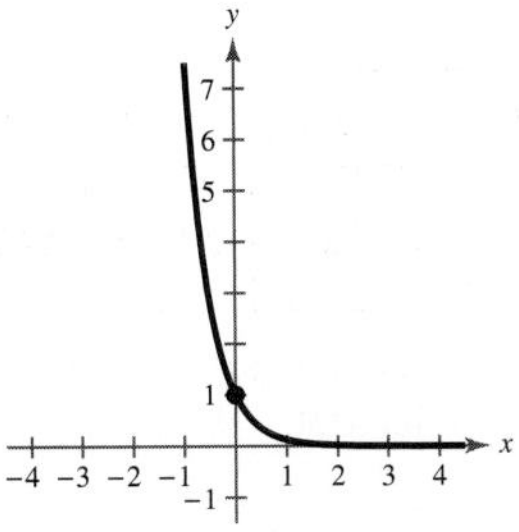

Horizontal asymptote: $y = 0$
y-intercept: $(0, 1)$
Decreasing on $(-\infty, \infty)$

13.

x	0	1	2	3	4
$h(x)$	0.37	1	2.72	7.39	20.09

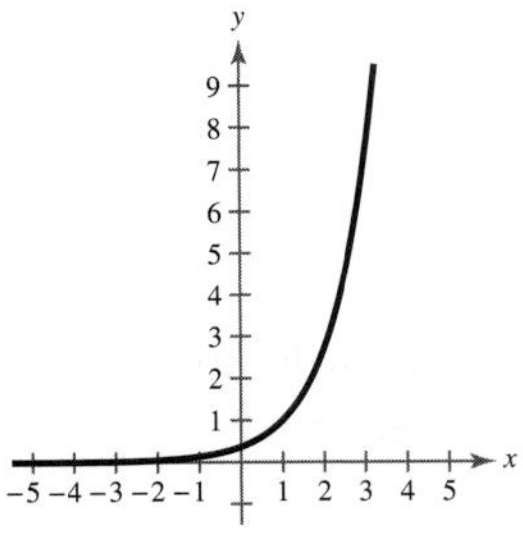

Horizontal asymptote: $y = 0$

15.

x	-2	-1	0	1	2
$h(x)$	2.86	2.63	2	0.28	-4.39

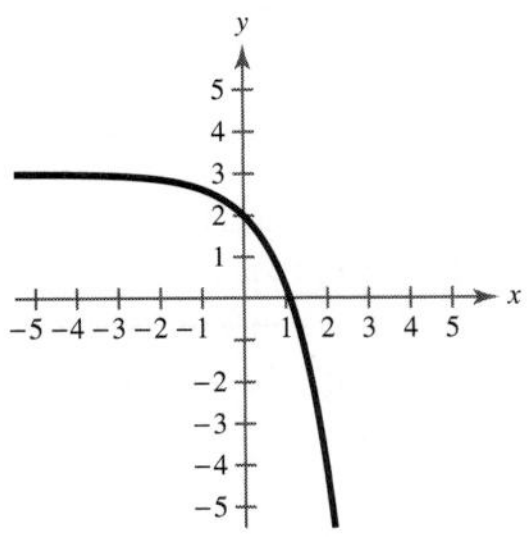

Horizontal asymptote: $y = 3$

17.

x	-1	0	1	2	3	4
$f(x)$	6.59	4	2.43	1.47	0.89	0.54

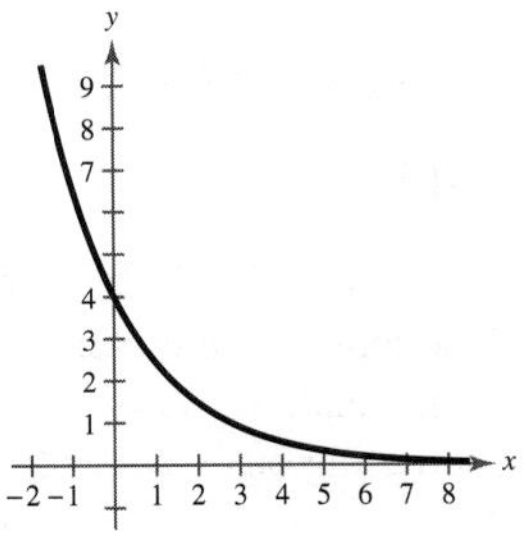

Horizontal asymptote: $y = 0$

19.

t	1	10	20
A	\$10,832.87	\$22,255.41	\$49,530.32

t	30	40	50
A	\$110,231.76	\$245,325.30	\$545,981.50

21. (a)

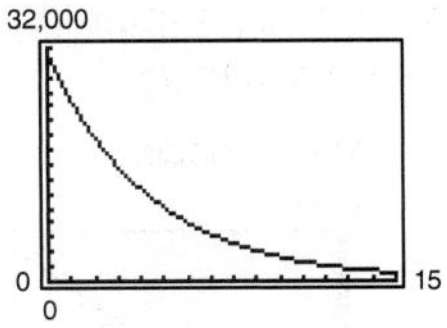

(b) \$19,708.80
(c) When it is first sold; yes; Answers will vary.

23. $5^4 = 625$ **25.** $64^{1/6} = 2$ **27.** $\log_4 64 = 3$
29. $\ln 24.532\ldots = 3.2$ **31.** $\log_{1/2} 8 = -3$
33. 3 **35.** -1

37. Domain: $(0, \infty)$
Vertical asymptote: $x = 0$
x-intercept: $(10{,}000, 0)$

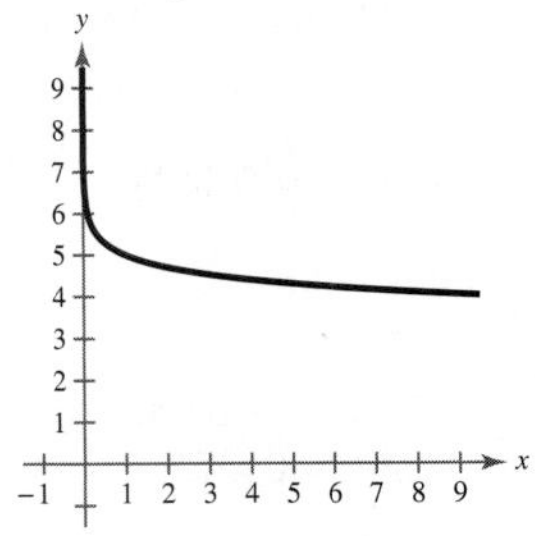

39. Domain: $(1, \infty)$
Vertical asymptote: $x = 1$
x-intercept: $(1.000001, 0)$

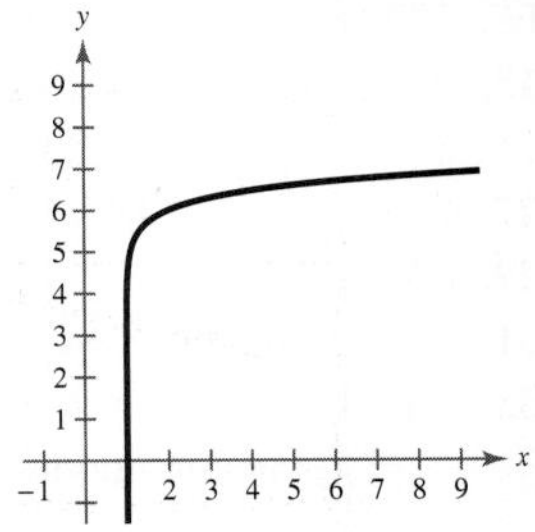

41. 3.068 **43.** 0.896

45.

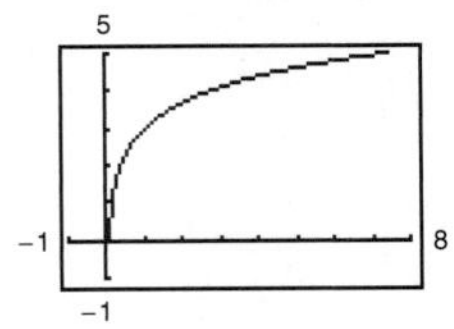

Domain: $(0, \infty)$
Vertical asymptote: $x = 0$
x-intercept: $(0.05, 0)$

47.

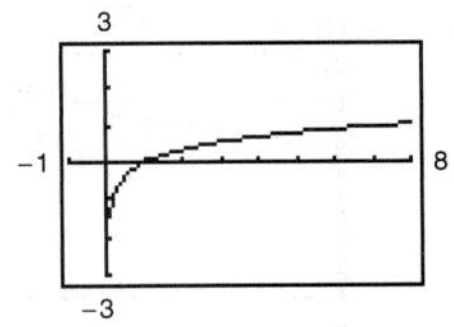

Domain: $(0, \infty)$
Vertical asymptote: $x = 0$
x-intercept: $(1, 0)$

49. (a) $0 \le h < 18{,}000$
(b)

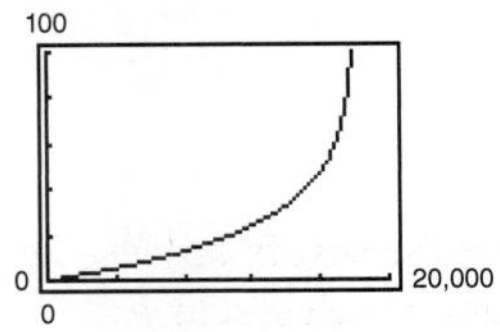

Asymptote: $h = 18{,}000$
(c) The time required to increase its altitude further increases.
(d) 5.46 min

51. 1.585 **53.** 1.041

55.

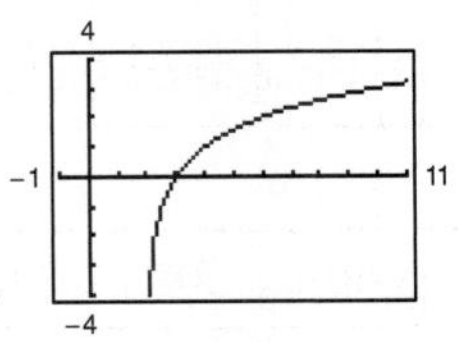

57.

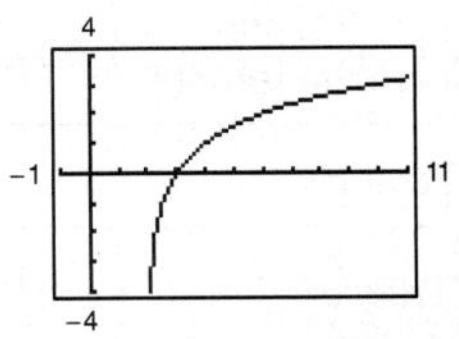

59. 1.13 **61.** 0.41 **63.** $\ln 5 - 2$ **65.** $2 + \log_{10} 2$
67. $1 + 2\log_5 x$ **69.** $\log_{10} 5 + \frac{1}{2}\log_{10} y - 2\log_{10} x$
71. $\ln(x + 3) - \ln x - \ln y$ **73.** $\log_2 9x$
75. $\ln\dfrac{\sqrt{2x - 1}}{(x + 1)^2}$ **77.** $\ln\dfrac{3\sqrt[3]{4 - x^2}}{x}$

79. (a)

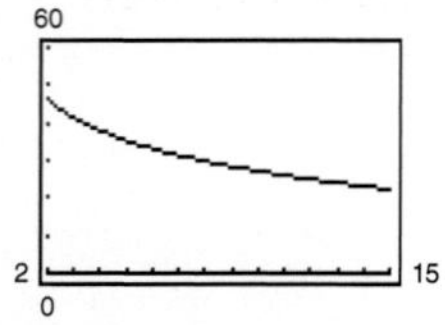

(b)

h	4	6	8	10	12	14
s	38	33	30	27	25	23

(c) The decrease in productivity starts to level off.

81. 4 **83.** -3 **85.** -5 **87.** 4096 **89.** 9
91. e^4 **93.** $e^2 + 1$ **95.** -0.757 **97.** 4.459
99. 1.760 **101.** 3.916 **103.** 1.609, 0.693
105. 200.615 **107.** 36.945 **109.** 53.598 **111.** 4
113. 0.9 **115.** -1 **117.** 0.368 **119.** 10.05 yr
121. e **122.** b **123.** f **124.** d **125.** a **126.** c
127. $k = 0.0184$; 11,645,427
129. (a) 11.08 weeks (b) 24.16 weeks
131. Logistic model
133. (a) Linear model: $N = 40.4t + 2728$; $r^2 \approx 0.9836$
Exponential model: $N = 2737.170(1.0135)^t$; $r^2 \approx 0.9832$
Power model: $N = 2688.8t^{0.0645}$; $r^2 \approx 0.8748$

(b) Linear model:

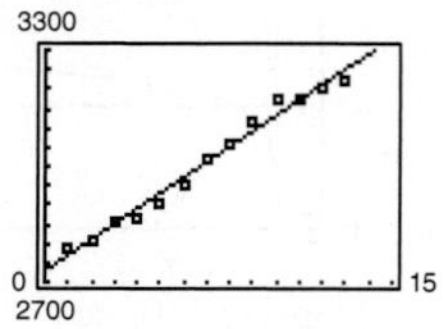

Exponential model:

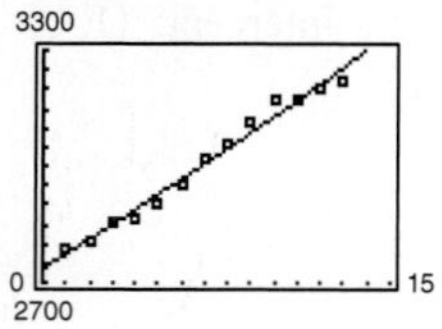

Power model:

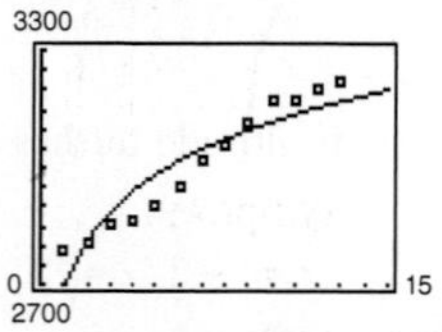

(c) The linear model is the best fit because its coefficient of determination is closest to 1.

(d) 2023–2024

135. True. $e^{x-1} = e^x \cdot e^{-1} = \dfrac{e^x}{e}$ **137.** False. $x > 0$

139. Because $1 < \sqrt{2} < 2$, then $2^1 < 2^{\sqrt{2}} < 2^2$.

Chapter Test (page 392)

1.

x	-2	-1	0	1	2
$f(x)$	100	10	1	0.1	0.01

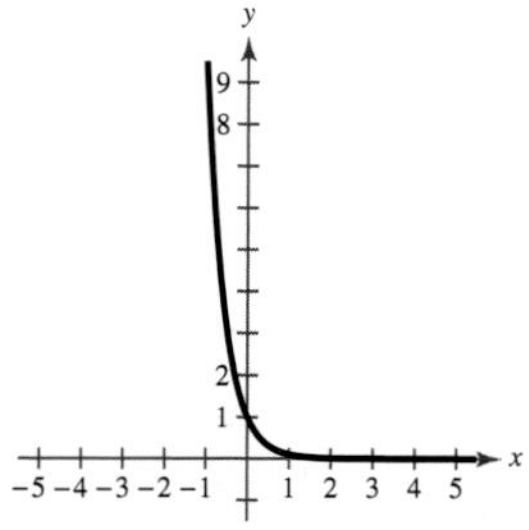

Horizontal asymptote: $y = 0$
y-intercept: $(0, 1)$

2.

x	0	2	3	4
$f(x)$	-0.03	-1	-6	-36

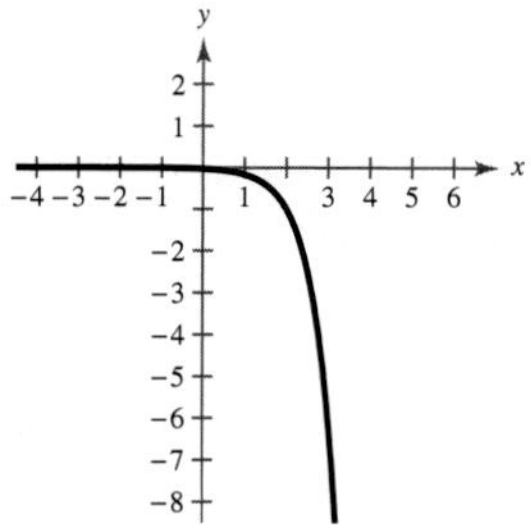

Horizontal asymptote: $y = 0$
y-intercept: $\left(0, -\frac{1}{36}\right)$

3.

x	-2	-1	0	1	2
$f(x)$	0.9817	0.8647	0	-6.3891	-53.5982

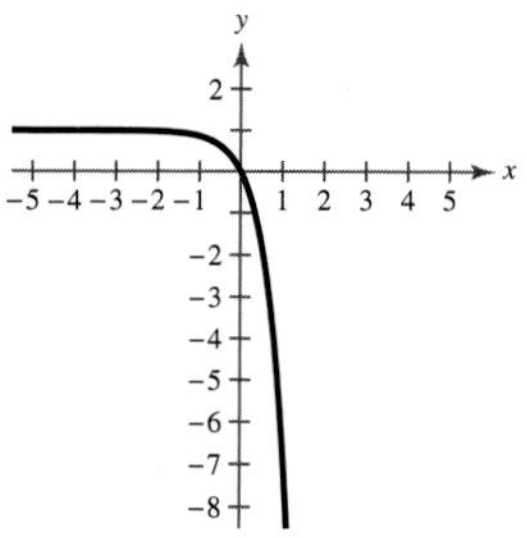

Horizontal asymptote: $y = 1$
Intercept: $(0, 0)$

4. -0.89 **5.** 9.2 **6.** 2

7.

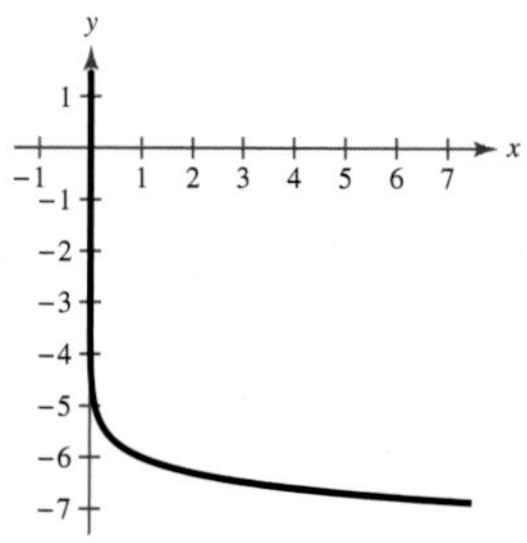

Domain: $(0, \infty)$
Vertical asymptote: $x = 0$
x-intercept: $(10^{-6}, 0)$

8.

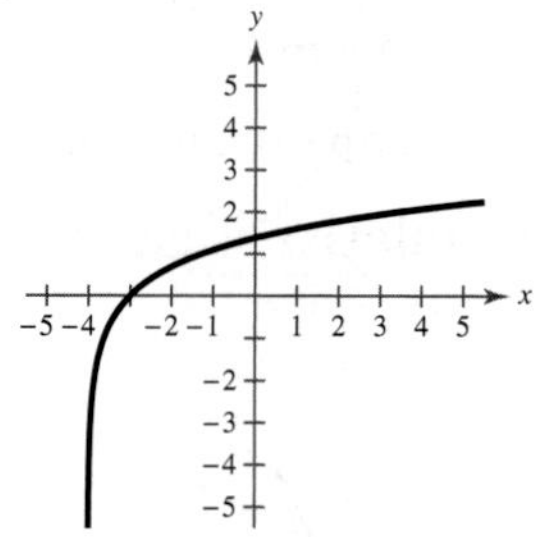

Domain: $(-4, \infty)$
Vertical asymptote: $x = -4$
x-intercept: $(-3, 0)$

9.

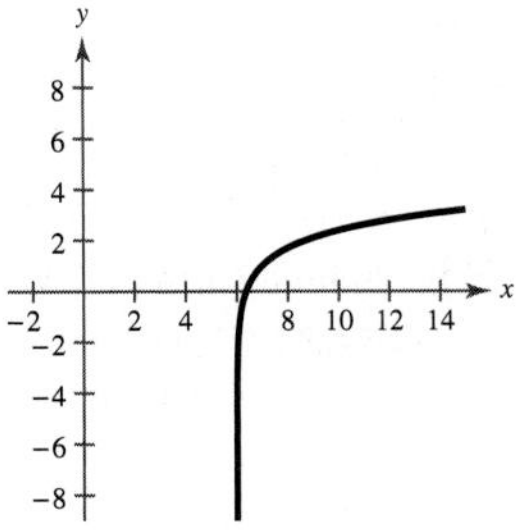

Domain: $(6, \infty)$
Vertical asymptote: $x = 6$
x-intercept: $(6.37, 0)$

10. 1.945 **11.** -0.286 **12.** 1.674
13. $\log_2 3 + 4 \log_2 a$ **14.** $\ln 5 + \frac{1}{2} \ln x - \ln 6$
15. $\ln x + \frac{1}{2} \ln(x + 1) - \ln 2 - 4$ **16.** $\log_3 13y$
17. $\ln \dfrac{x^4}{y^4}$ **18.** $\ln \dfrac{x(2x - 3)}{x + 2}$ **19.** 4
20. 2.431 **21.** 117,649 **22.** 100,004 **23.** 1.321
24. 1 **25.** 1.597 **26.** 1.649 **27.** 62.34%
28. (a) Logarithmic model: $R = 81.2970 \ln t - 144.646$
Exponential model: $R = 2.006(1.3266)^t$
Power model: $R = 0.099t^{2.5639}$
(b) Logarithmic model:

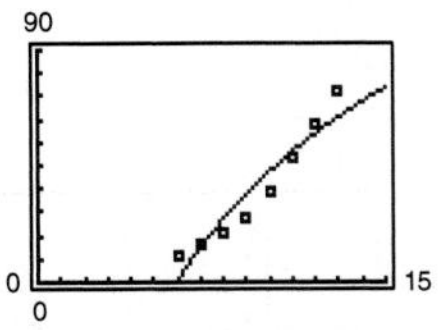

Exponential model:

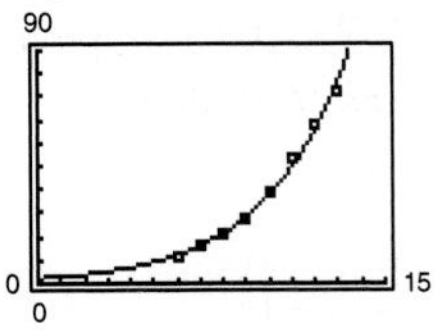

Power model:

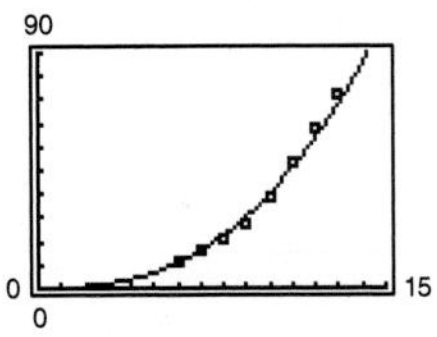

(c) Exponential model; $430.9 billion

Cumulative Test for Chapters 3–4 (page 393)

1.

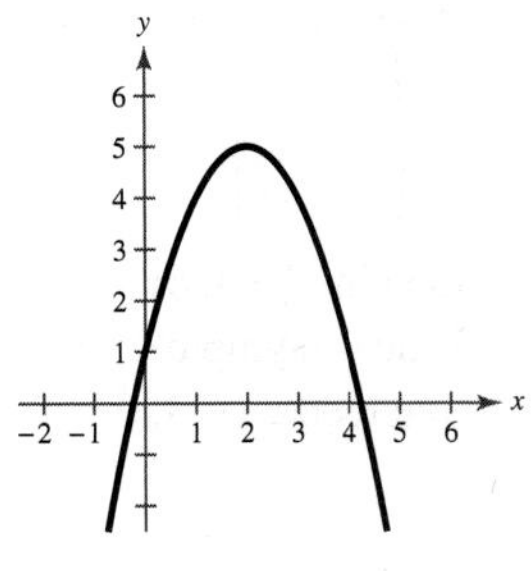

2.

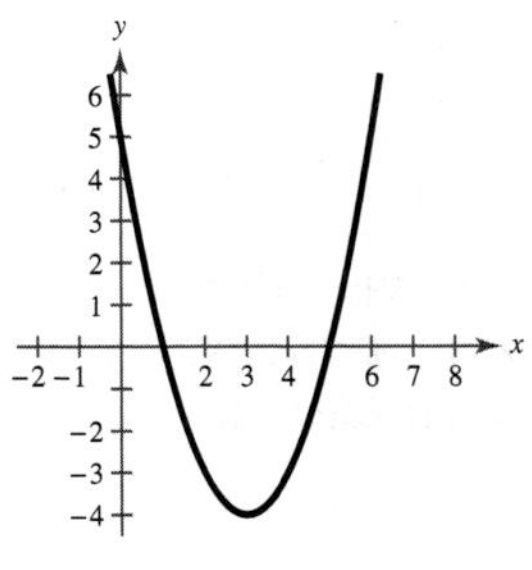

3.

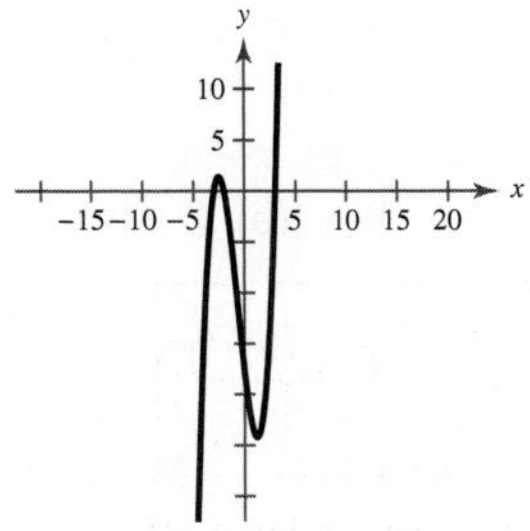

4. (a) 3 or 1 positive real zero(s), no negative real zeros
(b) $\pm 1, \pm 2, \pm 3, \pm 6, \pm 9, \pm 18$
(c)

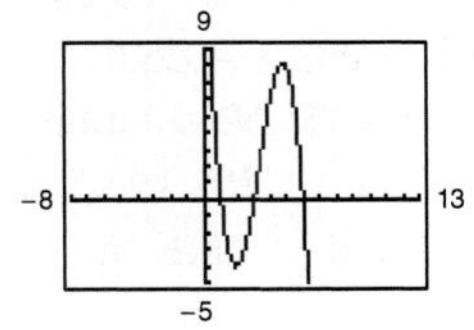

(d) 1, 3, 6

5. (a) 1 positive real zero, 2 or 0 negative real zero(s)
(b) $\pm 1, \pm 2, \pm 5, \pm 10, \pm \frac{1}{3}, \pm \frac{2}{3}, \pm \frac{5}{3}, \pm \frac{10}{3}$
(c)

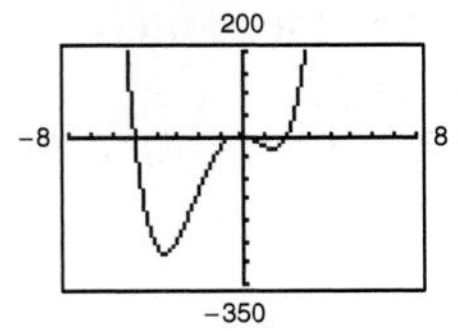

(d) $-5, -\frac{1}{3}, 0, 2$

6. $-2, \pm 2i$ **7.** 1.424 **8.** $4x + 2 - \dfrac{15}{x + 3}$

9. $2x^2 + 7x + 48 + \dfrac{268}{x - 6}$ **10.** $x^3 + 6x^2 - x - 6$

11. $x^4 - 4x^3 + 8x^2 - 8x$

12.

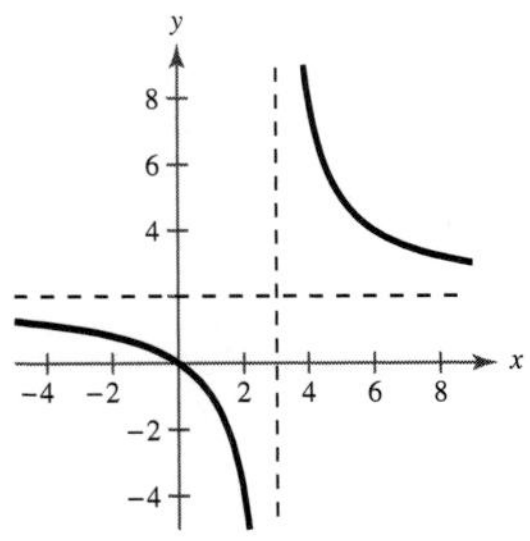

Asymptotes:
$x = 3, y = 2$

13.

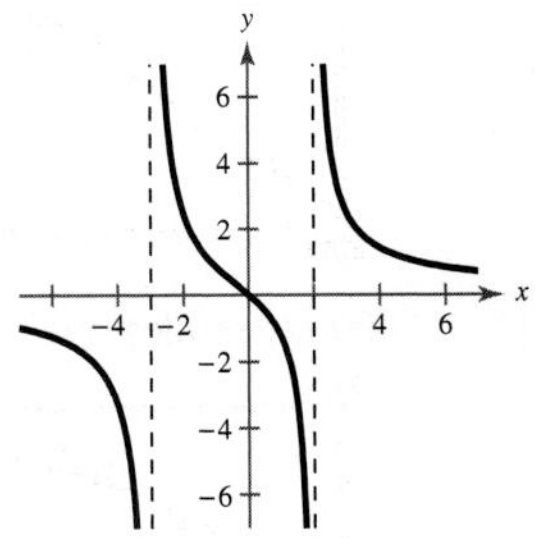

Asymptotes:
$x = -3, x = 2, y = 0$

14.

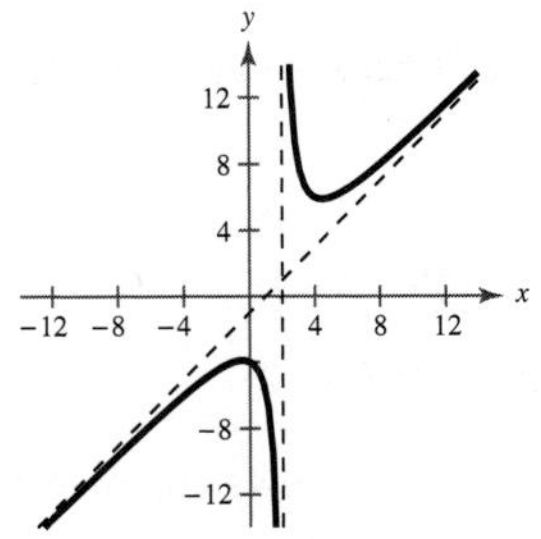

Asymptotes:
$x = 2, y = x - 1$

15.

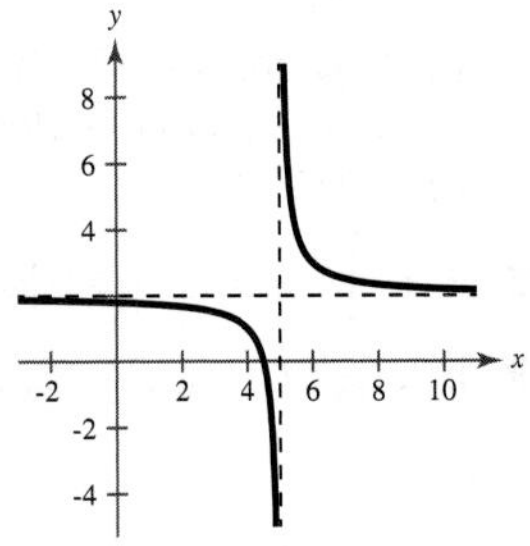

Horizontal shift 5 units to the right, vertical shift 2 units upwards. Zeros: $x = \frac{9}{5}$

16. Answers will vary. Sample answer: $y = \dfrac{4x^2}{x^2 + 1}$

17. Horizontal shift one unit left, vertical shift five units down

CHAPTER 4

18. Reflection in the x-axis, horizontal shift three units left
19. (a) All real numbers x (b) $(0, 10)$ (c) $y = 2$
20. (a) All real numbers x (b) $(0, 2)$ (c) $y = 0$
21. (a) All real numbers x such that $x > 0$ (b) $(1, 0)$
(c) $x = 0, y = 0$
22. (a) All real numbers x such that $x > \frac{1}{2}$ (b) $\left(\frac{7}{12}, 0\right)$
(c) $x = \frac{1}{2}$
23. -4 **24.** 10 **25.** -3 **26.** 1.723 **27.** 0.872
28. -7.484 **29.** -3.036
30. $\ln(x + 2) + \ln(x - 2) - \ln(x^2 + 1)$ **31.** $\ln\frac{x^2(x + 1)}{(x - 1)}$
32. $x \approx 0.973$ **33.** $x \approx 6.073$ **34.** $x = 25.6$
35. $x \approx 152.018$ **36.** $x = 0, 1$ **37.** No solution
38. \$50, \$9.50, \$5.45; \$5
39. (a) \$1204.52 (b) \$1209.93
40. (a) 300 (b) 570 (c) 9 yr
41. (a) $A = x(273 - x)$
(b)

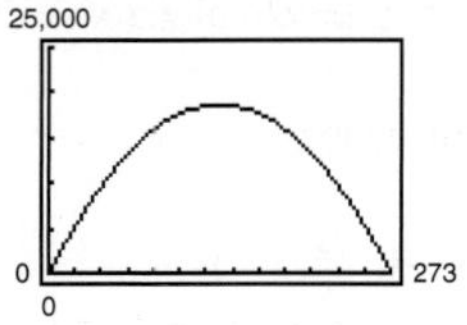

$0 < x < 273$
(c) 76.23 ft × 196.77 ft
42. (a) Quadratic model:
$S = -1.1693t^2 + 40.2326t + 147.853$; $r^2 \approx 0.9905$
Exponential model: $S = 229.067(1.0652)^t$; $r^2 \approx 0.9486$
Power model: $S = 160.681t^{0.4294}$; $r^2 \approx 0.9921$
(b) Quadratic model:

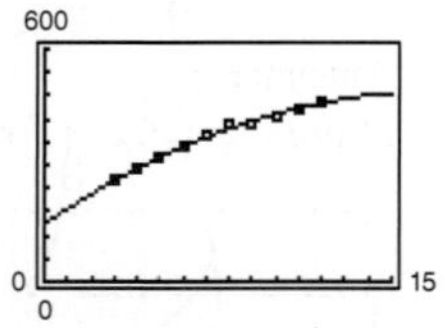

Exponential model:

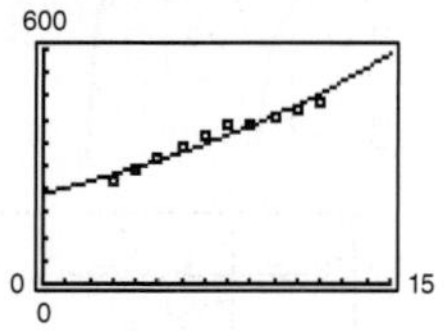

Power model:

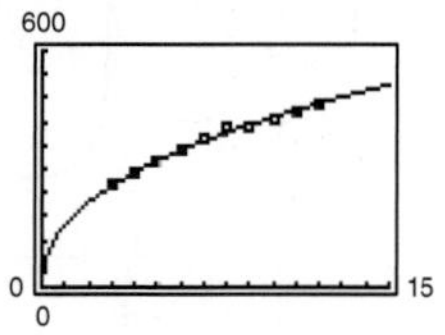

(c) The power model is the best fit because its coefficient of determination is closest to 1.
(d) \$581.603 billion; Answers will vary.

Chapter 5

Section 5.1 (page 404)

1. system, equations **3.** substitution **5.** Break-even point
7. (a) No (b) No (c) No (d) Yes
9. (a) No (b) Yes (c) No (d) No
11. $(2, 2)$ **13.** $(2, 6), (-1, 3)$
15. $(0, 2), \left(\sqrt{3}, 2 - 3\sqrt{3}\right), \left(-\sqrt{3}, 2 + 3\sqrt{3}\right)$ **17.** $(4, 4)$
19. $(10, -10)$ **21.** $\left(\frac{1}{2}, 3\right)$ **23.** $(1, 1)$ **25.** $\left(\frac{20}{3}, \frac{40}{3}\right)$
27. No solution **29.** \$2000 at 4%, \$16,000 at 2%
31. \$3500 at 5.6%, \$14,500 at 6.8% **33.** $(-2, 0), (3, 5)$
35. No real solution **37.** $(0, 0), (1, 1), (-1, -1)$
39. $(-3, 1)$ **41.** $(1, 2)$ **43.** No real solution
45. $(-3, 0), (2, -5)$ **47.** $(4, -0.5)$ **49.** $(8, 3), (3, -2)$
51. $(\pm 1.540, 2.372)$ **53.** $(0, 1)$ **55.** $(2.318, 2.841)$
57. $(2.25, 5.5)$ **59.** $(0, -13), (\pm 12, 5)$ **61.** $(1, 2)$
63. $(-2, 0), \left(\frac{29}{10}, \frac{21}{10}\right)$ **65.** No real solution **67.** $(0.25, 1.5)$
69. $(0.287, 1.751)$ **71.** $(0, 1), (1, 0)$ **73.** No real solution
75.

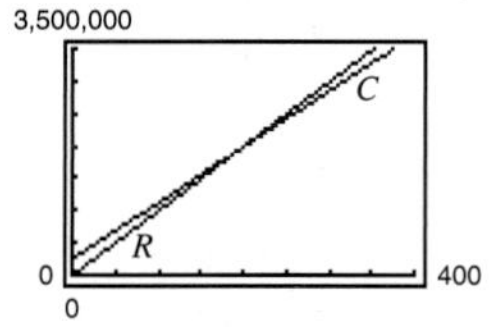

192 units; \$1,910,400

77.

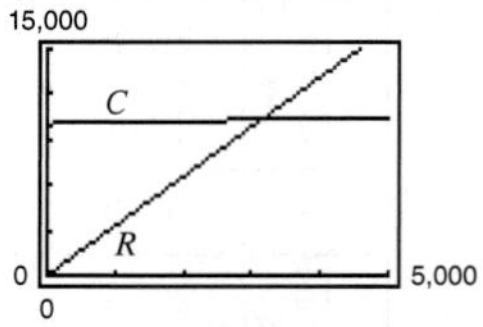

3133 units; \$10,308

79. 6 m × 9 m
81. (a)

Week	Animated	Horror
1	336	42
2	312	60
3	288	78
4	264	96
5	240	114
6	216	132
7	192	150
8	168	168
9	144	186
10	120	204
11	96	222
12	72	240

(b) and (c) $x = 8$
(d) The answers are the same.
(e) During week 8, the same number of animated and horror films were rented.
83. (a) $C = 9.45x + 16,000$
$R = 55.95x$
(b) 344 units

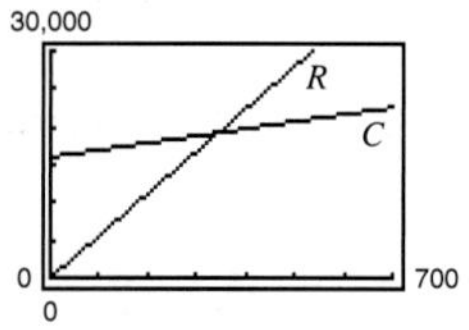

85. 8 mi × 12 mi

87. (a) $\begin{cases} x + y = 20{,}000 \\ 0.055x + 0.075y = 1300 \end{cases}$

(b)

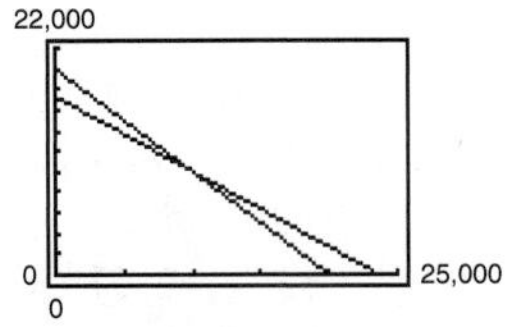

(c) \$10,000. The solution is (10,000, 10,000).

89. (a)

t	Year	Colorado	Minnesota
13	2013	5265	5417
14	2014	5339	5450
15	2015	5413	5483
16	2016	5487	5516
17	2017	5561	5549
18	2018	5635	5582
19	2019	5709	5615
20	2020	5783	5648

(b) 2017

(c)

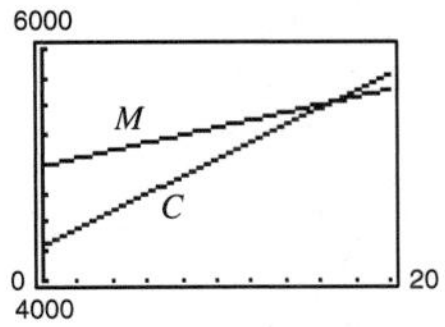

(16.71, 5539.34)

(d) (16.71, 5539.34)

(e) At one point in 2016, the populations of Colorado and Minnesota are equal.

91. False. You can solve for either variable before back-substituting.

93. For a linear system, the result will be a contradictory equation such as $0 = N$, where N is a nonzero real number. For a nonlinear system, there may be an equation with imaginary roots.

95. (a) $\begin{cases} 3x + y = 3 \\ 3x + y = 5 \end{cases}$ (b) $\begin{cases} 3x + y = 4 \\ 2x + y = 2 \end{cases}$ (c) $\begin{cases} 6x + 3y = 9 \\ 2x + y = 3 \end{cases}$

97. (a)

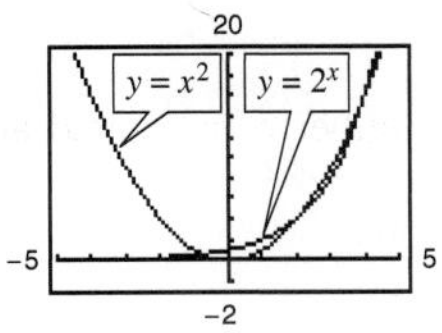

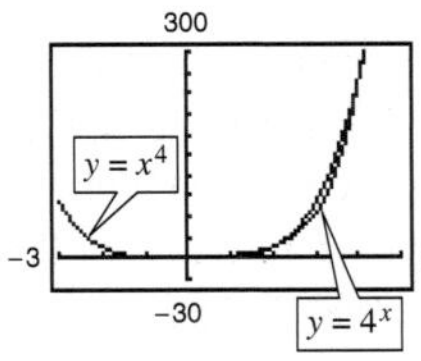

(b) There are three points of intersection when b is even.

99. $y = \frac{2}{7}x + \frac{22}{7}$

101. $x = 4$; A vertical line cannot be written in slope-intercept form.

103. Domain: All real numbers x except $x = 6$
Asymptotes: $y = 0, x = 6$

105. Domain: All real numbers x except $x = \pm 4$
Asymptotes: $y = 1, x = \pm 4$

107. Domain: All real numbers x
Asymptote: $y = 0$

Section 5.2 (page 413)

1. method, elimination **3.** Inconsistent **5.** Yes

7. (2, 1)

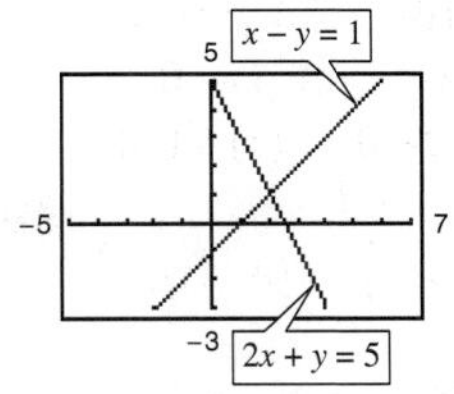

9. $(1, -1)$

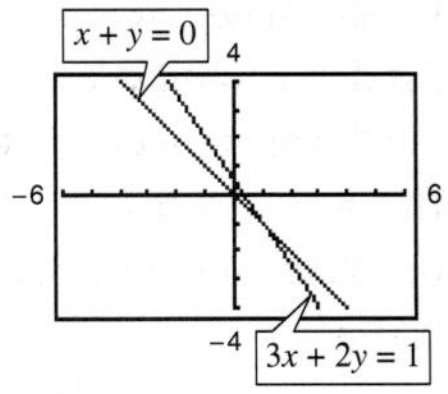

11. Inconsistent

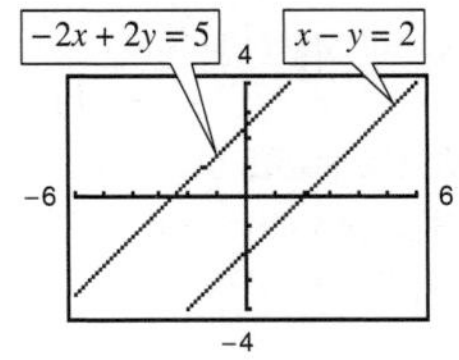

13. $\left(2, \frac{1}{2}\right)$ **15.** (3, 4) **17.** $\left(-3, \frac{3}{2}\right)$ **19.** $\left(\frac{12}{7}, \frac{18}{7}\right)$

21. $\left(\frac{5}{2}, 0\right)$ **23.** Inconsistent **25.** $\left(\frac{2}{9}, \frac{4}{3}\right)$

27. b; One solution, consistent

28. a; Infinitely many solutions, consistent

29. d; No solutions, inconsistent

30. c; One solution, consistent **31.** $\left(\frac{3}{2}, -\frac{1}{2}\right)$

33. Inconsistent **35.** All points on $6x + 8y - 1 = 0$

37. $\left(\frac{7}{2}, -\frac{1}{2}\right)$ **39.** All points on $-5x + 6y = -3$

41. Inconsistent **43.** (101, 96) **45.** $(-1, 1)$

47.

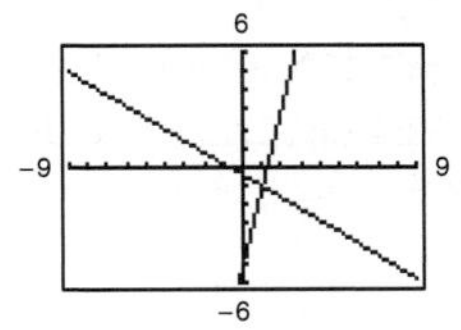

Consistent; $(1, -1)$

49.

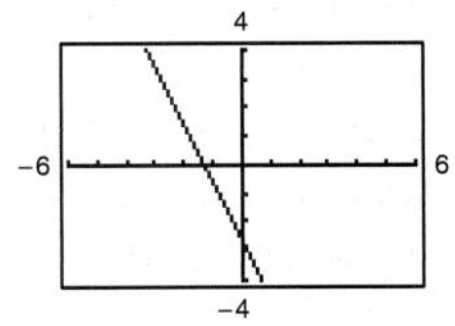

Consistent;
all points on $6x + 3y = -8$

51.

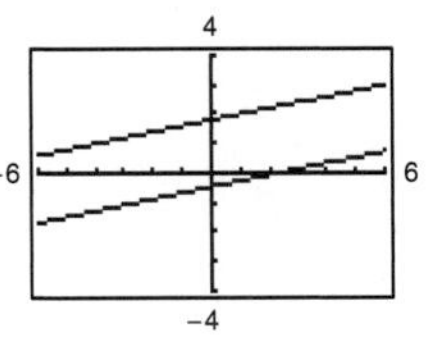

Inconsistent

53.

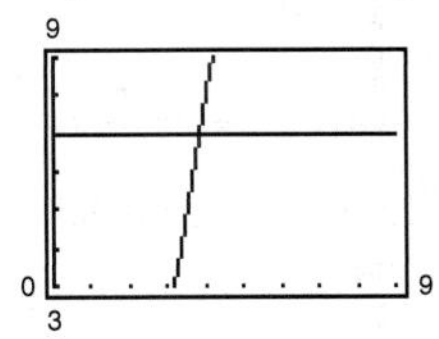

(3.833, 7)

55.

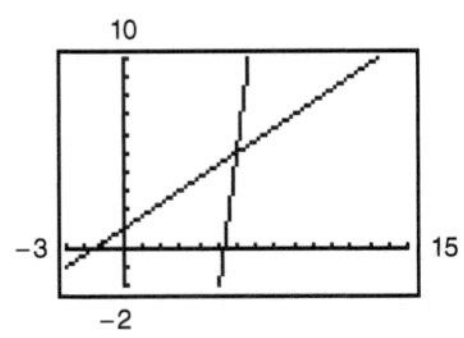

(6, 5)

57.

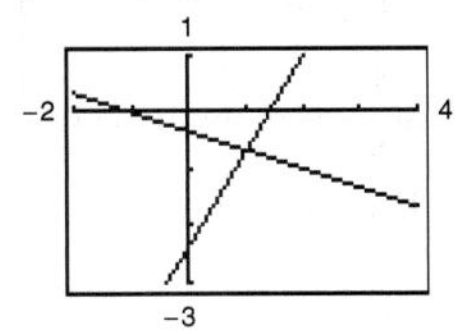

$(1, -0.667)$

59.

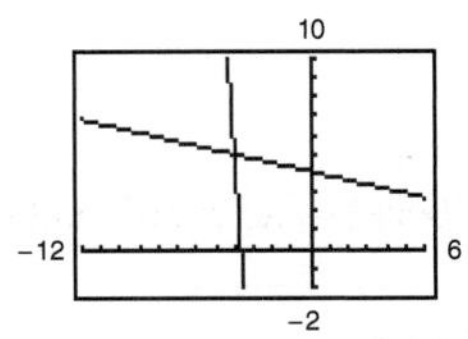

$(-4, 5)$

61. (4, 1) **63.** $(2, -1)$ **65.** $(6, -3)$ **67.** $\left(\frac{49}{4}, \frac{33}{4}\right)$

69. $\begin{cases} \frac{1}{5}x + y = 1 \\ 3x - 2y = 15 \end{cases}$ **71.** $\begin{cases} 2x + \frac{1}{4}y = 4 \\ -4x - 3y = 2 \end{cases}$

CHAPTER 5

73. (240, 404) **75.** (2,000,000, 100)

77. Plane: 550 mi/h, wind: 50 mi/h

79. (a) $\begin{cases} 15A + 12C = 16{,}275 \\ A + C = 1175 \end{cases}$

(b) $A = 725$, $C = 450$; Answers will vary.

(c) $A = 725$, $C = 450$

81. 9 oranges, 7 grapefruits

83. (a) $S \approx 1761.89$, $t \approx 8.98$; Answers will vary.

(b) In 2018, both retailers had sales of $1761.89 million.

(c) The coefficient is the annual increase in sales for each retailer.

(d) The retailers would never have the same amount of sales for any year.

85. $y = 0.97x + 2.1$ **87.** $y = -0.58x + 5.39$

89. (a) and (b) $y = 14x + 19$

(c)

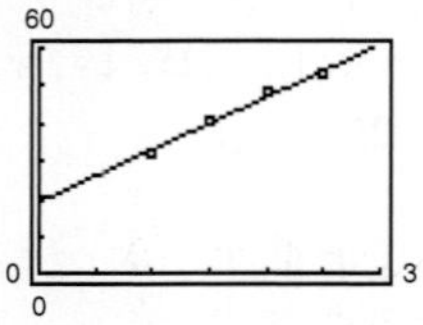

(d) 41.4 bushels per acre

91. True. A linear system can have only one solution, no solution, or infinitely many solutions.

93. False. Sometimes you will be able to get only a close approximation.

95. (a) $k = 18$ (b) $k \neq 18$ **97.** $u = -x \ln x$; $v = \ln x$

99. $x \leq -\frac{22}{3}$

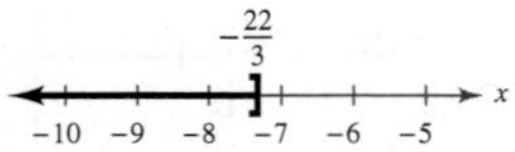

101. $-2 < x < 18$

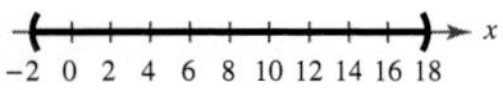

103. $-5 < x < \frac{7}{2}$

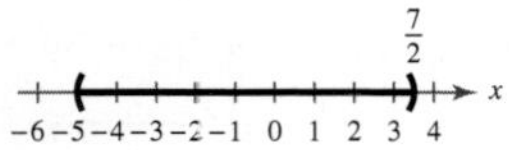

105. $\ln 6x$ **107.** $\log_9 \dfrac{12}{x}$ **109.** $\ln \dfrac{x^2}{x + 2}$

111. Answers will vary.

Section 5.3 (page 427)

1. row-echelon **3.** Gaussian **5.** three-dimensional

7. Independent

9. (a) No (b) Yes (c) No (d) No

11. (a) No (b) No (c) Yes (d) No

13. (2, −2, 2) **15.** (3, 10, 2) **17.** $\left(\frac{11}{4}, 7, 11\right)$

19. $\begin{cases} x - 2y + 3z = 5 \\ y - 2z = 9 \\ 2x - 3z = 0 \end{cases}$

It removed the x-term from Equation 2.

21. (1, 2, 3) **23.** (−4, 8, 5) **25.** (2, −3, −2)

27. Inconsistent **29.** $\left(1, -\frac{3}{2}, \frac{1}{2}\right)$ **31.** $(-a + 3, a + 1, a)$

33. Inconsistent **35.** (−1, 1, 0) **37.** Inconsistent

39. $(2a, 21a - 1, 8a)$ **41.** $(9a, -35a, 67a)$

43. $\begin{cases} x + y + z = 1 \\ 2x + y + z = 4 \\ x + y - 3z = -7 \end{cases}$ **45.** $\begin{cases} x + y + 2z = -10 \\ -x + 12y + 8z = -14 \\ x + 14y - 4z = -6 \end{cases}$

47. $\begin{cases} x - z = 0 \\ -x + y = 4 \end{cases}$

49.

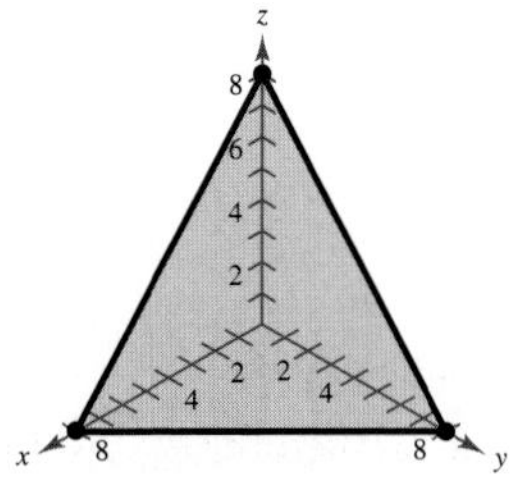

Sample answer: (8, 0, 0), (0, 8, 0), (0, 0, 8), (4, 0, 4)

51.

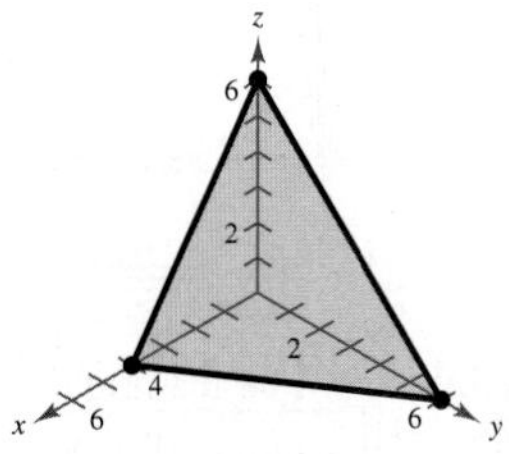

Sample answer: (4, 0, 0), (0, 6, 0), (0, 0, 6), (2, 3, 0)

53. $\dfrac{A}{x} + \dfrac{B}{x - 14}$ **55.** $\dfrac{A}{x} + \dfrac{B}{x^2} + \dfrac{C}{x - 10}$

57. $\dfrac{A}{x - 5} + \dfrac{B}{(x - 5)^2} + \dfrac{C}{(x - 5)^3}$

59. $\dfrac{A}{x} + \dfrac{Bx + C}{x^2 + 1} + \dfrac{Dx + E}{(x^2 + 1)^2}$ **61.** $\dfrac{1}{x} - \dfrac{1}{x + 1}$

63. $\dfrac{3}{2x - 1} - \dfrac{2}{x + 1}$ **65.** $-\dfrac{3}{x} - \dfrac{1}{x + 2} + \dfrac{5}{x - 2}$

67. $\dfrac{3}{x} - \dfrac{1}{x^2} + \dfrac{1}{x + 1}$ **69.** $2x - 7 + \dfrac{17}{x + 2} + \dfrac{1}{x + 1}$

71. $x + 3 + \dfrac{6}{x - 1} + \dfrac{4}{(x - 1)^2} + \dfrac{1}{(x - 1)^3}$

73. $-\dfrac{1}{x - 1} + \dfrac{x + 2}{x^2 - 2}$

75. $\dfrac{3}{x} - \dfrac{2}{x - 4}$

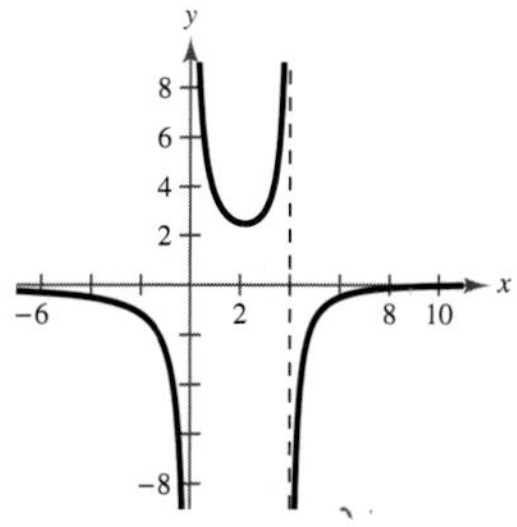

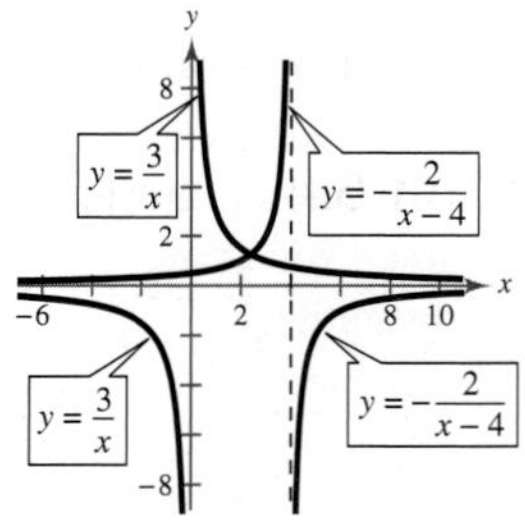

The vertical asymptotes are the same.

77. $s = -16t^2 + 144$ **79.** $s = -16t^2 - 32t + 400$

81. $y = x^2 - 3x$ **83.** $y = -2x^2 - 7x - 4$

85. $x^2 + y^2 - 10x = 0$ **87.** $x^2 + y^2 + 6x - 8y = 0$

89. $20,000 at 4%, $2500 at 6%, and $7500 at 8%

91. $187{,}500 + s$ in certificates of deposit

$187{,}500 - s$ in municipal bonds

$125{,}000 - s$ in blue-chip stocks

s in growth stocks

93. 22 two-point baskets, 13 three-point baskets, 10 free throws

95. $I_1 = 1, I_2 = 2, I_3 = 1$ **97.** $y = -\frac{5}{24}x^2 - \frac{3}{10}x + \frac{41}{6}$

99. $y = x^2 - x$

101. (a) $\begin{cases} 900a + 30b + c = 55 \\ 1600a + 40b + c = 105 \\ 2500a + 50b + c = 188 \end{cases}$

$y = 0.165x^2 - 6.55x + 103$

(b) 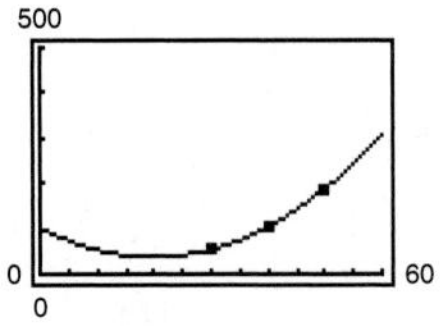

(c) 453 ft

103. $\dfrac{60}{100 - p} - \dfrac{60}{100 + p}$

105. False. The leading coefficients are not all 1.

107. The student did not work the problem correctly. Because $\dfrac{x^2 + 1}{x(x - 1)}$ is an improper fraction, the student should have divided before decomposing.

109. No. There are two arithmetic errors. The constant in the second equation should be -11 and the coefficient of z in the third equation should be 2.

111. $\begin{cases} 2x + y - 5z = 3 \\ -4x - 2y + 10z = 7 \end{cases}$ **113.** $x = 5, y = 5, \lambda = -5$

115. (a) $-4, 0, 3$
(b) 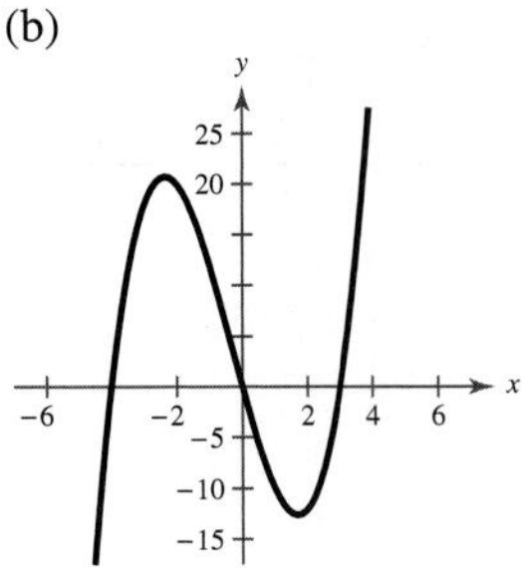

117. (a) $-4, -\frac{3}{2}, 3$
(b) 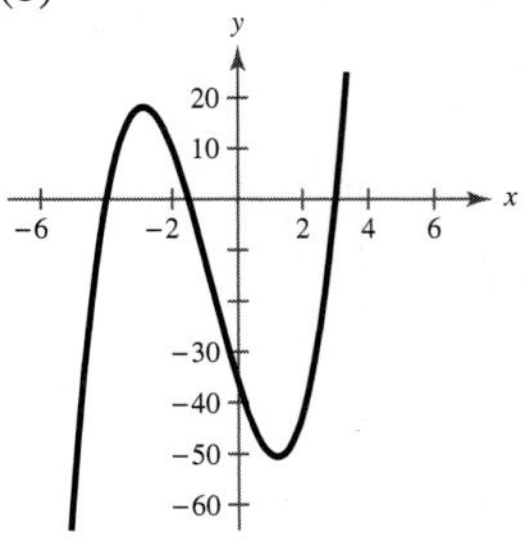

119. 9 **121.** Answers will vary.

Section 5.4 (page 441)

1. matrix **3.** Gauss-Jordan elimination **5.** No
7. 1×2 **9.** 3×1 **11.** 2×2

13. $\begin{bmatrix} 4 & -3 & \vdots & -5 \\ -1 & 3 & \vdots & 12 \end{bmatrix}$ 2×3

15. $\begin{bmatrix} 1 & 10 & -2 & \vdots & 2 \\ 5 & -3 & 4 & \vdots & 0 \\ 2 & 1 & 0 & \vdots & 6 \end{bmatrix}$ 3×4

17. $\begin{bmatrix} 7 & -5 & 1 & \vdots & 13 \\ 19 & 0 & -8 & \vdots & 10 \end{bmatrix}$ 2×4

19. $\begin{cases} 3x + 4y = 0 \\ x - y = 7 \end{cases}$ **21.** $\begin{cases} 12y + 3z = 0 \\ -2x + 18y = 10 \\ x + 7y - 8z = 43 \end{cases}$

23. Multiply R_1 by $-\frac{1}{4}$. **25.** Add 5 times R_1 to R_3.

27. $\begin{bmatrix} 1 & 4 & 3 \\ 0 & 2 & -1 \end{bmatrix}$

29. $\begin{bmatrix} 1 & 1 & 4 & -1 \\ 0 & 5 & -2 & 6 \\ 0 & 3 & 20 & 4 \end{bmatrix}, \begin{bmatrix} 1 & 1 & 4 & -1 \\ 0 & 1 & -\frac{2}{5} & \frac{6}{5} \\ 0 & 3 & 20 & 4 \end{bmatrix}$

31. (a) i) $\begin{bmatrix} 3 & 0 & \vdots & -6 \\ 6 & -4 & \vdots & -28 \end{bmatrix}$ ii) $\begin{bmatrix} 3 & 0 & \vdots & -6 \\ 0 & -4 & \vdots & -16 \end{bmatrix}$
iii) $\begin{bmatrix} 3 & 0 & \vdots & -6 \\ 0 & 1 & \vdots & 4 \end{bmatrix}$ iv) $\begin{bmatrix} 1 & 0 & \vdots & -2 \\ 0 & 1 & \vdots & 4 \end{bmatrix}$
(b) $x = -2, y = 4$ (c) Answers will vary.

33. i) 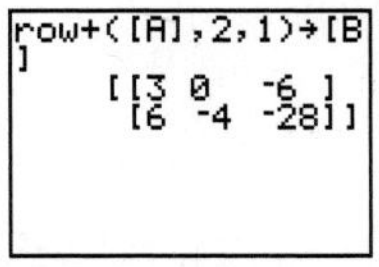

ii) 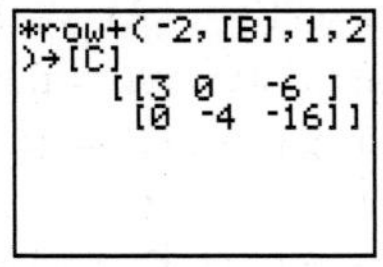

iii) 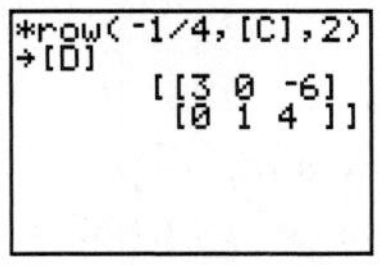

iv) 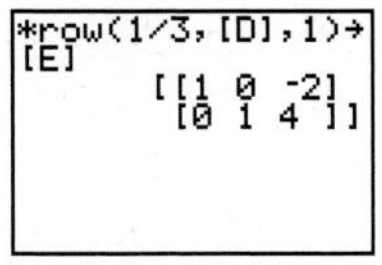

35. Reduced row-echelon form **37.** Not in row-echelon form

39. Row-echelon form **41.** $\begin{bmatrix} 1 & -3 & 3 \\ 0 & 1 & -2 \end{bmatrix}$

43. $\begin{bmatrix} 1 & 2 & -1 & 3 \\ 0 & 1 & -2 & 5 \\ 0 & 0 & 1 & -1 \end{bmatrix}$ **45.** $\begin{bmatrix} 1 & 0 & 0 \\ 0 & 1 & 0 \\ 0 & 0 & 1 \end{bmatrix}$

47. $\begin{bmatrix} 1 & 0 & -\frac{3}{7} & -\frac{8}{7} \\ 0 & 1 & -\frac{12}{7} & \frac{10}{7} \end{bmatrix}$

49. $\begin{cases} x - 2y = 4 \\ y = -3 \end{cases}$ **51.** $\begin{cases} x - y + 4z = 0 \\ y - z = 2 \\ z = -2 \end{cases}$
$(-2, -3)$ $(8, 0, -2)$

53. $(7, -5)$ **55.** $(-4, -8, 2)$ **57.** $(3, 2)$
59. Inconsistent **61.** $(-4, -3, 6)$ **63.** $(3, -2, 5, 0)$
65. $(4, -3, 2)$ **67.** Inconsistent **69.** $(2a + 1, 3a + 2, a)$
71. $(5a + 4, -3a + 2, a)$ **73.** $(0, -6, 2)$
75. $(-5a, a, 3)$ **77.** Yes; $(-1, 1, -3)$ **79.** No
81. $y = x^2 + 2x + 5$ **83.** $y = 2x^2 - x + 1$
85. $f(x) = -9x^2 - 5x + 11$ **87.** $f(x) = x^3 - 2x^2 - 4x + 1$
89. $I_1 = \frac{13}{10}, I_2 = \frac{11}{5}, I_3 = \frac{9}{10}$

91. $\begin{cases} x + 5y + 10z + 20w = 95 \\ x + y + z + w = 26 \\ y - 4z = 0 \\ x - 2y = -1 \end{cases}$
15 \$1 bills, 8 \$5 bills, 2 \$10 bills, 1 \$20 bill

93. $\dfrac{8x^2}{(x - 1)^2(x + 1)} = \dfrac{2}{x + 1} + \dfrac{6}{x - 1} + \dfrac{4}{(x - 1)^2}$

95. (a) $y = 24t^2 + 67t + 1831$
(b) 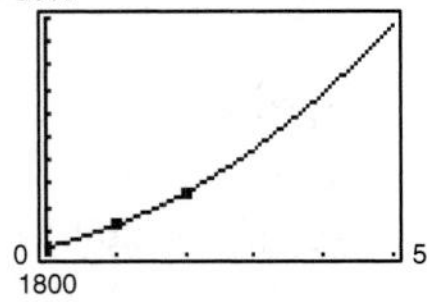

(c) 2015: \$2766
2020: \$4901
2025: \$8236
(d) Answers will vary.

97. (a) $x_1 = s, x_2 = t, x_3 = 600 - s,$
$x_4 = s - t, x_5 = 500 - t, x_6 = s, x_7 = t$
(b) $x_1 = 0, x_2 = 0, x_3 = 600, x_4 = 0, x_5 = 500,$
$x_6 = 0, x_7 = 0$
(c) $x_1 = 500, x_2 = 100, x_3 = 100, x_4 = 400,$
$x_5 = 400, x_6 = 500, x_7 = 100$

99. True. See Example 7.

101. $\begin{cases} x + y + 7z = -1 \\ x + 2y + 11z = 0 \\ 2x + y + 10z = -3 \end{cases}$
(Answer is not unique.)

103. No; Answers will vary.

105.

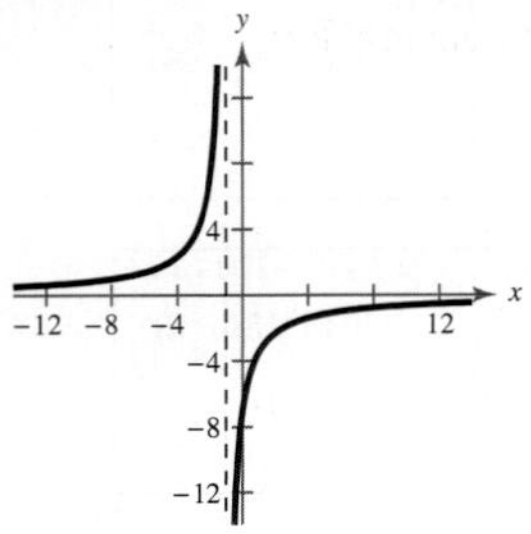

Asymptotes: $x = -1, y = 0$

107.

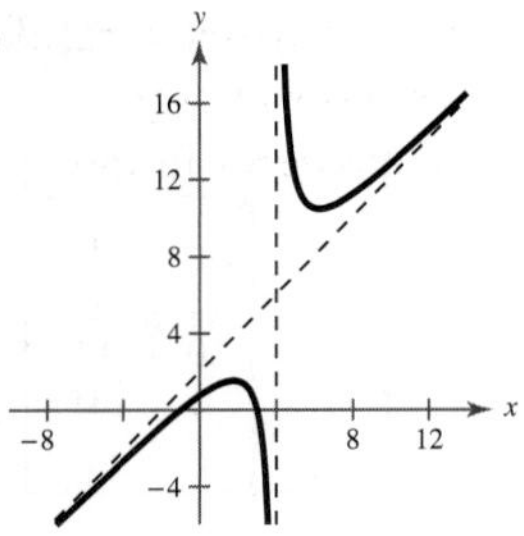

Asymptotes:
$x = 4, y = x + 2$

Section 5.5 (page 455)

1. equal **3.** zero, O

5. (a) iii (b) i (c) iv (d) v (e) ii

7. No, not in general. **9.** $x = 5, y = -8$

11. $x = -1, y = 4, z = 8$

13. (a) $\begin{bmatrix} 8 & -1 \\ 1 & 7 \end{bmatrix}$ (b) $\begin{bmatrix} 2 & -3 \\ 5 & -5 \end{bmatrix}$

(c) $\begin{bmatrix} 15 & -6 \\ 9 & 3 \end{bmatrix}$ (d) $\begin{bmatrix} 9 & -8 \\ 13 & -9 \end{bmatrix}$

15. (a) $\begin{bmatrix} 9 & 5 \\ 1 & -2 \\ -3 & 15 \end{bmatrix}$ (b) $\begin{bmatrix} 7 & -7 \\ 3 & 8 \\ -5 & -5 \end{bmatrix}$

(c) $\begin{bmatrix} 24 & -3 \\ 6 & 9 \\ -12 & 15 \end{bmatrix}$ (d) $\begin{bmatrix} 22 & -15 \\ 8 & 19 \\ -14 & -5 \end{bmatrix}$

17. (a) $\begin{bmatrix} 5 & 5 & -2 & 4 & 4 \\ -5 & 10 & 0 & -4 & -7 \end{bmatrix}$

(b) $\begin{bmatrix} 3 & 5 & 0 & 2 & 4 \\ 7 & -6 & -4 & 2 & 7 \end{bmatrix}$

(c) $\begin{bmatrix} 12 & 15 & -3 & 9 & 12 \\ 3 & 6 & -6 & -3 & 0 \end{bmatrix}$

(d) $\begin{bmatrix} 10 & 15 & -1 & 7 & 12 \\ 15 & -10 & -10 & 3 & 14 \end{bmatrix}$

19. (a) Not possible (b) Not possible

(c) $\begin{bmatrix} 18 & 0 & 9 \\ -3 & -12 & 0 \end{bmatrix}$ (d) Not possible

21. $\begin{bmatrix} -8 & -7 \\ 15 & -1 \end{bmatrix}$ **23.** $\begin{bmatrix} -3 & -\frac{1}{3} & 1 \\ -4 & \frac{8}{3} & -5 \end{bmatrix}$

25. $\begin{bmatrix} -17.143 & 2.143 \\ 11.571 & 10.286 \end{bmatrix}$ **27.** $\begin{bmatrix} -16.463 & -33.373 \\ 3.706 & -26.654 \\ 2.167 & 18.382 \end{bmatrix}$

29. $\begin{bmatrix} -\frac{2}{3} & \frac{8}{3} \\ \frac{7}{3} & 0 \\ -3 & -\frac{7}{3} \end{bmatrix}$ **31.** $\begin{bmatrix} 2 & -5 \\ -5 & 0 \\ 5 & 6 \end{bmatrix}$ **33.** $\begin{bmatrix} 11 & 1 \\ -11 & 5 \\ 54 & -6 \end{bmatrix}$

35. Not possible **37.** $\begin{bmatrix} 2 & 0 & 0 \\ 0 & -1 & 0 \\ 0 & 0 & -\frac{1}{3} \end{bmatrix}$

39. $\begin{bmatrix} 10 & -40 & 20 \\ -6 & 24 & -12 \\ 8 & -32 & 16 \end{bmatrix}$

41. (a) $\begin{bmatrix} 0 & 15 \\ 6 & 12 \end{bmatrix}$ (b) $\begin{bmatrix} -2 & 2 \\ 31 & 14 \end{bmatrix}$ (c) $\begin{bmatrix} 9 & 6 \\ 12 & 12 \end{bmatrix}$

43. (a) $\begin{bmatrix} 0 & -10 \\ 10 & 0 \end{bmatrix}$ (b) $\begin{bmatrix} 0 & -10 \\ 10 & 0 \end{bmatrix}$ (c) $\begin{bmatrix} 8 & -6 \\ 6 & 8 \end{bmatrix}$

45. (a) $\begin{bmatrix} 7 & 7 & 14 \\ 8 & 8 & 16 \\ -1 & -1 & -2 \end{bmatrix}$ (b) $[13]$ (c) Not possible

47. $\begin{bmatrix} 124 & -70 \\ 228 & 452 \\ 192 & -72 \end{bmatrix}$ **49.** Not possible **51.** $\begin{bmatrix} 5 & 8 \\ -4 & -16 \end{bmatrix}$

53. $\begin{bmatrix} -4 & 10 \\ 3 & 14 \end{bmatrix}$

55. (a) No (b) Yes (c) No (d) No

57. (a) No (b) No (c) No (d) Yes

59. (a) $\begin{bmatrix} -1 & 1 \\ -2 & 1 \end{bmatrix}\begin{bmatrix} x_1 \\ x_2 \end{bmatrix} = \begin{bmatrix} 4 \\ 0 \end{bmatrix}$ (b) $\begin{bmatrix} 4 \\ 8 \end{bmatrix}$

61. (a) $\begin{bmatrix} -2 & -3 \\ 6 & 1 \end{bmatrix}\begin{bmatrix} x_1 \\ x_2 \end{bmatrix} = \begin{bmatrix} -4 \\ -36 \end{bmatrix}$ (b) $\begin{bmatrix} -7 \\ 6 \end{bmatrix}$

63. (a) $\begin{bmatrix} 1 & -2 & 3 \\ -1 & 3 & -1 \\ 2 & -5 & 5 \end{bmatrix}\begin{bmatrix} x_1 \\ x_2 \\ x_3 \end{bmatrix} = \begin{bmatrix} 9 \\ -6 \\ 17 \end{bmatrix}$ (b) $\begin{bmatrix} 1 \\ -1 \\ 2 \end{bmatrix}$

65. (a) $\begin{bmatrix} 1 & -5 & 2 \\ -3 & 1 & -1 \\ 0 & -2 & 5 \end{bmatrix}\begin{bmatrix} x_1 \\ x_2 \\ x_3 \end{bmatrix} = \begin{bmatrix} -20 \\ 8 \\ -16 \end{bmatrix}$ (b) $\begin{bmatrix} -1 \\ 3 \\ -2 \end{bmatrix}$

67. (a) $\begin{bmatrix} 7 & -2 & 5 \\ -6 & 13 & -8 \\ 16 & 11 & -3 \end{bmatrix}$ (b) $\begin{bmatrix} 7 & -2 & 5 \\ -6 & 13 & -8 \\ 16 & 11 & -3 \end{bmatrix}$

The answers are the same.

69. (a) $\begin{bmatrix} 25 & -34 & 28 \\ -53 & 34 & -7 \\ -76 & 30 & 21 \end{bmatrix}$ (b) $\begin{bmatrix} 25 & -34 & 28 \\ -53 & 34 & -7 \\ -76 & 30 & 21 \end{bmatrix}$

The answers are the same.

71. (a) $\begin{bmatrix} 26 & 11 & 0 \\ 11 & 20 & -3 \\ 11 & 14 & 0 \end{bmatrix}$ (b) $\begin{bmatrix} 26 & 11 & 0 \\ 11 & 20 & -3 \\ 11 & 14 & 0 \end{bmatrix}$

The answers are the same.

73. (a) and (b) $\begin{bmatrix} 3 & -6 & 0 \\ 3 & 3 & 0 \end{bmatrix}$

75. Not possible, undefined **77.** Not possible, undefined

79. (a) and (b) $\begin{bmatrix} 5 & 16 & -13 \\ -8 & 5 & 0 \end{bmatrix}$ **81.** $\begin{bmatrix} -4 & 0 \\ 8 & 2 \end{bmatrix}$

83. $\begin{bmatrix} 85 & 102 & 51 & 34 \\ 119 & 136 & 170 & 68 \end{bmatrix}$ **85.** $\begin{bmatrix} 84 & 60 & 30 \\ 42 & 120 & 84 \end{bmatrix}$

87. (a) [\$3.50 \$6.00]

(b) [\$1037.50 \$1400.00 \$1012.50]

The entries represent the total profits made at the three outlets.

89. $\begin{bmatrix} \$23.20 & \$20.50 \\ \$38.20 & \$33.80 \\ \$76.90 & \$68.50 \end{bmatrix}$ The entries represent labor costs at the two plants for the three boat sizes.

91. $\begin{bmatrix} 0.40 & 0.15 & 0.15 \\ 0.28 & 0.53 & 0.17 \\ 0.32 & 0.32 & 0.68 \end{bmatrix}$ P^2 represents the proportion of changes in party affiliations after two elections.

93. True. To add two matrices, you add corresponding entries.
95. Not possible **97.** 3×2 **99.** Not possible
101. 2×3 **103 and 105.** Answers will vary.

107. $AC = BC = \begin{bmatrix} 2 & 3 \\ 2 & 3 \end{bmatrix}, A \neq B$

109. (a) $A^2 = \begin{bmatrix} -1 & 0 \\ 0 & -1 \end{bmatrix}, A^3 = \begin{bmatrix} -i & 0 \\ 0 & -i \end{bmatrix}, A^4 = \begin{bmatrix} 1 & 0 \\ 0 & 1 \end{bmatrix}$

The entries on the main diagonal are i^2 in A^2, i^3 in A^3, and i^4 in A^4.

(b) $B^2 = \begin{bmatrix} 1 & 0 \\ 0 & 1 \end{bmatrix}$

B^2 is the identity matrix.

111. (a) No, because they have different dimensions.
(b) No, because they have different dimensions.
(c) Yes, because the number of columns of A is equal to the number of rows of B.

113. $\ln \dfrac{64}{\sqrt[3]{x^2+3}}$

Section 5.6 (page 466)

1. inverse **3.** No **5–9.** Answers will vary.

11. $\begin{bmatrix} \frac{1}{2} & 0 \\ 0 & \frac{1}{3} \end{bmatrix}$ **13.** $\begin{bmatrix} -3 & 2 \\ -2 & 1 \end{bmatrix}$ **15.** Does not exist

17. Does not exist **19.** $\begin{bmatrix} 1 & 1 & -1 \\ -3 & 2 & -1 \\ 3 & -3 & 2 \end{bmatrix}$

21. $\begin{bmatrix} -\frac{3}{2} & \frac{3}{2} & 1 \\ \frac{9}{2} & -\frac{7}{2} & -3 \\ -1 & 1 & 1 \end{bmatrix}$ **23.** $\begin{bmatrix} -12 & -5 & -9 \\ -4 & -2 & -4 \\ -8 & -4 & -6 \end{bmatrix}$

25. $\dfrac{5}{11}\begin{bmatrix} 0 & -4 & 2 \\ -22 & 11 & 11 \\ 22 & -6 & -8 \end{bmatrix}$ **27.** Does not exist

29. Does not exist **31.** $\dfrac{1}{59}\begin{bmatrix} 16 & 15 \\ -4 & 70 \end{bmatrix}$ **33.** $\begin{bmatrix} \frac{5}{13} & -\frac{3}{13} \\ \frac{1}{13} & \frac{2}{13} \end{bmatrix}$

35. $k = 0$ **37.** $(5, 0)$ **39.** $(-8, -6)$ **41.** $(3, 8, -11)$
43. $(2, 1, 0, 0)$ **45.** $(2, -2)$
47. Not possible, because A is not invertible. **49.** $(-4, -8)$
51. $(-1, 3, 2)$ **53.** $(0.3125t + 0.8125, 1.1875t + 0.6875, t)$
55. $(5, 0, -2, 3)$
57. \$4000 in AAA-rated bonds, \$2000 in A-rated bonds, and \$4000 in B-rated bonds
59. \$30,000 in AAA-rated bonds, \$90,000 in A-rated bonds, and \$180,000 in B-rated bonds
61. $I_1 = \frac{1}{2}$ ampere, $I_2 = 3$ amperes, $I_3 = 3.5$ amperes
63. 100 bags for seedlings, 100 bags for general potting, 100 bags for hardwood plants

65. (a) $\begin{cases} 2.5r + 4l + 2i = 300 \\ -r + 2l + 2i = 0 \\ r + l + i = 120 \end{cases}$

(b) $\begin{bmatrix} 2.5 & 4 & 2 \\ -1 & 2 & 2 \\ 1 & 1 & 1 \end{bmatrix}\begin{bmatrix} r \\ l \\ i \end{bmatrix} = \begin{bmatrix} 300 \\ 0 \\ 120 \end{bmatrix}$

(c) 80 roses, 10 lilies, 30 irises

67. True; $AA^{-1} = I = A^{-1}A$.
69. False; Only square matrices are nonsingular.
71. Answers will vary.
73. (a) Answers will vary.

(b) $A^{-1} = \begin{bmatrix} \frac{1}{a_{11}} & 0 & 0 & 0 & \dots & 0 \\ 0 & \frac{1}{a_{22}} & 0 & 0 & \dots & 0 \\ 0 & 0 & \frac{1}{a_{33}} & 0 & \dots & 0 \\ \vdots & \vdots & \vdots & \vdots & \dots & \vdots \\ 0 & 0 & 0 & 0 & \dots & \frac{1}{a_{nn}} \end{bmatrix}$

75. $\ln 3 \approx 1.099$ **77.** $\dfrac{e^{12/7}}{3} \approx 1.851$

Section 5.7 (page 473)

1. determinant **3.** -5 **5.** 4 **7.** 32 **9.** -24
11. -9 **13.** 7.76
15. 11.998
17. (a) $M_{11} = -5, M_{12} = 2, M_{21} = 4, M_{22} = 3$
(b) $C_{11} = -5, C_{12} = -2, C_{21} = -4, C_{22} = 3$
19. (a) $M_{11} = 10, M_{12} = -43, M_{13} = 2, M_{21} = -30, M_{22} = 17, M_{23} = -6, M_{31} = 54, M_{32} = -53, M_{33} = -34$
(b) $C_{11} = 10, C_{12} = 43, C_{13} = 2, C_{21} = 30, C_{22} = 17, C_{23} = 6, C_{31} = 54, C_{32} = 53, C_{33} = -34$
21. (a) -75 (b) -75 **23.** (a) 170 (b) 170
25. -58 **27.** 0 **29.** -9 **31.** -168 **33.** 412
35. -336 **37.** 566

39. (a) 0 (b) -1 (c) $\begin{bmatrix} -2 & -5 \\ 4 & 10 \end{bmatrix}$
(d) 0; $|AB| = |A| \cdot |B|$

41. (a) -23 (b) 1 (c) $\begin{bmatrix} -9 & 4 & 1 \\ -5 & -10 & 6 \\ 4 & -1 & -1 \end{bmatrix}$
(d) -23; $|AB| = |A| \cdot |B|$

43. (a) -25 (b) -220 (c) $\begin{bmatrix} -7 & -16 & -1 & -28 \\ -4 & -14 & -11 & 8 \\ 13 & 4 & 4 & -4 \\ -2 & 3 & 2 & 2 \end{bmatrix}$
(d) 5500; $|AB| = |A| \cdot |B|$

45–49. Answers will vary. **51.** $x = \pm 2$ **53.** $x = \pm\frac{3}{2}$
55. $x = -1, 3$ **57.** $x = -4, -1$ **59.** $x = 1, \frac{1}{2}$
61. $x = 3$ **63.** $8uv - 1$ **65.** e^{5x} **67.** $1 - \ln x$
69. True. Expansion by cofactors on a row of zeros is zero.
71. Answers will vary. Sample answer:

$A = \begin{bmatrix} 1 & 0 & -3 \\ 6 & -2 & 7 \\ 9 & 5 & -1 \end{bmatrix}, \quad B = \begin{bmatrix} 3 & 1 & 5 \\ -8 & 1 & 0 \\ -7 & 6 & -2 \end{bmatrix}$

$|A + B| = -328, |A| + |B| = -404$

73. (a) 6 (b) $\begin{bmatrix} \frac{1}{3} & -\frac{1}{3} \\ \frac{1}{3} & \frac{1}{6} \end{bmatrix}$ (c) $\frac{1}{6}$ (d) They are reciprocals.

75. (a) 2 (b) $\begin{bmatrix} -4 & -5 & 1.5 \\ -1 & -1 & 0.5 \\ -1 & -1 & 0 \end{bmatrix}$ (c) $\frac{1}{2}$
(d) They are reciprocals.

77. (a) Columns 2 and 3 are interchanged.
(b) Rows 1 and 3 are interchanged.

79. (a) 3 is factored from the second row.
(b) 2 and 4 are factored from the first and second columns, respectively.

81. (a) 15 (b) -75 (c) -120
The determinant of a triangular matrix is the product of the numbers along the main diagonal.

83. Answers will vary.

85. $(2y-3)^2$ **87.** $(2, -4)$

Section 5.8 (page 484)

1. Cramer's Rule **3.** $-\frac{1}{2}$ **5.** $\frac{5}{2}$ **7.** $\frac{33}{8}$ **9.** 24

11. $x = 0, -\frac{16}{5}$ **13.** Collinear **15.** Not collinear

17. $x = 3$ **19.** $(-3, -2)$ **21.** $(-1, 2)$

23. Not possible **25.** (a) and (b) $\left(0, -\frac{1}{2}, \frac{1}{2}\right)$

27. (a) $y = 0.143t^2 + 1.417t + 57$
(b)
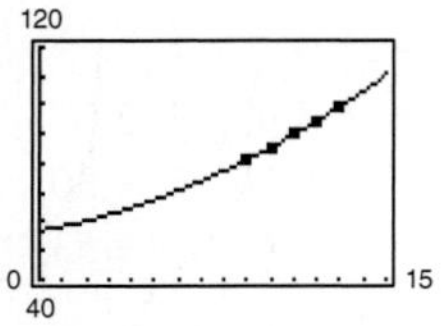

The model fits the data well.
(c) \$110.43 billion

29. (a) [3 15], [13 5], [0 8], [15 13], [5 0], [19 15], [15 14]
(b) 48 81 28 51 24 40 54 95 5 10 64 113 57 100

31. (a) [3 1 12], [12 0 13], [5 0 20], [15 13 15], [18 18 15], [23 0 0]
(b) -68 21 35 -66 14 39 -115 35 60 -62 15 32 -54 12 27 23 -23 0

33. HAPPY NEW YEAR **35.** CANCEL ORDERS SUE

37. True. Cramer's Rule divides by the determinant.

39. Answers will vary. **41.** $x + 4y - 19 = 0$

43. $2x - 7y - 27 = 0$

Review Exercises (page 488)

1. $(1, 1)$ **3.** No solution **5.** $(0.25, 0.625)$ **7.** $(5, 4)$

9. $(0, 0), (2, 8), (-2, 8)$ **11.** $(2, -0.5)$

13. $(0, 0), (4, -4)$ **15.** $(4, 4)$ **17.** 800 plants

19. 96 m × 144 m **21.** $(-8, 1)$ **23.** $\left(\frac{5}{2}, 3\right)$

25. Infinitely many solutions **27.** $(0, 0)$ **29.** No solution

31.
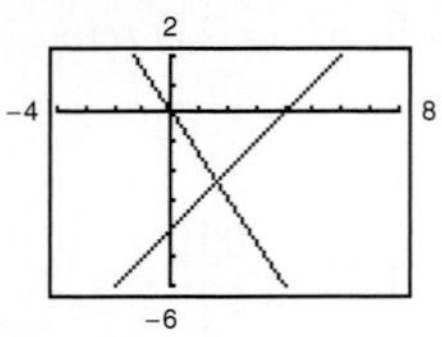

Consistent: $(1.6, -2.4)$

33.
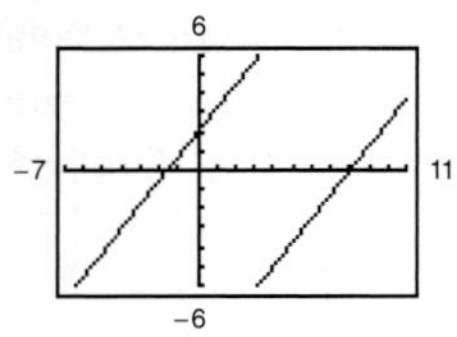

Inconsistent

35.
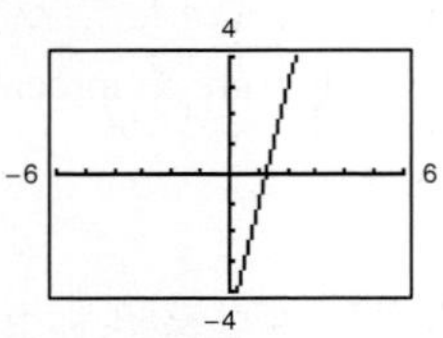

Consistent; All points on the line $8x - 2y = 10$

37. $\left(\dfrac{500{,}000}{7}, \dfrac{159}{7}\right)$ **39.** 218.75 mi/h; 193.75 mi/h

41. $(2, -4, -5)$ **43.** $\left(-\frac{257}{16}, -\frac{5}{16}, \frac{11}{16}\right)$

45. $(3a + 4, 2a + 5, a)$ **47.** $\left(-\frac{19}{6}, \frac{17}{12}, \frac{1}{3}\right)$

49. $(a - 4, a - 3, a)$

51.
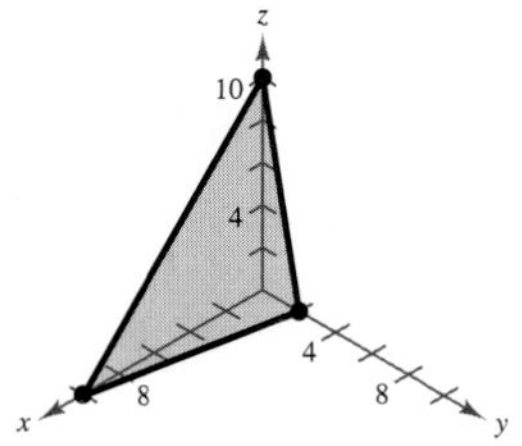

Sample answer: (10, 0, 0), (0, 2, 0), (0, 0, 10), (2, 1, 3)

53. $\dfrac{3}{x+2} - \dfrac{4}{x+4}$ **55.** $\dfrac{1}{2}\left(\dfrac{3}{x-3} - \dfrac{3}{x+3}\right)$

57. $\dfrac{1}{2}\left(\dfrac{3}{x-1} - \dfrac{x-3}{x^2+1}\right)$ **59.** $y = 2x^2 + x - 5$

61. 4 par-3 holes, 10 par-4 holes, 4 par-5 holes

63. 3×1 **65.** 1×1 **67.** $\left[\begin{array}{rr|r} 6 & -7 & 11 \\ -2 & 5 & -1 \end{array}\right]$

69. $\left[\begin{array}{rrr|r} 3 & -5 & 1 & 25 \\ -4 & 0 & -2 & -14 \\ 6 & 1 & 0 & 15 \end{array}\right]$

71. $\begin{cases} 5x + y + 7z = -9 \\ 4x + 2y = 10 \\ 9x + 4y + 2z = 3 \end{cases}$ **73.** $\begin{bmatrix} 1 & -\frac{5}{3} & 3 \\ 0 & 1 & \frac{9}{7} \end{bmatrix}$

75. $\begin{bmatrix} 1 & 1 & 1 \\ 0 & 1 & 2 \\ 0 & 0 & 1 \end{bmatrix}$ **77.** $\begin{bmatrix} 1 & 0 \\ 0 & 1 \end{bmatrix}$ **79.** $\begin{bmatrix} 1 & 0 & 0 \\ 0 & 1 & 0 \\ 0 & 0 & 1 \end{bmatrix}$

81. $(10, -12)$ **83.** Inconsistent **85.** $\left(\frac{1}{2}, -\frac{1}{3}, 1\right)$

87. $(3a + 1, -a, a)$ **89.** $(a - 3, -2a, a)$

91. $(2, -3, 3)$ **93.** Inconsistent **95.** $(1, 2, 2)$

97. $(3, 0, -4)$ **99.** Inconsistent **101.** $x = 12, y = -7$

103. $x = 1, y = 11$

105. (a) $\begin{bmatrix} 17 & -17 \\ 13 & 2 \end{bmatrix}$ (b) $\begin{bmatrix} -3 & 23 \\ -15 & 8 \end{bmatrix}$
(c) $\begin{bmatrix} 14 & 6 \\ -2 & 10 \end{bmatrix}$ (d) $\begin{bmatrix} 37 & -57 \\ 41 & -4 \end{bmatrix}$

107. (a) Not possible (b) Not possible
(c) $\begin{bmatrix} -22 & 32 & 38 \\ -14 & -4 & 2 \end{bmatrix}$ (d) Not possible

109. $\begin{bmatrix} -13 & -8 & 18 \\ 0 & 11 & -19 \end{bmatrix}$ **111.** $\begin{bmatrix} 17 & -39 \\ 37 & 6 \end{bmatrix}$

113. $\begin{bmatrix} 21 & -\frac{45}{4} & -\frac{159}{8} \\ \frac{93}{8} & \frac{327}{8} & \frac{291}{8} \end{bmatrix}$ **115.** $\begin{bmatrix} -4 & -2 \\ \frac{7}{2} & -\frac{17}{2} \\ -\frac{17}{2} & -1 \end{bmatrix}$

117. $\dfrac{1}{3}\begin{bmatrix} 6 & 2 \\ -4 & 11 \\ 10 & 0 \end{bmatrix}$ **119.** $\begin{bmatrix} 14 & -2 & 8 \\ 14 & -10 & 40 \\ 36 & -12 & 48 \end{bmatrix}$

121. Not possible

123. $\begin{bmatrix} 14 & -22 & 22 \\ 19 & -41 & 80 \\ 42 & -66 & 66 \end{bmatrix}$ **125.** $\begin{bmatrix} 1 & 17 \\ 12 & 36 \end{bmatrix}$

127. (a) [\$99 \$112]
(b) [\$12,400 \$23,080 \$22,820]
The entries represent the total prices of all tires at each factory.

129. Answers will vary. **131.** $\begin{bmatrix} 4 & -5 \\ 5 & -6 \end{bmatrix}$

133. $\begin{bmatrix} \frac{1}{2} & -1 & -\frac{1}{2} \\ \frac{1}{2} & -\frac{2}{3} & -\frac{5}{6} \\ 0 & \frac{2}{3} & \frac{1}{3} \end{bmatrix}$ **135.** $\begin{bmatrix} \frac{1}{5} & \frac{1}{5} \\ \frac{1}{10} & -\frac{1}{15} \end{bmatrix}$

137. Does not exist **139.** $\begin{bmatrix} -3 & 6 & -5.5 & 3.5 \\ 1 & -2 & 2 & -1 \\ 7 & -15 & 14.5 & -9.5 \\ -1 & 2.5 & -2.5 & 1.5 \end{bmatrix}$

141. $\begin{bmatrix} 1 & -1 \\ 4 & -\frac{7}{2} \end{bmatrix}$ **143.** Does not exist **145.** $\begin{bmatrix} 2 & \frac{20}{3} \\ \frac{1}{10} & \frac{1}{6} \end{bmatrix}$

147. $(36, 11)$ **149.** $(2, -1, -2)$ **151.** Not possible
153. $(-3, 1)$ **155.** $(1, 1, -2)$ **157.** -23 **159.** -42
161. 550
163. (a) $M_{11} = 4, M_{12} = 7, M_{21} = -1, M_{22} = 2$
(b) $C_{11} = 4, C_{12} = -7, C_{21} = 1, C_{22} = 2$
165. (a) $M_{11} = 30, M_{12} = -12, M_{13} = -21, M_{21} = 20,$
$M_{22} = 19, M_{23} = 22, M_{31} = 5, M_{32} = -2, M_{33} = 19$
(b) $C_{11} = 30, C_{12} = 12, C_{13} = -21, C_{21} = -20,$
$C_{22} = 19, C_{23} = -22, C_{31} = 5, C_{32} = 2, C_{33} = 19$
167. 130 **169.** -24 **171.** -27 **173.** 279
175. 16 **177.** 1.75 **179.** 48 **181.** Not collinear
183. Collinear **185.** $(1, 2)$ **187.** Not possible
189. $(-1, 4, 5)$ **191.** $(0, -2.4, -2.6)$
193. (a) and (b) $\left(\frac{53}{33}, -\frac{17}{33}, \frac{61}{66}\right)$
195. (a) [12 15 15], [11 0 15], [21 20 0], [2 5 12], [15 23 0]
(b) −21 6 0 −68 8 45 102 −42 −60 −53 20 21 99 −30 −69
197. I WILL BE BACK **199.** THAT IS MY FINAL ANSWER
201. False. The solution may have irrational numbers.
203. Elementary row operations correspond to the operations performed on a system of equations.

Chapter Test (page 494)

1. $(4, -2)$ **2.** $(0, -1), (1, 0), (2, 1)$
3. No solution **4.** $(3, -4)$ **5.** $\left(\frac{7}{13}, -\frac{85}{26}\right)$
6. $(1, 0, -2)$ **7.** $y = -\frac{1}{2}x^2 + x + 6$
8. $\frac{5}{x-1} + \frac{3}{(x-1)^2}$ **9.** $\frac{1}{x} + \frac{2}{x^2} - \frac{1}{x^2+1}$
10. $(-2a + 1.5, 2a + 1, a)$ **11.** $(5, 2, -6)$

12. (a) $\begin{bmatrix} 1 & 0 & 4 \\ -7 & -6 & -1 \\ 0 & 4 & 0 \end{bmatrix}$ (b) $\begin{bmatrix} 15 & 12 & 12 \\ -12 & -12 & 0 \\ 3 & 6 & 0 \end{bmatrix}$

(c) $\begin{bmatrix} 7 & 6 & 12 \\ -18 & -16 & -2 \\ 1 & 10 & 0 \end{bmatrix}$ (d) $\begin{bmatrix} 36 & 20 & 4 \\ -28 & -24 & -4 \\ 10 & 8 & 2 \end{bmatrix}$

13. $(-2, 3, -1)$ **14.** 67 **15.** -2
16. 30 **17.** $\left(1, -\frac{1}{2}\right)$
18. $x_1 = 700 - s - t, x_2 = 300 - s - t,$
$x_3 = s, x_4 = 100 - t, x_5 = t$

Cumulative Test for Chapters 3–5 (page 495)

1.

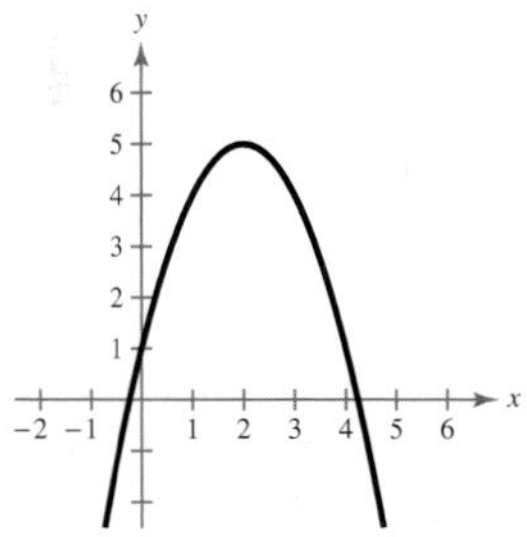

2.

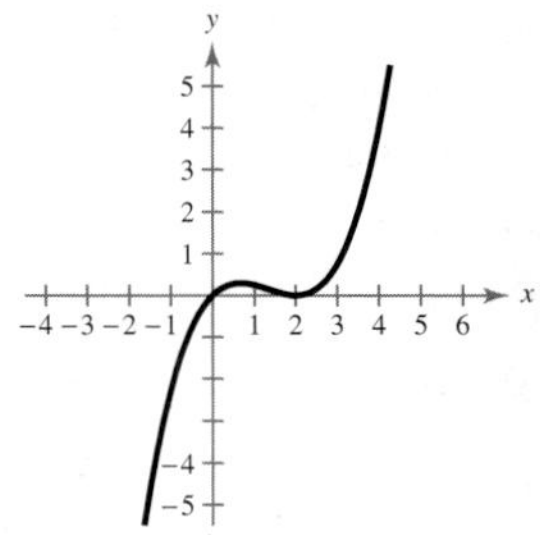

3. $(x + 2)(x - 2i)(x + 2i); -2, \pm 2i$ **4.** -5.151
5. $2x^2 + 7x + 48 + \frac{268}{x-6}$ **6.** $f(x) = x^4 + x^3 + 18x$

7.

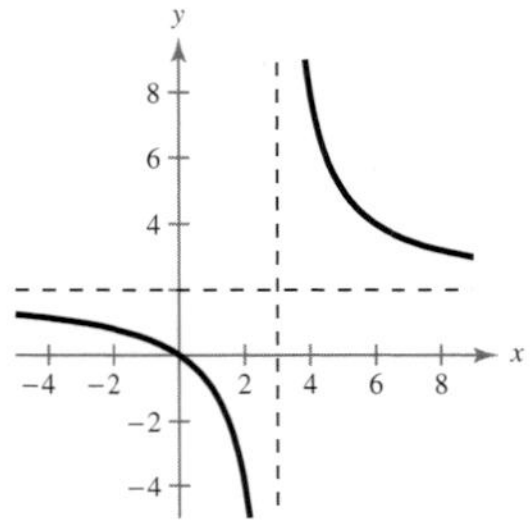

Asymptotes:
$x = 3, y = 2$

8.

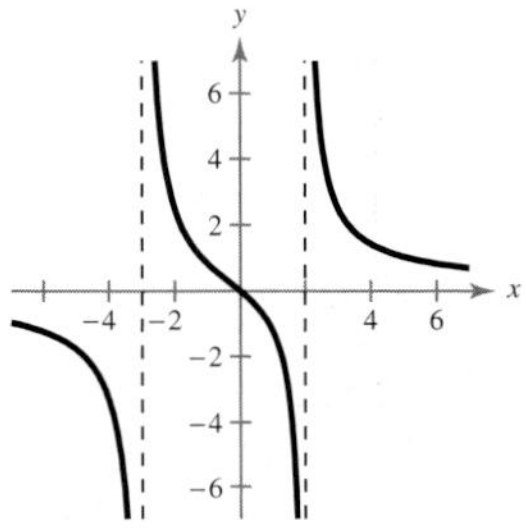

Asymptotes:
$x = -3, x = 2, y = 0$

9.

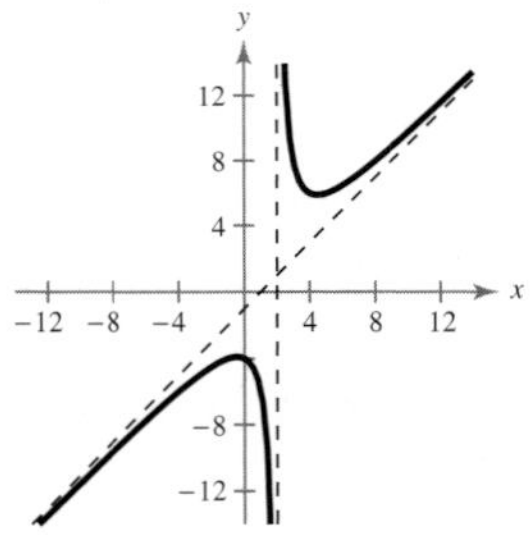

Asymptotes: $x = 2, y = x - 1$

10. Horizontal shift one unit to the left and vertical shift five units downward
11. Reflection in the axis and horizontal shift three units to the left
12. (a) All real numbers x (b) $(0, 10)$ (c) $y = 2$
13. (a) All real numbers x (b) $(0, 2)$ (c) $y = 0$
14. (a) All real numbers x such that $x > 0$
(b) $(1, 0)$ (c) $y = 0, x = 0$
15. (a) All real numbers x such that $x > \frac{1}{2}$
(b) $\left(\frac{7}{12}, 0\right)$ (c) $x = \frac{1}{2}$
16. -4 **17.** 10 **18.** -3 **19.** 1.723
20. 0.905 **21.** 0.585
22. $\ln(x + 2) + \ln(x - 2) - \ln(x^2 + 1)$ **23.** $\ln \frac{x^2(x+1)}{x-1}$
24. $x = 3$ **25.** 1.459 **26.** $x = 3$ **27.** 105.966
28. $x = 0, 1$ **29.** No solution **30.** $(11, 3)$
31. $(8, 4), (2, -2)$ **32.** $\left(-\frac{1}{2}, -\frac{2}{3}\right)$
33. $\left(-\frac{13}{6}a - \frac{1}{3}, \frac{5}{6}a + \frac{2}{3}, a\right)$ **34.** $(1, 2, 0)$ **35.** $(-2, 1, -3)$

36. $\begin{bmatrix} -7 & -10 & -16 \\ -6 & 18 & 9 \\ -12 & 16 & 7 \end{bmatrix}$ **37.** $\begin{bmatrix} -18 & 15 & -14 \\ 28 & 11 & 34 \\ -20 & 52 & -1 \end{bmatrix}$

38. $\begin{bmatrix} 3 & -31 & 2 \\ 22 & 18 & 6 \\ 52 & -40 & 14 \end{bmatrix}$ **39.** $\begin{bmatrix} 5 & 36 & 31 \\ -36 & 12 & -36 \\ 16 & 0 & 18 \end{bmatrix}$

40. 22 **41.** -214 **42.** -14 **43.** 2.8

44. $\begin{bmatrix} 1 & \frac{3}{2} & -\frac{18}{7} \\ 0 & -\frac{1}{2} & \frac{4}{7} \\ 0 & 0 & \frac{1}{7} \end{bmatrix}$ **45.** (a) \$1204.52 (b) \$1209.93

46. $k \approx 0.0106$; 4,112,067

47. (a) Quadratic model:
$S = -1.16963t^2 + 40.2326t + 147.853$; $r^2 \approx 0.9905$
Exponential model: $S = 229.067(1.0652)^t$; $r^2 \approx 0.9486$
Power model: $S = 160.681t^{0.4294}$; $r^2 \approx 0.9921$

(b) Quadratic model:

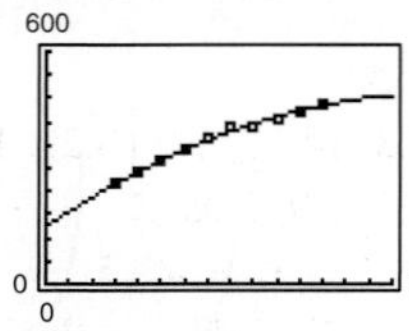

Exponential model:

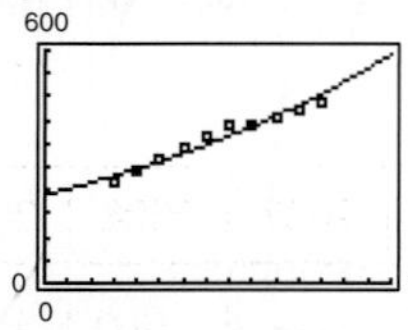

Power model:

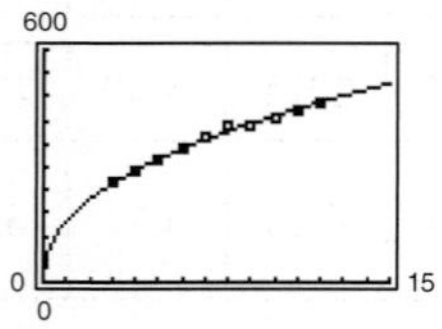

(c) The power model is the best fit because its coefficient of determination is closest to 1.

(d) \$581.603 billion; Answers will vary.

48. (a) $C = 59.95x + 150{,}000$
$R = 200x$

(b)

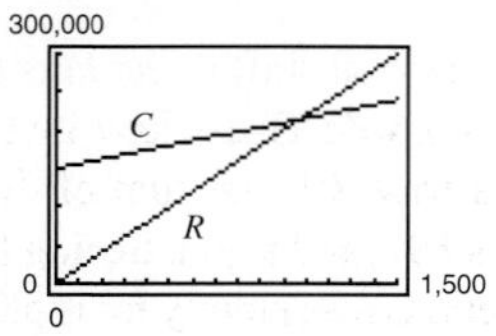

1071 units

49. 16 m × 22 m

Chapter 6

Section 6.1 (page 507)

1. terms **3.** index, upper limit, lower limit

5. (a) Finite sequence (b) Infinite sequence

7. $-3, -1, 1, 3, 5$ **9.** 3, 9, 27, 81, 243

11. $-\frac{1}{2}, \frac{1}{4}, -\frac{1}{8}, \frac{1}{16}, -\frac{1}{32}$ **13.** $2, \frac{3}{2}, \frac{4}{3}, \frac{5}{4}, \frac{6}{5}$ **15.** $\frac{1}{2}, \frac{2}{5}, \frac{3}{10}, \frac{4}{17}, \frac{5}{26}$

17. (a) 0, 1, 0, 0.5, 0 (b) $0, 1, 0, \frac{1}{2}, 0$

19. (a) 0.5, 0.75, 0.875, 0.938, 0.969 (b) $\frac{1}{2}, \frac{3}{4}, \frac{7}{8}, \frac{15}{16}, \frac{31}{32}$

21. (a) 1, 0.354, 0.192, 0.125, 0.089 (b) $1, \frac{1}{2^{3/2}}, \frac{1}{3^{3/2}}, \frac{1}{8}, \frac{1}{5^{3/2}}$

23. (a) $-1, 0.25, -0.111, 0.063, -0.04$
(b) $-1, \frac{1}{4}, -\frac{1}{9}, \frac{1}{16}, -\frac{1}{25}$

25. (a) and (b) 3, 15, 35, 63, 99

27. 9, 15, 21, 27, 33, 39, 45, 51, 57, 63

29. $3, \frac{5}{2}, \frac{7}{3}, \frac{9}{4}, \frac{11}{5}, \frac{13}{6}, \frac{15}{7}, \frac{17}{8}, \frac{19}{9}, \frac{21}{10}$ **31.** 0, 2, 0, 2, 0, 2, 0, 2, 0, 2

33. $\frac{100}{101}$ **35.** -73 **37.** $\frac{256}{129}$ **39.** $a_n = 5n - 2$

41. $a_n = 6n + 1$ **43.** $a_n = n^2 - 1$ **45.** $a_n = \frac{n+1}{n+2}$

47. $a_n = \frac{(-1)^{n+1}}{2^n}$ **49.** $a_n = 1 + \frac{1}{n}$

51. $a_n = (-1)^n + 2(1)^n = (-1)^n + 2$ **53.** 28, 24, 20, 16, 12

55. 3, 4, 6, 10, 18 **57.** 1, 3, 4, 7, 11

59. 6, 8, 10, 12, 14; $a_n = 2n + 4$

61. 81, 27, 9, 3, 1; $a_n = \frac{243}{3^n}$

63. (a) 1, 1, 0.5, 0.167, 0.042 (b) $1, 1, \frac{1}{2}, \frac{1}{6}, \frac{1}{24}$

65. (a) 0, 0.5, 0.667, 0.375, 0.133 (b) $0, \frac{1}{2}, \frac{2}{3}, \frac{3}{8}, \frac{2}{15}$

67. (a) 1, 0.5, 0.042, 0.001, 2.480×10^{-5}
(b) $1, \frac{1}{2}, \frac{1}{24}, \frac{1}{720}, \frac{1}{40{,}320}$

69. $\frac{1}{12}$ **71.** 495 **73.** $(n + 3)(n + 2)(n + 1)$

75. $\frac{1}{2n(2n + 1)}$ **77.** c **78.** b **79.** d **80.** a

81.

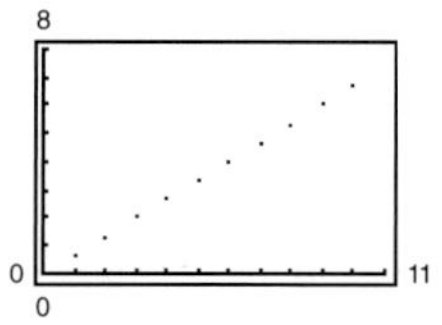

83.

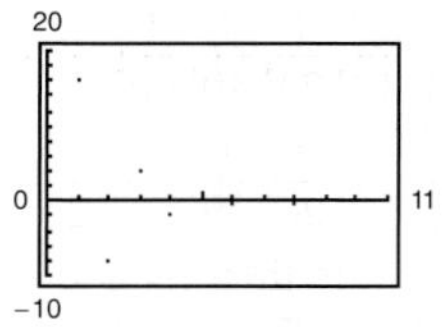

85.

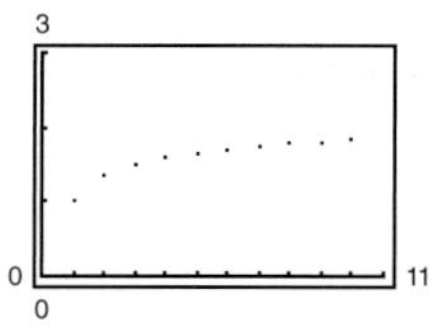

87. 35 **89.** 364 **91.** $\frac{124}{429}$ **93.** 40 **95.** 186

97. 31 **99.** 81 **101.** $0.6\overline{3}$ **103.** $\sum_{i=1}^{9} \frac{1}{3i} \approx 0.94299$

105. $\sum_{i=1}^{8} \left[2\left(\frac{i}{8}\right) + 3\right] = 33$ **107.** $\sum_{i=1}^{6} (-1)^i 3^i = 546$

109. $\sum_{i=1}^{20} \frac{(-1)^{i+1}}{i^2} \approx 0.821$ **111.** $\sum_{i=1}^{5} \frac{2^i - 1}{2^{i+1}} \approx 2.0156$

113. $\frac{1092}{625}$ **115.** $-\frac{3}{2}$ **117.** (a) $\frac{3333}{5000}$ (b) $\frac{2}{3}$

119. (a) $\frac{7777}{10{,}000}$ (b) $\frac{7}{9}$

121. (a) $A_1 = \$5037.50, A_2 = \$5075.28,$
$A_3 = \$5113.35, A_4 = \$5151.70,$
$A_5 = \$5190.33, A_6 = \$5229.26,$
$A_7 = \$5268.48, A_8 = \$5307.99$
(b) \$6741.74

123. (a) $p_0 = 5500, p_n = (0.75)p_{n-1} + 500$
(b) 4625, 3969, 3477, 3108; These values represent the number of trout in the lake for the years 2016 through 2019.
(c) 2000 trout; Answers will vary. Sample answer: As time passes, the population of trout decreases at a decreasing rate. Because the population is growing smaller and still declines 25%, each time 25% is taken from a smaller number there is a smaller decline in the number of trout.

125. True by the Properties of Sums

127. 1, 1, 2, 3, 5, 8, 13, 21, 34, 55, 89, 144;
$1, 2, \frac{3}{2}, \frac{5}{3}, \frac{8}{5}, \frac{13}{8}, \frac{21}{13}, \frac{34}{21}, \frac{55}{34}, \frac{89}{55}$

129. 1, 1, 2, 3, 5

131. $a_{n+1} = \frac{1}{2}a_n + \frac{(1+\sqrt{5})^n + (1-\sqrt{5})^n}{2^{n+1}}$
$a_{n+2} = \frac{3}{2}a_n + \frac{(1+\sqrt{5})^n + (1-\sqrt{5})^n}{2^{n+1}}$

133. $x, \frac{x^2}{2}, \frac{x^3}{6}, \frac{x^4}{24}, \frac{x^5}{120}$

135. $-\frac{x^3}{3}, \frac{x^5}{5}, -\frac{x^7}{7}, \frac{x^9}{9}, -\frac{x^{11}}{11}$

137. $-\frac{x^2}{2}, \frac{x^4}{24}, -\frac{x^6}{720}, \frac{x^8}{40{,}320}, -\frac{x^{10}}{3{,}628{,}800}$

139. $-x, \frac{x^2}{2}, -\frac{x^3}{6}, \frac{x^4}{24}, -\frac{x^5}{120}$

141. $x + 1, -\frac{(x+1)^2}{2}, \frac{(x+1)^3}{6}, -\frac{(x+1)^4}{24}, \frac{(x+1)^5}{120}$

143. $\frac{1}{4}, \frac{1}{12}, \frac{1}{24}, \frac{1}{40}, \frac{1}{60}; \frac{1}{2} - \frac{1}{2n+2}$

145. $\frac{1}{6}, \frac{1}{12}, \frac{1}{20}, \frac{1}{30}, \frac{1}{42}; \frac{1}{2} - \frac{1}{n+2}$

147. Yes, if there is a finite number of integer terms, you can always find a sum.

149. (a) $\begin{bmatrix} 2 \\ 6 \end{bmatrix}$ (b) $\begin{bmatrix} -10 \\ -15 \end{bmatrix}$ (c) Not possible
(d) Not possible

151. (a) $\begin{bmatrix} -3 & -7 & 4 \\ 4 & 4 & 1 \\ 1 & 4 & 3 \end{bmatrix}$ (b) $\begin{bmatrix} 8 & 17 & -14 \\ -12 & -13 & -9 \\ -3 & -15 & -10 \end{bmatrix}$
(c) $\begin{bmatrix} -2 & 7 & -16 \\ 4 & 42 & 45 \\ 1 & 23 & 48 \end{bmatrix}$ (d) $\begin{bmatrix} 16 & 31 & 42 \\ 10 & 47 & 31 \\ 13 & 22 & 25 \end{bmatrix}$

Section 6.2 (page 516)

1. $a_n = a_1 + (n-1)d$

3. A sequence is arithmetic when the differences between consecutive terms are the same.

5. Arithmetic sequence, $d = 2$

7. Arithmetic sequence, $d = -\frac{1}{2}$

9. Not an arithmetic sequence

11. 21, 34, 47, 60, 73
Arithmetic sequence, $d = 13$

13. 25, 21, 17, 13, 9
Arithmetic sequence, $d = -4$

15. 3, 6, 11, 20, 37
Not an arithmetic sequence

17. 1, 5, 1, 5, 1
Not an arithmetic sequence

19. $\frac{1}{4}, \frac{1}{4}, \frac{1}{4}, \frac{1}{4}, \frac{1}{4}$
Arithmetic sequence, $d = 0$

21. $a_n = 6n - 5$ **23.** $a_n = -7n + 50$

25. $a_n = -\frac{3}{4}n + \frac{15}{4}$ **27.** $a_n = 9n - 14$

29. $a_n = 103 - 3n$ **31.** 5, 11, 17, 23, 29

33. −2.6, −2.4, −2.2, −2.0, −1.8 **35.** −2, 2, 6, 10, 14

37. 22.45, 20.725, 19, 17.275, 15.55

39. 15, 19, 23, 27, 31; $d = 4$; $a_n = 11 + 4n$

41. $\frac{3}{5}, \frac{1}{2}, \frac{2}{5}, \frac{3}{10}, \frac{1}{5}$; $d = -\frac{1}{10}$; $a_n = -\frac{1}{10}n + \frac{7}{10}$

43. 59 **45.** −10.2

47.
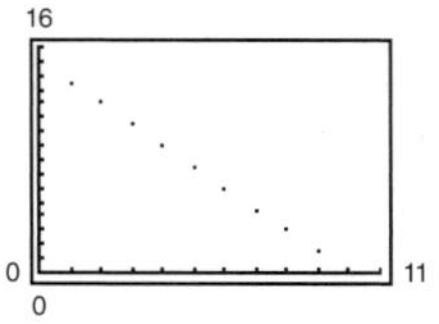

49.
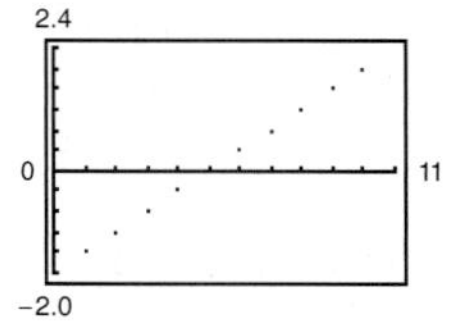

51. −1, 3, 7, 11, 15, 19, 23, 27, 31, 35

53. 19.25, 18.5, 17.75, 17, 16.25, 15.5, 14.75, 14, 13.25, 12.5

55. 1.55, 1.6, 1.65, 1.7, 1.75, 1.8, 1.85, 1.9, 1.95, 2

57. 110 **59.** −25 **61.** 5050 **63.** −4585 **65.** 620

67. 36 **69.** 4000 **71.** 1275 **73.** 355 **75.** 129,250

77. 440 **79.** 2777.5 **81.** 14,268 **83.** 405 bricks

85. $200,000

87. (a) $a_n = 0.39x + 1.89$
(b)

Year	2006	2007	2008	2009
Sales (in billions of dollars)	4.23	4.62	5.01	5.40

Year	2010	2011	2012	2013
Sales (in billions of dollars)	5.79	6.18	6.57	6.96

The model fits the data well.
(c) $44.76 billion (d) $69.72 billion; Answers will vary.

89. True. Use the recursion formula, $a_{n+1} = a_n + d$.

91. $x, 3x, 5x, 7x, 9x, 11x, 13x, 15x, 17x, 19x$

93. Given a_1 and a_2, you know that $d = a_2 - a_1$. So, $a_n = a_1 + (n-1)d$.

95. $S_n + 5n$; Answers will vary.

97. Answers will vary. Sample Answer: Gauss saw that the sum of the first and last numbers was 101, the sum of the second and second-last numbers was 101, and so on. Seeing that there were 50 such pairs of numbers, Gauss simply multiplied 50 by 101 to get the summation 5050.
$a_n = (n+1)\left(\frac{n}{2}\right)$, where n is the total number of natural numbers.

99. (1, 5, −1) **101.** Answers will vary.

Section 6.3 (page 525)

1. geometric, common **3.** geometric series **5.** $|r| < 1$

7. Geometric sequence, $r = 3$ **9.** Not a geometric sequence

11. Geometric sequence, $r = -\frac{1}{2}$

13. Geometric sequence, $r = 0.1$ **15.** Not a geometric sequence

17. 6, 18, 54, 162, 486 **19.** $1, \frac{1}{2}, \frac{1}{4}, \frac{1}{8}, \frac{1}{16}$

21. $5, -\frac{1}{2}, \frac{1}{20}, -\frac{1}{200}, \frac{1}{2000}$ **23.** $9, 9e, 9e^2, 9e^3, 9e^4$

25. 64, 32, 16, 8, 4; $r = \frac{1}{2}$; $a_n = 64\left(\frac{1}{2}\right)^{n-1}$

27. 9, 18, 36, 72, 144; $r = 2$; $a_n = 9(2)^{n-1}$

29. $6, -9, \frac{27}{2}, -\frac{81}{4}, \frac{243}{8}$; $r = -\frac{3}{2}$; $a_n = 6\left(-\frac{3}{2}\right)^{n-1}$

31. (a) and (b) 15.227 **33.** (a) 44.949 (b) $\frac{32{,}768}{729}$

35. (a) and (b) -8192 **37.** (a) and (b) 633.568

39. $a_n = 7(3)^{n-1}$; 45,927 **41.** $a_n = 5(6)^{n-1}$; 50,388,480

43. $\frac{1}{128}$ **45.** $-\frac{2}{9}$

47.

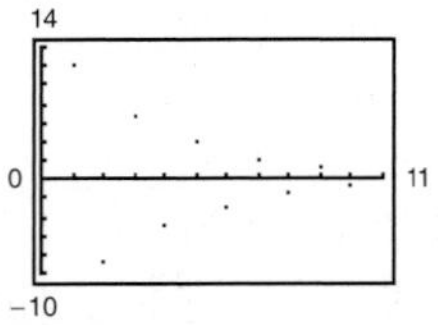

49.

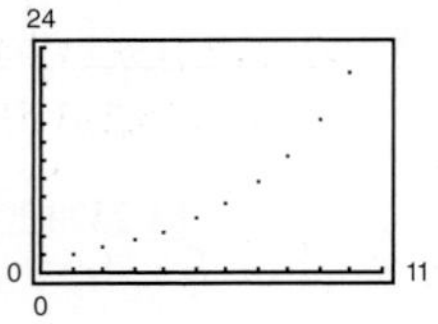

51. $8, 4, 6, 5, \frac{11}{2}$

53.

n	S_n
1	16
2	24
3	28
4	30
5	31
6	31.5
7	31.75
8	31.875
9	31.9375
10	31.96875

55. 511 **57.** 43 **59.** 2700.31 **61.** 6.4

63. 2092.60 **65.** $\sum_{n=1}^{7} 5(3)^{n-1}$ **67.** $\sum_{n=1}^{7} 2\left(-\frac{1}{4}\right)^{n-1}$

69. 50 **71.** $\frac{10}{3}$

73. Series does not have a finite sum because $\left|\frac{7}{3}\right| > 1$.

75. $\frac{1000}{89}$ **77.** $-\frac{30}{19}$ **79.** 27

81. Series does not have a finite sum because $\left|\frac{5}{2}\right| > 1$.

83. $-\frac{49}{9}$ **85.** $\frac{4}{11}$ **87.** $\frac{113}{90}$

89. Geometric; $r = 2$; 262,136

91. Geometric; $r = \frac{1}{3}$; 135 **93.** Arithmetic; $d = 6$; 720

95. Geometric; $r = 0.8$; 28.944 **97.** \$6480.83

99. Answers will vary.

101. (a) \$26,198.27 (b) \$26,263.88

103. (a) \$153,237.86 (b) \$153,657.02 **105.** 126 in.^{2}

107. (a) $T_n = 70(0.8)^n$ (b) 18.4°F; 4.8°F

(c)

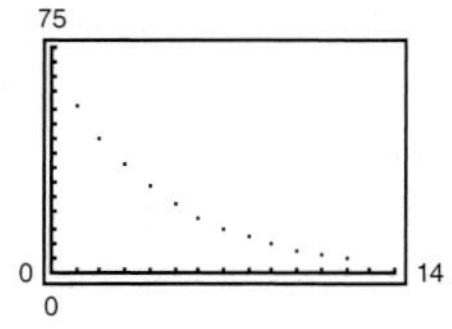

3.5 h

109. (a) $S = \sum_{n=0}^{\infty} 4\pi$ (b) $V = \sum_{n=0}^{\infty} \frac{4}{3}\pi\left(\frac{1}{3}\right)^n$

(c) Infinite surface area. Finite volume: 2π cubic units

111. 42 ft **113.** True. The terms all equal a_1.

115. $3, \frac{3x}{2}, \frac{3x^2}{4}, \frac{3x^3}{8}, \frac{3x^4}{16}$ **117.** $100e^{8x}$

119. (a)

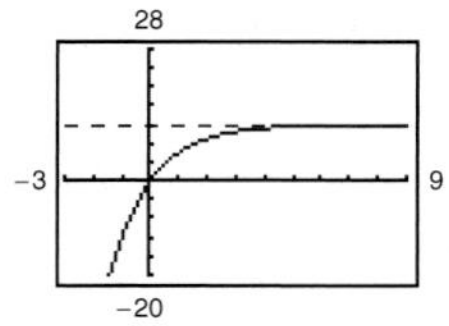

Horizontal asymptote: $y = 12$

Corresponds to the sum of the series

(b)

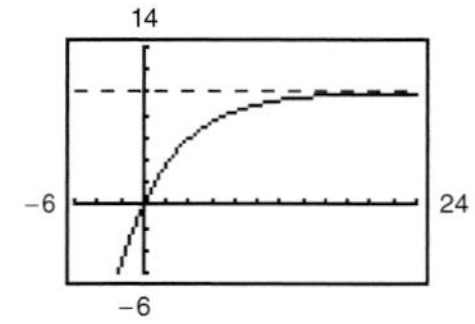

Horizontal asymptote: $y = 10$

Corresponds to the sum of the series.

121. Divide the second term by the first to obtain the common ratio. The nth term is the first term times the common ratio raised to the $(n - 1)$th power.

123. -38 **125.** Answers will vary.

Section 6.4 (page 534)

1. ${}_nC_r$ or $\binom{n}{r}$ **3.** Binomial Theorem, Pascal's Triangle

5. 21 **7.** 15,504 **9.** 14 **11.** 1 **13.** 210

15. 4950 **17.** 749,398 **19.** 230,300 **21.** 31,125

23. $x^5 + 5x^4 + 10x^3 + 10x^2 + 5x + 1$

25. $a^3 + 9a^2 + 27a + 27$ **27.** $y^3 - 12y^2 + 48y - 64$

29. $x^5 + 5x^4y + 10x^3y^2 + 10x^2y^3 + 5xy^4 + y^5$

31. $32x^5 - 80x^4y + 80x^3y^2 - 40x^2y^3 + 10xy^4 - y^5$

33. $64y^3 - 144y^2 + 108y - 27$

35. $64r^6 - 576r^5s + 2160r^4s^2 - 4320r^3s^3 + 4860r^2s^4 - 2916rs^5 + 729s^6$

37. $x^8 + 8x^6 + 24x^4 + 32x^2 + 16$

39. $-x^{10} + 25x^8 - 250x^6 + 1250x^4 - 3125x^2 + 3125$

41. $x^8 + 4x^6y^2 + 6x^4y^4 + 4x^2y^6 + y^8$

43. $729x^{18} - 1458x^{15}y + 1215x^{12}y^2 - 540x^9y^3 + 135x^6y^4 - 18x^3y^5 + y^6$

45. $\frac{1}{x^5} + \frac{5y}{x^4} + \frac{10y^2}{x^3} + \frac{10y^3}{x^2} + \frac{5y^4}{x} + y^5$

47. $\frac{16}{x^4} - \frac{64y}{x^3} + \frac{96y^2}{x^2} - \frac{64y^3}{x} + 16y^4$

49. $-512x^4 + 576x^3 - 240x^2 + 44x - 3$

51. $3x^5 + 15x^4 + 34x^3 + 42x^2 + 27x + 7$

53. $-4x^6 - 24x^5 - 60x^4 - 83x^3 - 42x^2 - 60x + 20$
55. $61{,}440x^7$ **57.** $360x^3y^2$ **59.** $1{,}259{,}712x^2y^7$
61. $-4{,}330{,}260{,}000x^3y^9$ **63.** 1,732,104 **65.** 180
67. $-489{,}888$ **69.** 210 **71.** 35 **73.** 6
75. $3125y^5 + 6250y^4 + 5000y^3 + 2000y^2 + 400y + 32$
77. $32x^5 + 240x^4y + 720x^3y^2 + 1080x^2y^3 + 810xy^4 + 243y^5$
79. $81t^4 - 216t^3v + 216t^2v^2 - 96tv^3 + 16v^4$
81. $27x^{3/2} + 135x + 225\sqrt{x} + 125$
83. $x^2 - 3x^{4/3}y^{1/3} + 3x^{2/3}y^{2/3} - y$
85. $3x^2 + 3xh + h^2, h \neq 0$
87. $6x^5 + 15x^4h + 20x^3h^2 + 15x^2h^3 + 6xh^4 + h^5, h \neq 0$
89. -4 **91.** $161 + 240i$ **93.** $2035 + 828i$
95. $-115 + 236i$ **97.** $-23 + 208\sqrt{3}i$ **99.** 1
101. $-\frac{1}{8}$ **103.** 1.172 **105.** 510,568.785
107.

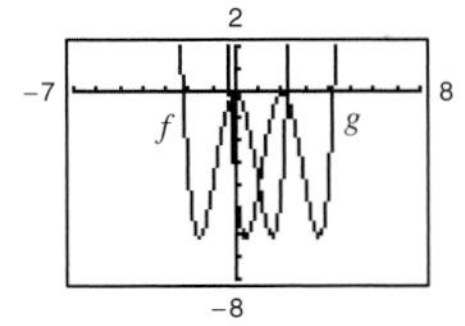

g is shifted two units to the right of f.
$g(x) = x^4 - 8x^3 + 19x^2 - 12x - 4$

109.

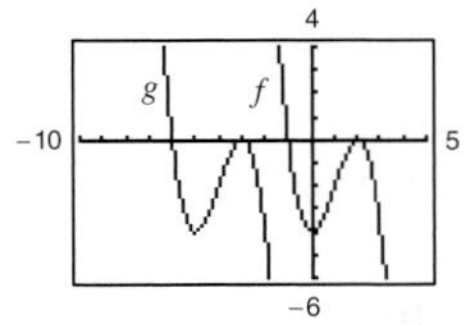

g is shifted five units to the left of f.
$g(x) = -x^3 - 12x^2 - 45x - 54$

111.

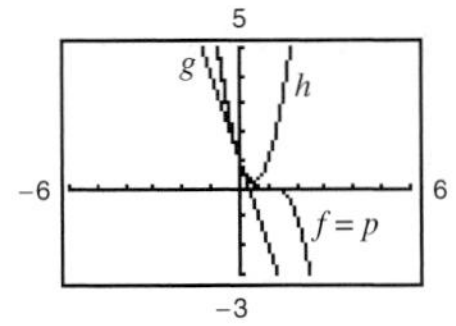

$p(x)$ is the expansion of $f(x)$.

113. 0.273 **115.** 0.171
117. (a) $g(t) = 0.005t^2 - 0.27t + 22.5$

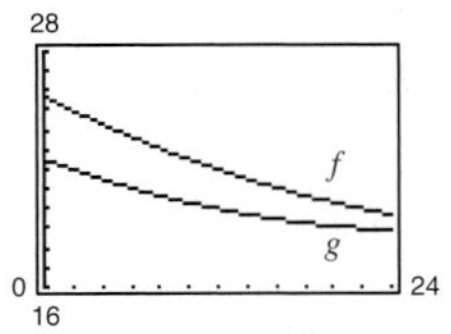

119. True. Pascal's Triangle is made up of binomial coefficients.
121. The first and last numbers in each row are 1. Every other number in each row is formed by adding the two numbers immediately above the number.

1 8 28 56 70 56 28 8 1
1 9 36 84 126 126 84 36 9 1
1 10 45 120 210 252 210 120 45 10 1

123. and 125. Answers will vary.
127. $5(2x)^4(-3y)^1 = -240x^4y$; Answers will vary.
129. $\begin{bmatrix} -1 & 6 \\ -1 & 5.5 \end{bmatrix}$

Section 6.5 (page 543)

1. Fundamental Counting Principle **3.** Permutation
5. 8 **7.** 6 **9.** 11 **11.** 3 **13.** 200 **15.** 1024
17. (a) 900 (b) 648 (c) 180 **19.** 24,000,000
21. (a) 100,000 (b) 20,000
23. (a) 10,000 (b) 9000 **25.** (a) 720 (b) 48
27. 120 **29.** 362,880 **31.** 24 **33.** 336 **35.** 720
37. 427,518,000 **39.** 197,149,680 **41.** 11,880
43. 1,092,624
45. ABCD, ABDC, ACBD, ACDB, ADBC, ADCB, BACD, BADC, CABD, CADB, DABC, DACB, BCAD, BDAC, CBAD, CDAB, DBAC, DCAB, BCDA, BDCA, CBDA, CDBA, DBCA, DCBA
47. 420 **49.** 2530 **51.** 10 **53.** 4 **55.** 1
57. 40,920 **59.** 850,668
61. AB, AC, AD, AE, AF, BC, BD, BE, BF, CD, CE, CF, DE, DF, EF
63. 4.42×10^{16} **65.** 5,586,853,480 **67.** 175,223,510
69. 462,000 **71.** 3744 **73.** 5 **75.** 20
77. $n = 5$ or $n = 6$ **79.** $n = 10$ **81.** $n = 3$
83. $n = 2$
85. False. Because order does not matter, it is an example of a combination.
87. For some calculators, the answer is too large.
89. ${}_nP_r$ represents the number of ways to choose and order r elements out of a collection of n elements.
91 and 93. Answers will vary. **95.** $(-2, -8)$

Section 6.6 (page 552)

1. sample space **3.** mutually exclusive **5.** $0 \leq P(E) \leq 1$
7. $P(E) = 1$
9. $\{(H, 1), (H, 2), (H, 3), (H, 4), (H, 5), (H, 6), (T, 1), (T, 2), (T, 3), (T, 4), (T, 5), (T, 6)\}$
11. {ABC, ACB, BAC, BCA, CAB, CBA}
13. $\frac{3}{8}$ **15.** $\frac{7}{8}$ **17.** $\frac{3}{13}$ **19.** $\frac{3}{26}$ **21.** $\frac{5}{36}$ **23.** $\frac{1}{6}$
25. $\frac{3}{100}$ **27.** $\frac{9}{25}$ **29.** $\frac{1}{5}$ **31.** $\frac{2}{5}$ **33.** 0.25 **35.** $\frac{1}{3}$
37. 0.88 **39.** $\frac{7}{20}$ **41.** $\frac{4}{13}$ **43.** $\frac{2}{13}$ **45.** $\frac{11}{32}$ **47.** $\frac{5}{8}$
49. $\frac{3}{32}$ **51.** $\frac{1}{8}$ **53.** $\frac{27}{1000}$ **55.** $\frac{2}{125}$
57. $P(\{\text{Taylor wins}\}) = \frac{1}{2}$
$P(\{\text{Moore wins}\}) = P(\{\text{Perez wins}\}) = \frac{1}{4}$
59. (a) 24.00 million (b) 0.317 (c) 0.881 (d) 0.415
61. (a) $\frac{1}{120}$ (b) $\frac{1}{24}$ **63.** (a) $\frac{14}{55}$ (b) $\frac{12}{55}$ (c) $\frac{54}{55}$
65. (a) 0.1024 (b) 0.32 (c) 0.4624
67. (a) $\frac{1}{15{,}625}$ (b) $\frac{4096}{15{,}625}$ (c) $\frac{11{,}529}{15{,}625}$
69. (a) $\frac{\pi}{4}$ (b) Answers will vary.
71. True. The sum of the probabilities of all outcomes must be 1.
73. (a) As you consider successive people with distinct birthdays, the probabilities must decrease to take into account the birth dates already used. Because the birth dates of people are independent events, multiply the respective probabilities of distinct birthdays.
(b) $\frac{365}{365} \cdot \frac{364}{365} \cdot \frac{363}{365} \cdot \frac{362}{365}$ (c) Answers will vary.
(d) Q_n is the probability that the birthdays are *not* distinct, which is equivalent to at least two people having the same birthday.

CHAPTER 6

(e)

n	10	15	20	23	30	40	50
P_n	0.88	0.75	0.59	0.49	0.29	0.11	0.03
Q_n	0.12	0.25	0.41	0.51	0.71	0.89	0.97

(f) 23; From the table in part (e), $Q_n = 0.51$ when $n = 23$.

75. (a) No. $P(A) + P(B) = 0.76 + 0.58 = 1.34 > 1$. The sum of the probabilities is greater than 1, so A and B cannot be mutually exclusive.
(b) Yes. $A' = 0.24$, $B' = 0.42$, and $A' + B' = 0.66 < 1$, so A' and B' can be mutually exclusive.
(c) $0.76 \le P(A \cup B) \le 1$

77. 70 **79.** 165

Review Exercises (page 558)

1. $\frac{2}{3}, \frac{4}{5}, \frac{8}{9}, \frac{16}{17}, \frac{32}{33}$ **3.** $a_n = 5n + 2$ **5.** $a_n = \dfrac{2}{2n - 1}$

7. $9, 3, 1, \frac{1}{3}, \frac{1}{9}$ **9.** $\frac{1}{380}$ **11.** $(n + 1)(n)$ **13.** 30

15. $\dfrac{205}{24}$ **17.** $\displaystyle\sum_{k=1}^{20} \frac{1}{2k} \approx 1.799$ **19.** $\displaystyle\sum_{k=1}^{9} \frac{k}{k + 1} \approx 7.071$

21. (a) $\frac{1111}{1250}$ (b) $\frac{8}{9}$ **23.** (a) 0.0404 (b) $\frac{4}{99}$

25. (a) 3015.00, 3030.08, 3045.23, 3060.45, 3075.75, 3091.13, 3106.59, 3122.12
(b) \$3662.38

27. Arithmetic sequence, $d = -2$

29. Not an arithmetic sequence **31.** 3, 7, 11, 15, 19

33. 1, 4, 7, 10, 13

35. 35, 32, 29, 26, 23; $d = -3$; $a_n = 38 - 3n$

37. $a_n = 103 - 3n$; 1600 **39.** 110 **41.** 6375

43. (a) \$57,564 (b) \$277,160

45. Geometric sequence, $r = 2$

47. Not a geometric sequence **49.** $4, -1, \frac{1}{4}, -\frac{1}{16}, \frac{1}{64}$

51. $9, 6, 4, \frac{8}{3}, \frac{16}{9}$ **53.** $120, 40, \frac{40}{3}, \frac{40}{9}, \frac{40}{27}$; $r = \frac{1}{3}$; $a_n = 120\left(\frac{1}{3}\right)^{n-1}$

55. (a) $-\frac{1}{2}$ (b) -0.5 **57.** 19,531 **59.** 9831

61. 32 **63.** Not possible because $\left|\frac{3}{2}\right| > 1$.

65. (a) $a_t = 130{,}000(0.7)^t$ (b) \$21,849.10 **67.** 45

69. 126 **71.** $x^4 + 20x^3 + 150x^2 + 500x + 625$

73. $a^5 - 20a^4b + 160a^3b^2 - 640a^2b^3 + 1280ab^4 - 1024b^5$

75. 21 **77.** 70 **79.** 10

81. (a) 216 (b) 108 (c) 36 **83.** 95,040 **85.** 5040

87. $n = 3$ **89.** 28 **91.** 479,001,600 **93.** $\frac{1}{9}$

95. (a) 0.416 (b) 0.8 (c) 0.074

97. True. $\dfrac{(n + 2)!}{n!} = \dfrac{(n + 2)(n + 1)n!}{n!} = (n + 2)(n + 1)$

99. (a) Each term is obtained by adding the same constant (common difference) to the preceding term.
(b) Each term is obtained by multiplying the same constant (common ratio) by the preceding term.

Chapter Test (page 561)

1. $3, 2, \frac{4}{3}, \frac{8}{9}, \frac{16}{27}$ **2.** 12, 16, 20, 24, 28

3. $-x, \dfrac{x^2}{2}, -\dfrac{x^3}{3}, \dfrac{x^4}{4}, -\dfrac{x^5}{5}$ **4.** $-x^3, -\dfrac{x^5}{2}, -\dfrac{x^7}{6}, -\dfrac{x^9}{24}, -\dfrac{x^{11}}{120}$

5. 7920 **6.** $\dfrac{1}{n + 1}$ **7.** $2n$ **8.** $a_n = n^2 + 1$

9. $a_n = 5100 - 100n$ **10.** $a_n = 4\left(\frac{1}{2}\right)^{n-1}$

11. $\displaystyle\sum_{n=1}^{12} \frac{2}{3n + 1}$ **12.** $\displaystyle\sum_{n=1}^{\infty} 2\left(\frac{1}{4}\right)^{n-1}$ **13.** 57.17 **14.** 189

15. $\frac{25}{7}$ **16.** $81a^4 - 540a^3b + 1350a^2b^2 - 1500ab^3 + 625b^4$

17. 84 **18.** 1140 **19.** 328,440 **20.** 72 **21.** $n = 3$

22. 26,000 **23.** 12,650 **24.** $\frac{3}{52}$ **25.** $\frac{1}{924}$

26. (a) $\frac{1}{4}$ (b) $\frac{121}{3600}$ (c) $\frac{1}{60}$

Chapter 7

Section 7.1 (page 573)

1. conic section **3.** circle, center

5. The standard form of the equation of a circle; (h, k) represents the center of the circle, r represents the radius of the circle.

7. $x^2 + y^2 = 16$

9. $(x - 3)^2 + (y - 7)^2 = 53$

11. $(x + 3)^2 + (y + 1)^2 = 10$

13. Center: $(0, 0)$
Radius: 6

15. Center: $(5, 0)$
Radius: 3

17. Center: $(-1, -6)$
Radius: $\sqrt{19}$

19. $x^2 + y^2 = 4$
Center: $(0, 0)$
Radius: 2

21. $x^2 + y^2 = \dfrac{3}{4}$
Center: $(0, 0)$
Radius: $\dfrac{\sqrt{3}}{2}$

23. $(x - 1)^2 + (y + 3)^2 = 1$
Center: $(1, -3)$
Radius: 1

25. $\left(x + \frac{3}{2}\right)^2 + (y - 3)^2 = 1$
Center: $\left(-\frac{3}{2}, 3\right)$
Radius: 1

27. Center: $(0, 0)$
Radius: 4

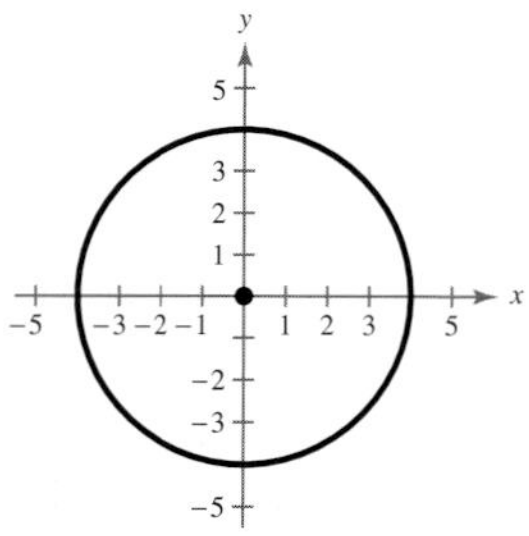

29. Center: $(-4, -1)$
Radius: 3

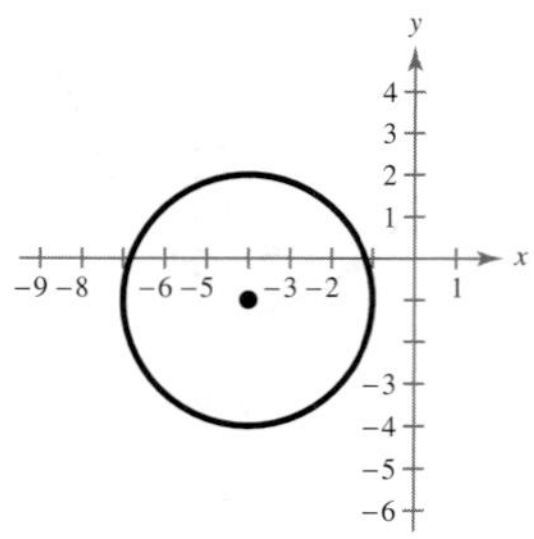

31. Center: $(7, -4)$
Radius: 5

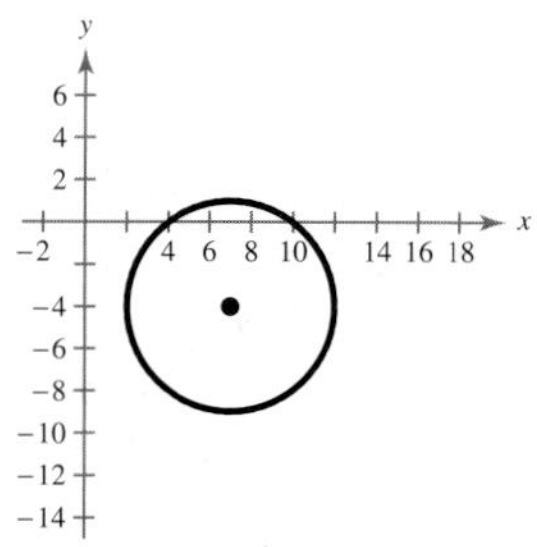

33. Center: $(-1, 0)$
Radius: 6

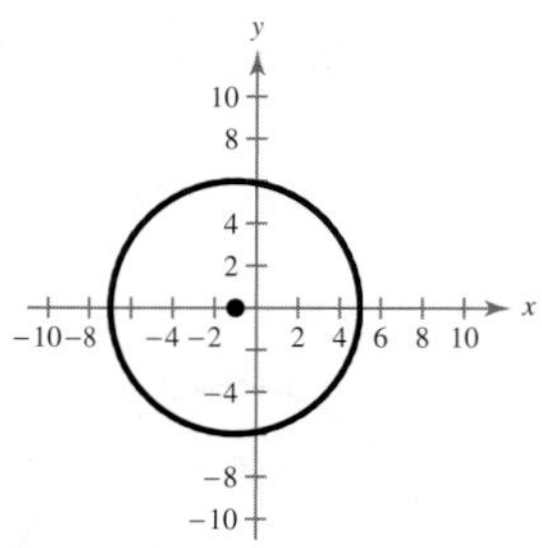

35. x-intercepts: $(-1, 0), (-9, 0)$
y-intercept: $(0, 3)$

37. x-intercepts: $\left(6 \pm \sqrt{7}, 0\right)$
y-intercepts: none

39. x-intercepts: $\left(1 \pm 2\sqrt{7}, 0\right)$
y-intercepts: $(0, 9), (0, -3)$

41. (a) $x^2 + y^2 = 2704$ (b) Yes (c) 2 mi

43. e **44.** b **45.** d **46.** f **47.** a **48.** c

49. $x^2 = \frac{3}{2}y$ **51.** $x^2 = 8y$ **53.** $y^2 = -2x$

55. $x^2 = -16y$ **57.** $y^2 = 8x$ **59.** $y^2 = 3x$

61. Vertex: $(0, 0)$
Focus: $\left(0, \frac{1}{2}\right)$
Directrix: $y = -\frac{1}{2}$

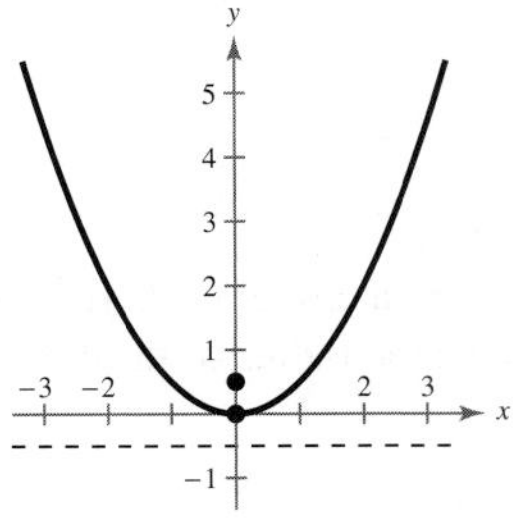

63. Vertex: $(0, 0)$
Focus: $\left(-\frac{3}{2}, 0\right)$
Directrix: $x = \frac{3}{2}$

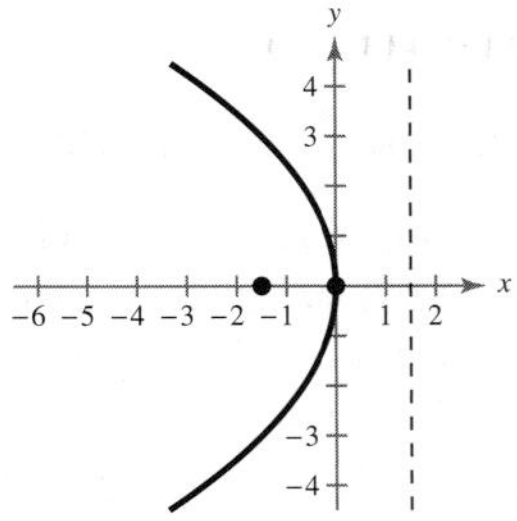

65. Vertex: $(0, 0)$
Focus: $\left(0, -\frac{3}{2}\right)$
Directrix: $y = \frac{3}{2}$

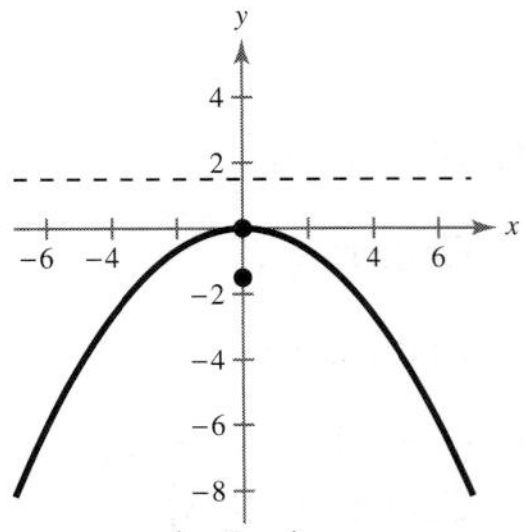

67. Vertex: $(-1, -2)$
Focus: $(-1, 0)$
Directrix: $y = -4$

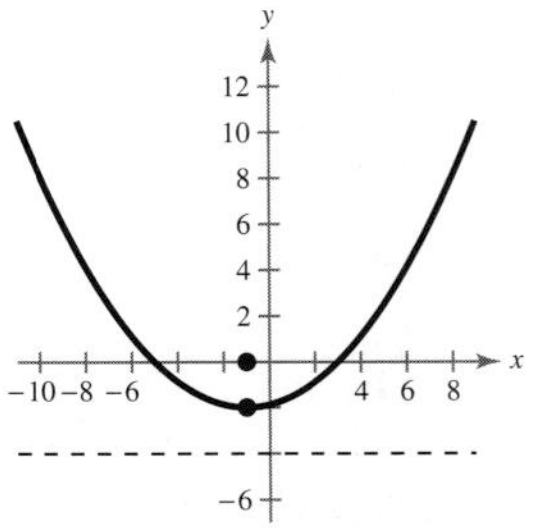

69. Vertex: $\left(-\frac{3}{2}, 2\right)$
Focus: $\left(-\frac{3}{2}, 3\right)$
Directrix: $y = 1$

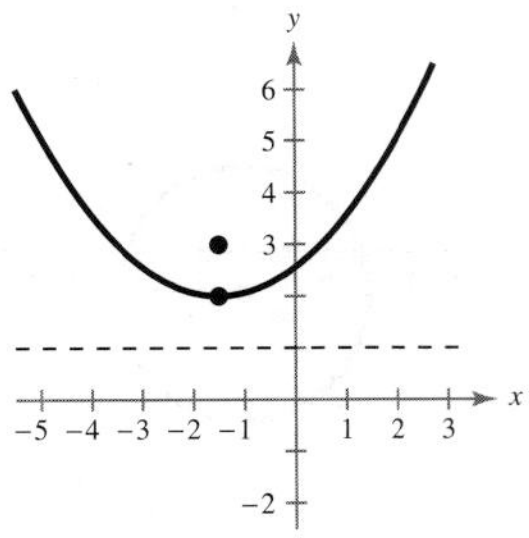

71. Vertex: $(-2, -3)$
Focus: $(-4, -3)$
Directrix: $x = 0$

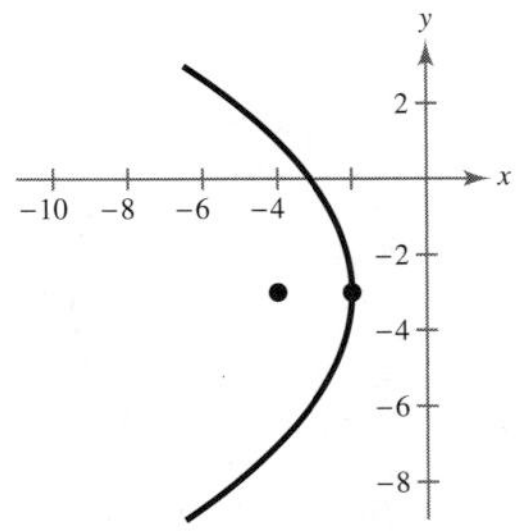

73. Vertex: $(-2, 1)$
Focus: $\left(-2, -\frac{1}{2}\right)$
Directrix: $y = \frac{5}{2}$

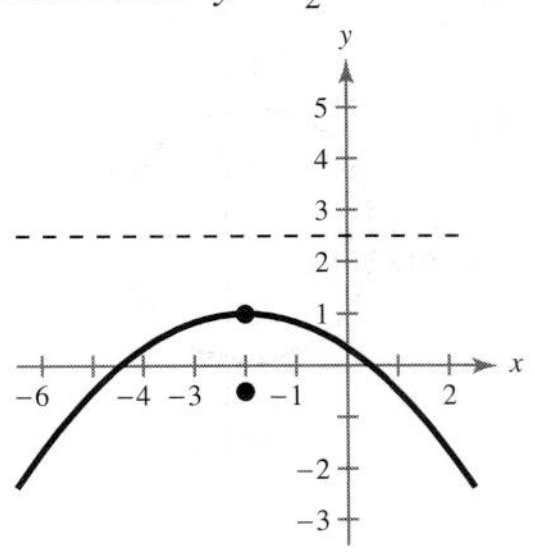

75. Vertex: $\left(\frac{1}{4}, -\frac{1}{2}\right)$
Focus: $\left(0, -\frac{1}{2}\right)$
Directrix: $x = \frac{1}{2}$

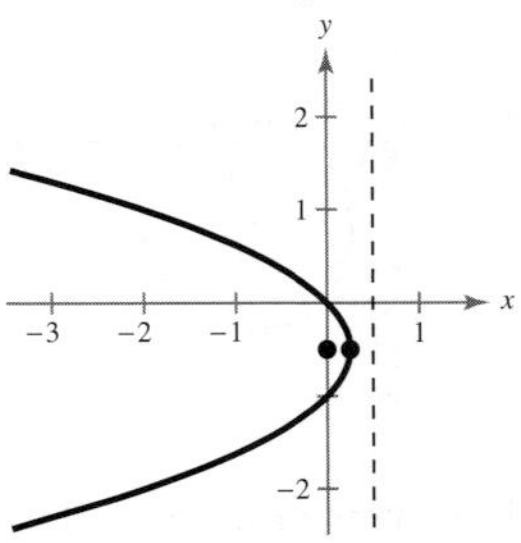

77. Vertex: $(1, 1)$
Focus: $(1, 2)$
Directrix: $y = 0$

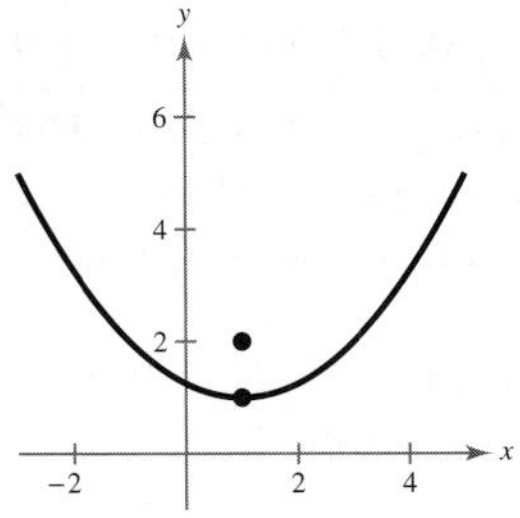

79. $(x - 3)^2 = -(y - 1)$ **81.** $y^2 = 4(x + 4)$

83. $y^2 = 2(x + 2)$ **85.** $(y - 2)^2 = -8(x - 5)$

87. $x^2 = 8(y - 4)$ **89.** $(y - 2)^2 = 8x$

91.

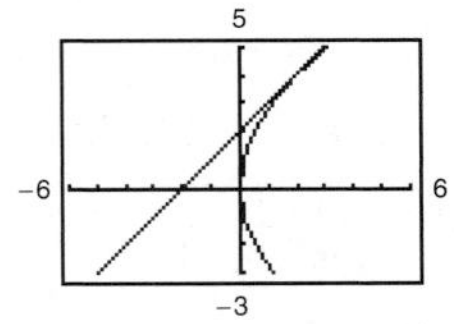

$(2, 4)$

93. $4x - y - 8 = 0$ **95.** $4y - x + 2 = 0$

97. $x^2 = -4(y - 4)$ **99.** $x^2 = -\frac{25}{4}(y - 48)$

101. (a)

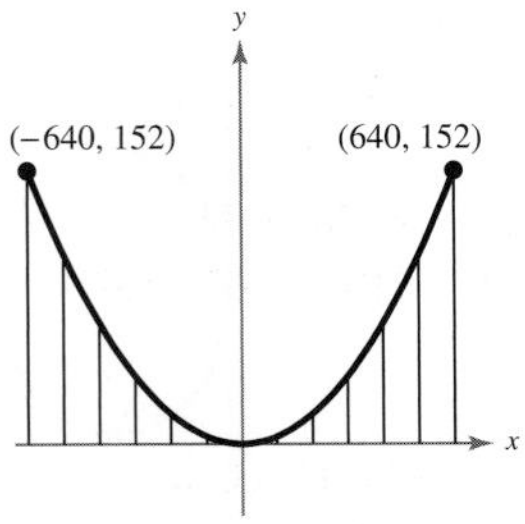

(b) $x^2 = \dfrac{51{,}200}{19}y$

(c)

x	0	200	400	500	600
y	0	14.844	59.375	92.773	133.59

103. (a) $y^2 = 6x$ (b) 2.67 in.

105. (a) $x^2 = -49(y - 100)$ (b) 70 ft **107.** $y = \frac{3}{4}x - \frac{25}{4}$

109. $y = \dfrac{\sqrt{2}}{2}x - 3\sqrt{2}$ **111.** $y = -\dfrac{3}{4}x + 12$

113. False. $x^2 + (y + 5)^2$ represents a circle with its center at $(0, -5)$ and a radius of 5.

115. False. A circle is a conic section.

117. False. A parabola does not touch its directrix.

119.

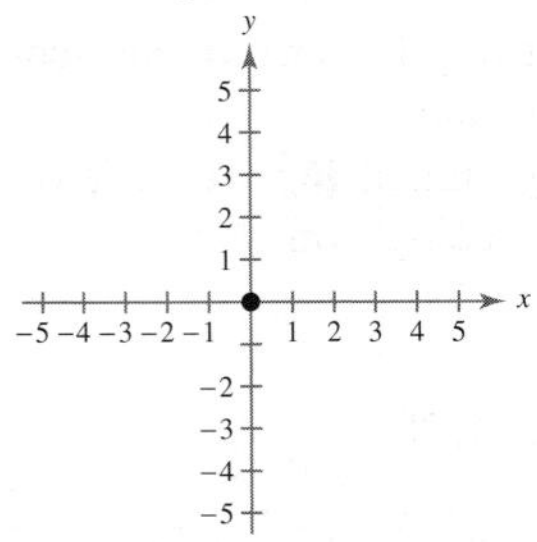

The intersection results in a point.

121. $y = \sqrt{6(x+1)} + 3$
123. Minimum: $(0.67, 0.22)$; maximum: $(-0.67, 3.78)$
125. Minimum: $(-0.79, 0.81)$

Section 7.2 (page 583)

1. ellipse, foci **3.** minor axis **5.** Vertical **7.** 4
9. b **10.** a **11.** c **12.** d

13. $\frac{x^2}{4} + \frac{y^2}{16} = 1$ **15.** $\frac{x^2}{9} + \frac{y^2}{5} = 1$ **17.** $\frac{x^2}{24} + \frac{y^2}{49} = 1$

19. $\frac{5x^2}{48} + \frac{y^2}{16} = 1$ **21.** $\frac{(x-2)^2}{4} + \frac{(y+1)^2}{1} = 1$

23. $\frac{(x-4)^2}{16} + \frac{(y-2)^2}{1} = 1$ **25.** $\frac{x^2}{308} + \frac{(y-4)^2}{324} = 1$

27. $\frac{(x-3)^2}{36} + \frac{(y-2)^2}{32} = 1$

29. Center: $(0, 0)$
Vertices: $(\pm 8, 0)$
Foci: $(\pm\sqrt{55}, 0)$
Eccentricity: $\frac{\sqrt{55}}{8}$

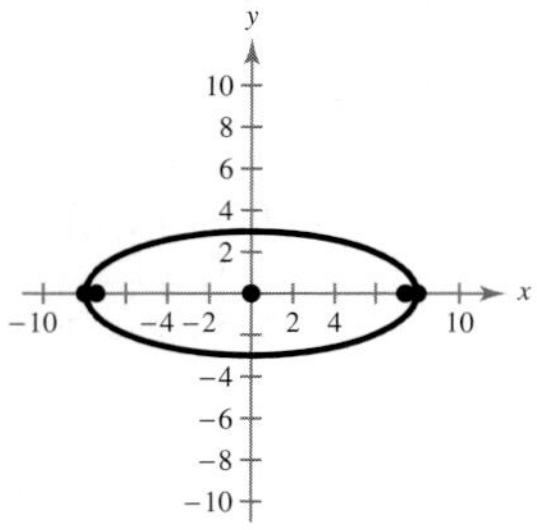

31. Center: $(4, -1)$
Vertices: $(4, 4), (4, -6)$
Foci: $(4, 2), (4, -4)$
Eccentricity: $\frac{3}{5}$

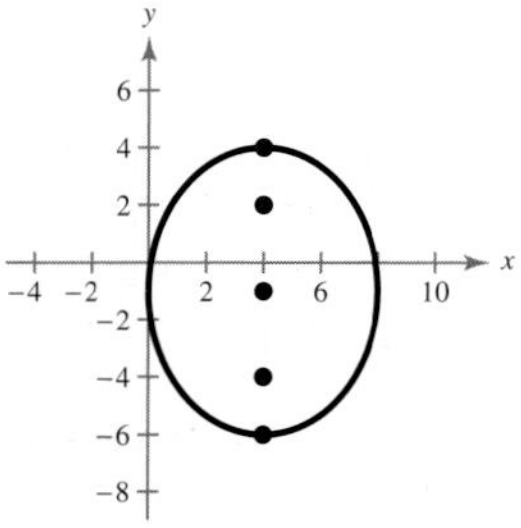

33. Center: $(-5, 1)$
Vertices: $\left(-\frac{7}{2}, 1\right), \left(-\frac{13}{2}, 1\right)$
Foci: $\left(-5 \pm \frac{\sqrt{5}}{2}, 1\right)$
Eccentricity: $\frac{\sqrt{5}}{3}$

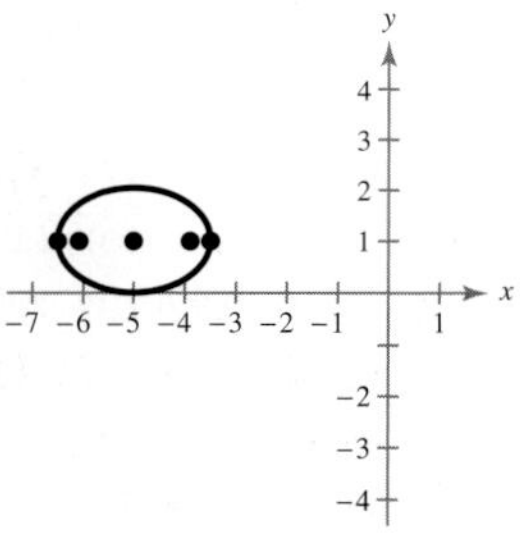

35. (a) $\frac{x^2}{36} + \frac{y^2}{4} = 1$ (c)
(b) Center: $(0, 0)$
Vertices: $(\pm 6, 0)$
Foci: $(\pm 4\sqrt{2}, 0)$
Eccentricity: $\frac{2\sqrt{2}}{3}$

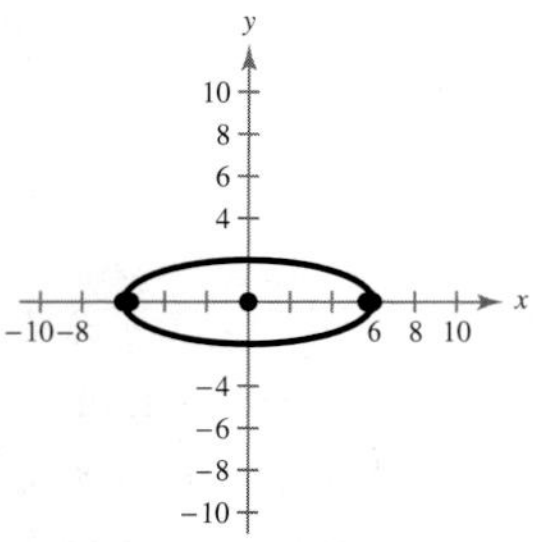

37. (a) $\frac{x^2}{4} + \frac{y^2}{49} = 1$ (c)
(b) Center: $(0, 0)$
Vertices: $(0, \pm 7)$
Foci: $(0, \pm 3\sqrt{5})$
Eccentricity: $\frac{3\sqrt{5}}{7}$

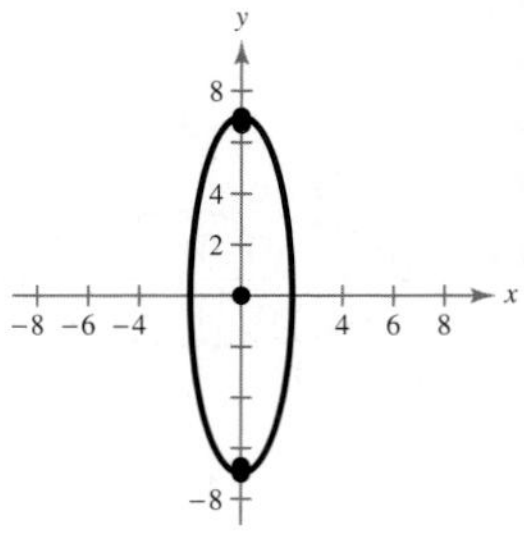

39. (a) $\frac{(x+2)^2}{4} + \frac{(y-3)^2}{9} = 1$ (c)
(b) Center: $(-2, 3)$
Vertices: $(-2, 6), (-2, 0)$
Foci: $(-2, 3 \pm \sqrt{5})$
Eccentricity: $\frac{\sqrt{5}}{3}$

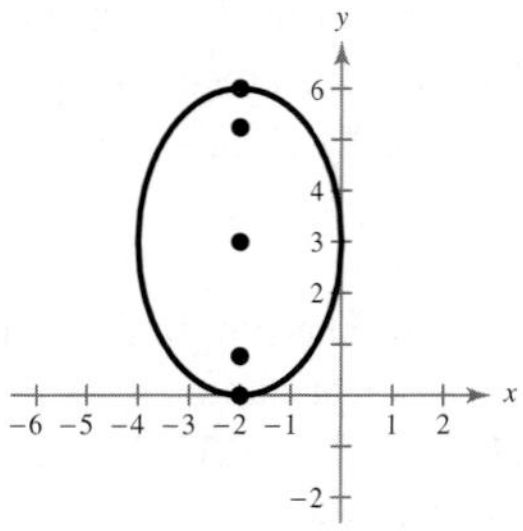

41. (a) $\frac{\left(x+\frac{3}{2}\right)^2}{4} + \frac{\left(y-\frac{5}{2}\right)^2}{12} = 1$
(b) Center: $\left(-\frac{3}{2}, \frac{5}{2}\right)$ (c)
Vertices: $\left(-\frac{3}{2}, \frac{5 \pm 4\sqrt{3}}{2}\right)$
Foci: $\left(-\frac{3}{2}, \frac{5}{2} \pm 2\sqrt{2}\right)$
Eccentricity: $\frac{\sqrt{6}}{3}$

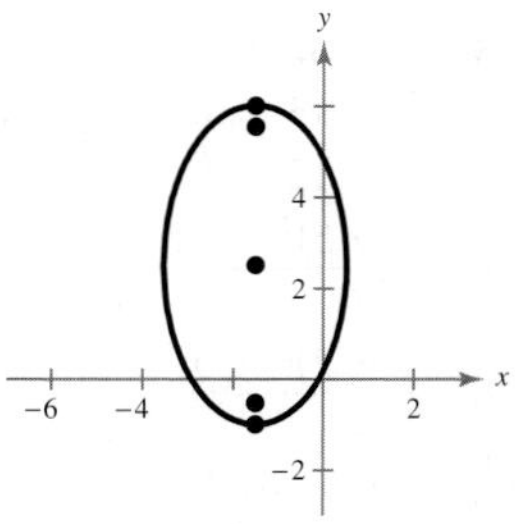

43. (a) $\frac{\left(x-\frac{1}{2}\right)^2}{5} + \frac{(y+1)^2}{3} = 1$
(b) Center: $\left(\frac{1}{2}, -1\right)$ (c)
Vertices: $\left(\frac{1}{2} \pm \sqrt{5}, -1\right)$
Foci: $\left(\frac{1}{2} \pm \sqrt{2}, -1\right)$
Eccentricity: $\frac{\sqrt{10}}{5}$

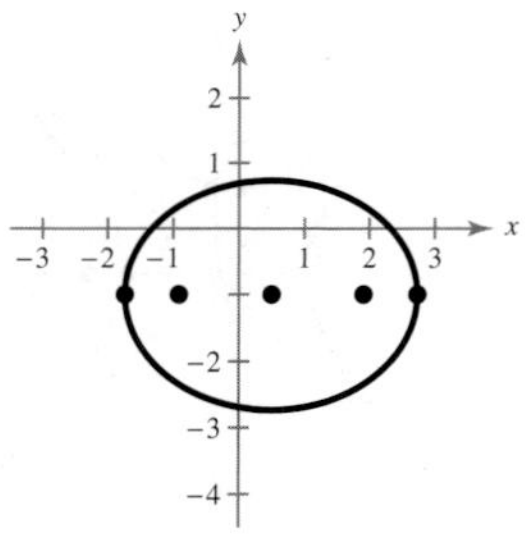

45. $\frac{\sqrt{5}}{3}$ **47.** $\frac{2\sqrt{2}}{3}$ **49.** $\frac{x^2}{25} + \frac{y^2}{16} = 1$

51. $\frac{(x-1)^2}{45} + \frac{(y-7)^2}{81} = 1$ **53.** $\frac{x^2}{318.09} + \frac{y^2}{18.80} = 1$

55. $40\sqrt{33}$ mm ≈ 229.8 mm **57.** 40 units

59.

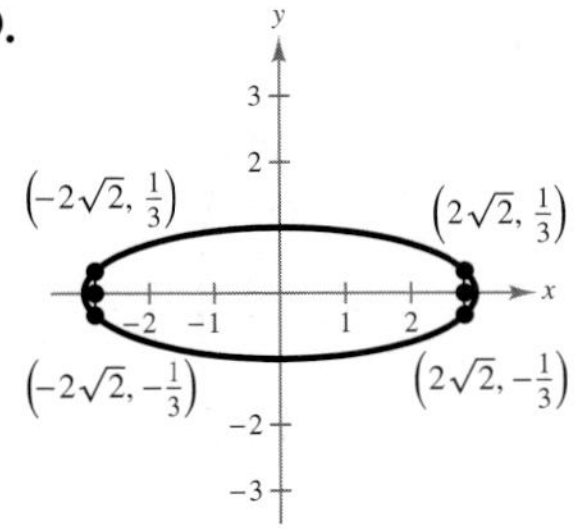

61.

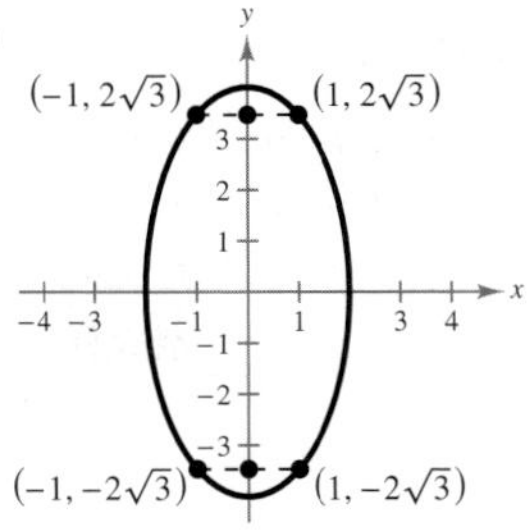

63. True. The ellipse is more elongated when e is close to 1.

65. No. The equation represents an ellipse.

67. Nearly circular because its eccentricity is about 0.055, which is close to zero.

69. (a) $2a$

(b) The sum of the distances from the two fixed points is constant.

71. Proof **73.** Geometric **75.** Arithmetic

77. 15.0990

Section 7.3 (page 593)

1. hyperbola **3.** (h, k) **5.** Center

7. a **8.** d **9.** b **10.** c

11. $\frac{y^2}{4} - \frac{x^2}{12} = 1$ **13.** $\frac{(x-4)^2}{4} - \frac{y^2}{12} = 1$

15. $\frac{(y-5)^2}{16} - \frac{(x-4)^2}{9} = 1$ **17.** $\frac{y^2}{9} - \frac{4(x-2)^2}{9} = 1$

19. $\frac{(y-2)^2}{4} - \frac{x^2}{4} = 1$

21. Center: $(0, 0)$
Vertices: $(\pm 1, 0)$
Foci: $(\pm\sqrt{2}, 0)$
Asymptotes: $y = \pm x$

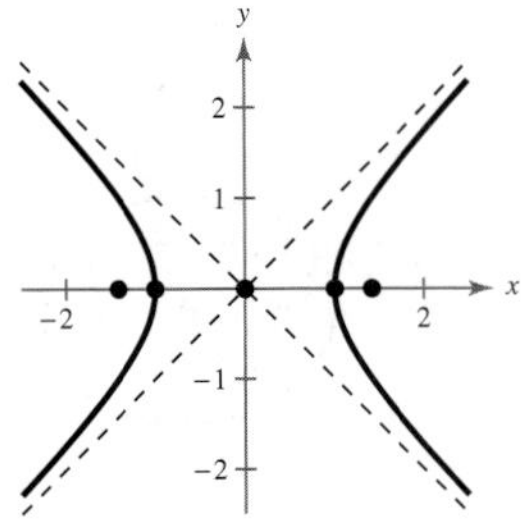

23. Center: $(0, 0)$
Vertices: $(0, \pm 1)$
Foci: $(0, \pm\sqrt{17})$
Asymptotes: $y = \pm\frac{1}{4}x$

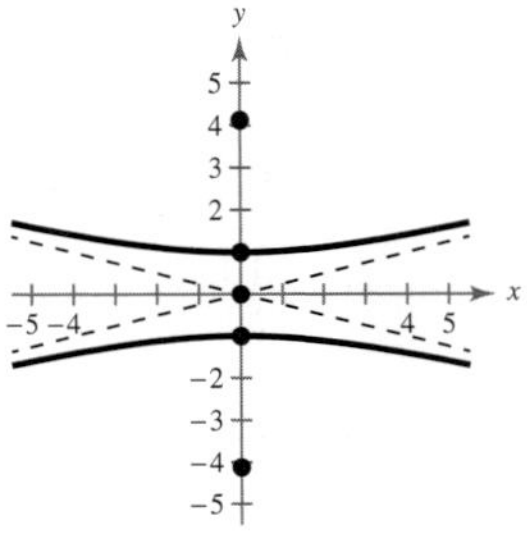

25. Center: $(0, 0)$
Vertices: $(0, \pm 4)$
Foci: $(0, \pm 2\sqrt{5})$
Asymptotes: $y = \pm 2x$

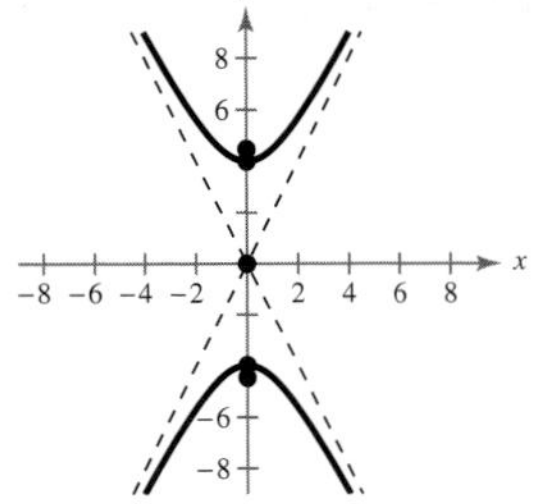

27. Center: $(3, 1)$
Vertices: $(0, 1), (6, 1)$
Foci: $(3 \pm \sqrt{10}, 1)$
Asymptotes:
$y = 1 \pm \frac{1}{3}(x - 3)$

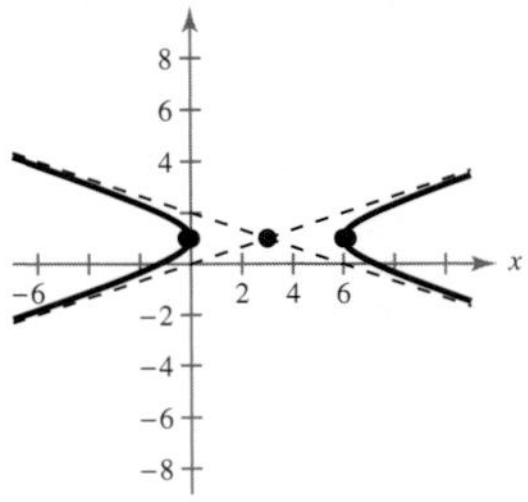

29. Center: $(2, -6)$
Vertices: $(2, -5), (2, -7)$
Foci: $\left(2, -6 \pm \frac{\sqrt{17}}{4}\right)$
Asymptotes: $y = -6 \pm 4(x - 2)$

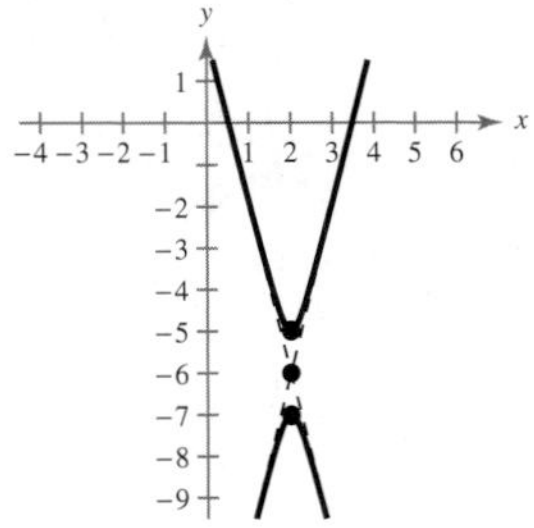

31. (a) $\frac{x^2}{9} - \frac{y^2}{4} = 1$

(b) Center: $(0, 0)$
Vertices: $(\pm 3, 0)$
Foci: $(\pm\sqrt{13}, 0)$
Asymptotes: $y = \pm\frac{2}{3}x$

(c)

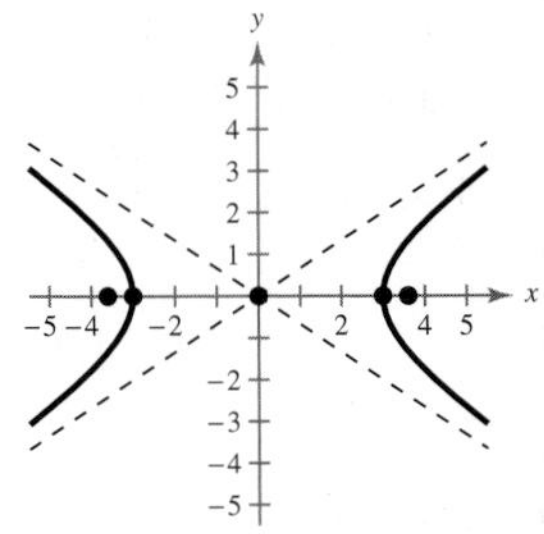

33. (a) $\frac{y^2}{4} - \frac{x^2}{8} = 1$

(b) Center: $(0, 0)$
Vertices: $(0, \pm 2)$
Foci: $(0, \pm 2\sqrt{3})$
Asymptotes: $y = \pm\frac{\sqrt{2}}{2}x$

(c)

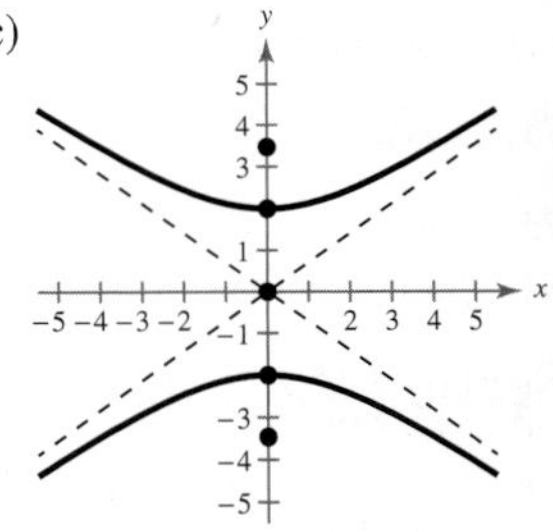

35. (a) $(x-2)^2 - \frac{(y+3)^2}{9} = 1$

(b) Center: $(2, -3)$
Vertices: $(3, -3), (1, -3)$
Foci: $(2 \pm \sqrt{10}, -3)$
Asymptotes:
$y = -3 \pm 3(x - 2)$

(c)

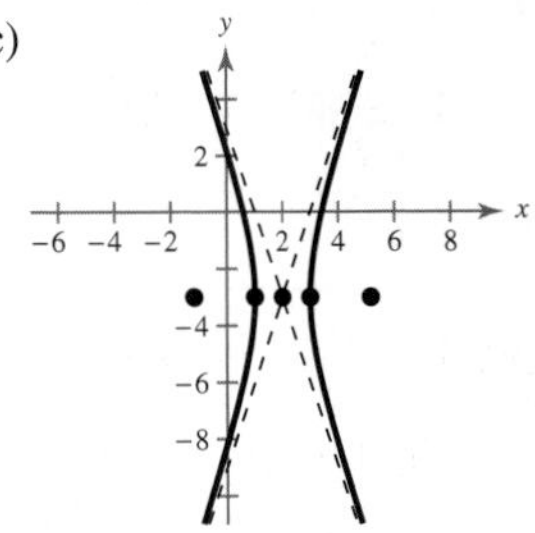

37. (a) $\frac{(x+4)^2}{7} - \frac{y^2}{2} = 1$

(b) Center: $(-4, 0)$
Vertices: $(-4 \pm \sqrt{7}, 0)$
Foci: $(-7, 0), (-1, 0)$
Asymptotes:
$y = \pm\frac{\sqrt{14}}{7}(x + 4)$

(c)

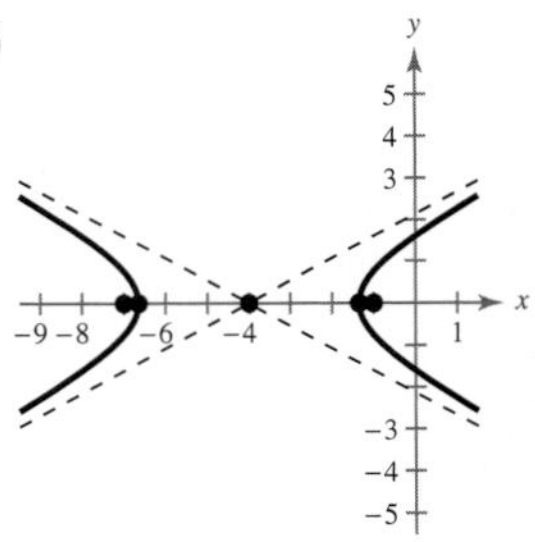

39. (a) $\dfrac{(y+3)^2}{2} - \dfrac{(x-1)^2}{18} = 1$

(b) Center: $(1, -3)$

Vertices: $\left(1, -3 \pm \sqrt{2}\right)$

Foci: $\left(1, -3 \pm 2\sqrt{5}\right)$

Asymptotes:

$y = -3 \pm \frac{1}{3}(x - 1)$

(c)

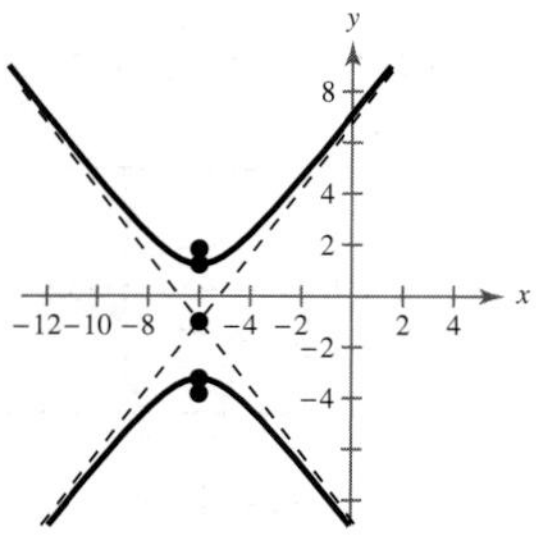

41. $\dfrac{x^2}{1} - \dfrac{y^2}{25} = 1$ **43.** $\dfrac{17y^2}{1024} - \dfrac{17x^2}{64} = 1$

45. $\dfrac{(x-2)^2}{1} - \dfrac{(y-2)^2}{1} = 1$ **47.** $\dfrac{(x-3)^2}{9} - \dfrac{(y-2)^2}{4} = 1$

49. $\dfrac{(x-4)^2}{16} - \dfrac{(y-3)^2}{9} = 1$

51. $\dfrac{x^2}{98{,}010{,}000} - \dfrac{y^2}{13{,}503{,}600} = 1$

53. (a) $x^2 - \dfrac{y^2}{27} = 1$ (b) 1.89 ft = 22.68 in.

55. $5 + \sqrt{41}$ **57.** Ellipse **59.** Hyperbola

61. Parabola **63.** Circle **65.** Parabola

67. True. For a hyperbola, $c^2 = a^2 + b^2$. The larger the ratio of b to a, the larger the eccentricity of the hyperbola, $e = c/a$.

69. False. If $D = E$ or $D = -E$, the graph is two intersecting lines. For example, the graph of $x^2 - y^2 - 2x + 2y = 0$ is two intersecting lines.

71. Answers will vary. **73.** $\dfrac{(x-6)^2}{9} - \dfrac{(y-2)^2}{7} = 1$

75. Proofs **77.** $x(x + 4)(x - 4)$ **79.** $2x(x - 6)^2$

81. $2(2x + 3)(4x^2 - 6x + 9)$

Section 7.4 (page 601)

1. plane curve, parametric equations, parameter

3. Eliminate the parameter.

5. c **6.** d **7.** b **8.** a

9. (a)

t	0	1	2	3	4
x	0	1	$\sqrt{2}$	$\sqrt{3}$	2
y	2	1	0	-1	-2

(b)

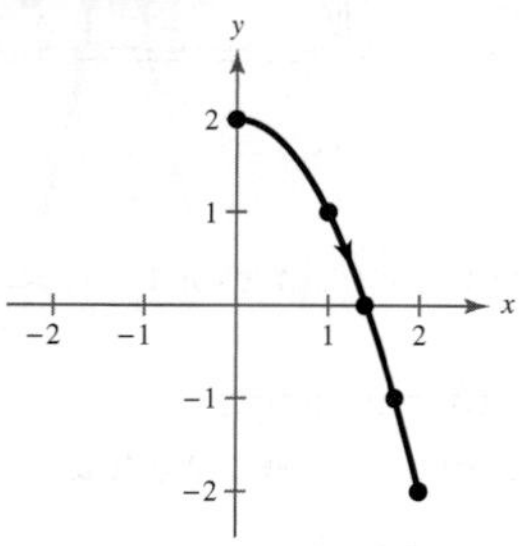

The curve starts at (0, 2) and moves along the right half of the parabola.

(c)

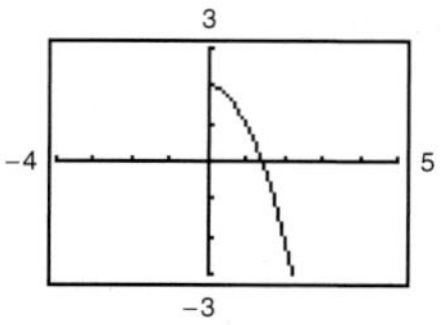

(d) $y = 2 - x^2$

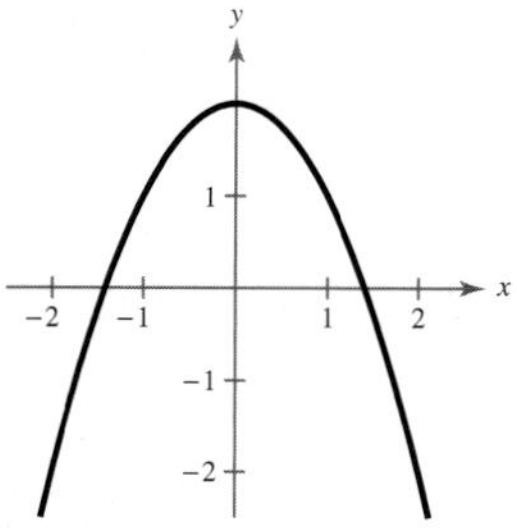

The graph is an entire parabola rather than just the right half.

11. b

13. (a)

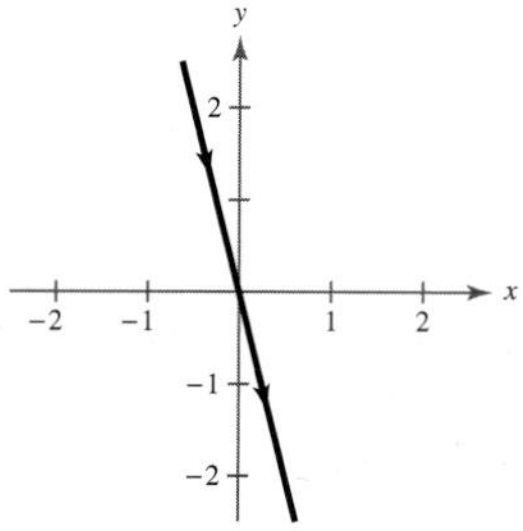

(b) $y = -4x$

15. (a)

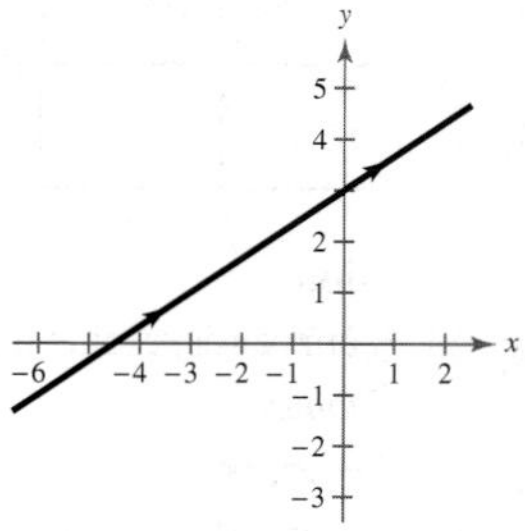

(b) $y = \frac{2}{3}x + 3$

17. (a)

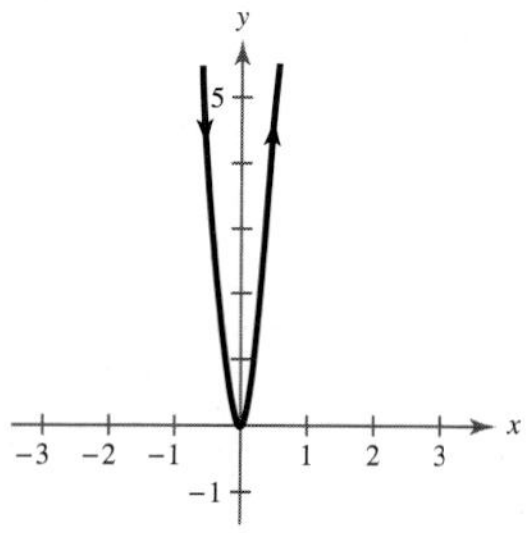

(b) $y = 16x^2$

19. (a)

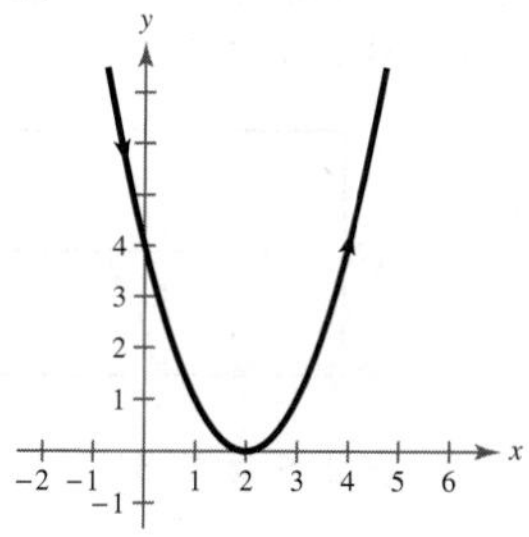

(b) $y = (x - 2)^2$

21. (a)

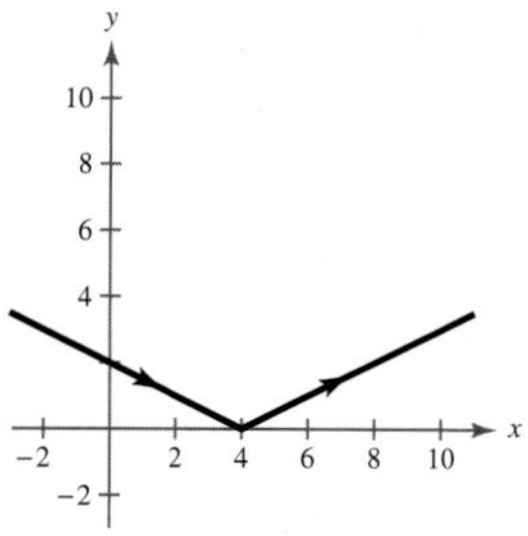

(b) $y = \frac{1}{2}|x - 4|$

23. (a)

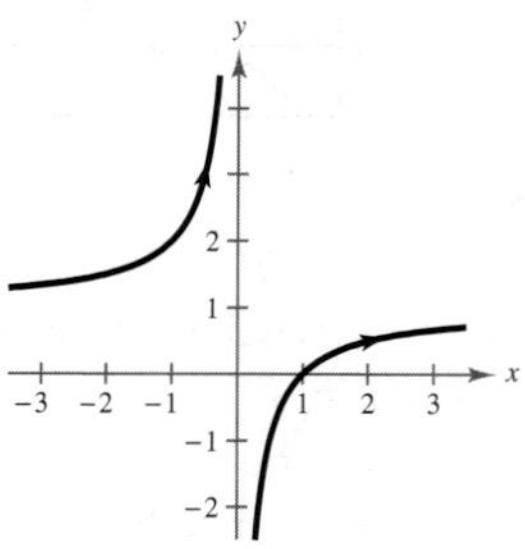

(b) $y = \dfrac{x - 1}{x}$

25. (a)

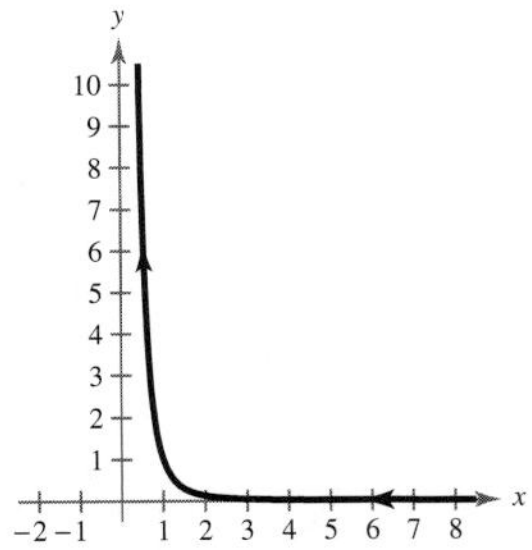

(b) $y = x^{-3}, x > 0$

27. (a)

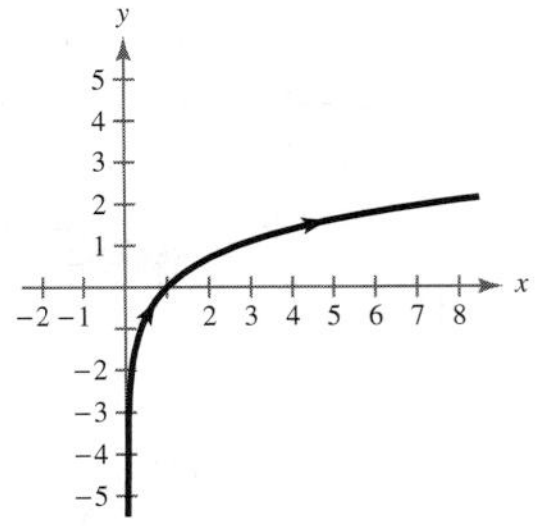

(b) $y = \ln x$

29.

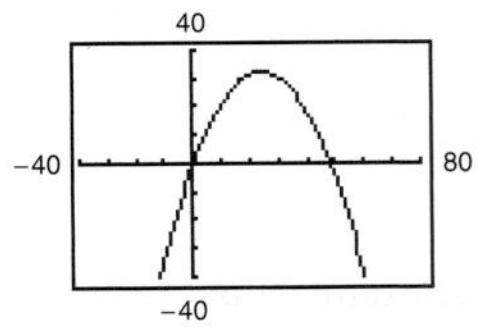

Function

31.

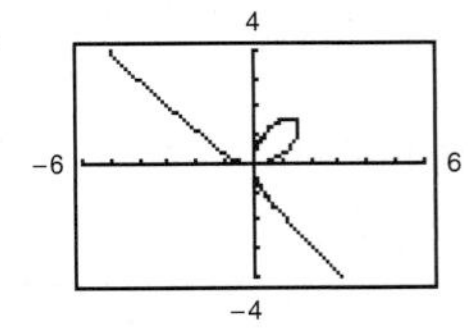

Not a function

33.

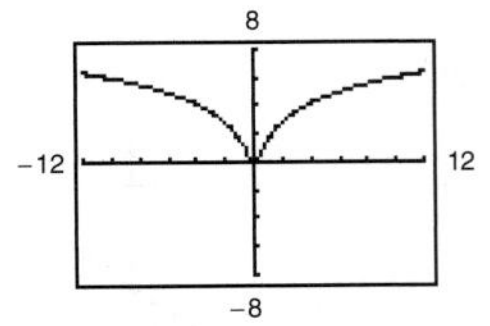

Function

35. Each curve represents a portion of the line $y = 2x + 1$.

	Domain	*Orientation*
(a)	$(-\infty, \infty)$	Left to right
(b)	$(-\infty, 0), (0, \infty)$	Right to left
(c)	$(0, \infty)$	Right to left
(d)	$(0, \infty)$	Left to right

37. Each curve represents a portion of the curve $y = x^3 - 1$.

(a) $0 \le x \le 1$

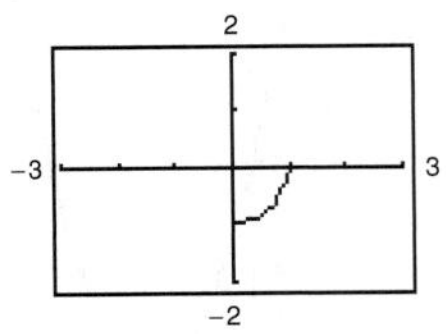

(b) $0 \le x \le 3$

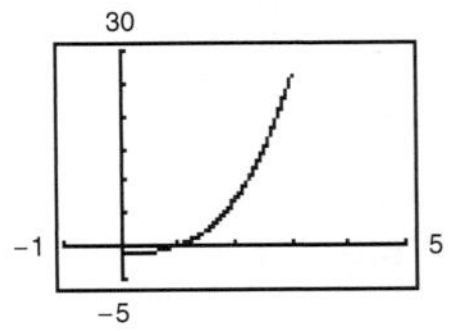

(c) $-2 \le x \le 3$

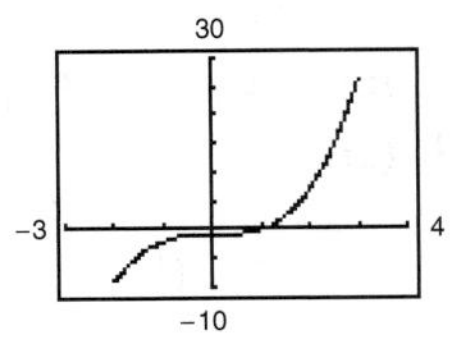

(d) $-3 \le x \le 3$

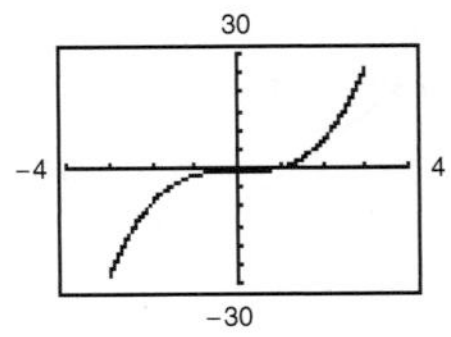

39. $x = 5t$
$y = 4 - 6t$

41. $x = -2 + 11t$
$y = 3 - 13t$

43. Answers will vary. Sample answers:

(a) $x = 3$
$y = t$
$-1 \le t \le 5$

(b) $x = 3$
$y = -t$
$-5 \le t \le 1$

45. (a) $x = t, y = -2t + 1$ (b) $x = 2 - t, y = 2t - 3$

47. (a) $x = t, y = \dfrac{3}{t}$ (b) $x = 2 - t, y = \dfrac{3}{2 - t}$

49. (a) $x = t, y = 4t^2 + 5$
(b) $x = 2 - t, y = 4t^2 - 16t + 21$

51. (a) $x = t, y = e^{-t}$ (b) $x = 2 - t, y = e^{t-2}$

53. (a) $x = 4 - t^2, y = t$
(b) $x = 4 - \dfrac{t^2}{4}, y = \dfrac{t}{2}$

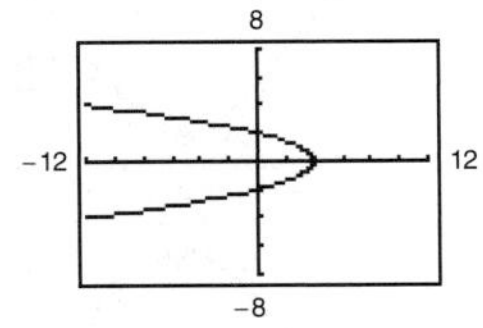

Answers will vary.

55. (a) $x = 44\sqrt{2}t, y = 44\sqrt{2}t - 16t^2$

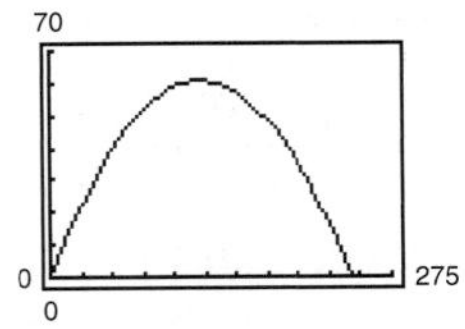

Maximum height: 60.5 ft
Horizontal distance: 242.0 ft

(b) $x = 66\sqrt{2}t, y = 66\sqrt{2}t - 16t^2$

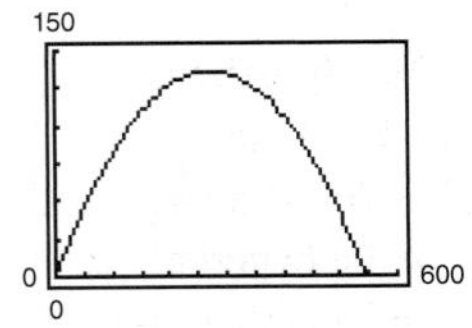

Maximum height: 136.1 ft
Horizontal distance: 544.4 ft

(c) $x = 44\sqrt{2}t, y = 30 + 44\sqrt{2}t - 16t^2$

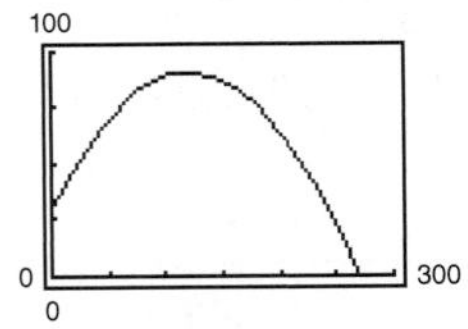

Maximum height: 90.5 ft
Horizontal distance: 269.0 ft

57. (a) $x = \dfrac{v_0\sqrt{2}}{2}t, y = 3 + \dfrac{v_0\sqrt{2}}{2}t - 16t^2$

(b)

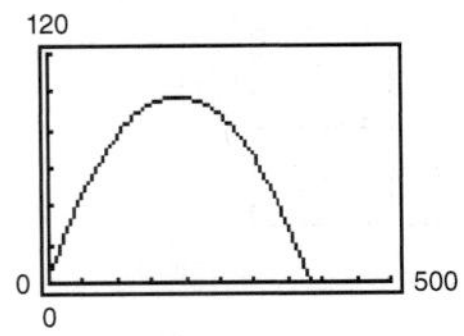

No

(c)

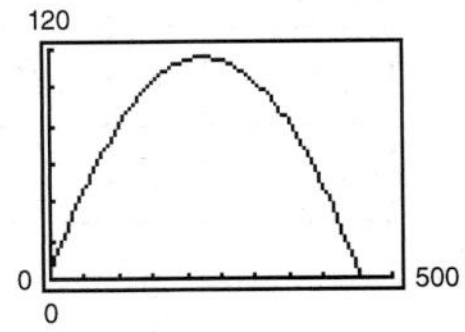

Yes

(d) 114.827 ft/sec

59. True. Both sets of parametric equations correspond to $y = x^2 + 1$.

61. False. The equations have different ranges.

63. Answers will vary. **65.** Even **67.** Neither

Review Exercises (page 606)

1. Hyperbola **3.** $x^2 + y^2 = 25$

5. $(x - 2)^2 + (y - 4)^2 = 13$

7. $x^2 + y^2 = 36$
Center: $(0, 0)$
Radius: 6

9. $\left(x - \frac{1}{2}\right)^2 + \left(y + \frac{3}{4}\right)^2 = 1$
Center: $\left(\frac{1}{2}, -\frac{3}{4}\right)$
Radius: 1

11.

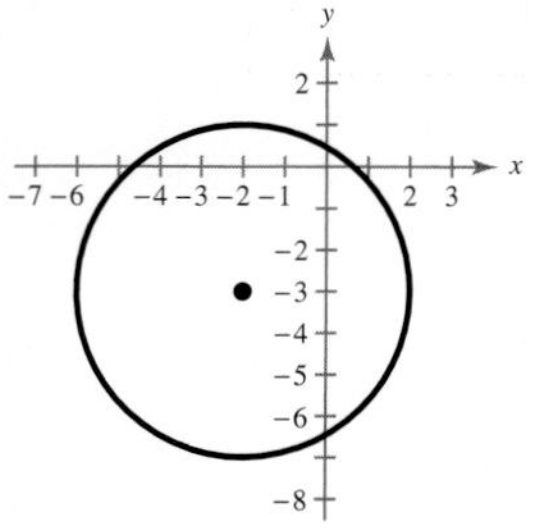

Center: $(-2, -3)$
Radius: 4

13. $\left(0, -4 \pm \sqrt{5}\right)$

15. Vertex: $(0, 0)$
Focus: $(0, -2)$
Directrix: $y = 2$

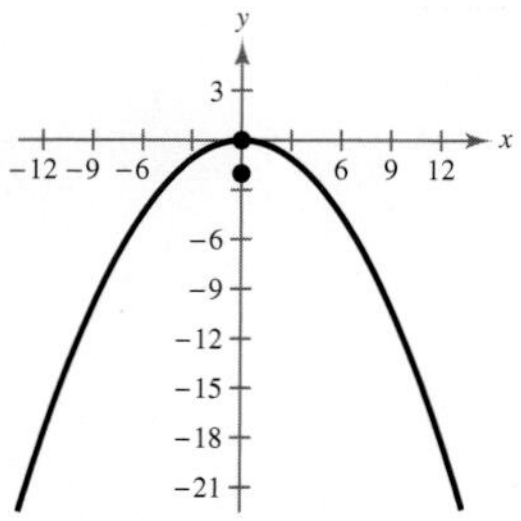

17. Vertex: $(0, 0)$
Focus: $(-9, 0)$
Directrix: $x = 9$

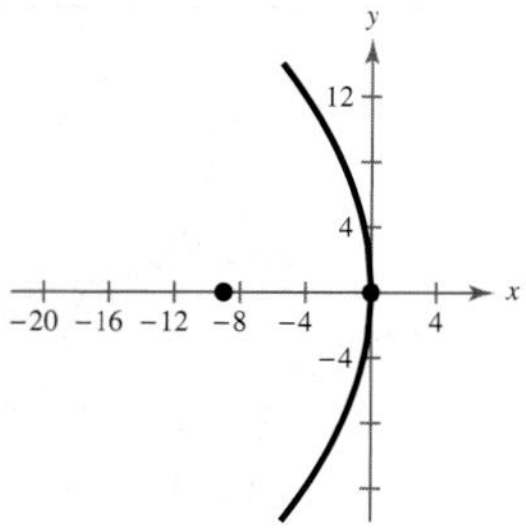

19. $x^2 = 20y$ **21.** $(x + 6)^2 = -9(y - 4)$

23. $(y - 2)^2 = 12x$ **25.** $2x + y - 2 = 0$ **27.** $(0, 50)$

29. (a) Parabola: $y = 4 - \frac{1}{9}x^2$
Circle: $x^2 + \left(y + 4\sqrt{10}\right)^2 = 196$

(b)

x	0	1	2	3	4
d	2.65	2.57	2.35	1.97	1.45

31. Center: $(0, 0)$
Vertices: $(0, \pm 6)$
Foci: $\left(0, \pm\sqrt{11}\right)$
Eccentricity: $\dfrac{\sqrt{11}}{6}$

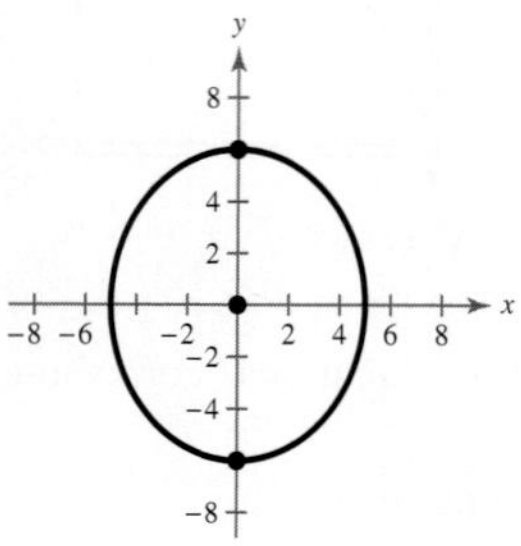

33. Center: $(3, 2)$
Vertices: $(9, 2), (-3, 2)$
Foci: $\left(3 \pm 4\sqrt{2}, 2\right)$
Eccentricity: $\dfrac{2\sqrt{2}}{3}$

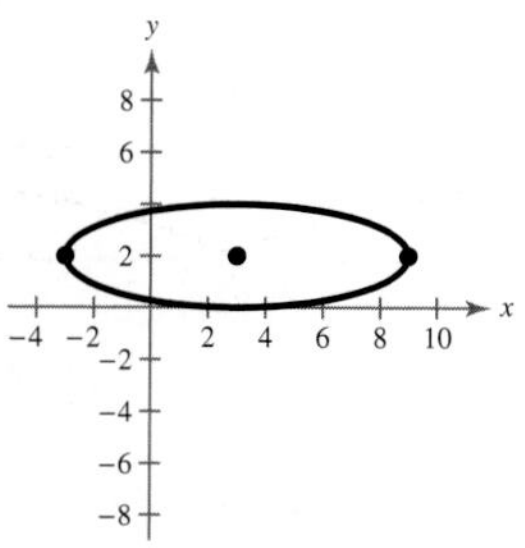

35. (a) $\dfrac{(x - 1)^2}{9} + \dfrac{(y + 4)^2}{16} = 1$
(b) Center: $(1, -4)$
Vertices: $(1, 0), (1, -8)$
Foci: $\left(1, -4 \pm \sqrt{7}\right)$
Eccentricity: $\dfrac{\sqrt{7}}{4}$
(c)

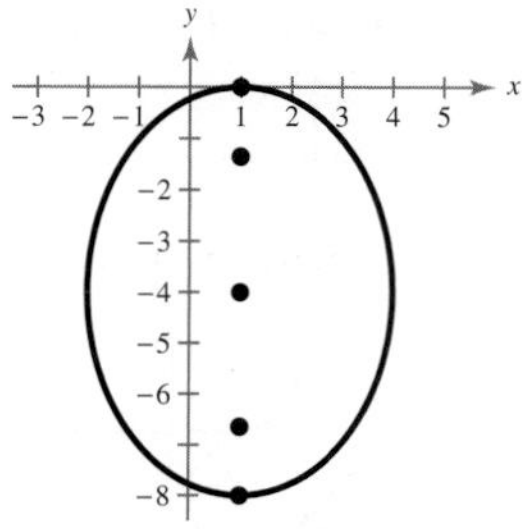

37. (a) $\dfrac{(x + 2)^2}{1/3} + \dfrac{(y - 7)^2}{1/8} = 1$
(b) Center: $(-2, 7)$
Vertices: $\left(-2 \pm \dfrac{\sqrt{3}}{3}, 7\right)$
Foci: $\left(-2 \pm \dfrac{\sqrt{30}}{12}, 7\right)$
Eccentricity: $\dfrac{\sqrt{10}}{4}$
(c)

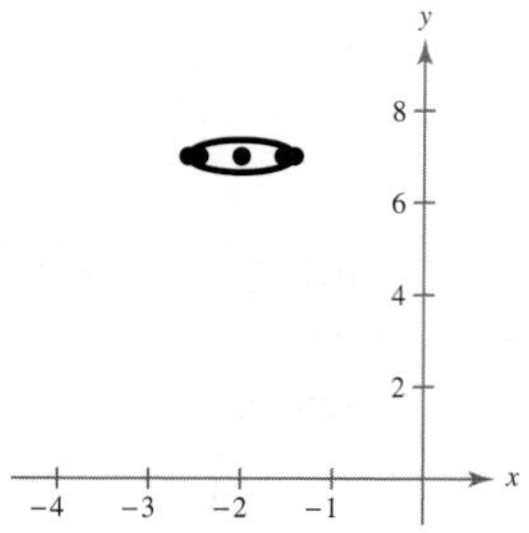

39. $\dfrac{(x - 5)^2}{25} + \dfrac{(y - 3)^2}{9} = 1$ **41.** $\dfrac{x^2}{25} + \dfrac{y^2}{9} = 1$

43. $\dfrac{(x - 2)^2}{25} + \dfrac{y^2}{21} = 1$ **45.** $\dfrac{(x - 2)^2}{4} + (y - 1)^2 = 1$

47. The foci should be placed 3 feet on either side of center at the same height as the pillars.

49. $e \approx 0.0539$ **51.** $x^2 - \dfrac{y^2}{4} = 1$ **53.** $\dfrac{x^2}{16} - \dfrac{y^2}{20} = 1$

55. $\dfrac{(y - 3)^2}{9} - \dfrac{x^2}{9} = 1$ **57.** $\dfrac{(x - 4)^2}{16/5} - \dfrac{y^2}{64/5} = 1$

59. (a) $\dfrac{y^2}{4} - \dfrac{x^2}{5} = 1$
(b) Center: $(0, 0)$
Vertices: $(0, \pm 2)$
Foci: $(0, \pm 3)$
Asymptotes: $y = \pm\dfrac{2\sqrt{5}}{5}x$
(c)

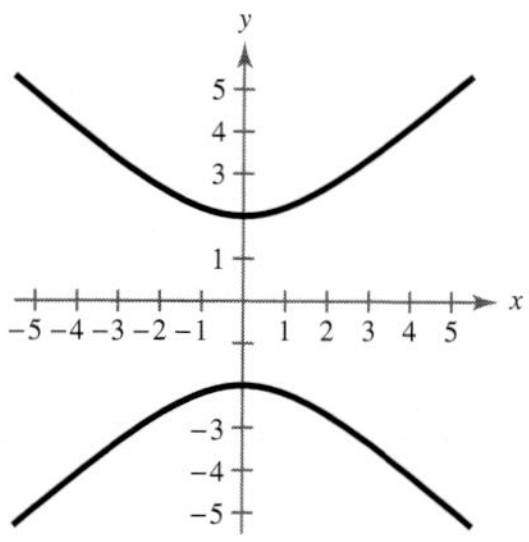

61. (a) $\dfrac{(x-1)^2}{16} - \dfrac{(y+1)^2}{9} = 1$

(b) Center: $(1, -1)$
Vertices: $(5, -1), (-3, -1)$
Foci: $(6, -1), (-4, -1)$
Asymptotes: $y = -1 \pm \dfrac{3}{4}(x - 1)$

(c)

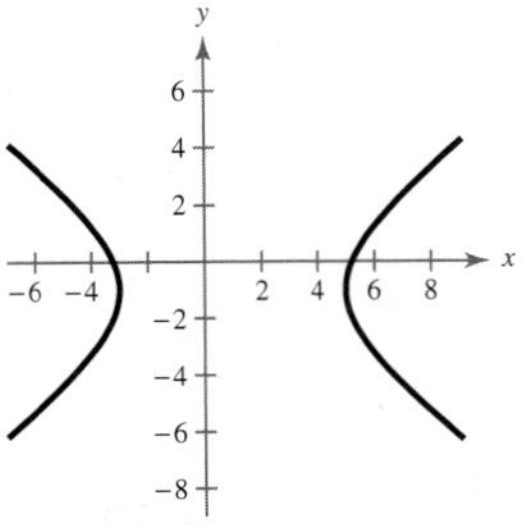

63. (a) $\dfrac{(x+6)^2}{101/2} - \dfrac{(y-1)^2}{202} = 1$

(b) Center: $(-6, 1)$
Vertices:
$\left(-6 \pm \dfrac{\sqrt{202}}{2}, 1\right)$
Foci: $\left(-6 \pm \dfrac{\sqrt{1010}}{2}, 1\right)$
Asymptotes:
$y = 1 \pm 2(x + 6)$

(c)

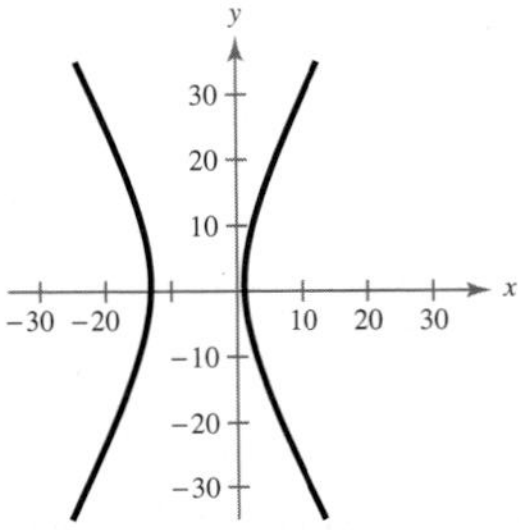

65. Center: $(0, 0)$
Vertices: $(\pm 6, 0)$
Foci: $(\pm\sqrt{61}, 0)$
Asymptotes: $y = \pm\frac{5}{6}x$

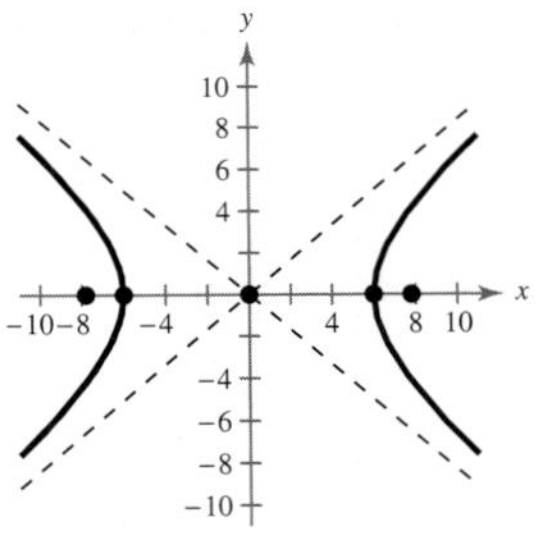

67. Center: $(3, 2)$
Vertices: $(3, 5), (3, -1)$
Foci: $\left(3, 2 \pm \sqrt{13}\right)$
Asymptotes:
$y = 2 \pm \frac{3}{2}(x - 3)$

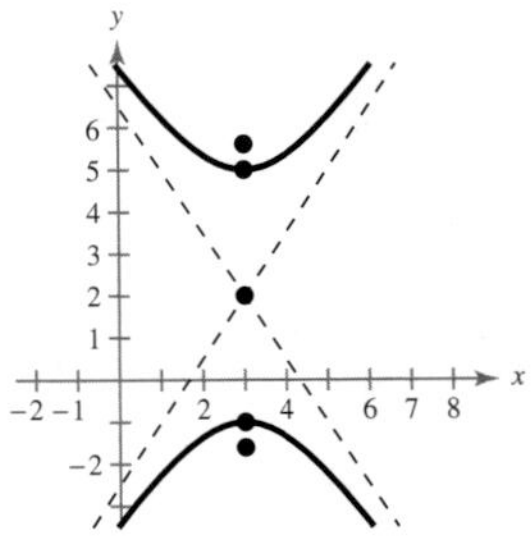

69. 72.2 mi **71.** Parabola **73.** Circle **75.** Ellipse

77.

t	-2	-1	0	1	2	3
x	-8	-5	-2	1	4	7
y	15	11	7	3	-1	-5

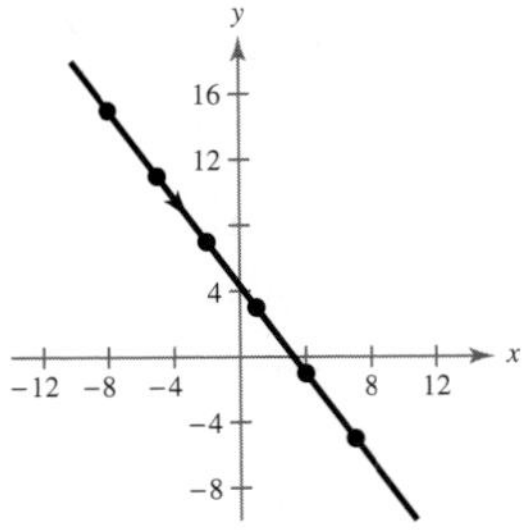

79.

t	-2	-1	1	2	3	4
x	-3	-6	6	3	2	$\frac{3}{2}$
y	2	3	5	6	7	8

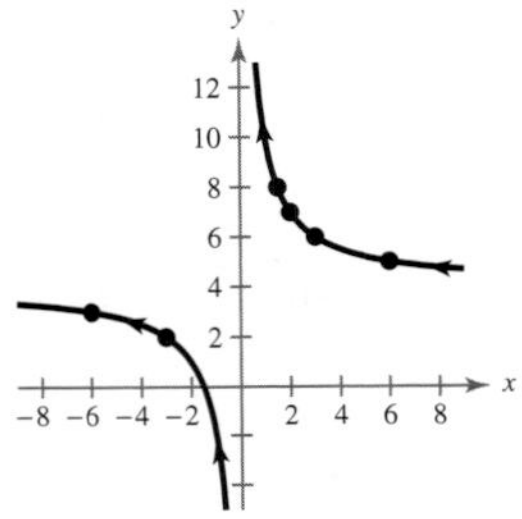

81. a **82.** c **83.** d **84.** b

85.

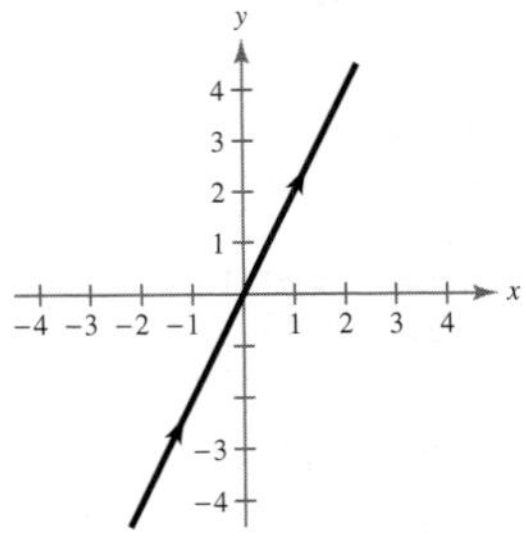

$y = 2x$

87.

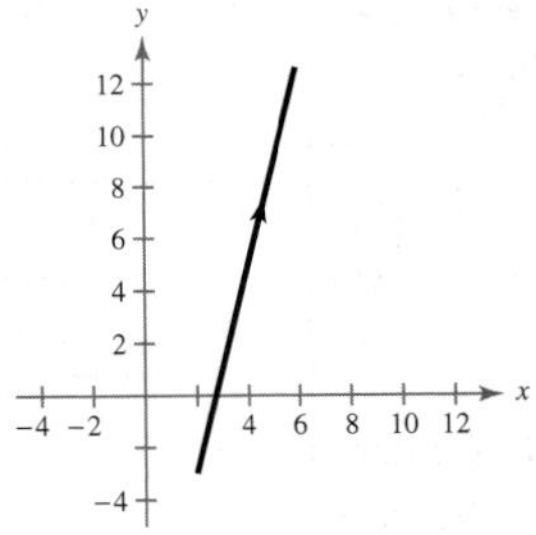

$y = 4x - 11, x \geq 2$

89.

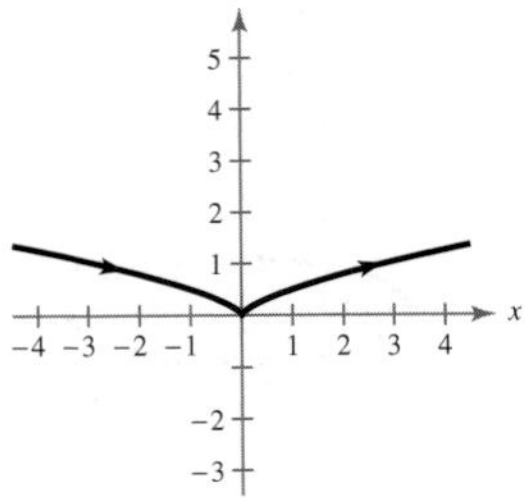

$y = \dfrac{1}{2}x^{2/3}$

91.

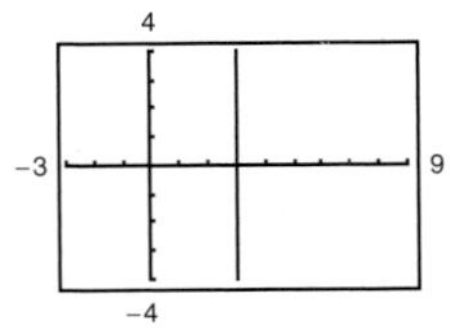

Not a function

93.

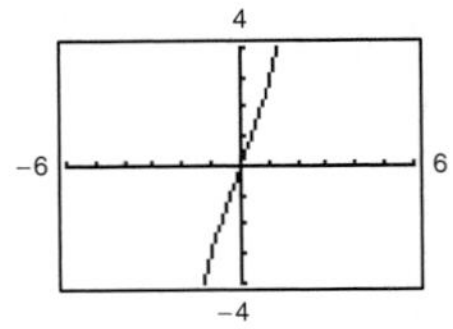

Function

CHAPTER 7

95.

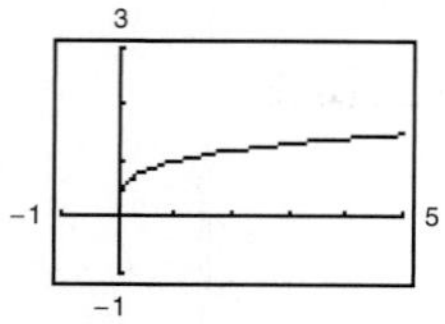

Function

97.

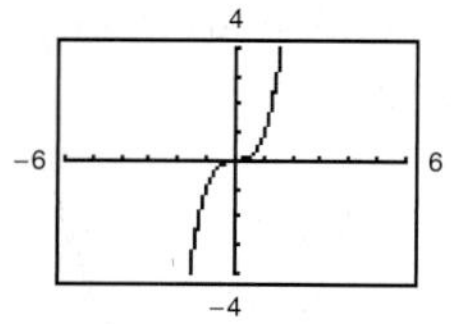

Function

99.

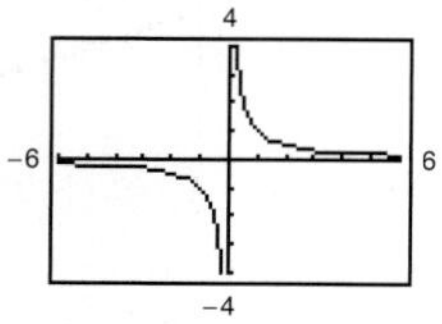

Function

101.

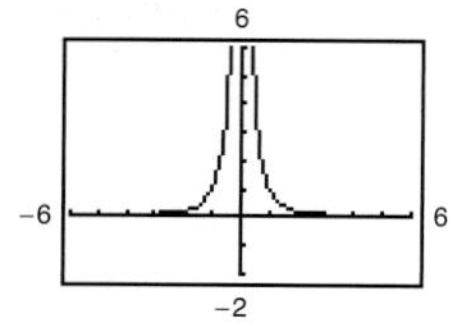

Function

103. (a) $x = t, y = 6t + 2$ (b) $x = 1 - t, y = 8 - 6t$

105. (a) $x = t, y = t^2 + 2$ (b) $x = 1 - t, y = t^2 - 2t + 3$

107. $x = 3, y = t$ **109.** $x = -5t, y = 4 + 3t$

111. $x = \frac{v_0\sqrt{2}}{2}t, y = 7 + \frac{v_0\sqrt{2}}{2}t - 16t^2$

113.

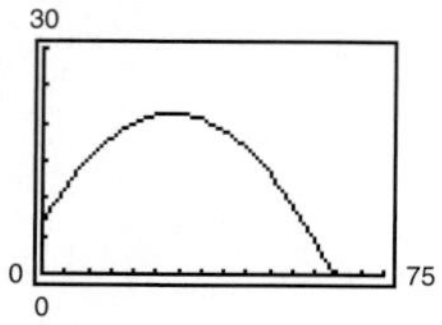

21.29 ft

115. False. If the graph intersected the directrix, then there would exist points nearer the directrix than the focus.

117. True

119. (a) Major axis is horizontal (b) Circle
(c) Ellipse is flatter. (d) Horizontal shift

Chapter Test (page 610)

1.

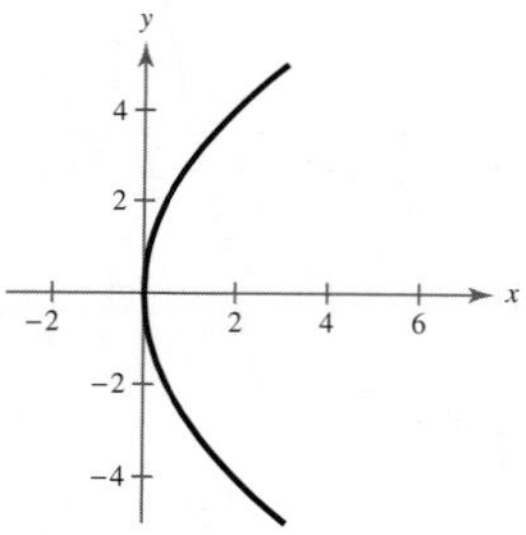

Vertex: $(0, 0)$
Focus: $(2, 0)$

2.

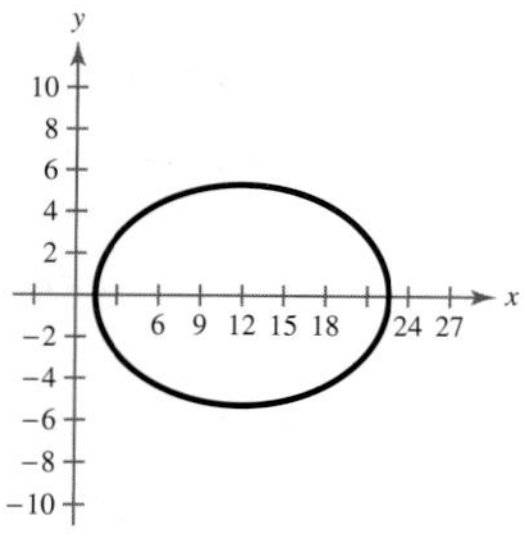

Vertices: $(12 \pm 4\sqrt{7}, 0)$
Foci: $(12 \pm 2\sqrt{21}, 0)$

3.

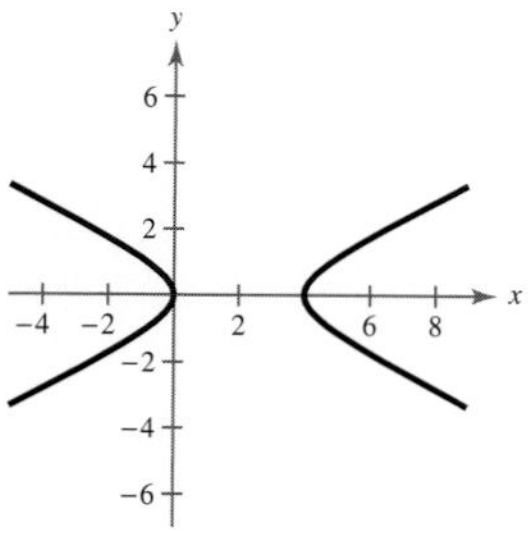

Vertices: $(0, 0), (4, 0)$
Foci: $(2 \pm \sqrt{5}, 0)$

4. $x^2 + y^2 = 64$

5. $(x + 3)^2 + (y + 3)^2 = 50$

6. $(y - 7)^2 = -\frac{49x}{8}$

7. $(y + 2)^2 = 8(x - 6)$

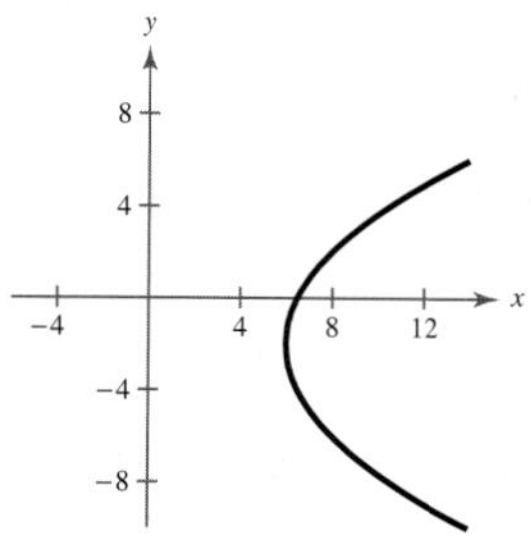

8. $\frac{(x - 4)^2}{16} + \frac{(y - 2)^2}{4} = 1$

9. $\frac{(x + 6)^2}{16} + \frac{(y - 3)^2}{49} = 1$

10. $\frac{y^2}{9} - \frac{x^2}{4} = 1$

11.

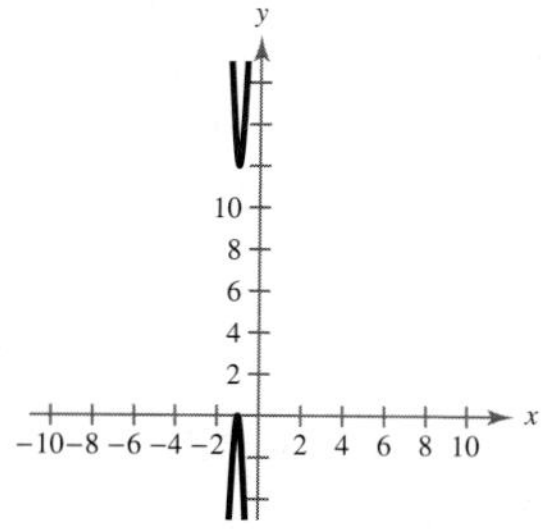

12.

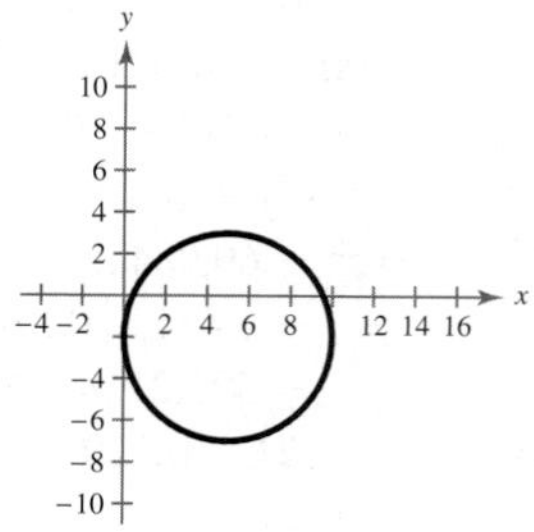

13.

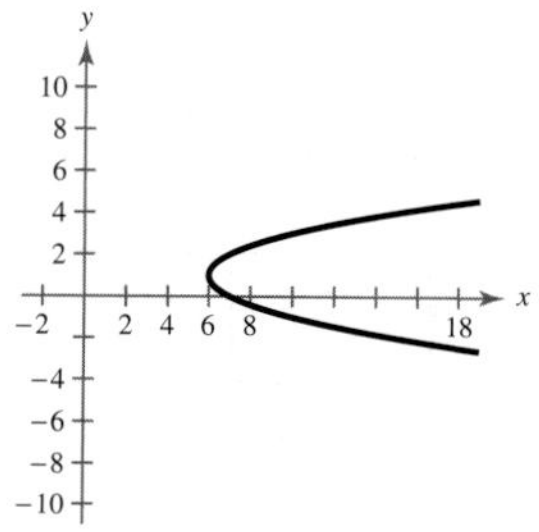

14.

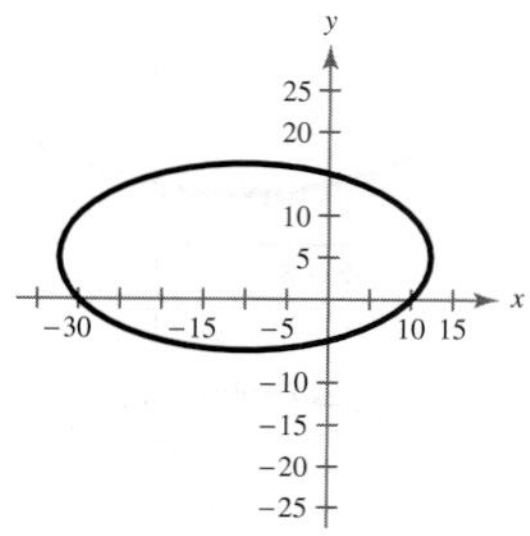

15.

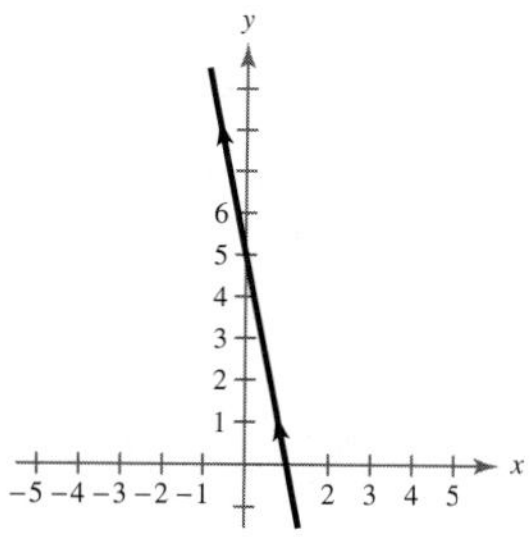

$y = 5 - 5x$

16.

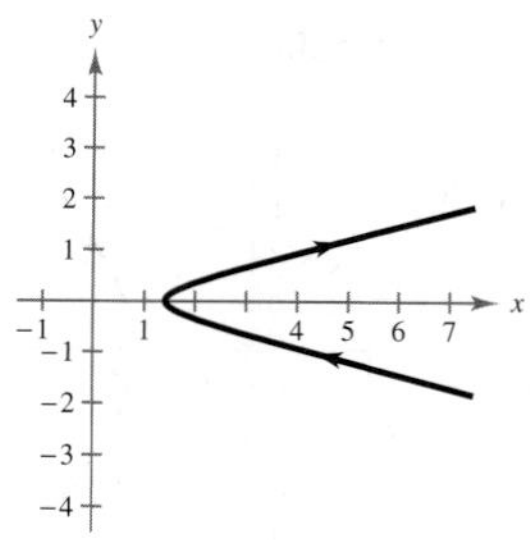

$y = \pm\frac{\sqrt{x^2 - 2}}{4}, x \geq \sqrt{2}$

17.

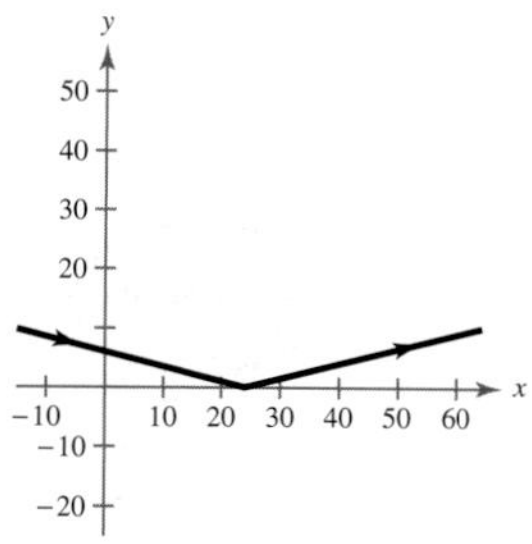

$y = \left|\frac{x}{4} - 6\right|$

18.

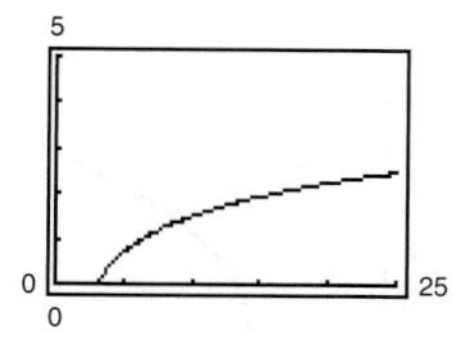

Function

19. (a) $x = t, y = 7t + 6$ (b) $x = 4t, y = 28t + 6$

20. (a) $x = t, y = \frac{8}{t}$ (b) $x = 4t, y = \frac{2}{t}$

21. (a) $x = t, y = \frac{4 - t^2}{2}$ (b) $x = 4t, y = 2 - 8t^2$

22. (a) $x = t, y = \frac{3}{t}$ (b) $x = 4t, y = \frac{3}{4t}$

23. About 33.9 m **24.** $\frac{x^2}{8556.25} + \frac{y^2}{11{,}990.25} = 1$

Cumulative Test for Chapters 6–7 (page 611)

1. $\frac{1}{5}, -\frac{1}{7}, \frac{1}{9}, -\frac{1}{11}, \frac{1}{13}$ **2.** 3, 6, 12, 24, 48 **3.** 2, 5, 11, 23, 47

4. $-x, \frac{x^2}{2}, \frac{-x^3}{6}, \frac{x^4}{24}, \frac{-x^5}{120}$ **5.** 110,544 **6.** 15,120

7. $\frac{n+1}{2}$ **8.** $\frac{1}{2n+1}$ **9.** 135 **10.** 3,937,563,750

11. 2,050,000 **12.** 96 **13.** About 34.480 **14.** 80

15. (a) 190 (b) 190 **16.** $x^4 + 12x^3 + 54x^2 + 108x + 81$

17. $32x^5 + 80x^4y^2 + 80x^3y^4 + 40x^2y^6 + 10xy^8 + y^{10}$

18. $x^6 - 12x^5y + 60x^4y^2 - 160x^3y^3 + 240x^2y^4 - 192xy^5 + 64y^6$

19. $6561a^8 - 69{,}984a^7b + 326{,}592a^6b^2 - 870{,}912a^5b^3 + 1{,}451{,}520a^4b^4 - 1{,}548{,}288a^3b^5 + 1{,}032{,}192a^2b^6 - 393{,}216ab^7 + 65{,}536b^8$

20. 120 **21.** 90,720 **22.** 420 **23.** 50,400

24. Hyperbola

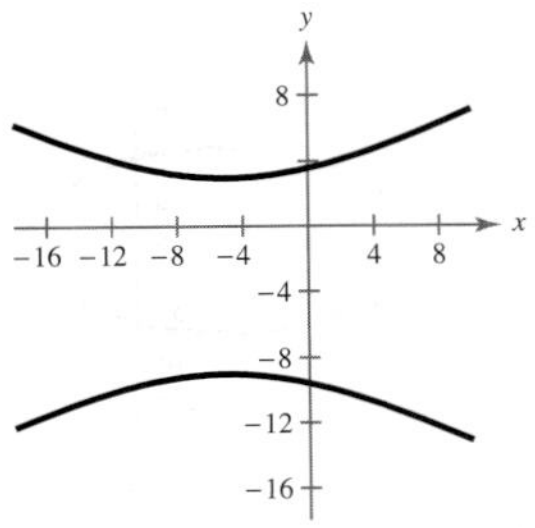

25. Ellipse

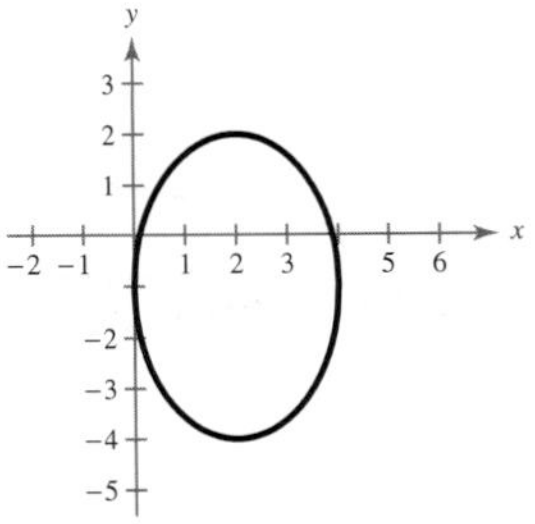

26. Hyperbola

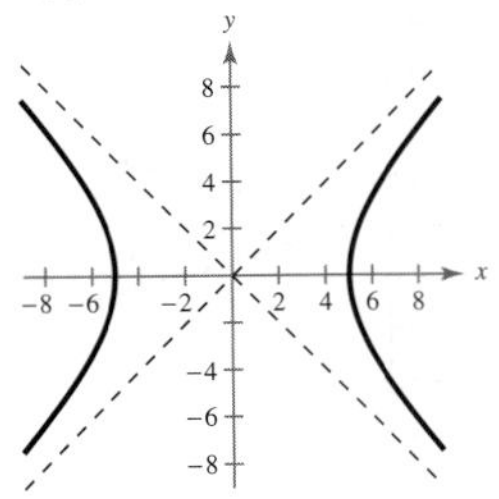

27. Circle

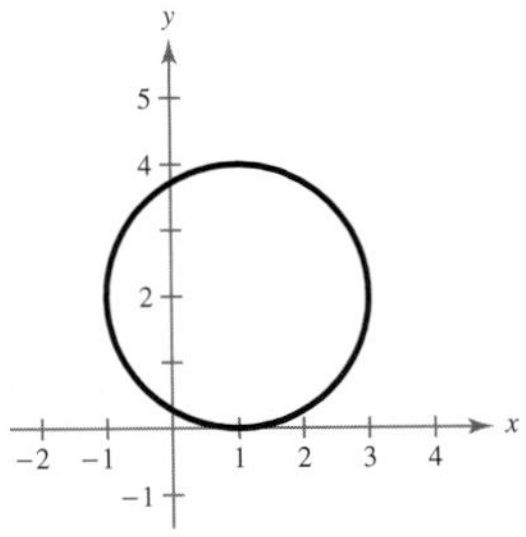

28. Parabola

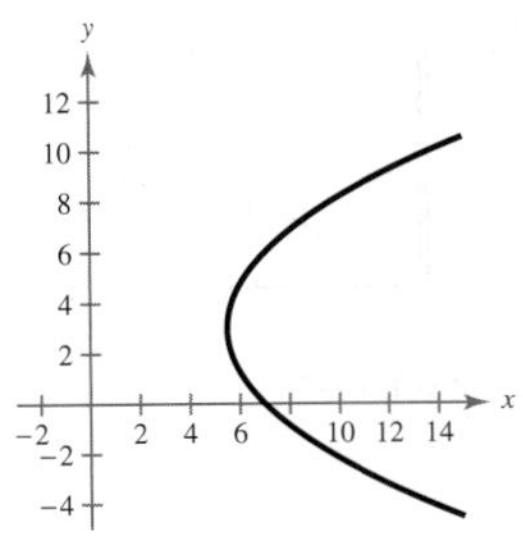

29. Ellipse

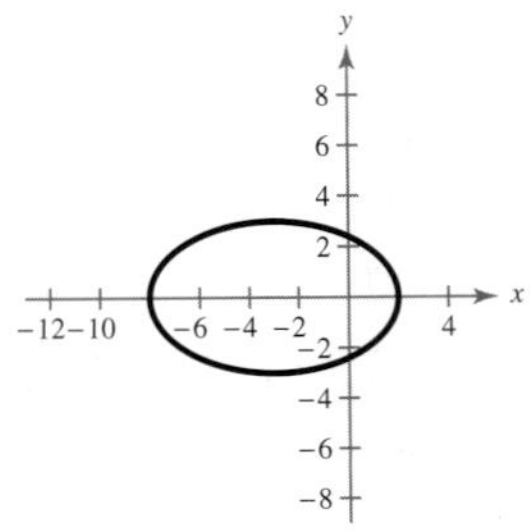

30. $(x - 2)^2 = -\frac{4}{3}(y - 3)$ **31.** $\frac{(x-1)^2}{25} + \frac{(y-4)^2}{4} = 1$

32. $\frac{(y+4)^2}{4} - \frac{3x^2}{16} = 1$ **33.** $(x + 2)^2 + (y - 4)^2 = 73$

34. $\frac{5(y-2)^2}{4} - \frac{5x^2}{16} = 1$ **35.** $\frac{(x-2)^2}{16} + \frac{(y-3)^2}{1} = 1$

36. $(y - 2)^2 = 8(x + 6)$

37. (a) and (b)

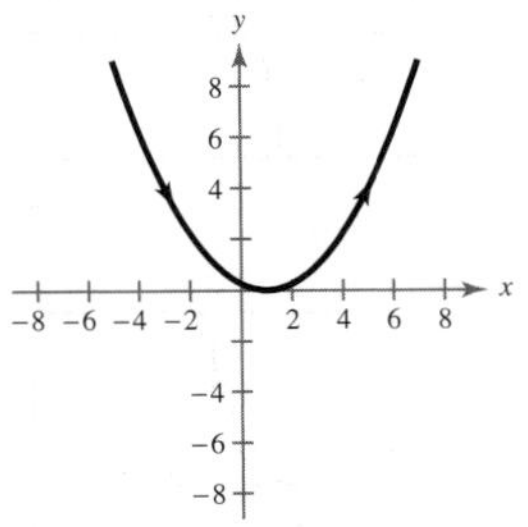

(c) $y = \left(\frac{x-1}{2}\right)^2$

38. (a) and (b)

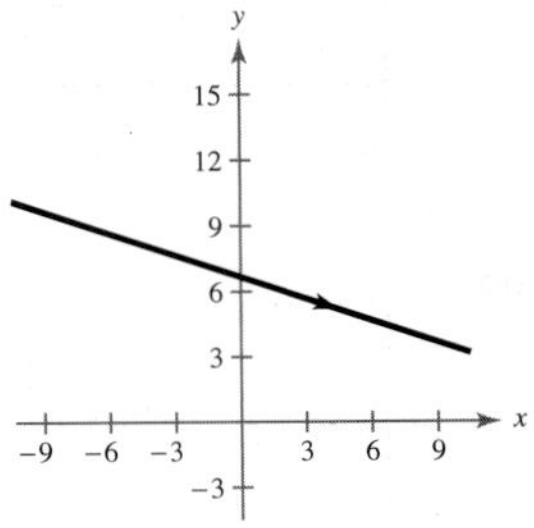

(c) $y = 4 - \frac{x-8}{3}$

39. (a) and (b)

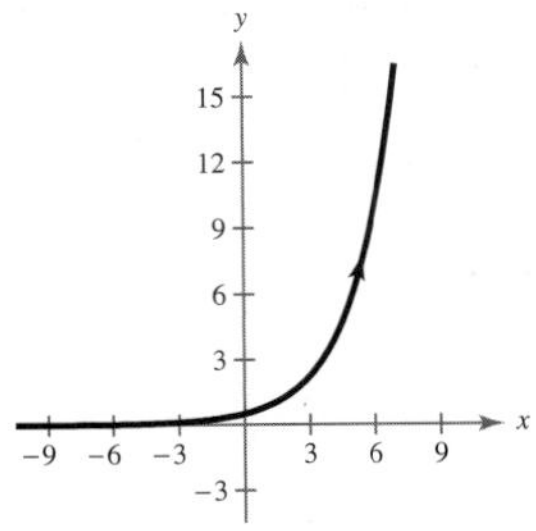

(c) $y = \frac{1}{2}e^{x/2}$

40. (a) and (b)

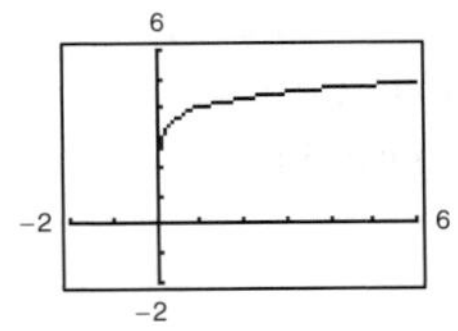

(c) $y = \frac{1}{2}\ln x + 4, x > 0$

41. (a) $x = t, y = 3t - 2$ (b) $x = \frac{t}{2}, y = \frac{3}{2}t - 2$

42. (a) $x = t, y = t^2 - 16$ (b) $x = \frac{t}{2}, y = \frac{t^2}{4} - 16$

43. (a) $x = t, y = \frac{2}{t}$ (b) $x = \frac{t}{2}, y = \frac{4}{t}$

44. (a) $x = t, y = \frac{e^{2t}}{e^{2t}+1}$ (b) $x = \frac{t}{2}, y = \frac{e^t}{e^t+1}$

45. $x = t, y = \frac{7}{4}t - \frac{13}{2}$

46. (a) 2008: \$21.125 billion
2009: \$24.406 billion
2010: \$30.023 billion
2011: \$37.974 billion
2012: \$48.261 billion
2013: \$60.882 billion
(b) \$222.671 billion
(c) 2014: \$75.838 billion
2015: \$93.128 billion
2016: \$112.754 billion
2017: \$134.714 billion

47. \$604,464.70 **48.** $\frac{1}{4}$ **49.** $\frac{x^2}{49} + \frac{y^2}{64} = 1$ **50.** No

Index of Selected Applications

Index

D

E

G

FORMULAS FROM GEOMETRY

Triangle

$h = a \sin \theta$

$\text{Area} = \frac{1}{2}bh$

Laws of Cosines:
$c^2 = a^2 + b^2 - 2ab \cos \theta$

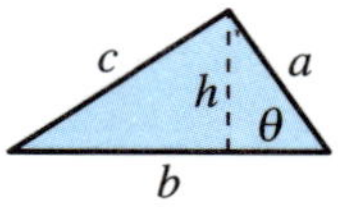

Right Triangle

Pythagorean Theorem:
$c^2 = a^2 + b^2$

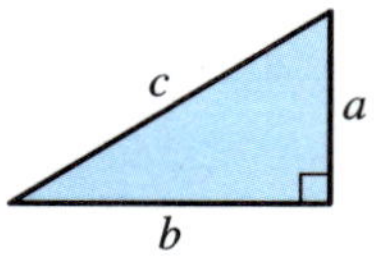

Equilateral Triangle

$h = \frac{\sqrt{3}s}{2}$

$\text{Area} = \frac{\sqrt{3}s^2}{4}$

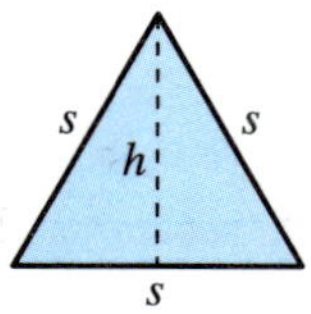

Parallelogram

$\text{Area} = bh$

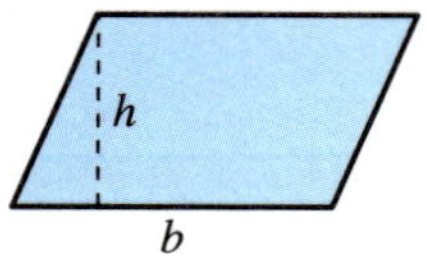

Trapezoid

$\text{Area} = \frac{h}{2}(a + b)$

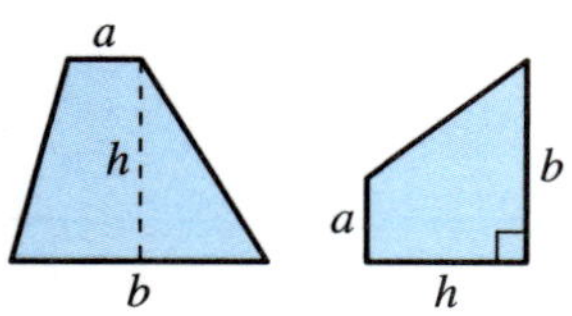

Circle

$\text{Area} = \pi r^2$
$\text{Circumference} = 2\pi r$

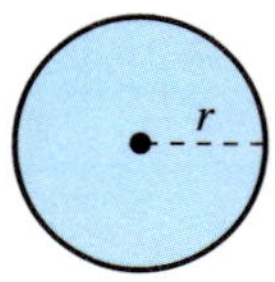

Sector of Circle

$\text{Area} = \frac{\theta r^2}{2}$

$s = r\theta$

(θ in radians)

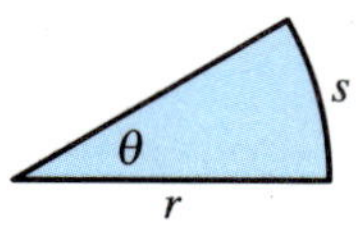

Circular Ring

$\text{Area} = \pi(R^2 - r^2)$

$= 2\pi pw$

(p = average radius,
w = width of ring)

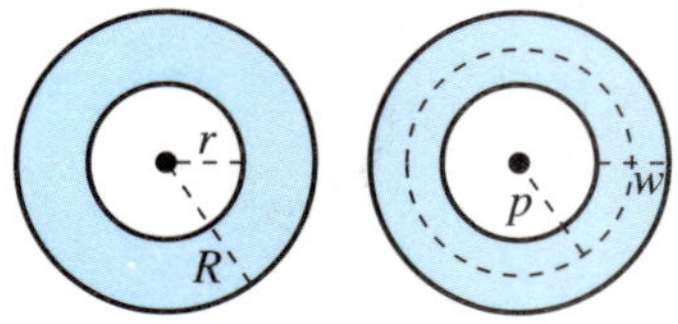

Ellipse

$\text{Area} = \pi ab$

$\text{Circumference} \approx 2\pi \sqrt{\frac{a^2 + b^2}{2}}$

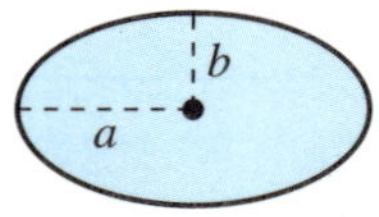

Cone

(A = area of base)

$\text{Volume} = \frac{Ah}{3}$

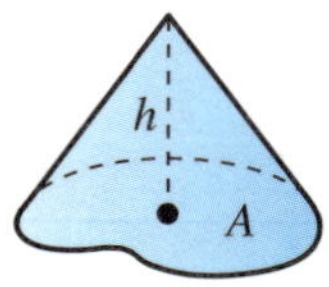

Right Circular Cone

$\text{Volume} = \frac{\pi r^2 h}{3}$

$\text{Lateral Surface Area} = \pi r \sqrt{r^2 + h^2}$

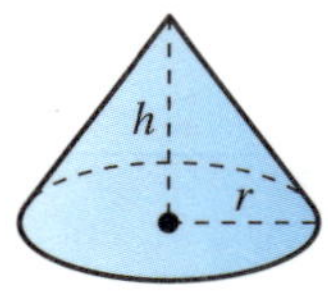

Frustum of Right Circular Cone

$\text{Volume} = \frac{\pi(r^2 + rR + R^2)h}{3}$

$\text{Lateral Surface Area} = \pi s(R + r)$

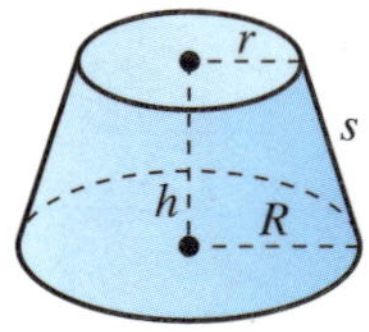

Right Circular Cylinder

$\text{Volume} = \pi r^2 h$

$\text{Lateral Surface Area} = 2\pi rh$

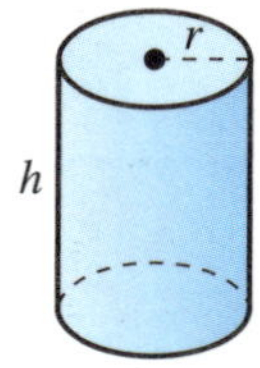

Sphere

$\text{Volume} = \frac{4}{3}\pi r^3$

$\text{Surface Area} = 4\pi r^2$

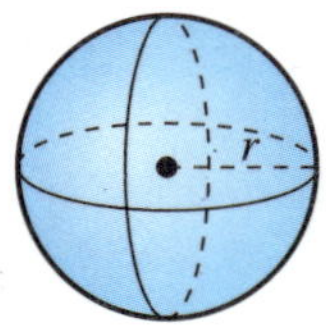

Rational Function ***(p. 298)***

$f(x) = \frac{1}{x}$

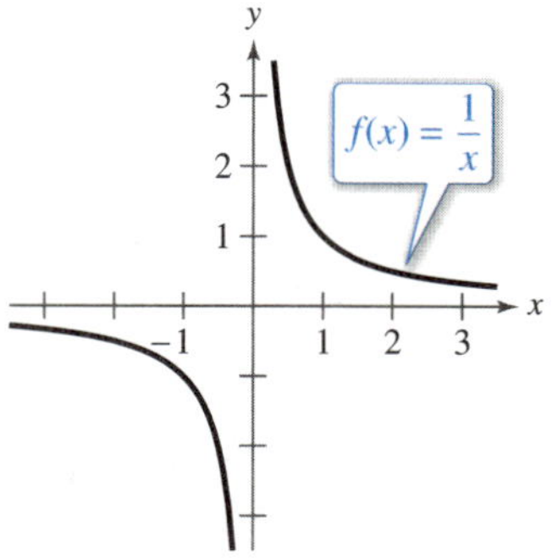

Domain: $(-\infty, 0) \cup (0, \infty)$
Range: $(-\infty, 0) \cup (0, \infty)$
No intercepts
Decreasing on $(-\infty, 0)$ and $(0, \infty)$
Odd function
Origin symmetry
Vertical asymptote: y-axis
Horizontal asymptote: x-axis

Exponential Function ***(p. 326)***

$f(x) = a^x, a > 1$

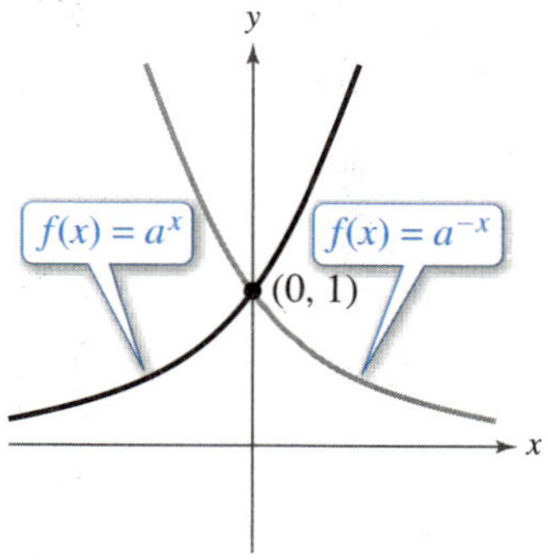

Domain: $(-\infty, \infty)$
Range: $(0, \infty)$
Intercept: $(0, 1)$
Increasing on $(-\infty, \infty)$
for $f(x) = a^x$
Decreasing on $(-\infty, \infty)$
for $f(x) = a^{-x}$
x-axis is a horizontal asymptote
Continuous

Logarithmic Function ***(p. 339)***

$f(x) = \log_a x, a > 1$

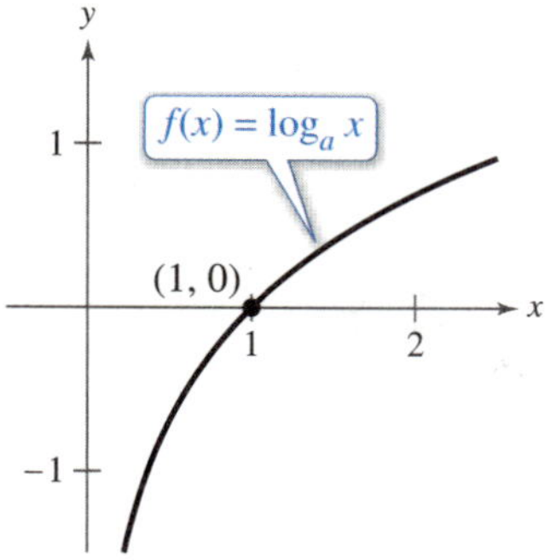

Domain: $(0, \infty)$
Range: $(-\infty, \infty)$
Intercept: $(1, 0)$
Increasing on $(0, \infty)$
y-axis is a vertical asymptote
Continuous
Reflection of graph of $f(x) = a^x$
in the line $y = x$